P9-EDC-448

PERGAMON INTERNATIONAL LIBRARY
of Science, Technology, Engineering and Social Studies
The 1000-volume original paperback library in aid of education, industrial training and the enjoyment of leisure
Publisher: Robert Maxwell, M.C.

PLANKTON AND PRODUCTIVITY IN THE OCEANS

Second Edition

Volume 2 – Zooplankton

By the same author

PLANKTON AND PRODUCTIVITY IN THE OCEANS, 2nd Edition,

Volume 1 Phytoplankton

Other titles of interest

CORLISS
The Ciliated Protozoa, 2nd Edition

McLUSKY & BERRY
Physiology and Behaviour of Marine Organisms

NAYLOR
Cyclical Phenomena in Marine Plants and Animals

PARSONS, TAKAHASHI & HARGRAVE
Biological Oceanographic Processes, 2nd Edition

PURCHON
The Biology of the Mollusca, 2nd Edition

VAN WINKLE
Proceedings of a Conference on Assessing the Effects of
Power-plant Induced Mortality on Fish Populations

WOLFE
Fate and Effects of Petroleum Hydrocarbons
in Marine Organisms and Ecosystems

PLANKTON AND PRODUCTIVITY IN THE OCEANS

Second Edition

Volume 2—Zooplankton

JOHN E. G. RAYMONT[†]

Department of Oceanography in the University of Southampton

PERGAMON PRESS

Oxford · New York · Toronto · Sydney · Paris · Frankfurt

U.K.	Pergamon Press Ltd., Headington Hill Hall, Oxford OX3 OBW, England
U.S.A.	Pergamon Press Inc., Maxwell House, Fairview Park, Elmsford, New York 10523, U.S.A.
CANADA	Pergamon Press Canada Ltd., Suite 104, 150 Consumers Rd., Willowdale, Ontario M2J 1P9, Canada
AUSTRALIA	Pergamon Press (Aust.) Pty. Ltd., P.O. Box 544, Potts Point, N.S.W. 2011, Australia
FRANCE	Pergamon Press SARL, 24 rue des Ecoles, 75240 Paris, Cedex 05, France
FEDERAL REPUBLIC OF GERMANY	Pergamon Press GmbH, Hammerweg 6, D-6242 Kronberg-Taunus, Federal Republic of Germany

First edition 1963

Second edition 1983

British Library Cataloguing in Publication Data

Raymont, John E. G.
Plankton and productivity in the oceans. – 2nd ed.
Vol. 2: Zooplankton
1. Marine plankton
I. Title
574.92 QH91.8.P5

ISBN 0-08-024404-1 Hardcover
ISBN 0-08-024403-3 Flexicover

Typeset by Macmillan India Ltd., Bangalore.

Printed and bound in Great Britain at The Camelot Press Ltd, Southampton

Contents

Introduction

The manuscript for this volume of *Plankton and Productivity in the Oceans* dealing with zooplankton and secondary production, was almost complete in August 1979. Had it not been for the help and encouragement of my late husband's colleagues I do not think I would have had the audacity to attempt to finalize the work. To them, and to the many other friends at the Institute of Oceanographic Sciences and elsewhere, I owe a deep debt of gratitude.

In particular I would like to thank Professor Charnock, F.R.S. and Professor Lockwood for allowing me office and library facilities in the Department of Oceanography, Southampton University; Sir Frederick Russell, F.R.S. who read Chapter II; Mrs. Gathergood who completed the typing even during retirement; and the editorial staff of Pergamon Press. Dr. Burton, whose task of reading the text of both volumes was made doubly onerous, has been of invaluable aid, ever ready with helpful criticisms and suggestions.

There has been an inevitable delay between the production of the two volumes; I have endeavoured to check any dubious points, and although I cannot aspire to my husband's standard, I know it would have been his wish that this volume should be completed and I hope I have not failed him.

—To John, who patiently taught me all the Biological Oceanography I shall ever know.

August 1982

Brigit Raymont.

Chapter 1
Introduction

General Considerations and Definitions

The floating and drifting animal population known as the zooplankton cannot be precisely delimited from another great pelagic community in the marine environment—the nekton. However, the plankton is usually regarded as being distinct in that it has little or no power of independent horizontal migration. The zooplankton inhabits all layers of the ocean down to the greatest depth sampled (cf. Banse, 1964; Vinogradov, 1962, 1968, 1972).

It has long been recognized that numerous zooplankton species drawn from many phyletic groups have powers of vertical migration, sometimes performing very extensive vertical movements. Some of the more powerful swimmers in the zooplankton, such as pelagic decapods and the larger euphausiids, may also occur in swarms which appear, at times, to move horizontally, not entirely due to the drift of the water. Although it is doubtful whether this could be called a horizontal migration, the precise difference between the nektonic freely swimming and migrating pelagic community and the plankton is difficult to judge. More recent observations on living plankton in the ocean have suggested that some of the larger medusae and siphonophores are capable of strong swimming movements, and certainly it is not easy to distinguish between such strongly moving plankton and the very small members of the nekton, (mainly small fishes and cephalopods, frequently classed as the micronekton), whose major movements are often essentially vertical migrations resembling those of the zooplankton. Although a very large number of zooplankton animals drawn from a wide variety of taxa (e.g. copepods, ostracods, sagittae, siphonophores, euphausiids, ctenophores, heteropods, pteropods, salps, etc.) remain planktonic throughout their whole existence, a large array of animals occur in the plankton during only a part of their lives. These are known as meroplanktonic animals, in contrast to those holoplanktonic forms which remain permanently in the plankton.

Meroplankton and Holoplankton

The meroplankton includes the various types of trochophore and veliger larvae of benthic worms and molluscs; different kinds of nauplii and zoeas of bottom-living crustaceans; cyphonautes larvae; ascidian and tornaria larvae; the several types of echinoderm larvae; larval squid; the planulae of Cnidaria, as well as medusae of the hydromedusan type. Also included in the meroplankton are the eggs and larval stages of most fishes which when adult are part of the nekton. These meroplanktonic forms are dealt with in more detail in Chapter 2.

As with the phytoplankton (Volume 1), a more or less coastal zooplankton extending from inter-tidal zones to the edge of the continental shelf, and thus very approximately delimited by the 200 m depth contour (the neritic zooplankton), can be distinguished from the oceanic zooplankton, which ranges offshore from beyond the continental slope across the whole extent of the oceans. While neritic and oceanic zooplankton populations are not sharply separated, there are fairly obvious differences. Neritic plankton would tend to have a larger proportion of meroplankton; oceanic meroplankton must be comparatively long-lived (*vide infra*). The meroplankton, particularly in more temperate waters, may also show marked fluctuations in species abundance with seasonal breeding, so that the population is more variable than offshore. Despite the importance of meroplankton in inshore waters, the holoplankton—often represented by a limited number of species—may be extremely plentiful and dominate the population over much of a year.

Harvey, Cooper, Lebour and Russell (1935) showed that certain, mainly neritic, copepods were dominant off Plymouth through most of the year. Of the meroplankton, cirripede nauplii were moderately abundant only over March, and polychaete larvae and rotifers early in the year. Similar investigations in inshore coastal waters have generally confirmed the overwhelming importance of holoplanktonic copepods in the total zooplankton, with meroplanktonic larvae being of minor significance over limited periods of the year (e.g. Wiborg, 1954; Deevey, 1956; Lie, 1967). Details are included in the chapter on seasonal cycles.

In estuarine waters and in certain seas characterized by low salinity (e.g. the Baltic), a considerable variety of meroplankton often can occur in high densities, at times even dominating the plankton population. A very low diversity of holoplankton is typical, but the few species represented may be present in extraordinarily high densities (cf. Jeffries, 1967). For example, Ackefors (1965) found that off Stockholm—where the salinity was about 7‰—copepods, especially *Acartia bifilosa*, dominated the catches; *Eurytemora* was also important. Cladocera (e.g. *Podon polyphemoides*) were plentiful from June to November. Meroplankton included rotifers and bivalve larvae (*Mytilus*, *Macoma*, *Mya arenaria* and *Cardium lamarcki*).

Certain genera and species of copepod are characteristic of estuarine waters. The genus *Acartia* includes a number of species (e.g. *A. bifilosa*, *A. tonsa*, *A. discaudata*) found widely in temperate estuaries while other species are typical of warmer waters. Tranter and Abrahams (1971) found seven species of *Acartia* and the closely related *Acartiella* in brackish Indian waters, in contrast to *A. erythraea*, a euryhaline marine, and *A. negligens*, a stenohaline marine species found outside estuarine waters. Other copepods typical of estuaries are *Eurytemora* spp., *Labidocera wollastoni*, *L. aestiva*, *Tortanus discaudatus*, *Centropages hamatus*, *Pseudocalanus elongatus*, *Paracalanus crassirostris*, *Pseudodiaptomus*, the harpacticoid *Euterpina acutifrons* and certain cyclopoids belonging to the genus *Oithona*, particularly *O. nana* and *O. brevicornis*. Other coastal copepods (*Paracalanus parvus*, *Temora longicornis*, *Isias*) commonly occur in estuaries. Although estuarine holoplankton is often dominated by a single species of copepod (*Eurytemora* or *Acartia*) (cf. Jeffries, 1964, 1967), other taxonomic groups may be significant, for example, chaetognaths (*Sagitta setosa*, *S. crassa*, *S. hispida*, *S. nagae*), the appendicularian *Oikopleura dioica*, ctenophores (*Pleurobrachia*, *Mnemiopsis*, *Bolinopsis*) and, amongst protozoans, a number of tintinnids. In warm waters the decapod *Lucifer faxoni* is a typically neritic estuarine species, as is *Acetes*, though this might be regarded as partly bottom-living—and as are several mysid species which are typical of, and may be very abundant in, estuarine waters. Although not holoplankton, a few hydromedusans (cf. Chapter 2)—*Podon polyphemoides* and *Penilia* among Cladocera, as well as the scyphozoan *Aurelia aurita*—may become enormously plentiful seasonally in estuaries.

In very shallow, particularly estuarine waters, this contribution from animals living partly on the bottom can be significant. The term tychozooplankton is sometimes given to such animals which, whether as young or adults, spend part of the 24-hour day on the bottom, but rise and spend part of their day, often the dark hours, as planktonic organisms. The distinction between tychoplankton and benthos may be difficult, since during storms, especially in shallow estuaries, a number of small bottom organisms may become artificially swept up into the water, but may be able to exist, and indeed to flourish for a time, in the intermediate layers. A genuine tychozooplankton, however, seems to exist, the animals spending more or less regularly a portion of their time every day as planktonic or benthic organisms. Such animals include a number of mysid species, amphipods, harpacticoid copepods, cumaceans, isopods, shrimps, prawns and other crustaceans. Phyla other than the Crustacea may contribute; some protozoans are tychoplanktonic, though these are probably more generally mechanically swept up from bottom deposits. Metamorphosing larvae and post-larvae of a variety of animal groups, as well as some rotifers and occasionally swarming benthic species (e.g. sexually mature polychaetes), may be part of the tychoplankton.

Outside specifically estuarine waters, but in the coastal areas, the zooplankton is much more obviously made up of holoplanktonic species, particularly copepods. Many of the species already listed for estuaries are found, though some very typical estuarine species such as *Eurytemora* hardly exist outside the mouths of rivers. The variety of copepods is considerably increased, however, although diversity is still usually greater in oceanic waters, especially warm oceans. In temperate seas, *Acartia clausi* with *Paracalanus parvus*, *Pseudocalanus elongatus*, *Temora longicornis and Oithona* spp., including *O. similis*, *Corycaeus* and *Oncaea*, become important, with increasing proportions of *Calanus finmarchicus* or *Calanus helgolandicus* in slightly more offshore waters. *Calanus* rapidly becomes an outstandingly dominant species over boreal continental shelf areas.

In warmer neritic waters, there is a greater diversity of copepods, but certain forms are typical of the coastal areas and are not found abundantly in more open oceans. Descriptions of geographical distribution, including distinctions between neritic and oceanic copepods, are included in Chapter 2.

The coastal neritic waters show a greater variety of other holoplankton. A very few euphausiids (*Nyctiphanes simplex*, *N. capensis*, *N. couchi*, *N. australis*, *Pseudeuphausia latifrons*) appear to be limited to coastal waters. Some chaetognaths are more typically inshore (e.g. *Sagitta frederici*) in addition to those listed for estuaries. In warmer waters *Muggiaea kochii* and *M. atlantica* appear to be somewhat coastal, and a few pteropods (e.g. *Creseis acicula*) can flourish over continental shelves. *Limacina retroversa* occurs in colder, mixed coastal waters. For the distribution of these and other taxa, as well as for copepods, the details given in Chapter 2 should be consulted.

A problem in delimiting neritic and oceanic forms arises with some very widely distributed species, because races probably exist which, though physiologically different, are not morphologically distinct (e.g. *Liriope*, *Oithona similis*). Sometimes there is evidence for the existence of two very similar species. Wiborg (1954), for example, believes that *Pseudocalanus* is represented by *P. minutus* living in oceanic regions, whereas *P. elongatus* is an inshore form. However, with some species (e.g. *Acartia clausi*, *Corycaeus anglicus*) long term and wide ranging investigations, such as those carried out by Continuous Plankton Recorder surveys (Glover, 1967), strongly suggest separate oceanic and neritic populations, but morphological separation into distinct species is not possible. Some cladocerans, though strictly meroplanktonic, appear not only to flourish near coasts but to be able to spread out over ocean depths, at least for periods. Some holoplanktonic appendicularians also appear to be able to range very widely over neritic and oceanic provinces. Sharp distinctions between oceanic and neritic species are therefore sometimes not possible and cosmopolitan species (e.g. *Scolecithricella minor*, *Oithona similis*, *Beroe cucumis*, *Tomopteris ligulata*) are well known, but the general distinction between oceanic and neritic zooplankton faunas is well founded.

Although neritic zooplankton is normally dominated by a few holoplanktonic species, there is a considerable variety of meroplanktonic larvae of benthic animals. Meroplanktonic larvae would be expected to be generally more abundant in coastal waters approximately corresponding to the limits of continental shelves, in part because of the greater density of the benthos producing the larvae in shallower seas and in part in relation to the comparatively small distance from the bottom, where the larvae are produced, to the surface waters. Moreover, many deep-sea benthic animals tend to have direct development, so few larvae may be expected throughout the enormous depth of open ocean.

Thorson (1964) suggests a total of 140,000 species of marine bottom fauna. Populations of perhaps 80 % of these species live, at least in part, at depths of less than 200 m, many at much shallower levels. Thorson estimates that 80 % of all the marine shallow water invertebrates studied, which could approximate to 90,000 species altogether, have planktonic larval stages. Not only are planktonic larval stages rare in the life cycles of deep-sea benthic invertebrates, but meroplanktonic larvae are not typical amongst Arctic or Antarctic benthos. The few bottom invertebrates at high latitudes (including larval stages in the life history) breed during the short summer. The vast majority of meroplanktonic larvae inhabit coastal neritic areas in temperate and warm seas. Even in such limited depths the great majority of larvae, according to Thorson, tend to be in the near-surface waters, largely due to their fairly strong positive phototactic responses. Such responses are, however, modified by temperature and salinity, so that the immediate surface is avoided by many species. Older larvae may later become photonegative and be distributed in the near-bottom layers prior to settlement and metamorphosis. Planktonic larvae of inter-tidal benthic species appear to remain photopositive throughout larval life except immediately before settlement. This may be an adaptation to maintaining the stock in inter-tidal areas. A list of meroplanktonic larvae with an indication of their light responses is included in Thorson's (1964) review.

The average period of planktonic life for the larvae of benthic species is believed to be 2–4 weeks. The short period should assist in maintaining a fairly large proportion of the meroplankton in the inshore plankton population despite the enormous mortality, mostly attributable to predation. Data from different authorities emphasize a heavy mortality: *Ostrea edulis* larvae in Dutch waters lose 14 % at each tidal cycle; of 3 million eggs spawned by a pair of *Mya arenaria* only about 0.001 % settle per year; of 13,000 nauplii released by one *Balanus balanoides* over its total life, only 0.2% survive to settlement; one adult crab (*Paralithodes*) gives rise to 200,000 first stage zoeas, but only 7000 on average, survive even to the glaucothoe stage. The success of meroplanktonic life despite this mortality, and the general restriction of the larvae to nearshore waters is evident from the long term stability of the benthos population over continental shelves, in spite of the marked fluctuations known for some species (Stephen, 1931, 1938; Thamdrup, 1935; Jones, 1950, 1956; Smidt, 1951; Sanders, 1956; Thorson, 1957). Such fluctuations in year classes of certain meroplanktonic benthic species are largely due to the relative success or failure of a brood, reflecting environmental factors which existed during the period of larval life.

Only about 5% of Arctic benthos produce planktonic larvae, and the proportion is even smaller in the Antarctic. In temperate latitudes, apart from the very much greater proportion of planktonic larvae, benthos are usually more abundant chiefly in spring and summer, though some larvae occur at all times of the year. Inshore plankton may thus change qualitatively with the changing meroplankton component. Some discussion of periods of meroplankton abundance is included in Chapter 4.

Mileikovskiy (1968) demonstrated that the distribution of a variety of planktonic larvae belonging to such different benthic adults as gastropods, nemertines, bryozoans, decapods, cirripedes, polychaetes, ophiuroids, echinoids, lamellibranchs, holothurians and asterioids, followed closely the shallower areas around the coasts of Norway and Russia. The distribution of larval density was more or less in accordance with the distribution of the breeding adults and was less influenced by currents, at least over areas of maximal larval density. Where more or less circular current systems were present, so that larvae were maintained in relatively shallow waters over extensive areas, as in the South Barents Sea, larvae could be really abundant even 200–300 miles from land. Elsewhere they were restricted to the fairly narrow continental shelves; less than 10% of the larvae were dispersed more than 50 miles offshore.

Scheltema (1975) emphasizes that water circulation, especially the coastal hydrography, and the behaviour patterns of meroplanktonic larvae both contribute to the extent to which larvae spread beyond the edge of the continental shelf where they are spawned. Even with a drift of about 100 miles offshore there is a good chance of return. The length of larval life is another factor in distribution. Scheltema cites the stability of the centre of the xanthid crab population along the Californian coast as evidence of a hydrographic mechanism for retaining the zoeas close to shore. Compensation currents, especially in concavities of the coastline, probably prevent too many larvae from being swept southwards by the current.

Along the Atlantic coast of North America between Florida and Cape Hatteras, zoeas of the crab *Callinectes sapidus* were widely distributed up to 60 miles offshore but were absent beyond the axis of the Gulf Stream. The density of larvae decreased with distance offshore, the early stages being nearer the coastline. Although surface drift disperses the larvae, there is evidence of a shoreward drift of bottom water along the continental shelf of North America, especially in the region off Delaware and Chesapeake Bays, thus providing a good chance for the return of older decapod larvae. Further south, along the Atlantic seaboard, surface and bottom waters both show considerable flow towards shore. Over the whole coastline Scheltema calculated that around 10% of the larvae may be returned shorewards by passive drift. The current pattern shows seasonal changes, however, and the spawning times of several decapod species vary with the area, promoting the maximum retention of the larvae.

Makarov (1969), describing the distribution of decapod larvae on the continental shelf of Kamchatka, agrees with the importance of the spawning area and pattern of nearshore currents in distribution. Larvae were most abundant in shallow waters (35–75 m); samples from stations exceeding 150–200 m rarely contained any decapod larvae. However, the distribution showed an increase in density on proceeding offshore from very shallow depths (*ca.* 30 m) to reach a maximum at 75 m depth, after which the density then declined again. What is described as a "larval belt" was found along the length of the shelf at medium depth.

The direction of offshore currents and the presence of compensatory currents and eddies are listed by Makarov as important factors in retaining meroplanktonic larvae, so that usually only an insignificant proportion passed to the open ocean. The influence of longshore and strong tidal currents was especially important over the central part of the Kamchatka shelf. The larval belt was much narrower in the central area, due largely to the surface current being restricted to a narrow band closely parallel to shore. At the northern and southern ends of the shelf, the surface currents were much wider and more diffuse, and the larvae were more widely distributed. However, they were more abundant in the northern and southern areas, since the strong longshore current in the central part of the coast carried them away from that region.

Although Makarov found that various decapod families followed this overall distribution pattern, there were specific differences. Among the Paguridae and Maiidae, although all the larvae were mainly in the larval belt, they showed depth preferences. The Crangonidae displayed greater differences; while *Crangon septemspinosa* and *C. dalli* larvae were mostly closer to shore (especially the former species) and were not abundant in the larval belt, *Mesocrangon intermedia* and *Crangon communis* larvae were mainly in the middle of the larval belt. Makarov attributed these varied distributions largely to the varying behaviour patterns of the different species. Vertical migration between the water layers, especially in response to light, salinity and pressure, along with tidal changes, could cause the larvae to encounter strata with compensatory or reversed flows and eddies, bringing the species into the particular depth favoured for settlement. Return migration, with shoreward currents, probably operates with many other meroplanktonic crustaceans. For instance, post-larval *Pandalus goriurna* were found to approach shallower and shallower waters, although the young larvae were found in deeper waters. While this might represent an active horizontal migration, in the vast majority of examples distribution results from vertical migration utilizing different current patterns.

A longer larval period may well result in a greater degree of dispersal of larvae offshore, but the use of returning currents may still operate. Johnson (1940) found that the larvae of *Emerita* (the sand crab) from the Californian coast may be swept out by the California Current with its offshore drift to a distance approaching 130 miles from the mainland. This wide dispersal is accompanied by rather a long larval life of up to 4 months. It was presumed that a heavy loss of larvae was inevitable but that the species inhabiting a fairly narrow continental shelf coastline must be adapted to the heavy mortality. Efford (1970) believes, however, that over the 4-month period many of the larvae are returned inshore by the bottom reversed currents.

Another example concerns the distribution of the phyllosoma larvae of *Palinurus interruptus* along the coasts of southern California and Baja California (Johnson,

1960). Although the larvae were widely dispersed over the hatching area and to some distance offshore, they were seldom found far to north or south. Swirls, eddies, including transient ones encountered during the study, and counter-currents retained the larvae for recruitment, avoiding any great loss particularly to the south with the prevailing Californian Current. This strong current might be expected to carry huge numbers of larvae outside their normal area of metamorphosis. In Johnson's words: "It is amazing to note how well the larval population appears to remain within the area throughout the floating larval period."

The seasonal Davidson Current in late autumn and winter may also return older larvae which have drifted south. Thus, despite the long larval life, the population is retained, although Johnson emphasizes that the circulation is not of the semi-enclosed circular pattern. Bigelow (1926) described the larvae of *Sebastes* as being largely retained in the Gulf of Maine, whereas the majority of fish larvae tended to drift out with the near-surface anticyclonic circulation. He attributes the retention of the *Sebastes* larvae to the location of the larval area in the northern part of the Gulf, and to the comparatively deep habitat which they occupy, which protects them from the superficial currents.

Particular behavioural patterns may hold for estuarine zooplankton to enable species, both holo- and meroplankton, to maintain a population against the constant flushing out to sea, or to return a substantial proportion of the population from the more open sea to the estuary. For the cladoceran, *Podon polyphemoides*, Bosch and Taylor (1973) suggest that, as for many other estuarine plankton, a change in level occurs. The cladocerans are mostly in the upper 4-m layer by day and, assuming a typical two-layer estuarine circulation, will be mainly transported seawards. At night they are in the deeper strata, probably due to passive sinking, and thus are returned to some extent by the inwardly-flowing, more saline deep layer. Changes in the level of a typical estuarine copepod, *Acartia tonsa*, in relation to different developmental stages, also appear to be responsible for retaining a proper proportion of the population from being flushed to sea (cf. Chapter 5).

The studies of Grice and Gibson (1975, 1977) have also confirmed the existence of overwintering eggs of some neritic copepods (e.g. *Labidocera aestiva*, *Pontella meadi*) which remain in the bottom deposit in shallow waters and hatch when conditions are favourable, giving rise to a new population. Kasahara, Uye and Onbe (1974) and Kasahara and Uye (1979) cite six calanoid species which produce "resting eggs" found in sediments of the Seto Inland Sea. The eggs of warm temperature species, e.g. *Tortanus forcipatus*, *Acartia erythraea* and *Calanopia thompsoni*, were most abundant in October–November when the animals declined in numbers in the plankton hauls, while the reverse was true for the temperate and winter plankton member *Acartia clausi* for which large numbers ($3.4 \times 10^6/m^2$) of resting eggs occurred in June. Madhupratrap and Haridas (1975) have similarly found resting eggs of copepods in the shallow brackish areas of the Cochin Backwater. If such resting eggs occur generally in estuaries and can accumulate in the bottom mud, they can be an important factor in the renewal of the copepod population of estuaries, especially in those subject to strong seasonal changes. Onbe (1977) has already conclusively demonstrated the presence of large numbers of resting eggs of the cladoceran *Penilia* in shallow waters. Presumably other cladocerans will use overwintering eggs for the re-establishment of populations. Other mechanisms may be employed; for example, some hydromedusans have a resting stage during the monsoon period at Cochin, but rapidly produce generations of medusae when conditions become favourable.

The adults or advanced developmental stages of some estuarine zooplankton species maintain a portion of the population against flushing by sheltering along the sides of the estuary in topographic irregularities, particularly in deeper pockets out of the main stream, so that breeding populations can be re-established (cf. Bakker and de Pauw, 1975). In any event, during the time when extensive populations of zooplankton are present in an estuary, there must be some loss to sea, and a balance is struck between this reduction and the breeding rate (for holoplankton) or spawning intensity (for meroplankton).

Some meroplankton (e.g. oyster larvae) make use of a behaviour pattern which, it is believed, assists retention inside an estuary. The larvae drop to the bottom on the ebb tide and become relatively inactive, so largely maintaining their position. On the flood tide however, they become active, rise in the water, and are carried landwards to some extent in the deeper inflow.

Changes in vertical distribution and swimming activity in some larval species may augment retention. Bousfield (1954a, b) demonstrated an effective pattern for cirripede larvae, mainly *Balanus improvisus*, for returning larvae in an estuary in eastern Canada. The early nauplii (N I and II) were found in great density in the estuary and at the head of the bay near the area of release. The different naupliar stages, however, exhibited a difference in vertical distribution. Young nauplii (I and II) were mostly in the upper 2 m; N III and IV were at 2.5–3.0 m, corresponding to the approximate level of no motion in the two-layered estuarine circulation; older nauplii (V and VI) were deeper (4 m), with some Nauplius VI accompanying the cyprid stage at levels exceeding 5 m.

Although from the exchange ratio in the estuary, about 15% of the 10,000 larvae produced per adult might be expected to be flushed out to sea at each tide—so that after the 18 days of larval life, only about 0.3% would have been retained inside the estuary—a much greater proportion, some 10% of the larval population, remained. Predation losses have not been included in calculating the proportion of larvae retained. The pattern of estuarine circulation as related to the larval behaviour appears to have been responsible for the much greater retention. The early larvae were transported seawards by the residual surface current and were found mainly along the southern side, near the river exit to the bay. Nauplius III and IV stages drifted progressively seawards, though less rapidly, mainly along the southern side, but N V larvae inhabiting the deeper layers were partly returned to near the river exit and were clearly in mid-channel. The N VI centre of distribution was even more landward and was on the northern side. Thus the young nauplii in the surface waters on the south side merge into the plankton of the outer bay at about mid-depth but are then returned to the estuary as older larvae ready for metamorphosis, approximately to the centre of the original population.

Some larvae of benthic invertebrates are swept away from continental shelf areas into the open sea. The great majority of such "strays" die, mostly from predation, but their transport to the ocean has led them away from their effective and viable settlement area. Those with a relatively long larval life may, however, have some chance of survival should they reach a suitable shallower region. Moreover, larvae from a variety of invertebrate benthic taxa can prolong their larval existence to some extent if conditions for metamorphosis are unfavourable. These matters are discussed in Chapter 4.

Neuston

A particular assemblage of zooplankton is now well recognized as being closely associated with the immediate surface film in the ocean. This community, known as the neuston, has its true members which live permanently in this habitat, but numerous other planktonic animals may also be found temporarily at the surface, usually migrating there at night. Some authorities class these temporary members of the neuston as the "facultative" neuston. Larger animals such as fish may also come up to the surface as a transient population. The facultative neuston will not be discussed in any detail.

Nets which have been specially designed to skim the ocean surface may catch animals which, although aquatic, live with their bodies simultaneously in air and water. These animals form the pleuston. Zaitsev (1971) differentiates between the neuston and the pleuston, while admitting that the distinction may be somewhat blurred. Both populations are associated with the surface film. That part of a pleuston animal which projects above the water surface can withstand prolonged desiccation and exposure to direct sunlight. Pleuston organisms tend to be of medium to large size and are usually dispersed by wind. Banse (1975) has reviewed the somewhat confused terminology, and has proposed the term hyponeuston for those organisms beneath, but *attached* to, the surface film. Among the few marine representatives, examples might be quoted from the Craspedophyceae (cf. Volume I). Banse further suggests that animals not attached but living within the uppermost decimeter be termed endopleuston. Zaitsev's broad distinction will, however, be retained in the following general description.

Pleuston animals are usually characterized by some sort of float; familiar examples are *Physalia*, *Velella* and *Porpita*. Apart from these siphonophores, *Ianthina* (a prosobranch mollusc) with its float of bubbles, and the nudibranch, *Glaucus*, which apparently contains bubbles of gas within its body cavity, might almost also be regarded as pleuston animals. David (1965) lists free floating anemones in the pleuston; these are members of the Minyadidae, a tropical family where the pedal disc is modified as a float (cf. Hyman, 1940). Possibly some of the stalked barnacles, which make a bubble float apparently surrounding some sort of foreign fragment, should be included in the pleuston, but precise limitation is obviously difficult.

Some surface organisms may make use of the pleuston as a source of food, or as a substratum. Laursen (1953) found that *Ianthina* would feed on *Velella* and other siphonophores, but stomach contents indicated that the gastropod would also prey on copepods and fairly generally on zooplankton. Bayer (1963) showed that *Velella* and *Porpita* were regularly eaten by *Ianthina*; *Physalia* was also taken. The nudibranch *Fiona* was also observed to feed on *Velella*; Bayer found that *Fiona* characteristically maintained a position on top of the siphonophore. Hyman (1951) quotes all three pelagic nudibranchs (*Fiona*, *Glaucus* and *Phyllirhoe*) as feeding on siphonophores. The diet of the three siphonophores *Velella*, *Porpita* and *Physalia* has also been investigated by Bieri (1966, 1970) (cf. Chapter 6).

According to Zaitsev, true neuston organisms, as distinct from the pleuston, may be aerial forms living on the air side of the surface film (the epineuston) or aquatic organisms inhabiting the water side of the film (the hyponeuston). Neuston organisms are of small to medium size and include organisms of all taxa from algae, bacteria and protozoans to fish fry. The epineuston is extremely limited taxonomically, being represented almost solely by the insect *Halobates*. *Halobates* appears to be very buoyant because of the film of air trapped on the hairs of the body (David, 1965). This insect feeds on the body fluids of animals in the surface film, such as the cnidarians of the pleuston (Cheng, 1973). *Halobates* may be eaten by small fish and, according to Cheng, by certain seabirds.

The assemblage of organisms inhabiting the water side of the surface film—the hyponeuston—is marked by a far greater array of taxa. Apart from representatives of most major zooplankton groups, micro-organisms are present. Autotrophic organisms are apparently less important and abundant than heterotrophic forms, though more recent observations of such algal groups as the Pterospermaceae, which inhabit the surface film (cf. Volume I, Chapter 4), indicate that any algae in the neuston are lost with the usual collecting techniques. A few algae which have been identified, including blue-green forms, flagellates and occasionally dinoflagellates, appear to be associated with foam at the surface. They tend to be nanoplankton forms, presumably remarkably adapted to high surface light intensity. The bacteria present, according to Zaitsev, are especially plentiful in the uppermost 2 cm in the Black Sea. At a depth of only 50 cm, for example, the density is orders of magnitude lower. What is generally termed the "bacterioneuston" is perhaps one hundred times richer than the

Table 1.1. Density of tintinnids (specimens/m³) of various areas at the surface of the Black Sea (Zaitsev, 1971)

	Species	Tube sampler 0–5 cm layer	Juday net 0–10 m layer
	Tintinnopsis karajacensis	82,000	0
	T. beroidea	4,000	0
	T. tubulosa	6,000	0
(a)	*Helicostomella subulata*	22,000	52
	Stenosemella ventricosa	8,000	17
	Coxliella helix	34,000	65
	C. annulata	0	1
	Stenosemella ventricosa	16,000	108
	Coxliella helix	16,000	97
	Helicostomella subulata	4,000	0
(b)	*Favella ehrenbergii*	2,000	0
	Tintinnopsis campanula	12,000	0
	T. compressa	2,000	0
	T. meunieri	0	1
	Favella ehrenbergii	26,000	119
	Stenosemella ventricosa	44,000	50
	Coxliella helix	50,000	32
(c)	*Helicostomella subulata*	32,000	0
	Stenosemella nucula	10,000	0
	Metacylis mereschkovskii	4,000	0
	M. ehrenbergii	2,000	0

bacterioplankton in the layers beneath (cf. also Tsyban, 1971, for Black Sea and Pacific Ocean). The species diversity of micro-organisms is also much greater in the 0–2-cm layer in the Black Sea.

Of the animal population, protozoans are an important group in the Black Sea, and the bacteria with protozoans probably form a major food source for the neuston invertebrates. *Noctiluca* can be especially plentiful in the upper 5 cm (>15,000/m^3), and tintinnids reach their peak abundance and variety of species in the same layer (cf. Tables 1.1, 1.2). Other protozoans include radiolarians (*Spumellaria*) and for-

Table 1.2. Abundance of tintinnids in various layers of the Black Sea at various times of the year (Zaitsev, 1971)

Layer (cm)	Species	Specimens
	August 1963	
0–5	*Tintinnopsis meunieri*	276
	T. tubulosa	44
	Stenosemella ventricosa	52
5–25	*T. meunieri*	3
	T. tubulosa	1
	S. ventricosa	44
	Coxliella helix	70
25–45	*S. ventricosa*	10
45–65	*S. ventricosa*	19
	July 1966	
0–5	*T. urnula*	16
	T. cylindrica	240
	T. kofoidi	28
	T. meunieri	84
	C. helix	164
	Helicostomella subulata	104
5–25	*T. tubulosa*	22
	T. meunieri	9
	T. kofoidi	3
	H. subulata	16
25–45	*T. cylindrica*	18
	T. beroidea	1
45–65	*T. cylindrica*	9
	T. meunieri	17
	T. tubulosa	1
	C. helix	8
	H. subulata	5
	October 1966	
0–5	*T. meunieri*	52
	T. beroidea	28
	T. karajacensis	8
	C. helix	102
5–25	*T. karajacensis*	28
	T. meunieri	15
	T. compressa	1
25–45	*T. meunieri*	47
	T. tubulosa	2
	C. helix	11
45–65	*T. karajacensis*	6

aminiferans; a few holotrich ciliates and Sarcodina have also been reported. In the Black Sea, rotifers form a major constituent of the metazoan population in the neuston. Numerous meroplanktonic larvae (larval polychaetes; molluscan veligers; cirripede nauplii; large numbers of zoeas, megalopas and other decapod larvae, euphausiid and stomatopod larvae) are usually present in the hyponeuston.

Of other larval forms, fish eggs and fry are an important neuston component. The eggs of certain families of fishes (Engraulidae, Mugilidae, Carangidae, Mullidae, Callionymidae, Pomatomidae, Pleuronectidae, Soleidae) exhibit very high buoyancy so that the eggs are found in the extreme upper layer. Even the larvae of many of these families tend to congregate in the upper 0–5-cm layer. Fish eggs which are typical of the neuston are said to develop more rapidly than other types of eggs from the same region. According to Zaitsev, the times of major spawning bursts avoid periods of stormy weather. Fish larvae from other families (Blennidae, Bellonidae, Exocoetidae, Atherinidae, Labridae, Ammodytidae, Gobiidae, Syngnathidae) add to the neuston population. Although fish eggs tend to be dispersed by moderate waves, the fry appear to maintain their level, and the eggs also reform just below the surface when calm conditions recur. Figure 1.1 (Zaitsev, 1971) illustrates the position of anchovy and mullet eggs at the surface. Fish larvae developing in the hyponeuston almost uniformly have air bladders even though these are lacking in the adults.

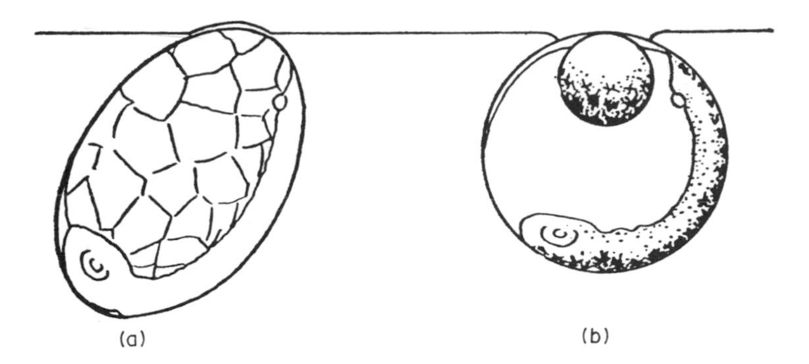

Fig. 1.1. Position of (a) anchovy, (b) mullet eggs at sea surface (Zaitsev, 1971).

Certain planktonic copepods, euphausiids and hyperiids which perform diurnal migrations can become transient members of the neuston. In the open ocean, *Planes minutus* and *Portunus portunus* can also be present at the surface, associated with floating *Sargassum*. In shallow waters, some species of normally benthic amphipods, cumaceans, isopods, mysids and occasionally other taxa, can migrate to the extreme surface and thus add to the transient hyponeuston.

Generally, the neuston of higher latitudes is less well known and is less abundant than in warm oceans. Zaitsev (Table 1.3) found that in summer in the northern North Pacific quite substantial populations of copepods (especially *Calanus tonsus*), certain hyperiids and other amphipods, as well as young stages of crustacean meroplanktonic species may be in the uppermost 5 cm. Only one pontellid copepod was present. Most of the plankton, however, including the copepods, is richer in the 5–25 cm layer than at

Table 1.3. Composition and average density (specimens/m³) of organisms at the surface of the Bering Sea and the neighboring Pacific waters (Zaitsev, 1971)

Organism	Microhorizon (cm) 0–5	5–25
Chaetognatha	7.90	9.20
Pteropoda	3.71	4.57
Cephalopoda, juv.	0.39	0.15
Calanus tonsus	77.30	123.20
C. cristatus	13.90	19.10
Eucalanus bungii	4.50	6.90
Epilabidocera amphitrites	4.80	0.12
Decapoda (crustacea), zoea	2.80	8.20
Brachyura, megalopa	5.60	1.10
Cumacea	9.10	1.20
Amphipoda, without Hyperiidae	18.60	1.60
Hyperiidae	362.40	140.50
Isopoda	3.10	0.60
Euphausiacea	3.30	2.10
Pisces, larvae	7.60	1.40

the immediate surface. Abundance is greatly affected by season, especially surface temperature.

While neritic neustonic species including meroplanktonic forms may be abundant in shallow waters, there is some indication of a decrease in the density of neuston with nearness to shore. A number of holoplankton animals are moreover characteristic of the permanent neuston over wide areas of ocean. Small copepods (e.g. *Oithona minuta*, *Centropages ponticus*, *Acartia clausi*, *Paracalanus parvus*), often mainly nauplius stages, are typical, together with *Sapphirina* in warmer waters; cladocerans, especially *Evadne tergestina*, may be abundant. However, probably the most typical holoplanktonic representatives are the pontellid copepods (Champalbert, 1971). Species of *Pontella*, *Pontellopsis* and *Labidocera* are particularly characteristic of the neuston of warm and warm/temperate waters, with *Anomalocera patersoni* and *Epilabidocera amphitrites* typical of higher latitudes.

Heinrich (1971) commented on the relative abundance of pontellid copepods in the neuston of the tropical Pacific. She found that certain species were typical of particular water masses and that species diversity varied with different communities. The highly productive oceanic equatorial area had a low diversity, whereas diversity was highest off New Guinea (a moderately productive area) and in the Pacific northern central area, a region of low production. Apart from the relative abundance of pontellids and the importance of fish eggs and larvae and of crustacean larvae in the neuston, David (1965) lists a few decapods including *Parapeneus longipes*, some salps, squids such as *Teleotheuthis* and certain fish and medusae as representatives of the hyponeuston.

There are indications of certain characteristic behaviour patterns amongst neuston animals. Hartman (1972) describes observations on sixteen species of neuston animals which he obtained alive in Mediterranean waters. They included the young of *Belone*, larvae of two other fishes, *Blennius* and *Atherina*, crustacean larvae including *Maia*, *Nyctiphanes* and two species of stomatopods, a species of the copepod *Sapphirina* and several pontellid copepods: *Anomalocera* and species of *Pontella*, *Pontellopsis* and

Labidocera. During day-time the animals obviously "preferred" the extreme upper layer down to a few centimetres depth. He also noted the characteristic jumping out of the water which is typical of many pontellids and possibly represents an escape reaction from predators. Many of the species actively glided under the surface, even knocking against the surface film. They would also exhibit "fright reaction" by fleeing to the surface and only very occasionally would they leave the immediate surface for a deeper layer. Some of the animals fed at the surface: *Labidocera* was observed to catch chaetognaths and copepods; *Pontella mediterranea* caught fish larvae and copepods, and one of the fishes was observed feeding on neuston. David (1965) also observed the surface-living fish, *Mupus*, feeding apparently preferentially on small animals trapped in the surface film. He believes, however, that some neuston animals appear to be herbivorous. Presumably they live on the comparatively few algae which are restricted to the immediate surface layer.

Hempel and Weikert (1972) emphasize, with regard to the neuston of the sub-tropical and boreal north-east Atlantic, that very little phytoplankton was present in the immediate surface layer. An indication of the thinness of that phytoplankton layer is their observation that even more phytoplankton was captured in the lower neuston net than in the surface one. Hempel and Weikert thus claim that the neuston is mainly omnivorous or carnivorous and suggest that its food, other than phytoplankton, might include buoyant plankton bodies as well as floating insects and fish scales in addition to bacteria. In shallow seas (e.g. the Black Sea) there is a considerably richer potential food supply at the surface including pollen, seeds, insects and foam bubble aggregates with bacteria colonies. In the open sea, terrigenous material is little represented and despite some greater concentration of food particles in the surface skin, the food supply at the surface will be very limited, especially in clear tropical waters. However, the surface habitat provides excellent light for visual feeders and the bacteria in the neuston may be grazed on by copepods and other "particle" feeders.

In these investigations in the sub-tropical Atlantic the density of neuston organisms was low (fish larvae, for example, averaged only 10/100 m^3). Species diversity was also low, though higher in sub-tropical than in temperate neuston. The vast majority (94 %) of the invertebrate neuston fauna were crustaceans, of which about half were copepods, especially pontellids, and one third ostracods, with *Conchoecia spinirostris* especially abundant. Some molluscs (heteropods, pteropods and a few cephalopods) occurred. The "facultative neuston", which concentrated at the surface mainly at night because of its vertical migration, was considerably richer and included some pelagic mysids and isopods. Though post-larval and juvenile fish occurred in the neuston, eggs and young larvae of fishes were generally less plentiful in the immediate surface. Considerable numbers of these, however, apparently occurred right at the surface in the turbid waters of the West Africa coastal upwelling area. About 30 fish species and some unidentified myctophids were found—not a high diversity for a sub-tropical region. The total number of species in the neuston was inversely related to the biomass. Generally in the warmer seas, the immediate surface was less rich than the microlayer beneath, which varied from 20 to 45 cm depth, but fish post-larvae were more abundant at the surface. Young stages of invertebrates, however, preferred the layers beneath the surface.

Hartmann (1970), in a detailed study of neuston of the north-east sub-tropical Atlantic, found that larvae of *Scombresox saurus*, about 6–29 mm length, were abundant in the immediate surface throughout the whole 24 hours and were much more plentiful in the 0–10-cm microlayer than in succeeding strata. The eggs and very young larvae were not so abundantly represented. The eggs and very young stages possibly cannot withstand the intense solar radiation but the young larvae may benefit by avoiding predators by living right at the surface. Older larvae (<50 mm length) were less plentiful at the surface and migrated there at night. Hartmann found that the food of *Scombresox* larvae could be classified into four categories according to the growth stages of the fish larvae: larvae up to 15 mm, protozoan–gastropod; 16 to 30 mm, gastropod–entomostracans; 31 to 45 mm, entomostracans; 46 to 113 mm, crustaceans. The crustacean food included cladocerans, calanoids, ostracods and the cyclopoids *Sapphirina*, *Oncaea* and *Corycaeus*.

Parin (1967) also described numerous larvae of fishes caught in the neuston in open waters of the tropical Indian and Pacific Oceans. Apart from myctophids, where the adults migrate from deeper layers, the other families represented were near-surface oceanic fishes. According to Parin, all the larvae were usually more frequent in the hyponeuston by day than by night. He attributes this to active visual feeding during daylight hours. Kawaguchi (1969) found certain myctophid fish migrated to the surface at night. Tuna larvae have been reported as comparatively rare in the neuston during the day by previous workers, but were said to come up possibly to the immediate surface at night. In a later study in the tropical Pacific, Parin, Gorbunova and Chuvasov (1972) collected numbers of larvae and juvenile fish with a neuston trawl. Larvae and juveniles (except myctophids) were more numerous in the neuston by day than at night, although there was a greater variety of species at night. For four stations they quoted densities of larvae (excluding myctophids) ranging from 5 to 85 individuals/1000 m^3 by day and from 1 to 60 individuals/1000 m^3 by night (cf. density quoted by Hempel and Weikert, 1972, for the Atlantic). Further details of the distribution of fish larvae are included in Chapter 2.

Of the advantages and disadvantages of neustonic existence, reduction of predation pressure by day may be a major benefit, though at night predation is markedly increased. A disadvantage would appear to be the little food available to particle feeders, despite the reported aggregation of micro-organisms. Bacteria and aggregates may not compensate for the paucity of phytoplankton, but in addition to organic aggregates, pteropod mucus webs, old larvacean houses etc., can be used as platforms as well as providing nutrition. However, Harvey (1971) claims that larger quantities of nanoplankton, surface active materials and non-living particulate matter were found in surface layers of the sea than at 10 cm depth. Another problem facing the permanent animal neuston is the required ability to withstand the intense radiation at the surface. Possibly related to this problem is the repeated observation on the colour of the neuston fauna. A blue coloration is exceedingly common, whilst not normally observed in other planktonic animals.

Herring (1967, 1977) emphasizes that the neuston, particularly of warmer seas, is not transparent but is predominantly blue to purple due to the possession of carotenoid proteins apparently similar to the pigment of the lobster. He has now shown that the carotenoid portion of the pigment is derived from astaxanthin and occurs in pontellids such as *Labidocera* and the cnidarians *Velella* and *Porpita*. However, the blue colour of *Physalia* is due to a bilin–protein complex and the pink and green colours seen in the float may arise from free bilins. (Bonnett, Head and Herring, 1979)

Herring (1972) suggests this may be the result of differences in the diet of the siphonophores; *Physalia* is mainly a fish eater and would derive bile pigments from its food, while the supply of astaxanthin to the two chondrophores could be provided by their mainly crustacean diet, high in these carotenoids. David (1965, 1967) points out that some colours, possibly for instance in *Sapphirina*, may not be due entirely to pigment, but to interference.

Herring's analysis, which is confined to pigment either in the cuticle (of crustaceans) or in the tissue beneath, shows that not only *Velella* and *Porpita*, but also *Sagitta robusta* and *S. regularis* (from the Indian Ocean) and some salps in the neuston, all show some degree of blue coloration; the blue carotenoid pigment of *Salpa cylindrica* is not astaxanthin but is similar to fucoxanthin (Herring, 1978). The pontellid copepods exhibit a particular preponderance of blue colour. According to Herring, every species of *Labidocera* and *Pontella* which was examined alive showed a pale blue colour in the chitin and a deeper colour in the hypodermis. Other copepods such as *Isias*, *Calanus arcuicornis*, *Temora discaudata* and species of *Acartia* showed some colour. Certain cyclopods (*Corycaeus*, *Corycella*, *Sapphirina*), as well as zoeas and other larvae of decapods also included blue coloration, with a smaller number of red chromatophores. The neuston squid, *Onychia caribaea*, revealed a blue structural coloration, as did some fish larvae and fishes. For *Ianthina*, *Atlanta*, *Creseis acicula* and *Euclio* among the neuston molluscs, however, while purple pigment was apparent, this was not a carotenoid protein.

The function of the blue pigment is extremely difficult to identify. It may have a protective function since the animals will be exposed to the intense rays of the sun, including ultra-violet radiation. However, Herring suggests that the spectral composition of the pigment does not seem to give any obvious protection. Possibly either the carotenoid or the protein part of the complex may be more protected by occurring in a compound in this form. Another often repeated suggestion concerns cryptic coloration, i.e. that animals with this colour are to some extent less visible and therefore less vulnerable. Though such ideas have been regarded as tentative, the work of Hairston (1975) showed that a lake copepod having high concentrations of carotenoids had a markedly greater survival rate than those with less pigment. He suggests that the carotenoid pigments in *Diaptomus* have both a photoprotective and a cryptic function. Other species rely on counter-shading (*Glaucus*) or reflection (*Idothea*).

The neuston, apart from serving as a horizon where many fish larvae feed, serves as a feeding layer for a number of adult pelagic fish, though they feed also in the deeper strata. Zaitsev (1971) also points to a number of birds (skimmers, shearwaters, fulmars, petrels, kittiwakes) which apparently skim the surface when feeding or pick particles from the surface, even though some, such as kittiwakes, also dive for food beneath the surface.

The Standing Crop of Zooplankton—Sampling Methods

The zooplankton is characterized not only by having representatives of almost every taxon of the animal kingdom but by including a very great range in body size. The well known classification of Dussart (cf. Table 1.4), one of several which divides the zooplankton into size groups, shows that zooplankton organisms may range from a few microns in diameter for instance small flagellates and ciliates similar in size to the

Table 1.4. Classification of zooplankton size groups (Dussart, 1965)

Zooplankton	Size range (μm)
Nanoplankton (*sensu lato*)	
—nanoplankton (*sensu stricto*)	< 2
—ultra microplankton	2–20
"Net" plankton (filterable plankton)	
—microplankton	20–200
—mesoplankton	200–2000
—megaplankton	> 2000

nanoplankton of the primary producers but lacking chlorophyll, to animals such as large jellyfishes which may measure a metre or more across. Apart from these exceptionally large medusae, some other zooplankton—certain ctenophores, siphonophores, medusae and a few deeper-living mysids, decapods, euphausiids and amphipods, amongst crustaceans—may be of appreciable size.

With such a range of size in the zooplankton, it is not possible for any single sampler to estimate the total plankton quantitatively. Almost all the sampling methods rely upon the catching power of relatively fine mesh nets, originally made of bolting silk, now usually made of nylon or other synthetic material. Somewhat coarser material is used for young fish trawls, employed for sampling certain types of the larger zooplankton (e.g. krill). Mesh size must be a critical factor in selecting organisms. While the finer silk nets may capture fairly effectively the young stages and eggs of planktonic animals, as well as some of the smaller appendicularians and similar animals, such nets filter too slowly and filter too small a volume of water to sample quantitatively the larger and more active zooplankton (e.g. euphausiids, chaetognaths and large pelagic decapods). On the other hand, comparatively large and coarse nets, suitable for sampling what is usually called the macroplankton, must allow a large though unknown fraction of the small microplankton to pass through the meshes. Between the finest and the coarsest plankton net there is a great variety in size and mesh. In attempting to estimate total plankton, a combination of nets of different mesh is frequently employed, but estimating the extent of overlap of the size ranges of zooplankton captured is extremely difficult. Moreover, the quantity of zooplankton which according to its size should pass through a net is variable, depending on many factors—elasticity of the net, speed of towing, the amount of clogging, the shape of the animals and other characteristic features of the species such as possession of spines and projecting appendages.

The desire to sample a specific body of water, either in terms of a section of the vertical column or a particular longitudinal column at some required depth in a horizontal tow, introduced the need for closing devices. Thus, early on, in addition to the appearance of different designs of plankton net for improving catching efficiency, closing nets were devised, using throttling mechanisms or lids operated by messengers—for example, the Apstein and Nansen closing nets. The use of vertically hauled closing nets greatly increased our knowledge of regional and seasonal quantitative variations in the zooplankton. However, with the design of different nets, complications arose concerning the precise volume of water filtered in relation to mesh size and speed of tow, the effect of closing devices and the depth of sampling. Clogging, which is itself variable with space and time, must be a factor affecting the volume filtered. Avoidance of the plankton net is another variable which is difficult to quantify.

It is obvious that with nets of comparatively small diameters—especially if towed slowly—many of the larger more active zooplankton can escape capture, and the design of large coarse mesh nets, towed at comparatively high speeds, can capture such animals more effectively, it is extremely difficult to assess how far a standard type of plankton net, towed normally, permits even less active zooplankton to escape. The degree of rarity of particular zooplankton species must be an important factor in choosing net size. In broad terms, the less common the animal, the greater the quantity of water that must be sampled, but this becomes impracticable unless the species is also fairly large and high speed coarse nets can be employed.

Even the very finest nets cannot sample the very smallest organisms. Samples may be collected with a large water bottle, as with nanoplankton primary producers (cf. Volume 1), and sedimentation techniques employed, followed by counting of the animals. The method is time-consuming and laborious. An alternative method for sampling the very small zooplankton is to pump monitored quantities of water from known depths and to filter the water through a very fine mesh gauze. Investigations, particularly by Beers and his colleagues in California (e.g. Beers and Stewart, 1969), demonstrated the very large numbers of some protozoans which can be recovered by such collecting methods. As Beers himself points out, however, even with the finest mesh (usually $< 30\,\mu m$), some organisms of very small size pass through the mesh in unknown quantity, their numbers depending partly on their size, shape, and compressibility and partly on the degree of clogging of the gauze used.

This so-called pump-and-hose method can be employed also for more general sampling of zooplankton, apart from its use for quantitative determinations of the very small organisms. Water can be pumped and measured by a pump situated on the deck of the ship or by the use of a submersible pump. Designs of pumping apparatus are now available which to a large extent avoid damage to organisms. Icanberry and Richardson (1973) describe a filter-pump system which, by comparison with plankton net sampling, appears to be effective for quantitatively sampling the zooplankton. In any pump method, speed of pumping is important. It is also uncertain how far the more active organisms avoid the mouth of the pump. Moreover, the method is obviously limited to comparatively shallow depths. The use of a very long hose subjected to sub-surface and deeper currents leads to obvious difficulties of operation, as also does the movement of the ship under open sea conditions. The pump method may, however, be employed as a check on the operation of the filtering power of nets. It is useful also for shallow inshore sampling, provided it is recognized that the larger more active animals will probably not be taken, certainly not quantitatively. In reviewing some of the problems related to sampling the zooplankton, substantial reference has been made to the UNESCO Monograph (1968).

One of the chief problems in quantitative sampling is to obtain accurate knowledge of the amount of water passing through a net. The quantity is affected by many factors—shape of net, material, speed of towing and so forth. Clogging is again a factor, as with the problem of size of retention of plankton, especially as its effect can become increasingly, though unpredictably, greater as towing continues and more and more plankton organisms are caught. Quantitative assessment is especially problematical if sampling coincides with a period of high production of phytoplankton. An early modification of the plankton net to attempt to standardize filtration was the introduction of a non-filtering cylinder in front of the filtering conical net, but despite such modifications the need for measuring actual flow through the net became imperative.

A number of flow meters have been designed. They should be robust, but small enough so as not to interfere with the flow of water through the net. In some sampling apparatus, such as the Gulf III sampler, the filtering equipment is housed inside a container and the water outflow measured beyond the net. Measurement of the quantity of water filtered was achieved with the Clarke–Bumpus plankton sampler. In effect, this consists of a cylindrical metal non-filtering section, which includes a rotating impeller acting as flow meter, and a filtering nylon gauze net attached to the metal cylinder. The net section is shaped as a long cone, is relatively small and interchangeable, so that the apparatus can be used with different mesh sizes. Alternatively, a number of samplers of different meshes can be used simultaneously, sometimes at different depths.

Even when a suitable flow meter has been designed, its position in the mouth of the net can be critical. In latter years, the use of experimental chambers and flumes where standard flow velocities can be accurately monitored has achieved much in permitting more accurate assessment of the performance of metered nets. Mahnken and Jossi (1967), for example, describe tests in an experimental flume in which they measured the distribution of flow across the mouth of a typical conical plankton net 1 m in diameter. Nets of three different mesh sizes were tested: 0.281, 0.221 and 0.111 mm. With bridled nets, flow was not uniform. Maximal velocity was found near the edge of the mouth and minimum velocity at the centre of the net. Unbridled nets tested in the flume showed comparatively uniform velocity of flow across the net mouth. The filtration efficiency of the net was relatively high for all the mesh sizes used, provided normal towing speeds were maintained. The precise position of the flow meter in a normal bridled net is thus critical. Flow data may be considerably affected by the position of the meter and flow data are recognized as often difficult to interpret.

Mahnken and Jossi quote earlier work from Nishizawa and Anraku which indicated also that there was an apparently smaller water flow through a conical net if the flow meter was mounted more or less centrally. Tranter and Smith (1968) also comment on the great importance of positioning of flow meters and the marked variation in flow with varying position from the centre. During their tests in the experimental flume, Mahnken and Jossi observed that the first part of a typical conical net formed a rather loose billowy cylinder, whereas the hindpart took up the shape of a more precise cone. They suggested that these two parts represented low and high filtration rate sections, respectively. At high channel velocities, equivalent to high towing speeds, they suggest that any net cannot filter completely and some water must be rejected at the mouth.

Tranter and Smith (1968) examined the filtration performance of plankton nets. Towing a net disturbs the water immediately in front of the entrance to the net as well as to some extent upstream; there are also disturbances due to the towing bridles and other obstructions. Some of the turbulence in front of a net can be reduced by towing from the top of the net and adding a type of depressor below.

With a normally towed net there is an acceleration front arising from the resistance of the net to the filtration through the gauze. Such disturbances, though they may extend only a short distance, are strong at the periphery and could warn animals of the approach of a net, thus leading to avoidance. The stream line ahead of a net should be modified as little as possible by the towing. Tranter and Smith, examining flow patterns with simple basic types of plankton net, demonstrated effects on the stream line of the common practice of using a mouth-reducing cone in front of the net. The effect is dependent on whether the cone is made of non-porous or porous material. Although water is accelerated and displaced at the mouth of a net as a consequence of the resistance to flow, the mouth reduction cone, besides increasing filtration efficiency, creates a low pressure area so that a column of water wider than the reduced mouth is drawn into the sampler.

Nets should be designed to get a fairly even flow of water along the length of the filtration area. Since organisms caught by the filtration gauze could be damaged by the pressure drop across the meshes, lowering the speed of tow or reducing the mouth area might improve the condition of the catch. Mouth-reducing cones, if used, must be of non-porous material. Any suggested change in speed of tow must be related to the problem of avoidance and escape by organisms. Although there has been a tendency to reduce the size of opening, permitting some increase in towing speed, the smaller opening itself raises particular problems.

The efficiency of filtration of a net depends on its porosity and on the surface area of filtration. The ratio between the porous (or filtration) area of the net and the area of the mouth, often termed the open area ratio, is critical in filtration efficiency. Below a ratio of about three, the efficiency declines progressively. For simple conical nets, however, the filtration efficiency, in the absence of other complicating factors, is fairly high—for open area ratios exceeding three it approximates to 85 %, and for ratios exceeding five, the filtration efficiency may be as high as 95 %.

While mesh size is so closely related to size of organisms retained and to filtration efficiency, mesh size itself is subject to several factors causing variability. Although metal gauzes may be used in rigid nets, particularly those enclosed in some type of container, silk, nylon, or other synthetic fibre is used for the great majority of nets and its quality is all-important. Type of weave and the ability to withstand wear is also important. Worn fabric may cause organisms to stick against the meshes so that the catch is not properly recovered. Distortion of the mesh is obviously undesirable. Imperfect net pattern and incorrect towing speed can cause distortion, all of which can lead to variability in catching power. When a new net is used and the catch subsequently recovered, a certain amount of organic material is fixed on the meshes. This may be considerably reduced by careful washing, but there is a tendency for accumulation of material with age. Older nets also may show frayed fabric. The smoothness, roughness or stickiness of the surface of an animal in relation to the net surface also affects retention.

The avoidance of a net by plankton organisms may be in response to visual stimuli, to pressure changes, acceleration or turbulence, or to actual contact with the towing apparatus. Various species of zooplankton also respond in different ways: by varying their pattern of swimming, by acceleration or cessation of movement or by change of direction. The most obvious factors related to net avoidance are size of net and speed of towing. Equipment concerned with towing should be reduced as far as possible and surfaces should not be shiny.

Apart from avoidance, with any net there must be loss of organisms through the meshes. Vannucci (1968) defines net mesh selection as the capacity of a net to select organisms from the population in the water that is passed through the net. But there is also some degree of loss of organisms of a size such that the smallest diameter is greater than the mesh size. This has been proved both in field and laboratory experiments. To some extent there may be active squeezing of living plankton through the net. Some experiments, for instance, have shown small differences in the amount of escape between living and dead copepods with the same mesh. More generally, however, escape is believed to be largely passive and the result of the pressure drop across the meshes, the bodies of plankton being to some degree compressible. Heron (1968) has argued, however, that with most gauzes there is considerable variability in mesh size and that escape of organisms is largely due to their being carried along the net until they find an aperture in the gauze (because of mesh variation) large enough for them to escape. Whatever the explanation, it is important to recognize in all quantitative sampling of zooplankton that some proportion of those organisms which are somewhat larger than the average mesh size will be lost.

Clogging may become a significant factor affecting filtration efficiency when sampling the smaller zooplankton, especially in the presence of dense phytoplankton. To some extent this can be monitored by using two flow meters, one inside and one mounted outside on the net. However, in view of the great importance of the position of flow meters in assessing flow, the proper positioning of the two flow meters is obviously critical. To some extent clogging can be partly reduced (cf. Clutter and Anraku, 1968) by including a porous cylindrical collar in front of the main filtering cone. Oscillatory movements of the collar occur during towing, probably caused by eddies, and this acts to some extent as a cleaning mechanism. A most important factor in avoiding or reducing clogging is to limit the distance of tow, but the volume of water filtered should not be too small for quantitative assessment to be made.

The depth of sampling must be known for most types of investigations. Somewhat crude estimates of the depth of tow may be made by relating the length of towing wire to the angle of tow. Difficulties occur with great lengths of towing wire, however, especially in the presence of strong currents. The determination of maximum depth of tow may be made by means of a Bourdon tube mounted in the tow net. With a type of recorder set up in addition to the tube, some trace of depth path is possible. Modern methods for determining depth more accurately and continuously usually involve telemetering, either acoustic or electrical.

The need to tow nets at specific depths and for definite periods of time not only demands satisfactory opening and closing devices, but also reliable triggering mechanisms. Although the well-known types of simple messenger are often satisfactory and multiple messengers may be used for taking a series of divided hauls, they are somewhat unreliable. Electrical triggering devices are often now employed. Since some closing devices were suspected of causing loss of a portion of the plankton catch near the front of the net, certain nets have devices for closure at the cod end. A net described by Foxton (1969) is of particular interest since the path to the cod end, in the form of an inverted Y, furnishes two cod ends which may be closed separately. This type of device is particularly useful for some of the larger plankton nets. Adaptations have been proposed to provide more than two chambers. The need to sample a series of strata in one haul is important, particularly in examining vertical distribution of zooplankton. Although a series of nets or of Clarke–Bumpus samplers operated by multiple messengers has been successfully employed, the apparatus can be difficult to operate. A device such as that of Bé (1962) in which a number of nets, previously folded and lowered together, may be subsequently opened and closed in turn at different depths is an interesting development for multiple sampling.

As Clutter and Anraku (1968) have repeated, there is no single instrument which can sample all the components of the zooplankton. Not only must the various taxa, which differ so much in size and mobility, be sampled adequately, for an investigation of the population age structure of a zooplankton species, but it is necessary to sample the different stages in development which also differ in size. The more rare, including the larger carnivorous zooplankton, must be quantitatively investigated, as well as the widespread more abundant herbivorous species. Error due to avoidance by the larger forms may become greater and might seriously affect estimates of abundance, patterns of vertical migration, indices of diversity and the spatial isolation of populations. Clutter and Anraku emphasize that avoidance has been proved by direct observation, by comparison of the catches of different types of nets, and by day and night sampling comparisons.

High speed samplers have been designed partly to sample the more rare and more mobile organisms as well as to effect a possible saving in ship time. It is still not clear whether such high speed samplers have better catching power. High speed samplers include the Gulf and Kinzer type, the Isaacs–Kidd mid-water trawl, and the Hardy Plankton Recorder. Such samplers may be of considerable size and weight. It is also necessary when towing at high speeds to ensure that the net is carried to the required depth without adding abnormally heavy weight. A number of depressors have been designed to assist towing at the required depth. A very useful catalogue is supplied by Tranter and Smith (1968) of types of nets, classified under low speed and high speed and in coarse, medium and fine gauze categories, with some indication of the type of investigation for which they have been successfully used. In selecting appropriate nets, however, Tranter and Smith emphasize that testing of all equipment is still essential.

The Continuous Plankton Recorder designed by Hardy has proved to be an excellent tool for sampling very large areas of ocean, with the advantage that it may be used by ships of opportunity. Apart from one or two obvious disadvantages such as the poorer condition of organisms captured and the shallow depth of sampling, a repeated criticism was the small size of mouth, possibly increasing the amount of selection. The modification by Longhurst, Reith, Bower and Seibert (1966), which consisted essentially of placing a fairly large, more or less normal pattern, plankton net in front of the Recorder, produced an instrument, the Longhurst–Hardy Plankton Recorder, which is widely used for plankton surveys. The apparatus can be towed horizontally, obliquely or vertically.

Haury (1973) pointed out that the residence time of animals in the Longhurst–Hardy Recorder net would be all-important in relation to their distribution on the gauze of the Recorder. Any permanent or temporary sticking of organisms against the net, or any build-up of plankton in front of the Recorder due to variation in flow, could influence the determination of zooplankton distribution. Haury injected plastic pellets, representing artificial zooplankton and also preserved *Calanus* adults (both dyed), into the net of a vertically hauled Recorder. Over 80 %, on average, of the injected plankton was recovered on the gauze. However, the residence time, (i.e. period for the first appearance of the dyed plankton) varied considerably (minimum *ca.* 13 sec; maximum

60 sec). Both temporary and more permanent sticking of plankton against the net was experienced, so that clusters of natural and of dyed plankton were washed onto the Recorder gauze. Intermittent holding back of the organisms could obviously register artificial patterns of abundance by the Recorder. While increased towing speed tended to reduce residence times, the increased filtration pressure could cause more permanent sticking of plankton against the net, and some animals might be lost from the mouth. With horizontally or obliquely towed Longhurst–Hardy Plankton Recorders, factors affecting flow through the net may act differently. Gravity might increase the clustering of animals, especially as some zooplankton (pteropods and ostracods) apparently cease swimming on capture and would sink.

In addition to errors due to the many factors involved in operating sampling nets, the natural variability of zooplankton, often referred to as "patchiness", must be responsible for much of the variation in net catches. Indeed, with sound sampling methods and in the absence of clogging, many planktonologists would accept that most of the variation between net catches is due to patchiness. Increasing the number of replicates when sampling should reduce error, but this demands additional time for taking samples (often not possible on oceanic cruises). Replicates also add enormously to the labour of counting zooplankton samples. Size of sample is an obvious factor in reducing variation due to patchiness. Wiebe (1971) studied the effects of the spatial structure of zooplankton on the error of replicate net hauls, using a computer model. Patches of variable size and number, variable position, and varying in density of organisms were simulated. The passage of a simulated net haul was examined and the volume of water filtered and the number of individuals captured was estimated for each patch encountered, the individuals in a patch being assumed to be randomly distributed. Wiebe's findings suggested strongly that increased size (diameter) of net as well as lengthening of tow both increased the precision of sampling. The length of haul was particularly important. Apparently this was not due only to the volume of water filtered, but was related to the number of patches sampled during a tow and probably to a suitable number of individuals of a species being captured. With tows of the same length, an increase in net diameter from 25 to 100 cm showed an increase in accuracy and precision of estimate for the larger net; this also increased with a net of 200 cm diameter. Presumably, with larger nets, each haul would take more or less the same number of patches, the patches having an approximately random distribution of individuals. A large net will presumably also perform better with regard to avoidance.

The methods and gear employed for sampling zooplankton must be greatly influenced by the particular aim of the investigation. For example, the study of the life history of a single species, an investigation of the horizontal distribution of plankton communities, or a study of the vertical distribution and diurnal vertical migration of a zooplankton taxon will demand very different approaches. One of the most difficult problems is to estimate the *total* zooplankton in any body of water. Consideration of the range in size and mobility of the organisms might suggest that at least four categories of zooplankton could be separately investigated: the microzooplankton, less than about 200 μm, by a pump or water-bottle method; the small zooplankton, ranging from about 200 μm to several millimetres length, by hauls using a net of about No. 3 mesh (0.33 mm aperture); the larger zooplankton, with a stramin or similar coarse material net (mesh *ca.* 1 mm aperture) hauled at fairly high speed; the fast-moving macroplankton and micronekton with a high speed sampler of the Isaacs–Kidd mid-water trawl type. The recommendations of Aron, Ahlstrom, Bary, Bé and Clarke (1965) review processes for measuring the productivity of plankton standing stock and related oceanic properties.

Apart from selection of the correct plankton samplers for estimating total zooplankton biomass, there is the problem of the overlap of catch, since, in theory, a net will select the lower size of animal captured, but cannot discriminate for the upper size limit. Bé (1968) has proposed a scheme for eliminating overlapping upper-sized fractions from each haul which are duplicated in the samples from the coarser mesh nets. A series of graded sieves, with meshes equivalent to those of the standard nets employed, is used to filter each catch through a separatory column and is divided into as many size classes as there are standard samplers. Figure 1.2 indicates Bé's model for four standard samplers. The fractions A, B, C and D should contain those plankton animals which would be collected ideally by each of the samplers, and their sum, corrected for volume of water filtered, would give an estimate of *total* plankton biomass.

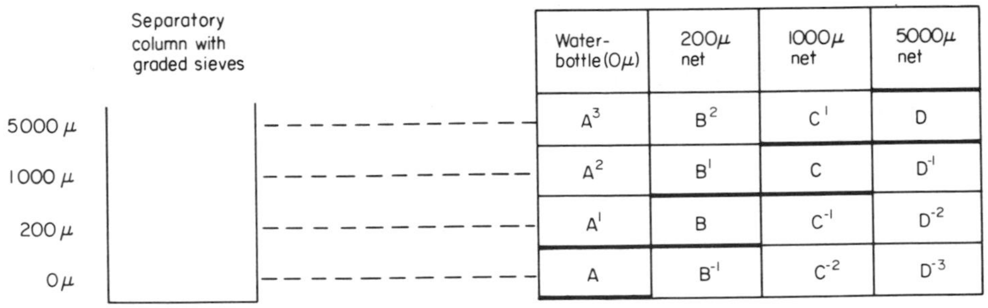

Separatory column with graded sieves

	Water-bottle (0μ)	200μ net	1000μ net	5000μ net
5000 μ	A^3	B^2	C^1	D
1000 μ	A^2	B^1	C	D^{-1}
200 μ	A^1	B	C^{-1}	D^{-2}
0 μ	A	B^{-1}	C^{-2}	D^{-3}

Fig. 1.2. Sampler model for estimation of total zooplankton biomass (Bé, 1968).

The Biomass of Zooplankton

The total amount (weight) of zooplankton is clearly an important parameter, relevant amongst other problems to regional comparisons and seasonal fluctuations and to considerations of secondary production, especially with regard to transfer of energy and matter from one trophic level to another. With the great range of zooplankton and the difficulties of quantitative sampling, there have been comparatively few analyses of *total* biomass. Whatever the choice of net, and even if several nets of different catching power are simultaneously employed, some degree of selection in sampling is inevitable. Brief reference to methods for attempting a total assessment of zooplankton is included on page 27. Since, however, many of the earlier studies used only one type of net and were limited to one period of the year, comparisons of biomass between regions, especially where studies have been separated by a considerable lapse of time and different methods employed, must be viewed with caution.

The whole zooplankton, in view of its great diversity, cannot be considered as a single entity—feeding, assimilation, growth, reproduction and life history, vertical distribution and role in higher trophic relationships, to mention only a few factors, can differ greatly from taxon to taxon and even between species of the same genus. Some biologists, especially in regions where one species is clearly dominant, have attempted to avoid this problem by concentrating attention on that species, but there are obvious dangers in this. Seasonal changes in overall population, increased diversity and a tendency to liken the behaviour of lesser known species to the few well-studied forms could lead to serious errors.

Many of the major zooplankton surveys have been followed by the laborious task of numerical assessment of the various animals. The results from some investigations, for example, the IIOE, have then permitted the publication of atlases showing the varying numbers of total zooplankton and also densities of the major taxa over the whole ocean.

The task of counting all the plankton organisms in a tow, or of counting an aliquot, is not only laborious and time-consuming; number of the different plankton animals may not be a satisfactory index to the crop, since the size and weight of various species and stages vary so greatly. A widely used method, therefore, has been to estimate the magnitude of zooplankton crop in terms of volume, expressed either as volume of organisms/m^3, or as the quantity under a column of water of unit area (volume/m^2). Although, with the degree of selection due to the type of net employed, samples cannot be regarded as reflecting the *total* zooplankton, for many purposes a comparison of plankton volumes as obtained from tows can be a useful first approximation. Moreover, the method is simple and rapid. A widely used method to measure total volume of zooplankton is by a displacement technique, after proper drainage of the catch to reduce extraneous water. A less satisfactory method is an estimate of the amount as a settled volume. This could be more subject to error due to the interstitial water between the organisms. Any volume measurement, however, is open to error in that the catch may contain variable proportions of the more compact and of the more gelatinous and generally larger zooplankton. Many investigators attempt to reduce this error by first removing the larger animals, including salps, medusae and siphonophores, and estimating these separately.

Total weight of plankton would appear to be a more satisfactory measurement, but weighing the total amount of wet zooplankton is not a very easy or accurate operation. Determination of dry weight would be more accurate, but the whole catch is then destroyed and taxonomic and other work is impossible. In many investigations the plankton catch has been divided using one of the many types of splitter and an aliquot dried and weighed, while the rest has been used for taxonomic and other forms of examination. Determination of the weight of organic matter or of some chemical constituent such as carbon, nitrogen or phosphorus may be used as an estimate of crop instead of dry weight. Comparative analyses have also been carried out; Beers & Stewart (1969) give figures for numbers of microzooplankton organisms and their related estimated volumes (Table 1.5a). Cushing, Humphrey, Banse and Laevastu (1958), for example, published a table showing conversion factors for estimating total plankton in terms of displacement volume, biomass (wet wt), dry weight, carbon, nitrogen and phosphorus content, etc. (Table 1.5b, c). These equivalents are related to zooplankton in general. Bé, Forns and Roels (1971) quote some biomass ratios and show the very great variation in, for example, wet and dry weight for different zooplankton groups. Wiebe, Boyd and Cox (1975) have also commented on the discrepancies involved in determining biomass, especially by displacement volume and by wet weight. They suggest that the chief error between the procedures is due to variable amounts of interstitial water adhering to the zooplankton. The bias introduced varies with the size of sample, being greater in small samples. However,

Table 1.5a. The numerical abundance and estimated volumes of the Protozoa, Metazoa, and total micro-zooplankton averaged over the euphotic zone in the 3 size classes of samples collected at Stations I to V (Beers and Stewart, 1969)

Station	Protozoa				Metazoa				Total micro-zooplankton			
	−35[a]	+35[b]	+103	Total	−35[a]	+35[b]	+103	Total	−35[a]	+35[b]	+103	Total
	Number/m³				Number/m³				Number/m³			
I	580,000	13,000	1,300	600,000	8,400	6,000	5,300	20,000	590,000	19,000	6,600	620,000
II	600,000	15,000	320	620,000	5,500	8,900	1,700	16,000	610,000	24,000	2,000	630,000
III	330,000	9,400	740	340,000	2,800	4,600	3,300	11,000	330,000	14,000	4,000	350,000
IV	430,000	10,000	300	440,000	1,300	14,000	1,500	17,000	430,000	25,000	1,800	460,000
V	170,000	24,000	400	190,000	270	7,500	2,100	10,000	170,000	31,000	2,500	200,000
	Volume (mm³/m³)				Volume (mm³/m³)				Volume (mm³/m³)			
I	2.0	1.4	0.4	3.8	0.6	3.0	15.3	18.9	2.6	4.4	15.7	22.7
II	1.7	0.8	0.1	2.6	0.3	5.2	4.5	10.0	2.0	6.0	4.6	12.6
III	4.0	0.5	0.2	4.7	0.1	2.4	9.3	11.8	4.1	2.9	9.6	16.6
IV	1.6	0.6	0.1	2.3	0.1	7.2	3.4	10.7	1.7	7.8	3.5	13.0
V	0.4	0.9	0.1	1.4	<0.1	3.1	6.0	9.1	0.4	4.0	6.1	10.5

[a] −20 at Station V.
[b] +20 at Station V.

Table 1.5b. Zooplankton equivalents

	Carbon (1 mg)	Nitrogen (1 mg)	Phosphorus (1 mg)	Plankton biomass (1 mg)	Dry plankton (1 mg)	Displacement vol. (1 ml)
Carbon, (mg)	1	6.0	75	0.12	0.60	96
Nitrogen, (mg)	0.17	1	13	0.020	0.10	16
Phosphorus, (mg)	0.013	0.078	1	0.0016	0.008	1.3
Plankton biomass, (mg)	8.3	50	620	1	5.0	800
Dry plankton, (mg)	1.7	10	130	0.20	1	160
Displacement vol., (ml)	0.010	0.060	0.75	0.0012	0.0060	1

Note: Dry organic matter in *zooplankton* is very nearly equal to dry plankton.

Table 1.5c. Grouping of plankton by size (Cushing, Humphrey, Banse and Laevatsu 1958)

Name of the size class	Approximate size limits	Characteristics by species composition
Megaloplankton	>1 cm.	Large organisms like squids, salps, etc.
Macroplankton	1 mm–1 cm.	Large zooplankton organisms
(Mesoplankton)	0.5–1 mm.	Small zooplankton organisms and their naupli, large diatoms
Microplankton	>60 μ.	Most phytoplankton
Nannoplankton	> 5 μ.	organisms
Ultraplankton	< 5 μ.	Bacteria, small flagellates

they emphasize the need to use past data given as displacement volumes or wet weight and to use non-destructive techniques on present plankton samples to estimate zooplankton biomass in terms of dry weight, especially of carbon. They have found reliable relationships, provided a constant technique is used, between displacement volume, wet weight, dry weight and carbon. For zooplankton samples from various areas and over different seasons, they determined carbon as 31–33 % dry weight. The bias introduced by variable amounts of interstitial water does not, however, permit a simple percentage relationship between displacement volume (or wet weight) and dry weight (e.g. dry wt is *ca.* 5 % displacement volume for small biomass/m^3; and *ca.* 13 % displacement volume for large biomass/m^3).

One of the earlier attempts to relate plankton volumes to dry weight and organic content is that of Curl (1962), who studied the biomass of plankton south of New York. Curl obtained data for wet and dry weights and for carbon, nitrogen and phosphorus content for various taxa of zooplankton (cf. Fig. 1.3). He suggests that organic weight is the best unit for estimating total biomass, but for the regular assessment of zooplankton crop this would appear to be rather difficult. Beyer (1962) has commented on a possible error inherent in volume (i.e. displacement) measurements: at least some zooplankton, including crustaceans, may show significant changes in water content so that increase in volume may not be a true representation of increased organic matter. Determinations of total crop in terms of some common

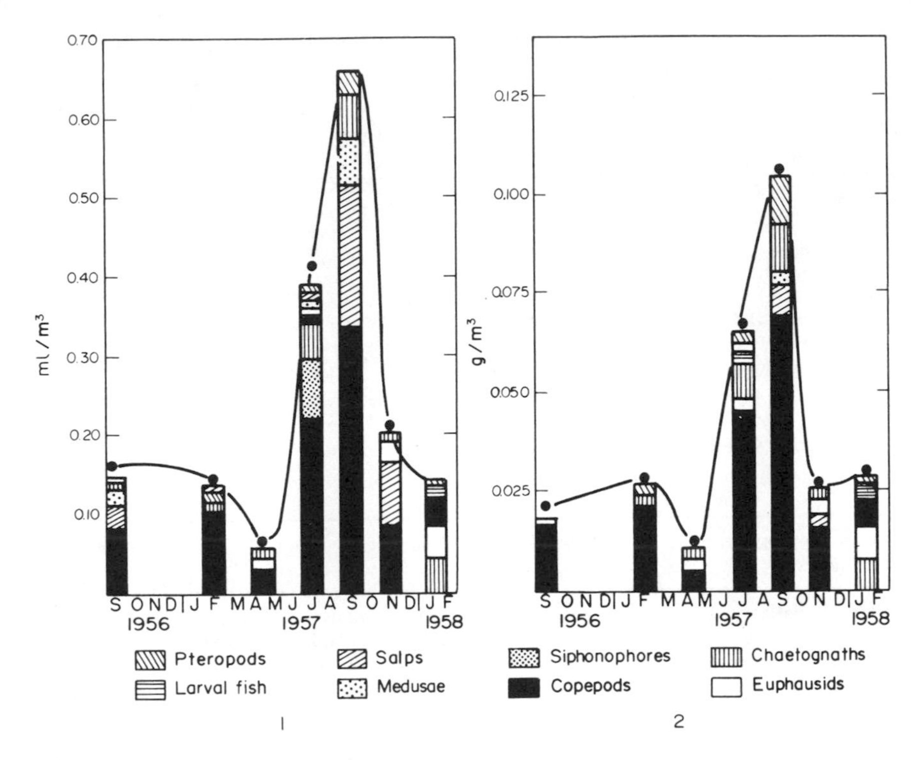

Fig. 1.3. (1) Average volumes and (2) average weights of zooplankton on the continental shelf (Curl, 1962).

biochemical constituent have been attempted. Since the proportion of carbohydrate is usually very small and the amount of lipid very variable (cf. Chapter 7) protein as measured by the biuret method has sometimes been used as a measure of standing stock. Since determinations must often be carried out on preserved samples, the use of other constituents such as ATP is precluded.

Since large numbers of data on zooplankton are recorded in terms of numerical counts of organisms, it is helpful to be able to compare counts with estimates of biomass as displacement volumes. Moreover, since the size range of various taxa of plankton and stages in development is so wide, it is important from many points of view, including trophic relationships, to be able to relate counts to weight of material. A number of investigators, including Wiborg (1954) have contributed to our knowledge of the relative volumes of different species and stages of plankton organisms. Wiborg related the volumes of some common species of copepod to that of *Calanus finmarchicus*, Copepodite V. Since data are also available for the dry weights of certain copepods, it is possible to estimate the contribution of the more important copepod components in terms of weight. As Matthews and Hestad (1977) have pointed out, many planktonologists believe that estimates of the biomass of zooplankton can best be made by using size/weight relationships as applied to numerical data from preserved plankton samples, but unfortunately few data for size/weight ratios of various species and stages exist. They quote Kamshilov's and Pertsova's data, but point out that these values apply to certain areas only (see also Ikeda and Motoda,

1978). It may be necessary to re-examine size/weight relationships especially for widespread species. Matthews and Hestad have derived size/weight ratios for a number of planktonic decapods, *Thysannoessa* and some other euphausiids, *Tomopteris*, *Aglantha* and *Eukrohnia*.

Most of the earlier information dealing with the range of plankton biomass over the oceans has been obtained, however, either by settlement or displacement techniques. Wiborg (1954) cites values per unit of surface for some North American and north European waters, some of the data being sufficient to indicate seasonal changes in volume. Wiborg also quotes mean monthly displacement volumes of zooplankton (as ml/m^3) in some European waters for the upper layers of the sea (cf. Table 1.6). (Some of the data have been obtained by recalculation from settlement volumes or dry weight.)

Discussion of the seasonal changes in zooplankton abundance in different regions of the world appears in a later chapter. It is obvious, however, that at high latitudes there is a marked fluctuation between winter and summer and therefore the determination of mean weight or volume (biomass) of zooplankton must be greatly influenced by seasonal changes, especially by the seasonal breeding of zooplankton. The pattern is also considerably changed by seasonal vertical migration (see later). Despite these fluctuations, however, the zooplankton biomass at high latitudes tends to be greater than that in warm waters, and at any latitude neritic plankton tends to be of greater biomass than offshore zooplankton.

It is not helpful to cite numerous biomass data from a wide variety of seas, but some data will be now discussed to illustrate some of the major differences. Despite the reserve with which data on biomass obtained with different apparatus and methods (and frequently employing conversion factors) must be treated, there appears to be general agreement that standing crops in warmer seas are usually smaller and exhibit less obvious seasonal changes. Jespersen (1924) demonstrated, however, that considerable variations in crop may appear apart from latitude. In the sub-tropical and tropical North Atlantic the volume of zooplankton, mainly macroplankton, appeared to fall sharply from areas off the American coast towards Bermuda. Considerably greater quantities were found north of the Azores and again towards the Spanish coast, where the volume was some eight times that found near Bermuda. There was also a sharp fall in the amount of zooplankton in the Mediterranean, especially from western to eastern regions. With regard to latitude, however, Jespersen emphasizes the very much greater standing crop of macroplankton found in typically boreal waters. Compared with volumes ranging from 200–2000 ml/hr found in different tropical and sub-tropical regions, very much greater volumes were typical towards the mouth of the English Channel (Fig. 1.4). West of Ireland, volumes approached 7–8 l/hr and in the colder seas round the Shetland and the Faroes relatively huge volumes of 18–19 l/hr were obtained.

In later investigations Jespersen (1935) confirmed the poverty of macroplankton in the Sargasso Sea as compared with the temperate North Atlantic. He suggests that by volume temperate areas may be some seven times richer. However, in the vicinity of the Cape Verde Islands and off the African mainland, in waters of greater nutrient content, rich hauls of zooplankton were obtained. In the Pacific Ocean, Jespersen similarly found that whereas high densities of zooplankton were found off the Galapagos Islands in the nutrient-rich waters of the Peru Current, in the central southern Pacific,

Table 1.6. Monthly mean volumes of zooplankton in ml/m³ in the upper layers of the sea in some European waters (Wiborg, 1954)

Month	Southwest. North Sea (*Savage* 1931) Av. depth. 65 m	Iceland coast (*Jespersen* 1940) 25–0 m	Ona (author) 50–0 m	St. M (author) 100–0 m	Eggum (author) 50–0 m	Barents Sea (*Manteufel* 1941) 25–0 m	Murman coast (*Pchelkina* 1939) 25–0 m
Jan.	—	—	0.01	0.01	0.05	—	—
Feb.	—	—	0.03	0.01	0.04	—	—
Mar.	—	0.04	0.04	0.04	0.04	—	0.05
Apr.	—	—	0.12	0.13	0.14	0.4–0.8	0.17
May	—	0.49	0.15	0.23	1.48	2.0–20.0	—
June	0.38	0.09–0.45	0.26	0.34	0.62	—	—
July	—	0.25–0.74	0.09	0.17	0.45	0.01–0.08	3.06
Aug.	0.33	0.15	0.10	0.09	0.11	0.4	0.58
Sept.	0.33	—	0.08	0.06	0.09	—	—
Oct.	0.33	—	0.06	0.05	0.14	—	—
Nov.	0.32	—	0.08	0.02	0.07	—	—
Dec.	—	—	0.03	0.01	0.12	—	—
Max. found	2.30	1.45	0.39	0.68	5.40	36.00	—

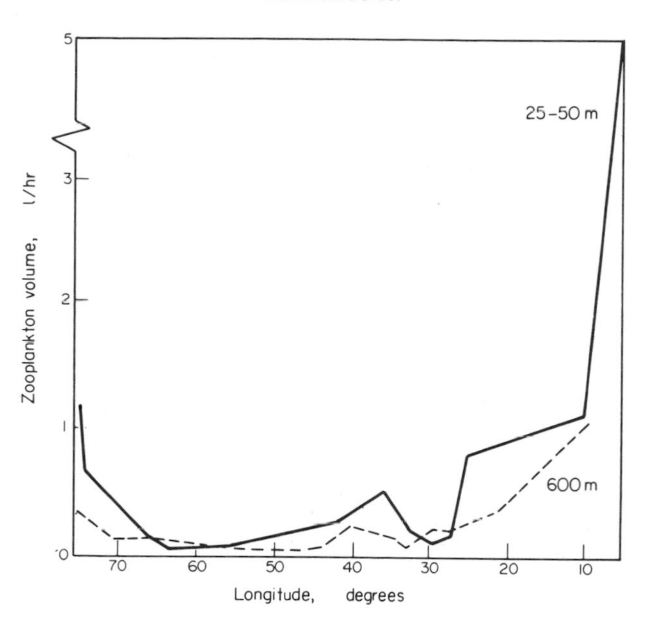

Fig. 1.4. Variations in the abundance of zooplankton with longitude in the North Atlantic at two depths. The line of stations runs from off Norfolk (U.S.A.) eastwards towards Bermuda and the Azores, then northwards towards the mouth of the English Channel (redrawn from Jesperson, 1924/25).

especially near Samoa and Fiji, the zooplankton was very poor. In the warm waters of the Indian Ocean large quantities of zooplankton were occasionally taken at certain stations, but these were associated with large concentrations of salps. Jespersen's investigations confirmed that in the tropical waters of the Sulu Sea and Celebes Sea a distinct seasonal change in the abundance of zooplankton occurs with an appreciable increase from spring to summer. Nevertheless, it is clear that the quantities of macroplankton generally present at any time in tropical waters do not approximate to the rich hauls obtained by Jespersen in the North Atlantic. He states that the results of the German Plankton Expedition suggest that for open oceans the average crop of zooplankton in the cold northern waters is about eight times that of tropical seas. He attributes this largely to the greater crop of phytoplankton typically found in more northern waters.

Not only are there substantial latitudinal differences in standing crop; the reduction in density of zooplankton with depth must be recognized. Thus, in comparisons of biomass, depth of sampling can be critical (*vide infra*). Although the exact depth at which the amount of plankton declines varies in different oceans in relation to the different hydrological conditions, especially of the sub-surface layers, it is characteristic of all oceans for the surface to be richest and for the density of zooplankton to fall sharply beneath the surface. Figure 1.5 shows the decline in biomass with depth in different oceans, as calculated by Mauchline (1972) from equations suggested by Johnston from various estimates of plankton abundance at different depths. The sharp reduction below depths of 500 m is not only typical of different water masses but holds for the abundance of zooplankton determined as number, weight or as volume.

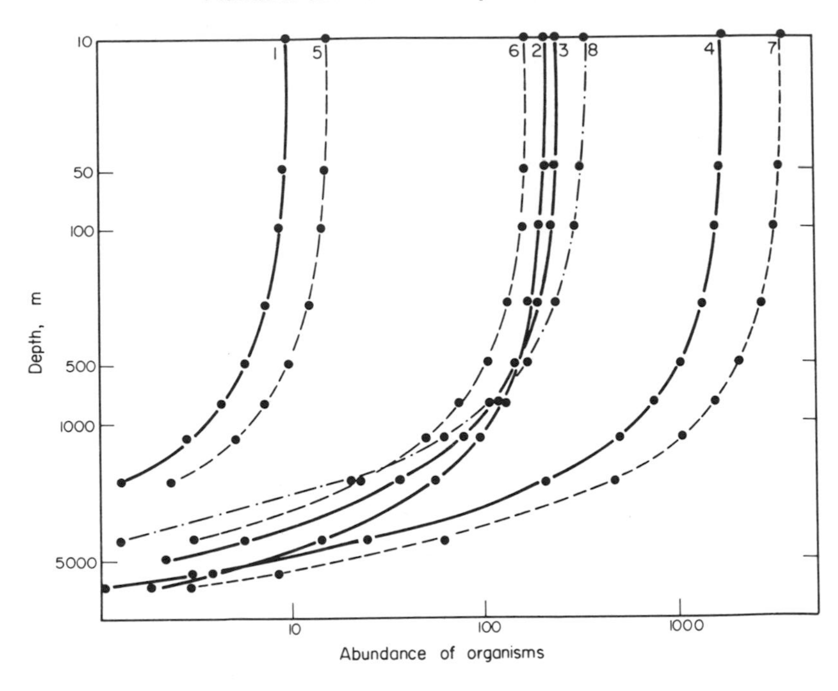

Fig. 1.5. Abundance of plankton in the water column.
Dry weight (mg/m³) (1) Plankton of North Equatorial and Canary Currents (Yashnov); (2) Tropical plankton (Bogorov); (3) Kurile Kamchatka Trench (Zenkevitch and Birstein); (4) Boreal plankton (Bogorov);
Number of organisms (5) Number of copepods per haul in Bay of Biscay (Farran); (6) Number of organisms per zone in Norwegian Sea (Ostvedt); (7) Number of copepods in North West Pacific (Brodsky);
Displacement volume (8) Volume of plankton in the West Atlantic (Leavitt) (from Mauchline, 1972).

Foxton (1956) examined data from the Discovery collections on plankton volumes in the Southern Ocean. While accepting the difficulty of comparing values on zooplankton biomass from other surveys, he points to the marked similarity of the results of all expeditions in showing maximal quantities of zooplankton at higher latitudes and minimal quantities at lower latitudes. There is great variation in the ratio, however, with a very high value (14:1) from the Meteor expeditions as against only 4.5:1 from Discovery expeditions. The ratio is largely dependent not only on the time of year sampled, but on the different types of organisms caught. The Meteor tows were, to a considerable extent, increased by phytoplankton caught in the net. Foxton points out that while there is a general tendency for lower zooplankton crops to occur nearer the equator (cf. Fig. 1.6) the smallest catches are not necessarily very close to the equator, since equatorial divergences, partly due to the Equatorial Undercurrent, lead to increased primary production and frequently to increased zooplankton in those areas. Few data can be cited for total standing crop of zooplankton at southern latitudes apart from the results of the Discovery expedition. Values obtained by Sheard showing somewhat greater volumes of zooplankton than those recorded by Discovery might be due to the inclusion of many gelatinous organisms and to the fact that the collections were taken in summer. Foxton's own results indicate that from the sub-tropics to the

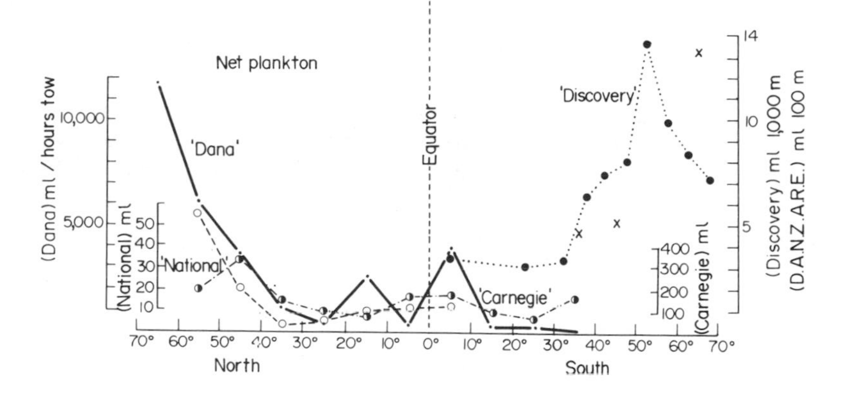

Fig. 1.6. The volumes of "zooplankton" ("net plankton") at different latitudes according to the results of various expeditions (Foxton, 1956).

high Antarctic there is a gradual increase in total zooplankton volume, with a maximum at about 50°–55°S.

At very high southern latitudes the volume of zooplankton tends to decrease. This may be partly an effect of the nets used, which failed to take any large euphausiids, particularly *Euphausia superba*, which can occur in very high density and could considerably influence total volume and biomass. The higher volume of zooplankton in latitudes 50°–55°S might be in part, however, an effect of passive drift of zooplankton from high latitudes. Not only do low latitude stations show small levels of standing crop, but at higher latitudes the crop shows a very great degree of variability, above all in latitudes between 50° and 55°S. (cf. Fig. 1.7).

Philippon (1972) has estimated the biomass of zooplankton, in terms of dry weight, for a neritic area of the Antarctic at the Kerguelen Islands. The average value for winter months was 46 mg/m^3, and a very marked seasonal increase occurred, with a summer average of 362 mg/m^3. Philippon has recalculated data from other authors in

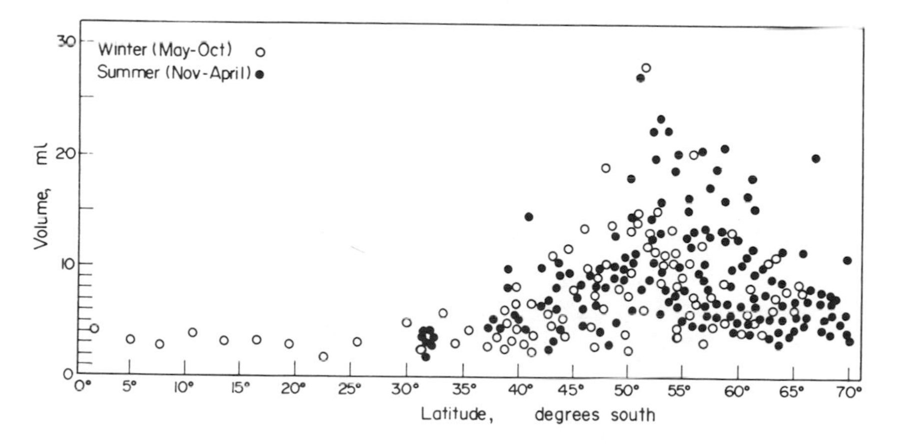

Fig. 1.7. Some individual volumes (0–1000 m) at each station plotted according to the latitude of the station (Foxton, 1956).

terms of dry weight, using data only for the more superficial layers since the depth off Kerguelen (about 50 m) was small. She estimates Foxton's data as approximating to 55 and 56 mg/m^3 for the Antarctic and sub-Antarctic zones respectively, and states that Voronina found a range of only 10–50 mg/m^3 for waters below 100–200 m depth. However, near the Antarctic Convergence Voronina obtained much higher values (*ca.* 300 mg/m^3) and around the Divergence about 80 mg/m^3. Vladimirskaya (1976) also noted for the Scotia Sea an increased biomass (2–3-fold) in areas south of the Convergence and where the Weddell Sea and Circumpolar Current mix. For the more superficial waters the maximal biomass could exceed 100 mg/m^3.

The high summer biomass reported for the Kerguelen Islands may reflect the coastal character of the station and the fact that the Islands are situated approximately on the Antarctic Convergence. Foxton (1956) suggests that the standing crop of zooplankton in the Antarctic is at least four times that of tropical southern regions, even though the Antarctic plankton may not have been adequately sampled. In the northern hemisphere, however, at the very highest latitudes the biomass of zooplankton appears to be greatly reduced. Zenkevitch (1963), summarizing the quantitative distribution of zooplankton in northern seas, believes that even allowing for the zooplankton in very cold seas possibly being richer in the lower layers, the total biomass of zooplankton is very low in the Polar basin. North of Siberia the markedly lower salinity may affect the biomass, but the general low productivity of Polar waters is reflected in the low zooplankton crop.

Figure 1.8 illustrates the remarkable rise in zooplankton density from low numbers in the warmer Pacific seas, through high numbers in boreal waters, and the very sharp reduction in Arctic areas. As regards the Arctic and sub-Arctic seas of the U.S.S.R., Zenkevitch points out that the south-western Barents Sea is by far the most productive. A zooplankton biomass up to 2000 mg/m^3 was recorded in the summer in some areas, but even for the whole Barents Sea the mean annual biomass amounted to 140 mg/m^3. The biomass in the colder seas is much lower; thus for the Laptev Sea and the eastern

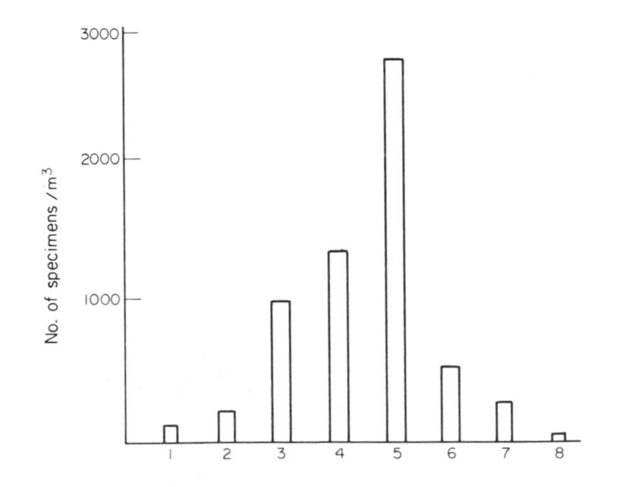

Fig. 1.8. Change in number of specimens per m^3 from (1) tropical part of Pacific Ocean, (2 and 3) through northern part of the Pacific, (4 and 5) Bering Sea, (6 and 7) Chukotsk Sea, and (8) Arctic Basin (Zenkevitch, 1963).

Siberian Sea Zenkevitch quotes Jashnov to the effect that even in the summer, the biomass does not exceed 72 mg/m^3 on average. In the more central Arctic basin there seems to be a very significant decrease in zooplankton. Zenkevitch quotes only 12 mg/m^3 in the upper 100 m with a rise to nearly 30 mg/m^3 in an intermediate layer, but only some 7 mg/m^3 in the deep layer below 800 m. Johnson (1963) also refers to the low standing crop of zooplankton in the Arctic basin. This agrees with the very low primary production. These standing crops might be compared with data given more recently for Arctic seas by Hopkins (1969).

In the central Arctic Ocean Hopkins estimated an exceedingly low value of 0.62 mg/m^3 (dry wt) for the standing crop in the surface layer (0–200 m). For the succeeding layer (200–900 m) of Atlantic water the amount was 0.14 mg/m^3, and for the Arctic deep water (>900 m) about 0.04 mg/m^3. Though there was some seasonal change, Hopkins points out that the increase during the summer was comparatively small. A value for biomass of 9 mg/m^3 dry weight for the upper 500 m (found by Hopkins and quoted by Bé, Forns and Roels (1971) for the East Greenland Current), though still low, suggests—as other authors have indicated—that zooplankton increases considerably outside the Arctic Basin itself. By comparison, Philippon (1972) quotes data obtained by Vinogradov for the Bering Sea indicating zooplankton biomass (as dry wt) of 250–1000 mg/m^3 in spring and summer, the periods of high primary production.

At high latitudes the seasonal changes in biomass especially in surface and sub-surface layers are remarkable. Vinogradov (1970) quotes the following data for Station P in the Pacific, dealing with the upper 100 m. During winter (November to March) the biomass is low (of the order of 10 mg/m^3); an increase in April results in comparatively high quantities from May to July, averaging 200 mg/m^3; by September/October the biomass is of the order of only 50–60 mg/m^3. At higher latitudes in the Pacific, even larger biomasses of zooplankton approaching 1000 mg/m^3 may be found during spring, but the seasonal effect is very strong and is obviously affected by seasonal vertical migration. Figure 1.9 shows averaged quantities of plankton over much of the north-west Pacific including the Bering Sea, and illustrates the changes in the 0–100-m layer and the influence of variations in biomass at deeper levels to 500 m. Figure 1.10 similarly gives seasonal changes in the biomass, expressed as displacement volumes,

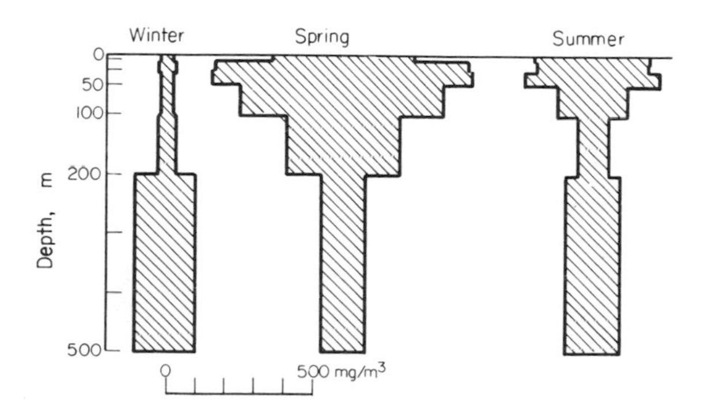

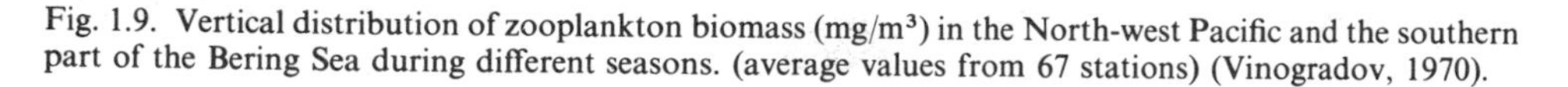
Fig. 1.9. Vertical distribution of zooplankton biomass (mg/m^3) in the North-west Pacific and the southern part of the Bering Sea during different seasons. (average values from 67 stations) (Vinogradov, 1970).

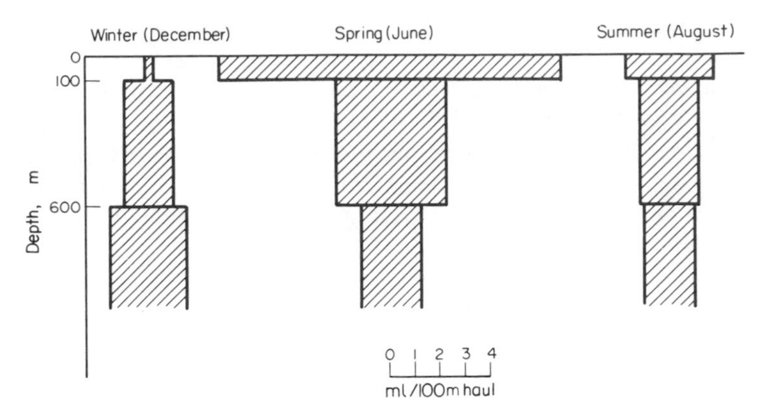

Fig. 1.10. Vertical distribution of the zooplankton biomass in the Norwegian Sea at different seasons (Vinogradov, 1970).

for the Norwegian Sea. The spring increase in surface with both vertical migration and reproduction is clear.

Bé, Forns and Roels (1971) calculated the seasonal variation for the upper 100 m from Wiborg's data as equivalent to a winter volume of 10 ml/1000 m^3, summer volume of 340 ml/1000 m^3, and spring volume of 230 ml/1000 m^3. Pavshtiks (1975) found that the numerical abundance of zooplankton in part of Davis Strait could vary seasonally by several hundred-fold, and that the variation in biomass, though not so remarkable, was still large. For the upper 200 m, in the West Greenland Current, for example, total biomass varied from an average in April of 15 mg/m^3 to 293 mg/m^3 in July; presumably a lower biomass might have occurred in mid-winter. The biomass varied considerably with water mass as well as with season in Davis Strait. For the very cold Canada Current the moderately large biomass was reached only in September (140 mg/m^3). The biomass of *Calanus* alone in the upper 50 m also showed considerable variation, ranging from a few mg/m^3 in April to several hundred mg/m^3 in different areas in summer. Especially high values were found in July in West Greenland waters (961 mg/m^3) and in Atlantic waters (1200 mg/m^3).

As regards the biomass of zooplankton in more temperate regions, a summary of volume as measured by displacement is given by Bigelow and Sears (1939) for areas of the northern Atlantic. Seas around Iceland, the Faroes and the north of Scotland appear to be especially rich. Bigelow and Sears suggest that the continental shelf waters off the north-east coast of U.S.A. rank next in quantity for the northern Atlantic. There is a considerable seasonal fluctuation, but the summer maximum (0.7–0.8 ml/m^3) compares favourably with that of the North Sea and is considerably higher than the volume of zooplankton in the English Channel (cf. Table 1.7).

Nagasawa and Marumo (1975) examined the biomass of zooplankton in Suruga Bay, Japan. In the surface layer the main Bay showed a greater mass of zooplankton (25–63 mg/m^3 wet weight) than the waters slightly further offshore (2–18 mg/m^3). The overall mean (19 mg/m^3) would correspond to a plankton volume of 20 ml/1000 m^3. The bottom shelves steeply in the Bay from around 20 m to nearly 250 m; the quantity of zooplankton in the near bottom layers showed marked variation (7–119 mg/m^3), with the greater biomass at the shallower stations. The average quantity for the bottom

Table 1.7. Comparison of volumes of zooplankton from different oceanic and coastal regions (Raymont, 1966)

Region	Plankton volume (ml/1000 m³)		Authority
	Winter	Summer	
Northern Atlantic—Iceland Faroes, Greenland		840[a]	
Coastal shelf, north-eastern U.S.A.	200–300;	700–800	Bigelow and Sears (1939)
North Sea		500–600	
English Channel		100	
Laptev and eastern Siberian Sea		70	
Central Arctic Basin		10	
Sea of Japan	30–500;	1000	Zenkevitch (1963)
Okhotsk Sea		1000–3000	
Bering Sea		1500–2500	
	Yearly Mean		
Barents Sea	140		Zenkevitch (1963)
Australian oceanic waters	20–50		
Australian shelf waters	100		Tranter (1962)
Australian inshore waters	>200		
Subtropical Pacific	<50		
Equatorial Pacific	100		Russian data quoted by Tranter (1962)
Boreal northern Pacific	500–1000		
New York—Bermuda			
Coastal waters	1070		
Slope waters	270		Grice and Hart (1962)
Sargasso waters	20		

[a] Most of smaller copepods not included owing to mesh size.

layers was 54 mg/m³, equivalent to a volume of 50 ml/1000 m³ (cf. Table 1.7). The surface layer was dominated by chaetognaths, with copepods, fish and sergestids giving 83% of biomass. In the deeper layers, chaetognaths and copepods were approximately equal (cf. Fig. 1.11). The size of the dominant zooplankton organisms must influence biomass and this may be relevant to comparisons of inshore and offshore populations. Kozasa (1974) investigated zooplankton in the East China Sea and found that small species of copepods including young stages were dominant. Numerically the zooplankton was richer nearer the coast than offshore, but this difference was not reflected in biomass as measured by settlement volume and even less by displacement volume (Table 1.8). Possibly the inshore samples contained a larger proportion of the smaller stages.

Sameoto (1977) examined biomass in the Gulf of Maine, particularly over the Scotia shelf and entrance to the Bay of Fundy. Although weights are given for copepods and

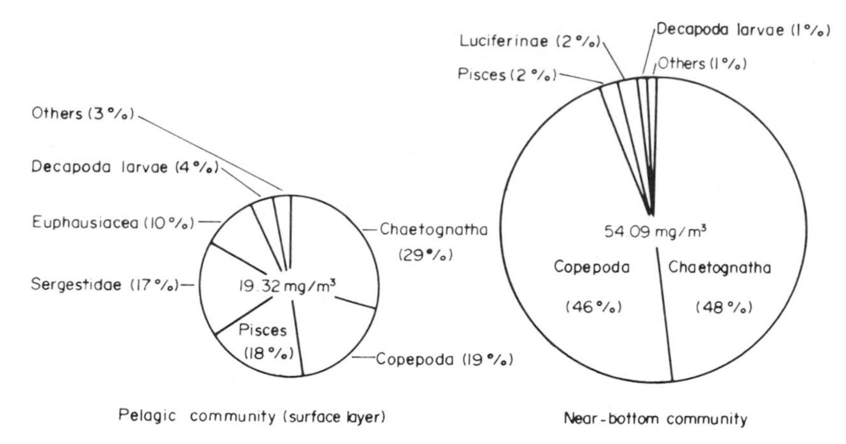

Fig. 1.11. Composition of zooplankton communities in Suruga Bay (Japan) (Nagasawa and Marumo, 1976).

Table 1.8. Biomass of zooplankton in the East China Sea (Kozasa, 1974)

Area	Copepods number/m^3	Settlement volume ml/m^3	Displacement volume ml/m^3
Coastal, west of Kyushu	4798	2860	650
Tsushima warm current	2604	2960	720
Mixed water	1541	2010	700
Yellow Sea cold water	1057	1140	520

chaetognaths and the biomass and some minor components are not listed, the variation over different areas is enormous, despite all sampling being conducted over one month (August). When values for copepods (by far the most dominant group) and chaetognaths are converted to volumes, the minimum is about 50 ml/1000 m^3 (occasionally even lower) and the highest values are between 200–300 ml/1000 m^3, a few exceeding 300 ml/1000 m^3. The highest quantities appeared to be near the entrance to the Bay of Fundy, though Sameoto found no obvious explanation for the difference in crop with area.

Tranter (1962) has studied the biomass of zooplankton in Australian seas. Oceanic regions, except where upwelling occurs, have a low crop—normally <25 mg/m^3 (wet biomass weight). Continental shelf areas had a higher crop, and inshore shallow stations, for example at Port Hacking, had even larger amounts. At times the Port Hacking plankton included large numbers of salps, but if these are excluded the biomass (>200 mg/m^3) is still higher than offshore. Tranter uses a conversion factor to compare his data with plankton volumes recorded by Bigelow and Sears. He also includes values of the biomass of zooplankton already quoted as weight/m^3, and obtained by Russian workers from the Pacific and converts these data to plankton volumes (Table 1.7). Tranter's values for oceanic Australian waters (20–

50 ml/1000 m^3) are similar to those of the Russian investigations in the sub-tropical Pacific. The richer equatorial Pacific area (100 ml/1000 m^3) is approximately equivalent to Australian shelf waters, but in large areas of the boreal northern Pacific, zooplankton volumes are very much greater, reaching even 1000 ml/1000 m^3 in summer and being maintained through the year at amounts exceeding 500 ml/1000 m^3. Plankton volumes during the summer in the Bering Sea and Sea of Okhotsk also exceed 1000 ml/1000 m^3 (Table 1.7) and high values are found in the south-western Barents Sea, but as already indicated certain very cold seas, (e.g. Laptev and East Siberian) have very low biomass, only 70 ml/1000 m^3 as displacement volumes.

Values for the biomass of zooplankton in warmer seas include those for the Mediterranean found by Arellano and Lennox and quoted by Philippon (1972). Standing crops of approximately 9 mg/m^3 were found in the warm period of the year, and 16.5 mg/m^3 in the cold season, with a maximum during two months varying from about 27–30 mg/m^3. Data for the Black Sea, where one would expect a distinctly larger crop, indicate a biomass of the order of 51 mg/m^3 in June, 17.8 mg/m^3 in May and 24 mg/m^3 in October.

Details are given in a later chapter (cf. Chapter 3) on the biomass of zooplankton, measured as volumes per m^2 for the upper 200-m water column and published as maps from the results of the IIOE (cf. Fig. 1.12a, b). It is sufficient to note here that volumes were generally low over the open ocean, some extensive regions not exceeding 10 ml/m^2, (cf. Fig. 1.12a, b). On the other hand, particularly in upwelling areas, very rich crops, even exceeding 50 ml/m^2, were found (e.g. off southern Arabia, Somalia, Oman and Kerala). The biomass of the most productive areas varied with the monsoon seasons.

Lenz (1973b) divided the catch for some samples from the Indian Ocean. Part of the preserved material was analysed for dry weight and part for organic carbon and the results compared with data for displacement volume determined as a measure of biomass. Lenz comments on the great difficulties in inter-converting settling volumes, displacement volumes, wet and dry weights, dry organic matter and carbon content. The average ratio, displacement volume to dry weight, for the samples was 12.3, an unusually high figure which Lenz attributes to the considerable numbers of salps and medusae. He points out, in agreement with other workers, that high biomass of plankton was found in the Gulf of Aden area, in the northeastern part of the Arabian Sea, especially in the Gulf of Oman and along the Somali coast. The average dry weight of standing crop for the upper 200 m of water varied between 2 and 51 mg/m^3, or, reckoning for the column of water, between 0.4 and 10.2 gm/m^2. These values may also be compared with those given by Tranter (1963) for the Indian Ocean during the month of August, when with different nets standing crop in terms of dry weight varied between 40 and 56 mg/m^3 (i.e. within the upper limit suggested by Lenz).

The effects of distance offshore and of the monsoonal climate can be seen from values by Gapishko (1971a) for zooplankton biomass from the Indian Ocean. High values of 500 mg/m^3 were found for the Gulf of Aden during the summer monsoon, and even up to 1000 mg/m^3 during September following upwelling, in the central part of the Gulf. Tjuleva (1971, 1973) also refers to seasonal changes due to the monsoon but quotes a peculiar instance off the west of the Indian sub-continent where the biomass increased somewhat at a distance of about 150 miles offshore. Biomass values were listed—Cape Diu, 520 mg/m^3; slope waters, 320 mg/m^3; deep sea (*ca.* 150 miles),

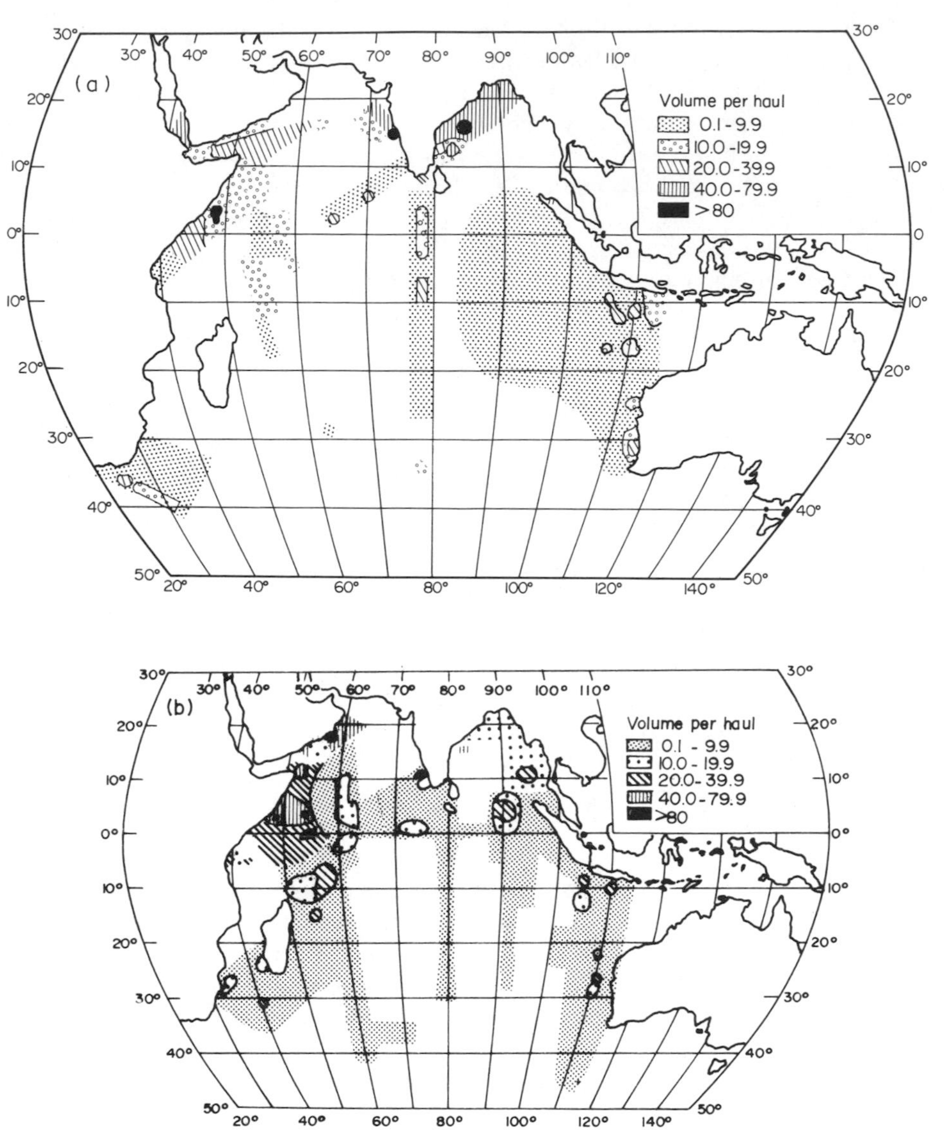

Fig. 1.12. Zooplankton biomass distribution in ml/m² for 200-m column in the Indian Ocean (a) December–February (North-east monsoon), (b) July–September (South-west monsoon) (Rao, 1973).

460 mg/m³. This might be another example of upwelling effects on the zooplankton drifting offshore. On the other hand, during the south-west monsoon, the difference between the large biomass inshore and over the deep ocean is well exemplified by the following data: Shelf (Gulf of Kutch): 153 mg/m³; off Cape Comorin: 1200 mg/m³; deep sea: 30–70 mg/m³.

Vinogradov (1970) has summarized information on the biomass volume of zooplankton in tropical oceans. In the upper 100 m where biomass is largest, volume

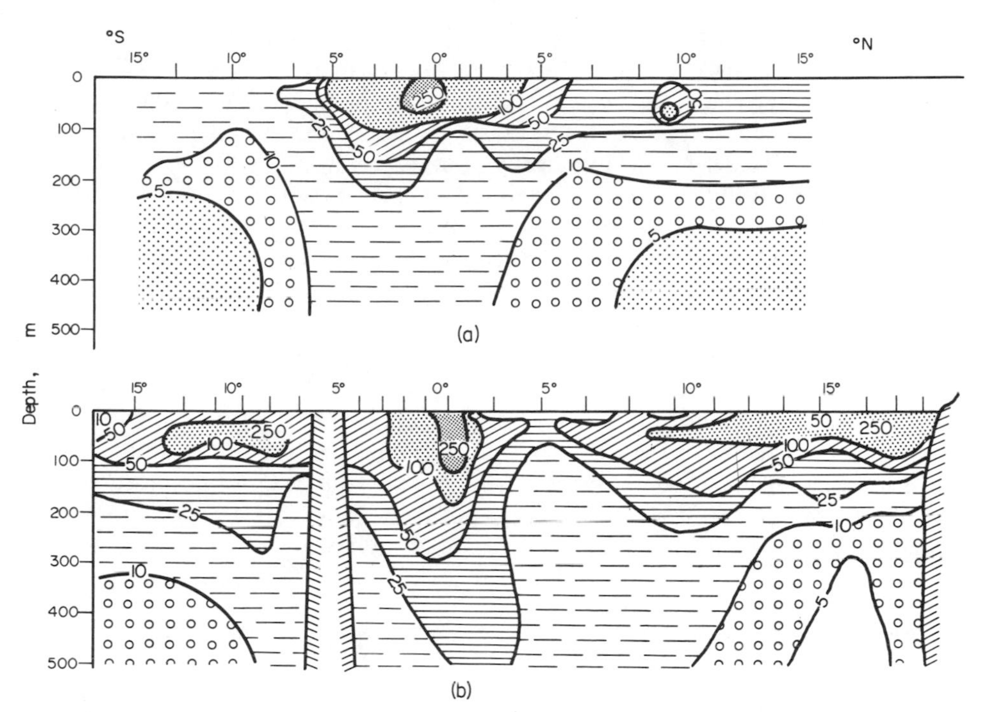

Fig. 1.13. Distribution of planktonic biomass (in ml/1000 m^3) in meridional sections through the tropical waters of the Pacific (a) and Indian (b) Oceans (Vinogradov, 1970).

ranged from 15–20 ml/1000 m^3 in depleted central waters to 250–300 ml/1000 m^3 in areas of intensive divergence. At sub-surface (200–500 m) levels the corresponding volumes of zooplankton were reduced to 5–10 ml/1000 m^3, and 10–25 ml/1000m^3. Figure 1.13 indicates the variation in biomass of zooplankton as volumes in sections ranging from 15°N to 15°S in the Pacific and Indian Ocean as determined by Vinogradov and Voronina. The small amounts of plankton typical of the nutrient-poor central tropical waters, the marked increases in volumes at divergences and the great reduction of zooplankton biomass with depth are obvious.

A comparison by Wickstead (1968) of variations in standing crop of zooplankton at different latitudes is included in Chapter 3 dealing with seasonal change. It is worth noting, however, that Wickstead, using displacement volume and dry weight determinations, found that mean values for the English Channel were below that for a neritic tropical area off Zanzibar. Volumes were around 100 and 180 ml/1000 m^3 for the English Channel and Zanzibar, respectively: dry weights were 19 and 42 mg/m^3. Although Wickstead accepts that tropical plankton often has a large proportion of gelatinous salps, medusae and siphonophores, the dry weight data for Zanzibar suggest a rich crop.

In more open oceans, however, there is considerable evidence for larger volumes of zooplankton at higher latitudes where mixing of upper waters occurs at some period of the year, and at any latitude where current patterns promote the mixing of surface layers. This is well exemplified by the investigations of Bé, Forns and Roels (1971) in the North Atlantic. They studied geographic variations in the biomass of zooplankton from approximately 60°N to near the equator, using various estimates of biomass, displacement volume, wet, dry and ash-free (i.e. organic matter) dry weight. Generally there was good agreement between the distribution of biomass over the ocean as determined by the different methods (cf. Figs 1.14a, b). Altogether 342 samples, taken with a 202 μm aperture net hauled vertically or obliquely from 300 m to the surface, at all seasons over several years were analysed so that seasonal variations, vertical migration and patchiness should not have significantly affected the overall pattern of biomass. Some bias may have been introduced since more of the sampling was conducted over summer and autumn, when crops, in particular at higher latitudes, tend to be seasonally larger.

Zooplankton abundance followed the pattern of being high in sub-Arctic and cold-temperate regions, but with marked and seasonal change. Volumes exceeding 50 ml/1000 m^3 were usually found north of 45°N, approximately within an area from Cape Hatteras to the Bay of Biscay, with highest values between May and October, often exceeding 100 ml/1000 m^3 (cf. Fig 1.14a). A peak volume in June almost reached 1000 ml/1000 m^3. In June 1962 along a transect from New York to Scotland, values often exceeded 100 ml/1000 m^3 (cf. Fig. 1.15). Such values appear to be typical of the

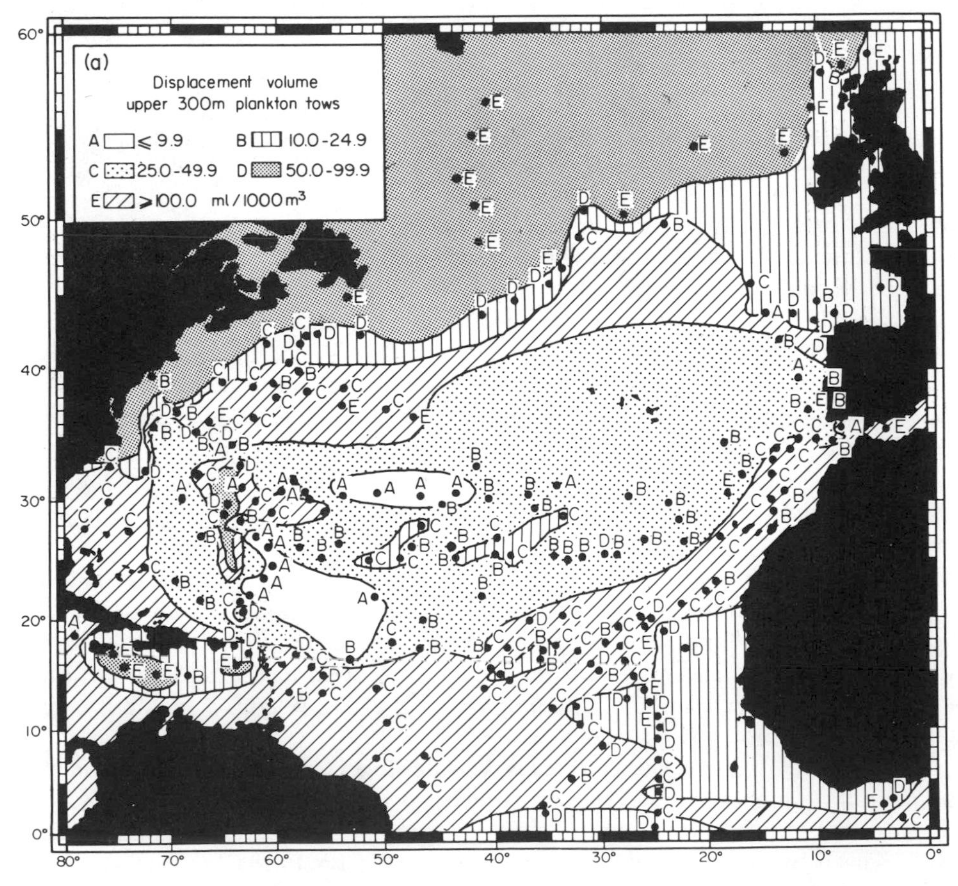

Fig. 1.14a

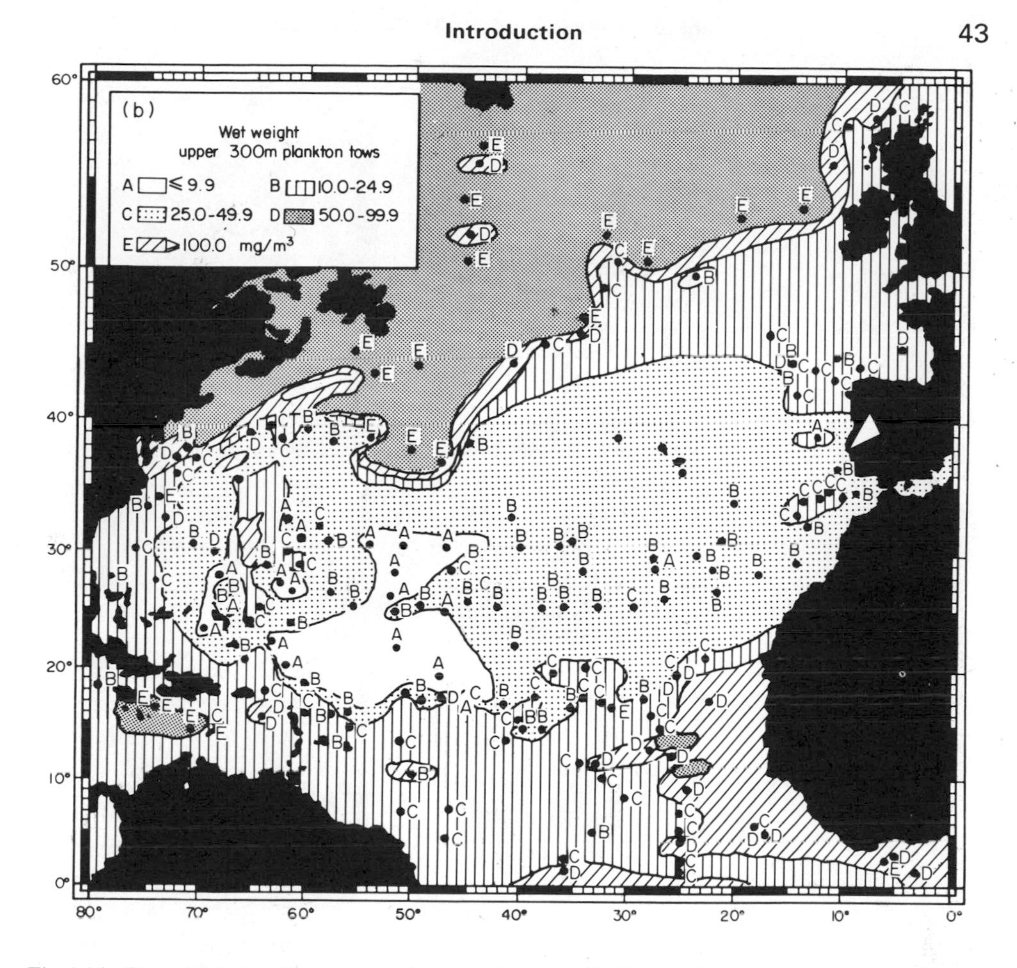

Fig. 1.14. Geographical variations in zooplankton biomass in the North Atlantic Ocean, based on plankton tows in the upper 300 m of water. (a) Displacement volumes, (b) wet weight (Bé, Forns and Roels, 1971).

boreal Atlantic. Volumes reported by Bé *et al.* (1971) from the work of Kusmorskaya for the upper 200 m of water in the North Atlantic between Newfoundland and the British Isles reached between 100–500 ml/m^3. Even higher amounts (1000–7000 ml/1000 m^3 from April to October; < 200 ml/1000m^3 from November to March) were found for the central Labrador Sea by recalculating data obtained originally by Kielhorn. In terms of wet weight, the cold-temperate and sub-Arctic North Atlantic areas yielded weights frequently exceeding 50 mg/m^3 (Fig. 1.14a, b), especially in the warmer months. Greater biomasses are reported from other investigators, for example, 250–500 mg/m^3 for the sub-Arctic North Atlantic, with higher values (300–700 mg/m^3) over April and May.

Regions with strong boundary current (Gulf Stream and North Atlantic Drift, Canaries Current, Antilles Current) were marked by fairly high displacement plankton volumes (25–50 ml/1000 m^3) according to Bé *et al.* (1971). Very high volumes were commonly found in the slope and continental shelf waters off the north-eastern coast of America.

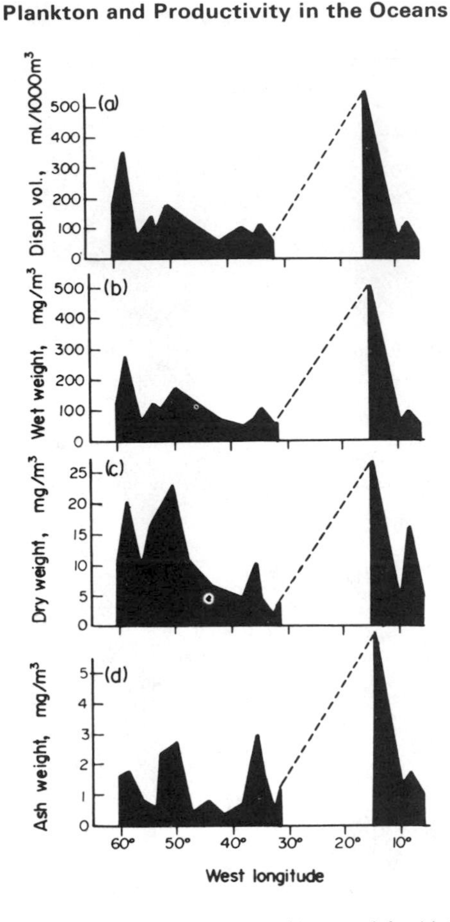

Fig. 1.15. Variation in plankton displacement, (a) volume; (b) wet weight; (c) dry weight; and (d) ash weight in the mid-latitudes of the North Atlantic (Bé, Forns and Roels, 1971).

In the central waters of the North Atlantic between about 40° and 20° N, the biomass of zooplankton was generally low ($<25\,\text{ml}/1000\,\text{m}^3$) with particularly small stocks ($<10\,\text{ml}/1000\,\text{m}^3$) in the south-western Sargasso Sea (Fig. 1.14a). The low biomass was thus characteristic of oligotrophic waters which do not show replenishment of the very low concentration of nutrients for phytoplankton growth from deeper layers. Where some degree of temporary winter enrichment occurs, a brief increase in plankton volume during spring may be observed. In the Bermuda region, while the average volume for the year was $<25\,\text{ml}/1000\,\text{m}^3$, very short periods of increased volume over spring amounted to six-and four-times (in one year even twenty-one-times) the minimum volume according to Menzel and Ryther (1961), and to ten-times the minimum volume according to Deevey (1962).

Seasonal changes in zooplankton were generally, however, very small in the Sargasso Sea, in contrast to slope and continental shelf waters off New York. Bé *et al.* noted fairly high volumes of zooplankton in tropical and sub-tropical regions where divergence and upwelling occurred. For example, a biomass equivalent to $50\,\text{ml}/1000\,\text{m}^3$ was noted off West Africa, extending partway across the Atlantic in the zone of divergence between 5° and 20°N (cf. Fig. 1.14a). In general, therefore, in the North Atlantic higher biomass was characteristic of regions where nutrient-rich mixed layers existed for at least some period of the year, as in sub-Arctic, cold-temperate and upwelling regions, and along oceanic margins, in contrast to central stable oligotrophic areas where biomass was low.

The very high quantities of zooplankton frequently recorded for boreal latitudes are to some degree repeated from results obtained by Barkhatov, Volkov, Dolzhenkov and Karedin (1973) in studies in Antarctic and sub-Antarctic seas. They collected macroplankton using an Isaacs–Kidd Midwater Trawl from the upper 100-m layer. South of the Antarctic Convergence the highest biomass in summer amounted to about 5–10 mg/m^3; some areas gave 12 mg/m^3 but huge swarms of *Euphausia superba* were specifically excluded. The biomass inside a swarm would have been much greater. At divergences, biomasses of <1–5 mg/m^3 were found. In sub-Antarctic waters off southern Chile higher quantities (20 mg/m^3 maximum) were found, but by far the greater biomass was found in the New Zealand area. In the Tasman Bay >50 mg/m^3 was recorded, but to the east of South Island, New Zealand, a comparatively enormous crop (>500 mg/m^3) was found associated with a huge mass of salps. One exceptionally large catch in this general region was estimated as 3.7 g/m^3! In general, the larger catches of macroplankton over the whole investigation were associated with large numbers of medusae, salps and euphausiids. Salps occurring in swarms can give exceptionally large biomass data.

Comparisons of biomass may be expressed in terms of the quantity of zooplankton in the water column beneath a unit area (m^2). Consideration is often restricted to a limited depth (e.g. 500, 1000, or 2000 m) since accurate sampling to the bottom over oceanic depths is difficult, and with the vastly reduced density of zooplankton in great depths, comparisons over total depth might be misleading. But while, as Vinogradov (1961) indicates, tropical surface waters have only approximately a tenth to a twentieth the crop of zooplankton of surface waters from high latitudes, and there is a great reduction in biomass with depth at all latitudes, the decrease is much sharper at low latitudes (cf. Fig. 1.16). The overall pattern of biomass also may be affected at greater depths where deep water masses from higher latitudes invade warmer oceans (e.g. the southward movement of deep water in the north-west Pacific and the movement of Antarctic waters at depth in the western Indian and South Atlantic Oceans). Thus precise knowledge of the vertical distribution of the zooplankton is essential to comparisons of the biomass for the whole water column.

Reference has already been made to the small seasonal changes in biomass in the Sargasso Sea as compared with waters off the American continental shelf. Clarke (1940) investigated the waters south-east of New York and found them very much richer than areas between the continental slope and Gulf Stream and the Sargasso Sea to the south-east. Clarke's data indicated plankton volumes for the coastal, slope and Sargasso Sea in the proportion of 16 : 4 : 1. Mean volumes were 194, 52 and 12 ml respectively for a standard 10-m tow. However, the seasonal fluctuations were also very different. Coastal zooplankton volumes were twenty- to forty-fold greater in summer than in winter; slope plankton was ten-fold larger in summer, and the plankton in the Sargasso Sea showed negligible seasonal change.

Riley, Stommel and Bumpus (1949) also compared the crop beneath a square metre of sea surface for coastal areas south of New York, for the semi-tropical Sargasso Sea,

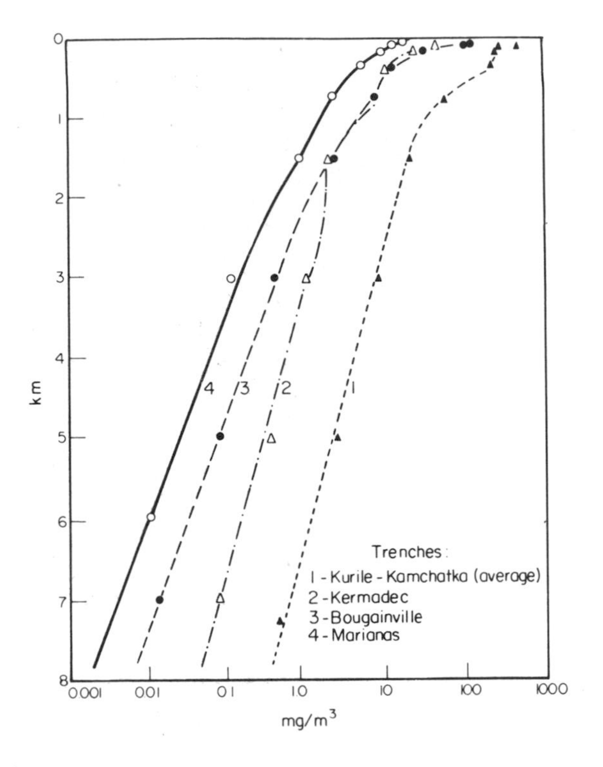

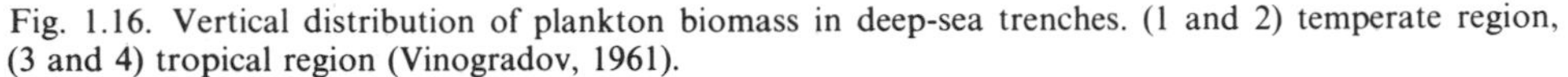

Fig. 1.16. Vertical distribution of plankton biomass in deep-sea trenches. (1 and 2) temperate region, (3 and 4) tropical region (Vinogradov, 1961).

and for the mixed continental slope water between. They found that the inshore water showed its typical greater richness; the mean ratios for the volumes of zooplankton of the three areas, coastal, "slope", and tropical waters, were approximately 10:4:1. Despite the greater depth of oceanic areas and the much greater crop fluctuations near shore, especially in coastal boreal waters, this investigation generally confirmed the view that tropical waters are usually poorer in zooplankton biomass than temperate waters, especially inshore areas. Oceanic plankton also appears to have a generally lower organic content (cf. Table 1.9 and Table 1.10).

Grice and Hart (1962), investigating the zooplankton of the upper 200 m of water between New York and Bermuda, emphasized the richness of the coastal areas. In numerical abundance, the ratio between coastal, "slope" and Sargasso Sea waters averaged some 22:4:1. By volume, the same comparison yielded ratios of approximately 50:3:1. Their results confirmed the very much greater seasonal fluctuations in neritic waters (approximately thirty-fold) compared with the sub-tropical Sargasso Sea waters (approximately four-fold). Although in coastal regions virtually the whole column of water from surface to bottom was sampled, whereas only the upper 200 m in oceanic stations was sampled, the greater richness of coastal waters is still evident. The volume in the coastal area exceeds that of most boreal northern Atlantic regions. The very low quantity found for the Sargasso Sea is similar to the values suggested for sub-tropical Pacific and Australian oceanic regions (Table 1.7). Deevey and Brooks (1977) recently confirmed the marked decrease in density and

Table 1.9. Volume of zooplankton in ml per 30 minute haul. Average of all stations (Riley, Stommel and Bumpus, 1949)

Area	Depth (metres)	Total Zooplankton (ml/30 min)	Crustacean Plankton (ml/30 min)
	25–0	118 (120)[a]	41
Coastal Area	Bottom–25	220 (119)[a]	50
	Mean	194 (120)[a]	50
	25 or 50–0	77	18
Slope Water	275–25 or 50	50	18
	Mean	52	18
	50–0	29	17
	100–50	27	17
Sargasso Sea	500–100	8	5
	900–500	5	3
	Mean	12	7

[a] Numbers in parentheses are recomputed averages omitting a series of stations on July 9–17, 1938, when salps were unusually abundant.

Table 1.10. Summary of regional plankton comparisons (Riley, Stommel and Bumpus, 1949)

	Coastal water	Georges Bank May	Georges Bank June	Slope water	Gulf Stream	Sargasso Sea
Herbivores, 10^{-9} g C/ml	15.5	72.0	150.0	17.2	5.2	2.9
Carnivores, 10^{-9} g C/ml	1.1	12.6	3.7	2.8	0.4	0.1
Both, 10^{-9} g C/ml	16.6	84.6	153.7	20.0	5.6	3.0
Depth Range, metres	0–50	0–50	0–50	0–200	0–400	0–400
Zooplankton, g C/m^2	0.83	4.2	7.7	4.0	2.2	1.2
Carbon, % of wet wt	5	5	5	2.4	2.4	2.4
Zooplankton, g wet wt/m^2						
Calculated	17	85	154	167	92	50
Observed	21	72	81	122	85(?)	45
Plant Pigment Units:						
Surface (per m^3)						
Calculated	1650	6760	5250	1340	1650	1050
Observed	2500	4490	8660	1840	800	860
Total (per m^2)						
Calculated	163,600	501,250	239,000	444,200	272,400	330,000
Observed	122,400	458,800	315,900	267,300	295,000	310,000

biomass of zooplankton between the temperate neritic waters off New York and the oligotrophic open Sargasso Sea. They point to the dominance of small nanoplankton, especially *Coccolithus huxleyi*, as the main primary producer in the Sargasso Sea as contrasted with the rich diverse net phytoplankton in more coastal waters. The zooplankton of the Sargasso Sea also tends to be of small body size; few are probably exclusively herbivores, and Deevey and Brooks believe that an omnivorous habit is widespread.

In an earlier report Deevey and Brooks (1971) found maximum displacement volumes in April for total zooplankton in the Sargasso Sea near Bermuda, for the upper 500 m layer. A second maximum, in numbers, occurred in October but this was

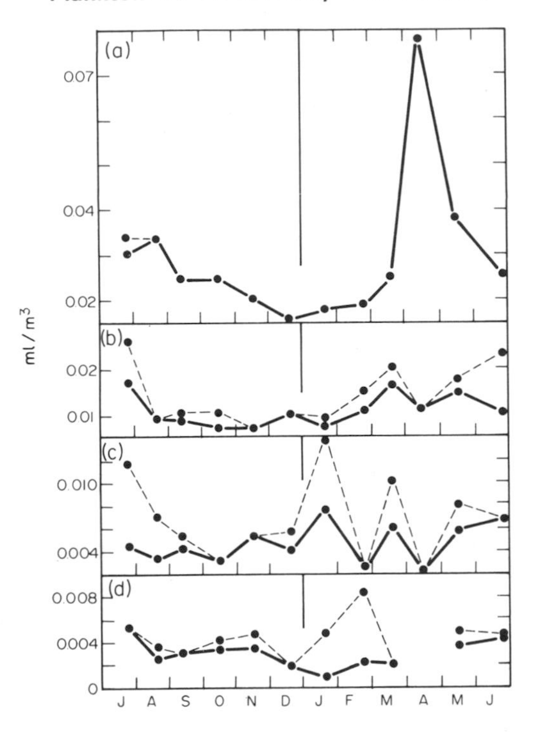

Fig. 1.17. Zooplankton displacement volumes in ml/m³ from July 1968–June 1969. Minimum (————), maximum (– – – –). (a) 0–500 m, (b) 500–1000 m, (c) 1000–1500 m, (d) 1500–2000 m (Deevey and Brooks, 1971).

made up of small copepods and was not reflected in the volume. Over the next depth interval (500–1000 m) volume maxima were found in July, March and May; from 1000 to 1500 m maxima were found during July, January, March and May and at the deepest level sampled (1500–2000 m) they occurred in July and June (cf. Fig. 1.17). The minimum volumes refer to smaller plankton; maximal volumes include all organisms. Crustaceans increased in importance and diversity below 500 m despite the sharp decrease in biomass with depth. Although "seasonal" fluctuations are apparent, the amplitude is comparatively small. This is also reflected in changes in numerical abundance (cf. Fig. 1.18).

The total biomass for the upper 500 m was 14.75 ml/m², excluding large zooplankton, or 14.9 ml/m² for total zooplankton. Over succeeding depths the biomass declined to 5.4 ml/m², (500–1000 m), 2.3 ml/m² (1000–1500 m) and 1.35 ml/m² (1500–2000 m), all values being for zooplankton excluding the large forms. Vinogradov and Sazhin (1978), for plankton tows down to 3400 m in the Sea of Japan, found that the greatest biomass occurred above 1500–2000 m and thereafter decreased with depth.

Mullin (1969) illustrates the effect of length of the water column from investigations of Menzel and Ryther (1961) in the Sargasso Sea. The standing crop of zooplankton was estimated from collections in the upper 500 m and from dry weight determinations, but from numerical comparisons of the zooplankton in the 0–500 m layer and from

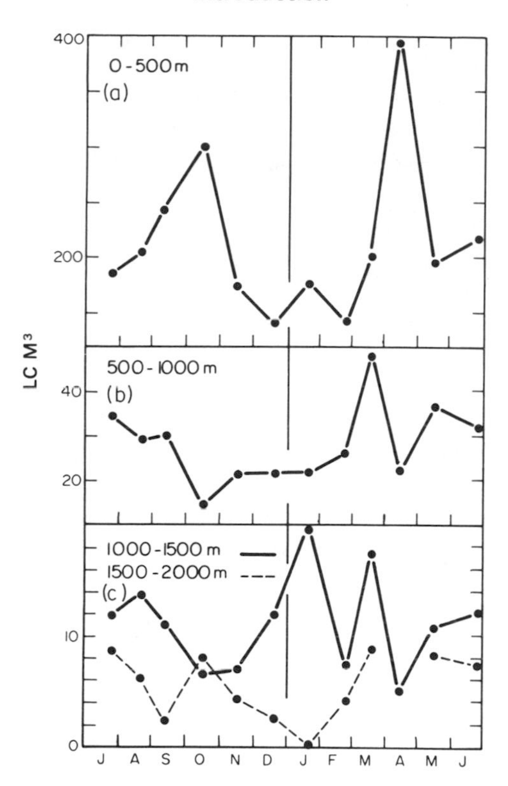

Fig. 1.18. Total zooplankton numbers/m³ taken as for Fig. 1.17. (a) upper 500 m, (b) 500–1000 m, (c) 1000–2000 m (Deevey and Brooks, 1971).

0–2000 m on a few occasions, an assessment was then made for the whole 0–2000 m. Except for one or two irregular spring peaks, the biomass was comparatively stable and ranged between 600 and 1000 mg C/m². This appears to be a very high biomass when compared, for example, with investigations by Smayda (1966) for another warm-water but comparatively shallow area, the Gulf of Panama. Though estimates were made as displacement volume, a conversion factor (carbon = 5 % wet wt) yielded a biomass for the Gulf of Panama from bottom to surface of 240 mg C/m² in the productive upwelling period (January–April), and of 135 mg C/m² for the rest of the year.

In temperate areas estimates in neritic waters do not at first sight appear to give such very high crops of zooplankton per unit of surface. Thus for the English Channel Harvey (1950) calculated 2.0 g/m² as the biomass of zooplankton, expressed as organic matter between April and October, omitting a marked seasonal peak. For the North Sargasso Sea from July to September he quotes a near comparable value of 1.4 g/m². A calculated value for the Gulf of Maine is distinctly higher—omitting the April peak the summer value was around 10 g/m². Bearing in mind that these data for cold-temperate areas are taken over the main productive season, whereas in the Sargasso Sea the difference between seasons is slight (*vide supra*), it might appear that the total production in a warm ocean like the Sargasso is substantial. However, in contrasting the huge depth of water column in an ocean with a shallow temperate neritic area like the Channel, or a tropical shallow region such as the Gulf of Panama, it is all-important to consider the vertical concentration of the zooplankton. The rich standing crops of shallow neritic seas are concentrated and can be suitably exploited by higher trophic levels. A crop of plankton widely dispersed vertically in a deep ocean cannot be so effectively utilized and comparison of a crop beneath a unit of surface can be misleading in relation to the whole ecosystem. Therefore comparisons between areas of similar depth are important. Harvey (1950) gives the mean annual quantity of zooplankton as 1.5 g/m^2 and 2.0 g/m^2 organic matter for shallow areas of the English Channel and Long Island Sound, respectively. In terms of wet weight (biomass) of zooplankton this would approximate to 9 g/m^2 and 12 g/m^2.

Comparisons of biomass must also be critically examined with reference not only to seasonal changes and length of water column but to the type and especially size category of zooplankton studied. LeBrasseur and Kennedy (1972) examined the micro-zooplankton of coastal (Strait of Georgia) and oceanic (ocean Station P) areas of the Pacific Ocean. The samples were filtered through 44 μm mesh nets, i.e. distinctly finer filtration than is normally used. The report then grouped all the organisms which were identified and counted into twelve size categories, starting from 1.25 μm and going up to more than 1000 μm.

The species composition was essentially similar in the Strait of Georgia and at ocean Station P, but naturally meroplankton as well as *Cladocera*, *Oikopleura* and also euphausiids were more abundant inshore. Though there are very substantial differences in the two areas in size and numbers of the zooplankton categories, the annual mean was approximately equal in the two areas with about twelve organisms per litre. The most significant difference is that coastal waters showed maximum numbers over the summer, roughly June–September, and the minimum over November–January. In contrast the oceanic waters showed a maximum in winter, especially December–January, and minimum plankton numbers in June. There were also size differences; the "smaller" organisms less than about 225 μm were about three times more numerous at the oceanic station than in coastal waters; whereas the 225–550 μm, category were more numerous in the coastal waters. The larger organisms, from 500 up to 1000 μm, were approximately equal in the two areas. The smallest size group, less than 125 μm, was particularly common in January in the oceanic waters. The largest organisms were only about one-third as numerous in the inshore stations in January, but on the other hand in June the inshore population had increased by five times.

The microzooplankton (in which LeBrasseur and Kennedy include everything less than 1 mm long axis normally running about 0.5 mm or less and passing the 350 μm net but retained by the 44 μm net) at the open ocean station consisted primarily of protozoans (*Colisphera* and *Globidurina*) with some dinoflagellates especially in March/April, and small copepods particularly *Serracalanus*, *Oithona* and *Oncaea*. Of nauplii, *Oithona* was overwhelmingly important. There is a suggestion that some of the large copepod species such as *Calanus plumchrus*, which develop through the winter at a time when primary production is small, may have to use microzooplankton as a food source. LeBrasseur and Kennedy also examine the question of the microzooplankton which is retained by the very small mesh used (44 μm), and how far this adds to the annual zooplankton biomass estimates which they have made with larger nets. They compute that it is less than 5 % for coastal areas and about only 14 % for oceanic areas. This still leaves open the important problem that, using chemical methods to estimate

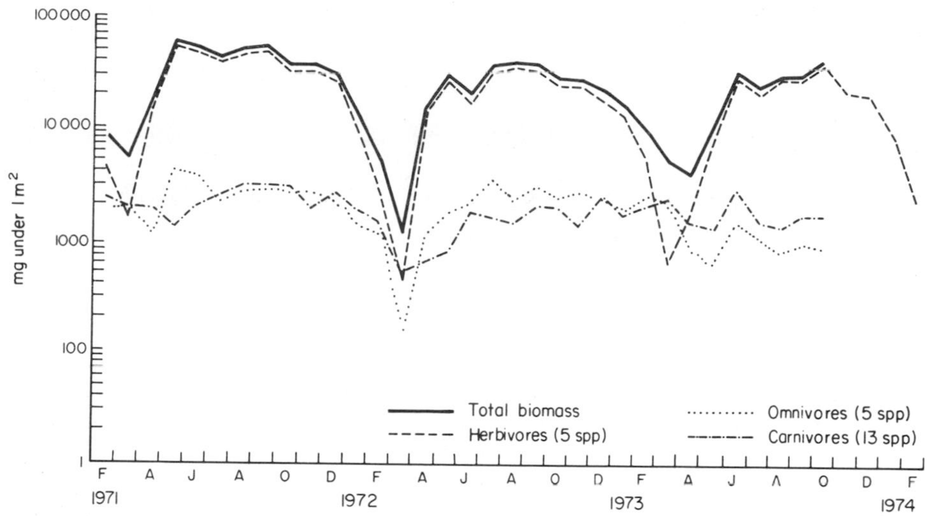

Fig. 1.19. Monthly biomass (mg dry weight/m² column of 690 m) plotted on a logarithmic scale (Matthews and Bakke, 1977).

non-plant protein, earlier workers such as Banse and others have estimated some 18 mg of nitrogen/m³. Despite the addition of this collection of microzooplankton by using 44 μm mesh nets there is still some 85% of the total non-plant protein unaccounted for in seawater. Does this arise from animal detritus, or is there still a very large population, for example, of small ciliates, which bypasses the 44 μm mesh and contributes substantially to total zooplankton biomass?

Whereas the values for biomass in the English Channel and Long Island Sound relate mainly to zooplankton of comparatively small size (small copepods were dominant in both areas), a very different zooplankton fauna has been investigated recently by Matthews and Bakke (1977). They give information on changes in biomass over the year for twenty-three species of mero- and macroplankton in a 690 m column at Korsfjorden (Norway). There is presumably no information on the smaller zooplankton. A marked seasonal change in biomass was described, with a minimum in March/April, rising sharply in June to a more constant quantity over summer. A maximum was attained by the end of summer or early autumn. The biomass beneath 1 m² varied between 1 and 60 g (dry wt). Figure 1.19 illustrates changes in the biomass of the zooplankton and of different trophic levels. Such an investigation is not only of great intrinsic interest, but has yielded important suggestions on the trophic relationships of many of the zooplankton fauna. It is difficult, however, to compare the standing crop with that for other neritic areas in which macroplankton is sparse or represented by other taxa, for example by ctenophores. Values for biomass in communities where the macroplankton makes a substantial contribution can hardly be compared with those where the smaller zooplankton comprises almost the whole of the biomass. It would also be difficult to contrast the biomass of plankton at Korsfjorden with that of high latitude open oceans.

Estimates of total zooplankton biomass, therefore, while of considerable value, may complement but cannot replace the detailed analysis of the taxonomic groups. The next chapter introduces the range of taxa in the zooplankton.

Chapter 2
The Major Taxa of the Marine Zooplankton

The enormous diversity of animals in the plankton is well recognized. There follows a general description of the major taxa, including some indication of the horizontal (geographical) and vertical distribution of certain of the better-known species.

Protozoa

Of the great variety of protozoans, foraminiferans and radiolarians were early recognized as important components of the plankton, their shells being recovered from surveys over vast areas of the oceans. Other shelled protozoans (e.g. tintinnids) are also well known, but, in contrast, our knowledge of the shell-less forms, which are easily destroyed during collection, is very limited.

Foraminifera

Almost all foraminiferans are marine. The cell is surrounded by a calcareous many-chambered shell, pierced by pores through which the ectoplasm passes. This layer gives rise to a network of very fine anastomosing pseudopodia. Although planktonic foraminifera can be extremely numerous so that their calcareous shells can contribute to a very great extent to deep-sea deposits leading to characteristic foraminiferan ooze, the number of living species of planktonic forms is extremely limited, in contrast to the numerous benthic species. Bé (1966; 1967) and Bé and Tolderlund (1971) quote 27–30 species of planktonic foraminiferans drawn from only two families.

As with many other plankton animals, the Foraminifera seem fairly clearly limited in their distribution to particular water masses. Temperature is an important factor, and they appear to be absent from water of low salinity. Bé and Tolderlund suggest that twenty-two species of foraminiferans are warm water forms, whereas only five are cold species (Arctic, Antarctic or sub-polar). These include *Globigerina quinqueloba*, *Globigerina bulloides*, *Globorotalia bradyi*, *Globorotalia scitula* and *Globigerina pachyderma*, but only the last species appears to be abundant at very high latitudes. It is the only species found in the Arctic Ocean and is dominant both north of the Arctic Circle and south of the Antarctic Convergence. This high latitude distribution is, however, largely confined to the left coiling variety of the species; right coiling *Globigerina pachyderma* is typical of sub-polar waters, both in the northern and southern hemispheres. *Globigerina bulloides* is usually the dominant foraminiferan of sub-Arctic and sub-Antarctic areas. In the sub-Antarctic the species can spread to a limited extent south of the Antarctic Convergence. Saidova (1972) found foraminiferans in the Kurile Kamchatka Trench, mainly *Globigerina pachyderma*, *G. bulloides* and *Globorotalia inflata*. Live specimens occurred to 200–250 m with a peak density of 4–7/m^3 in the 50–200 m layer, and only about 1/m^3 in the uppermost 50 m. Tests with some cytoplasmic remnants, however, were found at considerable depths. Belyayeva (1969) found *G. pachyderma* and *G. bulloides* in Antarctic waters in plankton and in sediments. Tests were less plentiful in shelf and deep-water deposits, owing it is believed, to dilution from land run-off for the shallow waters and to dissolution of calcium carbonate at greater depths for the deep deposits. Hada (1972) reported a new foraminiferan, *Globigerinella oyashiwo*, a cold-water form in Japanese waters, as the first known cold-water species of the genus.

In the transition zones marking the borders of warm and cold currents, such as the Gulf Stream and Labrador Current boundary, and the Kuroshio and Oyashio Current boundary, cold-water and warm-water species of foraminiferans overlap to some extent. *Globorotalia inflata*, however, appears to be a foraminiferan species which is indigenous to transition zones, both north and south. One species, *Globigerinita glutinata*, appears to be a eurythermal cosmopolitan form.

Of the twenty-two warm-water foraminiferan species, a dozen are mainly sub-tropical, being typically found in central oligotrophic water masses, roughly limited north and south by the 40° latitude. Some of these foraminiferans are characteristic of the edges of the central waters and occur in upwelling regions and in smaller numbers in transition zones. Though the sub-tropical foraminiferans may be quite common in tropical waters, some ten of the warm-water species (e.g. *Globigerinoides sacculifer*, *Pulleniatina obliquiloculata*, *Globorotalia menardii* and *Globorotalia tumida*) are more confined to the tropical waters of all three oceans (cf. Fig. 2.1). Three of the tropical species are restricted to the Indo-Pacific area.

Berger (1969) suggests that Foraminifera are most abundant in the upper 100 m and that numbers decrease rapidly with depth. The mean density recorded for 160 μm net captures was about 10/m^3, but enormous variations were found (less than 1 – 100,000/m^3), though change in size of mesh of the net appears to be partly responsible. The importance of net mesh size was confirmed by Bé and Tolderlund (1971). Generally higher densities were obtained with finer nets (cf. Chapter 1). Berger correlated high densities of Foraminifera with areas of high fertility; he implies that the protozoans are essentially algal feeders.

Bé and Tolderlund confirmed that most Foraminifera are surface or sub-surface dwellers, but while the majority of the spinose species live in the upper part of the euphotic zone (for example, all species of *Globigerinoides* and many species of *Globigerina*), the majority of the non-spinose species live at depths exceeding 50 m, some deeper than 500 m. All species of *Globorotalia* have a considerable range in depth; the young stages live in the euphotic zone but older foraminiferans, both of this and other genera, tend to live more deeply. *Hastigerina*, a spinose form with a transparent test, is also fairly deep living. While a very few foraminiferans may approach 1000 m depth, none appear to be really deep-living forms. Some of the deeper-living species, as with other taxa, appear to have a somewhat wider geographical range than near-surface species. Bé and Tolderlund's results confirm that areas of high primary productivity tend to be regions with richer Foraminifera populations. Oligotrophic waters may have foraminiferan abundance below 1 specimen/m^3. Figure 2.2 shows the association of high density Foraminifera areas with regions of current gyrals and

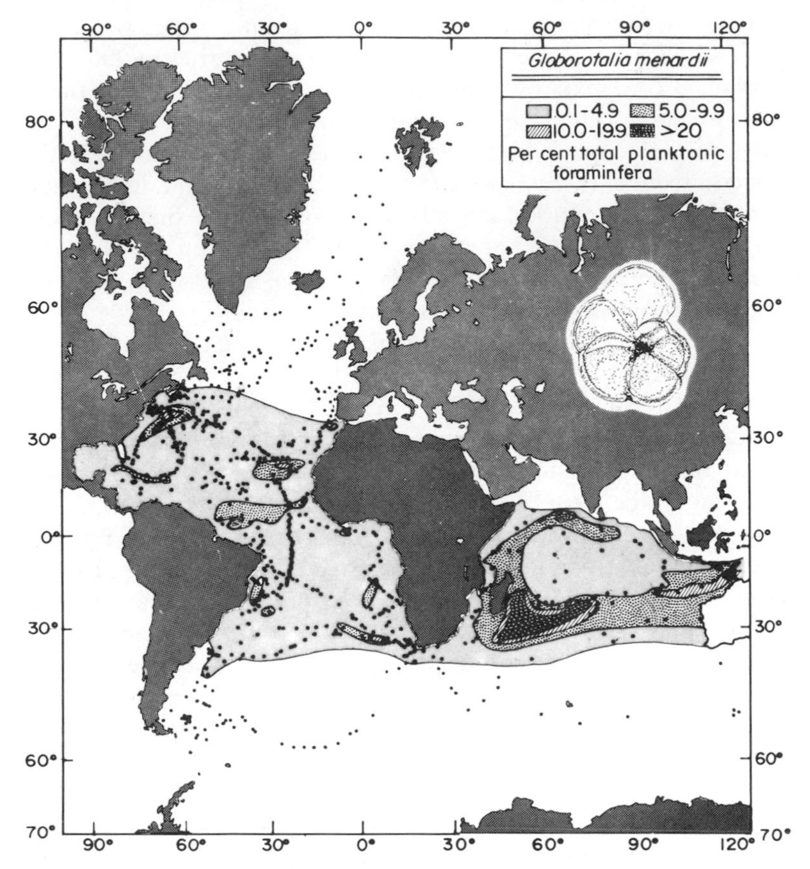

Fig. 2.1. Distribution of relative abundance of *Globorotalia menardii* in surface waters (0–10 m) (Bé and Tolderlund, 1971).

upwelling areas; these may have 100 or more Foraminifera/m^3. Bé, Vilkes and Lott (1971) also demonstrated the association of different foraminiferan faunas with the various water masses of the northern sub-tropical and southern sub-tropical Sargasso Sea and further differences between these warm assemblages and the sub-Arctic foraminiferans separated by the "slope" waters.

Bé, Harrison and Lott (1973) found that cold-water species of planktonic Foraminifera are usually smaller than sub-tropical or tropical species. Thus the polar and sub-polar *Globigerina pachyderma* and *G. quinqueloba* contrast with such warm-water forms as *Globorotalia menardii* and *Globigerinoides conglobatus*. *Orbulina universa*, however, one of the commoner Foraminifera in middle and lower latitudes, shows considerable variation in the form and size of the test. This species, while most abundant in relatively cool sub-tropical waters, has a considerable range from the transition zone between sub-tropical and sub-Antarctic waters throughout the sub-tropical and tropical zones. The species is unique among planktonic Foraminifera in that there are two stages in the life cycle, the trochospiral of the young stage being succeeded by a single-chamber spherical stage. No further growth occurs once the single chamber has been established, but the size of adult test shows considerable variation, particularly with latitude. Test diameter ranges from 210 μm to >1000 μm

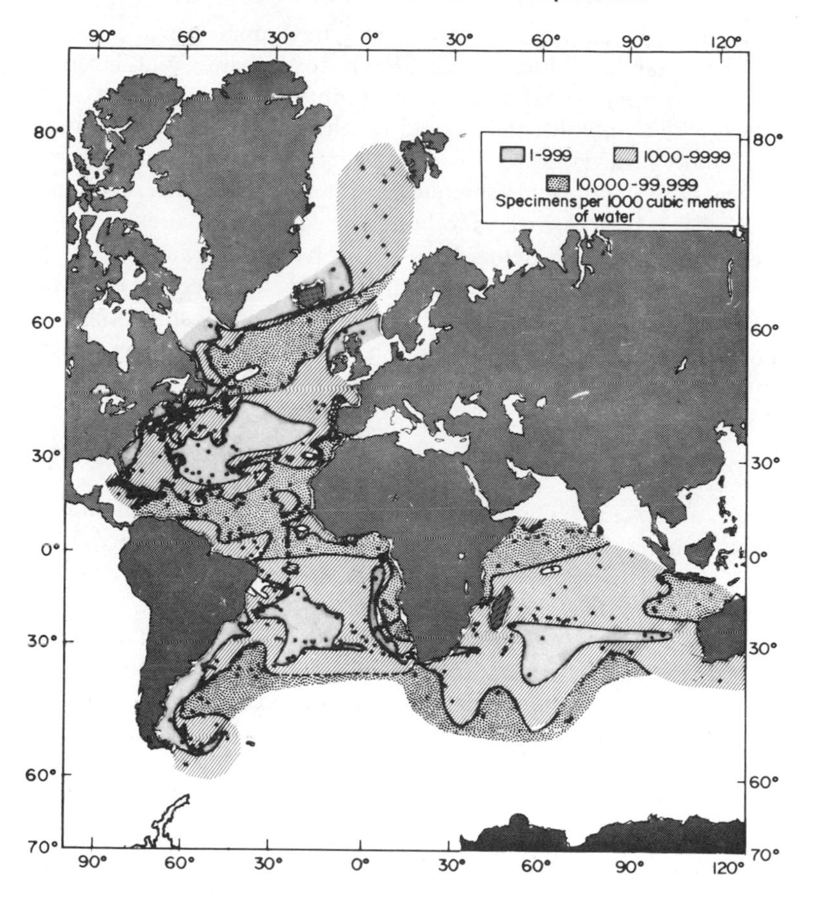

Fig. 2.2. Absolute abundance of total planktonic foraminifera in surface waters (0–10 m) (Bé and Tolderlund, 1971).

diameter, increasing at lower latitudes and directly with surface water temperature. In the Indian Ocean, for example, the larger shells (diameter 600–800 μm) occur in tropical and sub-tropical waters, medium sized shells (450–600 μm) in mid-latitudes (about 23°–32° S), and the smaller shells (<450 μm) from about 32°–45° S. Bé, Harrison and Lott (1973) and Hecht, Bé and Lott (1976) also demonstrated differences in shell diameter, variations in thickness of the wall of the test and differences in porosity of the test with latitude. The last characteristic, porosity, refers to a character of pelagic foraminiferans that the test wall is perforated by pores roughly in two size categories, small and large. Both the number of pores and especially the diameter of the smaller category of pores is variable. In *Orbulina universa* the lower the latitude, the greater the porosity (increase in diameter and in number of small pores). To some extent the differences in porosity appear to be true of some other foraminiferans.

Many planktonic Foraminifera, at least in the younger stages, are characterized by possessing long and delicate spines as well as the mass of pseudopodia. The calcareous spines presumably aid flotation. Bé has described a planktonic form, *Hastigerina pelagica*, which, though of considerable size with spines up to 7 mm long, is able to float by means of a bubble-like capsule, some 2 mm in diameter, made up of gas-filled chambers. The chambers house several photosynthetic dinoflagellate species (*Dissodinium pyrocystis*), apparently living symbiotically. Although surface-living foraminiferans with symbiotic algae could synthesize photosynthetic products, the majority are usually regarded as herbivorous feeders. Bé has, however, seen a copepod captured by the pseudopodia and subsequently digested by *Hastigerina pelagica*. Some Foraminifera might, therefore, have to be regarded as omnivores.

As regards reproduction, gametogenesis has been demonstrated for two planktonic foraminiferan species, so presumably sexual reproduction occurs. Huge numbers of flagellated gametes are released after the parent cell, having lost the long spines of the test, has sunk into somewhat deeper layers. Whether the reproductive cycle is similar to that of benthic foraminiferans has not yet been proved.

Radiolaria

Radiolarians are exclusively marine and, in contrast to foraminiferans, almost all are planktonic organisms. They are characteristically oceanic; indeed, it is doubtful whether there are any neritic species, though occasionally they can drift into bays in considerable numbers. The radiolarian cell can be of large size and considerable complexity. It is divided into an inner and outer layer of cytoplasm (sometimes claimed to be similar to the endoplasm and ectoplasm of other protozoans) by an organic membrane—the central capsule—containing a relatively large nucleus and certain other inclusions such as oil drops, granules and crystals. The extra-capsular cytoplasm is divided into a mainly assimilative layer where food is digested and a highly vacuolated frothy outer region called the calymma, with a further limiting outermost layer. The calymma is believed to assist in flotation of the cell; the size of the fluid-filled vacuoles or alveoli is subject to considerable change.

Numerous pseudopodia radiate from the cell, branch and anastomose. Some fine unbranched pseudopodia resembling axopodia are also present, though Mackinnon and Hawes (1961) state that the true axopodia do not occur in Radiolaria. A skeleton made of spicules is characteristic of the group. Radiolarians inhabit all oceans and some live in the various layers probably over the whole great depth range; differences in vertical distribution are seen, however, amongst the various species.

If Radiolaria are classed as an order of the Rhizopoda (Sarcodina), one sub-group, the Acantharia, is now regarded by most authorities as an entirely separate order. Acantharians apparently lack the perforated membrane of the central capsule typical of radiolarians, and have true axopodia as well as reticulated pseudopodia. The skeleton in Acantharia consists of a regular arrangement of spicules. Basically there are twenty spicules, arranged in five planes and meeting at the centre of the cell. They are composed of strontium sulphate instead of silica; typically a lattice skeleton is also present. A type of hydrostatic apparatus is present (Massera Bottazzi and Nencini, 1969). Myonemes arranged where the spicules leave the outside of the cell can extend the ectoplasm. Casey (1971) states that acantharians are capable of changing depth by means of the hydrostatic apparatus, for example, leaving the surface in stormy weather. Zooxanthellae are present in the endoplasm of the central capsule area. Acantharians are chiefly surface-living forms, mainly in the warmer oceans (cf. Haeckel, 1887a; Vinogradov, 1970).

Massera Bottazzi, Schreiber and Bowen (1971) report that while acantharians are rarely found in shallow seas or over continental shelves, they are reasonably abundant among the zooplankton organisms of open oceans and where the coast slopes sharply to deep water. Surveys indicate that in no part of the open Atlantic do Acantharia amount to less than 1% by number of the total zooplankton. In some areas (e.g. North Equatorial Current, South Atlantic Gyre, Eastern Caribbean) they usually comprise 10–20% of total zooplankton numbers, though with their generally small size, their contribution to biomass must be small.

In one North Atlantic cruise Massera Bottazzi *et al.* (1971) found that out of twenty-five plankton tows, nineteen showed Acantharia exceeding 10 specimens/m^3, whereas Foraminifera and Radiolaria exceeded 10/m^3 in only eight and six plankton tows, respectively. Moreover, the highest density for radiolarians was 28 specimens/m^3, and for foraminiferans 39/m^3, whereas acantharians exceeded 40/m^3 in eight plankton tows. The comparatively low densities probably reflect the failure of the plankton nets to retain smaller protozoans, but the data suggest that acantharians were the most abundant of the protozoans studied. Cifelli and Sachs (1966) found both foraminiferans and radiolarians (*Spumellaria* and *Nasellaria*) present throughout the whole ocean in a traverse in the North Atlantic from Nova Scotia to the Caribbean, but suggest that radiolarians were slightly less abundant (maximum 27/m^3; minimum 0.6/m^3).

Massera Bottazzi, Vijayakrishnan Nair and Balani (1967) have also described appreciable numbers of acantharians occurring in the upper 200 m in the Indian Ocean.

The distribution of acantharians in relation to water masses in the North Atlantic was investigated by Massera Bottazzi and Vannucci (1964, 1965a, 1965b). Many of the abundant species in warm and warm/temperate areas (e.g. *Acanthocolla cruciata*, *Amphibelone anomala*, *Phyllostaurus siculus*) are found in the three major oceans and the Mediterranean. Other abundant species, such as *Amphilouche elongata*, are described as cosmopolitan. Few acantharians were found in one cruise over continental shelf or slope regions, and the North Equatorial Current and Counter Current areas were richer (average numbers: 140 and 147/m^3) than the Canaries Current and Sargasso Sea (average numbers: 12 and 14/m^3). Another cruise indicated marked abundance of species and individuals in the South Sargasso and Caribbean and, as before, in the North Equatorial Current; few were found in the North Sargasso and Gulf Stream regions. The cold Labrador Current had exceedingly few species (*Acanthochiasma rubescens*, *A. fusiforme*, *Heteracon biformis*, *Gigartacon* spp.) and these were widely distributed over warmer waters also.

A sharp reduction in acantharian populations was experienced on nearing extensive shores. A considerable seasonal change in population was discovered, however, in continental slope areas. Slope waters were very poor in acantharians in spring/summer, but by early winter (November) adults were comparatively abundant (224/m^3). Shelf waters remained virtually destitute of Acantharia. The Gulf Stream populations were similar to those of the slope waters in November. Richness in species and abundance could sometimes occur in spring. The variations may reflect seasonal fluctuations in the course of the Gulf Stream. Little is known of the reproduction of acantharians but there is a suggestion that they reproduce at deeper levels, and the appearance of many juvenile forms in North Sargasso waters in spring and summer might indicate a rise of deeper waters.

Massera Bottazzi and Nencini (1969) describe the geographical distribution of three acantharians in the Atlantic Ocean. *Acanthochiasma rubescens* has a very wide distribution in the north and south Atlantic, including the Labrador Current and continental slope areas off North America, though it is not found on the continental shelf. The species also occurs in the Mediterranean. Another apparently widespread species is *Acanthochiasma fusiforme*, recorded for the North Sea and the Labrador Current as well as throughout most of the warm waters of the north and south Atlantic and the Mediterranean. *Acanthocyrta haeckeli*, on the other hand, appears to occur mainly in the warm tropical and sub-tropical parts of the North Atlantic and Mediterranean, including the Gulf Stream and Canaries Currents.

The Radiolaria may be distinguished from the Acantharia by certain morphological features. In radiolarians the central capsule is always perforated by pores. Zooxanthellae, present in many of those species inhabiting the upper layers of the sea, are situated in the extra-capsular protoplasm, mainly in the calymma. Though some species lack a skeleton, spicules of silica are present in most radiolarians, forming a skeleton of remarkable diversity. The spicules may form an irregular network or be present as more or less isolated structures. Most frequently, they are united as complex latticework structures characteristic of the species, sometimes as a sphere or as a series of spheres, sometimes as a helmet-shaped or cup-shaped chamber or series of chambers. Radiating spines are developed in many radiolarians and there may be other ornamentation (hooks, spines and teeth). While most radiolarians are approximately spherical, a number show a kind of bilateral symmetry and a few are of irregular form.

Campbell (1954) describes radiolarians as occurring in all seas, with many species having a cosmopolitan distribution. Similar faunas are frequently typical of the three oceans, but Arctic, Antarctic and temperate species may be distinguished from typical warm-water forms. As with many other zooplankton taxa, there is a greater diversity of species in warm oceans, but the few cold-water forms may occur in considerable densities. In general, the Atlantic fauna appears to be less rich than the Pacific. Radiolarians show a considerable range of form. The sub-groups also exhibit some differences in their patterns of vertical distribution, so that a brief description of the sub-orders is required. The classification of Radiolaria is artificial (cf. Haeckel, 1887b; Riedel, 1971) and a more natural classification is being developed.

Three sub-orders of Radiolaria may be distinguished: Nassellaria, Spumellaria and Phaeodaria. The first two are frequently grouped together as polycystine radiolarians (cf. Riedel, 1971). In the Nassellaria (Monopylea) the central capsule is roughly egg-shaped, with a flat base which is the only perforated pole. The skeleton is based on a tripod of three rods, a ring, or a lattice-type shell, and may become exceedingly complex; in many it is of the helmet-type (cf. Campbell, 1954). The Spumellaria (Peripylea) have the central capsule pierced by numerous fine pores; the skeleton frequently takes the form of a spherical lattice enclosing the central capsule or a series of concentric spheres. In more modified forms the shell becomes ellipsoid or disc-shaped and may be elongated along one or more axes. A few lack the skeleton (e.g. *Thalassicola*); some are colonial with many central capsules. Haeckel (1887b) stated that some of these latter species can be extremely numerous on the surface at times. The shells of the Nassellaria and Spumellaria contribute to the radiolarian ooze which is found over many oceanic areas. As Casey (1971) points out, deeper-living forms tend to be larger and laterally compressed and their skeletons are more massive. The Acantharia do not contribute much to oozes since the strontium skeleton dissolves.

The third sub-order of radiolarians, the Phaeodaria (Tripylea) have a central capsule with a double membrane. There is one major apical pore and typically a very few, usually two, other pores. A characteristic feature is a mass of pigmented material, the phaeodium, which lies just outside the central capsule and eccentric to the pore field. There are no zooxanthellae; their absence may be related to the deep-living habit of many of the order. Campbell (1954) emphasizes the numerous deep-living species, especially in the southern hemisphere. Vinogradov (1970) summarizes the vertical distribution of the sub-orders—the Nassellaria are typically in the surface waters of warmer oceans, with only a few species reaching deeper layers. While the Spumellaria have a fairly similar vertical distribution, more species extend to deep waters (species of *Stylotrochus*, *Heliodiscus*, *Trochodiscus* and *Plegmosphaera*), with some living at 2000 m or even exceeding 4000 m. Only the sub-order Phaeodaria is very well represented at deep ocean levels, as well as occurring in shallower depths.

Some Phaeodaria are of considerable size for protozoans; Haeckel (1887b) describes the majority as 1–2 mm, some exceptionally reaching 20 mm or more in diameter. Some of the families are typical of the deep ocean fauna. For example, the Challengeriidae, including *Challengeria*, *Challengeron* and *Pharyngella*, are well known in deep waters (Fig. 2.3). In this family the skeleton forms a tracery of fine hexagonal meshes and usually has a mouth opening at one end, often fringed with teeth. The shell may also have spines and horns. The Tuscadoridae have solid oval or spindle-shaped cells with a mouth; sometimes the shell has leg-like extensions (e.g. *Tuscadora* = *Tuscarora*). The Aulacanthidae have a skeleton with many hollow radial tubes, often branched externally and usually a network of spicules in addition; typical genera are *Aulacantha*, *Auloceros*, *Aulospathis* and *Aulographis* (Fig. 2.4). The Conchariidae have a bivalve shell; the two valves are relatively thick and may be unequal. In the Medusettidae there is an oval or cap-like shell with hollow articulated leg-like structures surrounding a wide open mouth (e.g. *Medusetta*, *Gorgonetta*) (Fig. 2.3). Other well known genera include *Haeckeliana*, with a spherical shell and unbranched radial spines, and *Castanea* with a spherical shell and large mouth armed with teeth. In general, in Phaeodaria the skeleton is made up of hollow rods or tubules, but some families have a more solid network. Although so many deep-sea species are known, their skeletons are comparatively few in deep-sea oozes. Reshetnjak (1971) believes that this is associated with the chemical nature of the skeleton. Although composed of siliceous organic material as in all true radiolarians, the precise composition of Phaeodaria apparently differs from that of the polycystines.

Little is known about reproduction in the radiolarians. Binary fission may occur, at least in some species; multiple fission, producing flagellated "swarmers", is known for a number of radiolarians but the details of reproductive cycles are still uncertain. Renz (1976) states that Nassellaria and Spumellaria can multiply by simple division, with one daughter cell remaining in the shell and the other forming a new skeleton, and also by a sexual process where, after repeated nuclear division, biflagellate swarmers are produced which fuse to form a zygote developing a new shell. He also refers to "young" individuals, distinguished as beginning to form the apical chamber or second shell-joint in Nassellaria or the medullary or cortical shell in Spumellaria.

A large proportion of small, young radiolarians was captured in central Pacific

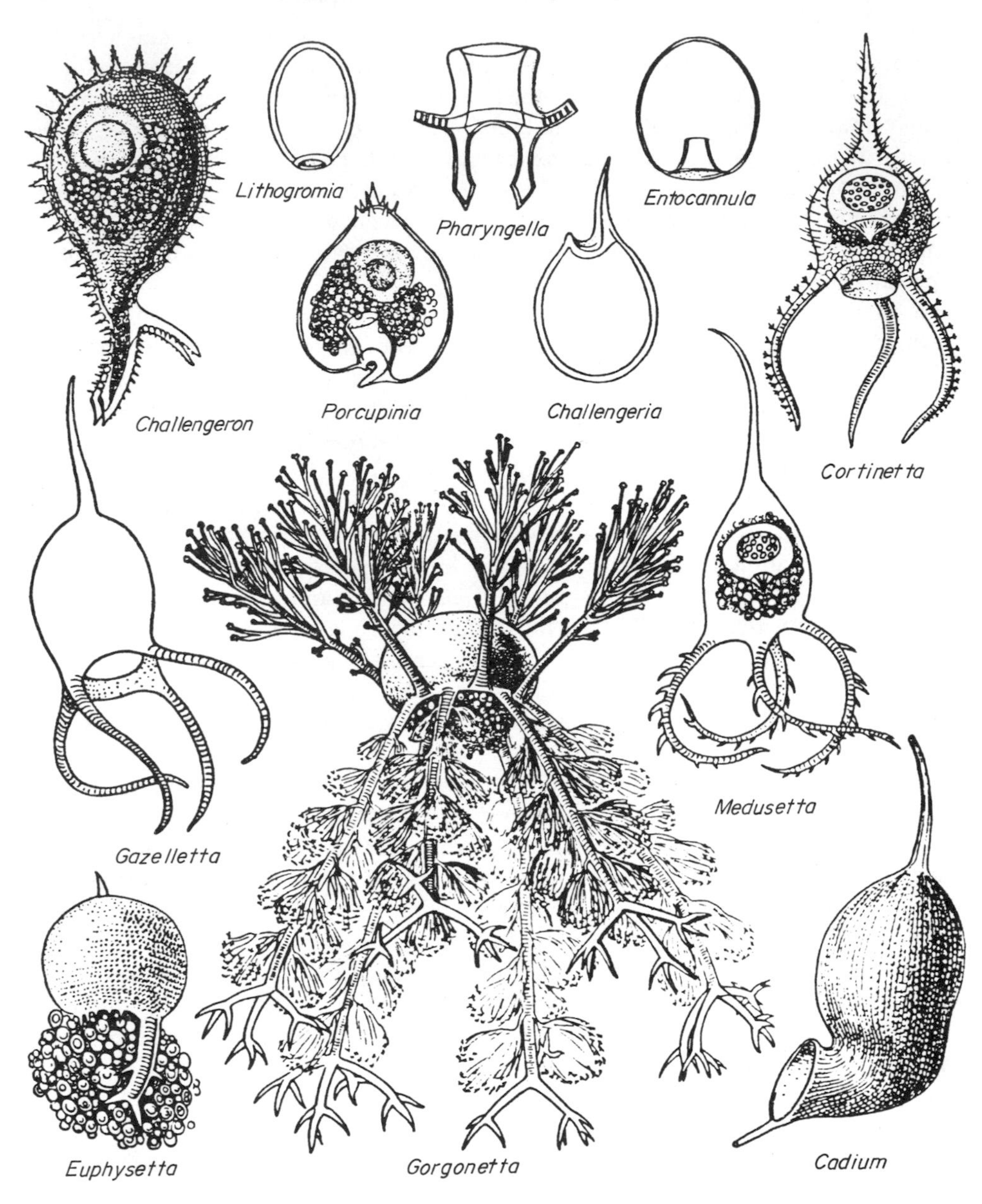

Fig. 2.3. Challengeriidae, Cadiidae, Medusettidae (Campbell, 1954).

waters. These measured >35 μm but <103 μm in diameter and are not taken in normal net samplings. Haeckel (1887b) states that some radiolarians can be so abundant as to play a large part as food of other planktonic and abyssal animals; for instance, medusan, salp, crustacean and pteropod guts can be a rich source of species. Vinogradov (1972b) describes zooplankton hauls from the Kurile–Kamchatka Trench at 5000–6000 m and deeper as always containing a large proportion of radiolarian debris, with a trace of phaeodia, and consisting mainly of Aulacanthidae,

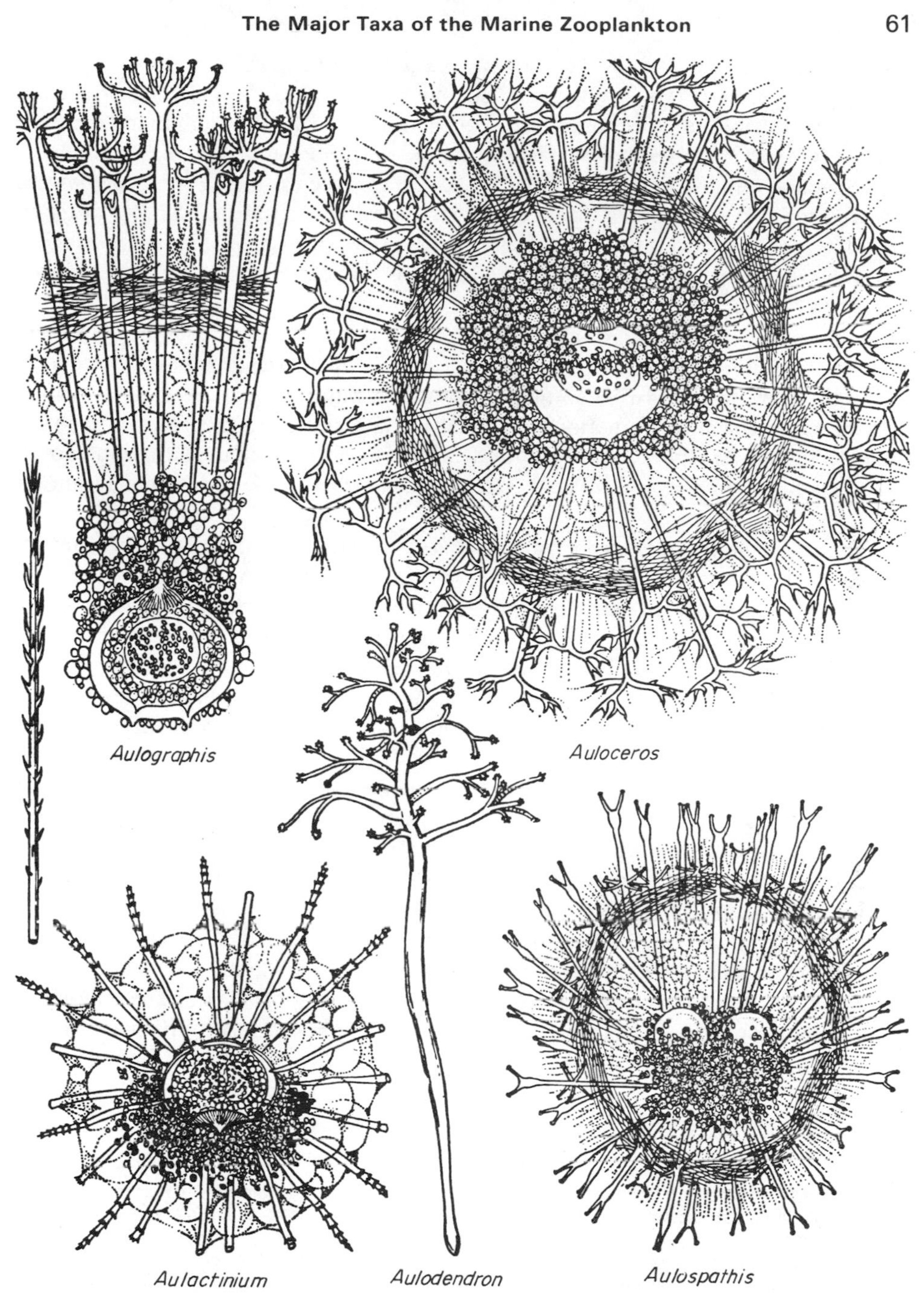

Fig. 2.4. Aulacanthidae (Campbell, 1954).

together with *Challengeria*, *Trochodiscus*, *Haeckeliana* and others. The intestinal contents of some of the deep-sea Metazoa (mainly gammarids and copepods) contained traces of radiolarian phaeodia, indicating that this is a major component of their diet.

Although Haeckel (1887a) described the geographical distribution of many radiolarian species taken during the Challenger Expedition, he held that a general analysis of distribution was premature. For the Phaeodaria, although many are deep living, there are a number of surface forms. Some genera (*Aulacantha, Castanella*) include many common and widely distributed surface species. *Aulacantha scolymantha* is described as the commonest surface-living species and is cosmopolitan, *Aulographis pandora* is also cosmopolitan but other surface species (e.g. *Aulodendron antarcticum* and *Aulosphaera bisternaria*) are described as occurring only in the Antarctic. Haeckel lists a few Arctic species (e.g. *Aulosphaera multifurca* and *Sagenoscena spathillata*). Among the Challengeriidae, *Challengeria naresii* is described as a cosmopolitan deep-living radiolarian. In contrast, *Challengeron swirei* and *C. richardsii* are regarded as Antarctic forms found fairly commonly in surface waters.

Even a superficial examination of Haeckel's data indicates the great number of radiolarian species, especially polycystines, in the warm waters of the North and South Atlantic and North and South Pacific Oceans, in contrast to the rather small numbers of cold-loving species.

From an examination of recent radiolarians from material from the tops of sediment cores at higher southern latitudes, Hays (1965) found that the Antarctic Convergence marked the limit between fairly sharply contrasted radiolarian faunas to north and south. Seven species were considered to be characteristic of the Antarctic fauna (cf. Table 2.1). Three were endemic (*Lithelius nautiloides, Triceraspyris antarctica, Peromelissa denticulata*); *Helotholus histricosa* and *Botryopyle antarctica* were bipolar; the remaining two species ranged outside Antarctic waters, though the abundance of *Spongoplegma antarcticum*, especially near the Convergence, was characteristic, as Haeckel (1887a) had described earlier. The warmer-water radiolarians north of the Antarctic Convergence were mostly cosmopolitan species, but in still lower latitudes the number of species increased, the fauna including tropical radiolarians.

Reviewing the distribution of polycystine radiolarians, Casey (1971) also refers to the Antarctic Convergence as causing a break in distribution. Species in Antarctic waters included a few endemic forms, some eight bipolar as well as thirteen cosmopolitan species. The latter tend to be somewhat deeper-living (>200 m), whereas the species restricted to very cold waters are mainly in the upper 200 m. The radiolarian population in near-surface Antarctic waters is often sparse; most individuals seem to inhabit the 200–400 m layer. Arctic waters may also apparently show a limited surface

Table 2.1. Species representative of the two faunas of the southern seas (Hays, 1965)

Antarctic Species	Warm-water Species
Spongoplegma antarcticum Haeckel	*Cenosphaera nagatai* Nakaseko
Lithelius nautiloides Popofsky	*Echinomma leptodermum* Jörgensen
Triceraspyris antarctica (Haecker)	*Axoprunum stauraxonium* Haeckel
Helotholus histricosa Jörgensen	*Heliodiscus asteriscus* Haeckel
Peromelissa denticulata (Ehrenberg)	*Androcyclas gamphonycha* Jörgensen
Theocalyptra davisiana (Ehrenberg)	*Calocyclas amicae* Haeckel
Botryopyle? antarctica (Haecker)	*Lamprocyclas maritalis* Haeckel
	Stichopilium annulatum Popofsky
	Eucyrtidium tumidulum? Bailey

population and a more widely distributed deeper radiolarian fauna.

Casey (1971) believes that the major zones in the geographical distribution of polycystine radiolarians over the North and South Pacific correspond approximately to the recognized hydrological divisions described for other zooplankton taxa. Although other faunal zones may exist for deeper-living radiolarians, corresponding to deep-water masses, many species are eurybathic, so that the distribution pattern is less certain.

The geographical distribution suggested by Petrushevskaya (1971a) follows a broadly similar outline with the main faunal divisions defined as: Antarctic species; bipolar species, found in the Antarctic and in the North Pacific and/or North Atlantic; cosmopolitan species distributed across the tropics but submerged below the warm surface waters; warm-water tropical species; species inhabiting slightly cooler waters, approximately sub-tropical. Petrushevskaya lists species typical of particular water masses, for example, *Euchitonia elegans*, *Botryocyrtis scutum* and *Pterocanium praetextum* inhabit surface tropical waters of all three oceans; *Lithelius nautiloides*, *Antarctissa strelkovi* (= *Helotholus histricosa*) and *Spongotrochus glacialis* are more or less confined to Antarctic waters; *Antarctissa denticulata* (= *Peromelissa denticulata*) and *Saccospyris antarctica* (= *Botryopyle antarctica*) are usually low Antarctic species (cf. Hays' results).

Petrushevskaya (1971b), investigating the warm-water Spumellaria and Nassellaria from the central Pacific, found a rich assemblage of species and individuals in the very warm surface waters to about 100 m depth (temperature 20°–28°C), with densities ranging from 5000 to 15,000/m^3. The richest populations occurred in the equatorial region; some species were found *only* in the upper layers and in the tropical regions (e.g. *Pseudocubus obeliscus*, *Zygocircus archicircus*). These are presumably stenobathic and stenothermal. Another less rich radiolarian assemblage, including *Cornutella verrucosa* and *Dictyophimus clevei*, occurred somewhat deeper in the central Pacific, where the temperature was lower (7°–19°C). These species are absent from the surface; they occur from about 100 m and are abundant at around 300 m. Their density is also much lower than surface forms; usually only 25–50 specimens/m^3 of any species were found. These cool-water species are, however, found widely in temperate areas of the oceans (cf. Fig. 2.5). A third faunal assemblage of polycystines, living below 500 m, had very low densities (*ca.* 10–20/m^3), but details were not studied (cf. Fig. 2.5). Dead skeletons of radiolarians were abundant in the whole water column; in deep water they exceeded the density of living forms (cf. Fig. 2.6).

Renz (1976) also investigated the distribution of radiolarians in warm central Pacific waters. He described a total of 110 species of which sixty-seven were common to the three domains, northern tropical, equatorial and southern tropical. While the radiolarians were abundant in richer upwelling areas, about 0°–10°N latitude, central gyre areas were poor. Although there was a gradual diminution in abundance to north and south from the equatorial zone, Renz obtained much higher densities for the whole region than those recorded by many earlier workers, viz., 242–18,370 individuals/m^3. Renz, however, analysed pump samples, filtered through mesh as fine as 35 μm. The high density of the smaller radiolarians is significant. Thus, the number of radiolarians captured by the 35 μm mesh was five times that obtained by the 103 μm mesh and fifty times that for the 363 μm mesh. One of the commonest species collected, *Psilomelissa calvata*, measured only 50–60 μm.

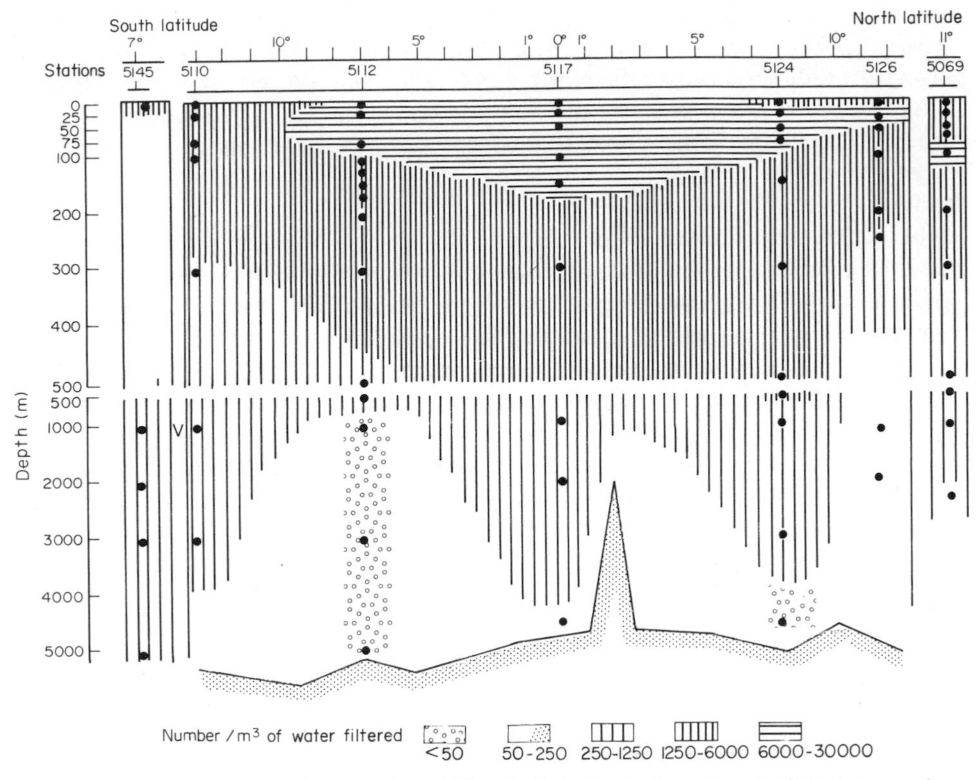

Fig. 2.5. Distribution of living Spumellaria and Nassellaria in the plankton along 154° W (Petrushevskaya, 1971).

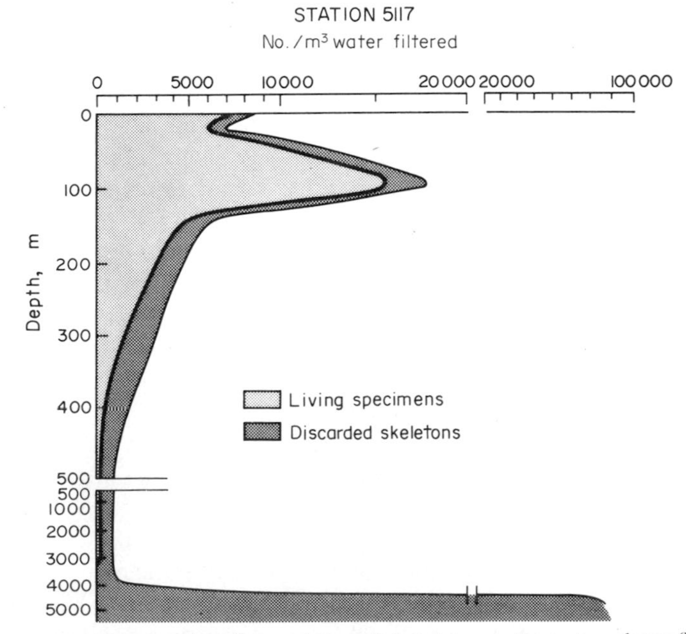

Fig. 2.6. The quantity of living radiolarians and discarded skeletons in the water column (Petrushevskaya, 1971).

As with other zooplankton, it is impossible to separate consideration of horizontal distribution of species from their vertical distribution. There are some indications of temporary depth changes in polycystine radiolarians, though the evidence is much less clear than for Acantharia. Casey (1971) cites polycystines coming into the immediate surface during winter in the Mediterranean and changes in depth in the upper 200 m off California associated with upwelling.

The vertical distribution of radiolarians has been described by many investigators. Haeckel (1887) divided the fauna into five zones. Haecker's (1907) classification indicated a colloid zone (0–50 m); challengerid zone (50–350/400 m); tuscadorial zone (350/400–1000/1500 m) and a pharyngellid zone (1000/1500–5000 m). The deeper forms were largely Phaeodaria, with the three deeper levels having, as characteristic species, numbers of Challengeriidae, Tuscadoridae and *Pharyngella* respectively. Casey (1971) points out that some Sphaerellaria and Cyrtellaria occur in deeper waters, from 400 to 5000 m. Reshetnjak's (1966) classification, quoted by Vinogradov (1970) and by Renz (1976), for the Kurile–Kamchatka area, indicated five zones: 0–50 m; 50–200 m; 200–1000 m; 1000–2000 m; 4000–8000 m (abyssal). In most of such classifications frequency of occurrence of the commoner species was the main criterion. Individual species were also held to be more or less confined to a particular depth stratum (stenobathic) (e.g. *Pandora* spp. from 400–700 m; *Aulospathis* spp. about 1000 m; *Aulographis* spp. mainly 400–1000 m). Vinogradov (1970) believes, however, that among the Phaeodaria at least half the species have a very considerable vertical range of about 2000 m and that only about 2 % of the species are really deep-sea stenobathic forms.

For shallower layers, Casey (1971) concluded from work in Californian waters that three vertical zones existed to about 1000 m, the deepest level sampled, and that these three upper zones corresponded to some degree with thermocline and pycnocline levels.

Almost all investigators agree that maximum radiolarian densities occur at the surface or at about 100 m in some areas (e.g. Indian Ocean tropical waters), and that a marked decline in density occurs with increasing depth. Renz (1976) found, for the upper 200 m of central Pacific waters, that a change in fauna occurred at around 75 m depth, but that rather few species were absolutely restricted by depth. Of a total of 137 species, twenty-six were thus restricted by the 75 m depth limit, about half the remaining species occurred mainly either above or below the 75 m level and some (31) had a range throughout the water column. While sampling below 200 m was much less complete, Renz suggests that only five species were restricted to depths exceeding 200 m. All other species found at depths greater than 200 m could also occur at the surface. Samples taken at four deeper stations revealed an usual sharp fall in density with depth. Samples at 1200 m varied from 111 and 48.6 individuals/m^3, as compared with only 1.4 radiolarians/m^3 at 4000 m.

Although it is accepted that many Radiolaria are eurybathic with a wide depth range as great as 50–2000 m, or even to 8000 m, Reshetnjak suggests that some 45 % of the Phaeodaria in the Kurile–Kamchatka Trench are stenobathic. Vinogradov (1970), however, from an examination of data for the northwestern Pacific, points out that of 103 species, sixty-three exhibited a remarkable eurybathic distribution and only twenty-eight species, living always deeper than 200 m, were restricted to a specific stratum (200–1000 m; 1000–2000 m; or 4000–8000 m). He alludes to differences in the vertical distribution of radiolarians according to Haecker's and Reshetnjak's data and believes that this supports the view that most radiolarians, essentially Phaeodaria, are widely eurybathic. For the Kamchatka area, if surface species to a depth of 200 m be excluded, eighty-two of the remaining species (41 %) have a range of more than 2000 m, while twenty-six species have a vertical range exceeding 4000 m. Only two stenobathic deep-sea species, less than 1 % of the fauna, live deeper than 2000–4000 m. Although most species of the families Tuscadoridae, Challengeriidae, and Thalassothamnidae (Spumellaria) along with a few other genera (*Circocastanea*, *Haeckeliana*, *Aulographis*) descend to 2000 m and even 4000 m, almost all these species reach the 50–200 m layer.

Even markedly eurybathic radiolarians do not have a uniform vertical distribution but show a clear tendency to prefer one water layer, where they may be found in great abundance. Not only may they constitute a considerable part of the total mass of plankton near the surface in certain regions, but they sometimes appear in masses at other depths. The rich fauna of Phaeodaria in the surface layer (0–50 m) in the North Pacific sharply differs from that at greater depths (Table 2.2). The fauna of the deeper layers reaches its greatest diversity in the 200–1000 m layer, and because of the eurybathic nature of most of the deeper-living species, faunistic changes are fairly gradual (Fig. 2.7). Vinogradov's data show that below the near surface (200 m) level the fauna becomes slowly impoverished, as other investigators have also found.

Table 2.2. The number of species of Radiolaria (Phaeodaria) occupying various depths[a] (Vinogradov, 1970)

Depth (m)	0–50	50–200	200–500	500–1,000	1,000–2,000	2,000–4,000	4,000–8,000
Total number of species	186	136	163	154	104	56	26
Number of "appearing" species		87	43	12	13	5	1
% of total number of species at this depth		64	26	8	13	9	4
Number of "disappearing" species	137	11	26	63	53	31	
% of total number of species at this depth	74	8	15	41	51	55	

[a] K. A. Brodskii (1957) and N. G. Vinogradova (1958) have shown that vertical changes in the faunistic composition are especially obvious on comparing the number of species having an upper or lower boundary of their distribution in each of the depth intervals sampled. For the sake of brevity these species will be designated as "appearing" or "disappearing" in a certain layer.

Flagellate protozoans and amoebae

Flagellates occur widely in the seas. Many of these are photosynthetic organisms and form part of the phytoplankton (cf. Volume 1). However, a number of colourless flagellates which are saprozoic or holozoic occur in the oceans. Many are minute forms and rarely estimated quantitatively; they are sometimes counted as part of investigations of the nanoplankton. Sorokin (1977) recorded large concentrations of the flagellate *Bodo* during an investigation in the Sea of Japan. Larger colourless flagellates

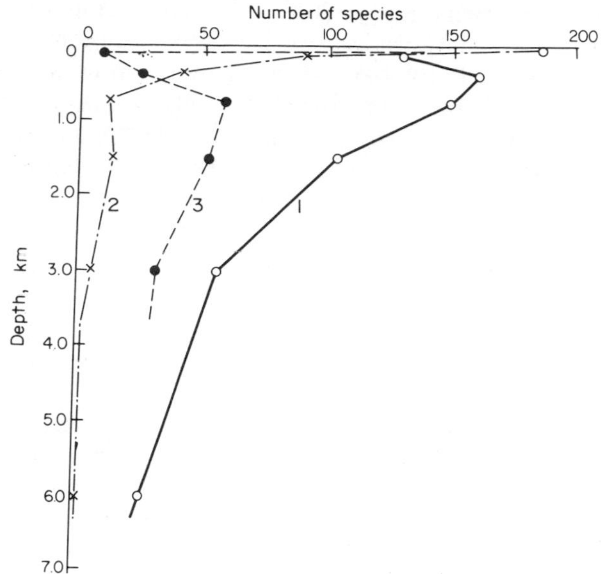

Fig. 2.7. Variation in number of species of Radiolaria (Phaeodaria) with depth. (1) total number of species, (2 and 3) number of species "appearing" or "disappearing" in each of the sampled layers (Vinogradov, 1970).

exist, amongst which are certain dinoflagellates. One of the best known is *Noctiluca scintillans* (*miliaris*), a relatively large species, which can occur in very great numbers at times in various marine communities. Petipa, Pavlova and Mironov (1970), for example, point to its importance in the Black Sea. Other genera of dinoflagellates in which chromatophores are lacking include *Oxyrrhis*, *Amphisolenia* and *Diplopsalis*. Certain species of *Phalacroma*, *Cochlodinium* and *Spirodinium* also lack chromatophores and nutrition must be holozoic or possibly, to a limited extent, saprozoic.

Even less well known are some amoebae which have been described from marine waters. Though many of these are undoubtedly bottom species which may be taken in shallow waters, some amoeboid forms almost certainly live in the open oceans, though they are normally very difficult to capture in identifiable form. There is also very little information on densities of amoebae in marine environments. Protozoa, however, are of considerable significance in the seas. With substantial populations and their potentially high rate of reproduction, they can have an important role in marine ecosystems (cf. Volume 1). Moreover, apart from herbivorous species which will feed on the phytoplankton, other protozoans can feed on material which is utilized by comparatively few metazoans. Some small flagellates and possibly a few other protozoans may be able to absorb dissolved organic matter, comparatively abundant in the deep oceans, but the majority feed on bacteria and detritus.

Ciliates

Ciliates are sometimes found in considerable abundance in seawater and their rate of turnover, as with other protozoans, tends to be high.

Many of the marine species feed mainly on bacteria and detritus but at least one important group, the tintinnids, are to a considerable extent phytoplankton feeders. The investigations of Gold (e.g. 1968, 1969, 1970) on culturing tintinnids, strongly suggest that these ciliates subsist largely on algae. Both Campbell (1954) and Zeitzschel (1966) describe tintinnids as utilizing bacteria and detritus but state that the food consists of naked flagellates, peridinians, coccolithophorids, diatoms and silicoflagellates. Tintinnids serve as food for a wide variety of zooplankton, e.g. copepods, euphausiids, Cladocera, salps, fish larvae and chaetognaths.

The cilia are greatly restricted on the tintinnid body; the major structure is the peristome with its twenty-four large feathery membranelles of fused cilia serving locomotion and feeding, the latter assisted by the pumping action at the base of the peristome. One of the most characteristic features of the tintinnids is the trumpet- or flask-shaped lorica, secreted as a fairly delicate test by the cell, but typically strengthened with foreign particles (Fig. 2.8). Tintinnids are readily recognized by the lorica (cf. Kofoid and Campbell, 1929, 1939), and more information is available on their distribution than for some other protozoan groups, since they can be more readily captured and identified. Tintinnids are often also the most abundant of the protozoans. Hada's (1972) studies of the protozoan fauna, in Akkeshi Bay, Japan, for example, indicated that apart from *Noctiluca scintillans*, most of the flagellates were autotrophic dinoflagellates. Though a few rhizopods, *Amoeba* sp., three foraminiferans, an acantharian, *Acanthometron pellucidum*, and one radiolarian, *Helotholus histricosa*—were reported, by far the commonest protozoans, and indeed the most abundant of the zooplankton organisms after copepods, were the ciliate group, Tintinnida. Hada states that tintinnids dominate the protozoan plankton, especially of colder Japanese waters. Of a total of forty-nine species of tintinnids recorded, no fewer than seventeen belonged to the genus, *Tintinnopsis*, a well known and common neritic tintinnid. Other species found were *Stenosemella nivalis*, *Codonellopsis frigida* and *C. borealis*, *Helicostomella fusiformis* and *H. sublata*, and species of *Favella*, *Parafavella*, *Ptychocylis*, *Acanthostomella* and *Tintinnus*.

About 800 species of tintinnids are known; almost all are planktonic and essentially marine, less than 2% being freshwater. There is general agreement that tintinnids are found in the surface layers of the ocean or very close to the surface (cf. Campbell, 1954). Hedin (1975) found all species, with one exception, in the uppermost 10 m off the west coast of Sweden. Zeitzschel (1967), sampling in the North Atlantic, referred to populations in the upper 50 m although, quoting from Vitiello, he points out that in the Mediterranean there are small differences in the vertical distribution of tintinnids with season, the maximum for the period November to March being at 30 m, but from May somewhat exceeding 50 m. He also describes tintinnids as exhibiting small diurnal vertical movements.

Many tintinnids have a very wide distribution and have been described as cosmopolitan. Campbell (1954) points out that North Atlantic forms are similar to those from the North Pacific, and that the tintinnid fauna of the equatorial Pacific is similar to that of the equatorial Atlantic. However, Antarctic and Arctic tintinnids are distinct. *Parafavella* and *Ptychocylis* are Arctic genera, in contrast to genera such as *Cymatocylis*, *Protocymatocylis* and *Laackmanniella*, which are confined to Antarctic waters. Marshall (1969) points out that *Cymatocylis*, though essentially Antarctic, has been found in the North Atlantic.

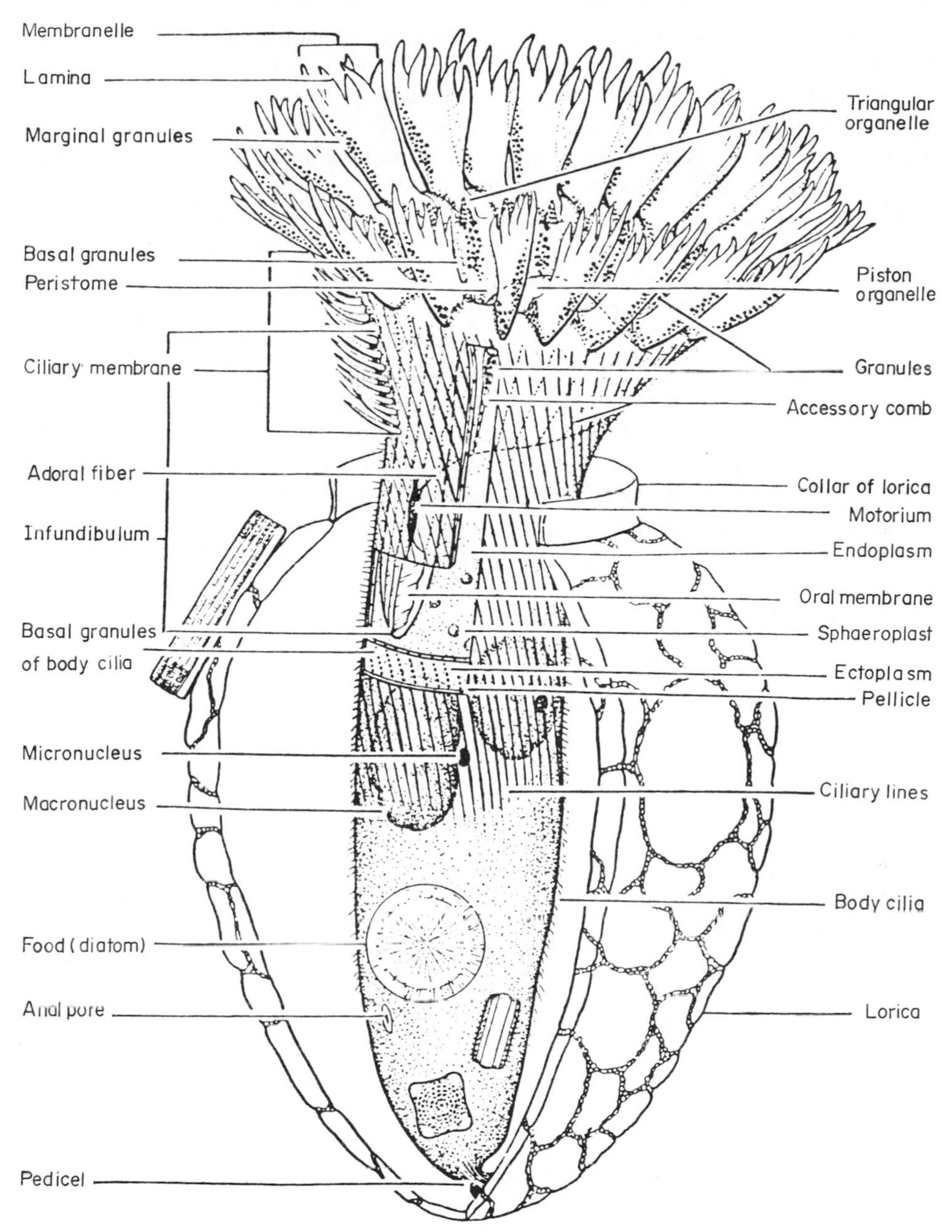

Fig. 2.8. Morphological features of a typical modern tintinnine *Stenosemella nivalis* from shallow Pacific waters off California. Part of the lorica is cut away in order to show soft parts of the organism inside. (× 1000) (Campbell, 1954).

A number of tintinnids, though widespread, are essentially neritic. Marshall (1969) describes *Tintinnopsis* as a mainly neritic and temperate genus, which is often very common in inshore waters and represented by a large number of species. Hedin (1975) found that out of a total of twenty-five tintinnids off the Swedish west coast, eleven species belonged to *Tintinnopsis*, *T. baltica* being extremely abundant and the other species of lesser importance. Burkill (1978) found six species of the genus *Tintinnopsis* out of a total of thirteen tintinnids in Southampton Water; *T. beroidea* was an extremely abundant species, present thoughout the year. *T. turbo* was also found throughout the year, though in somewhat smaller numbers. *T. beroidea* appears to be very widespread indeed in temperate neritic waters. Konovalova and Rogachenko (1974) quote it as the major species among the tintinnids of Amur Bay, where it was the main contributor to two large seasonal peaks of tintinnid abundance observed in spring and summer. The maximum density ($20–30 \times 10^6/m^3$) occurred in April/May.

A large number of other species of *Tintinnopsis* was found; the tintinnids were mainly in the upper 10 m. Another mainly neritic genus is *Stenosemella*, *S. nivalis* and *S. ventricosa* being extraordinarily widely distributed in the North Atlantic, including a range from Arctic seas to warm-temperate areas (Marshall, 1969). It is also known in the Mediterranean and can be abundant off the coast of British Columbia. In Hedin's (1974) investigations *Stenosemella nivalis* was the most abundant tintinnid in the area investigated, and occurred through the whole year. Species of *Helicostomella* are also neritic in character (cf. Hada, 1972). *H. subulata* is recognized as a relatively abundant neritic form. It was found by Hedin over the summer months in Gullmar Fjord and, though not very abundant, was present from April to October in Southampton Water (Burkill, 1978). Marshall (1969) gives an extremely wide geographical distribution for this neritic form in the Atlantic. A number of species belonging to other genera are mainly coastal in character. *Codonellopsis lagenula* is described by Gaarder (1946) as a neritic species; forms such as *Favella serrata* and *Favella ehrenbergii*, described by Marshall as having a very wide distribution and by Hedin as occurring in inshore waters, should be regarded as neritic species. Similarly, *Steenstrupiella steenstrupii* occurs widely in temperate warm waters in all three oceans. Jorgensen (1924), describing the tintinnids of the Mediterranean, quotes *Tintinnopsis beroidea, Stenosemella* spp., *Favella ehrenbergii* and *Favella serrata* among a number of neritic forms. Posta (1963), who investigated the tintinnids in the Bay of Villefranche, lists *Stenosemella ventricosa*, *Tintinnopsis* spp., *Codonella galea*, *Undella clarapedei*, *Codonellopsis schabi* and *Dictyocysta elegans* as among the six most important species in what must be an essentially coastal area. Marshall (1934) describes *Tintinnopsis*, *Codonellopsis* and *Leprotintinnus* spp., as confined to the neritic areas of the Great Barrier Reef.

The geographical distribution of oceanic tintinnids is related largely to temperature. Distinctions can be drawn between cold-loving, temperate and warm-water oceanic forms, despite the occurrence of many widespread species and the fact that many species exhibit considerable change of form in different parts of the geographical range. Jorgensen (1924), discussing the occurrence of tintinnids in the Mediterranean, divides them according to their probable latitudinal origin. Thus a sub-Arctic grouping includes the genus *Ptychocylis* with *Parafavella denticula* and *Tintinnus* (*Salpingella*) *acuminatus*, though Marshall (1969) gives a somewhat wider distribution for the latter species. Zeitzschel (1966) also cites it as a eurythermal form more or less uniformly present in warm and cold areas of the Atlantic. Jorgensen describes warm temperate species as normally extending in the North Atlantic to a latitude of about 50°–53°N, and he lists *Codonella galea*, *Codonellopsis orthocras*, *Dictyocysta mitra*, *Undella clarapedei* and many others. He groups a number of species together as sub-tropical and tropical (*Climacocylis scalaria*, *Codonella amphorella*, *C. cistella*, *C. nationalis*, *Cyttarocylis cassis*, *Undella* spp., *Xystonellopsis paradoxa*). Gaarder points out that several warm-water species of *Dictyocysta* occur in all three oceans. Marshall (1934, 1969) describes the genus *Parundella* as consisting mostly of tropical and temperate forms of small size, and the family Undellidae as mainly a warm-temperate and tropical family, with *U. hemispherica* widely distributed in tropical areas. Genera with mainly warm water species are *Epiplocylis*, *Epiplocycloides* and *Climacocylis* (Marshall, 1969). Gaarder (1946) quotes *Xystonellopsis cymatica* and *Climacocylis longa* as occurring in the three warm oceans (cf. Marshall, 1934).

As contrasted with these mainly warm-water forms, a number of tintinnids appear to be essentially cold-loving. Marshall (1969) supports the views of Jorgensen and of Campbell that *Ptychocylis* is a northern cold water genus. She lists some ten species, most of them spreading across the northern North Atlantic, including the Norwegian Sea, North Sea and the Newfoundland Banks. *P. cylindrica* and *P. glacialis* are, however, specified as occurring in Arctic seas. *Leprotintinnus pellucidus* is another Arctic and sub-Arctic form (cf. also Hada's collections). Konovalova and Rogachenko (1974) found *Leprotintinnus pellucidus* and also *Ptychocylis obtusa* mainly in very reduced tintinnid populations occurring under the ice during the winter in Amur Bay. Several species of *Coxliella* appear to be very cold forms. *C. ampla*, while found in the Arctic and having a mainly northern distribution, spreads a little into warmer areas, but *C. intermedia*, *C. meunieri* and *C. tubularis* are essentially Arctic species (Marshall, 1969). The genus *Acanthostomella* also has cold-loving species: *A. elongata* is found in the Arctic and *A. norvegica* more widely over the northern North Atlantic. Some species, however, appear to be entirely warm water. *Parafavella*, another cold-water genus, has certain species more or less confined to the Arctic but more are widely distributed over colder regions of the North Atlantic (cf. Hada's records for the North Pacific).

In relation to tintinnid abundance, Zeitzschel (1966) found that the maximum density of tintinnids in the North Atlantic was mainly confined to 1 or 2 months and that there was a correlation with the time of the water temperature maximum. Thus the date of the maximum became later in the year in the eastern North Atlantic, e.g. from May in the Bay of Biscay to October near the Shetlands. The distribution of several important tintinnid species was also partly related to temperature (Marshall 1934). *Parafavella* is regarded as a cold-water genus (cf. Marshall) but whereas *P. gigantea* is a more neritic tintinnid and was abundant off the Newfoundland Banks, west of Greenland, off Scotland and the Norwegian coasts and in the North Sea, it was rarely taken in the open Atlantic. *P. parumdentata* (and the closely related *P. edentata*), by contrast, were broadly distributed over the open North Atlantic, mainly in the colder waters north of latitude 40°N, reaching even beyond Spitzbergen. However, as a stenohaline species, it is confined to oceanic water; *P. gigantea* is sufficiently euryhaline to occur in waters of as low salinity as 12 ‰.

Zeitzschel's analysis of the distribution of *Acanthostomella* also recalls Marshall's summary. *A. norvegica* was found to be distributed widely as a cold water tintinnid over the boreal Atlantic and was only exceptionally found in the warmer North Atlantic Drift. It was not usually found in Arctic waters. By contrast, some *Acanthostomella* spp. are exclusively tropical. As an example of a characteristic warm water tintinnid, Zeitzschel cites *Eutintinnus fraknoi*, which occurs widely in tropical waters but is found only in those areas of higher latitudes in the North Atlantic where it has been transported by the Gulf Stream.

In a further study Zeitzschel (1967) found, at Weather Station India (60°N) in cold Atlantic waters, that the maximum development of tintinnids—occurring about May to July with a small burst in November—was due mainly to the cold species *Parafavella parumdentata*, *Parafavella edentata* and *Ptychocylis minor*. The maximum density corresponded approximately to the bloom period of phytoplankton. At Weather Station Juliet (latitude 53°N) situated in rather warmer waters of the North Atlantic Drift, tintinnid abundance did not vary much throughout the year. The density was substantially lower and the number of species was greater. The species included warmer-water forms (e.g. *Dictyocysta elegans* and *Eutintinnus fraknoi*).

Posta (1973), examining the changing tintinnid populations, including several neritic forms, through the year in the Bay of Villefranche, also suggests that changes in abundance are chiefly related to seasonal variation in temperature. Hedin (1975), discussing the same problem for tintinnids off the Swedish west coast, points out that colder-loving species (*Codonellopsis pusilla*, *Ptychocylis minor* and *Leprotintinnus pellucidus*) were abundant in the winter/early spring months, with maxima in October/November, December/January and March/April, respectively. *Ptychocylis urnula* replaced *P. minor* by April/May; *Parafavella gigantea*, usually regarded as a cold species, reached a peak in late spring about May/June when the waters in Gullmar Fjord were still relatively cold. During the warmer months, *Helicostomella subulata*, a very abundant species, *Tintinnopsis campanula*, *Favella ehrenbergii*, *Coxiella helix* and *Favella serrata* were plentiful, with maxima occurring from June to August, approximately in that order.

Temperature is believed to be the major factor controlling this succession of species, though the changes are complicated by the immigration of species from outside waters. Although some species such as *Stenosemella nivalis* and *Tintinnopsis baltica* were very abundant and more or less continuously present, these also showed changes in population so that there was marked fluctuation in total abundance. The extreme density range was $200/m^3$ to $91{,}000/m^3$, peak populations occurring between April and August and occasionally in November. Burkill (1978) similarly found in Southampton Water that numbers of tintinnids were generally low in winter, rising to a maximum in May ($>10^6/m^3$), with a secondary peak in October. While certain species were present all through the year, some were restricted to particular seasons. *Tintinnopsis campanula* occurred only in the summer; *T. parva* only in the autumn period. *Stenosemella nivalis* and *Parafavella obtusangula*, while occurring in the autumn, winter and early spring, avoided late spring and summer. *Favella serrata* was present mainly in the late spring and autumn. Burkill is inclined to ascribe the changes mainly to seasonal variations in temperature in Southampton Water.

Hedin (1975) quotes mean densities for the tintinnid populations for the year as $10{,}300/m^3$ inside Gullmar Fjord and $15{,}600/m^3$ outside the fjord. Tintinnid densities recorded by investigators for other areas can be considerably greater, e.g. $30 \times 10^6/m^3$ (maximum density) by Vitiello (1964) and $2 \times 10^6/m^3$ by Travers and Travers (1970) for Mediterranean areas. For one coastal area of the North Sea a population, mainly of *Tintinnopsis*, is given as $2.5 \times 10^6/m^3$. For the Arabian Sea a mean density of $19{,}000/m^3$ is comparable to that in Gullmar Fjord. Despite the relatively small biomass and the marked regional and temporal fluctuations in density of tintinnids, their potentially rapid rate of multiplication makes them an important part of the marine food web. The studies by Heinbokel (1978a, b) reinforce this view; he obtained good correlation between field and laboratory feeding behaviour observations in species from neritic south Californian waters which showed high ingestion rates and assimilation efficiency. In later work, Heinbokel and Beers (1979) calculate that tintinnids took less than 20 % of primary production and that certain species may be able to switch size categories of nanoplankton food particles (Rassoulzadegan and Etienne 1981), thus competing with small crustaceans. Various types of tintinnid cyst have been described (cf. Reid and John, 1978) but their role is uncertain.

Though far less obvious and less easily recognized than the tintinnids, the smaller naked ciliates may possibly be of even greater significance in food chains in the ocean. Few data exist on the identity, abundance and distribution of specific ciliates (apart from tintinnids) in the oceans. Hamburger and von Buddenbrock (1911) describe a number of species and Bock (1967) lists a few from Atlantic, North Sea and Baltic waters. Since naked ciliates preserve poorly, culturing is usually necessary to establish identification. Burkill (1978), for example, found chiefly oligotrichs, with a few small hypotrich and holotrich ciliates over a year in Southampton Water. These naked ciliates were more plentiful in summer, with a peak population in May. Identification was difficult, with poor preservation, but *Euplotes charon* and *Uronema marinum* were cultured. The latter euryhaline species fed on a wide variety of bacteria.

Zaika and Averina (1968) found reasonable densities of infusorians in the admittedly shallow waters at the entrance to Sevastopol Bay. "Small" ciliates (20–55 μm) showed a mean density over four summer months of 1600/l and "large" species (>55 μm) showed some 350/l; the biomass amounted to 36 and 40 mg/m^3, respectively. Rather higher densities of ciliates were found further inshore. The genera identified included *Strombidium*, *Askenasia*, *Lohmanniella* and *Mesodinium* of the "small" forms, and *Lohmanniella*, *Didinium*, *Chilodonella*, *Prorodon* and *Uronema* of the "larger" ciliates. Though a form such as *Didinium* is carnivorous, there is little indication of the feeding of the smaller infusorians. On the other hand, Lighthart (1969) working in the Puget Sound area, was able to isolate a few species of Protozoa (amongst which were flagellates and ciliates) including some from the water column, and he demonstrated that some were bacteriovorous.

In a quantitative survey off California in which sampling was confined to the upper 100 m, Beers and Stewart (1969) found high total numbers of Protozoa, ranging from approximately 600,000/^{3} in slope waters, to 190,000–440,000/^{3} at two oceanic stations. The number of protozoans may be compared with approximately 20,000/m^3 and 10,000–17,000/m^3 for total metazoans at the same stations. More important, however, is the proportion of ciliates in the protozoan population. In the finest samples (less than 35 μm) the ciliate groups together (chiefly small, non-loricate species) comprised 95 % or more of the total Protozoa. In the next size category, the larger tintinnids were much more important, but in all the samples together, tintinnids did not amount to more than about 20 % of the total ciliate population. Although the contribution of ciliates, especially small species, to total biomass is admittedly very small, with their great rapidity of reproduction, they could form an effective link in the food chain (cf. vide supra for tintinnids).

Zaika (1972) investigated the microzooplankton in the Mediterranean and north-west Atlantic using water bottle samples, which were subsequently filtered and counted for live animals. In order of decreasing abundance, infusorians were first, followed by nauplii, copepodites, radiolarians, appendicularians, etc. The number of infusorians was under-estimated, since some were almost certainly lost or destroyed. Of the infusorians, small forms (25–45 μm), including *Strombidium*, were common; larger species (125–150 μm) were rarer, and tintinnids usually amounted to only 10–20 % of the total. For the Mediterranean, the maximum numbers of ciliates rarely exceeded 100/l; fewer occurred right at the surface and a maximum tended to occur at about 20–30 m depth. In the north-west Atlantic densities were generally comparable, though a few samples even exceeded 300 infusorians per litre; a shallow and a second deeper maximum at about 50–75 m were present. In contrast, in both areas the density of radiolarian protozoans rarely exceeded 5/l.

Margalef (1963) has emphasized the importance of ciliates (mostly small species) in the food chain in western Mediterranean waters. The average density found was nearly 900 000/m^3 (900/litre) of which more than 90 % of the ciliates were oligotrichs and only 5 % tintinnids. The smallest population was 50/l and the largest 12,000/l. Although maximal density occurred at the surface, a considerable density was found with relatively little change with depth between 5 and 75 m. Denser populations appear to follow the phytoplankton maximum, but seasonal fluctuations were not very great. Margalef considers that the ciliates feed mainly on bacteria and small flagellates, and that in turn they are preyed upon by the net zooplankton, especially by nauplii, fish and invertebrate larvae. He believes that the biomass could approximate to that of the net zooplankton. In a further estimate of ciliate numbers in the western Mediterranean, Margalef (1969) found that at the end of July densities ranged from approximately 100–500/l. It is not only these comparatively large densities, but the rather high turnover rate and suitability as a food source of these ciliates, which emphasizes their significance.

An investigation of the ciliate population in the Mediterranean off northern Spain was of particular interest in that sampling was continued to a depth of 500 m. Naked ciliates, mainly oligotrichs, were again the dominant component of the population, tintinnids amounting to only 5 % of the total. The greatest ciliate population found over the whole study was 13,400/l, observed for a surface population in February 1966. Mean densities for the period ranged from 1692/l at the surface, to 242/l at 100 m, with a further reduction to 42 ciliates/l at 500 m. (Table 2.3). Mean numbers of phytoplankton cells, as a ratio to one ciliate, ranged from 25 to 142. Larger populations of ciliates were observed during the first part of the year, but seasonal fluctuations were smaller than for the phytoplankton.

Beers and Stewart (1970) in investigations off California, again emphasize the importance of protozoans which comprised > 95 % of the total microzooplankton (i.e. that retained by 35 μm mesh screen) (cf. Volume 1, Chapter 8). Ciliates dominated the protozoan fraction, and while tintinnid ciliates were a very substantial part of the fauna in the upper eutrophic layer, there were few at deeper levels, where smaller naked ciliates were significant. Some of the small ciliates passed through the 35-μm screens. Sorokin (1977) also mentions ciliates as part of an abundant fauna in the upper thermocline region during late summer in the Sea of Japan, but heterotrophic flagellates were far more plentiful; tintinnids were also present in significant densities.

Hada (1972) reports, for the Japanese coast, that non-tintinnid ciliates occurred, including *Didinium* spp. and *Tiarina*, both of which he records also as Antarctic genera.

Table 2.3. Density of ciliates and ratio of phytoplankton cells per ciliate with depth in the western Mediterranean (Margalef, 1968)

Depth (m)	Ciliates per litre (mean numbers)	Number of phytoplankton cells for each ciliate (mean)
0	1692	45
5	983	71
10	502	142
20	808	69
30	675	83
50	733	57
75	475	49
100	242	49
150	150	47
200	134	45
250	158	25
300	100	30
400	67	45
500	42	68

Stentor, *Prorodon*, a few suctorian ciliates and *Mesodinium rubrum*, described as a red tide species, were also found. *Mesodinium rubrum* is a well known ciliate occurring in inshore waters (cf. Parsons and Blackbourn, 1968), which may bloom at high densities producing red tide occurrences as, for example, off the Dutch coast, British Columbia, Britain, etc. *Cyclotrichium meunieri* is possibly identical; this ciliate may also bloom as, for example, in Peruvian waters. The presence of photosynthetic pigments (chlorophyll and carotenoid) in these ciliates is well authenticated and a fuller account is given in Volume 1. The pigments differ from partly decomposed photosynthetic pigments which would be expected if the algae were being consumed. Blackbourn, Taylor and Blackbourn (1973) believe that the pigments represent part of a peculiar endosymbiosis between degenerated algae and the ciliate, though they admit the possibility that the pigments could be derivatives of repeatedly ingested algae. Holm-Hansen, Taylor and Barsdate (1970) reported a ciliate red tide in Alaskan waters; the species, although not *Mesodinium*, also apparently contained chlorophyll. Blackbourn *et al.* (1973) identified two other ciliates (*Prorodon* sp. and *Strombidium* sp.) each of which had intact algal chloroplasts. How much certain ciliate species regularly possess photosynthetic pigments is not known, but species with endosymbiotic algae could act to some extent as primary producers.

Cnidaria (Coelenterata)

This phylum, formerly known as the Coelenterata, contains a number of groups which are important members of the marine plankton. There are comparatively few accurate estimates of the density of medusae and siphonophores since some of these are of relatively large size and are not easily captured in closing nets. More recent observations by Scuba diving, especially in oceanic waters (cf. Hamner, Madin, Alldredge, Gilmer & Hamner, 1975; Hamner, 1977), suggest, that these cnidarians, in common with some other "gelatinous" zooplankton, may be far more abundant than is generally envisaged and that they constitute a very significant carnivorous component in the zooplankton food web.

Hamner quotes earlier observations from submersibles (such as those of Beebe), indicating the comparative abundance of cnidarians, including siphonophores and medusae. Mackie and Boag (1963) also speak of the abundance of siphonophores in the ocean and the fact that they appear to compete with Hydromedusae in many marine food chains. Many medusae occur at times in vast numbers forming huge shoals. Zelickman, Gelfand and Shifrin (1969) report great densities of *Rathkea* and *Tiaropsis* in one Russian bay. *Aurelia*, *Liriope* and a number of species of medusoid stages of hydroids like *Leuckartiara* and *Tima* have also been reported in great numbers cast up by the tide on beaches. Dense swarming of other more oceanic medusae is quoted by Kramp (1968a, b). Thus the Narcomedusan *Solmissus incisa* which mainly occurs in deep water was encountered in the Peru Current system nearer the surface, apparently in dense shoals. In further cruises of the Galathea Expedition, swarms of *Crambionella orsini*, a surface tropical scyphomedusan, were encountered. Stiasny (1940) cites *Linuche unguiculata*, a common surface-living scyphomedusan of the tropical Atlantic, as another swarming species. Bigelow and Sears (1973) discuss the numbers of siphonophores present in the Mediterranean; *Velella* is mentioned as one species subject to swarming. The importance of planktonic cnidarians is enhanced by their occurrence in all seas and at all depths (cf. Vinogradov 1970; Barham, 1963).

Hydrozoa—Hydromedusae

Hydromedusae have a fixed hydroid generation giving rise to a free-living planktonic medusoid generation. This hydroid stage of many hydromedusae is not known but it is presumed that they all have a fixed benthic stage. These medusae thus form part of the meroplankton. A few brief details of structure are given in Chapter 4, which also includes some account of the distribution of species, especially in relation to latitude and time of appearance. An account of some typical Arctic and boreal species is given by Zelickman (1972).

Although hydromedusans are normally found reasonably near to coasts, forming part of the neritic plankton, some species occur over considerably deeper water, near the edge of continental shelves. For several of these species it is known, or frequently presumed, that the hydroid stages live near the bottom of the continental slope. Some medusae appear to be truly oceanic in that they are regularly found very far from the coasts. Kramp (1959) advances several reasons why such apparently meroplanktonic forms can be found over open oceans. Many hydromedusans are capable of asexual budding, giving rise to numbers of similar medusoid forms which finally become sexually mature or alternatively, as in *Turritopsis*, eggs may develop to planulae *in situ*. Kramp believes that some of these medusae can bud sufficiently rapidly for normal currents to transport these forms regularly over oceanic depths. He quotes *Turritopsis nutricula*, which is found, for example, in coastal areas around the British Isles, in the North Sea, in the English Channel and off the New England coast of North America, but also widely in open oceans. Indeed, it has been recorded from all three oceanic regions.

Another anthomedusan, *Bougainvillea niobe*, found in the Atlantic along the east coast of America, appears to be carried out to the Sargasso Sea, and probably is able to spread to these areas by its prolific budding. A related species, *Bougainvillea platygaster*, is widely distributed in tropical and sub-tropical areas of the Atlantic Ocean and it is also found in the Indian Ocean and the west Pacific (Fig. 2.9a). It appears to have remarkable powers of asexual propagation, as has *Cytaeis tetrastyla*, another surface-living anthomedusan very widely distributed in all warm oceans and as far south as the Sub-Tropical Convergence. In these latter two species the budding is unusual in character, though it is known in certain other medusae, in that the stomach wall gives rise not to daughter medusae but to small polyps.

There are certain other species of hydromedusae, such as *Phialopsis diegensis* and *Euphysora furcata*, which are often found at great distances from coasts; but little is known about their propagation and it is not clear why they are able to spread out to oceanic regions. One possible mechanism is that the hydroid stage of these hydromedusans may normally be attached to floating objects which are widely dispersed over oceanic areas. One proven example is the anthomedusan *Pandea conica*. This medusa has long been recognized as largely oceanic, with a very wide distribution (Fig. 2.9b). It is now known that the hydroid stage is attached to shells of the oceanic pteropod, *Cleodora* (*Clio*) *cuspidata*; the medusoid stage is correspondingly widely distributed. Tesch (1946) quotes observations from Kramp and other authorities that a few pteropods carry hydroid colonies other than *Pandea conica* ("*Campaniclava cleodorae*"), including *Clio* (*Euclio*) *balantium* bearing *Campaniclava clionis*, and *Diacria trispinosa* and *Cuvierina columnella* carrying *Laomedea striata* colonies.

Most of the species mentioned so far tend to live near the surface, as indeed do most hydromedusae in neritic areas. Several medusae, however, with hydroid stages living at depth on the continental slope, not only frequent more offshore waters but often tend to live in deep water. Such medusae include the large *Octophialucium funerarium* found in the Mediterranean and north-west Europe, some species of *Calycopsis*, *Ptychogena* and *Amphinema*, *Leuckartiara grimaldii* and *Russellia mirabilis* (Kramp, 1959). Compared with the large number of shallow-living neritic medusae, the number of deep-living slope species is small; Kramp (1959) quotes only sixteen altogether. Moreover, many of these deeper-living species have a very wide vertical distribution and may approach the surface. Vinogradov (1970) lists *Calycopsis geometrica*, *C. nematophora* and *C. chuni* among the few species which descend as deep as 2000 m but he points out that they are distributed from that depth almost to the surface. Kramp (1959) also quotes *Calycopsis chuni* as occurring in deep water but occasionally reaching the superficial layers, and since this species is generally fairly near coasts, he presumes that the polyp occurs deep on the continental slopes. The related *Calycopsis borchgrevinki* is also eurybathic but is confined to Antarctic waters. *Euphysora furcata* is a species with a very wide vertical distribution, though avoiding the immediate surface, and associated with colder water. *E. bigelowi*, by contrast, is neritic and epipelagic (cf. Kramp, 1965). Some anthomedusae, however, e.g. *Tiaranna*, *Pandea rubra*, *Chromatonema*, *Annatiara* and *Bythotiara murrayi*, are probably truly bathypelagic, occurring in deep and intermediate layers, although their precise depth distribution is unknown, mainly owing to the difficulty of capturing them in closing nets. Vinogradov (1970) lists only one species, *Meator rubratra*, an anthomedusan related to *Calycopsis*, but occurring only in the north-west Pacific, as restricted to deeper layers, probably from 500 to 2000 m. It seems likely that even the most truly bathypelagic medusae avoid the very deepest layers.

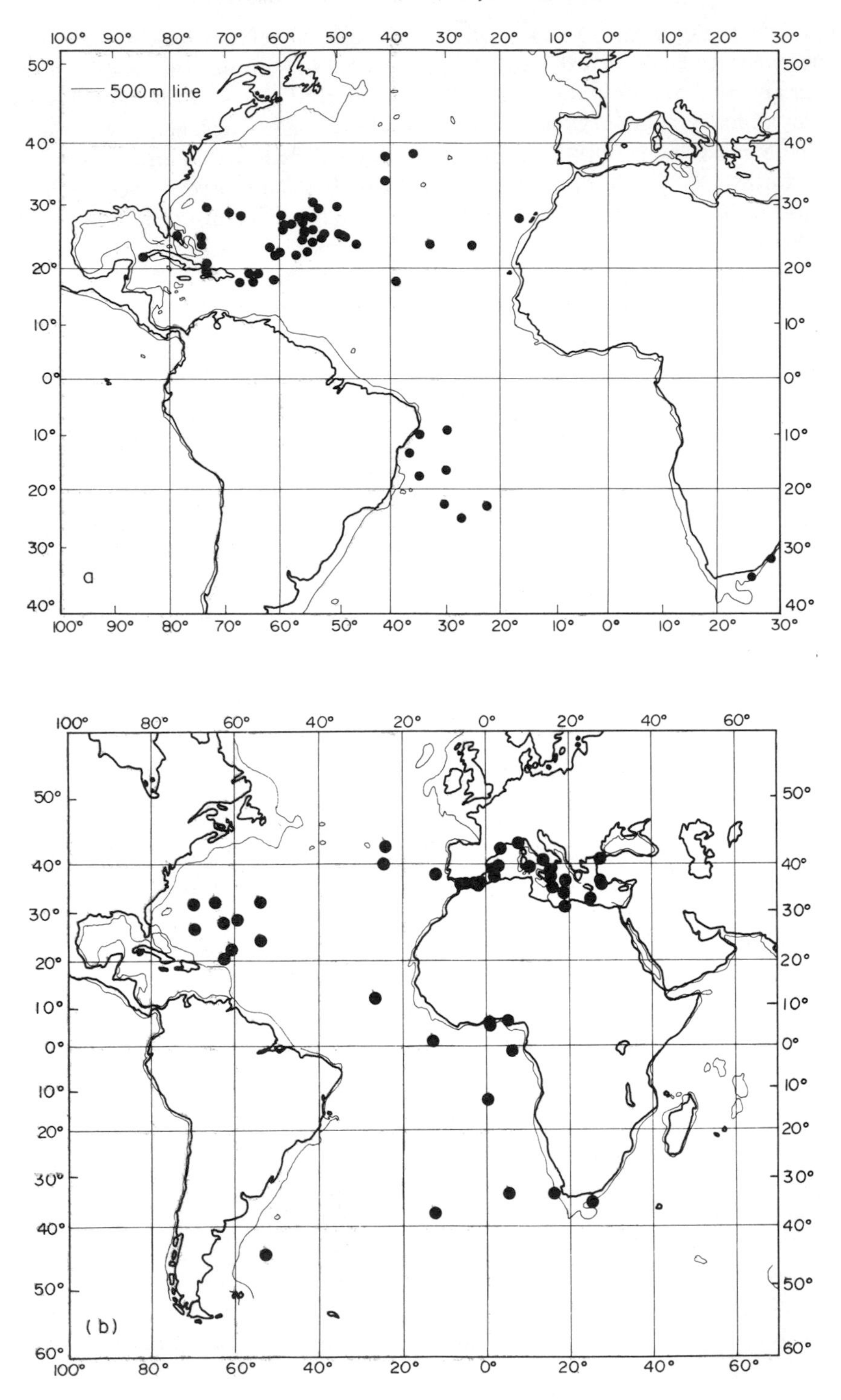

Fig. 2.9. (a) Distribution of *Bougainvillea platygaster* (b) Distribution of *Pandea conica* (Kramp, 1959).

The vast majority of Hydromedusae clearly live relatively near to the coast, mostly in the near surface layers. However, Hydromedusae occur in oceanic regions, and some may be found throughout the whole column of water. Though several hydromedusan species can reach even 2000 m, the great majority of oceanic medusae are Trachymedusae or Narcomedusae. Although the development of many species is unknown, the life history of those Narcomedusae and Trachymedusae which have been followed is direct and does not involve a fixed stage. Most probably this applies to all species.

In Trachymedusae the egg develops into an actinula larva which grows into the medusa. Narcomedusae include some asexual budding of the larva, at least in some forms (cf. Rees, 1966). These medusae may therefore occur widely over oceanic depths. Although they are usually of real importance in oceanic areas, the number of species is relatively small as compared with those of the hydromedusae. For example, Russell (1953) lists some seventy Hydromedusae and only four Narcomedusae and ten Trachymedusae occurring round the British Isles. Admittedly Russell's area of work is mostly over shallow water, but he includes a number of deep-living forms. The preponderance of hydromedusae is also emphasized from collections of deep-sea expeditions. Trachymedusae and Narcomedusae may of course be found in areas of continental slopes; indeed, they may penetrate into shallow coastal waters but there they are normally overwhelmed in numbers by hydromedusae. One of these medusae, *Ptychogastria polaris* (Fig. 2.10), is perhaps rather exceptional in that it might be classified as neritic. The species is typical of the cold, high latitude seas of both Arctic and Antarctic, but it spends part of its life attached by suckers to the sea bottom. Kramp has divided these oceanic medusae into epipelagic species, distributed from the surface to 200–250 m, and the deeper-living bathypelagic species. As Vinogradov (1970) points out, the much greater variety of species is generally in the surface waters, but a definite fauna exists in the deeper layers. Moreover, as with many zooplankton taxa, the cold surface waters of high latitudes tend to show few species of all types of cnidarian medusae, whereas surface waters of the warm seas show a great variety of species.

Amongst the epipelagic Trachymedusae there are some relatively common forms generally distributed in warm/temperate oceans. These include *Rhopalonema velatum* (Fig. 2.12), *Geryonia proboscidalis* and *Liriope tetraphylla*, (Fig. 2.11), the last being one of the commonest and most widely distributed of oceanic Trachymedusae, mainly in seas between 40° N and 40° S. *Aglaura hemistoma* is another epipelagic species, living in the same warm/temperate waters, but with a somewhat wider vertical distribution. Compared with these mainly warmer-water forms, *Aglantha digitale* is one of the commonest Trachymedusae in Arctic, sub-Arctic and boreal seas of both the Atlantic and Pacific Oceans. It is a widely distributed form, but, though mainly epipelagic in the North Pacific, it seems to have a wider vertical range and perhaps avoids the immediate surface layers in the Atlantic.

The Narcomedusae include a number of mainly epipelagic species (Fig. 2.11). *Solmundella bitentaculata* is very common, especially in the southern hemisphere, including the Antarctic area. The genus *Solmaris* includes two mainly coastal epipelagic forms, but other species are oceanic. Two other Narcomedusae, *Solmissus marshalli* and *Aeginopsis laurentii*, can occur at the surface but also are found in deeper layers; the former is widely distributed; the latter is confined to Arctic seas. Several

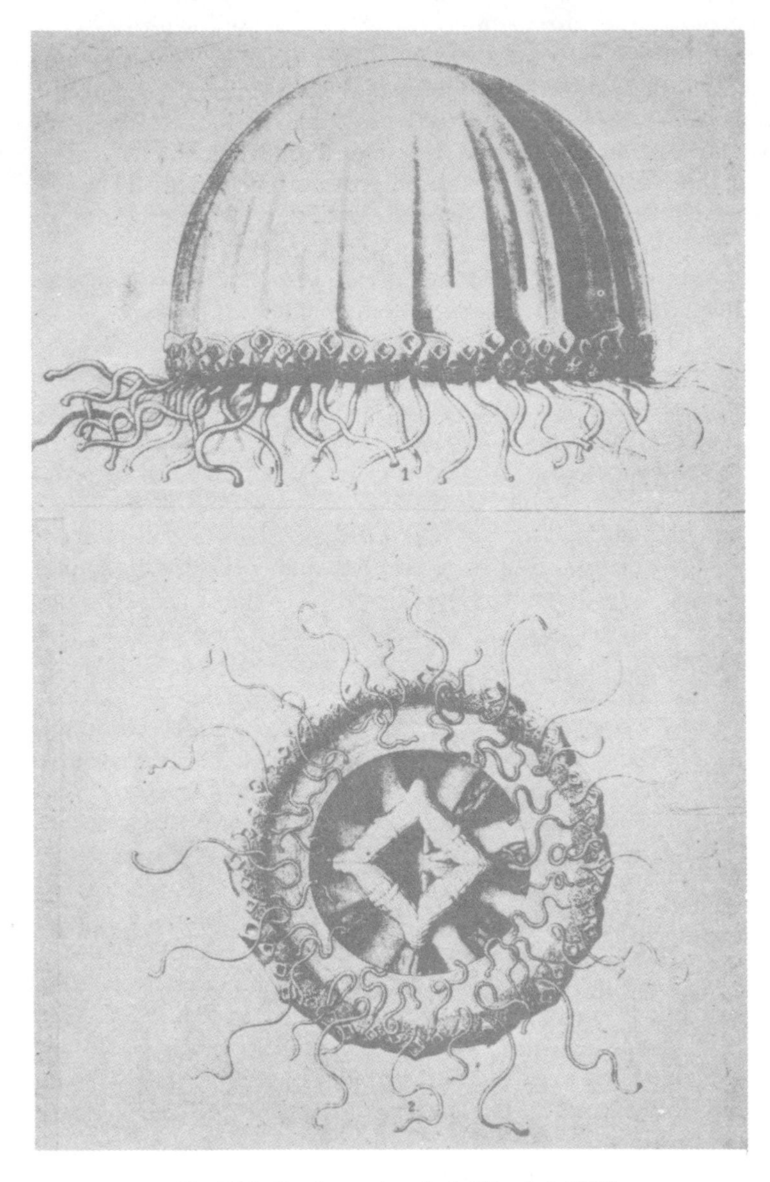

Fig. 2.10. *Ptychogastria polaris* (Haeckel, 1881).

other species have a very wide vertical distribution, even though some of them may appear more commonly in deeper layers. *Aegina citrea*, a narcomedusan, is very widely distributed in warm/temperate regions but occurs mostly in intermediate and deep cold layers and is also tolerant of low oxygen, as in the Indian Ocean (cf. Vannucci and Navas, 1973b). Casanova (1977) found it rarely above 600 m in the north-east Atlantic.

Species of Trachymedusae (Fig. 2.12) regarded as essentially bathypelagic by Kramp (1959, 1965) include *Halicreas minimum* which, while having a great depth range, hardly ever appears in the superficial layers, *Haliscera racovitzae*, a bathypelagic

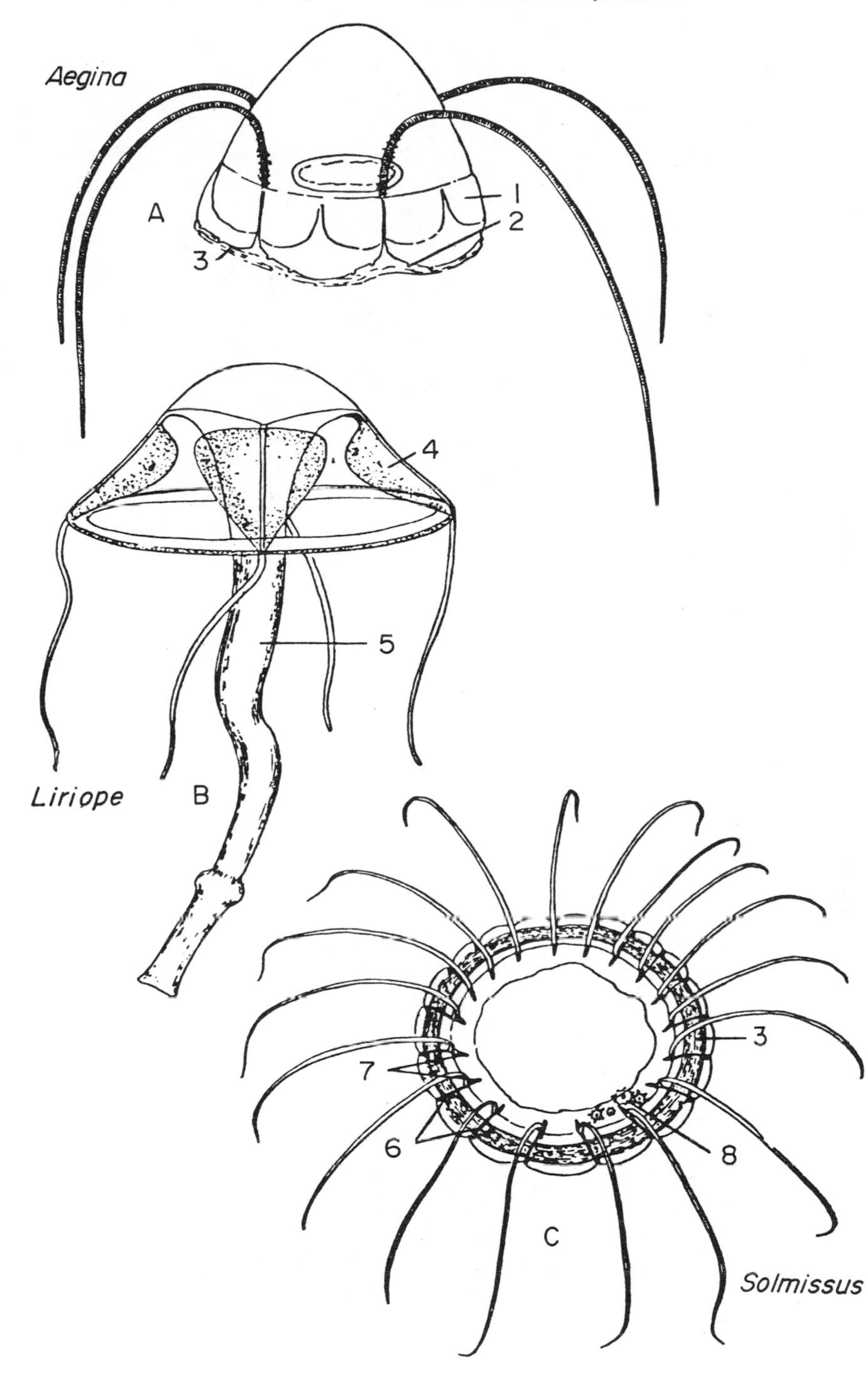

Fig. 2.11. Narcomedusae (A) *Aegina*; (B) *Solmissus* and Trachymedusa; (C) *Liriope*; (1) gastric pouches; (2) lithostyles; (3) velum; (4) gonad; (5) pseudomanubrum; (6) peronia; (7) tentacle roots; (8) budding medusae in floor of gastric cavity (Hyman, 1940).

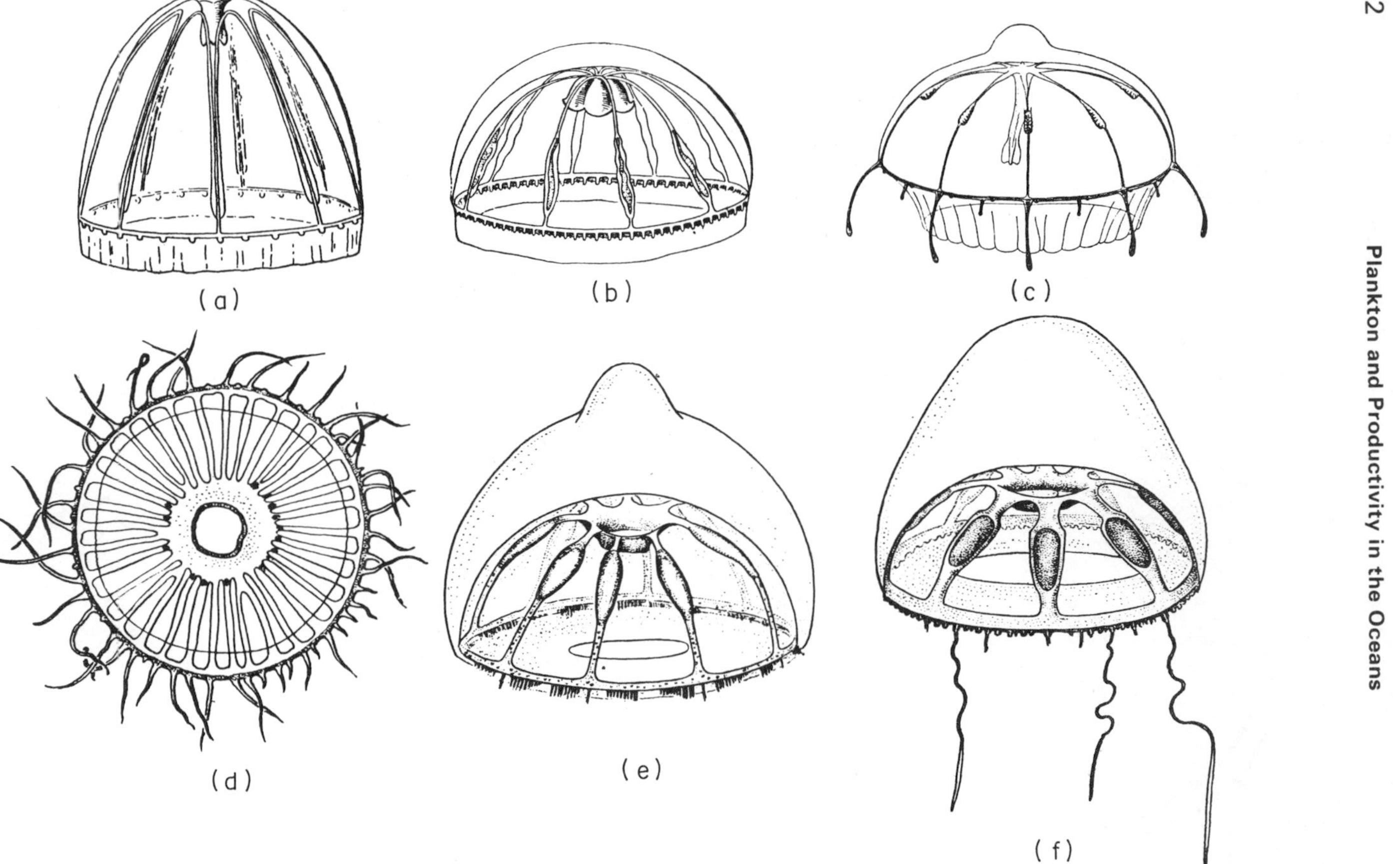

Fig. 2.12. Various Trachymedusae (a) *Colobonema sericeum*; (b) *Pantachogon haeckeli*; (c) *Rhopalonema velatum*; (d) *Halitrephes maasi*; (e) *Botrynema brucei*; (f) *Haliscera bigelowi* (Kramp, 1959).

form in Antarctic water, and *Botrynema brucei*, a bathypelagic inhabitant of most seas except the Arctic, where it is replaced by the allied species, *B. ellinorae*. *B. brucei* is carried into the deeper layers of the Indian Ocean by the Antarctic Intermediate Water, and while it is a cold Antarctic species, it has a considerable depth range. The genus *Halitrephes* occurs at deep levels in almost all warm/temperate seas. *Pantachogon haeckeli* is very widely distributed in deep regions of the Atlantic, apparently from the Arctic to the Antarctic slopes. It is found also in the Pacific and Indian Oceans, but in the Antarctic it is replaced by *P. scotti*. A generally bathypelagic genus is *Crossota*; it is widely distributed with different species in different areas: *C. norvegica* in the Arctic; *C. rufobrunnea* in the boreal Atlantic/Pacific; *C. brunnea* in the three main oceans (Fig. 2.13).

Vinogradov (1970), while emphasizing the difficulty of knowing precise limits for deep-living species, points to the decrease in the variety of species at deeper levels, but

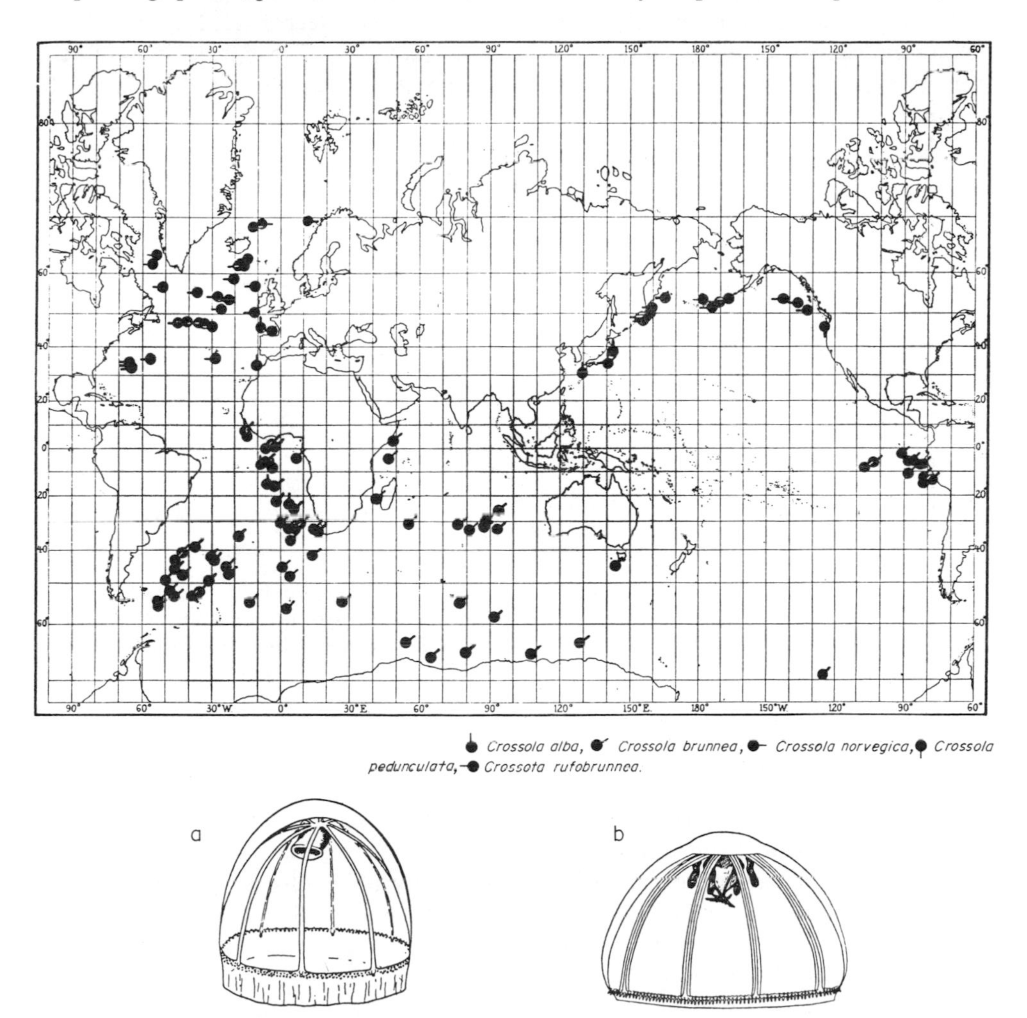

Fig. 2.13. Geographical distribution of all species of *Crossota* and (a) *Crossota brunnea* (b) *Crossota rufobrunnea* (Kramp, 1959).

lists species of *Halicreas*, *Botrynema*, *Pantachogon*, *Colobonema* (cf. Fig. 2.12) and *Crossota* as typical below 500 m. Casanova (1977) describes *Colobonema sericeum* as among the more abundant of the deeper-living Trachymedusae of the north-east Atlantic, with maxima between 600 and 1500 m. Though widely ranging, it does not reach Arctic or Antarctic waters and is more limited to the southern hemisphere. The narcomedusan *Aeginura grimaldii* also seems to be a true bathypelagic and widely distributed form (cf. Kramp, 1965; Vannucci and Navas, 1973b; Casanova, 1977).

Some collections from the north-west Pacific Ocean, using closing nets, permit a more precise interpretation of depth ranges. Very few of the bathypelagic species inhabit the whole water column from 500 to 5000 m. Thus, *Aglantha digitale* is essentially a cold surface form; *Crossota brunnea* has a very narrow depth range, being practically confined to the 500–1000 m layer; *Pantachogon haeckeli* mostly inhabits the same level, but some specimens descend as deep as 3000 m; *Botrynema brucei* occupies about the same depth range, but may extend somewhat deeper (cf. Fig. 2.14). The maximum density for *Aglantha*, quoted by Vinogradov, is 1–3 specimens/m^3, while the density maximum of *Crossota* at 500–1000 m is smaller, as expected; it may amount to 1/10 m^3, a remarkable population for medusae at that level.

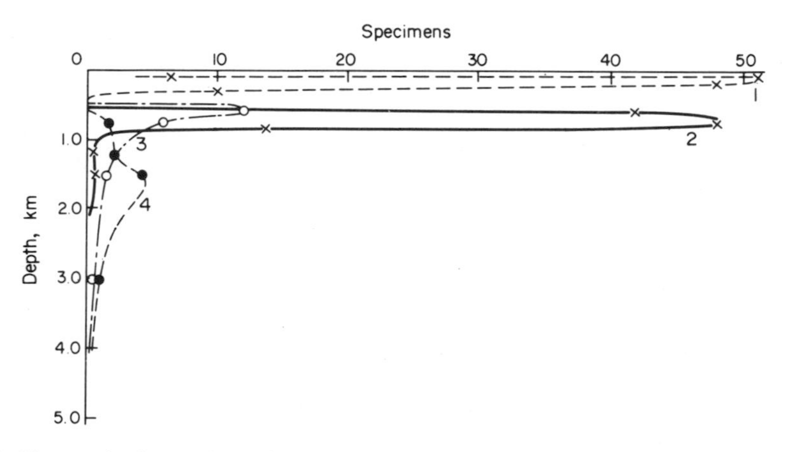

Fig. 2.14. Changes in the numbers of certain Hydromedusae with depth in the waters of the Kurile-Kamchatka Trench (average for 15 stations). (1) *Aglantha digitale* (spec./40 m^3); (2) *Crossota brunnea* (spec./1000 m^3); (3) *Pantachogon haeckeli* (spec./1000 m^3); (4) *Botrynema brucei* (spec./1000 m^3) (Vinogradov, 1970).

Hydrozoa—Siphonophora

The siphonophores appear to be distributed through all the layers of the sea and, as holoplanktonic forms, extend throughout the world's oceans. Barham (1963) speaks of divers commonly seeing siphonophores at a variety of depths in the ocean during bathyscaphe descents. Swarming is known for some species. There is a suggestion that they are less abundant in the cold waters of very high latitudes. According to Vinogradov (1970), concentrations of siphonophores can occur even at very great depths in the ocean, but most species have a wide vertical range. Totton (1954) considers that there are about 150 species of Siphonophora and that they occur abundantly in most seas.

Pugh (1974), from a survey of the literature, suggests that siphonophores, while seldom numerically abundant in comparison with zooplankton such as copepods or ostracods, are often important constituents of the macroplankton. However, usual collecting methods tend to damage or destroy many specimens, so that their density has probably been much underestimated. As an abundant and widespread carnivorous group of zooplankton, siphonophores graze effectively on herbivorous copepods and other taxa. Mackie and Boag (1963) described methods of capturing the prey. In *Nanomia* the stem and tentacles are fully extended only when the animal is quiescent and suspended in comparatively quiet water, the total expanded surface of the siphonophore forming an effective fishing net. Mackie and Boag estimated that the combined length of all the fishing filaments amounted to fifty times the length of the stem. When swimming the filaments are contracted. *Physalia*, in contrast, elongates the fishing tentacles, which can extend to 8–10 m or longer when swimming. *Muggiaea* has a complex co-ordinated feeding movement involving the extended tentacles. Some siphonophores perform more or less regular lengthening and shortening of the stem over long periods.

While no detailed morphological description of siphonophores and of the sub-orders will be given, it is helpful to appreciate some of the major characteristics distinguishing the main divisions of the group. In the Physonectae, the colony of polymorphic components ("individuals") is budded from an elongate stem, at the apex of which is a comparatively small gas-filled float chamber (pneumatophore). Further along the stem is a series of muscular swimming bells (nectophores or nectocalyces) below which is a zone occupied by numerous cormidia. Each cormidium typically consists of a protective or buoyant bract, a feeding gastrozooid with a long branched tentacle, one or more dactylozooids (palpons) with an unbranched tentacle and clusters of gonophores of both sexes which are set free as medusae and bear the gonads (e.g. *Halistemma*, *Agalma*) (Fig. 2.15). In some Physonectae the stem is much shortened, the float enlarged and the cormidia crowded beneath: members of the Rhodalini are now known to be benthic (Pugh, personal communication).

The Calycophorae also have an elongate stem but the summit of the colony has one, two or more large nectophores. A swimming bell conceals the apex of the stem, which has a somewhat enlarged canal, often with an oil droplet (the somatocyst). Below the bell, the stem carries numerous cormidia budded from a growing zone, so that the lowermost cormidium is the oldest. Each cormidium typically consists of a protective and buoyant bract, a gastrozooid and one or more gonophores (of the same sex), commonly acting also as swimming bells. There are no dactylozooids. The cormidia in most Calycophorae can break free, when they are termed eudoxids, and they can lead an independent existence, finally liberating the gonophores which can be replaced (Totton and Bargmann, 1965) (eg. *Diphyes*, *Muggiaea*) (Fig. 2.16).

The small sub-order Cystonectae is characterized by the possession of a comparatively large float filled with gas; the stem may be stout or so shortened, as in *Physalia*, that it is essentially a budding zone. No nectophores or bracts are present. *Physalia* has numerous gastrozooids lacking tentacles, large dactylozooids with very long tentacles and smaller dactylozooids and clusters of gonophores forming a mass of individuals beneath the bladder.

The Chondrophorae, represented by *Velella* and *Porpita*, now separated from the Siphonophora, are highly modified. The budding zone forms a compact flat

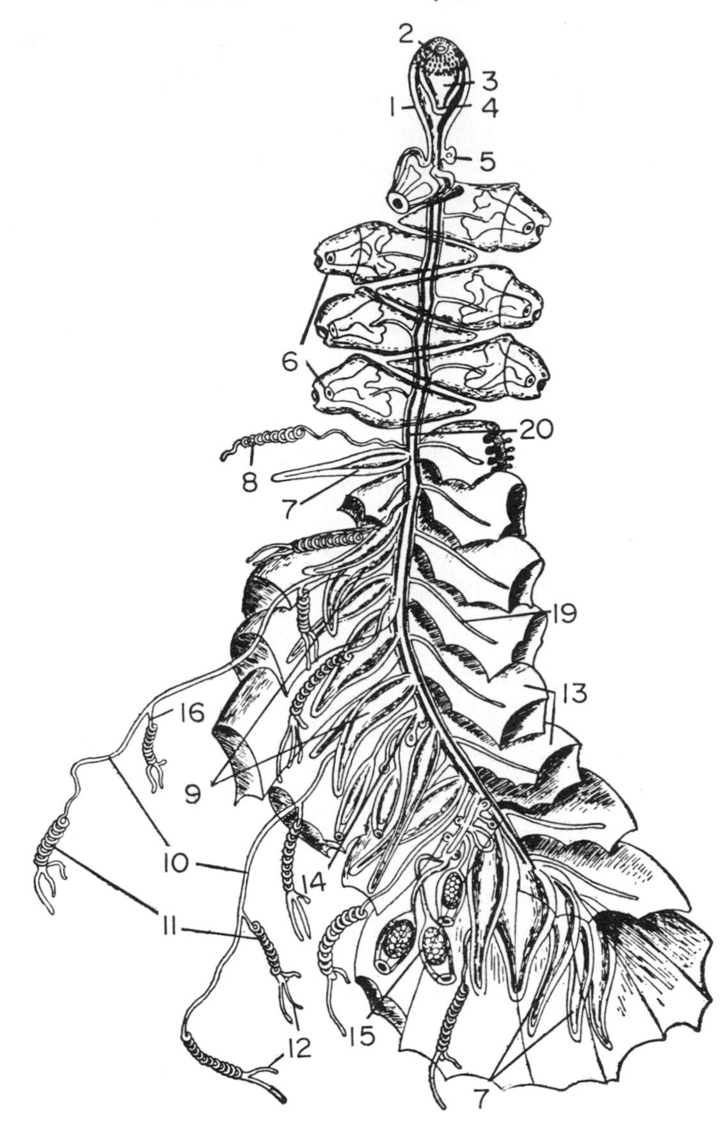

Fig. 2.15. *Agalma* (Physonectae) entire colony. (1) Float; (2) pigment; (3) air sac; (4) funnel; (5) budding zone of nectophores; (6) nectophores; (7) dactylozooids; (8) tentacles of dactylozooids; (9) gastrozooids; (10) tentacles of gastrozooids; (11) cnidoband; (12) end filaments; (13) bracts; (14) cluster of male and (15) female gonophores; (16) tentilla; (19) gastrovascular canals; (20) stem (Hyman, 1940).

disc of relatively tough consistency with a very large float with many air chambers above. Below the disc a central large gastrozooid and circles of gonophores and tentacle-like dactylozooids occur. Nectophores and bracts are absent. The gonophores of Chondrophorae are liberated as medusae.

In all siphonophores the fertilized egg develops into a planula, which then directly forms a larval structure which may resemble a swimming bell or may be more like a pneumatophore or a gastrozooid, depending on the group. Carré (1967), describing

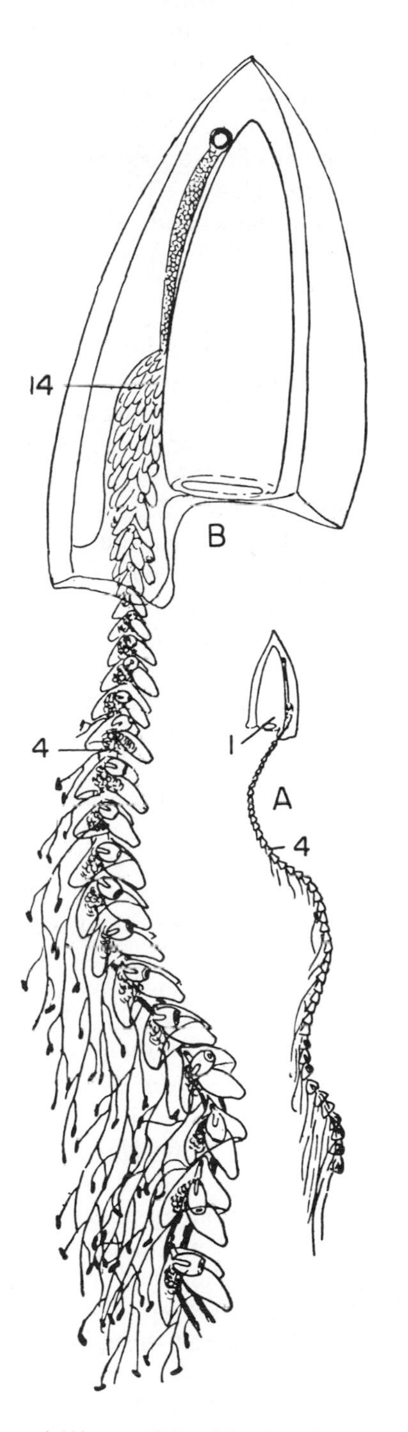

Fig. 2.16. *Muggiaea* (Calycophorae) (A) natural size; (B) enlarged, showing single swimming bell and long stem with cormidia. (1) Swimming bell; (4) stem with cormidia; (14) growing region of stem (Hyman, 1940).

the development of *Abylopsis tetragona*, found that the larva was unusual amongst Calycophorae in that the first nectophore persisted as the adult anterior bell. More generally the larval bell is replaced. Thus he described the nectophore of the larva of *Lensia conoidea* as deciduous. In the development of the physonectid larva of *Forskalia edwardsi* a pneumatophore is first formed. Budding from the larva occurring in open warm seas gives rise to the colony.

The abundance of the Calycophorae is undoubtedly increased by their ability to bud off eudoxids. Some authorities speak of an alternation of generations in the Calycophorae, since the colony (polygastric phase) buds off the eudoxids asexually and the eudoxids, after their independent existence, reproduce sexually. The liberated medusae produce eggs which after fertilization develop into a planula, which ultimately buds to form the polygastric generation. In other siphonophores the sexual phase is reduced at most to liberation of the medusae, releasing eggs and sperm. Calycophoran eudoxids appear to be present over most of the year. It is uncertain whether their release has any seasonality.

A number of siphonophore and chondrophore genera have only a single species, for instance *Porpita*, *Physalia*, *Physophora*, *Bassia* and *Rhodalia*; others have only two species, such as *Rhizophysa*, *Chelophyes*, *Chuniphyes* and *Abylopsis*. Alvarino (1971) points out that where such species pairs exist, one is often typical of warmer seas and the other found in colder waters, e.g. *Chelophyes appendiculata* is widely found in temperate cooler waters as well as in warmer areas, while *Chelophyes contorta* is confined essentially to tropical waters. As with most zooplankton groups, the greatest number of species is found in warmer waters and some, indeed many, species are confined to the tropics (e.g. *Velella*, *Porpita*) where they are typically widely distributed in the three major oceans. A few siphonophores have much broader distribution. *Physophora hydrostatica* is almost cosmopolitan, and is very common in the North Atlantic Drift (cf. also Totton and Fraser, 1955) and the Mediterranean; apparently it is less common in the Indian Ocean. It avoids the coldest areas of polar currents and it does not occur as far north at high latitudes in the Pacific; the records of Alvarino (1971) indicate this much more limited distribution there. Bigelow has quoted a temperature range of 7°–21°C for the species, but it has been recorded at temperatures from 3° to even 26.6°C. Zelickman (1972) records it in the waters of Barents Sea.

Another species with a very wide distribution, though occurring much more commonly in colder waters, is *Dimophyes arctica*. Zelickman quotes it as very common in the Barents Sea and comments that, though it is stenohaline, as indeed are most Siphonophora, it is eurythermal and eurybathic. He believes that it does not reproduce at temperatures exceeding 10°C and it is most abundant at temperatures of 1°–2°C. Its widespread distribution in the North Atlantic is apparent from the records of Totton and Fraser (1955). *Dimophyes arctica* is a cosmopolitan species found in all oceans and in the epiplankton of high latitudes, submerging to considerable depths in warm regions (bipolar) (Alvarino, 1971). Its distribution over most seas with its common occurrence in Antarctic waters is in contrast to that of *Diphyes antarctica*, which is confined to the Antarctic.

Pugh (1974) suggests that these contrasted distribution patterns could be related to the eurybathic nature of *Dimophyes* and its apparent ability of resorting to different strata (and thus different environmental temperatures) at different stages of the life history. Thus, off the Canaries, Pugh found that polygastric individuals of *Dimophyes* were mostly fairly shallow (200–250 m), whereas eudoxids were deeper (>600 m). Mackintosh (1934) repeatedly refers to the occurrence of *Dimophyes arctica* in cold Antarctic waters. According to Totton (1954) it has a very wide tolerance in all oceans; in the Indian Ocean he did not find it north of 6°S latitude, but it was breeding to 7°S, and its entire temperature range may approximate to −1°–13°C. Bigelow and Sears (1937) had earlier reported 2°–14°C as the range in the Atlantic Ocean, with a maximum around 3°C. The species, as with so many other siphonophores, appears to have a very wide depth range, though it is usually more abundant below the first 100 m. It is especially common to about 1000 m, but, at least in some hauls with closing nets, has been taken at depths of 2000–3000 m. Totton quotes its occurrence in the North Atlantic (latitude 44°N) at a depth of about 1000 m. Figure 2.17 shows its numerical abundance and distribution in relation to temperature. The peak distribution at relatively low temperatures is obvious, but its high thermal tolerance is clear. The species is apparently not known in the Mediterranean, despite its eurythermal character (cf. Casanova, 1977).

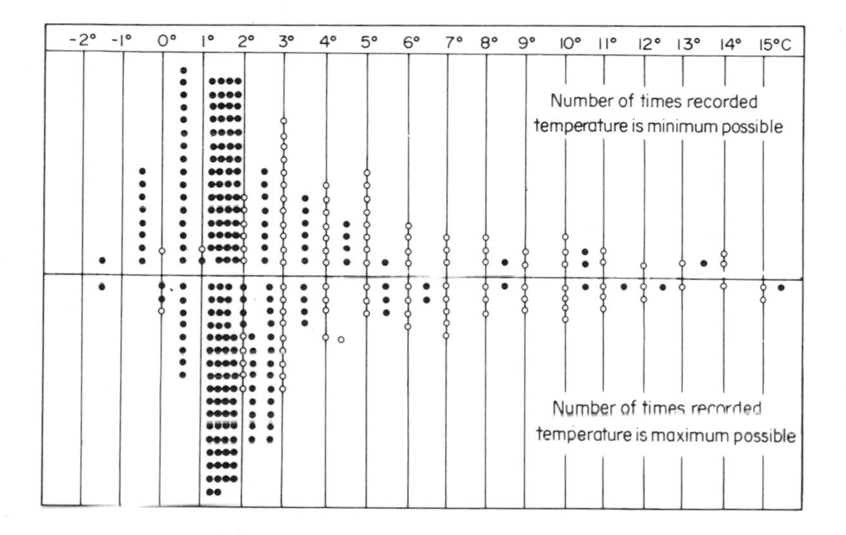

Fig. 2.17. Frequency of capture of *Dimophyes arctica* in water of known upper and lower temperature limits. Solid circles (●) for "Discovery" closing net hauls, open circles (○) for "Meteor" closing net hauls (data from Bigelow and Sears, 1937 (Totton, 1954)).

A few other siphonophores, while avoiding the very cold and the warmest waters, are relatively widespread over warm and temperate areas. *Nanomia* (*Stephanomia*) *cara* is found in the temperate Atlantic, including Norwegian and British waters. Totton and Fraser (1955) speak of it as a cosmopolitan species in the North Atlantic, occurring in the Faroe–Iceland and Faroe–Shetland Channels, and the Norwegian Sea, with a more doubtful record for the Barents Sea, as well as in less cold waters such as the North Sea, English Channel, south and west of Ireland and in the open Atlantic. Margulis (1972) lists it as a typical boreal species among the Physophorae and Calcycophorae of the North Atlantic Ocean. He states that it may reach the Davis Strait and even as far north as 70°–72°N off the Norwegian coast. Its southern limits

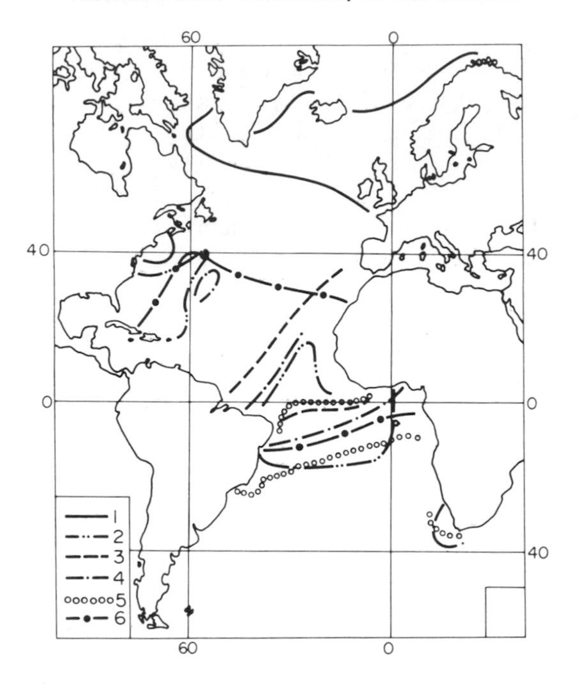

Fig. 2.18. Distribution of Siphonophores in the Atlantic Ocean. (1) *Nanomia cara*; (2) *Ceratocymba dentata*; (3) *Lensia hostile*; (4) *Maresearsia praeclara*; (5) *Nectropyramis natans*; (6) *Melophysa melo* (Margulis, 1972).

are the "slope" waters off the north-eastern United States and to the west of Spain (Margulis 1972) (Fig. 2.18).

In contrast, *Nanomia bijuga* is essentially a warm species found in the tropical Atlantic, West Indies and Mediterranean as well as in the Indian Ocean and warmer areas of the Pacific. *Halistemma rubrum* is a warm-water siphonophore distributed also through the warm oceans, including the Mediterranean and Red Sea. *Agalma elegans*, another species with a fairly wide distribution in warm/temperate waters, has a range in the north-east Atlantic, according to Totton and Fraser's (1955) records, generally similar to that for *Nanomia cara*. Margulis (1972) lists it with those generally warm-water siphonophores which extend to transitional zones, but which move considerably further north along the north-eastern side of the Atlantic. The related species, *Agalma okeni*, is a more tropical organism: Margulis states that it avoids transitional waters and rarely penetrates north of 40°N latitude, even in the north-east Atlantic (cf. Fig. 2.19). Similarly, in the Pacific Alvarino (1971) gives its limits as 40°N to 43°S.

Rosacea plicata is another widely distributed species, especially in the southern hemisphere, though occurring at least as far north as the latitude of Ireland in the northern hemisphere. It may appear in the uppermost layers but is another siphonophore with a very wide depth distribution; it is found throughout the upper 500 m, and frequently also at greater depths. It has been reported relatively commonly in Antarctic waters (Totton, 1954). Leloup's (1955) records for siphonophores from the Michael Sars Expedition confirm the generally wide distribution of such species as *Rosacea plicata*, *Chelophyes appendiculata*, *Hippopodius hippopus* and *Agalma elegans*.

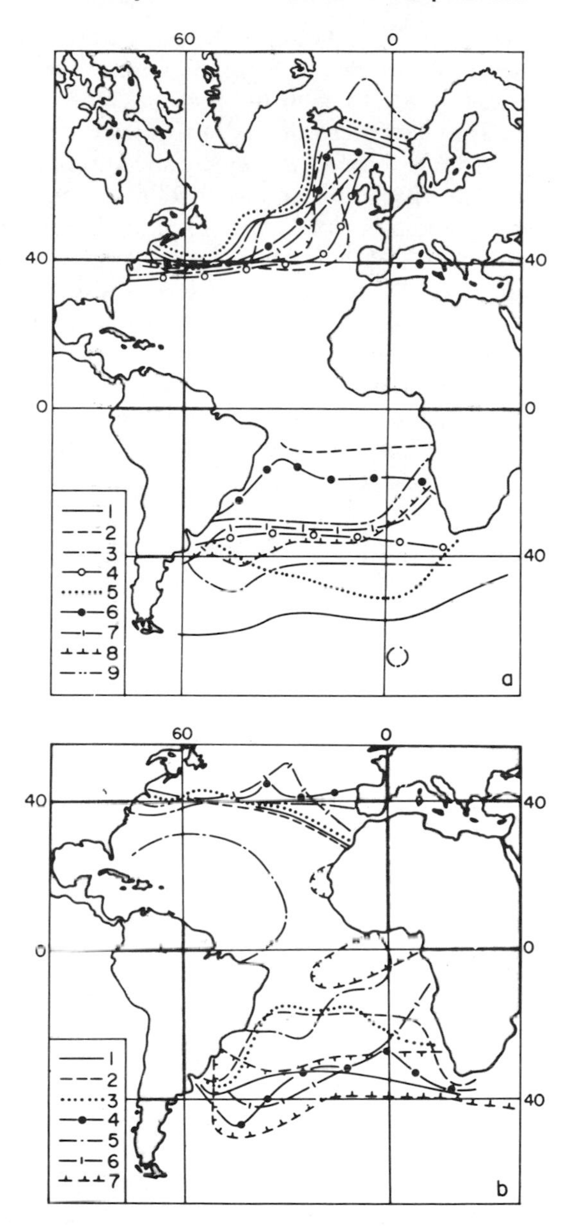

Fig. 2.19. Distribution of Siphonophores in the Atlantic Ocean. a: (1) *Rosacea plicata*; (2) *Agalma elegans*; (3) *Lensia conoidea*; (4) *L. subtilis*; (5) *Hippopodius hippopus*; (6) *Enneagonum hualynum*; (7) *Bassia bassensis*; (8) *Chelophyes appendiculata*; (9) *Ceratocymba sagittata*. b: (1) *Abylopsis eschscholtzii*; (2) *Eudoxoides mitra*; (3) *Diphyes bojani*; (4) *D. dispar*; (5) *Abyla trigona*; (6) *Agalma okeni*; (7) *Lensia hardy* (Margulis, 1972).

A few siphonophores, while being essentially warm or warm/temperate in distribution, are suggested by Totton as being relatively more abundant in neritic areas. These include *Muggiaea atlantica* and *M. kochi*, *Lensia subtilis*, *L. subtiloides* and *Diphyes chamissonis*. The last two species were the only abundant siphonophores in the Great Barrier Reef Lagoon.

Little mention has been made of the few cold water siphonophores, apart from *Dimophyes arctica*. *Marrus orthocanna* is a comparatively rare Arctic bathypelagic species. Pugh (1974) states that although the previous most southerly record was 55° N latitude, the species has now been taken in fairly deep water (700–800 m) at both 18° and 28° N. It may, therefore, occur more widely and improved sampling methods may necessitate a revision of geographical distribution. Zelickman (1972) lists only *Dimophyes* and *Marrus* in addition to *Physophora* in the Barents Sea. In the North Atlantic, however, van Soest (1973) indicates that several siphonophores may occur at moderately high latitudes (*ca.* 60° N). His list of species also suggests differences in distribution of those species broadly ranging over warm-temperate areas, and those more strictly tropical and sub-tropical (cf. Table 2.4).

The Antarctic has a larger number of siphonophores. According to Totton (1954), fifteen species have been identified: *Pyrostephos vanhoeffeni*, *Marrus antarcticus*, *Stephanomia (Moseria) convoluta*, *Diphyes antarctica*, *Muggiaea bargmannae*, *Dimophyes arctica*, *Lensia havock*, *L. achilles*, *Rosacea plicata*, *Vogtia serrata*,

Table 2.4. List of species of siphonophores collected in the North Atlantic Ocean. The stations at which species occurred (latitudinal position only) are indicated by crosses (van Soest, 1973)

Station position	66°	62°	59°	52°	49°	46°	45°	43°	42°	39°	34°	33°	32°	25°	22°
Siphonophora															
Thalassophyes crystallina		×													
Muggiaea kochi		×													
Dimophyes arctica	×	×													
Crystallophyes amygdalina		×					×								
Chuniphyes multidentata		×	×	×			×			×					
Apolemia uvaria		×			×		×								
Nanomia cara		×	×												
Physophora hydrostatica		×	×							×					
Lensia conoidea		×	×				×	×	×	×					
Lensia multicristata			×				×								
Lensia fowleri				×			×								
Rosacea plicata				×	×	×	×		×	×					
Vogtia pentacantha			×			×	×			×					
Vogtia spinosa					×	×	×	×	×	×					
Hippopodius hippopus					×		×	×	×	×				×	
Velella velella					×		×								
Eudoxoides spiralis					×	×	×	×	×	×					
Chelophyes appendiculata							×			×	×	×		×	
Halistemma rubrum						×		×	×	×					
Nectopyramis thetis							×			×					
Nectopyramis diomedeae							×								
Vogtia serrata										×					
Rosacea cymbiformis										×	×				
Lensia subtilis											×	×			
Diphyes dispar										×	×	×		×	×
Diphyes bojani											×	×			
Ceratocymba sagittata										×	×				
Bassia bassensis											×	×			
Abylopsis tetragona											×			×	
Abylopsis eschscholtzi											×	×			
Eudoxoides mitra												×		×	
Sulculeolaria monoica														×	

Thalassophyes crystallina, *Crystallophyes amygdalina*, *Heteropyramis maculata*, *Chuniphyes multidentata* and *C. moseri*. The first five of these are confined to the Antarctic and Antarctic Intermediate Water (Pugh, personal communication). Records for the North Atlantic include *Thalassophyes crystallina*, *Crystallophyes amygdalina* and *Chuniphyes multidentata*, apart from the cosmopolitan *Diomophyes* (cf. Table 2.4).

The warmer-water siphonophores (belonging to the Physophorae and Calycophorae) found in the Atlantic Ocean have been divided by Margulis (1972) into three rough groupings according to their distribution. Species which extend into transitional waters, while limited to about 40°N in the western Atlantic, reach much further north in the north-east Atlantic, approaching Iceland and southern Norway (e.g. *Rosacea plicata*, *Agalma elegans*, *Lensia conoidea*, *Chelophyes appendiculata*, *Hippopodius hippopus*, *Bassia bassensis* (cf. also Totton and Fraser, 1955). A group of more definitely tropical/sub-tropical siphonophores, including *Diphyes dispar*, *D. bojani*, *Agalma okeni* and *Eudoxoides mitra*, is restricted to latitudes less than 40°N particularly in the western North Atlantic, though there may be a slightly greater spread in the east (cf. Casanova, 1977). In the southern hemisphere the limit is the Sub-Tropical Convergence, or, for some of the species, only about 20°S. A third group (*Ceratocymba dentata*, *Lensia hostile*, *Nectopyramis natans*, *Maresearsia praeclara*) are held to be truly tropical species, keeping essentially to equatorial waters (cf. Fig. 2.18).

A number of siphonophores are found comparatively close to the surface in warmer seas, and some of them rank amongst the commonest species. Pugh (1974) lists *Chelophyes appendiculata* as relatively shallow-living and probably the commonest species. Other fairly shallow and abundant forms include *Eudoxoides mitra*, *Eudoxoides spiralis*, *Hippopodius hippopus*, *Diphyes dispar*, *Diphyes bojani*, *Bassia bassensis*, *Abylopsis eschscholtzii* and *Abylopsis tetragona* (cf. Casanova, 1977). Most of these may show a considerable depth range, however, some involving diurnal vertical migration. Neto (1970) cites a similar group of species for the Cape Verde Islands. Michel and Foyo (1976) list seven species as the commonest siphonophores in the Caribbean, and although found at shallow depths they display a marked depth range, apart from *Diphyes dispar* (cf. Table 2.5). An obvious absentee from the list of common species for the Caribbean is *Bassia bassensis*, which is apparently rarely found

Table 2.5. Relative abundance and vertical distribution of siphonophores collected in the Caribbean Sea and adjacent areas (Michel and Foyo 1976)

Species	Depth range (m)	Range of maximum numbers (m)	Total estimated numbers collected
A. eschscholtzii	0–1272	0–50	13,040
A. tetragona	0–1665	0–100	15,608
C. appendiculata	0–2500	0–100	7,482
D. bojani	0–996	0–100	14,667
D. dispar	0–50	0	4,213
E. mitra	0–822	0–250	4,250
E. spiralis	0–2500	0–100	1,700

there. Alvarino (1971) lists most of the same species as found from about 40–45°N to 40°S latitude in the Pacific, with *Chelophyes appendiculata* as a most abundant epiplanktonic form.

Several species of *Lensia* are known. Pugh (1974) found that while *L. multicristata* could occur through the whole water column (to 1000 m) off the Canaries, other species, perhaps somewhat unusually for siphonophores, were much more restricted in depth. *L. cossack* and *L. campanella* were very shallow, in approximately the upper 50 m; *L. fowleri* over the upper 100 m; *L. subtilis* ranged between 40 and 150 m. Other depth ranges were *L. hotspur* 85–150 m; *L. meteori* 150–400 m; *L. exeter* 400–450 m; *L. achilles* 500–625 m; *L. havock* 580–660 m; *L. lelouveteau* >700 m; and *L. hostile* 800–900 m.

Despite the narrow depth ranges indicated for some species of *Lensia*, Vinogradov (1970) stresses the widely held view that many siphonophores exhibit broad depth ranges. He suggests, however, that the Physophorae do not generally show as wide a range as many Calycophorae. Such species as *Nectopyramis thetis*, *N. spinosa*, *Lensia multicristata* and *Crystallophyes amygdalina* may extend from the surface to a depth of even 2000–3000 m. Only a few forms, such as the physophorans *Erenna* and *Dromalia alexandri*, are never found near the surface of the ocean (indeed, the latter is now thought to be benthic). *Clausophyes ovata*, another essentially deep-sea form, can sometimes appear, according to Vinogradov, in shallower layers. Pugh (1974) found that both polygastric and eudoxid stages were bathypelagic. The genus *Chuniphyes* is characteristic of great depths, with captures in hauls from 4000–8000 m (Vinogradov, 1970). In the Atlantic both Pugh (1974) and Casanova (1977) found *Chuniphyes multidentata* mostly at about 600–700 m, but there are shallower records, especially for younger forms.

Leloup (1955) describes several siphonophores taken in the Michael Sars Expedition as bathypelagic but usually with a wide vertical range.

Totton (1954), discussing the vertical distribution of species, mentions several of the siphonophores cited by Vinogradov as distributed to 2000 or 3000 m. But, whilst recognizing the difficulties of assessing precise depth limits he includes other forms as essentially deep-living. Examples are the Arctic and Antarctic species of *Marrus*, a new species, *M. orthocannoides*, from tropical deep waters, *Thalassophyes crystallina* and possibly *Maresearsia*, *Vogtia glabra* and *Muggiaea bargmannae*.

With their very wide vertical and horizontal distribution and marked carnivorous habits, siphonophores are significant tertiary producers.

Scyphozoa

The Scyphomedusae may be conspicuous members of the plankton population. Great shoals of jellyfish, particularly of *Aurelia aurita* can appear in coastal areas; *Pelagia* and *Linuche* occur further offshore (cf. Stiasny, 1940). But the variety of species of Scyphomedusae is far less than for Hydromedusae. Scyphozoa, moreover, are more conspicuous than many other cnidarians in that numerous species are of relatively large size for zooplankton. Common species such as *Aurelia*, *Chrysaora* and *Cyanea* may measure several inches to a foot across the bell. Some Scyphozoa are relative giants; Russell and Rees (1960) and Russell (1967) described the deep-water forms named *Stygiomedusa* and *Deepstaria*. Kramp (1968) also makes some reference to very large

Scyphomedusae. *Periphyllopsis galatheae* was said to have a diameter of 38 cm; a specimen of *Versuriga annadyomene* belonging to the Rhizostomeae was recorded with a diameter of the umbrella of 60 cm. A damaged specimen of a medusa, possibly a new species of *Cassiopea*, was discovered—a gigantic form with relatively enormous oral arms. Kramp (1961) lists about 160 known pelagic Scyphomedusae in the whole world, of which only six are known round the coasts of the British Isles.

There are four orders of planktonic Scyphomedusae: Cubomedusae, Coronatae, Semaeostomeae and Rhizostomeae. The latter two orders have a number of coastal species, all of which probably have a fixed stage in the life history. The free-swimming planktonic larva, the planula, settles to produce a hydroid-like form, the scyphistoma, which when fully developed, undergoes a type of transverse fission producing asexually a number of ephyrae. These are liberated and grow into medusae. The scyphistoma can increase by budding, developing a stolon and by forming resistant podocysts (cf. Chapman, 1966). The transverse asexual division to form ephyrae is called strobilation; the marked ability for asexual reproduction is a major factor in producing the great bursts of jellyfish sometimes encountered.

Russell (1970) has summarized some of the salient points in the life history of *Aurelia*. The adults normally liberate planulae about the end of the summer, and by December and through the next 2 or 3 months active strobilation of the polyps occurs. The average duration is some 70 days in winter, but there can be a second shorter summer period of perhaps 25 days. There are two peaks, therefore, in the production of the ephyrae, though apparently strobilation can go on throughout the year. In Britain temperature, light and food supply all possibly act as factors to which the reproductive cycle is adjusted. Presumably, in seas with little or no seasonal change, strobilation may proceed almost continuously.

Russell points out that certain developmental stages of scyphozoans are present all the year round along British coasts, with young ephyrae appearing early in the year in the plankton (cf. Chapter 4). The adult jellyfish, while common in the late spring and summer, tend to disappear in autumn, although capture of the adults at some distance from the coast during winter suggests that some of the population overwinter in deeper water.

Kraeuter and Setzler (1975) record a somewhat different seasonal cycle for Scyphozoa from Georgia estuaries. As suggested for British waters, some species (e.g. *Chrysaora*) persisted offshore in winter. *Aurelia* and *Pelagia* occurred in spring and early summer, *Chrysaora* and *Stomolophus* in spring, summer and autumn. The latter species was by far the more dominant form, generally accounting for half of the total jellyfish population, with densities reaching even $>90/1000\,m^3$.

The occurrence of species of *Chrysaora*, *Aurelia* and *Rhizostoma* near the coasts can be understood since in most Semaeostomeae and Rhizostomeae the scyphistoma stage is attached to the bottom in relatively shallow water. On the other hand, the genus *Pelagia*, another member of the Semaeostomeae, has direct development; it is not surprising that this species has an oceanic distribution.

Although Semaeostomeae and Rhizostomeae are more typical of coastal and offshore waters, they may spread to very considerable distances in more open oceans. The Coronatae, however, undoubtedly represent the most typical open sea Scyphozoa, except for *Pelagia*, which is widely distributed, particularly in warmer seas. Stiasny (1940) illustrates from records of scyphozoans caught on the Dana cruises how much

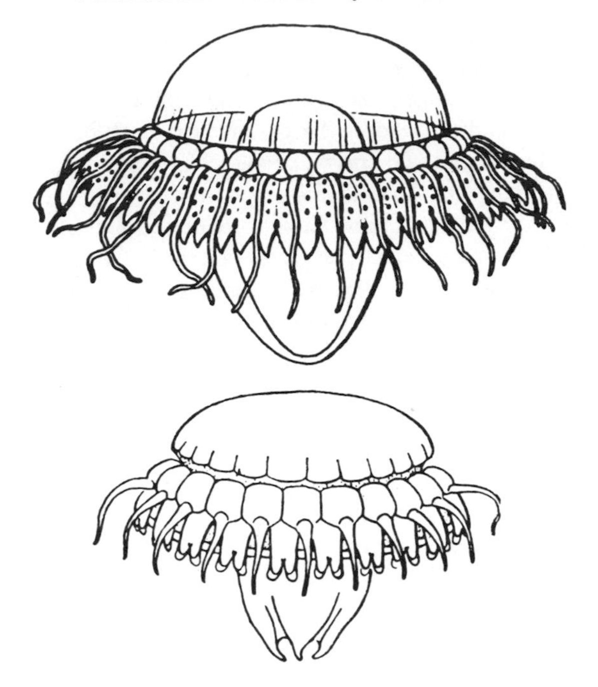

Fig. 2.20. *Atolla chuni* (upper) and *Atolla wyvillei* (lower) (Mayer, 1910).

the Coronatae dominated the catch. Two-thirds of the total were Coronatae, the vast majority being *Periphylla hyacinthina* (= *periphylla*) and *Atolla wyvillei* in equal proportions (Fig. 2.20). There were very few *Nausithoe punctata*. Of the remaining jellyfish, *Pelagia noctiluca* was extremely numerous. Kramp (1968) found from the Galathea collections, that *Nausithoe punctata*, rare in Stiasny's collections, was somewhat coastal in all warm seas, as was *Linuche unguiculata*. *Pelagica noctiluca* was confirmed as an oceanic holoplanktonic species, circumglobal in warm-temperate areas, forming dense surface swarms at times. Kramp suggested that while its breeding was restricted in the less warm parts of its range, it probably bred at all times in tropical seas.

The Coronatae can inhabit all depths. While *Nausithoe* is essentially a surface dweller and *Linuche unguiculata* is very common in surface waters of the tropical Atlantic, genera such as *Atolla* and *Periphylla* have a very wide range in the deep sea and are among the most typical oceanic deep sea plankton forms. According to Vinogradov (1970), deep-sea Scyphozoa inhabit all depths of the oceans from 500 m to at least 5000 m. As with other deep-sea genera, there are problems in separating species and in defining their geographical limits. Two species of *Atolla*, *A. wyvillei* and *A. vanhoeffeni*, are probably worldwide species (Fig. 2.20).

Kramp (1968) considers *Periphylla periphylla*, a monotypic genus, cosmopolitan and bathypelagic, apparently occurring in all seas including fringes of the Antarctic. It does not penetrate to the deep Arctic basins. Though *Periphylla* may occur even down to about 7000 m and is essentially bathypelagic, it is not so strictly limited in its vertical distribution and may occur in somewhat shallower waters in colder seas. Russell (1970) considers its depth of maximum abundance as about 400–1500 m, but in the Mediterranean it mostly exceeds 1000 m and can occur as deep as 2700 m. Stiasny (1934) found it in Antarctic waters as well as generally throughout the world oceans, and states that it can occur at any depth from 0–2500 m. It was often at the surface in the Arctic and sub-Antarctic but less often so superficial in warm waters.

Atolla wyvillei is cosmopolitan, according to Kramp, occurring in all seas including the Antarctic and Arctic (cf. also Stiasny, 1934). It is bathypelagic, mainly in intermediate and deep waters, occurring from 2000 to 7000 m in Kramp's records. Russell (1970) suggests its maximum occurrence as 500–1500 m; Casanova (1977) gives it as below 600 m with maximum around 1000 m in the European Atlantic. *Atolla vanhoeffeni* was found by Kramp in the Pacific and Indian Oceans, as well as being widely distributed in the Atlantic. It occurred at depths rarely less than 1000 m and extending to 5000 m. Casanova, however, suggests depths shallower than 1000 m off north-west Africa and southern Europe. Russell quotes a third species, *A. parva*, occurring in the Atlantic possibly as widely as the other two deep-sea species. According to Casanova (1977) there may be two races, the more southerly being of larger size. *A. chuni*, another bathypelagic form, apparently inhabits only southern latitudes, and *A. russelli* occurs off south-west Africa. Of other deep-living Coronatae, *Paraphyllina ransoni* is a rare deep-water species occurring in the Atlantic.

Although the life history of many of these typical deep-sea Scyphozoa is unknown, a type of scyphistoma stage named a "stephanoscyphus" has been described. This is a fixed polyp stage (a fairly typical scyphozoan character) of which the earlier specimens, described by Komai, had branching horny tubes, the whole colony being embedded in a sponge. Other species are now known which have solitary polyps (e.g. *Stephanoscyphus simplex* and *S. corniformis*, Kramp, 1959a). Though some species appear to have a wide distribution in relatively shallow waters, Kramp points out that *Stephanoscyphus simplex* has a very great depth range but frequents deeper water to perhaps 7000 m.

Russell (1970) summarizes our knowledge of the development of these Coronatae; the original description by Komai referred to the life history of *Nausithoe punctata*, a relatively shallow water scyphozoan. Werner has now reared a deep-sea species, *Atorella vanhoeffeni*, from a living stephanoscyphus brought back from the Indian Ocean. A detailed description of the ephyra produced from the stephanoscyphus is known. Other species of *Nausithoe* are deep-sea forms. Two species, *N. atlantica* and *N. globerifera*, occur in deeper waters to the west and north of the British Isles. Possibly *Stephanoscyphus simplex* is the polyp of one of the deeper-living species.

The small order of scyphozoans, the Cubomedusae (Cubozoa) have cuboidal bells with four flattened sides; from each corner springs a tentacle or tentacle group. A rhopalium is present in the middle of each side in a notch just above the margin (Fig. 2.21). As Mayer (1910) pointed out, Cubomedusae can occur in open oceans, but are usually confined to shallower waters and occur only in warmer seas. Of the total of 2254 scyphozoans taken in the Dana cruises, only fifteen were Cubomedusae. Kramp (1968) also confirms from the Galathea collections that the two relatively common Cubomedusae, *Carybdea alata* and *Tamoya haplonema*, were mainly coastal and warm-water species, *Carybdea* being essentially circumtropical. Kraeuter & Setzler (1975) found two cubomedusans, *Tamoya* and *Chiropsalmus*, to be reasonably common in the shallow waters of Georgia estuaries. *Chironex fleckeri* is another neritic

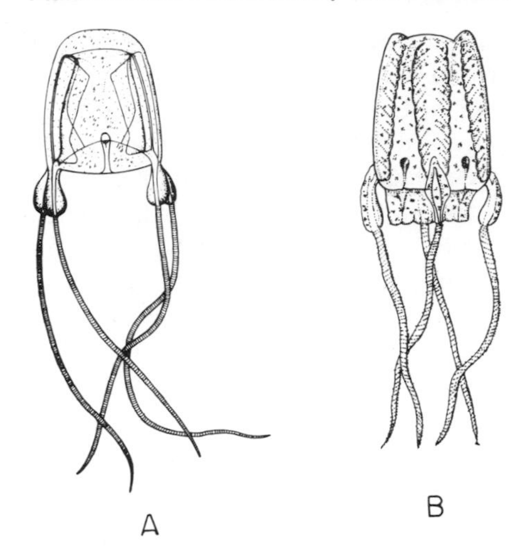

Fig. 2.21. Cubozoa species: (A) *Carybdea rastonii* and (B) *Tamoya bursaria* (Yamazi, 1973).

species, found all round Australian coasts to about latitude 26°S. Barnes (1966) believes that its appearance is associated with lower salinity water from river run-off. Many cubomedusans, especially *Chironex fleckeri*, are highly toxic animals.

Little is known of the life cycle of Cubomedusae (Hyman, 1940), though *Tripedalia* and *Charybdea* are said to produce a planula which settles to form a polyp stage with four tentacles (cf. Thiel, 1966). Werner (1971) confirmed the formation of a planula from the egg of *Tripedalia cystophora* within the gastral pocket of the medusa. After release the planula forms a small fixed polyp which can bud off daughter polyps from the base. Werner states that the polyps are more similar to those of Hydromedusae than Scyphomedusae. Both primary and secondary polyps develop directly into single medusae; the polyp tentacles are reabsorbed and the adult tentacles represent new developments. At a temperature of 28–29°C development from planula to medusa takes 10–12 weeks, with a similar further period for the medusa to reach sexual maturity.

Ctenophora

The phylum Ctenophora, at one time associated with the Cnidaria, is now usually separated, since the former lack typical stinging cells. Though this group has relatively few genera, it is extremely important as a holoplanktonic phylum whose representatives are widely known in all seas from the Arctic to the Antarctic and throughout the tropics. Although as many as 200 species have been suggested, many of these are likely to be varieties and some species, for example of *Beroe* and *Pleurobrachia*, are almost cosmopolitan. As holoplanktonic organisms, ctenophores are well known throughout the oceans but they are perhaps more common in coastal areas. Some of the species (*Cestum*, *Ocyropsis*, *Mnemiopsis*) can be of relatively large size.

In the least modified ctenophores like *Pleurobrachia* the globular or pear-shaped body has considerable bulk due to the thick mesogloea. In the middle of the upper pole is the apical sense organ. There is a pair of typically branched large tentacles, which can be withdrawn into sheaths. A narrow slit-like mouth at the opposite pole, leads into a compressed wide stomodaeum ("stomach"), from which eventually leads a ramifying system of tubes conforming to a strict plan. Prominent in this system are eight meridional canals, one running beneath each of the comb rows (ctenes), by which the animal swims. Two paragastric canals run along the stomodaeum. Ctenophores are exclusively carnivorous, feeding by trapping other plankton animals on sticky mucus from the tentacles or oral lobes (*vide infra*). Specialized colloblasts on the tentacles release filaments superficially resembling nematocysts, but these are sticky and not penetrating, and help to entangle food organisms.

Four orders of planktonic ctenophores are known, apart from the peculiar creeping Platyctenidae. The first three orders all have tentacles, at least at some stage, and are sometimes grouped as Tentaculata. The Cydippida have the two long tentacles throughout life. The meridional and paragastric canals end blindly. Genera include *Pleurobrachia*, *Hormiphora*, *Euchlora* and *Tinerfe*. In the Lobata (e.g. *Bolinopsis*, *Mnemiopsis*, *Leucothea*, *Ocyropsis*, *Eurhamphaea*), the body is slightly flattened and two large oral lobes are developed, two of the meridional canals extending into each oral lobe and fusing (Fig. 2.22). Four ribbon-like projections (auricles) are present close to the mouth and the four remaining meridional canals extend along them (cf. Mayer, 1912).

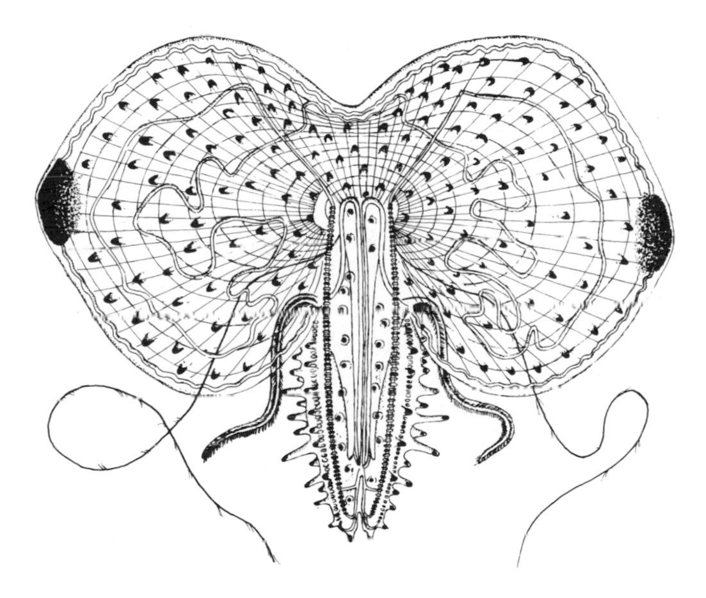

Fig. 2.22. *Leucothea ochracea* (× 1.5) (Mayer, 1912).

The Lobata pass through a cydippid larval stage, but in the adult the large tentacles are reduced. Typically, a row of small simple tentacles lies on each side of the mouth, with one larger feathered tentacle in the middle. *Ocyropsis* lacks tentacles altogether (cf. Mayer, 1912) (Fig. 2.23). Usually Lobata swim with the mouth forward; the winglike large oral lobes trap food, assisted by the auricles. In Cestida (*Cestum*, *Velamen*), the body is very greatly flattened so that it becomes ribbon-like. A whole row of small tentacles lies along the oral edge and an oral groove receives prey which is

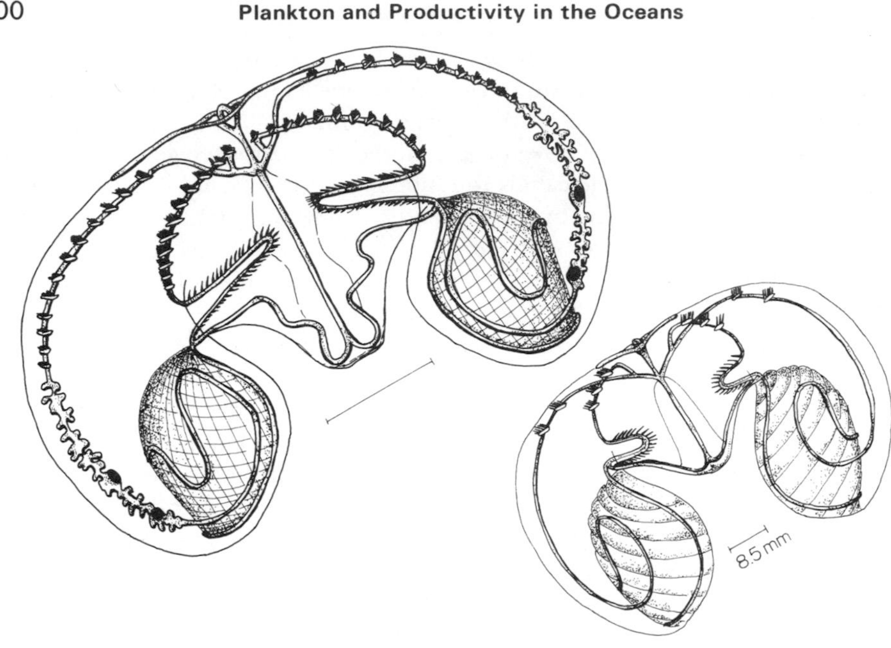

Fig. 2.23. *Ocyropsis crystallina* (young) —— denotes life size (Mayer, 1912).

carried to the mouth (Fig. 2.24). There is a cydippid larva. The Beroida (*Beroe*) (Fig. 2.25) possess a roughly cylindrical but somewhat flattened body. The stomodaeum is very wide and the meridional and paragastric canals give off side branches which anastomose. There are no tentacles whatsoever, not even in the larva.

The Cydippida feed with the long sticky tentacles serving as a fishing net. Harbison, Madin and Swanberg (1978) state that *Lampea pancerina* feeds exclusively on salps, and while large individuals can completely engulf a salp, the small ctenophores appear more like parasites and are probably identical with the organisms previously identified as *Gastrodes parasiticum*. The same authors, describing feeding mechanisms in Lobata, point to the importance of the oral lobes in all lobate ctenophores. Typically, particles

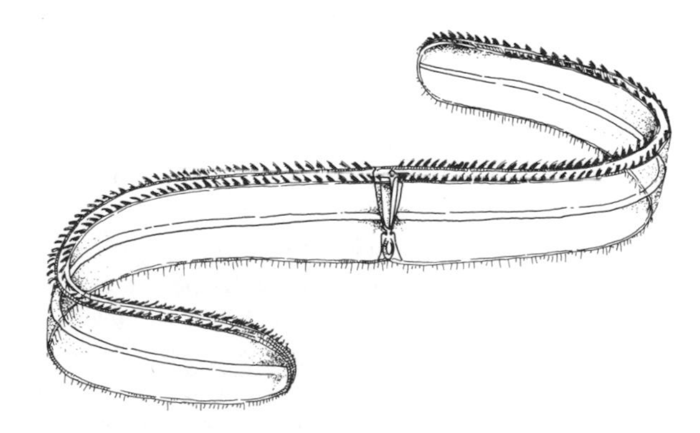

Fig. 2.24. *Folia paralellela* (Cestida) natural size (Mayer, 1912).

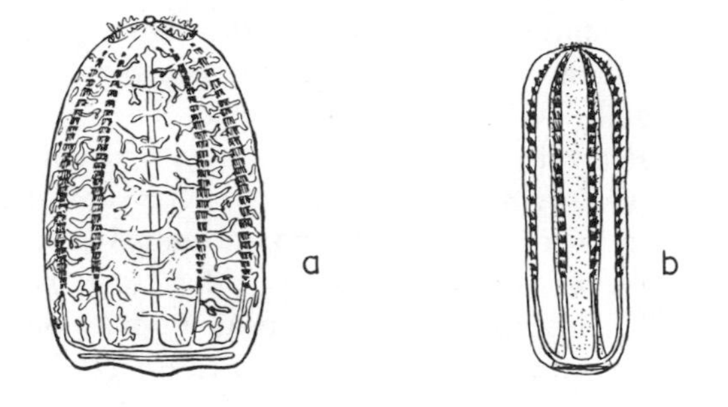

Fig. 2.25. *Beroe cucumis* (a) and *B. gracilis* (b) (Liley, 1958).

trapped in the mucus are passed to the mouth by cilia; the auricles between the upper edges of the lobes can assist. In *Leucothea*, which consumes prey mainly of copepod size, there are greatly enlarged oral lobes, each functionally divided into halves and capable of wrapping round prey sticking in the mucus. The auricles, which are long and slender, are constantly waving, producing a water current between the oral lobes. *Ocyropsis* has extremely muscular oral lobes (Fig. 2.23), and it can swim by a muscular flapping as well as more slowly by the ctenes. The feeding is entirely due to the muscular lobes which can be held widely open, but after contact with relatively large actively moving prey (euphausiids, small fish, etc.) the lobe curls over and directly transfers food to the mouth. Mucus plays no part in the process.

Many ctenophores may be extraordinarily abundant, occurring at times in huge dense shoals (Fraser, 1970) and in such numbers that they make serious depredations on the rest of the zooplankton (Zelickman, 1969). A sudden incursion of ctenophores in temperate latitudes is often associated with a marked decline of the crustacean zooplankton (cf. Bigelow, 1926). Mayer (1912) suggests that swarms of a single species are less common in the tropics.

Fulton (1968) summarizes observations in coastal waters off British Columbia by stating that ctenophores were the most important invertebrates in the euphotic zone in the Strait of Georgia from late June to October. *Pleurobrachia* comprised more than 25% of the surface plankton biomass in summer and autumn. *Bolinopsis* was also abundant in spring and *Beroe* was found through the year but in much lower density. Hirota (1974) demonstrated the abundance of post-larvae of *Pleurobrachia bachei* off California at a comparatively small distance from the shore. The maximum was found at about 3 miles. As a result, in part of the local circulation a high abundance of adults occurred inshore in the upper layers (mostly in the top 15 m) by day.

Field observations showed that growth from a body diameter of 1.5 mm to 6.5 mm could occur in around 30 days, with a somewhat slower initial rate of development and with slower growth for larger animals. Hirota showed, from rearing experiments using high feeding rates, that growth could be more rapid in the laboratory and was also accelerated between 15° and 20°C, though with considerably increased mortality. He concluded that at the high growth rate the population could double in 35 days. This might be compared with an observation quoted by Fraser (1970) that *Bolinopsis* off Monterey developed to maturity in three weeks.

Pleurobrachia is a seasonally dominant carnivore off California, preying selectively on small crustaceans and at times regulating their abundance. The ctenophores transfer a considerable quantity of organic matter through the food chain off La Jolla. The period of high abundance is from May to October, with a seasonal maximum usually in July or August; low densities are found from November to early spring. The changes may be associated with the Davidson Counter Current shoaling over winter off California. Numbers of *Beroe* also increase over the summer, but persist at low density during the winter, when a secondary maximum may occur. Hirota compares the seasonal changes for *Pleurobrachia bachei* with that for *P. pileus* in other regions. Thus, in the North Sea and English Channel peak populations were found about May/June, with a secondary maximum in autumn. Fraser's (1970) observations for the North Sea indicate a first peak in June with the main peak in November, resulting from breeding of the summer population, though this can vary from year to year. Off New Zealand the maximum occurs in July/August, i.e. in the austral winter.

Pianka (1974) believes that in temperate waters ctenophores like *Pleurobrachia* may be generally annual, but there appears to be considerable variation in spawning times. Off California, large adults overwinter and may spawn in spring; in Friday Harbour, though spring spawning occurs, some individuals continue to survive and grow and there appears to be summer spawning and possibly spawning in autumn. On the east coast of North America, south of Cape Cod, breeding appears to be in late spring and summer, but occurs in late summer and autumn north of Cape Cod. In the Mediterranean breeding may occur all the year, with a spring/summer peak, but in the Adriatic it is limited to summer. Reeve and Walter (1978) found no clear seasonal breeding pattern for *Mnemiopsis mccradyi* in Florida waters. They believe that a high density of zooplankton of suitable prey size is a major factor in stimulating reproduction. Breeding can occur at all sizes of ctenophore and paedogenesis has been generally recognized (cf. Pianka, Hirota, Reeve and Walter). An adult *Pleurobrachia* can produce 1000 eggs per day.

The comparatively sudden and massive increases in ctenophore population, typical of many inshore areas and often lasting only a few weeks, are due to a very few species. In Southampton Water, *Pleurobrachia pileus* increases in density very abruptly about May/June and high populations, with consequent reduction of the crustacean plankton, last for about six weeks only. Only small numbers of fairly large individuals are found over most of the remainder of the year (Raymont and Carrie, 1964). Greve (1972) describes *Pleurobrachia* as one of the most abundant holoplankton organisms in the North Sea, with a remarkable range of salinity (12–45‰) and temperature (–1°–26°C) tolerance. Burrell and Van Engel (1976) observed very sudden increases of *Mnemiopsis leidyi* over summer in the estuarine waters of Chesapeake Bay. The ctenophore can occur inshore in an environmental salinity of 15‰ in winter and even as low as 6‰ during the vast summer increase. The extreme temperature range is given as 1.3°–28.8°C.

Miller (1974) found very large populations of the same ctenophore in the Pamlico River estuary, North Carolina, during spring and autumn, but very low densities over summer. Its marked tolerance to very wide fluctuations in temperature and salinity is a factor enabling the species to inhabit the estuary throughout the year. Miller recorded its occurrence at salinities as low as 6‰ in winter and even less than 3.5‰ in summer. Kremer and Nixon (1976) also comment on the prevalence of this ctenophore in estuaries along the Atlantic and Gulf coasts of North America over summer. They point to the occurrence of *M. mccradyi* as a co-species in the southern states. In

Narragansett Bay, *M. leidyi* was observed to become suddenly abundant in July and remained as an important predator until October. The increase over summer was extremely rapid and the decline in autumn was also sudden. Often densities of >10 ctenophores/m^3 were found over summer, whereas only a remnant of the population (1–2 individuals per 10,000 m^3) occurred over winter, i.e. over 1 or 2 months there was a population change of five orders of magnitude. The very brief outburst in Narragansett Bay may be related to the species being near the northern limit of its occurrence at that latitude.

A longer period of high density occurrence appears to be true of more southern estuaries, with, in some areas (e.g. Maryland), a bimodal occurrence and even more or less continuous presence in some regions. Kremer and Nixon found that comparatively large individuals (almost all >4 cm) occurred over winter, but early in the main period of increase only about 20 % were of this size; small individuals were dominant over summer. The maximum densities indicated for ctenophores, bearing in mind their comparatively large size for zooplankton, are considerable. In Narragansett Bay, *Mnemiopsis* could reach 100/m^3, equivalent to 3.7 g/m^3 (dry wt); in Maryland, similar densities (50–100/m^3 maximum) have been recorded. *Bolinopsis* is reported in densities of 100–400/m^3 in the Barents Sea.

The decline of such populations of ctenophores inshore has been recognized in several areas as due primarily to the depredations of the ctenophore, *Beroe*. Some medusae, a few fish and certain amphipods (e.g. *Hyperoche mediterranea*—cf. Hirota 1961) feed on ctenophores, but Kamshilov (1955) in particular drew attention to the almost exclusive preying of *Beroe* on tentaculate and lobate ctenophores in Russian waters. Swanberg (1974) confirmed the feeding of *Beroe* on such ctenophores as *Bolinopsis*. Greve (1970, 1975b) states that *Beroe gracilis* feeds exclusively on *Pleurobrachia*, and *B. cucumis* selects *Bolinopsis*, though Swanberg (1974) believes the diet is less specific. In any event *B. ovata* will feed on many other ctenophores. Most workers believe that the diet consists exclusively of other ctenophores.

Burrell and Van Engel (1976) noted that although *Chrysaora* fed on *Mnemiopsis* in Chesapeake Bay there was no obvious effect on population density, but that *Beroe* caused a vast decrease in its numbers. On the other hand, Miller (1974) observed a great reduction in *Mnemiopsis* during spring in a North Carolina estuary, due to the voracious feeding of *Chrysaora* on the ctenophore.

While much attention has been devoted to the massive populations of ctenophores which can be built up in inshore waters, it is only fairly recently that the great significance of oceanic populations of ctenophores has been emphasized. Only a few cydippids and *Beroe* are sufficiently resistant to be taken in nets, and hauls can give only very doubtful estimates of density. Even inshore species are rapidly cut to small shreds by plankton nets. Hamner (1977) and his colleagues have pointed to the enormous difficulty of capturing many oceanic ctenophores. Many offshore species in particular are so delicate and fragile that even minor turbulence will destroy them, but diving techniques have now been employed which indicate the widespread occurrence and abundance of many species.

With medusae and siphonophores, ctenophores now rank as an extremely important carnivorous element in oceanic plankton food webs. Harbison, Madin and Swanberg (1978) believe that previous estimates of populations of ctenophores from open ocean are utterly misleading. From their Scuba observations they suggest that ctenophores may occur almost as plentifully in open oceans as inshore, where they are accepted as enormously abundant. The considerable diversity of feeding biology suggested by collections from the North Atlantic and Indian Oceans indicates that ctenophores represent a very significant component in the ecosystem of the upper waters of open oceans. Harbison *et al.* (1978) collected ctenophores in 75% of 250 dives, mostly in the North Sargasso Sea and, over summer, in the temperate and slope waters of the North Atlantic. The diversity of ctenophores other than those few seen in inshore waters is indicated by the list of species in their report. Included in their findings is a large very fragile ctenophore taken in slope water off New England (*Thalassocalyce inconstants*). It is believed to be a new species, somewhat intermediate between cydippids and lobates and probably representing a new family. The ctenophore has a medusa-like body, lacking oral lobes and auricles and with two tentacles without sheaths (cf. Madin and Harbison, 1978b).

Although a greater variety of species is found in warm waters, ctenophores may occur in high density in very cold seas. Large swarms of *Pleurobrachia* and *Bolinopsis* were recorded by Fraser (1970) from the Arctic; *Bolinopsis* is reported as reaching densities of $400/m^3$. *Mertensia ovum* is an Arctic species which drifts south in the Labrador Current. Digby (1953) recorded ctenophores occurring all the year in East Greenland waters, with large quantities of gelatinous material apparently derived from ctenophores in late summer. Zenkevitch (1963) reported ctenophores in cold Russian seas and Zelickman (1972) found *Pleurobrachia*, *Bolinopsis* and *Beroe* in the Barents Sea. The density of *Bolinopsis* was dependent on the supply of zooplankton food and especially on the depredation of *Beroe*. Hardy and Gunther (1935) identified both *Beroe* and *Pleurobrachia* from the Antarctic, though they point out that the abundance of ctenophores could not be estimated since many were destroyed in net capture. Most ctenophores are epipelagic and though there is little evidence of diurnal vertical migration they will leave the surface in rough weather. Though most ctenophores are surface dwellers, some may occur in deep waters, even to depths approaching 3000 m (Bayer and Owre, 1968). Vinogradov (1970) also believes that ctenophores are mainly confined to surface and sub-surface waters to depths of 200–500 m. Although very few ctenophores occur in deeper waters, a species of *Beroe* characteristic of the North Pacific was found from below 200 m to 2000 m.

Harbison, Madin and Swanberg (1978) suggest that the genus *Mnemiopsis* is mainly neritic, as also is *Pleurobrachia*. According to Fraser (1970), *P. pileus* is found widely in more coastal areas of the temperate Atlantic and in the Arctic (cf. Greve, 1975). The common species found in coastal areas of the North Pacific is probably a separate species, *P. bachei*. *Bolinopsis*, also neritic, is represented by *B. infundibulum* in cold-temperate waters as well as in the Arctic. Lenz (1973a) found that it was much more widely distributed in the Baltic Sea than was once considered and was commoner than *Pleurobrachia*.

The ctenophore is so delicate that it had been frequently missed in field collections. Harbison *et al.* (1978) found *B. vitrea* as a neritic form in tropical and sub-tropical waters and Mayer (1912) also describes it as abundant in warm coastal areas of the Caribbean. *Beroe cucumis* is a very widely distributed ctenophore found in Atlantic, Indian and Pacific Oceans as well as in Arctic and Antarctic waters. It is probably a truly cosmopolitan form, but a number of races, sometimes regarded as separate species, are known. *B. ovata* appears to be a more warm-water neritic species, found for example in warm Atlantic and West Indian waters and in the Mediterranean (cf. Mayer, 1912; Swanberg, 1974). Greve (1975a) lists *B. gracilis* in the southern North Sea and around the British Isles. Harbison, Madin and Swanberg (1978) refer to a number of other ctenophores which are oceanic and mainly tropical and sub-tropical (e.g. *Lampea pancerina*, *Leucothea* spp.) or tropical (e.g. *Eurhamphaea vexilligera*, *Ocyropsis crystallina* and *O. maculata*). *Callianira bialata* appears to be more temperate and sub-tropical in distribution and *Cestum veneris*, though mainly tropical, has been taken in temperate waters. *Velamen parallelum*, as a more temperate and sub-tropical species, appears to inhabit somewhat cooler waters than *Cestum*.

The widespread occurrence and abundance of ctenophores emphasizes the need to estimate their densities in various seas more precisely. Reeve (1977) has indicated from experiments on *Mnemiopsis* that their capacity for feeding is enormous. The ctenophore appeared to be satiated only when the gut was clogged. It may be, therefore, that the grazing potential of this carnivorous group of zooplankton is even greater than is now believed.

Nemertea

The planktonic nemertines all belong to the so-called "armed" order Hoplonemertea, sub-order Polystylifera. They are found in all the major oceans but never occur in shallow waters. They are bathypelagic, occurring at depths exceeding 500 m, and normally in waters of between 1000 and 3000 m. These pelagic nemertines thus differ sharply from the great majority of the phylum which are benthic animals. The planktonic group is comparatively small; Coe (1954), describing some new species from the Pacific, suggests that there are altogether some sixty-seven species belonging to eleven families.

The body is fairly firm but rather gelatinous, and, unlike benthic nemertines, is horizontally flattened so that it appears relatively broad. The musculature is comparatively feeble and the animals tend to float, maintaining their level at intermediate ocean depths. Coe suggests that an appearance of nemertines in upper layers may be a consequence of upwelling. Although some genera (e.g. *Nectonemertes*) have a fin around the tail, in some species even extending laterally, their weak musculature suggests that they are not capable of active swimming. Whereas a few species are only a few millimetres long, the great majority are some 20–40 mm and some range even up to 100 mm length. One individual of *Dinonemertes investigatoris*, a species found in the Indian and Atlantic Oceans, exceeded 200 mm. Coe (1945) describes it as the giant among the nemertines.

Exoskeletal remains of crustaceans have been found occasionally in the guts of the few planktonic nemertines examined, suggesting that pelagic Nemertea are carnivorous like the benthic representatives. Probably the long eversible proboscis can be coiled around the prey, assisted by a mucus secretion; the proboscis also has numerous minute stylets which may rasp the tissues of the prey.

A few species show clear sexual dimorphism; in *Nectonemertes*, for example, the males have clasping tentacles which possibly assist in associating with the female (Fig. 2.26).

Very few specimens of most planktonic nemertine species have been taken in oceanic expeditions, and since it is unlikely that they are active enough to escape nets, there is general agreement that they are represented in the oceans by an exceedingly sparse

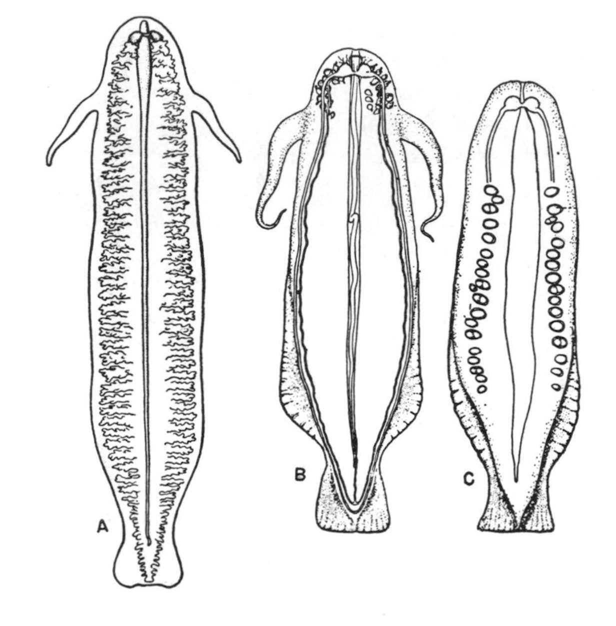

Fig. 2.26. *Nectonemertes mirabilis* (A) Male with relatively short tentacles showing lobed intestinal diverticula; (B) male showing spermaries, also proboscis and sheath; (C) female with 16 pairs of ovaries (Coe, 1956).

population (cf. Coe, 1945, 1954). A species with an apparently wide horizontal distribution is *Nectonemertes mirabilis*, reported for the Atlantic Ocean and also occurring in the Pacific off Japan and California (Brinkmann, 1917). Later records by Coe (1945, 1954) suggest that this is one of the more abundant pelagic Nemertea. It occurs east and west of Greenland, in latitude about 64°N, and throughout the North Atlantic, across the equator and widely in the South Atlantic to about 24°S. In the Pacific it is also found relatively far north, off the Kurile area and in the Bering Sea. In depth it always exceeds 500 m and mostly occurs at the 1000–2000-m level. Another fairly broadly distributed species is *Protopelagonemertes hubrechti*, found widely in the North and South Atlantic and in the North Pacific. One of the commonest, best known and most widely distributed species is *Pelagonemertes rollestoni* (Fig. 2.27) known from the Indian and Pacific Oceans and generally in the South Atlantic even to Antarctic waters. Wheeler (1934) gives one record as 50°S and describes the species as relatively common. In the North Atlantic, however, it is not known much beyond the equator; Coe (1935) gives its northern limit as 7°N.

Although a few pelagic nemertines have a wide distribution, many appear to be limited to one ocean. Coe (1954) suggests that of the sixty-seven species, twenty-nine occur in the Pacific but of these, twenty are found in the Pacific Ocean alone. Of some forty species recorded for the Atlantic, over thirty appear to be limited to that ocean. The geographical distribution may be considerably extended, however, when more Nemertea are captured. The very small number of records of this exceedingly sparse bathypelagic fauna implies that there are considerable difficulties in delimiting both its geographical and vertical distribution.

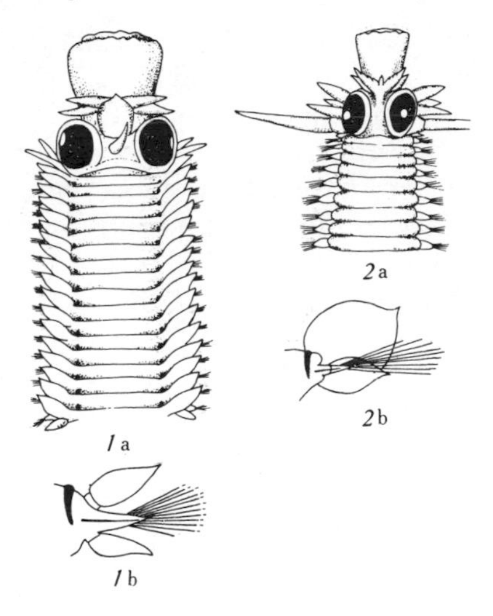

Fig. 2.27. *Pelagonemertes moseleyi* (Yamaji, 1973).

Annelida

The Annelida are represented in the plankton as meroplanktonic trochophore and post-trochophore larvae of benthic polychaetes and echiuroids. In shallow waters, adult polychaetes (e.g. some syllids) may sometimes be found living a planktonic life for part of their existence, i.e. they are tychopelagic. Epitokous polychaetes may also swarm temporarily in inshore waters, at times appearing quite conspicuously in the plankton. The somewhat peculiar annelid, *Poebius* (Heath, 1930), is a holoplanktonic species found in the North Pacific; views differ as to whether this worm is an aberrant echiuroid or a much modified polychaete. The distribution of the species appears to be confined mainly to sub-Arctic water masses of the North Pacific (cf. McGowan, 1960). Many authors have referred to adult polychaetes drawn from a variety of families which may occasionally be found in the plankton even over oceanic depths. Some of these may be truly holopelagic species.

In addition to these more or less isolated examples of holopelagic forms, however, there are a few families of polychaetes which provide a substantial number of well-known holoplanktonic worms. Among these the Aphroditidae have a very few pelagic species, these being mostly post-larval and bathypelagic forms (cf. Monro, 1930; Stop-Bowitz, 1948). Though some of these species have been recovered repeatedly in plankton hauls, there is still some doubt as to their precise habitat and whether this is truly planktonic. Certainly some of the Aphroditidae juveniles are only planktonic in the younger stages, and Wesenberg-Lund (1939), for instance, suggests that some of these may be very slowly descending to take up eventually a deep benthic existence. Of the other families, there is no doubt concerning their genuine permanent pelagic existence. They include a sub-family of the Phyllodocidae, the Lopadorhynchinae, now given full family rank, the Lopadorhynchidae, by most authorities. Two other extremely small groups are usually also given separate family rank: the Isopilidae is a very small family of holoplanktonic polychaetes including the important genus

Phalacrophorus. *Pontodora pelagica* is a single planktonic species assigned to a separate family, the Pontodoridae (cf. Dales and Peter, 1972). Three other families of polychaetes are exclusively planktonic: the Alciopidae and Tomopteridae have a number of representatives which are well known in the plankton; the Typhloscolecidae have only a few genera but some have a very wide geographical distribution.

In general pelagic polychaetes do not constitute a major part of the total zooplankton biomass, though Mileikovskiy (1969) emphasizes that they should not be regarded as a rare planktonic group; he quotes densities from data by Tebble of just under 1 polychaete/m^3 for the relatively rich sub-Arctic Pacific area. Tebble (1962) himself points out that over the North Pacific only three species occur in noticeably large, though variable, densities: *Tomopteris elegans*, *Tomopteris septentrionalis*, and *Typhloscolex muelleri*. The two tomopterids together approach densities of about 0.5 individuals/m^3. Mileikovskiy, however, states (from data by Dales) that higher densities approaching 5 polychaetes/m^3, even up to 20/m^3, occur in the Monterey area. His own data suggest densities slightly exceeding 1/m^3 for adults and young stages, both in the relatively shallow 200–500-m layer of the North Pacific and in the deeper 500–1000-m stratum. Since some of these polychaetes appear to be more or less cosmopolitan, their importance in particular water layers, especially as predatory species, should not be under-emphasized.

The Alciopidae are Errant Polychaeta, with a well developed prostomium bearing four to six antennae, and with an eversible proboscis. Most characteristic is a pair of relatively enormous eyes borne on the prostomium. The body itself is transparent with well developed uniramous parapodia, bearing setae, and fairly conspicuous foliaceous dorsal and ventral cirri – presumably an adaptation to planktonic life. Many of the species have patches of brown pigment ("segmental glands") at the bases of the parapodia. The Lopadorhynchidae are somewhat similar in that there is a prostomium with two pairs of short antennae, well developed uniramous parapodia bearing setae and a fairly conspicuous dorsal and ventral cirri which may be foliaceous. In contrast to the Alciopidae, however, the eyes are very small and inconspicuous; they may be very much reduced.

The Tomopteridae is an important family with one major genus, *Tomopteris*, with some forty species; some authorities subdivide *Tomopteris* into separate genera. The prostomium is produced into a pair of small but obvious lateral horns (the antennae) and though most species possess a pair of eyes, these are very small and inconspicuous. The parapodia are well developed, biramous paddles, lacking setae but expanded into foliaceous pinnules ("pinnae"), presumably also an adaptation to the planktonic existence (Fig 2.28). The second segment has a pair of very obvious long processes

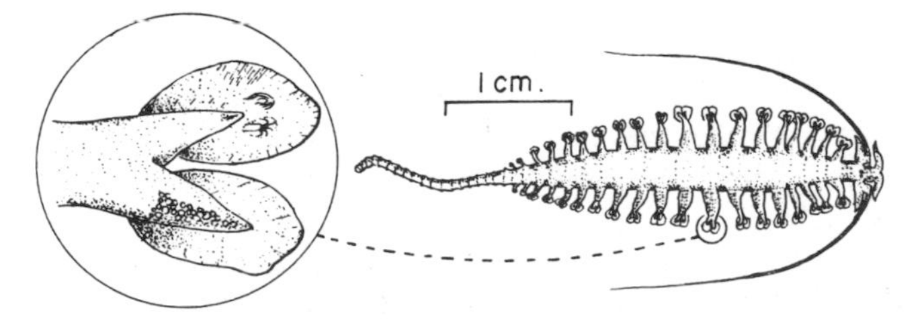

Fig. 2.28. *Tomopteris helgolandica* (with enlarged bi-lobed paddles) (Fraser, 1962).

("streamers") representing a pair of setigerous appendages; possibly the processes represent extended acicula. Some species have an additional pair of processes, the setigerous appendages of the first segment, between the streamers and the prostomium, but these are always very much smaller.

Several types of glands may be present on the parapodia and are useful in identification. Chromophile glands, large yellowish tubules, may occur, usually only on the neuropodium, opening by a single pore near its ventral border. Hyaline glands are relatively small, consisting of a mass of transparent tubules in either the notopodium or neuropodium, or in both; sometimes dark brown or reddish pigment may occur near the centre of the gland. Rosette glands are minute, apparently composed of a ring of only eight cells. They may be brownish or red in colour and situated on the trunk of the parapodium or on the noto- or neuro-podium. The fourth type, spur glands, are small glands, possibly accessory to the chromophile glands.

The Typhloscolecidae have a transparent, more or less spindle-shape or cylindrical body, with a very small conical and inconspicuous prostomium, which is not clearly marked off from the rest of the body. The prostomium lacks eyes and obvious sensory appendages, except for a pair of well-developed nuchal organs. The parapodia are rudimentary, and setae usually are recognizable only in the posterior segments, but there are comparatively well-developed dorsal and ventral cirri.

The number of species of pelagic polychaetes found in warmer waters of the world appears to be much higher than at high latitudes and Stop-Bowitz (1948), for example, calls attention to the few species which penetrate the North Atlantic. The great majority of polychaetes appear to be oceanic, many having a fairly wide distribution. Stop-Bowitz found three species of *Tomopteris* (*T. krampi*, *T. septentrionalis* and *T. messania*) and also *Travisiopsis lanceolata* with a very wide horizontal distribution in the North Atlantic. They also appeared to be eurybathic. This latter species is also listed by Wesenberg-Lund (1939) as having an extended range in the North Atlantic, penetrating further than most species to higher latitudes. *Naiades* (*Alciopa*) *cantraini* and *Vanadis formosa* are widely distributed in temperate and tropical areas of the Atlantic and Pacific. The records of Muus (1953) also suggest a fairly broad distribution in temperate and colder North Atlantic areas for such species as *Alciopa reynaudi*, *Pelagobia longicirrata* and several species of *Tomopteris*.

Unfortunately, in many plankton surveys, pelagic polychaetes have been taken in such low numbers and in rather poor condition of preservation, so that precise distribution patterns are difficult to deduce. The following summary of some of the recognized distributions is taken from reports including those of Dales (1957), Tebble (1960, 1962) and the investigations of Peter and Mileikovskiy, but particularly from the bibliography of Dales and Peter (1972).

Of the Lopadorhynchidae, *Pelagobia longicirrata* is a cosmopolitan species. Dales reports it as the most abundant and widely distributed species in the Pacific, occurring off the Aleutians, through the equatorial regions as far south as Chile. It is also found commonly in the Antarctic where it is probably the most abundant polychaete. Tebble (1960, 1962) suggests that it occurs in almost all water masses, with a wide depth range, throughout the South Atlantic including Antarctic areas, and, though possibly more patchily distributed, in sub-Antarctic waters. An earlier finding from 77°S latitude is probably the most southerly record for any adult planktonic polychaete. According to Peter (1973), the species is less common in the southern part of the Indian Ocean than in the warmer waters to the north, except near the Australian coast. Fauvel (1916) suggests that in the Atlantic it occurs at least from Greenland to the Antarctic as well as in the Mediterranean. Mileikovskiy (1969) confirms that it is common and widespread at high latitudes both in the North Atlantic and North Pacific. Ramirez (1977) found the species off the Argentinian shelf and considers that it is brought there from the sub-Antarctic in a deeper cold current.

Maupasia coeca, another member of the Lopadorhynchidae with a wide distribution, appears to be scattered throughout the North Pacific, the North and South Atlantic, including the Mediterranean, and the Indian Ocean. It occurs at high latitudes including Antarctic waters, as well as in sub-tropical and tropical areas, but is far less commonly taken in the Southern Ocean than *Pelagobia*. In contrast, several species of *Lopadorhynchus* appear to be essentially warmer-water organisms. *L. brevis*, *L. uncinatus* and *L. krohni*, according to Tebble, occur only in tropical and sub-tropical waters. These species may still be fairly widely distributed in warmer waters. For example, *L. brevis* occurs in the tropical and sub-tropical waters of the Atlantic, Mediterranean and Pacific, and *L. uncinatus* is widely distributed in the Pacific, North and South Atlantic and Indian Ocean (Dales and Peter, 1972). Peter (1977) records, amongst new pelagic polychaetes from the Indian Ocean, *Lopadorhynchus indica* as comparatively common in the surface waters of the eastern Indian Ocean where it appears to be confined to waters of high temperature. Tebble (1960) states that no species of *Lopadorhynchus* occurs south of the Sub-Tropical Convergence (approximately 40°S latitude in the Atlantic Ocean). Peter (1973) also quotes *L. nationalis* and *L. henseni* as common over the Indian Ocean but extending to somewhat cooler waters to latitude about 35°S. In the south-west Atlantic, it has a comparable distribution (cf. Ramirez (1977). Species of *Lopadorhynchus* are found north of the Secondary Polar Front in the Atlantic. *Phalacrophorus pictus*, in contrast, is a cosmopolitan species scattered throughout the world, though occurring perhaps more commonly at high latitudes including Arctic and sub-Arctic areas of the Atlantic and Pacific Oceans. Mileikovskiy (1969) states that it is comparatively abundant at high northern latitudes and there are some records from Antarctic waters.

Many more species and genera of Alciopidae occur in warm Atlantic and Pacific waters than at the higher latitudes of these two oceans. Species such as *Torrea candida* and *Niades* (*Alciopa*) *cantraini* are widely known from tropical and temperate Atlantic and Pacific Oceans, and were found abundantly in the Pacific off California by Dales, so that they might be described as cosmopolitan (Dales and Peter, 1972). Tebble (1960), however, reports that while common in the South Atlantic they do not reach the colder waters south of the Sub-Tropical Convergence, and equally do not penetrate the North Atlantic beyond the Secondary Polar Front.

Other species of Alciopidae show widespread distribution in the warmer waters of the world but apparently cannot penetrate the cold waters. For instance, *Alciopina parasitica* occurs widely in the North and South Atlantic, Mediterranean and the Indian Ocean; *Krohnia lepidota* and *Alciopa reynaudi* are two other species which, though they have a fairly wide distribution in all three oceans, according to Tebble, cannot stretch beyond the Sub-Tropical Convergence. Species of *Vanadis* also illustrate much the same pattern of geographical distribution. *Vanadis minuta* is found quite widely in the Pacific, Atlantic and Indian Oceans (Peter, 1974). In the South Atlantic, Ramirez (1977) found it extended farther into cooler waters than many other polychaetes, but it did not reach south of 38° S latitude. *Vanadis formosa* is a generally warm-water form in the tropical and sub-tropical Atlantic, the Mediterranean, and commonly in the Pacific, but Dales suggests that it barely penetrates waters below 12°C. It does not extend south of the Sub-Tropical Convergence in the Atlantic area (Tebble, 1960). *Vanadis crystallina* is a similar species, found widely in the North and South Atlantic, Pacific and Indian Oceans, but confined mainly to warmer waters and not proceeding beyond the Sub-Tropical Convergence. Three species of *Rhynchonerella*, *R. gracilis*, *R. moebii* and *R. petersi*, also seem to be largely confined to warmer waters of the oceans.

In contrast, a few Alciopidae can tolerate colder waters. *Vanadis longissima*, while found in the warmer waters of the Atlantic and Pacific Oceans, has a much broader range than many of the planktonic polychaetes. In the North Pacific, for example, it stretches to the border of the sub-Arctic water and Transition Zone (*ca.* 45° N latitude). In the South Atlantic it is found in sub-Antarctic waters but does not cross the Antarctic Convergence. According to Tebble (1960), it is the only species of pelagic polychaete which is limited in its southerly distribution by the Antarctic Convergence. South of the Antarctic Convergence is the very closely related species, *Vanadis antarctica*. Tebble (1960) raises the possibility that the two species might represent warm- and cold-water forms of a one-time single species.

The distribution of two other Alciopidae also relates to colder waters. *Rhynchonerella angelini* is a moderately widespread form found in the Atlantic, Pacific and Indian Oceans and though regarded mainly as a warm-water species not extending beyond the Sub-Tropical Convergence in the southern Atlantic, it can inhabit colder waters in the northern hemisphere. It is widespread in the Pacific and, according to Dales (1967), is found all over the world and is comparatively common in the rather cold waters of California. *R. angelini* appears to extend rather farther north than many Alciopidae in the North Pacific (cf. Tebble, 1960). In the North Atlantic it is not limited by the Secondary Polar Front. Muus (1953) records it in the Faroe–Iceland area. *Rhynchonerella bongraini* is an endemic Antarctic species, limited to Antarctic waters and in fact barely reaching the Antarctic Convergence in its most northerly distribution.

Of the family Tomopteridae, several species of *Tomopteris* (e.g. *T. mariana*, *T. nationalis*, *T. ligulata*, *T. apsteini*, *T. elegans*) are tropical and sub-tropical species or at least mainly limited to warmer waters. A number of warm water species are at present recorded from the Mediterranean only. However, a very few forms (*T. septentrionalis*, *T. planktonis*) are cosmopolitan, occurring widely in tropical, sub-tropical and temperate waters, as well as in colder waters at high latitudes. Both species are recorded from sub-Antarctic and Antarctic regions. Ramirez (1977) records *T. planktonis* off Argentina from the cold deep current from sub-Antarctic areas. Both species also reach far north in the North Atlantic to the Irminger Sea and Davis Strait, although *T. septentrionalis* is perhaps more widely distributed, including the Norwegian Sea, Iceland, and the North Sea (Muus, 1953). The species is found also in the cold northern Pacific at high latitudes, but *T. planktonis* does not extend into sub-Arctic waters in the North Pacific. On the other hand, *T. pacifica* is typical of the sub-Arctic Pacific waters and apparently is not found elsewhere.

Of other Tomopteridae, *T. helgolandica* is a widespread species in the North Atlantic only, where it can extend into high latitudes and into neritic areas including the North

Sea and English Channel. *T. krampi* is a relatively deep-living polychaete, found especially in the North and South Atlantic, and only sparingly recorded in the Pacific. Although it does not penetrate in the South Atlantic beyond the Sub-Tropical Convergence, in the North Atlantic it reaches high latitudes, even the Irminger Sea and Davis Strait areas (cf. Muus, 1953). *T. nisseni* is also found fairly broadly in the Atlantic Ocean but does not extent south of the Sub-Tropical Convergence (Tebble, 1960). In the North Atlantic its distribution extends far north and is similar to that of *T. krampi*. In the Pacific, *T. nisseni* is apparently restricted to the north and occurs in deeper waters. One species, *T. carpenteri*, is confined entirely to Antarctic waters. Tebble (1962), commenting on the considerable similarity of distribution of polychaetes—including Tomopteridae—in the Pacific and Atlantic Oceans, points to one marked difference. Whereas *T. nisseni* (as already noted) and *T. planktonis* stretch into cold boreal waters of the Atlantic Ocean, they do not extend to the sub-Arctic waters in the Pacific. Here they are replaced by *T. pacifica*.

Of the few species of Typhloscolecidae, at least two are fairly widely distributed. *Sagitella kowalevskii* occurs broadly over the Pacific, though Tebble (1962) regards it as mainly inhabiting warmer waters. The species has a considerable depth range. According to Peter (1973) *Sagitella* is very abundant and probably the most widespread polychaete in the Indian Ocean. It is found especially in the warmer waters extending south at least to about 30°S latitude, but in places can approach the Antarctic Convergence in small numbers. In the Atlantic it is also widely distributed, but does not reach colder waters to the south. It is not listed by Tebble (1960) from the Discovery collections. Ramirez (1977) has taken it in the South Atlantic at approximately 37°S latitude. According to Muus (1953), the species does not reach higher latitudes in the North Atlantic, though it is well recognized as far north as the area west of Ireland.

The second species, *Typhloscolex muelleri*, is cosmopolitan (Dales and Peter, 1972) and is widely distributed in the warmer waters of the three oceans, but is also well known at higher latitudes. In the southern hemisphere it extends south across the Antarctic Convergence and is well known in the Antarctic Ocean (cf. Tebble, 1960). It occurs in sub-Arctic and Arctic waters of the Atlantic, for instance in the Norwegian Sea and Barents Sea, and also in the sub-Arctic northern Pacific. It is found over a considerable depth range, though rarely near the surface (Mileikovskiy, 1969). Not only is it widely distributed but, according to Mileikovskiy (1969), it is one of the more important cold-water polychaetes of Atlantic and Pacific Oceans.

Of the species of *Travisiopsis*, *T. levinseni* is found generally in the north and south Atlantic, reaching fairly high latitudes, for instance, the Faroe–Iceland area and Davis Strait. It has also been found in colder waters of the Pacific, though records are rather scattered. Dales (1957) records it from somewhat deeper strata off California. Peter (1972) has recorded it from the Arabian Sea. It is probably a more or less cosmopolitan species since, in the southern hemisphere, it is also well known in Antarctic waters (cf. Tebble, 1960). *T. lanceolata* has a similar distribution in the North Atlantic, occurring in both warm and colder waters. However, although it is also well known in the South Atlantic, it does not extend past the Sub-Tropical Convergence (Tebble, 1960). The species has been taken by Peter (1973) in the Indian Ocean, where it appears to live at depth, perhaps in upwelled water. It apparently can survive a considerable range in temperature.

Tebble (1960, 1962) points out that many polychaetes typical of tropical and sub-tropical waters, are limited by cold barriers. For example, in the southern Atlantic some sixteen species do not extend beyond the Sub-Tropical Convergence. Most of these species are also restricted in the northern hemisphere by the much less defined Secondary Polar Front. In the North Pacific, where there is a rather clearer demarcation between the warmer waters and the colder sub-Arctic waters, of thirty-three species inhabiting the North Pacific twenty-one are limited by the Transition Zone. Some of these species may penetrate to some extent into transitional water, which may vary with season. A few cosmopolitan species may also spread to high latitudes, but there are apparently no endemic Arctic species. Only three species of polychaetes are endemic to the Antarctic (*Vanadis antarctica*, *Tomopteris carpenteri*, and *Rhynchonerella bongraini*), and they do not pass the Antarctic Convergence.

Mileikovskiy (1969) examined the depth distribution of adults and larvae of some common planktonic polychaetes at high latitudes. In the Norwegian Sea, adult *Pelagobia longicirrata* and *Typhloscolex muelleri* were infrequent in the uppermost 50 m and were most common at about 200–500 m depth, but larvae of these species were absent from the upper 500 m. Larvae of *Phalacrophorus pictus* were also found below 500 m. Breeding might, therefore, occur at greater depths in these three species. *Tomopteris septentrionalis* larvae, however, were plentiful in the upper layers to about 200 m.

In the Kurile–Kamchatka area of the North Pacific, on the other hand, *T. muelleri* was mainly in the 50–200-m layer, though extending to 500 m; *Tomopteris septentrionalis* was mainly distributed from 200 to 500 m, with some few specimens extending to 1000 m; *Pelagobia longicirrata*, by far the most abundant polychaete, was mainly in the 200 to 1000 m layer but had many representatives to 2000 m, and some specimens reached 4000 m. The species was not found shallower than 200 m. In general, the distribution of larval and young stages was fairly obviously correlated with the adult distribution. Thus, *Pelagobia longicirrata* larvae were lacking from the upper 200 m and were abundant in the 200–2000-m layer. This species is thus held to breed only in deeper water; the other three species breed at lesser depths, mainly 50–500 m.

Larval forms of Benthic Invertebrates

Some of the larval stages of the more important taxa of holoplanktonic forms will be mentioned in accounts of their life histories and breeding cycles (Chapter 3). No attempt is made to give a comprehensive description of the great variety of meroplanktonic larvae of benthic invertebrates, since good coverage exists in various textbooks and specialist keys. Apart from the brief summary which follows, however, a separate and somewhat fuller account of the meroplanktonic eggs and larvae of fishes is given later (Chapter 4), largely in view of their widespread occurrence and commercial significance.

The medusoid phase of hydromedusan cnidarians can be abundant in nearshore waters. Some Anthomedusae and Leptomedusae are released only for a comparatively short period from the hydroid stage, but many can live for at least several weeks, during which time the gonads mature. Newly liberated medusae are small (*ca.* 1 mm), and in species with only four marginal tentacles, there is little change, apart from an increase in size with age and the development of the gonads. Many species, however, possess more than the four tentacles and these medusae show various stages of development. In some there are also more than the typical four radial canals passing from the stomach to the ring canal, the extra canals developing during the free-swimming life of the medusa. Some species have branched canals and in others, certain canals are blind. Species of Hydromedusae possessing more than the basic four tentacles usually show an increasing number with age of development. Russell (1953) points out that the increase typically follows a regular pattern until the full complement for the species is achieved. In a few medusae (e.g. *Podocoryne carnea*) the gonads are fully developed at liberation and the length of life will be short. According to Russell (1953) the average life span is about 2 months in British waters although some, (e.g. *Aequorea*), may live for 6 months or more. In such temperate waters the majority of species occur from spring to autumn and the whole life history (medusa—egg—polyp—medusa) occupies about a year, though a few (e.g. *Obelia*) have more than one brood annually. More rapid cycles will occur in warmer waters (cf. Chapter 3).

Even when the medusa is free-living for several weeks (mainly in shallow water), because the hydroid stage is attached to the substratum, Hydromedusae will be essentially coastal, not normally spreading far beyond the continental shelf. Kramp (1957), however, points out that a few species appear to be truly oceanic, and are found regularly over deep oceans. Although these species have a benthic hydroid phase, usually in moderately shallow water, asexual budding by the medusae can permit a considerable spread over oceans. Asexual budding, more common amongst Anthomedusae, occurs normally from the stomach wall or base of the marginal tentacles. The reproductive cycle of some bathypelagic species (cf. Chapter 3) which appear to be fairly widely distributed is not known. Some species could have a deep-living hydroid stage (cf. also Russell, 1953).

Many hydroids (e.g. *Tubularia*, *Sertularia*, *Plumularia*) do not produce free-swimming medusae. The gonads develop in special gonothecae on the hydroid. In some species the fertilized egg develops directly into the larva before release. In a few (e.g. *Tubularia*), even the typical larva is not released but a more advanced actinula larva is formed. Typically, however, the fertilized egg develops into a planula, a larva usually of very elongated oval shape, uniformly ciliated, with a solid mass of cells internally. Fertilization may occur in some species on the medusa rather than free in the sea and some development of the fertilized egg takes place before release. *Turritopsis nutricola* develops to a full planula.

The ciliated planula larva is typical of the phylum Cnidaria. In the essentially oceanic Trachymedusae and Narcomedusae no hydroid phase occurs in the life history. A planula developing from the egg forms tentacles and is gradually transformed into the free-swimming medusa. The fertilized eggs of anemones develop into planula larvae before settling on the sea bottom. Over coral reefs, for example, enormous numbers of planulae may be released into the nearshore plankton. The release frequently shows an obvious periodicity. The prolonged planktonic larval stage of Cerianthidae is known as "arachnactis". In probably all coastal species of Scyphozoa the eggs are fertilized in the medusa and are liberated as planula larvae. In any event, the planulae soon attach to the substratum to form the relatively small polyp-like scyphistoma, which may bud and proliferate according to species and environmental conditions.

Strobilation gives rise to the ephyra larvae, varying from one to many, according to species and circumstances. The free-swimming ephyra is of characteristic flattened saucer shape, with eight lobes separated by deep incisions. Each lobe has a notch housing a developing marginal sense organ and bounded by a pair of lappets. A manubrium marks the developing gastro-vascular system with its canals. The development of the adult scyphozoan from the ephyra involves the formation of the full canal system, marginal tentacles and the thickening of the body to achieve the typical jellyfish form. Some oceanic scyphozoans can develop directly without a scyphistoma stage. The peculiar stephanoscyphus stage of coronate scyphozoans is referred to earlier (cf. page 97).

The trochophore type larva is characteristic of a number of invertebrate phyla. Typically the trochophore has a more or less spherical or ovoid body, with a girdle of ciliated cells, the prototroch, running approximately in an equatorial position in front of the mouth. Behind the mouth is another circle of locomotory cilia known as the metatroch. In the apical position there is a small group of cells with elongated sensory cilia forming an apical organ. The mouth opens into a stomodaeum situated in a fairly large body cavity, which also contains a simple stomach and intestine, opening posteriorly. The anus is surrounded usually by a small circle of cilia, the telotroch. Mesenchyme cells occur in the body cavity, which includes typically a pair of larval nephridia.

Many families of Polychaeta pass through a trochophore stage in their development. An important character in a typical polychaete trochophore is the possession of paired mesodermal blocks, which later multiply as a paired segmental series forming the main trunk of the adult worm. With this formation of the body segments, paired bundles of chaetae develop. Future body segments are also often marked by rings of cilia. At a stage when the mesoderm is beginning to segment, the larva is usually referred to as a post-trochophore. Many polychaetes hatch as post-trochophores rather than in the trochophore stage (cf. Fig. 2.29). The bundles of chaetae in trochophore and post-trochophore stages are often very long ("provisional chaetae"), and are characteristic of different polychaete families.

In Nemertine worms there is a larva which in overall pattern somewhat resembles a trochophore, possessing a large body cavity with the usual three-chambered gut and small apical sensory organ, but the ciliated major locomotory band lying at the base of the larva is drawn out into two large ciliated lobes (lappets), one on either side of the mouth. The larva is known as a pilidium. In those Bryozoa (Polyzoa) which include a larval stage—the cyphonautes—the general structure is again trochophore-like but the shape is approximately triangular, with a prominent locomotory ciliated band running along the base. Many sipunculids have a trochophore-type larva in their development. In shape the larva is elongate-oval or almost cylindrical, with a well marked telotroch. In several larvae there is considerable elongation of the body behind the mouth. In some sipunculids this stage is succeeded by a pelagosphaera larva (Rice, 1975).

Some pelagosphaera larvae are of large size and long lived. In other species, however, the trochophore is short-lived, the eggs are yolky and the development is shortened (cf. Fig. 2.30). In the small phylum of the Phoronidae the larva, of the general trochophore type, has a rather extended portion of the body behind the mouth and a fairly prominent telotroch, but there is no conspicuous apical sense organ. Most characteristic is the possession of a number of narrow ciliated blunt lobes forming arm-

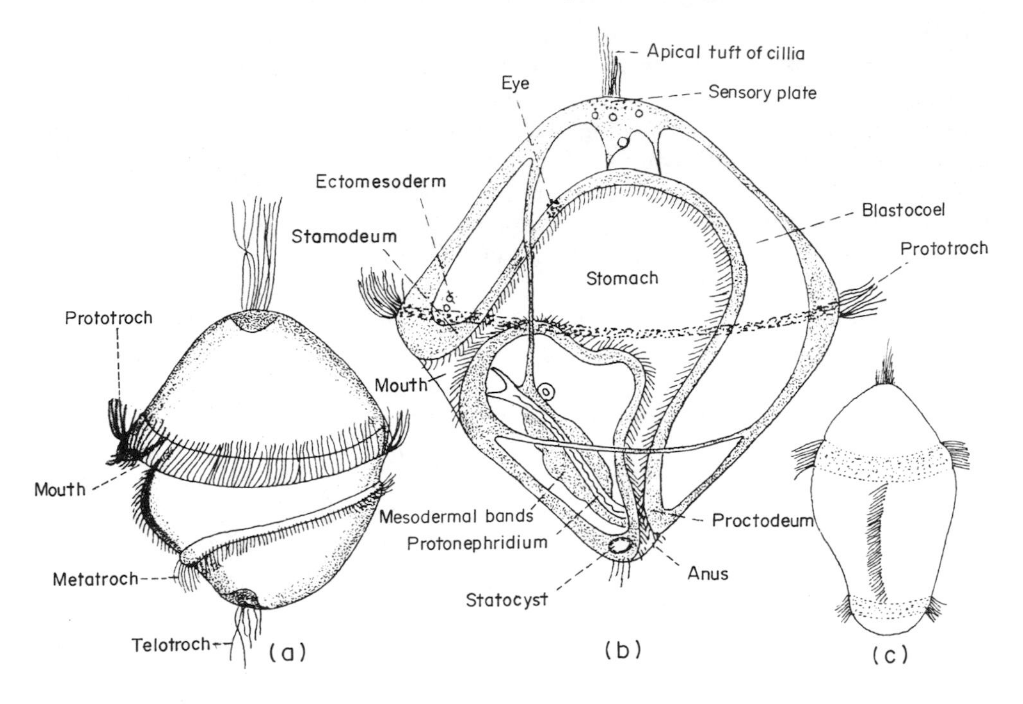

Fig. 2.29. Polychaete larval forms (a) Trochophore of *Polygordius*; (b) structure of an annelid trocophore; (c) Trochopore of *Marphysa* (Barnes, 1974).

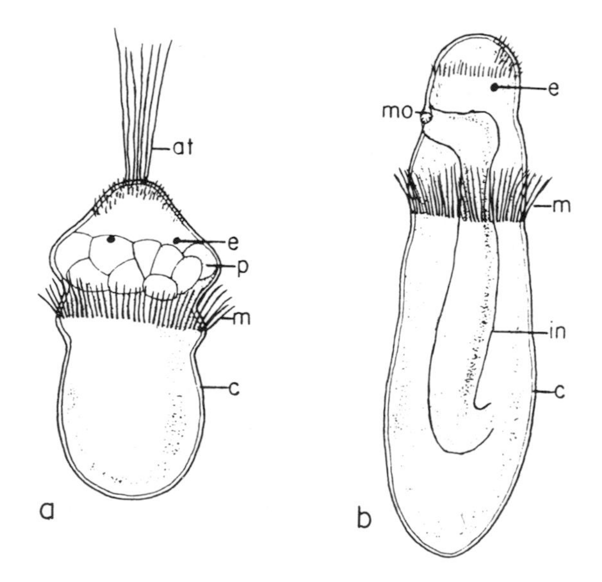

Fig. 2.30. *Golfingia vulgaris* (a) Metamorphosing trocophore, dorsal view; (b) Lecithotropic pelagosphaera larva (60 hours) lateral view. at = apical tuft; c = cuticle; e = eye; in = intestine; m = metatroch; mo = mouth; p = prototroch (Rice, 1975).

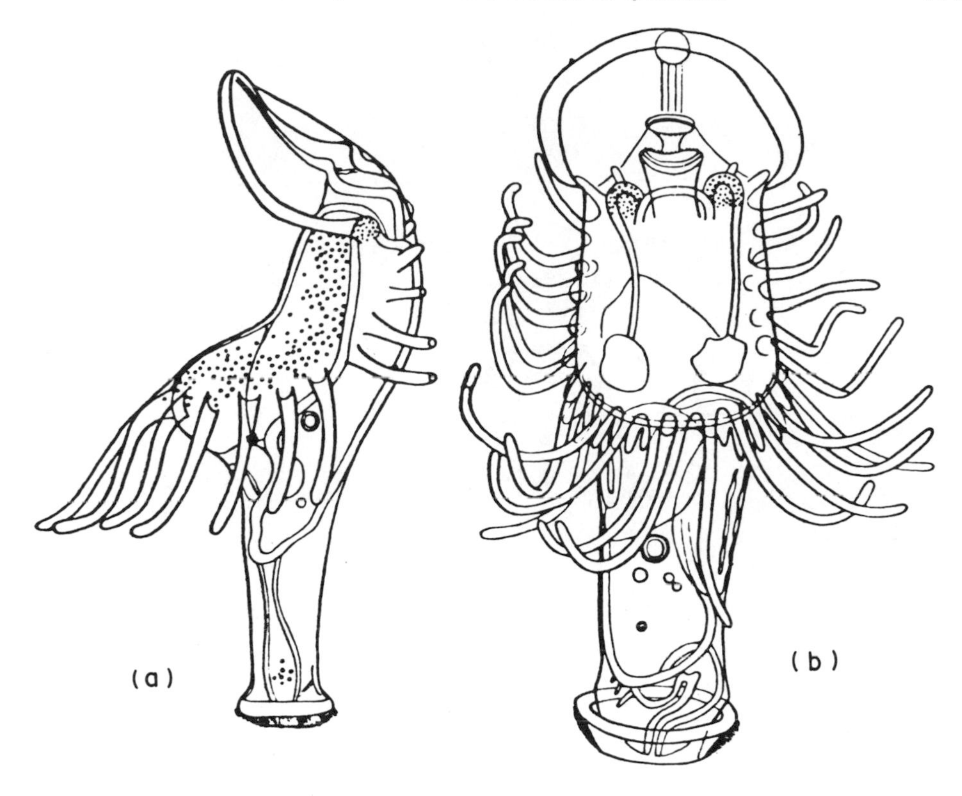

Fig. 2.31. Large size larva of phoronid. *Actinotrocha branchiata* (a) larva with 24 tentacles (left side); (b) fully developed larva with 28 larval tentacles and 24 definitive ones (Forneris, 1957).

like projections of varying length and apparently corresponding to the prototroch of the typical trochophore (Fig. 2.31).

The development of both lamellibranch and gastropod molluscs may include a free-swimming larva of more or less characteristic trochophore form. More typical of the Phylum Mollusca, however, is a free-swimming larva known as the veliger. Gastropods usually hatch as veliger larvae. The gastropod veliger has already developed a very fine thin shell of spiral shape, but the most characteristic feature is the presence of two large more or less semi-circular folds bearing long cilia (the velum), the main locomotory organ of the larva. The velum can be retracted, a suitable strong retractor muscle developing in more advanced veligers. In some species the velum may be extensively lobed and hypertrophied (Fig. 2.32).

The veliger is morphologically considerably more advanced than the trochophore. In the gastropod, a rudiment of the foot is already developed and there are paired eyes and tentacles, in addition to some complexity of internal organs. During the development of the gastropod veliger, a remarkable twisting of the orientation of the body (torsion) occurs. In the lamellibranch the first larva is usually a trochopore. The veliger, though generally similar to that of gastropods, never undergoes torsion so that the organs remain symmetrically arranged. Frequently, in bivalve development, when the veliger metamorphoses, the paired shell valves of the young lamellibranch are

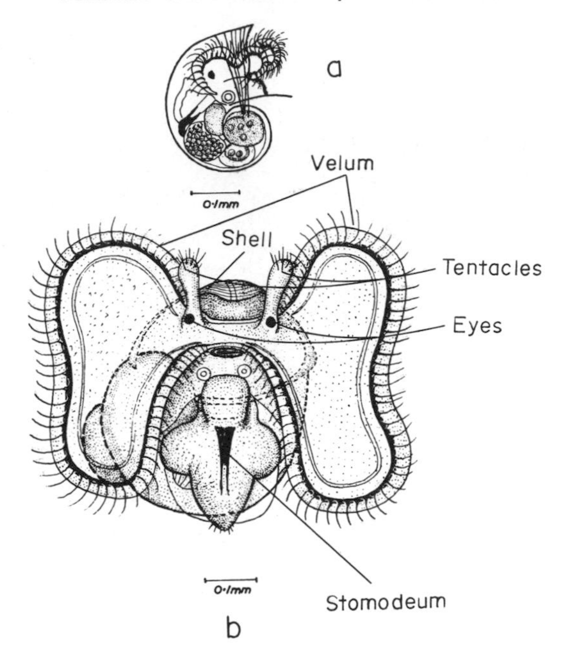

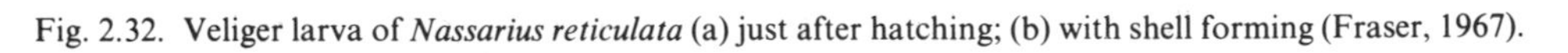

Fig. 2.32. Veliger larva of *Nassarius reticulata* (a) just after hatching; (b) with shell forming (Fraser, 1967).

extremely delicate and the larva remains planktonic for some time before descending to the bottom. Such juveniles can be very abundant on occasions in inshore plankton.

Though the larvae of the various classes of Echinodermata appear to differ considerably from one another, they have certain basic common characteristics. They are bilaterally symmetrical and the locomotory cilia are restricted to prominent bands, typically drawn out into symmetrical lobes or arms. There is a three-chambered gut, but both mouth and anus open ventrally. In the asterioids, the bipinnaria larva has the locomotor ciliated band drawn out into a clear pre-oral and larger post-oral loop. Each of these is further drawn out into paired blunt arms. The bipinnaria may be followed by a brachiolarian stage with additional paired longer arms and a fixation organ. The auricularia larva, characteristic of Holothuria, broadly resembles the bipinnaria, but the post-oral blunt locomotory band is never entirely separated from the pre-oral loop. The auricularia may be succeeded by a more or less oval or barrel-shaped larva, with five separate unfolded locomotory bands ("the pupa"). The larvae of Ophiuroidea and Echinoidea are known as plutei. They broadly resemble one another and differ sharply in shape from the auricularia and bipinnaria type. In the pluteus the locomotor ciliated band is drawn out into long symmetrical, relatively narrow, thin arms, supported by skeletal spicules. The precise disposition and relative lengths of the paired arms is characteristic of different plutei (Fig. 2.33).

The tornaria larvae of enteropneust Hemichordata to some degree resemble the fundamental pattern of echinoderm larvae, but possess, in addition to a folded anterior locomotory ciliated band, a separate unfolded prominent telotroch, just anterior to the anus, which is situated at the extreme posterior end of the larva. There is a fairly well

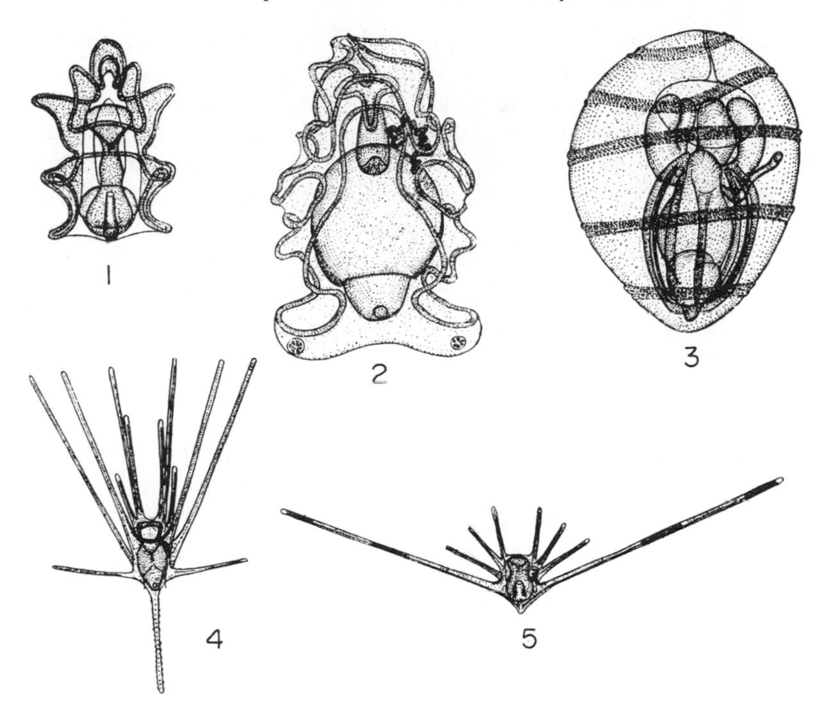

Fig. 2.33. Types of echinoderm larvae. (1) Bipinnaria of *Asterias glacialis* × 40; (2) Auricularia × 40 and (3) 'pupa' × 60 of *Synapta digitata*; (4) Echinopluteus of *Echinocardium cordatum* × 30; (5) Ophiopluteus of *Ophiothrix fragilis* × 175 (Johnstone, Scott and Chadwick, 1924).

developed ciliated apical tuft. Planctosphaera larvae are probably longer-lived larvae of certain hemichordates (cf. page 487).

Many species of fixed ascidians produce eggs which develop into free-swimming larvae, having a rather short period in the plankton before metamorphosis. The larvae are somewhat tadpole-like in body shape, and differ from other invertebrate larvae so far described in possessing gill slits, an endostyle and a post-anal tail, with muscles and notochord.

In the great class of Crustacea a whole series of larval types have developed and although these show numerous modifications and sometimes almost bizarre structure, they may generally be classified under the groupings of nauplius larva, zoea larva and juvenile or post-larval stage. The nauplius larva possesses three pairs of crustacean appendages: uniramous antennules, biramous antennae and mandibles. All three appendages are natatory. The nauplius has a typical median eye and a fairly prominent labrum in front of the mouth. Other developing paired appendages, posterior to the mandibles, may appear as small buds, gradually enlarging with successive moults. This development with the appearance of segmentation of the trunk, leads to a stage usually described as the metanauplius. Amongst crustacean benthic taxa, cirripedes are marked by having a series of usually six nauplius stages. Cirripede nauplii characteristically possess horns or spines, typically a lateral pair anteriorly and a single prominent posterior spine. The last nauplius stage in cirripedes is succeeded by the so-called cyprid stage which metamorphoses to the adult barnacle. The cyprid has a bivalve carapace; the antennules have formed a fixation organ, and six pairs of thoracic appendages have developed inside the carapace.

In some crustacean taxa the egg hatches as a nauplius, which is succeeded by a number of moults to give the more advanced zoeal stage. In other crustaceans (e.g. amphipods, isopods) development is more or less direct, however, the eggs being protected by some type of brood pouch formed by the female. In yet other crustaceans, with indirect development which includes a number of larval stages, the egg hatches directly as an advanced zoea larva. No free nauplius larva appears in the development. This is typical of the great majority of decapod crustaceans. In the typical zoea the head appendages are all well developed and the mouthparts (mandibles and maxillae) are functional as feeding appendages. There is a pair of compound eyes and certain thoracic limbs are developed, almost always three pairs of maxillipeds. These are largely setose appendages. Natatory exopods on some or all of the thoracic appendages effect the locomotion of the zoea (i.e. the propulsion is thoracic.) Some of the more posterior thoracic appendages are usually present only as small buds. The abdomen is typically well developed and segmented and ends in the telson, but pleopods are lacking or rudimentary (cf. Williamson, 1969).

Zoeas are usually ornamented with various spines, some of which may be very large. The spine pattern is characteristic, not only of families, but even of species. Most decapods pass through a large number of moults, with the zoeas gradually increasing in size and developing the thoracic appendages. The larvae of crabs tend to have a very large cephalothorax, with very prominent spines and large paired compound eyes (Fig. 2.34a, b). Brachyura have a final stage, before the metamorphosis to the adult crab, known as the megalopa larva. This generally resembles the adult crab, with fully developed thoracic limbs, but with an extended abdomen and completely developed pleopods. The final larval stage of pagurid anomurans, following a succession of zoeal stages, is also characteristic of the group. The larva (the glaucothoe) has the full development of head appendages and stalked eyes, a pair of chelae and the four pairs of legs on the thorax, and a large symmetrical abdomen with four pairs of biramous pleopods, as well as a tail fan, with flattened uropods. The larvae of the rock lobsters (Palinuridae) are the highly characteristic phyllosomas. They have a very broad depressed head and thoracic shield, very long biramous legs and an extremely small

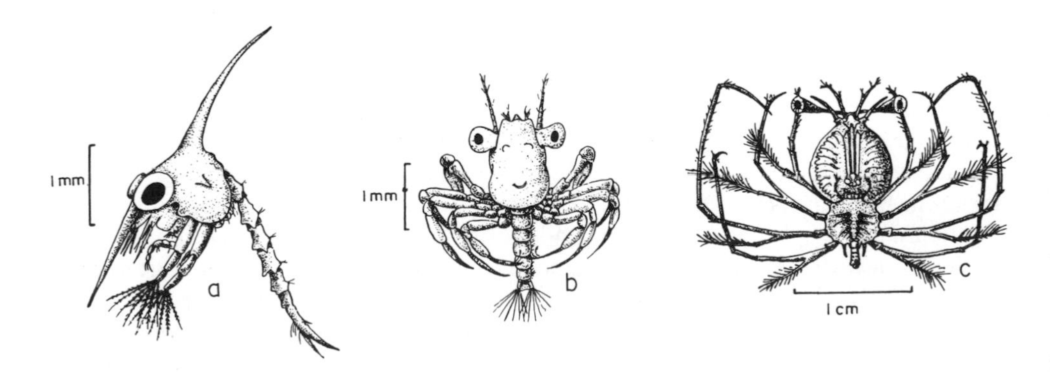

Fig. 2.34. Crustacean larvae (a) 3rd zoea and (b) megalopa of *Portunus puber*; (c) phyllosoma of *Palinurus* (Fraser, 1967).

abdomen (Fig. 2.34c). The advanced larvae of stomatopods, while generally resembling the body form of the adult, have a typical very large spiny carapace.

A number of special terms are used to describe the larvae of certain crustacean groups. An advanced type of larva, following the zoeal stages, with exopodites developed on all the thoracic limbs, is often described as a mysis larva, since it somewhat resembles adult mysids. Special names are also given to particular larval stages of euphausiids and of sergestids. The latter can be extremely peculiar. Brief descriptions are included in accounts of development of the holoplankton. Amongst the meroplanktonic larvae of certain benthic crustaceans, protozoeal stages may precede the zoea. In the protozoea the antennules and antennae are swimming appendages, but the mandibles are functional mouthparts and no longer natatory. The carapace is developed but may not be fused to the thorax, and the paired eyes may not be functional in the earlier stages. There are at least two pairs of maxillipeds, with setose natatory exopods. The abdomen is developed and is usually fully segmented. Williamson (1969) believes that protozoeal stages may be simply classed as early zoeas. Among the many excellent accounts of the remarkable variety of crustacean larvae, especially of decapods, may be mentioned those of Gurney (1942), Williamson (1957, 1967) and Pike and Williamson (1958).

Chaetognatha

The Chaetognatha form a small rather isolated phylum, confined to the marine environment. There are only some six genera, though the largest genus, *Sagitta*, has been sub-divided by some systematists. One genus (*Spadella*) is benthic; the rest are all holoplanktonic animals with worldwide distribution. It is surprising that a small, isolated group should assume such importance in the plankton; not only are chaetognaths found in all the oceans, they are usually relatively abundant, often ranking second or third in frequency after the ubiquitous copepods. In biomass they make a significant contribution to the total zooplankton.

The body form is remarkably uniform in the various genera. The bilaterally symmetrical elongate coelomate body is divided into three major segments, head, trunk and postanal tail. The phylum is one of the few in which all the representatives are hermaphrodite. The ovaries, occurring in the trunk segment vary greatly in size with the degree of maturity. The testes are found in the tail and lead through vasa deferentia, which end in fairly conspicuous seminal vesicles, the precise shape of which is used in systematic studies. Despite their hermaphrodite character, chaetognaths were often described as cross-fertilized and species generally exhibit protandry which might appear to obviate self-fertilization. In fact the degree of protandry seems to vary with the species and a process of transfer of sperm, subsequently effecting fertilization in the same individual, has been described (cf. Hyman, 1959). Cross-fertilization by mutual exchange of spermatophores has, however, been said to take place, the two sagittac lying in copulation, head to tail, in a manner reminiscent of earthworms. Dallot (1968) states clearly that reciprocal exchange of spermatophores does not take place; cross-fertilization may occur, but self-fertilization is possible, and Hyman (1959) regards it as typical.

Reeve and Walter (1972b) support the view from observation of cultures of *Sagitta hispida* that a type of copulation can occur, though not with mutual spermatophore transfer. Instead, one *Sagitta* can seize another and deposit a spermatophore more or less at random on the body surface. The sperm may penetrate the body cavity or, more probably, pass along the surface of the body to enter the opening of the oviducts. Development is direct; a small transparent juvenile hatches from the egg. Various growth stages can be distinguished (cf. Alvarino, 1965; Reeve, 1970a). When the animals are sexually mature, successive batches of eggs may ripen and be released, largely dependent, according to Reeve, on food availability.

Chaetognaths can move very actively; the body is provided with strong musculature and with one or two pairs of fins along the side of the body and a fin around the tail. The fins are supported by rays; they seem mainly for equilibration and assist flotation. Apart from gentle slow swimming, rapid darting vibratile movements can be made and a jack-knife type of flexing of the body is also possible. The capacity for rapid movement and the good co-ordination seem to relate to the marked ability of sagittae for capturing living prey. The head, which is very well developed, has two sets of teeth on each side, typically in four rows, with fewer teeth in the anterior rows. In addition there is a very well developed series of chitinous jaws (hooks) on each side, actuated by a strong musculature. The jaws are more or less sickle-shaped; there may be up to fourteen on each side, depending on the species. The whole is covered by a flap of skin, the hood, which can be pulled back so as to expose the hooks and teeth. The mouth is greatly expanded at the same time and prey seized by the jaws can be swallowed whole.

Sagittae appear to feed more or less indiscriminately on living zooplankton. Reeve (1966) found that dead prey did not evoke any feeding reaction. Minute stiff hair-like structures on the surface of sagittae are possibly stimulated by the prey's movements, the stimulus eliciting the feeding reaction. Feigenbaum and Reeve's (1977) experiments with a vibrating probe demonstrated the reaction to such stimuli. Bone and Pulsford (1978) describe epidermal ciliated cells, some forming ciliary fences which act as sensory receptors. Although copepods, as the commonest members of the zooplankton, are normally most frequently taken by sagittae, a great variety of zooplankton, including fish larvae, can be consumed; the range of size of prey is considerable. As an example of maximal size of prey, Reeve (1966) reported that one *Sagitta* of about 10 mm consumed a *Lucifer* nearly 15 mm length. The summary by Alvarino (1965) indicates the variety of prey taken by chaetognaths, including medusae, siphonophores, salps and other chaetognaths, apart from a wide range of crustaceans, and draws attention also to the relative large size of prey.

Apart from the sensory hair-like structures over the body, chaetognaths possess a pair of moderately well developed eyes situated on the head. Ducret (1975), in a recent detailed examination of the eyes of several species of chaetognaths, has demonstrated that *Sagitta* and *Eukrohnia* have different types of eyes, the latter genus possessing visual elements recalling the ommatidia of Arthropoda. In *E. hamata* the number of ommatidia decreases in deeper-living specimens from warmer waters, as compared with animals from shallower colder habitats. Despite the implication that chaetognath eyes have a real sensory function, Ducret doubts whether they are useful in prey capture.

The corona ciliata, a more or less folded band of epithelium on the dorsal surface of the head, is also generally believed to be sensory, possibly tactile or chemosensory. In *Pterosagitta* a pair of groups of elongated bristle-like hairs of unknown function is present at the sides of the body. Observations by Bieri (1966b) conclude that the sensory tufts detect water movements. Wings may be an elaboration of shorter tufts to improve reception and differentiate between various frequencies; there was no tactile response from touching tufts. A thickening of the surface of the body, approximately at the junction of head and trunk, known as the collarette, occurs in the majority of chaetognaths. It is especially well developed and extends along the body in *Pterosagitta.* The precise form of collarette and corona ciliata has been used to some extent in defining species, though Hirota (1961) suggests that in different forms of *Sagitta crassa* their structure is more related to temperature and salinity. Tokioka (1965) uses the form of the corona ciliata as a significant systematic feature. Figure 2.35 indicates some major differences in distinguishing species grouping. He suggests that the collarette may have a protective function or perhaps, with its often bladder-like cells, assist buoyancy, but is of little phylogenetic importance.

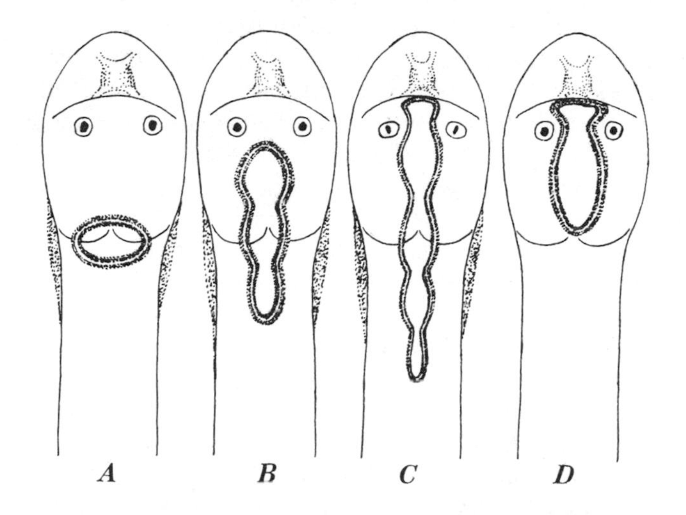

Fig. 2.35. Types of corona ciliata. (A) the most primitive state found in Spadellidae, Pterosagittidae and Krohnittidae; (B) the most primitive state in the Sagittidae (e.g. *Mesosagitta*); (C) the state common to *Sagitta, Serratosagitta*; (D) the most specialized state found in *Flaccisagitta* (Tokioka, 1965).

The gut in chaetognaths is a relatively simple straight tube, the major portion being the intestine, running from just behind the head segment to the ventral anus immediately in front of the tail. In a number of chaetognaths a pair of simple diverticula arise from the anterior end of the intestine and push forward against the head septum.

The identification and definition of some species of chaetognaths has proved a difficult problem. Difficulties include poor preservation of material and the separation of growth stages (cf. Tokioka, 1965; Alvarino, 1965, for detailed discussion). Another cause of confusion is the occurrence of polytypic species. Murakami (1959), considering the distribution of the neritic *Sagitta crassa*, particularly in relation to salinity, has suggested that several types exist, varying amongst other characteristics in the form of the collarette. The polytypes apparently relate only partly to salinity and in part to temperature. Other instances of polymorphism in chaetognaths include Ghirardelli's (1968) observation on *Sagitta enflata*, also noted by Furnestin (1957), and the identification by David (1955) of two forms of *Sagitta gazellae*, existing north and south respectively of the Antarctic Convergence.

While many chaetognaths are inhabitants of open oceans, certain species are neritic, with the ability to live in waters of somewhat lowered salinity (e.g. *S. setosa*, *S. friderici*, *S. bedoti*, *S. euneritica*, *S. crassa*, *S. euxina*). Studies by Reeve (1970) indicate that *S. hispida*, a species found in waters off the east Florida coast, as well as occurring more widely in the tropical Atlantic, according to Alvarino (1965), has the ability to withstand fluctuating salinities. Michel and Foyo (1976) found that *S. hispida* was a rare inhabitant of the open Caribbean and believe that the species is typical of inshore neritic waters. Alvarino (1965) includes a very detailed review of the precise distribution of oceanic and neritic chaetognaths. A summary by Tokioka (1965) lists twenty-two neritic species out of a total of about sixty, with seventeen in the Indo-Pacific, and eight in Atlantic areas. The range of neritic species varies considerably. *S. setosa* is the most northern form of the eastern Atlantic, typical of the North Sea and English Channel. *S. friderici* occurs in neritic waters from the French coast south to South Africa and in the Mediterranean on the eastern side of the Atlantic Ocean, and south of 15°S on western Atlantic coasts. *S. euneritica* occurs in the East pacific from 45°N latitude southwards to Baja California (cf. Alvarino, 1965).

S. bedoti is a neritic tropical species of the Indo-Pacific, found in the eastern Pacific from approximately 30°N to 15°S latitude, and in East Indies and coastal Indian waters. *S. crassa* is typical of waters of lowered salinity in the area of the Sea of Japan. *S. peruviana* and *S. popovicii* are species characteristic of Peruvian coastal waters.

Matsuzaki (1975) discusses the distribution of chaetognaths in the continental shelf areas of the East China Sea. Typical of the Kuroshio waters are warm-ocean species such as *Krohnitta subtilis* and *Sagitta hexaptera*; widespread in the mixed waters of the continental shelf are *S. bedoti*, *S. pacifica*, *S. enflata* and *Pterosagitta*, while in the coastal low salinity areas with temperatures below 15°C are such typical neritic forms as *S. crassa and S. nagae* (cf. Murakami, 1959). The importance of *S. nagae* in inshore waters is also stressed by Marumo and Nagasawa (1973) and by Nagasawa and Murumo (1975) (cf. Chapter 3).

Chaetognaths are found at all depths of the ocean. Some such as *Heterokrohnia*, species of *Eukrohnia* (*E. hamata*, *E. bathypelagica*, *E. fowleri*) and some species of *Sagitta* (*S. planctonis*, *S. zetesios*, *S. decipiens*, *S. macrocephala*, *S. maxima*) are generally regarded as bathypelagic or mesopelagic (Fig. 2.36). They are probably cosmopolitan and oceanic in distribution (cf. Alvarino, 1965 for details of depths), but some differences in distribution patterns appear. Thus *E. hamata* inhabits much shallower epiplanktonic waters at high latitudes, even reaching the surface. The shallow Arctic, sub-Arctic and Antarctic populations are continuous through the deeper-living temperate and even deeper tropical populations of the three major oceans. Thus, north of 50–60°N the species is mainly in the upper 200 m, with maximum numbers in the upper 100 m. In the tropics, however, hardly a specimen occurs above 1000 m. *E. hamata* is apparently absent from the Mediterranean, probably a depth limitation. To a limited extent, *E. fowleri* and *S. maxima* exhibit somewhat comparable depth changes in the Pacific at least.

Of the other deep-living forms, *S. decipiens*, a mesoplanktonic species, has a very wide geographical range (Fig. 2.37), approximately 50°N to 40°S in the Atlantic (including the Mediterranean), in the Pacific from the boundary of sub-Arctic waters in

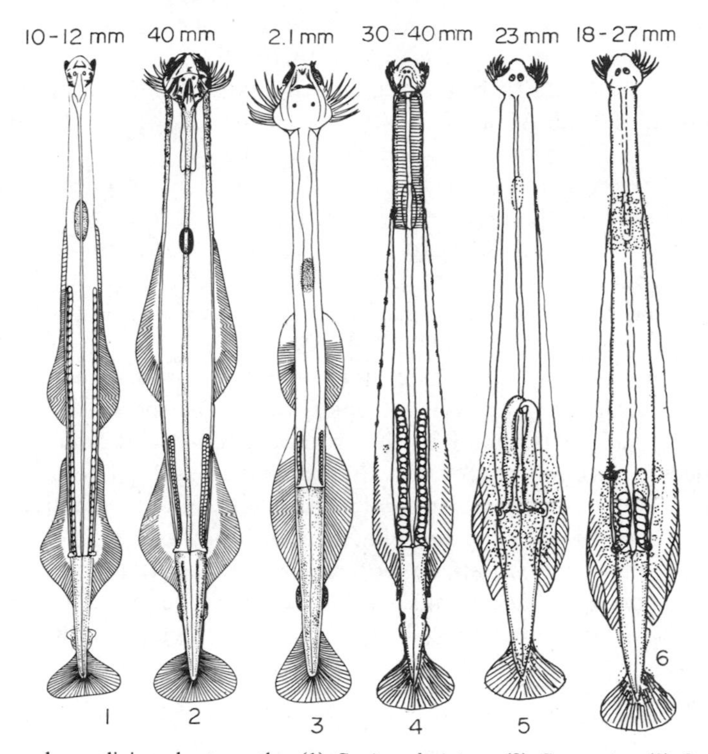

Fig. 2.36. Some deeper-living chaetognaths. (1) *Sagitta decipiens*; (2) *S. zetesios*; (3) *S. macrocephala*; (4) *Eukrohnia hamata*; (5) *E. bathypelagica*; (6) *E. fowleri*. Not to scale, average length of adult indicated (Yamaji, 1973).

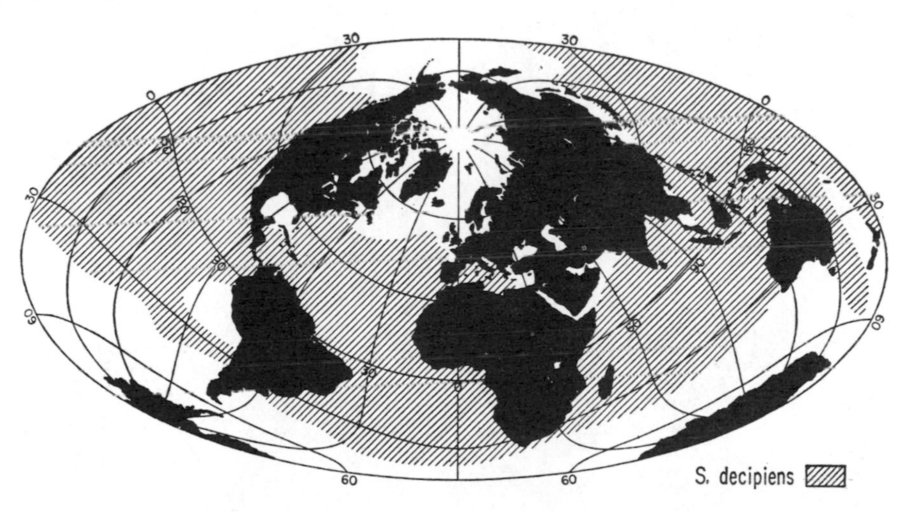

Fig. 2.37. World-wide distribution of *Sagitta decipiens* (Alvarino, 1965).

the north to the Sub-Tropical Convergence in the southern hemisphere, and throughout the Indian Ocean as far south as the Sub-Tropical Convergence. *S. zetesios* has a rather similar range, though perhaps not extending quite so far in the southern hemisphere. The mesopelagic *S. planctonis* occupies the southern parts of the three major oceans, extending to sub-Antarctic waters. *Eukrohnia bathypelagica* has been found in the Indian Ocean and also widely in the Pacific, mostly below 1000 m. *E. bathyantarctica* was formerly believed to be a circumpolar southern species, confined to deep Antarctic and sub-Antarctic waters, but is now known to occur elsewhere in deep water.

Apart from the deeper-living chaetognaths which are typically oceanic, surface and near surface species including *Krohnitta* spp., *Pterosagitta draco*, *S. pacifica*, *S. bipunctata*, *S. enflata*, *S. hexaptera*, *S. pulchra*, *S. regularis*, *S. ferox*, *S. serratodentata*, and *S. tasmanica*, are typical of the open oceans. Some of these may live slightly deeper than the immediate upper 200 m (e.g. *S. lyra*—cf. Furnestin, 1957) but essentially they are epiplanktonic (Fig. 2.38).

In general, epiplanktonic species are more abundant towards the tropics than at higher latitudes. *Krohnitta pacifica* is a tropical equatorial species found in all three oceans. *Sagitta pacifica* is found in the tropical Indo-Pacific and Red Sea, but apparently does not occur in the Atlantic. In the Pacific it spreads from about 40°N to 35°S latitude. Many other epiplanktonic chaetognaths are essentially tropical–equatorial in distribution in the Indo-Pacific, spreading out somewhat into more central waters of the Pacific Ocean, but not occurring in the Atlantic (e.g. *S. regularis*, *S. ferox*, *S. robusta*, also *S. bedoti* and *S. neglecta*, which are to some extent neritic in character). *S. pulchra*, another Indo-Pacific species, is roughly confined to latitudes

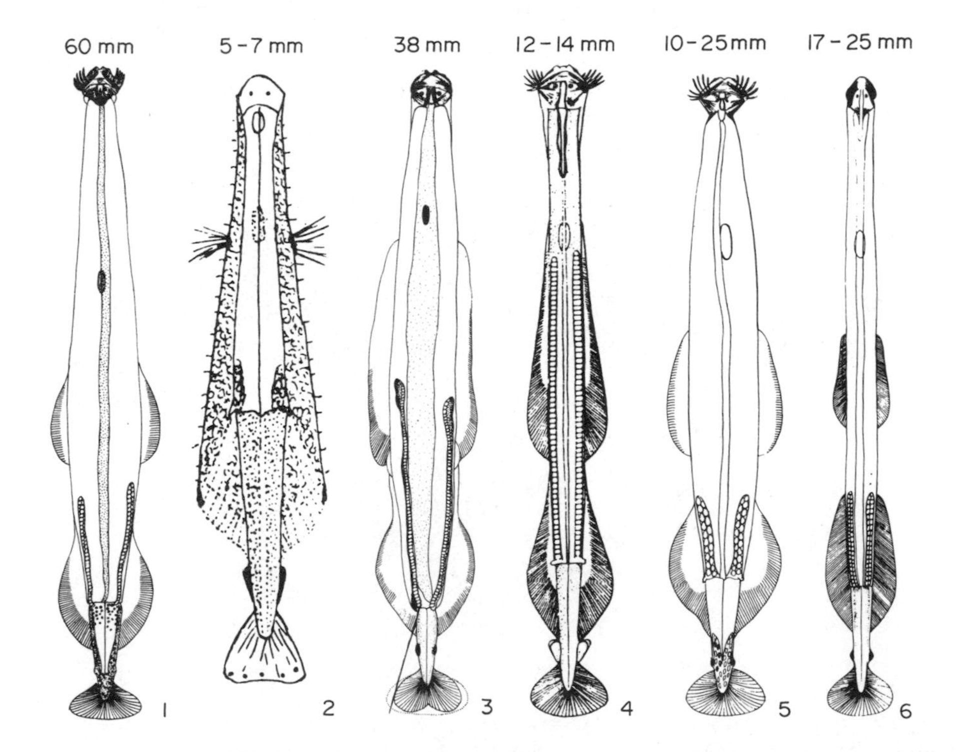

Fig. 2.38. Some warm and temperate-living chaetognaths. (1) *Sagitta hexaptera*; (2) *Pterosagitta draco*; (3) *Sagitta lyra*; (4) *S. bedoti*; (5) *S. enflata*; (6) *S. elegans*. Not to scale, average length of adult indicated (Yamaji 1973).

20°N to 10°S, except for some spread by the Kuroshio Current in the North Pacific and along the Great Barrier Reef in the South Pacific. Nair and Rao (1973) and Nair (1977) include several of these species as amongst the most plentiful in the Indian Ocean. In order of numerical abundance the species of chaetognath are: *Sagitta enflata* (36–80 % of total), *S. pacifica*, *S. bedoti*, *S. bipunctata*, *Pterosagitta draco* and *S. regularis*.

A considerable number of more or less surface-living oceanic chaetognaths are also typical of warm-temperate waters and are found in all three oceans, so that they are often described as of cosmopolitan (warm–cosmopolitan) distribution. Approximate ranges of latitude are from 40°–45°N, including the Mediterranean, to about the same latitude south in the Atlantic and the Pacific and widely in the southern Indian Ocean to about the Sub-Tropical Convergence. Such typical wide-ranging forms include *Krohnitta subtilis*, *Pterosagitta draco*, *Sagitta bipunctata*, *S. lyra*, *S. hexaptera* and *S. enflata*, which may also occur closer inshore in warm waters (cf. Matsuzaki, 1975). An extended distribution of several of these warmer species in the Gulf Stream region in the north-east Atlantic, even as far as the British Isles, is obvious, as for many other warm-temperate zooplanktonic forms. Michel and Foyo's (1976) analysis of chaetognaths from the Caribbean shows that the tropical surface waters and the sub-tropical waters immediately beneath are characterized by several of the same oceanic species. *Sagitta enflata* and *S. serratodentata* were the most abundant; *S. bipunctata* was less common; *Krohnitta pacifica*, *Pterosagitta draco* and *S. hexaptera* were also important, the latter two species being more usually associated with the sub-tropical water.

In cooler waters there are a few chaetognaths of oceanic distribution, some of these being somewhat restricted to particular water masses. Thus, according to Alvarino (1965), *Sagitta pseudoserratodentata* is apparently typical of Pacific Central waters and the Californian region, while *S. scrippsae*, another Pacific species, is found in the Transition region, including the Alaskan gyre and Californian Current, gradually disappearing as the Current fades to the south (Alvarino, 1965) (cf. Chapter 8). In the Atlantic, *S. serratodentata* has long been recognized as an oceanic species found mainly in tropical and warm-temperate areas, with approximate latitude ranges of 50°N to 45°S. Nair (1977) reports its occurrence at three stations off south-east Africa, the species presumably being carried into the borders of the Indian Ocean from the Atlantic.

The distribution of *S. serratodentata* may be somewhat confused by the occurrence of *S. tasmanica*, a species characteristic of sub-Antarctic waters and of the cold southernmost oceanic waters of the Pacific and Indian Oceans. According to Alvarino (1965), the Sub-Tropical Convergence represents the usual limit of northward extension of the species. However, in the Atlantic Ocean, *S. tasmanica* is also found in cold northern areas, approximately north of Cape Hatteras on the western side of the North Atlantic. It occurs as an immigrant species in the Gulf of Maine, probably not breeding there, though it may breed further east and was earlier widely reported as *S. serratodentata* (Bigelow 1926; Redfield and Beale, 1940). The reports of *S. serratodentata* in oceanic waters of the north-east Atlantic as far north as the Shetlands may also partly apply to *S. tasmanica* (cf. Fraser 1952, 1955). Furnestin (1970) lists *S. tasmanica* as a cold-water chaetognath from the west Mediterranean and North Atlantic. *S serratodentata* appears to be a more warm-loving species.

One of the best known cold-loving, shallow-living chaetognaths in northern waters is *Sagitta elegans*, a species typical of water bordering the continental shelf off north-west Europe—"mixed" water of Russell (1939) and Fraser (1939). It is found typically in the upper 100–150 m and has a clear northern distribution in Arctic, sub-Arctic and boreal waters of the Atlantic and Pacific Oceans. Fraser (1952) indicates temperature limits in Scottish waters of around 0° to 14°C and compares these with other species. In the Atlantic, the southernmost limits of *S. elegans* approximate to 46°N (East Atlantic) and 36°N (West Atlantic) and it has been regarded as another member of the "*Calanus* community" of the North Atlantic (cf. Chapter 8). In Arctic waters a variety of considerably greater size, *S. elegans arctica*, is found. Fraser (1952) also describes a Baltic form.

In considering the distribution of *S. elegans*, Alvarino (1965) suggests that Cape Hatteras (*ca.* 36°N) on the east coast of North America, is an approximate boundary for several Atlantic chaetognaths, with *S. elegans* to the north and *S. serratodentata* and *S. helenae* to the south. *S. elegans* is characteristic of the mixed shoal waters along the continental shelf, a habitat comparable to that described for the eastern North Atlantic (cf. Russell, 1939; Fraser, 1939, 1952). *S. helenae*, a species found off Florida and the Gulf of Mexico, spreads along the continental shelf to the south. In the North Pacific, *S. elegans* extends similarly in the cold Oyashio waters but is restricted to the south by the warm waters of the Kuroshio Current. Bieri (1959) states that it is typical of Pacific sub-Arctic waters and decreases in abundance in more Arctic waters near pack ice.

The distribution of chaetognaths, as with other taxa of zooplankton, may be complicated by the spread of deeper currents and their emergence in near surface waters. Fraser (1937, 1952, 1955) demonstrated how certain Lusitanian chaetognaths, carried in the Mediterranean outflow in the eastern Atlantic and reinforced by species of the North Atlantic deeper layers, may be distributed to the west of Ireland and Scotland.

In the colder waters of the southern hemisphere, *E. hamata* and *S. planctonis* occur as mesopelagic species, with *E. hamata* nearer the surface at high latitudes. *E. bathyantarctica* is found in deep water but certain other species are endemic to Antarctic waters. *S. marri* is mesobathypelagic and circumpolar in distribution; its northern limit is near to the Antarctic Convergence. *S. gazellae* is found in sub-Antarctic and Antarctic areas of all three oceans, extending as far as the Sub-Tropical Convergence. It is essentially a surface form but it exhibits pronounced seasonal vertical migrations. Details of its distribution are given by David (1955) and Mackintosh (1962) (cf. Chapter 8).

Species of chaetognaths are more numerous in the Indo-Pacific region than in the Atlantic. Tokioka gives fifty-five species for the Indo-Pacific, thirty-one being endemic, as opposed to thirty-one in the Atlantic region, of which eight are believed to be endemic.

Vinogradov (1970) considers that chaetognaths tend to divide into three groups according to their vertical distribution. The majority of species are surface-living, in the uppermost 50 or 100 m, and consist of both neritic and oceanic forms, mainly warmer-water species. Occasionally a very few specimens of these species may reach even deeper than 200 m. The second group includes more eurybathic species which can extend from the surface to reach depths of 1000 to 2000 m, sometimes even deeper. They include *Sagitta elegans*, *S. planktonis*, *S. zetesios*, *S. gazellae*, *S. marri*, *S. maxima*, *S. macrocephala* and *Eukrohnia hamata*. Young specimens of many of these species generally live closer to the surface, as with many other mesopelagic taxa (cf. Kotori, 1972; Fagetti, 1972). The third group consists of a few deep-sea species, always living below 500 to 1000 m, and includes three species of *Eukrohnia* (*E. fowleri*, *E. bathypelagica*, and *E. bathyantarctica*) and also possibly *Bathyspadella* and *Heterokrohnia*, if these are genuine species. The lower depth range for this third group is uncertain; they almost disappear at 4000–5000 m, though some small specimens of *Eukrohnia fowleri* have been found in the Kurile–Kamchatka Trench down to 6000–7000 m.

Kotori (1972), dealing with the vertical distribution of chaetognaths from the northern North Pacific and Bering Sea, has drawn attention to the considerable vertical range of *S. elegans*, normally in the upper 200 m but extending to 700 m, and *E. hamata*, with the maximal zone of abundance somewhat deeper than *S. elegans* but extending to nearly 1500 m. Meso- and bathypelagic species (*S. zetesios*, *S. macrocephala*, *E. bathypelagica*) also tended to live more shallowly at high northern latitudes than in more temperate and warmer seas, but in the Caribbean all the chaetognaths collected by Michel and Foyo (1976) had a wide depth range. *Krohnitta subtilis* and *Sagitta lyra* were more abundant in mesoplanktonic layers than in the epiplankton. *S. macrocephala*, though included amongst the four bathypelagic species, had an upper depth limit somewhat shallower than Vinogradov suggests.

Figure 2.39 compares the vertical distribution of the biomass of chaetognaths in the Kurile–Kamchatka Trench in the sub-Arctic Pacific with that in tropical areas of the Pacific Ocean. In both, the maximum biomass occurs in the near surface layers, but somewhat deeper (*ca.* 300–500 m) at high latitudes and nearer the immediate surface in the tropics. In both areas a secondary maximum appears, at about 2000–2500 m in the sub-Arctic, and at rather shallower depths in the tropical Pacific. Below about 3000 m the biomass is negligible in both regions, but in tropical areas this reduction takes place at somewhat lesser depths. According to Vinogradov, the biomass of chaetognaths is more important in sub-Arctic areas than in the tropics. In the sub-Arctic North Pacific, in the near surface layer and the 2000–2500 stratum, chaetognaths may contribute > 40 % of the total biomass. In Suruga Bay, Nagasawa and Marumo (1975) found that sagittae formed 48 % of the total biomass of the near bottom zooplankton. The maxima at higher latitudes can be identified with separate species: *S. elegans* in the upper 200 m, with *Eukrohnia hamata* somewhat deeper and *E. fowleri* in the secondary maximum.

Fagetti (1972) also identifies chaetognaths from three major depth levels (epi-, meso- and bathypelagic) off the coasts of Chile. *S. bipunctata*, *S. serratodentata* and *S. bierii* are near surface species, the former two restricted to the Central Pacific water mass, the latter to the coastal current. Mesopelagic species include *S. decipiens*, *E. hamata*, *E. fowleri*, *S. macrocephala*, *S. zetesios*, and *E. bathypelagica*, though there is some separation into upper and lower mesopelagic zones. Only *E. bathyantarctica* is truly bathypelagic in Chilean waters, being always deeper than 1000 m. This species has been reported also from deep layers of the Caribbean (cf. Owre, 1972; Michel and Foyo, 1976). The immature chaetognaths tend to live more shallowly (500–700 m).

Alvarino (1965) has suggested that the vertical distribution of some chaetognaths is related to the oxygen content of the water (e.g. *Sagitta decipiens* in the Sea of Cortez). The distribution of *S. elegans* in the North Pacific was believed to coincide with oxygen concentrations exceeding 6 ml/l. On the other hand, Kotori (1972) reports that

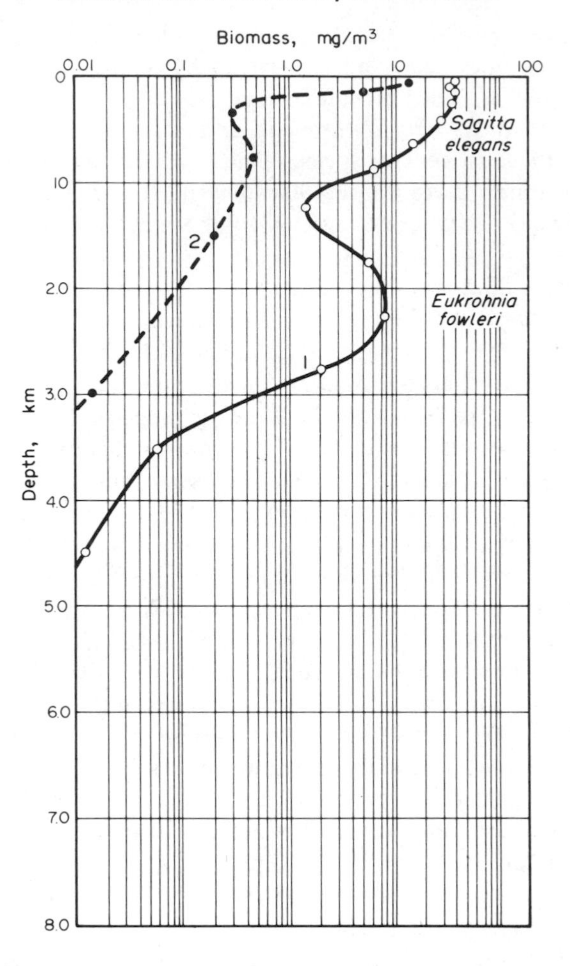

Fig. 2.39. Average biomass of Chaetognatha in (1) Kurile-Kamchatka Trench region; (2) tropical Pacific Ocean region. Dominant species in Trench area are listed (Vinogradov, 1970).

Marumo and Kitou both found numerous *S. elegans* in Japanese waters with an oxygen concentration below 4 ml/l, and he further stresses that in the northern North Pacific he collected many, apparently healthy, *S. elegans* from water containing only 1 ml O_2/l.

Furnestin (1970a, b) describes the distribution of the chaetognath fauna of the Mediterranean and warm-temperate areas of the north-eastern Atlantic. Boreal areas of the continental shelf with moderately high salinity at the entrance to the English Channel and Brittany coast are very rich, and characterized by *S. elegans*. *S. tasmanica* may also appear to a minor extent. In the more Channel-type waters (salinity <35‰) *S. setosa* is characteristic. In more oceanic regions to the south west of Ireland, in the Bay of Biscay and off the Portuguese coast were relatively broadly distributed species, including *S. bipunctata* as the abundant surface form, and *S. hexaptera*, *S. lyra* and *Eukrohnia hamata* as comparatively abundant species, rather deeper living, especially by day. Epiplanktonic warm-water chaetognaths, including *S. friderici* (typical of the

Moroccan coast) and *S. enflata*, appeared at some stations in areas with temperatures ranging from 18°–25°C, but they were never found where the winter surface temperature fell below 17°C. At somewhat deeper levels over these same areas *Pterosagitta*, *Krohnitta subtilis* and *S. serratodentata* were present.

In the Mediterranean, only a rather limited number of chaetognath species (17) were present, as compared with the African–Atlantic area (27). Omitting *S. setosa* (since this species, though locally abundant, is neritic) and *S. euxina* (characteristic of the Black Sea), only four species were numerically important. *S. enflata* was the most important. *S. bipunctata*, although a very constant constituent of the fauna, contributed far less and *S. hexaptera* and *S. lyra* were both significant at somewhat deeper levels. The two deeper-living species, *Eukrohnia fowleri* and *Sagitta macrocephala*, normally inhabiting depths >1000 m in the neighbouring tropical Atlantic, were not so depth limited in the Mediterranean, perhaps due to the degree of homothermy of the deep troughs.

Although, therefore, as with other zooplankton taxa, temperature, salinity, possibly oxygen, as well as suitable food availability and other factors, all play a part in determining both vertical and horizontal distribution of sagittae, the more precise limitations of particular species to water masses, frequently especially well typified by chaetognaths (see Chapter 8), may reflect subtler properties of the water mass.

Mollusca

Among the Mollusca certain whole taxa of gastropods—Heteropoda, Thecosomata, Gymnosomata—have adopted a holoplanktonic life, for which they are highly adapted. Apart from these taxa a few isolated gastropod genera have also become adapted for a holoplanktonic existence. *Fiona*, *Glaucus* and *Phyllirhoe* (cf. Bonnevie, 1931) are nudibranchs which show marked modifications for neustonic life, as also does the prosobranch genus, *Ianthina*.

In addition to the veliger larvae of such holoplanktonic forms, the developmental stages (veliger and post-veliger larvae) of many benthic molluscs may form, on occasions, a significant part of the plankton (cf. Chapter 4). Especially in inshore and estuarine waters, larval clams, mussels, oysters and other bivalves, as well as gastropod veligers, may appear in considerable numbers in the plankton.

The cephalopods represent a highly successful class of pelagic Mollusca.

Heteropoda

A group of holoplanktonic gastropods, most of which are very highly modified for planktonic existence, is the Heteropoda. Early on, these were associated with the pteropods since both groups were holoplanktonic molluscs. It is now appreciated that the heteropods are distinct; they are a section of prosobranchiate gastropods. A recent review by Thiriot-Quievreux (1973) mentions that the heteropods are mesogastropods, possibly near the Naticacea; Thiriot-Quievreux keeps the name Heteropoda as a super-family. Tesch (1949) lists only eight genera of Heteropoda and though at one time of the order of 100 species were described, he reduces the number to twenty-two. Although Thiriot-Quievreux mentions three new species, the group is a relatively small one. The numbers of specimens rarely seem to be really large, though

Aravindakshan (1977) states that *Firoloida desmaresti* was comparatively abundant in the Indian Ocean and that the species has been reported as swarming in the Mediterranean and is abundant in the Caribbean and Sargasso Sea. Heteropods would not usually, however, contribute significantly to the overall biomass of zooplankton.

It is almost impossible to quote densities since very few quantitative studies of distribution of Heteropoda have been made. Thiriot-Quievreux mentions that perhaps the greatest density of heteropods so far reported is off the edge of the continental shelf at Nosy Bé, Madagascar (Frontier, 1973). Numbers are quoted only per horizontal tow; they amounted to about 80 specimens for species of *Atlanta*—the commonest form. The maximum number per tow (1029) was of *Atlanta helicinoides*. Some idea of the very different numbers of heteropods and pteropods may be obtained by comparing this density with the 71,000 specimens of *Creseis acicula* taken at the same Station. Species of *Atlanta* are usually nevertheless the most abundant of the heteropods and this may be connected with their generally small size. Some may be as small as 1 or 2 mm in length and the largest species of *Atlanta* do not exceed about 10 mm. On the other hand, other genera of heteropods can be comparatively very large. Some reach 100 mm in length (*Pterotrachea*) and some may exceed 200 or 300 mm, even approaching 500 mm (*Carinaria*).

Little detailed information is available on the vertical distribution of heteropods, but undoubtedly most of them seem to inhabit the surface or near surface layers. Although deeper plankton hauls will occasionally catch a heteropod, the numbers are extremely reduced with depth; clearly they are essentially epiplanktonic. Their horizontal geographical distribution is also well defined. They are warm-water animals, confined to tropical and sub-tropical waters, including the Mediterranean, although the number of species there is greatly reduced. However, in warm seas many of the species are cosmopolitan (cf. Fig. 2.40 and Table 2.6). Species of *Pterotrachea* in particular occur in all three major oceans, together with *Atlanta peroni*, *A. lesueri* and *Firoloida desmaresti*, for example.

Even with the usual oceanic circulation, few heteropods are normally found beyond 40°N and 40°S, and some are much more strictly confined to a relatively narrow band of warmest tropical water. Tesch (1949) shows occasional findings of *Carinaria lamarcki* beyond 40° latitude, indicating, as for so many zooplankton animals, that specimens can be carried by currents outside their normal distribution. Dales (1957) gives some records of heteropods occurring in the north-east Atlantic, mostly between 30° and 40°N, but there are some records between 40° and 50° latitude. Furnestin (1964) lists several heteropods which she states are inhabitants of the open warm oceanic Atlantic waters and rarely approach the continental shelf. However, the invasion of tropical ocean waters along the south of the Moroccan coast occasionally transports a few individuals of the following species: *Oxygyrus keraudreni*, *Atlanta fusca*, *A. inflata*, *Carinaria lamarcki*, *Pterotrachea minuta* and *Firoloida desmaresti* (cf. also McGowan, 1967).

Seapy (1974) discusses the rather unusual distribution of the heteropod, *Carinaria japonica*, in the waters off California. He confirms earlier studies such as those of Dales (1953) demonstrating that this heteropod is one of the few species that can occur more in the transition zone of the North Pacific; such water may extend even to 45°N. Seapy's investigations allow some calculation of densities of *Carinaria*. Frequently, over much of the area, densities were below 1 heteropod/1000 m^3, though one or two

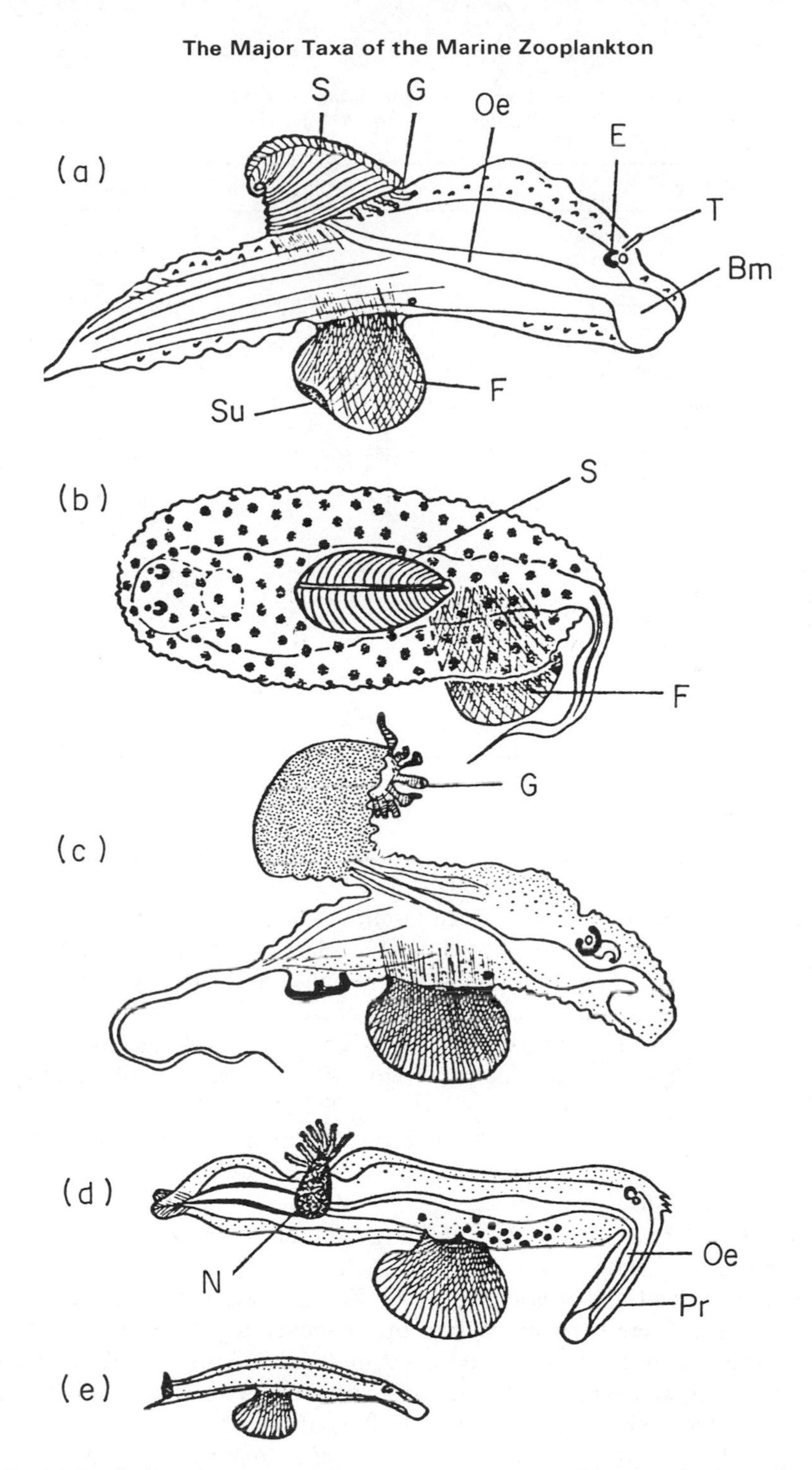

Fig. 2.40. Adults of various female heteropods. (a) *Carinaria lamarcki*; (b) *Pterosoma planum* (from above); (c) *Cardiopoda richardi*; (d) *Pterotrachea hippocampus*; (e) *Firoloida desmaresti*. Bm = buccal mass; E = eye; F = fin; G = gills; N = nucleus; Oe = oesophagus; Pr = proboscis; S = shell; Su = sucker; T = tentacle (Thiriot-Quievreux, 1973).

Table 2.6. The geographical distribution of the Heteropoda (Thiriot-Quievreux, 1973)

Species	Atlantic	Mediterranean	Indo-Pacific
Oxygyrus keraudreni	+[a]	+	+
Proatlanta souleyeti	+	+	+
Atlanta peroni	+	+	+
A. gaudichaudi	+	–[b]	+
A. inclinata	+	–	+
A. lesueuri	+	+	+
A. helicinoides	+	–	+
A. inflata	+	–	+
A. fusca	+	+	+
A. turriculata	–	–	+
Carinaria lamarcki	+	+	+
C. galea	–	–	+
C. cithara	–	–	+
C. cristata	–	–	+
Cardiopoda placenta	+	–	+
C. richardi	+	–	+
Pterosoma planum	–	–	+
Pterotrachea coronata	+	+	+
P. hippocampus	+	+	+
P. scutata	+	+	+
P. minuta	+	+	+
Firoloida desmaresti	+	+	+

[a] + = present
[b] – = absent

very exceptional hauls in relatively rich collections gave numbers exceeding 100 animals/1000 m^3; Dales found 20/1000 m^3.

The heteropods are divided into three families. Of these the Atlantidae possess a calcareous, transparent, flat, spiral shell, which is large enough to enclose the whole animal when retracted. The gastropod foot is modified as a laterally flattened fin-like structure, possessing an operculum and sucker. Tesch points out that the Atlantidae keep themselves up in the water only by means of continuous jerking movements of the fin. Hida (1957) reported fairly high, though sporadic, densities of *Atlanta* in the central Pacific.

The Carinariidae have a very much reduced shell, far smaller than the animal itself and enclosing only the visceral mass. There is a very well-developed flattened fin developed from the foot and this carries a sucker but no operculum. The body is generally much more elongated and has a large amount of thick jelly-like substance, giving ease of flotation.

In the Pterotracheidae there is no shell at all. The fin is well developed, flattened and rather elongated. There is no operculum but a sucker is present in the male. The gelatinous transparent body is more or less cylindrical and the animal is particularly well adapted to planktonic existence and is able to swim forwards and even backwards by undulatory movements of the body as well as of the fin.

In *Pterotrachea* the visceral mass is posterior to the swimming fin, which is succeeded by a distinct tail. Tesch points out that with the modified body form in the Carinariidae and Pterotracheidae the animal swims upside down, with the swimming fin upwards. Aravindakshan (1977) found that *Pterotrachea coronata* ranged from about 23°N to

39°S in the Indian Ocean and was moderately common in restricted localities, but *P. hippocampus* was mainly equatorial and was the commonest species of *Pterotrachea*. *Firoloida desmaresti* was the dominant species in the Indian Ocean, however, with a distribution approximately from 25°N to 40°S. Dales (1952) found *Firoloida* rare, but *P. hippocampus* comparatively plentiful, off the Pacific coast of North America. Michel and Foyo (1976) reported several species of *Atlanta*, *Cardiopoda placenta*, *Pterotrachea coronata* and *Firoloida desmaresti* from the Caribbean. *Atlanta peroni* and *Firoloida* were frequently caught, and *Atlanta inclinata* was numerous, but heteropods were not abundant in general.

Heteropods are very highly predaceous animals; the proboscis is very mobile and capable of being turned almost in any direction. With the nearly telescopic eye and the hooks of the radula, heteropods can sight and seize prey very actively. According to Tesch, the usual food includes ctenophores, salps and Hyperiidae, possibly with other crustaceans and sagittae. Thiriot-Quievreux (1973), however, points out that heteropods are very selective feeders. The Atlantidae use the rasping action of the radula to feed on pteropods, gastropod veligers and similar forms. The Carinariidae and Pterotracheidae tend to swallow their prey whole. Dales noted that *Carinaria japonica* seemed to be present in plankton which was rich in salps, cnidarians and chaetognaths, and Seapy's study indicated that the preferred prey of *Carinaria japonica* was salps and doliolids. Hamner, Madin, Alldredge, Gilmer and Hamner (1975) observed the heteropod *Cardiopoda placenta* feeding on *Salpa* in the field and believe that the prey was sighted. Other animals eaten in laboratory tests included *Phyllirhoe*, *Firoloida* and fish larvae. *Carinaria japonica* also fed on salps and *Pterotrachea coronata* was observed to eat physonect bracts in the sea.

Eggs of heteropods, deposited as strings or singly, develop into veligers which are well-recognized in the oceanic plankton (cf. Thiriot-Quievreux 1973).

Thecosomata

The shelled pteropods or Thecosomata are now generally recognized as opisthobranch molluscs having a separate origin from the Gymnosomata, although for convenience both groups are loosely classed as pteropods. The calcareous shell into which the animal can be withdrawn may be coiled as in the Limacinidae, or is symmetrical in the form of a more or less elongated cone; it may then be slightly curved, however, as in the Cavoliniidae to which the majority of the thecosomes belong. Limacinidae possess an operculum to the shell which is lacking in Cavoliniidae. Cavoliniidae are mostly larger than members of the Limacinidae and often occur in swarms.

Perhaps the most characteristic organ of the Thecosomata is the pair of wings. By their flapping movement pteropods are able to move in the water. A precise analysis of the swimming movement is given by Morton (1954). The wings are developed from the foot as in other holoplanktonic Mollusca. From the foot also arises a posterior foot lobe, ventral to the mouth and wings, and provided with ciliary and glandular cells. Although the posterior footlobe varies considerably (a triangular organ fixed to the base of the wings in *Limacina*; more broadly developed in *Cavolinia*), it is part of a ciliated feeding mechanism, which is completed by two lateral footlobes, projections of the wing base, which assist in carrying particles forward to the mouth opening.

Although the posterior and lateral footlobes and the wings are all modified parts of the foot, the footlobes form the food collecting mechanism; the wings are for locomotion. It has been suggested, however, that cilia at the very bases of the wings may assist in food collection (Yonge, 1926). In some pteropods such as *Limacina* these cilia aid in rejecting unwanted particles; these pass to the underside and upperside of the wings and are cast off. A short lobe on the right side of the mantle edge, the balancer, also seems to act as a rejection path.

The thecosomes are essentially herbivorous plankton feeders, grazing on dinoflagellates and diatoms of suitable size and shape. Possibly some pteropods can use nanoplankton flagellates. The feeding is a combination of ciliary transport with mucus secretion. Not only phytoplankton but occasionally Foraminifera and Radiolaria may be taken.

In the Cavoliniidae the footlobes act as a main food collecting mechanism. The ciliated mantle cavity may also play a role. The large pallial gland suggests active mucus secretion, though Morton (1954) suggested that it was difficult to see how any mucus food string from the mantle cavity could be handled. On the other hand, in *Limacina* the mantle cavity plays a very large part in the feeding process. Morton describes how, by the beating of cilia in the mantle cavity, a regular inhalent and exhalent current is maintained. Particles are trapped by the mucus secretion and the mucus string is carried to the mouth. Ciliary tracts on the footlobes also transport food (cf. Fig. 2.41).

Gilmer (1974), in a study of the feeding of *Cavolinia* spp., places considerable emphasis on the production of the feeding current by cilia of the mantle cavity, which is approximately U-shaped, with inhalent and exhalent openings. A ventral glandular thickening in the mantle cavity, the pallial gland, supplies mucus. Food strings produced in the mantle cavity exit at the exhalent aperture and pass by two pathways to the mouth. Most of the mucus food strings are drawn towards the dorsal side of the left wing from the exhalent opening and are then swept into the dorsal ciliated groove of the median foot lobe. The small lateral foot lobes also sweep food into the oral groove (cf. Fig. 2.42). Other food particles from the exhalent current are drawn into the ciliated lateral trough at the posterior margins of the wings. Excess food is rejected, partly by a flapping of the wings, which disengages a food mass in the oral groove, and partly by a ciliated rejection path along the anterior part of the wings.

Gilmer confirms that the food covers a whole range of phytoplankton, including microflagellates and also protozoans (tintinnids, foraminiferans and radiolarians). Organic aggregates, detritus and bacteria representing particles less than 5 μm are also believed to contribute a substantial part of the diet. The upper size limit is 200 μm. Gilmer believes that buoyancy is an important factor in the life of the thecosomes and probably considerably influences swimming speeds. Lobes of the mantle extending through clefts in the shell in *Cavolinia* and *Diacria* possibly assist buoyancy and the increase in surface area due to the appendages might be significant.

In Limacinidae the mantle cavity is on the dorsal side of the animal; in Cavoliniidae it is ventral. The Limacinidae have lost the typical mollusc gill, though a gill is present in the genus *Cavolinia*.

The thecosomes described so far form the Euthecosomata, as opposed to a few species (e.g. *Cymbulia*, *Corolla*) classed as the Pseudothecosomata. In these, instead of the typical calcareous external shell, there is a glassy jellylike cartilaginous pseudo-

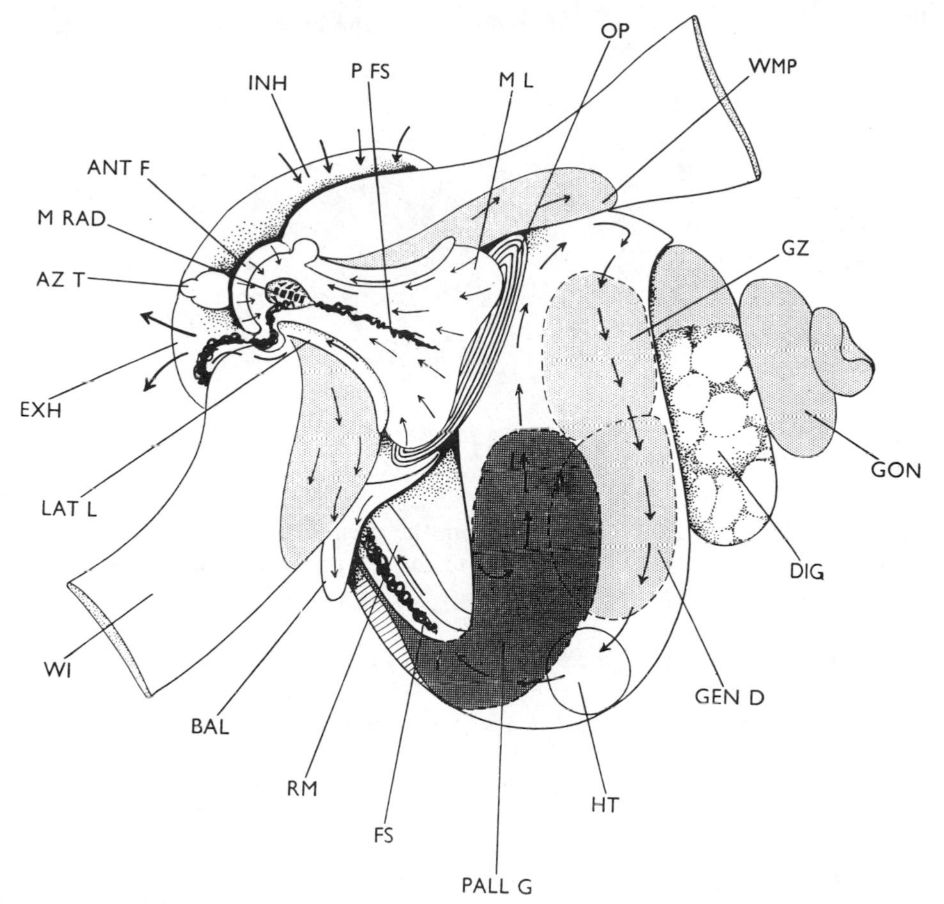

Fig. 2.41. *Limacina retroversa*, ventral view. Diagram showing the ciliary currents employed in food collecting. The surface of the foot and the ciliated portions of the wings are fully expanded, and the distal parts of the wings are omitted. ANT F, anterior horsehoe-shaped fold; AZ T, azygous tentacle; BAL, balancer; DIG, digestive gland; EXH, exhalant side of pallial cavity; FS, food string within the pallial cavity; GON, gonad, GEN D, genital duct; GZ, gizzard; HT, position of the heart; INH, inhalant side of the pallial cavity; LAT L, lateral lobe of the foot; M L, median lobe of the foot; M RAD, mouth, with the radula; OP, operculum; P FS, food strings formed on the median lobe of the foot; PALL G, pallial mucous gland; RM, rectum; WI, wing; WMP, ciliated "Wimperfeld" at base of wing (Morton, 1954).

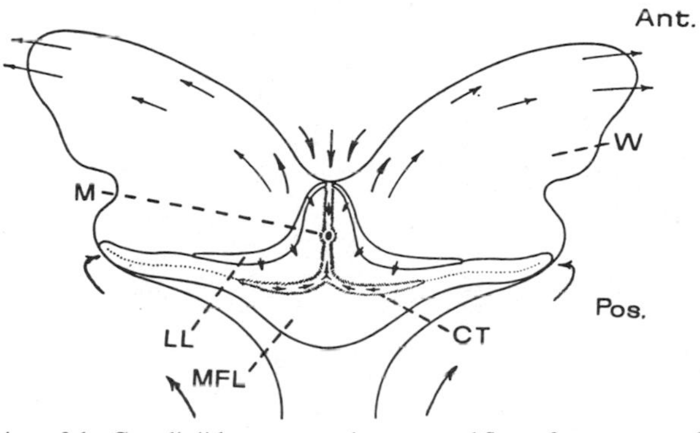

Fig. 2.42. Mouth region of the Cavoliniidae; arrows show general flow of water around the mouth (M), and indicate movement along anterior wing margins (Ant) as rejection currents for excess food; CT = ciliated tracts of the posterior wing margins; LL = lateral lobes of the foot; MFL = median unpaired foot lobe; W = wing (Gilmer, 1974).

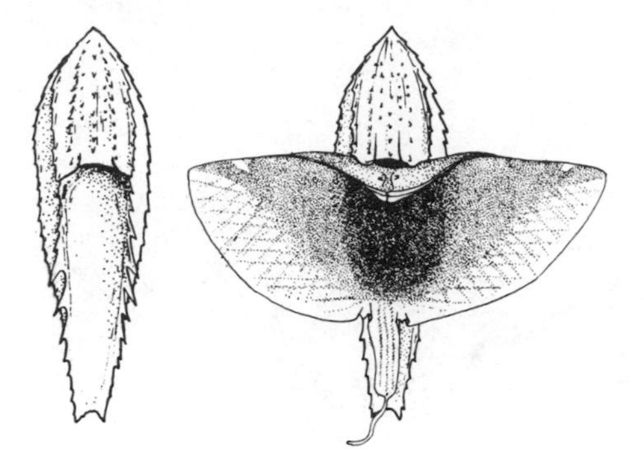

Fig. 2.43. *Cymbulia peroni* and *pseudoconcha* (50–60 mm) (Van der Spoel, 1972).

concha, attached to an exceedingly thin integument, so that the pseudoconcha is really internal. The animal is easily separated from the pseudoconcha (cf. Fig. 2.43). The lateral and posterior footlobes are united to form a proboscis with ciliated gutters opening to the mouth. The wings ("fins") are joined ventral to the mouth to form a large swimming plate.

Gilmer (1972, 1974) investigated the feeding of *Gleba* and *Corolla*, observing living animals by Scuba diving, as well as examining what particles were retained as food. *Gleba* and *Corolla* feed by means of a secreted large free-floating mucus net. The animal remains motionless with its mouth at the base of the net. A variety of phytoplankton and zooplankton is captured by the web, with a large proportion of very small particles including nanoplankton. Protozoans are also taken; tintinnids tend to dominate the diet. *Gleba* will take nauplii even up to 800 μm. The animal can swim rapidly after loosing from the net and may apply its mouth to a new position. A fresh net may also be formed. *Desmopterus* is a very peculiar form with a relatively large swimming plate and no pseudoconcha (cf. Fig. 2.44). McGowan even questions whether it could be an aberrant gymnosome. *Gleba*, *Corolla* and *Desmopterus* are essentially warm-water species, confined to the tropics (cf. Tesch, 1946), though McGowan (1967, 1968) finds the latter two species even to 40°N in the Californian

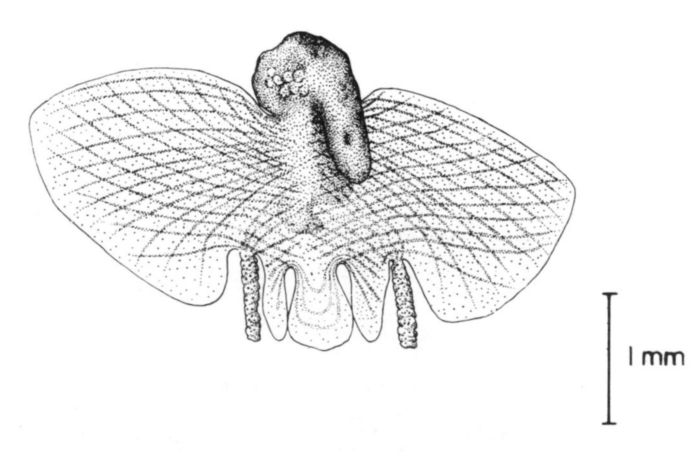

Fig. 2.44. *Desmopterus papilio* (soft parts only) (Van der Spoel, 1972).

Current area, especially in the central mixed zone. *Cymbulia* is more widely distributed (cf. *C. peroni* in the North Atlantic) (Tesch, 1947). The genus *Peraclis* is perhaps intermediate between Euthecosomata and Pseudothecosomata in that it has a calcareous shell rather resembling *Limacina*, but a proboscis-type mouth. It is a bathypelagic genus, with some species reaching a depth of about 1000–2500 m.

Thecosomes are mostly epipelagic, though they may not be strictly confined to the surface. Apparently all Limacinidae and some other species can perform diurnal vertical migration, to a maximum depth approaching 300 m (cf. McGowan, 1968; Moore, 1949; Chen and Bé, 1964). While their biomass does not rival that of copepods or euphausiids, McGowan states that on occasions densities can exceed 100/m^3.

The vast majority of thecosome pteropods, as with other zooplankton, are warm-water species. Bonnevie (1913) points out that the number of individual thecosomatous pteropods, as well as the number of species, was much higher in the southern crossing of the North Atlantic than in the northern crossing during the "Michael Sars" expedition. Many of the warmer-water species occur in all three oceans, some of them, however, fairly sharply confined to tropical waters. Species of *Limacina*, such as *L. bulimoides*, *L. inflata* and *L. lesueurii*, are essentially tropical or sub-tropical; *L. lesueurii* is possibly more bisub-tropical with *L. trochiformis* (Van der Spoel, 1967) (Fig. 2.45).

In contrast, certain species of *Limacina* are cold-loving. *Limacina helicina* occurs at relatively high latitudes in the Atlantic Ocean; Van der Spoel suggests temperature limits of −0.4°– +4.0°C. In the Pacific Ocean it does not reach such high latitudes; moreover, two forms of the species occur there, apparently forming separate populations (cf. Chapter 8). Another form of the same species, termed by some investigators *L. antarctica* appears at high Antarctic latitudes. At somewhat lower Antarctic latitudes and in the sub-Antarctic region it is replaced by *L. rangii*, probably another variety of *L. helicina* according to Van der Spoel (1967). *L. helicina*, therefore, appears to be a widely distributed species, at both northern and southern high latitudes, but present in a number of forms. (cf. Fig. 2.46).

Limacina retroversa is another cold-loving boreal species found in the North Atlantic (Fig. 2.47) with a somewhat more southerly distribution than *L. helicina* (cf. also Chen and Bé, 1964). Regarding the separation of *L. retroversa* and *L. balea* (cf. Bonnevie, 1913; Tesch, 1946), Van der Spoel suggests that these may be two forms of one species with a slight difference in geographical distribution. *L. balea* occurs generally in more northern waters and in the north-east Atlantic persisting even as far as 75°N. *L. retroversa* is a more typical boreal form reaching 65°N in the north-east Atlantic, and commonly found in the North Sea and in the entrance to the English Channel. Vane and Colebrook (1962) believe that, at least in certain years, separate populations of *L. retroversa* exist, over the oceanic waters west of Ireland and over the shallow waters of the North Sea. Van der Spoel (1967) suggests that shallow water is not favourable for most pteropods, but that this species may spread from deeper waters from time to time and persist at least for limited periods. The pteropod is sufficiently euryhaline to live at salinities down to 31‰.

Whereas *L. retroversa* can occur over both deep and shallow water, *L. balea* seems to be confined to deeper oceanic regions. Redfield (1939) describes invasions of juvenile *L. retroversa* from open North Atlantic waters into the Gulf of Maine. Growth occurs during the circulation of currents carrying these animals round the Gulf of Maine, the

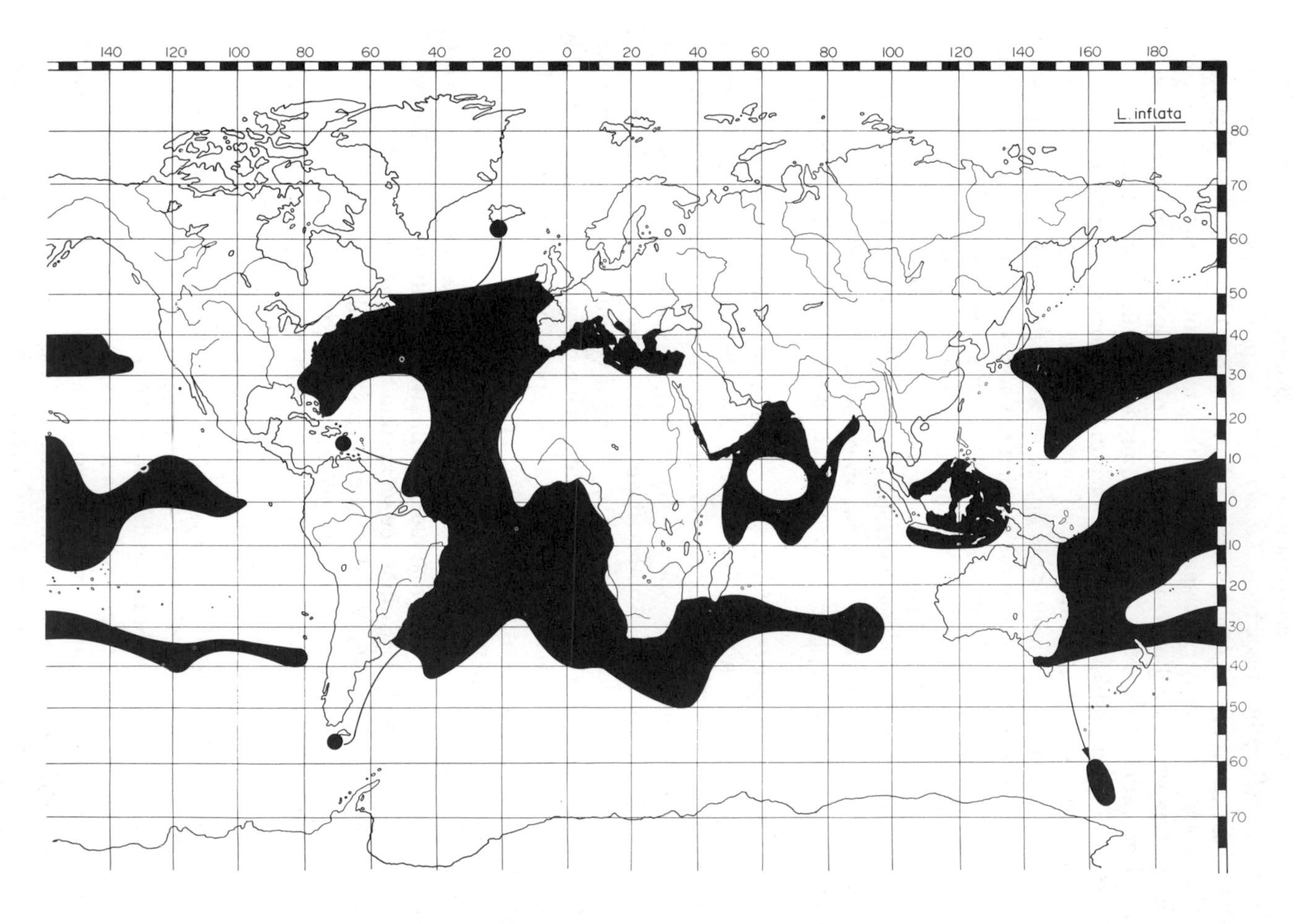

Fig. 2.45. Distribution of *Limacina inflata* (Van der Spoel, 1967).

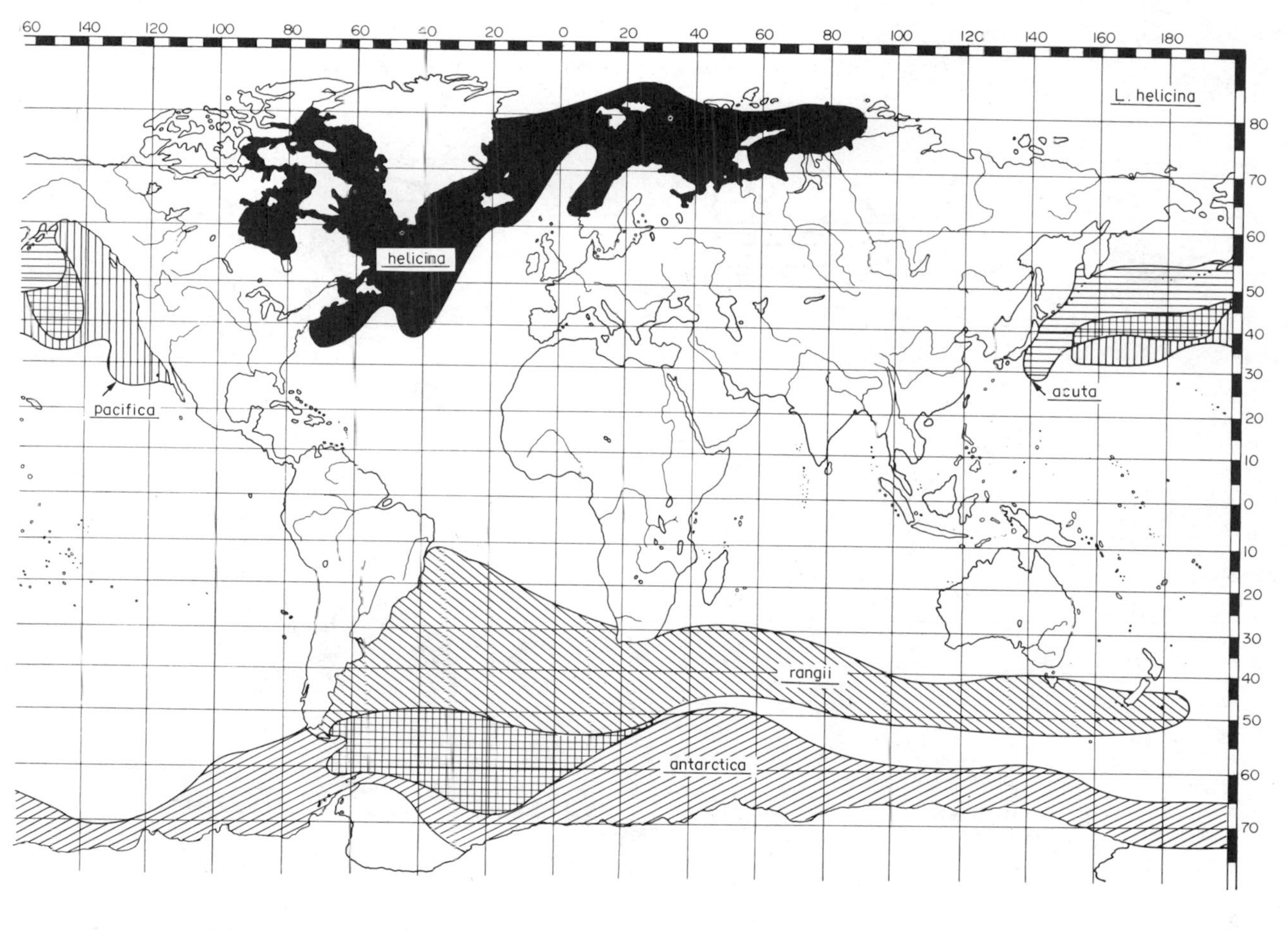

Fig. 2.46. Distribution of *Limacina helicina* (Van der Spoel, 1967).

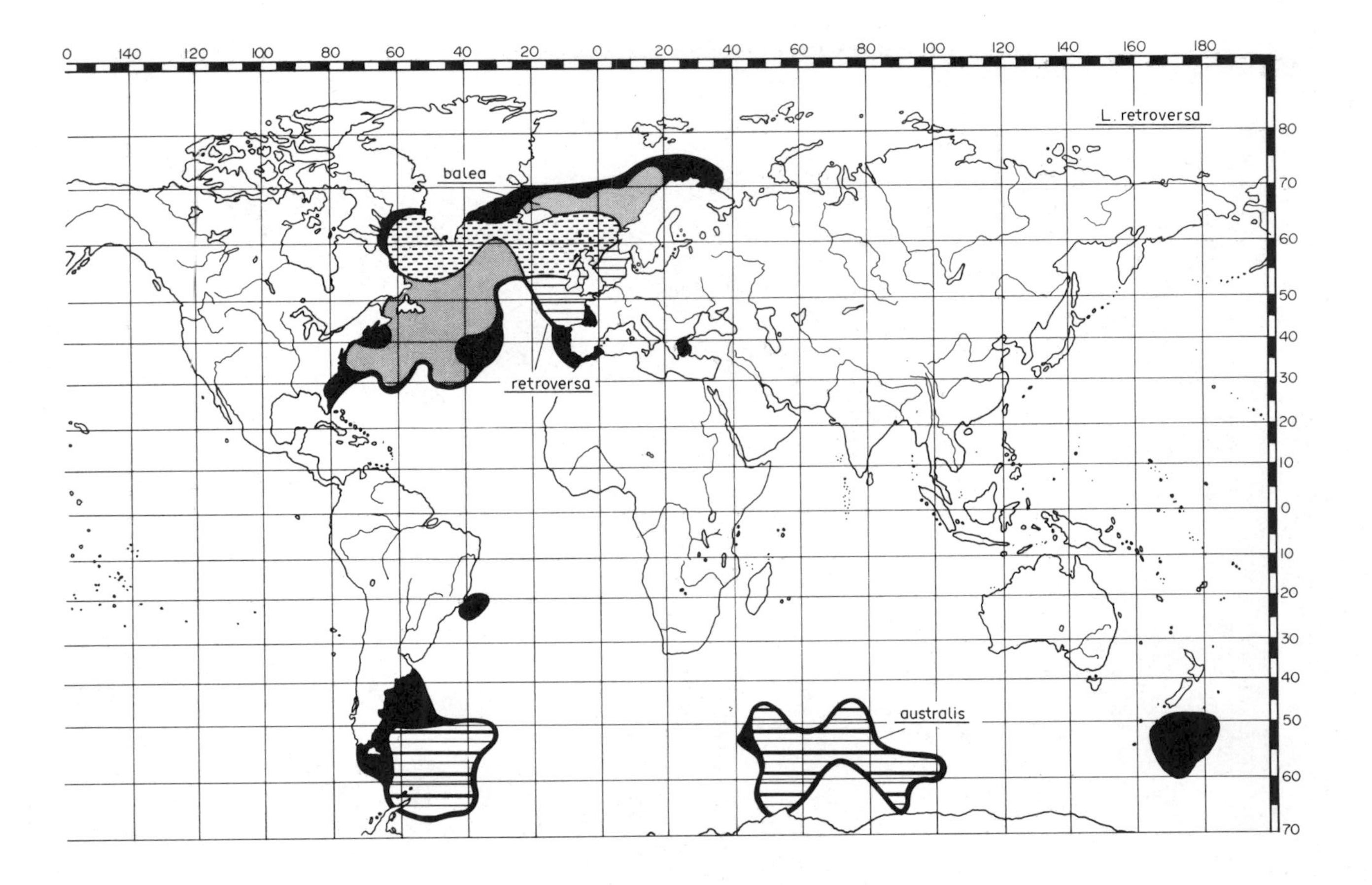

Fig. 2.47. Distribution of *Limacina retroversa* (Van der Spoel, 1967).

rate of growth being dependent particularly on temperature. Three successive invasions occurred—in December, April and June. Chen and Bé (1964), studying the thecosomes in the North Atlantic north of 30°N latitude, described the overwhelming dominance of *L. retroversa* in sub-Arctic regions of the western North Atlantic. The species made up almost the entire population, with considerably high densities (maximum $10-18/m^3$ in night tows at Weather Stations B and C). Maximal concentrations were found at temperatures around 8°–10°C.

In the southern hemisphere *L. retroversa* may be represented by the pteropod *Limacina australis*, probably a form of *L. retroversa balea*. It occurs most frequently between latitudes 50°S and 60°S, seldom in the coldest Antarctic waters (Fig. 2.47).

Limacina helicoides is relatively unusual among thecosomes in being a bathypelagic species. Tesch (1946) records it as never shallower than 500 m in the Atlantic Ocean and frequently at much greater depths. Tesch (1948) later noted its presence in the Pacific, but so far it has not been taken from the Indian Ocean. *L. helicoides* is a viviparous species.

A number of warm-water shelled pteropods occur, usually in all three major oceans as tropical or sub-tropical forms, limited to a latitude of about 45°N in the Atlantic and often being somewhat more abundant in latitudes about 20° north and south of the equator. These include such species as *Diacria quadridentata*, *Hyalocylis striata*, *Cavolinia uncinata* and *C. globulosa*, the latter only in the Indo-Pacific. In the warm-temperate western North Atlantic, Chen and Bé (1964) described *Limacina inflata* and *L. bulimoides* as representative of the sub-tropical but cold-tolerant species, with the cosmopolitan *Clio pyramidata* and also *Styliola subula*, attaining high densities at temperatures of between 18° and 22°C. Sub-tropical warm-tolerant species listed were *Limacina trochiformis* and *Creseis acicula*, found in all oceans, but not beyond 40°N in the Atlantic, with *Cavolinia inflexa* and *Creseis virgula* (Fig. 2.48).

Haagensen (1976) suggests that "sub-tropical, cold-tolerant" is perhaps equivalent to sub-tropical, and "sub-tropical warm-tolerant" to tropical, in the descriptions of other workers on thecosome distribution. In his own investigations in the Caribbean, however, he distinguishes the depth distribution in relation to the water masses. A tropical surface water mass (temperature >27°C; salinity <36.25‰) was characterized by such species as *Limacina trochiformis*, *Creseis acicula acicula*, *C. virgula virgula*, *C. virgula conica*, *C. chierchiae*, and *Cavolina longirostris*, with *Corolla* and *Gleba cordata*. These species were present in the surface waters throughout and could be regarded as tropical. Other thecosomes lived in a sub-surface sub-tropical warm water (temperature 10° to 27°C; salinity >36.25‰) during the day, though there was a marked tendency to migrate into the surface at night. Such species included *Styliola subula*, *Hyalocylis striata Limacina lesueurii*, *Limacina bulimoides Cuvierina columnella*, *Diacria trispinosa* and *Cavolinia inflexa* (Fig. 2.49). *Clio cuspidata* appeared as a sub-tropical species in the sub-surface waters but did not migrate.

Limacina inflata was by far the commonest thecosome forming 39% of the Caribbean collections and found in both water masses by day and night, though the average level changed. Sakthivel (1973) described *Limacina inflata* as an abundant species in the Indian Ocean. Despite differences in day and night hauls, the species appeared to be plentiful in richer, especially upwelling areas, but central regions of the Ocean were sparsely inhabited. South of the equator *Limacina* occurred to ap-

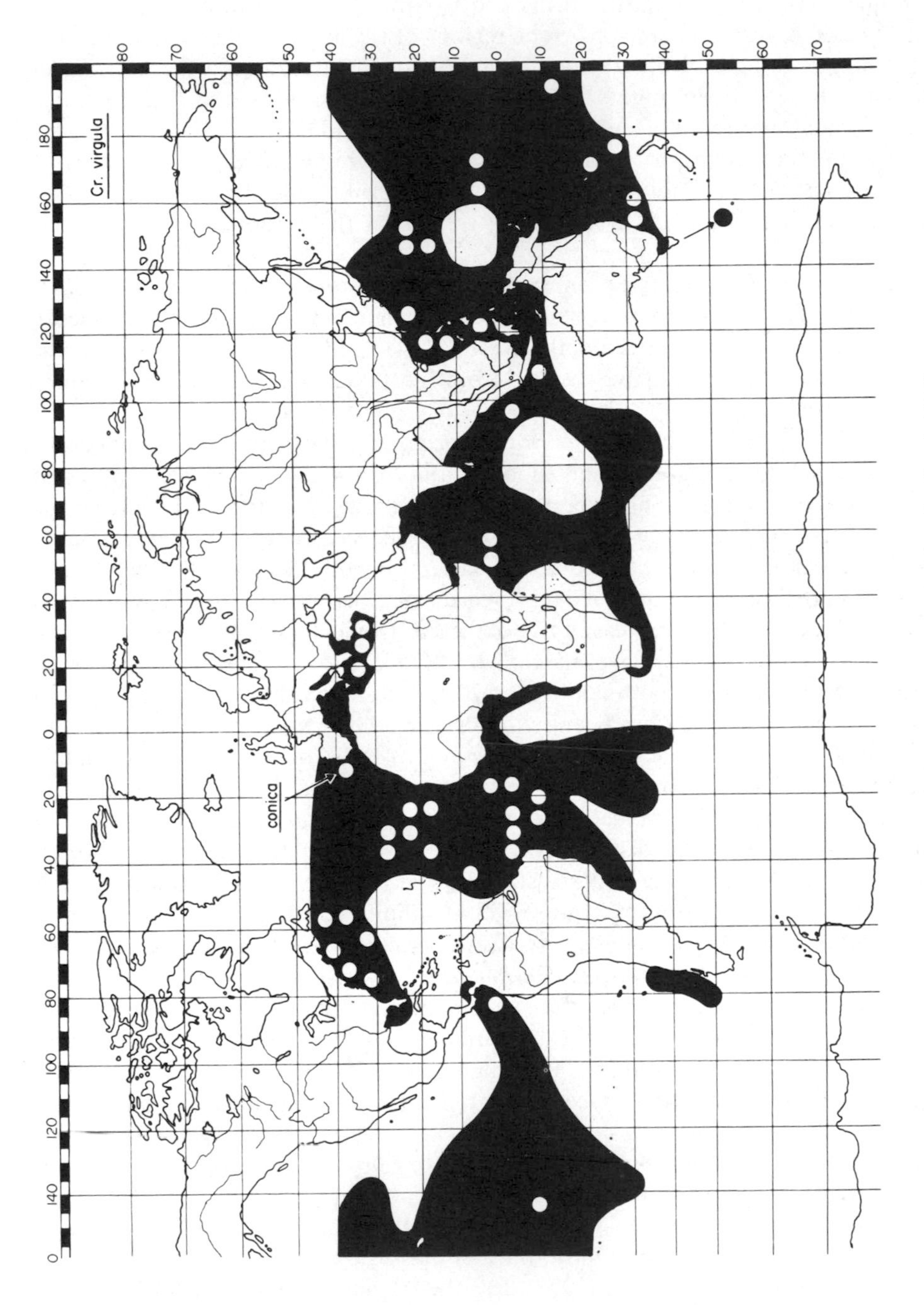

Fig. 2.48. Distribution of *Creseis virgula* (Van der Spoel, 1967).

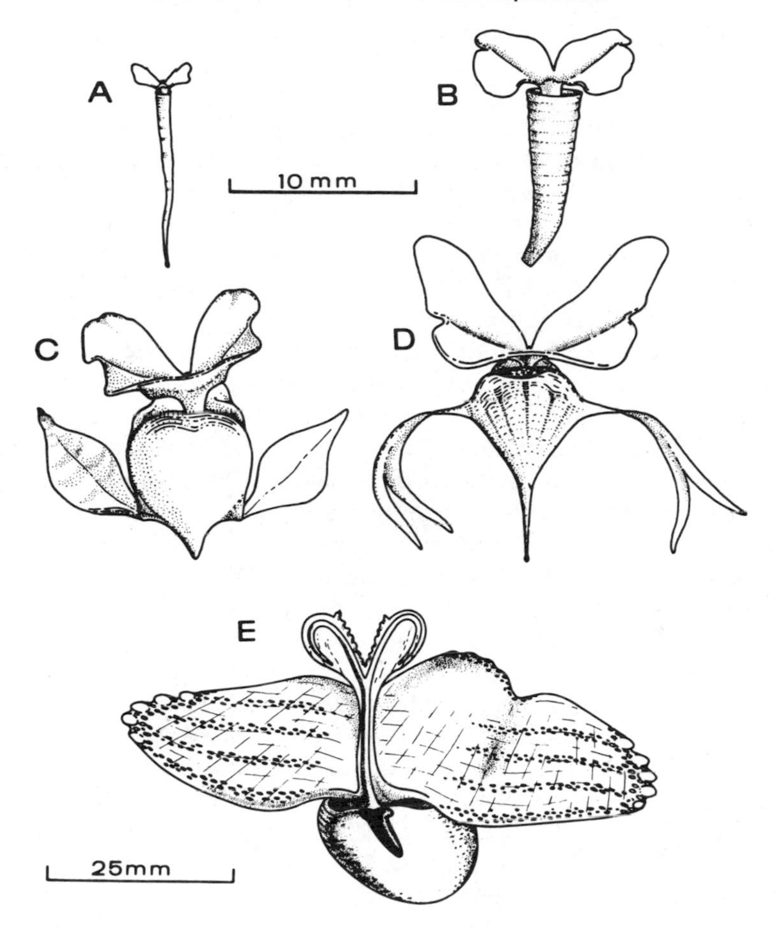

Fig. 2.49. Some adult thecosome species. (A) *Creseis acicula*; (B) *Hyalocylis striata*; (C) *Cavolinia tridentata*; (D) *Diacria trispinosa*; (E) *Gleba cordata* (Gilmer, 1974).

proximately latitude 32°S, but maximum abundance was around 11°S latitude and the occurrence fell to a minimum at 26°S.

Furnestin (1964), describing the warm saline oceanic waters off the Moroccan coast, lists several species of thecosomes (*Limacina inflata*, *L. bulinoides*, *Creseis acicula*, *C. virgula*, *Styliola subula*, *Hyalocylis striata*, *Cavolinia inflexa*, *C. longirostris*, *Diacria quadridentata*, and *Euclio balantium*) which she regards as warm-water, mainly tropical species. They are found more abundantly in the southern section of the Moroccan coast which is more subject to the influence of warm oceanic water. Casanova (1977) found *Limacina bulimoides* and *L. lesueurii* in surface waters of the Mediterranean, but only in the western region where the Atlantic inflow is significant. *Limacina inflata*, however, is widespread in the Mediterranean and is eurybathic in distribution.

Tesch (1946) mentions that some warm-water species may avoid the very warmest waters; he includes *Styliola subula* and *Cavolinia gibbosa*. A large group of warm-water pteropods are more widely distributed, spreading for instance in the Gulf Stream offshoots of the Atlantic Ocean, to somewhat higher latitudes, e.g. *Creseis acicula*,

C. virgula, *Cuvierina columnella*, *Cavolinia longirostris* and *C. inflexa*. *Creseis acicula* is often found nearer coasts and Haagensen (1976) describes it as the most neritic of thecosomes. Frontier (1973), for example, describes the species as occurring over the whole continental shelf off Nosy Bé, with dense populations at the end of the dry season. The only other thecosome listed for the coastal areas was *Creseis chierchiae*, found particularly in bays near Nosy Bé.

Clio (*Euclio*) *pyramidata* and *C. cuspidata* range particularly far. *Clio pyramidata*, with its varieties, has an enormous geographical spread, roughly from the boundary of the Antarctic to 65°N in the Atlantic Ocean, less far north (about 50°N) in the Pacific. It has an almost world-wide distribution, though avoiding really cold sub-Arctic waters. Bonnevie (1913) describes it as one of the commonest pteropods. *Diacria trispinosa* is another warm-water species which spreads considerable distances in more temperate seas, though it does not range nearly as far, especially to the south, as *Clio pyramidata*. Some thecosomes can be exceedingly abundant at times, forming swarms (e.g. *Styliola subula*, *Clio pyramidata*). It is not surprising that their calcareous shells may appear in sufficient quantities on the sea floor to form a typical pteropod ooze in some areas.

A few species of thecosomes are bathypelagic. *Limacina helicoides* is almost always deeper than 1000 m and can certainly extend to 2000 m. *Clio* (*Euclio*) *polita* is a bathypelagic species, apparently restricted to the Atlantic Ocean (but cf. Chen and Bé, 1964) where it is fairly widespread. Juveniles live in near surface waters. *Euclio chaptali* is also probably bathypelagic, but is found in all three major oceans, usually in tropical latitudes. The genus *Peraclis* has a number of deep-living species (*vide supra*). These are found in all three oceans (Tesch, 1946, 1948) as with many other bathypelagic forms. *P. reticulata* and *P. triacantha* can occur in somewhat shallower waters but have a very wide vertical distribution. Bonnevie (1913) speaks of swarms of *P. reticulata* in the Atlantic at depths of 50–250 m. Other species (e.g. *P. bispinosa* and *P. moluccensis*) are uncommon above 1000 m and may reach a depth of 3000 m. Haagensen (1976), however, records more moderate depths, though with a wide range, for such species as *P. apicifulva*, *P. bispinosa* and *P. reticulata* in Caribbean waters. The first two species are stated to be surface-living in Mediterranean waters and *P. reticulata* is described as eurybathic (Casanova, 1977). *Peraclis* (= *Procymbulia*) *valdivii* is another deep-living species found apparently in all three oceans.

Gymnosomata

This is a relatively small group of holoplanktonic molluscs highly adapted for pelagic life. They are opisthobranch Mollusca, as are the holoplanktonic Thecosomata. This latter group are the "shelled" pteropods; the Gymnosomata, totally without a shell, are often spoken of as the "naked" pteropods. The two groups, though both planktonic opisthobranch molluscs, appear to have very little in common except that they both swim by wing-like projections of the modified foot (cf. Morton, 1958). It is almost certain, as Tesch (1950) and many others have suggested, that the Thecosomata and Gymnosomata are quite separate in origin.

The body of a gymnosomatous pteropod is usually more or less streamlined, lacks mantle and mantle cavity and bears two pairs of relatively small tentacles on the head. The body lacks a shell and bears a pair of wings nearer to the head end by which it swims strongly; external gills may be present. In some species the body may be only a few millimetres in length; few exceed 30 mm. *Clione* may exceed 50 mm and even attain 70 mm in length in colder waters (Conover and Lalli, 1974).

The gymnosomes are highly predaceous animals. The mouth, at the anterior end, is surrounded by a variety of organs assisting in food capture; there is an array of hooks and teeth and also adhesive tentacles (cephaloconi) or tentacles bearing suckers (acetabula). For example, *Pneumoderma* and *Spongiobranchaea* both have sucker-bearing arms; *Clione* has three pairs of buccal cones and two hook sacs, but no sucker-bearing arms or jaws; *Cliopsis* has a long proboscis about three times the length of the cylindrical jelly-like body, but no buccal appendages (cf. Figs 2.50, 2.51). Bonnevie (1913) lists these characters for several species. Most of the buccal armature of gymnosomes is retracted when at rest but can be everted to seize food. A well developed radula, also partly eversible, assists in the taking of food. The capture of food is well described by Lalli (1970) for *Clione*. A recent description (Lalli, 1970) of the buccal armature of *Crucibranchaea macrochira* shows that the eversible proboscis has hook sacs, complex radula and jaws and hooks, and a pair of retractile sucker arms with separate suckers at their base.

The Gymnosomata is not a large group; according to Tesch (1950) there may be 20 genera with perhaps double that number of species. Many of the genera have only one or two species. Little is known about precise depth distribution but the group appears to have mostly epiplanktonic species; a few are bathypelagic. McGowan (1968) states that most gymnosomes are mesopelagic. Most of the species occur in warmer seas where they may have a fairly wide, almost cosmopolitan, distribution. Very few Gymnosomata are really abundant, but there are at least two exceptions. *Hydromyles globulosa* is an exclusively tropical Indo-Pacific form, but in that vast area it occurs so commonly that it forms more than 98 % of all the gymnosomes captured (Tesch, 1950). *Hydromyles* is an aberrant gymnosome, lacking proboscis, jaws, hook sacs and buccal appendages (cf. McGowan, 1968). The other remarkably abundant species is *Clione limacina*, a form which is typical of the boreal Atlantic though it occurs in the North Pacific also (cf. Lalli, 1970). In the North Atlantic it is so common that it amounts to about 80 % of the total gymnosome population and it is known as a food of the once plentiful Greenland right whale as well as being eaten by some fishes.

The gymnosome is believed to feed exclusively on the thecosome *Limacina*, *L. helicina* in high northern latitudes and *L. retroversa* in boreal waters (Conover and Lalli, 1972). For other examples of selective feeding see Chapter 6. The distribution of *Clione limacina*, while typically occurring everywhere in the boreal and Arctic Atlantic, is complex, since there may be a number of varieties. Probably the form occurring in the North Pacific is identical with *Clione limacina*. Tesch (1950) also believes that it is identical with the species which occurs in the Antarctic which has sometimes been given separate rank as *Clione antarctica*. In the warmer parts of the North Atlantic a form of *Clione limacina* exists which becomes sexually mature in more or less dwarf form. This dwarf variety undoubtedly is associated with somewhat warmer waters, even entering the English Channel, and extending to perhaps 40°N latitude. There is a suggestion that *Clione limacina* occurs in somewhat deeper layers to the south, but it is not a bathypelagic species.

A gymnosome which occurs only in the south is *Spongiobranchaea australis*, found in Antarctic and sub-Antarctic waters, probably not north of 35°S, and apparently a circumpolar southern species. *Pneumodermopsis ciliata* is a fairly widespread species,

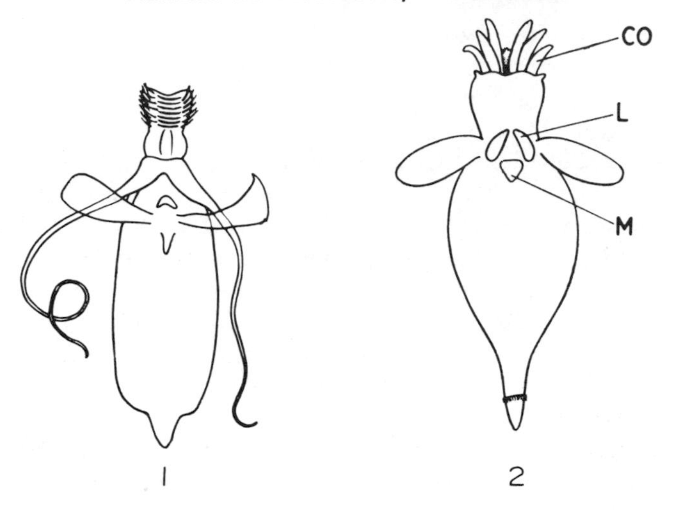

Fig. 2.50. (1) *Massya longecirrata*; (2) *Clione limacina*. CO = buccal cones; L = lateral; M = median lobes of foot (Morton, 1957).

occurring generally over the North Atlantic and recorded to the south of Iceland (cf. Morton, 1957) (Fig. 2.51). It is found in the warm waters of the Atlantic, Indian and Pacific Oceans, reaching as far north as at least 40° N latitude in the Pacific. Cooper and Forsyth (1963) found the species over oceanic depths only in the North Atlantic (always beyond the continental shelf) as far as the north of Scotland, but always in small numbers. The related *P.* paucidens occurs over shallow waters of the continental shelves on both sides of the North Atlantic, and according to Continuous Plankton Records it has now established itself in the North Sea (cf. Chapter 8). Of species found in warmer waters, *Notobranchaea macdonaldi* is essentially tropical/sub-tropical in all oceans and does not pass beyond about 40° N latitude; *Pneumoderma atlanticum* and *P. mediterraneum* are other tropical species in all three oceans.

Although most of the species are epipelagic, *Cephalobrachia macrochaeta* is a true bathypelagic species, usually exceeding 1000 m depth and probably with a very wide distribution. *Massya longecirrata* (Fig. 2.50) is another bathypelagic form (*ca.* 1000 m), and though so far found only in the North Atlantic, it is probably cosmopolitan (cf. Tesch, 1950; Morton, 1957). A third bathypelagic species, *Schizobrachium polycotylum*, is apparently widely distributed in southern oceans (cf. Tesch, 1950).

A small gymnosome, *Paedoclione doliiformis*, has been found to be regularly neotenous (Conover and Lalli, 1972, 1974). They ascribe the onset of neoteny to feeding on small-sized prey and believe that the dwarf forms of *Clione* also result from this habit.

Lalli and Conover (1973) describe the development of *Paedoclione*. Eggs are deposited in a string of mucus which is moulded into a ball floating on the surface. Hatching is apparently comparatively rapid and a shelled veliger with a thimble-shaped shell generally similar in shape to that of *Clione* (the only other gymnosome of which the development is known) is liberated. The veligers feed on phytoplankton and soon lose the shell; the velum becomes fragmented and is replaced by three ciliary bands as the larva lengthens (cf. Figs 2.52, 2.53). Further development is mainly an

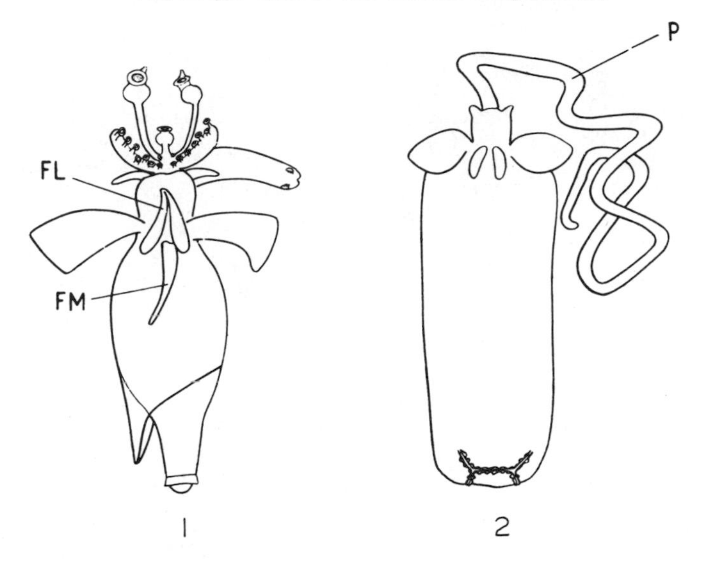

Fig. 2.51. (1) *Pneumodermopsis ciliata*; (2) *Cliopsis krohni*. FL, FM = lateral and median lobes of foot; P = everted proboscis (Morton, 1957).

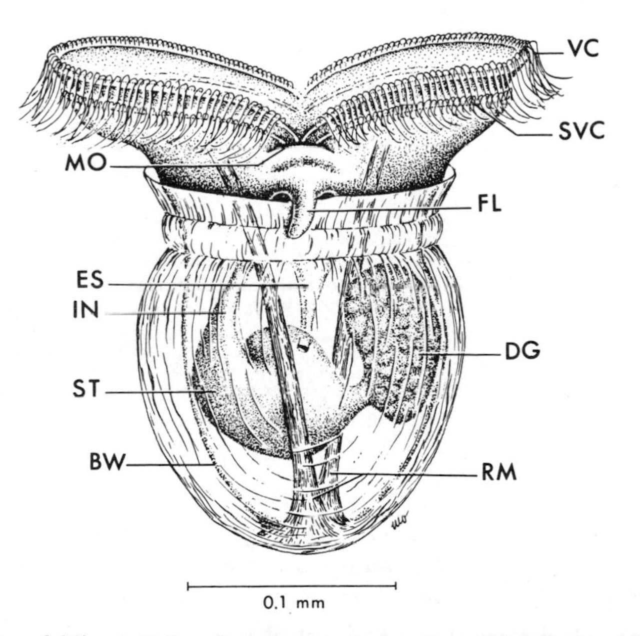

Fig. 2.52. *Paedoclione doliiformis*. Veliger. Retractor muscles have been slightly displaced from their normal mid-ventral and mid-dorsal positions. BW = larval body wall; DG = digestive gland; ES = oesophagus; FL = undifferentiated foot lobe; IN = intestinc; MO – mouth; RM = retractor muscles; ST = stomach; SVC = postoral cilia of subvelum; VC = preoral cilia of velum (Lalli and Conover, 973).

increase in size; polytrochous larvae feed on veligers of *Limacina*. In *Clione* these larvae slowly metamorphose to the adult, gradually elongating, developing the pair of wings and losing the ciliated bonds.

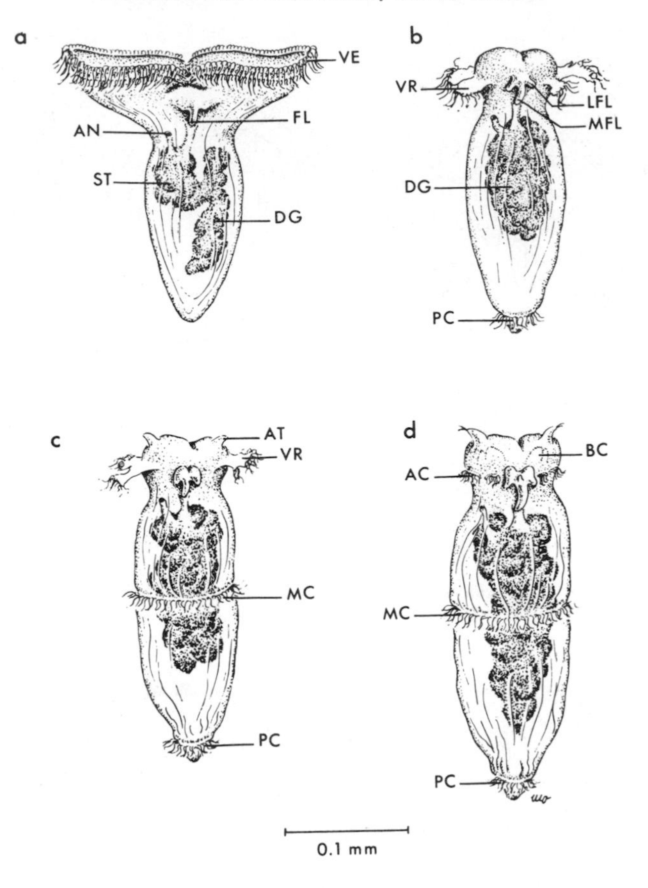

Fig. 2.53. *Paedoclione doliiformis*. Successive stages in metamorphosis from shell-less veliger to polytrochous larva. (a) Shell-less veliger larva; (b) larva with 1 ciliary band and velar rudiments; (c) larva with 2 ciliary bands and velar rudiments; (d) polytrochous larva with 3 ciliary bands and completely developed head structures. AC = Anterior ciliary band; AN = anus; AT = anterior tentacle; BC = retracted buccal cones; DG = digestive gland; FL = undifferentiated foot lobe; LFL = lateral foot lobe; MC = middle ciliary band; MFL = median foot lobe; PC = posterior ciliary band; ST = stomach; VE = velum; VR = velar rudiments (Lalli and Conover, 1973).

Cephalopoda

The cephalopods are pelagic molluscs, swimming by means of powerful and well co-ordinated muscles of the mantle, with its funnel assisted by the fins. The suckered mobile arms, together with the jaws, furnish a very efficient predaceous mechanism and the highly developed sense organs—especially the eyes—with the brain, provide excellent nervous co-ordination. Complex buoyancy mechanisms may exist for adjusting density to depth. The secretion of ink and the capacity for colour change are other adaptations to pelagic life.

The cephalopods can hardly be regarded as a predominantly planktonic group. Some species are bottom-living; other forms are members of the nekton—powerful swimmers in the depths of open oceans, some species achieving large size. The borderline between the smaller nektonic species (micronekton) and planktonic forms is not obvious, but some planktonic species exist, and these populations of markedly carnivorous holoplanktonic animals are reinforced by the planktonic larvae of nektonic and benthic squids and octopuses. Even some larger species may be essentially planktonic; the tropical *Argonauta* is an example.

Although the larvae of benthic and nektonic species would more properly be dealt with in the section concerned with meroplankton, it is convenient to describe briefly all the cephalopods in the plankton together. Some of the smaller species listed, for example, by Muus (1963) in his review of the cephalopods of the North Atlantic, may be captured occasionally in plankton hauls and some might be regarded as planktonic. No attempt will be made, however, to mention more than a few of these small adult cephalopods. McGowan (1967) examining the distribution of cephalopods in the Californian Current region, points out that the analysis relates to larvae only; no individuals exceeding 10 mm mantle length were captured.

In a later more detailed analysis, Okutani and McGowan (1969) describe the distribution and abundance of squid larvae. Since there are so few reports dealing with quantitative and distributional studies of cephalopod larvae, this investigation will be referred to in some detail. The study covers larvae collected over three years in plankton hauls. The distributional ranges for many of the species are, however, based largely on knowledge of the adults. Several species which are probably endemic to the Californian region were identified. The three most abundant species were *Abraliopsis felis*, *Gonatus fabricii* and *Loligo opalescens*. *Loligo opalescens*, a demersal spawner, is found from Puget Sound to around 27°N along the Mexican coast. *Abraliopsis felis* is probably also endemic to the California Current. Both adults and larvae are all from the area and it is not taken in sub-Arctic, central or eastern tropical Pacific waters. Okutani and McGowan suggest that its distribution might be compared with that of *A. scintillans*, known only in waters off Japan.

Meleagroteuthis heteropsis is possibly also endemic, with records from the Californian coast. Larvae were found of *Octopodoteuthopsis* sp. Although it is not clear whether this is a separate endemic species, it differs from *O. megaptera* from the New England coast and another species of the genus from African waters. Two species of squid larvae, which are believed to be widely distributed, were taken in the collections. *Pterygioteuthis giardi* is known in tropical and warm waters as well as in cool-temperate seas. It occurs in some tropical regions of both the eastern and western Pacific as well as in more temperate waters off California, Australia and New Zealand cf. also Nesis, (1972) for Peruvian waters. It is known in tropical waters of the eastern and western Atlantic, in boreal waters off Scandinavia and off New England and in warmer waters of the Bay of Biscay, the Azores and the Mediterranean (cf. Table 2.7). It has also been taken in the Indian Ocean.

Onychoteuthis banksii appears to be almost cosmopolitan, occurring in all three major oceans with a very wide distribution including tropical, sub-tropical and temperate waters and extending into boreal and sub-Antarctic regions. Clarke (1966) refers to the wide distribution of the adults, including cold waters, and Muus (1963) gives its range as extending into the Norwegian and Barents Seas. It is apparently excluded from the Arctic and Antarctic, but in the Atlantic extreme limits are given as latitude 71°N and 56°S. In the Pacific it occurs from the Bering Sea to New Zealand (cf. Nesis, 1972).

Other species identified with more limited ranges include *Helicocranchia pfefferi*

Table 2.7. General summary of distributional ranges of larval squid (Okutari and McGowan, 1969)

Region	Subregion (approximate positions)	Species													
		Loligo opalescens	*Abraliopsis felis*	*Pterygioteuthis giardi*	*Octopodoteuthopsis* sp.	*Onychoteuthis banksii*	*Gonatus fabricii*	*Meleagroteuthis heteropsis*	*Ctenopteryz sicula*	Rhynchoteuthion larvae	*Chiroteuthis veranyi*	*Pyrgopsis pacificus*	*Liguriella* sp.	*Teuthowenia megalops*	*Helicocranchia pfefferi*
Arctic							+								
West Pacific	Boreal (Kamchatka; northern Japan)					+	+								
	North temperate (Japan)					+				+		+			
	Tropical (East Indies; northern Australia)			+		+				+					
	South temperate (southern Australia)			+		+			+	+		+			
	Antiboreal (New Zealand)			+		+						+			
Mid-Pacific	Mostly tropical					+				+					
East Pacific	Boreal (Bering Sea; Alaska)					+	+								
	North temperate (*area under study*)	+	+	+	+	+	+	+	+	+	+	+	+	+	+
	Tropical (Central and northern South America)			+		+				+					
	South temperate (mid-South America)	(No information)													
	Antiboreal (Southern South America)	(No information)													
Atlantic	Boreal (Scandinavia; England; Davis Strait)			+		+	+							+	+
	North temperate (Bay of Biscay; Azores; Mediterranean; New England)			+		+	+		+	+	+			+	+
	Tropical (East Africa: Caribbean Sea)			+		+				+	+				
	South temperate (South Africa; Mid-South America)					+			+						
	Antiboreal (Southern South America)					+									
Indian Ocean	Mostly tropical to South temperate					+				+		+			
Antarctic							+								

from waters in the North Atlantic and now from California and found by Nesis (1972) off Peru, and *Chiroteuthis veranyi* from the Atlantic and Mediterranean. This larva is the curious doratopsid type (cf. Fig. 2.54) with a long "neck" and "tail", both of which are lost during development. Clarke (1966) gives records for adults also from the Pacific Ocean, and Nesis (1972) found larvae off Peru. Muus (1963) gives the distribution in the eastern Atlantic as south and west of Ireland but extending to the Faroe–Iceland area. A number of larval chiroteuthids, including the smallest, named *Planctoteuthis*, and larger ones, named *Doratopsis*, are also illustrated by Clarke (1966). Some have been taken in the Atlantic; others in the Indian Ocean and in Japanese waters, but it is uncertain at present to which adults the larvae belong.

Other squid larvae found by Okutani and McGowan (1969) off California with a fairly limited distribution (cf. Table 2.7) include *Ctenopteryx sicula*, a warm Mediterranean and Atlantic species, with limits 41°N to 32°S which is also reported off eastern Australia, and *Pyrgopsis pacificus*, an Indo-Pacific tropical and warm-temperate form found off Japan, Australia, the Banda Sea and African coasts. Clarke also quotes adults from the Atlantic and Mediterranean. *Teuthowenia megalops*, a member of the Taoniinae is known in the warmer areas of the Atlantic (Gulf Stream, Sargasso Sea, South Equatorial Current) and as far as the Faroes. Clarke (1966), however, states that generic names of Taoniinae have been used to some extent to describe

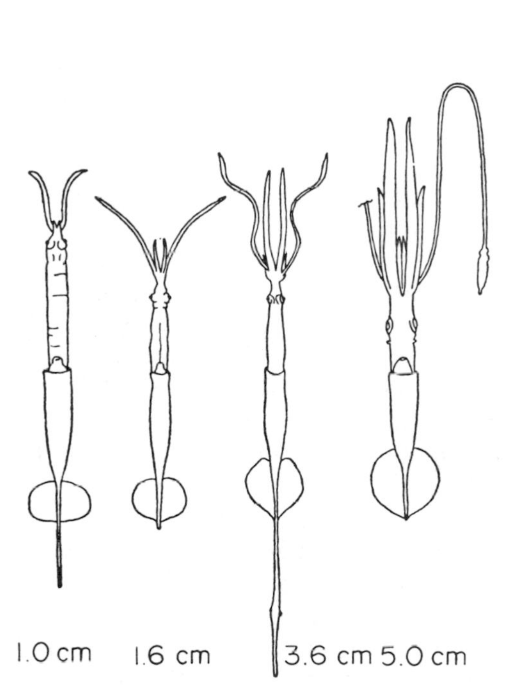

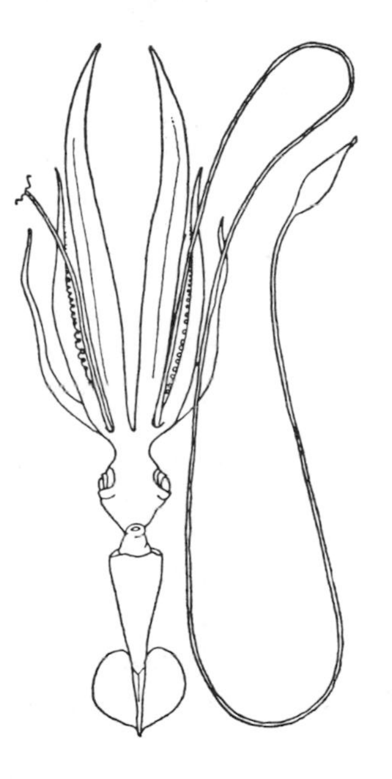

Fig. 2.54. Changes in form during growth of *Chiroteuthis veranyi* after various authors. The three smallest are referable to the larval "species" *Doratopsis vermicularis*. All from ventral side. Mantle lengths in cm (excluding the "tail") are indicated (Clarke, 1966).

developmental stages of these cephalopods. A series can be traced showing the gradual loss of larval characters, so that specific identification of many is uncertain. Nesis (1965) believes that this species is identical with *Desmoteuthis megalops*, a more cosmopolitan cephalopod described by Muus (1963) as reaching colder areas of the Atlantic, including the Faroe–Shetland area and the Norwegian Sea. A few larvae were found off the southern tip of Greenland and Nesis considers the species breeds throughout the year.

Okutani and McGowan (1969) also found larvae of the squid *Gonatus fabricii* off California. This species has a distribution very different from the others so far listed. It is found in Arctic and sub-Arctic areas of the Atlantic and Pacific Oceans. For example, in the Atlantic it is known from Greenland, Iceland, the Norwegian Sea as well as off north-east Britain and the New England and Nova Scotia coasts in the western Atlantic. Its southern limit is 36°N. In the Pacific there are records off Kamchatka, the Bering Sea, the Sea of Japan and the west coast of North America, with its southern limit reaching 54°N. It is known in Antarctic and sub-Antarctic waters, but some authorities believe this to be a separate species, *Gonatus antarcticus*.

Nesis (1965) cites its northern distribution as including Davis Strait and the Barents Sea. He states that larvae can be found in waters of between 0° and 7°C, mostly 5° to 7°C, avoiding the cold Labrador Current and true Arctic waters, though adults can occur there. Nesis (1965) quotes breeding in Biscay waters from February to June with a peak in April/May; in Labrador waters the peak is in May/June, though some breeding occurs throughout the year. The breeding is over great oceanic depths. The young are mainly in the 20–150 m layer in the Labrador Sea. Muus (1963) also indicates its high latitude range. Off California, *Gonatus fabricii* breeds chiefly in April, May and June (cf. Okutani and McGowan, 1969). These workers found *Abraliopsis felis* larvae mainly in May, June and July, and *Loligo opalescens*, a neritic spawner, spawning mostly in winter/spring. Some squid larvae occurred all through the year, but there were strong indications of seasonality.

For the north-eastern Atlantic region, Muus (1963) gives the distribution of some other cephalopods which as juveniles or as small species might be encountered in the plankton. Three species of *Sepiola* occur. *S. atlantica* is found widely in the North Atlantic and in the Norwegian Sea, the Faroe–Iceland area, south and west of Ireland, as well as in the North Sea, Skagerrak and in the English Channel. *S. aurantiaca* and *S. pfefferi* appear to be more limited to the North Atlantic, south and west of Ireland, although *S. pfefferi* is found in the Faroe–Shetland area and the northern North Sea.

Muus (1963) describes the distribution of other squid species whose juveniles might be expected to occur in plankton hauls. *Loligo* and *Alloteuthis* appear to be more coastal, being found round the British Isles, south and west of Ireland, in the North Sea and Skagerrak and reaching as far north as the Faroe–Shetland area. Muus (1963) also quotes two species of Octopoda which are planktonic as larvae. *Octopus vulgaris* occurs in the north-east Atlantic area off Ireland, the southern North Sea and English Channel; *Eledone cirrosa*, while found in the same areas, spreads farther north into the northern North Sea, Faroe–Shetland region and Norwegian Sea. *Octopus vulgaris* larvae are the most widely distributed of all benthic octopods according to Voss (1971). The larvae can maintain themselves in the plankton for considerable periods.

Of the cephalopods found typically in warm waters, *Onychia carribaea* is a surface-living squid that may be captured in nets. *Spirula* occurs in all three oceans in tropical and sub-tropical waters. Clarke (1966) states that the young forms may occur at depths ranging from 1000 m to 1750 m. The adults show marked vertical migration.

Nesis (1972) has reported on oceanic cephalopods in the Peru Current. The greater variety of species in warm as compared with colder waters is illustrated by the fact that, of thirty-seven oceanic species, thirty-four are tropical or sub-tropical, more than half occurring in all three major oceans. Two species (*Brachioteuthis riisei* and *Bathyteuthis abyssicola*) are widely cosmopolitan, while *Gonatus antarcticus* and *Todarodes angolensis* appear to frequent colder waters. Some three species were bathypelagic, being confined to cold deep waters. Unfortunately, nektonic and planktonic species cannot be precisely distinguished. However, "sluggish planktonic species" are listed—*Tremoctopus*, *Ocythoe* and *Argonauta* in the uppermost layers. Moreover, squid belonging to the Octopoteuthidae, Histioteuthidae, Chiroteuthidae, Cycloteuthidae and many Cranchiidae, and octopuses belonging to the Amphitetidae and Bolitaenidae, are described as deeper-living planktonic or semi-nektonic forms, occurring at medium depths in the ocean. Nesis (1965) points out that most of these cephalopods make extensive diurnal vertical migrations, the shallower-living reaching even the 100–200 m layer at night.

In general, the larvae of the medium depth species are in the surface or near-surface layers, while the adults are deeper living. For the common species, *Abraliopsis affinis*, spawning is in the upper layers and eggs are neutrally buoyant. The body length of the larvae increases, however, with depth. Larvae of such surface species as *Tremoctopus violaceus* and *Symplectoteuthis oualaniensis* remain near the surface, but *Argonauta* larvae may occur as deep as 500 m, though the majority are in the uppermost layer (see Table 2.8).

Table 2.8. Depth of occurrence of Argonauta larvae (Nesis, 1972)

Depth (m)	0–50	50–100	100–200	200–500
Number of larvae in equal 50 m interval	24	19	16	3

True bathypelagic species may not show any clear shallow distribution of larval stages, but *Japetella*, according to Nesis, reproduces at great depth. The larvae acquire neutral buoyancy so that larvae and juveniles are finally widely distributed in depth through the whole water column. A list of the cephalopod larvae and juveniles gives an indication of characteristic depth distribution (Tables 2.9, 2.10).

The larvae and juvenile stages of many cephalopods are not yet recognized. Clarke (1966) describes certain larvae which have been known as *Rhynchoteuthis* (or *Rhynchoteuthion*) and which are characterized by the tentacles being fused together. They are believed to be the larvae of ommastephrid squids, and some have been related to particular species (e.g. *Todarodes pacificus* in Japanese waters, *Nototodarus sloani* off eastern Australia, and two larvae taken in the Atlantic referable to *Ommastrephes caroli* or *O. pteropus*). There appears to be a wide depth range for the few larvae captured. One larva, believed to be that of *Symplectoteuthis oualaniensis*, was reported by Clarke (1966) as being captured in the Indian Ocean with a neuston net. This would agree with the conclusion of Nesis (1972) that the larvae of this squid in Peru–Chile waters remain at the surface.

Table 2.9. Average depths at which pelagic cephalopod mollusc larvae and early juveniles were caught (Nesis, 1972)

Species	Length of mantle (mm)	Average depth at which caught (m)
Abraliopsis affinis	≤10	87
Pterygioteuthis giardi	≤10	101
Onychoteuthis banksi	<13	109
Gonatus antarcticus?	4–6	74
Ctenopteryx sicula	<5	141
Brachioteuthis riisei	2–9	100
Histioteuthis sp.	2–3	112
Sympl. oualaniensis	1–17	27
Leachia pacifica	5–25	225
Helicocranchia pfefferi	3–15	161
Japetella diaphana	≤20	459
Octopus sp.	11–19	750
Tremoctopus gracilis	4–15	0
Argonauta sp.	1.4–4.6	96

Table 2.10. Distribution of *Japetella diaphana* by size and depth at which caught (Nesis, 1972)

Lower boundary of layer in which caught (m)	Length of mantle (mm) 0 10 20 30 40 50 > 50						Average length of mantle (mm)
0–100	7	—	—	—	—	—	6.5
101–200	3	1	—	—	—	—	6.8
201–500	1	9	7	2	—	—	21.6
501–1000	6	11	4	—	1	1	17.7
>1000	6	7	7	4	2	3	25.8

Crustacea

Among the great Class Crustacea, several predominantly benthic orders may have a few representatives in the marine plankton. The Leptostraca, for example, include one bathypelagic genus, *Nebaliopsis*. Two orders, the Cumacea and Isopoda, have very few planktonic forms (probably no cumacean is truly holoplanktonic) and will be discussed very briefly before examining crustacean orders of greater significance in the plankton.

Cumacea

Cumaceans are benthic crustaceans, mainly living in sand and mud, but although no species is known to be holoplanktonic, numbers of these crustaceans may sometimes appear in coastal waters, especially at night. The external form of a cumacean is characteristic. The head and thorax are very considerably enlarged; there is a very well developed carapace fused with the first three or four thoracic segments and prolonged in front as two lateral extensions to form a pseudorostrum. The abdomen is long and narrow, composed of six segments and carries elongated uropods (Fig. 2.55). Some

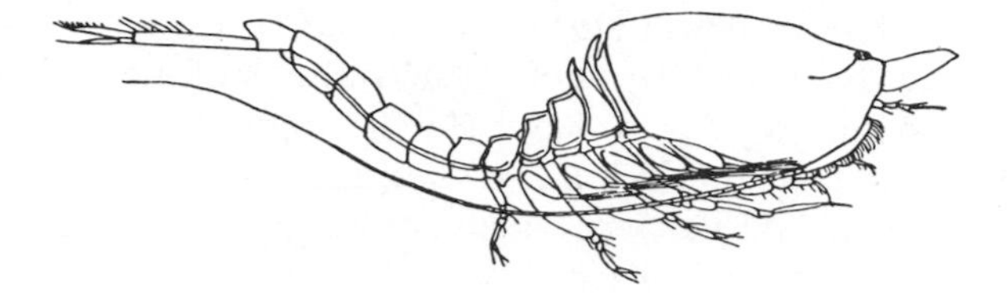

Fig. 2.55. *Campylaspis rubicunda*. Lateral view of male (Jones, 1957).

cumaceans appear to filter their food; other species scrape up detritus from the bottom.

The tychopelagic habit, true of some species, of spending part of their existence swimming in the plankton, is particularly characteristic of male cumaceans, which tend to swim towards the surface at night and may be attracted by artificial lights. Female cumaceans are more rarely pelagic. Such Cumacea include, amongst others, species of *Bodotria*, *Iphinoe*, *Lamprops*, *Campylaspis*, *Nannastacus*, *Cumella*, *Pseudocuma* and *Leptostylis*. The many species listed by Naylor (1957) are mostly found fairly widely in shallow waters off north-west Europe. These include *Iphinoe trispinosa*, *Pseudocuma similis* and *Bodotria scorpioides* in the waters around the British Isles, south and west of Ireland and in the North Sea, the latter species also occurring in the Belt Sea. *Nannastacus unguiculatus* appears to be more southerly in distribution, being found in the English Channel, southwest of Ireland and in the Bay of Biscay. *Campylaspis rubicunda* has a very wide spread, as with the first group of cumaceans, but also extends to the cold Norwegian, Barents, Kara and White Seas, and *Lamprops fuscata* appears to be restricted to these cold northern waters.

Fulton (1968) also reported that while cumaceans are usually regarded as benthic animals, males and females were found in surface tows off Vancouver Island. Altogether some seventeen species are known; though few in number they include *Diastylis pellucida*, *Lamprops carinata* and *Cumella vulgaris*. Gamo (1967) collected species of *Bodotria*, *Nannastacus*, *Cumella* and *Campylaspis* in shallow night tows in Japanese and south-east Asian waters. On the other hand, all the cumaceans taken in plankton hauls in IIOE collections were captured close to the coast and usually in small numbers. Kurian (1973) emphasizes that no cumaceans survived far from shore and that the group is benthic.

An area, however, where cumaceans have been reported even in considerable numbers is off south-west Africa (Jones, 1955). All stations where rich hauls were taken were near the coast and in water shallower than 150 m, and Jones regards all the five species taken as really bottom-living. *Iphinoe* was particularly abundant. Furnestin (1970a) also commented that Hart and Currie had reported cumaceans occasionally plentiful in the Benguela Current off West Africa and she quotes *Iphinoe fagei* as an important species.

Isopoda

Very few isopod crustaceans are strictly planktonic though many species may be found, even in some abundance, among floating weed or associated with floating timber. In shallow inshore waters bottom-living isopods may also temporarily swim in the overlying water and are taken in tow nets as tychopelagic forms (cf. Naylor, 1957a). Some species of *Idothea*, amongst other genera, are known inshore; *I. metallica* may be found more offshore but Tattersall (1911) points out that even this species is not genuinely planktonic. Baan and Holthuis (1969) also indicate the association of *Idothea* spp. with floating weed. Off Texel (Netherlands) they found four species amongst those listed by Naylor (1957b), *I. baltica* and *I. metallica* being moderately plentiful. They commented on the very wide distribution of the latter species in the Atlantic, Mediterranean and Indo-Pacific.

Tattersall listed *Munnopsis* as a bathypelagic genus. Amongst other characters, species of the genus have three pairs of posterior legs modified for swimming and the body tends to be somewhat laterally expanded. A revision of the genus recognized both *Munnopsis* and *Paramunnopsis*. *Munnopsis typica* is found offshore and is recorded at high latitudes in the Atlantic and in the Norwegian and Barents Seas (Naylor, 1957b) (Fig. 2.56). *Paramunnopsis oceanica* is a truly pelagic and deep-living species which may be found to about 3000 m depth. *Munnopsis murrayi* is also regarded as a pelagic species and is found to about the same depth. Naylor (1957b) reports it in the Faroe–Shetland area of the Atlantic.

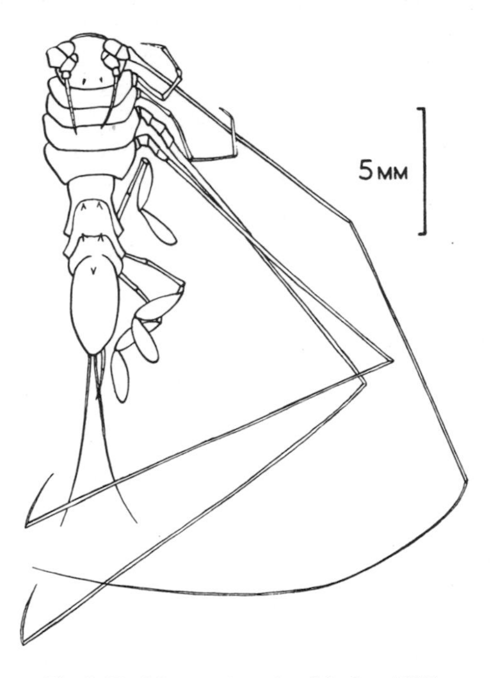

Fig. 2.56. *Munnopsis typica* (Naylor, 1957).

Wolf (1962), however, describing *Munneurycope* (= *Eurycope* = *Munnopsis*) *murrayi* as a cosmopolitan generally deep-living species (550–3000 m) taken in various areas in the three main oceans, believes that even this pelagic species seeks its food on the bottom. He states that the only truly planktonic asellote isopods are *Paramunnopsis* spp. (*P. oceanica*, *P. longicornis* and *P. spinifer*), possibly *Desmosoma chelatum* and some of the lesser known *Munnopsis* spp. may also be partly planktonic.

Another isopod genus which includes some planktonic species, mostly found inshore, is *Eurydice*. *E. truncata*, for example, is quoted by Tattersall (1911) as being widespread in the Atlantic, mainly at depths less than 200 m. *E. grimaldi*, also found widely in the eastern Atlantic and as far north as off Iceland and the Faroes, is apparently more oceanic in distribution and may occur to depths approaching 1000 m. Baan and Holthuis (1969) recorded *Eurydice spinigera*, listed by Naylor (1957b) as an inshore form, off Texel. They believe that it is planktonic for part of its life.

Cladocera

Of branchiopod crustaceans, the Cladocera are usually an important group in fresh waters. They are often regarded as being of comparatively little importance in the marine zooplankton since they are represented by only three genera, *Penilia*, *Podon* and *Evadne*. *Bosmina coregoni maritima* is reported from dilute estuarine waters and is endemic in the Baltic (Ackefors, 1971). Some species of marine cladocerans, however, have a remarkably wide distribution in the seas of the world, and at times, for short periods, their numbers can be exceptionally large, even exceeding the ubiquitous copepods. The importance of the Cladocera has been noted, among other areas, in Japanese waters (Onbe, 1977), in parts of the Indian Ocean (Della Croce and Venogopal, 1972), in the Mediterranean (e.g. Tregouboff, 1963; Casanova, 1968; Moraito-Apostolopoulou and Kiortsis, 1977) and in the Black Sea (Margineanu, 1963). As direct consumers of phytoplankton, with some species capable of feeding on very small particles, cladocerans can be important secondary producers.

The role of Cladocera in the marine zooplankton is, however, somewhat limited in two respects. They tend to be extremely seasonal, reaching high densities usually only over rather short periods. Thus, in temperate and warm-temperate regions they occur usually in the warmer period of the year. Cladocerans frequently show a marked species succession. For example, in the North Sea *Podon leuckarti* is succeeded by *P. intermedius* (Gieskes, 1971). Cronin, Diaber and Hulburt (1962) found that while *Evadne nordmanni* was usually the spring and early summer species found in Chesapeake Bay and in nearby areas of the continental shelf off the north-east coast of America, including Long Island Sound, it was succeeded by *Penilia avirostris* in high summer (mainly August/September).

In the Inland Sea of Japan, Onbe (1977) found that *Podon leuckarti* and *Evadne nordmanni* appeared in spring, and were followed by *Podon polyphemoides* and *Evadne tergestina*, with *Penilia avirostris* occurring in the high summer period. Water temperature is generally believed to be the chief factor associated with this succession. For example, the two congeneric species *Evadne nordmanni* and *Evadne tergestina* were found by Onbe (1977) to flourish at temperatures of 15–19°C and 24–28°C, respectively in the Inland Sea of Japan. For the north Aegean, Moraitou-Apostolopoulou and Kiortsis (1977) also found *Evadne nordmanni* only during the coldest period of the year, whereas *Penilia avirostris* reached maximal numbers only during the warm period; *Evadne tergestina* was abundant only between June and August, at a temperature of about 24°C.

A further factor limiting the significance of cladocerans is that, in the strictest sense, they are meroplanktonic organisms. Though they are capable of very rapid parthenogenetic reproduction, effecting the building of dense populations, they include a bisexual generation in the life history with a resting egg. They are, therefore, mainly neritic in distribution, and although many records testify to their capability of spreading out to considerable distances from the coast (e.g. Wiborg, 1955; Tregouboff, 1963; Calef and Grice, 1967; Urosa and Rao, 1974), they cannot contribute substantially to the zooplankton of open oceans. On the other hand, they may be extremely important in near-shore zooplankton. One species, *Podon polyphemoides*, is truly estuarine (cf. Jeffries, 1967; Gieskes, 1971; Bosch and Taylor, 1976).

Cladocerans are characterized by the body possessing a bivalve carapace fused to a few trunk segments, but leaving the head free. In *Penilia* the bivalve carapace covers the body and appendages, but in *Podon* and *Evadne* it forms a conspicuous dorsal chamber acting as a brood pouch. In *Podon* the brood pouch is almost hemispherical and is clearly demarcated by a groove from the head. In *Evadne* there is no obvious separation of the head from the large brood pouch and the body may be oval or triangular (Fig. 2.57). Development in cladocerans is direct. The brood pouch may contain resting eggs, usually a single one, with a comparatively thick wall (cf. Cheng, 1947), or (as with the characteristic parthenogenetic development) the brood pouch may be filled with embryos, which in some species can distort its shape. Of the appendages, the antennules vary in their development with the genus and sex. For example, they are rudimentary in *Podon* but are as long as the carapace in male *Penilia* (cf. Fig. 2.57). The chief locomotory appendage is the well developed biramous setose antenna, which lies behind a prominent compound eye in *Podon* and *Evadne*. The eye is comparatively small in *Penilia*. The trunk limbs number only four to six pairs.

Penilia is represented by a single species, *P. avirostris*, found in the tropical coastal waters of all the oceans and in warm-temperate inshore seas. Della Croce (1974) gives extreme records of 52°N and of 40°S latitude. Details of the reproductive patterns of *Penilia*, especially in relation to the geographical limits of the species, are included in Chapter 3.

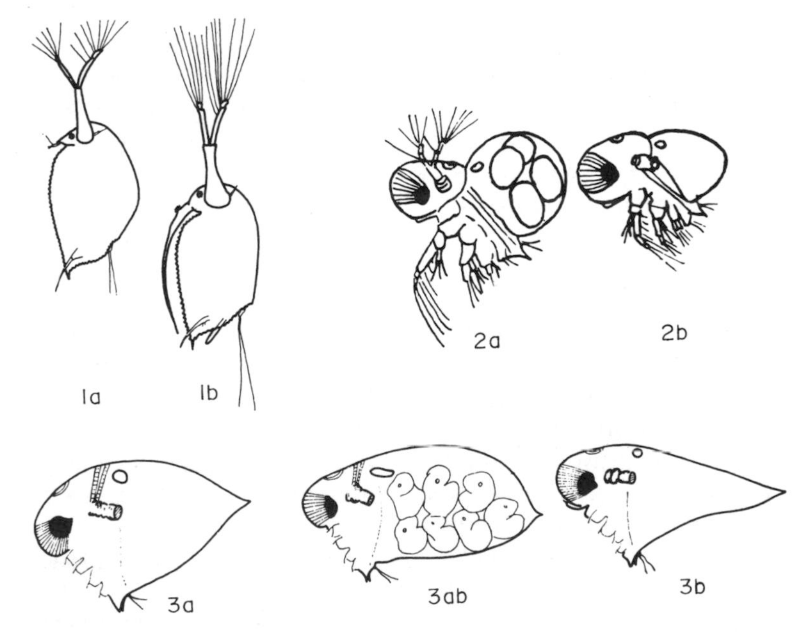

Fig. 2.57. Cladocera. (1) *Penilia avirostris*; (2) *Podon polyphemoides*, a = ♀; b = ♂; (3) *Evadne nordmanni* a = ♀, b = ♂, ab = ♀ with brood pouch (Della Croce, 1974)

Of the several species of *Podon*, *P. leuckarti* is distributed mostly north of 40°N and is found in coastal waters of the north-western and the north-eastern Atlantic, including the North Sea, Baltic and English Channel, and fairly abundantly, though close inshore, off Iceland and Greenland and in the Norwegian Sea, White Sea, Barents Sea, Kara and east Siberian Seas. It has been recorded at latitude 70°N off Greenland. In the Pacific Ocean the species has been found in inshore areas of the Bering Sea and the Sea of Japan and apparently also from Monterey Bay. Della Croce (1974) also notes its occurrence in the Black Sea. In the southern hemisphere it has been found on the western side of the Atlantic between 50° and 55°S.

P. leuckarti seems to be very much a neritic species. Gieskes (1971) points out that although it is recorded by Cheng off Plymouth, in the Clyde Sea area and in the Irish Sea, populations are small west of the British Isles. According to Gieskes the species does not occur where there is any inflow of oceanic water as in the north-western North Sea. In the southern North Sea off the Dutch coast *P. leuckarti* appears when the temperature reaches about 6.5°C; maximum populations tend to develop at temperatures between 10° and 13°C. A further rise in temperature, however, is associated with the development of the bisexual generation and the rapid disappearance of the species. In the southern North Sea its time of appearance is, therefore, mainly between April and June.

At temperatures of about 13°C, *P. leuckarti* tends to be succeeded by the related species *Podon intermedius* (Gieskes, 1971). This species seems to have a very slightly warmer distribution than *P. leuckarti*, and in the North Sea and north-eastern Atlantic area flourishes at about 13°–16°C. *P. intermedius* appears to be present only in the northern hemisphere. In the north-west Atlantic, it is found mainly in neritic waters in comparatively small numbers off the Newfoundland–Nova Scotia shelf. In the eastern North Atlantic Della Croce (1974) gives its distribution as the English Channel, North Sea, Baltic and Norwegian Seas as well as in seas off Ireland and Iceland. It is also recorded from the Mediterranean and Black Seas. Although a coastal species, it can spread out in the eastern North Atlantic to greater distances from the coast than in the western Atlantic. As a slightly warmer species, it is largely absent from the colder Icelandic coasts and the Norwegian Sea (Gieskes, 1971).

Podon polyphemoides appears to be somewhat similar in its distribution to *P. leuckarti* in that it is found mainly north of 40° latitude in the northern hemisphere, both in the north-western and north-eastern Atlantic. Apart from being distributed in the English Channel, North Sea, Baltic and Norwegian Sea, it is found in the Mediterranean and Black Seas, and occurs on both sides of the North Pacific. In the southern hemisphere it occurs in the east and west Atlantic between about 20° and 40°S, as well as being known off South Africa and New Zealand. While, like other cladocerans, it is mainly a coastal species, *Podon polyphemoides* is typically found in brackish waters. Gieskes quotes its salinity limits as between 5 and 30‰. It is the only cladoceran cited as truly estuarine by Jeffries (1967).

The distribution of *Evadne nordmanni*, a moderately cold-loving species, somewhat resembles that of *Podon leuckarti*. It is found mainly north of 40° in the northern hemisphere in both the Pacific and Atlantic. In the North Pacific it extends south as far as latitude 23°N. Plankton Recorder collections show that the species was plentiful in continental shelf waters, especially in the North Sea, and in more open waters of the eastern North Atlantic. Few were found in the Norwegian Sea, and the species

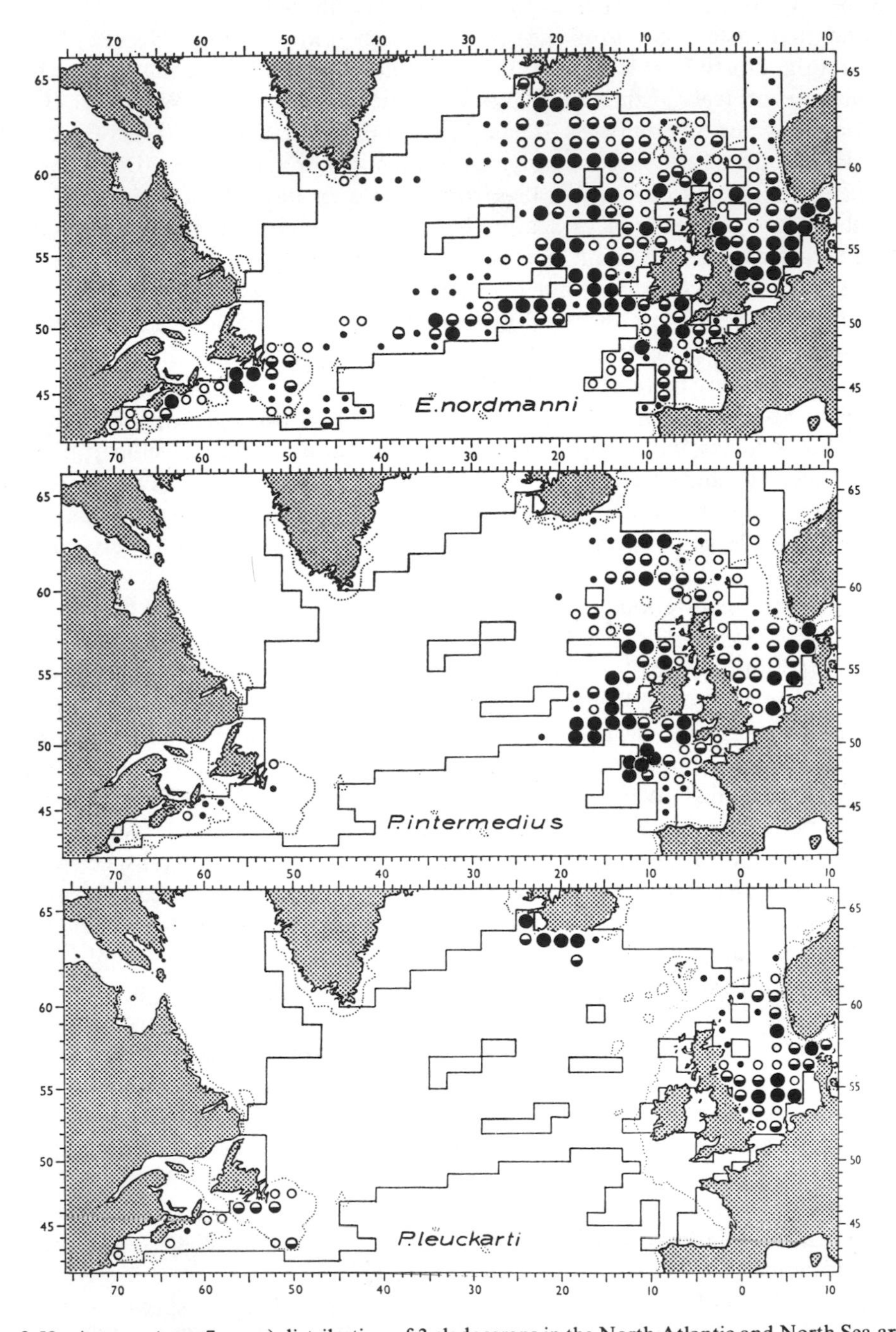

Fig. 2.58. Average (over 7 years) distribution of 3 cladocerans in the North Atlantic and North Sea areas. Numbers per sample (equivalent to $3\,m^3$ water filtered).
for *Evadne nordmanni*: ● = 20; ◒ = 15; ○ = 7; • = 2
Podon intermedius: ● = 20; ◒ = 5; ○ = 2.1; • = 1
P. leuckarti ● = 11; ◒ = 3; ○ = 1; • = 0.2 respectively (Gieskes 1971).

apparently did not occur in the cold waters of the Labrador and Irminger Current. It was abundant in the western North Atlantic in coastal waters off Newfoundland and Nova Scotia, but was more obviously confined to the coast (cf. Fig. 2.58). It occurs in the Mediterranean and Black Sea (Della Croce, 1974) and widely in the southern hemisphere, in the east and west Atlantic northwards to a latitude of about 18°S, and off south-east Australia and in the south-west Pacific. Like *P. leuckarti*, its first appearance in the southern North Sea is at a temperature of 6°–7°C and it is most abundant at a temperature of about 13°C. It therefore appears as early as March in the English Channel, usually about April or May in different areas of the North Sea and disappears by about October from British waters.

Evadne spinifera is a warmer-water cladoceran. It was found by Moraitou-Apostolopoulou and Kiortsis (1973, 1977) as the most abundant cladoceran in the Aegean area of the Mediterranean, and though most plentiful over the warmer months of the year it was absent from Greek waters only in one month (February). *Penilia* and *E. tergestina* occurred with *E. spinifera* over summer as warm-water species. *Podon intermedius*, found over summer in higher latitudes, was found only during winter in the warm Aegean. Further details of the seasonal cycles of cladocerans in the Black Sea and Mediterranean are included in Chapter 3. Della Croce (1974) gives the extreme distribution of *Evadne spinifera* in all seas as between 60°N and 40°S. As a warm-temperate species it occurred only in the southern part of the seas of north-west Europe (Gieskes, 1971), but it was found in the north-east North Sea and in the Skagerrak, possibly in relation to somewhat warmer inflowing water. Its appearance was limited to the three warmest months, July to September. The temperature range in northwest European waters approximated to 14°–18°C. *E. spinifera* is characteristic of more open waters. Gieskes' records were mostly beyond the 200 m contour and the species appears to avoid coastal waters of low salinity.

Evadne tergestina is a species found in tropical and warm-temperate waters of all oceans in both coastal and open seas. According to Della Croce (1974) its normal limits are between about 45°N and 35°S. Furnestin (1970), reviewing the cladocerans of the north-west coast of Africa, does not mention *E. Tergestina*, though *E. spinifera*, *E. nordmanni*, *Podon intermedius* and *P. polyphemoides* are all listed and could be abundant at certain seasons (cf. Chapter 8). Further south, along the west African coast, however, the species is well known. Bainbridge (1972) records it with *Penilia* off Lagos. Neto and Paiva (1966) found three cladocerans in coastal waters off Angola, *Penilia*, *Evadne spinifera* and *E. tergestina*, at times reaching high densities. *E. tergestina* was more abundant than *E. spinifera*. Thiriot (1977) makes several references to its occurrence, usually accompanied by *Penilia*, along the west African coast (Pointe Noire, off Dakar), but farther south the cladocerans are represented by *Podon polyphemoides* and *Evadne nordmanni*. *E. tergestina* similary occurred with *Penilia* off Nosy Bé, Madagascar (Frontier, 1974). Amongst many other tropical areas, it is also important off Brazil and Venezuela (Calef and Grice, 1967; Rao and Urosa, 1974).

The breeding of cladocera

Though strictly the breeding cycles of cladocerans should be part of the review of reproductive patterns of the meroplankton, it is convenient to include an account of some of the species with breeding of the holoplankton (Chapter 3). Certain features of the reproduction of Cladocera, however, now need discussion.

The life cycle of marine cladocerans is believed to include the production of comparatively thick-walled resting eggs (Fig. 2.59), which tide the species over an unfavourable period, as during winter in temperate climates. Hatching of the resting eggs produces females which are parthenogenetic and a series of parthenogenetic cycles occur, with a consequent rapid increase in population. A factor in the very rapid production of parthenogenetic females is that some females of *Evadne nordmanni* have been observed to be carrying advanced embryos which already had eggs in their embryonic brood pouches. Onbe (1977) has recently confirmed this observation for four species of *Podon* and *Evadne*.

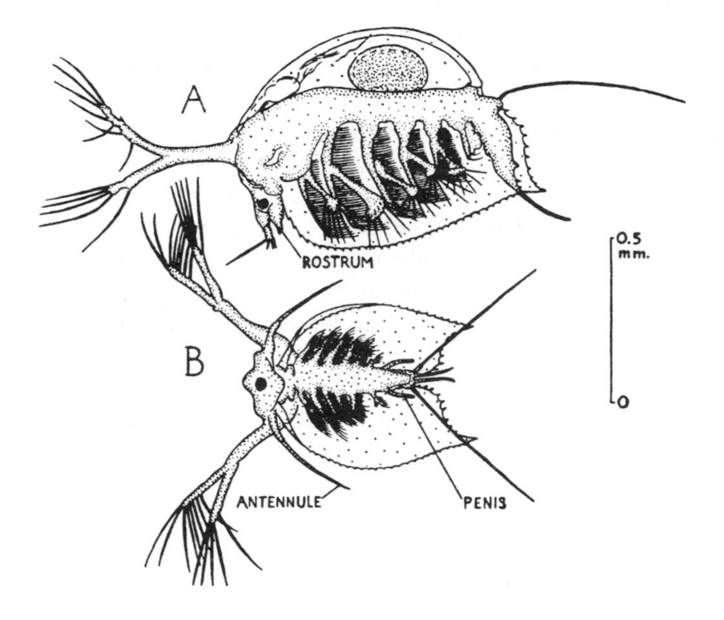

Fig. 2.59. *Penilia avirostris*. A = mature ♀ with resting egg in brood sac; B = male (Lochhead, 1954).

It is widely recognized that for various species of Cladocera, as the population peaks the fertility appears to decline. Cheng (1947), for example, found a marked reduction in fertility of *E. nordmanni* in the Clyde Sea area after July, corresponding to the time when the proportion of sexual individuals was increasing. He demonstrated an inverse relationship between the reproductive capacity of parthenogenetic females and the occurrence of bisexual individuals. With *Podon leuckarti*, Gieskes (1971) pointed out that in the North Sea, as temperatures rose beyond 13°C and the parthenogenetic breeding was less active, a bisexual generation developed. The time of appearance of *P. leuckarti* also seemed to vary between the eastern and western Atlantic, presumably in relation to temperature differences. Even in the North Sea the species reached its maximum and subsequently disappeared about a month earlier in the eastern region than in the somewhat cooler western region. A similar pattern was observed by Gieskes (1971) for *Podon intermedius*. This tendency for the appearance of a species to be later in the season and for the season of abundance to be shortened at higher latitudes, which has been noted for other zooplankton (cf. Robinson, 1965, Volume 1), appears to be particularly clear for *Evadne nordmanni*. The cladoceran appears earliest in shelf waters, even in March in the English Channel, mainly about June or July in the North Sea, but as late as September off south-west Greenland. As with other cladocerans, there is a strong tendency for the percentage of sexual individuals to increase with the passage of the season (cf. Fig. 2.60).

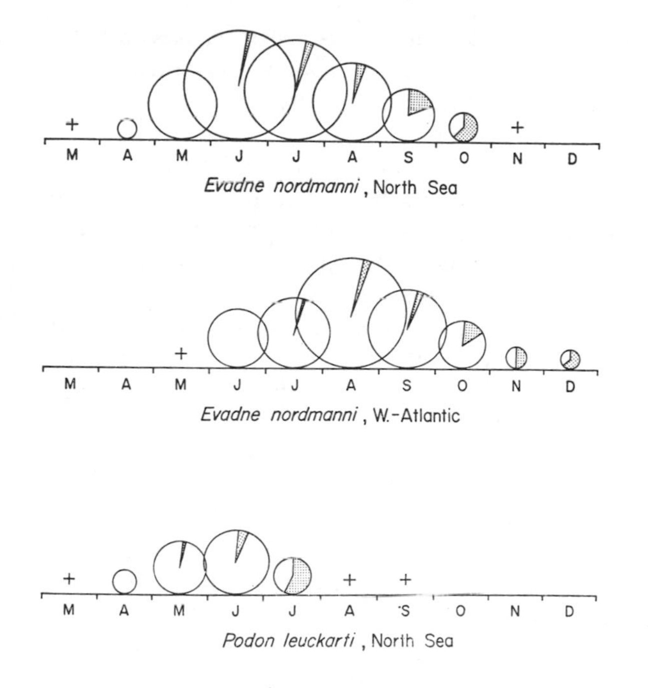

Fig. 2.60. Seasonal distribution of *Evadne nordmanni* and *Podon leuckarti* in two areas (mean numbers per 3 m^3 sample). Largest circle represents 100, smallest, one specimen. + indicates few individuals present; dotted segments indicate percentage of sexual individuals (Gieskes, 1971).

It was generally believed, but apparently without proof, that following sexual reproduction in cladocerans the resting eggs produced must sink to the bottom. Onbe (1977) found abundant resting eggs of all the cladoceran species occurring in the Inland Sea of Japan in the bottom mud. He also observed their hatching. A maximum density was recorded of no fewer than 122,000 resting eggs/m^2 of *Penilia* alone on the sea floor during September. The density of resting eggs in all four cladoceran species was highest just before the population disappeared from the plankton. Thereafter the density of resting eggs in the mud declined slowly. On hatching in the following year, a new brood of parthenogenetic females was produced, giving rise to the burst of intensive parthenogenetic reproduction. The whole breeding cycle is now clear.

For several cladoceran species, fertility, as judged by the number of advanced embryos in the brood pouch of parthenogenetic females, has been observed to change with season. For instance, in *Evadne nordmanni* the number of embryos usually ranged from about 14–16 during the period that the peak population was being established (Gieskes, 1971). After the peak had been reached there was a very marked fall, the number of embryos in the brood pouch ranging usually from one to three. Some details are given in Chapter 3 of seasonal cycles and the effect on brood size in *Penilia*. One of the interesting features in the breeding pattern of geographically widespread cladocerans is the variation in reproductive potential in different regions.

Copepoda

Of all the marine zooplankton this class of crustaceans is probably the most familiar. With few exceptions copepods are the dominant constituent of the plankton in every sea area, usually comprising at least 70 % of the plankton fauna. They must rank as the world's most abundant metazoans. A somewhat more complete description of the Copepoda than of the other classes is therefore justified.

The head and first thoracic segment in Copepoda are fused to form a cephalosome lacking carapace and compound eyes. The paired head appendages consist of uniramous first antennae, biramous second antennae (except in cyclopoids which lack the exopodite), mandibles with palps and two pairs of maxillae; typically the maxillae are more or less setose forming a filtering mechanism for a current produced mainly by the second antennae, assisted by the mandibular palps. Included in the mouthparts are the appendages of the first fused thoracic segment which are modified as maxillipeds.

Following the cephalosome are typically five free thoracic segments, though some may be fused together so that only four or three segments are distinguishable. This region is usually termed the metasome. Each of the five segments is usually provided with a pair of biramous swimming legs, each pair of appendages coupled at the base. With few exceptions the last (fifth) pair of legs tends to be reduced. In cyclopoids they are usually similar in males and females; in calanoids they are dissimilar in the two sexes, the appendages being modified as copulatory organs in the male. There are up to five abdominal segments, the first bearing the genital openings and the last bearing the caudal furca, but the abdomen lacks appendages (Fig. 2.61).

The development normally includes a series of twelve stages of which the first six are nauplius, referred to as N I to N VI. Very rarely among calanoid Copepoda are there fewer nauplius stages, but *Chiridius armatus* has only four (cf. Lovegrove, 1956; Matthews, 1964). The following six instars are copepodite stages, referred to as C I to C VI, sexual maturity being restricted to the last stage (C VI), which cannot moult further.

The Copepoda have been divided by taxonomists in several ways into major orders or sub-orders. A number of orders are non-planktonic and mostly parasitic, at least in some stage in the life history. Though a few of these essentially parasitic copepods, such as Monstrilloida and *Caligus*, may appear briefly in the marine plankton, there are only three important planktonic groups, classified here as orders: Calanoida, Cyclopoida and Harpacticoida. The latter two orders are often grouped as the Podoplea. They have in common a characteristic division of the main body regions. A major articulation in the trunk occurs immediately *anterior* to the last free thoracic segment, so that the last body region (urosome) consists of one segment, bearing a usually rudimentary pair of swimming legs, the genital segment, and the abdominal segments ending in the furca. In the Harpacticoida, however, the width of body segments changes relatively little between thorax and abdomen, so that the body appears generally elongate and the trunk division is much less obvious than in the Cyclopoida,

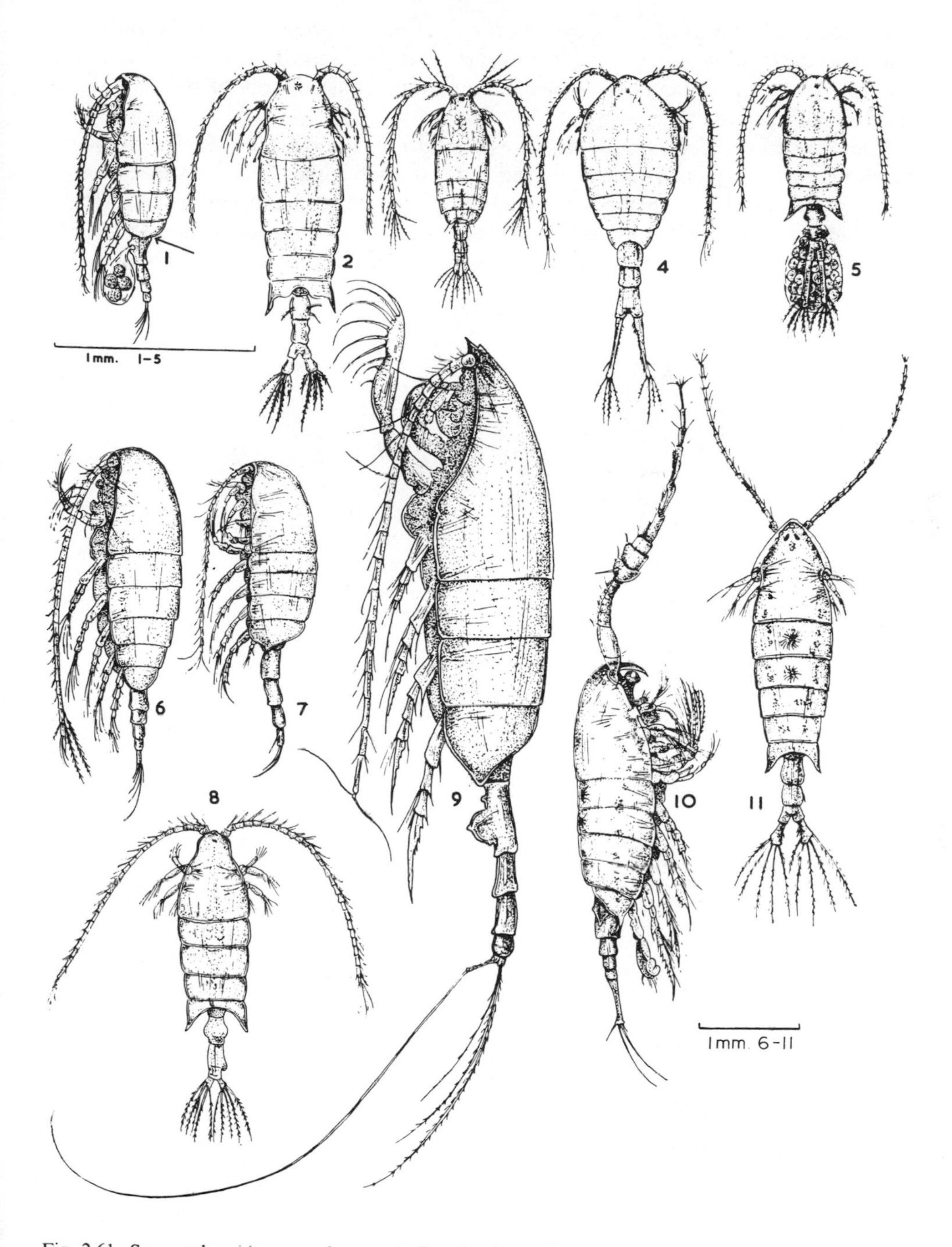

Fig. 2.61. Some calanoid copepod types 1, *Pseudocalanus elongatus*; 2, *Centropages typicus*; 3, *Acartia longiremis*; 4, *Temora longicornis*; 5, *Eurytemora hirundoides*; 6, *Calanus finmarchicus*; 7, *Metridia lucens*; 8, *Candacia armata*; 9, *Pareuchaeta norvegica*; 10, *Anomalocera patersoni*, male. II, female (modified from Fraser, 1962).

where the anterior part of the body is more or less pear-shaped. In the Calanoida, also known as the Gymnoplea, the major trunk articulation occurs immediately *posterior* to the last free thoracic segment, so that the urosome consists only of the genital and abdominal segments (Figs. 2.62, 2.63).

There are other less obvious differences between the main groups. In Calanoida the first antenna has typically twenty-five segments, though some fusion may take place in some species; rarely are there fewer than twenty distinct segments and in a number of calanoid genera one of the first antennae (nearly always the right one) is prehensile in the male. In Cyclopoida the segments of the first antenna are fewer in number, usually

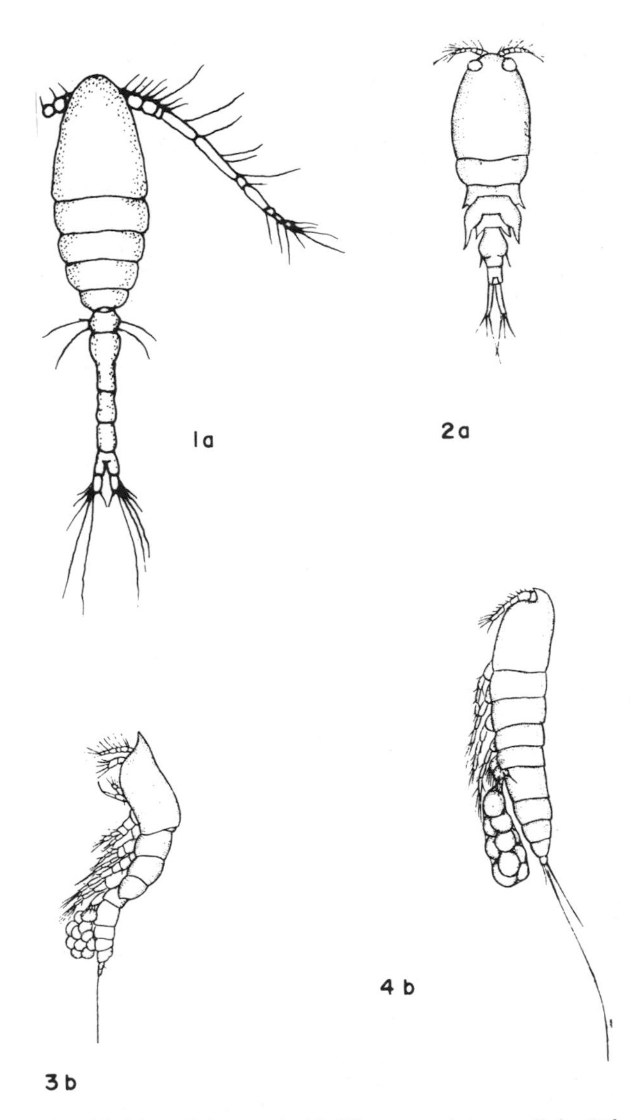

Fig. 2.62. Some cyclopoid (a) and harpacticoid (b) copepod types (1a) *Oithona similis* ♀ × 125; (2a) *Corycaeus anglicus* × 40; (3b) *Euterpina actifrons* × 50; (4b) *Microsetella norvegica* (Johnstone, Scott and Chadwick, 1924).

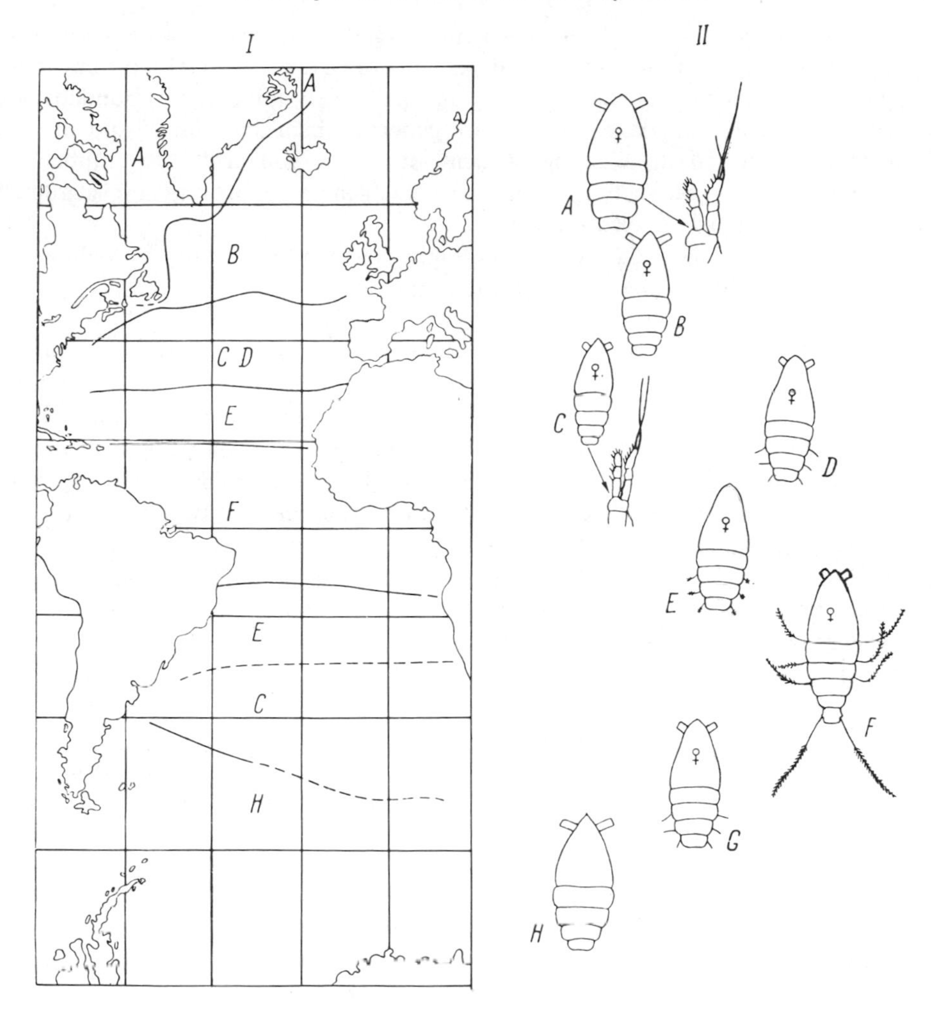

Fig. 2.63. Distribution and size variability of infraspecific forms of *Oithona atlantica* (A, B, C, H) with 4th pair of swimming feet of large and small forms; and *O. plumifera* (D, E, F, G) indicating changes in hairiness and length of setae on some appendages (Shuvalov, 1975).

eight to eighteen; the males of some genera have prehensile antennae but both of the pair are modified. In Harpacticoida the first antennae are usually short, normally three to nine segments, and both are prehensile in the male. The genital segment in male Calanoida has only one opening, usually on the left side; in Podoplea, by contrast, the male genital openings are paired and symmetrical. In the Podoplea also, the exopodite of antenna II is reduced (it is lacking in cyclopoids).

Although the three orders have each well over 1000 species, there are very marked differences in their importance in the plankton. Cyclopoida and Harpacticoida have far fewer marine planktonic species. The Cyclopoida are largely benthic copepods found among algae and in bottom sands and also in association with other animals; a number of species are parasitic or semi-parasitic. Freshwater forms are well recognized, *Cyclops* and closely related genera being well known freshwater planktonic animals. Amongst the marine representatives there are a few—perhaps a dozen—truly holoplanktonic genera, but they include some extremely important, abundant and widespread genera. The Harpacticoida are generally benthic animals and may be extraordinarily abundant as a constituent of the meiobenthos; a number are freshwater, but the majority are marine. Six genera, however, are genuinely holoplanktonic; many are tychopelagic.

In contrast, the Calanoida are outstandingly holoplanktonic, and though a very few genera (e.g. *Diaptomus*) are freshwater, the vast majority are marine, several genera contributing to the mass species of the world's oceans. Generally calanoids greatly outclass the cyclopoids in species and in numbers, but small cyclopoids can be exceedingly plentiful and diverse in warmer seas. Of 193 species of copepods identified by Farran (1936) from the Great Barrier Reef, 127 were calanoids, sixty cyclopoids and six harpacticoids. Deevey and Brooks (1977) listed 326 species of copepods from the warm waters off Bermuda; 84% were calanoids, 13% cyclopoids and 3% harpacticoids. Moore and Sander (1977) found 141 copepod species off Barbados, of which 66% were calanoids, 27% cyclopoids and 7% harpacticoids. Seven of the cyclopoid copepods and one harpacticoid, *Macrosetella gracilis*, were among the major species as regards numerical abundance. Michel and Foyo's (1976) studies in the Caribbean also demonstrated that of twenty of the most abundant copepods, two cyclopoids (*Oithona plumifera* and *Farranula carinata*) were most plentiful, with the harpacticoid *Microsetella rosea* next in abundance. Although the number of calanoid species greatly outnumbered cyclopoid species, the abundance of individuals was only slightly greater—49% as against 45%. Michel and Foyo suggest that the importance of cyclopoids, and perhaps harpacticoids, has been underestimated in the past, especially in tropical seas.

Cyclopoida

The total number of planktonic species (about seventy) represents approximately 5% of cyclopoids. Of the more important species, *Oithona similis* has been described by Bigelow (1926) as the most abundant and ubiquitous copepod in the world. Wilson (1942) found it the most abundant and widely distributed copepod in the "Carnegie" collections and Rose (1929, 1933) also speaks of its very wide distribution. Shuvalov (1975) comments that *O. similis* has one of the largest ranges of all planktonic organisms, its distribution including the North Atlantic, North Pacific, Arctic, Mediterranean, Black Sea, Red Sea and tropical oceanic regions, with a wide range in the southern hemisphere (Atlantic, Indian and Pacific Oceans) and in the Antarctic. Though the species occurs in temperate and tropical latitudes, it is especially abundant at high latitudes (cf. Digby, 1953; Grainger, 1959, 1965; Hansen, 1960; Grice, 1962; Pavshtiks, 1975). Zenkevitch (1963) comments on its particular abundance in the north-west Pacific, especially the Sea of Japan. Deevey (1960) refers to its common occurrence in Delaware Bay and the Gulf of Maine; in Block Island Sound it was one of the commonest copepods. She suggests a breeding temperature range of $-1°$ to $+20°C$, and points out that it is one of the copepods occurring all the year round in the Arctic.

Possibly, as Shuvalov's investigations indicate, *O. similis* is represented by a number of sub-species each having a more restricted geographical range. *Oithona* (*Oithonina*) *nana*, another widespread species, is perhaps more characteristic of coastal waters. Wilson (1932) says that it is less common off the north-east coast of America and more typically southern in distribution; he found it widespread over the Pacific and Atlantic (Wilson, 1942). It occurs abundantly also in the Black Sea. *Oithona plumifera* is found widely in all three major oceans, mainly in warmer waters, but ranging into temperate regions. In the eastern North Atlantic it occurs around the British Isles and the North Sea, even penetrating into colder sub-Arctic areas. It is reported from the Cape of Good Hope as well as from the Mediterranean and the Red Sea. However, Wilson states that it is chiefly tropical in its distribution.

There has been considerable argument as to the distinction between *O. plumifera* and *O. atlantica*, but Shuvalov (1975) demonstrates that the two are valid well-defined species without intermediates in the distinguishing morphological features and that each has its own geographical range. *O. atlantica* is a boreal form inhabiting the colder and temperate regions of the North Atlantic, the Mediterranean and adjacent areas, with a smaller variety in the southern oceans. *O. plumifera* is found in all tropical oceans, though currents may transport it to areas normally inhabited by *O. atlantica* (cf. Fig. 2.63). Farran (1949) lists *O. plumifera* as the second most abundant copepod in the whole inshore plankton from the Great Barrier Reef, and in Michel and Foyo's (1976) collections from the Caribbean it was the most numerous copepod.

Some *Oithona* species are more or less restricted to warmer regions, for example, *O. hebes* (Wilson, 1942) and *O. brevicornis*, which appears to occur more inshore in salt lagoons and tidal pools (cf. Wilson, 1932; Deevey, 1960).

Another cyclopoid genus, *Oncaea*, is similarly plentiful in most of the world's oceans. Some species (*O. venusta*, *O. minuta*, *O. media*) are found in all the oceans in warm and temperate areas, but avoid really cold waters (cf. Wilson, 1942). Deevey (1960) regards the two latter species as southern forms off the north-east United States coast. Other species of *Oncaea* range even more widely, *O. notopa* and *O. conifera* extending into Antarctic and Arctic Seas as well as occurring in temperate and tropical latitudes of all three major oceans (cf. Rose, 1933; Wilson, 1942), though Sewell (1947) throws some doubt on the Arctic distribution. Ostvedt (1955) found *O. conifera* extending through the whole water column to 2000 m in the cold Norwegian Sea. Some species (e.g. *O. curta*, *O. tenella*) are apparently found only in warm waters. On the Great Barrier Reef the wide ranging species *O. venusta* and *O. conifera* were both recorded, but the commonest and indeed the third most abundant copepod for the Inner Reef area was *O. clevei*, a species typical of warm seas (cf. Sewell, 1947).

By contrast, *Oncaea borealis*, found in the Arctic Sea and extending south along the Norwegian and British coasts as well as being recorded in the Gulf of Maine, appears to be a genuine cold water species which avoids warmer waters (cf. Wilson, 1932). Zenkevitch (1963) records it in the northern part of the Sea of Japan. Grice (1962) found it in the Polar Sea and gives its southern limit as south Labrador on the western side of the Atlantic. Sewell (1947) suggests that this is the only true Arctic species of *Oncaea*.

Three other genera of cyclopoids, *Corycaeus*, *Sapphirina* and *Copilia*, are important and characteristic constituents of the copepod fauna of warm-water plankton. *Corycaeus* includes one species, *C. anglicus*, which inhabits colder northern temperate waters, being well known in the North Sea and North Atlantic. Sars (1918) records it as an occasional visitor to the west coast of Norway, regarding it as a North Atlantic oceanic form. Colebrook, John and Brown (1961) consider that *Corycaeus anglicus* might be looked upon as a neritic species in the North Sea, but populations also exist in the more oceanic North Atlantic and the two groups may be ecologically distinct, though morphologically inseparable. Rose (1929, 1933) lists the Atlantic coast off north-west Europe, including round Great Britain, as typical for *C. anglicus* but does not include the species for the American side of the North Atlantic. Wilson (1942), however, recorded *C. anglicus* from a few localities in the western Atlantic and Caribbean and in certain areas of the central Pacific. From analysis of the "Carnegie" collections he regards *C. crassiusculus* as the most widely distributed and abundant species, found everywhere in the Pacific and Atlantic Oceans except at far northern and southern stations.

Most species of the genus *Corycaeus* (sometimes divided into separate genera) are, however, typical of tropical and warm-temperate seas. Wilson mentions that *Farranula* (= *Corycaeus*) (= *Corycella*), including *F. carinata*, *F. curta* and *F. rostrata*, is essentially tropical, and though of very small size, can occur in high density in warm surface waters. Michel and Foyo (1976) record *F. carinata* as the second most numerous species in the Caribbean. *Corycaeus speciosus*, *C. latus*, *C. flaccus* and *C. furcifer* are listed by Rose as occurring in all three tropical oceans but spreading to temperate areas. The last species, *C. furcifer*, is recorded as far as the Antarctic by Farran (1929). Other species (*C. ovalis*, *C. agilis*, *C. carinatus*) are perhaps more restricted to tropical waters, but almost all the species are essentially inhabitants of warmer seas. Although Wilson (1932) lists *C. speciosus* and *C. ovalis* as occurring as far north in the Atlantic as 45–50°N, he regards these as stragglers transported by the Gulf Stream (cf. Deevey, 1960). For the Great Barrier Reef collections Farran (1949) quotes a large number of *Corycaeus* species, and further details of their distribution, especially in the Indian Ocean, are given by Sewell (1947).

While the numerical importance of certain species of *Oithona* and *Oncaea*, at higher northern latitudes, has already been noted, they may also be important at extreme southern latitudes. Vervoort (1951) refers to the abundance of *Oithona similis* in Antarctic waters and to the frequent occurrence of *O. frigida*, an Antarctic form, which may also be found somewhat north of the Antarctic Convergence. Vervoort also notes that *Oncaea conifera*, the widespread species somewhat similar to *Oithona similis* in its distribution, is of general occurrence in Antarctic waters. However, species of these small cyclopoids may also be of great numerical significance in the zooplankton of warm oceans. For example, Deevey (1971) records the genera *Oithona* and *Oncaea* with the calanoid *Clausocalanus* (*vide infra*) as the three most numerically important copepods in the Sargasso Sea near Bermuda, totalling about 47% of the total copepod fauna. Deevey and Brooks (1971, 1977), in further analyses, found, for zooplankton off Bermuda, that the average density of cyclopoids (captured with a No. 8 net) over the upper 2000 m was approximately one-third of the total copepod population, numbers being distinctly higher in the upper 500 m.

The maximum species diversity for cyclopoids also appears to be in the upper layers of tropical and sub-tropical waters, the number of species decreasing with depth. Cyclopoids, however, were found throughout the water column. Species of *Oithona* included *O. plumifera* and *O. setigera*; species of *Oncaea* included *O. conifera*, *O. media*, *O. mediterranea* and *O. venusta*, with *O. ornata* as a deeper-living species, but the authors state that some of the very small cyclopoids were not identified. While *Oithona* was clearly more important than *Oncaea* in the upper waters, the latter became much more abundant at depth, especially from 1500–2000 m. Many of the species occurred throughout the whole water column. The total numbers of *Oithona* and of *Oncaea* greatly exceeded that of any other copepod genus in Bermuda waters. The authors speak of the tiny cyclopoids *Oithona* and *Oncaea*, with *Conaea gracilis* in deeper waters, as being "ubiquitous throughout the water column below 500 m". But they were abundant in the upper waters with other epiplanktonic cyclopoids—*Corycaeus*, represented by eight species apparently found all the year; *Farranula* (three species); *Sapphirina* (six species) and *Copilia* (four species).

Farran (1949) also recorded four species of *Corycaeus* amongst the fourteen most abundant copepods from the Great Barrier Reef. In waters off Barbados, Moore and Sander (1977) found that *Farranula gracilis* was numerically the most important species and that six cyclopoids, including several listed by Deevey off Bermuda, ranked among the first twelve major copepods.

The various species of *Sapphirina* and *Copilia* are all warm-water inhabitants. Sewell (1948) records the worldwide distribution of *Sapphirina* in the warm oceans, although stragglers may be carried by warm currents such as the Gulf Stream and Kuroshio, sometimes even to fairly high latitudes. Wilson's reports of *Sapphirina* off Woods Hole suggest that the copepod is tropical but may be spread by warmer currents. As with so many other zooplankton taxa, many of the species occur in all three warm oceans.

The body form of *Sapphirina* is unusual, being dorso-ventrally flattened and extended laterally into almost foliaceous projections, especially in the male. Division between the metasome and urosome is less obvious than is usual in cyclopoids, and the furca is leaf-like. The urosome of the male is very broad and there is a fairly pronounced sexual dimorphism, with the males being markedly iridescent. Most species are associated with pelagic tunicates (salps, etc.), though they may occur free in the water, and the precise relationship with the host is sometimes not clear. Heron (1973), commenting upon the frequent association of some copepods and salps, drew attention to the particular association of high numbers of *Sapphirina* with dense swarms of *Thalia democratica* in south-east Australian waters. He demonstrated that *Sapphirina angusta* in its young stages could grow inside the salp, living as a parasite, but that during feeding the juveniles could cause fatal damage to their hosts, especially by puncturing the blood system. Older stages of *Sapphirina* acted as predators, rapidly consuming the tougher tissues of the salp. The female copepod must leave the salp to mate, but once fertilized she could return, and after further voracious feeding egg production can occur in a very short time.

The growth rate and reproductive cycle of the copepod is very rapid, in conformity with the very high growth rate of the salp. The male *Sapphirina* is too large to remain inside the salp for its final moult; probably adult males do not feed. Heron comments that the second antennae and maxillipeds of *Sapphirina* are modified for attachment to the host and for tearing soft tissues of the prey, compared with the seizing and grasping type of appendages typical of carnivorous cyclopoids such as *Oncaea* and *Corycaeus* (see Fig. 2.64).

In *Copilia* the male, as in *Sapphirina*, has a very flattened leaf-like, almost transparent body, whereas the female is of more usual cyclopoid shape; the sexual dimorphism is extremely pronounced. The various species are essentially tropical, though some may stray into temperate waters.

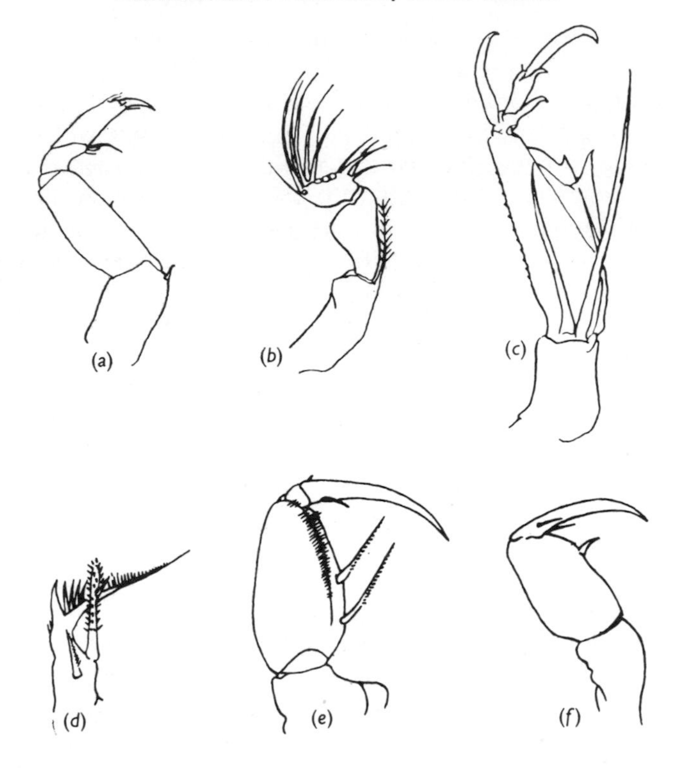

Fig. 2.64. Comparison of appendages of predatory cyclopoids second antennae: (a) *Sapphirina angusta* ♀; (b) *Oncaea venusta* ♀; (c) *Corycaeus furcifer* ♀; Maxillipeds; (d) *Sapphirina angusta* ♀; (e) *Oncaea venusta* ♀; (f) *Corycaeus furcifer* ♀ (Heron, 1973).

Harpacticoida

Among the few planktonic harpacticoid genera are certain species which may be very abundant and have a wide geographical range. *Euterpina acutifrons* is a neritic harpacticoid which can be seasonally abundant. Well known in warmer seas in all three major oceans, it spreads far into temperate waters, though it apparently avoids really cold areas (cf. Wells, 1970). Sars (1921) classes it as an occasional visitor off the Norwegian coast; it is widely known from the eastern side of the Atlantic, in the southern North Sea, the English Channel and off Ireland. Off the Atlantic coast of America, Deevey (1960) regards *Euterpina* as a southern stray copepod in the Delaware Bay region; although not recorded for the Woods Hole area, it is common south of Cape Hatteras. Deevey (1960) indicates a breeding range from 15°C to more than 30°C (cf. Deevey, 1960; Fig. 2.65a, b). *Euterpina* was found in small numbers inside the Great Barrier Reef by Farran (1936). In the plankton of Southampton Water *Euterpina* can become abundant in waters of somewhat reduced salinity in the late summer/autumn when temperatures are relatively high.

Macrosetella gracilis is an oceanic harpacticoid well known in the warmer waters of all three oceans (cf. Wilson, 1942; Wells, 1970; Michel and Foyo, 1976). To some extent it spreads into temperate waters, and it is recorded from the Antarctic (cf. Sewell, 1947). Farran records it as the most plentiful harpacticoid off the Great Barrier

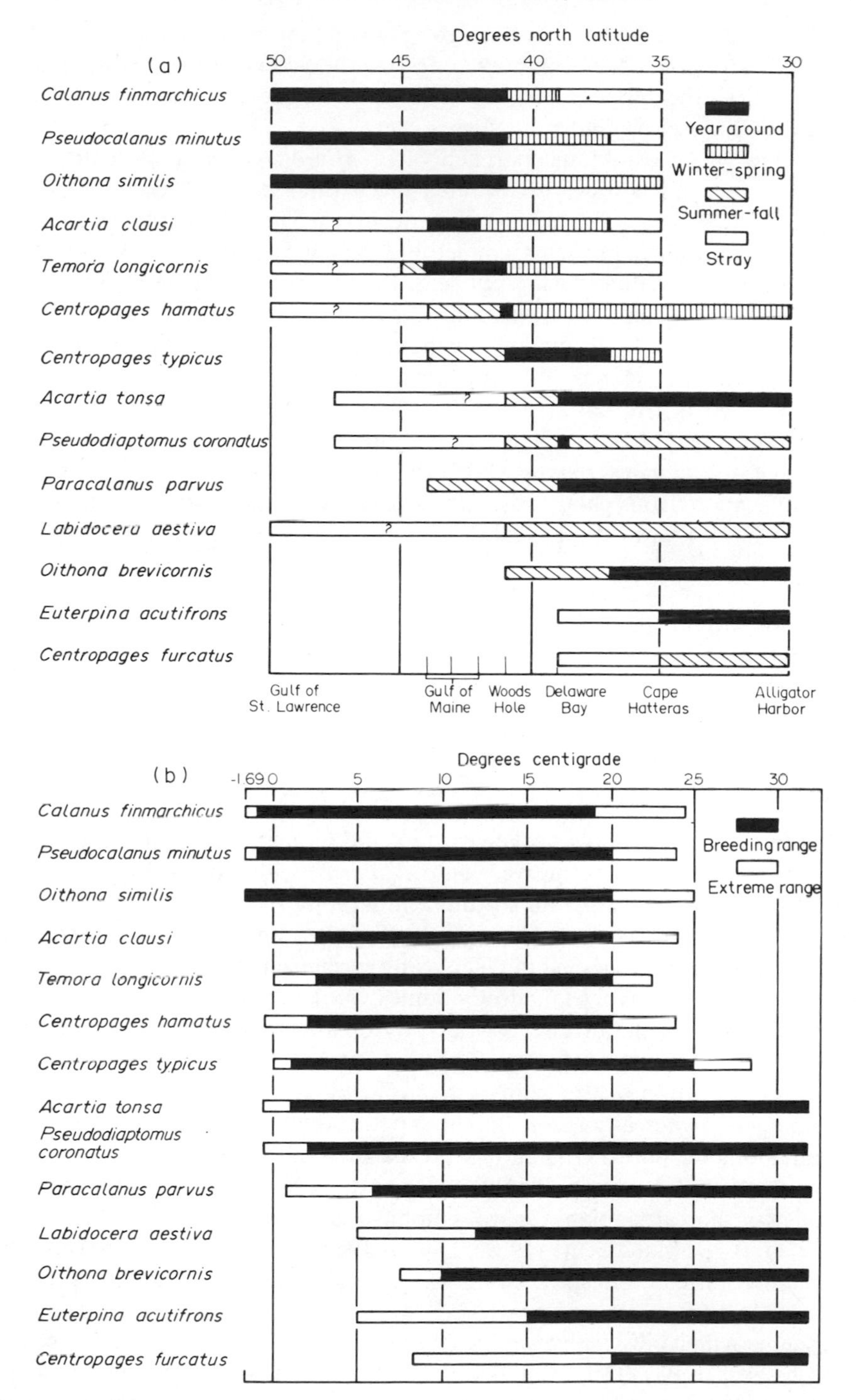

Fig. 2.65. Latitudinal and seasonal occurrence in eastern American waters of (a) certain copepods; (b) known temperature ranges of the copepods in the Delaware Bay area; the breeding ranges are shown in black (Deevey, 1950).

Reef. Calef and Grice (1966) found a strong correlation between the distribution of *Macrosetella* and *Trichodesmium* (*Oscillatoria*) in the waters off the north-east coast of South America; the filaments of the blue–green algae seemed to be used as a substratum for the crawling larval stage.

Of the few other planktonic harpacticoids, *Microsetella* is represented by two species. *M. norvegica* is widely distributed in all three oceans according to Wilson (1932). Klie's (1943) records show its wide occurrence in the Atlantic and in coastal seas such as the Baltic, English Channel and North Sea. It is also known at high latitudes including Arctic and Antarctic seas (Farran, 1929; Rose, 1933; Sewell, 1947). Its presence in the Indian Ocean was confirmed by Sewell (1947). The second species, *Microsetella rosea*, while also having a wide range in all the oceans and occurring in temperate areas even as far as 50°S, appears to be more typical of warmer waters. It is commonly taken in tropical areas (cf. Michel and Foyo, 1976). *Clytemnestra* and *Aegisthus* are other broadly-ranging planktonic harpacticoid genera. Wells (1970) suggests that *Aegisthus mucronatus* and both species of *Microsetella* have a worldwide distribution, apart from polar seas. *Miracea efferata* can be an abundant species, especially in warm seas (Moore and Sander, 1976; Deevey and Brooks, 1977).

Calanoida

Calanoida are outstandingly holoplanktonic copepods and tend to dominate the plankton in most seas. It is in the cold-temperate regions of the world, however, that apart from one or two important cyclopoids, the calanoids, especially Calanidae and Pseudocalanidae, become overwhelmingly dominant. Of these copepods, the genus *Calanus*, inhabiting the world's oceans, surpasses all others in abundance.

The genus Calanus

Using functional morphological characters and biometric characteristics, Brodsky (1975) postulates that the primitive type of *Calanus*, with "average" characteristics, originates in high and temperate latitudes, inhabiting the Arctic, North Pacific and North Atlantic boreal regions. He suggests that the number of species increases in warmer waters, and that they exhibit increasing specialization towards the tropics. The evolutionary process involves hydrodynamics, probably correlated with water viscosity, differences in the precise type of phytoplankton food, and increased competition, especially in tropical areas.

A new classification of the genus *Calanus sensu stricto* proposes ten species. Of these, *Calanus finmarchicus* is a boreal North Atlantic mass species particularly abundant along the borders of continental shelves at high latitudes (cf. Glover, 1967; Matthews, 1969; Edinburgh Plankton Records, 1973). The domination of this copepod in the boreal zooplankton was emphasized by Bigelow (1926). Wilson (1932), Fish (1936), Bigelow and Sears (1939), Digby (1954) and Grainger (1963) are among the numerous authors who speak of the vast abundance of this calanoid in the North Atlantic (cf. Chapter 3).

The southern limit of *C. finmarchicus* in the western North Atlantic would appear to be about Chesapeake Bay (Deevey, 1960). Over the eastern North Atlantic region many investigators (Somme, 1934; Ostvedt, 1955; Wiborg, 1955; Hansen, 1960; Matthews, 1969) speak of the overwhelming importance of *Calanus finmarchicus* in the boreal Atlantic region, including the Norwegian Sea and areas south of Iceland and

Greenland. It is one of the three most numerically important copepods in the Barents Sea (Zenkevitch, 1963). Marshall and Orr (1955a) remark on the widespread occurrence of *C. finmarchicus* over the North Atlantic and suggest a temperature range of −1.8° to +22°C. The distribution of *C. finmarchicus* is, however, complicated by the occurrence of other species of the genus.

Calanus hyperboreus was early recognized (cf. Sars, 1903; Bigelow, 1926) as a distinct Arctic circumpolar species occasionally found extending farther south in deeper water, for instance in deep Norwegian fjords (Somme, 1934), the deeper parts of the Norwegian Sea and an area north of Cape Cod (Conover, 1964). Brodsky gives its mean breeding temperature as −0.3°C. It is very doubtful whether this species occurs in northerly Pacific areas (cf. Grice, 1962).

Grainger (1962, 1963) and Jaschnov (1970) distinguished another relatively large *Calanus* at high latitudes as a separate Arctic species, *C. glacialis*. Brodsky describes *C. glacialis* as being confined to Arctic water with a circumpolar distribution; where Arctic and sub-Arctic waters meet there can be a mixture of *C. glacialis* and *C. finmarchicus*. Prygunkova (1968) records the species in the White Sea and Pavshtiks (1975) describes the changing proportions of *Calanus finmarchicus*, *C. hyperboreus* and *C. glacialis* in the Davis Strait. Jaschnov (1970) illustrates the characteristic Arctic distribution of *C. glacialis*, with its transport by Arctic currents along the coast of Greenland, off the north-east coasts of Japan and off Newfoundland (cf. Fig. 2.66a, b). Matthews (1966), however, found that temperature was inversely related to size in the development of *Calanus* and that some of the morphological characters used to separate *C. finmarchicus* and *C. glacialis* were temperature-dependent. Aurich (1966), studying the distribution of *Calanus* in the Irminger Sea and other North Atlantic areas, also held that intermediates existed between *C. glacialis*, *C. finmarchicus* and also *C. helgolandicus* (*vide infra*). If the view be accepted that *Calanus glacialis* is a

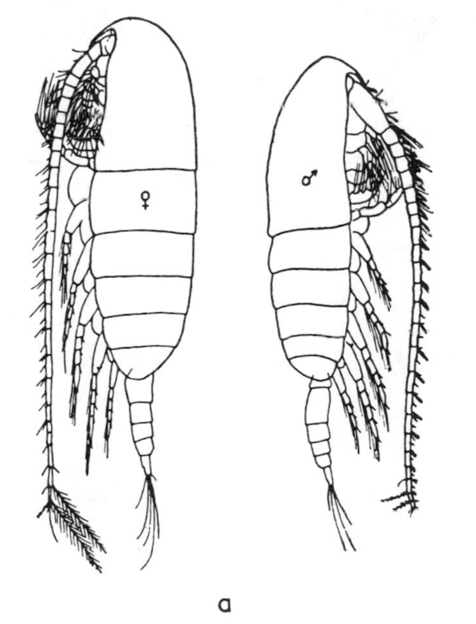

Fig. 2.66. (a) *Calanus finmarchicus* ♀ 2.7–5.4 mm; ♂ 2.4–3.6 mm (Rose, 1933).

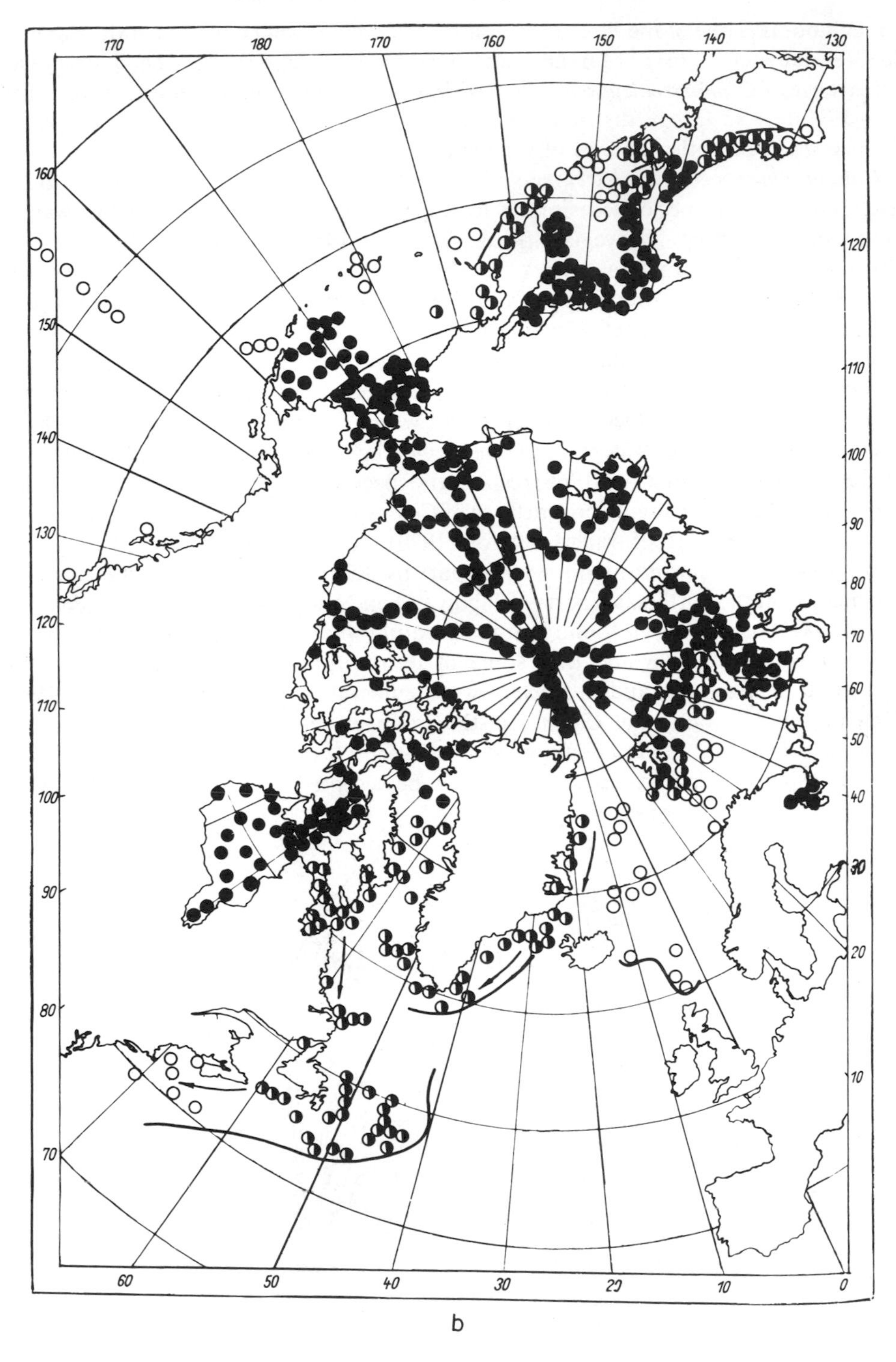

b

Fig. 2.66. (b) Distribution of *Calanus glacialis* in Arctic areas (Jaschnov, 1970).

distinct Arctic oceanic species usually inhabiting waters below 0°C, *C. finmarchicus* (*sensu stricto*) would appear to be excluded from the highest latitudes and is typical of the Atlantic boreal region (cf. Grainger, 1963; Jaschnov, 1970). Nevertheless, it is an abundant and widespread species; its breeding range, according to Brodsky, approximates to 4°–11°C.

A fourth species, *Calanus helgolandicus*, has long been the subject of debate as to its taxonomic ranking. Marshall and Orr (1955a) described it as a more southern, warmer-loving copepod than *C. finmarchicus*. Brodsky lists it as a separate species with a breeding temperature of about 12°C, its distribution covering such waters as the English Channel, southern North Sea, the area south-west of Ireland and the Mediterranean (cf. Brodsky, 1965; Matthews, 1967). Roe (1972a) found it as an abundant mid-water copepod off the Canary Islands.

The distributions of the four species are particularly well shown in the Edinburgh Plankton Records (1973) (Plates 150 to 153). Although *Calanus helgolandicus* was said not to be present at lower latitudes on the western side of the North Atlantic, it is now known to occur. Grice (1963) reported very few individuals in deeper water from the Sargasso Sea and Deevey and Brooks (1971) found it in deeper layers near Bermuda. Jaschnov (1970) described its abundance in the Mediterranean and Black Sea and its occurrence over the warmer areas of the eastern North Atlantic, probably largely due to transport from the deeper Mediterranean outflow (cf. Fig. 2.67a, b).

Matthews (1967) believes that *C. finmarchicus* and *C. helgolandicus* have sufficient morphological discontinuity and other differences to be regarded as separate species and he regards *C. glacialis* as an Arctic sub-species, though the separation of *C. finmarchicus* and *C. glacialis* into two ecological populations is clear. Brodsky (1975) also divided *Calanus finmarchicus* into two sub-species, *Calanus finmarchicus finmarchicus* and *Calanus finmarchicus glacialis*. Fleminger and Hulsemann (1977) confirmed the occurrence of *C. helgolandicus* in small numbers in shelf and slope waters of the western North Atlantic, the species tending to be in the warmer temperate waters compared with the distribution of *C. finmarchicus*. They also emphasize that an actively reproducing population of *C. helgolandicus* exists off eastern North America, the species spreading from this centre into the North Atlantic Drift and being known off the Grand Banks. Thus *C. helgolandicus*, *C. finmarchicus* and *C. glacialis* represent species typical of areas of progressively colder waters, but there is some overlap of the species distribution.

Fleminger and Hulsemann have distinguished pore patterns, which appear to be distinct, on the urosome of females of the different species. The pores may mark sexual glands, the secretions of which assist in preventing cross-mating between species. The populations of *C. helgolandicus* in the Atlantic and West Wind Drift are believed to arise from the centre in the western Atlantic rather than from the Mediterranean, which may, however, give rise to members of the species in the eastern Atlantic.

Outside the North Atlantic, several species of *Calanus* exist, some restricted to the North Pacific Ocean where, as in the Atlantic, they may be extraordinarily abundant and occur as mass species. The statement from the examination of the "Carnegie" collections that these North Pacific calanoids did not swarm as in the Atlantic probably merely reflects the time and particular areas visited.

Calanus cristatus inhabits the cold waters of the northern North Pacific, being more or less restricted to the south by the warm Kuroshio Current. Brodsky records an

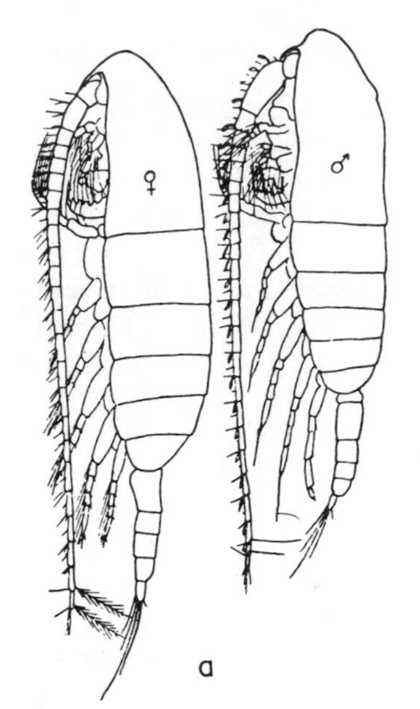

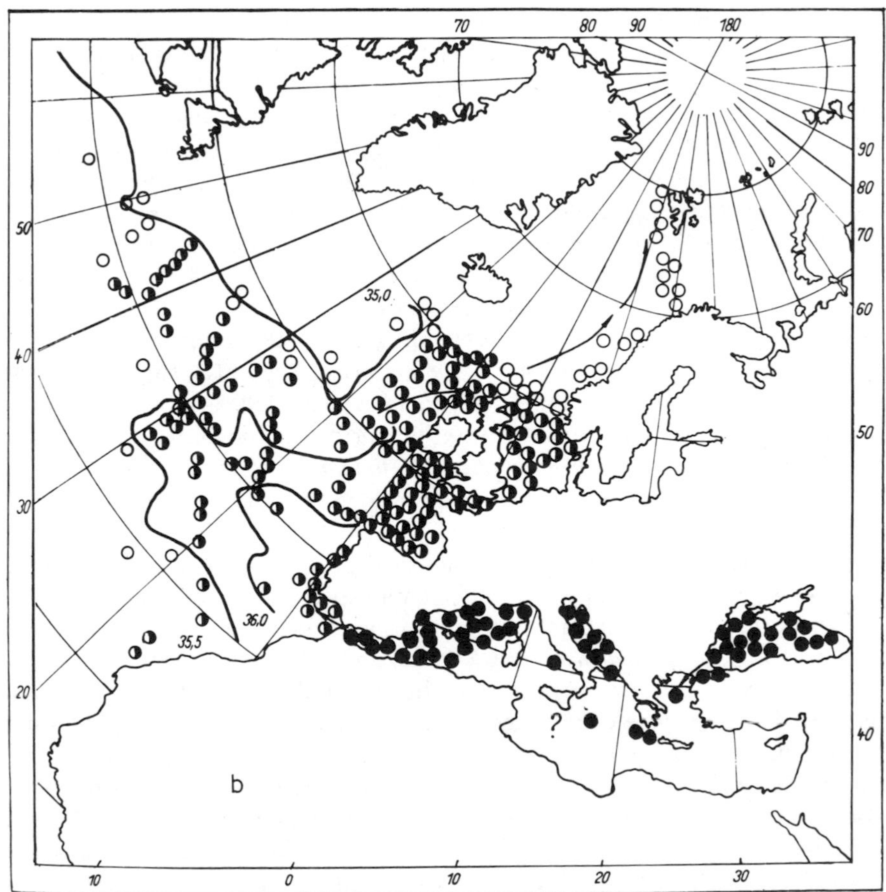

Fig. 2.67. (a) *Calanus helgolandicus* ♀ 3mm; ♂ 2.8 mm; (Rose, 1933); (b) Distribution of *C. helgolandicus* in Northern Hemisphere (Jaschnov, 1970).

average breeding temperature of 5°C. Johnson and Brinton (1963) list the species as one of the three common calanoids typical of sub-Arctic Pacific waters (roughly north of latitudes 40°–45°N; cf. Chapter 8). Omori (1967) describes this species as a typical sub-Arctic (boreal) copepod and one of the most important members of the zooplankton of the Bering Sea and North Pacific area, breeding in Japanese waters in the cold Oyashio Current, mainly in the upper 100 m. However, broods of *C. cristatus* can be transported, though in greatly diminished numbers, south in deeper waters (400–1000 m), where the colder Oyashio water spreads underneath the warm Kuroshio strata even to about 30°N latitude (cf. Fig. 2.68). The specimens drifting south, especially the C V copepodites, were considerably smaller (Omori, 1970).

The great abundance of *Calanus cristatus* in the cold open waters of the North Pacific is noted by Zenkevitch (1963). Both Morioka (1972) and Minoda (1972) describe the copepodite stages as widespread, especially in the upper 100 m; adult *C. cristatus* tend to be much deeper (>700 m), extending to considerable depths. Vinogradov (1972) points out that in the Kurile–Kamchatka area herbivorous copepods, mainly five species including *C. cristatus*, make up the great biomass in the upper 50–100 m, but they predominate even down to about 500 m and are represented in reduced densities at considerably greater depths.

Considerable interchange and vertical migration promotes the effective use of the phytoplankton food produced in the shallow euphotic zone (cf. Table 2.11). Heinrich (1962a) also points out the importance of this copepod in the North Pacific and its importance is implicit in the writings of LeBrasseur (1965), Parsons and LeBrasseur (1968) and Geynrikh (1968).

Calanus plumchrus, frequently reported as *C. tonsa*, is another of the very few calanoids with a vast geographical range in the North Pacific, more or less equivalent

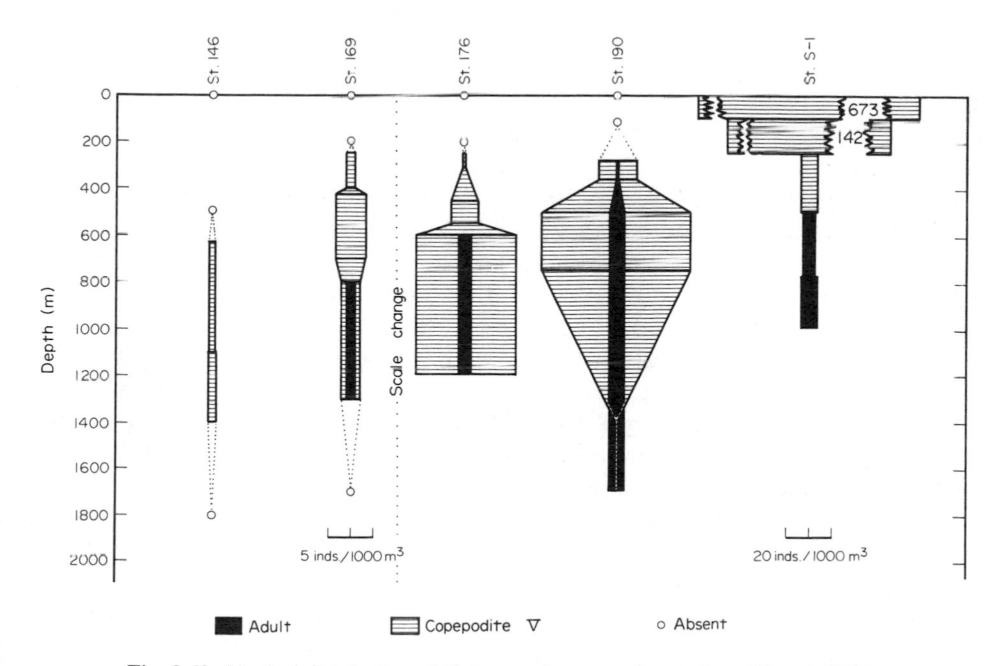

Fig. 2.68. Vertical distribution of *Calanus cristatus* at five stations (Omori, 1967).

Table 2.11. Biomass of the interzonal filter-feeding copepods (Calanus cristatus, C. plumchrus, Eucalanus bungii, Metridia okhotensis, M. pacifica) and their contribution to the total biomass of zooplankton in the Kurile–Kamchatka region. Average values for six stations (Vinogradov, 1972)

Depth (m)	Biomass (mg/m^3)	Per cent of total biomass in these hauls
0–50	555	75.5
50–100	67.1	51.0
100–200	41.1	37.6
200–500	151	62.1
500–750	41.9	42.6
750–1000	15.8	38.2
1000–1500	11.3	42.0
1500–2000	8.2	42.3
2000–2500	5.0	16.0
2500–3000	0.7	13.7
3000–4000	0.13	8.1
> 4000	0	0

to that of *C. cristatus*. It is another sub-Arctic species and whilst it is oceanic, it can be abundant closer to shore on both east and west Pacific coasts, for example, in Japanese seas (Zenkevitch, 1963) and in the Strait of Georgia (Parsons, LeBrasseur and Barraclough, 1970). Its importance together with *C. cristatus* is emphasized by LeBrasseur, Parsons and Geynrikh (*vide supra*). Morioka (1972), Minoda (1972) and Vinogradov (1972) all treat *C. plumchrus* with *C. cristatus* as very abundant, the copepodite stages especially so in the epiplankton and the adults descending more deeply. Kos (1975) states that the species spreads from as far north as the southern part of the Chukchi Sea (latitude *ca.* 70° N) southwards to 35° N latitude, where it lives more deeply. Off the Kuriles it forms, with *Calanus cristatus* and *Eucalanus bungii bungii*, about 80 % of the plankton biomass, but it is exceptionally abundant in the Sea of Japan and can contribute as much as 75–90 % of the biomass in north-western Japanese Sea areas. Even in the more coastal Japanese waters it can be abundant in deeper layers.

Calanus plumchrus is now separated as a North Pacific copepod from the closely allied *Calanus tonsus*, found in the southern hemisphere. Brodsky (1975) gives its breeding range as 3°–10°C, but Kos (1975) comments on great differences in surface temperature regimes in the various parts of its huge range. Kos employs statistical biometric analysis on the material to demonstrate that *C. plumchrus* may be separated into several morphologically distinct populations, some of which overlap geographically. He concludes that the species is polymorphic, with at least three intra-specific forms living in the extreme north, the north-east and the north-west of the Pacific respectively (cf. Fig. 2.69). This analysis does not extend to populations over vast open North Pacific areas.

Of the other species of *Calanus* in the Pacific, *C. pacificus* is a very common calanoid of the boreal North Pacific, not usually distributed so far north as *C. plumchrus*. Brodsky classifies it as "neritic–oceanic" and quotes very wide breeding temperatures (3°–20°C). Johnson and Brinton (1963) regard it as typical of the "transition zone", i.e. the relatively narrow area between sub-Arctic and sub-tropical water of approximately 38°–45°N latitude. The extension of these waters, especially in the

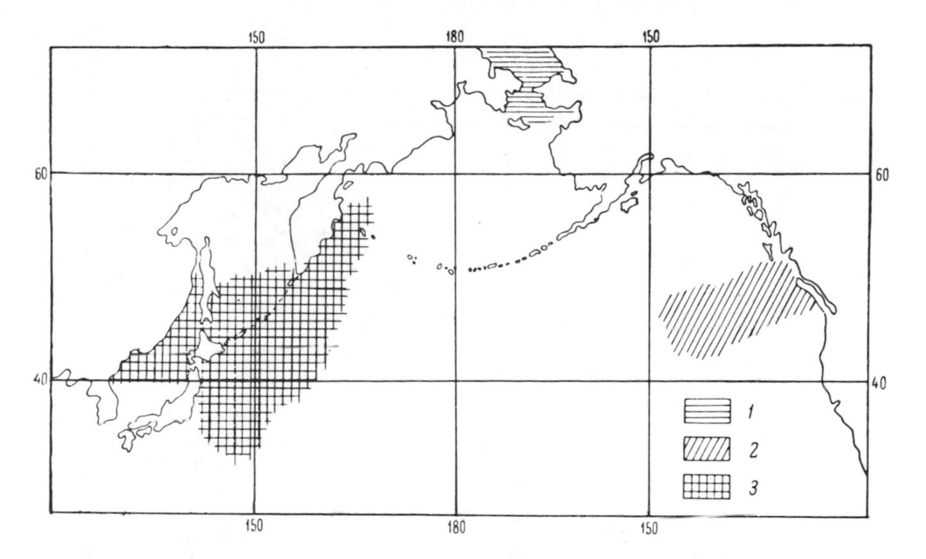

Fig. 2.69. Ranges of the infraspecific forms of *Calanus plumchrus*: (1) Northern form; (2) North-eastern form; (3) North-western form (Kos, 1975).

comparatively cold California Current, is another recognized habitat for the species. Jaschnov (1970) summarizes its distribution in Fig. 2.70. Earlier records were quoted as *Calanus finmarchicus* and more recently as *C. helgolandicus*, but both species are confined to the Atlantic.*

Brodsky (1961, 1965, 1975) has demonstrated the existence of three populations of *C. pacificus*: *C. pacificus pacificus* to the far west, including more northern areas of the Sea of Japan; *C. pacificus oceanicus* slightly more to the east of Japan and south of Kamchatka, with some overlapping areas; and *C. pacificus californicus* in the easternmost parts, off the coasts of Oregon and California. The retention of a stock of this copepod in the Californian Current area is partly affected by changes in their vertical distribution. Longhurst (1967a) describes an especially interesting pattern off Southern and Baja California where resting copepodites, mainly C V, exist as fairly dense pópulations in the oxygen-deficient water below 150 m. During the season of local coastal upwellings these copepods become abundant near the surface, feeding on the rich diatom blooms.

The abundance and importance of *C. pacificus* is noted by LeBrasseur (1965), Geynrikh (1968), Parsons *et al.* (1970) and Minoda (1972), amongst other authors. *C. pacificus* is now distinguished from a closely related form, *C. sinicus*, found off southern Japan and south China including the Gulf of Tonkin. This more neritic calanoid is regarded as a tropical or sub-tropical species with a breeding temperature of *ca.* 15°–20°C. A new species, *C. marshallae*, has been described by Frost (1974) from the eastern boreal North Pacific.

In the southern hemisphere, *Calanus tonsus* is an abundant oceanic and antiboreal ("notal") copepod, breeding, according to Brodsky, between 10° and 18°C. Jillett

* Many workers (eg. Mullin and Brooks, 1970) now use the specific name *C. helgolandicus* for *C. pacificus* when reporting experimental work (see Chapter 6).

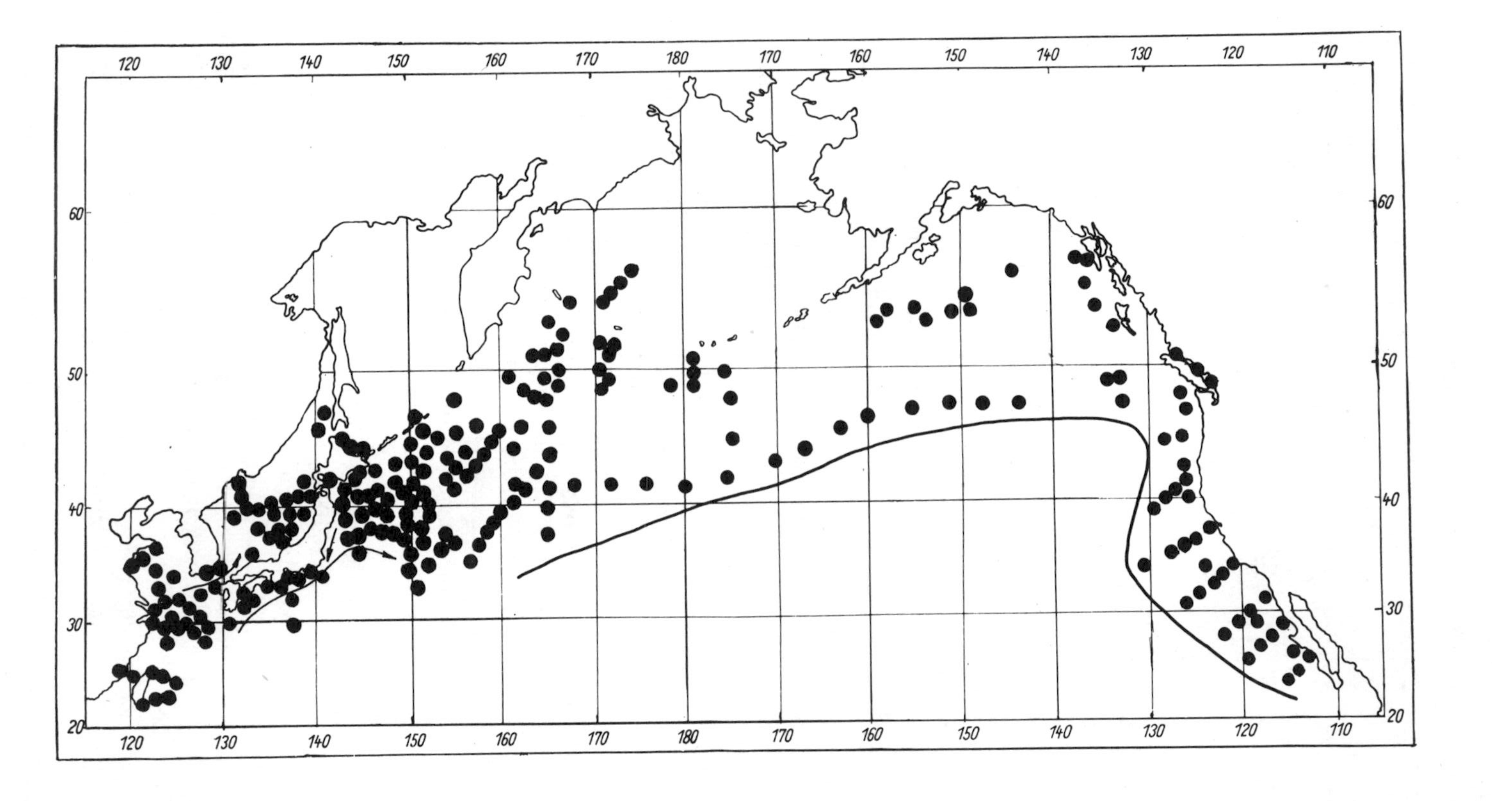

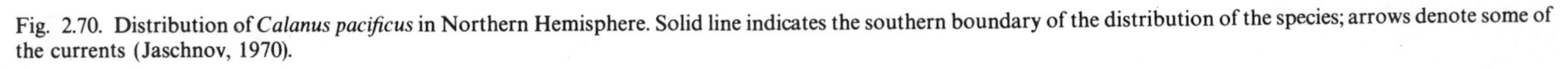
Fig. 2.70. Distribution of *Calanus pacificus* in Northern Hemisphere. Solid line indicates the southern boundary of the distribution of the species; arrows denote some of the currents (Jaschnov, 1970).

(1968) describes this species as being mainly distributed in the sub-Antarctic—from New Zealand across the southern Pacific, south of the Sub-Tropical Convergence—but not entering the true Antarctic. Occasional captures farther north are presumably due to a northward flow of colder deeper water. *C. plumchrus*, confined to the North Pacific, and *C. tonsus*, restricted to the Southern Ocean, thus appear to form a species pair. Off the east coast of New Zealand Jillett found *C. tonsus* abundant over the edges of the continental shelf and slope, somewhat less plentiful over oceanic depths, but avoiding neritic conditions inshore. In the areas of abundance, great swarms, especially of CV copepodites, occurred, Geynrikh (1973) found that *C. tonsus* penetrates up to 23°S along the coast of Chile, where it is carried north by the Peru Current.

In examining the distribution of *Calanus* spp. in the waters of the southern hemisphere, Brodsky (1961, 1965) distinguished two further species endemic to the south and distinct from *C. tonsus*. *C. australis* is found off the south of the three continents. Off S. America, Geynrikh (1973) reports a high biomass of this species occurring in coastal water from 5°–22°S associated with upwelling, giving high latitude conditions. *Calanus chilensis* appears to be much more restricted in distribution, being found off the coast of southern Chile with *C. australis* but not extending so far north or south. According to Brodsky both are neritic antiboreal species with a breeding temperature range of 12°–14°C for *C. chilensis* and 7°–15°C for *C. australis*. Jillett (1971) believes that northern New Zealand is the approximate northerly limit for *C. australis* (formerly recorded as *C. finmarchicus*), the species avoiding sub-tropical waters. The copepod is plentiful in the New Zealand area inside the 100 m depth contour as part of the shelf plankton, though it is not abundant close inshore.

In Brodsky's classification two species of *Calanus*, *C. simillimus* and *C. propinquus*, are transferred to a sub-genus *Neocalanus*, which includes *C. tenuicornis* and *C. gracilis* (cf. Sars, 1925; Rose, 1929). *C. simillimus* and *C. propinquus* are both well recognized high latitude southern calanoids. *C. propinquus* is an Antarctic oceanic form with a narrow breeding range (−1.8°–+2°C). *C. simillimus* ranges more widely over more northern Antarctic and sub-Antarctic oceanic areas, with a breeding range of 1°–8°C (cf. Chapters 3, 4). Mackintosh (1934) speaks of *C. simillimus* as a normal inhabitant of sub-Antarctic water, the view also of Vervoort (1951a) who points out that the species may occur near the Convergence and is to some extent carried south of the Convergence in particular areas. *C. propinquus*, on the other hand, is found everywhere in the true Antarctic region and is a typical Antarctic member of the epiplankton, occasionally being carried north of the Convergence (cf. Mackintosh, 1934; Vervoort, 1951a).

The two other species of *Calanus* (*Neocalanus*), *C. gracilis* and *C. tenuicornis*, are relatively warm-water calanoids. Brodsky describes the first as tropical (mean breeding temperature 19°C) and the second as widespread in tropical and warm-temperate waters of the Atlantic and Pacific, with a slightly lower minimum temperature range than *C. gracilis*. Rose (1933) suggests that *C. gracilis* has a fairly wide distribution in Atlantic (including the Mediterranean) and Pacific waters, and Wilson (1932), while describing *C. gracilis* as present in all three tropical oceans, suggests (1950) that it is more widely distributed and abundant than *C. tenuicornis* (cf. Sewell, 1947). Both species were found all the year round in the epiplankton of the Sargasso Sea by Deevey (1971). Roe (1972a) found *C. tenuicornis* as an abundant surface copepod off the Canaries and *Neocalanus gracilis* also common, but somewhat deeper. Jillett (1971) found *C. tenuicornis* off North Island, New Zealand, but seldom south of the Sub-Tropical Convergence.

C. robustior (sub-genus *Tropocalanus* of Brodsky) is another warm-water species of *Calanus* with a fairly wide range in Atlantic and Pacific waters (cf. Rose, 1933; Wilson, 1942). Sewell (1947) records it for all three major oceans in essentially tropical waters, apparently spreading to more temperate areas at greater depths.

Two species, *C. pauper* and *C. minor*, are proposed by Brodsky (1975) as separable to a genus *Canthocalanus*. *C. pauper* is a warm-water, mainly tropical form; the temperatures given by Brodsky as normal breeding limits (18°–28°C) are high. Wilson (1942) found it occasionally in Pacific collections; Sewell (1947) points out that it has not been recorded from the Atlantic (cf. also Deevey, 1971), though it has been found in the Mediterranean, perhaps transported via Suez. On the other hand, *Canthocalanus minor*, more generally known as *Nannocalanus*, is a widely known and relatively common form in warm and temperate seas. Wilson (1932, 1942, 1950) describes it as brought by the Gulf Stream to the region off Woods Hole and as being widespread in the warmer waters of all the oceans.

Brodsky gives a minimum breeding temperature as low as 12°C, which would agree with the species' wide geographical range. Sewell (1947) confirmed its wide occurrence in tropical and temperate regions of all three oceans: it can range far south (about 50°S) in the Pacific. Though usually surface-living, it may have a wide depth range. Jillett (1971) described *C. minor* as being fairly common off the north of New Zealand. Deevey (1971) found it all the year round and reasonably common in the Sargasso Sea, where it is one of the more abundant epiplanktonic calanoids (Deevey and Brooks, 1977). Colebrook, John and Brown (1961), examining the distribution of some oceanic copepods in the north-east Atlantic, state that in the warmer waters to the south *Calanus gracilis* occurred, and in slightly less warm areas, *C. minor* was found. Roe (1972c) described the latter as one of the very common surface-living copepods from the Canaries and it is repeatedly referred to as a warm-water surface form off west Africa (cf. Gaudy and Seguin, 1964; Stander and DeDecker, 1969; Bainbridge, 1972; Thiriot, 1977).

According to Brodsky's (1975) revision, the genus *Calanoides* includes only a single species, *C. patagoniensis*, an antiboreal species also reported by Sewell (1947). The last group of three species of *Calanus* is separated by Brodsky as a new sub-genus, *Carinocalanus*. *C. macrocarinatus* is a new species identified by Brodsky in temperate and neritic waters off South Africa. *C. carinatus* is a tropical and sub-tropical species described by Rose (1933) under the name of *C. brevicornis* for the warmer part of the Atlantic (including the eastern Mediterranean) as well as for the Indian Ocean. Under the name *Calanoides carinatus* the copepod has been widely recognized as a most important species in upwelling waters off the west African coast. Over periods of upwelling it may dominate the zooplankton. Details are included in Chapters 3 and 8. *C. acutus*, the third species, is a well known form confined to the Antarctic, where it is probably the most abundant copepod. It is found almost everywhere throughout surface Antarctic waters, sometimes in huge numbers, but is rare north of the Antarctic Convergence (cf. Mackintosh, 1934; Hardy and Gunther, 1935). Mackintosh's later (1964) summary confirms its restriction to the Antarctic; the few reports north of the Antarctic Convergence are at deeper levels, presumably carried in the Antarctic Intermediate Water (cf. Vervoort, 1951a). Brodsky gives its temperature breeding range in the Antarctic as $-1.8° - +2°C$ (cf. also Chapter 3).

Although this account of *Calanus* is so much more detailed than that for other zooplankton genera, the vast numbers and range of many of the species and their significance in marine food chains warrant the extensive treatment.

Calanoids from colder seas

A few calanoids, in addition to species of *Calanus*, are typical and abundant in colder high latitude seas. Some are very widespread copepods but are conveniently referred to here with cold-loving forms. Some of the species may live more deeply, even at high latitudes and therefore may be discussed again in a subsequent section.

In the northern North Atlantic *Metridia longa*, *Pseudocalanus minutus* and *Microcalanus pygmaeus* are among the characteristic species. Zenkevitch (1963) found these three species over much of the North Pacific and Sea of Japan.

Pseudocalanus is a regular high latitude copepod (cf. Ussing, 1938; Digby, 1954). Brodsky (1967) gives its distribution as the north Siberian seas and generally over the cold North Pacific, including the Sea of Okhotsk, Bering Sea and Sea of Japan, and over the North Atlantic Ocean including the Barents, Norwegian and Greenland Seas. He describes it as a mass species of cold waters. Fish (1936b) and Fish and Johnson (1937) commented on its importance off the north-east coast of the United States and it was one of the three most plentiful copepods in the Norwegian Sea (Ostvedt, 1955; Hansen, 1960). Grice (1962) described it as the most abundant copepod in Arctic seas (cf. also Grainger, 1965: Pavashtiks, 1975). There is some disagreement over the range of *Pseudocalanus*, since certain forms exist which are regarded by some authorities as distinct species. *P. elongatus* is suggested as somewhat more coastal by Hansen (1960), *P. minutus* as being more typical of open oceans. *P. gracilis* is apparently restricted to very high latitudes. *P. major* is another Arctic but neritic species, recorded for example by Chislenko (1975) for the fauna of Dikson Bay. Corkett & McLaren (1978) discuss the possible subdivision of the species. Rose (1929) stated that *P. minutus sensu lato* was comparatively rare south of the Bay of Biscay and was not normally found in the Mediterranean, though it is recorded for the Black Sea. On the American side of the North Atlantic it barely reached Cape Hatteras, and was a winter visitor to Delaware Bay (cf. Deevey, 1956).

Microcalanus pygmaeus has a somewhat similar range, with a higher latitude distribution resembling that of *Pseudocalanus*. It is especially abundant in Arctic seas but unlike *Pseudocalanus* it occurs in the Antarctic. Another calanoid at high latitudes, but also as a widespread species occurring in warmer areas of all three oceans, is *Scolecithricella minor*. This species can penetrate Arctic waters but probably exists in several different ecological forms. Zenkevitch (1963) states that it is one of the more important calanoids in eastern seas. Despite earlier records for Antarctic waters (cf. Hardy and Gunther, 1935), it is not reported there by either Vervoort (1951b) or Brodsky (1967), though *S. ovata*, widely distributed in deeper areas of the three major oceans, occurs in the Antarctic and *S. gracilis* is also a typical Antarctic shallow-living species.

Metridia longa, a typical Arctic copepod, is found in the extreme north of the Atlantic and in the Norwegian Sea, tending to live in the deeper layers and spreading a little south with colder currents both off Norway and the north-east coast of North America. Hansen (1960), Grice (1962), Johnson (1963), Grainger (1965), Dunbar and

Harding (1968), and others record it, with *Calanus hyperboreus*, as typical of surface Arctic waters. Brodsky (1967) states that it reaches the Chukchi Sea to the east in northern polar seas, but he questions whether it occurs in the coldest areas of the North Pacific Ocean, the species *M. okhotensis*, separately identified from the Sea of Okhotsk, apparently taking its place.

Metridia lucens is a very abundant oceanic boreal–temperate copepod, widespread in the North Atlantic and Norwegian Sea. Vervoort and Farran (1948) include a somewhat doubtful reference to its occurrence in the Barents Sea, but the species does not appear to extend far north beyond sub-Arctic waters. It is well known in the North Sea (cf. Colebrook, Glover and Robinson, 1961) and to the west of Ireland and extends into the western part of the English Channel. To the south, it reaches warm-temperate areas as far as the Bay of Biscay and the Mediterranean (cf. Rose, 1933; Brodsky, 1967). There is evidence that separate populations of *M. lucens* may exist in the Atlantic and on the continental shelf (cf. Colebrook, 1964, and Chapter 8).

There are numerous records of the species in the North Pacific, but it is probable that the widespread mass species in that ocean is in fact *M. pacifica* (cf. Zenkevitch, 1963; Brodsky, 1967) though it is recorded off Chile and Peru by Geynrikh (1973). With *Calanus plumchrus* and *Calanus cristatus*, *Metridia pacifica* is one of the common species in the North Pacific. Its southern limit would appear to be the warm Kuroshio water. Minoda (1972) describes *M. pacifica* with *Calanus* spp. and *Eucalanus bungii* as dominant in the North Pacific herbivorous plankton, *Metridia* being abundant to considerable depths. Vinogradov (1972) refers to the same four calanoids, with the addition of *M. okhotensis*, as the major copepods contributing to the biomass in the Kurile–Kamchatka Trench to a depth of about 1500 m.

Another rather less abundant calanoid typical of boreal waters is *Pareuchaeta* (*Euchaeta*) *norvegica*, a member of the North Atlantic *Calanus* community, but normally inhabiting somewhat deeper levels. Colebrook, Glover and Robinson (1961) class *Pareuchaeta norvegica* with the northern oceanic group of copepods from the north-east Atlantic. In the North Pacific, *Pareuchaeta japonica* seems to be an equivalent species.

Another temperate North Atlantic copepod, usually found off continental shelf areas, is *Centropages typicus*. Farran (1948) classes this as a euryhaline copepod, though it rarely comes into areas of markedly reduced salinity, in contrast to *Centropages hamatus*, which is common in more estuarine areas. Deevey (1956) states that *C. typicus* in the western North Atlantic does not penetrate north of New Brunswick; it is found off Delaware Bay and as far south as North Carolina. She gives a breeding range of 1°–24°C. Colebrook, Glover and Robinson (1961) and Colebrook (1964) associate *C. typicus* with southern intermediate copepod species in the eastern Atlantic (cf. Fig. 2.71).

Centropages hamatus, typical of areas of reduced salinity, is widely known from the temperate Atlantic waters, including the North Sea, Baltic Sea and English Channel as well as at higher latitudes in inshore areas of the Norwegian, Barents and Kara Seas (Zenkevitch, 1963). Wilson (1933) records the species widely on the American side of the Atlantic, including the Gulf of Maine. Deevey (1956) states that it occurs from Belle Isle to the Gulf of Mexico. In the North Pacific the genus is represented in inshore areas by *Centropages mcmurrichi*, which is common in bays and inlets in the Sea of Japan, Sea of Okhotsk and the Bering Sea (cf. Brodsky, 1967).

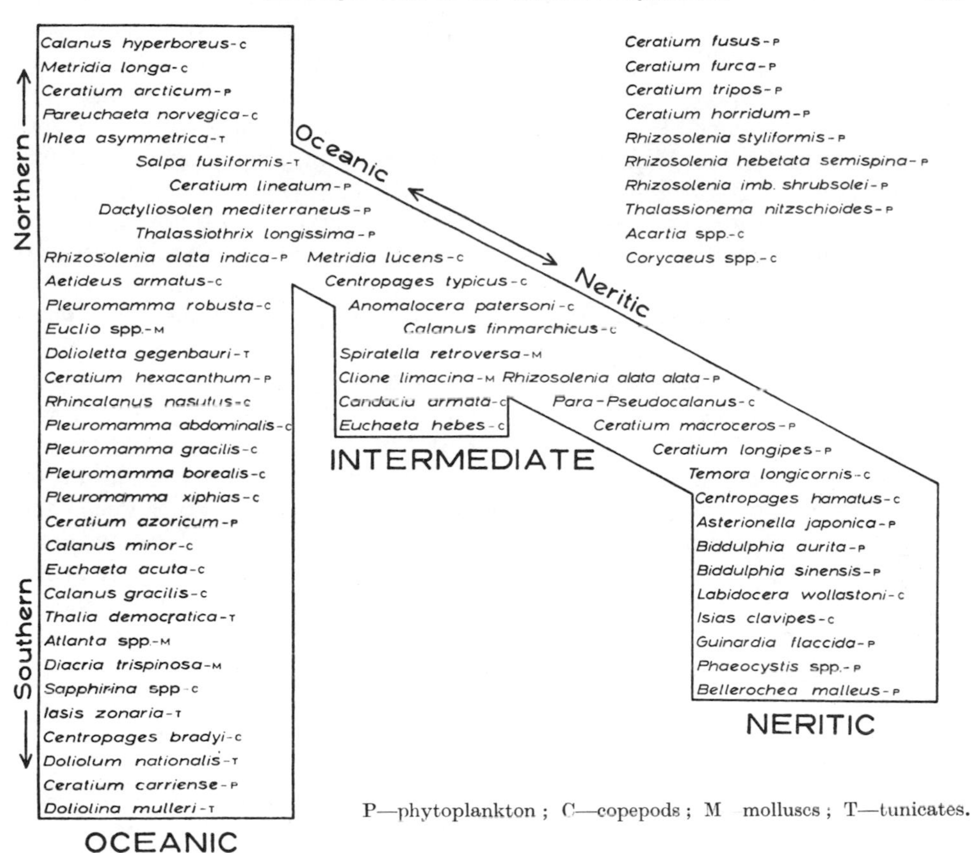

Fig. 2.71. A distribution series of some phyto- and zooplankton species. They are arranged in such a way that the distribution of each organism is most similar to those of the neighbouring organisms listed (Colebrook, Glover and Robinson, 1961).

Included in the zooplankton of colder seas are two small calanoids, *Acartia longiremis* and *A. clausi*. Both are described by Farran (1948) as euryhaline, and while they may be encountered over oceanic areas, they perhaps attain higher densities closer to shore. Another calanoid, *Temora longicornis*, is rather similar in its distribution. All three are widely distributed in the eastern coastal Atlantic region including the English Channel, the coastal North Sea, the Baltic and the Norwegian Sea (cf. Farran, 1948). *A. longiremis* is more northerly in distribution, however, not being recorded from the south and west of Ireland. Colebrook's (1964) investigations demonstrate the abundance of *A. clausi* in the north-east Atlantic and indicate the presence of discrete populations (cf. Chapter 8). Zenkevitch (1963) includes *Acartia* (mainly *A. longiremis*) and *Temora* in the Kara and Barents Seas, and *A. longiremis* is an important copepod in plankton from the Okhotsk and Bering Seas (cf. also Erikkson, 1973a).

Rose (1933) distinguishes *A. longiremis*, inhabiting polar and temperate areas, from *A. clausi*, which, though very widely distributed in temperate and warmer parts of the three oceans and the Mediterranean, is absent from the Arctic (cf. Brodsky, 1967). *Temora longicornis* was not recorded for the northern Pacific by Rose, Brodsky or Wilson (1932), though the latter author records it as one of the most abundant species off Woods Hole. Deevey (1952, 1956) describes *Temora longicornis* as a boreal–temperate, neritic species found in the western Atlantic from the Straits of Belle Isle to Cape Hatteras. This species, together with *Centropages*, *A. clausi* and *Pseudocalanus*, is a very important copepod in the Long Island, Block Island and, to a lesser extent, Delaware Bay area, though *Acartia tonsa* is generally of greater significance (cf. Conover, 1956).

Another species with a remarkably wide distribution and frequently very abundant in parts of the temperate North Atlantic is *Paracalanus parvus*. Though plentiful inshore, it can be common in surface oceanic waters. Rose (1933) and Wilson (1932) describe its wide occurrence, extending to all three major oceans, the Black Sea and Mediterranean and including warm as well as temperate waters. It is repeatedly described as abundant off regions of the west African coast, particularly during upwelling periods (cf. Thiriot, 1977). *Paracalanus parvus* does not occur in very high latitudes either in the northern or southern hemispheres. Farran and Vervoort (1951) describe its distribution in the north-east Atlantic as extending to the Faroe Iceland area and Norwegian Sea but not to the Barents Sea. In the North Pacific the species spreads from warmer waters to become abundant in the Sea of Japan, where it sometimes becomes a mass species, and it can extend to the southern part of the Okhotsk Sea (Zenkevitch, 1963; Brodsky, 1967). In the eastern Atlantic it is abundant off southern Norway, in the North Sea and in the English Channel. The combined copepod group *Paracalanus—Pseudocalanus* was among the commonest of the North Sea copepods identified by Colebrook *et al.* (1961). Deevey (1960) gives its northern limit off the American Atlantic coast as the Gulf of Maine, but states that it is probably more significant to the south.

The zooplankton fauna of temperate regions changes towards lower latitudes, with the inclusion of warmer species and a marked tendency to greater species diversity (cf. Deevey, 1971). Conversely, in proceeding to higher latitudes there is usually a reduction in the number of species, at least in the surface layers. Vinogradov (1970) suggests that about twenty species of copepods are found commonly in the surface layers of the North Atlantic north of 50°N, and for the North Pacific (the Bering Sea and the Sea of Okhotsk) only about ten calanoid species occur in the superficial layers. At much higher northern latitudes there is an increasing preponderance of very few species, *Calanus finmarchicus* (*glacialis*), *C. hyperboreus* and *Metridia longa*, especially in the Polar Basin.

Johnson (1963) points out that the calanoids typical of the sub-Arctic North Pacific (*C. cristatus*, *C. plumchrus*, *Metridia pacifica*, *Eucalanus bungii bungii* and *Acartia longiremis*) gradually die out as the Polar Basin is approached. Their place as the most abundant calanoids with greatest biomass, especially in near surface layers, is taken by the same four calanoids listed by Zenkevitch for the high latitude Atlantic. Apparently very few of the small copepods normally abundant at high latitudes (*Oncaea*, *Oithona*, *Microcalanus* and *Pseudocalanus*) were taken. In somewhat deeper layers, a greater variety of species was found, but in small numbers. *Scaphocalanus* spp., *Spinocalanus* spp. and *Temorites brevis* occurred fairly regularly.

Grice's (1962) investigations across the Polar Basin confirmed the wide distribution and frequent occurrence of *Calanus finmarchicus* (*glacialis*), *C. hyperboreus* and *Metridia longa*, with *Pseudocalanus* as the most abundant copepod; *Oithona similis* and *Microcalanus* were also taken. Other broadly ranging and usually deep-living species included *Scaphocalanus magnus*, *Spinocalanus magnus* and *Spinocalanus abyssalis*; such species may occur more shallowly in Arctic waters. Two other species, *Heterorhabdus norvegicus* and *Chiridius obtusifrons*, are essentially Arctic, but while they may spread south to a limited extent in the North Atlantic, inhabiting deeper-water layers, they are apparently excluded from the Pacific. Grainger (1965) lists amongst other copepods, *Gaidius tenuispinus*, *G. brevispinus*, *Heterorhabdus* and *Spinocalanus magnus* in sub-surface Arctic waters (cf. Dunbar and Harding, 1968). Table 2.12 and Fig. 2.72 summarize the relative frequency of the eighteen copepod species and their distribution in the Atlantic and Pacific.

Several of the calanoids found at high latitudes in the northern hemisphere have also been noted as occurring at high southern latitudes. Amongst the species restricted to Antarctic or sub-Antarctic seas, apart from the species of *Calanus* already described, two copepods are of great significance numerically, *Rhincalanus gigas* and *Metridia gerlachei*. Some account of *Rhincalanus* is included in Chapters 3 and 8. Mackintosh (1934) reported large numbers of *Rhincalanus gigas* almost everywhere in the slightly warmer parts of the Antarctic. Vervoort (1951a) found *Rhincalanus* in all seas round the Antarctic continent to the Convergence, mostly in the 50–250 m layers, with a few specimens passing north of the Convergence in summer. Near the Falkland Islands there may be some overlap between *R. gigas* and *R. nasutus* at its extreme southern limit. Voronina (1970) emphasized the marked seasonal depth migration of *R. gigas*, which takes place at somewhat different times according to latitude, and described the species as occurring throughout Antarctic and sub-Antarctic areas (cf. Ommanney, 1936). *Metridia gerlachei* is a cold-water form found from the Convergence to the highest southern latitudes and with some tendency to form shoals. Hardy and Gunther (1935) also found *Metridia gerlachei* south of the Convergence, but it was rarely in the immediate surface layer (cf. Vervoort, 1951b).

Table 2.12. Frequency of occurrence and abundance of the species in the 90 collections (Grice, 1962)

Species	Frequency	Number of specimens found
Metridia longa (Lubbock)	58	661
Calanus hyperboreus Kroyer	56	868
C. glacialis Jaschnov	45	273
Pseudocalanus minutus Kroyer	18	8,537
Scaphocalanus magnus (Scott)	15	21
Chiridius obtusifrons Sars	13	17
Ectinosoma finmarchicum Scott	12	43
Calanus finmarchicus (Gunnerus)	9	18
Oithona similis Claus	9	36
Acartia longiremis (Lilljeborg)	6	17
Paraeuchaeta glacialis (Hansen)	5	12
Oncaea borealis Sars	3	3
Microcalanus pygmaeus (Sars)	2	2
Spinocalanus magnus Wolfenden	2	2
S. abyssalis Giesbr. *var. pygmaeus* Farran	2	3
Heterorhabdus norvegicus (Boeck)	1	1
Temorites brevis Sars	1	1
Lubbockia glacialis Sars	1	1
Unidentified and mutilated specimens	15	30

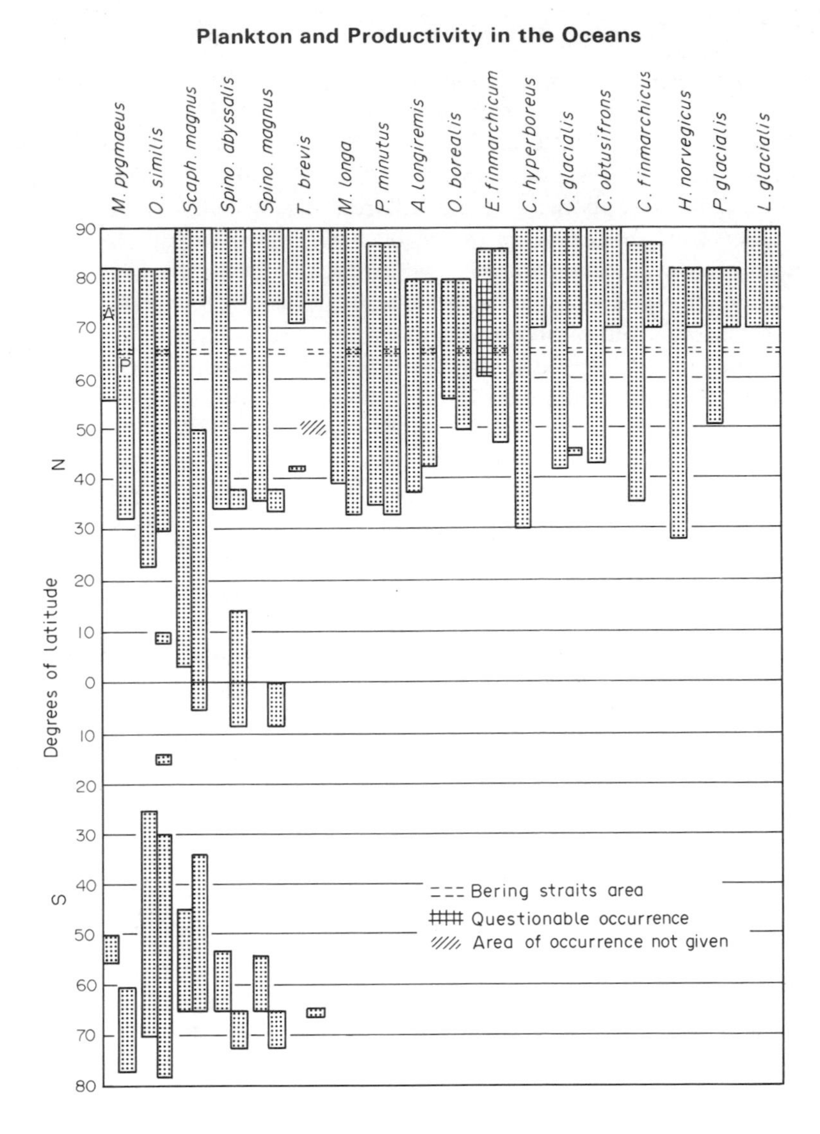

Fig. 2.72. Latitudinal distribution of copepod species in the Atlantic (A) and Pacific (P) Oceans (Grice, 1962).

A number of other calanoids (e.g. *Aetideopsis minor*, *Euchaeta antarctica*, *Haloptilus ocellatus*, *Gaetanus antarcticus* and *Heterorhabdus farrani*) are limited to Antarctic seas. Many are moderately deep-living and are mentioned later.

Deeper-living copepods

In boreal–temperate regions, south of these high Arctic latitudes, a deep-living zooplankton exists distinct to some extent from the shallower *Calanus finmarchicus* community of the North Atlantic and from the *Calanus cristatus – Calanus plumchrus – Metridia* community of the North Pacific. In both oceans, however, these more superficially living calanoids perform seasonal vertical migrations and have an extended vertical range (cf. Ostvedt, 1955; Vinogradov, 1972), so that although they diminish in numbers in deeper strata, they can make a substantial contribution to the deeper zooplankton fauna. A variety of calanoids are, however, more or less confined to the deeper layers and these become of greater significance with increasing depth. A few species (e.g. *Chiridius obtusifrons*, *Scaphocalanus* spp., *Heterorhabdus norvegicus*), while occurring at depth in the temperate and warmer seas, live at shallower depths at the highest latitudes. Sewell (1948) includes a list of such species which, he believes, have extended their distribution from high Arctic latitudes to the North Atlantic, mostly then inhabiting deeper layers.

Many deep-living species, however, despite an extensive vertical range, rarely (if ever) reach surface layers in any geographical area. Vinogradov (1972) defines for zooplankton generally a series of vertical ecological groupings: the surface (epipelagic zone) to about 200 m, the mesopelagic zone from 200–700 m (or 1000 m), the bathypelagic zone from 750 (1000) to 2500 (3000) m and the abyssopelagic zone, extending deeper than 3000 m. Russian planktonologists, especially Vinogradov himself, have contributed greatly to our knowledge of the vertical distribution of zooplankton, but the precise vertical (and horizontal) distribution of many deeper-living species is still not well established. Often in the following summary, therefore, the adjective deep-living is employed in place of the more precise terms mesopelagic, bathypelagic or abyssal.

The problem of defining vertical ranges of zooplankton, not only for calanoids, is complicated by the fact that closing net hauls were seldom employed in earlier surveys, and the possibility of contamination with species living in different strata is a matter for considerable debate. Vertical ranges for species almost certainly also vary with region. It is frequently true that many deeper-living calanoids exhibit a very wide range (cf. Michel and Foyo, 1976) (Table 2.13); some may show little evidence of a depth maximum, but for a large number of species a fairly obvious stratum can be demonstrated for the majority of individuals. Occasionally a species shows two clear maxima and very commonly the juvenile stages of a calanoid live at a different depth, often more shallowly, than the adults.

One of the most thorough recent investigations of the vertical distribution of calanoids is that due to Roe (1972, a, b, c) in an area near the Canary Islands. Closing net hauls were used to an upper depth limit of 40 m (day) and to a maximum depth of 950 m. Roe (1972c) found that the majority of calanoid species were scarce over the whole water column. Thus by day only about 25 of the total number of 203 species contributed more than 1 % of the total calanoid population through the column; only two (*Metridia venusta* and *Pleuromamma abdominalis*) amounted to more than 5 %. However, at a particular depth the contribution of a species might be very significant; for example, *Pleuromamma borealis* formed nearly 50 % of the calanoids at a depth of 350 m. A selection of genera of calanoids may be made for three major strata:

(i) Surface and near-surface: *Nannocalanus*, *Calanus*, *Clausocalanus*, *Acartia*, *Paracandacia*
(ii) 200–400 m: *Pleuromamma* (overwhelmingly important), *Haloptilus*, *Gaetanus*
(iii) > 700 m: *Metridia*, *Spinocalanus*, *Eucalanus* (cf. Fig. 2.73)

One family, the Metridiidae, represented almost entirely by *Pleuromamma* and Metridia, was by far the commonest, making up 23 %, by day, of the total calanoids.

Table 2.13. Relative abundance and vertical range of copepods collected in the Caribbean Sea and adjacent areas (Michel and Foyo, 1976)

Species or Group	Total Range (m)	No. of Samples	Total Numbers	%
Acrocalanus longicornis	0–2454	105	28,856	0.78
Clausocalanus furcatus	0–3594	233	189,191	5.11
Euchaeta marina	0–2660	147	82,029	2.21
Haloptilus longicornis	0–3560	244	50,990	1.37
Lucicutia flavicornis	0–4350	239	54,219	1.46
Mormonilla minor	27–5200	508	159,076	4.30
M. phasma	100–3560	329	14,270	0.38
Paracalanus aculeatus	0–4350	144	70,543	1.90
Rhincalanus cornutus forma *atlantica*	0–3602	394	53,206	1.43
Scolecithrix danae	0–2500	87	22,414	0.60
Undinula vulgaris	0–1000	83	66,514	1.79
Other Calanoida	0–7500	794	2,905,723	78.59
Total Calanoida			3,697,036	99.93
Aegisthus aculeatus	0–4350	173	1,694	0.43
Macrosetella gracilis	0–6220	343	80,563	20.83
Microsetella rosea	0–7500	532	270,068	69.83
Other Harpacticoida	0–6220	329	34,422	8.90
Total Harpacticoida			386,747	99.99
Conaea gracilis	0–5200	420	45,690	1.34
Farranula carinata	0–5200	243	412,740	12.14
F. gracilis	0–2500	62	12,895	0.37
Oithona plumifera	0–5200	601	491,616	14.46
Oncaea mediterranea	0–3144	301	56,584	1.66
O. venusta	0–5200	329	124,782	3.67
Other Cyclopoida	0–7500	783	2,254,849	66.33
Total Cyclopoida			3,399,156	99.97

To some extent families of calanoids also show differences in their vertical range. Grice and Hulsemann (1965), for the north-east Atlantic between 30° and 60°N latitude, found Centropagidae, Pontellidae, Acartiidae and Paracalanidae, though not represented by a large number of species, all in the upper 200 m (epipelagic zone). The Eucalanidae, the Calanidae and Pseudocalanidae were also more generally shallow-living, with more than half the 25 species in these three families occurring in the upper 200 m. On the other hand, almost all of the few species found of Phaennidae and Bathypontiidae were deep-living, below 1000 m. Although several other families had a very great bathymetric range, with some individual species exhibiting a huge vertical spread (*Spinocalanus abyssalis* and *S. magnus* from 180 to 5000 m), approximately half the species of Spinocalanidae and Lucicutiidae inhabited depths exceeding 1000 m and more than half of the species belonging to the families Aetideidae, Scolecithricidae and Augaptilidae were moderately deep-living, but were normally above 1000 m.

Vinogradov (1970) summarized the relative abundance of calanoid families as percentages of copepod biomass for the Kurile–Kamchatka region of the North Pacific. Calanidae and Eucalanidae were important throughout the upper 500 m, but

Pseudocalanidae were really significant in the deepest strata (4000–> 7000 m), as were Aetideidae, though the latter family was also important at intermediate depths (below 500–750 m). The Euchaetidae and Metridiidae might be regarded as of chief importance in bathypelagic and upper abyssal depths (see Table 2.14). With regard to the Calanidae and Eucalanidae, Deevey and Brooks (1971) found, in the Sargasso Sea, that some members of these families occurred throughout the whole water column (sampled to 2000 m) and that these species could be numerically important below 1000 m. Although some Calanidae, (e.g. *Nannocalanus minor*, *Calanus tenuicornis*) were epiplanktonic, other species (e.g. *Calanus finmarchicus*) were deep-living. Members of the Eucalanidae could be deep-living; *Eucalanus* was believed to occur in the shallower layers only during the spring. Outside these two families Deevey and Brooks (1971) found that a number of calanoids (*Haloptilus*, *Pleuromamma*, *Phyllopus*, *Heterorhabdus* and *Lucicutia*) could occur throughout the whole 2000-m water column, though they usually were present in considerably higher densities over a narrower depth range.

Table 2.14. Changes with depth in the ratios of some copepod families in the Kurile–Kamchatka region (in per cent of the total average biomass of copepods; the interzonal filter-feeding calanids not feeding at the depths of catch are not taken into account) (Vinogradov, 1972)

Depth (m)	Calanidae	Eucalanidae	Pseudo-calanidae	Aetideidae	Euchae-tidae	Metridiidae
0–50	94	5	1	—	1	1
50–100	39	53	1	1	1	7
100–200	24	61	1	1	3	9
200–500	39	46	1	2	3	8
500–750	1	—	1	21	51	6
750–1000	1	1	3	18	38	7
1000–1500	2	1	3	17	38	6
1500–2000	1	1	4	14	37	10
2000–2500	1	1	5	15	25	14
2500–3000	—	—	5	9	23	22
3000–4000	—	—	7	21	29	16
4000–5000	—	—	14	36	20	16
5000–6000	—	—	26	45	6	13
6000–7000	—	—	27	45	5	2
	—	—	41	46	8	5

No attempt will be made to provide a complete list of bathypelagic or abyssal copepods, since the list is formidable and the range not well known. However, some of the characteristic deep-living calanoids will be described, more particularly in the North Atlantic, but to some extent as regards world distribution. As an illustration of the large number of species, Sewell (1948) listed 357 species of deep-living Copepoda in the North Atlantic Ocean, though some are now known to occur in superficial as well as deeper layers. Examples of records of deep-living calanoids are drawn from the works of Sars (1925), Rose (1933), Lysholm and Nordgaard (1945), Sewell (1947, 1948), Vervoort (1951a, b, 1952, 1963, 1965), Ostvedt (1955), Grice and Hulsemann (1965), Deevey and Brooks (1971, 1977) and Roe (1972a, b, c). A list of species collated from these works by Grice and Hulsemann (1965) is included (Table 2.15).

Table 2.15. Calanoid copepod species in the north-east Atlantic (Grice and Hulseman, 1965)

Calanidae

1. *Calanus finmarchicus* (Gunnerus, 1765)
2. *C. glacialis* Jaschnov, 1955
3. *C. helgolandicus* (Claus, 1863)
4. *C. hyperboreus* Krøyer, 1838
5. *C. tenuicornis* Dana, 1849
6. *Nannocalanus minor* (Claus, 1863)
7. *Neocalanus gracilis* (Dana, 1849)
8. *N. robustior* (Giesbr., 1888)

Eucalanidae

9. *Eucalanus elongatus* (Dana, 1849)
10. *Mecynocera clausi* Thompson, 1888
11. *Rhincalanus cornutus* (Dana, 1849)
12. *R. nasutus* Giesbr., 1888

Paracalanidae

13. *Calocalanus contractus* Farran, 1926
14. *C. pavo* (Dana, 1849)
15. *C. plumulosus* (Claus, 1863)
16. *C. styliremis* Giesbr., 1888
17. *Paracalanus aculeatus* Giesbr., 1888
18. *P. nanus* Sars, 1907

Pseudocalanidae

19. *Clausocalanus arcuicornis* (Dana, 1849)
20. *C. furcatus* (Brady, 1883)
21. *C. paululus* Farran, 1926
22. *C. pergens* Farran, 1926
23. *Ctenocalanus vanus* Giesbr., 1888
24. *Microcalanus pygmaeus* Sars, 1900
25. *Pseudocalanus minutus* Krøyer, 1842
26. *Spicipes nanseni* gen. n. et sp. n.

Spinocalanidae

27. *Mimocalanus cultrifer* Farran, 1908
28. *M. inflatus* Davis, 1949
29. *M. nudus* Farran, 1908
30. *Monacilla tenera* Sars, 1907
31. *M. typica* Sars, 1905
32. *Spinocalanus abruptus* sp. n.
33. *S. abyssalis* Giesbr., 1888
    S. abyssalis var. *pygmaeus* Farran, 1926
34. *S. angusticeps* Sars, 1920
35. *S. hirtus* Sars, 1907
36. *S. magnus* Wolfenden, 1904
37. *S. ovalis* sp. n.
38. *S. polaris* Brodsky, 1950
39. *S. spinosus* Farran, 1908
40. *S. validus* Sars, 1905
41. *Tanyrhinus naso* Farran, 1936

Aetideidae

42. *Aetideopsis multiserrata* (Wolfenden, 1904)
43. *A. rostrata* Sars, 1903
44. *Aetideus armatus* (Boeck, 1872)
45. *Bradyetes inermis* Farran, 1905?
46. *Chiridiella brachydactyla* Sars, 1907
47. *C. macrodactyla* Sars, 1907
48. *C. subaequalis* sp. n.
49. *Chiridius obtusifrons* Sars, 1903
50. *Chirundina streetsi* Giesbr., 1895
51. *Euaetideus actus* (Farran, 1929)
52. *E. giesbrechti* (Cleve, 1904)
53. *Euchirella amoena* Giesbr., 1888
54. *E. bella* Giesbr., 1888
55. *E. curticauda* Giesbr., 1888
56. *E. intermedia* With, 1915
57. *E. messinensis* (Claus, 1863)
58. *Gaetanus minor* Farran, 1905
59. *G. pileatus* Farran, 1903
60. *Gaidius brevispinus* Sars, 1900
61. *G. tenuispinus* Sars, 1900
62. *Pseudaetideus armatus* (Boeck, 1872)
63. *Pseudeuchaeta brevicauda* Sars, 1905
64. *Pseudochirella fallax* (Sars, 1907)
65. *Undeuchaeta major* Giesbr., 1888
66. *U. plumosa* (Lubbock, 1856)

Euchaetidae

67. *Euchaeta acuta* Giesbr., 1892
68. *E. marina* (Prestandrea, 1833)
69. *E. media* Giesbr., 1888
70. *Paraeuchaeta farrani* (With, 1915)
71. *P. glacialis* (Hansen, 1886)
72. *P. gracilis* (Sars, 1905)
73. *P. norvegica* (Boeck, 1872)
74. *P. sarsi* (Farran, 1908)
75. *Valdiviella insignis* Farran, 1909

Phaennidae

76. *Amallophora typica* (T. Scott, 1894)
77. *Heteramalla dubia* (T. Scott, 1894)
78. *Xanthocalanus difficilis* sp. n.
79. *X. paraincertus* sp. n.

Scolecithricidae

80. *Amallothrix emarginata* (Farran, 1905)
81. *A. falcifer* (Farran, 1926)
82. *A. propinqua* (Sars, 1920)
83. *A. robustipes* sp. n.
84. *Racovitzanus antarcticus* Giesbr., 1902
85. *R. levis* Tanaka, 1961
86. *Scaphocalanus bogorovi* Brodsky, 1955
87. *S. brevicornis* (Sars, 1903)
88. *S. curtus* (Farran, 1926)
89. *S. echinatus* (Farran, 1905)
90. *S. elongatus* A. Scott, 1909
91. *S. longifurca* (Giesbr., 1888)
92. *S. magnus* (T. Scott, 1894)
93. *S. major* (T. Scott, 1894)
94. *S. medius* (Sars, 1907)
95. *S. robustus* (T. Scott, 1893)
96. *S. subbrevicornis* Wolfenden, 1911
97. *Scolecithricella abyssalis* (Giesbr., 1892)
98. *S. dentata* (Giesbr., 1892)
99. *S. laminata* Farran, 1926
100. *S. lobata* Sars, 1920

Table 2.15 (Contd.)

101. *S. minor* (Brady, 1883)
102. *S. ovata* (Farran, 1905)
103. *S. tenuiserrata* (Giesbr., 1892)
104. *S. unispinosa* sp. n.
105. *S. valida* (Farran, 1908)
106. *S. vittata* (Giesbr., 1892)
107. *Scolecithrix bradyi* Giesbr., 1888
108. *S. danae* (Lubbock, 1856)
109. *S. fowleri* Farran, 1926
110. *Scottocalanus helenae* (Lubbock, 1856)
111. *S. persecans* Giesbr., 1892

Tharybidae

112. *Undinella simplex* (Wolfenden, 1906)

Temoridae

113. *Temoropia mayumbaensis* T. Scott, 1893

Metridiidae

114. *Pleuromamma abdominalis* (Lubbock, 1856) f. typica Steuer, 1932
115. *P. borealis* (Dahl, 1893)
116. *P. gracilis* (Claus, 1863) f. minima Steuer, 1932
117. *P. piseki* Farran, 1929
118. *P. robusta* (Dahl, 1893)
119. *P. xiphias* (Giesbr., 1888)
120. *Metridia brevicauda* Giesbr., 1889
121. *M. discreta* Farran, 1946
122. *M. longa* (Lubbock, 1854)
123. *M. lucens* Boeck, 1864
124. *M. macrura* Sars, 1905
125. *M. princeps* Giesbr., 1889
126. *M. venusta* Giesbr., 1889

Centropagidae

127. *Centropages typicus* Krøyer, 1849
128. *C. violaceus* (Claus, 1863)

Lucicutiidae

129. *Lucicutia magna* Wolfenden, 1903
130. *L. clausi* (Giesbr., 1889)
131. *L. curta* Farran, 1905
132. *L. flavicornis* (Claus, 1863)
133. *L. gaussae* Grice, 1963
134. *L. gemina* Farran, 1926
135. *L. grandis* (Giesbr., 1895)
136. *L. anomala* Brodsky, 1950
137. *L. longiserrata* (Giesbr., 1889)
138. *L. ovalis* (Giesbr., 1889)
139. *L. parva* sp. n.
140. *L. intermedia* Sars, 1905
141. *L. tenuicauda* Sars, 1907

Heterorhabdidae

142. *Disseta minuta* sp. n.
143. *D. palumboi* Giesbr., 1889
144. *Hemirhabdus latus* (Sars, 1905)
145. *Heterorhabdus abyssalis* (Giesbr., 1889)
146. *H. clausi* (Giesbr., 1889)
147. *H. compactus* (Sars, 1900)
148. *H. norvegicus* (Boeck, 1872)
149. *H. papilliger* (Claus, 1863)
150. *H. spinifrons* (Claus, 1863)
151. *Mesorhabdus brevicaudatus* (Wolfenden, 1905)

Augaptilidae

152. *Augaptilus cornutus* Wolfenden, 1911
153. *A. glacialis* Sars, 1900
154. *Euaugaptilus bullifer* (Giesbr., 1892)
155. *E. elongatus* (Sars, 1905)
156. *E. facilis* (Farran, 1908)
157. *E. filiger* (Claus, 1863)
158. *E. gracilis* (Sars, 1905)
159. *E. hecticus* (Giesbr., 1889)
160. *E. humilis* (Farran, 1926)
161. *E. longimanus* (Sars, 1905)
162. *E. longiseta* sp. n.
163. *E. nodifrons* (Sars, 1905)
164. *E. palumboi* (Giesbr., 1892)
165. *E. sarsi* sp. n.
166. *E. squamatus* (Giesbr., 1892)
167. *Haloptilus acutifrons* (Giesbr., 1892)
168. *H. longicornis* (Claus, 1892)
169. *H. ornatus* (Giesbr., 1892)
170. *Pseudaugaptilus longiremis* Sars, 1907

Arietellidae

171. *Phyllopus helgae* Farran, 1908
172. *P. impar* Farran, 1908

Candaciidae

173. *Candacia bipinnata* (Giesbr., 1889)
174. *C. longimana* (Claus, 1863)
175. *Paracandacia bispinosa* (Claus, 1863)
176. *P. simplex* (Giesbr., 1889)

Pontellidae

177. *Pontellina plumata* (Dana, 1849)

Bathypontiidae

178. *Bathypontia minor* (Wolfenden, 1906)
179. *Foxtonia barbatula* Hulsemann and Grice, 1963.
180. *Temorites brevis* Sars, 1900
181. *T. discoveryae* sp. n.
182. *Zenkevitchiella atlantica* sp. n.

Acartiidae

183. *Acartia danae* Giesbr., 1889
184. *A. negligens* Dana, 1849

Incertae sedis

185. *Disco inflatus* gen. n. et sp. n.
186. *D. longus* sp. n.
187. *D. minutus* sp. n.

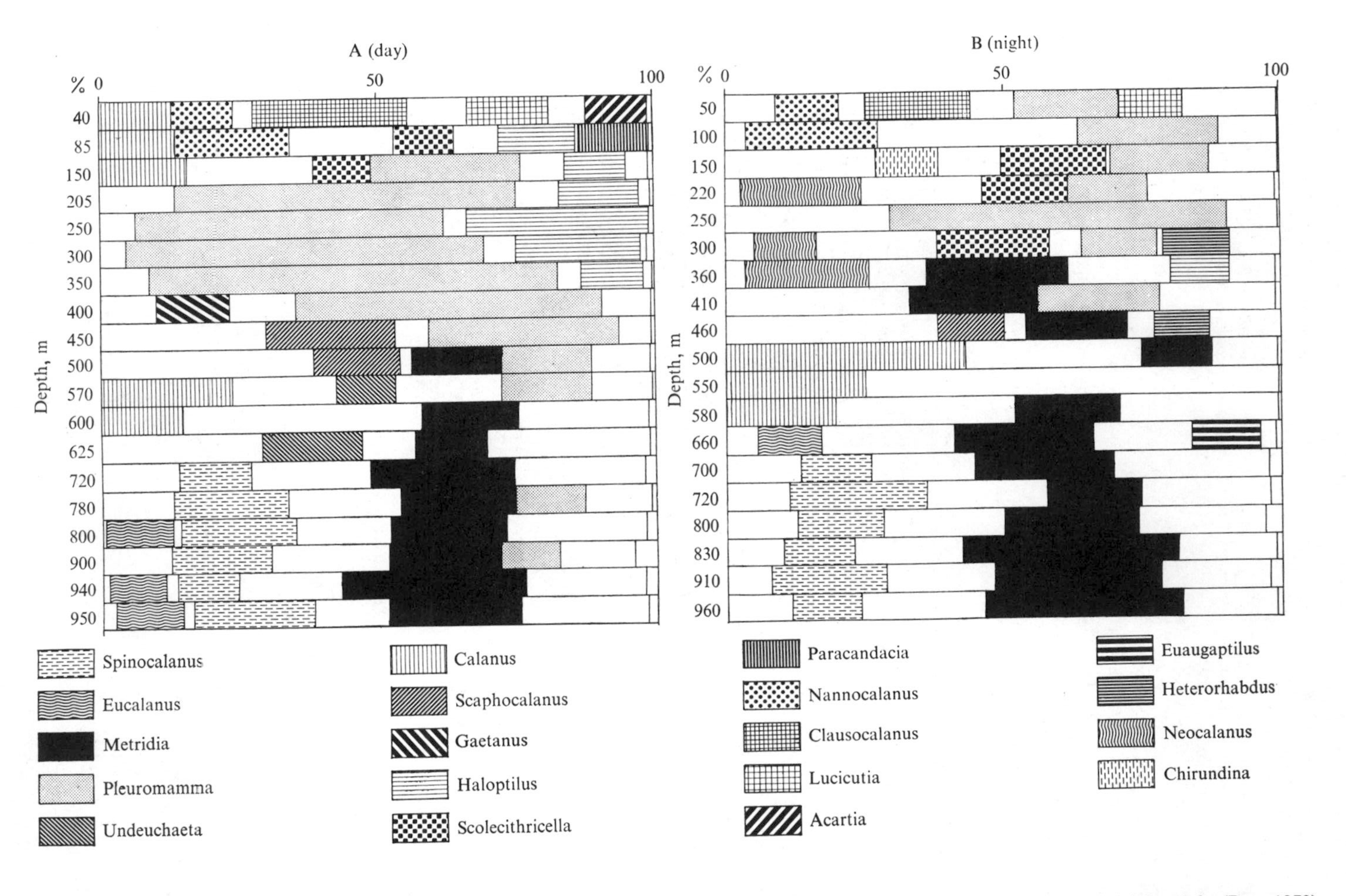

Fig. 2.73. The percentage of the estimated total number of specimens at each depth formed by each genus (A) by day and (B) by night (Roe, 1972).

The distribution of species of the genus *Calanus sensu lato* has already been described. *Calanus finmarchicus* in the Atlantic and *Calanus cristatus* and *Calanus plumchrus* in the Pacific tend to live in deeper layers in the southern parts of their ranges. *Calanus hyperboreus* tends to be markedly bathypelagic outside the Arctic. *Neocalanus gracilis* is occasionally deeper-living, though mainly epiplanktonic (cf. Vervoort, 1963; Roe, 1972a). Grice and Hulsemann's (1965) results indicate a range from 0–1000 m for *Neocalanus gracilis* and *N. robustior* and an even greater depth for *C. finmarchicus* and *C. hyperboreus* in the North Atlantic (cf. Deevey and Brooks, 1977).

The family Megacalanidae includes some relatively very large, truly bathypelagic calanoids. *Megacalanus princeps* is found fairly widely in the North and South Atlantic (Owre, 1962; Vervoort, 1963), the Indian Ocean (cf. Sewell, 1947) and the Pacific (Sars, 1925). *Bathycalanus* includes three species. *B. bradyi* has a distribution more or less similar to *Megacalanus princeps* and is found in the North Pacific, widely in the North Atlantic including equatorial regions, in the Indian Ocean and the Antarctic (Brodsky, 1967). *Bathycalanus princeps* and *B. richardi* are apparently confined to the Atlantic as is another deep sea species, *Bradycalanus sarsi*. Three of the species of Megacalanidae extend north to the Faroe–Iceland area (cf. Vervoort, 1951a, 1963).

Several species of *Eucalanus* appear to have a very great geographical range; though more typically living in warm and temperate seas, they may be distributed to higher latitudes in relatively small numbers. The account of the distribution of some species, however, may be somewhat confused since a number of sub-species have been reported. More recently, Fleminger (1973) has divided the genus into four groups comprising seventeen species and has listed the geographical distribution. Though many are broadly epiplanktonic, a number, especially of the *elongatus* group, are upper mesoplanktonic. Thus *Eucalanus elongatus* is found widely in the North and South Atlantic, the North and South Pacific, the Indian Ocean, Red Sea and Mediterranean. Sewell (1947) states that its vertical range is as great as from the surface to 3000 or 4000 m. It may show a narrower vertical extent in some areas of its geographical distribution; for example, in the Indian Ocean, it is mostly in the upper 500 m. In the North Atlantic it is carried north, presumably in warmer deep currents, to the Shetlands and south of Iceland. Vervoort (1963) emphasizes its particularly wide distribution (as *E. elongatus hyalinus*) in the Atlantic and its occurrence in deeper layers. Roe (1972a) refers to its abundance in deeper hauls with maximum abundance as 625 m.

Eucalanus attenuatus has a rather similar wide distribution in all three oceans, perhaps characteristically in warm-temperate waters, extending less far to higher latitudes. Brodsky (1967) describes its distribution as global. Its depth range is from the surface to more than 5000 m though, according to Sewell, it rarely exceeds about 1200 m in the Indian Ocean.

Eucalanus crassus similarly occurs in all three oceans, extending from the surface to 1000 or 1500 m. Vervoort (1963) speaks of its greater preference for more superficial layers, but Roe (1972a) found two populations, one from 500–950 m and the second very much shallower. Roe also found *Eucalanus subcrassus* and *E. pileatus* (which were not separated) with a very wide depth range but mainly around 500 m. Vervoort (1963) considers *E. pileatus* as a tropical species over the three oceans, and while mainly epiplanktonic, it may penetrate to deep layers. A number of other *Eucalanus* species live mainly near the surface and are mostly warm water. For the north-east Atlantic Vervoort (1951a) quoted all three species, *E. elongatus*, *E. attenuatus* and *E. crassus*, as occurring from the surface to deep water. Deevey and Brooks (1977) stress a seasonal migration of *Eucalanus* in the Sargasso Sea between the surface and 1500–2000 m. *E. hyalinus* (= *E. elongatus*) and *E. sewelli* (= *E. attenuatus*) were the most abundant and were deepest from summer to autumn, rising in winter and spring.

Rhincalanus cornutus and *R. nasutus* have somewhat similar distribution patterns to species of *Eucalanus*, being widely distributed but more abundant in the warm waters of all three major oceans. They penetrate considerably far north and south in the Atlantic (Vervoort, 1963). They exhibit a wide depth range from the surface to 1000 or 1500 m in the Indian Ocean, but to lesser depths in the Atlantic and Pacific (cf. Sewell, 1947). Grice and Hulsemann, however, found both *Rhincalanus nasutus* and *R. cornutus* extending to depths exceeding 1000 m in the warm-temperate north-east Atlantic (cf. Deevey and Brooks, 1971). Vervoort (1951a) reports *R. nasutus* reaching such high latitudes as the Faroe–Iceland area and the Norwegian Sea. Deevey and Brooks (1977) found a similar seasonal depth migration as with *Eucalanus*, but *R. cornutus* was mainly between 500 and 1000 m. Roe (1972a) states that both species are mostly mid-water (> 500 m), but can be nearer the surface.

Amongst the Pseudocalanidae, although *Pseudocalanus minutus* has been described as an extremely common copepod at high latitudes in superficial waters, it has a great depth range and a marked seasonal migration. Ostvedt (1955) found it throughout the water column at Weather Station M, but more plentifully at deep levels. At somewhat lower latitudes, its great depth range was confirmed by Grice and Hulsemann (1965) in the north-east Atlantic. *Ctenocalanus vanus*, a well known copepod with a great geographical range, is found widely in the Antarctic, mainly as epiplankton, but also at considerable depths (1000–2000 m). It is well known from tropical, sub-tropical and boreal parts of both Atlantic and Pacific Oceans. It is much more clearly bathypelagic outside the Antarctic according to Vervoort (1951a, 1963), but can occur at shallower depths; Grice and Hulsemann's records extend only to 200 m. Another usually deep-water genus is *Drepanopsis*, with *D. oblongus* an oceanic species found in the temperate Atlantic.

The genera *Spinocalanus*, *Mimocalanus* and *Monacilla*, are now usually separated as a family, the Spinocalanidae, which is essentially bathypelagic, although one or two species may live at somewhat shallower depths at highest latitudes (Rose, 1933). *Spinocalanus magnus* and *S. abyssalis* are both widely distributed, being found in Atlantic and Pacific Oceans and in Antarctic and Arctic waters (Sars, 1925; Vervoort, 1951b; Johnson, 1963, Brodsky, 1967). Although normally living in deep waters, *S. abyssalis* is reported by Vervoort (1951a) as epiplanktonic in the Arctic. Brodsky (1967) claims that *S. abyssalis* is restricted to the Pacific and that the polar species is *S. longicornis*. Other species of *Spinocalanus* appear to have a more restricted geographical distribution. For instance, *S. spinosus* is an oceanic deep-water form found in the temperate Atlantic; Grice and Hulsemann (1965) found it in their collections with a large vertical range of around 450–4000 m. Brodsky had described *S. polaris* as being confined to the Arctic Ocean, though Grice and Hulsemann (1965) found it later in the north-east Atlantic at depths of 1500–5000 m.

While the enormous depth range of *S. abyssalis* and *S. magnus* in the north-east Atlantic (*ca.* 200–5000 m) has already been noted (cf. page 194), other species recorded by Grice and Hulsemann (1965) are more limited to deep water (e.g. *S. ovalis*: 1500–5000 m; *S. validus*: 1500–4000 m; *S. hirtus*: *ca.* 3000–4500 m). Brodsky distinguishes some six species of *Spinocalanus* which are found only in the North Pacific area, all described as abyssal and mostly with a depth range of 1000–4000 m. In the other two genera—*Monacilla* and *Mimocalanus*—included in this family, the species are apparently exclusively bathypelagic or abyssal (cf. Vervoort, 1951a). Deevey and Brooks (1977) found *Spinocalanus* spp. contributing substantially to the calanoids in the Sargasso Sea below 500 m depth, with higher numbers between 1000 and 2000 m. *Monacilla* spp also occurred at depth. Roe (1972a) found *Spinocalanus magnus* and *S. spinosus* as the commoner species, with maximum numbers at 780 m and 950 m, respectively. Grice and Hulsemann's records indicate that *Mimocalanus cultrifer* has a great vertical spread (*ca.* 250–4500 m) in the north-east Atlantic, but that other species (*M. inflatus, M. nudus*) are mostly deeper than 1500 m and reach 2500 or 3500 m.

Species of the very large family, the Aetideidae, have been described by Rose (1933) as all living at sub-surface levels, some at very considerable depths, but some of these copepods are known to approach the surface, at least temporarily. The great vertical spread of some members of the Aetideidae has already been noted (cf. page 194/5 and Vinogradov, 1972). *Aetideus armatus* is an example of a fairly shallow-living species with only moderate depth range. It is a widely distributed calanoid recorded from all three oceans, though Brodsky regards the Pacific forms as probably belonging to a separate species. In the North Atlantic it extends to the Faroe–Shetlands area and to the Norwegian Sea. Ostvedt (1955), though finding it more typically in the 100–600 m level, records it as extending to the surface (cf. Farran and Vervoort, 1952). Grice and Hulsemann give the range as about 200–500 m, and Roe (1972a) describes it as a mid-water calanoid, with maximum abundance at about 600 m (cf. also Deevey and Brooks, 1977). *Euaetideus giesbrechti* is another relatively shallow species, (*ca.* 100–500 m) (Grice and Hulsemann, 1965), found in warm-temperate waters of the three oceans and the Mediterranean (cf. Lysholm and Nordgaard, 1945). Rose (1933) quotes depths of 100–250 m, but states that the copepod may reach the surface at night. Deevey (1971) found it in fairly small numbers all the year in the Sargasso Sea.

Aetideopsis rostrata also has a wide geographical range, being distributed in the Atlantic and Pacific Oceans as well as in the Malay Archipelago. It extends north to the Arctic Ocean and is a deeper-living calanoid, apparently always inhabiting layers below 500 m and normally deeper than 1000 m (cf. Ostvedt, 1955; Brodsky, 1967). The range given by Grice and Hulsemann (1965) for the north-east Atlantic is some 800–1300 m. The related *Aetideopsis minor* is another fairly deep-living copepod, ranging between 1000 m and the surface, but confined apparently to the Antarctic.

With regard to those Aetideidae normally living beneath the surface, Table 2.16 gives some indication of the diversity of species; only limited reference is made to their geographical and vertical distribution, but some details of several genera follow. A member of this family, *Chirundina streetsi*, is a true bathypelagic calanoid, taken in many parts of the three major oceans. It is a characteristic component of the deep calanoid fauna of the Atlantic; it has been recovered rarely at the surface (Vervoort, 1963). It was abundant in deeper hauls taken by Roe off the Canaries, with a wide depth range very similar to two other bathypelagic common species, *Undeuchaeta major* and *U. plumosa* (Table 2.16). Roe (1972a) suggests that the three species are more or less closely related.

Table 2.16. Distribution of some deeper-living species of Aetideidae

Species	Geographical Region	Depth Range[a] (m)	Authority
Aetideopsis rostrata	Atlantic, Pacific, Malay Archipelago, Arctic	>500–>1000	Ostvedt, 1955; Brodsky, 1967
	North Atlantic	800–1300	Grice & Hulsemann, 1965
A. minor	Antarctic	0–1000	Vervoort, 1951b
Chirundina streetsii	All three Oceans	450–1000	Vervoort, 1963; Grice & Hulsemann, 1965
	Atlantic Canaries, Sargasso	570–950	Sewell, 1947; Roe, 1972a; Deevey & Brooks, 1977
Chiridiella macrodactyla	Warm and warm/temperate Atlantic and Pacific	600	Vervoort, 1963
	Norwegian Sea	1000–2000	Ostvedt, 1955
	North Atlantic	2000–5000	Grice & Hulsemann, 1965
C. brachydactyla	North Atlantic	2000–3000	Grice & Hulsemann, 1965
C. subaequalis	North Atlantic	3000–4000	Grice & Hulsemann, 1965
Pseudochirella fallax	North Atlantic	2000–3000	Grice & Hulsemann, 1965
P. obtusa	Indian, North Atlantic, Pacific	0–1000–4800	Sewell, 1947; Vervoort, 1963
P. notacantha	Arabian Sea	0–1500	Sewell, 1947
	Indian, North Atlantic, Antarctic	770–1000	Vervoort, 1957
Undeuchaeta major	North Atlantic	500–2000	Grice & Hulsemann, 1965
	Atlantic Canaries	625	Roe, 1972a
	Atlantic, Pacific, Indian	0–550–2000	Rose, 1933; Sewell, 1947
U. plumosa	North Atlantic	500–1000	Grice & Hulsemann, 1965
	Sargasso	0–1500	Deevey & Brooks, 1977
Chiridius obtusifrons	North Atlantic	450–750	Grice & Hulsemann, 1965
	Norwegian Sea	200–800	Ostvedt, 1955
	Arctic areas	< 200	Grice, 1962; Johnson, 1963
	North Pacific	0–1500	Farran & Vervoort, 1952;
C. poppei	Mediterranean	200	Farran & Vervoort, 1952; Vervoort, 1963
	Tropical and temperate Atlantic, Indo-Pacific	> 100	Rose, 1933
	Atlantic Canaries	150–940	Roe, 1972a
	Sargasso	500–1500	Deevey & Brooks, 1977

Gaidius brevispinus	North Atlantic	450–1100	Grice & Hulsemann, 1965
	Arctic, Norwegian Sea	1000–2000	Ostvedt, 1955
	north-west Pacific, Bering Sea	> 200	Brodsky, 1967
G. tenuispinus	Atlantic and Polar Seas	> 200	Lysholm & Nordgaard, 1945
	North Atlantic	250–1000	Grice & Hulsemann, 1965
	Atlantic Canaries	ca. 500	Roe, 1972a
	North Atlantic, north-west Pacific, Arctic Ocean and Antarctic	> 200–4000	Brodsky, 1967
Gaetanus miles	Atlantic Canaries	250–350	Roe, 1972a
G. minor and *G. miles*	Sargasso	0–1000	Deevey & Brooks, 1977
Euchirella curticauda	North Atlantic	500–1000	Grice & Hulsemann, 1965
	Sargasso	500–1500	Deevey & Brooks, 1977
E. messinensis	North Atlantic	500–1000	Grice & Hulsemann, 1965
	Atlantic Canaries	570–950	Roe, 1972a
	Sargasso	0–1500	Deevey & Brooks, 1977
E. amoena	North Atlantic	100	Grice & Hulsemann, 1965
	Sargasso	1000–1500	Deevey & Brooks, 1977
E. intermedia	North Atlantic	1500–3000	Grice & Hulsemann, 1965
E. pulchra	Sargasso	1000–1500	Deevey & Brooks, 1977
	Atlantic Canaries	450–950	Roe, 1972a
	Central Arabian Sea	0–200	Sewell, 1947
E. rostrata	Sargasso	0–1000	Deevey & Brooks, 1977

[a] Depth of maximal abundance underlined. Records from Roe (1972) refer to maximum depth sampled of 950 m.

The deep-living genus *Chiridiella* (Sars, 1925; Rose, 1933) is represented in the Sargasso Sea by nine species all living more deeply than 500 m (Deevey and Brooks, 1977). Other species of *Chiridiella*, described by Brodsky (1967) as confined to the North Pacific, were not found above 1000 m. The genus *Pseudochirella*, according to Rose (1933), includes all deep-living species, though their vertical ranges are little known. Vervoort (1952) records six species of the genus in the temperate and northern Atlantic, some species reaching fairly high latitudes in the Faroe–Iceland area. In the Sargasso, of seven species found by Deevey and Brooks (1977), six exceeded depths of 1000 m. According to Brodsky (1967), species of *Pseudochirella* inhabiting the north-west Pacific may occur to a depth of 4000 m. Two other genera, *Bradyetes* and *Bryaxis*, are deep-living Aetideidae; North Atlantic species may live close to the bottom in moderately deep water. Indeed, Matthews (1964) suggests that some of this family may be benthic/planktonic forms. Amongst these he lists a species of *Chiridius*, *C. armatus*, a boreal Arctic species widely distributed in the Arctic Ocean, the North Atlantic and the North Pacific, and which can be carried south in cold currents. Rather surprisingly, Vervoort (1963) reports it from the Mediterranean and from the Gulf of Guinea. Bakke and Valderhaug (1978) regard it as an important carnivorous component of the deep community in a Norwegian fjord. The distribution of other species of *Chiridius* is given in Table 2.16 and Fig. 2.74.

Three genera of Aetideidae are generally recognized as typical of the strata more immediately beneath the upper layers, though some species may be deep-living. Many of the species are also very broadly distributed geographically. All species of *Gaidius* are described by Rose (1933) as bathypelagic. Vervoort (1952) lists all four species from the north-east Atlantic as deep-water (cf. also Deevey and Brooks, 1977, for the Sargasso Sea) but several of Brodsky's records for the north-west Pacific include species living at a few hundred metres depth. The two other genera, *Gaetanus* and *Euchirella* (Fig. 2.74), are among the most characteristic calanoids found in the sub-surface layers of many seas, though usually not in great abundance. Though both genera have a broad geographical range, many species of *Euchirella* are perhaps more characteristic of warmer and temperate seas (cf. Table 2.16).

According to Sars (1925) and Rose (1933) all known species are bathypelagic but Deevey and Brooks (1977) include *Gaetanus* among those genera decreasing in abundance from depths of 500 m in the Sargasso Sea. Although most species of *Gaetanus* live at considerable depths, certain species may be taken at the surface, for example, *G. miles* normally occurs in deeper water, even to 4000 m according to Sewell, but it may be found in surface waters. It is distributed in all three oceans in tropical, sub-tropical and temperate waters, but can penetrate far north in the Atlantic, for instance to the south of Iceland (Vervoort, 1963). *G. minor*, also found in all three oceans, has been taken at the surface but normally it is deeper-living. Ostvedt (1955) found it as far north as the Norwegian Sea, where it occurred at intermediate depths. *G. pileatus* appears always to live deeper than 200 m and can extend to considerable depths; in the North Atlantic it is more abundant at medium levels (*ca.* 300–600 m). Ostvedt confirmed its occurrence at intermediate depths in the Norwegian Sea. The species, found in all oceans, reaches to the south of Greenland in the North Atlantic. *G. latifrons* has a rather similar distribution extending to high latitudes in the North Atlantic. It is relatively common in the warm-temperate Atlantic and has been taken in the Indian and Pacific Oceans (Sewell, 1947). Another Atlantic species which appears

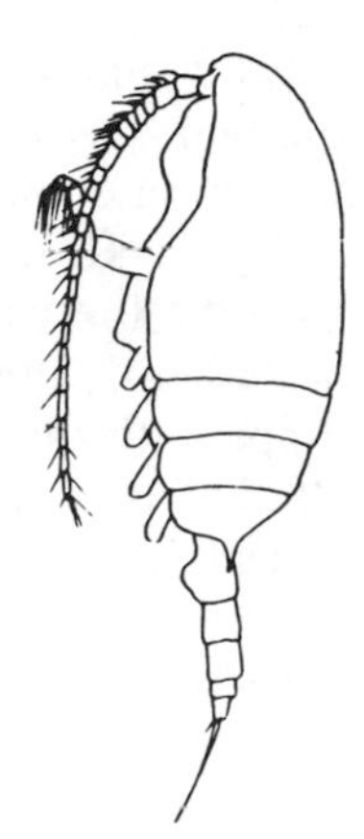

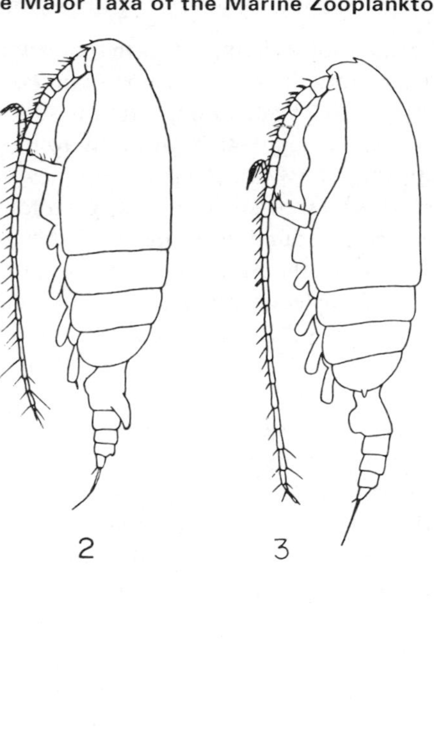

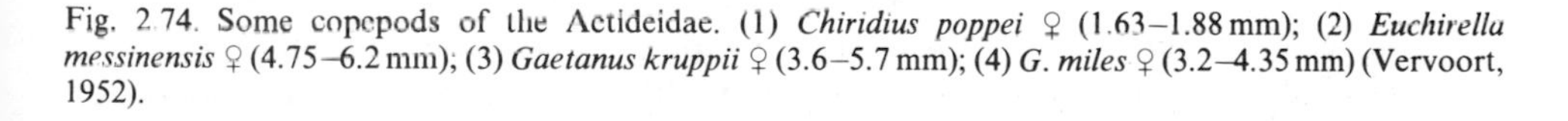

Fig. 2.74. Some copepods of the Actideidae. (1) *Chiridius poppei* ♀ (1.63–1.88 mm); (2) *Euchirella messinensis* ♀ (4.75–6.2 mm); (3) *Gaetanus kruppii* ♀ (3.6–5.7 mm); (4) *G. miles* ♀ (3.2–4.35 mm) (Vervoort, 1952).

to reach high latitudes is *G. curvicornis*, which Sewell has recorded from the Indian Ocean. Vervoort (1952) states that it occurs in the deep waters of the Atlantic and in the Arctic Ocean.

According to Sars (1925) and Sewell (1947), the commonest species, and indeed one of the most widely taken of all bathypelagic copepods, is *Gaetanus kruppi*, found in warm and temperate waters of all the oceans and in the Mediterranean. It reaches far north in both the Atlantic and Pacific (cf. Vervoort, 1963). The calanoid can live at considerable depths (more than 2000 m) but is often above 1000 m; however, it apparently never approaches the surface. All eight species of *Gaetanus* listed by Deevey and Brooks (1977) for the Sargasso Sea, apart from *G. miles* and *G. minor*, exceed 500 m and may extend to 2000 m. *G. antarcticus*, a species apparently confined to the Antarctic, was identified by Sewell from deep water of the Indian Ocean, presumably carried there in the cold Antarctic Drift. Several species apparently exclusive to the Pacific are noted by Brodsky (1967).

Many species of *Euchirella*, though living beneath the surface, generally favour only moderate depths (*ca.* 500 m). As with so many deeper-living zooplankton, some species show a considerable depth range. Vertical migration is sufficiently marked for some species to reach surface layers at night, though rarely found there by day (cf. Rose, 1933). Most *Euchirella* appear to be warm-temperate species (cf. Brodsky, 1967, for the Pacific), though in the North Atlantic some species reach fairly high latitudes. Of the various species, *E. messinensis*, found over most of the year but apparently not abundantly by Deevey (1971) in the Sargasso, is reported by Sars (1925) from the North and South Atlantic, the Mediterranean and the Pacific Ocean, including the Malay Archipelago. *E. rostrata* and *E. curticauda* are found in broadly similar areas but have been reported by Sars as far as 65°N in the Atlantic Ocean. Brodsky (1967) found *E. rostrata* in the north-west Pacific. Wilson (1932) records it as a visitor from deep water off Woods Hole. Ostvedt (1955) has identified both species in the Norwegian Sea at intermediate depths. Vervoort (1952) records altogether eight species of *Euchirella* for the North Atlantic, two species, *E. bitumida* and *E. curticauda*, being found as far north as the Faroe–Iceland area.

All species are said to live in deep or moderately deep water but they may occasionally appear at the surface, mainly at night. However, *E. maxima* and *E. intermedia* are never found at the surface. Sewell (1947) believed that *E. bitumida* could not be distinguished from *E. galeata*, a form found widely along the Pacific coast of North and South America, in the Malay Archipelago, as well as extensively in the Indian Ocean and in both the North and South Atlantic. Brodsky (1967), however, continued to separate *E. galeata*, giving its distribution as 35°N and 35°S in the Pacific Ocean at depths of from 600 m to the surface. Vervoort (1963) stated that *E. bitumida* is quite distinct as a species occurring over the North Atlantic and the whole Indo-Pacific. It seems to prefer moderately deep water. *E. pulchra*, a species with a fairly wide distribution in all three oceans appears to be an essentially warm-water form (cf. Sewell, 1947; Vervoort, 1963). Wilson (1932) records it occasionally from the Woods Hole area. The Edinburgh Plankton Records (1973) show rare occurrences of this species (and *E. messinensis*) in their North Atlantic studies, but *E. rostrata* by contrast, was frequently observed. *E. rostromagna* is perhaps unusual among species of *Euchirella* in being a cold-water species found in many Antarctic areas in moderately deep water (cf. Vervoort, 1951b).

The genus *Valdiviella*, sometimes assigned to the family Euchaetidae, is represented by only four species. These are deep-living calanoids mainly from temperate and warm waters. *V. insignis*, a large copepod reaching 12 mm length, has, however, been recorded by Sewell (1947) as far north as 61°N in the Atlantic. It occurs also in temperate and tropical areas and is found in the Indian Ocean. Grice and Hulsemann's (1965) records are from 1000 to 2000 m depth for the North Atlantic. Brodsky (1967) describes *V. brevicornis* at considerable depths of 2000–2500 m from temperate and tropical areas of the Atlantic and Indian Oceans and at even greater depths (*ca.* 4000 m) from the north-west Pacific. Deevey and Brooks (1977) give depth ranges for the two species as 1500–2000 and 1000–1500 m respectively.

The genus *Euchaeta* is often divided into *Euchaeta* and *Pareuchaeta*; species of the restricted genus *Euchaeta* are mainly epiplanktonic calanoids, widely distributed in temperate and warm waters but normally not reaching high latitudes. *E. marina* is perhaps the most broadly distributed and commonest species found in all three oceans and in the Mediterranean. Michel and Foyo (1976) regard it as one of the most important epipelagic calanoids of warm seas (Table 2.13). *E. acuta*, *E. pubera* and *E. spinosa* are also taken in warm-temperate areas of all oceans, though the last species is relatively uncommon. Lysholm and Nordgaard suggest that it may be somewhat deeper-living—a view supported by Vervoort (1963). Roe (1972b) found *E. marina* mainly at sub-surface levels, and *E. media* and *E. acuta* both relatively plentiful at mid-depths.

In contrast to *Euchaeta*, *Pareuchaeta* is a deep-living genus; according to Rose (1933) and Sars (1929) all species are bathypelagic. Grice and Hulsemann's (1965) findings for the north-east Atlantic support the general depth distinction between the two genera. *E. marina* was surface-living, though it can have a considerable range (Vervoort, 1963); *E. acuta* extended from the surface to a few hundred metres; *E. media* was a little deeper. The various species of *Pareuchaeta*, however, occurred mainly below 500 m: *P. farrani* to about 1500 m; *P. norvegica* to 2000 m; *P. sarsi* from 2000 to 3000 m. Many species of *Pareuchaeta* inhabit very cold seas at high latitudes. *P. glacialis* is an Arctic species found in the Atlantic; *P. polaris* is apparently confined to the Polar Basin (cf. Brodsky, 1967); *P. bradyi* is a North Atlantic species found as far north as 68°N. Ostvedt (1955) found both *P. glacialis* and *P. bradyi* in the deepest layers of the Norwegian Sea, but *P. glacialis* ranged into intermediate depths. *P. farrani* is also found in the Norwegian Sea and in the Greenland and Spitzbergen areas living in deep waters.

Pareuchaeta norvegica is widely distributed in deep water over the boreal Atlantic; in the Norwegian Sea the majority lived at intermediate levels (100–600 m). Its distribution, which is similar to that of *Calanus finmarchicus*, is in sharp contrast to the more southern North Atlantic ranges of *Euchaeta* spp. (*E. acuta*, *E. marina* and *E. pubera*) (cf. Edinburgh Plankton Records, 1973). *E. hebes*, found over the tropical and temperate North Atlantic, occurs in the Biscay area and off the mouth of the English Channel. It is common in the Mediterranean but sparingly distributed in the Pacific (Vervoort, 1963).

Other North Atlantic species of *Pareuchaeta* occurring in fairly high latitudes include *P. hanseni*, *P. scotti* and *P. barbata*, all found in deep water (Sewell, 1947). Deevey and Brooks' (1977) observations also show deeper levels for (*Par*) *euchaeta barbata*, *P. hanseni*, *P. norvegica* and *P. scotti* compared with such forms as *E. marina*, *E. media* and *E. acuta*. Brodsky (1967) cites a number of species of *Pareuchaeta* restricted to the North Pacific. *P. antarctica* is a high latitude southern species, found throughout the Antarctic area always south of the Convergence; though a deep-water species, it migrates towards the surface at night (cf. Hardy and Gunther, 1935; Vervoort, 1951b).

The Phaennidae includes *Xanthocalanus*; all the species of the genus are described by Rose (1933) as living in deep water. The two species identified by Grice and Hulsemann in the Atlantic were both confined to deep layers, one reaching 5000 m. Brodsky (1967) lists *X. polaris* as confined to Arctic polar waters, but *X. borealis* is spread over the North Atlantic, reaching high latitudes including Arctic Seas (cf. Johnson, 1963). Many species are taken in the temperate and boreal Atlantic, *X. pinguis* reaching 61°N. Sewell (1947) has found *X. greeni* in the Arabian Sea in addition to its broad range over the Atlantic. Brodsky describes five species apparently confined to deep waters of the North Pacific.

Onchocalanus and *Cornucalanus* are related genera, apparently all deep-living calanoids (cf. Owre and Foyo, 1964; Vervoort, 1965; Deevey and Brooks, 1977). *O. magnus* is a widely distributed species occurring in the three oceans and in the Antarctic (Vervoort, 1951b).

The family Scolecithricidae comprises several sub-surface deeper-living calanoids. The cosmopolitan species *Scolecithricella minor* usually lives at comparatively shallow depths. *S. ovata*, another fairly widely ranging species in all three oceans, is also well known from the Antarctic, and although it reaches far north it does not extend to the Arctic Ocean. Grice and Hulsemann (1965) obtained several species of *Scolecithricella* from the north-east Atlantic showing a gradation in vertical range, e.g. from *S. tenuiserrata* living just beneath the surface to a few species (*S. lobata* and *S. valida*) occurring at really deep levels (cf. Table 2.17).

Eleven species of the related genus *Amallothrix*, from temperate and warm areas of the North Atlantic and some other areas, are all described by Rose (1933) as deep water forms. Most of the species do not penetrate very far north in the Atlantic, though *A. emarginata* is reported from 65° N (Sars, 1925) and Sewell (1947) describes a wide geographical spread of this species in the three major oceans. Although Deevey and Brooks (1977) record *Amallothrix emarginata* as shallowly as 500 m (Table 2.17), all other species occurring in their Sargasso Sea collections are said to range at least deeper than 1000 m, including *A. valida*, another wide ranging species in the Atlantic, Pacific and Malay Archipelago and Antarctic Seas, according to Brodsky. *Scottocalanus persecans* and *Scottocalanus securifrons*, found in all three oceans in temperate and especially warm seas, live mainly bathypelagically but can approach the surface (Vervoort, 1965). Sewell (1947) records *S. securifrons* as far north as the Faroe–Iceland area. Roe (1972b) found *S. helenae* abundant in mid-water hauls, but the species was fairly sharply restricted to depths of about 500 m (cf. Deevey and Brooks, 1977).

One of the most important deep-sea genera of Scolecithricidae is *Scaphocalanus*. Grice and Hulsemann (1965) list eleven species from their north-east Atlantic collections; all live deeper than 200 m but the vertical ranges differed (Table 2.17). Deevey and Brooks (1977) list some nine species from the Sargasso Sea and describe the genus as among the important deeper-living calanoids. Roe (1972b) found *S. echinatus* as the commonest *Scaphocalanus* living in mid-water. Other comparatively common species were *S. magnus* and *S. major*. All species were essentially deep-living in the temperate Atlantic, the latter species exclusively (Table 2.17). *Scaphocalanus magnus* has a worldwide distribution (Sewell, 1947) in temperate and tropical areas of the three oceans, reaching very high latitudes, including the Arctic and Antarctic. Whereas in the Arctic it lives in both shallow and deeper layers, it is deep-living elsewhere. *S. brevicornis* also may live at intermediate depths in the Antarctic, but can occur at the surface at night (cf. *S. subbrevicornis* Table 2.17). According to Ostvedt (1955), *S. brevicornis* is a permanent member of the plankton in the Norwegian Sea at depths exceeding 1000 m. *Lophothrix* is another mainly deep water genus of Scolecithricidae (Owre and Foyo, 1964; Vervoort, 1965; Roe, 1972b; Deevey and Brooks, 1977). *L. frontalis* is one of the very broadly distributed species, ranging throughout temperate areas of the North and South Atlantic, Indian and Pacific Oceans, though not extending to Arctic or Antarctic areas. According to Sewell (1947), its vertical distribution is about 400–2000 m.

Other deep-living calanoids belonging to a relatively small family include

Table 2.17. Distribution of some deeper-living species of Scolecithricidae

Species	Geographical Region	Depth Range[a] (m)	Authority
Scolecithricella minor	Norwegian Sea	100–600	Ostvedt, 1955
S. ovata	Antarctic	500–1000	Vervoort, 1951
	Greenland, and Bering Sea	> 350	Brodsky, 1967
	Open Oceans	1000–3000	Brodsky, 1967
	Atlantic Canaries	350–940	Roe, 1972b
S. abyssalis	Atlantic, Pacific, Mediterranean	1000–4000	Brodsky, 1967
	North Atlantic	200–500	Grice and Hulsemann, 1965
S. vittata	Atlantic Canaries	85–<u>150</u>–570	Roe, 1972b
S. valiada	North Atlantic	2000–3000	Grice and Hulsemann, 1965
Amallothrix emarginata	North Atlantic	1500–2000	Grice and Hulsemann, 1965
	Sargasso	500–2000	Deevey and Brooks, 1977
Scaphocalanus curtus	North Atlantic	200–1000	Grice and Hulsemann, 1965
S. magnus	Sargasso	500–2000	Deevey and Brooks, 1977
	North Atlantic	500–2000	Grice and Hulsemann, 1965
	Atlantic Canaries	570–<u>800</u>–950	Roe, 1972b
S. subbrevicornis	North Atlantic	450–2250	Grice and Hulsemann, 1965
	Antarctic	0–1000	Vervoort, 1951b
	Pacific	1000–4000	Brodsky, 1967
S. elongatus	Sargasso	500–2000	Deevey and Brooks, 1977
	North Atlantic	1500–4000	Grice and Hulsemann, 1965
S. echinatus	Atlantic Canaries	400–<u>500</u>–950	Roe, 1972b
	Sargasso	500–1000	Deevey and Brooks, 1977

[a] Depth of maximal abundance underlined. Records from Roe (1972) refer to maximum depth sampled of 950 m.

Bathypontia and *Temorites*. *T. brevis* appears in the Mediterranean, Arctic, Antarctic and North Pacific (cf. Brodsky, 1967). According to Grice (1962) it is a widespread but generally deep-living species. Apart from recording the occurrence of *Temorites brevis* and *Bathypontia minor* to depths of about 1000 and 2000 m respectively in the north-east Atlantic, Grice and Hulsemann (1965) note three new species of Bathypontidae (*Foxtonia*, *Zenkevitchiella* and *Temorites discoveryae*) which may occur to great depths (4000–5000 m). *Temorites brevis* and *Bathypontia* spp. are all deep-living species recorded by Deevey and Brooks (1977) for the Sargasso, as are various species of *Undinella*. This rather isolated, mainly deep-sea genus includes *U. oblonga*, a widespread calanoid in the North Atlantic, North Pacific and Arctic Ocean (cf. Johnson, 1963), with a depth range of 500–1000 m in the North Atlantic (Grice and Hulsemann, 1965).

The family Metridiidae is an important deeper-living calanoid group. Only one of the seven species listed by Farran (1948) of the genus *Metridia* (Fig. 2.61), *M. lucens*, is not described as either deep-water or bathypelagic. Most species are oceanic. Deevey and Brooks (1971, 1977) record the same species as rare above 500 m and mostly occurring below 1000 m. The genus was one of the very few which was dominant below 1000 m depth. *M. curticauda* was also found in the Sargasso Sea, mainly between 1000 and 1500 m. Roe (1972b) described *M. venusta* as one of the commonest calanoids at mid-water and deep levels, with maximum abundance at around 500 m. *M. curticauda*, as an exclusively deep species, had its maximum occurrence at 940 m and *M. brevicauda* was also plentiful in intermediate and deep layers. Roe found Metridiidae to be by far the commonest calanoid family, with especial significance in deep layers. Vinogradov (1970) states that at very great depths (6000–8000 m) in the North Pacific the most abundant families are Metridiidae and Scolecithricidae, together with the genus *Spinocalanus*.

A stepwise arrangement of species with regard to depth, including some species with fairly restricted vertical spread and others with a very wide vertical range, is typical of deep-sea families such as the Spinocalanidae, Aetideidae, Scolecithricidae, Metridiidae, Lucicutiidae, Heterorhabdidae and Augaptilidae (cf. Grice and Hulsemann, 1965) (Fig. 2.75). As regards the Metridiidae, *M. princeps* is widespread in seas including the North temperate Atlantic, Indian Ocean, North and South Pacific and Antarctic. According to Brodsky it never occurs above 1000 m, though Grice and Hulsemann give its range as 500–2000 m. They suggest that *M. brevicauda* (a species found in Atlantic and Pacific Oceans) has about the same vertical range in the North Atlantic. *M. curticauda*, another wide-ranging form in all three major oceans, including tropical areas, and in the Antarctic, is said to occur at depths of 1000–4000 m in the Atlantic, but at lesser depths (minimum *ca.* 600 m) in Antarctic seas (cf. Brodsky, 1967). Some further examples of the stepwise distribution of species of *Metridia* are from Grice and Hulsemann's reports that *M. macrura* extends from 1500 to 3000 m and *M. discreta* from 100 m to nearly 5000 m.

The genus *Pleuromamma* includes several sub-surface species with a fairly wide depth range and at least one species, *P. xiphias*, which is bathypelagic, living at 200–1000 m, rarely deeper (Brodsky, 1967). Another species, *P. gracilis*, is also not usually found at the surface but reaches only moderate depths (about 500 m) (cf. Vervoort, 1965). Both species, with others (e.g. *P. abdominalis*), tend to be widely distributed in all three oceans, mostly in tropical, sub-tropical or warm-temperate areas (Wilson,

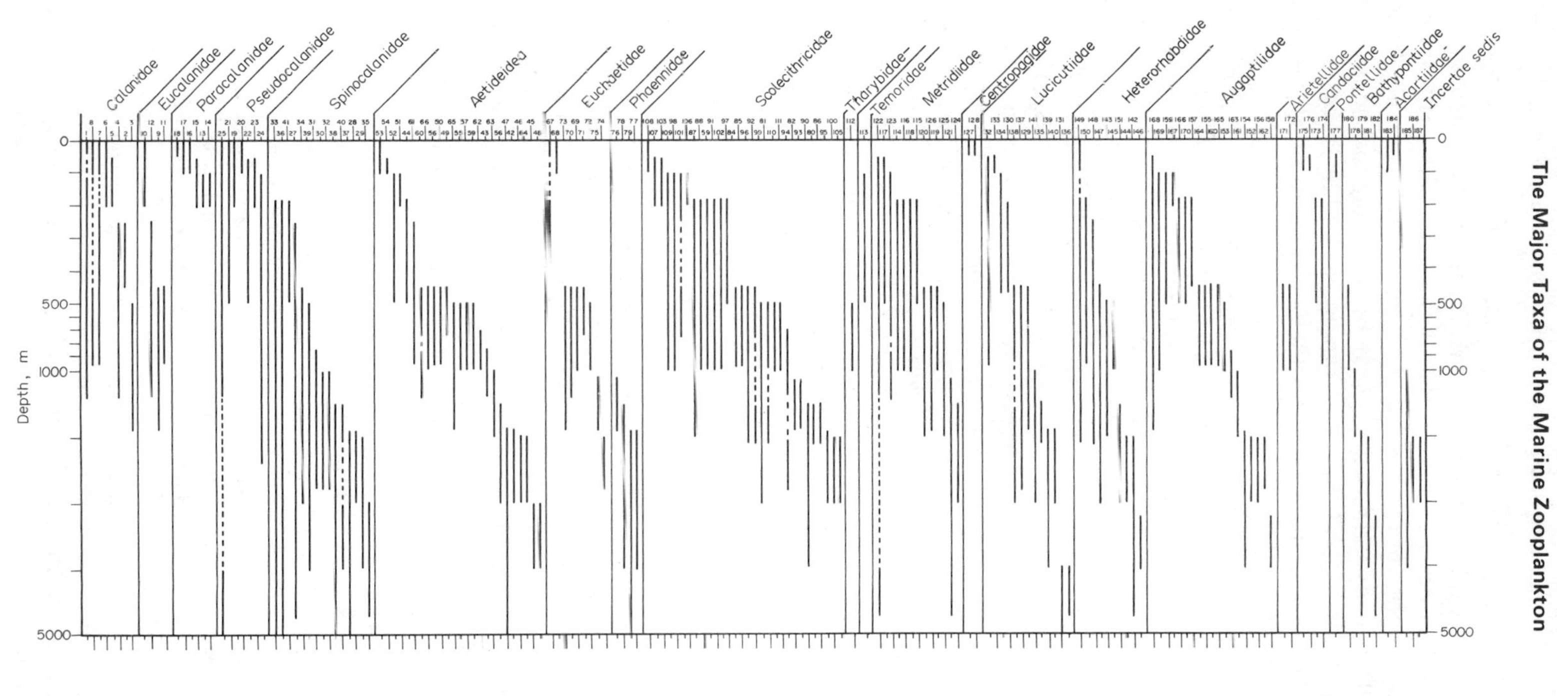

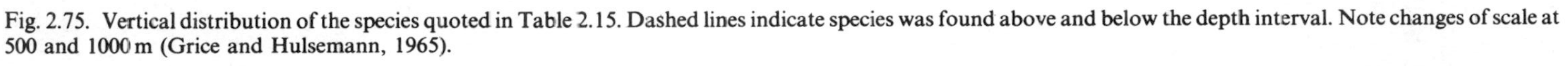
Fig. 2.75. Vertical distribution of the species quoted in Table 2.15. Dashed lines indicate species was found above and below the depth interval. Note changes of scale at 500 and 1000 m (Grice and Hulsemann, 1965).

1932; Sewell, 1947). *P. abdominalis* reaches as far north as near Iceland, however, presumably carried there by warmer currents (cf. Sars, 1925). *P. robusta* appears to have a more northerly extension in the North Atlantic; both Farran (1948) and Ostvedt (1955) record it for the Norwegian Sea. *P. robusta* also occurs in the Antarctic (Mackintosh, 1934; Hardy and Gunther, 1935), but probably does not go much south of the Antarctic Convergence. It is well known in the South Atlantic. It can occur at any depth to about 1000 m, mostly around the 600 m level, and shows pronounced vertical migrations.

Deevey (1971) found *P. abdominalis*, *P. xiphias*, *P. gracilis* and *P. piseki* all the year round in the Sargasso Sea; she also quotes the results of Fish (1954) that *Pleuromamma* was an important copepod at Station "E". Deevey and Brooks (1971, 1977) found that *Pleuromamma* could occur at all depths in the 2000-m water column, but it was important in the upper 500 m in the Sargasso and was one of the dominant genera at depths between 500 and 1000 m. Whereas *P. gracilis* and *P. piseki* were dominant in the upper waters, *P. abdominalis* and *P. xiphias* were most abundant from 500 to 1000 m. Roe (1972b) found *Pleuromamma* to be an outstandingly important genus in waters off the Canaries. *P. abdominalis* was one of only two calanoids to contribute more than 5% to the whole population of the water column. The water layer from approximately 200–400 m was the level where *Pleuromamma* was overwhelmingly dominant. Roe obtained a remarkably clear grading of maximal levels of abundance: *P. gracilis* (200 m); *P. piseki* and *P. borealis* (350 m); *P. robusta* and *P. abdominalis* (400 m); *P. xiphias* (570 m). Grice and Hulsemann's (1965) results also suggest that *P. abdominalis*, *P. gracilis* and *P. robusta* are found broadly at depths of 200–1000 m; *P. piseki* occurs rather more shallowly and *P. xiphias* is somewhat more restricted to the 450–1000-m layer.

Of other Metridiidae, *Gaussia princeps* (=*scotti*) is an abyssal copepod found in all three oceans, mainly in warmer areas. Saraswathy (1973) has described a separate species, *G. sewelli*, apparently confined to the northern Indian Ocean.

Lucicutia, with a relatively large number of species, is described by Brodsky as essentially a deep-water genus. Deevey and Brooks (1971, 1977) state that *Lucicutia*, like *Pleuromamma*, can occur at all depths in the Sargasso Sea, though mostly in the upper 500 m. The genus includes surface-dwelling as well as bathypelagic species. Of the deeper forms, *L. curta* was present in the Sargasso between 1000 and 2000 m; *L. ovalis* and *L. wolfendi* occurred mostly below 1000 m and *L. pera* and *L. longiserrata* from 500 to 1500 m. Owing to the abundance of surface species, however, the total numbers for the genus fell below 500 m. Roe (1972c) also found maximum numbers of *L. curta* comparatively deep (800 m); *L. longiserrata* and *L. magna* had maximal abundance at approximately 625 m. *Lucicutia flavicornis* is a widespread species in all three oceans. The stepwise pattern of vertical distribution already described is well illustrated with regard to species of *Lucicutia* as recorded by Grice and Hulsemann.

Deeper-living species with a very large vertical range included *L. longiserrata* and *L. ovalis* (approximately 450–3000 m); those with relatively narrow ranges included *L. grandis*—1500–2000 m; *L. intermedia*—2000–3000 m; *L. curta* and *L. anomala*—4000–5000 m. As regards the geographical spread, *L. grandis* is widely found north and south in the temperate Atlantic as well as in the Pacific and Indian Oceans, including tropical areas, and usually in the Antarctic, according to Brodsky. *L. bicornuta*, with a similarly wide distribution in all three oceans and the Antarctic, usually inhabits deep layers below 1500 m (cf. also Lysholm and Nordgaard's records). Two species, *L. polaris* and *L. anomala*, are Arctic organisms (cf. Johnson, 1963; Brodsky, 1967), again found at fairly considerable depths, though the latter species has been described by Grice and Hulsemann from the North Atlantic. Several other species are recorded from Antarctic waters (e.g. *L. maxima*, found also in the Atlantic and Pacific; *L. frigida*, mainly in intermediate layers; *L. curta* and *L. grandis*, from deep Antarctic waters) (cf. Farran, 1929; Hardy and Gunther, 1935).

Five genera included in the family Heterorhabdidae may all be described as deep-living, although some species inhabit intermediate depths and may even approach the surface; many have a great vertical range. Vinogradov (1970) lists only eleven species belonging to the Euchaetidae, Lucicutiidae and Heterorhabdidae for the upper 500 m of the Kurile–Kamchatka area, but a doubling of the number of species in the 500–1000 m layer and relatively little further change until the 2000–4000 m level, when more than forty species were present. Several species were taken at depths exceeding 6000 m. All three families had a broad eurybathic range, but this was especially true of the Heterorhabdidae; more than half the species had a vertical range exceeding 6000 m. For example, *Heterorhabdus compactus*, taken by Brodsky in a 6000–8500-m haul, was found by Geptner at shallower levels. Grice and Hulsemann (1965) found it ranged from 3000–4500 m at various North Atlantic stations. Roe (1972c) found a small day maximum at 940 m but he suggests that most of the population lived more deeply and that the species is confined to deep waters, though showing a wide range. Geographically the species ranges through the temperate North Atlantic and North Pacific and into the Arctic Ocean where it appears to live more shallowly (Johnson, 1963). It is found at intermediate depths in the Antarctic (Hardy and Gunther, 1935).

One of the best known members of the Heterorhabdidae is *Disseta palumboi*, a deep-living oceanic copepod widely distributed in temperate and warm areas of North and South Atlantic and in the Pacific and Indian Oceans. It has been taken even off South Georgia. It is one of the most abundant of the deep-living calanoids (Sars, 1925). Roe (1972c) found it abundant in his deepest hauls but concluded that the majority of the population was deeper—the species is confined to deeper water. Grice and Hulsemann indicate a range of 500–2000 m, with a deeper habitat (2000–4750 m) for *Disseta minuta*, a new species from the North Atlantic. *Heterorhabdus norvegicus* is widely recognized over the temperate and northern North Atlantic and in the Arctic and is common in deeper waters with a depth range of 100–2000 m (Sewell, 1947), or even deeper, according to Brodsky. Roe (1972c) found it from the deepest levels to about 400 m, but it lives more shallowly in the Arctic (Johnson, 1963). The species is not found in the Pacific Ocean (Grice, 1962).

H. abyssalis is found at generally similar depths in all three oceans, but not at polar latitudes according to Brodsky. *H. spinifrons*, found in all three oceans and the Mediterranean in temperate and tropical regions, has a very wide depth range from just below the surface in some areas to more than 1000 m in other oceans (cf. Sewell, 1947; Grice and Hulsemann, 1965; Roe, 1972a). While several other species of *Heterorhabdus* are deep-living, some are near-surface forms but with a wide depth range. *H. papilliger* is found in temperate and warm oceans at only moderate depths (cf. page 224). Although the genus *Heterorhabdus* was represented at all depths and was among the more numerous of the carnivorous copepods in the Sargasso, the greatest contribution was in the 500–1000-m level, and while *H. papilliger* was more abundant in the upper layer, *H. compactus*, *H. clausi* and *H. norvegicus* were common between 500 and 2000 m (Deevey, 1971; Deevey and Brooks, 1977).

Other genera of Heterorhabdidae, mostly deep-living, include *Heterostylites*, *Hemirhabdus* and *Mesorhabdus*. *Heterostylites major* is found widely in the Atlantic (including high latitudes) as well as in the Indian and Pacific Oceans and Antarctic. According to Farran and Vervoort (1948) it has a wide depth range, but Brodsky (1967) records it mostly in the 1500–3000-m layer. *Hemirhabdus grimaldi*, a large calanoid, is found at great depths in the Atlantic and Pacific (Rose, 1933).

The Augaptilidae, another important mainly deep-sea family is described by Vinogradov (1970) as one of the five families of calanoids with maximum species diversity at depths exceeding 1000–2000 m, especially in tropical and sub-tropical parts of the Pacific and Indian Oceans. In the North Pacific it is one of the more abundant families in the 500–2000 m strata. Characteristic genera include *Haloptilus*, many species of which inhabit moderate depths. *H. longicornis* is known from all three major oceans in warm and temperate areas; although described as bathypelagic by Rose (1933), it tends to inhabit deeper sub-surface waters (Table 2.18). *H. acutifrons*, a species with a wide distribution in all the oceans, including warm seas, is also found in the Arctic living more shallowly (Johnson 1963). Hardy and Gunther (1935) record it from intermediate depths in high southern latitudes. *H. oxycephalus*, living in the Antarctic both north and south of the Convergence, is found mostly in the epiplankton. Outside the Antarctic it is reported from the Mediterranean and the Atlantic, Pacific and Indian Oceans at moderate depths (Table 2.18) (cf. Sars, 1925; Sewell, 1947; Wilson, 1950). The characteristic Antarctic species *H. ocellatus* is, however, rarely taken north of 60°S. Deevey (1971) includes *Haloptilus* among the commoner genera in the Sargasso Sea collections with several species represented, though only *H. longicornis* occurred all through the year. Although the majority occurred in the upper 1000 m some lived more deeply (Table 2.18).

Of the many species of *Augaptilus*, *A. longicaudatus* is described by Rose (1933) and Brodsky (1967) as a bathypelagic species taken relatively frequently in temperate, sub-tropical and tropical waters of the North and South Atlantic, the Mediterranean, and the Pacific and Indian Oceans (cf. Sewell, 1947; Vervoort, 1965). *A. glacialis* is a deep species distributed broadly over the temperate and northern North Atlantic; off Greenland it occurs at depths of more than 1000 m but the species may sometimes be found more shallowly in the Arctic Ocean (cf. Johnson, 1963). Vervoort (1951b) comments on the very wide distribution of *A. glacialis* from the Arctic Ocean and Antarctic waters through temperate and sub-tropical to tropical areas, where it lives exclusively in the cold deep strata. The species appears to be confined to the Atlantic. In contrast, *A. cornutus* is found, according to Brodsky (1967), only in the north-west Pacific and in the Antarctic, but Grice and Hulsemann (1965) report it from deep layers of the Atlantic.

Euaugaptilus, usually regarded as a separate genus, also has a large number of species, at least forty according to Brodsky (1967). Deevey and Brooks (1977) list no fewer than nineteen species from the Sargasso Sea and, with only two exceptions, they are stated to live below 500 m, with many occurring down to the lowest depth sampled (2000 m). The commonest species were *E. bullifer*, *E. filigerus* (a shallower form), *E. magnus* and *E. palumboi* (Table 2.18). The last species was by far the most common

Distribution of some deeper-living species of Augaptilidae

Species	Geographical Region	Depth Range[a] (m)	Authority
Haloptilus longicornis	Atlantic Canaries	40–250–780	Roe, 1972c
	Sargasso	0–1000	Deevey and Brooks, 1977
	North Atlantic	50–2000	Grice and Hulsemann, 1965
	Sub-tropical/temperate oceans	0–175	Vervoort, 1965
H. acutifrons	Sargasso	0–2000	Deevey and Brooks, 1977
	North Atlantic	100–200	Grice and Hulsemann, 1965
	Atlantic Canaries	40–150–625–720	Roe, 1972c
	Arctic Ocean, Greenland	400–>1000	Brodsky, 1967
H. oxycephalus	Antarctic	0–500	Vervoort, 1951b
	Sargasso	0–1500	Deevey and Brooks, 1977
H. fons	Sargasso	1500–2000	Deevey and Brooks, 1977
Augaptilus glacialis	Arctic Ocean	200–2500	Brodsky, 1967
	North Atlantic	600–1000	Grice and Hulsemann, 1965
	Sargasso	500–2000	Deevey and Brooks, 1977
A. longicaudatus	Sargasso	500–2000	Deevey and Brooks, 1977
	All Oceans, Mediterranean	>200	Brodsky, 1967
A. cornutus	Antarctic	3000	Brodsky, 1967
	north-west Pacific	1000–4000	Brodsky, 1967
	North Atlantic	2000–3000	Grice and Hulsemann, 1965
Euaugaptilus palumboi	Sargasso	500–2000	Deevey and Brooks, 1960
	Atlantic Canaries	500–625–940	Roe, 1972c
	Pacific–Suruga	200–500	Brodsky, 1967
	North Atlantic	500–1250	Grice and Hulsemann, 1965
E. bullifer	Sargasso	500–2000	Deevey and Brooks, 1977
	North Atlantic	2000–4000	Grice and Hulsemann, 1965
E. filigerus	Sargasso	0–1500	Deevey and Brooks, 1965
	North Atlantic	200–450	Grice and Hulsemann, 1965
E. hyperboreus	Arctic Ocean	>200	Brodsky, 1967
E. longimanus	Sargasso	1000–1500	Deevey and Brooks, 1977
	North Atlantic	1000–2000	Grice and Hulsemann, 1965
	Gulf of Guinea	170–600	Vervoort, 1965
E. facilis	North Atlantic	2000–3000	Grice and Hulsemann, 1965
	Sargasso	1500–2000	Deevey and Brooks, 1977
	Gulf of Guinea	600	Vervoort, 1965
E. gracilis	Sargasso	1500–2000	Deevey and Brooks, 1977
	North Atlantic	3250–4000	Grice and Hulsemann, 1965
E. hecticus	North Atlantic	100–500	Grice and Hulsemann, 1965

[a] Depth of maximal abundance underlined. Records from Roe (1972) refer to maximum depth sampled of 950 m.

species of *Euaugaptilus* in Roe's (1972c) collections off the Canaries. While some species from the Pacific Ocean are described by Brodsky (1967) as doubtfully bathypelagic, other new species from the north-west Pacific (e.g. *E. parabullifer*, *E. similis*, *E. mixtus*) are listed as abyssal (1000–4000 m).

Some of the better known species of *Euaugaptilus* include *E. magnus*, generally found in the temperate North and South Atlantic and Pacific. It is also recorded from the Indian Ocean by Sewell (1947). Several species (*E. bullifer*, *E. filigerus*, *E. oblongus*, *E. facilis* and *E. nodifrons*) are known from temperate and warmer waters of all three oceans. *E. laticeps*, described from the temperate Atlantic and from the Mediterranean and Pacific, appears to be another relatively common species (cf. Sars, 1925). It was also recorded by Sewell (1947) from the Indian Ocean and apparently from Antarctic waters. *E. hyperboreus* is described by Brodsky (1967) as a deep-living form confined to the Arctic Ocean (Table 2.18) (cf. Johnson, 1963). Grice and Hulsemann (1965) give a detailed analysis of the depth distribution of *Euaugaptilus* in the Atlantic. The vertical spread of the species varies, as with other deep-living families, from the comparatively shallow-living *E. hecticus* to the deeper-living *E. gracilis* (Table 2.18), a species recorded from all three oceans. A few species appear to be more restricted in geographical range. For example, *E. clavatus* is reported by Vervoort (1965) from the Atlantic only, and *E. indicus* from the Indian Ocean (Sewell, 1947).

Centraugaptilus, a related genus of Augaptilidae, includes *C. rattrayi* and *C. horridus*, both classed as moderately deep-living (Lysholm and Nordgaard, 1945). Rose (1933) describes *C. rattrayi* as occurring in temperate and warmer parts of the Atlantic Ocean at depths as shallow as 200 m. Hardy and Gunther (1935) record the species occasionally at intermediate depths both north and south of the Antarctic Convergence. According to Sewell, *C. horridus* occurs in all three oceans, widely in the North Atlantic and in tropical and sub-tropical areas of the Indo-Pacific (cf. Vervoort, 1965). Other generally deep-sea genera of Augaptilidae, with relatively few species, include *Pontoptilus* and *Pachytilus*. Deevey and Brooks (1977) record species of these genera and of *Centraugaptilus* all from depths exceeding 500 m.

A calanoid family which includes many relatively deep-living copepods is the Arietellidae. *Arietellus setosus*, a warm-temperate species recorded widely in all three warm oceans and in the Mediterranean, reaches at least to 63° N in the Atlantic. It is a bathypelagic form, being found at more or less moderate depths of about 100–700 m, but also ranging from 0–1000 m (Deevey and Brooks, 1977). *A. simplex*, also widely distributed in all three oceans tends to reach high latitudes in the north-west Pacific and North Atlantic. According to Brodsky, the species is abyssal with records between 1000 and 4000 m, but a range of 0–1500 m is given for some localities. Sewell (1947) quotes rather shallower depths for the Indian Ocean.

Another genus, *Phyllopus*, is also mainly bathypelagic. *P. impar*, found in the North Atlantic (mostly in temperate and tropical latitudes but ranging to 60°N) as well as occurring off Malaya in the Indian Ocean, is given a vertical range by Grice and Hulsemann of 500–1000 m. They also quote *P. helgae* over the same depth range; this would agree with Deevey and Brooks' (1971) analysis for the Sargasso Sea, where the genus occurred throughout most of the year but rarely outside the 500–1000 m level. *P. helgae* is found in all three oceans; Vervoort (1965) states that in the Atlantic it stretches from off Iceland in the northern hemisphere to as far south as South Georgia, but that it is much less frequently found in the Indo-Pacific. *P. bidentatus* occurs fairly widely and is bathypelagic. *P. integer*, however, is recorded only from off California and at moderate depths, according to Brodsky (1967). Off the Canaries, Roe (1972c) found *Phyllopus helgae* as abundant in mid-water, with maximum numbers at approximately 600 m; *P. impar* occurred sparingly in deep water. Deevey and Brooks (1977) described *Phyllopus* as occurring throughout the water column off Bermuda, but being an important genus between 500 and 1000 m. Six species of *Arietellus* are mostly reported as below 500 m, though *A. plumifer* as well as *A. setosus* reach the upper layer.

Of other families, *Mormonilla* was an important genus, both *M. minor* and *M. phasma* being abundant between 500 and 2000 m off Bermuda. The former species inhabited the more superficial layers.

Deep-sea zooplankton is often said to include a number of cosmopolitan species. Brodsky (1967), however, claims that the bathypelagic calanoid fauna of the North Pacific, which is rich and diverse, comprising around 230 species, exhibits a considerable degree of endemism. He believes that while many recently identified endemic species are related to Atlantic forms, a number of the previously known species are more closely linked to Antarctic calanoids. Deevey and Brooks (1977) list some Pacific calanoids, however, now known from the Atlantic also. Brodsky further claims that the calanoid abyssal fauna is even more specific to the northern Pacific, with not only endemic species but endemic genera. Among genera found in other oceans but with many endemic Pacific species are *Lucicutia*, *Heterorhabdus*, *Pareuchaeta*, *Bathypontia*, *Bathycalanus*, *Spinocalanus*, *Pachyptilus* and *Heteroptilus*.

Wheeler (1970), reviewing the problem of endemism and density of copepods in deep ocean waters, finds for the Atlantic that both species and number of individuals increase with decline in latitude. This may reflect the species gradient in the upper layers (i.e. the deeper faunas may be derived mainly from the upper strata rather than by a slow colonization from north or south at depth). Earlier data for the Atlantic fauna of adult calanoids indicated a decrease in population density with depth: for the 10–50 m depth interval—6–22 individuals/m^3; 200–500 m—9 individuals/m^3; >2000 m—approximately 0.1–0.3 individuals/m^3. Wheeler obtained good agreement between these earlier data and his own findings, which indicated fewer than 0.3 adult calanoid/m^3 at 2000–4000 m in the North and South Atlantic. Calanoids were the most abundant metazoans at depths exceeding 2000 m; moreover, a number of cosmopolitan species occurred at depth. Table 2.19a gives a list of ten deep-water calanoids with an indication from Table 2.19b of their wide occurrence. According to Wheeler, the list could be doubled if the 1000–2000-m stratum were also included. Table 2.20 lists the species taken by Wheeler and includes a considerable number of deep sea calanoids already described in earlier pages. Although a majority of the species have a very extensive vertical range more or less throughout the 1000–4000-m water column, two calanoids, *Foxtonia barbutula* and *Temorites discoveryae*, described from the north-east Atlantic and subsequently from the Indian Ocean, are apparently limited to depths of more than 2000 m.

Somewhat in contrast to Brodsky's findings, Wheeler emphasizes the number of cosmopolitan deep-living calanoids. Although a considerable endemism is revealed in the North Pacific, some species have been identified more recently from other areas, and further study might reveal an increased geographical spread for some of these forms.

Table 2.19a. Ten deep-water calanoid species (Wheeler, 1970)

Species	Region[a]
Mimocalanus cultrifer	2, 3, 4
Spinocalanus abyssalis	1, 2, 3, 4
S. magnus	1, 2, 3, 4, 5, 6
Scaphocalanus brevicornis	4, 5, 6
S. magnus	3, 4, 5, 6
Metridia brevicauda	1, 4, 5
M. princeps	1, 3, 4
Lucicutia curta	1, 2, 3, 4
Heterorhabdus compactus	2, 3, 4, 5
Haloptilus longicornis	1, 4, 5

[a] See Table 2.19b.

Table 2.19b. Occurrence of the ten calanoid species given in Table 2.19a (Wheeler, 1970)

Region	Number species total	Number species in common
1. Bay of Biscay (Farran, 1926) 4000–2000 m	16	8
2. North-east Atlantic (Grice and Hulsemann, 1965) 4000–2000 m.	55	22
3. Antarctic (Vervoort, 1965b) "abyssal species".	48	14
4. Indian Ocean (Grice and Hulsemann, 1967) 4000–2000 m.	67	24
5. Pacific Ocean (Brodsky, 1957) "abyssal calanoid fauna"	113	14
6. Polar Basin (Johnson, 1963) 2000–1000 m	7	4

Epiplanktonic warm water copepoda

Some description has been given of epiplanktonic copepods from colder seas and a few incidental references to near-surface calanoids which may have a wide vertical range have been included in the discussion of deeper living forms. The diversity of essentially epiplanktonic species in tropical and sub-tropical waters, however, is so great that a short summary is essential. A division between the predominantly surface-living Copepoda and the deep-living species of warmer seas appears in a summary by Farran (1936) (Table 2.21) for the Great Barrier Reef collections. Though the list is incomplete and the "deep" fauna is deduced from hauls not exceeding 600 m, many species mentioned in subsequent pages appear in Farran's epiplankton.

Species of *Calanus sensu lato* found in relatively warm shallow layers have already been described (cf. pages 185–86). *Undinula* is a genus of Calanidae typical of warm

Table 2.20. List of adult calanoid copepod species (excluding contaminants) from Atlantic deep water (Wheeler, 1970)

EUCALANIDAE

1. *Eucalanus pileatus* Giesbrecht, 1888
2. *Rhincalanus cornutus* (Dana, 1849)

PSEUDOCALANIDAE

3. *Microcalanus pygmaeus* Sars, 1900
4. *Farrania friqida* (Wolfenden, 1911)

SPINOCALANIDAE

5. *Mimocalanus cultrifer* Farran, 1908
6. *Mimocalanus sulcifrons* new species
7. *Monacilla tenera* Sars, 1907
8. *M. typica* Sars, 1905
9. *Spinocalanus abyssalis* Giesbrecht, 1888
   S. abyssalis var. *pyqmaeus* Farran, 1926
10. *S. angusticeps* Sars, 1920
11. *S. magnus* Wolfenden, 1904
12. *S. spinosus* Farran, 1908
13. *Spinocalanus* species
14. *Teneriforma naso* (Farran, 1936)

AETIDEIDAE

15. *Aetideopsis retusa* Grice and Hulsemann, 1967
16. *Bradyetes brevis* Farran, 1936
17. *Euaetideus giesbrechti* (Cleve, 1904)
18. *Euchirella pulchra* (Lubbock, 1856)
19. *E. rostrata* (Claus, 1866)
20. *Gaetanus curvicornis* Sars, 1905
21. *G. minor* Farran, 1905
22. *G. pileatus* Farran, 1903
23. *Gaidius tenuispinus* Sars, 1900
24. *Pativella naporai* new species
25. *Pseudochirella fallax* (Sars, 1907)

EUCHAETIDAE

26. *Euchaeta marina* (Prestandrea, 1833)
27. *E. media* Giesbrecht, 1888
28. *Paraeuchaeta gracilis* (Sars, 1905)

SCOLECITHRICIDAE

29. *Scaphocalanus brevicornis* (Sars, 1903)
30. *S. echinatus* (Farran, 1905)
31. *S. longifurca* (Giesbrecht, 1888)
32. *S. magnus* (T. Scott, 1894)
33. *Scolecithricella dentata* (Giesbrecht, 1892)
34. *S. laminata* Farran, 1926
35. *S. timida* Tanaka, 1962
36. *Scolecithrix danae* (Lubbock, 1856)

THARYBIDAE

37. *Undinella gricei* new species

TEMORIDAE

38. *Temoropia mayumbaensis* T. Scott, 1893

METRIDIIDAE

39. *Metridia brevicauda* Giesbrecht, 1889
40. *M. discreta* Farran, 1946
41. *M. longa* (Lubbock, 1854)
42. *M. princeps* Giesbrecht, 1892
43. *M. venusta* Giesbrecht, 1892
44. *Pleuromamma abdominalis abdominalis* (Lubbock, 1856)
45. *P. borealis* (Dahl, 1893)
46. *P. gracilis gracilis* (Claus, 1863)
47. *P. piseki* Farran, 1929
48. *P. xiphias* (Giesbrecht, 1888)

LUCICUTIIDAE

49. *Lucicutia curta* Farran, 1905
50. *L. flavicornis* (Claus, 1863)
51. *L. intermedia* Sars, 1905
52. *L. longiserrata* (Giesbrecht, 1889)
53. *L. magna* Wolfenden, 1903
54. *L. ovalis* Giesbrecht, 1898

HETERORHABDIDAE

55. *Heterorhabdus abyssalis* (Giesbrecht, 1889)
56. *H. compactus* (Sars, 1900)
57. *H. papilliqer* (Claus, 1863)
58. *Heterostylites major* (Dahl, 1894)

AUGAPTILIDAE

59. *Euaugaptilus bullifer* (Giesbrecht, 1892)
60. *E. rostratus* (Esterly, 1906)
61. *Haloptilus fons* Farran, 1908
62. *H. longicornis* (Claus, 1863)
63. *H. mucronatus* (Claus, 1863)
64. *H. spiniceps* (Giesbrecht, 1892)

BATHYPONTIIDAE

65. *Bathypontia sarsi* (Grice and Hulsemann, 1965
66. *Bathypontia* species
67. *Foxtonia barbatula* Hulsemann and Grice, 1963
68. *Temorites brevis* Sars, 1900
69. *T. discoveryae* Grice and Hulsemann, 1965
70. *Zenkevitchiella tridentae* new species

Table 2.21. List of Great Barrier Reef collections of Copepoda (Farran, 1936)

Coastal	Open sea		Deep water
	CALANOIDA		
Calanus pauper	*Calanus tenuicornis*	*Euchaeta russelli,*	*Spinocalanus abyssalis*
Undinula vulgaris	*C. minor*	*Scolecithrix bradyi*	*Chiridius gracilis*
Eucalanus subcrassus	*C. gracilis*	*Centropages calaninus*	*Pseudotharybis zetlandica*
Paracalanus aculeatus	*Undinula darwini*	*Temora turbinata*	*Gaetanus miles*
P. parvus	*Eucalanus elongatus*	*Lucicutia flavicornis*	*G. pileatus*
Acrocalanus gibber	*E. attenuatus*	*L. gemina*	*G. minor*
Clausocalanus arcuicornis	*E. mucronatus*	*L. clausi*	*Undeuchaeta plumosa*
Calocalanus pavoninus,	*E. crassus*	*L. ovalis*	*Xanthocalanus squamatus,*
Undinopsis tropicus	*Rhincalanus cornutus*	*Heterorhabdus papilliger*	*Racovitzanus antarcticus*
Euchaeta concinna	*R. nasutus*	*Haloptilus spiniceps*	*Scolecithrix ctenopus*
Scolecithrix danae	*Mecynocera clausi*	*H. acutifrons*	*Scolecithricella dentata*
Centropages furcatus	*Paracalanus denudatus*	*H. mucronatus*	*S. ovata*
C. gracilis	*Acrocalanus longicornis*	*H. longicornis*	*S. profunda*
C. orsinii	*A. gracilis*	*Euaugaptilus palumboi*	*S. vittata*
Temora discaudata	*A. monachus*	*Candacia curta*	*S. tenuiserrata*
Candacia aethiopica	*Clausocalanus furcatus*	*C. bispinosa*	*S. nicobarica*
C. discaudata	*C. farrani*	*C. simplex*	*Scaphocalanus echinatus*
Calanopia elliptica	*C. paululus*	*C. truncata*	*Lophothrix latipes*
Labidocera acutifrons	*Calocalanus pavo*	*C. catula*	*Scottocalanus longispinus*
L. acuta	*C. styliremis*	*C. longimana*	*Scottocalanus sedatus,*
L. laevidentata	*C. contractus*	*Calanopia aurivillii*	*S. australis,*
L. sp.?	*C. plumulosus*	*Labidocera minuta*	*Scolecocalanus galeatus,*
L. detruncata	*Ctenocalanus vanus*	*Pontella securifer*	*S. lobatus,*
Pontella cristata	*Tanyrhinus naso,*	*Pontellina plumata*	*Macandrewella asymmetrica,*
P. fera	*Aetideus armatus*	*Acartia pietschmanni*	*M. sewelli,*
P. danae	*Euaetideus acutus*	*Acartia australis,*	*M. mera,*
Pontellopsis regalis	*E. giesbrechti*	*Acartia danae*	*Temoropia mayumbaensis*
P. krameri	*Euchaeta longicornis*	*A. negligens*	*Metridia venusta*
P. macronyx	*E. media*		*Pleuromamma abdominalis*
Acartia pacifica	*Euchaeta consimilis,*		*P. xiphias*

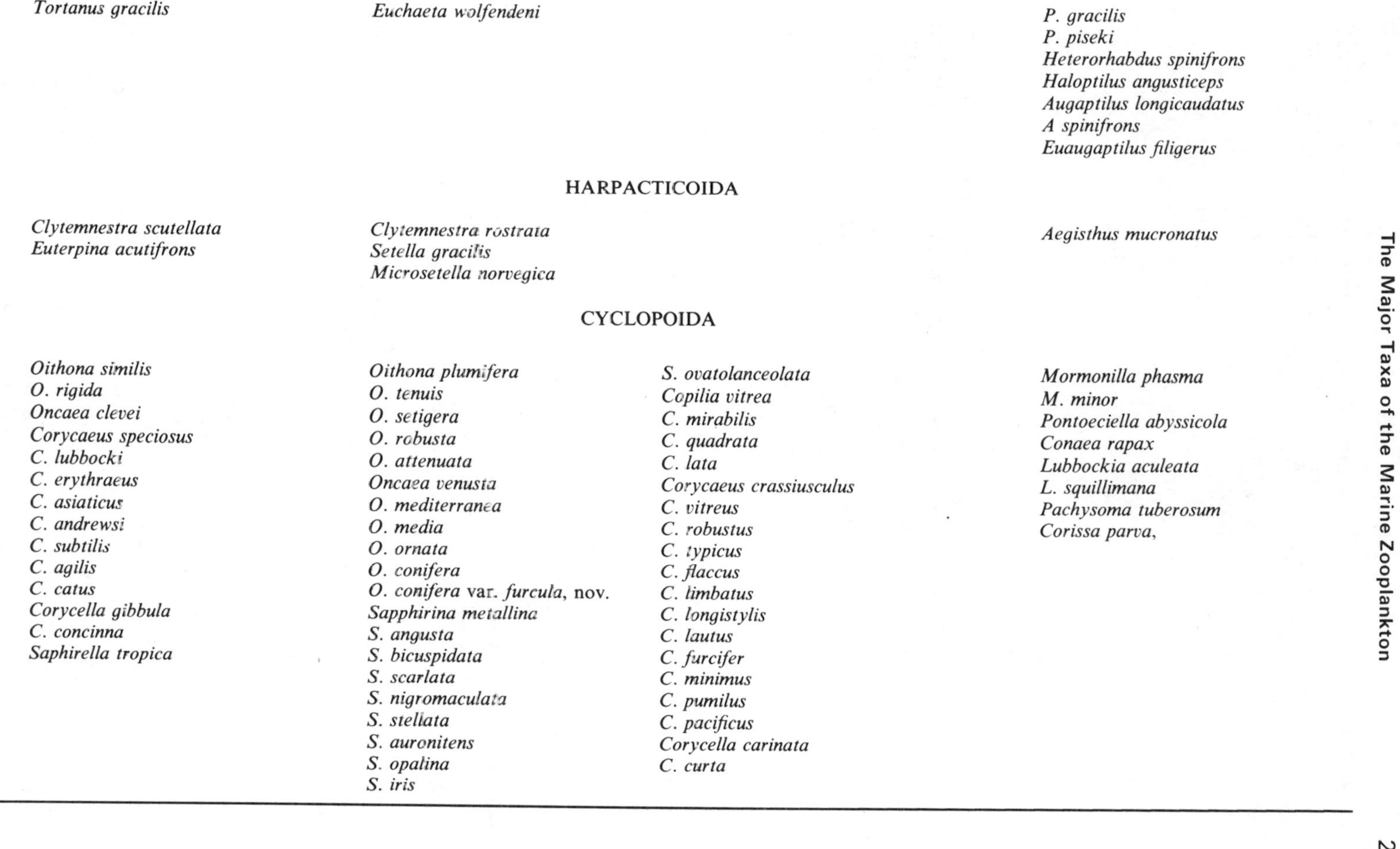

Tortanus gracilis	*Euchaeta wolfendeni*		*P. gracilis*
			P. piseki
			Heterorhabdus spinifrons
			Haloptilus angusticeps
			Augaptilus longicaudatus
			A spinifrons
			Euaugaptilus filigerus
	HARPACTICOIDA		
Clytemnestra scutellata	*Clytemnestra rostrata*		*Aegisthus mucronatus*
Euterpina acutifrons	*Setella gracilis*		
	Microsetella norvegica		
	CYCLOPOIDA		
Oithona similis	*Oithona plumifera*	*S. ovatolanceolata*	*Mormonilla phasma*
O. rigida	*O. tenuis*	*Copilia vitrea*	*M. minor*
Oncaea clevei	*O. setigera*	*C. mirabilis*	*Pontoeciella abyssicola*
Corycaeus speciosus	*O. robusta*	*C. quadrata*	*Conaea rapax*
C. lubbocki	*O. attenuata*	*C. lata*	*Lubbockia aculeata*
C. erythraeus	*Oncaea venusta*	*Corycaeus crassiusculus*	*L. squillimana*
C. asiaticus	*O. mediterranea*	*C. vitreus*	*Pachysoma tuberosum*
C. andrewsi	*O. media*	*C. robustus*	*Corissa parva,*
C. subtilis	*O. ornata*	*C. typicus*	
C. agilis	*O. conifera*	*C. flaccus*	
C. catus	*O. conifera* var. *furcula*, nov.	*C. limbatus*	
Corycella gibbula	*Sapphirina metallina*	*C. longistylis*	
C. concinna	*S. angusta*	*C. lautus*	
Saphirella tropica	*S. bicuspidata*	*C. furcifer*	
	S. scarlata	*C. minimus*	
	S. nigromaculata	*C. pumilus*	
	S. stellata	*C. pacificus*	
	S. auronitens	*Corycella carinata*	
	S. opalina	*C. curta*	
	S. iris		

waters. Both *U. darwini* and *U. vulgaris* are well known from all three major oceans, with the latter spreading somewhat into warm-temperate areas. Wilson (1932) states that it extends as far north as 41°N in the Atlantic Ocean, but Deevey and Brooks do not include it among their more important epiplanktonic genera. *U. vulgaris* occurs mainly nearer coasts in the tropical and sub-tropical Atlantic (Vervoort, 1963), suggesting slight neritic tendencies. It is a very common surface calanoid. Moore and Sander (1977) found it always present and comparatively plentiful off Barbados, where it was the second most abundant copepod. In Caribbean waters it was characteristic and abundant, but restricted to tropical surface and sub-surface waters. Binet (1973) refers to it as a superficial warm water species off the Ivory Coast.

The Paracalanidae are surface or near-surface dwellers. *Paracalanus parvus* has already been described as a widespread species in temperate and boreal waters, though not spreading to really high latitudes. But, the copepod is common also in surface layers of warm seas; Rose (1933) and Brodsky (1967) report its occurrence in Pacific, Indian and Atlantic warm waters as well as in the Mediterranean, Black Sea and Red Sea (cf. also Sewell, 1947). Probably several geographical races exist. Vervoort (1963) describes it as cosmopolitan in all temperate and warm oceans. It can be very abundant in surface waters off West Africa (Bainbridge, 1972) and in shallow waters off Florida (cf. Chapter 8). Farran (1936) records *P. parvus* for the Great Barrier Reef collections, but it was far less plentiful than the related *P. aculeatus*, the most common copepod in Reef hauls. Rose (1933) and Sewell (1947) found this latter calanoid widely distributed in all warm and temperate regions of the three oceans, including the Red Sea; though relatively common in surface waters of both the tropical and South Atlantic, it has not yet been recorded from European coasts in the North Atlantic (Farran and Vervoort, 1951). It is one of the more abundant surface species off African coasts (cf. Vervoort, 1963; Bainbridge, 1972).

Deevey (1971) took both species of *Paracalanus* in the Sargasso Sea, but only *P. nanus* occurred all the year. Grice and Hart (1962) recorded the same three species in their survey from New York towards Bermuda, but only *P. parvus* was encountered throughout all areas. In the Caribbean, *P. aculeatus* was one of the abundant calanoids in the surface tropical and sub-tropical waters (Michel and Foyo, 1976) (Table 2.13) and it was a major species inshore and more offshore at Barbados (Moore and Sander, 1976).

The genus *Acrocalanus*, another member of the Paracalanidae, includes five species of often abundant warm-water surface copepods found in the Indo-Pacific with one species, *A. longicornis*, in the tropical Atlantic (cf. Wilson, 1942; Sewell, 1947). Deevey and Brooks (1977) record it for the Sargasso Sea, though it does not appear to be abundant. Though a considerable vertical range is reported for the Caribbean, it was abundant only in surface tropical waters (Michel and Foyo, 1976). *A. andersoni* has also been described from the warm Atlantic (Vervoort, 1963).

Also typical of the epiplankton of warmer seas is *Calocalanus*. *C. pavo* is a tropical copepod widely distributed in all oceans in warm and warm-temperate areas; the species is occasionally found in the north-east Atlantic (Farran and Vervoort, 1951). Other species, *C. styliremis* and *C. contractus*, occur in the warm-temperate parts of the Atlantic. Bernard (1960) gives details of the distribution of these species in warm-temperate seas. Deevey (1971) and Deevey and Brooks (1977) mention these species as being found all the year in the Sargasso and found *Calocalanus* as important amongst the epipelagic copepods. The same species, with *C. tenuis*, are recorded as surface forms by Grice and Hulsemann (1965). Moore and Sander (1976) describe *C. pavo* as one of the more plentiful calanoids in inshore waters at Barbados and Jamaica. In addition to *C. pavo*, *C. plumulosus* inhabits all three major oceans and the Mediterranean, but according to Sewell (1947) it is more sub-tropical and tropical in distribution.

Mecynocera clausi is a surface form common in the tropical and sub-tropical areas of the three oceans (Vervoort, 1963). Grice and Hart (1962) state that it was numerically important in Gulf Stream and Sargasso samples, being amongst the six most common epiplanktonic copepods. Deevey (1971) and Deevey and Brooks (1977) record it as abundant all the year and an important epiplanktonic species in the Sargasso Sea (cf. also Moore, 1949). Roe (1972a) found it at the surface off the Canaries.

Clausocalanus is another mainly surface-living genus, though some of the species can extend their vertical range to about 500 m depth (cf. page 211); indeed, Farran and Vervoort (1951) describe two species as more deep-water forms. Most species are tropical or warm-temperate, and Rose (1933) indicates a wide distribution in all three oceans, the Mediterranean and Red Sea for *Clausocalanus arcuicornis*, and *C. furcatus*. Sewell (1947) suggests that *C. arcuicornis* may be found far to the south in the Pacific Ocean and north to 59°N in the Atlantic, but that *C. furcatus* is more restricted to warm-temperate waters. Deevey (1971) lists three species, *C. arcuicornis*, *C. furcatus* and *C. paululus*, occurring all the year in the Sargasso Sea.

Although the depth range from the surface is considerable, Deevey and Brooks (1977) rank *Clausocalanus* as a major constituent of the epiplankton. Moore (1949) also found *Clausocalanus* relatively abundant in the Sargasso; in collections by Fish (1954) from the western warm Atlantic *Clausocalanus* formed 16 % of the copepod population. Grice and Hart (1962) found *C. arcuicornis* and *C. furcatus* widely distributed in their collections south from New York to Bermuda, the latter species being especially common in Gulf Stream and Sargasso waters; *C. paululus* was apparently restricted to the Sargasso Sea. Further details of the distribution of *C. paululus* in the three oceans are given by Vervoort (1963). *Clausocalanus furcatus* was the most abundant calanoid taken in the Caribbean by Michel and Foyo (1976). Although it could extend in depth, almost the whole population inhabited the upper 100 m. *Clausocalanus* was also a most important calanoid in Roe's (1972c) collections from off the Canaries. It was entirely a surface genus forming more than 25 % of the copepods at that level. It is abundant also off West Africa (cf. Bainbridge, 1972).

Species of Eucalanidae, living in warmer seas, frequently exhibit a very great depth range. Some may approach the surface layer at least for part of the year and make significant contributions to the epiplankton (cf. Deevey, 1971; Bainbridge, 1972; Deevey and Brooks 1977). Although many species of *Lucicutia* are deep-living, at least two species, *L. flavicornis* and *L. gaussae*, are usually found at the surface, though with a wide vertical range. Grice and Hart (1962) found the former to be one of the commonest copepods in warm waters. In the north-east Atlantic it occurred from just beneath the surface to around 1000 m (Grice and Hulsemann, 1965). It was a sub-surface species, migrating at night to the surface in the Aegean (Moraito-Apostolopoulou, 1971). Roe (1972c) found *L. flavicornis* and *L. gaussae*, the first species being very abundant in surface waters, and although Michel and Foyo (1976) believe that *L. flavicornis* avoids the immediate surface during the day in the Caribbean (as does *Euchaeta marina* and *Acrocalanus*), the majority are in the upper 100 m stratum. The geographical spread of the species is extremely wide, covering tropical, sub-tropical and temperate areas of all three oceans.

The Euchaetidae includes some warm-water, comparatively shallow-living species, some of which are important and plentiful. *Euchaeta marina*, known from all three major oceans, has a very wide distribution in warm and temperate Atlantic waters; Vervoort (1963) suggests at least 52°N to 36°S. It is also broadly distributed over the sub-tropical and tropical Indo-Pacific. The species, which is a very characteristic component of surface plankton, is described by Wilson (1942) as the most widely distributed of the genus from the Carnegie cruise collections. In the Caribbean, Michel and Foyo (1976) report it as the third most abundant calanoid, almost entirely present in the upper 100 m. Roe (1972a) describes it as typically epiplanktonic. Farran (1936) lists species of *Euchaeta* from relatively shallow waters off the Great Barrier Reef.

Several calanoids occur fairly plentifully in surface waters of the warm oceans, although the great majority of species from the families to which these calanoids belong are generally bathypelagic. The Scolecithricidae, for example, have a very substantial number of deep-living species (*vide supra*), but *Scolecithrix danae* and *S. bradyi*, both of which are broadly ranging through the tropical and sub-tropical waters of all oceans and the Mediterranean, are surface dwellers or sub-surface/surface species (Vervoort, 1965). *S. danae* was plentiful and consistently in the upper 100 m in the Caribbean (Michel and Foyo, 1976); Roe (1972b) found both species as surface or sub-surface in distribution. He also recorded a marked stepwise distribution of species of *Heterorhabdus* off the Canary Islands, with *H. papilliger* abundant at the surface, though with small numbers extending down (Roe, 1972c). *H. papilliger* is well known in warm seas and mainly considered as a sub-surface calanoid, sometimes reaching the surface.

By contrast, most Temoridae would appear to be surface or near-surface calanoids (Fig. 2.61). *Temora longicornis*, already described as a widespread species of temperate and boreal regions, tends to be restricted by slightly warmer waters. *Temora stylifera*, on the other hand, occurs as a widely distributed epiplanktonic species in tropical, sub-tropical and temperate seas. Grice and Hart (1962) record it for all the areas from New York to Bermuda, and Moore and Sander (1976) list both *T. stylifera* and *T. turbinata* as common for inshore waters at Jamaica and Barbados. *T. stylifera* occurs abundantly in the North and South Atlantic and the Mediterranean, especially nearer coasts (cf. Bernard, 1970; Moraitou-Apostolopoulou, 1978). It is well known over the Indian Ocean and over part of the sub-tropical and temperate North Pacific (Vervoort, 1965). Bainbridge (1972) found it abundantly both inshore and offshore along the West African coast as a surface copepod; *T. turbinata* was also present.

Sewell (1947) does not record *T. stylifera* from the Indian Ocean, but lists *T. discaudata* and *T. turbinata*. Both species are widespread in tropical, sub-tropical and temperate surface waters of the Pacific and Indian Oceans, *T. discaudata* occurring also in the Red Sea and Mediterranean (cf. Fleminger and Hulsemann, 1973). *T. turbinata* is recorded for the North Atlantic as a straggler even off Woods Hole (cf. Wilson, 1932). The species is moderately plentiful in Florida inshore waters (cf. Reeve, 1964). Owre (1962) suggests that it occurs in both coastal and oceanic warm waters, and Owre and Foyo (1964) found it abundantly in the Caribbean. *Temora* spp. appear to flourish in waters nearer coasts.

The Centropagidae are also mainly surface dwellers. *Centropages typicus*, described for temperate waters (cf. Chapter 8) can extend to moderately low latitudes. It was one of the commonest calanoids in the neritic shelf waters off New York, though greatly reduced in numbers in "slope" waters (Grice and Hart, 1962). The species is also found as far south as the Mediterranean. Other more warm-temperate species include *C. bradyi* found in the Atlantic, the Mediterranean and the Pacific Ocean. Brodsky (1967) describes it as tropical in distribution, but spreading into temperate waters; for example, it is carried a considerable distance north in the Atlantic and Pacific by the Gulf Stream and Kuroshio Currents respectively (cf. Edinburgh Plankton Records, 1973). Roe (1972a) found it only at the surface.

Other warm-water surface Centropagidae include *Centropages violaceus*, found widely in the three oceans and Mediterranean. Vervoort (1965) suggests that its usual limits are 40°N and 35°S. *Centropages orsinii* and *C. gracilis* inhabit warm surface waters of the Indian and Pacific Oceans (Sewell, 1947). The latter has also been identified in the Atlantic. Farran (1936) described *C. gracilis*, *C. orsinii* and *C. furcatus* as coastal species from the Great Barrier Reef; *C. furcatus* was the commonest. *C. furcatus* is widely distributed in the Pacific, often occurring in large numbers (Wilson, 1950) and it is well known in the tropical and sub-tropical areas of the Atlantic and Indian Oceans (Vervoort, 1965; Fish, 1962; Bainbridge, 1972). Owre (1962) regards it as epiplanktonic in both coastal and oceanic areas of the Caribbean.

The Pontellidae may be regarded as an epiplanktonic family, though a very few species extend to some depth from the surface. Few Pontellidae (e.g. *Labidocera wollastoni*, *Anomalocera pattersoni*, *Epilabidocera amphitrites*) live in boreal waters; the great majority inhabit tropical, sub-tropical or warm-temperate surface waters (Fig. 2.61). The genus *Calanopia* includes *C. elliptica*, found in the warm waters of the Pacific and Indian Oceans, the Red Sea and Mediterranean (Sewell, 1947). Both this species and *C. aurivilli* are noted by Farran (1936) for the Great Barrier Reef. Moore (1949) described *C. americana* as one of the most common copepods inside the reefs off Bermuda; Reeve (1964) identified it from shallow waters off Miami, and Moore and Sander (1976) from inshore at Jamaica and Barbados.

The genus *Labidocera* includes several widely-distributed surface species. *L. acutifrons* is found in the warm waters of all three major oceans (cf. Vervoort, 1965); Wilson (1932) believes its occurrence off Woods Hole is mainly due to transport by the Gulf Stream. *L. acuta* similarly is reported from all three oceans, particularly the Indo-Pacific, extending from warm to temperate areas (Sewell, 1947). Fleminger (1975) outlines the distribution of the two species of *L. nerii*, found over much of the warm Atlantic. Some species (e.g. *L. minuta*) appear to be confined to Indo-Pacific warm water areas. *Labidocera* is typically an epiplanktonic warm water genus. Thus, according to Wilson (1942), the Carnegie cruises did not encounter any species of *Labidocera* north of 50°N in the Atlantic and 40°N in the Pacific. One of the commonest species in the more temperate regions off the north-east United States is *Labidocera aestiva*, described as a neritic warm-temperate species by Grice and Hart (1962).

Fleminger (1975) has discussed the biogeography of certain species of the genus

Labidocera in the coastal zones of the warm-temperate waters of the western hemisphere. Different hydrological conditions on the eastern and western sides of the American continent are believed to be partly responsible for considerable differences of distribution of neritic species of *Labidocera* on east and west coasts (cf. Fleminger, 1975) (Figs. 2.76, 2.77). Differences in sexual structures between the species presumably prevent crossing between genetically closely related species where some geographical overlap occurs.

Another essentially surface-dwelling genus is *Pontella. P. securifer* occurs in the tropical and sub-tropical regions of all three oceans, mainly between 35°N and 35°S. Vervoort (1965) doubts reports of its occurrence in cooler waters. Sherman (1963) comments on the abundance of pontellids in extreme surface waters of the central North Pacific (particularly *Labidocera acutifrons*, *L. detruncata*, immature *Labidocera* and *Pontella tenuiremis*). Two species (*Pontella securifer* and *P. princeps*) were common in the equatorial North Pacific (Heinrich, 1971) (cf. Chapter 8). Wilson (1950) draws attention to the very large number of epiplanktonic species of *Pontella* encountered in the Carnegie cruises. An exception to the generally warm-water habitat is *Pontella lobianco*, found in temperate latitudes of the Atlantic and Pacific. According to Rose, it extends from the Mediterranean to the English Channel and is found in the western Atlantic occasionally off Woods Hole (cf. Wilson, 1932).

The genus *Pontellina* was believed to have a single species, *P. plumata*, found generally in the warm waters of the world. Fleminger and Hulsemann (1974) have,

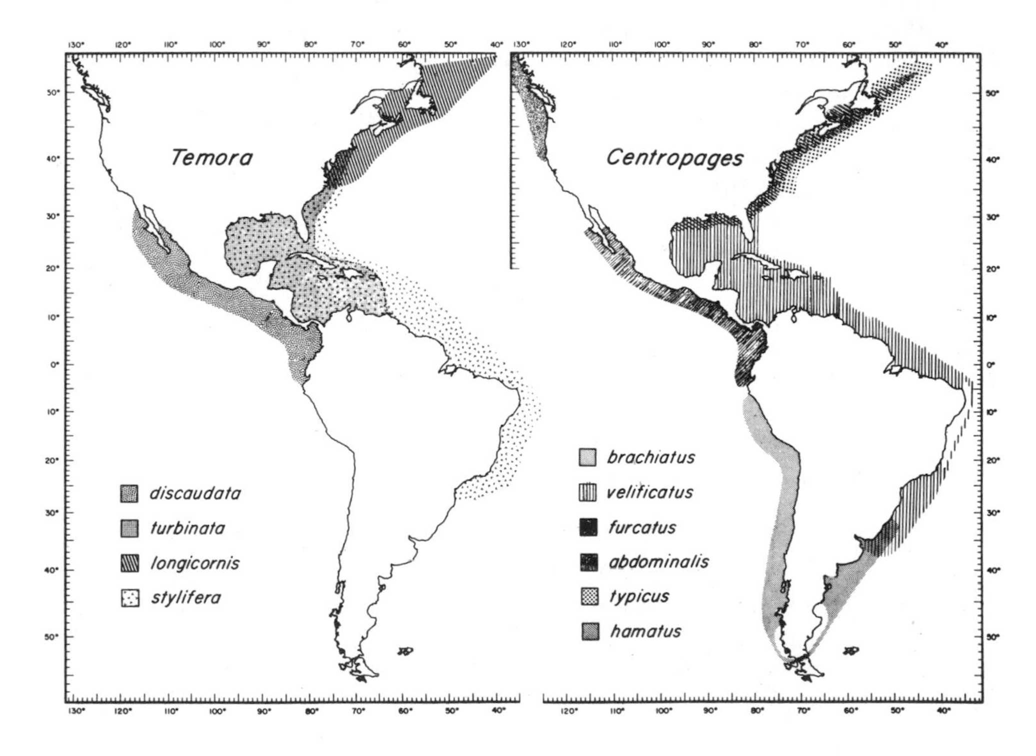

Fig. 2.76. Schematic distribution of species of the calanoid copepod genera *Temora* and *Centropages* in American neritic waters. (Fleminger, 1975).

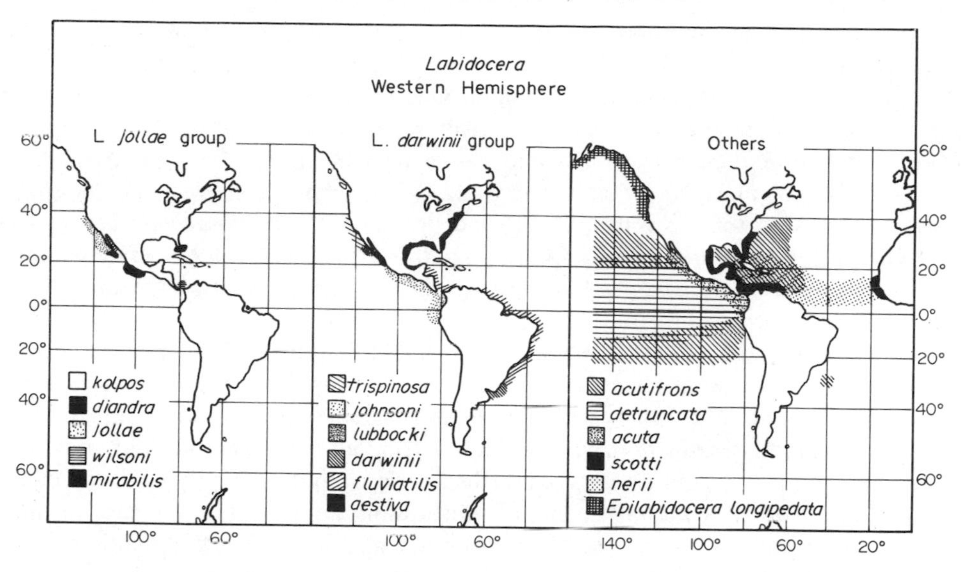

Fig. 2.77. Distribution of *Labidocera* in the Western Hemisphere. Coastal-zone species belonging to the jollae species group are on the left, those of the darwinii species group in the middle panel. *L. scotti* shown in the panel on the right belongs to the darwinii group. Distributions are based on the published records expanded by unpublished data (Fleminger, 1975).

however, distinguished four species. *P. plumata sensu stricto* is a warm-water circum-global species, roughly delimited north and south by the sub-tropical Convergences; though epipelagic, it can extend to sub-surface levels. The three other species, *P. platychela*, *P. morii* and *P. sobrina*, have a more restricted geographical range, the first in the equatorial Atlantic, the next in the Indo-Pacific and the last in the eastern tropical Pacific. All three are limited to the surface (the upper 50 m) and are truly tropical, being found from 20°N to 20°S approximately. Fleminger & Hulsemann suggest that whereas *P. plumata* occurs in oligotrophic areas, the other three species are confined to more eutrophic tropical regions. Other pontellid genera include *Pontellopsis*. *P. regalis* occurs in the warm waters of all three oceans. *P. perspicax*, found as a species in the tropical Atlantic to 13°N and in the Indo-Pacific, appears to be more restricted to tropical waters.

The majority of species of the genus *Candacia* (Fig. 2.61) are epiplanktonic (cf. Moore, 1949) and most are also warm-water copepods, though Brodsky quotes two species from deeper cold layers of the North Pacific. *Candacia armata* is a temperate rather than a warm-water species found over a wide area of the North Atlantic (including the Mediterranean) and reaching moderately high latitudes. Despite Rose's (1933) comment that *C. longimana* (and *C. tenuimana*) are deeper-living, Farran describes these as warm-water epiplanktonic species and Vervoort (1965) lists *C. longimana* as occurring in surface and sub-surface waters of the tropics and sub-tropics of all three major oceans, though extending in the temperate Atlantic.

Many species of *Candacia* appear to range into temperate regions while being mainly distributed in surface tropical or warm waters of all oceans (e.g. *C. curta*, *C. aethiopica*, *C. bipinnata*). Deevey and Brooks (1977) list nine species of Candaciidae, many near surface but some extending more deeply: *Paracandacia bispinosa* occurred moderately commonly. Roe (1972c) found two species, *Paracandacia simplex* and *P. bispinosa*, both relatively abundant in surface hauls. At depths of just less than 100 m the genus made a considerable contribution to the plankton (cf. Fig. 2.77). Moore (1949) found three epiplanktonic species near Bermuda. *Candacia aethiopica* occurred fairly regularly and *Paracandacia simplex* and *Paracandacia bispinosa* more sporadically. According to Wilson (1942) and Vervoort (1965), *Paracandacia simplex* is a very widely distributed surface species in both warm-water and temperate regions of all three oceans. *Paracandacia bispinosa*, though similarly broadly distributed, does not range so far into temperate waters.

The Acartiidae (Fig. 2.61) is a predominantly surface-living family of calanoids found over open oceans but to a marked degree near coasts; some species are typical of brackish waters. Some neritic species may be excessively abundant and dominate the inshore plankton. *Acartia tonsa* is a widely distributed calanoid found abundantly along the north-east Atlantic coasts of North America roughly from the Gulf of Maine to the Gulf of Mexico; it can become of outstanding numerical importance in that region in bays and inshore areas. It is well known also in the South Atlantic off the Brazilian coast, off the Pacific coasts of both North and South America, in the Indian Ocean and off Australia (Rose, 1933). Although Reeve (1964) found that *A. tonsa* was a very important copepod off Miami (cf. also Woodmansee, 1958), other Acartiidae (*A. spinata* and *A. bermudensis*) were probably commoner in that area. The two species were very important in nearshore Bermuda waters (Herman and Beers, 1969).

Of the many warm-loving species of *Acartia* (the genus is now divided into a number of separate genera), *A. danae* is described by Rose (1929, 1933) from the surface of the Mediterranean and warm-temperate areas of the three major oceans (cf. Farran, 1948). Wilson (1950) describes it as a very widely distributed surface-dwelling species from all warm seas. *A. negligens*, while generally similar in its broad distribution in surface warm waters of all three oceans as well as the Red Sea and Mediterranean, is, according to Vervoort (1965), more restricted to latitudes 35°N to 35°S, though it can be more abundant than *A. danae*. According to Moore (1949), it is the most specifically surface-living of the local copepods off Bermuda. Grice and Hulsemann (1965) describe both *A. negligens* and *A. danae* as surface-living forms in their north-east Atlantic studies. Off the Canaries, Roe (1972a) identified *A. negligens* and *A. danae*, both right at the surface where they made a quite significant contribution to the plankton (Fig. 2.77). Other epiplanktonic species are *A. latisetosa* from the temperate Atlantic and warmer areas off north-west Africa, the Mediterranean and the Black Sea, *A. plumosa*, a tropical neritic species in the Indo-Pacific, and *A. pacifica*, another Indo-Pacific species, described as sub-tropical and littoral by Brodsky (1967) and found off the Great Barrier Reef by Farran (1936).

As with many other planktonic taxa, the diversity of calanoids is obviously much higher in warm waters, and this is true particularly of epiplanktonic copepods. The contrast between high latitudes, where some five or six species of surface-living copepods may dominate and be overwhelmingly abundant over enormous areas of ocean, and oceanic low latitudes, characterized by a fascinating variety of copepod species, is very clear.

Ostracoda

Ostracods are not usually regarded as contributing significantly to the biomass of zooplankton. Their rather small size means that their abundance is often underestimated by coarser net hauls. Deevey (1968) found that off Bermuda, however, apart from Protozoa they ranked third in abundance after copepods and tunicates, with numbers ranging from a maximum of $36/m^3$ (March) to $11–15/m^3$ in early July and September–February. In the IIOE collections also, ostracods were next in numerical importance to copepods. They could occur in high density in some areas, though their distribution was patchy (cf. Chapter 3).

According to Vinogradov (1970), in warmer oceans the Ostracoda reach peak numbers in the upper 500 m, but even then they contribute little to total biomass, except for certain coastal tropical regions where they may become really abundant. Although numbers generally decrease with depth, in certain warm oceans some rise in density is evident in very deep layers. For example, in the Bougainville Trench a maximum occurs at depths exceeding 6000 m, due largely to *Archiconchoecia*. Poulsen (1969b) records *Archiconchoecia cucullata* as a bathypelagic species (500–2500 m) in the Atlantic to latitude 55° N. In warm waters the greatest concentrations appear to be almost at the surface, but in cold oceans surface abundance is often less than that in sub-surface layers, even down to 500 m (Vinogradov, 1970). Angel and Fasham's (1975) studies in the north-east Atlantic from 10.5° N to 60° N (longitude 20° W) also showed that the main population of ostracods occurred progressively deeper towards higher latitudes. Although epiplanktonic at low latitudes, they tended to be mainly shallow mesoplanktonic where seasonality was marked (cf. Fig. 2.78). The greatest density occurred at about 30° N. Figure 2.78 suggests that in that area the maximum density by day approached 10 ostracods/m^3. Fulton (1968) gives a density of *Conchoecia elegans* and *C. alata minor* in coastal waters off British Columbia exceeding $1/m^3$ at 50–200 m depth.

Vinogradov (1970) demonstrated that following the peak population at a few hundred metres depth, in a cold sea such as the Kurile–Kamchatka area the numbers of ostracods decreased again in deep layers, with very small populations remaining at depths exceeding 5000 m. Nevertheless, they contributed to the deep sea plankton. The peak biomass in the Kurile–Kamchatka area amounted to *ca.* 1.5–3 mg/m^3 at 100–500 m. At depths exceeding 3000 m it was only a few hundredths of a milligram. Angel

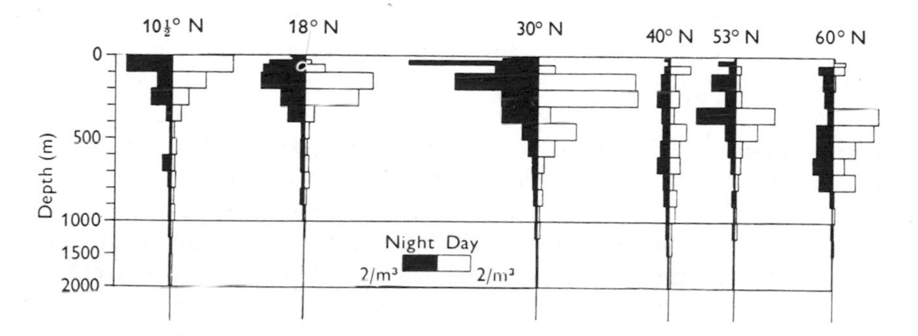

Fig. 2.78. Histograms of the estimated numbers of ostracods per cubic metre in each depth horizon sampled by day and by night at each of 6 positions (Angel and Fasham, 1975).

and Fasham (1975) suggest that the biomass of ostracods, which they estimate in the north-east Atlantic at a somewhat comparable latitude (53°N), is very similar to that quoted by Vinogradov. The number of ostracod species changes, according to Vinogradov (1970), relatively little to 2000 m and perhaps to 3000 m. Although a sharp reduction in species diversity follows in deeper strata, some ostracods are still found. Within 100 m of the seabed a further increase in species, many of which are undescribed, occurs.

Ostracods are easily recognized by the presence of the carapace as a bivalve shell, closed by an adductor muscle, all of the appendages being capable of being withdrawn inside the shell. The head is very little defined and there is no obvious body segmentation externally. There are well developed pairs of first and second antennae, the second typically biramous, and a pair of mandibles with a well developed palp, but the number of paired appendages posterior to the mandibles is markedly reduced. The first pair is usually termed the maxillae; at most, three pairs occur behind this.

The great majority of ostracods are benthic, but there are planktonic members, and one family, the Conchoeciidae or Halocypridae, is an important marine planktonic group. Cypridinids comprise numerous neritic species, especially in the Indo-Pacific. For example, *Cypridina dentata* is the dominant epipelagic species in the upwelling region of the South Arabian coast, and contributes markedly to luminescence in that area (cf. page 233). Both the Cypridinidae and the Conchoeciidae usually have a fairly prominent notch at the anterior end of each shell valve by which the antennae can be extruded. The second antennae are greatly developed and are the real swimming organs. Both families also have the full three pairs of thoracic appendages and both have lamelliform caudal rami; a heart is present (Fig. 2.79a, b).

In the Cypridinidae the shell valves are strong and calcified. There are normally two stalked compound eyes present; the organ of Bellonci, which is homologous to the ocellus in other Crustacea, may have a light receptor function. The second antenna has a very strong basal part and the main ramus (exopodite) forms a strong swimming limb. The mandibles are strong and end in claw-like setae. None of the appendages behind the maxilla takes the form of walking or swimming limbs: the first pair of limbs have vibratile plates at their bases; the last pair are slender, flexible vermiform appendages, possibly used for cleaning. The caudal rami are armed with a series of claws posteriorly, but there are none in the front of the ramus. The eggs are retained for a short time by the female in a cavity of the shell where they complete the first part of the larval development. In all ostracods the larva is a modified type of nauplius in which the bivalve shell is already developed; the appendages, however, are uniramous.

The Cypridinidae include a few neritic epibenthic planktonic species, though the majority are benthic. Deevey (1968) states that probably only two genera should be regarded as strictly planktonic—*Macrocypridina* and *Gigantocypris*. *Macrocypridina* (*Cypridina*) *castanea* is the only species of the first genus, but it has a very wide distribution. Poulsen (1962) records it from all three major oceans, ranging from near surface to at least 3000 m in the Atlantic, to 2000 m in the Indo-Pacific and to 2000 m in the Mediterranean; it is generally more frequent below a depth of about 1000 m. It is a eurythermal species occurring at moderate latitudes, mostly from about 40°N even to 53°S in the Atlantic and Indian Oceans; Poulsen's map (Fig. 2.80) gives a slightly greater range to perhaps 50°N (cf. also Deevey, 1968), but the species clearly does not reach high northern latitudes. Another genus, *Philomedes*, appears to be a partly

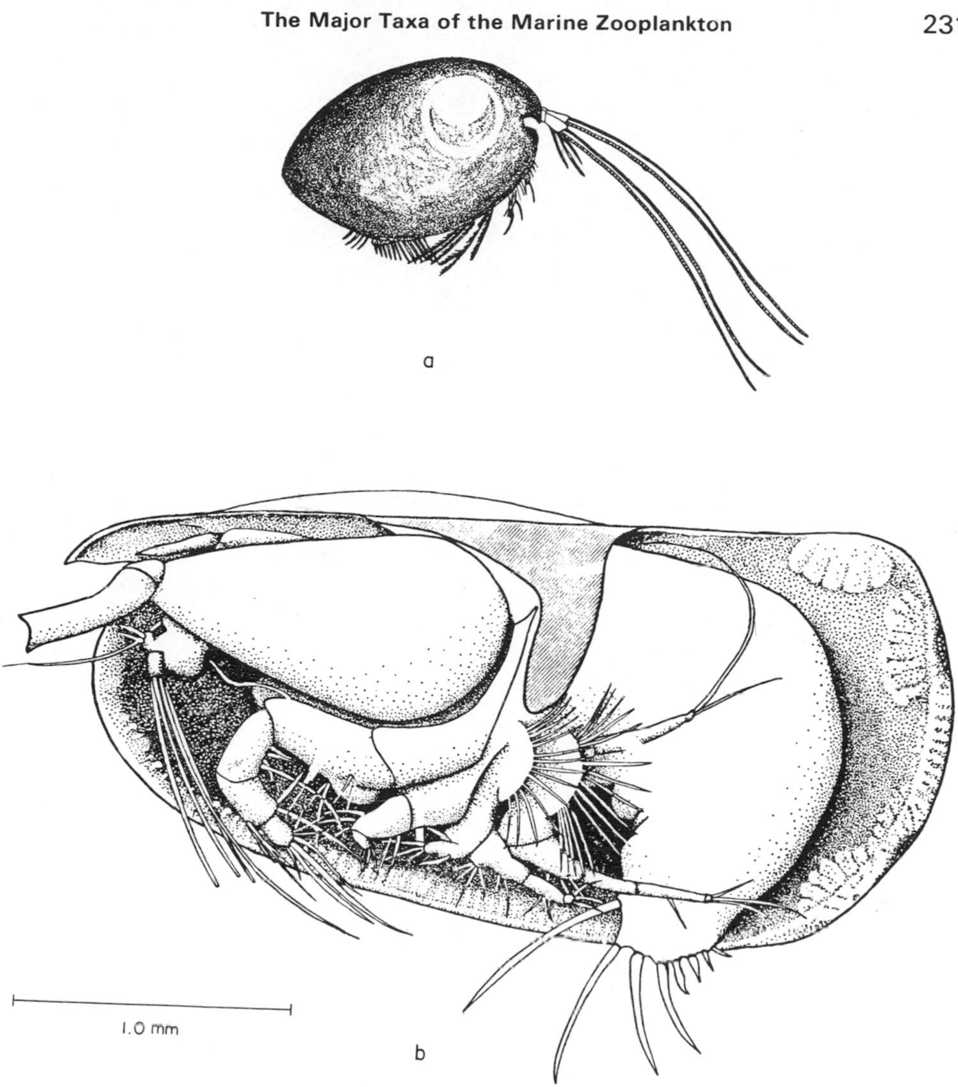

Fig. 2.79. (a) Myodocopid ostracod in lateral view showing antennal notches in valves (Barnes, 1974); (b) *Conchoecia borealis antipoda* ♀ with left valve of carapace removed (Isles, 1961).

planktonic form; one or two species are perhaps planktonic during the reproductive period. *P. globosa* is recorded by Tregouboff and Rose (1957) as appearing at night. Klie (1944a) records *P. globosa* as an Arctic–boreal species occurring widely in northwest Europe but also in the Mediterranean; *P. lilljeborgi* is also northerly in distribution, whereas *P. macandrei* is more southerly in the Atlantic. A few other essentially benthic Cypridinidae may occasionally be taken in deep hauls.

Shallow, mainly surface, hauls may capture certain predominantly coastal ostracods that appear to be largely planktonic. Vinogradov (1970) refers repeatedly to the importance of *Cypridina* in surface waters of marginal tropical seas and coastal areas. Poulsen (1962) lists several species of *Cypridina*, including *C. acuminata* and *C. sinuosa*,

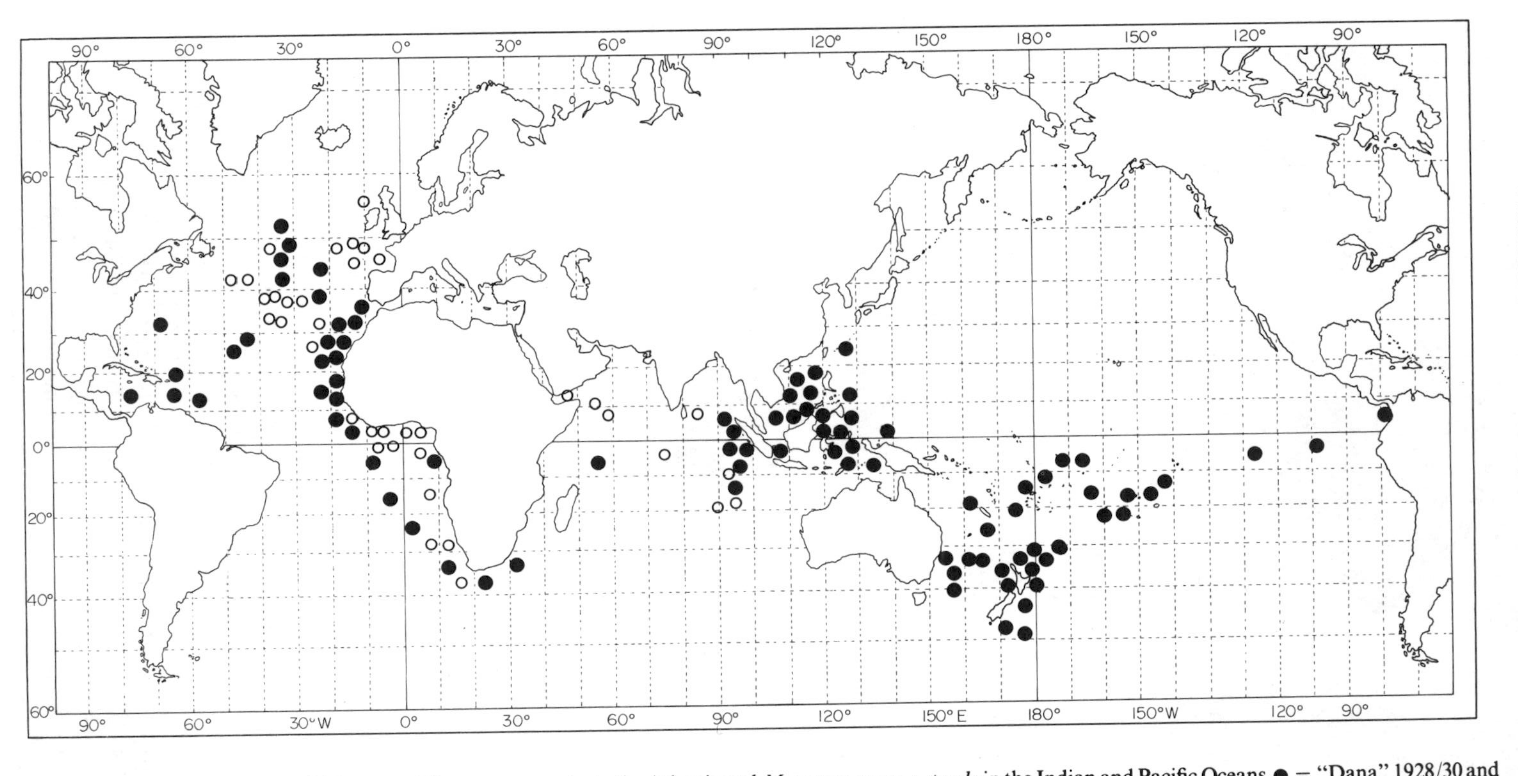

Fig. 2.80. Regional distribution of *Macrocypridina castanea* s. str. in the Atlantic and *M. castanea* var. *rotunda* in the Indian and Pacific Oceans. ● = "Dana" 1928/30 and "Galathea" 1950/51, ○ = earlier workers (Poulsen, 1962).

as shallow water ostracods, mainly in the western Pacific. Some species are now allocated to separate genera. Rudyakov (1972) deals particularly with the species *C.* (*Pyrocypris*) *sinuosa*, found in surface waters of mainly coastal regions round Fiji and other Pacific equatorial areas. Large numbers were also taken in more open tropical Pacific waters (Rudyakov, 1973) and their depth distribution to around 300 m analysed. Some appreciation of their importance may be gleaned from the observation that 3600 individuals were counted and that remnants apparently of the same ostracod species were recovered from the guts of the amphipod *Anchylomera*. George (1967) found ostracods were far more abundant in coastal waters, especially of the northern Arabian Sea, than in open oceanic areas of the Indian Ocean; *Cypridina dentata* was the most plentiful species, and though usually restricted to the coast, could occur plentifully offshore. In studies of the zooplankton of two atolls in the Laccadive Islands, Tranter and George (1972) found very dense swarms in night hauls in one lagoon (Kavaratti Atoll) of a luminescent ostracod belonging to the same genus, *Cypridina*; densities approached 1000 ostracods/m^3.

The other truly pelagic genus of Cypridinidae, *Gigantocypris*, is the largest of the planktonic ostracods, measuring in excess of 20 mm. According to Poulsen (1962), there are probably five species; all are bathypelagic and some reach very great depths. *Gigantocypris agassizi* from the Pacific, particularly the eastern and central areas, is usually deeper than 1400 m and may reach 2500 m. *G. mulleri* is an Atlantic species with a very wide distribution, probably at least 63°N–62°S, and may occur deeper than 2000 m, though the use of open nets makes some of the earlier records doubtful. *G. mulleri* juveniles are found normally at 800–1000 m, while females carrying eggs occurred at *ca.* 1200 m (Moguilevsky 1976). *G. danae* occurs in the Indian Ocean, mostly at depths of around 2000 m. Other bathypelagic forms are *G. australis* in the south-west Pacific and *G. dracontovalis* in the tropical Atlantic, Indian and West Pacific, but has been taken at 42–52°N at 2500 m, near the bottom (Fig. 2.81). All these species are strongly stenothermal.

Skogsberg (1920) states that *Gigantocypris* can live at considerably shallower depths, at least in the Sargasso Sea, and that the specific gravity of the seawater is matched by changes in the specific gravity of the ostracod. But many ostracods show only a moderate reduction of specific gravity, and swimming is an essential part of the maintenance of level in the water. Sinking rate in *Conchoecia* species was observed by Gooday and Moguilevsky (1975) to be related to fullness of gut in both sexes or size of egg mass in females; speed of sinking was usually increased in animals with fuller gut or egg content. Most species showed controlled sinking velocity by orientation with anterior end vertically downwards or at an angle of 45°. Truly planktonic ostracods, however, show a great reduction of calcium carbonate in the shell; deposition of fat may assist in reducing the specific gravity.

This shell reduction is particularly true of Conchoeciidae (Halocypridae), an essentially, probably entirely, planktonic family. Though they share with the Cypridinidae the characteristic notch in the shell, the valves in the Conchoeciidae are usually very thin and are modified in shape so that the dorsal margin is more or less straight. In contrast to the Cypridinidae there are no eyes, but a frontal tentacle which ends in a club-like dilation is present. The three pairs of limbs behind the maxilla, while variable, are more leg-like in form; the last pair is exceedingly small and never vermiform in shape. The caudal lamellae are rounded and there is one well-developed

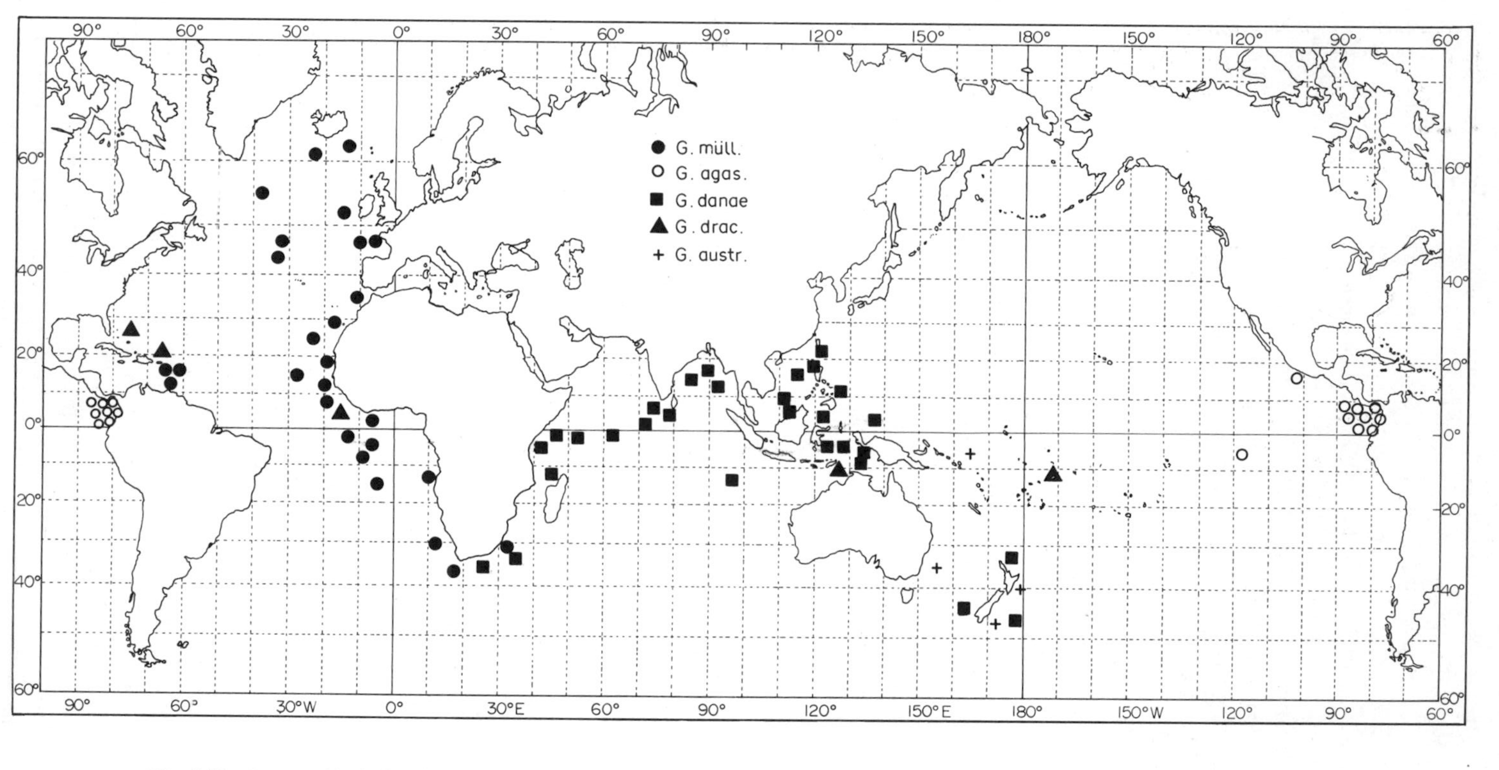

Fig. 2.81. Geographical distribution of species of the genus *Gigantocypris* from "Dana" and "Galathea" records (Poulsen, 1962).

claw-like seta on the anterior edge as well as claws on the posterior edge. In the typical genus *Conchoecia* the shell is generally somewhat elongated. With one exception the eggs are never retained, even for a short period, by the female (Fig. 2.82).

Some of the species of ostracods in this family can be quite abundant; some form swarms. Angel (1970) recorded a remarkable occurrence of *Conchoecia spinirostris*,

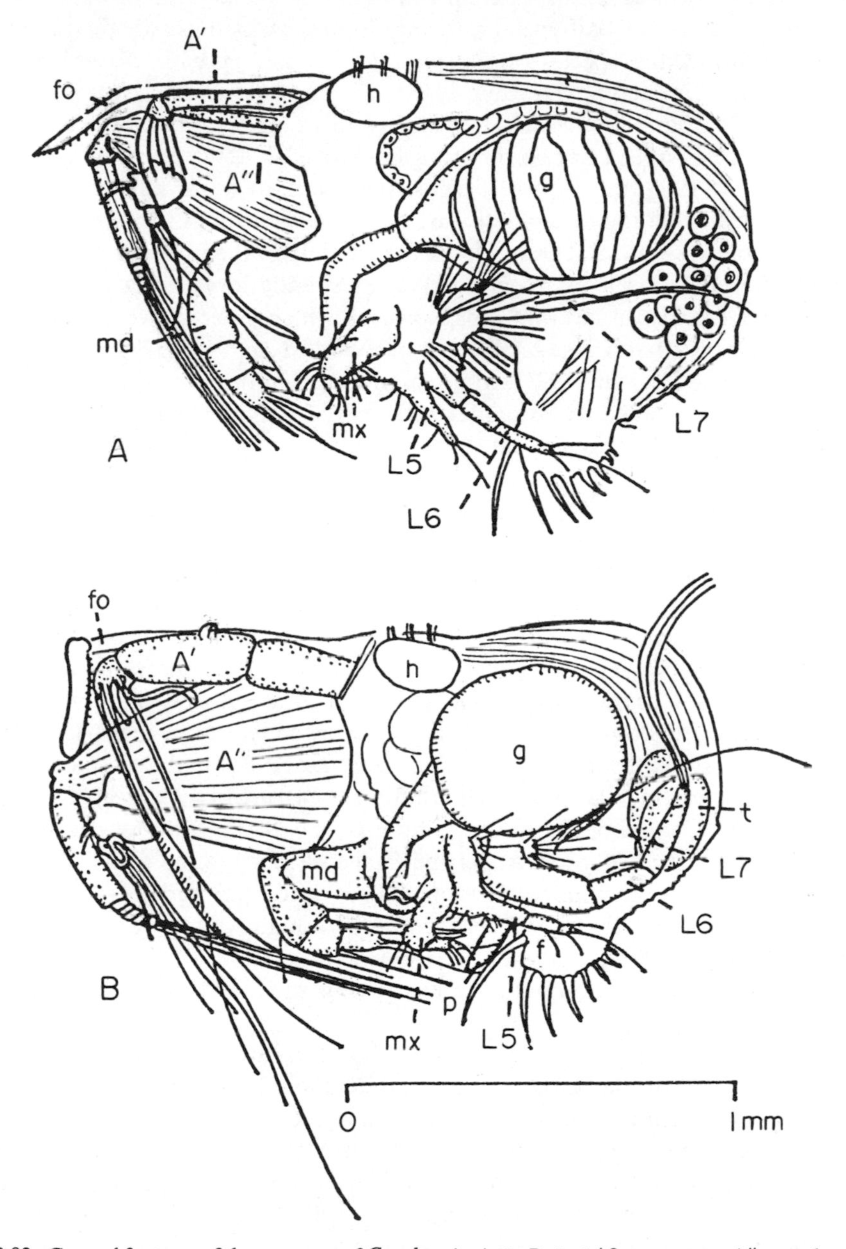

Fig. 2.82. General features of the anatomy of *Conchoecia*. A ♀; B ♂; A′ first antenna; A″ second antenna; fo: frontal organ; h: heart; md: mandible; mx: maxilla; L5, L6, L7: limbs; g: gut; t: testis; p: penis; f: furca (Deevey, 1968).

with enormous numbers of males taken in surface hauls at night by the neuston net; normally females preponderate in the upper layers. Moguilevsky and Angel (1975) found that this species comprised 88 % of total neustonic ostracods taken in Atlantic hauls. The family includes *Conchoecia*, divided by Poulsen (1973) into seventeen different genera (but cf. Angel and Fasham, 1975), *Halocypris*, and one genus, *Thaumatocypris*, with several species including many ancient geological forms and deep-living epibenthic or bathypelagic forms, often separated as a distinct family (cf. Poulsen, 1969b). Many species of *Conchoecia* have a very wide distribution both horizontally and in depth (Deevey, 1968). In the following account of species distribution, the original name for the whole genus *Conchoecia* is retained.

C. elegans is a widespread species, occurring from *ca.* 70° N–50° S latitude (or even more widely—Deevey, 1968) in all three oceans. It is perhaps more abundant at higher latitudes, though avoiding the coldest waters. While plentiful in the tropics, it avoids the warmest surface waters. The species occurs mainly in the upper layers, extending possibly to 500 m. Klie (1944b) refers to it as an essentially eurythermal and euryhaline form. The species exists in several forms, however, which differ in size, the larger form north of 60° N in the Atlantic, the smallest between *ca.* 40° N and 40° S (Poulsen, 1973). The study by Angel and Fasham (1975), while confirming that the large variety occurred in high latitudes, being taken for example, in the Norwegian Sea, showed that at lower latitudes the large form was present, but at greater depths and beneath the shallow smaller variety.

A few species appear to be restricted to colder waters. *C. obtusata*, according to Klie (1944), is a sub-Arctic or northern boreal form, found chiefly north of 60° latitude (50° according to Angel and Fasham, 1975) but appearing also in the Antarctic. It has a wide depth distribution occurring from surface layers to the bathypelagic zone. *C. chuni* is a southern species; according to Poulsen (1973) none have been recorded further north than 30° S. The species has a wide depth distribution, from the surface to more than 1000 m. *C. borealis* is a moderately cold-water species found in the North Atlantic in the more temperate Gulf Stream area as widely as from Bermuda to the Lofoten Islands at depths mainly more than 1000 m at 40°N and 500–700 m at 50°–60°N. *C. maxima* is a high latitude northern form occurring in these deeper layers, but mostly in the Arctic area of the Atlantic about 61–80°N. To some extent the species overlaps the range of *C. borealis*. Angel and Fasham (1975) found *C. borealis* in their samples from middle and higher latitudes. They also recorded *C. maxima* in the deepest hauls at latitude 60°N, and believe that, as a high Arctic species, it may have occurred in a deep Arctic outflow. *C. symmetrica* appears to be restricted to the temperate Southern and Antarctic oceans, living mostly deeper than 500 m (cf. Poulsen, 1973).

The great majority of species of *Conchoecia*, however, are widely distributed mainly in the warm oceans of the world. Thus, Poulsen (1969b) lists amongst other species *Conchoecia macrocheira*, *C. bispinosa*, *C. magna* and *C. hyalophyllum* as warm-water ostracods in all the oceans, limited in the North Atlantic to around 46° N latitude, although the last species was taken at 60°N, and *C. magna* may occur more commonly farther south at 30°N (Angel, 1979). Vertical and geographical distribution must also be considered, however, especially in relation to water masses. While *C. magna*, for instance, was abundant in the upper few hundred metres, an ostracod such as *C. spinifera* also occurs widely in the warm waters of all three major oceans. It is rare at

Table 2.22. List of species of some planktonic ostracods together with their zoogeographical grouping (or factor number) according to Angel and Fasham (1975) in daytime samples in the north east Atlantic (Angel, 1979)

Factor 1. Principally epipelagic species associated with North Atlantic Central Water	
1. *Conchoecia bispinosa*	6. *C. oblonga* Form B
2. *C. ctenophora*	7. *C. parthenoda*
3. *C. curta*	8. *C. procera*
4. *C. magna*	9. *C. secernenda*
5. *C. oblonga* Form A	10. *C. subarcuata*

Factor 2. Widespread shallow mesopelagic species	
11. *C. daphnoides*	16. *C. loricata*
12. *C. hyalophyllum*	17. *C. rhynchena*
13. *C. imbricata*	18. *C. spinifera*
14. *C. inermis*	19. *C. subedentata*
15. *C. lophura*	

Factor 3. Mostly shallow to deep mesopelagic species associated with Antarctic Intermediate Water	
20. *C. brachyaskos*	24. *C. pusilla* (northern population)
21. *C. dentata*	25. *C. pusilla* (southern population)
22. *C. macromma*	26. *C. valdiviae*
23. *C. mollis*	

Factor 4. Principally epipelagic species associated with South Atlantic Central Water	
27. *C. allotherium*	32. *C. microprocera*
28. *C. atlantica*	33. *C. nasotuberculata*
29. *C. echinata*	34. *C. porrecta*
30. *C. giesbrechti*	35. *C. pseudoparthenoda*
31. *C. kyrtophora*	36. *C. spinirostris*

Factor 5. Northern high latitude species	
37. *C. acuticosta*	39. *C. haddoni sensu strictu*
38. *C. borealis*	40. *C. obtusata*

Factor 6. Medium to high latitude deep mesopelagic to bathypelagic species	
41. *C. ametra*	44. *C. kampta*
42. *C. dichotoma*	45. *C. stigmatica*
43. *C. dorsotuberculata*	

Other species	
46. *C. alata*	50. *C. haddoni* (southern form)
47. *C. acuminata*	51. *C. macroprocera*
48. *C. mamillata*	52. *C. teretivalvata*
49. *C. plinthina*	

the surface and is present mainly from 500–1000 m, extending to some 2500 m according to Poulsen (1973). Deevey (1968) gives latitudinal limits for the species of 52°N to 35°S in the three oceans and found few in the upper layers. *C. inermis* is another tropical form found in all three oceans, but mostly in deeper waters; *C. hirsuta* and *C. aequiseta* are warm-water species also present at depth. *C. hirsuta* was found in the Antarctic (Angel, unpublished data). The two species tend to replace each other in the three oceans, with *C. aequiseta* distributed mainly in the Atlantic and Indian Oceans and *C. hirsuta* in the Pacific (Poulsen, 1973).

The importance of considering the vertical and horizontal distribution is also illustrated by the results of Angel and Fasham's (1975) survey in the north-east Atlantic (Table 2.22). Using factor and principal component analysis techniques, they grouped a large number of ostracod species. Thus *Conchoecia aequiseta*, which was not distinguished from *C. hirsuta*, occurred in maximal abundance with *C. macrocheira* at mesopelagic depths around 18°N; *C. inermis* and *C. spinifera* were shallow mesopelagic species, especially plentiful about 30° and 40°N; *C. curta, C. bispinosa, C. magna* and *C. spinirostris* were shallow (mainly 100–300 m) species, with greatest abundance at 30°N and 18°N. Poulsen (1973) also points to *C. acuminata* and *C. curta* as warm-water tropical forms found mostly from about 40°N to 45°S, (Fig. 2.83) but whereas *C. acuminata* is rare at the surface and widely distributed in deeper layers even to 3000 m depth, *C. curta* is mostly abundant at 100–300 m, although it can be neustonic (Table 2.22) (Fig. 2.84). *C. spinirostris* is another ostracod found in sub-tropical and tropical waters in all oceans and abundant in the upper 300 m according to Deevey (1968). It was one of the commonest species in the Bermuda area (cf. Angel and Fasham, 1975).

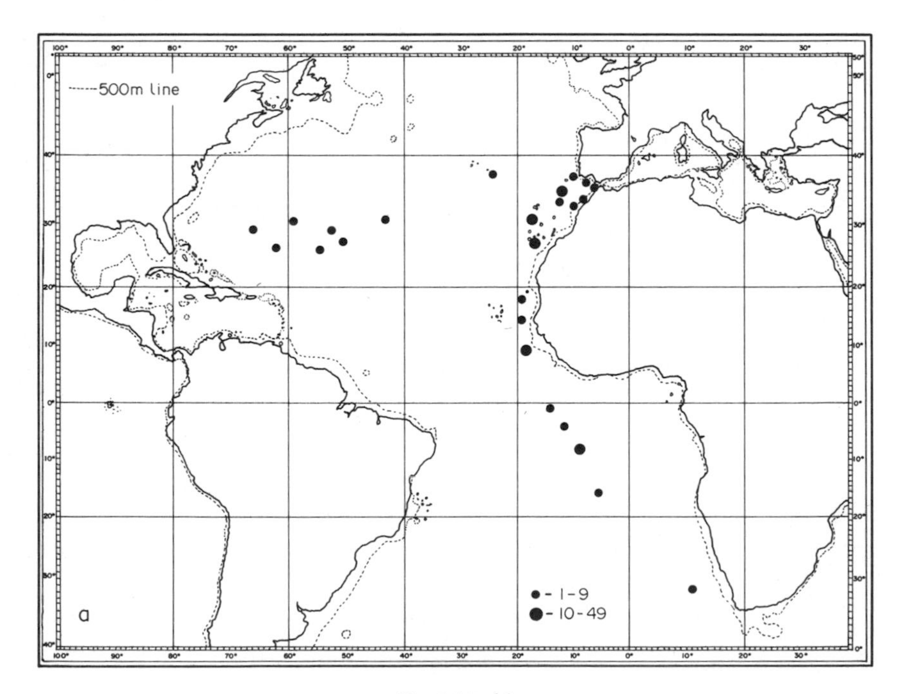

Fig. 2.83. (a)

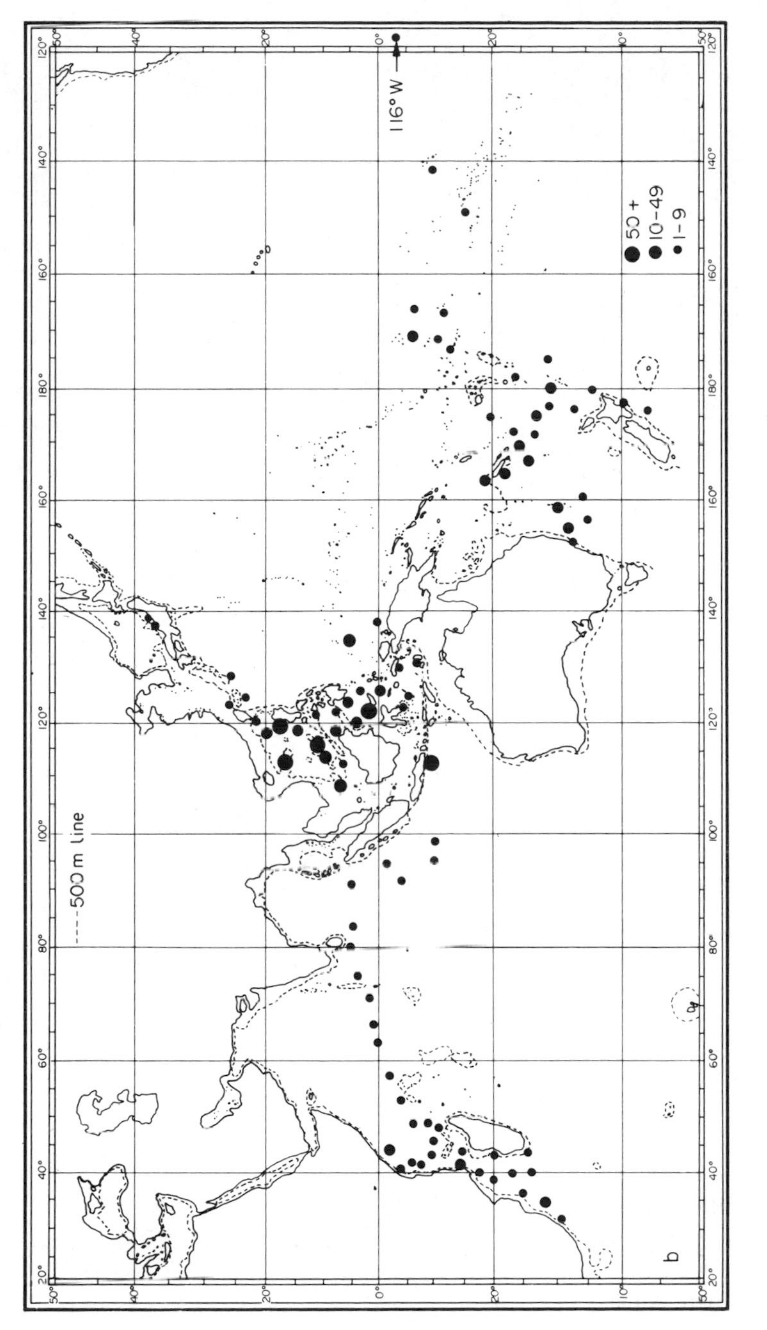

Fig. 2.83. (b) Distribution of *Conchoecia acuminata* (a) in Atlantic; (b) in Indo-Pacific Ocean (Poulsen, 1969).

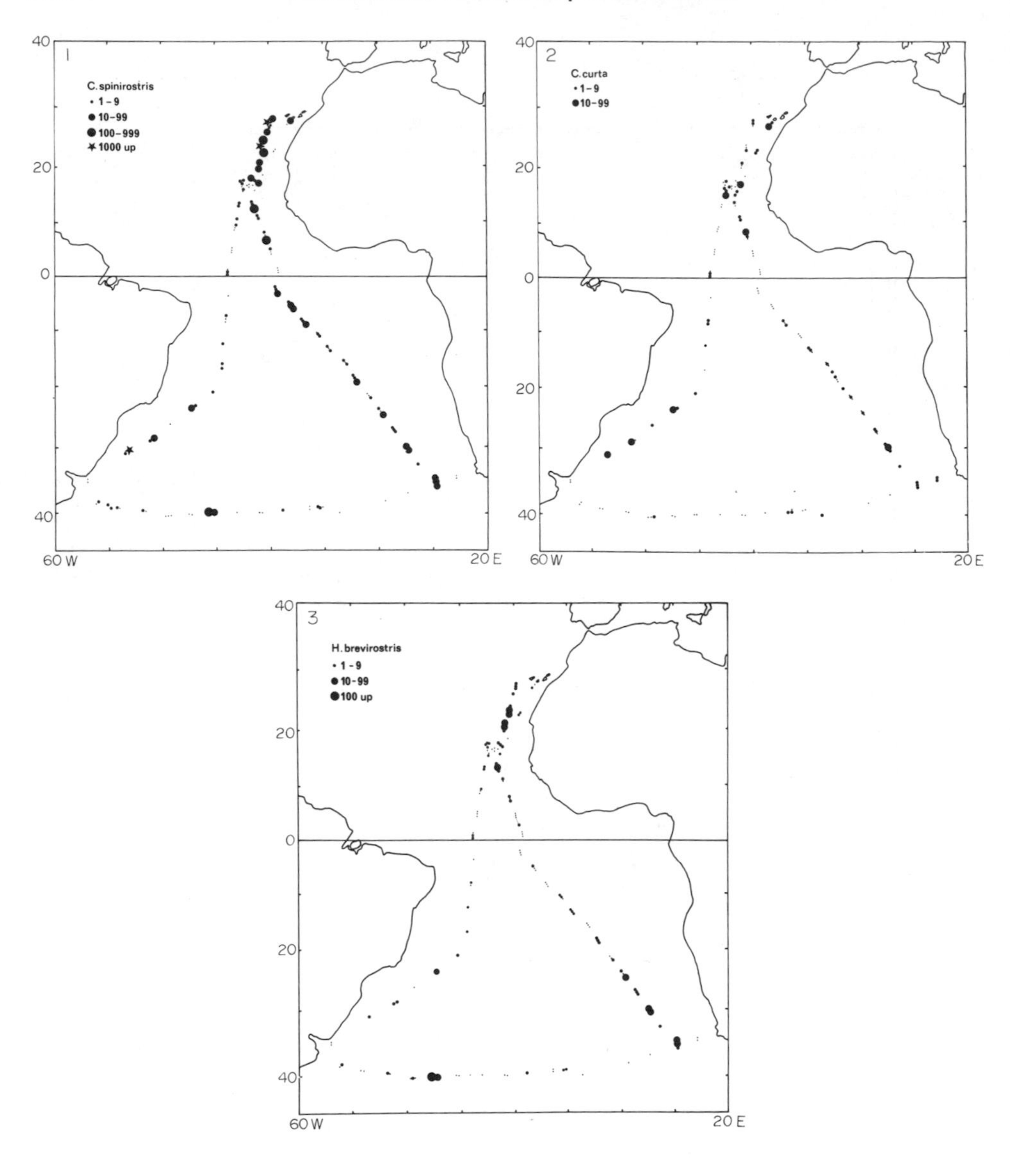

Fig. 2.84. Distribution of the 3 most abundant ostracod species in neuston net samples taken in the Atlantic Ocean. The cruise track is outlined and abundance in samples is given in orders of magnitude. (1) *Conchoecia spinrostris*; (2) *C. curta*; (3) *Halocypris brevirostris* (Moguilevsky and Angel, 1975).

C. haddoni differs in distribution, ranging from more than 60°N to more than 40°S and apparently avoiding the zones closest to the Equator. In vertical range it is said to reach at least 3000 m, though it is comparatively rare in the uppermost strata (Poulsen, 1973). According to Angel and Fasham (1975), *C. haddoni* is another ostracod in which different forms may be distinguished. The smaller "southern" variety occurs in the tropical Atlantic and maximum populations were observed off Cape Blanc. The typical larger form has a distribution at higher latitudes (53° and 60°N in the survey), and

Angel and Fasham include it in the northern group of ostracods, with *C. borealis*, *C. obtusata*, *C. daphnoides* and the higher latitude variety of *C. elegans*. *C. daphnoides* is one of the most widely distributed and abundant of ostracods in all three oceans, ranging from 61°N in the Atlantic to 47°S in the Pacific (Poulsen, 1973). It is more common in temperate and sub-tropical waters than in warmer areas, and becomes more abundant in sub-surface layers, extending to about 3000 m depth. *C. imbricata*, another widely distributed ostracod with a large vertical spread, ranges from the latitude of around Iceland (65°N) in the Atlantic to 47°S in the Pacific (cf. Poulsen, 1973), and has been recorded from Antarctic seas (Fig. 2.85).

Fasham and Angel (1975) emphasize that some of the faunal assemblages of ostracods in their north-east Atlantic survey may be related to specific water masses, but that the determination of horizontal and vertical boundaries are both essential. Moreover, very few species are restricted to one zone only. It is the relative numerical abundance between zones which is really significant (cf. Chapter 8).

As with many other groups of zooplankton, earlier records of depth distribution (e.g. Poulsen's "Dana data" 1969a, 1973), must be examined critically, since closing nets were rarely employed, and a very wide vertical range which is frequently quoted for species may not be accurate. Nevertheless, Vinogradov (1970) mentions *C. curta*, *C. daphnoides*, *C. elegans* and *Macrocypridina* spp., amongst other ostracods, as having a very wide vertical distribution. The majority of ostracods, however, are more limited in depth; relatively few species exceed 3000 m. Deevey's (1968) investigations give only limited information on deeper levels. Most ostracods were found in the upper 300–400 m by day, though some were present in the deepest samplings (2000 m). Few, apart from *C. spinirostris*, were abundant in the uppermost 100 m (but cf. Moguilevsky and Angel, 1975). Among species said to occupy typically a deeper habitat, *C. valdivae* is restricted to deeper water, being common from about 1500–3000 m. It is confined to tropical and near-tropical waters of all oceans. *C. leptothrix* is similarly a tropical and moderately deep-water form (500–2000 m).

The obvious richness of ostracod species in warmer waters is emphasized by the work of Angel (1969) and Angel and Fasham (1975). Off the Canaries, ninety-six species of ostracods were recorded, including thirty-two species of *Conchoecia*. Of a total of ninety-six species identified in their north to south survey along the 20°W meridian, the expected increase in number of species at lower latitudes was very clear, (e.g. thirty-seven species at 60°N, eighty one species at latitude 18°N). Angel (1969) points out that during the day, off the Canaries, there was a distinct layering of species. A zone rich in numbers and species (e.g. *C. porrecta*, *C. spinirostris*, *C. elegans*, *C. acuminata*, *C. oblonga*, *C. bispinosa*, and *C. daphnoides*) extended from the surface to about 250 m. A relatively poor zone occurred at about 300—400 m, with mostly *Macrocypridina*; another rich zone from 450–625 m included *C. imbricata*, *C. inermis*, *C. haddoni* and *C. alata*, and was succeeded by a zone of low density below about 720 m. Two species, *C. ametra* and *C. kampta*, had their maximum centre of abundance deeper than 800 m. In their later survey, Angel and Fasham (1975) include *C. kampta* with *C. mamillata* and *C. plinthina* as a deep mesopelagic group of ostracods ranging from about 18° to 40°N (Table 2.22). *C. ametra* was also mesopelagic ranging north to 60°.

Apart from *Conchoecia*, *Halocypris globosa* is very widely distributed in all oceans. In the Atlantic it occurs from about 65°N to at least the Cape of Good Hope; in the

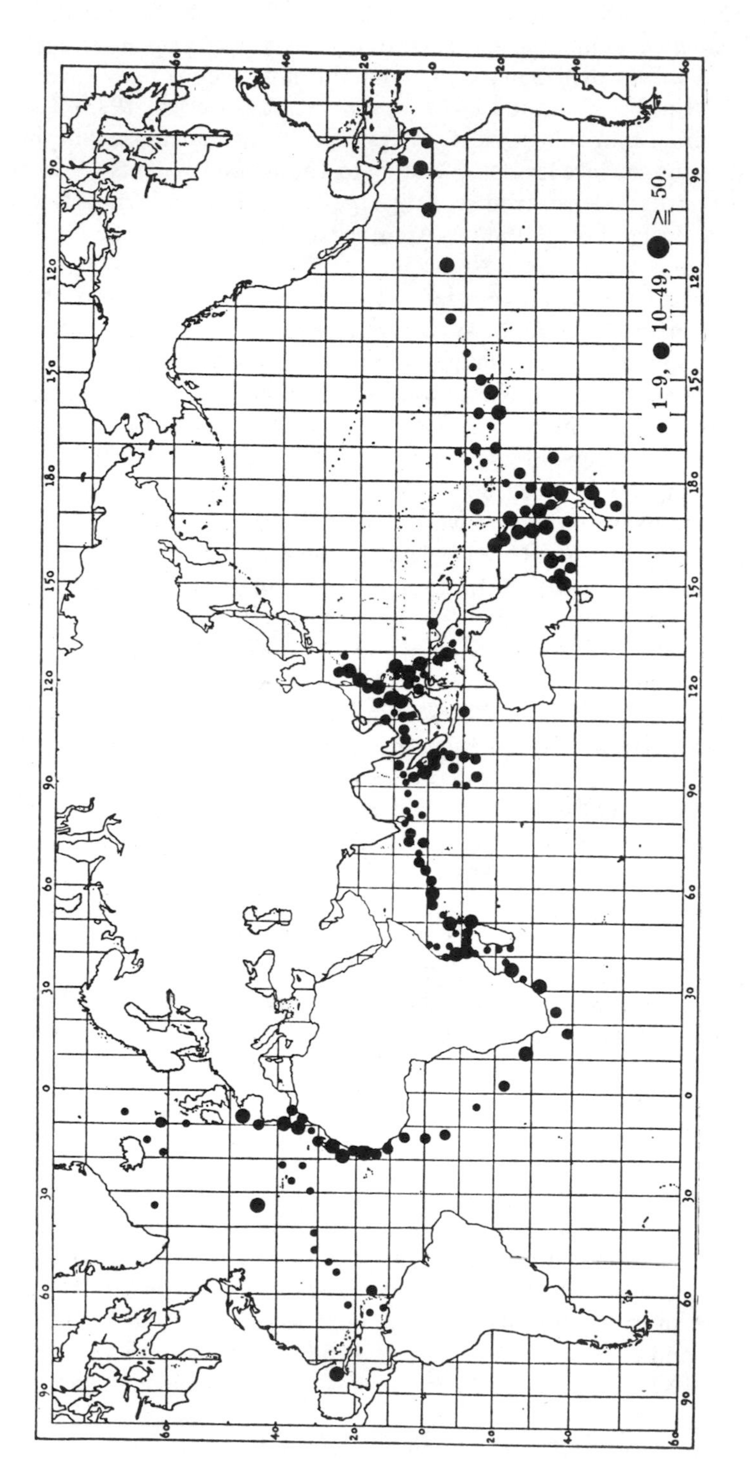

Fig. 2.85. Distribution of *Conchoecia imbricata* (Poulsen, 1973).

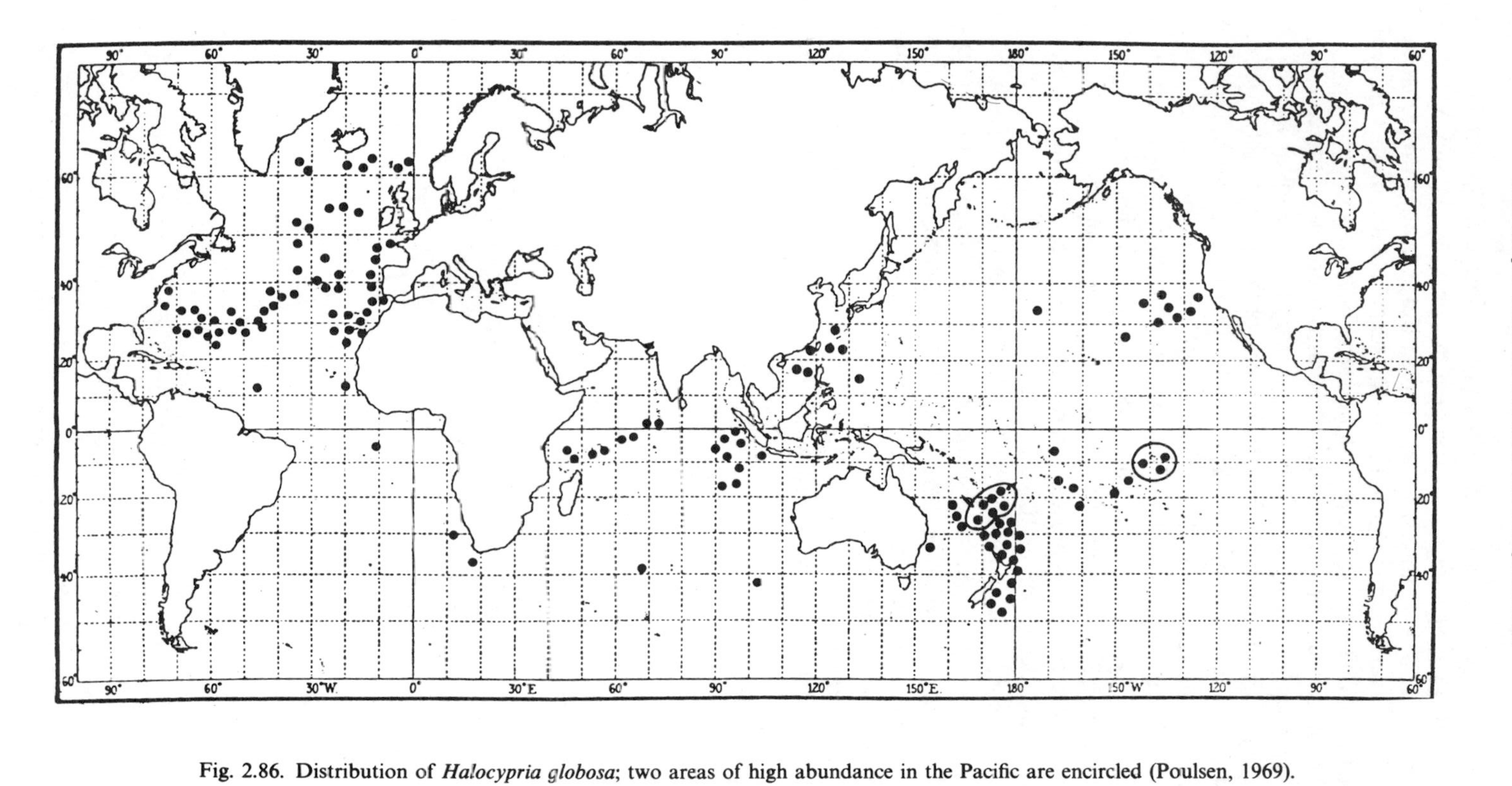

Fig. 2.86. Distribution of *Halocypria globosa*; two areas of high abundance in the Pacific are encircled (Poulsen, 1969).

Indian Ocean it extends to 40°S, and even farther south in the Pacific (cf. Poulsen, 1969) (Fig. 2.86). The species apparently does not occur in the sub-Arctic or farther south than about 50° latitude. It is not so plentiful in tropical regions, but has an enormously wide vertical distribution from the surface to about 3000 m, perhaps less deep in the Indian Ocean. It is clearly an extremely eurythermal species.

The other species, *Halocypris brevirostris*, is also found in all three oceans, but is a warm-water form occurring mostly in tropical and sub-tropical areas. The Dana material does not record it either north or south of latitude 45° (cf. Poulsen, 1969a) (Figs 2.84, 2.87, 2.88). Poulsen (1969a) shows the contrast in distribution between the two species *H. globosa* and *H. brevirostris*. *H. brevirostris* also displays a very considerable depth range with the maximum near the surface, numbers declining in deeper waters, but densities increasing at depths approaching 1500–2000 m and the range extending to 3000 m. Two other species, *Halocypris bicornis* and *H. cornuta*, are fairly obvious tropical species but are very rarely found in near surface layers and reach maximum abundance at about 1000–2000 m.

A few ostracods belonging to the families Cytheridae and Cytherellidae are occasionally encountered in the plankton. In these families, in contrast to the Cypridinidae and Halocypridae, there is a relatively thick shell without any notch; the second antenna has a reduced or vestigial exopodite; there is no heart and only a single nauplius eye. The caudal furca is reduced in the Cytheridae, but present in the Cytherellidae. Representatives of both families are really benthic, but specimens are occasionally collected, mostly in deep-water plankton hauls, according to Tregouboff and Rose (1957). They may approach the surface in more coastal waters in response to artificial illumination; they cannot, however, be regarded as regular constituents of the plankton.

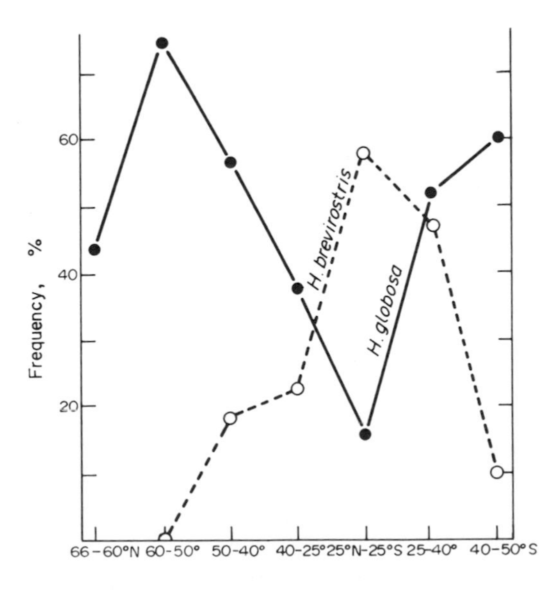

Fig. 2.88. *Halocypria globosa* and *Halocypris brevirostris* showing frequency percentage of occurrence at Dana stations in all oceans from North to South (Poulsen, 1969).

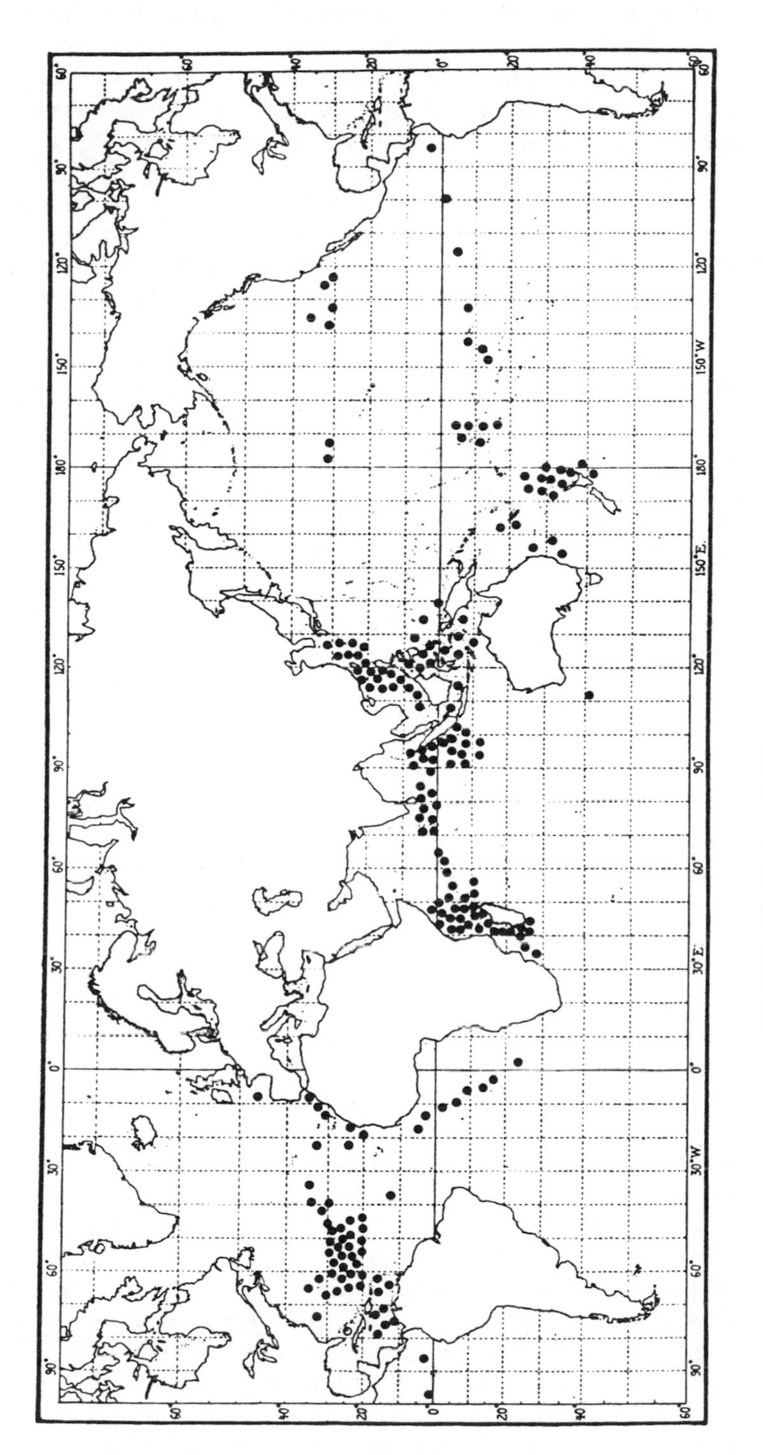

Fig. 2.87. Distribution of *Halocypris brevirostris* (Poulsen, 1969).

Though the ostracods are perhaps not such conspicuous members of the plankton, especially in surface tows at moderately high latitudes, the occurrence of many different species at all levels down to considerable depths, the wide depth range of some species, and their extensive horizontal distribution, render them a reasonably important element in the marine plankton, despite their small size.

Amphipoda

Although the majority of amphipods are benthic crustaceans belonging to the Gammaridea, the Hyperiidea (Hyperina), a smaller sub-order, are exclusively planktonic, but usually constitute a relatively small component of the total zooplankton. Hyperiids with very few exceptions are oceanic, and though different species and families may inhabit the whole water column in the ocean, at certain times a few species may swarm near the surface. *Parathemisto gaudichaudi* swarms in temperate waters. Nair (1972) reported a large dense swarm of hyperiids, mainly *Hyperia sibaginis*, off the Kerala Coast at depths down to more than 150 m.

With few exceptions the biomass of hyperiids is normally not very significant. Nair, Jacob and Kumaran (1973) found that planktonic amphipods showed higher concentrations in the northern part of the Indian Ocean, especially in areas subject to upwelling or land drainage. They stress the high densities of amphipods which can be found in productive tropical and sub-tropical waters. The IIOE collections showed by far the greatest numerical contribution to the total amphipods was due to Hyperiidae. Dunbar's (1942, 1964) studies, indicate that in Arctic or sub-Arctic seas *Parathemisto libellula* and *P. abyssorum* may be widespread and make an important contribution; *P. gaudichaudi* can be plentiful at somewhat lower latitudes. At a depth of some 4000–5000 m, hyperiids may also form a more important part of the zooplankton (cf. Vinogradov, 1973).

The body form of the Hyperiidea varies considerably. In some usually bathypelagic species, the shape may be almost globular, the musculature weak and the cuticle thin. In others the body is comparatively compact, even slender, and the muscles well developed. Most hyperiids have a somewhat swollen head and thoracic region (Fig. 2.89). Typically, in Physocephalata (*vide infra*) the head is globular, with most of the surface bearing a pair of enormously developed eyes, leaving the dorsal midline of the head free for muscle attachment of the gut. *Vibilia*, however, possesses only small eyes. In the Physosomata (*vide infra*), the head is comparatively short and the eyes are small or absent and situated at the sides of the head.

The coxae in hyperiids may be indistinct and fused to the thorax, or separated by a suture, but they are never enlarged as in many Gammaridea. The urosome consists of two segments only, the original second and third segment being fused, sometimes also with the telson. In the Anchylomeridae the uropods appear as single leaf-like segments. In Platyscelidae and Parascelidae the bases of the fifth and sixth thoracic legs are expanded to form opercula covering much of the ventral surface, the abdomen being very narrow and the telson and uropods fitting beneath the body (Fig. 2.90). The antennae of hyperiids lie either anteriorly or ventrally on the head. Antenna 1 never has an accessory flagellum and both antenna 1 and 2 are often reduced in female hyperiids. The flagellum of the antenna may be multi-segmented (e.g. in male Hyperiidae and Phronimidae), composed of one large segment with one or a few small

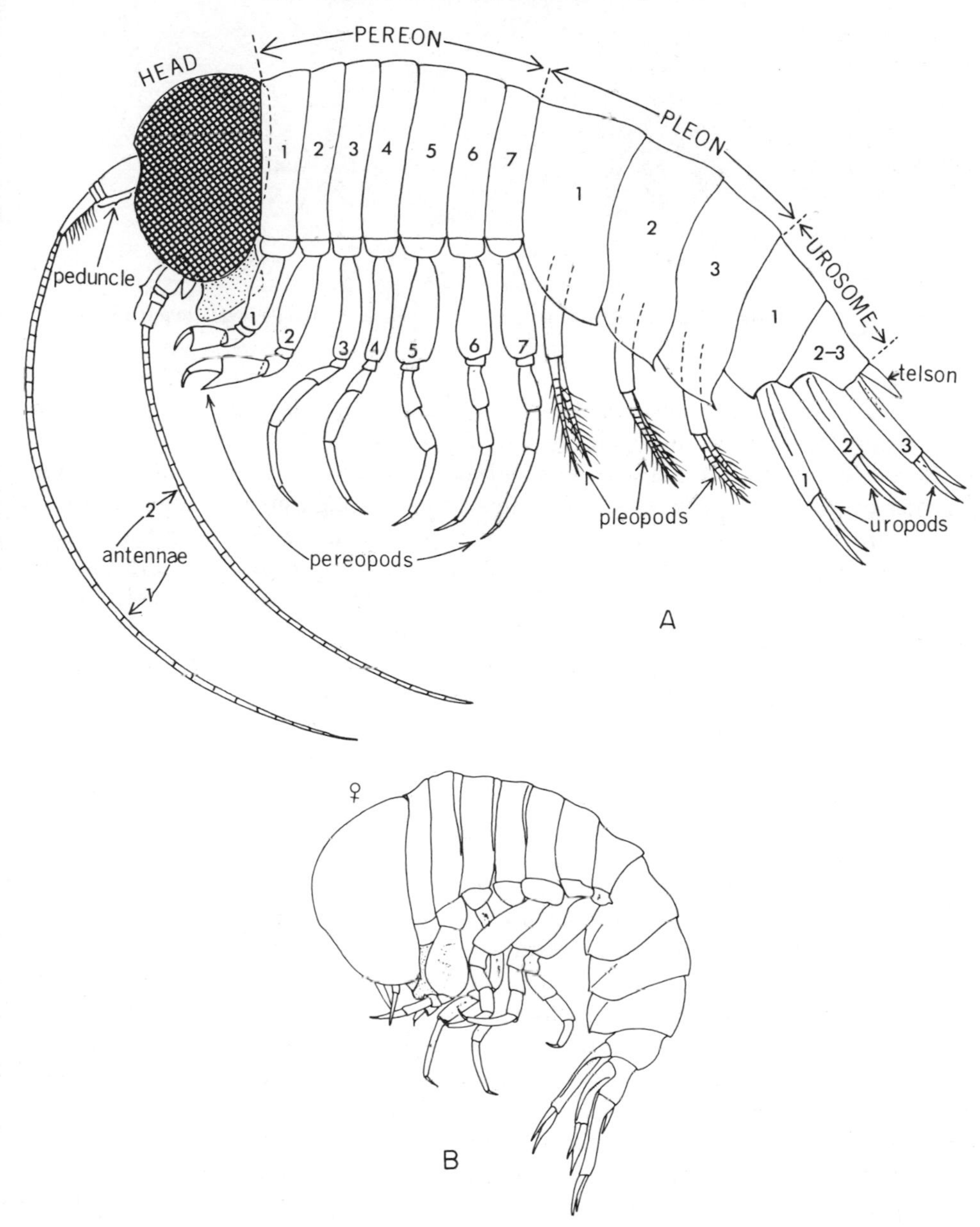

Fig. 2.89. Diagram of a hyperiid amphipod based on ♂ *Hyperia* (A); (B) ♀ *Hyperia* (Bowman and Gruner, (1973).

additional segments (e.g. male Lycaeopsidae), or formed of a few large segments folded upon each other (cf. Fig. 2.90) as in antenna 2 of males of Platyscelidae, Pronoidae and Lycaeidae. In the Oxycephalidae also the males have a well developed and elongated second antenna which is folded or angulated. This family is characterized, however, in that the head usually forms a well developed rostrum so that the eyes, though still large, do not cover the head. The mouthparts in hyperiids are less well developed than in

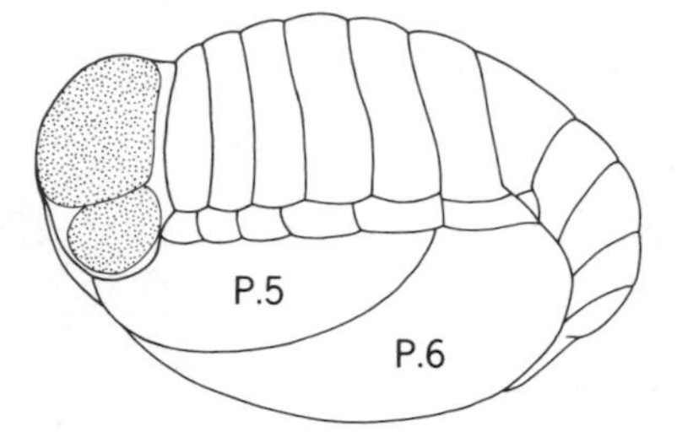

Fig. 2.90. Platyscelidae form (size range 2–24 mm) showing expanded P5 and P6 (pereopods) (Bowman and Gruner, 1973).

gammarids. The maxillipeds lack a palp and their inner lobes fuse in the Physocephalata but remain separate in the Physosomata.

Development is direct in amphipods and the young, released from the marsupium formed by oostegites, appear as miniature adults, though juvenile and adolescent stages are usually distinguished. In many hyperiids, however, the pleopods and uropods are not fully developed when the young stages are released and they are incapable of swimming (cf. Laval, 1965; Bowman and Gruner, 1973; Sheader, 1977).

In the Physosomata, hyperiids in which the head is small and not swollen, several families such as the Lanceolidae and Scinidae, are of generally normal body shape. But other, usually less well known, families (Archaeoscinidae, Mimonectidae, Microphasmidae) show a relatively great swelling dorsally of certain of the thoracic segments, giving an inflated appearance. The inflation is characteristic of the female only in the Archaeoscinidae and Mimonectidae (Fig. 2.91). Stephensen (1923, 1925) described the thin-skinned, almost colourless body, typical of many forms. These families are represented by very few species. They are comparatively rare animals, known from the deep sea. Stephensen (1923) grouped these families under the term "Primativa", and Pirlot (1929) classified them as a tribe, the Physosomata. Bowman and Gruner (1973), reviewing the classification of hyperiids, have retained the name Physosomata for the infra-order. All the remaining families (the Eugenuina of Pirlot) are included in an infra-order, the Physocephalata.

The Physosomata, apart from the Scinidae, are rather poorly known. Shoemaker (1945) found a number of specimens of *Microphasma agassizi* at depth off Bermuda, and several species of *Mimonectes*. *Mimonectes loveni* and *M. gaussi*, found between 1200 and >2000 m depth, were represented by a number of animals. These two species appear to have a wide distribution in deep waters of all three oceans. Vinogradov (1972b) recorded three species of *Mimonectes*, as well as *Microphasma agassizi*, in a wide depth range from a few hundred metres to 2000–3000 m. An excellent description of genera is included by Bowman and Gruner (1973).

The Lanceolidae, while also mainly deep sea, are somewhat better known. The small head, with small or rudimentary eyes, bears the first antenna which is moderately large and lanceolate. The second antenna is somewhat longer and slender, consisting of a few elongated segments (Stebbing, 1888) (Fig. 2.92). Some species may be comparatively large, even exceeding 60 mm. According to Bowman and Gruner (1973) the young stages show an inflated body but this is not characteristic of the adults. Stephensen (1923, 1925) describes the cuticle as thick and leathery. The genus *Lanceola* is typically

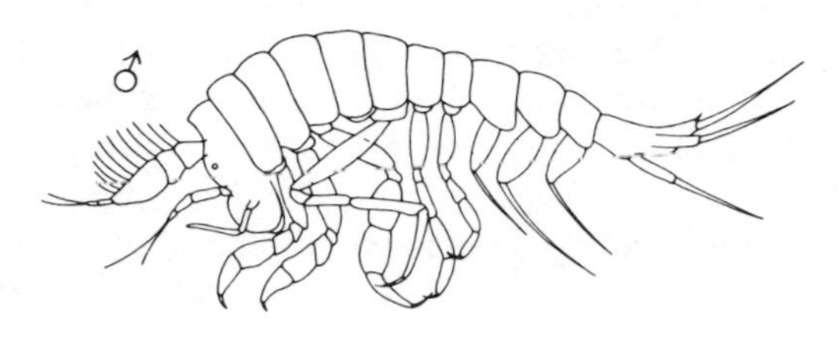

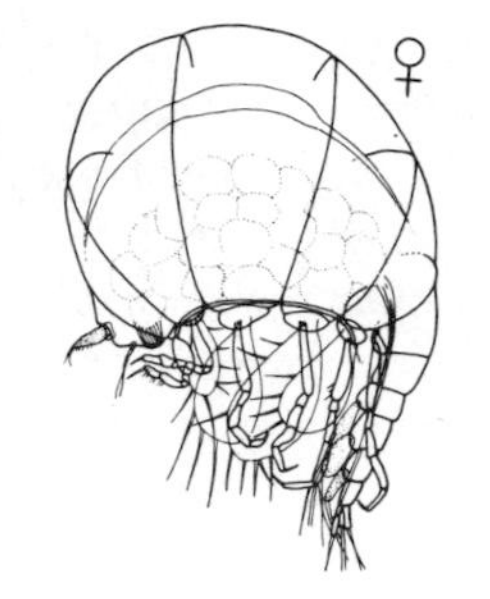

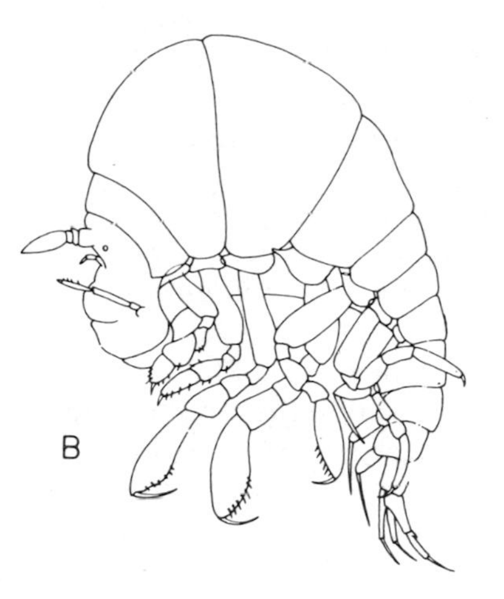

Fig. 2.91. (A) *Archaeoscina* ♂ and ♀; (B) *Microphasma* (Bowman and Gruner, 1973).

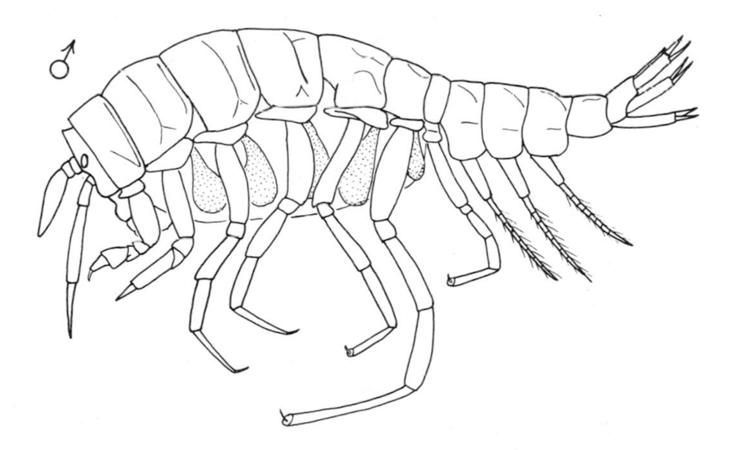

Fig. 2.92. *Lanceola* (Bowman and Gruner, 1973).

bathypelagic and often widely distributed (Barnard, 1932). For example, *Lanceola sayana* occurs in all three major oceans, with a distribution in the Atlantic from 56°N to 32.5°S; *Lanceola pacifica* is found in the North and South Atlantic and the Pacific; *Lanceola serrata* in the North Atlantic, including Davis Strait, to as far south as 41°N, and also fairly widely in the South Atlantic between 33° and 52°S, in comparatively deep water. An apparently more geographically restricted species, *Lanceola clausi*, is also a deep-living species (Chevreux, 1935) and is found in the Arctic (cf. Dunbar and Harding, 1968). Vinogradov (1972b), however, found all the four species, as well as *L. loveni* and *L. laticarpa*, in the Kurile–Kamchatka Trench and recorded three new species of the genus. *L. sayana* and *L. clausi* were comparatively abundant. The majority were more deep-living, mostly between 1000 and 2000 m.

In the genus *Scypholanceola* the eyes are developed more as strips at the side of the head and reflectors are present. Probably all the species should be referred to *S. vanhoeffeni*, a very widely distributed species, occurring at great depths (cf. Pirlot, 1939). Barnard (1932) records it widely in the Atlantic, as well as the Indian Ocean, and in the Antarctic as far as 64.5°S.

The Physosomata differ from the Physocephalata in their vertical distribution. Stephensen (1923, 1925) pointed out that most "Primativa" were deep-living. According to Vinogradov (1973), though Physosomata can inhabit the whole water column, few have colonized the sub-surface layers and they are the only hyperiids to occur in the deep layers. By contrast, the Physocephalata are distributed mainly in surface and sub-surface water. With the exception of the Cystisomatidae, containing the one genus *Cystisoma*, which occurs in the deeper layers to a depth of about 3000 m, few go deeper than about 500 m. *Cystisoma* is a comparatively large amphipod with a thin, almost transparent cuticle. The head is very large, rounded above, almost flat below and has a pair of oval eyes on the dorsal surface. Barnard (1932) confirms that *Cystisoma* is generally deep-living, normally exceeding 500 m and extending to about 3000 m. Shoemaker (1945) found *C. magna* to depths of >2000 m off Bermuda; the species is known from South Atlantic and Indian Oceans.

Although typically deep-living, a number of species of Physosomata have a very wide vertical distribution—a characteristic of deep-living members of other taxa. For example, no fewer than 15 species range from the surface or sub-surface layers to a depth of 2000–4000 m: *Lanceola sayana* from the surface to 3000–4000 m; *L. pacifica* from 200 m–4000, or even 8000 m; *L. clausi* from sub-surface layers to 5000 or 6000 m (Vinogradov, 1973) (cf. also Vinogradov, 1972b). Stephensen (1923, 1925) also cites *Lanceola sayana* as having a very wide range.

The hyperiids of the family scinidae have generally slender bodies (Fig. 2.93). The small head carries a long rod- or sword-shaped first antenna with an especially large second segment. This family is also deep-living and includes some species with a broad vertical distribution (e.g. *Scina marginata*, *Scina crassicornis*). *Scina borealis* is very widely distributed both vertically and horizontally, occurring generally in the North Atlantic, the Mediterranean, Indian Ocean and the North Pacific; it is recorded even to 80°N latitude (cf. Reid, 1955). Barnard (1932) also comments on its great depth and geographical range which extends to the Antarctic. Vinogradov (1972b) recorded it as the commonest species of the genus in the Kurile–Kamchatka Trench in depths to 2000 m, occasionally deeper. Several other species were found in the Trench, some at great depths.

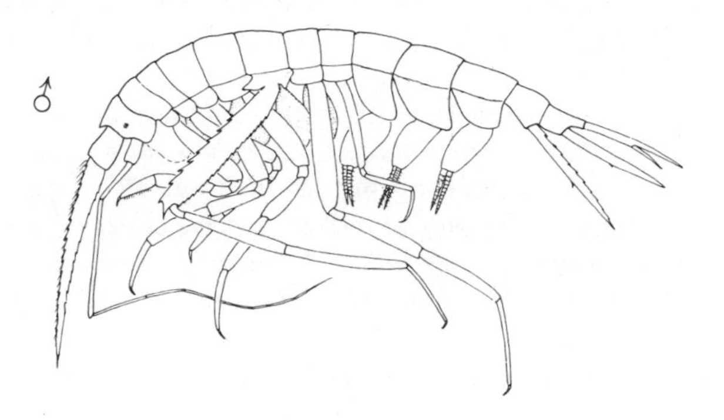

Fig. 2.93. *Scina* ♂ (Bowman and Gruner, 1973).

Thurston (1976b), investigating the planktonic amphipods off the Canaries to a depth of 950 m, also found that whereas Physosomata, like most pelagic gammarids, were relatively abundant at depths exceeding 500 m, Physocephalata were almost entirely confined to the surface. *Scina borealis* was one of the commonest hyperiids and showed a wide vertical range (40–950 m), agreeing with most observations from other geographical areas. Thurston describes the species as pan-oceanic and eurythermal. *S. crassicornis* also showed a wide vertical range (150–900 m). In the Kurile–Kamchatka Trench, despite the broad depth distribution of many Physosomata, a distinct vertical zonation is discernible. For example, *Lanceola loveni*, *Scina woltereckі*, *Scina wagleri*, and *Ctenoscina brevicaudata* do not appear much above 1500 m depth; *Scypholanceola agassizi*, *Ctenoscina spinosa*, and *Lanceola clausi var. gracilis* appear in the 3000–4000 m zone; only certain of the species (*L. clausi*, *Scina wagleri*) occur in the Trench itself below 6000 m (Vinogradov, 1973). In the north-east Atlantic, however, Thurston (1976b) found *Scina woltereckі* at much shallower depths (maximum abundance 570 m).

Of the Physocephalata, the family Vibiliidae is somewhat unusual in that of the two genera, *Vibilia* has a comparatively small head and the eyes are small or of moderate size. The first antenna is roughly conical. In *Cyllopus*, however, the head is swollen and bears large eyes.

Although the Physocephalata are mainly in surface waters, some have an apparently wide depth range (e.g. *Vibilia* spp., *Paraphronima* spp., *Primno macropa*) (cf. Chevreux and Fage, 1925; Reid, 1955). Stephensen (1925) lists *Vibilioides* (*Vibilia*) *alberti*, *Parathemisto abyssorum*, and *Hyperia medusarum* as other deeper-living Physocephalata, though the last two species can be abundant at the surface. Diurnal vertical migration is common amongst hyperiids, though the precise depth ranges are rarely known (cf. Thurston, 1976b).

Although the majority of species and families of Physocephalata are found in warmer seas, some can be relatively abundant at high latitudes and some have a wide distribution. Reid (1955) lists a number of species as 'cosmopolitan' (e.g. *Vibilia viatrix*, *Phrosina semilunata*, *Eupronoe maculata*) and a number as probably so (e.g. *Anchylomera blossvillei*, *Oxycephalus clausi*, *Streetsia porcella*). Chevreux and Fage (1925), and Stephensen (1923, 1924, 1925) also describe several of these species as cosmopolitan or very widely distributed and add *Euthamneus platyrrhunchus* and

Brachyscelus crusculum. Barnard's (1932) records for *Phrosina semilunata* and *Primno macropa*, including their occurrence in Antarctic waters, confirm their broad distribution.

In Thurston's (1976b) studies off the Canaries, the commonest Physocephalata were *Hyperioides longipes*, *Phrosina semilunata* and *Primno macropa*, the last species being by far the most abundant, comprising 44 % of the total pelagic amphipods. Thurston emphasizes that the great majority of the 86 species collected were represented by exceedingly few specimens. *Hyperioides longipes* is described as a widespread epiplanktonic species, the great majority of the population living in the uppermost 50 m, though a small number of a clearly distinguished larger form lived more deeply. *Phrosina semilunata* was also epiplanktonic, but with a few individuals extending to shallow sub-surface layers. *Phronima sedentaria*, though mostly at the surface, appeared to have a wide vertical range. *Primno macropa* was observed by Thurston to have two maxima, at near 40 m and 625 m, and many records in other areas suggest a vertical spread, from the surface to several hundred metres. The species is one of the most widespread, stretching from far south in the Southern Ocean to the Bering Sea, and as far as north of 60°N in the Atlantic. Thurston states that the species seems to be absent only from the Arctic Basin.

Although Thurston found comparatively low densities of hyperiids off the Canaries, Repelin (1970) found relatively large populations in the Indo-Australian Basin in the upper 200 m of tropical and sub-tropical waters. The major species was *Phronimella elongata* (54 % of total), with *Phronima atlantica* and *Phronima sedentaria* each comprising about 12 %. In sub-tropical waters, *Phronima atlantica* and *Phronimella elongata* were almost equally important (33 % and 30 %, respectively); *Phronima colleti*, which was not found in tropical waters, was also taken. Nair (1977) found *Paraphronima gracilis*, a species occurring in tropical and warm-temperate waters of all the oceans and Mediterranean, very widely and abundantly distributed in the Indian Ocean, particularly from 10° to 20°N and 10° to 20°S. *P. crassipes* appeared to be less abundant, though dominant in the Bay of Bengal.

A genus apparently typical of higher latitudes is *Parathemisto*, but there has been confusion in its relation to *Themisto* and *Euthemisto*. It is generally now considered that only one genus, *Parathemisto*, exists (Fig. 2.94) with species occurring in Arctic and boreal areas (cf. Stephensen, 1923), some being bi-polar in distribution.

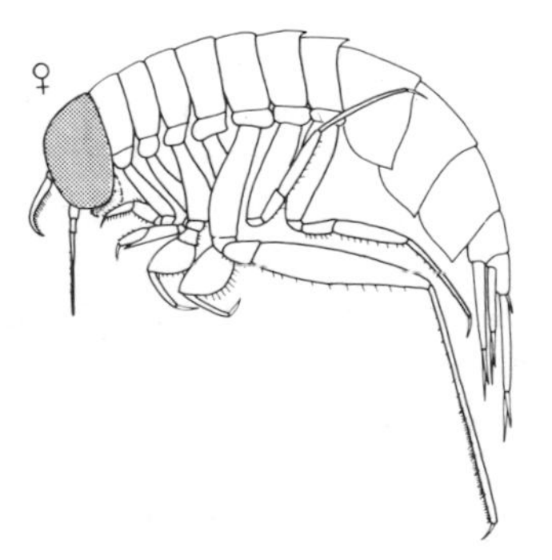

Fig. 2.94. *Parathemisto* ♀ (Bowman and Gruner, 1973).

Parathemisto libellula is an Arctic and sub-Arctic form, though it does not appear to be so abundant in the Arctic Ocean itself. Dunbar (1942) states that it was dominant in parts of the Canadian Arctic free from Atlantic influence. As a carnivorous species, *P. libellula* may be limited by the sparsity of food in the Arctic Ocean, breeding mainly taking place on the periphery (cf. Dunbar, 1964, 1957). It reaches south in the west Atlantic Ocean only as far as Belle Isle Strait, presumably in conformity with the flow of the cold Labrador Current.

Parathemisto abyssorum (Bowman, 1960; Dunbar, 1963, 1964) is a sub-Arctic or boreal form, occurring generally in Norwegian waters and along the eastern coast of Greenland. In the Arctic Ocean it occurs apparently only in the deeper warmer layers, though it is known from such cold areas as the Kara and Laptev Seas. *Parathemisto gaudichaudi* is described by Dunbar (1964) as sub-Arctic and boreal. It is well known in the Southern Ocean. Although it is abundant in some sub-Arctic waters, it is more limited in its range northwards in the northern hemisphere than *P. abyssorum.* It reaches south as far as the mixed water zone, and is not found in the coastal waters of E. Greenland, nor in Hudson Bay and the east Barents Sea. Dunbar states that this species is rare over open areas of the North Sea and it occurs only in the northern part in winter. Bowman (1960), in his account of the genus *Parathemisto* in the North Pacific, states that *P. pacifica* is widely distributed to the north of the Pacific sub-Arctic water boundary (about 36°N) but does not appear to enter the Arctic Ocean, nor does it occur west of Japan in the Japan Sea where it is replaced by *P. japonica*. *P. australis* is confined to warm water round the south-west coasts of Australia and New Zealand.

Sheader and Evans (1974), reviewing the taxonomic relationships of *Parathemisto gaudichaudi*, pointed out that *P. gaudichaudi* and *P. gracilipes* were believed to occur in sub-polar and temperate regions of both hemispheres, *P. gaudichaudi* being largely oceanic, as suggested by Dunbar, and *P. gracilipes* being described as neritic. Stephensen (1923) commented that the latter species was abundant at times, forming shoals. To some extent the distribution of the species was believed to overlap in intermediate habitats. The species were regarded as distinct, *P. gracilipes* being smaller, the body length maturing at 3–14 mm; *P. gaudichaudi* was larger, with a mature body length of 7–35mm.

A confusing condition was recognized in that both *P. gracilipes* and *P. gaudichaudi* could exist in a range of morphological forms, the extremes being known as the "compressa" and "bispinosa" forms, respectively. Sheader and Evans summarize the differences for *P. gaudichaudi*: the "compressa" form has the fifth and sixth legs approximately equal, whereas in the "bispinosa" form the fifth leg is about twice the length of the sixth. Their investigations have now demonstrated, however, that the two species are one and the same and that the original morphological characters, attempting to distinguish them, are not adequate. Both species can exhibit marked size variation and assume the "compressa" or "bispinosa" condition. Considerable variability results, especially as the morphological characters previously used to separate the two species depended on body size.

A particularly interesting experiment was that eggs and young removed from the marsupium of females of *P. gaudichaudi*, when kept at temperatures around 8°–9°C, matured at a smaller body size, and appeared to be identical to summer breeding specimens of *P. gracilipes*. Both *P. gracilipes* and *P. gaudichaudi* forms are bipolar and the authors propose that only a single species, *Parathemisto gaudichaudi*, is now acceptable, having a very wide distribution at relatively high latitudes. It is common in Antarctic waters, with a considerable vertical range off South Georgia, according to Hardy and Gunther (1935). Barnard (1932) records it as far south as 68°S, but mostly in the upper 100 m.

Although so many hyperiids are warm oceanic forms some species, in addition to *Parathemisto*, inhabit high latitudes. *Hyperoche medusarum* (= *H. kroyeri*) is a boreo-Arctic species, occurring mostly between 53° and 77°N in the Atlantic. It occurs also in the North and South Pacific and in association with ctenophores especially *Pleurobrachia spp.* mainly near the surface—a peculiar feature of many hyperiids (*vide infra*) (cf. Evans and Sheader, 1972; Floresco and Brusca, 1975). Thurston (1977) regards all species of *Hyperia* as parasitic on Scyphomedusae as larvae or at later stages. *Hyperia medusarum*, another high latitude species of the North Atlantic and Pacific, mostly north of 60°N, is essentially bathypelagic to about 1000 m, and occurs rarely at the surface (cf. Dunbar, 1942). It associates with *Cyanea* and *Aurelia* (Stephensen, 1923, 1924) and with other Scyphozoa (Sheader, 1974, Thurston, 1977). *Hyperia spinigera* is found within the body of the Coronate Scyphozoan *Periphylla periphylla*. *H. galba*, which is associated with the medusae *Pelagia*, *Aurelia* and *Chrysaora* and with the ctenophore *Beroe*, appears to be a very widely distributed hyperiid; Stephensen (1923) describes it as probably circumpolar. It is found generally over the North Atlantic, including the North Sea, and in the North Pacific, as well as in the Indian Ocean (off South Africa) and in the Antarctic. Dunbar (1963) reports it as far south as Bermuda in the North Atlantic, and Reid (1955) records it for the tropical Atlantic.

At high latitudes in Antarctic seas, even south of 60°S, apart from Physosomata (e.g. *Lanceola clausi*, *Scina borealis*, *S. antarctica*) and such cosmopolitan species as *Primno macropa*, certain Physocephalata are found which are typical of Antarctic waters (e.g. *Cyllopus lucasi*, *Vibilia antarctica*, *Hyperia macronyx*, *Hyperoche capucinus*, *Hyperiella antarctica*, *Cyllopus magellanicus*) (cf. Barnard 1930, 1932). This last species appears to be mainly sub-Antarctic in all three major oceans from around 30°S latitude; it occurs in the sub-surface layers to about 2000 m, though appearing at the surface at night. *Cyllopus lucasi*, by contrast, is more confined to the colder waters, not appearing north of 53.5°S. *Vibilia antarctica* is relatively common in both Antarctic and sub-Antarctic waters below 500 m by day. According to Hardy and Gunther (1935), it is next in abundance to *Parathemisto*.

Many hyperiids, however, are more or less confined to warm waters, with a number of such species being found in all three warm oceans. *Paraphronima crassipes* occurs in the Atlantic to about 40°N, as well as in the Mediterranean and tropical Pacific (cf. Stephensen, 1924; Chevreux and Fage, 1925). Reid (1955) records it in the Indian Ocean (cf. Nair, 1977—*vide supra*) and Red Sea, and fairly widely in the Atlantic. Chevreux and Fage (1925) list *Phronimopsis spinifera* in all sub-tropical seas. Stephensen reports *Hyperia schizogeneios* as a comparatively warm-water species found widely in the Mediterranean, and in the Atlantic, mostly between 40°N and 7°S, with some records off South Africa and in the warm waters of the Pacific. The related species, *H. hydrocephala* and *H. macrophthalma*, are reported as even more confined to warm waters. Reid (1955) cites under warm-loving species, *Vibilia pyripes* and *Lycaea pulex* in tropical and sub-tropical areas of all oceans. *Glossocephalus milne-edwardsii* appears to have a similar distribution (Stephensen, 1925; Chevreux and Fage, 1925). Barnard (1930, 1932) lists *Rhabdosoma armatum*, an oxycephalid reaching >150 mm length, for all tropical and sub-tropical oceans. *Pronoe capito*, *Rhabdosoma whitei* and *Lycaeopsis themistoides* would appear to be amongst many other typically warm-water species (cf. Stephensen, 1925; Reid, 1955; Thurston, 1976b).

Shih (1969) presents more precise zoogeographical limits for the family Phronimidae (cf. also Shih and Dunbar, 1963). According to Shih, all species, with the exception of *Phronima sedentaria* and *P. atlantica*, are restricted north and south by the Sub-Tropical Convergences (latitudes about 40°N and 40°S, respectively), and all the species occur in all three oceans. In general, in the western North Atlantic, *ca.* 35° to 40°N marks the limit of distribution. As with other taxa, with the occurrence of the North Atlantic Drift, these warm-water hyperiids range further north in the eastern Atlantic. Excepting *P. sedentaria* and *P. atlantica*, no Phronimidae reach as far north as the Bay of Biscay. In the South Atlantic the Sub-Tropical Convergence marks even more exactly the limit of distribution, and this holds for the Pacific and Indian Oceans.

Phronima sedentaria, the largest of the Phronimidae, with the female attaining about 40 mm, reaches about 45°N in the western North Atlantic, but in the eastern Atlantic there are captures from off the Shetland Isles and south west of Iceland. Shih suggests that the 15°C surface isotherm for August and the 10°C isotherm for February mark approximate northern limits. The species occurs widely in the warm waters of tropical and sub-tropical areas of all three oceans. It is very common in the Mediterranean. In the Pacific, it is widely recorded off Vancouver and has been found in the Gulf of Alaska, but does not extend much beyond 40° latitude in the western North Pacific. The species is found south of South Africa, Australia and New Zealand to approximately 50°–52°S.

Phronima atlantica has approximately the same wide distribution in the sub-tropical and tropical waters of all three oceans. It extends beyond the limits of the Sub-Tropical Convergence in the North Atlantic and North Pacific, though not as far north as *P. sedentaria*. In the southern hemisphere, *P. atlantica* reaches south of Australia and New Zealand, and penetrates Antarctic as well as sub-Antarctic waters, with a record even at 65°S latitude (cf. also Shih and Dunbar, 1963; Barnard, 1932). Although the precise vertical distribution of many Phronimidae is uncertain, *P. sedentaria* and *P. atlantica* are believed to extend by day from the surface to 400–500 m (cf. Thurston, 1976b).

One of the peculiar and characteristic features of hyperiid amphipods is that many species appear to live commensally, perhaps more parasitically, on other zooplankton, particularly cnidarians including medusae and siphonophores and pelagic tunicates such as *Pyrosoma*, salps and doliolids. For example, *Phronima sedentaria* has long been known to live in the "houses", the cellulose-like outer surface of tunicates (Fig. 2.95). Other well known examples include *Hyperoche medusarum*, *Hyperia medusarum*, and *Hyperia galba*, which are associated with medusae and ctenophores (cf. page. 254). Chevreux and Fage (1925) cite *Vibilia viatrix* and *V. jeangerardii* as often commensal with salps and Hardy and Gunther (1935) quote *Vibilia antarctica* as associated with *Salpa fusiformis* in Antarctic waters. Many other examples, including an association of *Lycaeopsis themistoides* with siphonophores (including *Diphyes*), are given by Stephensen (1925). An excellent recent list of the associations of various hyperiids with other invertebrates is given by Madin and Harbison (1977) and Harbison, Biggs and Madin (1977). (*vide infra*). Vinogradov (1973) suggests that deeper-living hyperiids may not be specifically confined to a particular host/prey species, but might change

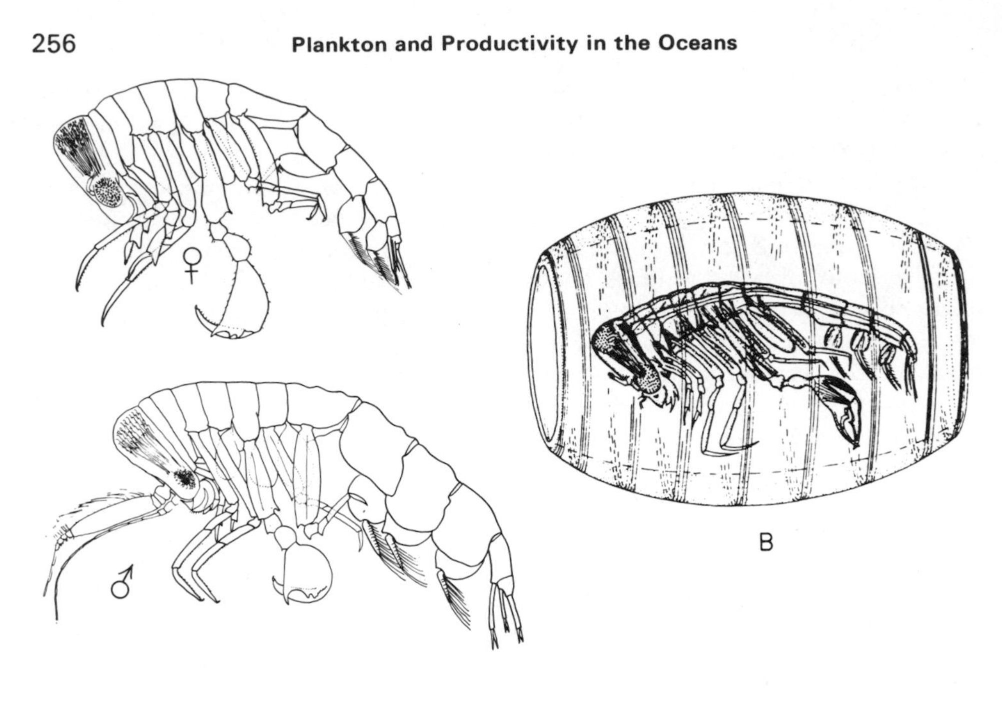

Fig. 2.95. (A) *Phronima* ♀ and ♂ (Bowman and Gruner, 1973); (B) *P. sedentaria* (3 cm long) within the tunicate "house" (Barnes, 1974).

their vertical distribution and live in or on bathypelagic cnidarians or tunicates as available.

With regard to *Phronima sedentaria*, many authorities relate the inhabiting of the house particularly to the egg-bearing females and to the housing of the young stages (cf. Stephensen, 1924). Batches of these young have been described attached to the inside of the house (cf. Barnard, 1932). Shih (1969) describes the development of the earlier stages as mainly within the houses ("barrels") of the tunicates. Barnard (1932) discusses the views of Dudich who believes that the young stages feed on the house material and that the distribution and occurrence of the amphipod appears to coincide with *Pyrosoma*. Both *P. sedentaria* and *P. atlantica* young are found inside the "barrels" of *Pyrosoma* salps.

Many hyperiids appear to feed on soft-bodied zooplankton such as salps and cnidarians. Deep-sea species may eat the tissues of deep-living medusae and siphonophores. Repelin (1970) confirmed that medusae and siphonophores formed a major part of the diet of hyperiids from Indo-Australian waters. Sheader and Evans (1975), while observing that *Parathemisto gaudichaudi* was a fairly general carnivore, found that several hydromedusae were readily eaten. *Aequorea* was a fairly common food at certain periods of the year. Details of feeding of amphipods are discussed in Chapter 6; it would appear that the widely recognized association of hyperiids and other planktonic invertebrates may be partly a trophic relationship (cf. Thurston, 1977).

In a detailed investigation, observing hyperiids at sea by means of Scuba diving, Madin and Harbison (1977) found that the amphipods *Lycaea* and *Vibilia* are specific and obligate symbionts of salps. *Vibilia* feeds on the material collected by the salp, whereas *Lycaea* feeds on the salp tissues, and this appears to be true for the various species investigated. When adult, *Vibilia* seems to occur as only one per salp, although many juveniles may occur together. With *Lycaea*, the number of juveniles and adults per salp is variable. The defence of the salp against the attacks is said by Madin and Harbison to be a rapid reproductive and growth rate. Other genera of hyperiids (*Brachyscelus*, *Parathemisto*, *Oxycephalus*, *Phronima*) are often associated with salps of several different species.

Harbison, Biggs and Madin (1977) believe that most hyperiids are associated with some species of gelatinous zooplankton, at least at some stage of their life history. For *Phronima*, the salps are both a source of food and of protective barrels. *Parathemisto* is probably free-living but juvenile states may have an association. Paraphronimidae, Lycaeopsidae, Pronoidae, Parascelidae and Platyscelidae are associated only with siphonophores; Hyperiidae and Brachyscelidae, primarily with medusae, some with colonial radiolarians; Oxycephalidae, often with ctenophores, but many are free-living. Whereas all appear to feed either on the food of their host, or on the host tissue, more or less opportunistically except for *Vibilia*, some associations seem to be highly specific, perhaps implying an approach to obligate parasitism.

While the Hyperiidea are entirely planktonic and the Gammaridea are regarded as benthic, as with so many other taxa, certain gammarids may appear temporarily in the plankton in shallow neritic waters. It is likely that a few gammarids are truly pelagic in shallow waters. Dunbar (1942) lists a few of those captured in Arctic regions as genuinely planktonic (e.g. *Apherusa glacialis*, *Pseudalibrotus glacialis* and *Pseudalibrotus nanseni*). But in addition to such essentially shallow water species, a number of oceanic planktonic forms have been recognized amongst different families of Gammaridea.

Shoemaker (1945) identified twenty-two species of pelagic gammarids from the Bermuda Expeditions oceanographic collections. Many of them were deep-living and apparently rarely encountered, but a few species were more common. *Cyphocaris* is one of the fairly well known genera, with several mainly bathypelagic species. For example, Shoemaker lists *C. challengeri* and *C. anonyx* at depths of about 480 to >2000 m. They are found in all three major oceans (cf. Barnard, 1937). *Eurythenes gryllus*, a species found at high latitudes in the North Atlantic, South Atlantic, North and South Pacific and Indian Oceans, was also captured in deep water (1200–>2000 m) off Bermuda in fair numbers. *Hyperiopsis tridentata* and *Parandania boecki* occur among other usually deep-living and rare species, listed by Shoemaker for the Atlantic and by Barnard for the Indian Ocean. *Eusiropsis riisei*, deep-living in tropical seas, occurred somewhat more commonly. One exception to the general deep habitat of planktonic gammarids is *Synopia ultramarina*, found in surface waters of the tropics and sub-tropics (Barnard, 1937; Shoemaker, 1945).

Russian investigators, in particular, have drawn attention to the existence of pelagic gammarids as a regular and significant constituent of the relatively sparse, deep-sea zooplankton. Vinogradov (1970, 1973), discussing the importance of these planktonic gammarids in the deep trenches of the North Pacific, points out that some 115 species of pelagic gammarids have been described. Some of the species occur as shallow as 500–1000 m, though with a wide depth range, but most are confined to great depths. While several species of *Cyphocaris* are typical of very deep ocean waters, with such gammarids as *Hyperiopsis*, *Halice*, *Eusirus* and *Eusirella*, other species of *Cyphocaris* (e.g. *C. anonyx*, *C. challengeri*) may apparently sometimes be found at sub-surface levels, and *C. richardi* with *Paradania boecki* are examples of gammarids with a very wide vertical distribution.

Studies by Thurston (1976a) off the Canaries, also showed *C. challengeri* and *C. anonyx* as mesoplanktonic, with depth ranges of 200–800 m and 625–720 m, respectively; *Eusiropsis riisei* lived more shallowly. In contrast, *Halice macronyx* was deeper-living (720–950 m), and is reported to occur at depths exceeding 1000 m in the Southern Ocean. Although *C. anonyx* was the most plentiful gammarid and was one of the more abundant amphipods, many pelagic gammarids were probably not sampled at the maximum fishing depth of 950 m.

Vinogradov states that of twenty-six species occurring in the surface and sub-surface layers, 11 were not found below 500 m. In the next stratum, from 500–3000 m, there was a rather uniform fauna with a fairly wide distribution. A comparatively rich fauna inhabited depths exceeding 5000 m, including at least twelve species of *Hyperiopsis*, *Hirondellea*, *Halice*, *Vitjaziana* and *Eusirus*. Not surprisingly, some of these gammarids kept extremely close to the bottom; some might be even partly benthic, but genuine pelagic species live in these great depths. There is a tendency for the few, truly planktonic, families to produce in the course of evolution a number of similar species in very deep waters. Vinogradov comments on the relatively small contribution of planktonic gammarids to total biomass except at maximum depths in deep sea trenches, although in the north-west Pacific, *Cyphocaris challengeri* makes a considerable contribution even above 500 m. Birstein and Vinogradov (1972) recorded forty-eight species of pelagic gammarids from the Kurile–Kamchatka area, but only *Cyphocaris challengeri* inhabited the sub-surface waters above 500–750 m.

Mysidacea

The Mysidacea share with the euphausiids the general superficial appearance of the decapod shrimps. Indeed, the three groups were previously loosely associated, and for many years euphausiids and mysids were grouped together as the Schizopoda. Although the Decapoda and Euphausiacea are often associated as Eucarida, the Mysidacea are separated by a number of basic characters and are regarded as a distinct Order of Crustacea.

The majority of the mysids range in length from about 5–25 mm but a very few species (e.g. *Gnathophausia*) may exceed 150 mm. While most mysids are benthic, a number of species, especially in shallow inshore waters, are tychopelagic, moving particularly at night from the bottom into the water layers above. Many mysids show well marked swarming behaviour and some (e.g. *Neomysis*, *Praunus*) are typical and very abundant estuarine species (cf. Cronin, Diaber and Hulburt, 1962; Ralph, 1965). In some estuarine areas of Japan, India and Thailand, mysids, chiefly *Neomysis* spp. and *Mesopodopsis*, occur so plentifully that they are used for food, sometimes mixed with the decapod, *Acetes*.

The speed and extent of the vertical movements of mysids from the bottom into the overlying layers varies (cf. Russell, 1931). Off the Great Barrier Reef the increase both in the number of mysid species and of individuals between day and night was especially striking although swarms have been observed off the rock faces during day by Scuba divers (cf. Tattersall, 1936a, b). Rather similar movements have been observed by Hobson and Chess (1976) in kelp beds off southern California, *Siriella pacifica*, amongst other species, moving from the bottom into more open water at night. Russell has described a particularly marked vertical movement in British waters for *Anchialina agilis*, the species appearing right at the surface at night. Hoenigman (1963) found this species with other mysids (*Gastrosaccus lobatus*, *Siriella thompsonii*) in the upper layers of the Adriatic in both day and night hauls. Fage (1942) had already noted the occurrence of several species in the superficial waters of the Mediterranean during dark hours.

The sporadic occurrence of mysids in the plankton, and their migrations, raise difficulties in distinguishing between some benthic and genuinely planktonic species. Many are rarely captured except close to the bottom in nets attached to trawls or dredges, typically in deep waters, so that their precise habitat is uncertain. Some records of mysids in the plankton are also due to the breeding movements of bottom living adults to more superficial layers (e.g. *Leptomysis gracilis*). Tattersall and Tattersall (1951) also cite females of the semi-benthic *Lophogaster typicus* as moving surfacewards to liberate their young. The younger stages of many mysid species live more shallowly than older stages. Certain mysids (e.g. *Neomysis americana*, *Mesopodopsis slabberi*, *Gastrosaccus spinifera*) undergo seasonal changes in depth distribution so that over a period they may appear more commonly in the plankton. An indication of the planktonic and partly pelagic species found in the zooplankton of north-east Atlantic waters is given by Nouvel (1950), but the distinction between benthic and planktonic species is often difficult.

A few genera of pelagic mysids, on the other hand (e.g. *Gnathophausia*, *Eucopia*) have long been recognised as amongst the most characteristic deep-sea zooplankton. In addition, certain members of mainly benthic genera have undoubtedly adopted a planktonic habit. *Boreomysis megalops* and *Arachnomysis leuckartii* are examples of bathypelagic species, while *Petalophthalmus oculatus* occurs more in midwater depths. Other mysids are epiplanktonic, e.g. *Siriella gracilis*, *S. thompsonii*, *Pseudanchialina pusilla*, and *Anchialina typica* (Pillai, 1973).

With few exceptions the cuticle of mysid crustaceans is thin and relatively soft. The shrimp-like carapace is fused with the head and part of the thorax, and extends laterally on each side of the thorax. The last four thoracic segments are not fused to the carapace, however. The more posterior of the free segments are obvious dorsally and there is some flexibility of the body between this region and the abdomen (Fig. 2.96A).

The head bears a pair of well developed, usually stalked eyes, biramous antennules and antennae with a well developed flagellum and the exopod as a large scale. The mandibles have a strong gnathobase (Fig 2.96B). The maxillae are well developed and assist in filter feeding. The eight biramous thoracic appendages form the main swimming limbs in the Mysida. The first two pairs are somewhat modified as maxillipeds. Some of the thoracic appendages may be greatly elongated (e.g. the fifth to seventh pairs in *Eucopia* (Fig. 2.97)). In the sub-order Lophogastrida (*Gnathophausia*, *Lophogaster*, *Eucopia*) thoracic epipodial gills are present on most segments. The Mysida do not possess gills.

A comparatively large brood pouch (marsupium) is formed below the thorax from two to seven pairs of oostegites. Eggs are shed directly into the brood pouch where fertilization occurs and where development is completed. Tattersall and Tattersall

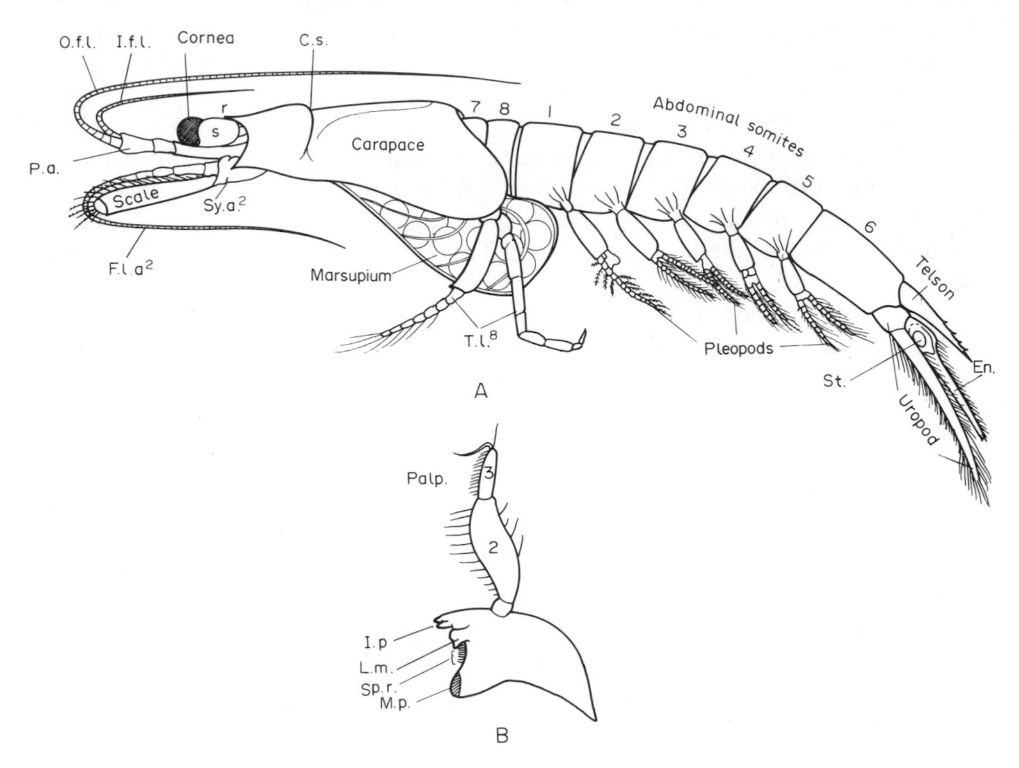

Fig. 2.96. Diagram of (A) typical ♀ mysid in side view; pd: peduncle of a 3-segmented antennule with outer (o.fl.) and inner (i.fl.) flagella; sy a: sympod of antenna; fla^2: flagellum; s: stalk of eye; r: rostrum; c.s: cervical sulcus; 7 and 8: 7th and 8th thoracic somites; t.l^8: 8th thoracic limb; en: endopod of uropod; st: statocyst: (**B**) mandible with 3 segmented palp; I.p.: incisor process; I.m.: lacinia mobilis; Sp.r: spine row: M.p.: molar process (Tattersall and Tattersall, 1951).

(1951) state that immediately following the release of a brood, which occurs at night, the female moults and that another copulation must follow almost at once. Clutter and Theilacker (1971) have confirmed this observation. The number of embryos, with a maximum usually approaching forty, varies with species, time of year, and the size of female. Mesopelagic and bathypelagic species seem to have fewer and relatively larger yolky eggs (cf. Mauchline, 1973).

The abdomen carries a pair of pleopods on each of the first five segments. In females of the sub-order Mysida these pleopods are very small simple plates. The males have larger biramous pleopods, but in many genera one or more pairs may be reduced in the males and the fourth or fifth pairs may be much elongated and used in copulation. In Lophogastrida, however, the pleopods are well developed in both sexes and are the main locomotory organs. The sixth abdominal segments bears a pair of large uropods forming a tail fan with the telson. In the family Mysidae each endopod of the uropods has an obvious statocyst.

Many mysids are omnivorous. Living and dead animal, as well as plant material being consumed. When feeding on relatively large particles mysids hold the material below the body with the thoracic appendages. Fine particulate matter, both detritus and phytoplankton, is also consumed using a filter-feeding process (cf. Lucas, 1936;

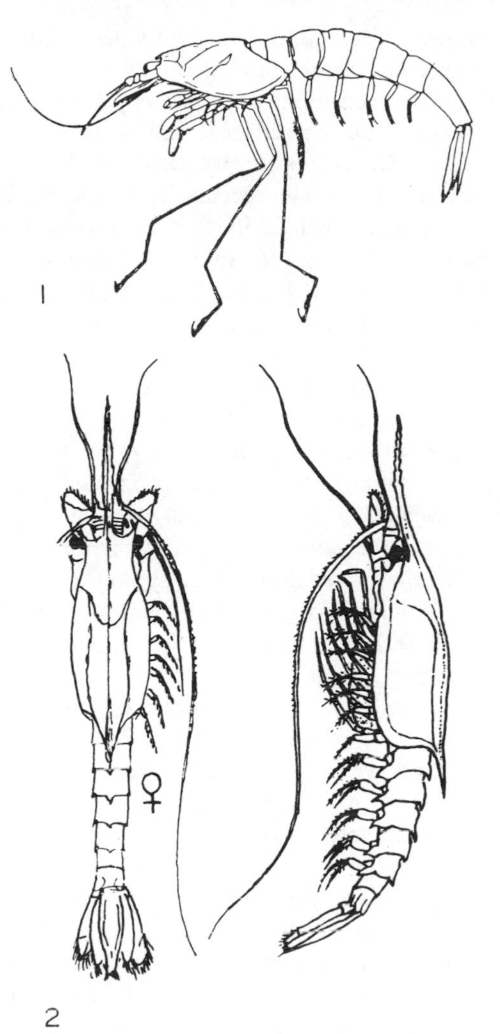

Fig. 2.97. (1) *Eucopia sculpticauda*; (2) *Gnathophausia zoea* (Nouvel, 1950).

Ferguson, 1973). The thoracic exopods produce a feeding current, aided by the maxillae, the setae of which also strain off the food particles. In Lophogastrida the feeding current is apparently due to the maxillae alone, the thoracic exopods largely maintaining a respiratory current over the gills.

Geographical and vertical distribution

The sub-order Lophogastrida, though small, includes important zooplankton genera. Members of the sub-order are regarded as more primitive in exhibiting certain characters: large gills on some or all of the thoracic appendages; well developed swimming pleopods in both sexes, without modification in males; a marsupium developed from seven pairs of oostegites; the endopods of the uropods lacking statocysts.

Gnathophausia, one genus of Lophogastrida, characteristically has a shield-like large carapace, which is partly calcareous and furnished with prominent spines. Tattersall and Tattersall (1951) regard it as the most primitive living mysid. Seven, possibly eight, species are recognized; all are deep-living and always planktonic, though Fage (1941, 1942) states that some species may breed near the bottom, the young migrating to somewhat shallower depths. *G. zoea* is known from all the oceans and is the most commonly recorded species. It is widespread in temperate and tropical waters, stretching from off Greenland and Iceland in the North Atlantic to the South Atlantic and from the North Pacific, including off Japan, across the mid-Pacific to the South Pacific, as well as in the Indian Ocean. The adults are bathypelagic but are not usually found much below 3000 m, the young live more shallowly (500–1000 m) (Tattersall, 1955).

Gnathophausia ingens appears to be a somewhat more warm-water species in all three major oceans. Fage (1941, 1942) gives its depth in the Atlantic and Pacific as about 500–1000 m, and Vinogradov (1970) gives a similar range for the north-east Pacific, but Tattersall (1955) suggests a somewhat greater depth range overall. Casanova (1977) quotes about 600 m for juveniles in the north-east Atlantic. *G. gigas* probably has the widest geographical range for the genus and is known from all three major oceans. In the Atlantic it stretches from off Greenland throughout the North and South Atlantic to reach high latitudes off South Georgia, with a record from the Bellingshausen Sea about 69°S (Tattersall, 1955). In the North Pacific it appears to reach farther north than other *Gnathophausia* spp., being known off Alaska and in the Bering Sea. It is recorded far south in the Indian Ocean, off Kerguelen Island. Its depth range is given as 650–4000 m; Vinogradov (1970) quotes a shallower depth (250 m) for the north-east Pacific, though mature adults were found only at depths exceeding 2000 m.

Species of *Lophogaster* live on or very near the bottom and are predominantly neritic, though found mostly in deeper areas of the continental shelf, or in continental slope waters (Fage, 1941, 1942; Casanova, 1977). Some species may be regarded as pelagic, for example, *L. spinosus* in the North and South Atlantic (Tattersall, 1955; Casanova, 1977) and *L. affinis* in the Red Sea (Pillai, 1973). The only British species, *L. typicus*, appears to be semi-benthic and is widespread in temperate waters including the Mediterranean. The liberated young of *Lophogaster* live in the upper water layers. *Charalaspidium* (= *Charalaspis*) *alatum*, another deep-living but essentially pelagic lophogastrid, has a wide distribution including the East Pacific, South Atlantic and Indian Ocean as far south as Kerguelen Island. It is a good swimmer with a large vertical depth range (*ca.* 1000–2500 m) (Tattersall and Tattersall, 1951).

Eucopia, the only genus of the family Eucopiidae, included in the Lophogastrida, is easily recognized by its very large membranous carapace. It resembles *Gnathophausia* in being an oceanic widespread bathypelagic genus. The adults have a considerable depth range, perhaps mostly at 1500–3000 m levels, though the young of some species live much more shallowly. *Eucopia hanseni* (= *unguiculata*) is the most common of deep sea mysids, and is found in all three major oceans and the Mediterranean, mainly in tropical, sub-tropical and temperate waters. In the Atlantic Ocean it reaches relatively high latitudes (about 64°N) off Greenland. Tattersall and Tattersall (1951) state that the species is commonest at depths of about 2000 m, though it may be found at depths exceeding 6000 m. In the north-west Pacific, Vinogradov (1970) gives its lower limit as about 3000 m. *E. hanseni* appears to have a very considerable vertical range and can live at much shallower depths than other species of *Eucopia*, at least to

500 m. In the Jabuka Trench area of the Adriatic, adult *E. hanseni* were found as shallowly as 250 m by Hoenigman (1963) and their small mean size indicated the existence of an isolated dwarf race (Nouvel, 1934; Casanova, 1977).

Eucopia grimaldii appears to have a generally similar wide distribution in all three oceans, though it does not penetrate the Mediterranean and appears to be somewhat more restricted by temperature and salinity. Its northern limit in the Atlantic Ocean is not clearly distinguishable from that of the previous species, *E. hanseni*. In the southern hemisphere *E. grimaldii* is found fairly far south, with some infrequent records to the north of South Georgia and in the Scotia Sea (Tattersall, 1955). Its commonest depth would appear to be about 1500–2000 m, but it can extend somewhat deeper. Its vertical range is considerable, and it may reach depths of less than 500 m, though it apparently cannot occur as shallowly as *E. hanseni*. The young live more shallowly. Vinogradov (1970) quotes its maximum depth as equal to that of *E. hanseni* in the north-west Pacific.

Eucopia australis is also recorded for all three oceans, being widely distributed in the Pacific and South Atlantic, somewhat less widely in the Indian Ocean. It is not known generally in the North Atlantic, although Fage (1941, 1942) records it as far as latitude 33°N. In the Southern Ocean and the Antarctic it is recorded from such areas as off South Georgia and in the Bellingshausen Sea, and the species extends far north in the North Pacific, reaching the Bering Sea. Vinogradov (1970) lists both *E. grimaldii* and *E. australis* for the Kurile–Kamchatka Trench and Okhotsk Sea, but the first species lives more shallowly at about 500–1500 m in the Trench area, whereas *E. australis* occurs mainly at depths of 2000–2500 m. Fage also states that both young and adults of *E. australis* tend to maintain a deep-living habitat, with their main depth distribution about 3000–4000 m. The three species, *E. grimaldii*, *E. hanseni* and *E. australis*, appear to be closely related.

Other less closely related species of *Eucopia* include *E. sculpticauda*, found in all three oceans, but, according to Fage (1942), more typically in warmer waters and always below 1000–1500 m, except for the Indo-Malaysian Archipelago and Sulu Sea areas, where it may occur at only 400 m depth. In the Atlantic, Fage states that as a species generally limited by cold waters, its distribution follows the Gulf Stream. The northern limit quoted by Tattersall and Tattersall (1951) off the British Isles is 57°46′N.

The sub-order Mysida contains the small family Petalophthalmidae, which is exceptional among Mysida in having seven pairs of oostegites and in lacking statocysts, has some planktonic species. *Petalophthalmus armiger* occurs in all three major oceans and though comparatively rare, it is apparently very widely distributed in deep waters. It is recorded as far north as the Bering Sea. *P. oculatus*, found in the Indian Ocean and in the Caribbean, is now known from the Pacific (Phillai, 1973). It is unusual in the genus in having well developed eyes and the mouthparts also suggest a predatory species living in mid-water (Fig. 2.98).

In the relatively large family Mysidae, *Boreomysis* is the sole genus of a somewhat primitive sub-family. Although some species of *Boreomysis* are bottom dwellers, some bathypelagic forms are important members of the zooplankton. The one pelagic British species, according to Tattersall and Tattersall (1951), is *B. microps*, an Atlantic species occurring widely over tropical, sub-tropical and temperate areas and extending to fairly high latitudes off Greenland and Iceland. It is found in deep water, never

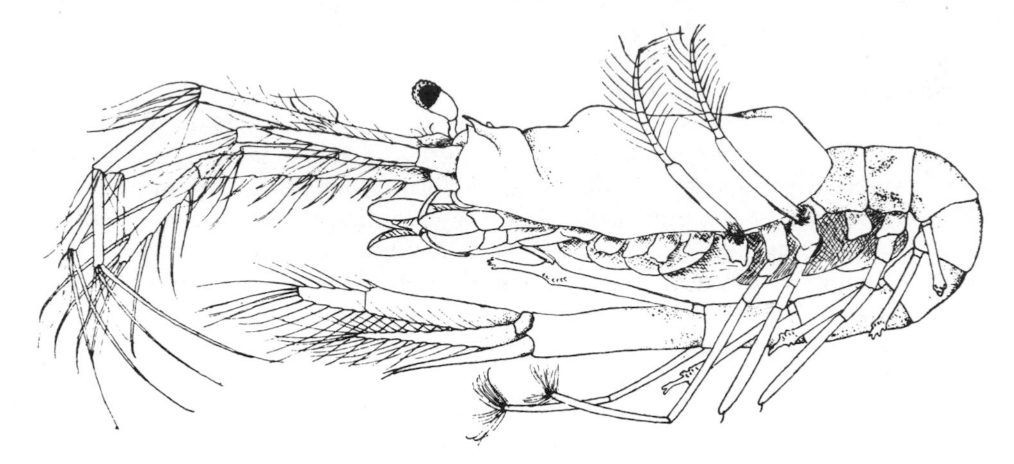

Fig. 2.98. *Petalophthalmus oculatus* (Pillai, 1973).

above 500 m. *B. inermis* is a rather shallower-living species found in the Pacific Ocean to high latitudes and in Antarctic waters (Tattersall, 1955). *B. rostrata*, a more mesoplanktonic species, is found in the southern parts of the three oceans and in Antarctic waters. *Boreomysis arctica* is a well known species occurring in very cold waters at high latitudes in the North Atlantic. Dunbar (1942) refers to it as a deep-living Arctic species. In the Korsfjorden, Norway, it is an important species in the deep water pelagic community (cf. Bamstedt, 1978).

Vinogradov (1970), referring to a number of mysids of importance in the zooplankton of the northern Pacific, includes several species of *Boreomysis*, which are eurybathic, or which migrate vertically over very considerable distances. *Boreomysis incisa* appears to be of particular significance. It is found by day from 4000–6000 m, on one or two occasions even to 7000 m, but does not appear to reach such great depths in the warmer parts of the Pacific. Figure 2.99 illustrates the important contribution which *Boreomysis incisa* makes to the relatively sparse fauna at about 5000 m depth in the Kurile–Kamchatka Trench. *Boreomysis incisa* is believed to make very substantial vertical migrations. Although occurring at very great depths, it feeds at sub-surface levels.

Vinogradov suggests that in tropical parts of the ocean, mysids may occur in fair numbers in the surface layer, whereas surface waters of cold oceans are inhabited only by a few species in small numbers (e.g. *Ceratolepis*, *Lophogaster typicus*, *Lophogaster multispinosus*, *Ceratomysis*). The very cold waters in the upper 200 m of the Pacific Ocean have particularly low numbers of very few species (e.g. *Meterythrops robusta*, *Ceratomysis vanclevei*). The importance of mysids and the species diversity appears to increase with depth. Thus in the north-west Pacific there are some ten species at depths of 200–500 m, whereas at depths of about 2000–4000 m a maximum diversity is reached with some 25 species. In tropical areas the increase in species diversity appears more shallowly, at about 500–1000 m. Closing net hauls in the north-west Pacific indicate that although the biomass of mysids there is about maximal at 500–1000 m depth, mysids contribute only about 4–6 % to the total plankton. However, at depths of 4000–6000 m, though the biomass of mysids has decreased, their contribution to the plankton may reach between 20 and 28 %.

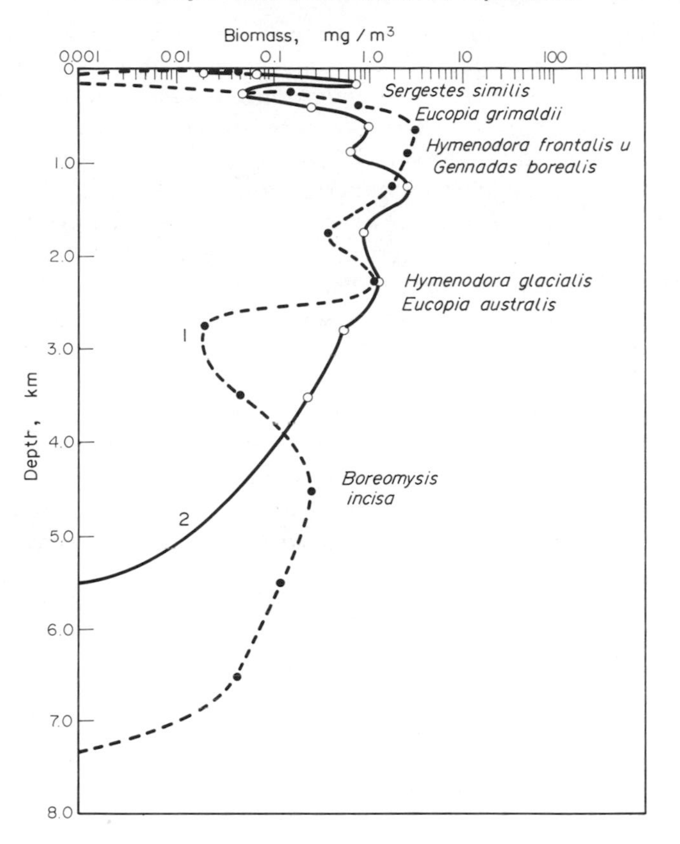

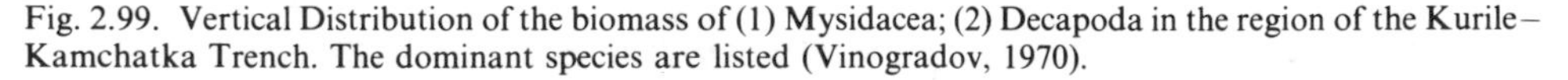
Fig. 2.99. Vertical Distribution of the biomass of (1) Mysidacea; (2) Decapoda in the region of the Kurile–Kamchatka Trench. The dominant species are listed (Vinogradov, 1970).

Of the planktonic Mysida which inhabit the surface and sub-surface layers, more especially of warmer waters, some species of *Siriella* may be comparatively plentiful. *S. thompsonii* (Fig. 2.100) is found in all three oceans, especially in tropical and subtropical areas, and is widely distributed both over oceanic areas and in more coastal regions. *S. gracilis* is confined to more open Indo-Pacific warm waters. In the Indian Ocean, Pillai (1973) found that three genera, *Pseudanchialina*, *Siriella* and *Hemisiriella*, tended to dominate the mysid fauna. They are also widely distributed over the Pacific. *Doxomysis quadrispinosa* and *Anchialina typica* also occurred generally at surface and

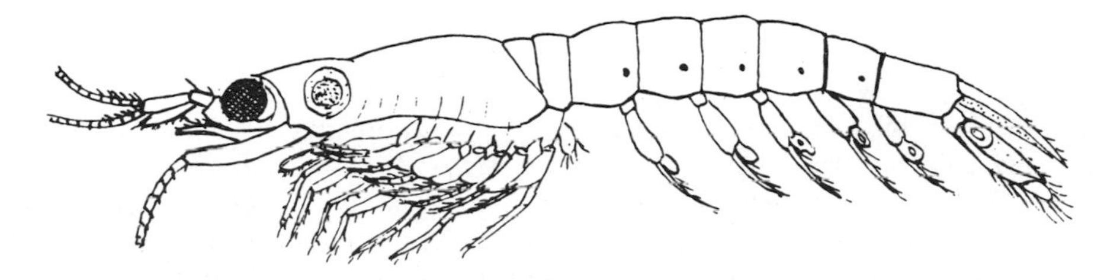

Fig. 2.100. *Siriella thompsonii* ♂ (Tregouboff and Rose, 1957).

sub-surface levels over the Indian Ocean, though they were somewhat less common. Apart from *Anchialina typica* and *Siriella thompsonii*, which occur in the Atlantic, Pillai believes that the other warm-loving epiplanktonic mysids are prevented from entering the Atlantic from the Indian Ocean by the comparatively cold waters off South Africa. Among other mysids living in superficial layers and found in warm-temperate areas including the Mediterranean, are *Anchialina oculata*, *Anchialina agilis*, *Gastrosaccus lobatus* and *Leptomysis gracilis*. Such species as *Siriella armata*, *Gastrosaccus spinifera*, *Schistomysis kervillei* and *Mesopodospsis slabberi* may be found in coastal waters from Norway to the Mediterranean.

In temperate and colder seas, at depth, several genera of mysids almost certainly include some planktonic species. *Meterythrops picta* is found in the Atlantic; *M. robusta* and *M. microphthalma* occur in such cold areas of the Pacific as the Bering Sea, the Kurile–Kamchatka Trench and the Okhotsk Sea. *M. robusta* lives in the upper 3000 m, at times even in the sub-surface layers. *M. microphthalma* is always found deeper than 3000 m. *Katerythrops oceanae* is also regarded as a deeper-living, probably oceanic, mysid (Holt and Tattersall, 1905). Some planktonic mysids appear, however, to be more mesopelagic, e.g. *Longithorax capensis*, a widely distributed species in the South Atlantic (Tattersall, 1955).

Antarctomysis maxima is notable in that, whereas many earlier findings of Antarctic planktonic mysids have been relatively isolated records, *Antarctomysis* has been repeatedly taken over many areas of the Antarctic. *Antarctomysis maxima* appears to be the commonest mysid in Antarctic waters and is found in relatively shallow layers, mainly 75–150 m. Although it has been regarded as an entirely Antarctic form, some rare records outside the Antarctic suggest that the species might occur farther north, migrating to Antarctic waters only during the spring/summer (Tattersall, 1955).

Mysids are of relatively large size for zooplankton and may therefore form a useful link in the food web from particulate matter (detritus and phytoplankton) to macroplankton. They are also consumed by fish and are well recognized as a constituent of the diet of several fishes in shallow waters. The tendency of certain more open ocean mysids to form swarms (e.g. *Siriella*), however, may suggest that they could contribute to the food of more open ocean fishes. Pillai (1973) suggests that *Pseudanchialina pusilla* is important in the diet of some ocean fishes. The well known and widespread occurrence of some mysids (*Eucopia*, *Gnathophausia*, possibly *Boreomysis*) in deeper layers of open oceans, where zooplankton tends to be scarce, might suggest that these deeper-living oceanic species are also important as food of oceanic fishes. They might share this role with deeper-living oceanic decapods, especially at depths below that inhabited mainly by euphausiids, which form such an important link in the food webs in oceanic environments.

Euphausiacea

The euphausiids form a major component of the zooplankton since some species, especially colder water inhabitants such as *Meganyctiphanes*, *Euphausia superba*, *E. pacifica*, *Nyctiphanes australis*, *Thysanoessa inermis*, and *T. raschii*, can be present in huge numbers over large areas and may show marked swarming behaviour. Komaki (1967) reports swarms of *Euphausia pacifica* in nearshore Japanese waters, mostly >20 mm length; he associates swarming with the approach of cold offshore waters.

Swarms of *E. superba* have been estimated as 60,000/m^3 and swarms of *Thysanoessa longicaudata*, in the Shetland Islands, as approaching 600,000/m^3 (Forsyth and Jones, 1966). In warmer waters, Baker (1970) reported surface swarming of *Euphausia krohnii*, estimated at 31,000/m^3.

Euphausiids are also of relatively large size compared with many zooplankton; the adult body length for many species ranges between 15 and 20 mm, with some smaller species being less than 10 mm, e.g. *Stylocheiron microphthalma* (6–7 mm); *Euphausia tenera* (8–9 mm). The larger species include *E. superba*, attaining 60 mm, and some species of *Thysanopoda*. For example, *T. cornuta* may measure 60–70 mm; specimens of this species and of *T. spinicaudata* have been taken exceeding 100 mm. The biomass of more abundant euphausiids, such as *E. superba*, may therefore be considerable, outweighing in some areas the biomass of the ubiquitous copepods. Euphausiids may be of outstanding importance in marine food chains (cf. Ponomareva, 1966), both in the Antarctic and in the northern hemisphere. According to Ponomareva, 28M tons of euphausiids are present as total biomass for the world's oceans, but this is a very conservative estimate. Marr's estimate for the Antarctic krill is very much higher (cf. Marr, 1962).

Euphausiids are of a special significance moreover in open oceans, for though a few such as *Nyctiphanes* and *Pseudeuphausia* are neritic, the great majority are oceanic forms. Their role in marine food chains is perhaps more obvious at high latitudes, where they can contribute very significantly to the food of whales. This is well known for the North Pacific Ocean and the North Atlantic and, above all, in Antarctic seas. Euphausiids may also be an important food of many fishes (e.g. herring, Pacific mackerel, capelin, Pacific salmon species, sardine, and Pacific pollack); in the Antarctic, fishes such as *Micromesistius* feed on euphausiids (cf. Permitin, 1970). Certain seals (e.g. *Phoca hispida*) prey on euphausiids, as do a number of oceanic birds (cf. Marr, 1962; Ponomareva, 1966). Although in the past considerable attention has been focused on the importance of euphausiids at higher latitudes, more recent investigations (e.g. Roger, 1973a, b, 1974, 1977) demonstrate that in tropical seas euphausiids may also play an important part in food chains, particularly with respect to oceanic fishes such as tunas.

Despite their abundance, euphausiids form a relatively small zoological taxon. Mauchline and Fisher (1969) list eighty-five species and Ponomareva (1966) lists eighty-two species. They are restricted to the marine environment, not occurring usually in water less than 28‰ salinity. The majority of species are relatively shallow living, especially in the upper 500 m; a number of species have a wider depth range through the upper 1000 m or even 2000 m. Very few species inhabit really deep layers (e.g. *Bentheuphausia amblyops*, *Thysanopoda cornuta*) and, according to Ponomareva (1966), these do not descend farther than 5000 m.

Euphausiids share with mysids and planktonic decapods a general shrimp-like appearance. The larger anterior part of the body, the cephalothorax, is covered with a thin carapace, fused to all the thoracic segments. Though the carapace extends laterally, the gills which are attached to the exopodites of the thoracic limbs, are exposed (cf. Fig. 2.101). The thoracic limbs form a paired series of similar biramous setose appendages. The first thoracic limb was formerly described as a maxilliped, but this term is not now usually employed, since none of the thoracic limbs is markedly modified for masticatory purposes (cf. Boden, Johnson and Brinton, 1955; Mauchline

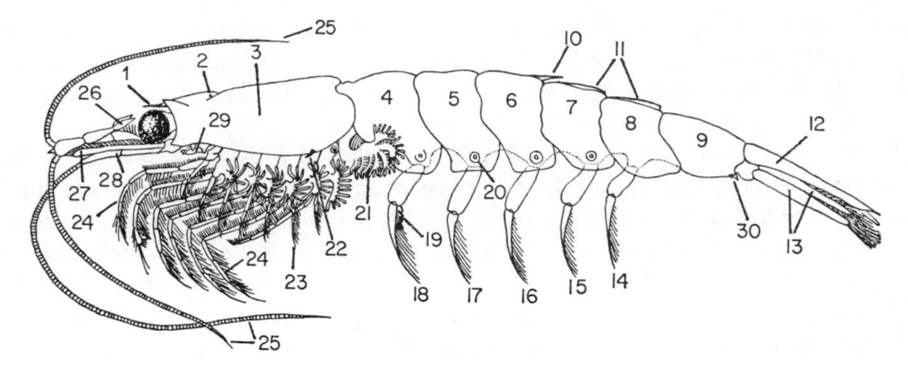

Fig. 2.101. The external morphology of an euphausiid. (1) Rostrum; (2) cervical groove; (3) carapace covering cephalothorax; (4–9) six abdominal segments; (10) dorsal spine; (11) dorsal keel; (12) telson; (13) uropods; (14–17) fifth to second pairs of pleopods; (18) first pair of pleopods modified in male to form; (19) the petasma; (20) luminescent organ of second abdominal segment; (21) gills; (22) lateral denticle of carapace; (23) exopodite of sixth thoracic limb; (24) endopodites of first and sixth thoracic limbs; (25) flagellae of first and second antennae (antennules and antennae respectively); (26) position of reflexed leaflets or of lappets on peduncle of antennule; (27) antennal scale; (28) peduncle of antenna; (29) mouthparts; (30) preanal spine of sixth abdominal segment (Mauchline and Fisher, 1969).

and Fisher, 1969). Some relatively minor modifications, however, may involve the first two, or even three pairs of the thoracic limbs, which are normally furnished with more plumose setae than the succeeding limbs. The setae assist in a filterfeeding process, for which the exopodites of the limbs provide a feeding current.

Each thoracic segment may carry a pair of limbs, but the full eight pairs are present only in the family Bentheuphausiidae, represented by a single species (*Bentheuphausia amblyops*). In all other euphausiids the last (eighth) pair, and in some genera also the seventh pair, of thoracic appendages is reduced.

The first five abdominal segments each carries a pair of pleopods—biramous setose swimming appendages. Their constant beating is responsible for a respiratory current over the ventral posterior region of the thorax so that the gills are irrigated. The last abdominal segment carries the telson and a pair of biramous, relatively flattened uropods, forming a tail-fan (cf. Fig. 2.101).

In the male the endopods of the first and second abdominal segments are modified as copulatory organs; in male Bentheuphausiidae, however, these pleopods are unmodified.

The head bears a pair of relatively large well developed compound eyes; these are divided in some genera (e.g. *Stylocheiron*, *Nematoscelis*), the ommatidia in such divided eyes frequently showing different structure in the upper and lower parts of the eye. Eyes are poorly developed in *Bentheuphausia*. Luminescent organs (photophores) are also typical of euphausiids, though lacking in *Bentheuphausia*. Except in *Stylocheiron*, which possesses only three, euphausiids have ten photophores: one on each eyestalk, a pair at the base of the second and seventh thoracic limbs, and a single photophore below each of the first four abdominal segments. Also carried on the head is a pair of antennules (Ant. I) each bearing two flagella, and a pair of antennae (Ant. II) carrying a flagellum and a scale.

The mouth parts consist of paired mandibles, having grinding cusps and well developed palps assisting in the feeding process, a pair of setose maxillules (first

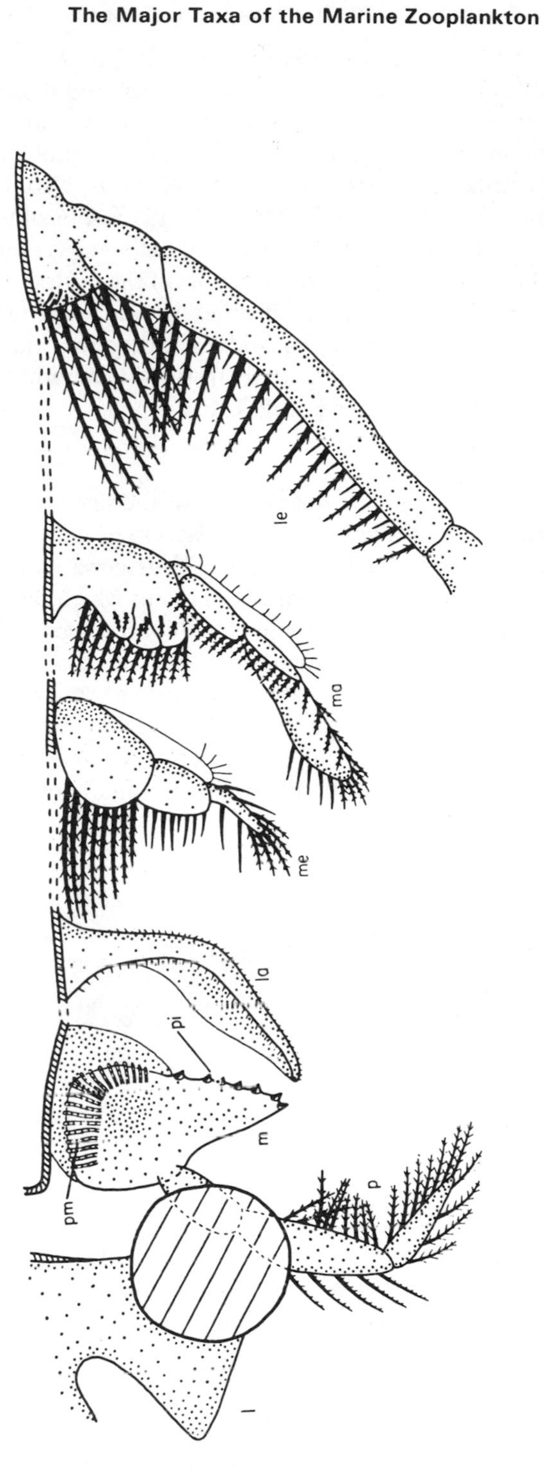

Fig. 2.102. Right-hand feeding appendages of an euphausiid, drawn from the left side. l = labrum; la = labium; le = first thoracic limb; m = mandible; ma = maxilla; me = maxillule; p = mandibular palp; pi = pars incisa; pm = pars molaris (Mauchline and Fisher, 1969).

maxillae), and paired setose maxillae (second maxillae) (cf. Fig. 2.102). The suspended food particles are drawn into a food groove below the head and thorax. The particles are strained off mainly by the first thoracic limbs and maxillae, and then passed by movements of the mouthparts to the grinding areas of the mandibles. In addition to phytoplankton and detrital particles, live zooplankton is also taken by many euphausiids (cf. Chapter 6). The thoracic limbs are mainly responsible for the capture of prey which is then held against the mouthparts, so that spinose setae can pierce the prey. Mauchline and Fisher (1969) give further details of the feeding process. Nemoto (1967) also discusses detailed differences in the feeding appendages of mainly herbivorous and mainly carnivorous euphausiids. He describes the patterns of the thoracic legs, the armature of the legs and their relative lengths, with particular reference to feeding habits (cf. Chapter 6, Fig. 6.10).

Life history

About two-thirds of the species shed their eggs into the sea but a few genera (e.g. *Nematobrachion*, *Nematoscelis*, *Nyctiphanes*, *Stylocheiron*) protect the eggs, which are attached to the posterior thoracic limbs until a nauplius larva hatches (Fig. 2.103).

The development involves a number of larval stages to which distinct names were originally given. The first stage is a nauplius, with three pairs of appendages—uniramous antennules, biramous antennae and mandibles, all lacking setae and unable to filter food; there is no mouth (cf. Fig. 2.104). In most species two similar naupliar stages occur; the second is larger and has a median eye. The subsequent stage is a metanauplius, lacking the mandibles, but with antennules and antennae as in the earlier nauplius stages; the body is distinctly larger with an abdomen appearing, but still no feeding occurs.

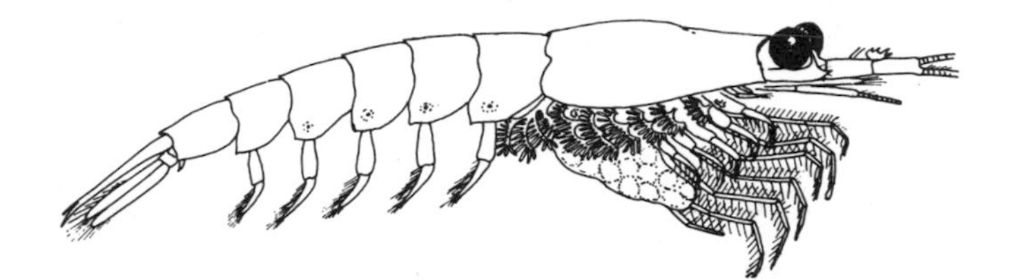

Fig. 2.103. *Nyctiphanes couchii* (ovig. ♀) (Mauchline, 1971).

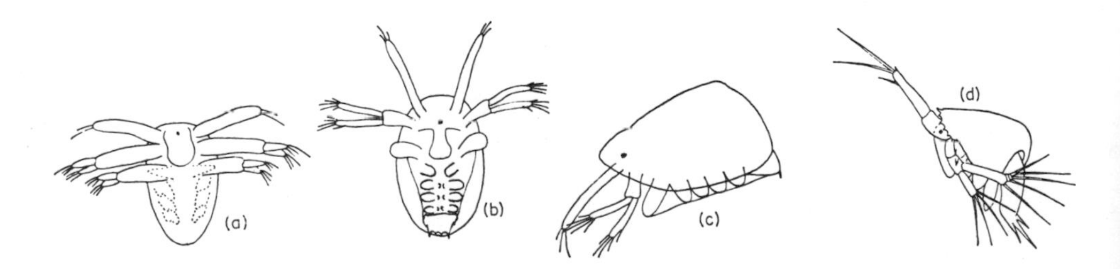

Fig. 2.104. (a) Free swimming first nauplius of *Meganyctiphanes norvegica*, ventral view; (b) ventral, and (c) lateral views of second nauplius of *Nyctiphanes simplex*; (d) lateral view of metanauplius of *M. norvegica* (Mauchline and Fisher, 1969).

Three calyptopis stages succeed the metanauplius, and these stages include the development of compound eyes and the abdomen. The calyptopis larvae are supplied with biting mandibles and setose maxillae (two pairs) so that the larvae can feed. Calyptopis I has a well developed carapace covering the front and sides of the cephalothorax. Eyes begin to develop but are unpigmented and a median eye is still present. The abdomen is usually unsegmented. In Calyptopis II the abdomen is five-segmented and uropods are appearing. The eyes are slightly better developed and the median (Nauplius) eye is lost. In Calyptopis III the abdomen has six segments and is more elongated. The eyes are better developed and globular.

The Calyptopis stages are followed by a series of larval furcilia stages, but there are substantial differences in the number of furcilia between the different species and even some variation in the precise development inside a species. Mauchline distinguishes four phases—during the first two the pleopods become functional; in the third and fourth stages the thoracic limbs, as well as the telson and the abdominal photophores develop. During the first phase some abdominal segments acquire non-setose pleopods. In many species the first furcilia has no pleopods and it is only at the next moult that one or more pairs of non-setose pleopods are acquired. But in other euphausiids the last calyptopis can moult directly to this later condition. In any event, more pleopods are added in subsequent moults until the full complement is achieved. A stage is often passed through in which the anterior pleopods are setose and the most posterior pleopods non-setose.

Whatever the precise course of development, the more anterior abdominal segments tend to possess fully developed setose pleopods earlier than the more posterior segments. The number of instars necessary to reach the final condition varies with the species, but finally a furcilia stage is reached when all five pairs of pleopods are fully developed and setose (Fig. 2.105). The variations in course of development in some species (e.g. *Nyctiphanes australis*, *Meganyctiphanes norvegica*, *Thysanoessa raschii*) are described by Mauchline and Fisher (1969).

Later furcilia show progressive reduction of spines on the telson and the development of the abdominal photophores. With the proper development of the thoracic appendages, the adolescent stage is attained; this involves also some small changes in certain mouthparts. Similar slight modifications may occur during growth from adolescent to adult, but essentially this change is in the maturing of the reproductive organs and the development of the secondary sex characters.

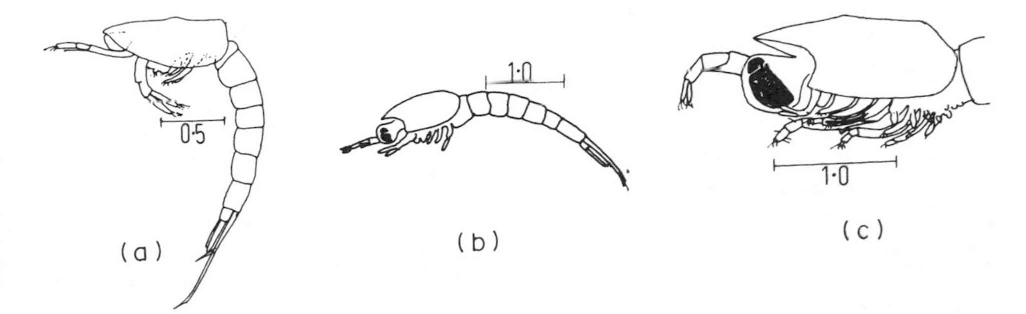

Fig. 2.105. Some stages in development of *Thysanoessa inermis* (a) Calyptopis III; (b) Furcilia I; (c) Furcilia IV (Scales in mm.) (Mauchline, 1971).

Geographical and vertical distribution

Bentheuphausia amblyops, the only representative of the Bentheuphausiidae, is also distinct from the great majority of euphausiids in being deep-living, normally exceeding 1000 m and reaching even 5000 m. It is widespread in warm and warm-temperate areas, more especially in tropical and sub-tropical regions. Mauchline and Fisher suggest limits of 46°N to 40°S for the Atlantic, though it is not found in the Mediterranean. Brinton (1962) indicates that the species is found more or less consistently in the Pacific Ocean from about 53°N to about 54°S latitude, provided the depth exceeds 1000 m, though in greater abundance in the eastern Pacific between 38° and 45°N and in eastern equatorial waters. Ponomareva (1966), however, does not believe that the northern distribution of *Bentheuphausia* is bounded by the Aleutians. The euphausiid can be taken in the Bering Sea, entering with deeper oceanic water through the Aleutian Straits. In the Indian Ocean, *Bentheuphausia* occurs at least between 10°N and 10°S latitude.

Of the Euphausiidae, epiplanktonic forms inhabiting near surface waters to depths of about 300 m or even 500 m are represented at high latitudes, as in several other taxa by a few species present in very considerable abundance. In near Arctic waters *Thysanoessa raschii* and *Thysanoessa inermis* occur in the Atlantic Ocean with, *Thysanoessa longicaudata* and *Meganyctiphanes norvegica*. In the Pacific Ocean the former two species are present with *Thysanoessa longipes* and *Euphausia pacifica*. Dunbar (1964) classes *Thysanoessa raschii* and *T. inermis* as sub-Arctic, occurring at somewhat higher latitudes than *T. longicaudata*, *T. longipes*, and *Euphausia pacifica*, which are boreal species penetrating sub-Arctic waters. *Thysanoessa longicaudata* is rather similar in its distribution in the Atlantic to the widespread and abundant *Meganyctiphanes norvegica*.

Mauchline and Fisher (1967) agree with Dunbar that *Meganyctiphanes* probably does not breed in the sub-Arctic. They point out that breeding does not normally occur in West Greenland waters or north of about 70°N latitude. *Meganyctiphanes* has a wide geographical range, however, stretching south in the western Atlantic as far as Cape Hatteras. It is abundant off the north east coast of U.S.A. and Canada, including the Gulf of Maine, Labrador and Baffin Island. It is also comparatively abundant off both west and east Greenland. In the Barents Sea it is found as far north as Spitzbergen and off the European coast it is particularly abundant in the Norwegian Sea. It is present in the North Sea, stretching south as far as the western Mediterranean and north west Africa (*vide infra*). Bigelow (1926) includes *Meganyctiphanes norvegica* as a member of the "*Calanus* community" (Chapter 8). *Meganyctiphanes* and *Thysanoessa inermis* are both mentioned as occurring in small numbers off the continental shelf, south of Cape Cod (Bigelow and Sears, 1939).

Thysanoessa raschii tends to be more neritic in character than *T. inermis* (Figs 2.106, 2.107), occurring in coastal regions of the northern North Pacific, including the Okhotsk Sea, Sea of Japan, Bering Sea and Gulf of Alaska. Its southern limit in the North Pacific is about the border between the sub-Arctic and transition water masses. In the Atlantic Ocean, though reported from 78°N, it probably does not breed beyond about 70°N; its southern limit in the west Atlantic is around 40°N. It is common off northern Scotland, along the Norwegian coast, around Iceland, and in the Barents and White Seas, and is the commonest euphausiid in west Greenland waters. As a relatively

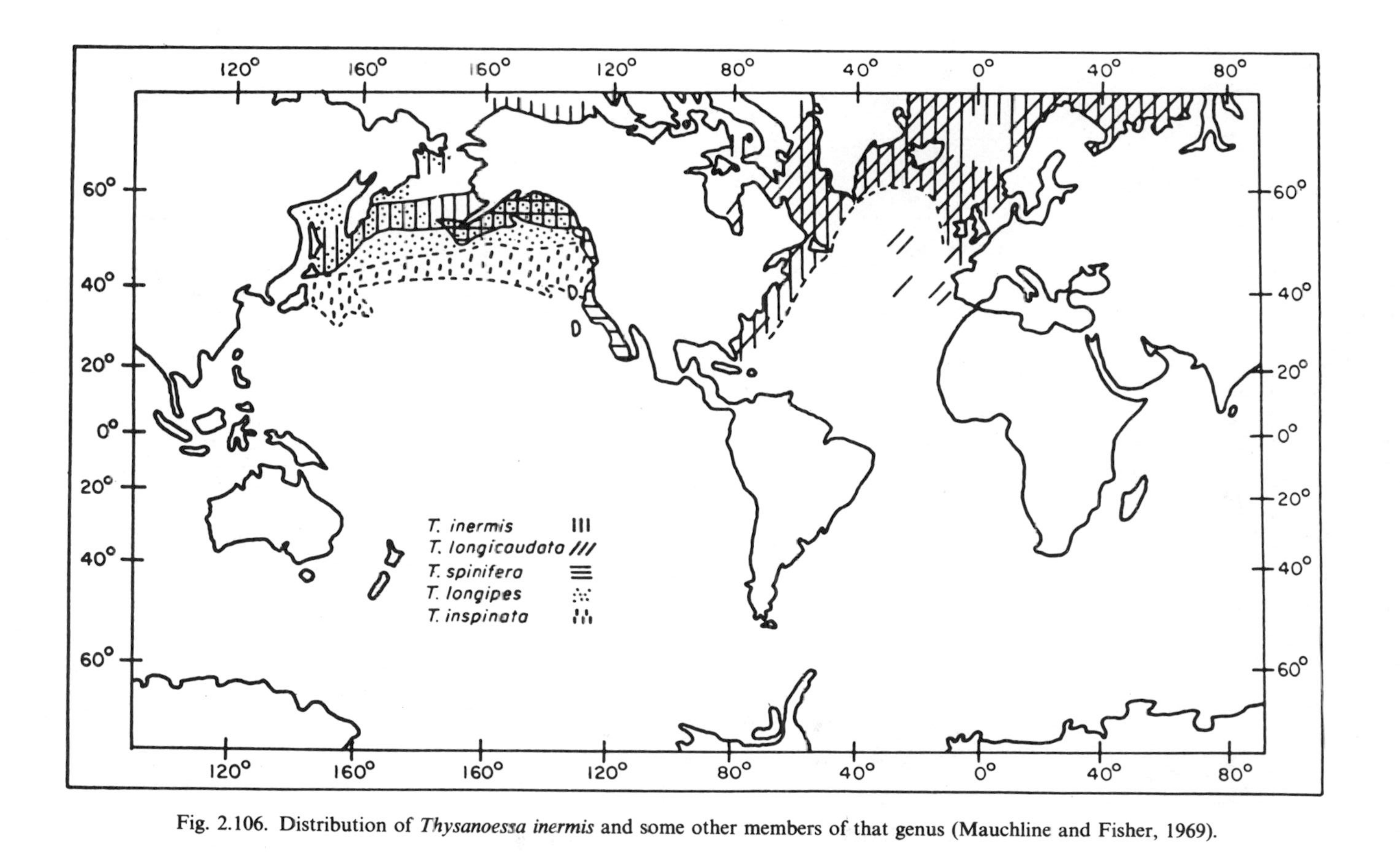

Fig. 2.106. Distribution of *Thysanoessa inermis* and some other members of that genus (Mauchline and Fisher, 1969).

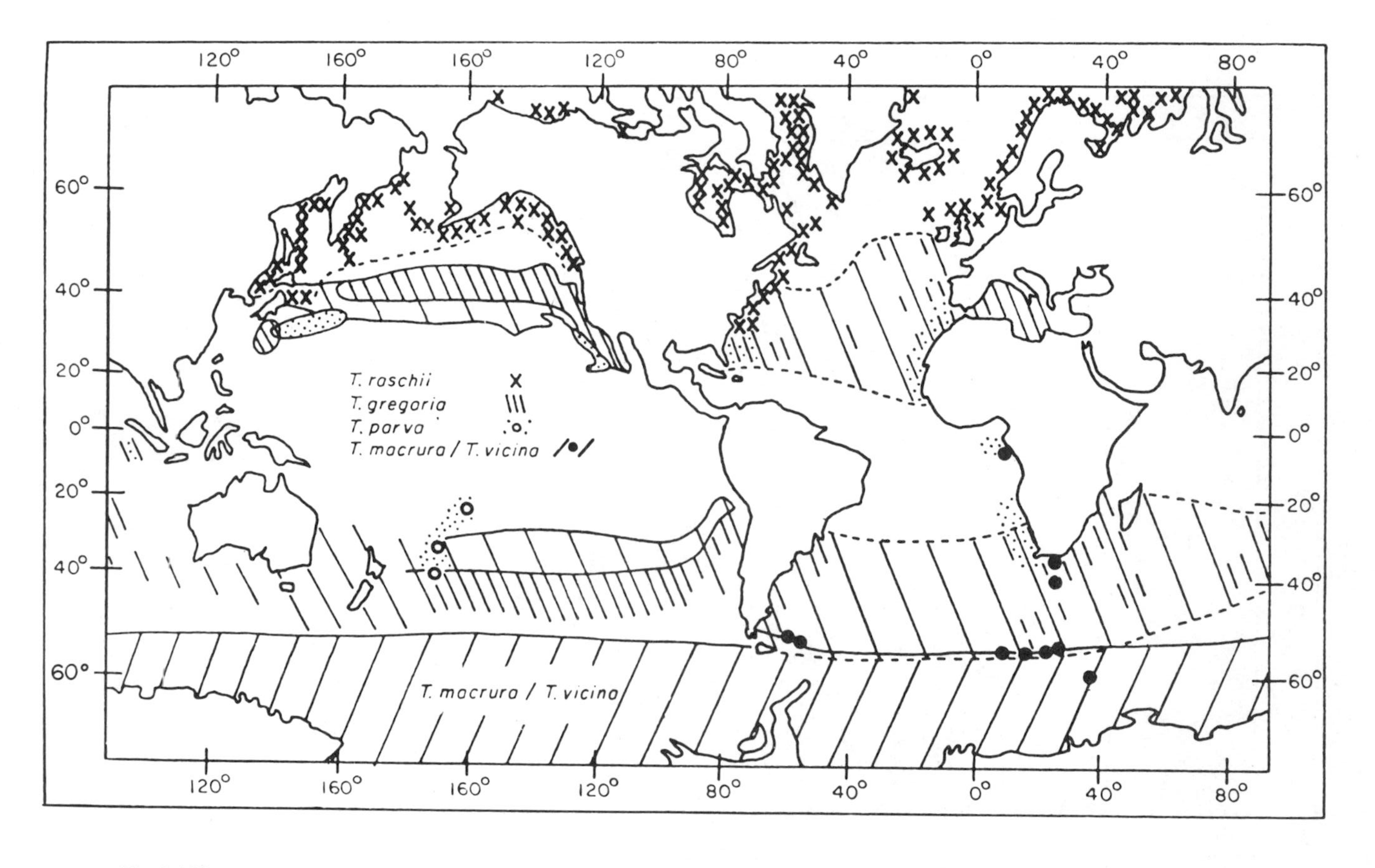

Fig. 2.107. Distribution of 5 *Thysanoessa* species (dots and crosses indicate records of occurrence) (Mauchline, Fisher, 1969).

shallow coastal species, it occurs mainly within the continental shelf, from about 200 m to near the surface (Nemoto, 1957; Wiborg, 1971).

Thysanoessa inermis is more oceanic, though it is relatively common off the coasts of the Gulf of Maine, Greenland, Norway, Iceland and the Barents Sea. It apparently does not breed north of 70°N in the North Atlantic, though it can be found at higher latitudes. In the Pacific Ocean the species does not reach so far north, but it is found abundantly in the Bering Sea. Its southern limit is about 43°N in the western Pacific and 50°N in the eastern Pacific; 40°N in the western Atlantic and 50°N in the eastern Atlantic (Boden, Johnson and Brinton, 1955; Brinton, 1962). Its depth range is from <300 m to the near surface.

Thysanoessa longicaudata, found only in the North Atlantic Ocean, may occur south as far as 35° or 40°N but extends even to 80°N off north-east Greenland. It is relatively abundant, however, only between about 55–70°N, and is an off-shore species. Its vertical range extends from the surface to somewhat deeper water than the two former species. *Thysanoessa longipes* is confined to the North Pacific where it inhabits the entire sub-Arctic water mass, but may reach a little farther south of 40°N by transport in the Californian Current (Fig. 2.106). It is well known in the Bering Sea and has been reported even beyond 70°N (cf. Boden *et al.*, 1955) (Fig. 2.108).

The species occurs in two varieties, the spined and unspined forms, the former apparently more truly sub-Arctic and of larger adult size. The depth distribution is from the surface to 500 m off the Alaskan Shelf (Nemoto, 1957), but it may reach deeper levels, especially in the Sea of Japan (Ponomareva, 1966). Zhuravlev (1977) states that the spined form of *T. longipes* is dominant in the Sea of Okhotsk central deep water.

The only species of *Euphausia* found in the northern North Pacific, *Euphausia pacifica*, is restricted to that great area (Fig. 2.108). Johnson and Brinton (1963) state that, though typical of sub-Arctic waters, the species is more wide ranging than other sub-Arctic euphausiids. It mostly occurs at depths <300 m. Though found in the southern part of the Bering Sea, the species is somewhat rare in that area and in the Sea of Okhotsk where it is confined to the southernmost area (Zhuravlev, 1977). It does not appear to breed there. It is plentiful in the Sea of Japan but is especially typical of the main part of the North Pacific Drift and the Californian Current areas. The geographical limit of *E. pacifica* is essentially the northern limit of the transition waters, but it can occur somewhat farther south, at least at greater depths, even to about 30°N latitude, according to Ponomareva. It appears to be somewhat more eurythermal than the other sub-Arctic species.

Thysanoessa spinifera, a Pacific boreal species, is much more restricted in its distribution. It is a neritic species, mainly in the eastern Pacific and does not exceed about 200 miles off shore (Boden *et al.*, 1955). It is abundant along the coast of the Gulf of Alaska, south to the Californian coast (Fig. 2.106).

Ponomareva distinguishes to some extent the euphausiid populations of the Far Eastern Seas from that of the main North Pacific. The Sea of Japan has a particularly rich euphausiid population notably in the spring, the main species being *Thysanoessa longipes* and *Euphausia pacifica*. In the Sea of Okhotsk the species are mainly *Thysanoessa longipes* and *Thysanoessa raschii*, and in the Bering Sea, *Thysanoessa inermis* and *Thysanoessa raschii* (Fig. 2.106).

Although the vertical extent of the epipelagic sub-Arctic euphausiids shows some

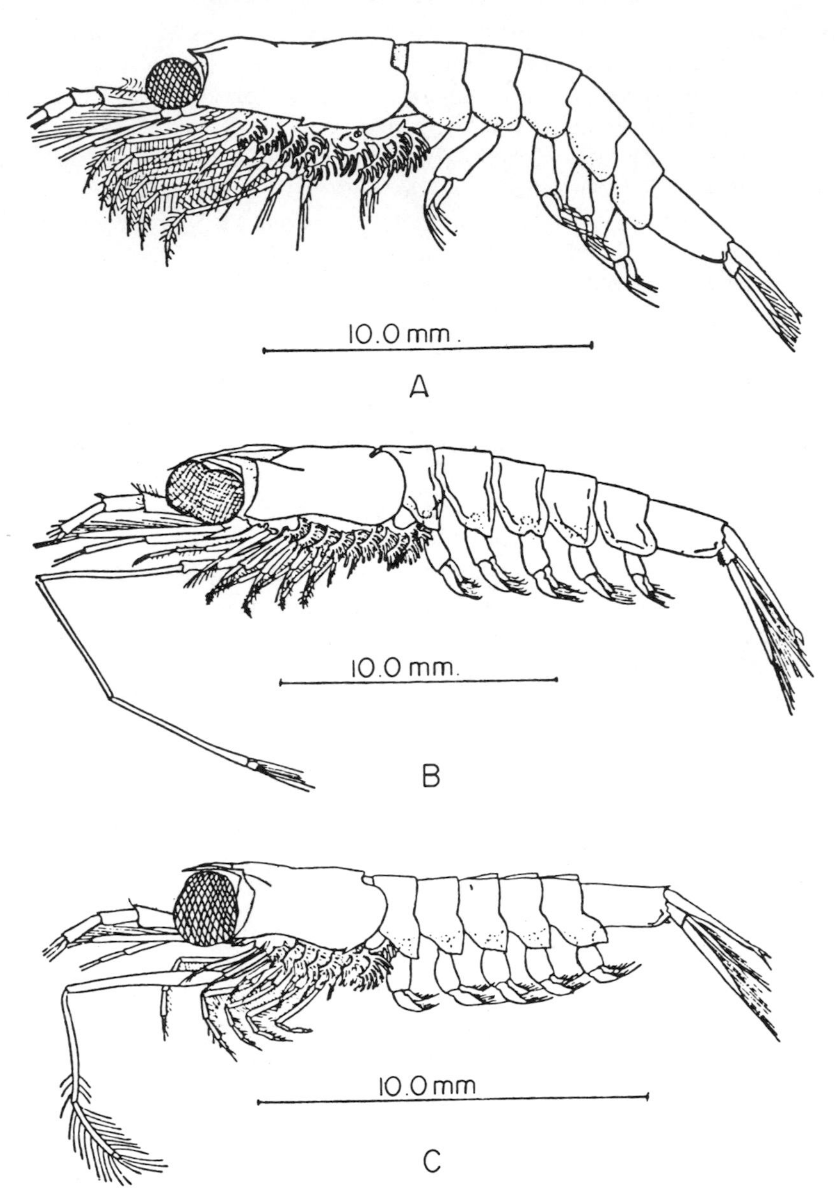

Fig. 2.108. Three euphausiid species. (A) *Euphausia pacifica* ♀; (B) *Nematoscelis difficilis* ♂; (C) *Thysanoessa longipes* (Boden, Johnson and Brinton, 1955).

specific differences, a general depth range for the five species in the northern North Pacific is 0–180 m (Brinton, 1962). The North Pacific is also inhabited by deeper-living species. As a cosmopolitan bathypelagic species, *Bentheuphausia amblyops* is taken generally over the North Pacific (*vide supra*). *Tessarabrachion oculatus*, a mesopelagic euphausiid, may be found at any depth between about the surface and 1000 m, and occurs from the latitude of the Aleutians southwards to about 35°N. According to Ponomareva (1966) it does not exceed 500 m in the North Pacific Ocean. Another

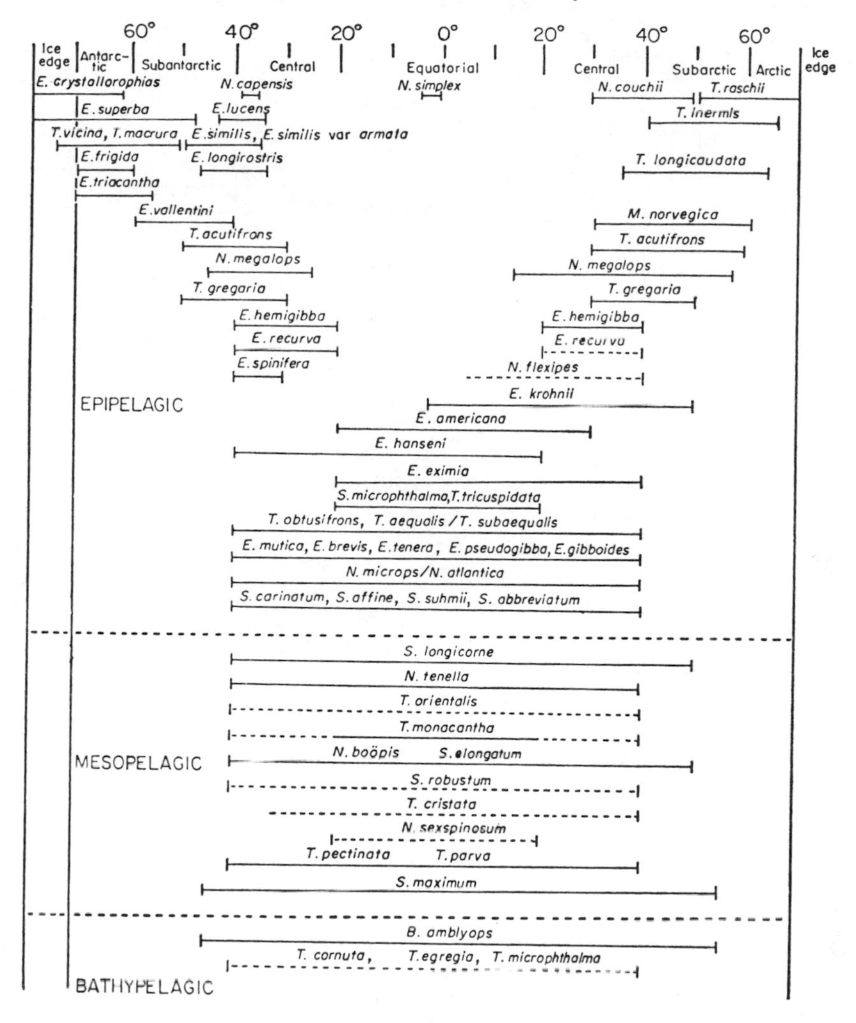

Fig. 2.109. The approximate horizontal and vertical distribution of euphausiids in the Atlantic Ocean (Mauchline and Fisher, 1969).

mesopelagic species found in the sub-Arctic Pacific, with a depth range of about 140–1000 m, is *Stylocheiron maximum*, a cosmopolitan form found in all oceans outside the Arctic (Fig. 2.109).

In the oceans in general, apart from *Bentheuphausia*, the only other bathypelagic euphausiids are species of *Thysanopoda*. *T. cornuta* and *T. egregia* are found in all three oceans, the first being more wide ranging, reaching about 50° N, while *T. egregia* is more restricted. Brinton (1962) suggests limits of about 40° N and 55° S latitude for the Pacific Ocean. *T. spinicaudata* is also bathypelagic but apparently lives only in the Pacific, while *T. microphthalma* appears to be confined to deep waters of the Atlantic and Indian Oceans (Figs. 2.109, 2.110). Captures of deep sea euphausiids are, however, somewhat rare and difficult and further deep sea collections may extend these geographical ranges. Though the immature stages of these species may be taken above

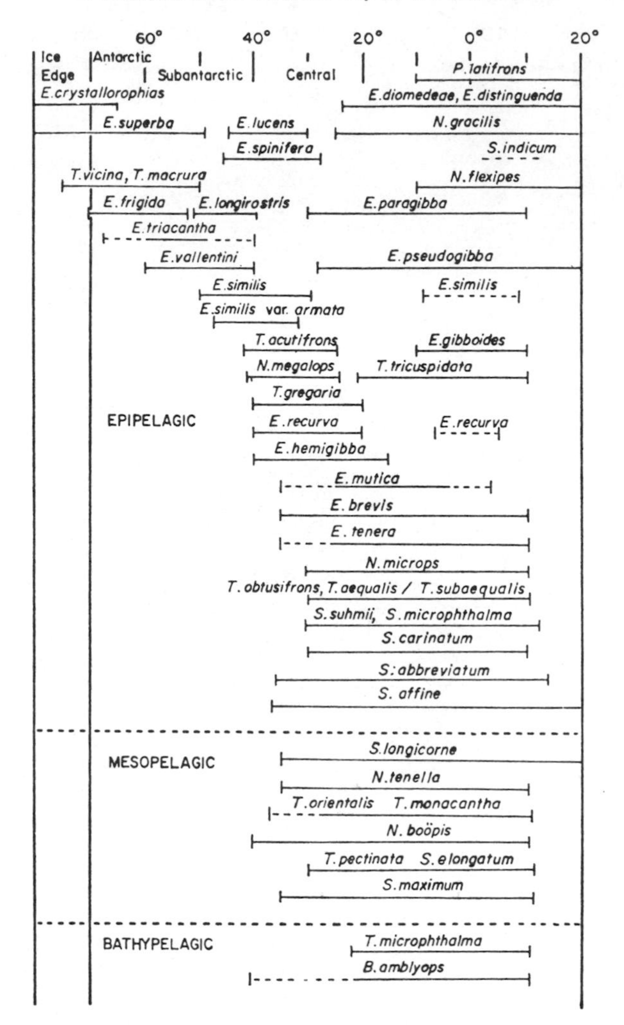

Fig. 2.110. The approximate horizontal and vertical distribution of euphausiids in the Indian Ocean (Mauchline and Fisher, 1969).

500 m depth, typically they are deeper-living. The adults are almost always living deeper than 1000 m, often exceeding 2000 m depth. The lower depth limit may even approach 5000 m according to Ponomareva.

According to Brinton (1962), mesopelagic euphausiids may live as shallowly as 140 m in sub-Arctic waters, but they are usually as deep as 400–500 m in warm seas. At their lower depth range they may extend to about 1000 m. They do not reach as far as the euphotic zone even with vertical migration. Mauchline and Fisher (1969) list a dozen euphausiids as definitely mesopelagic, including four species of *Thysanopoda* and four of *Stylocheiron*, all of which except *Stylocheiron robustum* occur in all three oceans. At least two other mesopelagic species also are known from Atlantic, Pacific and Indian Oceans so that their widespread distribution is obvious.

In the Pacific Ocean, Brinton (1962) points out that the mesopelagic euphausiids typical of the central water masses, such as *Stylocheiron elongatum*, *Nematoscelis tenella*, *Nematobrachion boopis*, *Thysanopoda orientalis*, and *Thysanopoda monacantha*, extend across the equatorial zone so that they range roughly from 40°N and 40°S latitude, this broad distribution being probably related to the more uniform temperature at intermediate depths in the tropics and sub-tropics. Some species are somewhat more restricted. Ranges in the Atlantic Ocean, and to a more limited extent in the Indian Ocean, are comparable. The more cosmopolitan *Stylocheiron maximum* has a greater geographical range (about 60°N to 60°S); *Stylocheiron longicorne* is also somewhat more wide ranging.

In the North Pacific, in the transition zone between sub-Arctic and central waters (*ca.* 45°N and 35°N latitude), Johnson and Brinton (1963) list *Nematoscelis difficilis* (Fig. 2.106) and *Thysanoessa gregaria* as typical epipelagic euphausiids (cf. Chapter 8). These species are found in the Californian Current, with the latter species in the Kuroshio area also. In the North Atlantic, at approximately similar latitudes, *Thysanoessa gregaria* also occurs, but *Nematoscelis difficilis* is replaced by *Nematoscelis megalops*. *Euphausia krohnii*, a species confined to the Atlantic Ocean, is also present in this region and appears to have a very wide range extending southwards even to equatorial regions of the Atlantic.

Thysanopoda acutifrons extends over a broad zone between sub-Arctic and central waters, and has a wide geographic range, from the Gulf of Alaska (about 55°N) generally over the North Pacific Drift, to reach about 35°N latitude. The species is also found in the Atlantic Ocean, even to about 70°N latitude. It is well known off Iceland and Greenland. To the south it occurs as far as the Gulf of Maine and in the eastern Atlantic to Gibraltar, but it is not found in the Mediterranean. Although classed generally as an epipelagic species, *T. acutifrons* is usually beneath the surface and can reach considerable depths: according to Ponomareva (1966), about 1000 m in the Pacific and even deeper in the Atlantic and Indian Oceans. Mauchline and Fisher give a vertical range of about 140–4000 m. *T. acutifrons* also occurs in the Southern Ocean; it is found in the South Pacific from about 35° to about 60°S, but not so far south in the Atlantic. It is not known from tropical regions. All other species of *Thysanopoda* are tropical or sub-tropical in distribution and are mostly mesopelagic or bathypelagic.

In the South Pacific a comparable transition zone exists between the sub-Antarctic and central waters, approximating to 45–35°S latitude. Apart from the presence of the bi-polar species, *Thysanopoda acutifrons*, *Thysanoessa gregaria* occurs and *Nematoscelis megalops* occupies the place of *Nematoscelis difficilis* of the North Pacific. These same three euphausiids are found in comparable water masses of the Indian Ocean and the South Atlantic (Fig. 2.106).

There are many suggestions that some euphausiids are limited in depth by particular isotherms. *Thysanoessa gregaria* may be bounded by the 7° and 11° isotherms in both hemispheres (Brinton, 1962, 1979).

The vertical ranges of many epiplanktonic and mesopelagic euphausiids and their diurnal migrations relate largely to the thickness of the upper water mass which may vary from about 200 m to nearly 1000 m. Salinity and especially temperature characteristics place a restriction on the vertical spread of a species. As Mauchline and Fisher (1969) suggest, a majority of tropical and sub-tropical species live mostly above 700 m at temperatures exceeding 10°C while some equatorial species do not extend below about 300 m, approximately the depth of the warm upper layer. On the other hand, sub-Arctic and sub-Antarctic epiplanktonic euphausiids live relatively shallowly because of the low temperature in the sub-surface deeper layers (Fig. 2.109).

Thus the warmer-water euphausiids may be roughly grouped into those resident in central water masses ("sub-tropical"), both north and south of the equator, and those typical of the equatorial zone. Although approximate latitudes have been quoted for the normal ranges of different euphausiids, the true distribution of any species, as with other zooplankton, is usually better described in relation to a water mass which serves as the centre of distribution. In Fig. 2.111, Brinton (1979) outlines some group associates in the eastern tropical Pacific and indicates the bio- and hydrographic conditions. Arctic, sub-Arctic transition, central and equatorial zones, and the corresponding zones in the southern hemisphere, may approximate to geographical latitudes but major surface currents may introduce irregularities. If these populations are to be retained inside their normal distribution areas, current systems must exist to prevent the dispersal of too large a proportion of the population as "expatriates" to unsuitable water masses where successful breeding is hindered or prevented. For neritic species local hydrographic conditions may serve to maintain the endemic population. In open oceans, gyres currents and counter-currents serve as the usual

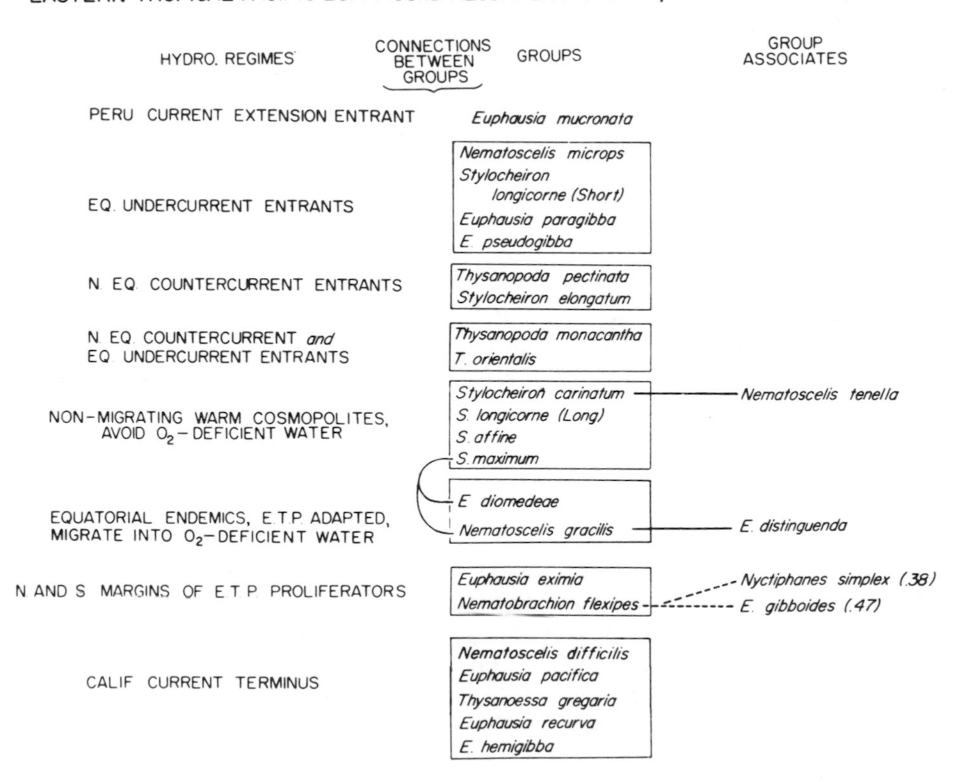

Fig. 2.111. Recurrent groups of euphausiid species and group associates in the Eastern Tropical Pacific. Affinity level > 0.50. Affinities of more weakly associated species are shown by broken lines (Brinton, 1979).

mechanisms. Some species may use depth migration to utlize currents assisting in the return of breeding stock (cf. Chapter 8).

As typical of the eastern and western central water masses of the Pacific Ocean, Brinton (1962) lists *Nematoscelis atlantica* and *Euphausia brevis*, found both north and south of the equator (bi-sub-tropical). *Thysanopoda obtusifrons* and *Stylocheiron suhmii* appear to be generally similarly distributed. *Euphausia hemigibba*, typical of the central water mass of the North Pacific, is replaced by *Euphausia gibba* in the central water mass of the South Pacific. Other species, *Euphausia recurva*, *E. mutica* and *Thysanopoda aequalis*, occur north and south of the equator, but their main populations are more at the periphery of the water masses. *Thysanopoda aequalis* is also practically limited to the eastern water mass in the North Pacific; in the western gyre it is replaced by *Thysanopoda subaequalis* (cf. Chapter 8). In the South Pacific both species occur in central waters but *Thysanopoda subaequalis* is more restricted in its range. Although the euphausiids listed for central waters have been described as epipelagic, many tend to live in the sub-surface layers. For instance *Euphausia hemigibba*, *Nematoscelis atlantica*, *Thysanopoda* sp., and *Stylocheiron* sp., are rarely shallower than 150 m by day. Many of these euphausiids approach the surface at night, but the genus *Stylocheiron* does not migrate, though it has an extensive vertical range.

In the Atlantic Ocean, the distribution patterns of euphausiids in relation to central and equatorial water masses are less clear than in the Pacific. The bi-polar distribution of *Thysanopoda acutifrons* to some extent overlaps the bi-anti-tropical pattern of *Nematoscelis megalops* and *Thysanoessa gregaria*. Isolated patches of *N. megalops* occurring in the Sargasso Sea are nearly all associated with cold core rings. (Haury, McGowan and Wiebe, 1978). *Euphausia hemigibba*, and possibly *E. recurva*, occur in central zones north and south of the equator, though the latter species has been identified in the North Atlantic only off North-west Africa (cf. Mauchline and Fisher, 1969). Thiriot (1977) has reported that off Mauretania, while *Nyctiphanes capensis* is dominant in the neritic zone, more offshore species such as *E. americana*, *E. tenera* and *E. gibboides*, occur in the southern part (11°N), and *E. krohnii*, *E. pseudogibba*, and *E. hemigibba* to the north (20°–24°N) (Fig. 2.109).

Although several euphausiid species are known from the oceanic zone, the maximum biomass is dominated by *E. krohnii*, with *Nematoscelis megalops* and *Thysanoessa gregaria* present in much smaller numbers (Fig. 2.108). Off the Moroccan coast, euphausiids are somewhat less abundant. *Nyctiphanes couchii* occurs inshore. Offshore *E. krohnii* is again dominant, but *Meganyctiphanes norvegica* occurs in reasonable quantities, although it is near the southern limit of its distribution. The latter species is a major contributor to the euphausiid biomass in the Gulf of Genoa. *Nematoscelis atlantica* and *Euphausia krohnii* are important species in the western Mediterranean (Boucher and Thiriot, 1972). Baker (1970) reports *E. krohnii* as one of the most abundant species taken off the Canaries; the daytime peak was about 500 m, but a dense surface swarm was sighted at night. In the Indian Ocean, *Euphausia hemigibba* and *E. recurva* are found south of the equator in central waters and *Thysanopoda acutifrons*, *Thysanoessa gregaria* and *Nematoscelis megalops* are present as transition zone species.

So far only the epiplanktonic euphausiids living mainly in central Pacific waters have been described. Many species inhabit the equatorial water mass of the Pacific Ocean. Some (*Euphausia diomedeae*, *E. paragibba*, *Nematoscelis gracilis* and *Stylocheiron microphthalma*) stretch almost across the equatorial Pacific. Others (*Euphausia distinguenda*, *E. eximia*, *E. lamelligera*, *Nyctiphanes simplex*) occur mainly in the eastern region while *Euphausia sibogae*, *E. fallax*, *Pseudeuphausia latifrons* and *Stylocheiron insulare* are found mostly in western tropical waters. Some equatorial species (*Euphausia tenera*, *Thysanopoda tricupsidata*) have a more widespread distribution from the equatorial water mass to the central waters in the West Pacific.

The importance of tropical euphausiids has been emphasized by the investigations of Legand Bourret, Fourmanoir, Grandberiun, Gueredrat, Michel, Rancurel, Rebelin and Roger (1972) and Roger (1974a) who found extensive euphausiid populations in open tropical waters of the western Pacific (*ca.* 10°–20°S). On average, euphausiids comprised 10 % of the zooplankton catch (Roger, 1977). Two genera formed the bulk numerically of the collections: 50 % approximately was contributed by *Euphausia*—*E. diomedeae*, and *E. tenera* near the equator and by *E. brevis* and *E. mutica* in subequatorial waters (Roger (1973a) refers to the great importance as biomass of *E. diomedeae*); 30 % of the catch came from *Stylocheiron*, with *S. abbreviatum* and *S. carinatum* as the major contributors.

As suggested by Brinton (1962), *Stylocheiron* tended to be more superficial in distribution and migrated little; *Euphausia* was more typically mesopelagic and exhibited more extensive diurnal vertical migrations. Thus, in the 0–200-m layer, during daytime, the epiplanktonic forms included particularly *Stylocheiron carinatum*, *S. affine* and *S. suhmii*, all very small euphausiids. The 200–400 m layer included some individuals of these more shallow-living species, but was characterized by *Euphausia diomedeae*, *E. fallax*, *Stylocheiron elongatum*, and *S. abbreviatum*. Among the deeper-living (400–800 m) mesopelagic species were several *Thysanopoda* species (*T. aequalis*, *T. tricuspidata*, *T. monacantha*), *Euphausia brevis* and the small *E. tenera* and species of *Nematoscelis* (*N. atlantica*, *N. tenella*, *N. microps*). By day, 40 % of the euphausiid population were in the upper 400 m, but owing to the dominance of small species (especially *Stylocheiron*), this comprised only a mere 20 % of the biomass. At night, 88 % of the population (72 % of the biomass) was present in the upper 200 m, indicating the pronounced diurnal vertical migrations of many of the larger deeper-living euphausiids. (Further details of diurnal migration are given in Chapter 5.)

In the Indian Ocean typical species found near the equator include *Euphausia diomedeae*, *E. distinguenda*, *E. paragibba*, *E. brevis* and *Thysanopoda subaequalis* (Fig. 2.110). In the Atlantic Ocean the distinction between euphausiids inhabiting near-surface equatorial and central waters is less clear than in the Pacific. A few species (e.g. *Euphausia americana*) are restricted to the Atlantic; this euphausiid extends from north of Bermuda across the equator to about 20°S latitude. But the majority of the warm-water surface species (*Euphausia eximia*, *E. mutica*, *E. tenera*, *E. pseudogibba*, *Stylocheiron microphthalma*, *Thysanopoda tricuspidata*) are well known also in the Indo-Pacific (Fig. 2.110). *Stylocheiron carinatum*, *S. abbreviatum*, and *S. suhmii* are among other euphausiids occurring at slightly deeper levels over a wide area in the warmer Atlantic. At about 30°N, Baker (1970) found *Stylocheiron affine* and *S. suhmii* in the upper layers, with *S. elongatum* living a little more deeply. *Nematoscelis*, especially *N. atlantica* and *N. microps*, was an important contributor to the total euphausiid population; *Euphausia* and *Stylocheiron* were also significant. In the Caribbean, Michel and Foyo (1976) found *Euphausia tenera*, *E. brevis*, and

E. americana, with *Stylocheiron affine*, *S. carinatum* and *S. longicorne*, as the most abundant euphausiids, the last three species living comparatively shallowly. There was a considerable measure of agreement between the more important species taken in the Caribbean and those reported by Roger.

The account of the warm-water near-surface euphausiids of the Indo-Pacific included a coastal species, *Pseudeuphausia latifrons*. This euphausiid is present in neritic areas of the tropical west Pacific and Indian Oceans and extends as far south as about 32°S latitude on the east coast of Australia. It is very abundant in northern coastal areas of the Indian Ocean, where *Euphausia distinguenda* appears to have a somewhat similar distribution (Brinton and Gopalakrishnan, 1973). *Pseudeuphausia latifrons* overlaps *Nyctiphanes australis* off south-east Australia. This latter neritic species also occurs along the coast of New Zealand and reaches about 43°S latitude. Another species of *Nyctiphanes*, *N. capensis*, is neritic in the waters off South and West Africa.

The genus *Nyctiphanes* appears to be typical of near coastal areas, especially where a transition occurs between warm and colder waters. *N. simplex* is found in coastal areas off California as far as 40°N latitude. In the southern hemisphere it occurs off South America to about 30°S latitude and in the Peru Current region as far as the Galapagos. The habitat of *N. couchii* is perhaps similar. It is a coastal form restricted to the eastern Atlantic, and though it reaches as far as 60°N, it stretches southwards to the North African coast and along northern coastal areas of the Mediterranean. It is relatively common in the North Sea and over the continental slope and shelf west of the British Isles. The distribution of *Euphausia mucronata* is somewhat peculiar: it is very abundant in the Peru Current region, but it does not penetrate westwards, as does *Nyctiphanes simplex*, towards the Galapagos. To the south it does not extend beyond about 40°S latitude, not reaching sub-Antarctic waters.

In the southern hemisphere, the extensive waters of the Southern Ocean south of the transition zone are inhabited by a rich population of euphausiids, but as in sub-Arctic waters, relatively few species are present. The species are typically circum-polar in distribution (cf. Chapter 8).

Euphausia crystallorophias is a neritic species, fairly closely associated with the coastline of Antarctica, living mainly under the ice at the highest southern latitudes. *Euphausia superba* is much more widely ranging between the continent and the Antarctic Convergence. Although circumpolar, it is particularly abundant at really high latitudes in the East Wind Drift area and at lower latitudes occurs particularly in the region of Bransfield Strait and the area round South Georgia (cf. Marr, 1962). *E. superba* is recognized as a relatively shallow-living euphausiid, which is very patchy in distribution and is liable to swarm at or close to the surface. The Antarctic species *Euphausia frigida*, extends into somewhat less high latitudes, typically ranging between the ice edge and the Antarctic Convergence. It spreads vertically to nearly 500 m during day time. *E. triacantha*, according to Baker (1959), has its maximum abundance at the Antarctic Convergence; smaller densities are found on each side of the Convergence. It may also reach about 500 m in depth. Figure 2.112 indicates the normal range and extreme distribution of these euphausiids, together with those of some sub-Antarctic and warmer-water species (cf. Mackintosh, 1964).

Two species of *Thysanoessa*, *T. macrura* and *T. vicina*, are also shallow sub-surface circumpolar Antarctic euphausiids, living mostly south of about 60°S latitude.

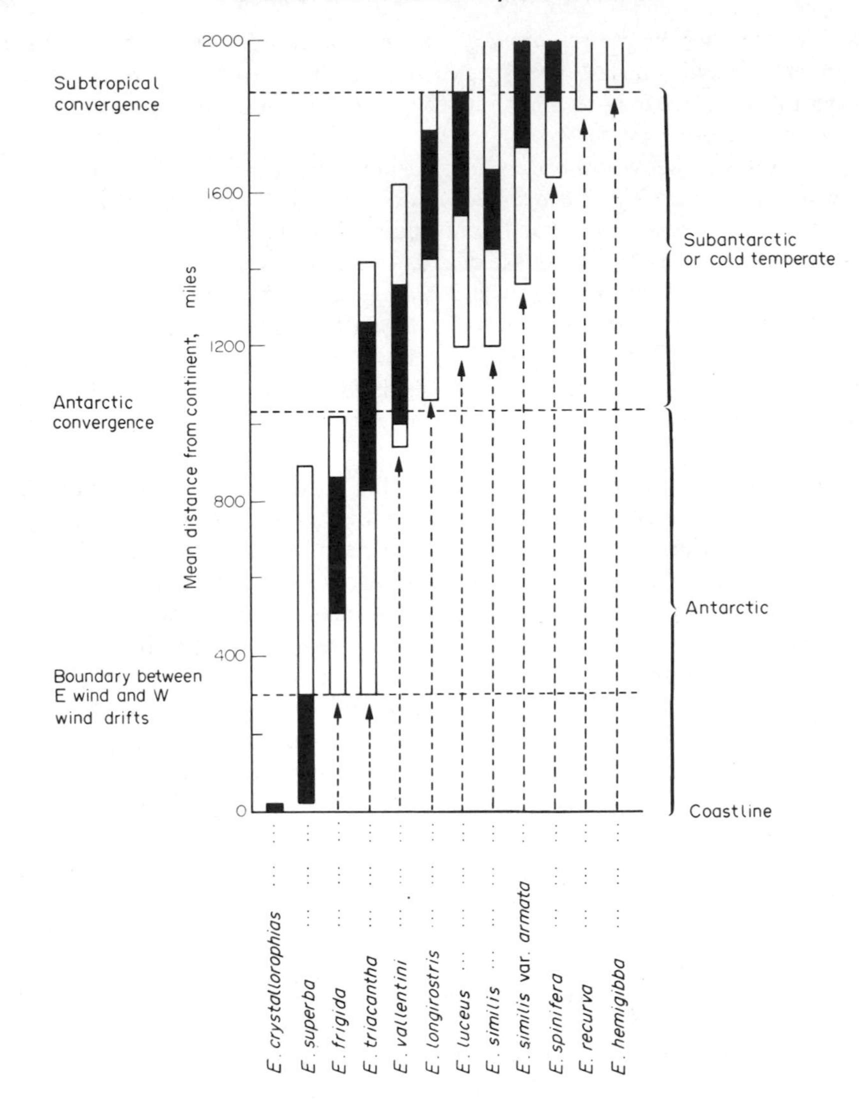

Fig. 2.112. Succession of euphausiid species from Antarctic northwards. Each species occupies a circumpolar zone, and range indicated is the estimated mean extension over the Antarctic and Subantarctic Zones. Black shaded parts indicate belt of concentration, outline denotes range occupied (Mackintosh, 1964).

Mauchline and Fisher (1969) indicate (Table 2.23) that *T. vicina* extends mainly from the ice edge to the Antarctic Convergence, while *T. macrura* ranges from the Antarctic continent to the Antarctic Convergence (Fig. 2.108).

Sub-Antarctic euphausiids also show a general circumpolar distribution. *Euphausia vallentini* is a typical sub-Antarctic form. Few individuals are found south of the Antarctic Convergence, but the species extends northwards towards the Sub-Tropical Convergence. *Euphausia longirostris* is also sub-Antarctic, but tends to have its main centre of distribution at slightly lower latitudes than *E. vallentini*. *E. similis*, though

Table 2.23. Twelve species of euphausiids that have circumpolar distributions in the Antarctic (Mauchline and Fisher, 1969)

Region	Species
Off Antarctic coast lines	*Euphausia crystallorophias*
Extending from Antarctic coast to Antarctic Convergence (50°–60°S)	*Euphausia superba*
	Thysanoëssa macrura
Extending from ice edge to Antarctic Convergence	*Euphausia frigida*
	Thysanoëssa vicina
Antarctic Convergence	*Euphausia triacantha*
Antarctic Convergence to Sub-Tropical Convergence	*Euphausia vallentini*
	E. similis
	E. longirostis
Southern regions of sub-tropical waters	*Euphausia lucens*
	E. similis var. armata
	E. spinifera

somewhat similar in distribution, certainly reaches to the Sub-Tropical Convergence. The variety *E. similis* var. *armata* apparently lives north of the Sub-Tropical Convergence though it crosses into sub-Antarctic waters (cf. Mackintosh, 1964; Roger, 1974a).

Many investigators, particularly Ponomareva, have emphasized that although euphausiids are of considerable significance everywhere in the shallower layers of the oceans, their biomass is greatest in the North Pacific, North Atlantic and Antarctic Oceans. Nemoto and others (e.g. Nemoto, 1967, 1970) have also pointed out that in those areas the euphausiid population is dominated by mainly herbivorous forms. The euphausiids make effective use of the relatively large standing crops of phytoplankton, especially diatoms, which occur at higher latitudes. Even such an obviously herbivorous species as *Euphausia superba*, however, may make very limited use of zooplankton during periods of phytoplankton scarcity. In the North Atlantic *Meganyctiphanes norvegica*, particularly older individuals, are partly carnivorous (cf. Chapter 6). Effective use of the phytoplankton at high latitudes is clear, whether this is utilized directly or in part through the euphausiid feeding on smaller zooplankton, which in turn consumes phytoplankton (cf. Parsons and LeBrasseur, 1970). Warm water oceans, with a large proportion of carnivorous euphausiids, might appear to be less productive. Nevertheless, the recent investigations of Legand *et al.* (1972) and Roger (1973, 1975, 1977) emphasize that the marked diurnal vertical migrations of much of the mesopelagic euphausiid population of open tropical and sub-tropical oceans lead to a substantial transfer of food to higher trophic levels.

Decapoda

Decapod crustaceans are correctly regarded as a most important group of benthic marine animals, although a number are holoplanktonic. The benthic species also contribute to the inshore plankton with the abundance of decapod zoea larvae, the megalopa larvae of crabs and the array of planktonic larvae of bottom-living shrimps, prawns and anomurans. An isolated group of pelagic decapods belonging to the galatheid Anomura is of significance in some seas in that the post-larval stages and adolescents of *Munida gregaria*, *Pleuroncodes monodon* and *Pleuroncodes planipes* can form vast swarms off New Zealand, Peru and Chile, the Patagonian coast including the Falklands, and the Gulf of California (cf. Longhurst, 1967).

The truly holoplanktonic decapods all belong to the Macrura Natantia, although Omori (1974) emphasizes the dominance of benthic habitat in this sub-order also; only about 210 out of 1940 Natantia species are pelagic. The pelagic members may be divided systematically into two groups, the Tribe (Division) Penaeidea and the Tribe (Division) Caridea (Fig. 2.113). The Penaeidea may be distinguished from the Caridea, amongst other characters, by the third thoracic legs being chelate (except where they are reduced) in Penaeidea and non-chelate in Caridea. The third maxilliped is seven-jointed in Penaeidea whereas it is four-to-six jointed in Caridea. The pleura of the second abdominal segment overlap those in front in Caridea but are not overlapping in Penaeidea, and the abdomen tends to show a sharp downward bend or hump in Caridea. The eggs are carried attached to the pleopods of female Caridea; Penaeidea shed their eggs into the sea.

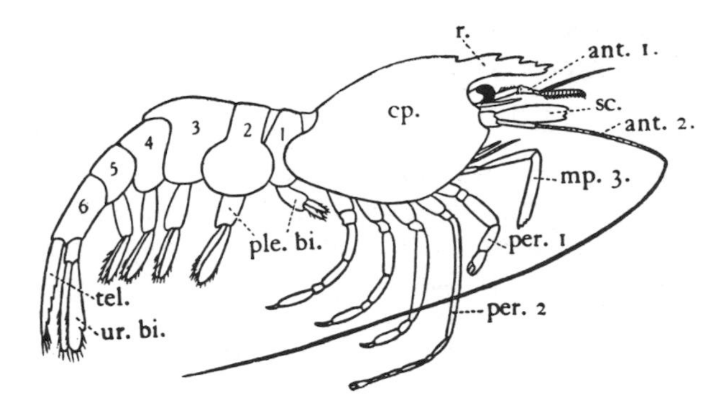

Fig. 2.113. Diagram of caridean decapod. Ant.1: antennule; ant.2: antenna; cp: carapace; mp.3: third maxilliped; per.1, per.2: first and second pereiopods; ple.bi.: biramose pleopod; r: rostrum; sc.: scale of antenna; tel.: telson; ur.bi.: biramose uropod; 1–6; abdominal segments (Eales, 1967).

The Tribe Penaeidea includes the family Penaeidae (Fig. 2.114), with a nearly complete series of gills, the first legs chelate and the fourth and fifth legs well developed. Among planktonic genera are *Funchalia* and a large genus, *Gennadas*. In the other family, the Sergestidae, the gill series is incomplete, the first legs are nonchelate and the fourth and fifth pairs are reduced in size or even absent. The genus *Sergestes* (with *Sergia*) includes a large number of important species. *Lucifer* is a highly modified genus lacking gills and with the front of the thorax greatly elongated; it is assigned to a separate sub-family (Fig. 2.114).

The Caridea include several planktonic genera (*Pasiphaea*, *Parapandalus*) in families which are mainly benthic, but one family is exclusively planktonic—the Oplophoridae—with a number of widespread genera (*Oplophorus*, *Acanthephyra*, *Notostomus*, *Hymenodora*, *Systellaspis*) and many species (Fig. 2.115). In both the Oplophoridae and the Pasiphaeidae exopodites are present on the thoracic legs.

Lucifer and *Acetes*, another sergestid genus, occur in neritic waters, *Acetes* swarming in very shallow inshore estuarine areas (Omori, 1974, 1977). Troost (1975) found

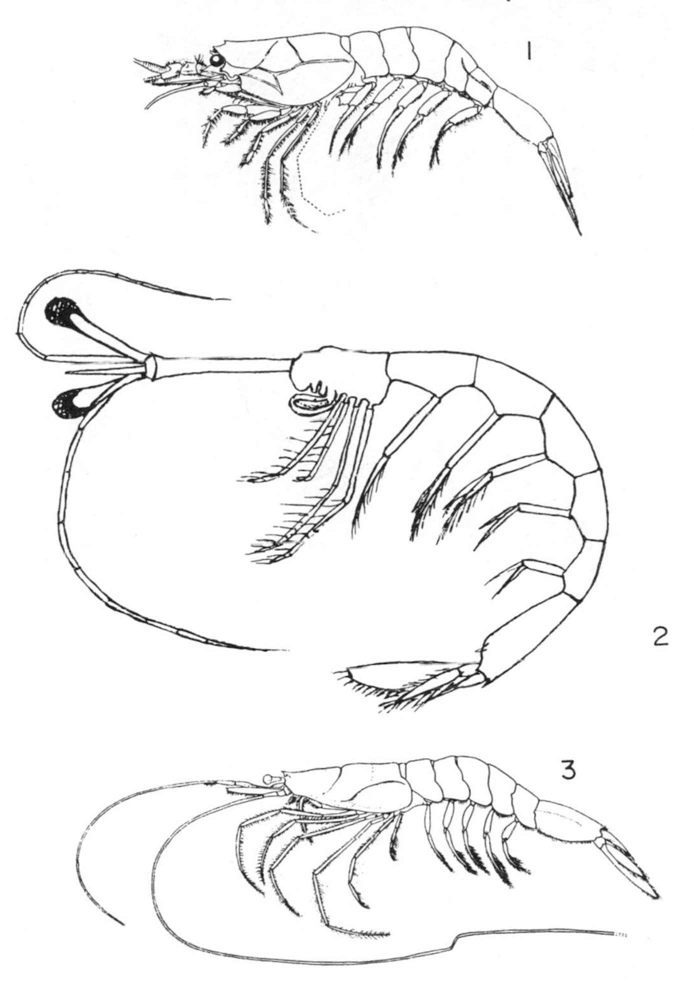

Fig. 2.114. Some penaeid species. (1) *Gennadas elegans* ♂ (*ca.* 45 mm.); (2) *Lucifer acestre*; (3) *Sergestes arcticus* ♀ (28–65 mm) (Lagardare, 1978; Rose, 1957).

Lucifer faxoni to be abundant in coastal waters off northern South America. A few species of *Funchalia*, *Sergestes* and *Pasiphaea* may sometimes be encountered over continental shelf regions, but the planktonic decapods are essentially oceanic in distribution. Some near-bottom species of *Pasiphaea*, *Aristaeus*, *Acanthephyra* and *Parapandalus* may spend at least part of their existence as benthic animals and may only doubtfully be characterized as planktonic (Omori, 1974). Matthews and Pinnoi (1973) found *Pasiphaea tarda* close to the bottom in a Norwegian fjord.

The majority of the pelagic decapods are relatively large for zooplankton (10–100 mm); some, including the near-bottom species (e.g. *Acanthephyra eximia*) can be of very considerable size, even up to 180 mm length. The thoracic legs and the pleopods are efficient swimming appendages, a significant feature in such large planktonic animals. Some species also have very prominent spines, which may retard sinking. A number of genera possess excessively long setose antennal flagella, well seen, for

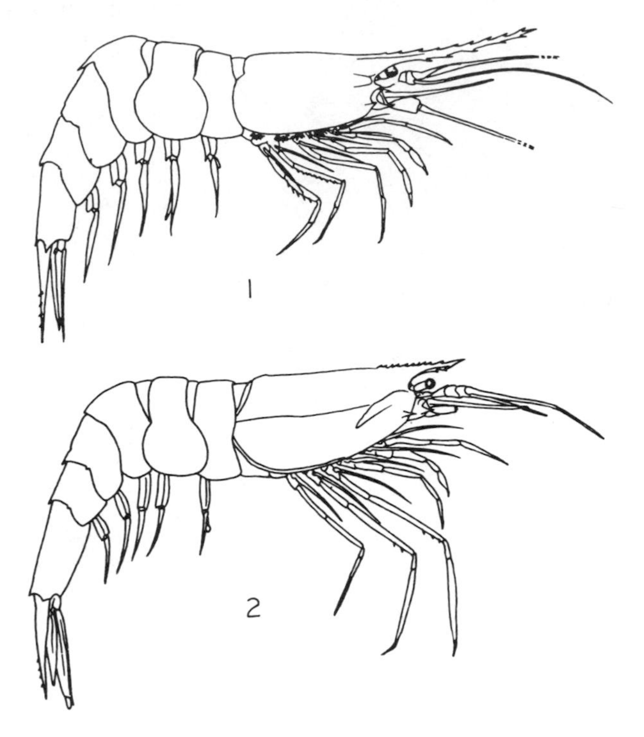

Fig. 2.115. Two caridean species. (1) *Acanthephyra purpurea* (147 mm); (2) *Meningodora vesca* (*ca.* 140 mm) (Rice, 1967).

example, in some sergestids. The exoskeleton is frequently rather poorly calcified. Upper water decapods are mostly transparent with relatively few red chromatophores; deeper water species are more or less uniformly red and heavily pigmented.

A few decapods (e.g. *Sergestes similis*, *Sergestes* (*Sergia*) *lucens*, *Acetes*) may occur in dense swarms; Omori (1974) states that *Sergestes* may then exceed 20 animals/m^3 in density, though much lower values are normally recorded. Such swarms of pelagic shrimps may comprise a substantial part of the total plankton biomass, though only over very restricted areas. Even when the biomass is large in inshore areas and over some continental slopes, especially where the bottom slopes steeply upwards and upwelling is common, a sharp decrease tends to occur with distance from shore (cf. Table 2.24).

Apart from swarming species occurring in neritic areas, the density of pelagic decapods in the open oceans appears to be very low. While such active animals may avoid nets, so that earlier records using the classical type of tow net may have given misleading impressions of density, the use of more efficient methods such as the Isaacs–Kidd mid-water trawl have confirmed the low decapod densities. Foxton (1970b) states that population densities of even the most abundant decapods in most open oceans are low as compared with other moderately sized planktonic animals such as euphausiids. Estimates of density are extremely speculative, but as against densities of euphausiids that may certainly exceed 10 individuals/m^3, maximum densities of decapods in the open ocean were one individual/500 m^3, or perhaps one

Table 2.24. Average biomass (dry wt) of pelagic shrimps in the oceans (Omori, 1974)

Area	Depth (m)	mg/1000 m³	Authority
Western Pacific Ocean			
Sagami Bay	0–1000	619	Aizawa, 1968
50°–40°N	0–1000	85	
40°–30°N	0–1000	283	
30°–20°N	0–1000	96	
20°–10°N	0–1000	37	
10°N–5°S	0–1000	127	
Eastern Pacific Ocean (30°N–25°S)			
less than 300 miles from the coast	0–90	180	Blackburn, 1968;
between 300 and 600 miles	0–90	78	Blackburn *et al.*,
more than 600 miles from the coast	0–90	15	1970
Eastern Indian Ocean			
9°–20°S	0–210	164	Legand, 1969
20°–32°S	0–210	67	
Western Mediterranean Sea			
North of the Balearic Islands	100–200	316	Le Gall and
	200–300	209	L'Herroux, 1971
South of the Balearic Islands	100–200	543	
	200–300	388	

individual/200 m³ (i.e. three orders of magnitude lower than the euphausiids). Donaldson's (1975) estimate that sergestids in the Sargasso Sea rarely exceeded 1 individual/2000 m³ would appear to be in reasonable agreement with this.

Planktonic decapods do not make a great contribution to the plankton, especially at high latitudes, though in the Kurile–Kamchatka Trench area species of *Hymenodora* may be relatively frequent and form a very significant part of the sparse plankton biomass at considerable depths (2000–4000 m) (Vinogradov, 1970, 1972). There appear to be very few records in the Antarctic. Omori (1974) includes *Petalidium* and *Hymenodora gracilis* at 67°S and a record of *Pasiphaea* in the Ross Sea. The contribution to the biomass made by pelagic decapods and the species diversity increase in sub-tropical and tropical regions. Over temperate and tropical areas Omori (1974) has proposed an average contribution to the biomass of about 10% for the layer 0–1000 m.

Pelagic decapods are distributed vertically from the surface to at least 6000 m depth. Neritic genera such as *Acetes* and *Lucifer* are relatively shallow-living, and a few species of *Sergestes* (e.g. *S. similis, S. lucens*) occur mostly in the upper 200 m, but the vast majority of decapods are mesopelagic or bathypelagic. While light intensity and temperature are regarded as important factors in the vertical distribution, as with many other taxa of zooplankton, Omori stresses the significance of food supply; decapods apparently concentrate in strata close to comparatively rich, more superficial layers (Fig. 2.116). Foxton (1970b) also suggests that the vertical stepwise distribution of decapods and their pattern of diurnal migration, with deeper-living species migrating to intermediate depths and shallower species almost reaching the surface, may in part be influenced by the necessity of distributing available food.

The great majority of decapods are characterized by a pronounced diurnal vertical migration that may extend to several hundred metres, though at least some of the

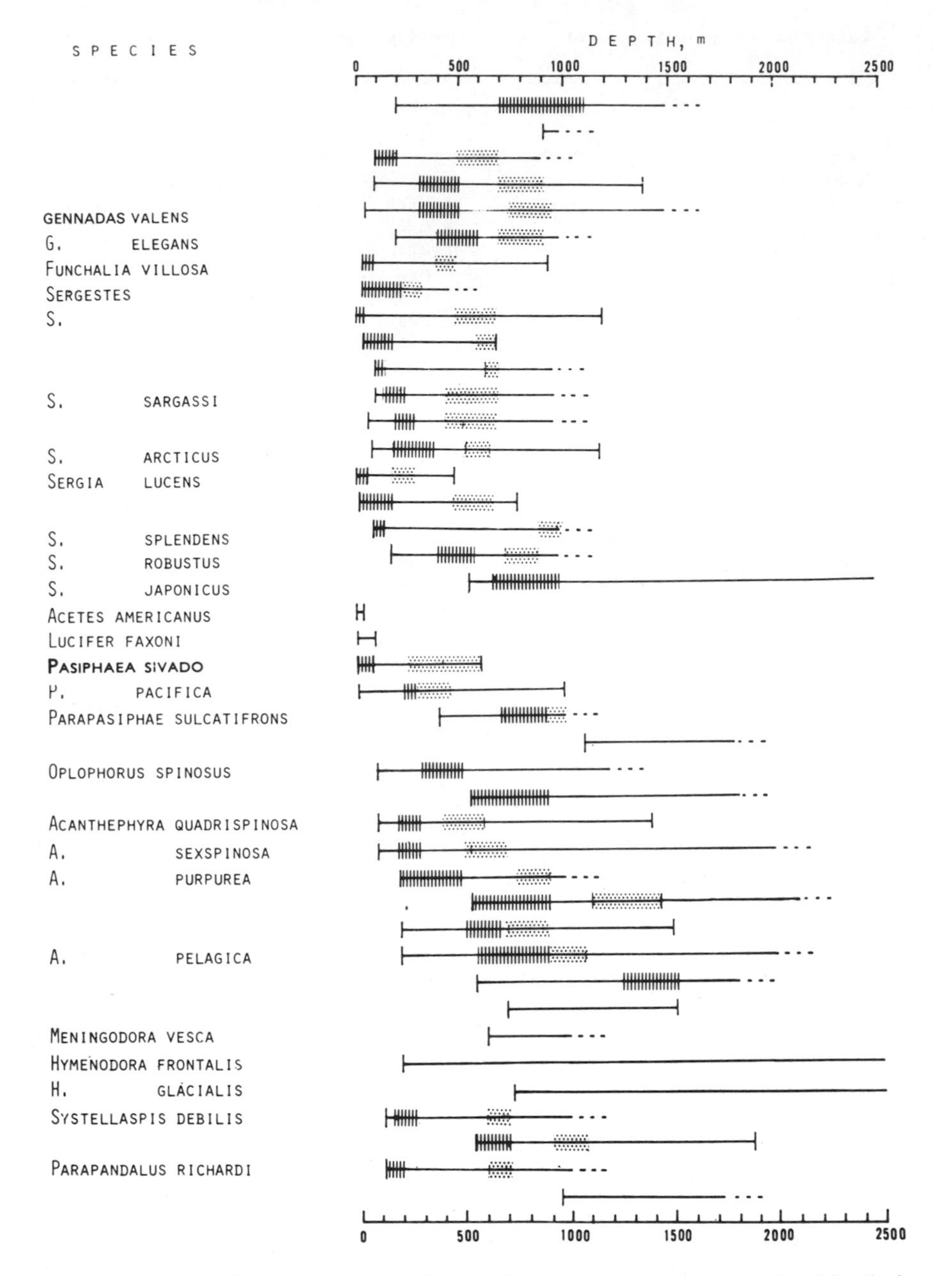

Fig. 2.116. Vertical distribution of pelagic shrimps. Solid lines indicate range. Dotted and hatched areas indicate depths of main day and night concentrations (modified from Omori, 1974).

bathypelagic forms appear to have a more limited migration (cf. Chapter 5). As with other zooplankton, the larval stages usually inhabit shallower layers. The diurnal migrations may lead to considerable vertical shifts in biomass. Omori points out that, for example, in Sagami Bay, where *Acanthephyra quadrispinosa* is the most abundant decapod, the maximum biomass changes with migration from about 400–600 m by day to 100–400 m at night, the deeper stratum then having rather low values (cf. Foxton, 1970a, b and Chapter 5 for further details).

Development

The Penaeidea are unique among decapods and resemble the euphausiids in that development includes a nauplius stage, with the three pairs of limbs, antennules, antennae and mandibles, all natatory. This is followed by a protozoea stage which possesses swimming antennules, biramous antennae, mandibles, maxillae and at least

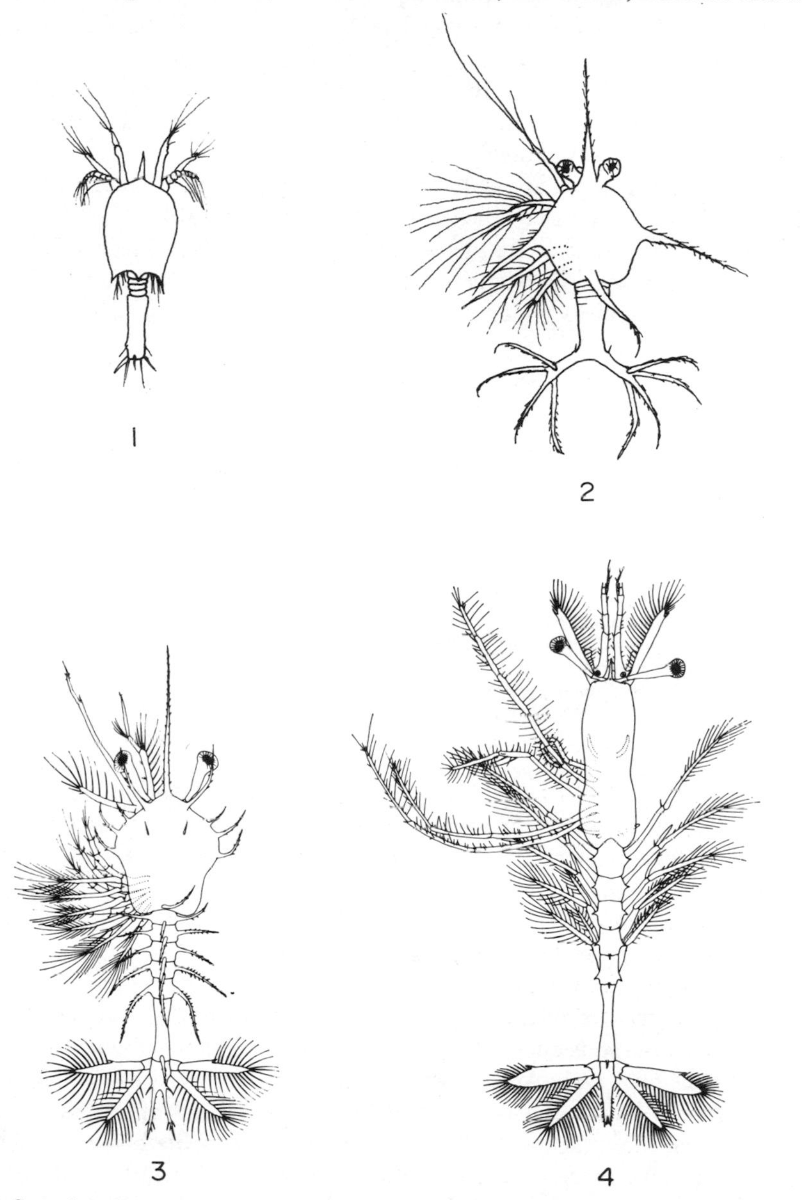

Fig. 2.117. Some larval decapod stages. (1) *Lucifer faxoni*, 1st protozoea; (2) *Sergia lucens* 2nd protozoea (elaphocaris); (3) 1st zoea; (4) 3rd post-larva (Omori, 1974).

two pairs of setose biramous maxillipeds. Paired eyes are present and the thoracic segments are not fused to the carapace. The protozoea of sergestids is often termed an elaphocaris. The protozoea stage is succeeded by a larva usually regarded as corresponding to the zoea of other decapods (Fig. 2.117). In sergestids this is the acanthosoma; in Penaeidae it is termed the mysis larva.

All the Penaeidae appear to have three protozoea stages, but the number of nauplius and zoea stages varies with genus and species, and sometimes, according to Omori (1974), even with environmental conditions. For example, in *Sergestes* (*Sergia*) *lucens*, Omori described two nauplius stages, of which the second grew gradually into a metanauplius. *Acetes* has three nauplius and one zoea stage; *Gennadas* has four zoea stages (cf. Table 2.25). In most of the species, especially the sergestids, protozoeal and zoeal stages are remarkably ornamented with spines and complex processes and the appendages may be highly setose.

In the Caridea, the eggs carried by the female hatch as protozoea or zoea larvae; there is little distinction between the various stages but probably all are best described as zoeas. Deep-living species frequently are characterized by large yolky eggs and abbreviated development. Herring (1974) draws attention to the extended incubation period of up to 10 months which may occur with such developments.

Table 2.25. Developmental stages of pelagic shrimps (Omori, 1974)

Gennadas: nauplius (?)-protozoea (3)[a]-zoea (4)-post-larva
Sergestes and *Sergia:* nauplius (2)-protozoea (3)-zoea (2)-post-larva
Petalidium: nauplius (?)-protozoea (3)-zoea (1)-post-larva
Acetes: nauplius (3)-protozoea (3)-zoea (1)-post-larva
Lucifer: nauplius (2)-protozoea (3)-zoea (2)-post-larva
Pasiphaea and *Parapasiphae:* zoea (4)-post-larva
Oplophorus, *Hymenodora* and *Systellaspis:* zoea (5)-post-larva
Acanthephyra: zoea (7 or more)-post-larva

[a] The number in parenthesis indicates the number of instars.

Geographical distribution

The deeper-living lower mesopelagic and bathypelagic decapods are frequently very widely distributed, as with many other zooplankton taxa, occurring in all three oceans (e.g. *Sergestes* (*Sergia*) *japonicus*, *Oplophorus spinicauda*, *Acanthephyra microphthalma*, *Systellaspis debilis* (cf. Holthuis, 1951). By contrast, the shallower oceanic species, and especially the neritic forms, are often confined not only to one ocean but to a specific area (e.g. *Sergestes lucens* in Suruga Bay; *Acetes chinensis* in the Yellow Sea and *Peisos petrunkevichi* off the south-eastern coast of South America).

Of the pelagic decapods occurring at high northern latitudes, *Hymenodora glacialis* appears to be the only truly Arctic species; it is found widely in the Atlantic east and west of Greenland, in the Norwegian and Barents Seas, and in the Pacific from the Bering Sea along the west coast of America to Ecuador (Stephensen, 1935). In the North Atlantic there are few records south of the Wyville Thompson Ridge, an area more generally typical of the related *Hymenodora gracilis*, a more southerly species (cf. Rice, 1967). Chace (1940) found only *H. gracilis* off Bermuda. Holthuis' (1951) records embraced the entire Atlantic Ocean, with scattered occurrences for the Indian Ocean and Pacific, but included both *H. gracilis* and *H. glacialis*. Later Sivertzen and Holthuis (1956) separated *H. glacialis*, an Arctic species occurring as far south in the North Atlantic to about 30°N latitude, from *H. gracilis*, occurring widely in the North Atlantic south of Greenland and in the South Atlantic as well as in the Indian Ocean; there is one record near Antarctica (68°S). Both species are normally bathypelagic; *H. glacialis* is found from 0 to 3000 m and *H. gracilis* from 500 to 5300 m.

Other decapods found at fairly high latitudes include species of *Pasiphaea*. Of the thirty-five species mentioned by Omori (1974), however, only three occur in high northern latitudes in the Atlantic Ocean—*P. tarda*, *P. sivado* and *P. multidentata*, all listed by Rice (1967) for the Norwegian and Barents Seas. Stephensen (1935) quotes *P. tarda* and *P. sivado* as occurring southwards from the Norwegian coast, Denmark Strait and Iceland areas in latitude 65–68°N. Although the species are described as bathypelagic, they are recorded in shallower seas (North Sea and Skagerrak). *P. tarda* extends south in the eastern Atlantic approximately to the Bay of Biscay and to the Carolinas in the western Atlantic. In the Pacific it is known from Alaska to Washington State and possibly off Ecuador. It appears to be more boreal than *P. multidentata* and *P. savido*, both of which extend to the Mediterranean. According to Casanova (1977) these two decapods are, however, limited to the cooler western Mediterranean, with *P. multidentata* the more numerous. In the western Atlantic, however, according to Sivertzen and Holthuis (1956), *P. multidentata* does not extend south of Cape Cod. *P. savido* is known from the Indo-west Pacific.

Parapasiphae sulcatifrons is another bathypelagic species (500–5400 m) found in the Atlantic from the Greenland, Iceland and Faroes area southwards to about 35°S off South Africa and in the southern part of the Indian Ocean (cf. Chace, 1940; Holthuis, 1951). Another of the few decapods to reach sub-Arctic waters is *Sergestes arcticus*. Stephensen (1935) describes its northerly limit in the North Atlantic as about 64°N, not usually beyond the latitude of Iceland and Baffin Land. Though there are occasional records farther north, this decapod is never truly Arctic. To the south it extends to about 35°N, being found in the Mediterranean, where it is a comparatively abundant decapod (Casanova, 1977). Lagardare (1978) quotes findings from South Africa and the Indo-Pacific. Sund (1932) describes it as relatively common and widely distributed over the North Atlantic, mostly deeper than 500 m (cf. also Hansen, 1922). Matthews and Pinnoi (1973) found *Sergestes arcticus* in Korsfjorden, West Norway, with the three species of *Pasiphaea*, *P. multidentata*, *P. sivado* and *P. tarda*, *P. multidentata* being the most common. All lived comparatively deeply, *S. arcticus* mainly between 200 and 450 m. The penaeid is also known from a few locations in the South Atlantic. Another sergestid, the large *S. robustus*, occurs to moderately high latitudes off New England and north of the British Isles (Sund, 1932).

In the North Pacific certain species extend to sub-Arctic waters also. Apart from *Hymenodora glacialis* (*vide supra*), Vinogradov (1970) reports *Sergestes similis* in the sub-Arctic, reaching about 50°N and occurring mainly as an epipelagic form. Farther south, in the warm Kuroshio waters off Japan (about 35°N latitude), the species is confined to somewhat deeper layers. *Bentheogennema* (*Gennadas*) *borealis* is a bathypelagic species, mostly occurring at depths greater than 700 m in temperate areas of the North Pacific, but occurring abundantly, according to Vinogradov, at rather more shallow depths (200–800 m) in sub-Arctic waters. *Hymenodora frontalis* is also

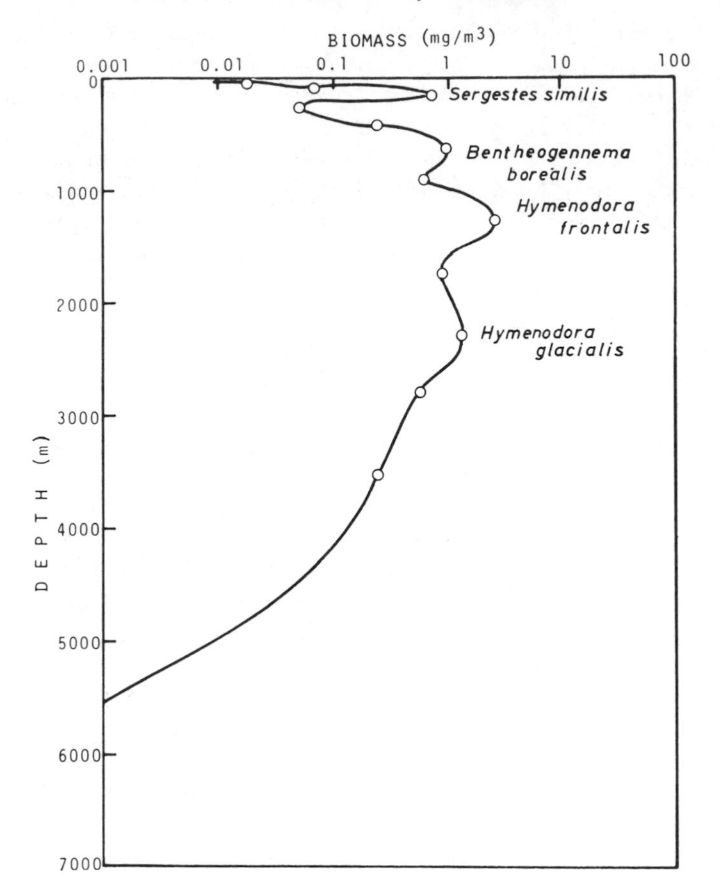

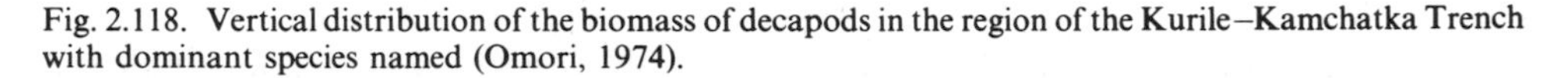
Fig. 2.118. Vertical distribution of the biomass of decapods in the region of the Kurile–Kamchatka Trench with dominant species named (Omori, 1974).

found in the north-west Pacific at depths of *ca.* 750–1500 m (cf. Fig. 2.118).

Farther south, Pearcy and Forss (1969) record *Sergestes similis* as the most abundant pelagic decapod of north-east Pacific waters and off the Oregon coast. *Pasiphaea pacifica* was plentiful off the Columbia River region, but other decapods, including *Hymenodora frontalis*, *H. gracilis* and *Bentheogennema borealis* were rarely taken. In the north-west Pacific off Japan, Omori (1974) lists, in addition to *Hymenodora frontalis* and *Bentheogennema borealis*, four decapods, *Acanthephyra quadrispinosa*, *Sergestes prehensilis*, *Gennadas parvas* and *G. incertus*.

Apart from an occasional doubtful record, no pelagic decapod appears to extend to very high latitudes in the southern hemisphere to Antarctic waters, though *Petalidium foliaceum* is found far south with *Hymenodora gracilis*.

The great majority of pelagic decapods inhabit the warm and warm-temperate oceans. Hansen (1922) listed only two species of sergestids for the North Atlantic (*Sergestes arcticus* and *S. robustus*) reaching cold northern waters, whereas sixteen species were described for warm-temperate, sub-tropical and tropical areas.

In the North Atlantic one of the most common species is *Gennadas elegans*

(Stephensen, 1935). It may reach from beyond 60°N southwards to the equator and is found sparingly in the South Atlantic (e.g. off South Africa). By day, it usually occurs at depths of *ca.* 1500–3000 m but is much nearer the surface at night. Burkenroad (1936) found that it was the most abundant species in the North Atlantic, especially in the north-east from about 60° to 30°N latitude, but that it was much less common at lower latitudes in the south-western North Atlantic. The species is apparently confined to the Atlantic Ocean. *Gennadas capensis* occurred in those areas where *G. elegans* was absent (Gulf of Mexico, Bahamas), as well as in the South Atlantic. Another relatively common species in Burkenroad's collections was *G. valens*, recorded from about 50°N to 36°S. Foxton (1970b) found *G. valens* to be by far the most abundant decapod in an area of the north-east Atlantic off the Canaries (about 28°N latitude). It is a deep mesopelagic species with a centre of distribution round 800 m. *G. elegans* was the third most abundant decapod in Foxton's catches, found mostly inhabiting depths greater than 800 m and probably extending considerably below 1000 m. Casanova (1977) describes *G. elegans* throughout the Mediterranean and as one of the most important decapods with *Sergestes arcticus*. *G. valens* has not colonized Mediterranean waters. Although these two species of *Gennadas* are Atlantic forms, other mainly warm-water species (*G. bouvieri* and *G. scutatus*) are found in all three major oceans.

Among the penaeids, *Funchalia villosa* is found widely distributed in warm tropical areas of the Atlantic and Pacific (cf. Burkenroad, 1936). Foxton (1970b) reported that the species was fairly common off the Canaries; while occurring at about 450 m by day, it was somewhat unusual among decapods in migrating almost to the surface at night. The species is also known from the Mediterranean.

Acanthephyra is one of the essentially deep-sea genera of Oplophoridae, containing twenty-seven species in all, some being widely distributed, usually in warm-temperate waters. *A. purpurea* is recorded by Sivertsen and Holthuis (1956) for the North Atlantic from about 20° to 53°N latitude; it does not occur in the Mediterranean. While records exist for depths of 0–1500 m, most of the adults taken during the Michael Sars' Expedition were inhabiting strata around 600 m. Off Bermuda, however, according to Chace (1940), where *A. purpurea* was the most abundant decapod, the centre of abundance was deeper (1200–1500 m). The related species *A. pelagica* (= *multispina* = *haeckeli*) appears to be deeper-living than *A. purpurea*. Sivertsen and Holthuis state that this species lives mainly below 700 m, with many individuals exceeding 1500 m depth. Chace (1940) found the centre of abundance of this species even deeper (> 1800 m) off Bermuda. *A. pelagica* also appears to be a more northerly distributed species in the North Atlantic (cf. Rice, 1967). From the collections of the Michael Sars' expedition in the more southerly section of the North Atlantic, of the two species combined, 86 % were *A. purpurea*; in the more northerly section, 79 % were *A. pelagica*. Apparently *A. pelagica* extends from 64°N (Davis Strait) even to 13°S presumably then occurring at considerable depths. It is found in the Mediterranean and again in the South Atlantic south of 23°S, as well as in southern parts of the Pacific and Indian Oceans even to 57°S. There is some evidence that *A. pelagica* takes 3 years to reach sexual maturity and *A. purpurea* only 2 years (Sivertsen and Holthuis, 1956). Foxton (1970b) found *A. purpurea* to be the second most common decapod in the Canaries region of the Atlantic; indeed, this species with the two species of *Gennadas* (*G. valens* and *G. elegans*) formed 60 % of the total decapod catch.

Other species of *Acanthephyra* in warm-temperate areas are *A. sexspinosa*, which is restricted to central and South Atlantic waters (about 17°N to 18°S), *A. acanthitelsonis*, apparently also mainly in the central and South Atlantic, and a number of species such as *A. brevirostris*, *A. eximia* and *A. microphthalma* found in all three major oceans (cf. Chace, 1940). All are relatively deep-living, *A. microphthalma* being apparently restricted to about 2000–4700 m. This latter species is found in the North Atlantic only as far as about 36°N (Sivertsen and Holthuis, 1956). Rice (1967) also points out that both *A. brevirostris* and *A. microphthalma* are rare in the north-east Atlantic area.

Oplophorus spinosus, another well known caridean, is a fairly common species in the warm-temperate North and South Atlantic as well as the Pacific and Indian Oceans. Casanova's (1977) records indicate the presence of *Oplophorus spinosus* off the coasts of north-west Africa, but that its northerly limit is approximately 40°N. It is somewhat unusual among pelagic decapods in that its vertical range is fairly wide (0–1800 m) but it exhibits little or no diurnal vertical migration (cf. Foxton, 1970b). Foxton found its range in the north-east Atlantic mostly about 100–500 m. A number of other Caridea found in all three oceans include *Oplophorus grimaldii*, *Meningodora vesca* (750–5300 m), found widely in temperate and warm areas of the Atlantic, with one record as far north as the south of Iceland; *Meningodora mollis*, found in the warmer waters south of about 40° in the Atlantic; and *Notostomus longirostris*, also from the warmer Atlantic, at depths from 600–3500 m (Holthuis, 1951; Sivertsen and Holthuis, 1956). Foxton (1970a) records last three species from the north-east Atlantic.

The genus *Systellaspis* includes *S. debilis*, found from about 650–1000 m by day in the north east Atlantic (Foxton, 1970a), but with a wide range of 25–3000 m according to Holthuis (1951). The species occurs widely in the Atlantic, in temperate waters south of Iceland and in sub-tropical and tropical areas as well as in the Indian and Pacific Oceans. It was the second commonest decapod in collections off Bermuda (Chace, 1940). *S. braueri* is also found in all three major oceans.

The predominantly benthic family Pandalidae includes the pelagic species *Parapandalus richardi*, which is relatively common in the warmer North Atlantic, usually not exceeding 37°N latitude and with a depth range of 25–1800 m (Sivertsen and Holthuis, 1956). Chace (1940) also regards *P. richardi* as a warm-water form.

Omori (1974) lists more than 50 species of *Sergestes* (including *Sergia*). *S. arcticus* and *S. robustus* have already been noted as occurring to fairly high latitudes in the North Atlantic (*vide supra*); *S. robustus* is very widely distributed. Sund (1932) recorded it over the North Atlantic, in Gulf Stream areas as well as off New England and the British Isles, and Casanova (1977) refers to its distribution from an even higher latitude in the North Atlantic to equatorial regions; Lagardare (1978) confirms these findings. It was comparatively common off north-west Africa and in the Mediterranean, especially in the eastern region. This large species is deep-living. Foxton (1970b) found it to be the most common sergestid off the Canaries and classes it as deep mesopelagic, with a maximum at around 800 m but probably extending to more than 1000 m. Earlier records suggest that it is comparatively deep-living, never appearing at the surface. Lagardare (1978) extends the depth range to 5000 m, with maximum concentration at night at 500–800 m and even up to 200 m on occasions. *S. japonicus* has a moderately extended northerly distribution; Sund (1932) recorded it (as *S. mollis*) off Rockall in the North Atlantic, but generally it occurs from off the

Table 2.26. Summarised data on the vertical distribution of twenty-five species in the Canary Islands area (Foxton, 1970b)

Species	Day: Vertical range (m) Upper	Day: Vertical range (m) Lower	Day: Depth of Maximum	Night: Vertical range (m) Upper	Night: Vertical range (m) Lower	Night: Depth of Maximum	Range of migration (m): Upper limit	Range of migration (m): Maximum	Range of migration (m): 50% level
PENAEIDEA									
Sergestes (Sergestes) sargassi	300	950	650	110	415?[a]	200	190	450	370
S. (S.) armatus	450	600?	580	110	?	200	340	·[b]	·
S. (S.) vigilax	560	930	650	110	?	110	450	·	·
S. (S.) pectinatus	560	700	650	70	110?	110	490	·	·
S. (S.) corniculum	600	930	700	70	?	250	530	·	430
S. (S.) atlanticus	·	·	·	·	·	·	·	·	·
Sergestes (Sergia) robustus	700	950+[c]	800	200	615?	565	500	235	280
S. (S.) splendens	775	950+	930	110	435?	110	665	320	720
S. (S.) japonicus	800	950+	950	565	925?	925	235	25	80
S. (S.) tenuiremis	930+	·	·	·	·	·	·	·	·
Gennadas valens	700	950+	800	200	925	415	500	385	400
G. elegans	775	950+	875	565	925	925	210	50	50
Bentheogennema intermedia	930+	·	·	·	·	·	·	·	·
Funchalia villosa	450	775?	450	+70[d]	900	70	380	380	430
CARIDEA									
Acanthephyra purpurea	700	950+	800	200	925	415	500	385	350
Systellaspis debilis	650	930?	650	115	925	150	535	500	490
S. cristata	875+	·	·	·	·	·	·	·	·
S. braueri	950+	·	·	·	·	·	·	·	·
Ophlophorus spinosus	95	495?	450	70	435?	435	25	15	20
Parapandalus richardi	560	700?	650	110	925	150	445	·	·
Ephyrina hoskynii	930+	·	·	925+	·	·	·	·	·
Meningodora vesca	875+	·	·	710+	·	·	·	·	·
M. mollis	·	·	·	925+	·	·	·	·	·
Notostomus longirostris	775+	·	·	750+	·	·	·	·	·
Pasiphaea hoplocerca	700+	·	·	200+	·	·	500	·	·
Eupasiphaea gilesii	800+	·	·	925+	·	·	·	·	·

[a] ? = lower limit provisional
[b] · = insufficient data
[c] Depth + = range probably deeper than indicated
[d] + Depth = range probably shallower than indicated

British Isles to Angola and is very rarely found in the Mediterranean (Casanova, 1977). It is a deep mesopelagic sergestid.

The great majority of sergestids inhabit warm seas and many are restricted to tropical and sub-tropical latitudes. Some, such as *S. tenuiremis* and *S. japonicus*, are found in all three major oceans. In the Atlantic, warm water species include *S. pectinatus*, reaching about 40°N latitude in the western Atlantic, and found in south African waters, off the Azores, Canaries (Foxton 1970b) and north-west Africa (Casanova, 1977). *S. sargassi*, *S. vigilax S. armatus* and *S. splendens* appear to have generally similar distributions in the North Atlantic, though the first species was described by Sund (1932) as reaching about 45°N in the western Atlantic. *S. splendens* may also extend slightly beyond 40°N in the west, but is restricted to the latitude of Gibraltar in the eastern Atlantic. All four species occurred off the Canaries (Foxton, 1970b) and off north-west Africa, with *S. sargassi* and *S. vigilax* occurring also in the Mediterranean (Casanova, 1977).

The genus *Sergestes* is now often divided into *Sergestes sensu stricto* and *Sergia*. Foxton (1970b) has shown that for the north-east Atlantic the species of *Sergestes* (*sensu stricto*) are mainly upper-water shallow mesopelagic forms. They are semi-transparent, with few chromatophores ("half red"), and with internal ventral photophores (organs of Pesta) which are probably of adaptive significance in assisting to render the animal inconspicuous in the low light intensity typical of their day depth. Bioluminescence seems to be important to species of *Sergestes*.

At depths greater than 650–700 m the sergestid population off the Canaries is represented by species of *Sergia*. These are lower-water deep mesopelagic forms and uniformly red in colour; photophores, if present, are dermal in position. Foxton also demonstrates that a progression may be observed in species depth distribution from the shallowest (*Sergestes sargassi*) to the deepest (*S. tenuiremis*). While the vertical

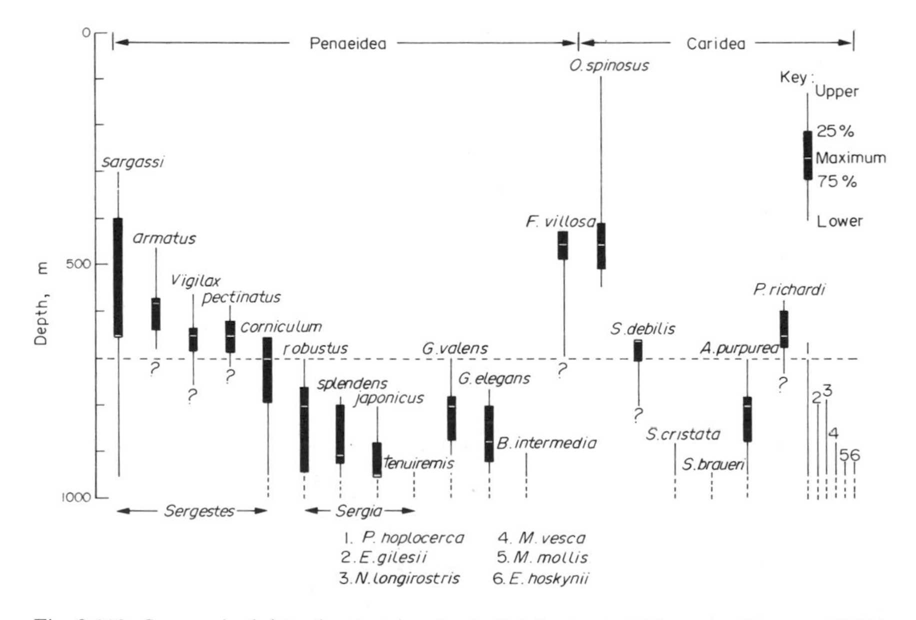

Fig. 2.119. Summarized data showing day depth distributions of 25 species (Foxton, 1970b).

distributions overlap to some extent and the vertical range of any one species may be large, the majority of the individuals of that species tend to be restricted to a stratum of 100–200 m thickness. Moreover, the diurnal migration patterns differ, though there is considerable overlap (cf. Table 2.26). Somewhat comparable patterns may be identified for penaeids (*Funchalia*, *Gennadas*, *Bentheogennema*) and for some Caridea (e.g. *Systellaspis* spp.) (cf. Fig. 2.119).

Donaldson (1975) reaches a generally similar conclusion from an investigation of the sergestids off Bermuda. Of fourteen species, the shallower-living *Sergestes* may be distinguished from *Sergia*, living generally more deeply, but each species tends to have a fairly narrow vertical distribution, ranging from *S. cornutus* and *S. sargassi* (shallowest, *ca.* 500 m) to *S. tenuiremis* and *S. japonicus* (deepest, the latter exceeding 1000 m). Though all species exhibit vertical migration, the extent varies (cf. Chapter 5). These differences, in addition to some degree of dietary preference (*S. splendens*, for example, includes more euphausiids in the diet and *S. japonicus* includes a substantial amount of detritus) may assist in reducing inter-specific competition.

Tunicata

The great phylum of chordate animals includes certain invertebrates, some of which are important in the marine plankton. Of the sub-phylum Tunicata, the best known representatives are probably the benthic ascidians or sea squirts, which are relatively common in coastal and some deep bottom waters. A large number of these ascidians produce larvae, usually known as tadpole larvae, which live in the plankton until metamorphosis. Two classes of Tunicata, however, the Appendicularia and Thaliacea, are holoplanktonic.

Appendicularia

These animals, known also as the Larvacea or Copelata, superficially resemble the tadpole larvae of ascidians. Their bodies consist of a small trunk, frequently not more than 1 mm in length, and a tail which is several times (usually not exceeding three times) the length of the trunk. The mouth leads into a pharyngeal cavity with two branchial openings sometimes called spiracles or gill slits, but there is no peribranchial cavity. Ventrally, the pharynx has a short endostyle, typically of a few very large mucus-secreting cells. The mucus flows along peripharyngeal bands which unite in the dorsal part of the pharynx. The cilia lining the bands carry mucus food strings back to the oesophagus. The cilia of the spiracles draw water with food particles into the terminal mouth, which is open permanently. The spiracular cilia can, however, reverse to eject any large particles blocking the mouth.

The stomach is roughly globular and may have lobes and is succeeded by a more or less oval intestine and rectum, ending at an anus which opens ventrally or to the side (Fig. 2.120). The animals are hermaphrodite, the gonads occupying a relatively large part of the body. The tail has a nerve cord from the cerebral ganglion running along its length, with inconspicuous ganglia supplying sensory and motor nerves. The axis of the tail, however, is at right angles to the plane of symmetry of the trunk, that is, it is twisted through 90°, and the junction of the tail with the trunk is constricted and shifted somewhat along the ventral side (Fig. 2.121). The typical chordate notochord

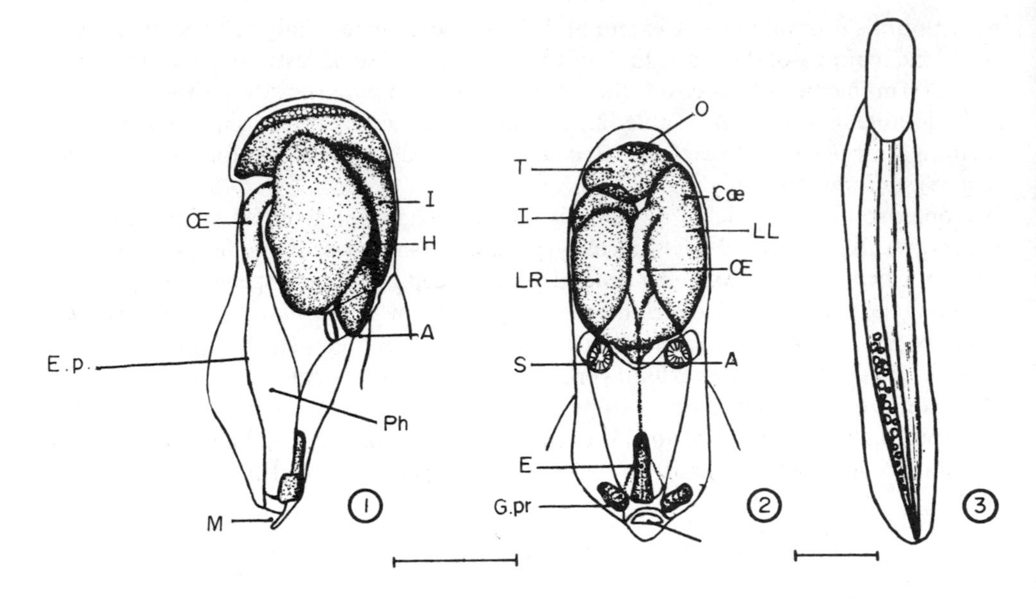

Fig. 2.120. *Oikopleura albicans* adult stage. (1) trunk seen from left side; (2) dorsal view of trunk; (3) whole animal. Scale 1 mm. A: anus; Cae: caecum; E: endostyle; E.P.: epipharyngeal band; H: heart; I: intestine; L.L., L.R.: left and right lobes of stomach; M: mouth; OE: oesophagus; O: ovary; Ph: pharynx; G.pr: prebuccal gland; S: spiracle; S.H: secretion of house; T: testis. (Fenaux, 1967).

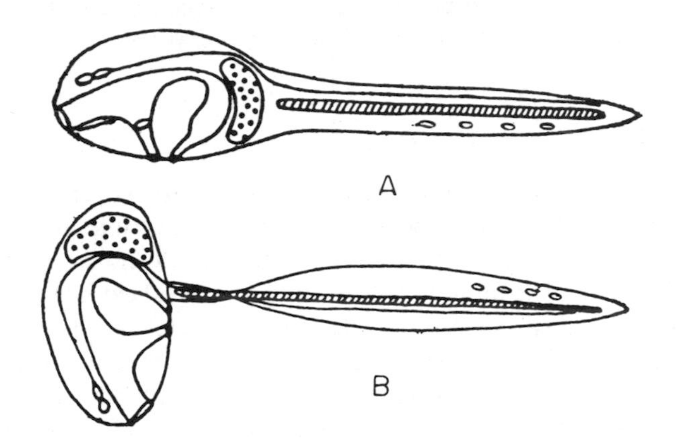

Fig. 2.121. Diagram of (A) before and (B) after twist and shift of tail in *Oikopleura*. Gut, gonads, nerve cord and ganglia, and notochord can be seen (Fenaux, 1967).

consists of a limited number (20) of very large cells and is found in the tail only, with the nerve cord on its left side. The tail has a fairly broad fin composed of the epidermal layer.

One of the most characteristic features of Appendicularia is their ability to secrete mucopolysaccharide to form a test or "house". The secretion is due to the oikoplasts, glandular epidermal cells often visible on the trunk and situated in particular positions on the body which are characteristic of the species. In life the appendicularian typically lives inside the house, which it may abandon; a new one is then secreted. The differentiation of the house, however, varies greatly with species. Even in one family—the Oikopleuridae, which is best known—the characteristics of the house vary. For example, in *Oikopleura longicauda*, it is very simple with one aperture in front for the entry of water and one behind for the exit of water. On the other hand, species like *Oikopleura albicans*, have a complicated house with a pair of special entrances for water inflow guarded by sieve mechanisms (cf. Lohmann, 1933/34). Whatever the degree of complexity, the houses are intimately connected with food collection.

The food consists of small phytoplankton and protozoan organisms, especially the minute nanoplankton, as first discovered by Lohmann. Over many years he contributed a series of papers on the biology of appendicularians; an excellent summary is included in Kukenthal and Krumbach's Handbook (1933–1934). The feeding process may best be described for appendicularians such as *Oikopleura albicans*. Much of this account is drawn from the monograph by Fenaux (1967) (cf. also Berrill, 1950). The constant beating of the tail of the appendicularian draws water through the house of *Oikopleura albicans*, which is provided anteriorly with a pair of sieves in the form of windows of cross fibres. The animal itself lies in a part of the house divided into a trunk chamber and a tail chamber. The current of water passes out of the house through an orifice behind the tail chamber. The force of the water expelled is strong enough to push the whole house with the animal forwards, but essentially the function of the water stream is to provide a feeding current. The window sieves exclude larger particles and debris.

The water passing into the house is directed to two filter food traps situated right and left of the tail chamber. Their fine filters collect the very minute nanoplankton, and they lead via a funnel (where the particles adhere to mucus secreted by the endostyle and are propelled by cilia) to the mouth of the appendicularian. These food filters are formed of two parallel leaflets separated by a prolongation of the house wall in such a way that the water has to pass all the way down and back along the leaflets, with a network of complex channels along the upper one providing a very effective filtration (Fenaux, 1967, Fig. 2.122). The upper channel of the leaflet opens to the dorsal chamber of the house, from which the water is expelled via the tail chamber.

Clogging of the filtering mechanisms usually causes the animal to vacate its house. According to Fenaux, a new secretion can take place in from 1.5–4 hours. It consists at first of a gelatinous covering of the trunk which is rapidly expanded in a matter of minutes to form the typical house. Alldredge (1976b) has also given data from observations both at sea and in the laboratory suggesting that houses can be vacated and new ones secreted very rapidly. The empty appendicularian houses may sometimes form a large part of plankton debris which may become an important food for other zooplankton, since it can collect bacteria and other organisms. In *Fritillaria* there is no house; a gelatinous vesicular apparatus, formed of two valves, hangs from the mouth as a temporary food collecting mechanism. When not in use, the structure is folded and appears as a small bubble-like structure near the mouth, but it can be expanded to be relatively larger than the houses of some oikopleurans (Lohmann, 1933/34). *Fritillaria* appears to be suspended from the filtering apparatus when feeding, with its body and tail free in the water (Fig. 2.123).

Of the three families of Appendicularia, the Oikopleuridae have a pair of gill clefts, opening to the exterior by tubular spiracles situated fairly far back from the mouth,

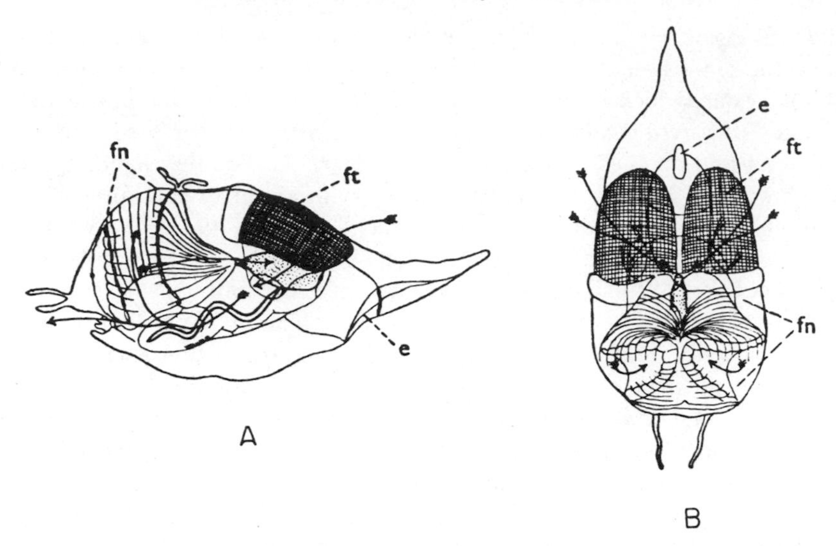

Fig. 2.122. *Oikopleura albicans* within house, showing filters, food net and direction of water currents. (A) from left side; (B) dorsal view, e: water exit (propulsion); fn: foot net; ft: filtering window (Berrill, 1950).

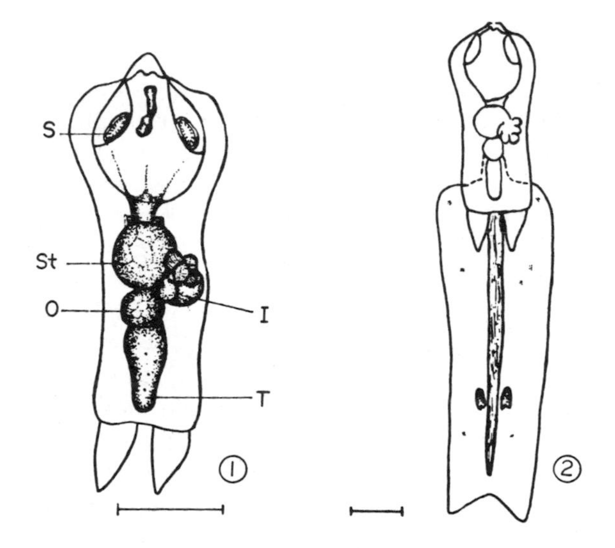

Fig. 2.123. *Fritillaria tenella* (1) Dorsal view of trunk; (2) whole animal; I: intestine; O: ovary; S: spiracle; St: stomach; T: testis. Scale 250 μm (Fenaux, 1967).

which has a prominent ventral lip. The animals are protandrous; *Oikopleura dioica* is exceptional in being dioecious. In the Fritillaridae the spiracles open anteriorly, direct to the exterior. The body is slender, often constricted at the abdomen, and the tail is relatively short and wide and normally indented. Fritillaridae are usually minute. The Kowalevskiidae have only one genus, *Kowalevskia*, with a globular trunk and a broad pharynx partly divided longitudinally and lacking endostyle and peripharyngeal bands. There is a long tail shaped like a willow-leaf (Fenaux, 1967). The house is a simple hemispherical structure lacking membranous food traps and with one orifice. Although there are thirteen genera of appendicularians, *Oikopleura* and *Fritillaria* have the great majority of species. Many have an ocean-wide distribution and may occur from the surface to approximately 2000 m (Berrill, 1950), but generally they are shallow-living. An exception is the comparatively large and bathypelagic *Bathochordaeus charon* (cf. Thompson, 1948). *Megalocercus abyssorum*, also large, is usually regarded as bathypelagic, but it has been taken on the surface (Fenaux, 1967, 1973; Lohmann, 1933/34).

There is a very sharp decrease in the density of appendicularian populations below the upper 200 m, even below 100 m, and exceedingly few are captured below 1000 m. (Lohmann, 1933/34). While maximum numbers of individuals are found in the extreme surface layers, a greater number of species may occur at a few hundred metres depth. Thus in the Mediterranean the 0–25-m layer had sixteen species, contributing 31 % of the catch, whereas in the 150–300 m layer there were twenty-one species, but contributing only 5.6 % (Fenaux, 1967).

Appendicularians are able to reproduce rapidly to form dense populations. Although Wyatt (1973) considered that *Oikopleura dioica* passed through three generations during December to June in the English Channel/Southern Bight area, indicating that the generation time was fairly long, Fenaux (1976) showed that the species could have a life cycle of three to five days at a temperature of 22°C. Although a reduction in temperature to 14°C markedly lengthened the generation time, it was still as brief as 10–12 days. The remarkable speed of development is indicated in Table 2.27. Formation of the first house follows soon after hatching, and even while the animal is in its first house, it secretes a new one in the form of a mucus sleeve which can swell in a few seconds when the old house is abandoned. At 22°C, for example, *Oikopleura dioica* can secrete up to eight houses in 24 hours. Reproduction of a new

Table 2.27. Principal events during life-cycle of *Oikopleura dioica*, with tail growth measurements (adapted from Fenaux, 1976)

Time		Mean length of tail (μm)		Main events	
Days	Hours	(14°C)	(22°C)	(14°C)	(22°C)
	0			Fertilization	Fertilization
	1				
	2				
	3		90		Hatching
	4		195		
	5		250		
	6	90		Hatching	
	8	180	370		Tail shift, 1st house
	18	370		Tail shift, 1st house	
3			1610		1st spawning
4		1500	2300		Main spawning
5		1800			All spawned
10		2310		1st spawning	
11		2350		Main spawning	
12				All spawned	

brood is also very rapid. At the same temperature (22°C), all the individuals of one brood of *Oikopleura dioica* had themselves spawned within five days. Clearly, very dense blooms of appendicularians can build up in a short space of time (cf. Seki, 1973).

The great majority of appendicularians are warm-loving and somewhat eurythermal, but some species have penetrated cold waters. Fenaux (1967) emphasizes the significance of seasonal influence on colder water populations. Data are quoted by Lohmann and Buckmann from the Antarctic suggesting that, whereas in the productive season a haul from 200 m to the surface might catch some 50,000 appendicularians, comprising almost entirely one species, *Fritillaria borealis typica*, in the winter a similar haul might be two or even four orders of magnitude lower. The seasonal effect is probably more related to phytoplankton production than to temperature change. Populations of warm water appendicularians may also show remarkable changes in density, mainly reflecting local variations in the crop of nanoplankton.

Distribution

Of the few cold-loving species, *Oikopleura gaussica* appears to be restricted to the very cold Antarctic. Three other closely related species, *O. drygalskii*, *O. valdiviae* and *O. weddeli*, have a similar distribution but may be varieties of *O. gaussica* (cf. Tokioka, 1964; Fenaux, 1967). Other Antarctic species are *Pelagopleura magna*, *P. australis*, and *Fritillaria antarctica*. Buckmann (1970), however, recorded *P. australis*, typical of the Antarctic and the Southern Ocean, in the North Atlantic off the Hebrides, and he questions whether *Sinisteroffia scrippsi*, regarded as an Antarctic species by Tokioka (1964), could be synonymous with this species. *Fritillaria borealis* may be very abundant in Antarctic waters. Several more eurythermal species can extend as far as the Antarctic (*vide infra*).

In exceedingly cold northern waters probably the only truly Arctic species is *Fritillaria polaris*. Two other appendicularians, *Oikopleura labradoriensis* and *O. vanhoeffeni*, are typical of high northern latitudes of the Atlantic and Pacific, but they can survive somewhat farther south in winter and are found in very cold southerly flowing currents. *O. vanhoeffeni* is perhaps the more cold-loving. It occurs to the north of Spitsbergen, even to latitude 81°N (Berrill, 1950), and is very abundant off Greenland and the Davis Strait. It is a typical member of the zooplankton in the Norwegian Sea and the Barents Sea, but occasionally is found off the Shetlands and in the northern North Sea. Buckmann (1970) regards *O. vanhoeffeni* as an Arctic species; during the spring of 1955, he found this species at only three stations in the North Atlantic, two of which had Arctic water with temperatures of around 0°C. Bigelow (1926) and Thompson (1948) associate the presence of *O. vanhoeffeni* with a flow of Arctic water in the Gulf of Maine and off Newfoundland, respectively. In the North Pacific, *O. vanhoeffeni* may occasionally be found even off the Californian coast, suggesting an influx of very cold water (Thompson, 1948; Lohmann, 1933/34),

Oikopleura labradoriensis occurs in the Atlantic very abundantly to around 70°N latitude, but can reach north of Spitzbergen and the Murmansk area; it is common in Davis Strait and the Labrador Current. Buckmann (1970) classed it as a boreal species and the most abundant of the cold water appendicularians. The other important boreal species in the northern North Atlantic is *Fritillaria borealis typica*, which can be enormously plentiful at times, especially in the Irminger Current and Davis Strait areas (Buckmann, 1970). As a bipolar species, it can be of great importance in Antarctic seas (*vide supra*). The cold water appendicularian fauna of the North Atlantic stretches south roughly as far as the Newfoundland area, the south of Greenland and the southern limit of the Norwegian Sea. However, the approximate border separating cold from warmer appendicularian faunas shifts north during the later summer, when the numbers of cold water species also increases (Buckmann, 1970).

All other appendicularians are more warm-loving or temperate forms, but a number can reach colder areas. *Oikopleura parva*, a eurythermal and mainly warmer species, can extend into cold near-Arctic waters, but it usually inhabits the deeper layers. The species may also reach Antarctic waters. Other appendicularians such as *Oikopleura longicauda*, *Stegosoma magnum*, *Folia gracilis* and a number of species of *Fritillaria* (*F. haplostoma*, *F. pellucida*, *F. tenella*, *F. venusta*) as well as *Kowalevskia tenuis* and *Appendicularia sicula* may also extend their distribution into cold Antarctic seas. Lohmann (1933/34) suggests that these visitors from warmer water may avoid the coldest surface waters (cf. Tokioka, 1964; Fenaux, 1967).

In the warm-temperate areas of the North Atlantic, Buckmann found that *O. longicauda*, a widespread species, was usually commonest, with *O. fusiformis* plentiful in both oceanic and coastal waters and *O. dioica* also in oceanic and coastal environments and extending somewhat into boreal waters. Although warm-temperate appendicularians normally flourish at temperatures exceeding 11°C, *O. dioica* is exceptionally eurythermal and euryhaline. According to Paffenhofer (1973), its habitat includes estuarine waters and it occurs widely in temperate and warm waters of all oceans. Fenaux (1967) gives its temperature limits as 3.2°–29.5°C and its salinity range as 11–38.3‰. It is known from all seas except the Antarctic, but is more abundant in temperate than tropical waters. Other fairly abundant appendicularians in the warmer-temperate areas of the North Atlantic include *Oikopleura albicans*, *O. rufescens*, *Fritillaria pellucida*, *F. haplostoma* and *F. borealis sargassi*.

Buckmann (1970) observed that, while over winter in the warm-temperate Atlantic *O. longicauda* was the most plentiful species, during late summer *O. dioica* became the more abundant species, even in the open ocean, and *O. fusiformis* increased markedly by the late summer. *O. fusiformis* also tends to be more important than *O. longicauda* with increasing latitude. *F. haplostoma* can be plentiful in restricted areas.

Fraser (1961) found that only three appendicularians were abundant in the zooplankton of the north-east Atlantic around Britain, particularly off Scottish coasts: the boreal species *Oikopleura labradoriensis* and *Fritillaria borealis*, and the widespread *Oikopleura dioica*, *O. fusiformis*, *Appendicularia sicula*, *Fritillaria pellucida* and *F. venusta* were occasionally taken. The latter species with *F. tenella*, appears to have an extremely wide distribution: while they are predominantly warm-temperate forms found in all oceans and in the Mediterranean, they are found also in cold seas, apart from the Arctic.

In sub-Arctic areas of the North Pacific Ocean, Tokioka (1960) found only two appendicularians, *Oikopleura labradoriensis* and *Fritillaria borealis*, the fauna thus being comparable to that in boreal areas of the North Atlantic. Where mixing of sub-Arctic and warmer waters occurred the two boreal species were plentiful and *Oikopleura longicauda* was abundant. In contrast to the North Atlantic, *Oikopleura fusiformis* was comparatively rare and was dominated by *O. longicauda*. In the mixed waters off California, Tokioka (1960) found *Fritillaria borealis typica* abundant, with *Oikopleura longicauda* and small numbers of warm species such as *Fritillaria borealis sargassi*. The neritic *Oikopleura dioica* was plentiful. Earlier records included a number of other warm-water species (*Oikopleura cophocerca*, *Fritillaria haplostoma*, *F. megachile*) with some cold-water forms.

Farther south, in the sub-tropical and tropical areas off central America and off Ecuador, Tokioka (1960) records thirty-six species of appendicularians. The commonest was *O. longicauda*, with *O. cophocerca*, *O. rufescens*, *O. fusiformis*, *Fritillaria pellucida*, and *F. borealis sargassi*. The warmer waters across the whole northern Pacific usually show a dominance of *O. longicauda*, with *O. fusiformis* next in abundance. Other common forms included *Fritillaria borealis sargassi*, *Oikopleura dioica*, *Fritillaria pellucida*, *Stegosoma magnum*, *Oikopleura graciloides* and *O. albicans*. In tropical regions *Megalocercus huxleyi*, a species confined to the Pacific and Indian Oceans, could be very common, along with *Stegosoma magnum*.

In south-east Australian Pacific waters, Thompson (1948) also demonstrated the outstanding dominance of *O. longicauda* (see Table 2.28). Next in order of importance were *Oikopleura fusiformis*, *Fritillaria pellucida*, *Oikopleura rufescens*, *O. dioica* and *Megalocercus huxleyi*. *Stegosoma magnum* could be plentiful.

In the Atlantic, Fenaux (1967) records forty-three species, the majority in the

Table 2.28. Proportional numbers of occurrence of species in south-eastern Australian waters (Thompson, 1948)

Species	Number	Species	Number
COPELATA.		DESMOMYARIA.	
Oikopleura longicauda	269,912	*Thalia democratica*	1,421,652
Oikopleura fusiformis	57,485	*Ihlea magalhanica*	132,144
Fritillaria pellucida	46,654	*Salpa fusiformis*	25,057
Oikopleura rufescens	45,718	*Traustedtia multitentaculata*	6,531
Oikopleura dioica	15,159	*Brooksia rostrata*	2,765
Megalocercus huxleyi	11,294	*Salpa maxima*	1,561
Stegosoma magnum	9,751	*Salpa cylindrica*	1,184
Oikopleura albicans	8,897	*Iasis zonaria*	914
Oikopleura cophocerca	3,597	*Pegea confoederata*	682
Oikopleura cornutogastra	3,483	*Cyclosalpa pinnata*	236
Fritillaria formica	1,580	*Cyclosalpa bakeri*	110
Oikopleura parva	1,536	*Thetys vagina* . .	91
Oikopleura intermedia	1,263	*Ritteriella amboinensis*	41
Fritillaria borealis f. *sargassi* . .	815	*Metcalfina hexagona*	2
Fritillaria haplostoma	300	*Cyclosalpa affinis*	2
Fritillaria fraudax	211	*Cyclosalpa floridana*	1
Fritillaria megachile	150	*Cyclosalpa virgula*	1
Tectillaria fertilis	143	*Ritteriella picteti*	1
Fritillaria bicornis	100		
Althoffia tumida	4		
Kowalevskia tenuis	3		
Bathochordaeus charon	1		
PYROSOMATIDA.		DOLIOLIDA.	
Pyrosoma atlanticum (colonies)	12,815+	*Doliolum denticulatum*	112,794
		Dolioletta gegenbauri	28,863
		Doliolidae (blastozooids)	13,717

equatorial region. The warmer areas, as in the Pacific, are dominated by *Oikopleura longicauda*, with *O. fusiformis*, *O. rufescens* and *O. cophocerca*. *O. longicauda* appears to be fairly limited southwards in the Atlantic, but to the north it reaches about 60°N latitude. *O. fusiformis* can extend in the North Atlantic even to the south of Greenland and in the South Atlantic to areas off the Cape of Good Hope and similar latitudes off South America. Other common species in the warm North Atlantic, apart from *O. dioica*, which is abundant in some coastal areas (e.g. Morocco), include *Fritillaria formica*, *F. pellucida*, *F. borealis sargassi* and, in more coastal areas, *F. haplostoma*.

In the Indian Ocean, including the Red Sea, of thirty-five species of appendicularians, *O. longicauda* is again the commonest, comprising 50–60% of the total. The fauna is rather similar to that in the warmer parts of the Atlantic and Pacific Oceans, with *Megalocercus huxleyi*, *Oikopleura rufescens* and *O. fusiformis* being the more important species after *Oikopleura longicauda*. The greatest biomass is found between 10° and 20°N, with a marked decline south of the equator (Fenaux, 1973).

Lohmann (1933/34) stresses the poverty of cold water appendicularian species, especially Arctic forms, as contrasted with the richness of warm water species, and the considerable similarity of the warm water appendicularian fauna in the three oceans. A marked exception is the distribution of *Megalocercus*. *M. huxleyi* is confined to the Pacific and Indian Oceans. *M. atlanticus* was believed to be confined to the Atlantic and *M. abyssorum* to the Mediterranean. *M. abyssorum*, however, is now known from the Pacific and Indian Ocean and the species cannot be distinguished from *M. atlanticus* (Fenaux, 1967, 1973).

Thaliacea

The other holoplanktonic class of the Tunicata, the Thaliacea, is made up of three orders: Pyrosomida, Doliolida and Salpida.

Pyrosomida

This order has only one genus, *Pyrosoma*, though it is sub-divided by some authorities. The species are primarily warm-water forms found in all seas except the Arctic Ocean, and they tend to be cosmopolitan, though found especially in tropical and sub-tropical areas. They are not normally present in coastal waters, and though they have been recorded to great depths, even 2850 m, according to Berrill (1950) they are typical of surface or near-surface layers.

All species take the form of cylindrical colonies closed anteriorly and composed of many zooids embedded in a gelatinous wall. The colony is propelled forwards by the passage of water through each zooid, and out via the posteriorly situated so-called cloaca. In each zooid the pharynx and atrium typical of chordate animals lie opposite to one another along the axis; between the two are the alimentary canal and gonads. The pharynx has an anterior buccal area separated by a pair of peripharyngeal bands from a hinder region, the branchial sac, with greatly elongated and numerous gill slits. An endostyle runs the length of the branchial sac. A heart and a stolon lie immediately behind the endostyle. Muscles are generally weak on the body, but there is an oral (branchial) and an atrial sphincter and paired cloacal muscles (cf. Fig. 2.124).

The life history is fairly complex. The egg develops into an individual of relatively small size called a cyathozooid, which has a stolon producing buds. A group of four

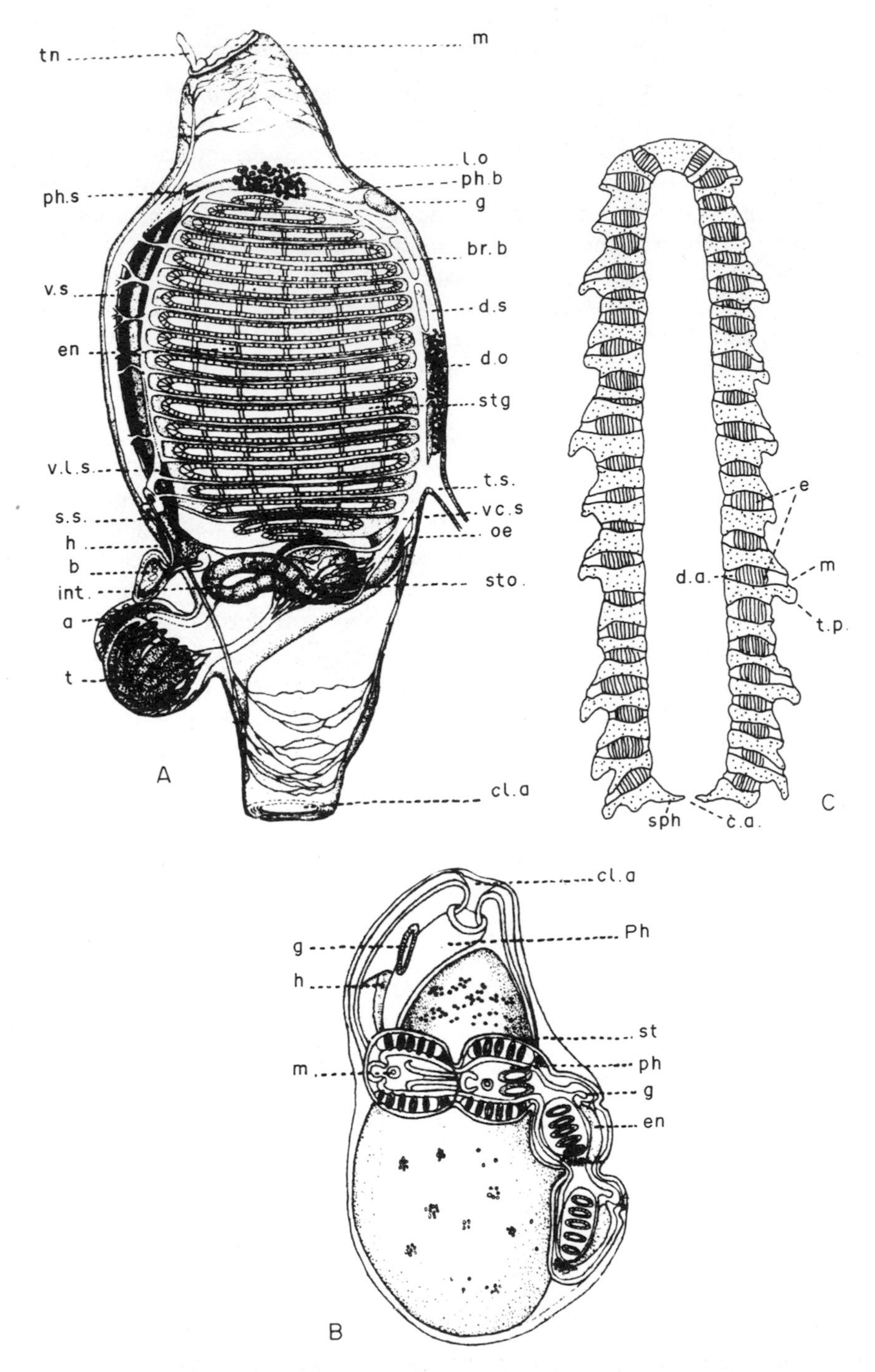

Fig. 2.124. *Pyrosoma atlanticum* (A) blastozooid, lateral view; (B) cyathozooid showing 4 primary buds; (C) colony in longitudinal section, 2 × natural size. b: bud; brb.: branchial bars; c.a.: colonial aperture; cl.a.: cloacal aperture; d.o.: dorsal organ; d.s.: dorsal sinus; en.: endostyle; g: ganglion; h: heart; int: intestine; l.o.: luminous organ; m: mouth; o: ovary; oe: oesophagus; ph.b.: peripharyngeal band; ph.s.: peripharyngeal sinus; sph: sphincter; s.s.: stolon sinus; st: stolon stigma; sto: stomach; t: testis; tn: tentacle; t.p.: test process; t.s.: tunic sinus; vc.s.: visceral sinus; v.l.s.: ventro-lateral sinus; v.s.: ventral sinus (A and C from Thompson, 1948; B from Tregouboff and Rose, 1957).

primary zooids develops from the stolon, the group being subsequently set free via the cloaca of the colony. By further budding, this young colony develops into a new individual. At least in some species, the numerous zooids are arranged in more or less orderly fashion.

Species of *Pyrosoma* include *P. atlanticum giganteum*, which is found in the Atlantic, commonly in the Gulf Stream and warmer waters, but occasionally penetrating even to the English Channel and approaches to the North Sea. The species occurs also in the Indian Ocean and Antarctic waters. According to Berrill (1950), the similar form found in the Pacific Ocean is a separate sub-species. It has been recorded off Tasmania and New Zealand (Thompson, 1948) as well as generally over warmer parts of the Pacific. The length of the colony may be from 12 up to 40 cm. Another common species in all three oceans, *Pyrosoma spinosum*, may also be very large; some colonies are said to exceed a metre in length. Thompson (1948) gives a list of the distributions of the species and the intergrading varieties. Godeaux (1962) found *Pyrosoma aherniosum* off the West African coast in addition to *P. atlanticum*. He describes the latter as being widespread in all warmer oceans, but *P. aherniosum*, as previously reported occurs in the Indian Ocean and Gulf of Oman. Godeaux (1977) records both species from the Atlantide Expedition, *P. aherniosum* being described as tropical and stenothermal, while *P. atlanticum*, living in warm seas, is a little more eurythermal. One of the most characteristic and well known features of *Pyrosoma* is its brilliant bioluminescence.

Doliolida

These planktonic typically oceanic animals have a barrel-shaped body with anterior branchial (oral) and posterior atrial apertures, each opening being surrounded with a series of small lobes. Muscle rings arranged as eight hoops pass around the body, the first and last being the oral and atrial sphincters, respectively. The gill slits, usually numerous, occur on the posterior wall of the pharynx, in the floor of which, somewhat more anteriorly, is an endostyle. Both pharyngeal and atrial cavities are very large. The test, typical of tunicates, is very thin and transparent. The heart lies just behind the endostyle, near the gut and the gonads, both ovary and testes (cf. Fig. 2.125).

The life history is exceedingly complex. The individual described represents the solitary sexual gonozooid. The egg from the gonozooid develops after fertilization to form a tailed larva, to some extent comparable with the ascidian tadpole larva. The oozoid stage develops from the larva. At first it is of small size, being developed from the trunk region of the tadpole larva. It differs when fully developed in some ways from the gonozooid; for instance, it has nine muscle bands and only four gill slits, and, of greatest significance. It has no gonads but carries just behind the endostyle, a ventral stolon which produces buds. On the dorsal side of the hind end of the body a small dorsal process, the cadophore, occurs. The oozooid is the asexual stage in the life history.

Numerous buds are formed by the stolon and these migrate around the body to anchor themselves to the dorsal process, which eventually elongates to carry three rows of functional individual zooids. Some of the buds remain permanently attached and are modified, having very large branchial openings; these so-called trophozooids act essentially as feeding zooids for the colony. The original parent loses the gut and other organs, but the musculature of the body becomes excessively developed so that this individual becomes the main propulsive unit. Some may persist in the plankton and are known as "old nurse" stages. Long before this degeneration, however, some of the

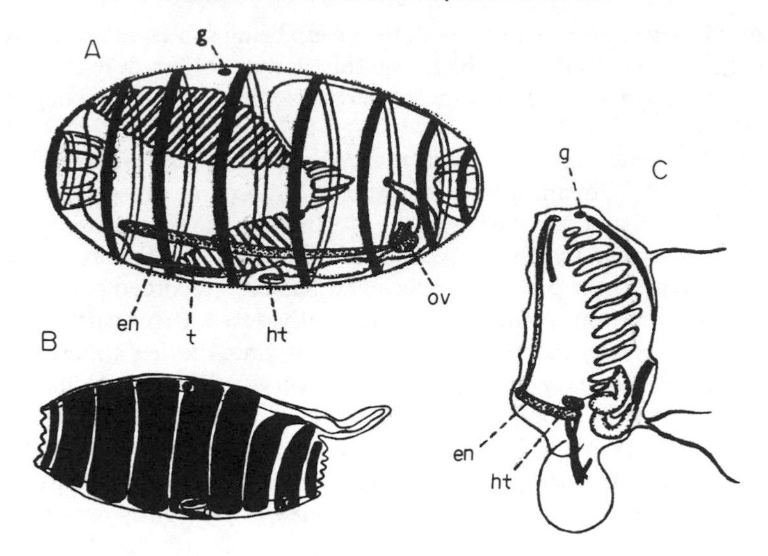

Fig. 2.125. *Doliolum nationalis* (A) gonozoid; *D. denticulatum* (B) "old nurse" (oozoid); (C) trophozoid; en: endostyle; g: ganglion; ht: heart; ov: ovary; t: testis (Berrill, 1950).

small buds in the mid-line on the cadophore develop differently and eventually break away to live independently as so-called phorozooids.

In body form the phorozooid generally resembles the oozooid and lacks gonads, but it is provided with a ventral process in the posterior part of the body. Before each phorozooid breaks away from the colony, small buds have settled on this ventral process; each bud finally develops into a gonozooid stage, the solitary sexual hermaphrodite individual which, when set free, completes the complex life history. An obvious alternation of sexual and asexual generations occurs and considerable numbers of individuals can appear relatively suddenly in the plankton. It is not surprising that on occasions doliolids can swarm in huge numbers.

Feeding of doliolids, like other tunicates, is by a ciliary mechanism; particles are trapped by the mucus secretion of the endostyle, the mucus strings being passed back to the gut. While most of the food is phytoplankton, Protozoa may be taken, and possibly very small metazoans. Doliolids are mainly warm-water species; many are warm–cosmopolitan. They are rarely found in shallow coastal waters.

Of the doliolids, *Doliolum denticulatum* is very widely distributed in the warmer waters of the Atlantic and Pacific Oceans and also occurs widely in the warm waters of the Indian Ocean (Tokioka, 1960). It was the commonest doliolid off the Great Barrier Reef (Russell and Colman, 1935). *Doliolum nationalis*, another very widely distributed species in warm waters, tends to be somewhat more abundant than *D. denticulatum* in waters near islands and continental slopes. *D. nationalis* does not appear so widespread in the South Atlantic, but it has been recorded off South Africa. To some extent it may alternate in its occurrence, possibly seasonally, with *D. denticulatum*, occurring more in the South Atlantic Ocean in winter, while *D. denticulatum* is more plentiful in spring/summer (Tokioka, 1960). Godeaux (1977), however, states that it was frequently caught off the West African coast and is locally abundant. He classes the species as neritic and able to live under variable temperature and salinity conditions. *D. denticulatum* is not found in the Sea of Japan or the Inland Sea, where *D. nationalis* is more common. Both species are recorded from tropical and sub-tropical waters off Central America and widely across the warmer waters of all three major oceans.

Doliolum (*Dolioletta*) *gegenbauri* somewhat resembles *D. nationalis* in distribution, but spreads a little farther north in the Atlantic, reaching the Faroe–Shetland area and even entering the North Sea. In the South Atlantic the species extends round the Cape of Good Hope and to the cool waters of the Benguela Current. Berrill (1950) considers it a somewhat cooler water species, though plentiful in warm seas. It was one of the more abundant thaliaceans found by Thompson (1948) in Australian waters. Fraser (1961) classed *Dolioletta gegenbauri* as cosmopolitan and able to reproduce over the warmer season in Atlantic waters off the north of Scotland, and *Doliolum nationalis* as a Lusitanian species. Hunt's (1968) Continuous Plankton Records of the North Atlantic demonstrated that *D. gegenbauri* was the most abundant doliolid, at times even the most abundant thaliacean. While a few specimens occurred over the continental shelf, the majority were found over deep water. The species was not found in collections from the North Sea (cf. Fig. 2.126).

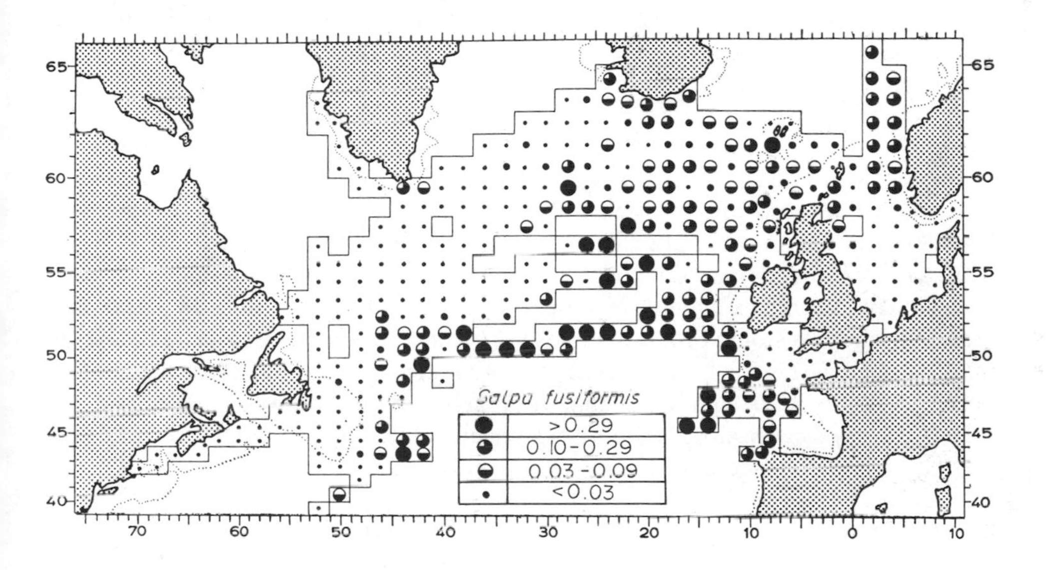

Fig. 2.126. Distribution of *Dolioletta gegenbauri* in the North Atlantic. Where only a part of the sampled area is shown, the organism represented was not observed in the excluded area (Hunt, 1968).

Off the British Isles, the northern limit of *D. gegenbauri* gradually shifted north, with numbers increasing to a maximum about June or July. In the western Pacific, Tokioka (1960) reports its distribution as rather similar to that of *D. nationalis*, including waters around islands and in Japanese seas and extending somewhat to cooler areas. *D. gegenbauri* is abundant in the cool waters of the Californian Current, along the length of the coasts of California and Oregon (cf. Berner, 1967). *D. denticulatum* was less abundant in the inshore collections but was plentiful offshore and to the south. Apart from spreading into slightly cooler waters, *D. gegenbauri* is well known from the warmer areas of all three major oceans. It is very abundant in certain parts of the Indian Ocean as well as being known from many regions of the warm Pacific—the Arafura Sea, off the Phillipines, off Central America and off the Great Barrier Reef. Godeaux (1977) classes it as eurythermal and cosmopolitan, occurring commonly off West Africa and generally over the warm Atlantic and Mediterranean.

Salpida

Salps resemble the doliolids in the generally oceanic habit and in possessing a transparent test enclosing the body, but the test may become thick, tough and hard and may have projections and keels. In some species, however, it remains soft. The body is cylindrical or more spindle-shaped, with the oral and atrial apertures at opposite ends of the body. Both pharynx and atrial cavity are exceedingly large; gill slits are absent except for a single very large opening on each side of the body. The reduction of the pharynx wall is so great that the dorsal lamina appears as an oblique band, often confusingly termed the "branchia".

The relative positions of endostyle, heart, gut and gonads, both ovary and testes, are generally much as in doliolids. The gut is very compact, often appearing as a prominent round mass in the body termed the "nucleus". An oral sphincter muscle, with muscles to a lip, and a complete atrial sphincter muscle are present. The other body muscles vary in number; frequently there are seven or nine. Though they virtually surround the body, they are incomplete; often they approximate dorsally but are interrupted ventrally. The muscles vary considerably with species, both in number and extent.

Feeding is by ciliary action, the secretion of mucus trapping particles much as in doliolids. While the majority of food is phytoplankton, Madin has shown that there is a considerable range in the size of particle filtered (cf. Chapter 6) (cf. also Foxton, 1966). Most salps are larger than doliolids; some may even exceed 80–100 mm.

Salps have a complex life history with an alternation of generations. In the solitary asexual oozooid, a stolon producing buds lies behind the endostyle. The buds develop into the aggregate sexual blastozooids (gonozooids) which differ from the solitary oozooids of the same species in that they are generally slightly smaller and are somewhat asymmetrical. The most significant difference, however, is that the blastozooid (= gonozooid) possesses gonads, both ovary and testes, but lacks a stolon (cf. Foxton, 1961 Fig. 2.127).

Eggs produced by the blastozooid are fertilized internally and the embryo develops inside the body of the mother in a pouch composed mainly of the lining of the atrium. A local thickening of parental tissue ventral to the embryo forms a so-called placenta. Many salps have only a single embryo, but in some species (e.g. *Thetys vagina*) several embryos develop simultaneously. There is no larval stage in the development. The salp is ultimately released from the parent as the solitary asexual oozooid stage, but even before release a stolon is developing. In the free oozooid, as the stolon develops further it emerges from the body; in many salps it forms a spiral growing out posteriorly. Its buds form chains of sexual blastozooids. The chain can be very long, but as the stolon develops the chain tends to divide into a series of blocks, each zooid of a block being at the same stage of development. As the blocks are released, they form aggregations of individuals which continue for a time as loose associations, finally separating into separate forms. This method of asexual reproduction can produce very large swarms of

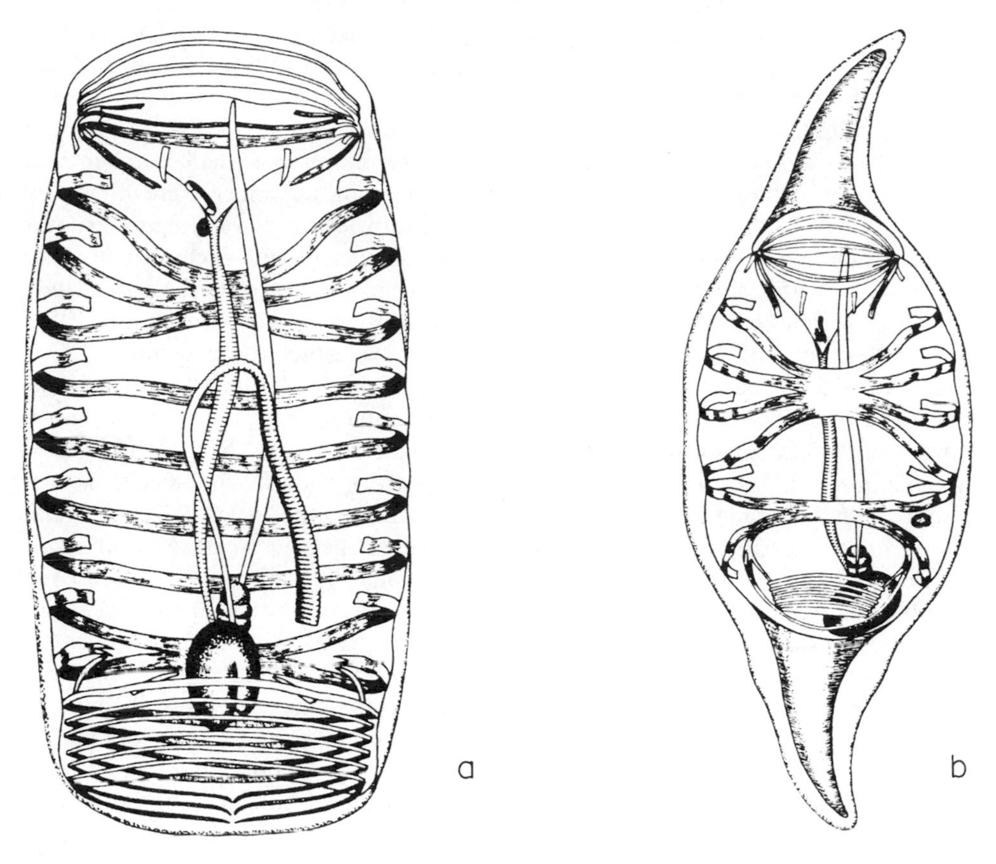

Fig. 2.127. *Salpa fusiformis* (a) Solitary form, ventral view (length 20 mm.); (b) Aggregate form, dorsal view (length 20 mm.) (Foxton, 1961).

salps over a short time in a particular area, a characteristic feature of this group of animals.

Foxton (1966) found huge swarms of *Salpa thompsoni*, an abundant salp of the Southern Ocean, consisting mostly of aggregates. In *S. thompsoni* the total number of buds per individual may approximate to 800. While small aggregates tended to mass near the surface, older aggregates with well developed embryos were mostly deeper than 100 m. Fraser (1947) believes that while most salps are near the surface in warmer oceans, they become more mesoplanktonic in colder waters. According to Berrill (1950), salps may extend even to 1500 m depth. Clearly a few solitary salps can give rise to an enormous swarm under favourable conditions; even warm-temperate species in cooler waters can produce swarms by asexual budding.

Distribution

Salpa (*Thalia*) *democratica* is one of the most widely distributed of all Thaliacea, and is probably the commonest salp in the world (Fraser, 1947, 1961). In the western Atlantic it occurs off New England and stretches across the Atlantic including the area of the Gulf Stream; it reaches the west of Britain and extends to the Norwegian Sea, possibly occasionally into the North Sea. It occurs at least to 40°S latitude in the South Atlantic (Berrill, 1950).

Hunt (1968), analysing Continuous Plankton Records over the North Atlantic, found that *Thalia democratica* was one of the three commoner thaliaceans in shelf waters west of Britain, though not in the North Sea. Highest numbers were found west of the continental slope over deeper water, and the salp occurred over deep water south of Newfoundland. West of Britain a seasonal northerly shift occurs in its main distribution, with maximum numbers attained in August or September. An area where it is known to swarm during summer is off the Moroccan coast (Furnestin, 1957). *Thalia democratica* also appears to be the most widespread and common of all thaliaceans in warmer Pacific waters (Berrill, 1950; Tokioka, 1960). Thompson (1948) described the species as the outstandingly dominant thaliacean in south-east Australian waters, and it was the commonest salp off the Great Barrier Reef (Russell and Colman, 1935), occurring sporadically in great shoals. The species is also abundant in the Indian Ocean (Tokioka, 1960). According to Heron (1972a, b), *Thalia democratica* was the third most common animal in the south-east Australia zooplankton. Its extraordinary rate of multiplication, especially the rapid asexual budding, and remarkably high growth, rate lead to massive swarms. Even with extensive mortality, there were sufficient survivors to permit the development of swarms when favourable conditions, especially dense phytoplankton blooms, occurred. Troost, Sutomo and Wenno (1976) have reported dense swarms of the tropical *Thalia sibogae* in the inner areas of Ambon Bay, where it dominated the zooplankton.

Salpa fusiformis is described by Berrill (1950) as a truly cosmopolitan species, being found in the Atlantic Ocean from 55°S to 60°N latitude, and extending in the Gulf Stream to the Norwegian Sea. It occurs widely in the Indian Ocean and across the Pacific and was believed to extend to the Antarctic to 65°–70°S, but this view is now modified (*vide infra*). The species ranges over the North Pacific; off Japan it reaches to about 50°N; in the Central Pacific its limits are approximately 40°S to 30°N, but in the eastern Pacific it extends from California to Alaska (Berrill, 1950). Fraser (1947) records this salp from the Norwegian Sea and states (Fraser, 1961) that being cosmopolitan in warm and temperate seas, it was the commonest species in the north-east Atlantic off Scotland. Hunt (1968) found that the distribution of *Salpa fusiformis*, one of the three most plentiful and widespread thaliaceans in the north-east Atlantic, appeared to follow the North Atlantic Drift from off Newfoundland to Iceland and European waters, including shelf areas off the British Isles. It was most abundant in mid-Atlantic and was fairly sharply limited to shelf waters to the west of France and south-west Britain. Its centre of abundance shifted north with the season (cf. Fig. 2.128). Tokioka (1960), while recording the salp from many warm areas of the Pacific Ocean, found it plentifully in cooler waters (e.g. off Japan).

Salpa fusiformis is the commonest thaliacean in those areas of the North Pacific where there is some mixture of sub-Arctic with warm water, and was abundant in the southern hemisphere, where there was mixed water. Thompson (1948) considered *S. fusiformis* to be one of the very common salps in Australian waters. Studies by Foxton (1961) demonstrated the widespread occurrence of *Salpa fusiformis*, certainly as far north as 60°N latitude, including round the North Cape and off Iceland. On the other hand, although the species (as *S. fusiformis aspera*) is the commonest salp of the

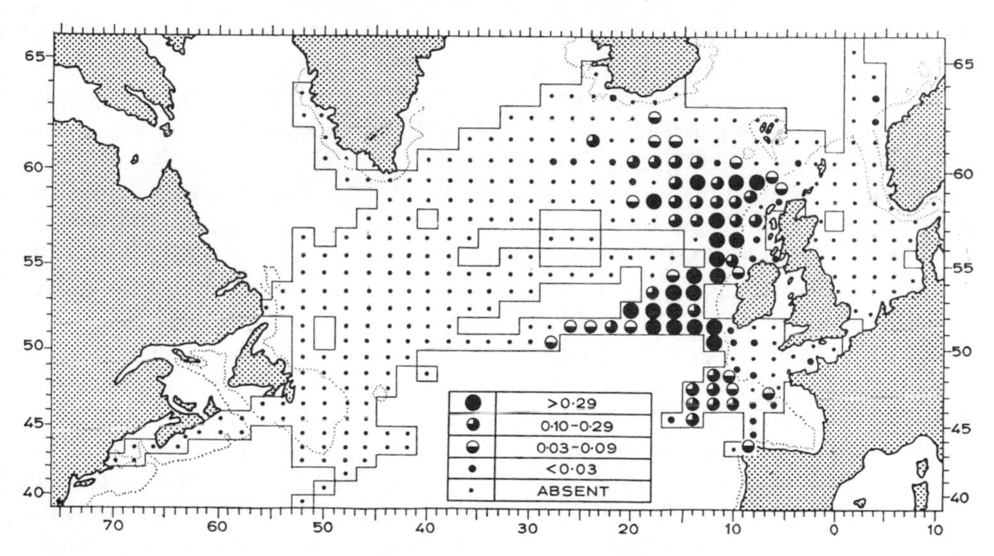

Fig. 2.128. Distribution of *Salpa fusiformis* in the North Atlantic. Symbols on key refer to mean number of organisms per continuous Plankton Recorder sample at 10 m. depth (Hunt, 1968).

Southern Ocean, it does not extend south beyond the Sub-Tropical Convergence. *S. aspera* should be regarded as a distinct tropical or sub-tropical form found in all three major oceans north of the Sub-Tropical Convergence, although its northern limits are uncertain. Godeaux (1977) states that it can extend to a considerable distance in the three oceans.

Salps, other than *Thalia democratica*, which appear to be relatively common in tropical and sub-tropical waters of the Pacific include *Salpa cylindrica*, *Cyclosalpa pinnata*, *Brooksia rostrata*, *Pegea confoederata* and *Iasis zonaria*. Berner (1967) indicates that while *S. fusiformis* is very common along the Oregon and Californian coasts, more offshore in the southern California region and farther south, the presence of *Salpa cylindrica* and other salps suggests a warmer-water fauna. Off the Great Barrier Reef, Russell and Colman (1935) found the tropical salps, *Salpa cylindrica*, *Cyclosalpa pinnata*, *Pegea confoederata*, and *Iasis zonaria* with the ubiquitous *Thalia*. These warm water salps are also found in the warm areas of the Atlantic and Indian Ocean (cf. Sewell, 1953; Tokioka, 1960; de Decker, 1973). Godeaux (1977) found *Salpa cylindrica* to be common among thaliaceans from West Africa and describes it as a tropical stenothermal species widespread through the whole intertropical belt. *Salpa maxima* and *Cyclosalpa affinis* are also widely found in the warm regions of all three major oceans; *Brooksia rostrata*, occurring in warm seas, is apparently more euryhaline. Two salps, *Iasis zonaria* and *Thetys vagina*, found in warm areas of the three oceans and Mediterranean, appear to be slightly more eurythermal and may occasionally reach somewhat higher latitudes. Fraser (1947, 1961) reports the species as rarely found in the cooler waters off Scotland; *Thetys vagina* occasionally reaches the mouth of the English Channel (Berrill, 1950).

Two other salps, *Ihlea symmetrica* (found in the warm North Atlantic) and *Cyclosalpa bakeri* (included by Godeaux (1977) as eurythermal species), are regarded as

Lusitanian species and may reach the moderately high latitude of the Faroe–Shetland area in the north-east Atlantic (Fraser, 1947, 1961).

The relative abundance of the major species of Thaliacea appears to be rather similar in the three oceans. While *Thalia democratica* is the commonest and *Salpa fusiformis* very abundant amongst the salps, the most important doliolids are *D. denticulatum*, occurring widely in all the warm-water regions, with *D. nationalis and D. gegenbauri* especially abundant in certain areas.

Although one or two salps, particularly *Salpa fusiformis*, have been occasionally found in the Norwegian Sea, no thaliacean is a regular member of the zooplankton at high northern latitudes. In the southern hemisphere, by contrast, a new species, *Salpa thompsoni*, is circumpolar but is confined to Antarctic and sub-Antarctic waters, not extending north of the Sub-Tropical Convergence (Foxton, 1961, 1966). Another salp, *S. gerlachei*, is found only at high southern latitudes and is confined to the Pacific sector of the Antarctic (Foxton, 1961, 1966). Two other salps have a circumpolar distribution in the Southern Ocean (Foxton, 1971). *Ihlea magalhanica* is a sub-tropical and northern sub-Antarctic form which does not reach the Antarctic Convergence. *Ihlea racovitsi* is a truly Antarctic species and is always south of the Convergence. *Salpa thompsoni*, as the most abundant salp of the Southern Ocean, probably overlaps both species in their distribution. *S. thompsoni*, however, is extremely patchy in its distribution. Many stations inside its area of occurrence had no salps; others produced 10,000–20,000 specimens, and swarms could measure 2–3 miles across. The patchiness in some ways parallels that of *Euphausia superba* in the same region. Since both animals are intensive phytoplankton grazers, the dense swarms may be in competition.

Fish Eggs and Larvae

Some account of fish eggs and larvae is essential in a text on zooplankton, since these can form an important fraction of the meroplankton. Apart from adult nektic fishes, including the great commercial fish populations over continental shelves, and those species which range widely over the world oceans, many fish of relatively small size, typically found in open oceans and collectively usually classified as part of the micronekton, contribute as eggs and larvae to the plankton. To attempt a description of such a very great variety of meroplankton and their distribution would be a task far beyond the scope of this general text. Proper treatment in any event would be better left to fisheries biologists. For completeness, however, some description of the contribution of fish eggs and larvae to the zooplankton is essential. This account, while dealing with distribution and other topics affecting fish larvae, must inevitably relate considerably to reproductive cycles. This aspect is reviewed here rather than in Chapter 4, which is concerned generally with the breeding of meroplankton.

The planktonic eggs of fishes are small, mostly of the order of a millimetre in diameter, transparent and practically spherical, except for those of the anchovy (*Engraulis encrasicolus*) which are oblong. The external surface of the egg is usually smooth. *Callionymus* eggs show hexagonal sculpturing. *Belone* eggs have fine tendrils. A spherical mass of yolk almost fills the egg in most species, leaving only a narrow perivitelline space. In a few species, however, such as *Sardina pilchardus*, the perivitelline space is large and surrounds a relatively small yolk sphere. Some fishes have one or more oil globules in the egg. As the fertilized egg segments, the blastoderm

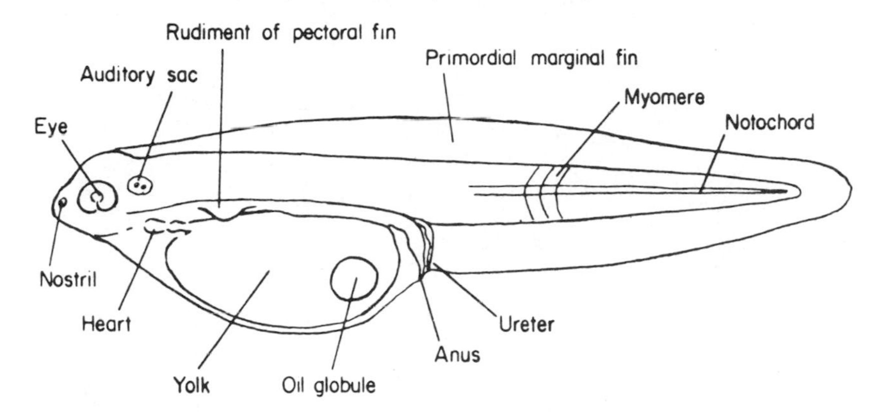

Fig. 2.129. Diagrammatic representation of a newly hatched larva from a pelagic egg (Russell, 1976).

gradually forms a coating over the yolk surface and during later development the embryo appears as a rod stretching across the surface of the yolk. Eyes, auditory sacs and muscle blocks develop in the older embryo, which extends farther across the yolk surface, until the developed tail lifts the embryo from the surface (Fig. 2.129).

Eggs are identified largely by their size, the characteristics of the yolk, the presence of oil globules and the pigmentation. The eggs of most north-west European fishes hatch in a period varying from 3 days to 3 weeks, the time varying with the size of egg, the amount of yolk and especially with the environmental temperature. Many tropical fishes can hatch much more rapidly.

The newly hatched fish larva is usually less than 4 mm in length, except for example, in the Clupeidae, which have a very slender and comparatively long larva. Apart from the well developed head (with a usually conspicuous eye), gut, heart, notochord and body muscles are normally visible. A marked characteristic of the young larva is the presence of a yolk sac anterior on the lower side. A fin, lacking finrays, runs along the length of the dorsal side of the body, around the tail and ventrally as far as the anus, situated in most species just behind the yolk sac (cf. Fig. 2.129).

The complete absorption of the yolk sac marks the commencement of the post-larval period. Pigment cells develop in patterns which are diagnostic for the different species of post-larvae. Amongst other characteristics used in their identification are the body shape and size and the size of the paired fins. During further post-larval stages the tail fin develops, followed by the differentiation of dorsal and anal fins. The vertebrae and rib bones ossify and with further changes in body pigmentation, the post-larva reaches the fish fry stage (Fig. 2.130).

Details of the feeding of some fish larvae are included in Chapter 6. Survival and growth of a brood of fish larvae are greatly dependent, among other factors, on the presence of sufficient density of food organisms of the right type and size (cf. Cushing 1975, and discussion in Volume 1). Numerous reports associate high densities of fish eggs and larvae with rich zooplankton patches (e.g. Thiriot, 1977).

Planktonic fish eggs are usually distributed more or less evenly in the near surface layers, though there is a tendency in calm weather for some to concentrate very close to the surface. Later embryonic stages may show some slight difference in vertical

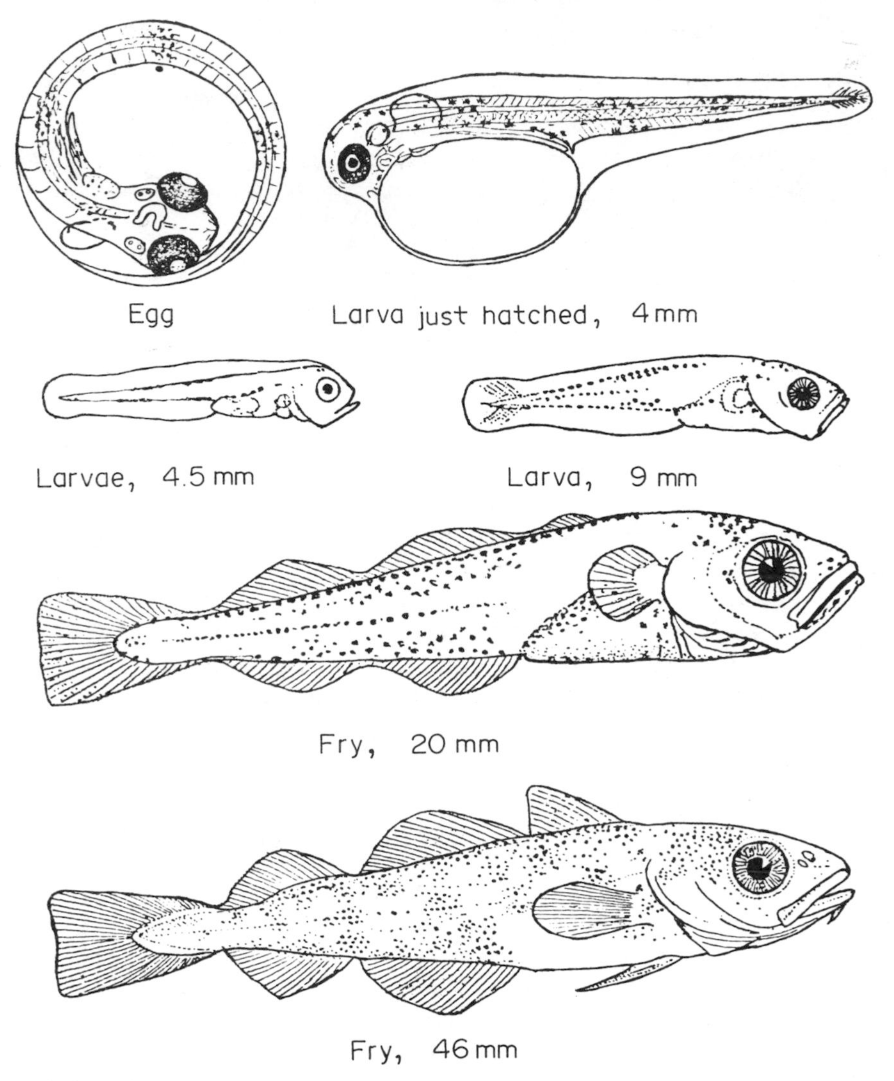

Fig. 2.130. *Gadus callarius* stages in early life history (Bigelow and Welsh, 1925).

distribution. There are also certain minor species differences. Thus, Russell (1976) noted that rockling (*Ctenolabrus*) eggs are very light, eggs of *Buglossidium* (*Solea*) *luteum*, *Limanda* and *Callionymus* arc relatively heavy, while eggs of *Sprattus*, *Scomber*, and *Scophthalmus* (turbot) are intermediate in specific gravity. Russell demonstrated that the larvae and post-larvae of fishes exhibited considerable differences in their daytime vertical distribution. Those which remained mainly in the upper 10 m in the shallow waters of the English Channel included pollack, Labridae, turbot and brill. At intermediate depths (mostly 15–30 m) whiting, dab, lemon sole, gurnard and hake larvae occurred, while close to the bottom were post-larval gobies. Sprat, pilchard and sand eel larvae were rather irregularly distributed but avoided the surface, although, according to Nellen and Hempel (1970), sprat, herring, cod and sand eel larvae may sometimes appear at the surface.

In temperate and higher latitudes the spawning of fishes and the appearance, therefore, of their larvae is usually markedly seasonal and to a large extent temperature dependent. Differences may appear from year to year; for example, unusually low temperatures in winter and spring may delay the appearance of larvae by about a month (Russell, 1976), or even 1–2 months (Kandler, 1950). For an area such as the English Channel a series of studies by Russell in the 1930s and subsequently, on the appearance of spring and summer spawned fish larvae showed a broad relation to the boreal or warm-temperate habitat of the adult fishes. Thus, Russell showed that flatfish such as plaice (*Pleuronectes platessa*), flounder (*P. flesus*) and the sole (*Solea vulgaris*), as well as fish like cod, haddock, whiting and pollack, produced very abundant eggs about March or April, with the larvae appearing in the plankton a few weeks afterwards. On the other hand, herring eggs which are, of course, demersal were deposited through the winter period; larvae occurred about January and February.

Callionymus spawned off Plymouth slightly later than the bulk of species in spring, with the post-larvae appearing in May and June. A group of fishes including the soles, *Solea lutea* and *S. lascaris*, *Trachinus*, *Caranx*, various gobies and *Mullus* are summer spawners, so that the post-larvae, according to Russell's investigations, were present in the plankton in July and August. Species of *Arnoglossus* were even slightly later, post-larvae being found during September. The studies by Qasim (1956) generally confirmed the times of abundance of different fish larvae in relation to their latitudinal distribution (cf. Table 2.29). Changes may, however, be also considerably affected by current patterns. Fives and O'Brien's (1976) studies on fish larvae found in Galway Bay demonstrated changes in relative abundance with season. For example, herring larvae were mostly found in late autumn and winter; *Sprattus*, *Merlangus*, *Trisopterus esmarkii* and *Pollachius*, mainly about April; *Callionymus* and *Gymnammodytes* in May; *Ctenolabrus* and *Arnoglossus* from July to September. The changes were related to the mixing of inshore littoral and sublittoral with offshore fish larvae, but also with incursions of offshore water, especially from deeper northern areas. An influx of offshore southerly water, occasionally experienced as during August of one year, brought more southerly distributed species.

Bigelow (1926) had earlier pointed to the importance of currents transporting fish eggs and larvae, but he emphasized the great distances to which larvae could be carried from the spawning ground, especially if the species had a fairly extended larval life. For example, he suggested that off Iceland both the American pollack (*Gadus virens*) and the cod, spawned about April and larvae drifted mainly westwards and northwards. However, while *G. virens* descended to the bottom comparatively near the spawning grounds, possibly due to drifting more slowly in deeper layers, cod could drift as much as 500 miles all around Iceland before metamorphosis. Haddock larvae could also drift considerable distances between May and July. In the Gulf of Maine, Bigelow suggests that eggs and larvae of fishes drift westerly and south-westerly round the periphery of the Gulf before being carried more offshore. Both *G. virens* and haddock larvae are believed to be pelagic in the Gulf of Maine for about 3 months; cod larvae for at least 2 months. There appears to be some difference in their vertical distribution in the Gulf of Maine, with the pollack being nearer the surface, the cod larvae rather deeper and the haddock larvae deeper still.

Table 2.29. Time of abundance of different fish larvae in relation to their latitudinal distribution (Qasim, 1956)

Fauna	Species	Time and duration of spawning in British waters											
		Dec.	Jan.	Feb.	March	Apr.	May	Jun.	July	Aug.	Sep.	Oct.	Nov.
Northern species (arctic-boreal)	*Centronotus gunnellus* (Gunnel)		+	+	+								
	Clupea harengus (Herring)	+	+	+									
	Cottus scorpius (Sea scorpion)	+	+	+	+								
	Gadus aeglefinus (Haddock)		+	+	+	+							
	Gadus callarias (Cod)			+	+	+							
	Pleuronectes platessa (Plaice)	+	+	+	+								
Southern species (mediterranean-boreal)	*Blennius pholis* (Shanny)					+	+	+	+	+			
	Clupea pilchardus (Pilchard)				+	+	+	+	+	+	+		
	Clupea sprattus (Sprat)				+	+	+	+	+	+			
	Gadus merlangus (Whiting)			+	+	+	+	+					
	Merluccius merluccius (Hake)					+	+	+	+	+			
	Scomber scombrus (Mackerel)				+	+	+	+	+				

The silver hake, *Merluccius bilinearis*, according to Bigelow and Welsh (1925), is mainly a summer spawner in the Massachusetts area, spawning roughly in July/August; the larvae are relatively deep-living at 40 m or more. The cod is a very early spring spawner, approximately from about February to April in the Gulf of Maine, and the haddock spawns slightly later from late February to May. The mackerel is a late spring spawner, from approximately the end of May to June or July in Massachusetts Bay and, as is frequently observed with other species, in the somewhat colder area of the St. Lawrence, it spawns later, from July to August. The American pollack (*Gadus virens*) is a late autumn/early winter spawner, roughly from November to January in the Gulf of Maine, but early in spring in European waters.

The spawning times of the European hake, *Merluccius merluccius*, in northwest European waters is to some extent associated with latitude. Wheeler (1969) suggests that, whereas off Ireland the hake, which tends to spawn in somewhat deeper water than many of the common gadoid fishes, spawns about April to July, off Scotland it is later, from about May to August, approximately the same time as Bigelow quotes for American silver hake in the Massachusetts area. However, Wheeler suggests that farther south, in European waters off Biscay, the hake can spawn in early spring. The fish also tends to spawn earlier over deeper water, gradually becoming later in somewhat shallower areas.

The Gadidae, as a major family extending from high latitudes to the sub-tropics, seems to be especially important in the cooler waters of the North Atlantic. The larvae in general have a relatively long planktonic life. Most of the commoner commercial species are fairly shallow-living, so that an abundance of larvae of such fishes as cod, haddock, whiting, saithe and pollack is found over spring and early summer in coastal waters of boreal temperate areas. In colder waters *Boreogadus saida*, a circumpolar species, spawns in the Barents Sea, from approximately November to February. Russell quotes Rass to the effect that larvae occur in the Barents Sea in February and March. This very cold water species undergoes its development at temperatures from <0 to 2°C, commencing under ice. Another widespread, cold-loving fish is the capelin, *Mallotus villosus*. It spawns during spring along the Murmansk coast, off the North Cape, Iceland, in the Barents Sea, along the Norwegian coast and off Newfoundland. Spawning occurs a little later in the White Sea, near Novaya Zemlya, and in the region of Hudson Bay. Spawning temperature, though somewhat variable, usually exceeds 2°C (cf. Prokhorov, 1968; Templeman, 1968).

The species of fish eggs and larvae considered so far are found chiefly over continental shelf areas in temperate boreal regions or, as with the hake, in slightly deeper waters offshore. Farther offshore in the north-east Atlantic, however, such species as the blue whiting, *Micromesistius poutassou*, can be extremely plentiful. Bailey (1974) reports on the spawning of this species in the north-east Atlantic. Earlier indications were that the species spawned mainly in depths of water greater than 1000 m, west of the British Isles, between latitudes of about 43°–63°N. It was also believed that spawning tended to become progressively later from south to north, so that, whereas in the Bay of Biscay spawning might be about February to April, south of Iceland it occurred about April or May and there were reports of larvae in the Norwegian Sea as late as June. Zilanov (1968) suggested a minimum spawning temperature of 8°–9°C. While he believes that the most intensive spawning occurs west of Ireland and the Hebrides, larvae are found in the Barents Sea and are derived from spawnings in the Norwegian Sea. Bailey found that in the area of Rockall, spawning occurred about March/April, approximately the time of the beginning of the main plankton increase in the area and was fairly sharply limited to a period of 3–4 weeks. The larvae were found mainly in April and May. The range of the species is considerable, also occurring south and west of Ireland and west of the Bay of Biscay (Fig. 2.131a). According to Bailey, the larvae have a fairly high growth rate as compared with other gadoids, about 0.15 % of the length per day, as compared with

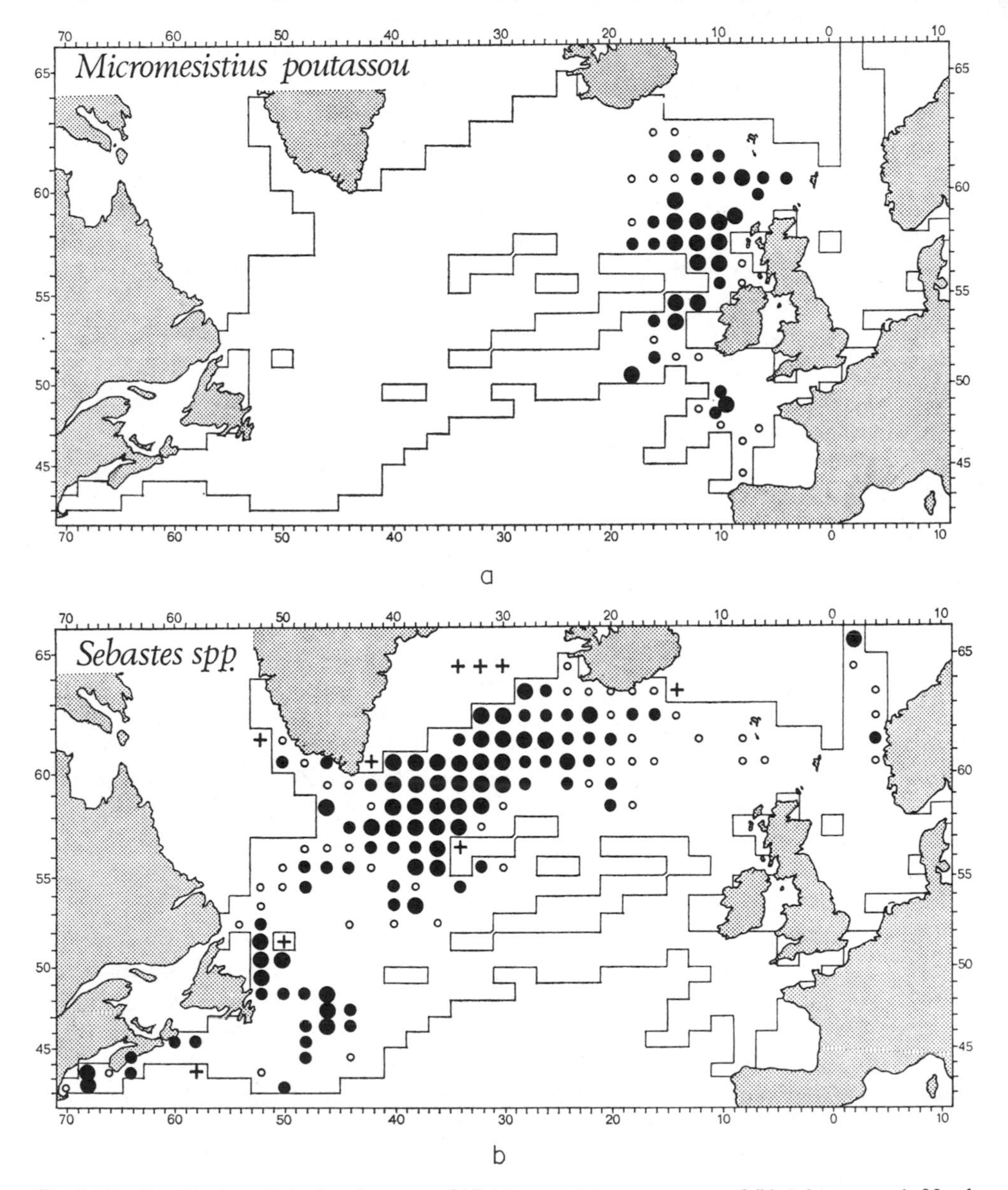

Fig. 2.131. Distribution of planktonic stages of (a) *Micromesistius poutassou* and (b) *Sebastes* spp. in North Atlantic. (a) ● = > 0.13; • = 0.03–0.13; ○ = <0.03; (b) ● = >0.14; • = 0.04–0.14; ○ = <0.04 for mean recorder sample category levels (Henderson, 1973).

estimates for haddock of 0.012%, and for cod of 0.011%. Gadoids appear to be generally slow growers, since estimates for other fishes suggest that the sardine, for example, may grow at a length of approximately 0.19% per day, similar to the Clyde herring and the mackerel at about 0.16% per day.

A species which seems to have an even more oceanic distribution in the young stages than *Micromesistius poutassou* is the redfish, *Sebastes marinus*. Templeman (1961) described the larvae and young of *Sebastes marinus* as occurring abundantly in the Norwegian Sea, south and west of Iceland, south of Greenland, off Newfoundland and Nova Scotia and in the Gulf of Maine. The larvae appear to grow slowly, with a pelagic life of several months in the upper water layers. A fishery began in the Gulf of Maine only in 1935, but it is now an important commercial undertaking. The adult fish are obviously northerly in distribution. Henderson has contributed several reports (e.g. Henderson 1964, 1966) from the Continuous Plankton Recorder analysis of the occurrence of *Sebastes* larvae.

Sebastes is a viviparous species, but according to Henderson, the main months for appearance of the larvae in the higher latitudes of the North Atlantic seem to be from about April to September. Figure 2.131b summarizes the distribution of the larvae. There is a suggestion of a separation of a population over the Nova Scotia shelf and Gulf of Maine from a more oceanic population, mostly north of 55°N in areas over the northern Atlantic off south-west Iceland. In the Gulf of Maine, Kelly and Barker (1961) suggest that the young larvae are close to the surface at about 10 m depth and that the release of the young appears to be quite late in spring, about June. The somewhat older larvae occurring in July/August are at about 10–30 m depth and by September they are found at a depth of 80 m or more. Kelly and Barker confirm the relatively long pelagic life of about 4 to 5 months, which allows the larvae to drift substantial distances. The considerable depth at which the older larvae are found may be related to Bigelow's (1926) observation that *Sebastes* larvae avoid being swept out of the Gulf of Maine.

In the Pacific the Scorpaenidae, the family to which *Sebastes* belongs, also includes the genera *Scorpaena* and *Sebastodes*, the latter genus being very closely related to *Sebastes*. *Sebastodes*, which has a large number of species, has a very wide distribution off the Pacific coast, approximately from the Gulf of Alaska to Baja California. Like *Sebastes*, *Sebastodes* is viviparous. Ahlstrom (1961) showed that the larvae of *Sebastodes* off California and southern California, occurred mostly in the upper mixed layer, 80% being found above a depth of 65 m. They appear to be released mostly during winter and spring in the area, and were more abundant from December to April. The larvae form an important part of the larval fish population off the Californian coast. During the years 1955 and 1956, for example, they ranked third in abundance, the other two important species in 1955 being the northern anchovy, *Engraulis mordax* (about 39%), and the hake, *Merluccius productus* (about 17%).

McHugh (1967) has emphasized the enormous importance of estuaries and shallow coastal waters off the Atlantic coasts of North America as breeding and nursery grounds for many fish species, some of great commercial significance. He points to the bay anchovy, *Anchova mitchilli*, as probably the most abundant of all Atlantic fishes, extending from Massachusetts to the Gulf. Other important fishes are the menhaden, *Brevoortia tyrannus*, which is probably next in abundance, certain Sciaenidae (e.g. *Micropogon* and *Leiostomus*), and also mullets like *Mugil cephalus*. Though some of the species spawn a little offshore, the young appear to be carried in, so that there is a great abundance of larval stages in estuarine and inshore waters. Postlarval menhaden are enormously abundant in Chesapeake Bay and other brackish waters.

Table 2.30 gives some idea of the population of larval fish, as taken by plankton net, in Lower Chesapeake Bay waters. In the estuaries and coastal areas of the New England States, representatives of the cod family, together with considerable numbers of some species of the herring family, are more important. Gunter (1967) suggests that in coastal American states in the Gulf of Mexico and off the Atlantic coast, larvae drift shorewards from offshore spawnings, so that the larval populations are extremely plentiful in the shallow waters. According to McHugh (1967) the situation in the Gulf and Atlantic coasts of the U.S.A. contrasts sharply with the Pacific coasts, where there are few estuaries of similar pattern to those on the Atlantic coast. Apart from the anadromous salmons and certain smelts, there is no mass movement of fish to spawning grounds off the Pacific coast, followed by a return of larvae into shallow coastal waters. One species which does appear to move offshore to spawn, with young larvae subsequently carried into shallow waters, is the Pacific halibut.

Table 2.30. Percentage catch in Lower Chesapeake Bay, taken by plankton net (McHugh, 1967)

Larval fish	Percentage of catch
Engraulidae	75.7
Clupeidae	2.4
Sciaenidae [a]	4.8
Sparidae [a]	0
Gadidae [a]	0
Gobiidae	12.5
All others	4.6
	100.0

[a] These appear as adult fish in other net collections.

This difference on the Pacific coast, the absence of shallow estuaries, the narrowness of the continental shelf, and the considerable population of deep water fish larvae close to shore, is illustrated particularly by the investigations of Ahlstrom. Despite the considerable range of fish larval species, Ahlstrom (1959) demonstrated that the great majority of larvae were in the upper mixed layer. Thus, twelve species were confined approximately to the upper 120 m, though each species had a rather narrower vertical distribution inside that limit (Figs 2.132, 2.133). Three species, however, *Merluccius productus*, *Bathylagus wesethi* and *Leuroglossus stilbius*, were deeper, extending to 200 m or more. This vertical distribution of fish larvae off California appeared to hold more widely in the North Pacific. In the upper 130 m approximately, the following were common: *Vinciguerria*, *Engraulis*, *Cyclothone*, and *Lampanyctus*. Of rather deeper-living larvae, *Argyropelecus* easily dominated the catch.

Ahlstrom (1965) suggests, from collections over four years, that twelve fish larval species made up the vast majority (over 90%) of the total larval catch off California, with a further twelve species adding approximately another 5%. The northern

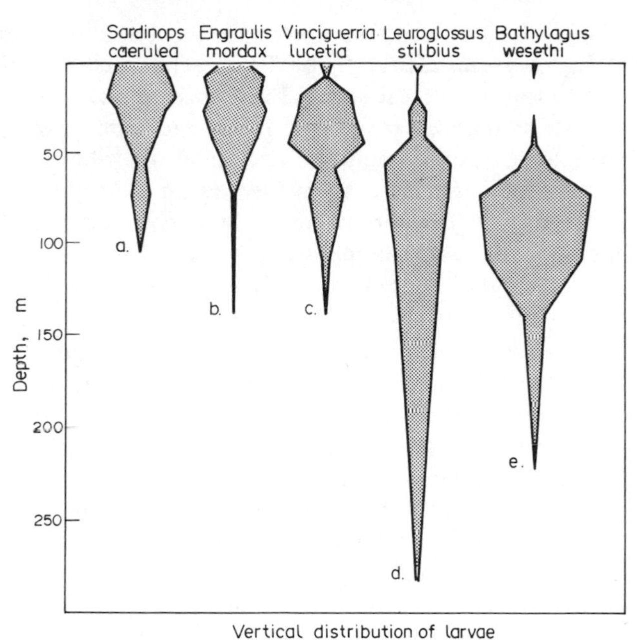

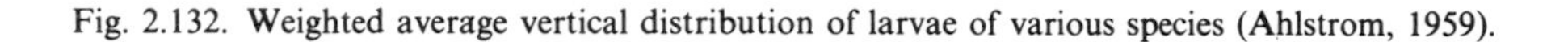
Fig. 2.132. Weighted average vertical distribution of larvae of various species (Ahlstrom, 1959).

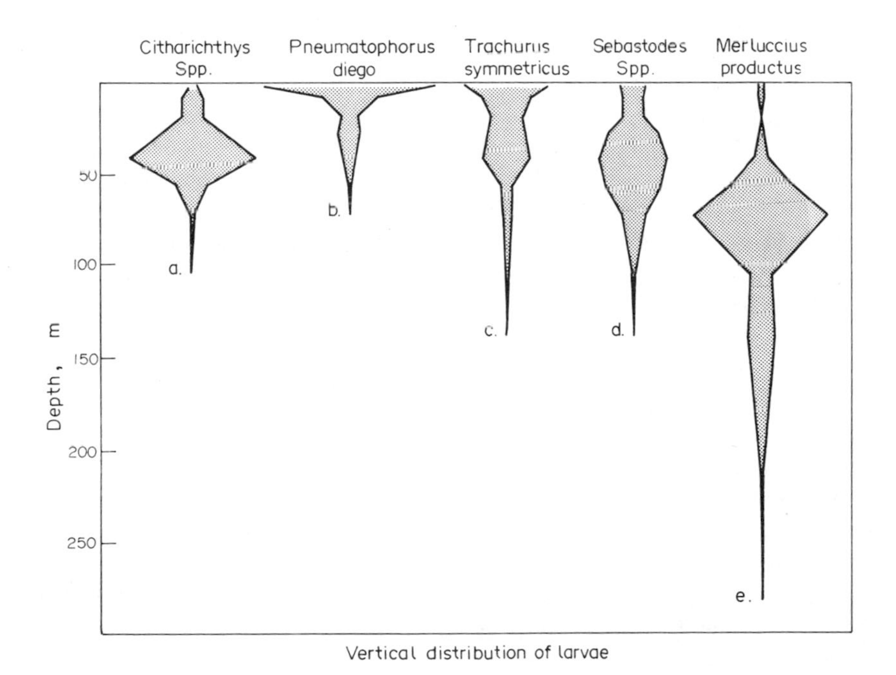

Fig. 2.133. Weighted average vertical distribution of larvae of various species (Ahlstrom, 1959).

anchovy, *Engraulis mordax*, appears always to be the most abundant species followed by the hake, *Merluccius productus* (cf. Table 2.31). The Pacific sardine, *Sardinops caerulea*, usually occurs amongst the first half dozen species, but Ahlstrom emphasizes the variability of the spawning areas for various fish species from year to year. The jack mackerel, *Trachurus*, appears to be important and widely distributed, extending more than 1000 miles from the coast, as a temperate oceanic North Pacific species. It probably does not reach farther southwards, but may extend farther north and seawards. The hake, *Merluccius*, probably has a similar distribution.

Apart from *Sebastodes* (*vide supra*), other larvae which appear in considerable numbers are the young of various myctophids (e.g. *Lampanyctus* spp., *Tarletonbeania crenularis* and *Diogenichthys*). Their occurrence emphasizes the importance of oceanic constituents of the ichthyoplankton off California. Ahlstrom points to the larvae of such species as *Tarletonbeania crenularis* and *Stenobrachius* (*Lampanyctus*) *leucopsarus*, which are temperate or sub-Arctic in distribution, being found mostly over winter and spring in the Californian area. In contrast, the larvae of *Triphoturus* (*Lampanyctus*) *mexicanus*, a sub-tropical species, occurred mostly over summer. *Diogenichthys laternatus* is an oceanic but even more tropical species, and the larvae only reach California in a warm year. In some years, according to Ahlstrom (1965, 1972) tropical and sub-tropical fish larvae extend north, at least as far as the latitude of Los Angeles. These visitors include *Vinciguerria* larvae, probably the commonest fish larva and widespread in the open tropical East pacific and *Triphoturus mexicanus* which can become abundant in some years and less plentiful in others when the temperate *Stenobrachius leucopsarus* occurs in larger numbers. Such variations stress the significance of ocean currents in the distribution of the ichthyoplankton. Although larvae of Bathylagidae are not nearly so plentiful, their presence underlines the contribution of deeper-living fishes to the larval population. The type of ichthyoplankton certainly contrasts with that off the Atlantic coast of the United States and off north-west Europe.

The region off Peru and northern Chile is also characterized by a very narrow continental shelf. Santander and Castillo's (1977a) investigations showed that the sardine, *Sardinops sagax sagax*, one of the important fishes in the area, spawned from August to about February over an area north of about 13°S latitude, but that maximum spawning was in August when more than 1000 eggs/m^2 were recorded. Spawning was maximal at temperatures around 19°–22°C and in water of 35‰ salinity. The sardine larvae could extend even several hundred miles offshore, but were mainly concentrated at about 30–50 miles distance from the coast. According to the year, maximum spawning occurred in different areas between *ca.* 3° and 13°S latitude, but in one year (1972) the distribution of the larvae extended southwards as far as 18°S. This extension of the spawning range and some lengthening of the spawning season may have been associated with a marked southward flow of sub-tropical waters (El Nino) raising temperatures along the coast by more than 3°C.

Santander and Castillo (1977b) examined the spawning of the anchoveta, *Engraulis ringens*, which appeared to be somewhat closer to the coast. Spawning off Peru (south of about 6°S latitude) was extended from about July to March, with a peak intensity in September (the southern spring). Anchoveta larvae were overwhelmingly dominant among the ichthyoplankton, averaging 92% over several years. Other larvae which were comparatively abundant included those of myctophids, especially *Lampanyctus*

Table 2.31. Comparison of relative abundance of fish larvae in vertical distribution series with their abundance in survey cruises of the California Cooperative Oceanic Fisheries Investigations in 1955 and 1956 (Ahlstrom, 1965)

Species	Fish larvae taken in vertical series			Total fish larvae taken in 1955			Total fish larvae taken in 1956		
	Number taken	Percent of total	Rank	Number taken[a]	Percent of total	Rank	Number taken[a]	Percent of total	Rank
Engraulis mordax	7,991	44.28	1	140,183	39.03	1	134,931	33.06	1
Merluccius productus	2,892	16.03	2	60,090	16.73	2	94,277	23.10	2
Sardinops caerulea	1,842	10.21	3	14,121	3.93	6	15,523	3.80	6
Sebastodes spp	1,267	7.02	4	29,344	8.17	3	29,144	7.14	3
Leuroglossus stilbius	838	4.64	5	15,111	4.21	5	18,620	4.56	5
Trachurus symmetricus	784	4.34	6	13,246	3.69	7	8,027	1.97	10
Stenobrachius leucopsarus	619	3.43	7	7,454	2.08	10	15,125	3.71	7
Triphoturus mexicanus	273	1.51	8	13,165	3.67	8	10,802	2.65	8
Pneumatophorus diego	217	1.20	9	1,950	0.54	14	1,520	0.37	20
Citharichthys spp	181	1.00	10	20,411	5.68	4	23,635	5.79	4
Lampanyctus ritteri	173	0.96	11	1,988	0.55	13	1,924	0.47	18
Vinciguerria lucetia	159	0.88	12	12,654	3.52	9	9,832	2.41	9
Diogenichthys laternatus	110	0.61	13	4,771	1.33	11	3,158	0.77	13
Bathylagus wesethi	84	0.47	14	3.245	0.90	12	2,146	0.53	17
Tarletonbeania crenularis	66	0.37	15	999	0.28	22	3,352	0.82	12
All others	549	3.04	—	20,423	5.69	—	36,124	8.85	—
Total	18,045	99.99	—	359,155	100.00	—	408,140	100.00	—

[a] Standard haul totals.

and *Diogenichthys*, mainly in spring, the widespread *Vinciguerria* larvae present over a fairly extended period, and Bathylagidae larvae, including *Leuroglossus*, with a spring peak. Among other moderately plentiful larvae were *Trachurus symmetricus murphyi* (August to February), *Scomber japonicus peruanus* (spring and summer) and *Merluccius gayi* (August to December).

A fairly extended spawning period appeared to be true of most fishes, though some showed increased intensity over spring. The larval fish population may be compared with that off southern California. The breeding of the Peru anchoveta may also be considered against that of the Argentina anchovy (*Engraulis anchoita*). Ciechomski (1971) found that *Engraulis anchoita* reproduced through the year, but with greatest intensity during the southern spring (October/November). The eggs and larvae dominated the ichthyoplankton at all seasons over the Argentina shelf, with two main spawning grounds, the greater off the La Plata River, but spawning did not occur south of about 47°S. The density of eggs and larvae of *Engraulis anchoita* exceeded that of *Engraulis ringens*. The only fish larvae in considerable density off Argentina, apart from *Engraulis*, were *Scomber japonicus marplatensis*. Off Peru and California, on the other hand, *Sardinops sagax* and *Sardinops caerulea*, respectively, may be in competition with the anchovy.

The larvae of fishes which as adults are rare visitors to north-west European coasts, particularly to Britain, and are normally present in ocean waters often farther south, would not usually be found in the plankton off the British Isles. Many of these fishes will spawn in the Mediterranean. Bigelow and Welsh (1925) similarly recorded several fishes (essentially warm water species) as occasional visitors to the Gulf of Maine, but not reproducing there. The saury pike, *Scombresox saurus*, is an oceanic fish found usually in warmer waters. The eggs and larvae are present in open oceans, normally not further than 40°N latitude. The John Dory, *Zeus faber*, apparently may spawn during the summer in western British waters, but is much more typically a spawner in Biscay and Mediterranean areas. The boar fish, *Capros aper*, also a summer spawner, is a fish more typically found in southern European waters and in the Mediterranean. Of the sea breams (Sparidae), essentially tropical and sub-tropical fishes, only one common species, *Pagellus bogaraveo*, occurs in northern European seas. It is found to the west of the English Channel and is a late summer spawner, but as with so many fishes that are more southern in distribution, spawns much earlier to the south.

Several other fishes, very rare visitors off Britain, never spawn off north-west Europe. *Xiphias gladius*, a fish typical of the tropical Atlantic, breeds only in waters exceeding 24°C. In the Atlantic, spawning occurs mainly from February to April. In the Mediterranean it takes place from June to August. *Thunnus thunnus* also breeds about June or July in the Mediterranean, whereas off Florida spawning is about May/June. Even in the Mediterranean, there may be variation in spawning times in different areas and according to the habitat (temperate, warm-temperate, tropical) of the adult fish species. Marinaro (1968), for example, working in the Bay of Algiers, distinguished the eggs and larvae of *Helicolenus dactylopterus* from certain other species such as *Scorpaena* spp. The investigations demonstrated variation in the spawning periods even in the warmer southern waters (cf. Table 2.32). Other species confined to warmer waters for spawning include the bonito (*Sarda sarda*) and *Thunnus alalunga*, which occurs as a summer spawner in the Mediterranean. The oceanic bonito, *Katsuwomus pelamis*, apparently spawns all through the year in tropical

Table 2.32. Fish spawning periods in the Bay of Algiers (Marinaro, 1968)

Species	Spawning period
Scorpaena porcus L.	May–August
Scorpaena scrofa L.	May–September
Scorpaena ustulata Lowe	May
Helicolenus dactylopterus Del.	January–March
Dactylopterus volitans L.	July

waters. Off North Africa, however, it mainly spawns from April to September.

The seasonal changes in hydrological conditions, especially upwelling, typical of much of the west African coast (cf. Chapter 8) are likely to be reflected in fish spawning times. For example, off Senegal, Boely and Champagnat (1970) found two periods of reproduction for *Sardinella aurita*, from December to February and from May to June. Zei (1967), however, found that off Ghana, where upwelling occurs, though some limited spawning of *Sardinella* spp. could take place through the year, the main period was from July to October. Eggs and larvae of *Anchoviella* were found over the same time, but the period extended beyond the time of upwelling. The richness was apparent in that up to 20,000 *Anchoviella* eggs per haul were obtained. *Anchoviella* eggs were more abundant offshore, *Sardinella* more abundant inshore. Postel (1955) quotes *Sardinella aurita* and *S. eba* as reproducing from June to August off Dakar. During September, larvae were observed in the coastal waters. Rossignol (1955) found, for the Bay of Pointe-Noire (Congo), that although spawning of both species could take place through the year, maximum periods were observed for *S. eba* in July/August and November/December and for *S. aurita* in May/June and again in December/January.

Furnestin (1955) quotes spawning of the sardine (*Clupea pilchardus*) as markedly seasonal off Morocco, being concentrated over winter and spring when the temperature was lowest in that area. No eggs were recovered over the summer months. For the anchovy (*Engraulis encrasicolus*), spawning was almost restricted to the summer, with maximal temperatures (15°–22°C). Both species are markedly coastal, eggs being found mostly at depths not exceeding 60 m. In European waters, the anchovy spawning period is prolonged, from April to August, or even to October along the Biscay coast (Wheeler, 1969). Kiliachenkova (1970) found that eggs and larvae of the horse-mackerel (*Trachurus trachurus*) occurred off northwest Africa over a wide area (8°–23°N). The species is believed to breed throughout the year, but maximum numbers of larvae occurred off Cap Blanc in December/January at 16°–18°C, and off Dakar in April (22°–23°C). Hotta and Nakashima (1971), investigating the allied species *Trachurus japonicus* in warm western seas off Japan, also found that although some spawning took place through the year, four geographically distinct populations were present, exhibiting different periods of maximal spawning, varying from mid-winter (January/February) to late winter, spring and summer (July/August).

Razniewski (1970), working off north-west Africa, observed another species of Carangidae, *Caranx rhonchus*, spawning in July/August south of Cap Blanc. Other fish species (e.g. *Vomer*, *Dentex*, *Hemirhamphus*) were also noted as being sexually mature and apparently forming pre-spawning shoals in June/July. According to Domanevsky (1970), *Dentex macrophthalmus*, abundant off north-west Africa, spawns from

October to April at the edge of the continental slope. *Dentex maroccanus*, however, is restricted to the months of May–August and to shallow waters. Of fishes typical of more open tropical waters, Frade and Postel (1955) described spawning periods of several wide ranging tropical Scombridae from collections of larvae ranging up to about 100 miles off Dakar (Table 2.33). For several species spawning is prolonged since they do not release all their eggs at one time. Moreover, fish of different ages show some variation in breeding times. These breeding periods compare well with those cited by Wheeler (1969), particularly for warm Mediterranean waters.

Table 2.33. Spawning periods for certain tropical Scombridae (Frade and Postel, 1955)

Habitat	Species	Spawning Period
Warm seas	*Sarda sarda* (bonito)	February/March
High seas, tropical	*Orcynopsis unicolor* (plain bonito)	end of May
Warm open seas	*Katsuwonus pelamis* (oceanic bonito)	May–September
Warm seas	*Neothunnus albacora*	May–September
Open seas, tropical	*Auxis thazard* (frigate mackerel)	June
Tropical	*Euthynnus alliteratus* (false albacore)	June–September/October

Many of the fishes characteristic of open warm waters of the three oceans appear to have very extended spawning periods, often all through the year (e.g. *Nesiarchus nasutus* a member of a typical ocean family, the Gempylidae) so that larvae may be found in most months. The larvae of oceanic warm water fishes also often appear in neuston catches (cf. Chapter 1). In the sub-tropical Atlantic, Hempel and Weikert (1972) recorded numerous post-larval fishes in the neuston, with eggs and young larval stages slightly below the immediate surface. Hartmann (1970) found the larvae of *Scombresox saurus* to be the most abundant species in the sub-tropical north-east Atlantic. Parin (1967) also found for sub-tropical waters of the North Pacific Ocean that *Scombresox saurus*, with certain others of the same family, were among the most plentiful of the fish larvae. In the tropical Pacific and the Indian Oceans in the near surface waters, larvae of Exocoetidae, Hemirhamphidae, Coryphaenidae and Gempylidae were found, myctophid larvae were less abundant and there were a few larvae of Thunnidae and Xiphiidae.

In a subsequent study of the near-surface ichthyoplankton in the tropical western Pacific, Parin, Gorbunova and Chuvasov (1972) collected fish larvae and juveniles belonging to sixty species, apart from leptocephali. Of the twenty-six families represented, one broad grouping (Hemirhamphidae, Exocoetidae, Coryphaenidae, Scombridae, Xiphiidae and Istiophoridae) was drawn from open ocean fishes, dwelling always in the upper levels. A second category comprised some Myctophidae and their juveniles—oceanic fishes living more deeply, but migrating surfacewards at night. Some other families living deeply but not migrating at night had larvae in the hyponeuston (Gonostomidae, Ceratoidei and other Myctophidae). A further group, mainly neritic in distribution (Holocentridae, Mullidae, Didontidae, Balistidae) were represented by their larvae which spend part of their life cycle in the open ocean. The larvae of some typically neritic fishes (e.g. Engraulidae, Clupeidae, Carangidae, Serranidae, Labridae, Blennidae, Gobiidae) were also found in the oceanic neuston, being transported from coastal areas by currents. Apart from the myctophids, which tended to be dominant, especially at night, the ichthyoplankton varied greatly in abundance and species, especially in relation to nearness of oceanic islands.

In investigations in the Indian Ocean, Nellen (1973) points to the overall importance of the larvae especially of Myctophidae and Gonostomatidae, both typically oceanic families. In the Red Sea and Gulf of Aden region, apart from these oceanic forms there is a considerable mixture of larvae of inshore families (Carangidae, Gobiidae, Synodidae). From Cape Guardafui to Mombasa, with the narrow shelf, the oceanic species greatly outnumbered coastal forms. Gonostomids and myctophids were present with a few tuna larvae. In the more open Indian Ocean from East Africa to South India, Gonostomatidae and Myctophidae with Labridae and Bothidae were frequently taken. Some other distinctions can be drawn between the distribution of larvae of tunas over coastal and oceanic areas. Suda (1973) described skipjack larvae as occurring over wide oceanic areas, whereas little tuna and frigate mackerel were mainly coastal. Nellen found that in the Gulf of Oman and also in some areas between Bombay and Karachi, where myctophids were especially abundant, a single species, *Benthosema pterota* made up the bulk of larvae.

Apart from qualitative differences, different areas of the Indian Ocean showed marked variation in the abundance of fish larvae. For the Red Sea and Gulf of Aden, a mean density of 75 larvae/m^2 occurred, 28 larvae/m^2 were found off Cape Guardafui and Mombasa, 42 larvae/m^2 along the west coast of India and about the same concentration in the Persian Gulf, but an exceptionally high density (695 larvae/m^2) occurred in the productive Gulf of Oman. Ali Kahn (1976) also found a high biomass of zooplankton along the coast of West Pakistan in waters south of Karachi, as far as the edge of the shelf, an area which appears to be a good spawning ground for *Sardinella sindensis*. Apart from rich hauls of *Sardinella* larvae, other commonly captured species included the myctophid *Benthosema* spp. (cf. Nellen, 1973), the engraulid *Amentum commersoni*, the gonostomatid *Vinciguerria* and *Diaphus* spp.

Seasonal variations in the ichthyoplankton in the Indian Ocean have been described by Menon and George (1977). Off the coast of south-west India, ichthyoplankton occurred through the year, but with a lower density over the winter and a comparatively long period of marked abundance from about March to September.

A detailed analysis of the seasonal distribution of fish eggs and larvae off the south-east coast of India in the estuarine waters off Porto Novo is given by Venkataramanujam and Ramamoorthi (1977). Their results are related to earlier investigations off the east coast of India. Some species appeared to breed almost throughout the year (e.g. *Cynoglossus* sp., *Ambassis commersoni*, *Caranx* sp.), though *Caranx* shows an apparently greater intensity of breeding over some months. Other fishes (e.g. *Mugil*, *Sphyraena jello*) appear to breed bi-annually, while several species (*Pomadasys hasta*, *Odontamblyopus rubicundus* and many others) breed only annually, either during the summer or during the post-monsoon period. A few fishes (e.g. *Grammoplites scaber*) appear to breed discontinuously, but at random (cf. Fig. 2.134).

Undoubtedly, in waters off the coast of India, the monsoons strongly affect breeding patterns. Heavy rainfall, due to the monsoon, exerts a considerable influence on breeding. For example, mullet in years of normal rainfall apparently breed during the pre- and post-monsoon months, but in one year, when there was a very low monsoon rainfall, they bred almost continuously. With many annual breeders, it is also obvious

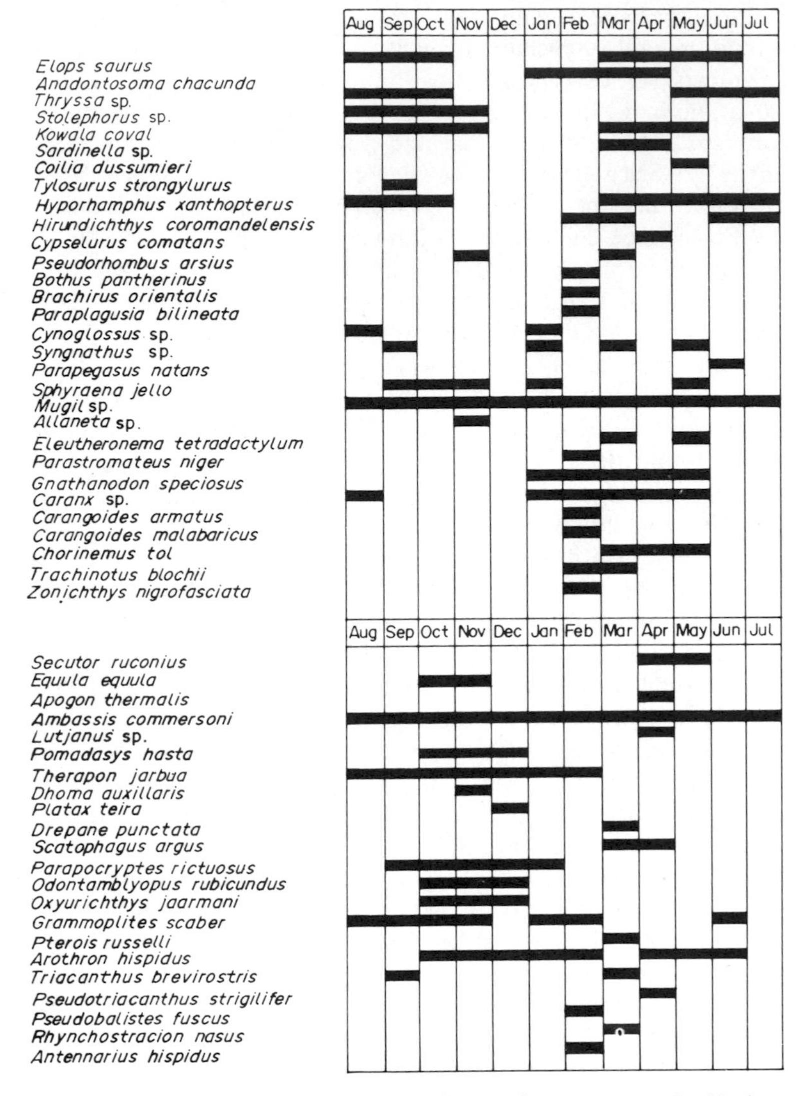

Fig. 2.134. Period of occurrence of larvae for some Indian south east coast species (Venkataramanujam and Ramamoorthi, 1977).

that breeding is either pre- or post-monsoon, but generally in normal years along the east coast of India, an intense breeding of fishes occurs immediately following the ending of the north-east monsoon. Many of the locally very abundant fish species off Porto Novo had very few larvae in the collections. Presumably such species breed farther offshore.

While monsoonal influences are marked in Indian waters, the breeding of fishes generally reflects the patterns typical of other warm oceans. Extended spawning periods with times of increased intensity, with more limited times for some species, are reminiscent of the different breeding patterns found for many invertebrate benthic taxa in tropical regions.

Chapter 3
Seasonal Changes and Breeding of the Holoplankton

Whatever the contribution of the meroplankton, even in shallow brackish warm areas where its density may be considerable, the zooplankton is dominated, with very few exceptions, by the holoplankton and especially by copepods. Seasonal breeding of the holoplankton must, therefore, play a very large part in the changes and abundance of the whole zooplankton. In examining seasonal patterns it is convenient to begin with coastal zooplankton, since quantitative surveys repeated at frequent intervals have been carried out in many inshore areas. However, surveys repeated sufficiently often to investigate seasonal cycles in oceanic areas are rare. The following review of seasonal changes will, therefore, deal mainly with neritic or near-shore surveys, but will include data, where available, from the open oceans of the same regions. The investigations will be very broadly divided into lower latitudes (essentially tropical and sub-tropical regions) and colder seas (temperate and high latitudes).

Lower Latitudes (Tropical and sub-Tropical Seas)

In the studies already cited by Herman and Beers (1969) carried out close to Bermuda, at the outer station copepods averaged 95 % of the total zooplankton and dominated the population over the year. The chief species at the outer station was *Paracalanus parvus*, averaging 52 % of the total copepods. Other reported species were *Acartia spinata* and *Calanopia americana* in approximately equal proportions. *Calanopia* showed main peaks of abundance in late summer and winter; *Acartia* in autumn and winter. *Paracalanus* was especially abundant in early summer and was most sparse in late autumn and winter. The total copepod population changes were rather variable over the two years, but generally stocks built up in spring and early summer and declined in autumn to a low density in winter. Even in winter, however, brief increases could occur. Young stages of copepods occurred almost throughout the year (according to the data of Herman and Beers) indicating that breeding cycles could be almost continuous, as might be expected in warmer waters.

In the more neritic station, where copepods averaged 77 % of the total zooplankton, *Paracalanus* contributed only 19 % of the copepods, though periods of abundance were similar to the outer station. *Calanopia* and *Acartia bermudensis* in approximately equal proportions together made up 50 % of the total copepods. However, *Calanopia* was most plentiful at this station in spring and late winter, almost reversing the patterns at the more open station. *Acartia* was most plentiful outside the summer season (Fig. 3.1a, b) The seasonal variation in total copepod population was not constant

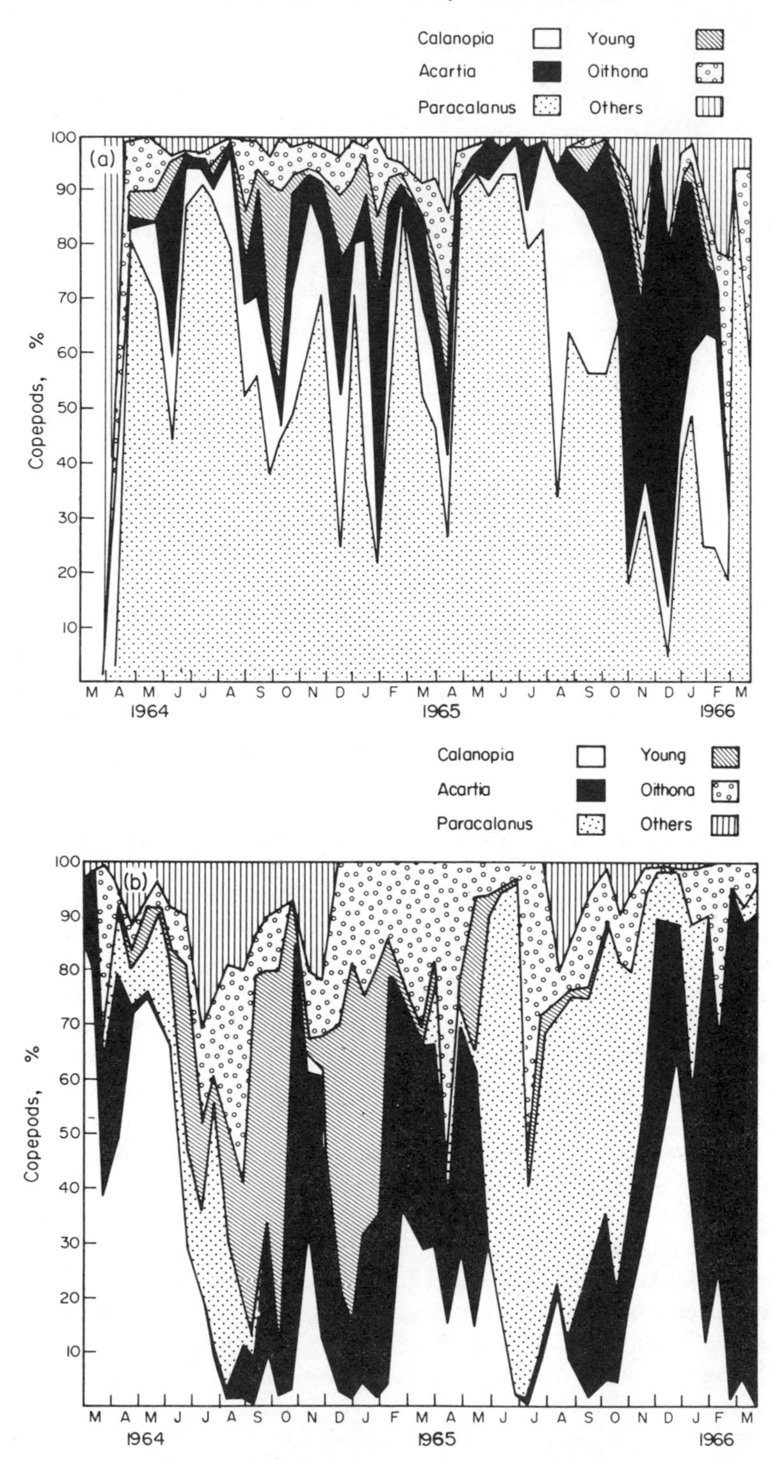

Fig. 3.1. Percentage distribution of copepods collected during the sampling period at (a) oceanic and (b) neritic stations. (Herman and Beers, 1969).

over the two years. For example, in the first year late summer densities were fairly high, but fell sharply in the second year. Herman and Beers, however, state that the cycle for *Acartia* at both stations resembled that described earlier by Woodmansee (1958) and by Reeve (1964) for very shallow waters off Miami.

Investigations by Woodmansee (1958) in Biscayne Bay showed that *Acartia tonsa* was the most important of the dominant copepods, being most abundant from about June to December, with the largest population in October. Both adult males and females were found throughout the year (unlike the seasonal pattern which holds for many major copepods at higher latitudes), suggesting that breeding was more or less continuous in Biscayne Bay. Woodmansee points out that Grice also considered *Acartia* as breeding continuously in Alligator Harbour, Florida. At the environmental temperatures in Florida, *Acartia tonsa* breeds fairly rapidly, about eleven generations occurring over a year; the period for one generation varied from about four to seven weeks, the duration decreasing with increase in temperature. In September the period was possibly as little as three weeks and a burst of reproduction produced the marked October peak density. The rather coarse mesh net employed in Woodmansee's investigations precluded an analysis of nauplius stages.

The second most important copepod was *Paracalanus parvus*, which occurred through the year, but most plentifully in winter and spring and thus to some extent alternating with the *Acartia* populations. *Labidocera* spp. (mostly *L. scottii*) copepodites were present through the year with adults in small numbers, except during July and August, suggesting again continuous breeding. Other copepods included the harpacticoid *Meta jousseaumei* and *Oithonina nana* in all months, though the latter small cyclopoid was not adequately sampled. *Calanopia americana* and *Temora turbinata* were occasionally fairly abundant. Woodmansee comments on the lack of a winter minimum and absence of a sharp spring rise in zooplankton density, typical of temperate regions. Presumably the higher environmental temperature of Florida permits year-round breeding, though phytoplankton is generally sparse in such warm seas, often even inshore. Although run-off from the land might promote primary production (cf. Volume 1), and Woodmansee believes that the zooplankton maximum in October, especially of *Acartia tonsa*, may have been related to an extensive run-off in the previous month, there was no correlation between run-off and zooplankton peaks in other months.

The reproductive potential of holoplanktonic forms such as copepods with a generation time of three or four weeks in tropical and sub-tropical waters is striking. Wickstead (1968) claims that some small tropical copepods can pass through a life cycle "in a matter of days". Amongst other taxa which can breed remarkably rapidly at high environmental temperatures with an increase in available phytoplankton food are salps and some appendicularians (cf. Volume 1). Although, as Wickstead remarks, breeding in tropical zooplankton does show some increase at one period of the year the number of generations possible and the breeding of some species in almost any month enables them to utilize excess food rapidly, thus avoiding the massive fluctuations which are so much more typical of high latitudes.

Reeve (1964) also studied the zooplankton of Biscayne Bay through the year. Peak populations were found for January, April, May, July and October with the latter month showing the maximum density. Though most plankton groups were represented in the collections, copepods were dominant, forming 65–85% of the total population, and their peaks followed the same density maxima with the greatest breeding peaks in July and October. Different species of copepods, however, reach their maximal densities in different months over the year, a feature which seems characteristic of many warm seas. Thus *Acartia*, represented mainly by *A. spinata* and *A. bermudensis*, and *Paracalanus parvus*, another important calanoid, had somewhat similar seasonal patterns with minimum densities at the end of July and August, but, Reeve points out that their peaks of abundance tend to alternate (e.g. *Acartia* early September—*Paracalanus* late November; *Acartia* January—*Paracalanus* February). Of other copepods, *Temora turbinata* was moderately abundant with several maxima, but its main peak in July/August was the minimum period for the dominant *Acartia* and *Paracalanus*. Cyclopoid and harpacticoid copepods, together with the calanoid *Calanopia*, showed sporadic bursts throughout the year (cf. Fig. 3.2). The various peaks of copepod abundance might indicate a generation period for different species of from four to seven weeks.

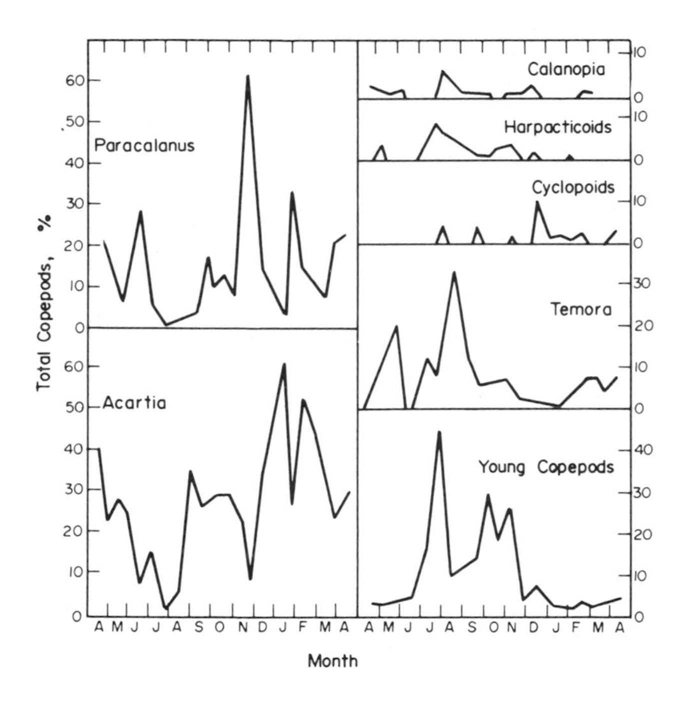

Fig. 3.2. Seasonal variation in the proportion of various elements of the copepod population (Reeve, 1964).

The other main zooplankton species found by Reeve was a neritic chaetognath, *Sagitta hispida*, which comprised 1–10% of the total population and was found in all stages of development throughout the year. Younger animals were more common from May to October, however, and older animals, including those showing mature ovaries, were much more plentiful from November to February (cf. Fig. 3.3). Breeding, as for copepods would, therefore, appear to be possible at any time of the year: Reeve quotes the views of other investigators that there is no well defined breeding period for the species in Florida or that repeated spawning times occur over the year.

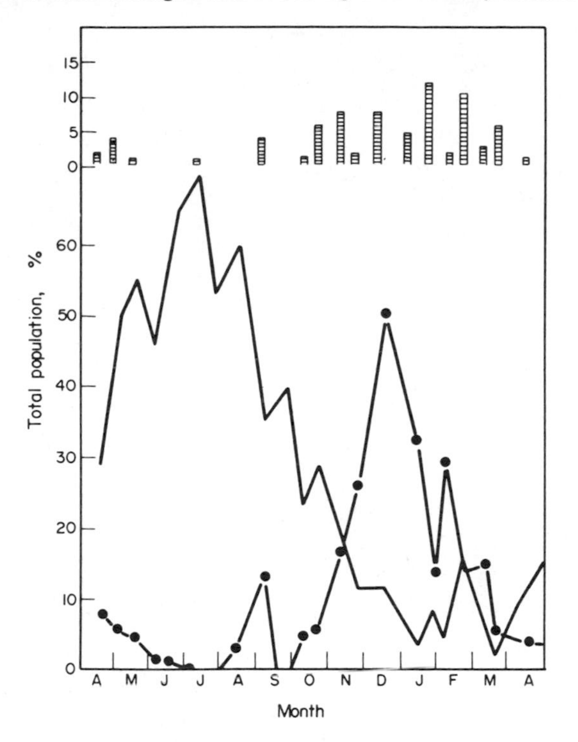

Fig. 3.3. Seasonal variation of *Sagitta* over 6 mm (closed circle line), and under 3 mm. Histogram represents percentage of animals with any development of ovaries (Reeve, 1964).

In a further study Reeve (1970b) confirmed the great significance of the copepods in Biscayne Bay. Among the remainder of the holoplankton only the carnivorous *Sagitta hispida* and ctenophores appeared to be of any importance. *Acartia tonsa* was the chief copepod. The large fluctuations from month to month again indicated repeated broods, but numbers were lower from June to September. Reeve cites Heinle (1966) who states that the generation time might be as little as seven days and that rapid changes, for instance in the available food, could trigger breeding. *Paracalanus parvus*, next in importance, was not abundant from January to May; *Temora turbinata* showed a peak in August. *Labidocera scottii* contributed to the biomass during winter inshore. *Oithona* (*Oithonina*) *nana* was not adequately sampled, but counts made from a very fine mesh net (35 μm) demonstrated a massive population, especially inshore. In numbers it comprised 80 % of the inshore copepod population. In biomass (dry wt) it formed 27 % and 17 % of the total weight inshore and in the mid-Bay, respectively (cf. Fig. 3.4). There were numerous peaks of abundance, the greatest being in October/November and most samples contained egg-sac-bearing females, indicating, as for other species, year-round reproduction. Counts of nauplii (cf. Fig. 3.5) from the fine net samples also showed many peak populations, especially in October, presumably of *Oithona*, and of *Acartia* in March and June/July, with lower numbers from December to February.

The importance of the breeding of a few copepods, such as species of *Acartia* and *Oithona*, is also clear from the investigations of Hopkins (1977) on the zooplankton

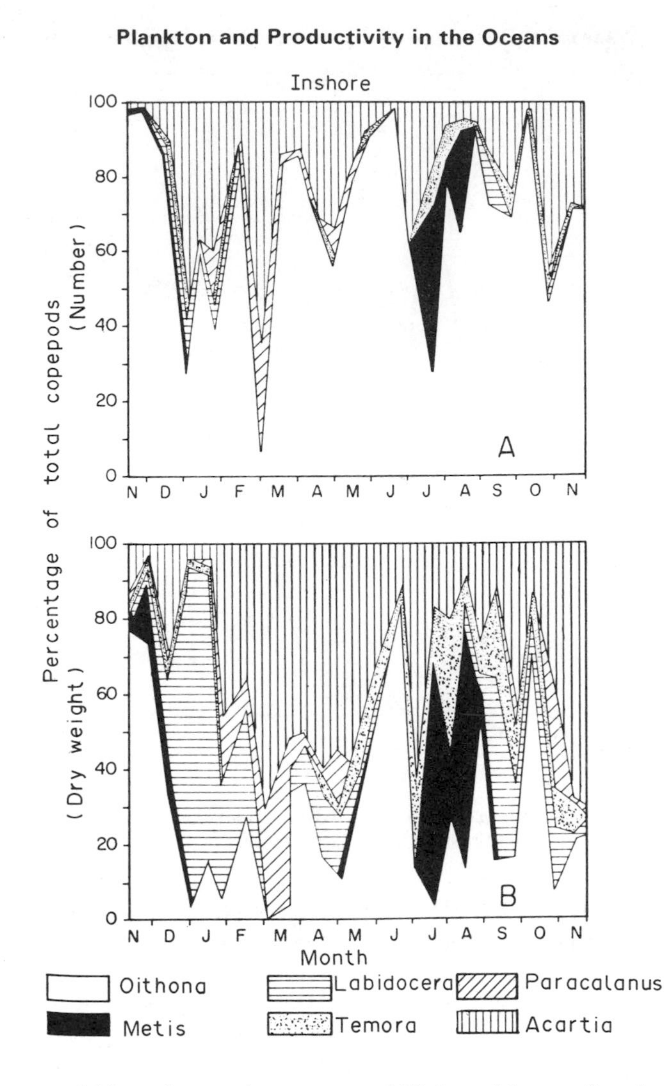

Fig. 3.4. Percentages of (a) numbers, and percentages of (b) dry weight contributed to the total by the constituent species of inshore copepods (Reeve, 1970).

of Tampa Bay, Florida. Although meroplankton was plentiful at times the holoplankton was far more important, with four species—three copepods (*Oithona brevicornis = colcarva, Acartia tonsa, Paracalanus crassirostris*) and the appendicularian *Oikopleura dioica*—forming 60 % of the average total biomass of zooplankton. *Oithona brevicornis* was by far the most common animal, though precise numbers could be misleading, since copepod nauplii, forming 29 % of the total plankton numbers, were not assigned to separate species. The numbers of all these species and of copepod nauplii were far lower in the winter season. However, the presence of fair numbers of nauplii during winter and the comparatively large populations over the other three seasons indicated fairly continuous reproduction, though of varying intensity.

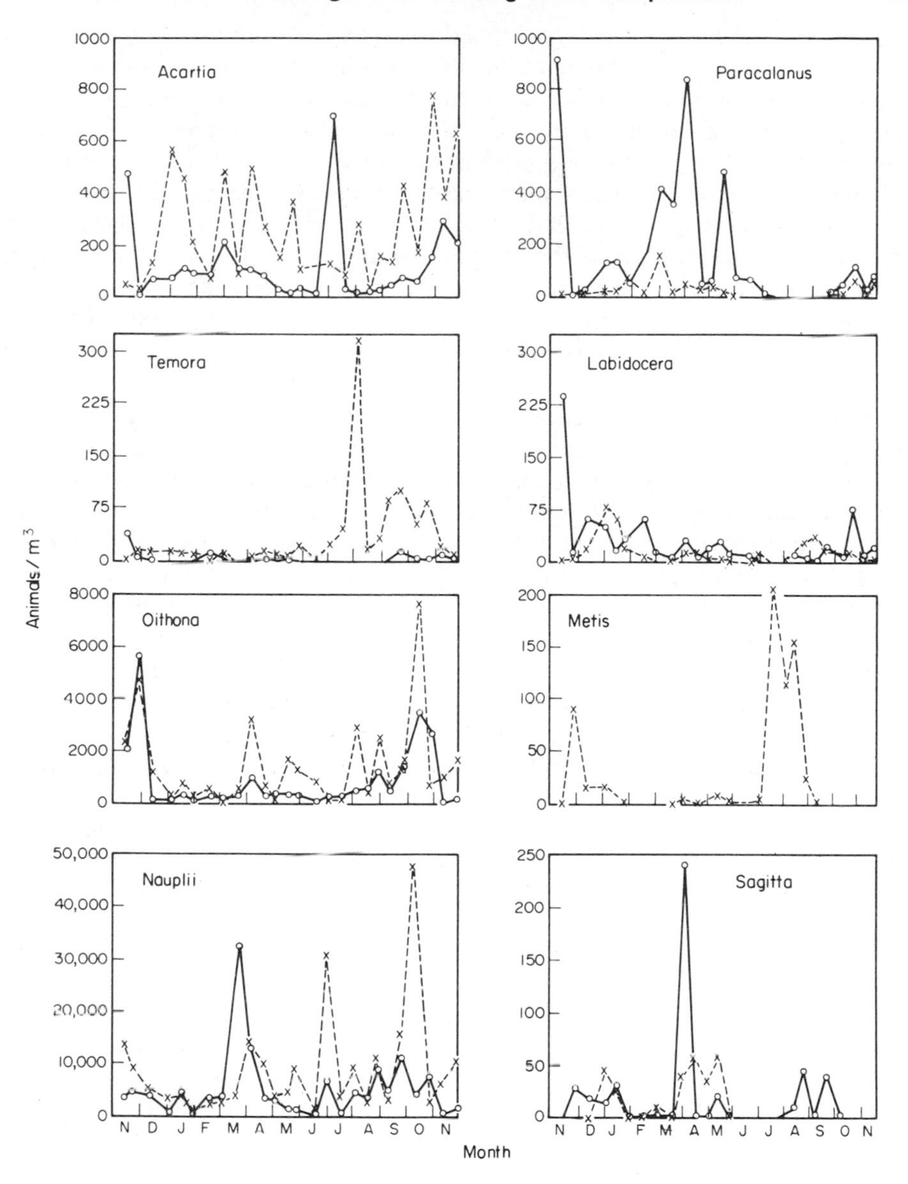

Fig. 3.5. Seasonal variation of holoplankton in numbers/m³ inshore (crosses and broken lines) and in midbay (circles and unbroken lines) (Reeve, 1970).

The cladoceran *Evadne tergestina* and four other copepods (*Oithona nana*, *Oithona simplex*, *Euterpina acutifrons* and *Labidocera aestiva*) together contributed on average 5 % to the total numbers of zooplankton. All had lower populations in winter, but each exhibited some period with a large population (exceeding 1000/m³). Most of the remaining 22 species, which occurred in comparatively low numbers, did not penetrate far into the bay. Rotifers were occasionally very abundant (63,000/m³) but were probably underestimated by the nets used (74 μm mesh). While only some of the fauna appear to be indigenous, the rich phytoplankton crop and production in these inshore waters presumably allows rapid reproduction of much of the fauna at different times. It appears that food supply is not the main factor regulating the zooplankton population. However, the total zooplankton showed a clear seasonal pattern: while over the three warm seasons average density and biomass were fairly similar (93,100–108,600/m^3; 32–54 mg/m^3), the density was seven to nine times greater than during winter.

In about the same latitude but in more open ocean water, Bsharah (1957) investigated the zooplankton of the Florida Current. A small seasonal fluctuation in density was observed with the larger quantities, in March to June, averaging two to three times that for the remainder of the year. Counts of the copepod population confirmed the presence of a spring peak with a density two to three times that of other months.

One of the few broad investigations of seasonal changes in zooplankton in warm waters of the North Atlantic is that of Moore (1949) for oceanic waters in the Bermuda region. Over the region as a whole there were areas of somewhat richer zooplankton. Thus, local variations in winds and currents carrying richer or poorer zooplankton past a sampling point could produce apparent seasonal changes at an individual station with typically denser populations in spring and autumn (cf. Fig. 3.6). However, no appreciable change in the average quantity of zooplankton for the upper 300 m over the whole region was evident. Mean values, reckoned as volume/plankton/haul, for December, May and July cruises were 13.2, 14.6 and 13.9 ml respectively. At the same time some species showed fairly defined breeding periods, though these occurred at different times of the year. For example, of the copepods, *Candacia aethiopica* appeared to have an autumn/winter maximum and *Acartia negligens* a peak about August, September or October with a smaller maximum at the end of June. *Occulosetella gracilis* had an autumn maximum, *Pleuromamma gracilis* and *Pleuromamma piseki* a winter or spring maximum, and *Calanus minor* a spring maximum.

For many of the species, however, including some of the commoner forms, there was

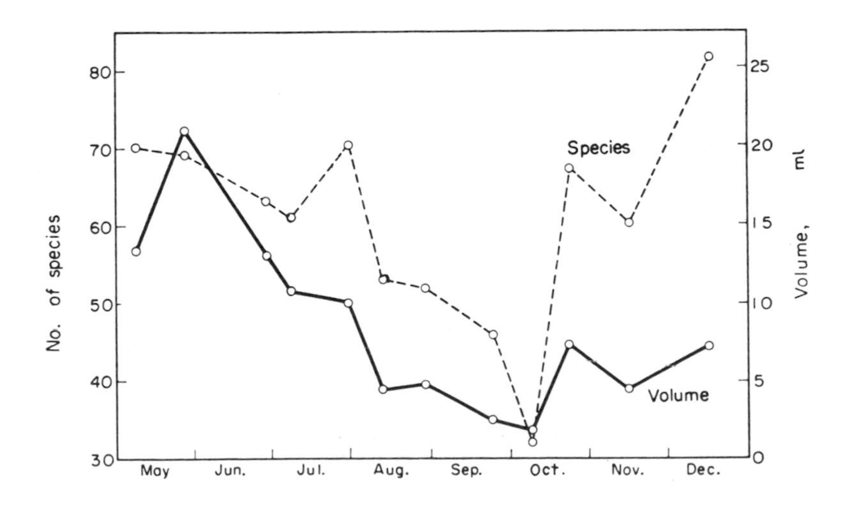

Fig. 3.6. Changes in the total volume of zooplankton and in number of species in waters at one station off Bermuda (Moore, 1949).

no apparent seasonal peak (e.g. *Mecynocera clausi*, *Clausocalanus* spp., *Euchaeta media*, *Oithona setigera*). Several species showing a period of maximal abundance also had other minor peaks. Taxa other than Copepoda showed similar variation in the times of abundance, for example of the euphausiids, *Thysanoessa gregaria* was probably a spring form. Three species of *Stylocheiron* (*S. carinatum*, *S. suhmi* and *S. longicorne*) had maxima in winter, autumn and summer respectively; several other euphausiids showed no clear period of maximal abundance. Of the Medusae, *Bougainvillea niobe* was maximal in winter, *Rhopalonema velatum* in late summer and *Liriope tertraphyllum* in spring. The pteropod *Limacina inflata* showed a more marked winter/spring maximum. *Creseis acicula* was more abundant in winter or in spring and autumn. Siphonophores showed the same variation in periods of abundance and presumably, therefore, breeding times. For example, *Lensia subtilis* and *Lensia fowleri* showed maxima in summer, *Diphyes bojani* in winter and *Bassia bassinensis* in winter and spring. Again, however, some siphonophores such as *Hippopodius hippopus* and *Chelophyes appendiculata* appeared to have no clear seasonal maximum. To understand the varying periods of abundance of holoplankton in warm oceans, data on the growth and life cycles of individual species are essential but as Moore (1949) has pointed out, such data are few and almost entirely restricted to the northern seas. With the high diversity of species in warmer waters, the elucidation of life cycles is especially difficult. Moore's own examination of cyclic changes in the proportions of the two generations in the life histories of some salps and siphonophores, however, led him to propose for Bermuda waters the following cycles per year: *Bassia bassinensis*—5; *Eudoxoides spiralis*—3; *Eudoxoides mitra*—5; *Thalia democratica*—3. Presumably similar cycles would be apparent for other species were sufficient material available for analysis.

Menzel and Ryther (1961) have also commented that zooplankton studies in warm oceans such as the Sargasso Sea have largely been concerned with providing data on the abundance of zooplankton at scattered times and places, but were not adequate for describing seasonal cycles. Their own investigations in ocean waters south-east of Bermuda were concerned with the biomass of zooplankton over the year and showed that there was a seasonal cycle in so far as a maximum occurred during the spring (cf. Fig. 3.7). Though this maximum seemed to occur in each of three years, the very high peak achieved in 1958 appears to be exceptional and probably not a true indication of zooplankton abundance. In any event, except for the spring period the standing crop of zooplankton is generally low and more or less constant, probably reflecting a series of breeding cycles for different animals spread over the year.

Deevey and Brooks (1971), studying changes in the Sargasso Sea (31° N), found that the total zooplankton showed a seasonal fluctuation with peaks in October and April for the upper 500 m, though for the biomass (as displacement volume) only the April maximum was apparent. Peaks at lower levels occurred in different months, for example at 500–1000 m in July, March and May, with the minimum in October, while at 1000–1500 m those volumes were practically reversed. The seasonal amplitude was comparatively small (*ca.* three-fold or four-fold) except at the deepest level (1500–2000 m), where the extremely small minimum ($0.4/m^3$) exaggerated the amplitude. Copepods formed 70% of the catch in the upper 500 m and were more dominant at deeper levels: ostracods were next in abundance. Other taxa made comparatively minor contributions to deeper fauna. A detailed study of the copepods (Deevey and

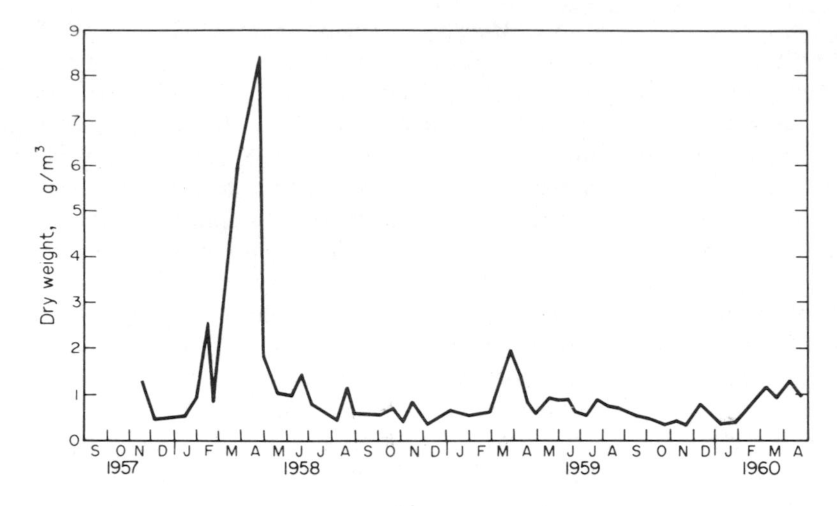

Fig. 3.7. Seasonal cycle in zooplankton biomass (dry weight) south-east of Bermuda (Menzel and Ryther, 1961).

Brooks, 1977) showed a clear seasonal cycle in the upper 0–500 m. A maximum occurred about March, following a brief phytoplankton bloom in February, but further maxima were found over summer or early autumn. Minimal populations were observed in late autumn and in May. The seasonal range in density was more than four-fold. Although total zooplankton decreased very sharply with depth (as other observers had found), the copepods became even more dominant, reaching >90 % of population.

Diversity, which was high, as is usual in warm waters, increased with depth; for calanoids the highest diversity occurred in the 500–1000 m stratum. Most species of cyclopoids, however, were found in the upper layers. Seasonal fluctuations, also studied in the deeper strata, showed more variability, but in general there was a greater abundance of copepods from spring to autumn and minimal densities in late autumn and for early winter (Fig. 3.8). The amplitudes in the different layers were: 0–500 m, 68–309 copepods/m^3; 500–1000 m, 6–37 copepods/m^3; 1000–1500 m, 2.3–17.4 copepods/m^3; and 1500–2000 m, 0.2–8.4 copepods/m^3 (Fig. 3.9). Deevey and Brooks suggest that although changes through the year in the deeper layers must be influenced by the seasonal changes in the upper stratum, variations in deeper water flows such as cold core eddies in the Gulf Stream may be a major influence and presumably introduce some irregularity into the yearly patterns.

While in the upper 500 m the number of species of copepods increased over summer, the changes were irregular in deeper layers. Characteristic depth patterns for some genera, illustrated in Fig. 3.9, suggest that the genus *Calanus* is represented by a near-surface concentration of *C. tenuicornis* and a deeper maximum for *C. finmarchicus* and *C. helgolandicus*, and that for the genus *Pleuromamma*, *P. piseki* and *P. gracilis* are nearer the surface than *P. abdominalis* and *P. xiphias*. There may be seasonal changes in depth for some copepods. *Eucalanus* spp., for instance, is generally deeper in summer and autumn, but rises towards the surface in winter/spring.

Lewis, Brundritt and Fish (1962) detected a greater volume of zooplankton near

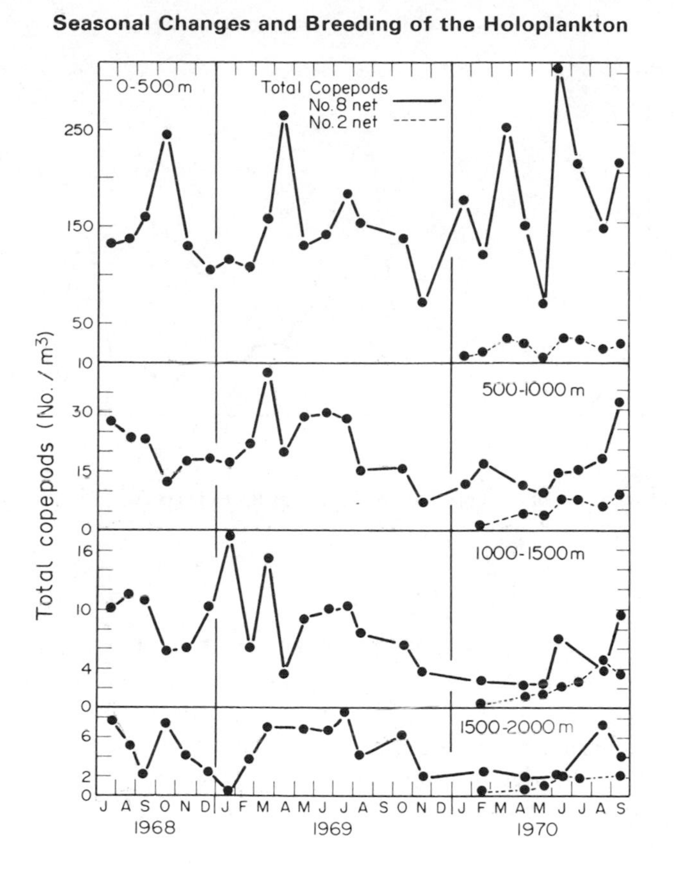

Fig. 3.8. Variations in total numbers/m³ of copepods over the four depth zones in the Sargasso Sea (Deevey and Brooks, 1977).

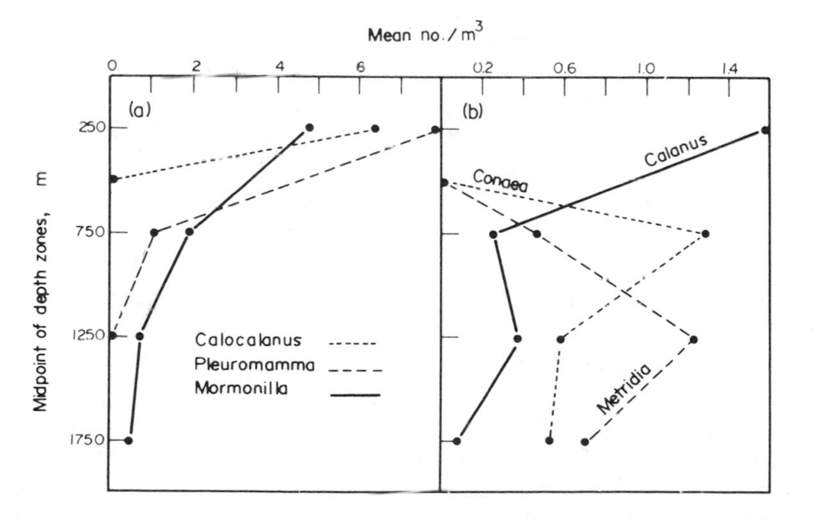

Fig. 3.9. Patterns of depth distribution of copepod genera, plotted as mean numbers/m³ at the midpoints of the four depth zones (Deevey and Brooks, 1977).

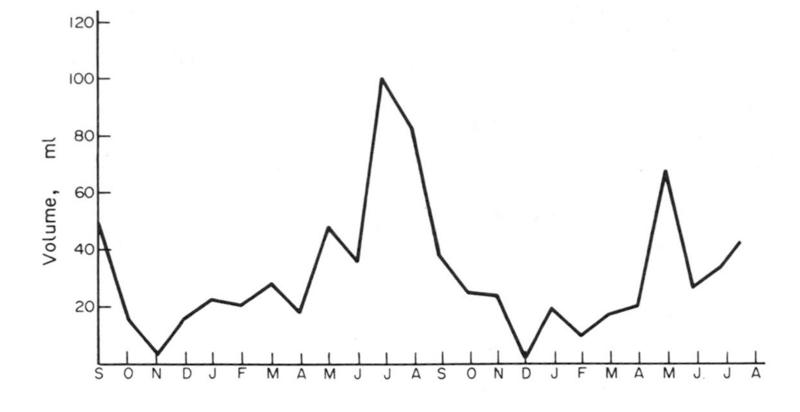

Fig. 3.10. Seasonal variation of total plankton by volume at Barbados (Lewis, Brundritt and Fish, 1962).

Barbados during the summer months (cf. Fig. 3.10), but Moore and Sander (1977) report that subsequent work by Lewis and Fish indicated that although some species seemed to follow a seasonal pattern, the variations appeared to be more associated with local hydrological conditions than with any regular seasonal change. The data of Lewis *et al.* (1962) also demonstrated that the pattern of fluctuations in copepods although the major group of zooplankton differed from the general pattern. In one year largest numbers were found in January–March and September/October, with a smaller peak in June (cf. Fig. 3.11). In the subsequent year marked fluctuations occurred with high numbers mostly from January to May with peaks in January, March and May. Series of peaks due to several different taxa also suggest bursts of reproductive activity over much of the year (Fig. 3.11).

Studies by Fish (1962), on the copepod population through the year at a station just west of Barbados are relevant. Seasonal environmental variations in this warm western Atlantic area are extremely small. The annual variation in total copepods showed a higher population from November to March with a peak in January and a second smaller rise in density from May to July, the peak being in June (cf. Fig. 3.12). Some of the species did not follow this seasonal pattern. Examples of breeding periods for particular species are as follows: *Neocalanus gracillis* occurred all through the year, with peaks during spring, early summer and winter (Fig. 3.13a). Female *Undinula vulgaris* were the most abundant copepods all through the year, but with peaks in spring, July and January/February (Fig. 3.13b). *Euchaeta marina* showed clear peaks in early spring and in winter (Fig. 3.13c) and *Candacia pachydactyla* in June and winter (Fig. 3.13d).

Fish believes that his data are rather similar to Moore's earlier results for those copepods which occurred in sufficient numbers to examine seasonal changes. He suggests that Moore's data show greatest numbers from November to January, with a smaller increase about April/May, but with low numbers in late summer. Bearing in mind, however, the marked tendency in warm waters for repeated or more or less continuous reproduction, the reasons for these apparent changes over the year are not obvious. Presumably hydrological changes during the year could be partly responsible.

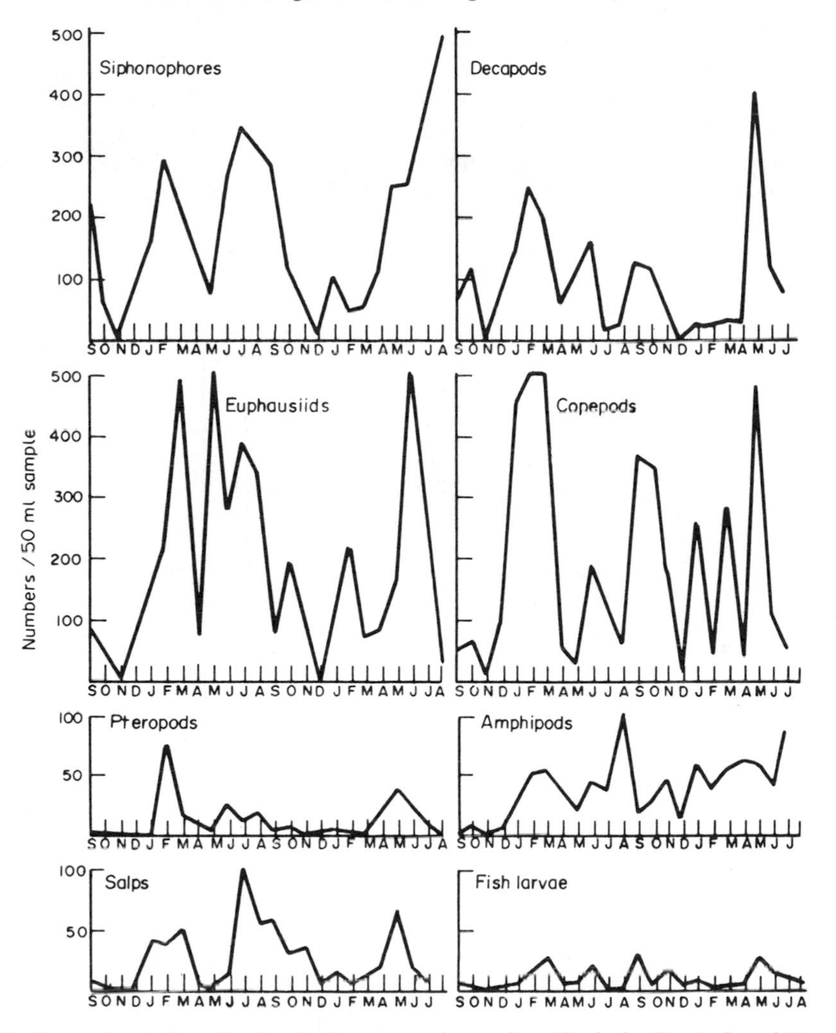

Fig. 3.11. Seasonal variation of major plankton groups by number at Barbados (Lewis, Brundritt and Fish, 1962).

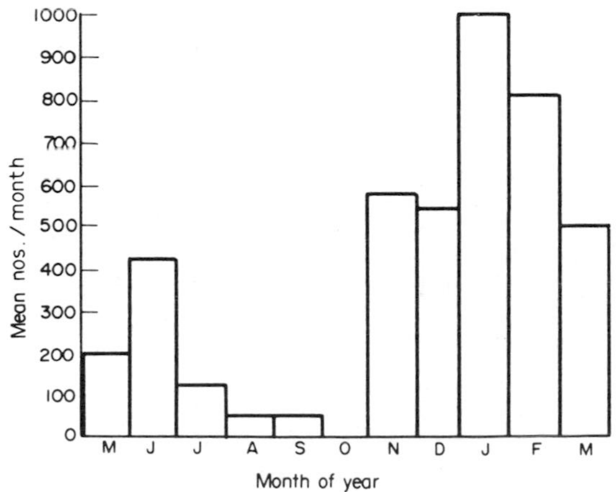

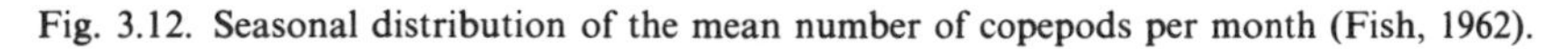

Fig. 3.12. Seasonal distribution of the mean number of copepods per month (Fish, 1962).

A number of investigations on zooplankton already noted for the west coast of Africa give valuable information on the changes in abundance of the holoplankton of inshore waters, though knowledge of the detailed breeding cycles for tropical species appears to be very limited. The fluctuation in seasonal abundance over large areas of the African west coast is made more complex owing to hydrological changes, especially to variations in the intensity and in timing of upwelling in different parts of the

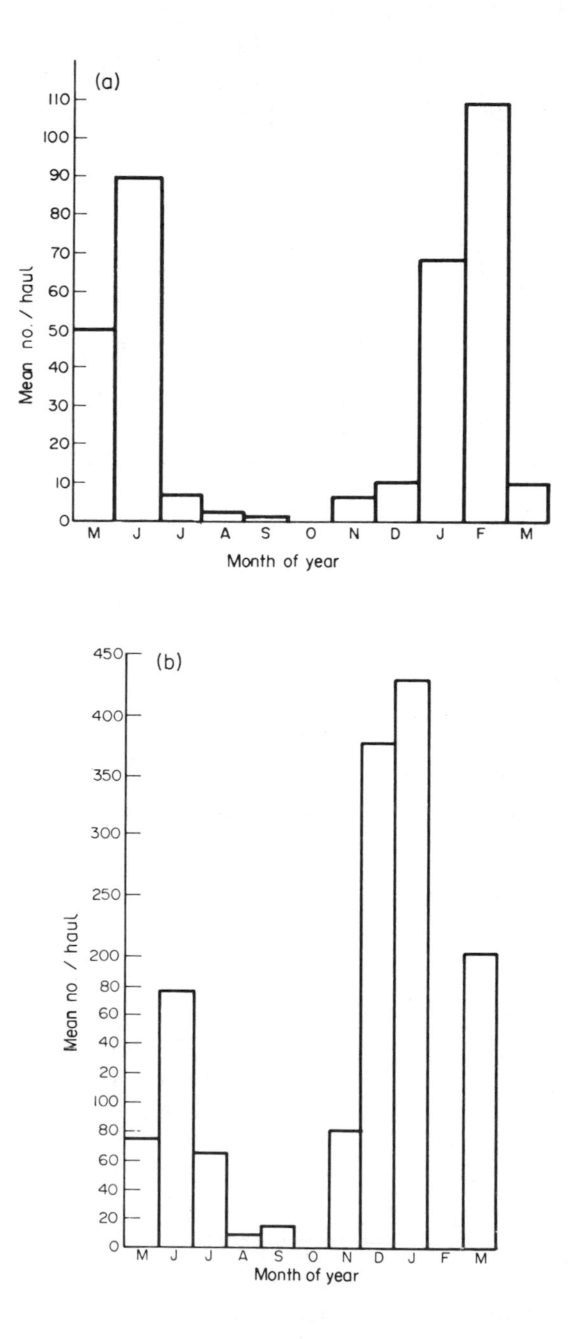

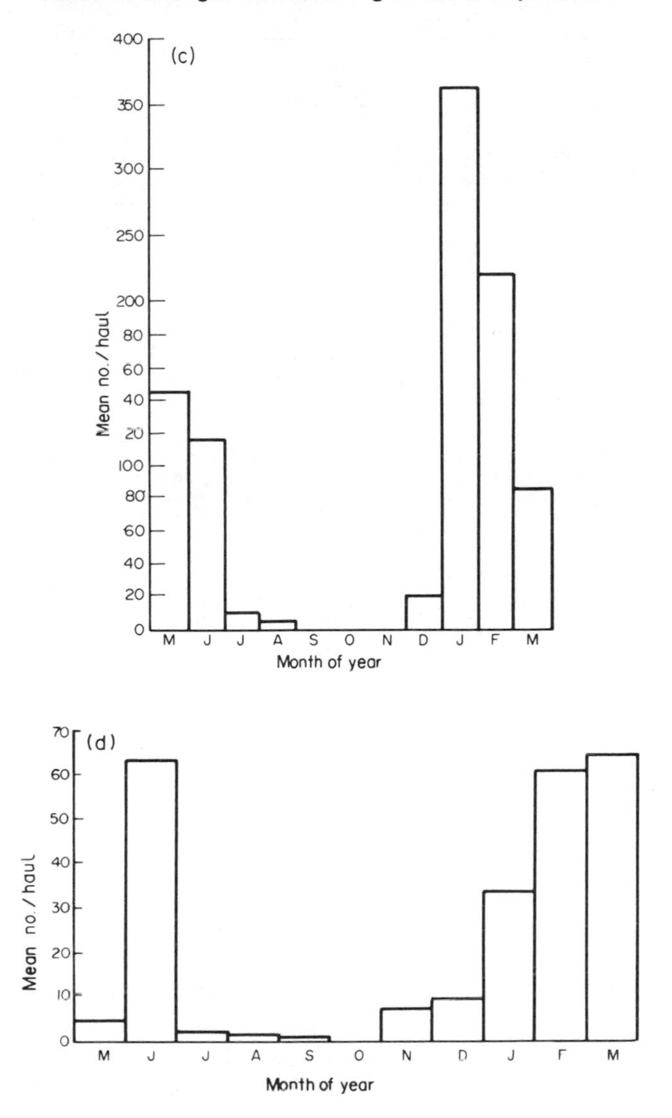

Fig. 3.13. Seasonal distribution of some copepod species (a) *Neocalanus gracilis* ♀; b) *Undinula vulgaris* ♀; c) *Euchaeta marina* ♀; d) *Candacia pachydactyla* (Fish, 1962).

continent (cf. Volume 1). The standing crop of zooplankton was shown by Mahnken (1969) to vary off Senegal and Guinea with seasonal hydrological changes. Rich crops were typical of February/March and were greater than, for example, during August. Changes in the density of zooplankton off Gabon and Congo were also observed and were believed to be correlated with seasonal changes in the temperature front in that area, with consequent effects on nutrient enrichment and plankton production. In more open ocean Atlantic waters, Mahnken found greater zooplankton populations between about 2°N and 4°S latitude during July and August, and more uniform concentration over February and March (cf. Fig. 3.14).

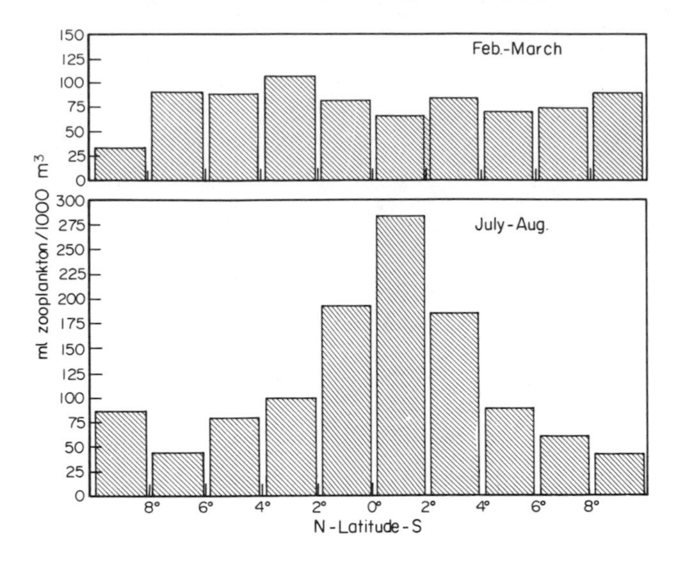

Fig. 3.14. Comparison of mean displacement volumes of zooplankton at the equator in ocean waters off West Africa in February–March, and July–August (Mahnken, 1969).

The area of Saint Helena Bay (about 30°S) although outside the tropics, has received considerable attention. Thiriot (1977) reviewed the results of several authors which showed that the whole plankton increases during the period of main upwelling of cold water (September to March). The zooplankton exceeded the phytoplankton in abundance during the southern autumn, due especially to the major copepods (*Calanus*, *Centropages*, *Paracalanus* and *Metridia*). Presumably their active reproduction follows the phytoplankton outburst. The eggs and larvae of euphausiids were found mostly from July to December, but the main period for adults and juveniles was from November to March with the overwhelming dominance of two species, *Euphausia lucens* and the neritic *Nyctiphanes capensis*. Whether these are annual breeders is not clear, though Boden (1955) gives detailed descriptions of their life histories. Salps and doliolids could also be plentiful in the Saint Helena Bay area, *Thalia democratica* in the warmer season, *Doliolum nationalis* in the colder period. Other holoplankton included chaetognaths, especially the neritic *Sagitta friderici*, *S. tasmanica* in somewhat colder waters and the hyperiids *Hyperia galba* and especially *Parathemisto gaudichaudi*. The latter species was characteristically very abundant from December to March in coastal waters. It may have a breeding cycle off South Africa similar to that described by Kane (1966) for animals from the Southern Ocean.

In the region of Walvis Bay, the southern limit of the tropical zone off West Africa, upwelling is more or less continuous through the year. Kollmer (1963) found that zones of different abundance and populations of zooplankton ran roughly parallel to the coast in the bay. The shallowest zone was dominated by *Acartia* with cladocerans (mostly *Pondon polyphemoides*), *Noctiluca* and meroplanktonic larvae. The next zone, the richest, had mainly the copepods *Calanoides carinatus* as dominant, *Metridia lucens* and *Centropages brachiatus* with a mixture of other species including euphausiids, (especially *Nyctiphanes capensis*), hyperiids (especially *Parathemisto*) and sometimes large densities of radiolarians in the mixed waters farther seawards. Farther offshore the zooplankton included salps, doliolids, siphonophores, ctenophores and pteropods, with *Liriope* sometimes in considerable numbers. Chaetognaths were represented inshore by *Sagitta friderici.*

Unteruberbacher (1964), working in the same region, found that the copepod abundance (for three major species listed) occurred during the cooler period June to January, though there was apparently not any clear regular seasonal fluctuation. The strong upwelling, however, presumably provided the best phytoplankton growth for intense reproduction of the copepods. *Calanoides carinatus*, although abundant at all times, was especially plentiful when the phytoplankton was rich in spring and early summer. Three other copepods, *Paracalanus parvus*, *Paracalanus crassirostris* (especially in the neritic zone) and *Oithona similis*, occurred throughout the year with the species found most typically in upwelled water (Table 3.1). *Paracalanus crassirostris* and *Paracartia africana* are truly neritic copepods, while the cooler water population has comparatively few species dominated by *Calanoides carinatus*, a species characteristic of upwelled water off West Africa but otherwise nowhere very plentiful except in spring off east and south Australia.

Table 3.1. Comparison of copepod numbers (percentages of total) in two areas of the coast of Walvis Bay, south-west Africa (Unteruberbacher, 1964)

Copepod	Offshore Area (%)	Coastal Area (%)
Calanoides carinatus	18.6	15.0
Paracalanus parvus	17.3	14.1
Metridia lucens	14.4	14.9
Centropages brachiatus	13.3	17.9
Oithona similis	9.4	7.6
Paracalanus crassirostris	4.6	12.9
	77.6	82.4

In the north-west a warm-water copepod fauna with many species (e.g. *Acartia danae*, *Centropages bradyi*, *Centropages chierchiae*, *Mecynocera*, *Eucalanus* spp., *Nannocalanus minor*, *Neocalanus gracilis*) is characteristic. Deeper-living species have also been recognized. Stander and De Decker (1969) compare the zooplankton over 1963, a much warmer year, with other years in the same area. Warmer water apparently extended into the region from the north-west. The volume of plankton forming the bulk of the catch was much reduced. Although *Calanoides carinatus* maintained its dominance over other species, its abundance was greatly diminished. *Metridia lucens*, *Paracalanus parvus* and *Centropages brachiatus* also exhibited marked decreases, though the latter showed some recovery over part of the area. What might be regarded as a group of neritic species for the region (*Paracartia africana*, *Oithona nana*, *Paracalanus crassirostris* with cladocerans and others, (cf. Unteruberbacher, 1964)) also suffered a reduction. In contrast, *Nannocalanus minor* showed an obvious rise in numbers and was accompanied by some other warm water species as listed by Unteruberbacher. Thus, as in many regions which may experience hydrological changes, seasonal changes and reproductive patterns of the major zooplankton may be greatly affected by unusual current flows.

Near the coast of West Africa, meroplankton (polychaete, echinoderm and caridean larvae) can become significant at some period of the year. Although pteropods, hyperiids and siphonophores, being essentially oceanic, were mostly near the shelf edge, *Muggiaea atlantica* could spread over the shelf and become very common. Salps could also be present in large concentrations, mainly during summer and autumn. Venter (1969), examining the chaetognath population, confirmed the neritic distribution of *Sagitta friderici* and found that it occurred, not in the major upwelling time, but mostly in the summer and autumn. *S. tasmanica*, though a more warm eurythermal type, occurred in mixed coastal and oceanic waters, especially in summer. *S. enflata* and *S. hexaptera*, with some other species, are more offshore forms; two species were more abundant in summer.

De Decker (1973) describes the plankton farther south on the Agulhas Bank off Cape Town. The phytoplankton crop is much less rich than off the west coast, though subject to seasonal change. Despite the lower crop, the volumes of zooplankton are generally higher than off the west coast, with large maxima in late spring/summer. De Decker explains that the apparent anomaly is due to massive and dense swarms of thaliaceans, mainly *Thalia democratica* and *Doliolum denticulatum*. He describes the swarms as sometimes extending for thousands of square miles and suggests that they may be hemmed in between an inshore water mass of cold upwelled water and a seasonally shifting offshore body of water. Four or five appendicularians (*Oikopleura longicauda*, *O. fusiformis*, *O. dioica* and either *O. cophocerca* or *O. rufescens*) made up 80% of total appendicularians. Though seasonal peaks for most of the species were evident, they occurred at different months and were not always repeated in both years. Of euphausiids, only two species occurred through the year, the neritic *Nyctiphanes capensis*, and *Euphausia recurva*, common in the Indian Ocean. *E. lucens*, which is abundant on the west coast, was found in winter only. *Nyctiphanes* repeatedly formed swarms over the Bank. Several cladocerans were found. The dominant species (*Penilia avirostris*) at times could outnumber all other zooplankton, especially in summer. Its reproduction includes a number of parthenogenetic cycles and a single sexual generation over the year on the Bank (Angelino and Della Croce, 1975). Copepods did not show a high level of species diversity. The main species important for dispersal and abundance are *Nannocalanus minor*, *Calanoides carinatus* and *Centropages brachiatus*.

The changes in population (cf. Table 3.2) must be strongly influenced by the

Table 3.2. Peaks of abundance of some copepod species for two separate years (De Decker, 1973)

Month	1960–61	1963–64
April	*Centropages chierchiae*	*Centropages chierchiae*
July	*Centropages brachiatus*	*Centropages brachiatus*
October	*Nannocalanus minor* *Calanoides carinatus*	none
January	*Rhincalanus nasutus*	*Rhincalanus nasutus* *Calanoides carinatus* *Nannocalanus minor*

hydrological movements, but the abundant copepods (*Centropages*, *Calanoides*) were observed to have their youngest stages concentrated in small areas, especially near upwelling centres. From these "nurseries", both inshore and along the shelf edge, they were dispersed.

Bainbridge's (1972) study of the seasonal changes off Lagos contrast with the investigations off south-west Africa, since apart from the latitudinal difference, the hydrological conditions off the coast of Nigeria are stable over most of the year, with a warm uppermost layer over the continental shelf disturbed by upwelling in some years during August and September. Generally, smaller zooplankton populations were found over deep water compared with shelf and slope areas. Both the density and volume of the zooplankton, however, show large increases during August and September except at the outermost stations (cf. Fig. 3.15).

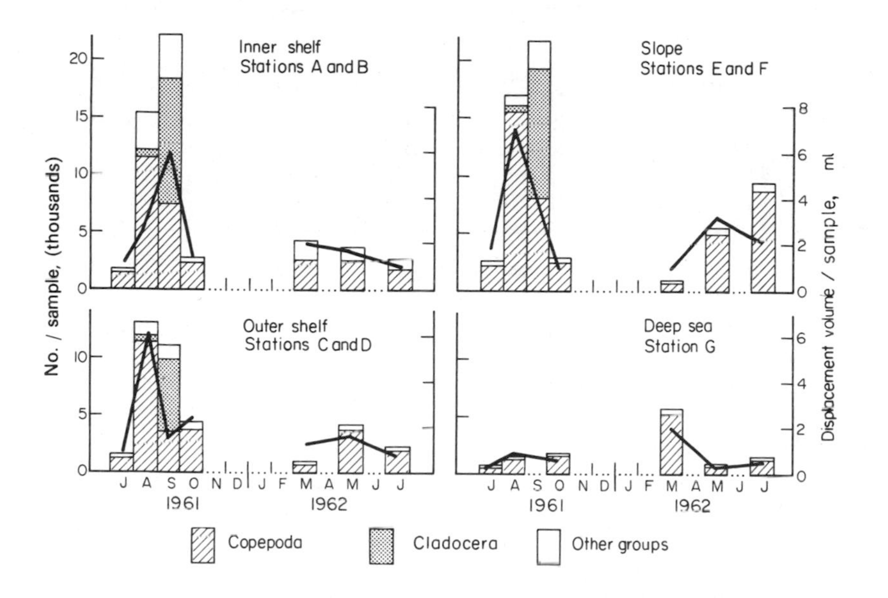

Fig. 3.15. Variations in the quantity and composition of the zooplankton along the Lagos transect. Displacement volumes are shown by line graph and the average numbers of all animals by histograms (Bainbridge, 1972).

Bainbridge illustrates the wide range of holoplanktonic copepods, including important species above and below the thermocline in shelf waters and species living more deeply in slope or oceanic waters. During the stable (major) period of the year *Clausocalanus furcatus* is usually commonest, but the zooplankton is relatively sparse, with high species diversity and a considerable proportion (20–40%) of carnivorous species (cf. Volume 1). In the upwelling period *Paracalanus parvus* is the most common copepod. The temperature change and the ability of the species to graze more effectively on the increased primary production may be factors enabling *Paracalanus* to compete more successfully. Other moderately common holoplankton during the upwelling period are *Oikopleura longicauda*, *Euconchoecia chierchiae*, *Lucifer faxoni* and *Sagitta enflata*. Cladocerans, however, chiefly *Penilia avirostris* and

Evadne tergestina, become important during upwelling (cf. Longhurst and Bainbridge, 1964), and briefly dominate the zooplankton (cf. Fig. 3.15).

Cladocera can be of great importance in inshore waters, but with their capacity for rapid parthenogenetic reproduction, they can spread from the coast and build up high density populations comparatively rapidly (cf. Bainbridge, 1972). Although cladocerans have resting eggs, their contribution to seasonal changes will be discussed when reviewing the holoplankton. Bainbridge found that during the upwelling the total zooplankton was more abundant, though with lower species diversity and with a decline in the proportion of carnivores. This applied chiefly to outer shelf stations; closer inshore, meroplanktonic larvae tended to obscure the effect (cf. Fig. 3.16). One copepod, *Calanoides carinatus* (and to a lesser extent *Eucalanus monachus*) is comparatively rare in Bainbridge's collections, occurring in deeper slope waters, but the species increased very markedly during the upwelling of water on the continental

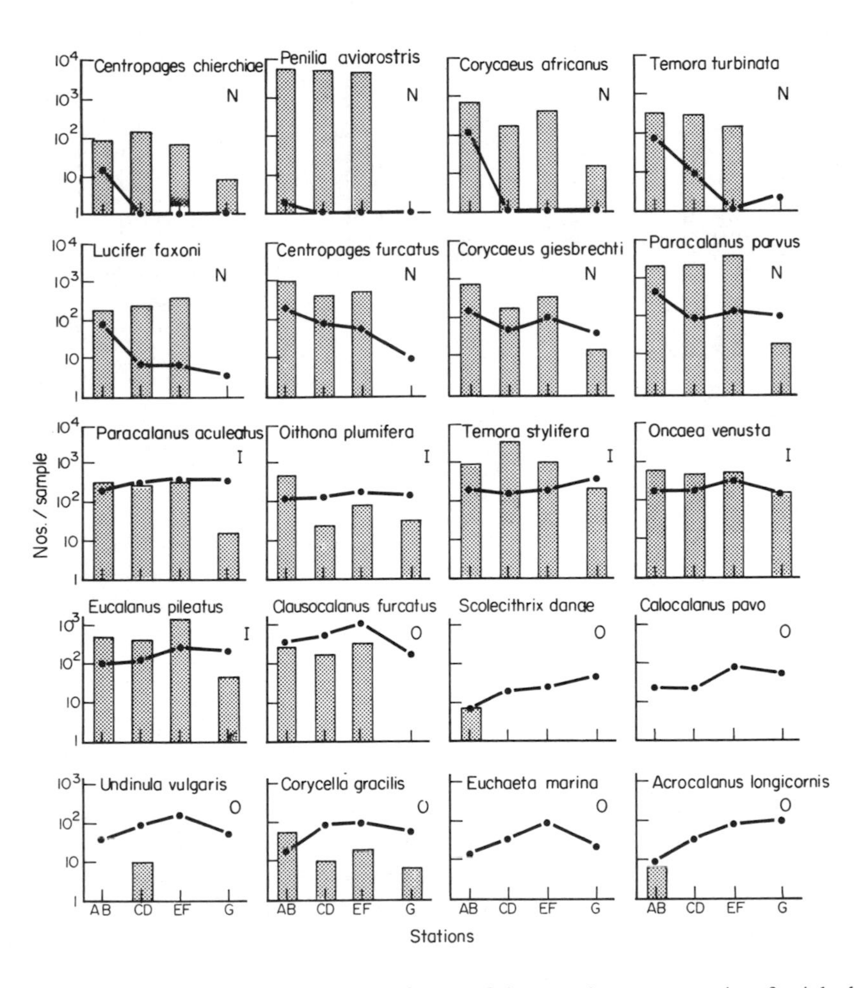

Fig. 3.16. Comparison of the average numbers of some of the more important species of epiplanktonic crustacea at stations off Lagos during months of thermal stability (line graphs) and during upwelling (histograms). Type of distribution during stable conditions indicated by (N) neritic; (I) intermediate; and (O) oceanic (Bainbridge, 1972).

shelf, as indeed in other regions off West Africa (*vide supra*). With more pronounced upwelling to the west of Nigeria, *Calanoides carinatus*, with *Temora turbinata*, comprised more than 90 % of the total copepods. In more offshore areas outside upwelling zones, the plankton is diverse, with a large proportion of gelatinous forms (salps, doliolids, medusae, siphonophores, etc) and lower in volume.

To the east of Nigeria, where there is more hydrological stability than off Lagos, the two most plentiful copepods were *Paracalanus* and *Clausocalanus* in a diverse zooplankton. Bainbridge comments on the change with upwelling from oligotrophic to eutrophic conditions and points to *Calanoides carinatus* as a species which seems to take special advantage of increased primary production. In tropical waters with upwelling, *C. carinatus* appears to fulfil a role similar to that of *Calanus finmarchicus* in the North Atlantic or *Calanoides acutus* in southern seas. The copepod may use the deeper oceanic or slope waters as a refuge, particularly for Stage V copepodites, and when carried into the continental shelf it appears as an "opportunistic" species, presumably passing through a fairly rapid reproductive cycle. Mensah (1966) refers also to maximum volumes and numbers of zooplankton over the coastal shelf waters off Ghana being found during the upwelling from about June to August, and to *Calanoides carinatus* being associated with the upwelling. There is comparatively little information concerning cycles of reproduction of zooplankton species in tropical environments. The contribution of Binet and Suisse de Sainte Claire (1975) on the breeding of copepods, particularly *Calanoides carinatus*, is therefore especially pertinent.

From samplings over five years at a shallow station on the continental shelf of Abidjan (Ivory Coast), Binet and Suisse de Sainte Claire studied the life cycle of *Calanoides carinatus*, a species which they recognize as being unusual among tropical planktonic copepods in its capacity to dominate the zooplankton. Although the copepod extends from 47°N to 37°S along the eastern Atlantic, it is comparatively rare and deeper-living, but can be very abundant in upwelled waters. Major development occurs only at temperatures below 23°C and it is limited along tropical regions to the cold upwelling season. They quote Mensah as finding three generations off Ghana and that C V stage of the last generation descended to deep water (1000 m) as a less active stock, remaining there for about nine months.

Binet and Suisse de Sainte Claire found that *Calanoides carinatus* appeared off Abidjan only in the latter half of June, with the upwelled water, in very small numbers (*ca.* 1/m^3). The population built up extremely rapidly with the cooling of the upwelled water reaching by August/September densities which could exceed 1000/m^3. In certain cases far higher densities were found in some years, and *Calanoides* formed 90 % of the copepods. A rapid decrease in October led to their practical disappearance towards the end of November (cf. Fig. 3.17). There is an inverse correlation between length of the female and temperature reckoned over the preceding fortnight, but increased body length is almost certainly also correlated with increased phytoplankton, which occurs with upwelling. During the cold season the vertical distribution of the six copepodite stages indicated that the younger copepodites were drifting eastwards in the surface Guinea Current; the later copepodites were rather deeper (> 15 m) and drifted westwards in the counter flow, maturation occurring in the deeper layers. Egg production near the upwelling source ensured that the youngest stages were in the upper 10 m. However, with the ending of the upwelling period, C. V are comparatively

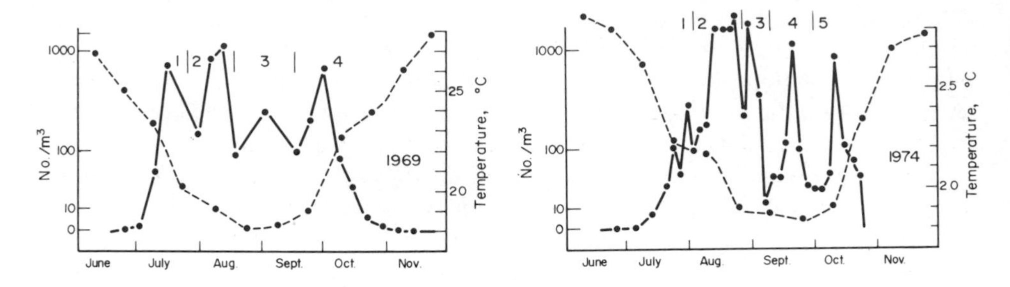

Fig. 3.17. Variations in numbers m^3 over two years, of *Calanoides carinatus* from sub-surface tows at a neritic station. The different generations are numbered and separated by a vertical line. The dotted line represents mean temperature at 10 m (Binet and Suisse de Sainte Claire, 1975).

abundant in deeper water along the continental slope. C. IV and some adults are present, but few other stages. The adults gradually disappear but the C. V descend and may be found at considerable depths (*ca.* 500 m). They appear to be a resistant stock, more or less equivalent to the "over-wintering" C. V *Calanus* of higher latitudes.

With the beginning of upwelling, the surviving C. V mature and adults appear in the surface layers. No reproduction occurs until the new upwelling season. Using the peaks of population and the succession of the copepodite stages as criteria, despite the tendency for the hydrological movements to mix populations, the authors were able to identify four to six generations during the upwelling period of reproduction (June to October). The males probably die after copulation. The succeeding generation is two to four weeks later.

Binet and Suisse de Sainte Claire (1975) contrast the reproductive cycle of *Calanoides* with that observed for other copepods on the Ivory Coast shelf. Most of the tropical neritic species reproduce all the year round, with about 12 generations. The smaller copepods, *Centropages* and *Temora*, have 13–14 and 10–11 generations respectively. Those of larger size have fewer generations according to Binet and Suisse de Sainte Claire. Thus *Eucalanus pileatus* has nine generations, *Undinula vulgaris* seven and *Euchaeta paraconcinna* six. These workers also liken the reproductive cycle of *Calanoides carinatus* more to that of *Calanus helgolandicus* (*pacificus*) in the Californian Current region. The species exploits the phytoplankton burst during the upwelling season. Thereafter it descends, mainly as C. V, into oxygen-poor deep waters (*ca.* 1000 m) as a "hibernating" stage (cf. Mensah).

Seguin (1970) studied the seasonal cycle of zooplankton off Abidjan, Ivory Coast, where a warm season from October to June alternates with a colder upwelling period from June to October. Although diversity is greatest in the warmer season, the maximum of zooplankton occurs during the colder upwelling period. Copepods dominated the population over the year, at times exceeding 90% of the total zooplankton. The contributions from other planktonic groups, while generally small, could show occasional outbursts, and though details of reproductive cycles could not be followed, it is probable that rapid generation cycles occurred at various periods. Thus chaetognaths became more important in September to November, very marked outbursts of thaliaceans, mainly *Thalia democratica*, occurred in December and July, and appendicularians were abundant in June and October. Cladocerans, though never as important, were moderately plentiful in September and January.

There were changes in the most abundant copepods. During the cold season, *Calanoides carinatus* was extremely plentiful, especially during August and September, with some contributions from *Eucalanus attenuatus* and *Euchirella rostrata*. During the warm but less saline period (October to December), *Paracalanus parvus* was very important. The warm but saline waters present from about December to March were marked by a variety of copepods. Some were present without any obvious peak densities (e.g. *Oncaea venusta*, *Copilia*, *Miracia*, *Candacia*). Others, while always represented, showed peak abundances (e.g. *Corycaeus* spp. in September and November; *Temora stylifera* in September, February, April, May). Other species were present only during a particular time (e.g. *Eucalanus elongatus* with a May maximum, *Mecynocera clausi* in February and May and *Euchaeta marina* with a maximum in February).

Since nauplii and copepodite stages were present in almost every month and there was little obvious seasonal pattern, and adults were found over different periods of the year, it would appear that reproduction could occur in any month. The increases in density of various species would then indicate the relative intensity of breeding. Although breeding cycles are not generally known for species, the studies of Gueredrat (1974) concerning copepods of the Ivory Coast include an analysis of the reproduction of *Euchaeta marina*, which appears to breed all the year. Gueredrat demonstrated, however, using criteria such as the proportions of males, of females bearing spermatophores and of egg-bearing females, that the intensity of reproduction varied during the year. For example, males were most abundant from August to November with a smaller peak in April/May. Females carrying spermatophores generally followed the same changes in abundance. Females bearing egg sacs were maximal in August/September, with a secondary small maximum from January to March. Stage V copepodites showed maxima in October/November and in March/April. The data suggest, as with many other tropical zooplankton species, that breeding can occur in almost any month, but that there are more intensive periods (*vide infra*). Figures 3.18 and 3.19 show clearly, however, not only the domination of the copepods as a whole, but the great significance of *Calanoides carinatus* during August and September in effecting a very great increase in the zooplankton population.

Binet (1972, 1974) and Le Borgne and Binet (1974) also draw attention to the importance of upwelling, especially to the contribution of certain deeper-living copepods, for the shallower continental shelf area of the Ivory Coast and Mauretania. The biomass of zooplankton follows the phytoplankton changes being deeper during the colder season and shallower in the warmer months. Young herbivorous stages of copepods tend to be nearer the surface. For successful survival, Binet suggests that zooplankton populations might remain in the near surface, whence they may well be transported to more oligotrophic waters, but they must reproduce more or less continuously, according to the food available. Likely examples are the copepods *Undinula vulgaris* and *Farranula gracilis*. Their reproduction would, therefore, be expected to occur at almost any period of the year (cf. Moore and Sander, 1976; *vide infra* for the Caribbean). Alternatively, species might change in depth, but would probably therefore need to interrupt their breeding cycles. *Calanoides carinatus* and *Eucalanus monachus* live normally fairly deeply (*ca.* 500 m) during the hot season, but migrate vertically nearer the surface over the cold productive months. During this period cohorts of *Calanoides* appear to follow in a succession of approximately three-week

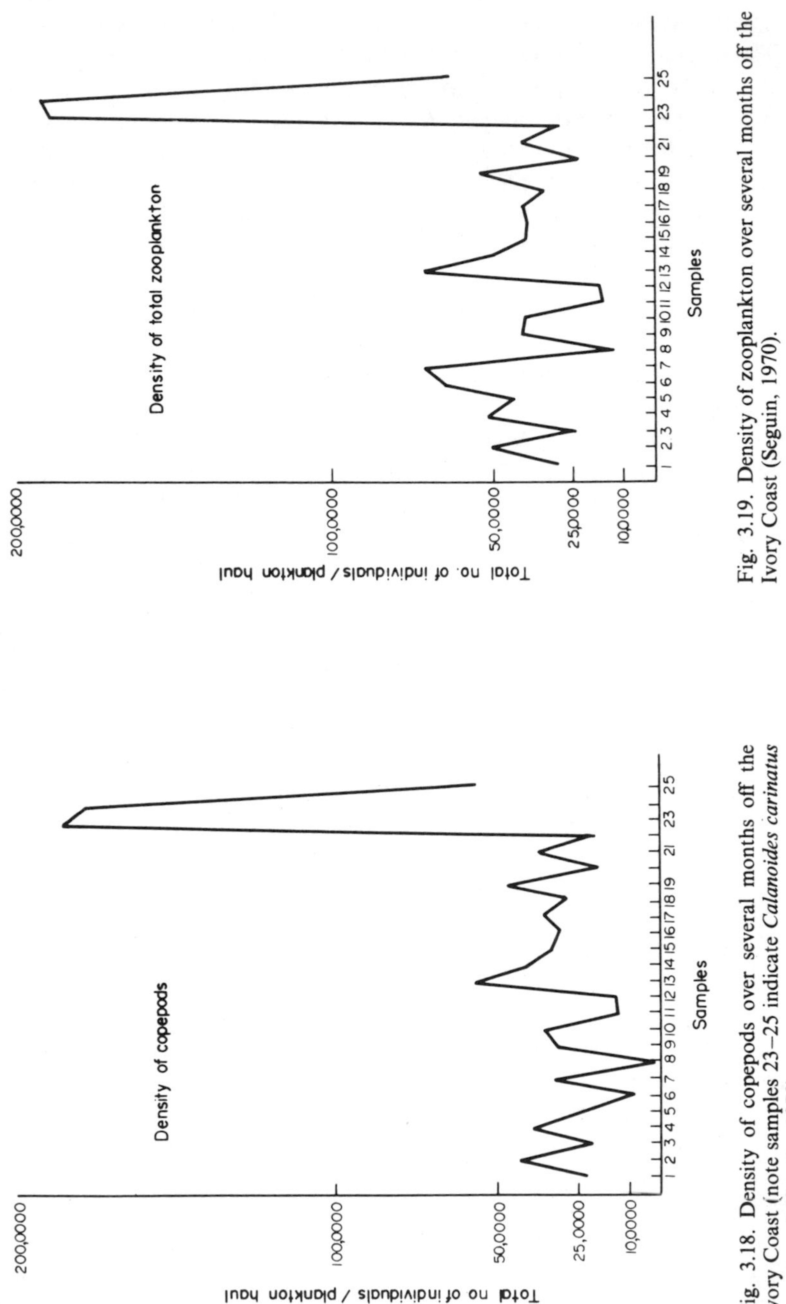

Fig. 3.18. Density of copepods over several months off the Ivory Coast (note samples 23–25 indicate *Calanoides carinatus* occurrence) (Seguin, 1970).

Fig. 3.19. Density of zooplankton over several months off the Ivory Coast (Seguin, 1970).

intervals, presumably effectively utilizing the more abundant food over that time.

Binet, also describing the Ivory Coast, points out that with seasonal upwelling and the comparatively sudden increase in phytoplankton, the herbivorous zooplankton continues to reproduce while remaining in the enriched waters, or, if transported to poorer waters, must interrupt the reproductive cycle. Thaliaceans and cladocerans can interrupt their cycles with the alternation of asexual or parthenogenetic cycles. The carnivore *Euchaeta concinna* is believed to go through fairly rapid reproductive cycles.

With the lessening of upwelling, however, diversification also increases, with maximum populations of *Centropages chierchiae*, *Temora turbinata*, salps, *Penilia* and chaetognaths. During the hot stable season (October to January) with the lower biomass and generally more oligotrophic conditions, two copepods, *Undinula vulgaris* and *Farranula gracilis*, are important. Cladocerans and thaliaceans effectively utilize richer phytoplankton by rapid cycles of asexual reproduction. Binet found, as has frequently been observed elsewhere, that with upwelling biomass increases and diversity is reduced. As hydrological stability is re-established, diversity increases with many new species appearing, usually in fairly low densities. Some (e.g. *Centropages chierchiae*) achieve their maximum populations about this time; chaetognaths also tend to be more abundant during this period. Carnivorous plankton can profitably feed on the increased density of herbivorous species if their reproductive cycles are so adjusted. *Euchaeta paraconcinna* appears to have the ability of undergoing rapid breeding cycles.

It appears that upwelling is not strong along the coasts of Liberia and Sierra Leone: the hydrological stability is preserved over the year, with surface waters being more or less oligotrophic. Off Senegal, however, two well-marked seasons occur, the cooler period with marked upwelling from December to May and the warmer period from June to November.

A clear demonstration of the seasonal effects of upwelling is that of Seguin (1966) from investigations in the coastal area (to about 30 m depth) off Dakar. The investigation, together with an earlier study by Gaudy and Seguin, suggests that several zooplankton maxima, expressed as volume, can appear during the year. The most obvious is about February/March; a less obvious one, according to Furnestin (1970) occurs in August to October. The chief maximum is correlated with the marked upwelling which occurs from about December /January to April. Upwelling promotes an extraordinary richness of phytoplankton production off Mauritania and Senegal, which in turn leads to dense zooplankton crops. The copepods were the dominant group.

As is so frequently true of warmer seas, the different taxa, and even species within taxa, can have their maxima at various times of the year. A greater intensity of reproduction is not confined to one season. According to Seguin, a number of species seem to be present all the year round, without any very obvious seasonal preference, (e.g. *Eucalanus subtenuis*, *Euchaeta marina*, *Temora stylifera*, *Centropages chierchiae*, *Candacia pachydactyla*, *Paracalanus parvus*, together with several species of cyclopoids and harpacticoids). Most of these are more typically neritic tropical species. With them occur other zooplankton types such as *Lucifer faxoni*, *Creseis acicula*, *Penilia avirostris*, chaetognaths and *Oikopleura longicauda*. During the colder season (approximately January to May) the copepod population becomes dominated by *Calanoides carinatus* in very considerable densities, with *Eucalanus attenuatus*,

Euchirella rostrata and *Euchaeta hebes* species more generally occurring in somewhat deeper waters. As in so many other areas off West Africa in which marked upwelling prevails, *Calanoides* is especially characteristic of the colder, upwelled water. In contrast, during the early part of the warmer season *Nannocalanus minor* and *Undinula vulgaris* are characteristic, with some pteropods, medusae and *Physalia*. In July and August pontellid copepods are relatively abundant, with siphonophores (*Muggiaea*, *Lensia* and *Chelophyes*), ctenophores and salps. Exceptionally large invasions of ctenophores (chiefly *Pleurobrachia* and *Beroe*) occur in October, and other siphonophores such as *Diphyes bojani* with doliolids and *Thalia democratica* may occur in considerable numbers. Of chaetognaths, *Sagitta enflata* occurred especially in August and *Sagitta friderici* and other species more in the cooler season. *Nannocalanus minor* can be regarded as the typical herbivore of the warmer periods as compared with *Calanoides carinatus* in the cooler months. The proportion of carnivorous zooplankton also tends to be somewhat lower during upwelling.

Casanova (1977) found very rich zooplankton off the coast of Mauritania which was particularly associated with upwelling water farther offshore along the continental slope—mainly deeper-living copepods (*Eucalanus*, *Pleuromamma*) and with euphausiids, especially *Euphausia krohnii*. In areas nearer the edge of the continental shelf these copepods occur with *Calanoides carinatus*, which became increasingly dominant, even comprising 90 % of the copepods in the cold upwelling season. The neritic euphausiid *Nyctiphanes capensis* could be very abundant, however, in the same area, even forming the major part of the total biomass. Very close to the coast, however, the zooplankton biomass was small and was dominated by *Paracalanus parvus*. According to Thiriot (1977), *Euphausia krohnii* is not distributed farther south along the African continent, but off the Mauritanian continental slope it can be very important. Off the coast of Morocco, near Cape Ghir in an area where a phytoplankton bloom was commencing, he also noted a burst of eggs and nauplii of euphausiids, chiefly *Euphausia krohnii*. He believes for more oceanic areas off the west coast of Africa that several regions marked by divergences, although not studied in detail, show an increase, especially in zooplankton such as copepods and euphausiids, which presumably have responded to the greater primary production in such areas. Kinzer (1966b) also comments that while particularly rich zooplankton is found in areas of upwelling between Cape Lopez and Pointe Noire, and in the central region of the River Congo outflow, even as far as 300 miles from the mouth, the open ocean has consistently low values except for zones of upwelling between 1°N and 4°S latitude.

In detailed investigations off the coast of Morocco, mainly within a depth of 200 m, Furnestin (1957, 1970) found that the waters were stable during winter, but that during spring, upwelling occurred reaching a maximum in the summer and gradually spreading from south to north. A major growth of phytoplankton occurred with the upwelling over summer. Although different regions of the Moroccan coast differ in their abundance of zooplankton, the northern regions being generally poorer than the south, and as is so frequently true, the abundance of zooplankton decreases farther offshore, there is also a clear seasonal effect. The volume of zooplankton is richer in autumn and winter than during the spring, and above all, than during the summer. Taking all the sectors together, the mean volumes were: autumn 77.5, winter 60.9, spring 53.0, summer 26.6. Furnestin contrasts this cycle strongly with that typical of north-west European and north-east American coasts, where a late spring or early summer increase is usual, more or less following the phytoplankton bloom. In such temperate areas there may often be two maxima.

Apart from the more neritic copepods such as *Paracalanus parvus* and *Centropages furcatus* in the coastal waters, the importance of *Calanoides carinatus* in the colder upwelling waters is again apparent. The cladocerans show a well-marked seasonal change off the Moroccan coast. Apart from a few *Podon intermedius* in winter/spring, three species were important. *Podon polyphemoides* accounts for more than one-third of total cladocerans and occurs throughout the year, but its density falls from autumn to spring and increases to an obvious summer maximum. *Evadne nordmanni* and *E. spinifera* are both well represented together forming more than 50 % of cladocerans, but whereas the latter tends to spread out from the coast and is present all the year, with a maximum in spring and a minimum in autumn, *E. nordmanni* is more strictly neritic, with a very clear seasonal cycle, occurring only in summer. Both appear to be able to take advantage of favourable nutritional conditions by rapid parthenogenetic cycles. Total cladoceran densities are maximal over summer when the total zooplankton is relatively poor.

Of appendicularians, *Oikopleura dioica* comprises about 70 % of the population, followed by *O. longicauda*. Both species are present all the year, but *O. dioica* is maximal in spring and autumn. It is likely that the appendicularians pass through very rapid reproductive cycles as in other warm regions. Salps can make massive invasions in this region, and at certain times nearly the whole of the zooplankton appears to be composed of these organisms. The massive numbers are due almost entirely to *Thalia democratica*. There is also an extremely clear seasonal cycle with outstanding summer abundance (cf. Fig. 3.20). While these great summer numbers are partly due to invasion from more offshore waters, it is very clear, according to Furnestin, that rapid reproductive cycles occur. Young solitary forms as well as aggregated chains appear, but the aggregated forms are three times as numerous during the summer invasions as in other areas (cf. Heron, 1972a, b; Troost, Sutuno and Wenne, 1976). Salps appear to

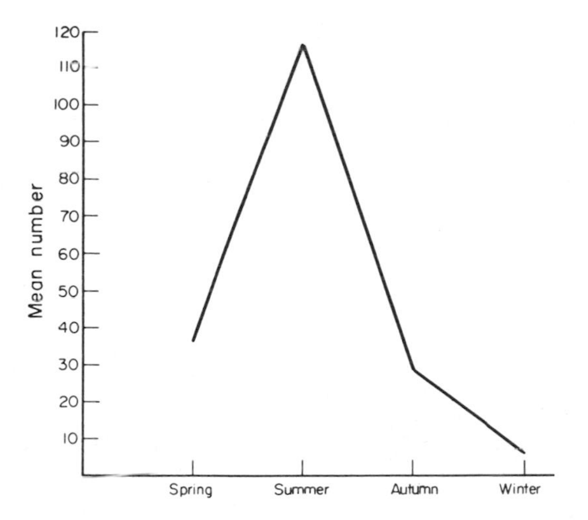

Fig. 3.20. Seasonal cycle of salps in the coastal zone off Morocco. Mean numbers of specimens/station over three years (Furnestin, 1957).

be able to utilize the massive blooms of phytoplankton which occur off the Moroccan coast over the summer and to reproduce exceedingly rapidly. Furnestin considers two alternative explanations for the paucity of zooplankton other than salps over the summer on the Moroccan coast. She points out that frequently the enormous blooms of phytoplankton have the characteristics of "red tide" phenomena, and this might inhibit most zooplankton. On the other hand, it may be that the massive invasions of salps feed so effectively on the phytoplankton that they compete very successfully and oust other zooplankton taxa.

Of other taxa, siphonophores were not very numerous but on average were rather more plentiful in spring and least abundant in summer. Of the four or five main species, *Lensia conoidea* (really an oceanic form) made up nearly 50% of the siphonophores and occurred chiefly in spring and summer. *Chelophyes appendiculata*, the second most abundant species, is mainly a summer form, with moderate numbers extending into the autumn. *Muggiaea atlantica* is regarded mainly as a coastal siphonophore and is third in abundance off Morocco. It is very clearly a spring species. Of the others, *Diphyes dispar* was entirely absent in summer and *Bassia bassensis* occurred mainly during autumn. These different seasons of abundance of various siphonophores probably mainly reflect movements of water masses towards and away from the coast, since most of the species live more offshore. The main periods of abundance of the more common species, however, may reflect major times of reproduction. The analysis by van Soest (1973) of the distribution of siphonophores in the North Atlantic illustrates the changing abundance of species with latitude. *Lensia conoidea* is described as being fairly common west of Great Britain to latitude 60°N; *Chelophyes appendiculata* is frequent south of 50°N latitude; *Bassia bassensis* and *Diphyes dispar* were dominant south of 35°N.

The chaetognath *Sagitta friderici* occurs throughout the year, but most plentifully in summer and autumn. The dominant stage over summer consists of juveniles and presumably a younger brood of smaller juveniles is found during the autumn. Reproduction probably occurs throughout the year, though with periods of increased intensity. Such periods are not necessarily the same each year, but probably reflect local hydrological factors. The life history is believed to occupy about a year. However, during spring and summer maturation occurs more quickly, so that individuals are smaller than over autumn/winter. The breeding of other warm water chaetognaths is described later.

In contrast to the variations in zooplankton populations in those African waters with marked seasonal upwelling, the more stable inshore areas and the more oceanic environments show comparatively little change throughout a year. Close to the island of Tenerife, for example, where there is little temperature variation (about 4°C), Thiriot (1977) quotes from studies by Corral Estrada, essentially on epiplanktonic copepods, that conditions remained oligotrophic in that area more or less throughout the year, with a large diversity of species but with comparatively few individuals occurring at any time. Various copepod species become relatively more important at different times, probably reflecting more intensive periods of reproduction. Some species, however, presumably reproduced more or less steadily.

Seasonal changes in relation to the breeding of zooplankton have been reported by Bernard (1967) in Mediterranean waters. The copepod *Temora stylifera* is found widely, occurring in dense swarms in some areas, especially in coastal but highly stratified waters. Bernard (1970) analysed the reproductive cycle of *Temora*. The species is scarce in February and March throughout the Mediterranean; maximal densities are found from July to December. Bernard believes that in many areas of the Mediterranean the copepod is capable of breeding through most of the year, unlike boreal species. Off Barcelona, Split, Marseilles and other coasts, *Temora* is scarce until May/June and a single maximum, sometimes with a minor secondary peak, is typical. In contrast, in the Gulf of Genoa peaks occur each month from June to September and in early November. In the Bay of Algiers peaks follow approximately monthly from May to December.

Bernard cites a study by Gaudy of *Temora stylifera* off Marseilles where the populations are more clearly separated. An examination of the changing densities of developmental stages (nauplii and copepodites) led Gaudy to claim that adult females spawn more or less synchronously, producing separate broods, as do many copepod species from high latitudes. Five generations were believed to occur annually. Bernard doubts whether females reproduce synchronously and attributes peaks in population of the copepod to favourable conditions, especially to increased primary production in addition to other factors such as lowered predation pressure, effecting an increased population density. She points out that the numerous peaks found for the Bay of Algiers may reflect, in contrast to some other Mediterranean regions, the considerable level of primary production over much of the year. In support of the view that *Temora* is able to breed throughout most of the year, she finds that females brought into the laboratory oviposited fairly freely in almost every month.

Both Gaudy and Bernard call attention to the competition which apparently exists between *Temora* and *Centropages typicus*. (*Centropages kroyeri* formed at times a considerable part of the population off Algiers.) *Centropages* forms overall the greater part of the population in most Mediterranean regions. As the colder-temperate copepod, it tends to increase to a small extent in the colder months of February and March when *Temora*, a warm-temperate species is scarce or absent. Between April to July, varying with the region, *Centropages* reaches main peaks then progressively declines, while *Temora* increases in density. During September and October *Temora* reaches its maximum importance, while *Centropages* densities are low, but in the succeeding three months *Temora* becomes reduced in numbers and *Centropages*, though not abundant, begins to assume its greater importance. The two copepods co-exist over much of the year, and while seasonal change in temperature is a major factor favouring the production of one species rather than the other, other factors (e.g. food, predators and unusual hydrographic conditions) may influence the composition of the zooplankton population. The ratio *Centropages*/*Temora* exceeds 20, and can be much greater in those areas where *Centropages* reaches its peak early in spring, but owing to competition with *Temora* the ratio falls below 10 in areas where its peak comes late. Results from other studies suggest other species of copepod (*Clausocalanus*, *Paracalanus* and *Oithona* spp.) found widely over the Mediterranean reach comparatively high densities over the late summer/autumn, when *Temora stylifera* is maximal. While reproducing like *Temora* over much of the year, they possibly show greater reproductive intensity over the later summer.

Moraitou-Apostolopoulou (1969) studied populations of six common copepods in the north Aegean approximately quarterly for a year. *Temora stylifera*, the commonest species, exhibited peak density in summer, diminishing in autumn to a low winter level, but reproduction continued throughout the year (cf. Bernard, 1970). *Centropages typicus*, while showing peaks in summer and autumn, showed a slight retardation in reproduction during the summer. *Centropages violaceus*, while rare in the Adriatic and western Mediterranean, was common in the Aegean and increased in summer to a peak over autumn. Population was minimal over winter and reproduction of this warm water species was confined to summer and autumn. *Calanus minor*, another common copepod, showed a peak population in autumn. *Calanus helgolandicus* and *C. tenuicornis*, both of which favour somewhat cooler waters, were common over winter, when reproduction took place. Summer and autumn population were low.

Bernard (1963) reared the planktonic harpacticoid *Euterpina acutifrons* in the laboratory, where the generation time was shorter than in nature. The average duration in the sea is probably from 6–8 weeks in Mediterranean waters. As with other harpacticoids, a succession of egg sacs is produced. Bernard quotes results from Della Croce indicating that reproduction can continue almost through the year in the Ligurian Sea, and this seems to apply also to the waters off Algiers.

Many studies of seasonal cycles in the Mediterranean have concerned population changes of Cladocera. Margineaunu (1963) reported on changes in the cladoceran population off the coast of Rumania (Black Sea) near the mouth of the Danube, where considerable fluctuations in salinity and temperature occurred over the year. *Podon polyphemoides* and *Evadne spinifera* appeared about April at temperatures of near 12°C and persisted in smaller numbers over the summer, when *Penilia avirostris* became dominant, with temperatures reaching about 21°C. *Penilia* also requires somewhat higher salinities (exceeding 17‰). A massive population of *Penilia* was typical during July. *Evadne tergestina* was generally similar in its requirements. These two species, following their rapid asexual breeding over the summer, disappeared first with the decline of temperature during autumn and were followed by the other cladoceran species. Margineaunu suggests that a few survived over the winter in deeper waters (cf. Della Croce and Venugopal, 1973).

In the Mediterranean itself, cladocerans were extremely abundant at times off Villefranche, but spread far out into more open waters (Tregouboff, 1963). *Evadne spinifera* apparently occurred more abundantly than other cladocerans over autumn and winter, when all the species tended to be deeper-living. *Podon intermedius* appeared mainly in early summer, and was followed by *E. tergestina* and *Penilia avirostris* in July. The species reached peak densities in August/September and disappeared during the autumn. Thiriot (1972) studied cyclic changes in cladocerans over five years off Banyuls in the western Mediterranean. *Evadne nordmanni*, the most coastal species, was present until the end of spring, the number of embryos carried by the parthenogenetic females varying with temperature. None of this species occurred over summer. *Evadne spinifera* was found over spring and during November/December. Over the summer the population declined, but a large increase followed during the autumn. *Podon polyphemoides* and *P. intermedius* occurred throughout the year and were neritic, though the former could be present in much diluted water. *P. intermedius* showed four maxima during the year. *Penilia avirostris* occurred over the summer period with a typical maximum.

Of other taxa, Braconnot (1974), working in the Mediterranean near Villefranche, described the breeding cycle of *Pyrosoma atlanticum* from an analysis over the year of changing densities of five growth stages (Classes I to V). Three cycles apparently occur over the year. Sexual reproduction of a fairly large population of mainly Class IV individuals during August/September produced numerous small (Class I) individuals which grew up during October and November. They reached Class IV and V by the end of that period. These in turn produced a new brood in December (85 % Class I) which was numerically the largest brood. Two cycles thus occurred between August and January. A third brood appeared in spring with numerous Class II and III being recorded by June.

The holoplankton of the warm waters of the western Atlantic Ocean was studied by Legare (1961) and Legare and Zoppi (1961). Their investigations in Venezuelan waters, mainly in the Cariaco region, demonstrated a greater richness of zooplankton inside the Gulf of Cariaco compared with waters only a short distance offshore over the Cariaco Trench. Although no clear pattern of changes in zooplankton abundance and composition at various stations in the Gulf was identified, comparatively poor crops of zooplankton occurred from July to November due to depleted offshore surface waters being carried by winds into the Gulf. When such a surface incursion was not evident the crop of zooplankton in the Gulf could be five-fold to eighteen-fold richer than in the Trench waters outside. Dense zooplankton was especially typical of the area near the outflow of the Manzana River.

The richness of the Gulf is to some extent increased by an inflow of deeper water. Rao and Urosa (1974) point out that with a westward drift of surface waters from the Gulf, some inflow of deeper richer water can occur towards the eastern end of the Gulf, promoting production. When conditions are more stable, productivity is lower. The results indicate therefore, that while a true seasonal effect is not generally evident for this tropical area, surface drift of water in or out of the Gulf causes fluctuations in the plankton (cf. Chapter 8). The Trench area over the deep water was relatively poor over most of the survey, a finding fairly typical for more open tropical waters. Copepods dominated the plankton of the Gulf (comprising 65 % on average), with Cladocera, especially *Evadne tergestina* and *Penilia avirostris*, appendicularians, chaetognaths, pteropods and meroplankton making up a further 33 %. Siphonophores, especially *Muggiaea*, and also doliolids could be moderately abundant. The copepods were mainly neritic and ubiquitous forms; *Temora*, *Oithona*, *Oncaea*, *Microsetella* and *Eucalanus* were the chief genera in that order of importance. Over the Trench area a different community with some deeper-living species occurred, and diversity was greater. Salps, ctenophores, the medusa *Liriope*, oceanic chaetognaths, pteropods and many euphausiids were also of greater significance offshore. In the Cariaco region 63 % of the chaetognaths were made up of only two species, *Sagitta hispida*, which was very abundant, and *Sagitta helenae*.

In a further study on the copepods over the Cariaco Trench, Zoppi (1964) described the fauna as a mixture of neritic and oceanic species, almost all of which were tropical. Four genera (*Clausocalanus*, with two very common species, *Paracalanus*, *Oithona* and *Temora*) made up 62 % of the total. There was little variation from month to month and no obvious season of maximal abundance. Different species tended to predominate at different periods, as has been frequently found for other tropical areas. Thus, in January the dominant species were *Labidocera scotti*, *Centropages furcatus* and *Clausocalanus* spp.; in May *Temora turbinata*, *T. stylifera* and *Labidocera acutifrons*; in July *Pontellopsis brevis* and *Oncaea conifera*; in August *Corycaeus* spp. and *Oncaea conifera*; in September *Acartia clausi*, *Clausocalanus* spp. and *Paracalanus* spp.; and in November *Centropages furcatus* and *Temora turbinata*. The more or less constant occurrence of adults of these various species might indicate that breeding of most of the copepods occurred throughout the twelve months, but that more intensive periods were typical of some species.

Off the east coast of Venezuela and more generally along the north-east coast of South America, Urosa and Rao (1974) pointed to the outflow of the Orinoco and Amazon rivers mixing with the Guinea Current waters, causing considerable changes in temperature and salinity. Earlier, Calef and Grice (1967) suggested a considerable influence of the River Amazon outflow on the production of zooplankton. The total volume of zooplankton during the wet season (May/June) was approximately three-fold greater than in the dry period (October/November) and the number of copepods showed a similar seasonal variation. Lenses of law salinity water even several hundred miles from the river mouth were identified with extensive populations of actively reproducing cladocerans (*Evadne tergestina*), *Lucifer faxoni* and some copepods.

Hydrological changes at the edges of the outflow might be influential in increasing the plankton production. Urosa and Rao (1974) also questioned whether the alternation of a wet period (May/June) and a dry season (October/November) could induce seasonal changes in the plankton. Results of a study of the chaetognath fauna, indicating the relative abundance of six major species (Table 3.3) at different periods, supported the suggestion of seasonal change, although the widespread species *Sagitta enflata* did not undergo any clear seasonal change. Total zooplankton biomass appeared to be greater in July/August than in the October/November (dry) or May/June (wet) season. Across the continental shelf, in the same region, a lower salinity band (32‰) occurred offshore, probably related to part of the Amazon outflow. Inshore, the salinity was greater (>36‰). Investigations demonstrated that the numbers of copepods, cladocerans and chaetognaths, as well as the biomass of total zooplankton, increased from the coast with the fall in salinity to reach a maximum in the mixed waters before the low salinity band was reached. Only Cladocera reached their maximum in the low salinity water itself (cf. Fig. 3.21). Any seasonal changes in such an area are presumably, therefore, considerably dependent on river outflow.

Much further south along the South American coast, studies at Cananeia (Sao Paulo, Brazil), at latitude about 25°S, have yielded data on the breeding and seasonal changes of certain warm-water inshore holoplankton. The decapod *Lucifer faxoni*, an important member of the neritic plankton in many tropical and sub-tropical regions, was shown by Lopez (1966) to be present both as adults and juveniles in all months of

Table 3.3. Order of abundance of the six major chaetognath species (Urosa and Rao, 1974)

Oct.–Nov. 1964	May–Jun. 1965	Jul.–Aug. 1968
S. enflata	*S. enflata*	*S. enflata*
S. serratodentata	*S. serratodentata*	*S. hispida*
K. mutabbii	*Pt. draco*	*S. serratodentata*
S. tenuis	*K. mutabbii*	*K. pacifica*
Pt. draco	*S. hexaptera*	*S. helenae*
S. hexaptera	*S. tenuis*	*S. tenuis*

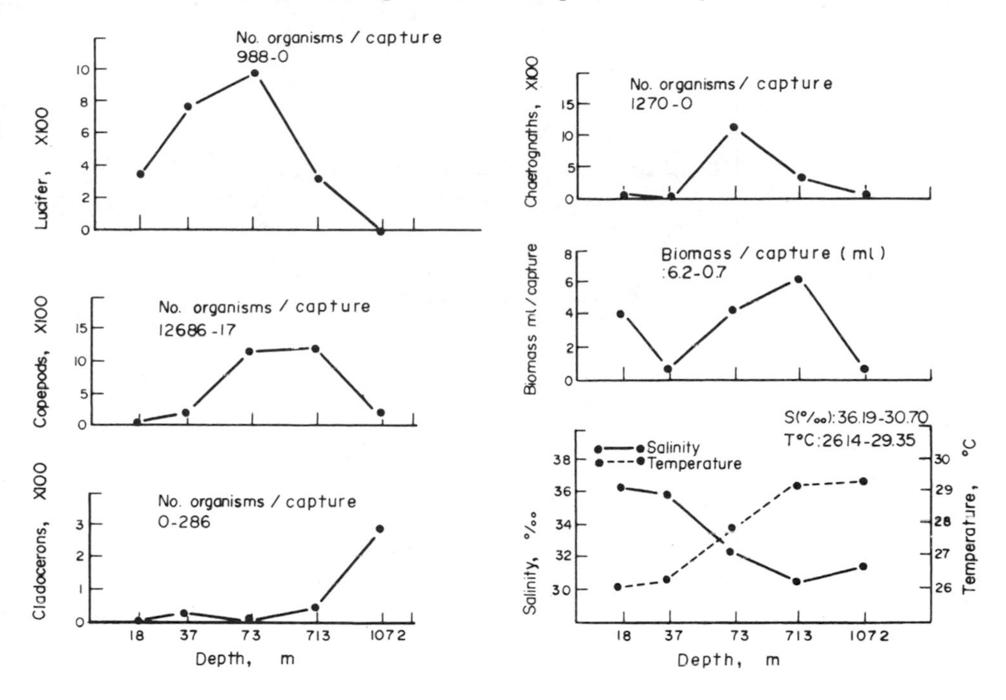

Fig. 3.21. Variations in temperature, salinity and biomass of some zooplankton groups for stations off Venezuela (the fluctuation in amount is given in each graph) (Urosa and Rao, 1974).

the year, with fully mature adults present as a lower percentage of population from April to September. Juveniles occurred as a maximum percentage of the population from December to April and again, as a much smaller maximum, in September/October. Adults were maximal from December to April (cf. Fig. 3.22). The data would suggest that breeding occurred all through the year, though with varying intensity.

In the same region, Tundisi (1970) investigated changes in appendicularians, both inshore and offshore. *Oikopleura longicauda*, the major species (71 % of the total zooplankton), showed varied changes in population over three years of study. Thus off Cananeia the maximum was in summer and autumn and the minimum in spring. Off Santos, maxima were observed in one year in March/April, September/October and December, but this did not hold for the previous year. The fluctuations thus appear to be irregular and were presumably connected with local sporadic hydrological changes rather than with any regular seasonal change. Tavares (1967) recorded invasions of the salp *Thalia democratica* in the Santos region. This occurred only during the cooler season, mainly April to July/August and during September/October. Doliolids, mostly *D. nationalis* and *D. gegenbauri*, however, were present more or less throughout the year.

The investigations of Russell and Colman (1934) and Russell (1934) on the zooplankton of the Great Barrier Reef have yielded considerable information on the fluctuations, abundance, seasonal cycles and relative importance of different species of tropical zooplankton. In considering the results, however, it must be remembered that

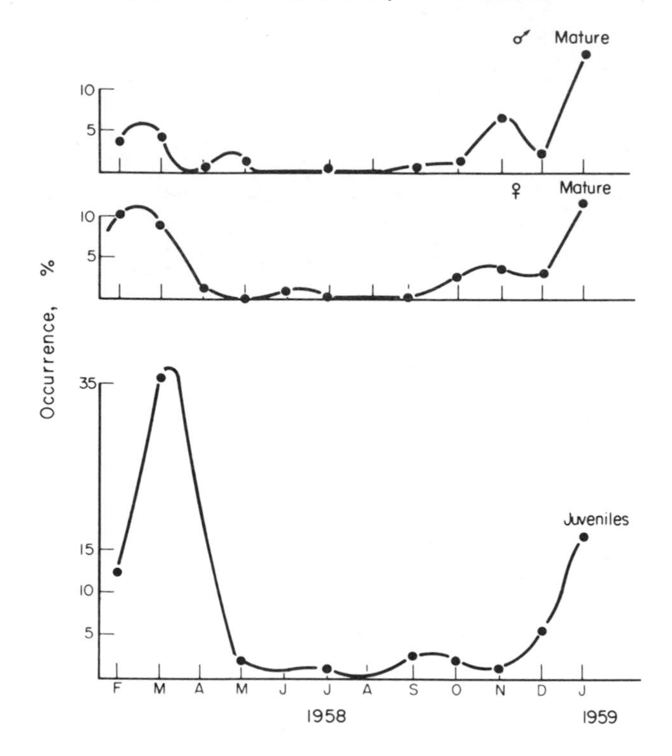

Fig. 3.22. Periods of abundance of mature males and females, and juveniles of *Lucifer faxoni* (Lopez, 1966).

the area is neritic, and seasonal changes may be very different in open tropical oceans.

Despite the variety of zooplankton characteristic of tropical waters, Russell and Colman found a preponderance of copepods on the Great Barrier Reef, a finding which appears to hold for the zooplankton of all seas. On the Barrier Reef they formed about 70% of the population. Other important holoplankton animals included tunicates, chaetognaths, pteropods, heteropods and siphonophores (Fig. 3.23). Although there were differences in the abundance of the zooplankton over the year, the variations were not nearly so obvious as in the seasonal fluctuations of the higher latitudes. For total zooplankton the amplitude was about 4; for copepods the amplitude exceeded 5 (Table 3.4). The amplitudes of seasonal change in various parts of shallow north European seas are much greater (cf. Table 3.4).

On the other hand, the plankton catches from week to week on the Great Barrier Reef showed marked variation in density, though apart from sudden massive invasions of salps, the plankton did not show periods with an overwhelming abundance of one or two species, as is so often true of higher latitudes. The average richness of the Great Barrier Reef zooplankton is approximately equal to that of higher temperate latitudes, and despite the high proportion of gelatinous type plankton typically found at lower latitudes, the volumes are greater even than for an inshore temperate area, for example off Plymouth (cf. also Wickstead, 1968). Copepods appeared to be less abundant on the Reef than in the English Channel, but in view of their generally smaller size, they may have been inadequately sampled.

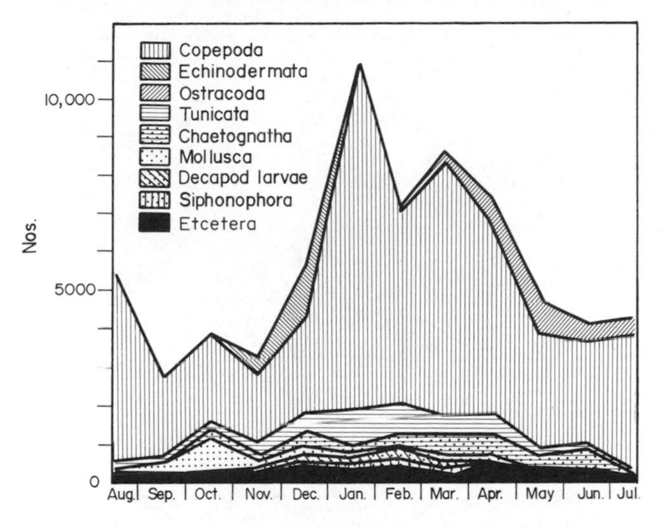

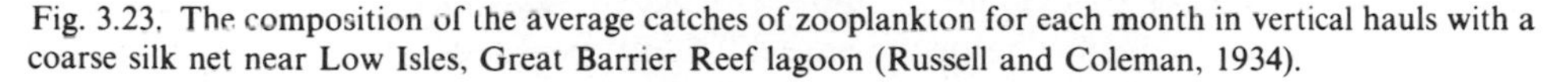

Fig. 3.23. The composition of the average catches of zooplankton for each month in vertical hauls with a coarse silk net near Low Isles, Great Barrier Reef lagoon (Russell and Coleman, 1934).

Table 3.4. Seasonal changes in the abundance of zooplankton from five European lightship stations, and the Great Barrier Reef. The figures represent the quotient of the largest monthly average catch divided by the smallest monthly average catch (Russell, 1934)

	All animals	Copepods
Anholt Knob[a]	10.6	18.7
	6.9	12.7
Smith's Knoll[a]	61.3	218.1
Borkumriff[a]	18.1	21.7
Varne[a]	218.5	421.0
Sevenstones[a]	47.2	46.7
G. B. R. Lagoon	4.1	5.3

[a] Lightships.

Despite the neritic character of the Reef, the temporary plankton amounted only to about 10 % of the whole zooplankton. Gastropod, lamellibranch, decapod and, for a limited period, echinoderm larvae were the most important constituents. Farran (1949) from seasonal differences in the relative abundance of the various copepods, suggested that there were differences in the breeding times of species. However, data for the immature stages were not generally available and precise breeding cycles are not known. The indications are that some species bred over only a limited period, though the times for different forms did not necessarily coincide. Some species (e.g. *Calanus pauper*, *Paracalanus aculeatus*, *Undinula vulgaris*, *Paracalanus parvus*) were present as adults and as immature stages almost throughout the year (cf. Fig. 3.24), but even these species showed changes in density. *Calanus pauper* showed a maximum in February,

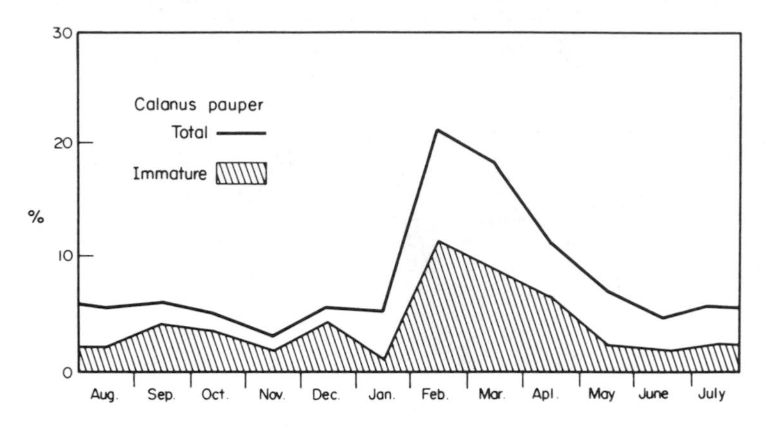

Fig. 3.24. Average catches of *Calanus pauper* with a coarse silk net (Farran, 1949).

Undinula from October to December, and *Paracalanus aculeatus* was more numerous from February to August. The occurrence of many peaks of abundance for many species probably indicates that a succession of broods occurred throughout the year with some intensification of breeding at particular periods. On the other hand, some copepods became abundant only over a restricted period of the year, for example, *Calanopia aurivilli* occurred from January to April and *Tortanus gracilis* showed a clear maximum about March/April (cf. Fig. 3.25).

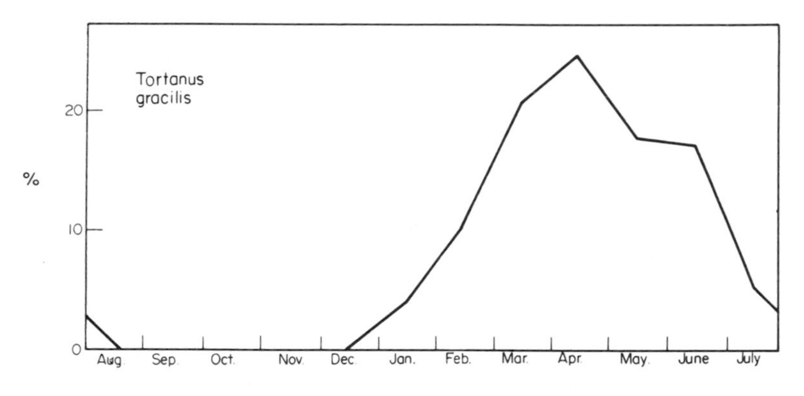

Fig. 3.25. Average catches of *Tortanus gracilis* with a coarse silk net at same location as Fig. 3.23 (Farran, 1949).

In such a neritic area changes in population may be caused to some extent by incursions of zooplankton from outside waters, but the major fluctuations appear to depend on local breeding. In *Labidocera acuta*, though the copepod was present only from December to July, rather sharp increases in populations occurred at intervals, in December, February, April and June (Fig. 3.26). However, sudden increases in *Acartia negligens* and *Acartia australis* are believed to be due to the drift of offshore waters onto the Reef: the species are open-sea forms.

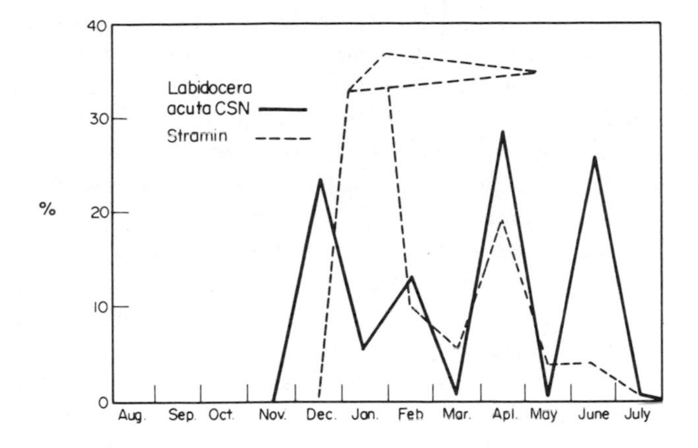

Fig. 3.26. Average catches of *Labidocera acuta* with two types of net (Farran, 1949).

The tendency for zooplankton species on the Great Barrier Reef to have extended breeding periods, which are intensified in certain months, appears to hold for zooplankton groups other than copepods. For appendicularians, Russell and Colman (1935) found considerable variations in abundance throughout the year with a maximum about January/February. Doliolids showed considerable fluctuations, with a long period of comparative abundance from December to June and a reasonable number of individuals during other months. Marked changes in density of salps were seen, though these were probably partly associated with swarms of individuals brought into the area. There were two periods of particular abundance for *Thalia democratica*, September to November and April/May.

Russell and Colman suggest that periods of highest reproduction may be associated with temperature changes, reproduction being most marked when the temperature ranged between 24° and 28°C. Of the pelagic pteropods, two species of *Creseis*, *C. virgula* and *C. acicula*, were by far the most abundant. There were clear indications of seasonal variation in density (Fig. 3.27), but these pteropods were relatively abundant over much of the year. Variations in the reproductive intensity of some

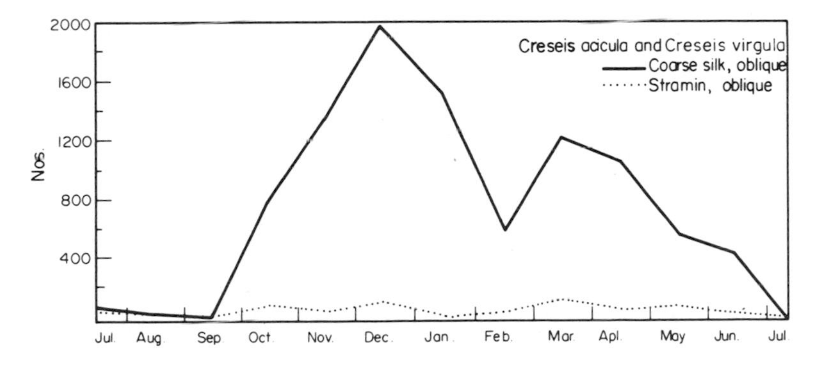

Fig. 3.27. Average catch of *Creseis acicula* and *C. virgula* combined for each month. Type of net is shown (Russell and Colman, 1935).

siphonophores occurred during the year, but the times of maximal reproduction varied from species to species. Thus *Diphyes chamissonis* had its maximal abundance about July or August, with numbers decreasing from January (Fig. 3.28a), whilst *Lensia subtiloides* was plentiful from October to March, with a maximum about December or January (Fig. 3.28b). *Abylopsis eschscholtzii* occurred mainly from August to November. *Eudoxoides spiralis* was clearly seasonal, being present only during three or four months of the late summer.

Another investigation of seasonal fluctuations in the zooplankton of tropical seas is that of Wickstead (1958), which also relates to neritic waters. Wickstead demonstrated that the zooplankton of the Singapore Straits exhibited changes simulating a spring and autumn maximum. The plankton population was greatly influenced, however, by the movement of water masses through the Straits due to the changing monsoons. Apart from qualitative differences in the plankton of the water masses during the change-over period between the north-east and the south-west monsoons, the variability in wind and the consequent effect on the water circulation tended to increase

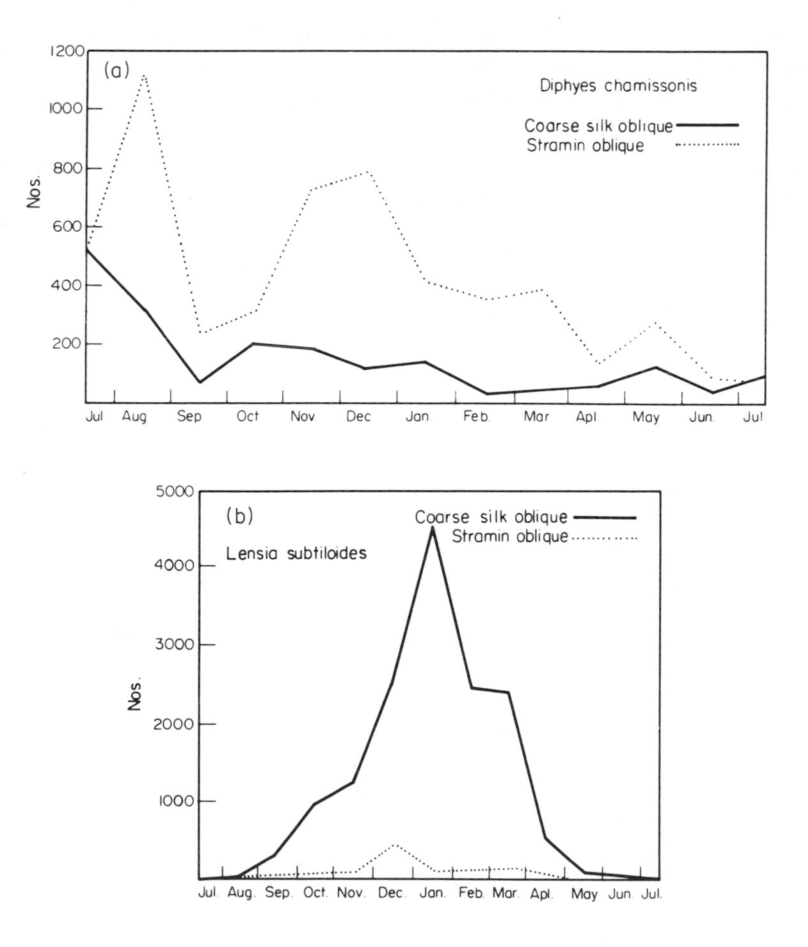

Fig. 3.28. Average monthly catches of two siphonophores, *Diphyes chamissonis* (a); *Lensia subtiloides* (b) Same location as Fig. 3.23. Type of net is shown (Russell and Colman, 1935)

the nutrient concentration in the water. There was, therefore, a richer crop of plankton after each switch of the monsoons, and the holoplankton of various taxa showed the same two seasons of relative abundance. Ctenophores were exceptional: from about January to September they exhibited fairly regular bimonthly periods of abundance perhaps reflecting a breeding cycle pattern, but they then declined. The meroplankton, mainly brachyuran with some caridean larvae and penaeids, also did not show spring and autumn maxima. Possibly the meroplankton was affected more by local spawning cycles than by major movements of water masses.

Wickstead followed this investigation with a study of the comparability of tropical and temperate crops of zooplankton (Wickstead, 1968). Similar samplings were made in the English Channel just outside Plymouth at L4, off Singapore and at two stations off Zanzibar for 13 months, so that any seasonal changes could also be examined. All the stations showed a seasonal cycle to some extent when density of zooplankton was reckoned as numbers/m^3. Wickstead calculated an Annual Plankton Fluctuation Factor by dividing the lowest monthly mean density into the highest monthly mean. The factor was greatest for the Channel Station (15.6); the two Zanzibar stations had factors of 5.3 and 6.7 respectively, and the Singapore station had the lowest factor (2.0). Figure 3.29 shows that the Zanzibar station was richest as regards mean zooplankton density and Singapore poorest. Displacement value measurements were available for the English Channel and Zanzibar stations only. The mean volume for the English Channel was below that for Zanzibar, but the Annual Plankton Fluctuation Factor was highest for L4. Similar results were obtained for dry weight comparisons.

Seasonal temperature fluctuation is, of course, far greater in the English Channel than in the two tropical regions. Although all three regions show annual cycles of zooplankton, the plankton composition in the English Channel remains fairly similar throughout the year. The fluctuations are due mostly to light and temperature changes. At Singapore the changes mainly reflect differences in the plankton population as a result of the monsoons and consequent water flow. Wickstead believes that the seasonal fluctuation at Zanzibar is also largely a result of a change of monsoon, though there is a somewhat higher rate of reproduction when temperature is maximal. Although the zooplankton crop in the English Channel would be expected to be lower in winter than in tropical inshore areas, rather surprisingly the spring crop in the Channel was not appreciably higher than that of tropical stations. Numerical density might not, however, have given a correct comparison; volumes might also be misleading, since tropical plankton often has considerable quantities of medusae, siphonophores and salps. But dry weight data showed that the average crop in the English Channel was about half that at Zanzibar. Though dry weight data were not available for Singapore, from comparisons of numbers and sizes of zooplankton Wickstead believes that the mean dry weight at Singapore would approximate to that of the English Channel. In view of the generally accepted view that for many plankton animals reproductive rate is higher in the tropics, though the crop is more stable than at higher latitudes (cf. the Annual Plankton Fluctuation Factor), turnover must be more rapid and the total annual *production* of zooplankton in such inshore tropical waters might well be greater than that of nearshore temperate seas.

Panikkar and Rao (1973) review investigations of seasonal changes in the zooplankton in a number of areas, mostly fairly close to the coast around India. Data

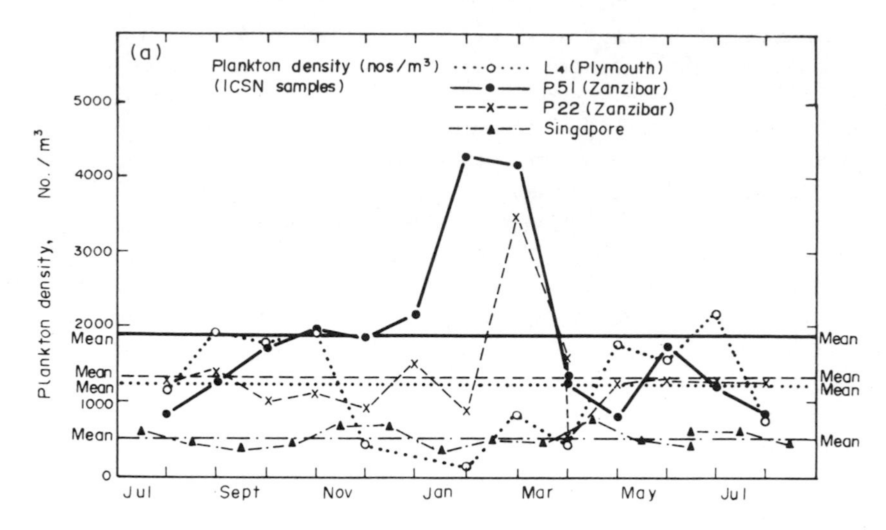

Fig. 3.29a. Plankton density (no./m^3) for three tropical and a temperate station. -----○---- Plymouth L4; ——●—— Zanzibar (close inshore); ----x-- Zanzibar (offshore) -·-▲-·- Singapore. Taken with Isaacs–Kidd sampling net (Wickstead, 1968).

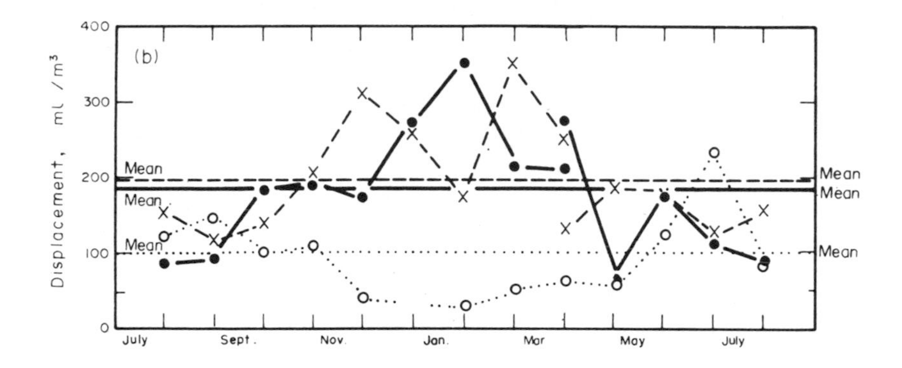

Fig. 3.29b. Plankton density (displacement volumes/m^3) for Plymouth and Zanzibar. Symbols as Fig. 3.29a (Wickstead, 1968).

for the volume of plankton (cf. Fig. 3.30) suggest that frequently there are two peak periods. For example, Prasad found a major peak in March (Gulf of Mannar) or April (Palk Strait) and a smaller peak in September and October. The precise period for the larger populations varies with the area, and there appears to be a north-to-south movement in the peak period along the east coast and a reverse south-to-north succession along the west coast. Panikkar and Rao relate this to monsoonal effects. On the west coast the south-west monsoon, from about May to September, not only reduces nearshore salinity but seems to be partly responsible for coastal upwelling. This upwelling sets in earlier in the south and gradually moves north. On the east coast, on the other hand, the run-off occurs mainly over the period September to November. Comparatively small temperature changes follow the monsoonal effects, so that the

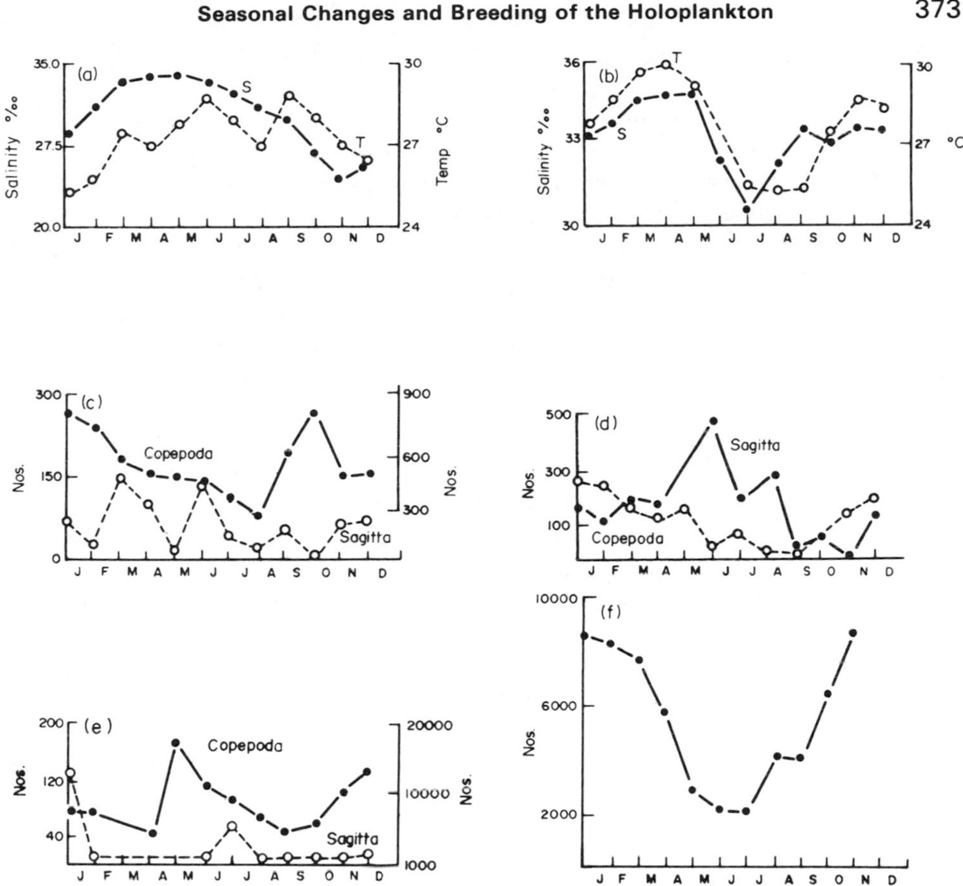

Fig. 3.30. Annual variation of zooplankton numbers/haul, temperature and salinity off the coast of India. a. Salinity and temperature off Waltair (East Coast); b. Salinity and temperature off Calicut (West Coast); c. Sagittae and copepods off Mandapam; d. Sagittae and copepods off Trivandrum (West Coast); e. Sagittae and copepods off Waltair (East Coast); f. Total (2 yr average) Copepods off Madras (Panikkar and Rao, 1973).

terms spring and autumn peaks are hardly appropriate for Indian coastal waters. In any event, Panikkar and Rao point out that daily or weekly fluctuations in local weather conditions may be considerable, sometimes even exceeding the more regular seasonal variations.

With the seasonal monsoon pattern, along the west coast the overall effect is that from about May to October phytoplankton tends to be abundant, but copepods are sparse. From about November to April the phytoplankton is generally lower; copepods become more abundant and chaetognaths, particularly *Sagitta enflata* and *Sagitta bedoti*, which begin to increase from about October, become an important group. For the east coast off Waltair, Ganapati and Rao's (1958) data indicate that the period August to December, with lowered salinity, is marked by only minor phytoplankton blooms, few chaetognaths and moderate densities of some tunicates. Between January and August, with higher salinities, there is rich phytoplankton, an abundance of appendicularians and considerable populations of chaetognaths.

The results of the International Indian Ocean Expedition not only provide data on the relative quantities of zooplankton in various parts of the ocean, but permit some examination of seasonal changes. Unfortunately, such observations apply to a few restricted areas, since much of the Ocean was visited only once during the Expedition. Generally the biomass of the total zooplankton is highest in certain regions of the Indian Ocean, particularly the southern coast of Saudi Arabia, Somalia, the south east regions of the Indian Ocean, the northern part of the Bay of Bengal and south of Java Reference to Volume 1 shows that most of these areas are also the regions of high primary production and large phytoplankton crop. Rao (1973) gives values for zooplankton biomass, expressed as displacement volumes (ml/haul), obtained with a standard IOSN net vertically hauled from 200 m to the surface. Off Somalia, volumes were about 15–35 ml/m^2, off Saudi Arabia, they even exceeded 50 ml/m^2, for the south-east Indian Ocean, the northern Bay of Bengal and the south of Java, they were about 15 ml/m^2. For most of the vast area of the remainder of the Indian Ocean the biomass was usually less than 15 ml/m^2. Off south-west Australia the biomass is generally very low, but over eight months there are two peaks, in February and August, though the maximum is only of the order of 8 ml/m^2.

Figures 3.31a, b illustrate the biomass of zooplankton during the north-east and south-west monsoons, respectively. During the south-west monsoon three centres of high biomass (about 50–60 ml/m^2) are present in the Arabian Sea off Somalia, Oman and Kerala. Fairly high biomass (20–40 ml/m^2) is also characteristic towards the head of the Bay of Bengal. A high biomass of zooplankton occurs off south-west India, and it is likely that this comparatively rich zooplankton is typical of most of the west coast of the Indian peninsula over the south-west monsoon, but unfortunately no data are available, as for much of the Indian east coast. During the period of the north-east monsoon, according to Rao (1973) the high zooplankton values typical of Kerala and Somalia decline and the high crop off the coast of Saudi Arabia tends to shift westwards. Figure 3.31b shows high values in the Gulf of Aden. A small area of high biomass appears off Goa and a similar small area off Madras. The east coast of India is probably generally rich in zooplankton during the north-east monsoon. Over the Indian Ocean as a whole there is an increase in the area of higher zooplankton biomass during the south-west monsoon, compared with the period of the north-east monsoon.

Panikkar and Rao (1973) found that copepods comprised 70–80% of the total zooplankton at most stations during the surveys, especially nearer the Indian coast. Moreover, in the areas of higher biomass it was during the winter monsoon that the greater development of the zooplankton population occurred. This probably reflects the development primarily of a largely herbivorous population, and might suggest that, following the phytoplankton increase during the summer monsoon, some lag is experienced before the zooplankton increases. Off the Malabar coast, for example, upwelling and the peak in phytoplankton population occurs about July/August, but the increase in zooplankton commences only about October/November. The area of increased zooplankton production, however, may extend for some hundreds of miles away from the major areas of phytoplankton crop, presumably indicating the drift of near-surface water (cf. Volume 1).

Haq, Ali Khan and Chugtai (1973) found very high zooplankton biomass off the coast of Pakistan. In the main area of the continental shelf the period of high abundance of zooplankton was from October to December, while March was a time of

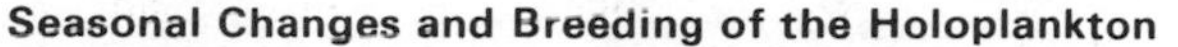

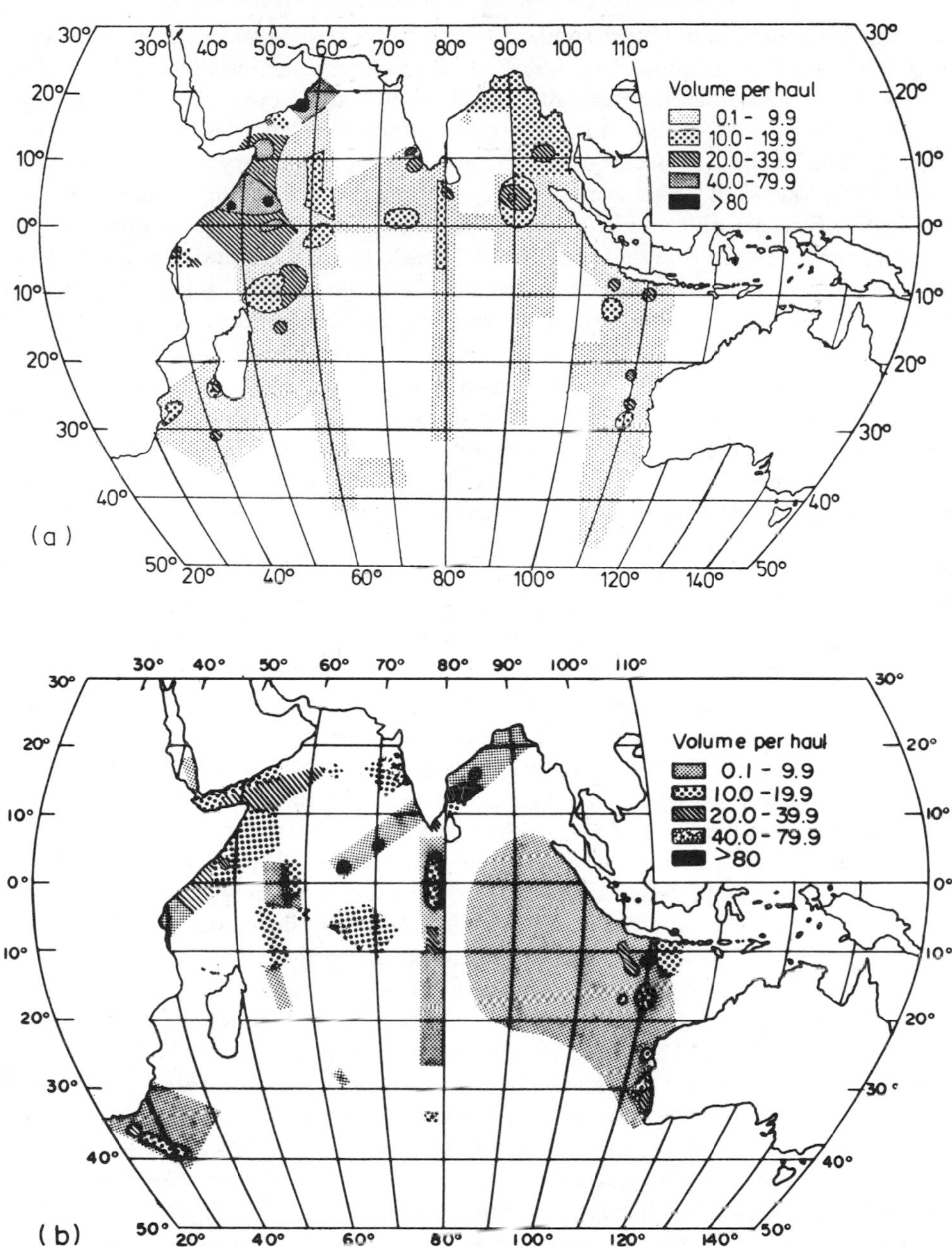

Fig. 3.31. Biomass distribution in the Indian Ocean in ml/m² for 200-m column; a) July–September (South west monsoon); b) December–February (North-east monsoon) (Rao, 1973).

very low density, though the whole period lies within the season of the north-east monsoon (October to April). While, according to Haq *et al.*, records from the IOBC suggest an average biomass for zooplankton for the north-east Arabian Sea of about 15–25 ml/m², data for the area off the coast of Pakistan suggest a much higher biomass (average 55 ml/m^2), values more similar to those productive areas of the Indian Ocean during the south-west monsoon. South of Karachi during December, Haq *et al.* reported a remarkable biomass of 166 ml/m^2 for a column of water 0–80 m. Particularly over November and December, when rich zooplankton crops seem typical, calm conditions occur over the area, with cool but oxygen-deficient water penetrating the shelf. This water is rich in nutrients and high phytoplankton production has been reported in the area over the period, presumably favouring a high crop of zooplankton. From these months till March there is a gradual restriction of this movement of sub-surface water, presumably associated with a generally poorer production. Surface copepods form the bulk of the zooplankton on the shelf, but deeper-living, more oceanic forms contribute to the biomass, especially *Pleuromamma indica* and *Euchaeta wolfendeni*. The former is particularly associated with water of much lowered oxygen content (apparently <0.1 ml O_2/l). Thus when oxygen-depleted sub-surface waters move towards the surface, *Pleuromamma indica* and certain other copepods become important. They appear to form the bulk of the copepod population in the slope waters and in some restricted shelf areas, especially over the period of high biomass towards the end of the year.

Paulinose and Aravindakshan (1977) also recorded exceedingly high zooplankton biomass from the north and north-eastern parts of the Arabian Sea. Off Kutch a volume of 560 ml/m^2 for a haul from 200 m to the surface was more than three times the highest volume of zooplankton recorded for any sample from the IIOE. In many of the collections ostracods were exceedingly abundant. At one station they outnumbered the normally abundant copepods a hundred-fold.

Over the Indian Ocean generally the major holoplankton groups follow the distribution pattern of the dominant copepods, with the chief areas of abundance off Saudi Arabia and the Somali coast. After copepods, in order of importance are ostracods, chaetognaths, euphausiids and thecosomes. While the percentage of copepods to total zooplankton changes little with season, ranging from 70–75% over some five months, the seasonal change with the monsoon in the density of copepods may be appreciable. Average populations are not meaningful with the geographical variations in density, but Rao (1973) quotes a Marsden square (5°) selected at random with the following populations of copepods: February, 75,629; March, 54,384; June, 29,683; August, 19,514; October, 55,456. The data illustrate the comparative low density over the summer monsoon (June and August) and the rise from October onwards. While not surprisingly the distributions of copepod density and zooplankton biomass show very much the same pattern, there is a fairly clear tendency, as shown by Kasturirangan, Saraswathy and Gopalakrishnan (1973) for the rich copepod areas to be much more extensive during the winter monsoon (cf. Figs 3.32a, b). The high density areas extend somewhat along the Saudi Arabian coast into the Gulf of Aden and also to the entrance of the Persian Gulf.

In contrast to the west coast of India, where the winter monsoon period appears to be richer, the northern Bay of Bengal seems to have an increased copepod population from April to October. More localized changes in density with season in the Java Sea and off western Australia may be due to small-scale upwelling during the July/August period. Figures 3.32a and b emphasize the low density of copepods in the more southern Indian Ocean, especially in the great central open ocean area, and though there is some change in distribution between the monsoons, change in relative

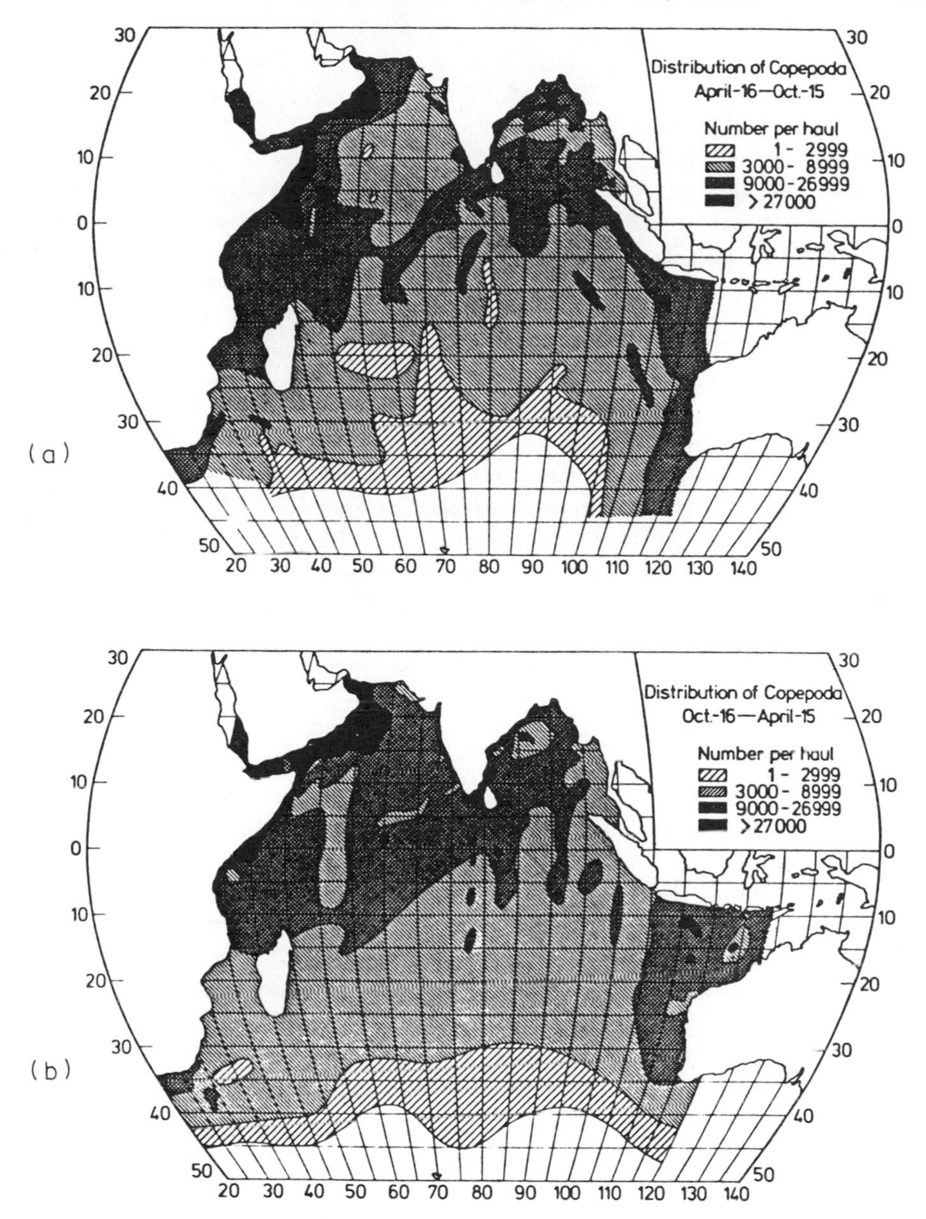

Fig. 3.32. Distribution of Copepoda in the Indian Ocean a) April to October; b) October to April (Kasturirangan, Saraswathy and Gopalakrishnan, 1973).

abundance is not obvious. A more detailed study of zooplankton changes in the Gulf of Aden is due to Gapishko (1971a,b), with particular reference to fluctuations in the copepod fauna over the monsoonal seasons. Biomass is moderately high in March, mainly due to *Calanus*, *Eucalanus*, *Rhincalanus*, *Paracalanus*, *Calocalanus* and *Clausocalanus*, but decreases to a minimum in May. During August, with the south-westerly monsoon, upwelling brings cooler-water species such as *Calanoides carinatus*,

Eucalanus attenuatus, *Eucalanus crassus* and *Rhincalanus nasutus*, and there is increased reproductive activity over August and September associated with the phytoplankton peak. *Calocalanus* and *Clausocalanus* are relatively less abundant. With the onset of the north-easterly monsoon and the flow of warmer water, other copepods (*Undinula vulgaris*, *Canthocalanus pauper* and *Nannocalanus minor*) become important by November, when another characteristic form is *Pseudodiaptomus* sp.

Although ostracods tended to be patchy in distribution, Panikkar and Rao (1973) found them in very high densities in the northern Arabian Sea, off southern Arabia and the south-west Indian coast. Other districts were not so rich, except for comparatively small isolated areas. However, during the winter monsoon period, ostracods appeared to spread as a high density band approximately from the Gulf of Aden along the coast of Saudi Arabia and across the northern Arabian Sea to the south-west coast of India. Euphausiids, like copepods, are in highest density along the north-east coast of Africa, the Arabian coast, south of Sri Lanka, off Madras and in an area in the Java Sea. Seasonal effects are again evident, however. During the south-west monsoon the high densities build up off Somalia, southern Arabia and Madras, and appear to extend offshore for hundreds of miles. During the north-east monsoon, while the areas of high density off Madras and Sri Lanka were reduced, a very wide region of moderately high density extended almost from the African to the Australian coast. Panikkar and Rao suggest that this may represent another example of the slower development of secondary production, following the increase in primary production during the summer monsoon.

An analysis of the distribution of Indian medusae by Vannucci and Navas (1973b) demonstrated that among the most abundant species are the trachymedusans, *Liriope tetraphylla*, *Aglaura hemistoma* and *Rhopalonema velatum*, and the narcomedusan *Solmundella bitentaculata*. Together these made up 61% of the total specimens. Whether their breeding times are affected by the seasonal monsoons is apparently unknown. In the Bay of Bengal considerable numbers of Anthomedusae and Leptomedusae are found, emphasizing the more neritic character of the area. In the open Indian Ocean, *Cytaeis tetrastyla* is abundant and widely distributed. Although this species has a benthic hydroid stage, it has a remarkable potential for vegetative reproduction, so that dense swarms may be encountered even over the high seas.

As we have already noted, in regions of the Indian Ocean experiencing pronounced upwelling with a corresponding increase in plankton density, the richness was often observed to spread offshore to a considerable distance. This spread was true of a number of taxa, including some meroplankton. Thus some decapod larvae could be found in considerable densities offshore. Although it would be appropriate to include data on larvae in the general section on meroplankton, it is more convenient to discuss these observations on meroplankton with the general review of the Indian Ocean zooplankton.

Menon, Menon and Paulinose (1967), describing decapod larvae other than phyllosomas and sergestids, found higher densities close to land, especially off the west coast of India, the south-west Arabian coast and the Somalia coast. At several stations caridean and brachyuran larvae were more abundant than those of other decapods. The distribution of penaeid and anomuran larvae was rather similar, but anomuran larvae were far less numerous. All decapod larvae were abundant around the southern tip of India. There was some evidence of greater abundance of penaeid larvae during the monsoon period from May to October in the Arabian Sea. According to Panikkar and Menon, the adult prawns are said to breed in these months. Menon and Paulinose (1973), apart from stressing the abundance of decapod larvae nearer costs (cf. Fig. 3.33), point out that the general pattern of abundance agrees with that of total zooplankton biomass, but particularly high concentrations of decapod larvae occur off Port Elizabeth, south of Java, and in three areas off the west coast of the Indian continent. In contrast, the central Indian ocean is extremely poor.

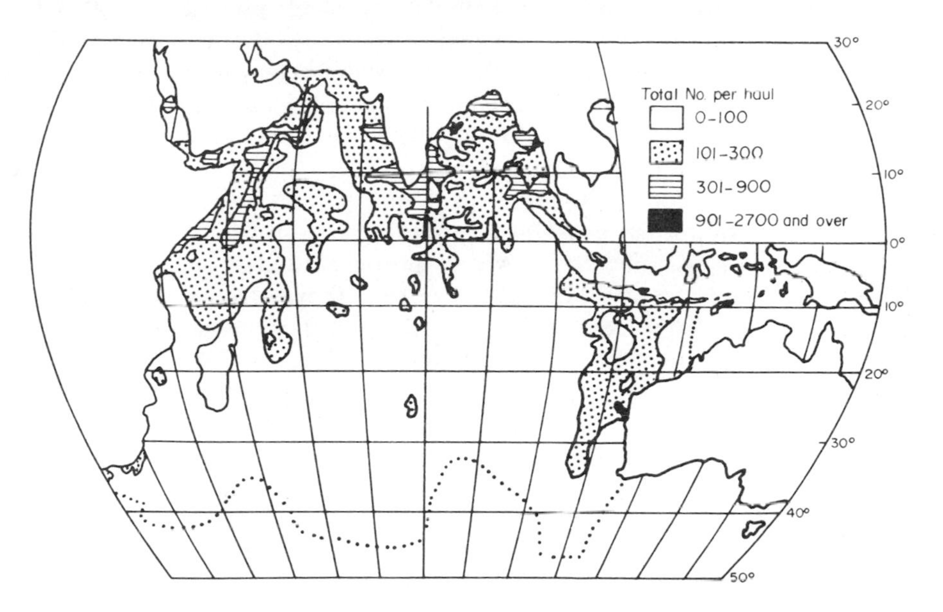

Fig. 3.33. Distribution of decapod larvae in numbers/haul (Menon and Paulinose, 1973).

During the south-west monsoon there was a wider distribution of decapod larvae and an extension of the high density areas, especially off Somalia, the Saudi Arabian coast, the west coast of India, most of the Bay of Bengal, south of Sri Lanka and west of Sumatra. On the other hand, an opposite effect (i.e. richer populations during the north-east monsoon) was observed in the Gulf of Aden, off Oman, around the south tip of India and near the outflow of the River Gascoyne off western Australia. While the rich populations off Port Elizabeth and Java were mainly sergestid and caridean larvae, much of the east coast of South Africa was poor in sergestid populations. Over the whole Indian Ocean survey, sergestid larvae dominated the collections, being present in 95 % of the samples, and their densities being five times that of other decapod larvae. Penaeids were moderately well represented. The developmental stages of one species, *Penaeopsis rectacuta*, occurred mainly in coastal waters and were particularly abundant off the Arabian and Somali coasts, along the west coast of India and most of the Bay of Bengal, and off north-western Australia. Brachyuran larvae occurred most regularly next to sergestid larvae, being found in 55 % of the collections. In certain areas they were the major decapod group.

There is some indication that apart from a monsoonal effect promoting production by mixing and local upwelling, excessive disturbance of coastal waters may reduce decapod larval abundance. For example, the Bay of Bengal is richer during the south-west monsoon, when the waters are generally less disturbed than during the north-east monsoon. On the other hand, the west coast, which is more sheltered during the north-east monsoon, is richer over that winter period than the Bay of Bengal. Increased production with pronounced upwelling at some distance from the coast appears to be true for decapod larvae. Thus larvae were not very plentiful close to the coast of southern Arabia, but at 200–300 miles offshore they could be extremely abundant. Rao (1973) also calls attention to the high density of decapod larvae (300–900 per haul) found adjacent to the coasts of Somalia, India and Burma, though off Saudi Arabia the rich populations were much farther offshore, especially over the south-west monsoon period. In general, higher populations of larvae are believed to be characteristic for the summer monsoon months for the ocean as a whole. Of other meroplankton, Rao quotes bivalve larvae as being dense along the Somali and south Arabian coasts, off south-west India, north of Sumatra and south of Java, with dense patches scattered over these areas. Bivalve larvae appear to be very sparse, however, along the coasts of Australia, South Africa and the east coast of India.

Investigations west of Australia in the eastern Indian Ocean along the 110°E meridian have demonstrated considerable differences in daily primary production with latitude, but also a distinct seasonal variation (cf. Figs 3.34, 3.35). According to Jitts (1969), the low period occurred from late November to early May ("summer") and the period of greater productivity from late May to early November ("winter"). Tranter and Kerr (1969) investigated the zooplankton of the upper 200 m over the same region and also found latitudinal differences, with the biomass being fairly uniform between about 16–27°S latitude but more than doubled to the north (9–12°S) and being reduced to less than half to the south. There was a seasonal variation, the biomass being greatest in September, with a second smaller maximum in March, and least in June (cf. Fig. 3.36).

Collections, using an Isaacs–Kidd midwater trawl, by Legand (1969) over the same region were studied for the larger zooplankton and micronekton. The organisms were

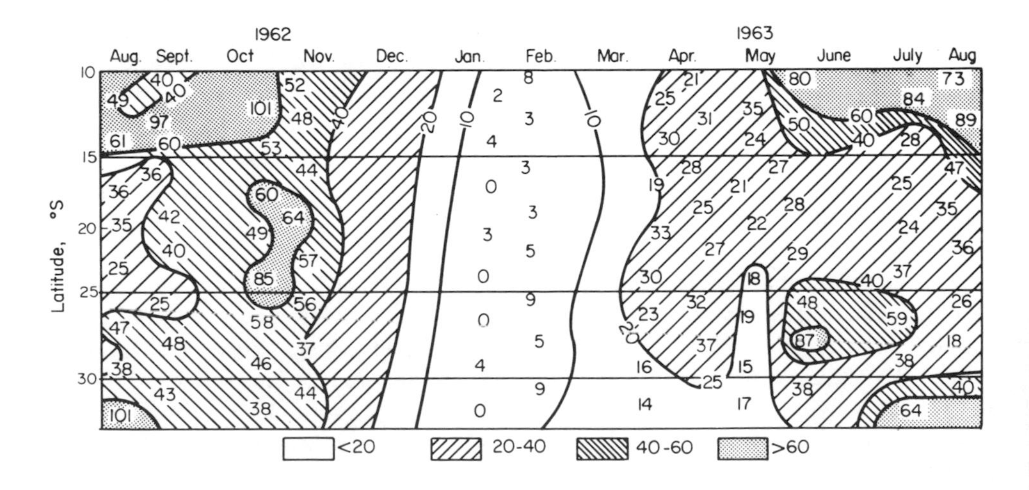

Fig. 3.34. Seasonal variation with latitude of the primary productivity, day samples in mgC/(hr m^2) of the column under 1 m^2 (Jitts, 1969).

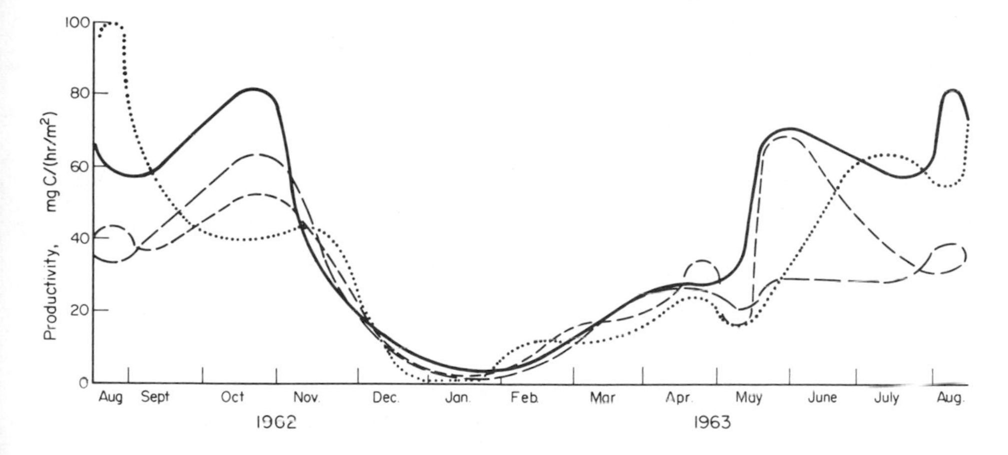

Fig. 3.35. Mean primary productivity under 1 m^2 in four latitudinal intervals. —9–15°S; ═══15–24°S; - - -24–30°S; • • • • • • 30–32°S (Jitts, 1969).

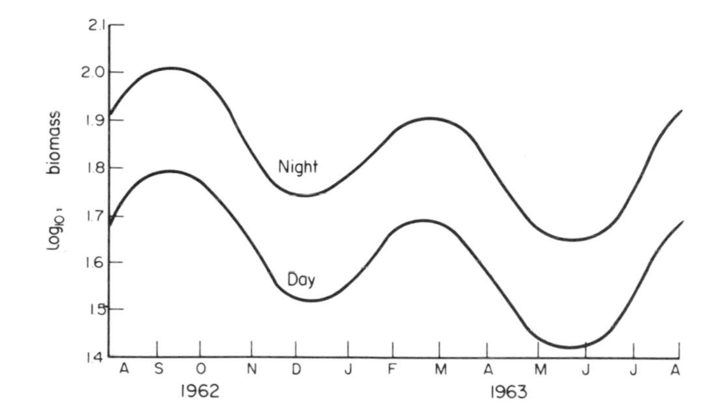

Fig. 3.36. Biomass distribution during the year. Average for the section, as estimated from the regression of $\log_{10}$ (biomass) (mg/m^3) on season. The SD varied between 0.19 and 0.25. IOSN vertical hauls 200–0 m (Tranter and Kerr, 1969).

sorted into four categories: gelatinous organisms, small plankton, macroplankton and micronekton. Latitudinal differences, reported by Jitts, Tranter and Kerr (*vide supra*), were also apparent in this study. The maximum for the plankton and macroplankton was in the north of the region (9°–12°S) except for the gelatinous organisms, which exhibited maxima at 24°S and 32°S. Seasonal differences were also observed, with marked maxima for all four categories of animals between August and October/November and minima between December and February. Less clear secondary maxima occurred between February and July (cf. Figs 3.37, 3.38a, b, c).

Consideration of the numerical seasonal variation of separate groups of zooplankton confirmed the main maximum in the August–November period, especially in the northern and southern regions of the meridian. The more variable time of the smaller secondary maxima, separated by 6–9 months from the main maximum, might reflect the seasonal cycles of dominant species, the young gradually growing at different rates according to their natural cycles. This view received some support from the observation that each group was of smaller average size in the August–November peak, and that average size increased in the secondary peak with growth of

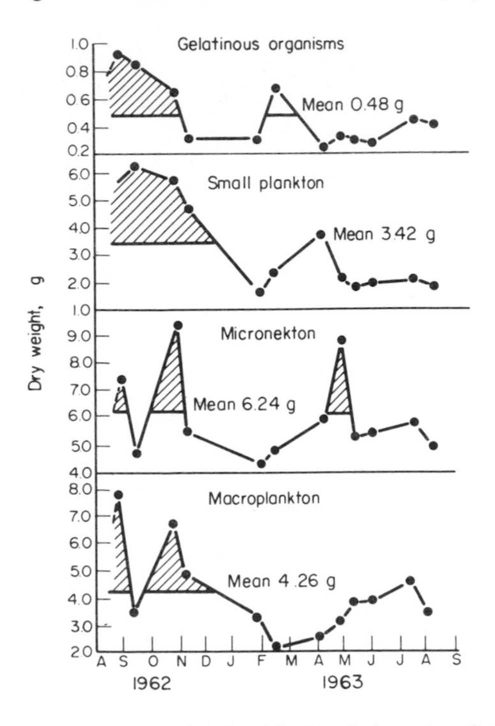

Fig. 3.37. Seasonal variations of mean dry weight/haul for the whole section (9–32°S). Hatched areas show values > the means for the whole period (Legand, 1969).

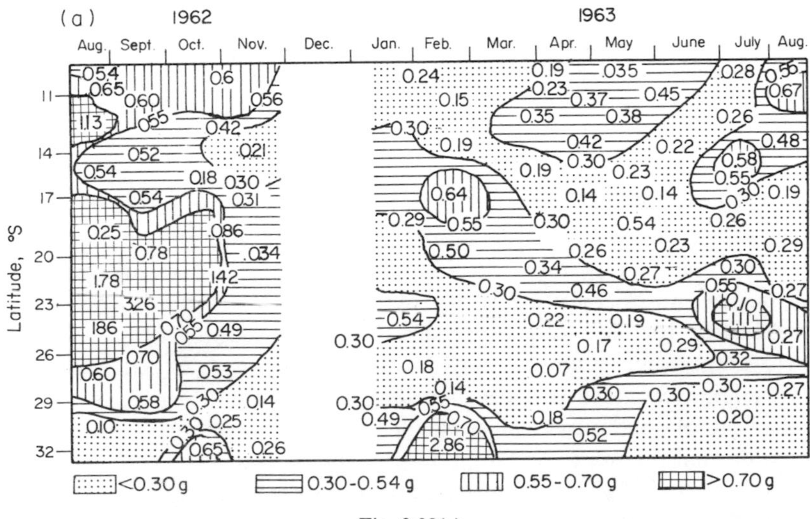

Fig. 3.38(a)

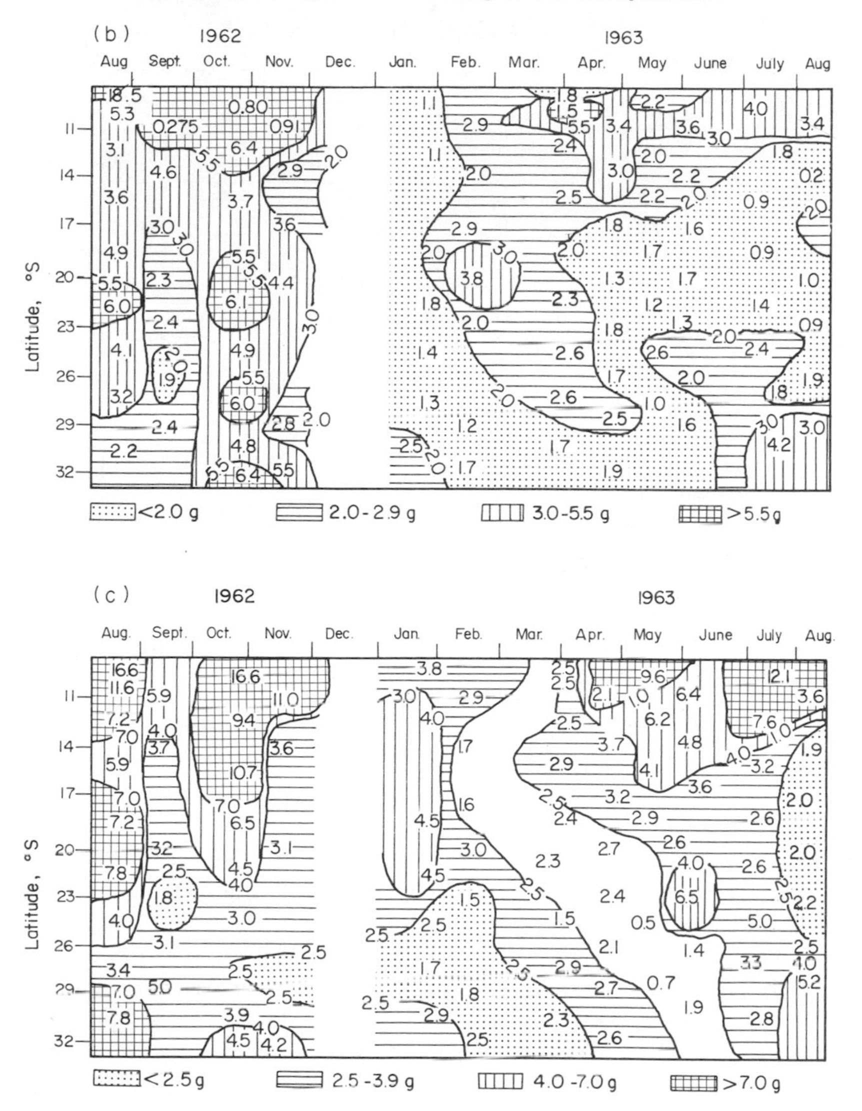

Fig. 3.38. Contours of dry weight/haul as a function of latitude and time a. gelatinous organisms; b. plankton; c. macroplankton (Legand, 1969).

populations. In the northern region, Legand observed that the maxima for the two size groups of plankton captured by the Clarke–Bumpus (smaller zooplankton including herbivores) and the Isaacs–Kidd midwater trawl (macroplankton), were separated by a period of 2–3 months. This could indicate, as Vinogradov and others have suggested, a time separation of different trophic groups dependent on life cycles, growth rates and speed of dispersal by water movement.

Tranter (1973), reviewing this same work in the area, pointed out that the Indian Ocean is more seasonal than other tropical oceans, largely because of the monsoonal influence. Expressing seasonal amplitude as the standard deviation of the monthly (or

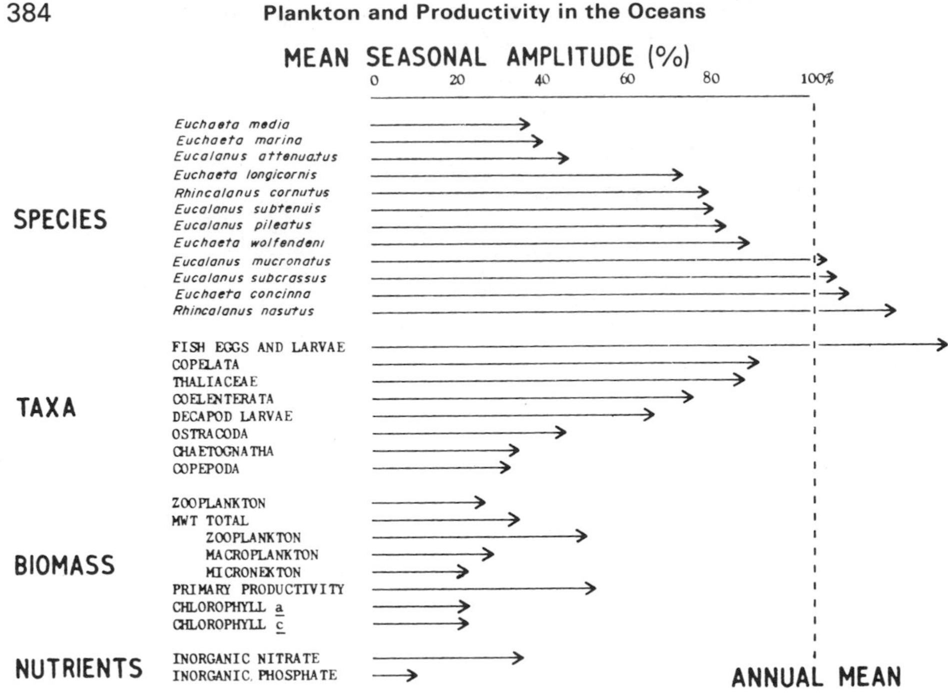

Fig. 3.39. Mean seasonal amplitude of species, taxa, biological stocks and nutrients in the eastern Indian Ocean (Tranter, 1973).

quarterly) means (as a percentage of the annual mean) showed that different zooplankton taxa exhibited differences in amplitude (Fig. 3.39). The greatest variation was for fish eggs and larvae (120 %); Appendicularia and Thaliacea showed an amplitude of around 80 %. Whereas the amplitude for the whole taxon of copepods was around 40 %, at the species level there was distinctly more variation. Tranter also examined differences in seasonal amplitude between the tropical and sub-tropical areas. For some copepods, (e.g. *Euchaeta marina* in the sub-tropics), seasonal differences were small. For others, for example *Rhincalanus nasutus*, there was a greater amplitude of change in the sub-tropics than the tropics, whereas the reverse was true of *Eucalanus subtenuis*. The seasonal amplitude for other areas, calculated as the coefficient of variation of monthly (quarterly) means and separated into coastal and oceanic regions is given in Table 3.5. Although the seasonal variation is small in the eastern Indian Ocean and is less than in the Sargasso Sea, it is greater than for the eastern Pacific. Tranter points out that the large amplitudes typical of high latitude areas may not involve the water column as a whole and may be strongly influenced by vertical seasonal migration (*vide infra*) (cf. Fig. 3.40 and Chapters 1, 5).

Regarding seasonal changes in other taxa in the Indian Ocean, Della Croce and Venugopal (1973) examined the distribution and reproduction of cladocerans, especially *Penilia avirostris*. The species occurs from Cape Agulhas to Zanzibar in a more or less continuous geographical distribution, but its seasonal appearances range from September to March. In the eastern Indian Ocean the species occurs from the Bay of Bengal to Singapore Straits, but its presence in the Bay of Bengal is extremely

Table 3.5. Seasonal variation in phytoplankton (P) and zooplankton (Z) species, taxa and biomass from different areas, calculated as the coefficient of variation of monthly means (Tranter, 1973)

Area	Numerical abundance				Biomass of Zooplankton
	Species Z	Taxa Z	Total P	Total Z	
OCEANIC					
Sub-arctic					
Stn "P" (North Pacific)					85
Stn "B" (Labrador Sea)	228		111	219	
Stn "M" (Norwegian Sea)	246		165		
Antarctic	110			93	41
Sub-antarctic					71
Sub-tropical					
Sargasso Sea		27			87
Gulf Stream		26			
Tropical					
Eastern Pacific					14
COASTAL					
Temperate					
Long Is. Sound			142	80	66
Str. Georgia (BC)	177				123
Plymouth		75		62	
Irish Sea	106				83
Sub-tropical					
Sydney (Aust.)		33	140		104
Tropical					
Singapore		56	130	56	
Cochin (S. India)		107	187		57

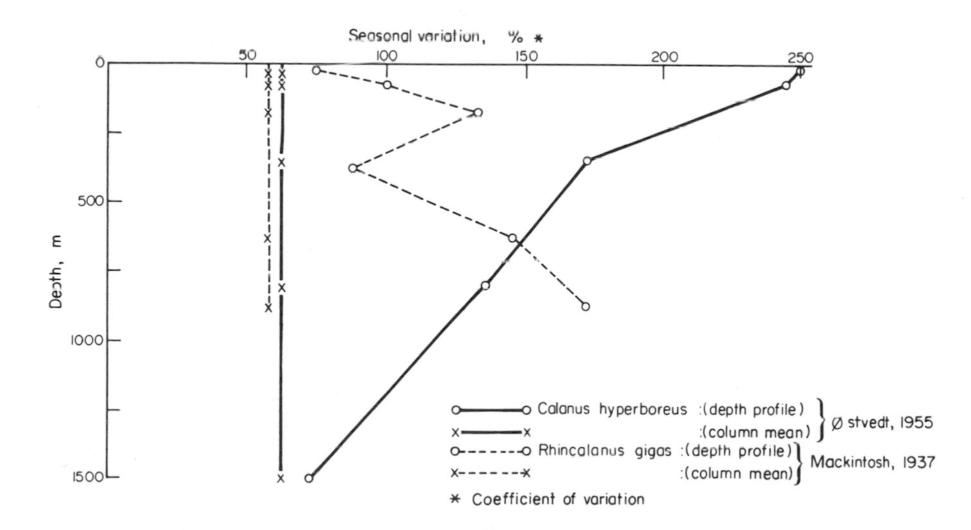

Fig. 3.40. Seasonal variation of zooplankton at high latitudes as a function of depth (Tranter, 1973).

seasonal, usually in March or April. It is characteristic of *Penilia* to show peak populations, for instance, off Madras in October, off Zanzibar in January, during the change-over of the monsoons in Singapore Straits in late September and early April, but apparently all the year round on the Agulhas Bank, with high densities from December to June and July/August off Madagascar. Although *Penilia* can range offshore, it tends to be mainly coastal (Frontier, 1973, 1974). The abrupt outbursts in population are due to the parthenogenetic females, which are derived from resting eggs, passing through a number of broods. These females appear to have six large successive broods, and these in turn give rise very rapidly to further parthenogenetic generations. During the establishment of the peak population, however, depression factors appear and some of the females form resting eggs and produce embryos which develop into males. Although other females continue parthenogenetic reproduction after the depression period, the broods are smaller, so the population declines.

It is convenient to examine the reproductive cycles of *Penilia* in other regions. *Penilia* populations appear fairly suddenly where the regular annual temperature cycles of more temperate latitudes, such as at Naples and Beaufort, North Carolina are typical. The population peaks rapidly but then gradually declines as the favourable conditions change. At higher latitudes (e.g. Narragansett Bay) or at lower latitudes (e.g. Zanzibar), the population appears and disappears very abruptly. Della Croce and Venugopal (1973) believe that a single population is typical of such areas and that hatching of the resting eggs occurs only after the old population has disappeared. In contrast, on the Agulhas Bank and also off Algiers and Sierre Leone, *Penilia* can be present all through the year. Though there is a temperature cycle, a favourable environment is more or less maintained, so that the cladocerans can be present in all months (cf. Fig. 3.41). At Singapore, two populations apparently occur corresponding to the two peaks in the temperature cycle. Breeding of both *Penilia avirostris* and *Evadne tergestina* is linked to plankton abundance following heavy rains at Nosy Bé Madagascar (Frontier, 1973, 1974). According to Della Croce and Venugopal, the breeding cycle has never been clearly demonstrated to be supported by par-

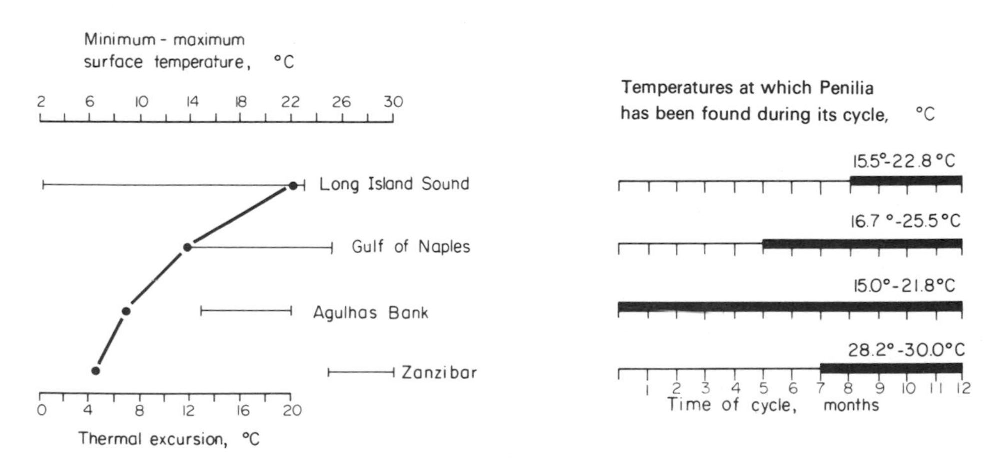

Fig. 3.41. Temperature values and length of seasonal and annual cycles in different areas for *Penilia avirostris* (Della Croce and Venugopal, 1973).

thenogenesis alone. Reproduction in *Penilia* can clearly be very rapid. Estimates by Wickstead suggest that a female may produce six broods in 36–40 days; a brood may bear young in less than 7 days. Though the size of the parent has some effect on brood size, the chief variation in reproductive potential is a cycle in brood size, commencing with high fertility, followed by a decline in the depression period and increasing fertility until the population abruptly disappears. The reproductive potential, however, differs between regions (cf. Fig. 3.42). Further investigations are required to establish those factors which cause the production of resting eggs not only in the Indian Ocean population, but elsewhere.

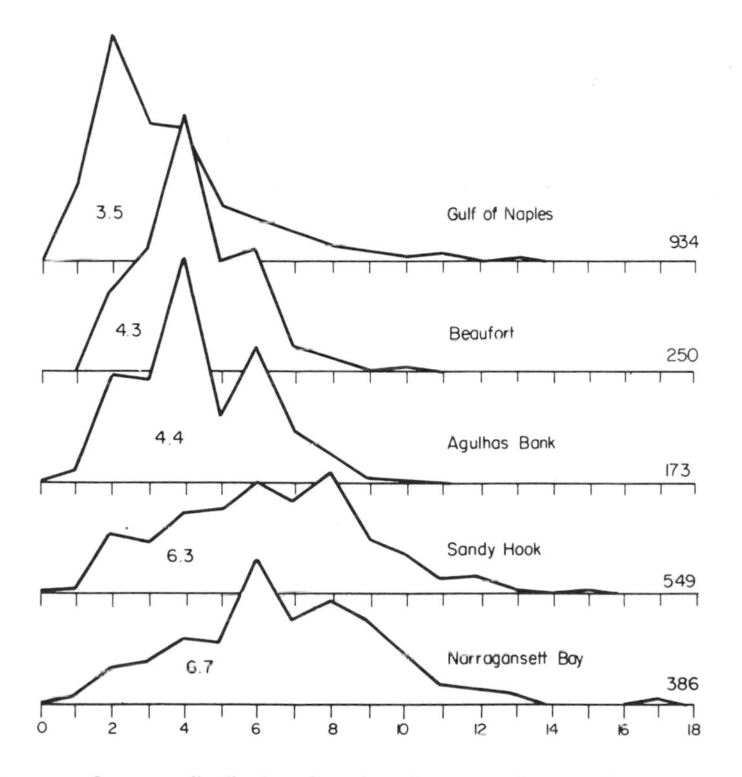

Fig. 3.42. Percentage frequency distribution of number of eggs or embryos per brood during seasonal and annual cycles in different areas for *Penilia avirostris*. For each area, mean number of eggs and embryos (L-hand side) and number of individuals (R-hand side) are given (della Croce and Venugopal, 1973).

Though the IIOE collections have not permitted the establishment of precise breeding cycles for many zooplankton taxa in the Indian Ocean, knowledge of reproductive patterns is available for some groups. Rao and Kelly (1962) examined the breeding cycle of the chaetognath *Sagitta enflata* over three years off Waltair. They found that immature stages and two maturing stages, classified as Stages I and II, existed almost throughout the year. The mature stage (Stage III) occurred in nine months of the period from October 1956 to December 1957, though apparently not in the months of January, May and September. Adults were few from September to November, but immature stages were plentiful. Rao and Kelly concluded that breeding was continuous over the year. The presence of adults in nine months, in addition to the younger animals, may imply at least nine broods from October 1956 to December 1957. Breeding was particularly intense over June and July, and the maximum density for all stages of chaetognath was from June to September. Considerable seasonal variation in size of the various stages of *Sagitta enflata* was observed, but temperature variation was too slight to account for the size differences which Rao and Kelly are inclined to attribute mainly to availability of food. Maximum size of the chaetognaths occurred at times of maximal copepod density.

Nair (1973) studied the breeding in the Cochin Backwaters of two species of chaetognath, *Sagitta bedoti* and *Sagitta enflata*, the same species studied by Rao and Kelly on the Indian east coast and by Furnestin off west Africa and in the Mediterranean. According to Furnestin (1957) and Alvarino (1965), in tropical waters off Africa the species has repeated breeding cycles (probably four) during the year. In the Mediterranean, which may represent the near northern limit of its range, it matures at a much smaller size and the breeding is reduced to one or possibly two cycles. Nair points out that at Aroor, a few miles inside the Cochin Backwaters, the summer monsoon greatly lowers the salinity; from June to November salinity can range from 0.4‰ to 22‰. From December to May the salinity is comparatively high (27.2–32.8‰).

The chaetognaths were also studied at Fairway Buoy, outside the mouth of the Backwaters, where during the south-west monsoon, salinity could range from 11.6–21.5‰. Although fully mature chaetognaths were comparatively rare in samplings, the occurrence of juveniles and of stages with developing gonads were used to indicate breeding intensity. At Aroor, chaetognaths were absent from July to November owing to the very low salinity. During the period when chaetognaths were present, *Sagitta bedoti* appeared to breed more or less continuously, with more intense breeding periods mainly in mid-January, mid-February and April/May. *Sagitta enflata* occurred less regularly in the Backwaters; two peaks of abundance of juvenile *S. enflata* appeared at Aroor in December and February. At the outside station (Fairway Buoy) the breeding cycles were affected by the changes in surface salinity and by some inflow of deeper water in certain months (e.g. September). From January to May *Sagitta bedoti* showed intense spawning. A peak of juveniles occurred in April, but heavy spawnings probably occurred several times, for example, in early February, February/March and in April. Several additional small peaks of abundance may be associated with inflowing water during July, and a further clear peak of juveniles and early maturing chaetognaths was observed in September. *Sagitta enflata* showed similar peaks of abundance from January to May as well as in July, September and December. Despite considerable variation in mean lifespan of both species, the marked overlapping of populations suggested repeated reproductive cycles. Nair believes that apart from the period of exceptionally low salinity at Aroor, breeding is more or less continuous for both species, agreeing with previous studies in other tropical areas. Even at the outside station, the chaetognaths, especially the adults, tend to be restricted to the deeper layers of water during the time of lowered salinity. Table 3.6 suggests that rich populations of copepods can support the sudden burst of chaetognaths.

It is convenient to compare these observations from the Indian Ocean with breeding in other regions. Matsuzaki (1975) found large numbers of *Sagitta enflata* in the East China Sea from July to January; the chaetognaths were absent in April. Adults were

Table 3.6. Average number of Chaetognatha and Copepoda at Aroor and Fairway Buoy during the year 1968 (Nair, 1973)

Month	Aroor		Fairway Buoy	
	Chaetognatha (no/100 m³)	Copepoda (no/100 m³)	Chaetognatha (no/100 m³)	Copepoda (no/100 m³)
January	17	171100	24	90200
February	1121	97800	122	46200
March	175	303600	501	21700
April	80	41400	4765	469500
May	129	55800	4311	101100
June	20	20100	129	45900
July	—	30100	1378	87600
August	—	16600	777	351400
September	—	28500	12843	399300
October	—	8600	141	25800
November	—	297200	—	31900
December	44	116900	474	101900

present from July onwards with maximum populations in that month. Breeding was believed to be confined to the summer, possibly in relation to the fall in temperature during winter. Owre (1960) studied the breeding of tropical chaetognaths in the Florida Current area of the Atlantic, and in particular examined the breeding of *S. enflata*. She points out that *S. enflata* is generally abundant in more coastal tropical areas, though not in salinities below 30‰. In the Florida Current east of Miami, the seasonal maximum for the species occurred in summer (early August). In contrast to chaetognaths at higher latitudes, which usually die after breeding, *S. enflata*, and possibly other species in warm waters, may reproduce more than once during their lifetime. Off Miami, *S. enflata* appears to show continuous spawning, though with periods of heavier spawning superimposed. Differences in the length at maturity, however, indicate that a more rapid maturing of the sagittae occurs during the warmer months. Owre suggests that broods were produced possibly in May and June; that these young chaetognaths themselves bred in August and September, and that their offspring in turn reproduced with October and November broods, but it may well be that maturation is even more rapid. A late major spawning occurs in about early November and another major brood in late February. It is uncertain whether a major spawning occurs during the winter.

Owre believes that another chaetognath, *Sagitta hispida*, breeds all through the year, though with possibly greater intensity in mid-winter and spring. The presence of very young stages indicated particularly heavy spawning in February, April, May, June, August and November. *Sagitta serratodentata* also appears to breed more or less continuously in the Florida Current area, but there are indications of intensification of spawning during autumn and spring. As with *Sagitta enflata*, the proportion of mature chaetognaths preceding these periods of exceptional juvenile abundance was above average, which is additional evidence for assuming intensification of spawning (cf. Figs 3.43a, b). *Pterosagitta draco* appears also to spawn throughout the year, with numerous juveniles occurring in almost every month, but with periods of more intense reproduction, for example, February, April, May, July and October. There is some evidence that individual *Pterosagitta* may also spawn more than once.

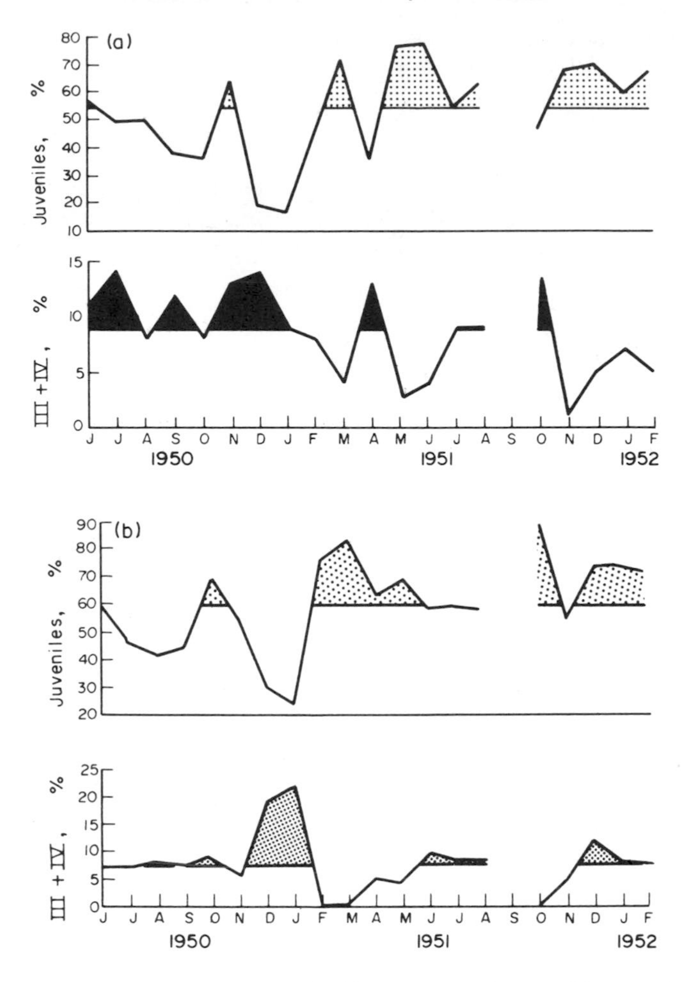

Fig. 3.43. (a) Percentage of juveniles compared with percentage of stages III and IV of *Sagitta enflata* (b) Percentage of juveniles compared with percentage of stages III and IV of *Sagitta serratodentata*. Darkened areas indicate values above the mean percentage (Owre, 1960).

Two chaetognaths occurring in the Florida Current area appear to be more limited in their times of reproduction. *Sagitta bipunctata* seems to be restricted to early spring and summer, although Furnestin (1957) has suggested reproduction throughout the year in warm waters. Breeding is not continuous with *Sagitta minima* off Miami. The reproductive periods are somewhat uncertain, but breeding is mainly during summer, with a fairly clear period in autumn (October/November) and possibly a winter spawning. Despite the very small environmental changes in the Florida Current area over the year, even for those chaetognath species which breed continuously, there are periods of more intense reproduction. Although there is such an extended breeding time, in contrast to the seasonal reproduction in colder waters, the total stock of chaetognaths off Florida is very low compared with that at higher latitudes. *Sagitta serratodentata*, according to Owre (1960), breeds in autumn and winter off Miami, whereas Deevey (1960) found in the cooler waters off Delaware Bay that spawning occurred in summer and late autumn. The lack of mature sagittae in surface waters in winter and spring indicates that no breeding takes place over that period. This is probably a reflection of lowered temperature. The population may need to be recruited from warmer offshore waters.

In the sub-tropical waters of the East China Sea, Matsuzaki (1975) reports on the breeding of *Sagitta nagae*, a species which tolerates cooler waters, now separated from *S. bedoti*, (cf. Nagasawa and Marumo, 1975). Dense populations with a large proportion of adults were present from January to July, indicating breeding during winter, but populations declined over the summer and disappeared during autumn. This cycle is, therefore, the reverse of the breeding pattern of *S. enflata* in the area (*vide supra*).

In the Pacific Ocean, Jespersen (1935) found high densities of macroplankton in such areas as the Galapagos Islands and the nutrient-rich waters of the Peru Current, whereas in the central South Pacific the zooplankton was generally very poor. No data were available on seasonal changes. On the other hand, in the tropical waters of the Sulu Sea and the Celebes Sea a seasonal change in the abundance of zooplankton was recorded, with an appreciable increase from spring to summer. The lack of seasonal change typical of open tropical oceans may therefore be modified through the year by hydrological change due to alternating monsoons, nearness to land, current patterns and other factors. In the open Pacific the well-known divergence close to the equator which is responsible for an increase in primary production is reflected in a greater crop of zooplankton (cf. Austin and Brock, 1959). As King and Demond (1953) and King and Hida (1957) have shown, however, despite greater concentrations of zooplankton near the equator in regions of divergence and upwelling in the Pacific, seasonal changes in abundance are relatively slight.

One of the very few investigations of seasonal changes in zooplankton in open tropical Pacific waters is due to Blackburn, Laurs, Owen and Zeitzschel (1970). In the region more regularly sampled (the western area), a small seasonal variation of the zooplankton in the upper 200 m was found. This was statistically significant for the day plankton collections though not for the night hauls (cf. Fig. 3.44). The maximum in the seasonal fluctuation was during the period April to September and the minimum between October and January, but the amplitude of change was very small (less than 2). A minor fluctuation in the phytoplankton crop might suggest that the seasonal change reflects the food available for the herbivorous zooplankton, but the reasons for any seasonal fluctuation are not obvious.

Of the somewhat larger pelagic crustaceans (apart from *Pleuroncodes planipes*, (which may be termed "crustacean micronekton"), euphausiids formed an important constituent. This crustacean micronekton showed considerable fluctuations with latitude and to some extent with season, but only the seasonal variation at the equator was significant. The maximum occurred in June/July and the minimum in December/January, but the amplitude of change (14.8) was greater than for the general zooplankton (Fig. 3.44). As in other areas, copepods were the dominant zooplankton group (54 %); protozoans amounted to 10 %; chaetognaths, appendicularians, euphausiids, tunicates, miscellaneous animals, ostracods and siphonophores followed in order of numerical importance, ranging from 6 % to 3 %. The very small seasonal change confirms the general view for open warm oceans. The similarity of the night zooplankton collections might be associated with diurnal vertical migration producing

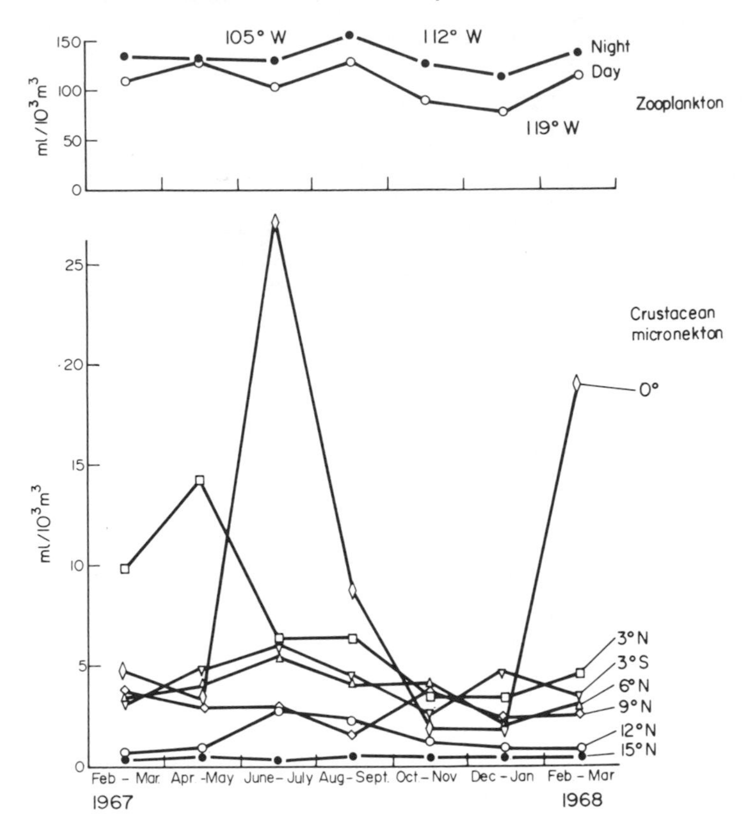

Fig. 3.44. Variation in standing stocks in the West Pacific area. Data are for all latitudes and longitudes combined except where shown otherwise (Blackburn *et al.*, 1970).

a substantial change in the night population. The change in the timing of maximal and minimal abundance of the crustacean micronekton might be related to the considerable carnivorous component of the micronekton. Conclusions must be tentative for the eastern part of the area, since investigations were confined to two periods of the year. Night zooplankton collections showed a significant seasonal effect, with the crop higher in February/March and lower during August/September, but the amplitude was very small (less than 2). At least one series of day zooplankton hauls (in longitude 82°W) also showed a small but significant seasonal change. In general, zooplankton stocks declined from east to west, and some areas showed no seasonal change whatsoever. There were considerable latitudinal variations in zooplankton crop. An obvious maximum close to the equator was presumably associated with the increased phytoplankton at the equatorial divergence, but a maximum of zooplankton was also evident at about 9°–10°N, where no phytoplankton increase was observed. A local convergence might be responsible. In the eastern region, areas close to the coast which experienced upwelling or which had a comparatively shallow thermocline had greater zooplankton concentrations.

King and Iversen (1962), from an analysis of mid-water trawls, found that night catches greatly exceeded day hauls, with faster-swimming organisms such as decapods being taken more at night. Rich catches were characteristic of the region of upwelling at the equator and in the Aleutian Current, and the poorest catches occurred in the South Equatorial Current south of 5°S, in the North Equatorial Current between 10° and 18°N latitude and in Hawaiian waters. In the central North Pacific no seasonal differences were observed in mean catch volumes between late summer and winter. Indeed, although the investigation was not designed to test such differences, King and Iversen remark on the general spatial and temporal uniformity of the collections.

Although such investigations do not yield data on the breeding of individual species in relation to seasonal change, Roger (1974a) provides details of the reproduction of some euphausiids in the tropical Pacific. Important species which could occur in very large swarms were *Euphausia diomedeae*, which at times formed nearly half the population, *Stylocheiron abbreviatum* and *Thysanopoda tricuspidata*. Roger found that although the seasonal variation was small, differences appeared in the breeding periods of the various species, including the three major forms. For example, *Nematoscelis microps*, *Euphausia paragibba* and *Stylocheiron carinatum* possibly had no distinct breeding period. Other species, though breeding over most of the year, showed periods of increased activity. Thus *Thysanopoda tricuspidata* bred mainly from October to December and also in March/April; *T. monacantha* mainly in September/October and January to March and *T. aequalis* appeared to breed all the year round, but the intensity was lower during July/August, a pattern rather similar to that of *Euphausia diomedeae*, whose reproduction was less intense during August/September. *Nematoscelis tenella*, *Nematoscelis boopsis* and *Bentheuphausia amblyops* bred all the year through (Table 3.7). There were also differences between the species in the length of time to reach maturity and in the amount of growth achieved in a month (Table 3.7). However, most mature in about a year, males a little more quickly. For at least two species it is known that females die after a single spawning.

Table 3.7. Breeding cycles of some tropical Pacific euphausiids (Roger, 1974a)

Species	Period to maturity (months)	Growth in one month (mm)	Main breeding period
T. tricupsidata	12–14	2.0	October–Dec. and March/April
T. monacantha	13–15	2.0	Jan–March and Sept/Oct
T. aequalis	10–12	1.2	All year (slow in July/Aug)
N. tenella	11–13	1.6	All year
E. diomedeae	11–13	1.0	All year (slow in Aug/Sept)
N. boopsis	15	1.4	All year
B. amblyops	24	1.0	All year

While most of these euphausiids appear to be largely mesoplanktonic in the tropical Pacific, according to Roger, a few species (e.g. *Euphausia tenera* and *Stylocheiron affine*) are more epipelagic and appear to have a shorter cycle, perhaps only 4–6 months, during which time they grow to a length of 9–12 mm. The data given for the growth of tropical euphausiids suggest, as expected, that growth is more or less continuous through the year. In contrast, most temperate or cold-loving euphausiids tend to have (at best) very slow growth during the winter but accelerated growth in spring and summer. For example, *Euphausia pacifica*, a northern North Pacific species, apparently grows twice as fast as these tropical forms during its main growth period (cf. Mackintosh, 1972, for *E. superba*). *Bentheuphausia amblyops*, a fairly widely distributed bathypelagic species, possibly has a shorter life (cf. Table 3.7) in tropical Pacific waters than in higher latitudes. If rates of growth are similar to that of other bathypelagic euphausiids, its length of life might be at least three years. (cf. Mauchline, 1972).

Along the border of the warm waters of the North Pacific in the Transition Zone, the characteristic euphausiid *Nematoscelis difficilis* has been described by Nemoto, Kamada and Hara (1972) as carrying eggs throughout the year off Japan, suggesting a long breeding period. In Sagami Bay (latitude 34°N), however, breeding appeared to be maximal in winter and spring, while off the Californian coast at a higher latitude (41°N), breeding was apparently concentrated later in the year, during summer.

A special area of the Pacific Ocean where seasonal change in the zooplankton is marked is in the Gulf of California. Manrique (1977) states that the temperature was minimal (16.5°C) in January, remained fairly low till May, but rose very abruptly in June to a maximum (29.7°C) in July. The warm season lasted to about the end of October. Manrique believes that temperature has a marked effect on the seasonal distribution of the zooplankton. Though copepods were the dominant group in number and species, cladocerans, hydromedusae, appendicularians, chaetognaths, thaliaceans and meroplanktonic forms could be important. Copepods were particularly abundant in January, April, August and December; cladocerans were dominant in February and were important in May and July; siphonophora, although occurring over most of the year, became dominant only in October. Generally the holoplankton dominated the meroplankton, but there was a greater proportion of meroplanktonic forms during the warm season, and during June meroplankton actually became the major group, with crustacean larvae, almost entirely brachyuran zoeas, constituting 62% of the biomass.

Although breeding periods of the zooplankton were not elucidated, the seasonal pattern of species, particularly of copepods, is believed to be closely related to temperature changes. For example, *Acartia tonsa* was very abundant in the cold months, but during the warm season tended to be replaced by *Acartia lilljeborgii*, this latter species being overwhelmingly important in August and September. Other species found only or mainly in the cold months included *Calanus helgolandicus*, a very important species, *Labidocera kolpos* and *Labidocera acuta*. *Labidocera johnsoni* and *Centropages furcatus* were found only during the warm season. These times of appearance presumably reflect the periods of active reproduction.

Though Manrique's studies were made in the somewhat warmer waters of the North Pacific and those of Jillett (1971) off New Zealand in the South Pacific, the latter investigations were in warm waters inside the Sub-tropical Convergence and may well be included here.

Jillett (1971) studied seasonal changes in zooplankton in Hauraki Gulf, New Zealand (latitude 36°S). Surface temperature varied from about 12° (August) to 22.7°C (February) at the inshore Harbour Station (20 m); the deeper Channel Station (*ca.* 50 m) had a smaller range (7°C). Surface salinity did not fall below 33‰. Holoplankton was less varied at the Harbour Station, with copepods forming more than 70% of the mean annual catch. Meroplanktonic larvae ranked next (22%); appendicularians and occasionally cladocerans were moderately abundant. In the Channel Station copepods formed 63% of the total, appendicularians 12% and meroplankton only 6%, but some oceanic species and groups of more offshore holoplankton contributed to the zooplankton, at least over part of the year. Figures 3.45a, b illustrate the relative abundance of the major zooplankton groups at the two stations.

At the inshore station, during the southern autumn (March to May), the most important copepods (*Temora turbinata* and *Corycaeus aucklandicus*) gradually declined and were succeeded by *Paracalanus parvus*, a winter form. Over winter (June to August), calanoids became even more important, due to two species, *Paracalanus* and *Acartia clausi*. *Pleurobrachia* was also an important winter animal. During the spring, *Paracalanus* and *Acartia* tended to decline as *Corycaeus* increased; *Temora* was also moderately important. Some meroplankton, including decapod and cirripede larvae, began to increase. Of the two major copepods characteristic of the summer, *Temora turbinata* was abundant and *Corycaeus* fairly plentiful. Cirripede nauplii dominated the meroplankton with maximum populations until about February. Other meroplankton (zoeas, larval polychaetes and molluscs, plutei) were also plentiful, as were the cladocerans *Podon polyphemoides* and *Evadne nordmanni*.

In the deeper Channel Station a succession of dominant copepods was also typical. Over autumn the population was very similar to that inshore, but *Oithona* (not found inshore) was a fairly important species. Doliolids, which were also taken, were never found inshore. Over winter the changes were similar, with *Paracalanus* and *Acartia* dominant, but there were significant numbers of oceanic species (*Ctenocalanus vanus*, *Clausocalanus arcuicornis*, *Oithona plumifera*). Over spring, *Paracalanus* and *Acartia* maintained their importance, and the euphausiid *Nyctiphanes australis* appeared. Copepods, though less abundant than in the shallower station during the summer, appeared as a series of species varying in relative importance. The cladoceran *Penilia* was far more abundant offshore. The changes in species composition at the Harbour Station have been related to the seasonal variations in temperature and salinity. *Acartia* and *Paracalanus* appeared as winter species; *Temora* and cirripede larvae when temperature and salinity were maximal. For the Channel Station, the population was more variable. Changing components were due more to incursions of oceanic water, with winter and late summer as times of strong inflows.

Although at the inshore station there was a tendency for a low density in winter, increasing to a peak at mid-summer, there was marked variability between consecutive samples and between years. At the offshore station there was a more recognizable seasonal pattern, with the minimum in May (*ca.* 2000 organisms/m^3) and maximum in January (*ca.* 35,000/m^3) (cf. Fig. 3.45c). Jillett suggests that these data indicate higher densities and a greater degree of seasonal variation for these sub-tropical latitudes than for areas nearer the equator. The results agree very well with those obtained by Wickstead for offshore waters at Zanzibar (cf. p. 371).

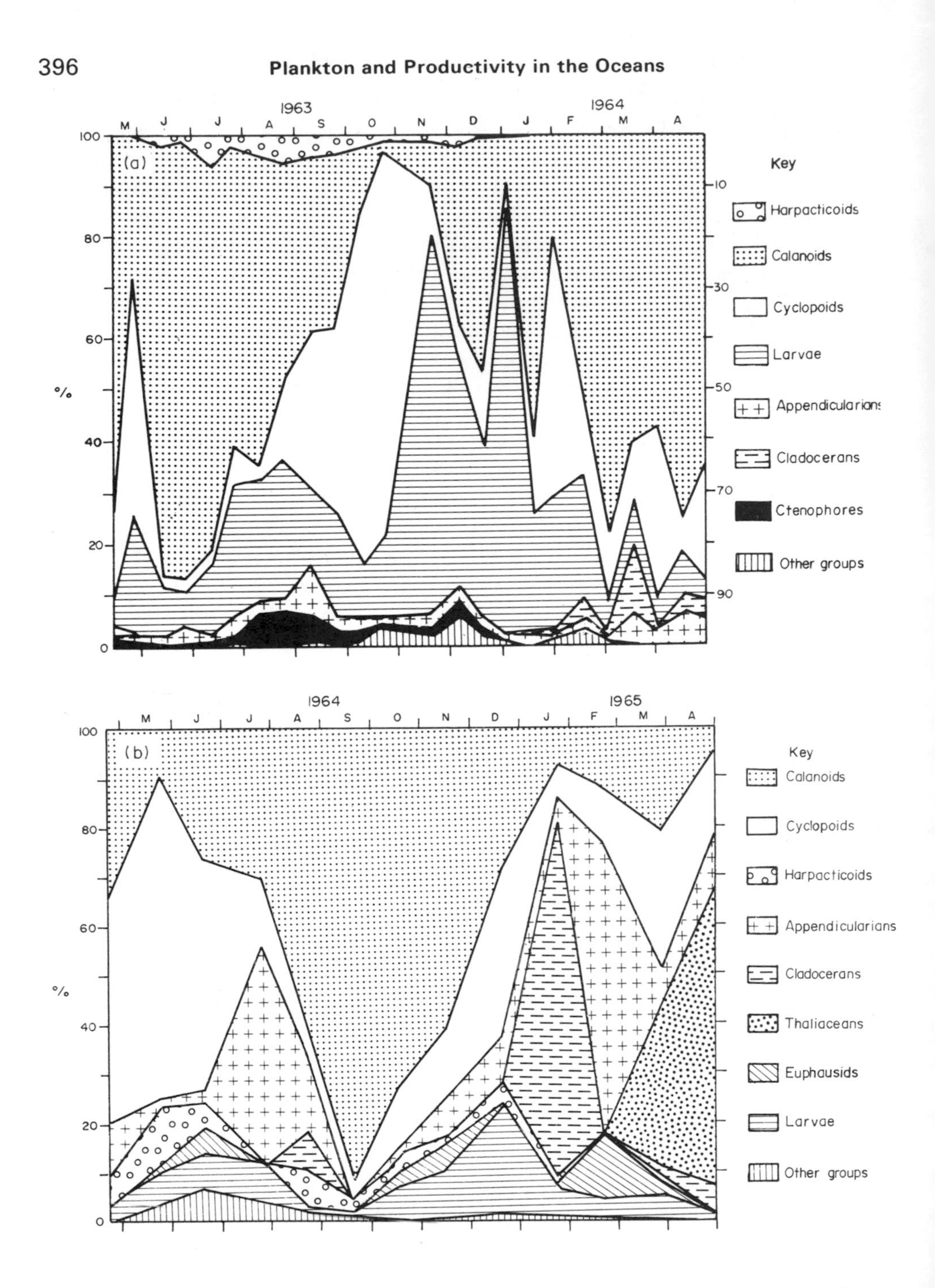

The overall review of the zooplankton of warm waters suggests, therefore, that in open tropical and sub-tropical oceans and even in neritic waters, the holoplankton tends to dominate, with copepods forming the major constituent for most of the year. Seasonal changes in total zooplankton may hardly be recognizable in many open oceans, though where there are considerable seasonal hydrological changes, in

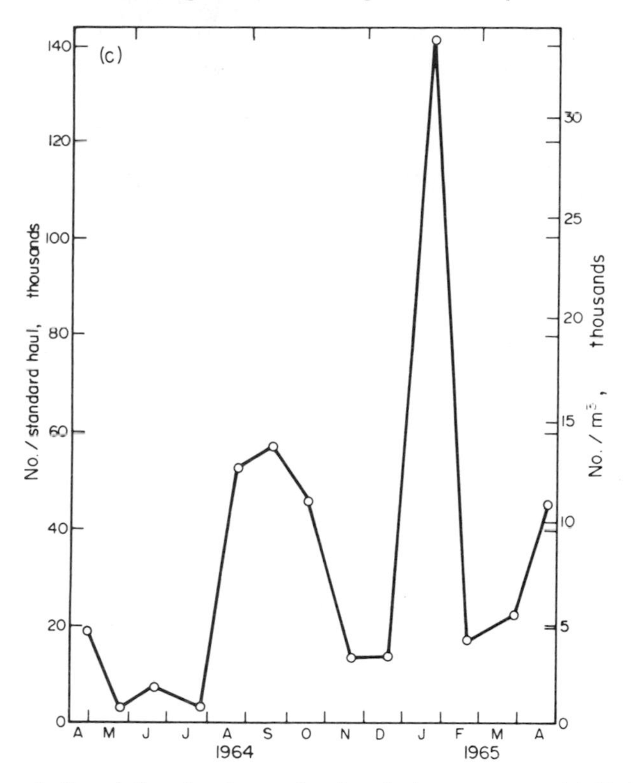

Fig. 3.45. Variations in the relative abundance of major plankton groups at (a) Harbour station and (b) Channel Station (c) Variations of zooplankton numbers at the Channel station (Jillett, 1971).

particular in regions subject to monsoons, seasonal effects may be apparent. Even where seasonal changes have been discovered, investigations have usually concerned total volume or numbers of particular taxa. The breeding cycles of separate species have not usually been studied, though there are exceptions, for example, the work of Binet and Suisse de Sainte Claire (1975). The study by Moore and Sander (1976) is therefore relevant, in that some consideration has been given to specific breeding periodicities in a study of the zooplankton in the Caribbean at Barbados and Jamaica. At Barbados numbers and biomass of zooplankton varied substantially, with many peaks in population, but there was no evidence of seasonality. Similarly, fluctuations appeared to be more or less random off Jamaica.

The neritic zooplankton was dominated by copepods, which formed 74 % and 76 % of the total zooplankton at Barbados and Jamaica respectively.

The contributions of other groups (Table 3.8) indicated considerable similarity in the general composition of the zooplankton between the two islands, though appendicularians were much more important off Jamaica. The number of copepod species was far larger at Barbados, but Table 3.9 shows that two herbivorous calanoids, *Clausocalanus furcatus* and *Paracalanus aculeatus*, made very similar contributions to both populations. Most of the species were widespread tropical and sub-tropical copepods.

Table 3.8. Average percentage contributions of the major zooplankton groups off Barbados and Jamaica (Moore and Sander, 1976)

Group	Barbados	Jamaica
Copepods	84.4	76.0
Larvae	5.0	3.6
Fish eggs	4.8	5.1
Chaetognaths	3.5	1.4
Larvaceans	0.8	11.7
Foraminiferans	*	1.4
Siphonophores	*	*
Thaliaceans	*	*
Pteropods	*	*
Meduase	*	*
Cladocerans	*	*
Euphausiids	*	*
Decapods	*	*
Amphipods	*	*
Ostracods	*	*
Ctenophores	*	*
Lancelets	*	–

* = < 1 %.

Table 3.9. Percentage contributions to total copepods of the ten most common species at the Barbados and Jamaica stations (Moore and Sander, 1976)

Barbados Copepod species	Mean	Jamaica Copepod species	Mean
Clausocalanus furcatus	34.26	*Paracalanus aculeatus*	36.75
Acartia spinata	17.84	*Undinula vulgaris*	12.97
Farranula gracilis	9.24	*Farranula gracilis*	6.57
Oncaea mediterranea	7.77	*Corycaeus latus*	3.51
Oncaea notopus	6.45	*Calocalanus pavo*	2.98
Calocalanus pavo	5.34	*Oncaea mediterranea*	2.66
Oithona setigera	5.03	*Oncaea venusta*	2.49
Macrosetella gracilis	3.90	*Corycaeus amazonicus*	1.86
Oithona oculata	1.53	*Temora turbinata*	1.65
Oithona plumifera	1.35	*Oithona nana*	1.50
Total	92.72		72.94

Moore and Sander used several criteria for distinguishing breeding copepods: the carrying of egg sacs, developed ovaries visible through the transparent body, pigmentation characteristic of ripe females and the bearing of spermatophores by ripe females. Such criteria were employed instead of attempting to trace the development of cohorts of young stages. Breeding of the various species appears to lack seasonal pattern or synchronization. Two breeding types are indicated:

(a) *Continuous breeders*, e.g. *Farranula gracilis* at both stations, *Oncaea mediterranea* at Barbados (cf. Fig. 3.46) and *Oncaea venusta* at Jamaica. *F. gracilis* also appeared to breed continuously further offshore near Jamaica, where it formed an important and sometimes dominant species. Despite being preyed on by larger copepods, chaeto-

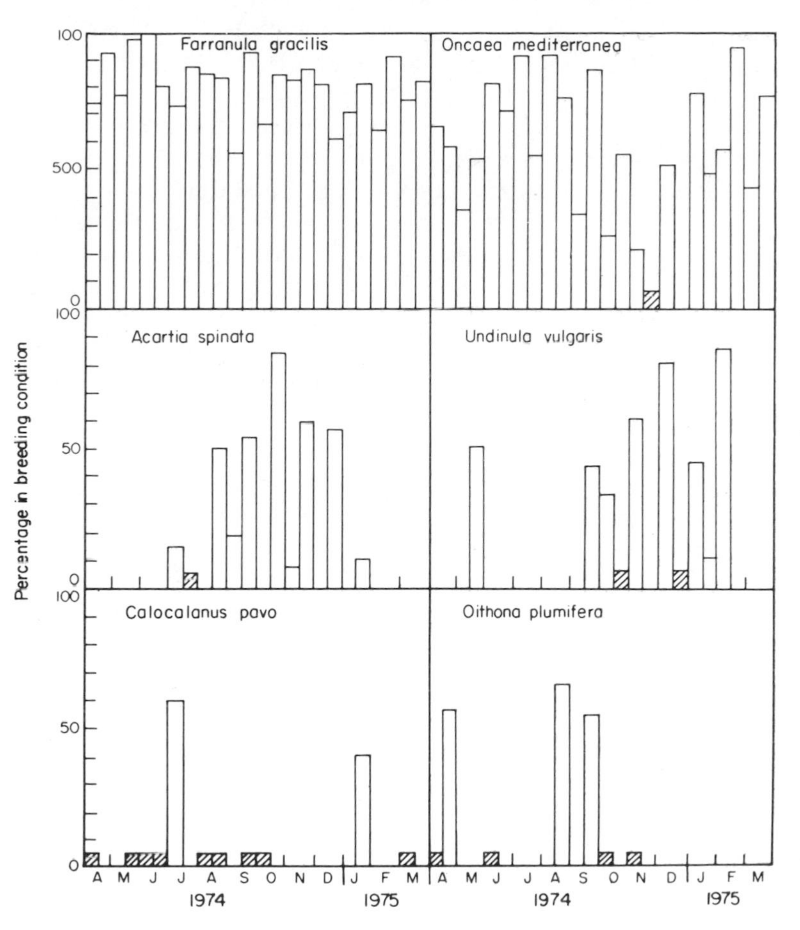

Fig. 3.46. Breeding activities of six copepods expressed as percentage of respective populations in breeding condition. Short hatched columns denote breeding condition observed in less than 1 % of adults (Moore and Sander, 1976).

gnaths and coelenterates, its high reproductive rate permitted the maintenance of variable but high populations.

(b) *Sporadic breeders*, some with protracted reproductive periods. Species which exhibited more intense periods of activity included *Acartia spinata* and *Undinula vulgaris* (Fig. 3.46). At Barbados, *Corycaeus speciosus* also showed an obvious series of intensive breeding periods, 100 % of the population breeding from April to early June, from mid-July to September, in the latter half of January and at the end of February. Although the most numerous species might be expected to breed continuously, *Paracalanus aculeatus* and *Clausocalanus furcatus*, dominant at Jamaica and Barbados respectively, appear to be sporadic breeders. For the former there is some evidence that the cycle of breeding may be of the order of a month and that a rapid series of broods occurs with perhaps occasional intensification. Such a pattern is reminiscent of Woodmansee's (1958) findings for *Acartia tonsa*.

Although the total number of species of copepod actively breeding at Barbados varied from four to twenty-one, no clear periodicity was apparent. For the harpacticoid *Miracia efferata*, a generation time of perhaps two months was indicated, with a fairly long adult life. At Jamaica possibly some degree of periodicity existed in that a very large percentage of the adults were egg-bearing. Despite an apparently high reproductive potential for the species, as estimated from the average number of eggs per egg-sac, the number of egg-sacs and mean generation time, the actual population was small. Some other copepods, (e.g. *Farranula* and *Oncaea*), also appeared to have a very high rate of reproduction. At both Jamaica and Barbados, however, the increase in zooplankton biomass normally expected in proceeding from oceanic to neritic areas was very small, perhaps reflecting the small increase in phytoplankton production. This might be a factor limiting the reproductive potential of the copepods. The differing abilities of small copepods to make rapid use of any available food may be particularly important. At Barbados, *Acartia spinata* showed very large fluctuations in abundance, at times forming 90 % of the zooplankton.

Moore and Sander (1977) also investigated fluctuations in abundance of zooplankton and breeding patterns in the open waters of the tropical western Atlantic off Barbados. Fish (1962) had previously suggested some seasonal change in the area (*vide supra*). Over a period of 28 months, however, Moore and Sander found no clear seasonal variation in zooplankton biomass, though in one year average biomass was greater during September and, during a succeeding year from January to August (cf. Fig. 3.47). The average volume and composition of the zooplankton was fairly comparable to that found more widely over the tropical western Atlantic. Copepods formed the major group both in numbers and in species. The contribution of other groups is illustrated in Table 3.10. Some copepods (e.g. *Undinula vulgaris*) did not show

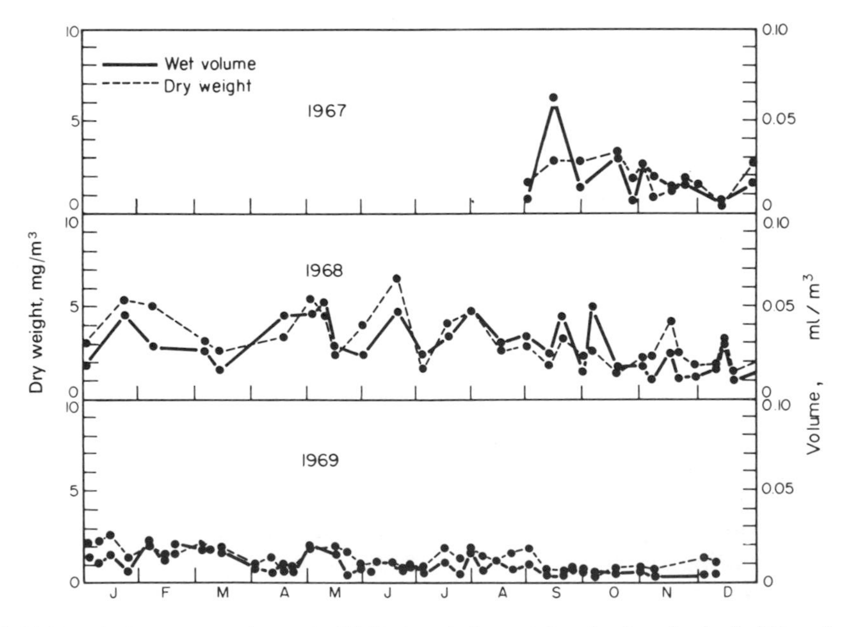

Fig. 3.47. Zooplankton wet volumes (solid line) and dry weights (broken line) of oblique hauls off Barbados. Daily collection values omitted (Moore and Sander, 1977).

Table 3.10. Average percentage contributions by the major groups to the zooplankton near Barbados (Moore and Sander, 1977)

Group	Surface	Oblique
Copepods	61.7	65.4
Fish eggs	19.7	12.5
Foraminiferans	9.6	6.6
Chaetognaths	2.8	4.6
Larvae	2.0	3.0
Ostracods	1.8	3.7
Larvaceans	1.7	2.4

marked fluctuations in abundance. Others (e.g. *Oncaea mediterranea*), while usually present, varied considerably in density, while certain species such as *Scolecithrix danae* occurred only sporadically but could reach high densities (cf. Fig. 3.48).

Altogether 82 species of copepods were taken in breeding condition at some

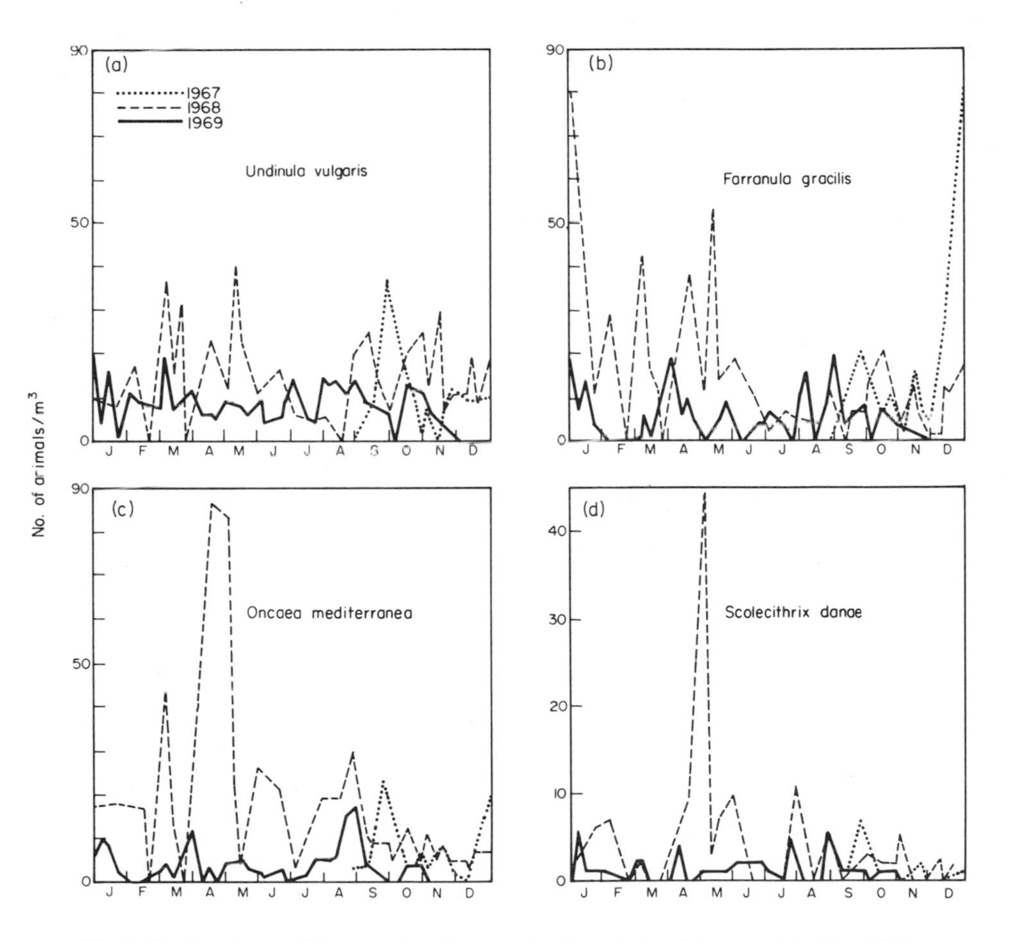

Fig. 3.48. Abundance of four species of copepods off Barbados (Moore and Sander, 1977).

period of the study, but no seasonal breeding pattern was evident either at species or at community level. Three categories of breeding could be distinguished:

(1) Some individuals of the species could always be found in breeding condition (e.g. *Farranula gracilis*, *Oncaea venusta*).
(2) Breeding was discontinuous, with protracted though random periods (e.g. *Undinula vulgaris*, *Macrosetella gracilis*).
(3) Breeding was discontinuous and sporadic and confined to brief periods only (e.g. *Paracalanus aculeatus*, *Corycaeus lautus*).

Since temperature and salinity fluctuations in the ocean off Barbados are very small, they are unlikely to have been serious factors affecting the breeding patterns. Small increases in phytoplankton production such as have been already recorded for warm waters by Menzel and Ryther (1961) in the Sargasso Sea may rapidly benefit the more continuous breeders, increasing brood size. Moore and Sander (1977) also suggest that a rise in phytoplankton above subsistence level might stimulate the more sporadic breeders to commence reproduction. The phytoplankton crop will, however, be so rapidly grazed down that fluctuations are not apparent. It is difficult to indicate factors which could be significant in the breeding of those species showing more discontinuous reproduction, but presumably, with such a diversity of species, very small differences in the total environment are sufficient to favour the breeding of a species.

The successful transport of breeding populations may also be a factor. Owre and Foyo (1964) earlier suggested that the relatively large numbers of copepodite stages of copepods which they found in open Caribbean waters might indicate the eastern Caribbean as a nursery area. With slower water movement, sexual reproduction of the copepods could be particularly efficient. (Owre and Foyo assumed that breeding in such tropical waters was probably year round, while admitting to the paucity of supporting data.) A study of the relative success of young stages of copepods transported to other areas, assuming such a nursery area exists, would be of great interest.

Colder Seas (Temperate and High Latitudes)

Cushing (1975) describes the production cycle for zooplankton, presumably the main period of egg and larval production, as varying between two and twelve months of the year. In most regions the period exceeds the duration of the phytoplankton cycle. The amplitude of seasonal change of the zooplankton also varies, from unity to more than 100, but is generally smaller than for phytoplankton. There are fairly clear indications that the zooplankton of inshore areas shows wider seasonal fluctuations than that typical of oceanic regions, but, according to Cushing, there appears to be no definite correlation between amplitude and latitude. Nevertheless, there is a marked tendency for crop size, and especially fluctuations in zooplankton crop of open warm oceans aside from upwelling areas, to be far less than those at high latitudes (cf. Menzel and Ryther, 1961; Wickstead, 1968; Mahnken, 1969; Sournia, 1969; Blackburn, Laurs, Owen and Zeitzschel, 1970; Jillett, 1971; Panikkar and Rao, 1973).

Total zooplankton production over the year at lower latitudes may be considerable, in view of the greater stability of the population, in contrast to the wide seasonal fluctuations and the lower winter density of plankton which is typical of most high latitudes (cf. page 46). Any examination of seasonal amplitude and production, however, must recognize the widespread tendency of the zooplankton, especially at higher latitudes, to vary its vertical distribution seasonally. Over winter, in very cold seas, much of the zooplankton migrates to greater depths. In coastal areas the extent of vertical migration tends to be limited. Wickstead (1968) suggests that although total annual production may be higher near coasts of tropical regions than in temperate waters, the Annual Fluctuation Factor (*vide supra*) is much larger in colder waters.

Many examples of marked seasonal variations in zooplankton biomass in temperate and higher latitudes can be described. Though spring and summer are the chief seasons of reproduction for most meroplankton, so that for brief periods in inshore waters larvae of benthic invertebrates and of fishes may appear to dominate the plankton, the seasonal changes in the holoplankton due to the breeding patterns of the major species dominate the fluctuations in the zooplankton even over continental shelves. For the Gulf of Maine, Bigelow (1926) described the total zooplankton as being minimal over February and March, but during April there was a marked increase, due almost entirely to the young stages of the copepod *Calanus finmarchicus*. By June, although the proportion of the young stages of *Calanus* had decreased, the zooplankton as a whole was at its maximum density and was almost homogeneous, with *Calanus finmarchicus* dominant everywhere.

However, other copepods and holoplankton, such as euphausiids, hyperiid amphipods, sagittae and *Limacina* reproduced over the later spring and summer months. Although a greater variety of zooplankton was typical of summer, the total quantity of plankton was less than in early June. Other holoplanktonic species, including the copepods *Centropages typicus* and *Temora longicornis*, became more important as the year advanced, and with continued reproduction a second augmentation occurred in September and October. Following this increase the total density of zooplankton began to fall, and a rapid decrease followed in the first months of the year to the winter minimum. Observations of Fish and Johnson (1937), in the Gulf of Maine, confirmed these conclusions (Fig. 3.49). The greatest changes were due to *Calanus finmarchicus*, though *Pseudocalanus minutus* and euphausiids were also important. The maximum quantity of zooplankton occurred in May or June.

At Kiel, Lohmann (1908) found that the zooplankton, of which copepods were the chief constituent, was fairly low and at a steady value in winter from January to early spring. A very sharp rise occurred in the late spring (May), to reach very high densities of zooplankton, though relatively low values occurred over summer. The late summer and autumn saw another large increase, leading to an autumn maximum. Thereafter there was a rapid fall to the winter minimum. Though meroplankton made some significant contribution to the summer populations, the major increases were attributable to the two copepods, *Acartia* (mainly in spring) and *Oithona* (in autumn) (Fig. 3.50). Harvey, Cooper, Lebour and Russell (1935) also demonstrated marked seasonal zooplankton changes in the English Channel. The minimum occurred about January, but in the early spring there was a large augmentation of zooplankton so that during April/May the volume of zooplankton was maximal for the year. Over the summer the numbers, though high, showed marked fluctuations with a second maximum about August/September (Fig. 3.51). This later maximum was the greatest for the year, if total number of animals be considered, though in volume the

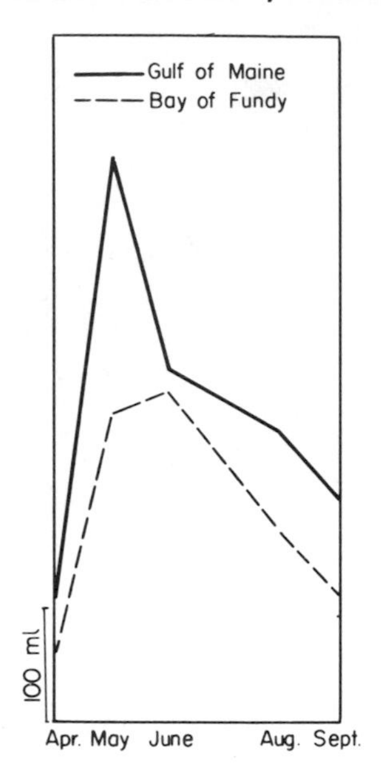

Fig. 3.49. Changes in the mean volume of zooplankton in the Gulf of Maine and Bay of Fundy (Fish and Johnson, 1937).

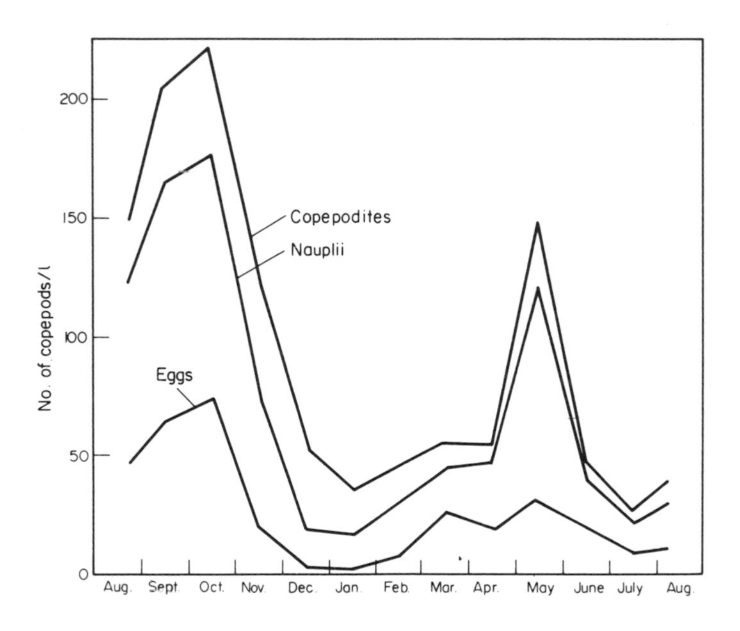

Fig. 3.50. Seasonal variation in the abundance of copepods at Laboe (redrawn from Lohmann, 1908).

zooplankton was slightly less than the spring peak. Harvey's results suggested that changes due to meroplankton were small apart from a contribution by cirripede larvae about March.

Almost the whole of the changes in the holoplankton were due to the fluctuations in

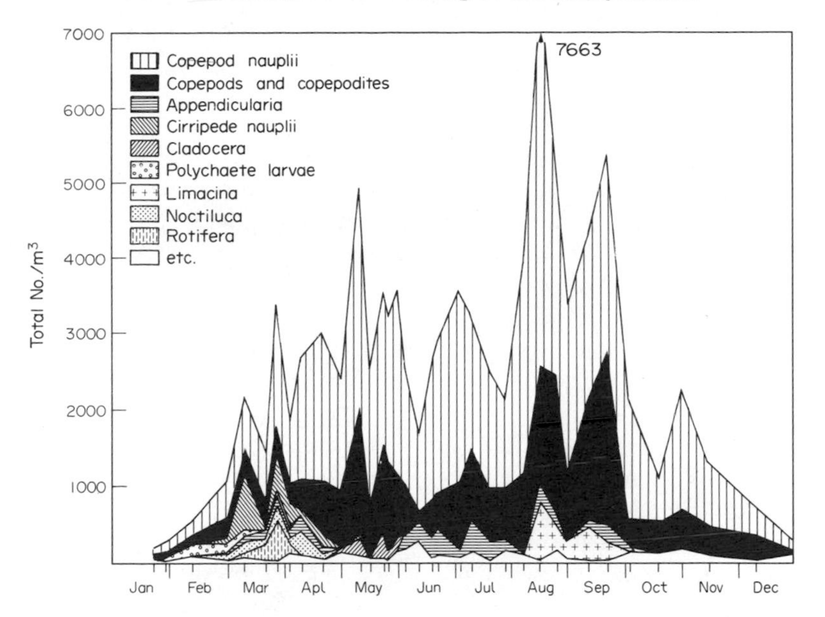

Fig. 3.51. The total number and composition of the zooplankton caught per m^3 at a station in the English Channel off Plymouth during 1934 from 45-m depth to surface (Harvey *et al.*, 1935).

copepods, although there were small increases of appendicularians and *Limacina* in summer (Fig. 3.51). Copepods, however, dominated the entire zooplankton. From April to autumn there were great numbers of copepod nauplii, with copepodite stages becoming more important towards the late summer. The changes were due largely to *Oithona*, *Pseudocalanus* and other small copepods, with a rather large number of *Temora* occurring in the first peak in May (Fig. 3.52). This pattern of zooplankton

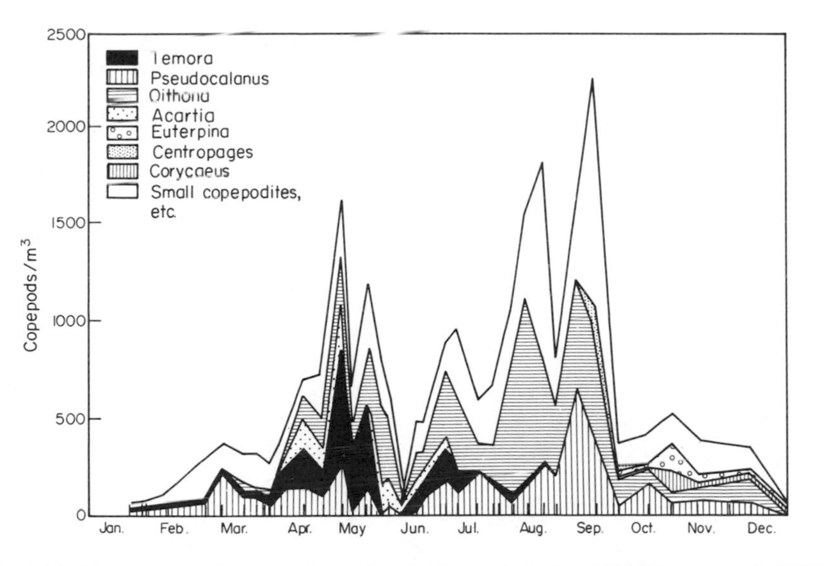

Fig. 3.52. The numbers of copepods caught off Plymouth during 1934 (Harvey *et al.*, 1935).

changes throughout the year off Plymouth has been confirmed by Mare (1940), who showed a very low zooplankton minimum in January/February, a sharp rise in April and May due to the reproduction of copepods, and a decline in summer, followed by a second maximum in the late summer.

The results of the investigations of Deevey (1956) on the zooplankton of the inshore areas of Long Island Sound and Block Island Sound show marked seasonal fluctuations in zooplankton densities, with maximum numbers appearing in summer. The main changes were again due to copepod reproduction, especially to the species *Acartia tonsa*, *A. clausi*, *Oithona* sp., *Temora longicornis* and *Paracalanus crassirostris* (Fig. 3.53). Wiborg (1954) also found that the contribution of meroplankton in coastal Norwegian waters was small: copepods were responsible for the overall population changes, contributing generally some 80–90 % of the total zooplankton. Of other holoplankton, appendicularians, cladocerans and euphausiid larvae were important at times (Fig. 3.54). The copepods *Calanus finmarchicus*, *Pseudocalanus elongatus*, *Microcalanus pusillus* and *Oithona similis* were most significant numerically over the year as a whole (Fig. 3.55). Two peaks occurred in the zooplankton population, the first during spring and early summer and the second between August and October. These peaks would appear to reflect chiefly the seasonal reproductive cycle of the more important copepods.

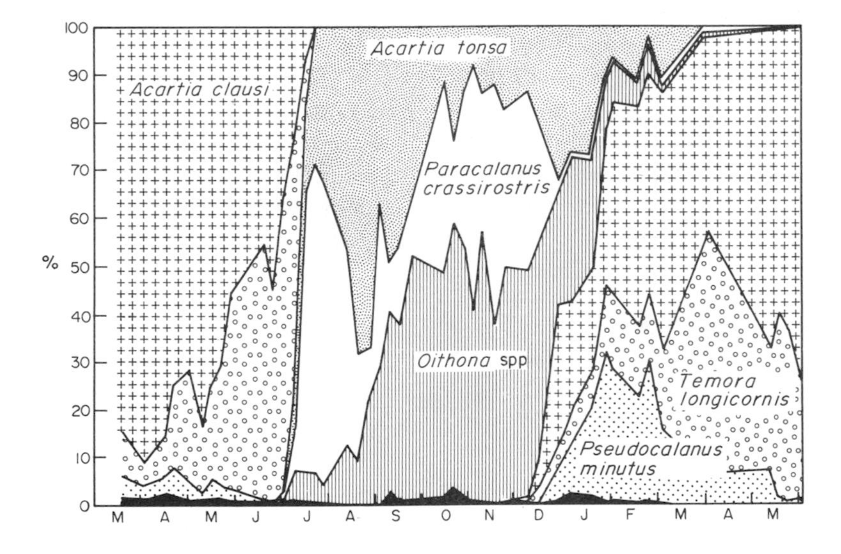

Fig. 3.53. Relative percentages of important copepods obtained from Long Island Sound over 15 months (Deevey, 1956).

The breeding cycle of *Calanus finmarchicus* is described in detail later, with that of some other copepods, but in support of the general view that copepod breeding is outstandingly important in the zooplankton cycles of temperate regions, it is significant that Marshall (1949) showed that the smaller copepods in Loch Striven had marked seasonal breeding, with maximum numbers of all species except *Temora* in July or August, though the precise date varied from one species to another. There was

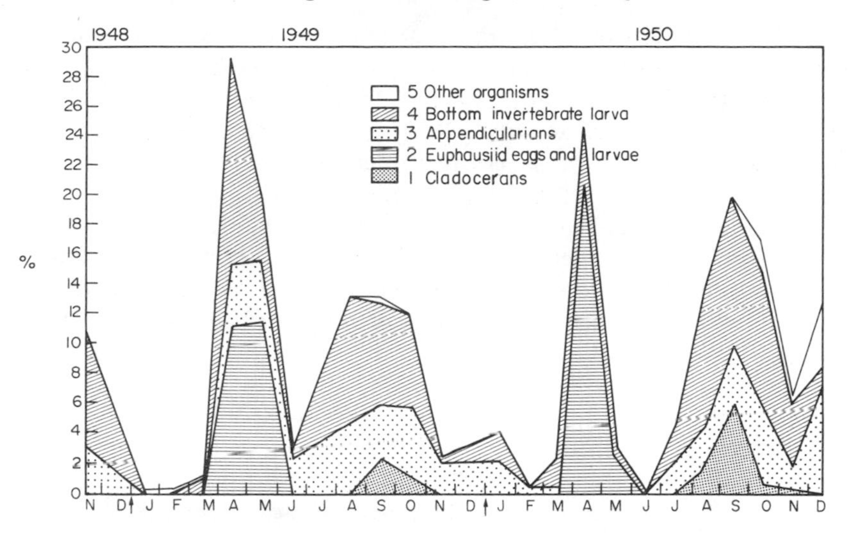

Fig. 3.54. The relative importance of "non-copepod zooplankton" at Eggum, Norway (Wiborg, 1954).

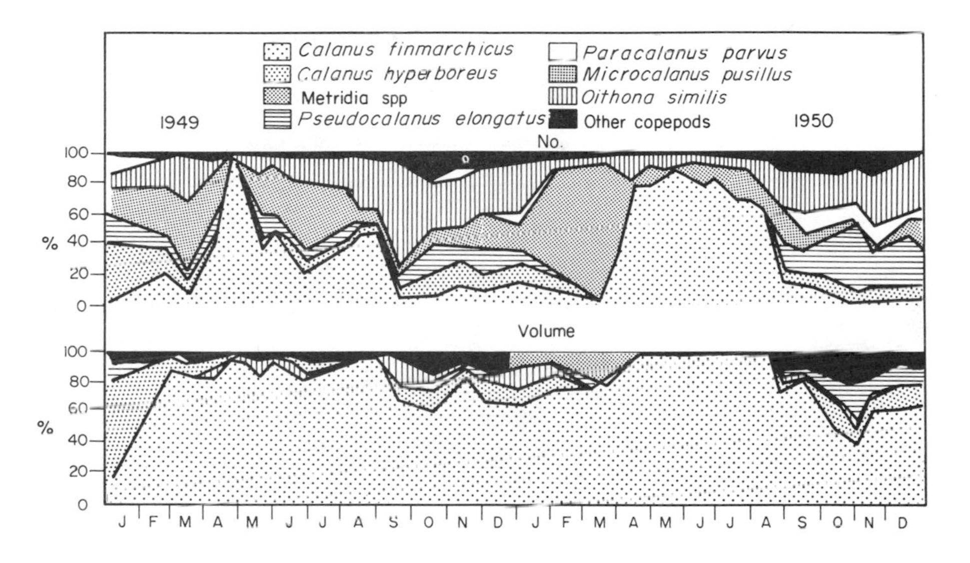

Fig. 3.55. Variations in the percentage composition of the copepods at Eggum, Norway (Wiborg, 1954).

usually a secondary, smaller and variable maximum in the spring about April or May; with *Temora* this peak was somewhat larger than the late summer density.

Of particular relevance to the observations of Harvey *et al.* (1935) on the zooplankton off Plymouth are studies by Bodo, Razouls and Thiriot (1965) on seasonal variations in zooplankton in the English Channel off Roscoff. Plankton collections were made at a shallow station (20 m depth) and at a deeper station (92 m) some 13 miles offshore, the latter being qualitatively similar to that off Plymouth. Inshore, though meroplankton was important, two copepods, *Centropages hamatus* and *Temora longicornis*, were the major species, with *Euterpina acutifrons* also fairly abundant. Offshore, though seasonal changes were similar, the major species were *Pseudocalanus elongatus*, *Acartia clausi* and *Oithona* spp., in that order (cf. Fig. 3.56a, b). At both stations, Bodo *et al.* demonstrated a very marked seasonal variation. The amplitude of change inshore, while not precisely determined, was of the order of 20-fold; offshore it was slightly greater.

The population, minimal in January/February, was made up largely of *Oncaea venusta*, with a few *Pseudocalanus*, *Acartia*, *Oithona* and *Euterpina*. In March–April the first generation of the spring/summer species (*Centropages hamatus*, *Temora longicornis*) occurred, the main spring burst of breeding being completed by June.

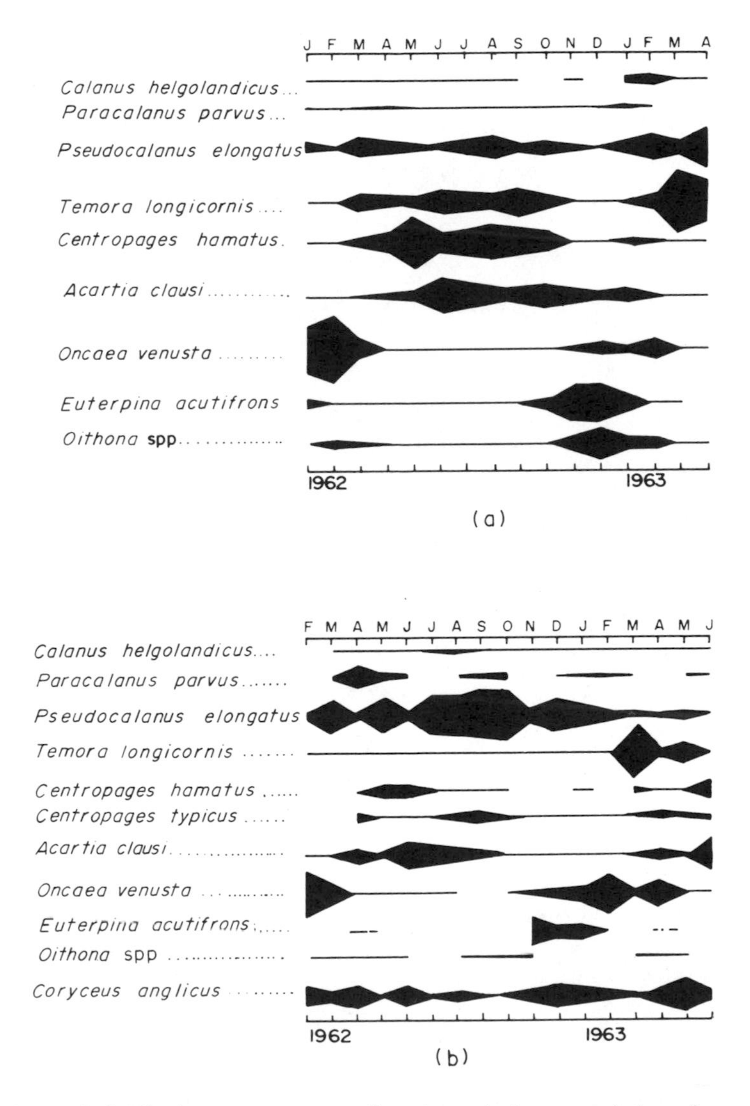

Fig. 3.56. Seasonal distribution as percentage of total zooplankton population of copepod species. (a) coastal station; (b) offshore station (Bodo, Razouls and Thiriot, 1965).

Whereas *Centropages* was the more important copepod in May, *Temora* reached its maximum in July. Cladocera were abundant only in late spring and summer. A second copepod peak occurred in August, with further broods by different species, but total biomass was maximal and more or less constant over the summer months. In September and continuing through October the population began to decline, with *Euterpina acutifrons*, *Oithona helgolandica* and *Oncaea venusta* (autumn/winter forms) succeeding the summer species.

Bodo *et al.* believed that temperature had a less marked effect on reproduction cycles compared with the seasonal variation in phytoplankton and possibly interspecific competition. For the major copepods (*Pseudocalanus*, *Temora*, *Centropages* spp., *Euterpina*) there were possibly five generations (four for *Acartia*), and presumably breeding does not entirely cease even over winter (cf. Fig. 3.56a, b). The results are similar to those of Digby (1953) for the breeding of small copepods off Plymouth. The usual inverse relationship between body size and temperature was observed for all the major copepods (*vide infra*). Eriksson's (1973a) studies off Goteborg, Sweden, also pointed to substantial changes with season in the zooplankton of shallow waters. Holoplankton species, especially copepods, were largely responsible for the changes. In neritic waters, the variety of copepod species was not large. *Oithona similis* was most abundant, followed in numerical importance by *Paracalanus parvus* and *Pseudocalanus elongatus*. Seven other species were important: *Acartia clausi*, *A. longiremis*, *Temora longicornis*, *Centropages hamatus*, *C. typicus*, *C. finmarchicus* and *C. helgolandicus* in that order, though *Calanus* densities, especially juveniles, were probably underestimated.

Seasonal changes in the holoplankton of colder waters, especially of temperate seas, must be greatly influenced by the seasonal variation in temperature, which will be more marked over shallow continental shelf areas. Over deeper areas the zooplankton may to some extent avoid seasonal temperature variation by vertical migration. Nevertheless, quite apart from the cycle of phytoplankton and other seasonal changes, which exert a profound influence on the pattern of the zooplankton population (cf. Volume 1), the temperature changes of the surface and sub-surface layers to some degree dictate the appearance and abundance of species. A limited species succession is often observed. The breeding cycle of holoplankton (number of broods, length of generation, size of individual) can also be considerably influenced by the seasonal temperature pattern, though other factors, particularly food supply, may be as important (*vide infra*).

In Eriksson's (1973a) investigations off Goteborg, for example, the colder species (*Calanus finmarchicus*, *Pseudocalanus elongatus*, *Centropages hamatus* and *Acartia longiremis*) occurred more in the early half of the year, and were followed by *Calanus helgolandicus*, *Paracalanus parvus*, *Centropages typicus* and *Acartia clausi* in the warmer half of the year. The development period may also be shortened during the summer; Eriksson estimates that for some small copepods it may be only three to four weeks. *Pseudocalanus* was usually dominant over the colder half of the year, though the pattern might be complicated by the possibility that it was represented by more than one species. *Pseudocalanus* was succeeded by *Paracalanus*, but an unusually cold spring could delay the reproductive cycle of *Pseudocalanus* so that the two species more or less coincided in their major period of abundance. While other species may alternate in their time of maximal density between colder and warmer seasons (cf. *Calanus finmarchicus* and *C. helgolandicus*; *Acartia longiremis* and *A. clausi*), this may also be related to the fact that some species are immigrant forms, whereas others are autochthonous. Thus *Pseudocalanus*, *Oithona*, *Temora*, *Centropages hamatus* and *Acartia* spp. are probably autochthonous copepods and formed the major part of the fauna off Goteborg.

Paracalanus parvus, despite being an immigrant form, was the second most abundant copepod, while *Calanus* spp., also immigrants, were much less plentiful. Periodic inflows of water may therefore be significant in seasonal changes in zooplankton (cf. Chapter 4). Temperature is in any event almost certainly not the only factor in succession. Food supply may be significant, affecting the size of copepods, but as Eriksson suggests, related species with very similar feeding requirements may tend to be separated in time in their reproductive cycles (e.g. *Acartia longiremis* and *A. clausi*). If food selected were more differentiated, the copepods might overlap to some degree (e.g. *Pseudocalanus* and *Paracalanus*). By contrast, *Centropages typicus* and *C. hamatus*, as omnivorous/carnivorous feeders, and *Oithona similis*, a carnivorous copepod at least when adult, would not compete with the herbivorous species. But, a species succession can appear between *Centropages hamatus* (winter) and *C. typicus* (summer).

The seasonal changes in zooplankton off Goteborg showed several peak densities in the shallow water (20 m) over spring and summer (April 14,200/m^3; July 12,500/m^3; September 6600/m^3). Farther offshore (>40 m), densities were not quite so great, though in the second summer high values were found in August (12,900/m^3) and September (9200/m^3). The minimum annual population occurred in January/February (*ca.* 200/m^3). Actual densities would have been considerably greater had copepod nauplii been adequately sampled. Eriksson (1973a) points to the high proportion of copepod nauplii (62 % of small copepod population) obtained by Digby (1950), using a net of 0.05 mm mesh in an English Channel survey, compared with a mean of only 13 % as nauplii found by Eriksson using a coarser net (0.16 mm mesh).

Modifications of the timing of the seasonal cycle of holoplankton in colder waters occur especially with latitude, essentially reflecting lower temperatures. Generally the spring burst of zooplankton and the attainment of a summer peak comes later on proceeding to high latitudes and the season of abundance is typically shortened (cf. Volume 1).

Steemann Nielsen (1937) suggests that in northern waters the augmentation of the plankton may be as late as June or July and that numbers are declining again from September onwards. Digby (1953), analysed the changes in biomass of the zooplankton in east Greenland waters and found that the maximum occurred in July and August and the minimum towards the end of March. The late summer biomass was almost twenty times that of the winter minimum. Medusae and ctenophores occurred all the year round, and during late summer a great quantity of gelatinous material occurred which was apparently derived largely from ctenophores. Appendicularians were also fairly abundant in late summer. Other organisms contributed very little to the total zooplankton apart from copepods, which were again responsible for the major changes in zooplankton abundance (cf. Fig. 3.57). Other workers have noted that relatively large increases in zooplankton may be due to bursts of medusae and ctenophores in Arctic waters, and to an abundance of salps in the rather warmer waters.

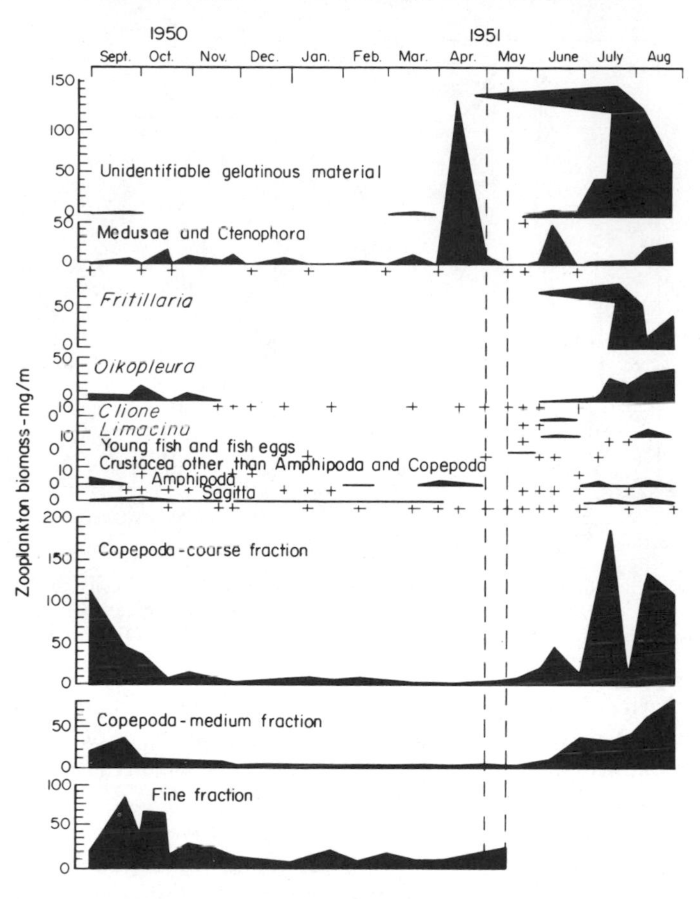

Fig. 3.57. Biomass of zooplankton organisms in upper 50 m at Scoresby Sound, Greenland. Vertical hatched lines represent the onset of spring phytoplankton production as shown by the appearance of diatoms and by plant pigment analyses (modified from Digby, 1953).

Breeding of copepods in the North Atlantic

There is general agreement that the major fluctuations in zooplankton abundance are due to copepods. Thus, apart from the observations already noted due to Bigelow, Lohmann, Harvey *et al.*, Wiborg, Deevey and others, Fish (1954) noticed the marked augmentation of the zooplankton in oceanic waters off Labrador associated with the rapid increase in calanoids, especially *Calanus finmarchicus*. This copepod dominated the changes in the total zooplankton, with a peak density in July and August (cf. Fig. 3.58). The major species appeared to have a single brood during the relatively short summer. Kielhorn (1952) had earlier drawn attention to the outstanding importance of *Calanus finmarchicus* in the central Labrador Sea, and suggested that the copepod had one main annual reproductive cycle with the main augmentation in numbers in late July. However, two copepods, *Oithona atlantica* and *Metridia longa*, bred in mid-winter.

The marked changes in zooplankton density obviously focus attention on the breeding cycles of holoplanktonic animals, especially copepods. Comparatively few

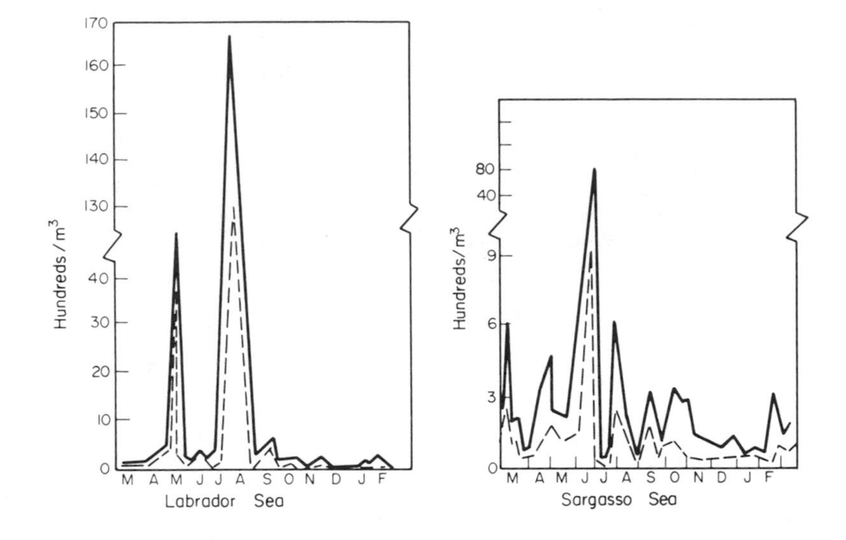

Fig. 3.58. A comparison of the annual cycle of zooplankton at an oceanic station in the Labrador Sea and at an oceanic station in the Sargasso Sea. Full line denotes total zooplankton; hatched line for Labrador Sea denotes *Calanus finmarchicus* and for Sargasso Sea, all copepods (redrawn from Fish, 1954).

species have been extensively studied, the best known being *Calanus finmarchicus*. As far as is known, all calanoid copepods show a similar pattern of development, passing through six nauplius stages and six copepodite stages, of which the last, Stage VI, is the adult (cf. Marshall and Orr, 1955a; Johnson, 1958).

The most extensive study of the reproduction of *Calanus finmarchicus* is that of Marshall, Nicholls and Orr (1933–35) for copepods of the Clyde Sea Area. They showed that in autumn the number of copepods was falling to reach a minimum density about February/March. The autumn and early winter was passed chiefly as the Stage V copepodite, but about the end of December those copepodites which had survived the early winter commenced their final moult to the adult condition. Thus by February, although the actual numbers of *Calanus* were very low, the percentage of adults found in the catches was at its maximum for the year, and these adults included a considerable proportion of males, which are normally scarce (cf. Figs 3.59, 3.60). This stock which had persisted through autumn and part of the winter now spawned, so that the eggs comprised a large percentage of the catch towards the end of February and beginning of March.

Although the percentage of *Calanus* ova was high, the number of eggs obtained from the small breeding population (Fig. 3.59) was not very large. The brood developed during March, and it was possible to follow a succession of naupliar and copepodite stages. This first brood attained maturity by about the beginning of April, and a second spawning occurred about April/May when the usual high proportion of eggs and nauplii was evident. The density of this generation was overwhelmingly greater at certain stations sampled, so that a very sudden rise in the overall population of *Calanus* was experienced in these areas. This second brood in turn grew up and became adult, and a third spawning occurred about June/July. This third generation passed through the various successive naupliar and copepodite stages as had earlier broods, and it is

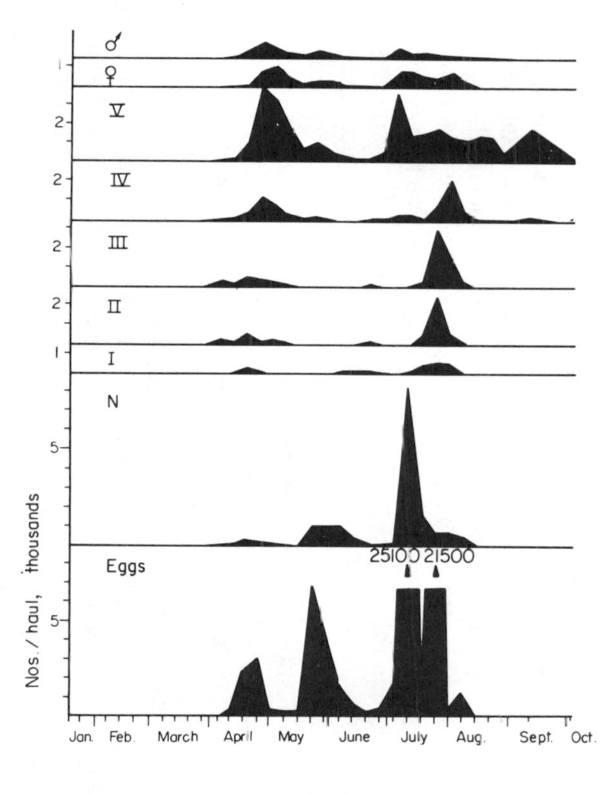

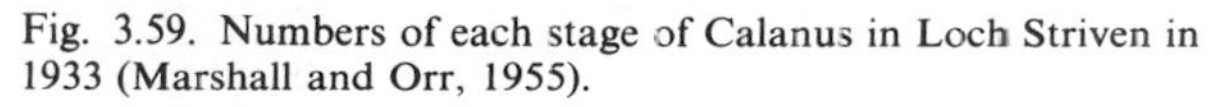
Fig. 3.59. Numbers of each stage of Calanus in Loch Striven in 1933 (Marshall and Orr, 1955).

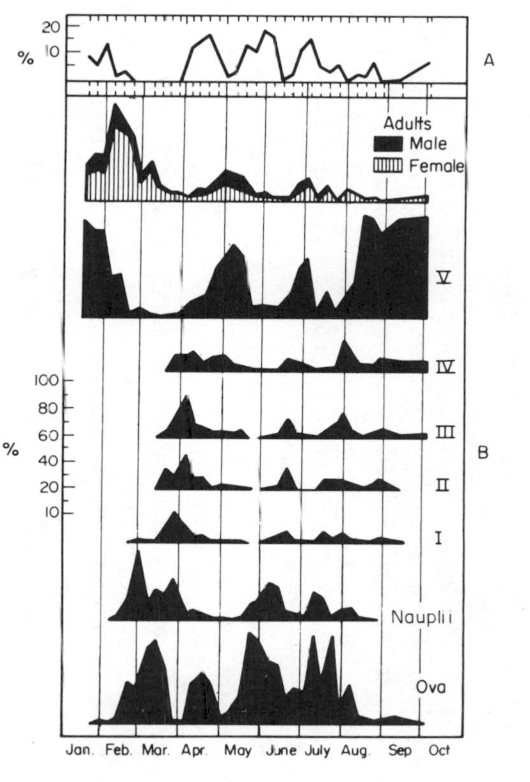

Fig. 3.60. Percentage composition of hauls of *Calanus* in Loch Striven in 1933. A = percentage of ♀♀ bearing spermatophores; B = succession of developmental stages through each breeding period (Marshall and Orr, 1955).

possible that a few passed through the final moult to attain the adult condition. But the main proportion of the third generation ceased moulting at the Stage V copepodite stage. As a result there tended to be an accumulation of Stage V copepodites over the late summer (August/September) (Fig. 3.60).

Nicholls and his fellow workers noticed on some occasions small numbers of eggs and nauplii during the August/September period, which suggested that there may have been some small-scale, irregular breeding after the third main generation of June/July (Fig. 3.60). This implied that a few of the ova from the June/July spawning developed to Stage V, but instead of persisting in this stage, moulted to give adult *Calanus* which bred to give rise to a very small number of ova. In any event, the offspring of these irregular small-scale spawnings grew up only to the Stage V copepodite condition and added to the general stock of Stage V overwintering forms. It appears, therefore, that in the Clyde Sea Area there are three major generations: February/March, April/May and June/July. Later work by Marshall and Orr (1955a, b) suggests that in such temperate areas the development from egg to adult takes about one month in *Calanus finmarchicus* and a further period of approximately one month is necessary for the ripening of the ova in the females. However, it appears that egg-laying can go on for at least two or three weeks, the release of ova being considerably affected by the density of phytoplankton available as food.

The total life span for a female *C. finmarchicus* of the main spring or summer stocks is of the order of two and a half months. On the other hand, approximately seven months is a reasonable estimate for the life span of the overwintering brood spawned about June/July. The length of life of male *Calanus* seems to be distinctly shorter than that for the females. Raymont and Gross (1942) suggested that males not only had a shorter adult life, but that they fed much less extensively, a point which was confirmed by Marshall and Orr (1955b).

Variations in the mean size of different generations of copepods have been noted for some time. Adler and Jespersen (1920) found the size of *Temora*, *Pseudocalanus* and *Calanus*, including copepodites and adults, to be minimal in summer and maximal in spring. They suggested that temperature was probably the most important factor, but differences existed even between North Sea and Kattegat populations. Marshall, Nicholls and Orr (1933–35) found that there was considerable size variation in the broods of *Calanus* which they studied; the maximum size and weight was normally typical of the first brood that matured about the end of March, i.e. those copepods which were developing in the coldest water. In contrast, the smallest size occurred in the summer/autumn brood, that is, amongst those copepods which developed in the warmest water. Nicholls also claimed that for any brood there was a decrease in mean size towards the end of the breeding period.

Though temperature has a marked effect on size, other factors must certainly play some part. Wiborg (1955) holds that the availability of food is responsible for considerable size differences between broods, an opinion which is shared by Ussing (1938) and Grainger (1959). Deevey (1960) suggested that a strong negative correlation existed between mean copepod length and environmental temperature where a marked seasonal fluctuation, probably exceeding 14°C, occurred. In such a locality there appears to be no relationship between the size of copepods and the amount of phytoplankton. In intermediate regions, however, with a moderate seasonal range in temperature (*ca.* 7°–14°C), the mean body length of successive generations of copepods appears to be almost directly correlated with the amount of phytoplankton during the month preceding the brood, but inversely correlated with temperature. In the Sargasso Sea, where the temperature range is extremely small, Deevey (1964) suggests that any copepod length/temperature relationship is extremely unlikely. With the very limited but varying phytoplankton crop, a correlation between copepod size and phytoplankton is probable; such a positive correlation was found for three species of *Pleuromamma*.

Anraku's (1964a) work on *Acartia* lends support to the view that temperature and other factors such as food supply may affect body size. He found that *Acartia tonsa* in Buzzards Bay (near Woods Hole) were smaller in summer but showed an increase in mean body size during autumn/winter to reach a maximum in late February. In the colder environment of Cape Cod Bay, *A. tonsa* was consistently smaller than in Buzzards Bay, though some seasonal change in body size occurred, with a maximum in October and a decline again in winter. He believes that temperature was more favourable to this somewhat warm water species in Buzzards Bay, but that since the winter temperature was the same in both areas, the greater food supply in Buzzards Bay became more significant.

The usual inverse relationship between body size and seasonal temperature was evident for the Buzzards Bay copepods. For the cold water species, *Acartia clausi*, maximal size was achieved in winter/early spring in Buzzards Bay and mean size was greater over that season than for specimens from Cape Cod Bay. Maximum body size was found for Cape Cod Bay animals during April, and over the summer mean body size was greater than in Buzzards Bay. Anraku suggests that the higher summer temperatures in Buzzards Bay probably restricted growth in *A. clausi*, but that over the colder part of the year the copepod there could benefit from the better food supply.

For British waters it appears, from the work of Russell (1935a) and Bogorov (1934) in the Plymouth area and Wimpenny (1937) in the North Sea, that the three main breeding periods suggested by Nicholls for the Clyde Sea Area for *Calanus finmarchicus* generally hold, the first brood occurring about March. Results from various investigations have also established that the overwintering Stage V passes into the deeper water layers and remains there during the late autumn/winter, during which time it appears to be more or less unresponsive to light and shows no diurnal vertical migration. After attaining maturity the adults rise towards the surface and spawn (Fig. 3.61). It is suggested that the overwintering copepods may be living economically, largely on their fat reserves, during this period when food is relatively scarce. A lowered metabolic rate appears to be also typical of this overwintering Stage V (cf. Chapters 6, 7).

The reproduction of *Calanus finmarchicus* has been studied in several other areas. The rapidity of development, the number of generations per year and the date of beginning of reproduction may vary considerably with the area. In general it would seem that reproduction in colder waters is slower and begins later and that there tend to be fewer broods during the year. For example, Fish (1936a) concluded that in the Gulf of Maine the first brood of *Calanus* occurred in March/April and a second spawning in June/July. It is possible that a third, smaller breeding took place in September. The overwintering stock was of Stage V copepodites which were derived partly from the June spawning and partly from the smaller later brood, when this occurred. Some of the copepodites must, therefore, have lived from June until April or May of the following year, a period of 10 or 11 months, most of the time existing as Stage V. Clarke

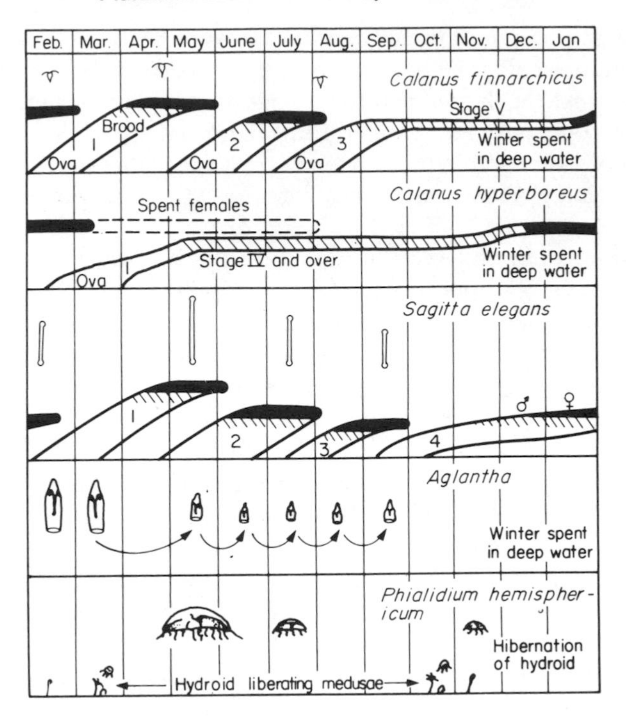

Fig. 3.61. Diagramatic representation of life histories of five zooplankton animals under boreal conditions. An indication of the different sizes of the various broods is included (Russell, 1935).

(1940) confirmed that off the north-east coast of North America there were two major generations of *Calanus*.

The work of Ruud (1929) suggested that only two spawnings of *Calanus* occurred off the coast of Norway: the first in March and the second about June. However, Wiborg (1954) has pointed out that Ruud's observations did not continue beyond July, and his own analysis suggests that there were four breeding periods for more southern regions of Norway: February/March, May/June, July/August and September/October. Wiborg's observations support the general contention advanced by Marshall and Orr that in more northern waters generations tend to be reduced, for in the Lofoten area Wiborg found only three: in April/May, June, and July/September.

In even colder areas, Paulsen (1906) recorded one main extended brood of *Calanus* lasting from approximately March to June with a possible second breeding period later in Icelandic waters. The observations of Fish also suggest one major brood for the Labrador area. Perhaps the most extreme case for the slowing of reproduction in *Calanus* at high latitudes comes from the work of Digby (1954) off Greenland. Most of the *Calanus finmarchicus* took one year for development, the nauplii occurring as a single brood spread out from May to October but mainly during the months of June, July and August. Development proceeded as far as the copepodite stages by the early autumn, but in those polar regions overwintering could occur apparently in any of the copepodite stages, Stages III–V. It had earlier been observed that even in more northern Norwegian waters, Stages IV and V both occurred as overwintering stages.

Digby found that at high latitudes moulting to the adult condition could take place about the beginning of the year in January, but that the single brood did not appear until June or July. However, Digby's results suggested that for some individuals of *Calanus* in the polar regions development was even more prolonged. Some of the yearly brood did not proceed nearly as rapidly with their development and thus were only very early copepodites by the beginning of autumn. These overwintered as Stage I or II copepodites and had only reached the typical Stage V overwintering stage by the following autumn. Thus they lived for almost two years before they spawned.

The closely related *Calanus hyperboreus* appears from the work of Somme (1934) and of Wiborg (1954) to have a clear single breeding period in the waters of the Norwegian Sea, where it is endemic. *C. hyperboreus* overwinters, much as *C. finmarchicus*, as a reduced stock mostly in the Stage V condition and it occurs very deep in the Norwegian fjords. It moults about January to the adult condition and a single spawning takes place about February or March. The eggs pass through their earliest development in deep water, but by April the young stages reach the surface layers and continue their development to reach the Stage V copepodite during the summer months. They then descend to deep water to complete the annual cycle (Fig. 3.61).

Other factors apart from latitude may affect the time of the zooplankton increase; warm and cold currents may be significant. The generally lower temperatures of North Atlantic waters, which are affected by the cold Labrador and Nova Scotia currents, in contrast with the eastern North Atlantic, warmed by the North Atlantic Drift, markedly affects the timing of zooplankton growth. Similarly, clear differences appear in the eastern and western Pacific. On a more restricted scale, cold waters in parts of the Baltic may retard the zooplankton increase. Eriksson (1973c) found in the shallow Oregrund waters off the east coast of Sweden that the summer increase in zooplankton was late due to very low temperature. Copepods, *Acartia bifilosa* and *Eurytemora* dominated the plankton in June and in July/August respectively, but although a very rich plankton built up in the summer, its duration was very brief.

In less severe conditions, the zooplankton maximum may occur somewhat later inshore than at the same latitude offshore. Lie (1967), sampling the plankton in Hardangerfjord, Norway, found the minimum in December–March and maximum in July–September, with the maximum tending to become later farther inside the fjord. As so frequently occurs in colder seas, analysis of seasonal change is greatly complicated by seasonal vertical migration. Lie demonstrated that many of the copepods descended to greater depths from September, the majority exceeding 100 m by November. Ascent of the population starts in January and $>70\%$ of biomass is in the upper 100 m from June to September. Densities showed great seasonal variation (e.g. 200 animals/m^3 in March and 13,500/m^3 in September at a mid-fjord station) and even greater variation at inner stations. Copepods dominated the collections (from 84%–99%) as with most boreal zooplankton. Some larger animals (e.g. euphausiids) are unlikely to have been captured. Seven copepod species (*Oithona similis, Paracalanus parvus, Calanus finmarchicus, Oncaea borealis, Acartia longiremis, Pseudocalanus elongatus* and *Microcalanus pusillus*) comprised on average 84% of total organisms. Some deeper-living species (e.g. *Calanus hyperboreus, Chiridius, Metridia*) may be more important in deep hauls.

Calanus finmarchicus, the dominant species, had three spawnings in Hardangerfjord: in February/March, May/June and September. In the coastal region outside the fjord the major spawning was in the spring and there was a rapid rise in biomass. Inside the fjord spawnings were more equal and the increase in biomass more gradual, leading to a late summer maximum.

Some comparison might be made with seasonal changes in the zooplankton of the Norwegian Sea, but in oceanic waters at Weathership M (latitude 66°N) Ostvedt (1955), analysing divided hauls taken to the bottom (2000 m), found that copepods were dominant at all depths. *Calanus finmarchicus*, *Pseudocalanus minutus* and *Oithona similis* were the chief species in the upper 1000 m (80–90 %). The first two copepods were also dominant for the whole collections, occurring at greater depths in winter; other species may be important at deeper levels (Table 3.11). *Acartia*, *Paracalanus* and *Centropages typicus*, found near the surface, were markedly seasonal, occurring mainly in October–November and disappearing from December–March.

Table 3.11. Frequencies of the most abundant copepod species in per cent of total number of copepods (Ostvedt, 1955)

Depth (m)	0–50	50–100	100–600	600–1000	1000–2000
Total number	481,605	146,181	386,396	524,681	511,188
Calanus finmarchicus	34.83	33.77	42.25	41.34	25.51
Calanus hyperboreus	0.38	0.90	1.40	1.15	9.25
Pseudocalanus minutus	28.41	16.75	7.07	20.40	39.63
Microcalanus pygmaeus	0.86	7.95	16.26	6.27	9.14
Metridia longa *Metridia lucens*	0.79	3.57	10.77	2.30	0.90
Oithona similis	26.81	26.81	9.35	3.91	1.93
Oithona spinirostris	0.74	4.58	6.84	1.83	0.27
Oncaea borealis	5.46	3.56	4.75	22.25	12.20
Total	98.28	97.90	98.70	99.45	98.84

The surface population showed a spring increase in April/May, especially of *Calanus* and *Pseudocalanus*, with a peak population in June (cf. Fig. 3.62). *Oithona* and *C. hyperboreus* also increased as minor components. By June/July, *Calanus* Stage IV and V copepodites were sinking into deeper water and nauplii were lacking. During summer some carnivorous species which preyed on the earlier increased copepod populations spawned in the upper layer. A second smaller spawning of *Calanus* took place during autumn. If this was not due to an immigrant stock, it appears that much of the spring brood descended into deep water without a second spawning. The deep overwintering stock consisted mostly of C. IV and V, and the ascent to the surface again did not take place until April, when maturation had occurred. From April to June the abundance of nauplii and young copepodites illustrated the active breeding period. *Pseudocalanus* similarly ascended about March/April; a single spawning occurred mainly over April and the stock (almost entirely C V) returned to deep levels by June or July. Ostvedt also confirmed the breeding cycle of *Calanus hyperboreus* (cf. Fig. 3.63).

The overwintering C III, IV and V, with some VI, are deep-living; a portion of the stock ascends in spring and spawns in April, and the copepods are all in deep water again by July. Hansen (1960) also believes that although there may be a second

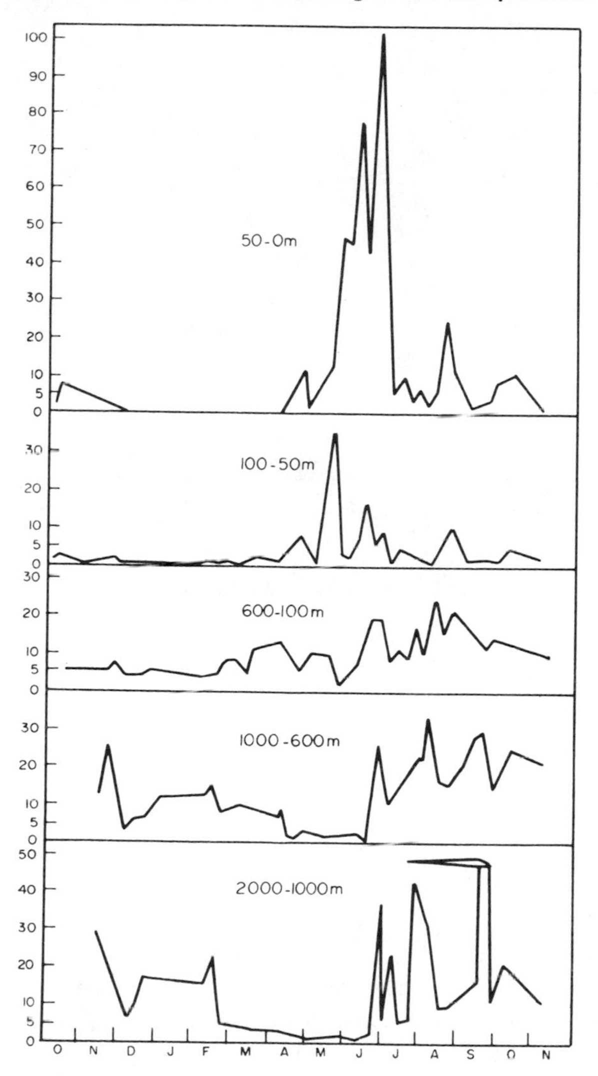

Fig. 3.62. Total numbers of copepods in thousands per haul at the different depths, Oct. 1948–Nov. 1949 (Ostvedt, 1955).

spawning of *Calanus finmarchicus* in the open Norwegian Sea in autumn, the species is essentially annual in that region. For *Calanus hyperboreus*, *C. finmarchicus* and *Pseudocalanus minutus* a seasonal vertical migration is characteristic, with overwintering stages present as copepodites IV and V. The average biomass of zooplankton over the upper 50 m in June was about 30 times that for the minimum period of November/December. The difference not only reflected reproduction but was partly due to seasonal vertical migration. In coastal waters, where some neritic copepods remained nearer the surface, the seasonal fluctuation was not so great. Hansen distinguishes between the oceanic *Pseudocalanus minutus*, essentially an annual

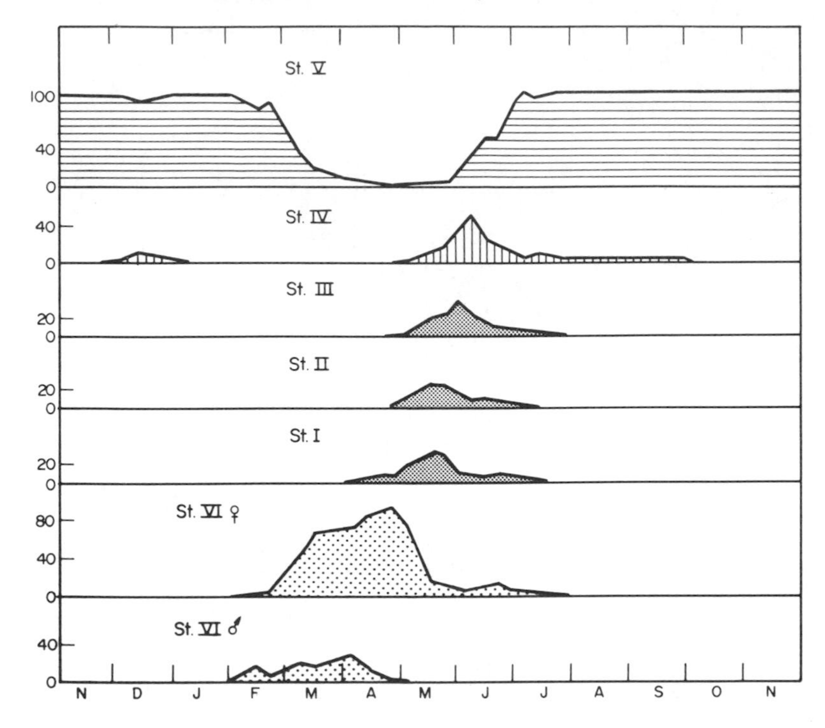

Fig. 3.63. Frequencies in percentage of the different stages of *Pseudocalanus minutus* calculated for the whole water-column from 2000 m to surface (Ostvedt, 1955).

reproducer, and *Pseudocalanus elongatus*, which is coastal in distribution with two broods per year.

Ostvedt (1955) found that apart from the main seasonal changes due to *Calanus* and *Pseudocalanus*, certain species regarded as southern immigrants (*Metridia lucens*, *Acartia clausi*, *Paracalanus parvus*) were important from August through the autumn, but were disappearing from the end of October. The minimal population at the surface over winter consisted mainly of *Oithona* and *Microcalanus*.

Intermediate and deeper layers were dominated by *Calanus finmarchicus* and *Pseudocalanus* over winter. Some of the stock, however, appeared to remain in deeper water when the normal spring migration towards the surface took place. Other species usually in the deeper layers included *Metridia longa*, *Pareuchaeta norvegica*, *Calanus hyperboreus* and *Microcalanus* with *Aglantha* and *Eukrohnia* among the fauna other than copepods. Ostvedt considers that apart from the marked changes with seasonal vertical migrations in the deeper layers the comparative constancy of conditions suggests less seasonal change compared with the superficial layers. The changes over the seasons at deeper levels are shown in Fig. 3.62.

Tentative observations were made on the breeding times of some of the deep-living copepods. *Pleuromamma robusta* (mainly in the 100–600-m layer) was more abundant with younger copepodites present, over February–April. Adults of *Metridia longa* were mostly deep-living and dominated the population of the species from January to June. From June to October, however, C IV and V were more important. Younger copepodites were shallower and occurred mainly over May/June. Probably only one spawning (April/May) occurred annually. A few deeper-living copepods appeared to breed in winter and summer (e.g. *Chiridius obtusifrons*) or all through the year (e.g. *Gaidius brevispinus*, *Pareuchaeta farrani*, *Pareuchaeta norvegica*). The last species was obtained in fair numbers; all stages, with mature females and males, occurred at all times. Although spawning was extended, it was more intense in winter/spring. A more extended breeding period appears to be typical of deeper, especially carnivorous, zooplankton. Thus, the seasonal pattern of deeper-water zooplankton as described by Ostvedt (cf. Fig. 3.62) to a large extent reflects seasonal migrations of copepods, but is probably influenced also by deeper ocean currents.

The breeding of some deeper carnivorous copepods has been examined by a few other workers. Bigelow (1926) stated that *Pareuchaeta* spawned throughout the year in the Gulf of Maine, and Wiborg (1954) believed that while breeding could occur in any month, there were periods of greater intensity. More recently, Bakke (1977) found that at Korsfjorden, Norway, breeding of *Euchaeta norvegica* could take place all the year round, but that there were two main generations with peaks in February and August. During winter the period from spawning to C I (November to February) was more extended (90 days) compared with the summer generation, spawned in June, in which development to C I took only 30 days. The total time was 5.5–6 months for the summer generation, compared with 6–7 months for the winter brood. The adult male matured later than the female, especially in autumn, and had a longer period as C V, possibly allowing a more extended feeding time and greater storage of food reserves, since the adult males cannot feed. The release of egg sacs by the females may also be related to periods when food is present in reasonable quantities.

Bakke and Valderhaug (1978) studied another carnivorous deeper-living copepod, *Chiridius armatus*, from the same area. Although two main reproductive periods occurred annually, the duration of breeding appears to be extended so that there is considerable overlapping of generations. For example, peaks of C V males and females were found in April, May, October, November and December, further peaks of adults in June and January; and all copepodite stages were found throughout the year in some numbers. The period of development from C III to C VI took 4–5 months (with one possibly shorter time) and the maximum period between generations was about 8 months. Since developmental stages younger than C III were not regularly sampled, precise times for development from egg to C III are not known.

Matthews and Bakke (1977) confirmed the domination of *Calanus finmarchicus* and the lesser contribution of *C. hyperboreus* and *Metridia longa* to the overall biomass of the water column at Korsfjorden, Norway. Apart from the investigations of breeding cycles in *Chiridius* and *Euchaeta* (*vide supra*), changes in some less common copepods (*Metridia lucens*, *Calanus helgolandicus*, *Candacia norvegica*) and deep-living species (*Heterorhabdus norvegicus* and *Scaphocalanus magnus*) may influence seasonal changes in biomass to a small degree. Cycles of some of the other carnivorous representatives of the Korsfjorden community will be discussed later, but there is a suggestion that *Metridia longa* may be a food source for *Chiridius*. There is a marked tendency for the cycles of various carnivores to be delayed in relation to the cycles of the herbivorous zooplankton (Fig. 3.64).

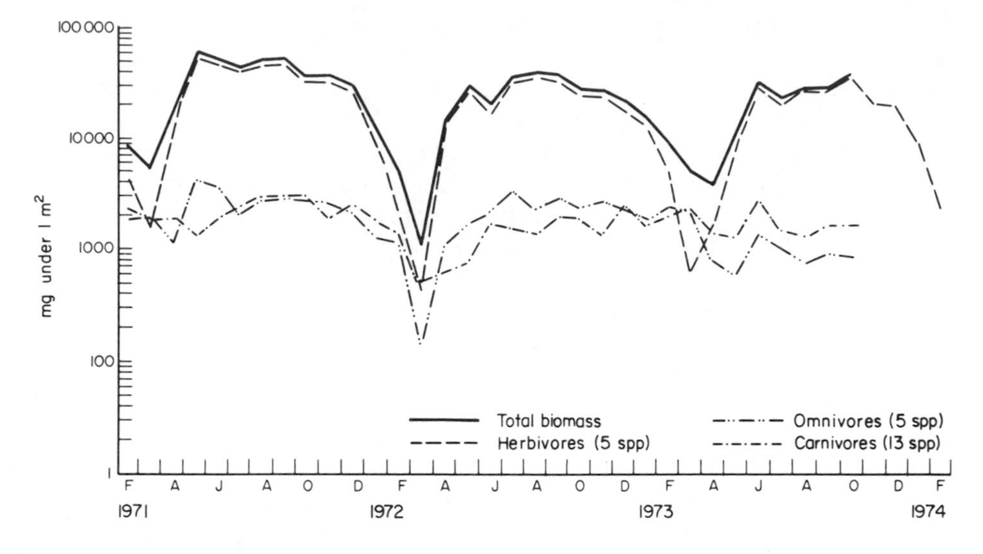

Fig. 3.64. Monthly estimates of total biomass and of biomass of herbivores, omnivores and carnivores separately, plotted on a logarithmic scale (Matthews and Bakke, 1977).

Seasonal changes in very cold waters

Very little is known of the breeding patterns of zooplankton in high Arctic regions. Hopkins (1969) demonstrated some increase in zooplankton biomass in Arctic surface water (0–100 m) during summer (July/August), though the difference in density between September and December was not large. Copepods, mainly *Calanus*, contributed by far the largest proportion to the zooplankton; in agreement with the results of Jashnov (1961), Johnson (1963), and Minoda (1967) in earlier studies, they formed more than 80 % of total biomass, and by number more than 90 %. Other groups of moderate importance included chaetognaths (mainly *Eukrohnia*), amphipods (*Parathemisto*, *Cyclocaris*), the decapod *Hymenodora*, appendicularians and hydromedusans. Radiolarians were also very abundant from July to the end of September, outnumbering metazoans.

Actual zooplankton numbers (taken with the No. 20 net) were small: for the 0–200-m layer the mean number was 56 organisms/m^3. Nauplii were found throughout the year, indicating that breeding of at least some of the high Arctic copepod species does not follow only the period of phytoplankton increase. Perhaps these nauplii are omnivorous forms, relying especially on detritus; some are presumably carnivores (Hopkins, 1969). Johnson (1963), dealing with copepods from the Arctic basin, also commented that the breeding period appeared to be very prolonged. Further evidence comes from Grainger's (1965) investigations, which suggest that most of the Arctic copepods breed over a considerable period of the year and are not confined to the brief period of phytoplankton increase (e.g. *Metridia longa*, *Microcalanus pygmaeus*). *Oithona similis* may have more than one brood, but with greater populations when food is more abundant. *Microcalanus* appears to take two years to reach maturity, overwintering mainly as C III and IV, but some individuals may reproduce when approximately one year old (cf. *Calanus glacialis*—*vide infra*).

Earlier investigations outside the Arctic Basin itself but in the very cold seas of the U.S.S.R. have been summarized by Zenkevitch (1963). In the Barents Sea Zenkevitch observed a domination of copepods, with three important species (*Calanus finmarchicus*, *Oithona similis* and the Arctic form *Metridia longa*) among the permanent inhabitants. Warmer-water species, including *Temora*, *Acartia*, *Centropages* and the pteropod *Limacina retroversa*, were found mainly along the western edge in less cold waters. The populations of these species tended to move eastwards in summer and returned during the cold winter, giving a kind of seasonal succession. They also showed higher densities in warmer years. In inlets, these species, with cladocerans and some meroplankton (cirripede larvae), were more abundant. A maximum in the more neritic areas occurred about May or June with *Balanus* larvae, *Calanus* and *Thysanoessa*; later the neritic copepods (*Temora*, *Pseudocalanus*), medusae and other forms became more important. There were some seasonal differences in distribution over the main oceanic region; the more Arctic species developed mostly in winter and spring, while during summer they were found more to the north or occurred in deeper water. Apart from the appendicularians *Oikopleura labradoriensis* and *O. vanhoeffeni* and some euphausiids, typical winter forms were chiefly the copepods *Euchaeta glacialis*, *Euchaeta norvegica*, *Calanus hyperboreus*, *Oithona* sp. and *Metridia longa*, apart from *C. finmarchicus*.

Copepods, especially *Calanus finmarchicus*, formed by far the greater proportion of total zooplankton biomass, especially over the open sea. The zooplankton maximum in the southern part of the region occurred about June, but could be as late as August in the north of the area. Whereas the maximum biomass was near the surface during summer, it was much deeper during winter. This, as Zenkevitch explained, quoting Kamshilov (1955), may be due to two broods of *Calanus finmarchicus* occurring over a year in the Barents Sea, the first in April and the second in June or July, with the animals overwintering in the deeper layers. Zenkevitch states that Jashnov, however, believed that the breeding of *Calanus* was monocyclic. The discrepancy might be related to geographical differences in the populations. Bogorov (1958) found population differences in the size of C V *Calanus* in the southern, central and northern areas of the Barents Sea. He also confirms the suggestions made earlier that in polar waters there is normally only a single generation of *Calanus* compared with two or three at lower latitudes. Indeed, for the Barents Sea he believes that all the large copepods are essentially annual reproducers. The pattern is comparable to that described by Heinrich for the Bering Sea, where all the large species except *Metridia pacifica* have one brood per year. Grainger (1959) found that *Calanus hyperboreus* required at least two years to complete a generation in the Arctic Ocean, while from observations made off an ice island at 84°N, Dawson (1978), suggests that it probably has a three-year cycle. The copepods seem adapted to living under ice cover, with C II and C III from a spring spawning overwintering in deeper water to become C IV and C V the following spring. Spawning may be delayed still further or a brood may fail if the phytoplankton is insufficient to sustain the youngest stages; it is even possible that recruitment of C IV from other areas may maintain the population.

Prygunkova (1968) described the life cycle of the Arctic copepod *Calanus glacialis* as being normally monocyclic, though Jashnov had suggested that a proportion of *Calanus glacialis* may take two years to reproduce (cf. Digby, 1953, for *Calanus* off Greenland). Grainger (1965) also described breeding being delayed till late summer in the Arctic, with C III as the main overwintering stage. The total breeding cycle thus appeared to occupy two years. Tidmarsh (1973) points to an acceleration of development in an area off north-west Greenland, where warmer ice-free conditions led to an earlier phytoplankton crop; *C. glacialis* spawned earlier and completed the life cycle in a year. Grainger (1965) draws attention to *C. glacialis* being somewhat unusual among Arctic copepods in that it requires growth of phytoplankton for breeding. In the White Sea Prygunkova found that the cycle for *C. glacialis* included a very short spawning in May. Rapid development in the upper layer proceeded as far as C IV, which then descended to rather deeper water, largely because of increasing temperature at the surface. The species overwintered mainly as C IV and moulted to C V by the following April or May after surfacing. They migrate again to deeper water over the summer and do not moult to the adult condition until the following autumn or winter, with oviposition delayed until the following April or May (i.e. a two-year cycle).

The main difference, according to Prygunkova, between *Calanus glacialis* populations from the White Sea and from other regions is that *all* the White Sea stock takes two years to achieve maturity. Unlike most stocks of *Calanus glacialis* living in superficial very cold waters, the species is forced to leave the surface during the summer months in the White Sea owing to the relatively high temperature there. Feeding is confined to the rich upper layers and is therefore restricted to a brief period, thus slowing the development. The seasonal environmental changes in the White Sea are reflected in the succession of zooplankton observed by Kolosova, Kokin and Vishenskiy (1977). Species characteristic of different major regions (Arctic, Arctic–boreal, boreal, and cosmopolitan) achieved their maximum densities between March and September. The Arctic species however, developed first in water still under ice. The less cold species increased later, more or less in succession, as the coldest plankton sank to greater depths, continuing in the very cold water.

An interesting comparison of the life cycle of *Calanus glacialis* in the White Sea is that described for *Oithona similis* by Shuvalov (1965, 1975) from the same area. Although *Oithona* has an extended breeding season from May to October, two major periods appear for nauplii with maxima in June and August. Shuvalov believes that two overlapping generations exist which differ in life span. A spring generation occurs, showing fairly rapid development over 2–3 months—mainly from June to August. The sexually mature adults overwinter till about April (7–10 months) and then spawn and die, so that they have a total life span of 9–12 months. At the end of summer some small females spawn to give rise to a slowly developing autumn generation. The development takes about 7–9 months, so that these become sexually mature about May. As adults they continue to exist from about May to October, giving a total life span of 12–15 months. He finds considerable variation in the adult size of *Oithona similis* from the White Sea, where temperature variation in the uppermost layer is large. Females appearing over spring/early summer are generally large; those developing over summer and found over autumn/winter are small. Pavshtiks (1975) regards this species as a summer breeder in the Davis Strait. The numbers begin to increase in July and abundance reaches a peak in August/September, while populations of *Calanus finmarchicus* show a decline.

In contrast to *Calanus glacialis*, *Oithona similis* appears to be more eurythermal or perhaps exists as different races with different temperature characteristics. According to Zenkevitch (1963), some of the larger, though less numerically important zooplankton crustaceans in the Barents Sea (some euphausiids, *Themisto*) may have a two-year breeding cycle.

So far discussion of seasonal changes and breeding of boreal and high latitude copepods has mainly concerned the North Atlantic, including the Arctic. With regard to reproductive cycles of the more abundant North Pacific copepods, Geynrikh (1968) examined the boreal oceanic community in the upper 500-m layer of the north-east Pacific and Bering Sea. Four major copepods, *Calanus pacificus* (*helgolandicus*), *C. plumchrus*, *C. cristatus* and *Eucalanus bungii bungii*—all essentially boreal in distribution—were considered over a three-month period, October, November and December.

For *Eucalanus bungii bungii* and *Calanus pacificus*, an overwintering stock, mostly of Stage IV and V copepodites, is formed during autumn. A slightly greater proportion of younger copepodites found in one area with somewhat higher temperature and possibly more autumn phytoplankton suggests a very limited extension of the breeding season. Essentially, however, there is no further breeding until the next year, when the presence of phytoplankton and a suitable temperature gives maturity and "spring" breeding. LeBrasseur (1965) found *Calanus pacificus* at Station P (50°N) mainly from June to December with adults predominant, suggesting that reproduction may occur over much of the year. Geynrikh (1968) states that reports indicate that three generations may be produced annually, beginning in April with other peaks in June/July and September/October. An overwintering stock composed mainly of C V is built up by November. The development of the overwintering stage (mostly C V) of *Eucalanus bungii bungii* to the adult occurs about May and there is an intensive period of reproduction over summer, about July. The copepodites grow up to form the overwintering generation by the autumn. South of latitude 40°–45°N where *Calanus pacificus* (*helgolandicus*) extends along the U.S.A. coast, the species shows a more continuous type of breeding with no overwintering stage. Mullin and Brooks (1967) and Brodsky (1967, 1975) have suggested that this is a different southern race.

With regard to the two other major species, *Calanus cristatus* and *C. plumchrus*, Geynrikh finds that both show a strong predominance of older copepodites (Stage V and VI) over the three months of the investigation in some areas, but of young copepodites (Stage I and II) in others, indicating that the copepodites may mature and breed during the autumn. A phytoplankton burst is apparently not necessary for maturation and reproduction. Growth occurred in the upper layers to reach late copepodite stages, mainly C V, by May, but the copepods then descended to deeper levels—some even below 500 m, where they remained till about October or November and then matured. *C. cristatus* is probably an annual species. After maturation of the stock of C V in early autumn, breeding commences probably about October or November. In the *very* cold extreme north-west Pacific breeding is much later, with young copepodites not appearing until after January.

Calanus plumchrus is probably also an annual species according to LeBrasseur (1965) and Geynrikh. C V, the dominant stage, remains at deeper levels from about May until October, but then matures and breeds, giving probably one generation a year. A race in the very cold Bering Sea multiplies later during winter and young stages are dominant about March. Beklemishev (1954) summarized earlier investigations on *C. plumchrus* and *C. cristatus*, stating that in both species a single annual brood occurred, the females spawning in comparatively deep water. The females are believed to oviposit without feeding, the eggs being produced at the expense of lipid reserves stored by the Stage V copepodites (cf. also Heinrich, 1962).

In the open oceanic North Pacific young *Calanus cristatus* and *C. plumchrus* are already present to graze upon the early spring phytoplankton. The other two species graze over the summer, so there is relatively little change in phytoplankton crop over the season. The partial temporal separation of the herbivorous species also reduces competition. Parsons and LeBrasseur (1968) found that around Station P in the North Pacific a massive increase in primary production due to increasing light and stability in the spring starts earlier and builds up less suddenly than in the North Atlantic. To some extent this is related to the mixed layer not extending so deeply in the North Pacific, but in addition the zooplankton grazers, dominated especially by *C. plumchrus*, effect a more stable relationship between phytoplankton and zooplankton (cf. Volume 1).

Parsons, LeBrasseur and Barraclough (1970) found that in the inshore waters of the Strait of Georgia, where the spring increase begins about February/March and peaks about May, the herbivorous zooplankton is also dominated by *Calanus plumchrus*. The nauplii appear first in February, developing through the copepodite stages in near-surface waters over the three months March, April and May. They then descend to deeper water (though not so deep as in the open ocean), where they remain without feeding. C V moults to the adult stage and oviposition follows about January/February, before the spring increase, for which a grazing population of young stages is already present. In the inshore waters two other copepods (*C. pacificus* and *Pseudocalanus minutus*) are also abundant. These produce several broods per year and are dependent on phytoplankton for breeding. The amplitude of change in zooplankton of coastal waters of British Columbia (dominated by changes in the copepod population) may, however, be many times that of the oceanic waters in approximately the same latitude.

Antarctic and sub-Antarctic breeding

Patterns of seasonal change and breeding of holoplankton at high latitudes have so far been concerned with the northern hemisphere. In the southern hemisphere *Calanus tonsus*, a species closely related to *C. plumchrus*, is found in sub-Antarctic waters. Previously there was confusion over the identity and distribution of the two species (cf. Chapter 2). Jillett (1968) found *C. tonsus* plentiful in the spring and early summer in surface waters off south-east New Zealand and farther offshore, though not in neritic waters. Its life history appears to be similar to that of *C. plumchrus*. Young copepodites, apparently the progeny of an overwintering population, first appeared in surface waters in September/October (southern spring). By November this generation had presumably grown to maturity; C V were abundant and females fairly plentiful, while young copepodites were sparse. A new brood represented by numerous young copepodites appeared in December, but by January this brood had developed to late copepodites, the great majority as C V. A few C V persisted in the surface waters in February, but thereafter *Calanus tonsus* disappeared until September; they apparently descended to overwinter at greater depths.

The rich lipid reserves in these copepodites emphasize their role as overwintering stages. While only C V were obtained in deeper waters during the southern autumn and winter, by July and August adult males and females were present in the deeper layers. C V had virtually disappeared by September and the great majority in deeper water were adults, mainly females. These presumably spawned to give the young copepodites of the first brood which migrate to the surface in September and October. The first spawning thus appears to be independent of the phytoplankton outburst (cf. Heinrich, 1962). The frequent dominance over much of the year of C V in very high densities leads Jillett to suggest that the adult female *C. tonsus* may be comparatively short-lived.

Calanus tonsus, though it occurs in the colder waters south of the Sub-Tropical Convergence, does not extend to the Antarctic. In true Antarctic waters south of the Antarctic Convergence, earlier studies, particularly those of Hardy and Gunther (1935) and of Mackintosh (1934, 1937) pointed to the vast importance of a comparatively few species of copepods (e.g. *Rhincalanus gigas*, *Calanoides propinquus*) and of euphausiids (*Euphausia superba*, *E. frigida*, *E. triacantha*) in the great seasonal changes of the zooplankton. Mackintosh indicated that a rapid increase in zooplankton occurred in Antarctic waters about November or December and that zooplankton was extremely abundant over the southern summer, with a peak value about February. By April the population was declining to reach low densities over the winter. Seasonal variations are to a considerable extent affected by changes in depth distribution which occur in the life histories of a number of the species (cf. the northern hemisphere). With the marked absence of meroplankton in the Antarctic, seasonal changes must relate closely to breeding and to vertical migration of the holoplankton.

Hardy and Gunther's (1935) investigations dealt with a limited line of stations over a few months of a year and could not therefore yield critical data on seasonal change, though they noted a general increase in spring and decline with the approach of winter. Most species of copepod increased over summer and declined in the late autumn. A few comments on other species suggested that *Parathemisto* was most plentiful in December and March, *Limacina helicina* chiefly in early summer and *L. balea* in autumn. The large-scale patchiness of some species (*Euphausia superba*, *Salpa*, *Parathemisto*, *Calanus proquinquus*) tended to obscure seasonal changes.

Mackintosh (1937) emphasized the marked effect of seasonal vertical migration, especially by the three species *Rhincalanus*, *Calanoides* and *Eukrohnia* on population changes, but changes in overall density due to breeding were also discernible. Thus, whereas during December the greatest concentration of zooplankton was near the surface, especially in sub-Antarctic waters, by March numbers had markedly increased and though the concentration was chiefly at the surface, it had spread considerably southwards past the Convergence. By September, while the greatest concentration was at intermediate depths, the whole distribution was greatly extended vertically, but there was also considerably less plankton than over summer. Early spring saw the gradual move towards the surface and some increase in numbers. Change in volume as well as numerical change was examined in the same area by Foxton (1956). Though in June (early winter) shallow hauls were poorer than in April, the deeper layers showed some increase. By August the bulk of the plankton was deep, but all levels showed a decrease. The rise to the surface began in October. The seasonal fluctuation was greatest at the surface; for the 0–50-m layer the amplitude was six- or seven-fold (cf. Fig. 3.65). The increase over spring and summer was also somewhat steeper in the surface layer. For the whole 0–1000-m column, however, the seasonal fluctuation in plankton volume in the Antarctic was slight. Though there appears to have been some

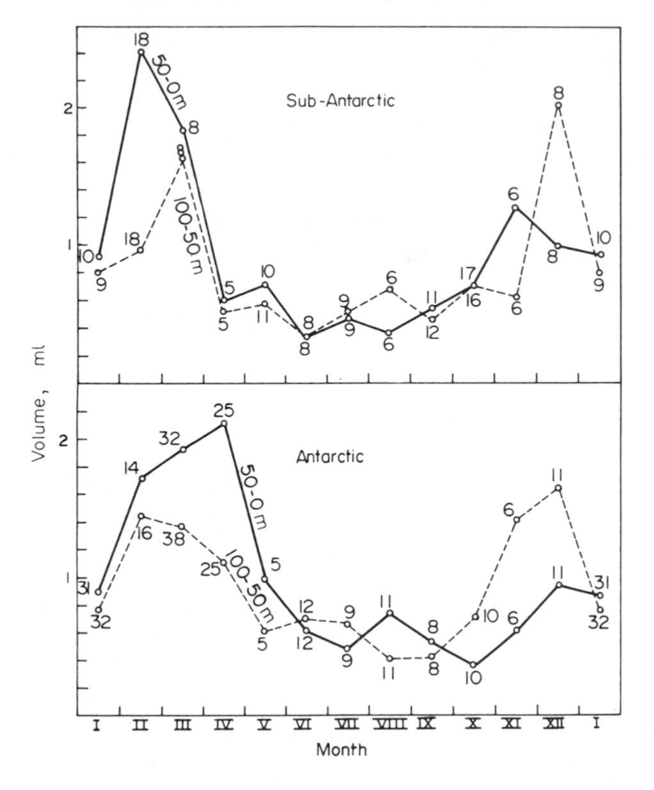

Fig. 3.65. Seasonal variation in the mean concentration of zooplankton in the 100–50 m and 50–0 m horizons of sub-Antarctic and Antarctic surface water. The numbers on each curve refer to the number of observations in each month (Foxton, 1956).

augmentation between December/January and April, followed by a decline in July, the volume was as large again as in April.

The expected seasonal fluctuation in zooplankton volume is a little more obvious in sub-Antarctic waters, but July (winter) again shows a small increase. The lack of a well-marked summer maximum and winter minimum for the whole water column in the Antarctic is largely explained by the occurrence of a winter concentration of zooplankton of greater depths (500–1000 m), i.e. in the warmer deep layers. As Foxton emphasizes, however, this crop is spread vertically during winter throughout a long water column; concentration is far less than over the summer. Moreover, the lack of phytoplankton production in the surface euphotic layer over the winter implies that zooplankton grazing is negligible and that production by the winter crop of zooplankton is at a minimum.

Examination of zooplankton volumes from the sub-tropics to the high Antarctic shows a gradual increase in total volume to a maximum at around 50–55°S, followed by a drop towards the ice edge. Very little variation occurs in winter in the surface layer, since the high latitude zooplankton migrates to deeper water. Figure 3.66, which includes a few near-equator (Indian Ocean) stations, however, shows that between 50° and 55°S high but greatly variable catches occur. Low and constant volumes are typical of low latitudes and the highest latitudes have small catches.

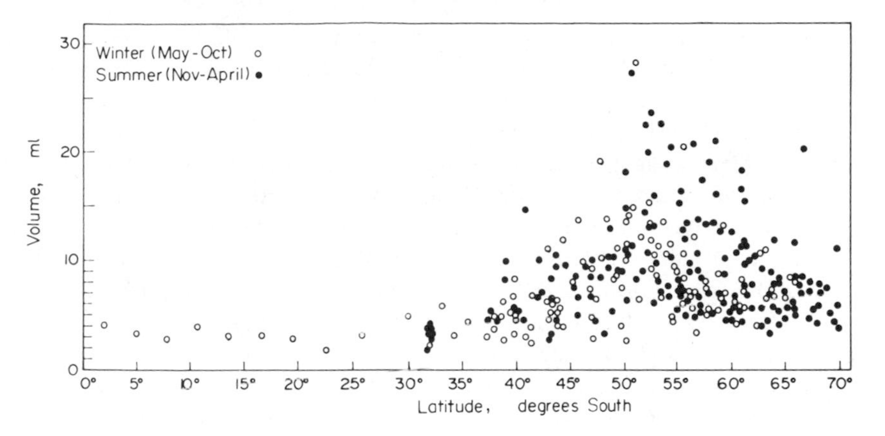

Fig. 3.66. Individual total volumes (1000–0 m) at each station plotted according to the latitude of the station. (A few observations had to be omitted for clarity) (Foxton, 1956).

In terms of temperature, the quantity of zooplankton in the whole column increases with fall in temperature and except for winter, that in the upper 100 m also increases. Maximum volumes in summer occur at approximately 4.0°–1.5°C—about six times those taken at temperatures 21.5°–19.0°C. These data on plankton volumes, however, take no account of the quantities of *Euphausia superba* at high latitudes, since it avoids the nets, except for small specimens. If this species was taken into account there might be an increased volume even up to the coldest (highest latitude) water. It is also possible that passive drift of zooplankton to 50°–55°S may be a major factor in the greater volume there. Figure 3.67 shows the increase in standing crop of zooplankton at all

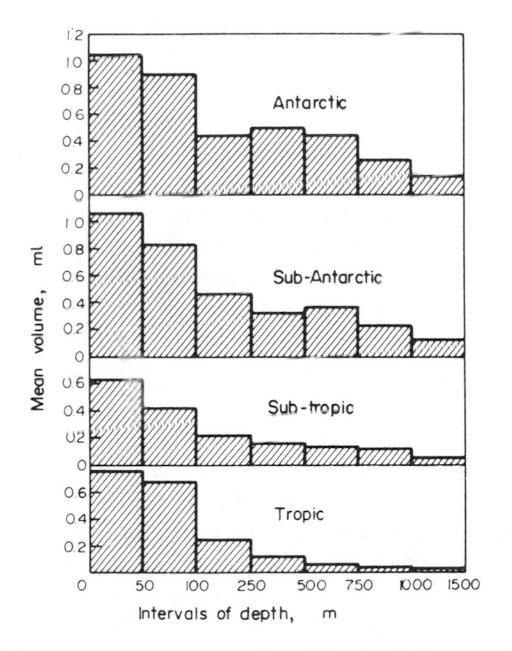

Fig. 3.67. Comparison of the mean concentration (volume/50 m haul) of zooplankton at seven intervals of depth in the Antarctic, sub-Antarctic, sub-tropic and tropic zones (Foxton, 1956).

depths in passing from the Antarctic to the tropics and that shallowest zones are always the major layers. Overall comparison gives the ratios: tropics/sub-tropics/sub-Antarctic/Antarctic as 1 : 1.3 : 2.7 : 3.3.

A contribution to our knowledge of seasonal change in Antarctic waters comes from Philippon's (1972) investigation of the zooplankton off the Kerguelen Islands, situated almost on the Antarctic Convergence. Philippon examined the zooplankton over a year at two shallow stations at approximately 35 m and 50 m depth. Copepods formed 90–99% of the population; amphipods, though far less abundant, came next in importance. The contribution of other taxonomic groups was insignificant, though there was a minor contribution due to brachyuran zoeas from December to March.

Even for such a high-latitude neritic area the variety of holoplankton was exceedingly limited. Only four species of copepods occurred in significant numbers. Of these, *Drepanopus forcipatus* and *Drepanopus pectinatus*, with *Oithona helgolandica*, occurred all the year round. *Drepanopus* showed considerable fluctuations in abundance, but copepodites were generally dominant, especially in June, July and December; adults formed the major proportion only during February. Although precise breeding cycles were not followed, there were possibly overlapping periods. Phillippon (1972) suggests that *Oithona* and the two species of *Drepanopus* reproduced to some extent throughout the year, but with marked intensity following the phytoplankton maximum. The chief phytoplankton increase was from September to November, with the maximum in October and a smaller second rise in March. Although the breeding of *Drepanopus* and *Oithona* seemed to follow the phytoplankton increase, the low temperature would slow development so that the maximum for zooplankton might be considerably later (cf. Fig. 3.68).

In contrast to these species, *Calanus simillimus*, the remaining important copepod, was markedly seasonal in appearance off Kerguelen. It was very abundant about August/September and in February and almost disappeared during June, July and November/December. The start of the breeding of *Calanus* does not appear to be dependent on the phytoplankton rise (cf. Volume 1). Although the breeding cycle was not completely elucidated, the few copepods occurring in June and July were mainly Stage V and Stage VI (adults): very few younger copepodites were present (cf. Fig. 3.69). In August and September, on the other hand, the population was dominated by young copepodites present in great numbers. This might suggest that a small, overwintering late copepodite stock had matured to produce a new brood about August/September. A considerable accumulation of late copepodites was recorded by Philippon in February/March; this might indicate, by analogy with other copepod cycles at high latitudes, the formation of an overwintering stock possibly living in deeper water.

The only other holoplankton taxon of significance, the Amphipoda, was represented by *Parathemisto gaudichaudi*. Few animals occurred before August, there was a clear November peak, and they disappeared by March. August/September appeared to be the only period of marked reproduction. The fluctuating densities of the four copepods and of *Parathemisto*, however, caused a very clear seasonal fluctuation with a peak of zooplankton over the summer. The minimum population occurred in July with 610 individuals/m^3 and the maximum in February with 7000 animals/m^3 (mean monthly values). Although a relationship can be drawn between the summer zooplankton maximum and near-maximum temperature (cf. Fig. 3.70), the major factor in the

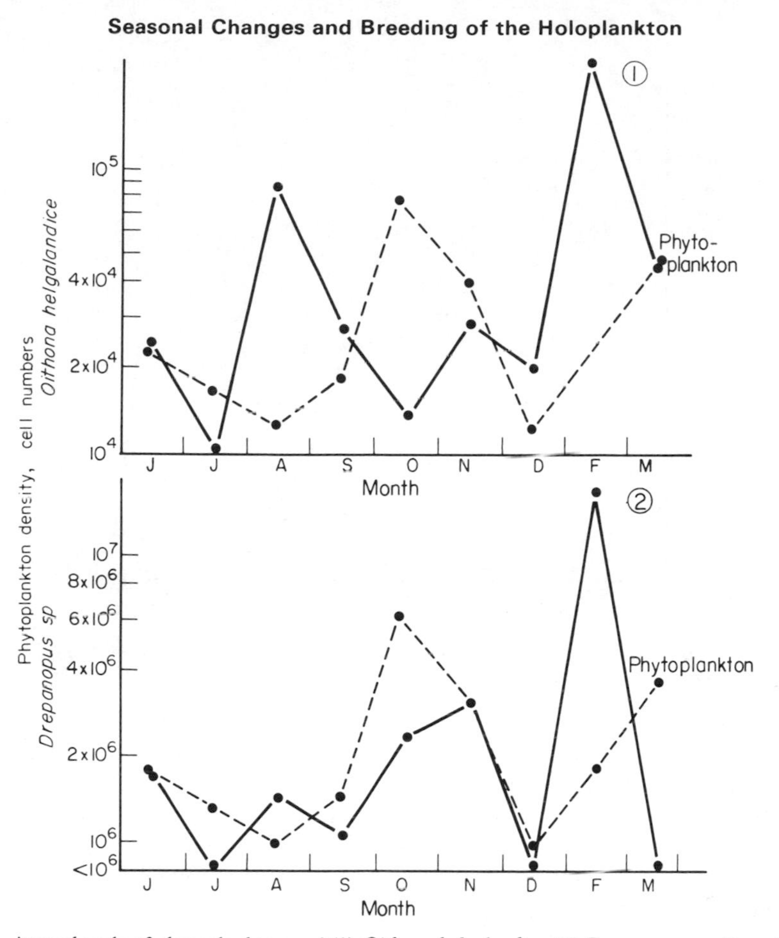

Fig. 3.68. Annual cycle of phytoplankton and (1) *Oithona helgolandica*; (2) *Drepanopus* sp. (from Philippon, 1972).

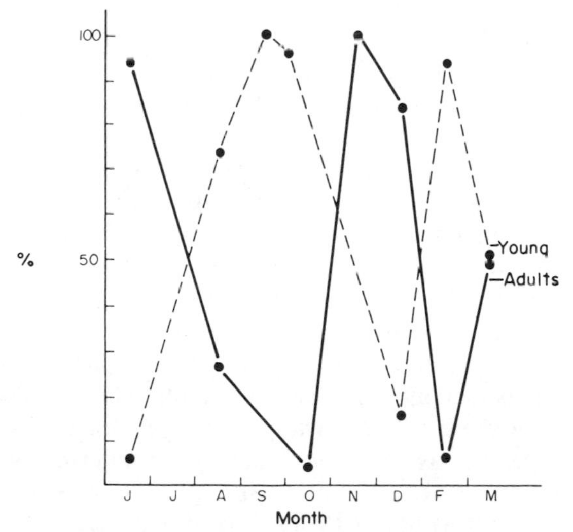

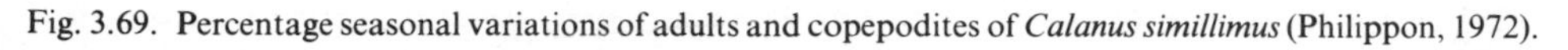
Fig. 3.69. Percentage seasonal variations of adults and copepodites of *Calanus simillimus* (Philippon, 1972).

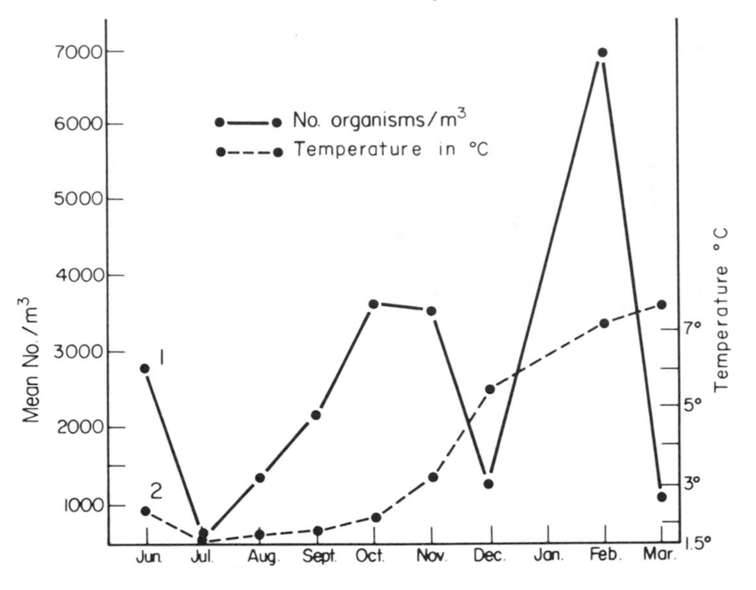

Fig. 3.70. Seasonal cycle of zooplankton abundance as a function of temperature (Philippon, 1972).

seasonal increase in zooplankton is undoubtedly the late spring growth of phytoplankton, itself partly conditioned by temperature change. Philippon points out that her results agree fairly well with the suggestions of Foxton that in Antarctic waters the phytoplankton burst occurs mainly in November, with sometimes a much smaller one in autumn (April), but that the maximum of herbivores is mainly in February, though a smaller one may occur in November. The biomass of zooplankton off Kerguelen varied from 22.5 to 557.7 mg/m^3.

The breeding of some Antarctic and sub-Antarctic copepods

Seasonal changes tend to become clearer when separate species are considered, though comparatively few life cycles of zooplankton have been studied in the Antarctic. For the few which are elucidated, the similarity to those of plankton at high northern latitudes is obvious.

For *Rhincalanus gigas*, the common copepod of Antarctic waters, Ommanney (1936) showed that there were two breeding periods during a year. The first spawning occurs in the Antarctic summer, beginning about mid-December and continuing throughout January, the somewhat warmer waters being the first to show the spawning. Most *Rhincalanus* spawn only in Antarctic waters ranging between *ca.* $+1°$ to $+4°$C. With such an extended spawning period, a mixture of copepodite stages persists throughout the Antarctic summer. However, about April this summer stock descends into deeper water and is then mostly present as adults. These adults spawn in deeper water about May/June, the second spawning for the year, and the offspring from this second spawning persist as a deep-living, overwintering stock, mainly as CIII, IV and V. By about October (early spring) these copepodites begin to rise towards the surface mainly as Stages IV and V and they mature late in November to produce the first spawning of the new year. Ommanney suggests that the overwintering stage appears to depend to some extent on the temperature: in the warmer sub-Antarctic waters the overwintering stages are predominantly C IV and V, but in colder regions there may be a considerable mixture of Stage III copepodites.

Voronina (1970) confirms for *Rhincalanus* the lagging of life cycle in higher latitudes farther south. The species is found in sub-Antarctic as well as Antarctic waters, with summer as the main period of abundance at the surface. In the sub-Antarctic the copepods migrate down earlier and have descended by April, but to the south they breed about April/May. This, according to Voronina, is Ommanney's second generation; she believes that in true Antarctic latitudes the species remains near the surface until June. Clearly, the zone of surface water inhabited by *R. gigas* moves from north to south.

Since the copepodites develop earlier in the north, the period of maximum summer biomass also changes from north to south. *Calanoides acutus* and *Calanus propinquus*, with *Rhincalanus gigas*, are three of the most important copepods making up 73% of total biomass of zooplankton. Breeding period, time of domination of each copepodite stage and the time of summer biomass is different for each species. First is *Calanoides acutus*, then follows *C. propinquus* and last is *R. gigas*. *R. gigas* extends farthest north, next is *C. propinquus* and most restricted is *C. acutus*. The differences in timing should make for a more efficient use of phytoplankton crop.

Of other studies on Antarctic copepods, Andrews (1966) investigated the life history of *Calanoides acutus*, which is found in Antarctic seas as a circumpolar species. As Voronina has indicated, its distribution to the north is very restricted. Although a few specimens are found in somewhat deeper layers north of the Antarctic Convergence, their transport in the Antarctic Intermediate Water is very limited. The life history of *Calanoides acutus* is intimately bound with seasonal vertical distribution. Andrews studied the copepodite stages only, since nauplii were not caught quantitatively. During the winter season (July to September) the entire population lives below 250 m and appears to consist almost solely of C IV and V; C III and VI amount to less than 4% of the population. From October, the overwintering C IV and V stock moults and there is a marked increase in the upper layers, with a seasonal vertical migration of the resulting C V and VI.

Almost all the adults found were female and although their spermathecal sacs were full of sperm, almost none carried spermatophores. One must assume that the moulting of the overwintering stock to Stage VI males, and copulation, must take place before the ascent. Males, as so often with calanoids, were presumably short-lived. Females migrating to the surface generally show immature ovaries, but during November egg-laying commences, reaching a maximum in December and January. By February and March oviposition is beginning to decline (cf. Fig. 3.71) and spent females are observed. From about May to September females are almost entirely absent from the catches (cf. Fig. 3.72). Young copepodites, presumably of the new brood, appear in the surface layers from November onwards throughout the southern summer, C I and II achieving maximal density during December (Fig. 3.72). Nauplii, though not regularly sampled, are also very abundant from November onwards. During the late summer, almost all the *Calanoides* population is represented by C III, IV and V, which gradually desert the surface until finally all are overwintering in deep water as C IV and V. *Calanoides acutus* appears, therefore, to have a single breeding cycle similar to that of *C. hyperboreus* and of *C. finmarchicus* at high northern latitudes.

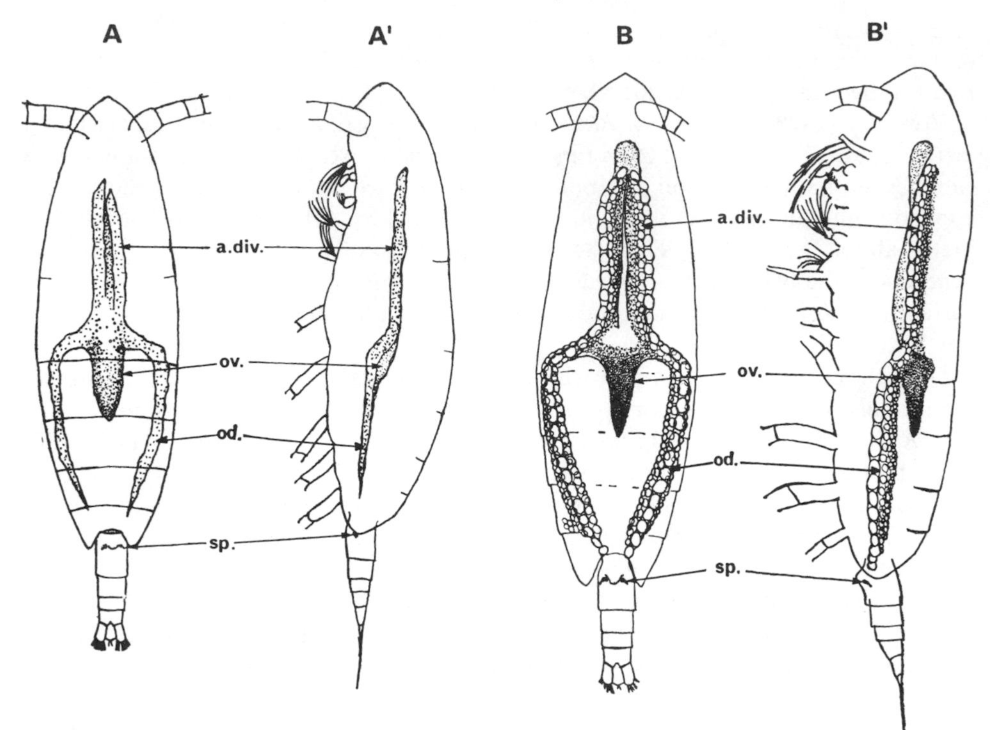

Fig. 3.71. Stages of development of the ovary and oviduct of the adult female. A. Maturity condition "A": dorsal view showing the ovary (ov.) with proliferating eggs migrating into the anterior diverticula of the oviduct (a.div.) A[1] a lateral view of the genital segment of the urosome of the same individual. B. Maturity condition "B": dorsal view showing the oviducts (od.) and their anterior diverticula (a.div.) packed with developing eggs. B[1]lateral view of the same individual showing the well-developed ventral protruberance of the genital segment of the urosome with spermathecae (sp.) packed with spermatozoa (Andrews, 1966).

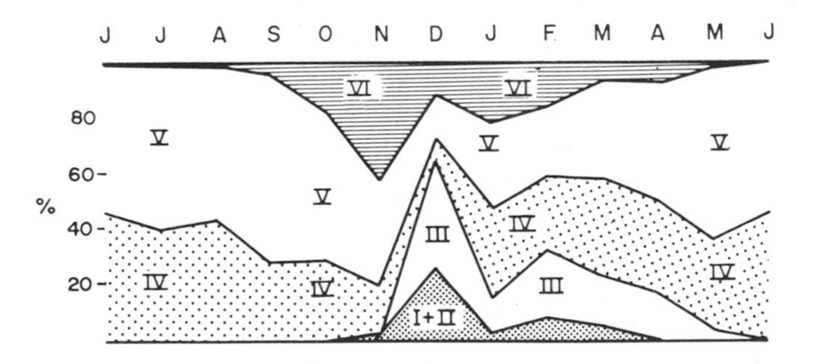

Fig. 3.72. The seasonal copepodite composition of the total population in the 1000–0-m layer. The abundance of each copepodite stage is shown as a percentage of the mean total population in each month. The samples used were taken throughout the Antarctic region although the majority were from the areas between 30°E and 30°W of the 0° meridian (Andrews, 1966).

Analysis of gut contents of the overwintering stock of CIV and V shows that copepodites are practically non-feeding. During the summer when feeding is very active an oil sac is well developed as a reserve for the overwintering period.

Reproductive cycles of chaetognaths

Although reproductive cycles of copepods have attracted most attention, breeding patterns of Chaetognatha have also been studied in some details (cf. Alvarino, 1965). Russell (1932, 1933) found that the chaetognath *Sagitta elegans* appeared to have four spawning periods in the English Channel: February, May, June/July and September. The offspring of the last September spawning apparently did not continue their development up to complete maturity during the autumn but overwintered in an immature stage. Moreover, there were indications that they descended to somewhat deeper levels during the overwintering period—a marked similarity to the behaviour of *Calanus*. By about December this overwintering chaetognath stock began to mature, the testes developing first, followed by the ovaries in the following month. The adults arising from the overwintering stock spawned about February to give the first brood (Fig. 3.61). Russell's results suggest that adult length is highest about May, so that these would appear to be the animals that arise from the February spawning and would be developing over the coldest period of the year. The mean length tends to fall during June, July and September. Temperature would thus again appear to be an important factor affecting size.

In the North Sea, Wimpenny (1937) concluded that *Sagitta elegans* had only three main generations. However, in the Irish Sea Pierce (1941) found that although *S. elegans* began to reproduce in February, breeding was continuous until the end of May and no further broods followed during the remainder of the year. In the Gulf of Maine, Redfield and Beale (1940) found that *S. elegans* had an extended breeding period over spring and summer. According to Clarke, Pierce and Bumpus (1943), however, over Georges Bank the chief breeding time was April or May, and a distinct second generation followed in late summer or autumn. This type of cycle is also confirmed by Zo (1973) for the shallow Bedford basin (Nova Scotia). In any event, it would appear that in the colder waters of the Gulf of Maine area the breeding of *S. elegans* exhibits a reduction in the number of broods compared with the English Channel. The breeding of *S. elegans* in the Gulf of Maine has been confirmed by Sherman and Schaner (1968). The density of the chaetognath in the Gulf of Maine is low in spring, high in summer and declining in the autumn and winter. This is related to the breeding cycle which is essentially one generation annually. Breeding continues through the spring to autumn, with immature stages dominant (60–95 %). Mature specimens were dominant (80 %) over winter but apparently died off after breeding in spring. The adults tended to be deeper-living. The population over Georges Bank and the more coastal populations along the Gulf of Maine appear to be more or less separate and are not dependent on immigration from other areas.

In the Norwegian Sea, Wiborg's (1954) observations suggest a spring spawning for the species; a peak of small individuals was found during April/May, though in one year a smaller peak occurred in September. Ostvedt (1955) also suggests late spring or early summer as the spawning time in the open Norwegian Sea (Weather Station M), and a similar time is indicated off eastern Greenland, though Alvarino (1965) quotes Kramp who states that the period extends in Greenland waters as long as from April/May to August/September, with early autumn as the main time. Only a single brood occurred. A similar extended season appears to be true for the Gulf of St Lawrence.

According to Dunbar (1940) *Sagitta elegans arctica* showed two distinct age groups at any one time in the Canadian Arctic. One brood had completed only about half its growth when the other brood was spawned. Probably only a single autumn spawning occurred. These *Sagitta* could, therefore, take two years to reach maturity in the very cold Arctic. Dunbar (1962) confirmed a two-year cycle in Arctic and sub-Arctic seas for *Sagitta elegans arctica*, the populations forming two distinct groups breeding in different years. The spawning period was long, probably from June/July to October or even as late as February. Spawning of *Sagitta* appeared to depend not so much on availability of food, as on the limitations of growth imposed by the low temperature. Thus Dunbar points out that at the head of one Arctic lagoon, where warmer conditions were typical, there was a marked acceleration of breeding. The cycle was annual and mature individuals were only about half the normal body size. Alvarino (1965) reports that in the Barents Sea *S. elegans* shows an annual breeding cycle spawning in summer, but according to Bogorov the cycle may occupy two years.

The neritic species *Sagitta setosa*, also studied by Russell (1932) off Plymouth, had a rather similar life history but with a greater number of breeding periods. Though the number of separate broods is not entirely clear, they appear to occur in February, April/May, June, July/August, September and October. The last (October) brood did not develop to maturity, but overwintered in deeper waters, testes developing in December and ovaries in January, the first brood thus being produced in February. Body length of the late summer and autumn broods was less than that of the average spring and summer broods.

In the southern North Sea the main breeding period for *Sagitta setosa* is from April to July, though eggs are found as late as September (Wimpenny, 1937). Off Flamborough there appear to be fewer broods than in the English Channel, and this appears to hold for the Irish Sea, where Pierce (1941) found two main breeding periods.

Murakami's (1959) studies of the neritic Japanese chaetognath, *Sagitta crassa*, a somewhat warmer water species, are possibly relevant to the variations noted for the breeding of *S. setosa*. Several polytypes of *S. crassa* exist. Three main polytypes have been reared experimentally by varying the conditions of salinity and temperature. The species was successfully reared even at temperatures below 7°C. Populations of the chaetognath in very shallow inlet Japanese waters must withstand low temperatures during winter, and the life cycle differs considerably from that of the same species in slightly deeper more offshore waters. Generally, three broods occur over a year: spring/summer, summer/autumn, with a life span of about three months, and autumn/winter, with a longer span (*ca.* 5 months). Apparently, however, some of the sagittae that are spawned in summer last until winter. In the slightly deeper areas the spring/summer brood is the largest, whereas in the shallow inlets, though breeding tends to commence earlier in the year, the most productive brood is in autumn/winter.

As in many other taxa, the breeding of chaetognaths appears to be very much related to temperature. Several species of *Sagitta* show a relationship between body size, maturity and temperature, with smaller body size generally coinciding with higher temperature during development. The number of broods in a year also tends to increase with higher habitat temperature in some of the widely distributed species (cf. Russell, 1935b; Reeve and Cosper, 1975). McLaren (1965) also suggested that in Arctic regions there is selection in chaetognaths for large size maturity and for high fecundity, and that slow development occurs where the food supply is limited and variable. Food and temperature may both be significant in the breeding pattern.

Sagitta maxima occurs mostly in deeper waters at higher latitudes. According to Alvarino (1965), breeding takes place off Greenland in the deep layers in summer and autumn, with a water temperature not much exceeding 0°C. Some of the population appears to delay till winter. From May to September juveniles and adults were found plentifully in Davis Strait, but mature specimens were taken only in deep water where spawning took place. The breeding cycle is certainly not less than annual; it could conceivably occupy a longer period.

The chaetognath *Eukrohnia hamata* occurred widely in the Norwegian Sea but was also a deeper-living species. Wiborg (1954) found highest numbers from April to August off the Norwegian coast, with peaks in May and July. He believed that breeding was chiefly in April/May. According to Ostvedt (1955) young individuals occurred more shallowly in June/July and spawning was believed to take place over summer. Kramp is reported by Alvarino (1965) as finding breeding over a large area (Davis Strait, Labrador Sea, Baffin Bay) stretching from spring through summer, but declining in autumn. Breeding, which takes place at rather deeper levels, occurs through much of the spring, peaking in summer and continuing in autumn. It is not entirely clear, however, whether breeding is an annual or a two-year cycle. In the Barents Sea, according to Bogorov (1958), the breeding is in autumn and there is a two-year life span. In the Bering Sea Heinrich (1962) found that breeding took place in the colder part of the year, but that there was a two-year span with much mixing of stages. On the other hand, in Norwegian waters the main numbers occurred from April to August, with small individuals chiefly in April/May and some in February/March. While this might be interpreted as an annual cycle, a recent study by Sands (quoted by Bamstedt (1978)) states that in the Korsfjorden *E. hamata* breed intensively from May to late October, but that although full body length is attained after one year, the adults do not become sexually mature until the second year.

In the southern hemisphere *Eukrohnia hamata* is the most abundant chaetognath in the Southern Ocean, being found in sub-Antarctic and Antarctic waters. David (1958) comments on the great variation in body size of the species, with the Antarctic Convergence appearing to separate two races, the smaller sized individuals to the north (cf. *Sagitta gazellae—vide infra*). In the Southern Ocean *Eukrohnia* generally inhabits surface and sub-surface levels from 0–500 m, with some extending to 1500 m, and the species descends to deeper water in winter. Breeding takes places in deep water, the mature stage (III and IV) being found only below 750 m. David's data suggest a very large increase in population by late summer (March); possibly this indicates one main annual brood, but details are not known. A spent form seems able to live for some time after spawning.

Sagitta maxima is found in the southern hemisphere, though individuals are of much smaller size in the Arctic. It is essentially a sub-Antarctic species, living mostly at depths of *ca.* 150–500 m. A few are found south of the Antarctic Convergence. The more mature stages of *S. maxima* (II and IV) occur at greater depths (below 750 m). David (1958) points out that all the plentiful Southern Ocean chaetognaths have the common characteristic of descending to deep waters for breeding. Thus few mature stages of *Sagitta marri*, which lives at about the same depth as *S. maxima*, but south of the Antarctic Convergence (cf. David, 1958; Mackintosh, 1964), were ever captured, but some Stage III were recovered and all were below 750 m.

Of all the Southern Ocean chaetognaths, the life cycle is best known for *Sagitta gazellae*, a species which occurs mainly in sub-Antarctic waters but also extends south of the Convergence. *S. gazellae* lives in the upper 200 m, though it may be a little deeper in winter. In the sub-Antarctic the abundant population is represented by an apparently distinct race of small-sized individuals. The Convergence has a reduced density, but another concentration in Antarctic waters consists of comparatively large individuals. The "northern small" race was studied in detail by David (1955). The chaetognaths descend to deep water to spawn; the mature Stage IV, for example, are mainly at 1000–1500 m. Spawning occurs over a long drawn-out period over spring, summer and autumn and possibly even to a very limited extent in winter. From eggs spawned, for example, in autumn (March/April), the young grow at a rate of about 5 mm/month and they reach the upper (0–100-m) layer by about May in considerable density. Growth continues, but more slowly in mid-winter, when the animals are somewhat deeper. By September the young have grown to a length of some 25–30 mm, but a more rapid growth occurs during the southern spring and summer to reach 55–60 mm—approximately full size—by the end of the summer.

With the extended spawning, however, there is a great overlapping of size groups as the year advances. The immature chaetognaths remain in the superficial layers till about February/March, but as they mature they descend as Stages II and III, to *ca.* 500–750 m. The mature Stage IV, which form the new breeding stock, are mainly below 1000 m, and average about 55–60 mm in length (i.e. little or no further growth in length takes place as the gonads mature). *S. gazellae* is, therefore, essentially an annual breeder. Though this cycle applies to the northern race, the southern race, which can reach 80 or even 90 mm in length, is also probably an annual. In the sub-Antarctic waters, unusually large individuals (*ca.* 90 mm) are encountered mostly in winter and spring in surface waters. They are, however, non-breeding individuals and do not take part in the seasonal breeding migration.

Low temperature, apart from other effects on breeding of chaetognaths, may delay maturity in species near the limit of their environmental range. Bigelow (1926) and Redfield (1939, 1941) described *Sagitta serratodentata* (probably *S. tasmanica*) populations which penetrate the relatively cold waters of the Gulf of Maine. The animals could persist there for quite a long period but did not attain maturity, though they reached large size. The effect may be primarily temperature-controlled, since outside the Gulf, in the warmer waters, populations of the chaetognath include mature and young forms.

The breeding of some cold water euphausiids

The euphausiids contribute substantially to the biomass of zooplankton at higher latitudes, frequently being second in importance to the great biomass of copepods. In certain regions, as in parts of the Antarctic, their biomass may temporarily exceed that of copepods, especially in the surface layers, since some euphausiids are well known to form dense swarms. Vinogradov (1970) quotes the biomass of Antarctic swarms as amounting to several hundred grams/m^3, even reaching kilograms/m^3. The breeding of temperate and high latitude euphausiid species can, therefore, be of great significance in seasonal change in biomass.

In the temperate waters of the English Channel, Lebour (1924) showed that the coastal species *Nyctiphanes couchii* reproduced mainly during the spring, with a peak of spawning about May. However, Lebour reported that some spawning appeared to occur all the year.

Many of the North Atlantic euphausiids are annual (Einarsson, 1945), though some take two years or even more to complete the life cycle. Moreover, for some euphausiids a lowering of temperature tends to delay maturity and slow development, an effect similar to that seen in many calanoid copepods. The deep-water euphausiid *Thysanopoda acutifrons* appears to take two years to mature throughout its range in the North Atlantic. Spawning occurs about May and the animals grow to a size of about 30 mm by the end of their second summer. They mature during the following winter and spawn at an average length of some 40 mm, after which the majority of the two-year-old stock dies off (cf. Fig. 3.73).

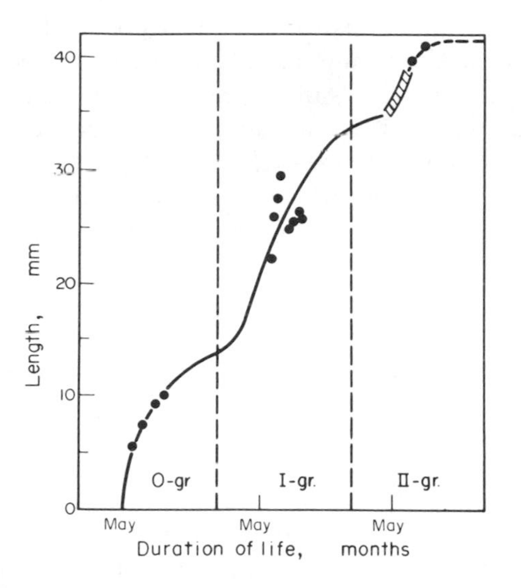

Fig. 3.73. The life cycle of *Thysanopoda acutifrons*. The doubled contoured part of the curve denotes the first spawning (Einarsson, 1945).

The effect of temperature variation on breeding is not seen in this species, probably owing to its living in deep water where conditions are more uniform than in the upper layers. According to Einarsson, *Meganyctiphanes norvegica* reaches maturity in one year in Atlantic waters. However, in the more southern part of its range spawning occurs from about February to April and most of the animals die at the age of one year. In more northern waters the spawning is somewhat later (March to July), and although *Meganyctiphanes* becomes mature in one year, a considerable portion of the stock appears to persist for a further year and spawns a second time (cf. Whiteley, 1948). This history of the breeding cycle was confirmed by the investigations of Mauchline (1959) for the Clyde. Spawning of *Meganyctiphanes* extended over spring and summer with peaks in April and at the end of June. Development to the adolescent stage, when the animals averaged 25 mm length, occupied 2–3 months. Sexual maturity was attained only in late winter, transference of spermatophores occurring in January to February.

Many of the males died at this time. The females spawned in spring, and a heavy mortality followed after spawning. However, a number of females persisted through the summer (when they exceeded 35 mm in length) and they survived for a second winter, spawning again in the following spring. Mauchline believes that a very small percentage of these females may persist even after the second year and might spawn for a third time.

Mauchline and Fisher (1969) also suggested that *Meganyctiphanes* might spawn twice in a season in temperate waters, giving peaks in April and June/July. Although a second spawning is not proved, the very protracted spawning period for several euphausiids (e.g. *Nyctiphanes australis* spawning over eight to eleven months, *Nyctiphanes couchi*, *Euphausia pacifica*) gives possible support to the suggestion. If such multiple spawning occurs, it probably varies greatly with latitude.

The life cycle for *Meganyctiphanes* was confirmed by Wiborg (1971) in the Hardangerfjord, Norway. There was a long spawning period from about April to June; maturity was reached in about nine months and transference of spermatophores occurred about February/March. Although some individuals are annual, Wiborg believes that some survive for two years and possibly a few even for three years; the stock was heavily reduced after spawning.

Matthews (1973) found that the same species breeds in the spring in the Korsfjorden and becomes mature at one year. Though the euphausiids can live for over two years, apparently only a small proportion survive to breed in a second year. There is great seasonal variation in rate of growth, with growth being greatly reduced over winter and slightly lowered over mid-summer. The slowing of growth over winter was not confirmed by Wiborg, but growth rate decreased during the second year of life in the Hardangerfjord.

According to Einarsson (1945), the life history of *Thysanoessa longicaudata* in cold northern (Greenland) waters appears to be somewhat similar in that the euphausiid matures in one year, but a portion of the stock persists for a longer time. However, in more southern areas the whole stock, after growing to maturity in a year, dies off.

A marked effect of temperature on reproduction occurs with two other species of *Thysanoessa*, *T. inermis* and *T. raschii*. Both spawn during the spring and may become sexually mature in one year, when they are comparatively small, measuring only about 14–17 mm. In the more temperate areas these euphausiids persist into a second year and reproduce again in the subsequent spring. By this time the animals exceed 20 mm in length. In sub-Arctic water (e.g. north of Iceland), however, these two species of euphausiids do not breed in the first year. Sexual maturity is delayed until the second spring, when they spawn at a length rather similar to those euphausiids of two years old in the more southern regions (Fig. 3.74). It appears that in the extremely cold Arctic waters of West Greenland, *Thysanoessa inermis* and *T. raschii* survive at least for a third year and spawn again (Fig. 3.74).

In general the body size of euphausiids occurring in very cold waters exceeds that of the same species in more temperate parts of the North Atlantic (cf. copepods). These observations are confirmed by the investigations of Wiborg (1971). Whereas *T. inermis* is an annual species in the Byfjord and the Hardangerfjord, spawning from April to June, with possibly a few individuals surviving for more than one year, at Station M in the Norwegian Sea the same species take two years to reach maturity. Nemoto (1957) claims that *T. inermis* is biennial in Aleutian Islands waters and may even reach a

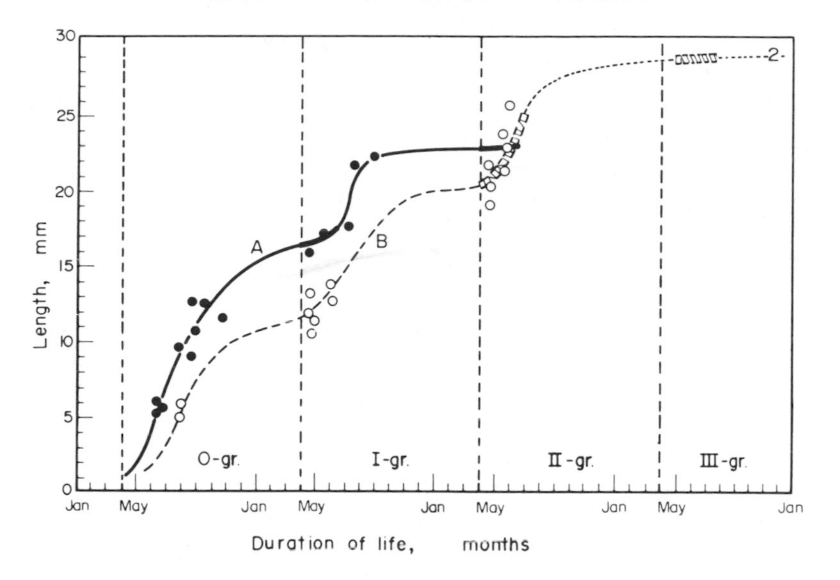

Fig. 3.74. The life cycle of *Thysanoessa inermis* (A) In southern localities of the North Atlantic; thickened parts of the curve denote first and second spawnings; (B) growth in north and east Icelandic waters (dashed curve) and further growth of the species in West Greenland (dotted curve). Spawning indicated by double contoured curve (Einarsson, 1945).

three-year cycle in the exceedingly cold waters north-east of the Aleutians. Nemoto's results for *T. longipes* for the north-west Pacific indicate a similar relationship. Whereas off the Aleutians a spineless form of the species matures and spawns at one year but may survive for a longer period, off Kamchatka maturity is not achieved until the second year at considerably greater body length, and the euphausiids may survive for three years and spawn. *T. spinifera* is also biennial, though some survive to a third year.

Euphausia pacifica, the common euphausiid of the boreal North Pacific, appears from the investigations of Nemoto (1957) to be an annual species, individuals of 15 mm length reaching maturity in a year and probably breeding once only. However, in the Aleutian Islands area spawning occurs in spring/summer of the second year at a length of about 22 mm.

Of the euphausiids so far described from cold waters of the northern hemisphere, only *Thysanopoda acutifrons* appears to achieve maturity regularly after only two years and to have a single breeding. This, however, appears to be the normal pattern of breeding of euphausiid species in the Antarctic. The enormously abundant southern krill, *Euphausia superba*, is believed to take two years to mature (cf. Fraser, 1936; Bargmann, 1945). About nine months is spent in the larval stage, but a period of more than a year follows as an adolescent before maturity is attained. Spawning may occur over a fairly long period from November to as late as March or April and the populations therefore tend to be very heterogeneous.

Burukovskiy (1967) examined krill from December to March in the surface water drifting north from the Weddell Sea. Fertilization was at its peak in December/January, declined in February and ceased in March, but reproduction occurred later on proceeding south. While there was an abundance of juveniles, which increased in average size from December onwards, there was a decline in population in March. Marr (1962) pointed out that eggs of *E. superba* were found comparatively rarely and in fairly deep water; he believed that the adults might descend to spawn. This paucity of eggs and young stages in collections even from depth in the Antarctic, despite the enormous density of the adults, was, however a puzzling feature in attempting to determine breeding time and location.

According to Mauchline and Fisher (1969), euphausiid eggs generally are more dense than sea water, and since *E. superba* spawns over much greater depths than many northern species the eggs will be at depth. They may sink at least to 500 m, but Mauchline and Fisher doubt whether the adults descend to spawn. Those eggs that are spawned near to the Antarctic shelf will reach the near bottom layers. The nauplii, metanauplii and calyptopes then have a long ascent towards the surface. Although adult krill were believed by Fraser to drift northwards towards the Antarctic Convergence and to spawn over deep water mainly in that region, the larvae and young stages drifting in the deeper layers mainly southwards towards the antarctic continent, it is more likely that spawning occurs far south, especially near the East Wind Drift divergence. Those eggs that are released over deeper oceanic areas may sink to 1000 or 1500 m, and after hatching the larval stages must climb to the surface. The furcilia and adolescent stages of *E. superba* are in the near surface Antarctic layer, which is moving roughly north-east in the Weddell area. There is little or no vertical migration. They may spend many months in the near surface layers at this period of their development, feeding on the phytoplankton.

Mackintosh (1972) has re-examined the life cycle, taking into consideration some material which included greater numbers of eggs and young stages of *E. superba*. Spawning appears to be mainly in the East Wind Drift in the Weddell area and in the Bransfield Strait and along the ridge of the Scotia Arc, all in regions where water of low temperature descends to comparatively great depth. Although spawning may occur from November to March (*vide supra*), Mackintosh found that the peak was in February or early March. He confirmed that growth to maturity takes two years, though an "intermediate" size group may appear in some years. Fevolden (1979) confirms that in the inner Weddell Sea (75°S) three years elapse before maturity is attained. Ivanov (1970), from a study of size composition curves, also found three size groups and concluded that krill might take three rather than two years to attain maturity, spawning only in the fourth year. Mackintosh, however, points out that the second of the three size groups corresponds to his "intermediate" group, and that more probably the group represents a population experiencing an abnormally high growth rate, typical of a warm year.

Growth rates of *E. superba* as determined by a number of investigators coincide fairly well, provided it is accepted that different regions of the Antarctic and different years may show considerable rate variations. For example, the Pacific sector appears to have a slower growth, perhaps reflecting a retarded increase in phytoplankton. The onset of growth of krill occurs later at colder higher latitudes, and the delay may amount to two months. The time of icemelt in any locality is a significant factor in the starting date. In all areas growth is rapid in spring and summer, with the rich phytoplankton crop; in winter growth almost ceases. Mackintosh points out that *E. superba* is approximately twice the body length (50 mm when adult) of *E. pacifica* and reaches maturity in twice the time (2 years). Since growth almost ceases over winter in *E. superba*, Mackintosh suggests that the growth rate over spring may approach twice that of *E. pacifica*. Chekunova and Rynkova (1974) described length/weight changes during the growth of *E. superba* in the Scotia Sea. During the spring following the year of larval development the young krill showed a very marked gain in weight (about 5 % per day). They suggest that limited feeding continues in winter and that at least one moult occurs over that period. All workers appear to accept that spawning in *E. superba* occurs over one breeding season only.

Euphausia triacantha occurs in Antarctic and sub-Antarctic waters and is abundant in the region of the Antarctic Convergence. Baker (1959) found that, as with *E. superba*, the normal life span from hatching to breeding takes two years. Spawning is again somewhat extended, commencing as early as September but occurring mainly in October/November and ending in December. In spring (September/October) three age groups may be distinguished—the earliest larvae (0-group), adolescents approximately one year old, and adults about to spawn. From October, the young 0-group shows a fairly steady growth, which Baker estimates as about 2–5 mm per month. By April this group is post-larval. Growth is much slower over the winter months, but by August they average 18–19 mm. As I-group euphausiids, growth continues again from about September at a more rapid rate, though not as fast as in the first year. By February the animals average about 30–32 mm and growth ceases. The two relatively rapid periods of growth correspond to the times of phytoplankton abundance. It appears that males mature about March but that few females are then mature; pairing is delayed till about June. Whether any of the females survive into a third year is uncertain.

The Antarctic euphausiid *E. frigida* appears to be circumpolar, as are several other Antarctic species, but it is limited to waters south of the Antarctic convergence (cf. Mackintosh, 1964). Makarov (1977) studied the life history of the species in the Scotia Sea and found comparatively abundant nauplius and metanauplius stages confined to the deeper waters, mostly below 500 m. Calyptopis stages were distributed throughout the whole water column, including the surface, indicating that the young larvae ascend, as has been suggested for other Antarctic euphausiids. Furcilias were taken in smaller numbers but were mainly in the superficial layers. The study was limited to only two months, December and January, and in the northern region spawning was already advanced, but it was clearly delayed to the south, nauplius and metanauplius stages contributing 67 % of the population in the south, compared with 29 % in the north. Furcilias amounted to less than 1.0 % in the south and 17 % in the northern area. Fevolden (1979) noted a protracted spawning season with berried females in late February as far north as Bouvetoya (*ca.* 55°S). Further investigations must determine whether *E. frigida* resembles *E. superba* and *E. triacantha* in its life cycle, but there are grounds for suggesting that several Antarctic euphausiids, *E. frigida*, *E. crystallorophias*, *Thysanoessa macrura* and *Thysanoessa vicina*, all take two years to reach maturity (Fevolden 1979).

Breeding cycles of some other cold water holoplankton

Information on breeding cycles at higher latitudes of holoplankton belonging to other taxa appears to be limited. Russell (1935a, 1953) suggested for the boreal trachymedusan *Aglantha digitale* that off the British Isles an overwintering stock of medusae spawned in March, producing a fairly large brood, and that two or three smaller broods were produced throughout the summer. The offspring of the last spawning, about August/September, formed an overwintering stock which tended to seek deeper water (Fig. 3.61). Russell suggests that *Aglantha* may show a similar pattern of life cycle to that suggested for *Calanus*, namely that while having several generations in more temperate seas, it will show a reduction in the number of broods and a slower development at higher latitudes. Ostvedt (1955) suggested late spring and summer as the spawning time in the Norwegian Sea for *Aglantha*. Wiborg (1954) recorded maximal numbers in June off Norway and Zenkevitch (1963) mentions summer spawning. Russell (1953) includes several observations suggesting that the first brood is of large-sized individuals; summer broods yielded specimens of smaller average size (cf. also Bigelow's (1926) observations for the Gulf of Maine).

Dunbar (1946) described the life cycle of the amphipod *Themisto libellula* in Canadian Arctic waters. This very important species appears to commence breeding about March, but the period is extended probably until July. During the summer Dunbar found two distinct size groups of *Themisto*, smaller "juveniles" and larger "adolescents", both sexually immature. The "juveniles" appear to be the products of the spring/summer breeding; the "adolescents" possibly become mature and breed in autumn. It is likely, however, that they delay breeding until the following spring, in which case the life cycle occupies two years instead of one. In either event, Dunbar points out that two distinct age-groups aside from new offspring exist at the same time. When newly spawned individuals are also present there are three distinct age-groups. Dunbar found this type of breeding cycle with other planktonic animals at high latitudes (e.g. *Sagitta elegans* and *Thysanoessa* spp.—*vide supra*). Ostvedt (1955) found ovigerous females of *Themisto abyssorum* in the Norwegian Sea in February, March, April and May. "Larvae" were most frequent from April to June, indicating spawning in spring or early summer. He believes that breeding may be connected with an annual vertical migration.

In the Southern Ocean, Kane (1966) observed that the common amphipod *Parathemisto gaudichaudi* reached maximum abundance in December, with almost 100% of the population occurring as small juvenile individuals. Breeding took place in the southern winter. Very large numbers of the *Parathemisto* caught in August were sexually mature and some release of the young from the brood pouch of the females occurred as early as September, but release continued to give a maximum in December. Growth of the young continued at a fairly steady rate in the Southern Ocean until March and immature animals persisted until July/August, when maturity was achieved. It is likely that during August/September mature females can produce more than one brood, but very few mature individuals survived after early summer and none survived after April. The life cycle of *Parathemisto* is therefore annual, but the presence of a few unusually large adults suggests that a few individuals might survive to a second year. Kane comments that this is unlikely for the populations which occur at lower latitudes off South Africa. Those populations were characterized by the unusually small size of the individuals. Philippon's (1972) observation of single annual breeding of *Parathemisto gandichaudi* during August and September in Antarctic waters off Kerguelen agrees with Kane's analysis.

Many of the larger crustaceans in boreal regions are believed to be annual breeders (cf. Fish and Johnson, 1937), though some of the larger pelagic decapods probably live longer (cf. Whiteley, 1948). There are few records of breeding times. Ostvedt (1955) notes that ovigerous females of *Hymenodora glacialis* were sometimes found in May and June in the Norwegian Sea and Wiborg (1954) mentions the very young specimens found in July. A more recent detailed study by Matthews and Pinnoi (1973) has shown that the decapod *Pasiphaea multidentata*, the commonest *Pasiphaea* in Korsfjorden, bred throughout most of the year, though with periods of greater intensity (e.g. May/June, October and even in winter). Maturity was achieved in about one and a half years and few survived for more than two years. At least two broods may be produced. Another pelagic decapod, *Sergestes arcticus*, bred in the spring and became mature in a year. The species probably lives for more than two years.

Although data on the breeding of planktonic decapods in the Mediterranean appear to be irrelevant to a discussion of breeding at high latitudes, the animals tend to inhabit fairly deep layers with lowered temperatures, so that records of their breeding are not entirely out of context. Seridji (1968) found that although few species of sergestids occurred off Algiers, numbers of larvae, mainly those of *Sergestes arcticus* and *S. robustus*, were comparatively abundant, mostly in winter and spring. *S. arcticus* larvae were found chiefly in November, December and January; *S. robustus* over the same months, but extending later, even to April. Larvae of *Gennadas elegans* were also taken over a long period, though more abundantly from winter until April. However, an indication of a succession of progressively older larvae was apparent, with protozoeas mostly in January and February, mysis I stages in March and April and even to July, and a very few older stages up to November.

Of other holoplankton, the pteropod *Limacina retroversa* was investigated by Redfield (1939) in the Gulf of Maine. The two main generations occurred over a year, with a possible third brood. The pattern was considerably influenced by the circulation in the Gulf. Thus in one year, Redfield showed that a population entered about December and was transported round the Gulf, growing up successfully during the following six months, until most of the individuals disappeared by the following June. However, two months before (in April), a second incursion of Limacina occurred in the eastern sectors of the Gulf; again, these spread round throughout the summer, though growing faster than the earlier immigrants. There was almost certainly some local reproduction in the Gulf from these immigrant populations, but despite this local reproduction over the summer months, *Limacina* does not seem to be able to maintain its numbers without constant incursions from offshore waters. The real centre for production of *Limacina* is presumably eastwards of the Gulf of Maine, and the population at any time inside the Gulf is the difference between immigration and any local reproduction on the one hand and mortality and drift on the other. Other studies of this widely distributed species in other areas would be of interest.

Salps have been widely recognized as being capable of very rapid growth and reproduction, causing dense swarms to appear suddenly. While swarming is often seen in warm-temperate regions, especially when cold and warmer water masses meet, swarms were early recognized at high latitudes in the Antarctic (cf. Hardy and Gunther, 1935). Swarms of salps and of *Euphausia superba* sometimes appear to occupy alternate patches in Antarctic waters and they may indeed compete to some extent for the rich crop of phytoplankton. The breeding of salps in the Antarctic is therefore of particular interest.

Foxton (1966) examined the breeding of the Antarctic and sub-Antarctic species, *Salpa thompsoni*. A huge rise in density occurred due to the budding of the aggregate forms, mostly in the season October to February/March: numbers could increase fifty-fold. Smaller fluctuations occurred in the solitary individuals, with a less clear seasonal effect, but a rise was discernible from about December to March. Both forms showed a sharp fall over winter, but solitaries were much fewer than aggregates in all months. The aggregate swarms are produced by the release of chains of buds. The winter aggregates consisted mostly of only a few small-sized individuals, the difference from the large aggregates typical of summer being related to the reduction in budding over winter. Different populations of solitary forms were observed in the shallow (0–100-m) and deeper (> 100-m) layers. In the shallow layer the smallest individuals measured less than 10 mm from November onwards and about 10–20 mm by March. This was probably the same group of individuals that attained a length of 50–60 mm by September, suggesting a growth rate of 6–8 mm per month. In the deeper layer smaller forms predominated even during March and April and probably represented newly released individuals. These smaller animals were present also in September and appear to be slower-growing overwintering animals. Foxton also observed that maximum catches of the salp, *Salpa gerlachei*, which is confined to very high southern latitudes, occurred in January. The swarming season probably coincided with the relatively short period of phytoplankton abundance typical of the very high Antarctic.

Breeding of Bathypelagic Zooplankton

Very little is known about breeding cycles of deep-living (bathypelagic) zooplankton. Mauchline (1972) has instanced a number of apparently unrelated observations on deeper-living marine animals which might suggest that bathypelagic plankton is comparatively long-lived compared with near-surface species. Whereas epipelagic forms tend to grow and mature rapidly, breed and then die, bathypelagic and deeper-living species are likely to grow comparatively slowly in view of the great reduction of food at depth in addition to the generally low temperature. Deep sea animals have been observed to be marked by an unusually small proportion of ovigerous females, which could indicate that breeding occurs less frequently than annually, as well as a slow attainment of sexual maturity.

The range in body length often characteristic of the comparatively few representatives of a deep-sea zooplankton species might also suggest that sexual maturity can be achieved well below maximal length and that continuous slow growth accompanies increased longevity. Mauchline has suggested that maturation might also be related to the process of mating for deep-sea forms; if mating does not occur, maturation might be delayed. He instances a deep-living euphausiid, *Thysanopoda cornuta*, in which a comparatively large female was still unmated (i.e. without spermatophore), whereas a smaller individual had already copulated, though the ovary was not yet ripe. With the low population density typical of deep sea zooplankton, opportunity for mating will be limited. Mauchline believes that the sex ratio in many deep sea forms is markedly in favour of females, probably with an extended breeding period, and that males above a minimum size are possibly mature throughout the year. Isolated individuals or very small groups may be unable to mate. The occurrence of occasional "giant-sized" individuals of deep-sea species might be related to such unmated individuals, which continue to grow but never achieve sexual maturity.

In considering the rate of growth of deep-living zooplankton species, the young stage emerging from the egg is usually comparatively larger than its epipelagic relative. Deep-sea species tend to have fewer eggs of larger size than similar epipelagic forms, so that the young are larger. Mauchline has attempted to relate body size to longevity and to test whether both are generally greater in deep-living zooplankton crustaceans. In his analysis he has used data from certain epipelagic species of a taxon to establish a relationship between age and length; from the body length of deep sea forms, he has then attempted to deduce their age by extrapolation. Data from higher latitude epipelagic species are more relevant, since while the larval period may not be much affected by the lower temperature at higher latitudes, maturity may be delayed compared with epipelagic species from warm seas. An example for one taxon (mysids) is illustrated in Fig. 3.75. In examining the body length of certain deep-living euphausiids and mysids, Mauchline has tended to employ the maximum length ("potential for growth or size") for a species rather than its average length, since the latter value might be much in error with the small number of deep-sea individuals usually captured. For these two crustacean taxa it appears that epipelagic species, which for the present purpose are regarded as from 0–500 m, are generally not so large in potential size as mesopelagic species, regarded as occurring from 300–1500 m, and bathypelagic species (>1000 m).

Potential for growth would appear to increase with depth for these two crustacean taxa. There are exceptions to the overall relationship. Thus among the euphausiids, *Meganyctiphanes* (about 40 mm) and *Euphausia superba* (about 60 mm) are of

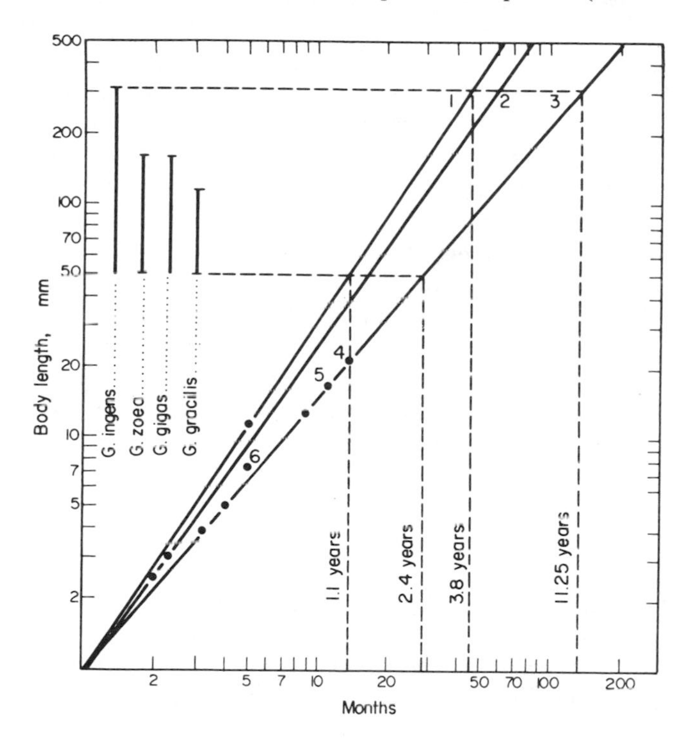

Fig. 3.75. Growth rates of epipelagic species of mysids extrapolated to estimate longevity of bathypelagic species of the genus *Gnathophausia*: 1. *Schistomysis kervillei*; *Neomysis integer*; 2. *S. spiritus*; 3. *Praunus inermis*; *Mysidopsis gibbosa*; 4. *P. neglectus*; 5. *P. flexuosus*; 6. *Erythrops serrata* (Mauchline, 1972).

relatively large size, though epipelagic. However, species of *Thysanopoda*, which is a typically deep genus, support the general theory in that they may exceed 100 mm in length. For copepods, deeper-living species are again generally larger than the epipelagic forms, but there is some difference in the views of authors regarding very deep-living species. Mauchline believes that the potential for size increases to a depth of *ca.* 1000–2000 m, but is less obvious at greater depths and may even tend to decline. A recent survey of the copepods of the Sargasso Sea by Deevey and Brooks (1977) demonstrated that the dominant species in the upper layers were of small size. Larger species were common below 500 m, but at depths greater than *ca.* 1500 m overall body size tended to decrease. Observations on chaetognaths suggest that they tend to be of large size in both deep and shallow waters, but that small species occur only at shallow depths.

In considering the growth rate of the more accessible epipelagic species for any taxon, there may be a considerable scatter of results. Faster and slower-growing surface species are well recognized. Nevertheless, a mean value can be deduced from which, by extrapolation, the potential body sizes of the bathypelagic species may be estimated and their probable age obtained (cf. Fig. 3.75). Rate of growth is assumed constant. For euphausiids and mysids, the deep-living species were estimated to be some three to seven times as long-lived as surface forms. Thus, many surface euphausiids live for a year, some perhaps for 2 or 3 years. Bathypelagic species may live for 3 to 5 years (e.g. *Thysanopoda egregia*) or some of the very largest (e.g. *Thysanopoda spinicaudata*) possibly up to 20 years. These species probably do not become sexually mature until they are 3 or 4 years old. Nemoto (1976) found mature *Thysanopoda cornuta* males and females of 70 mm living at >1000 m depth and suggested that breeding of these individuals might be continuous. He observed females of *T. egregia* and *T. spinicaudata* carrying from 200–800 eggs per individual. Bathypelagic mysids, represented by different species of *Gnathophausia*, can be 2 to more than 10 years old probably maturing after 1 or 2 years, compared with surface species of mysids which have a life span of from 2 to 12 *months*.

For copepods, there is obviously great variability in the life span and growth rate for epipelagic species of the order of 1–12 months. For bathypelagic forms (e.g. *Bradycalanus*, *Bathycalanus*, *Megacalanus* and *Valdiviella*), Mauchline calculates the longevity as from 3 to 58 months, an increase approaching five-fold.

Though very little information is available for chaetognaths, a two- to five-fold increase in longevity is indicated for bathypelagic species.

The greater size which may be achieved, despite the slow growth, for at least some bathypelagic zooplankton may be of advantage in accommodating greater stores of food in the body. Since many bathypelagic species are carnivorous rather than omnivorous, large size may be favoured in view of the greater mobility and capacity for ranging more widely for the limited food. The sharpest contrast is with the epipelagic zooplankton typical of tropical waters (e.g. numerous herbivorous species of very small copepods), which have a high metabolic rate and a very short life span, grazing rapidly on the small and brief increases of phytoplankton crop. They develop rapidly and mature quickly, giving rise to a new generation.

Chapter 4
Breeding of the Meroplankton

Temperate Latitudes

In middle and higher latitudes reproduction tends to be markedly seasonal. Fluctuations in temperature and in food supply, with the variation in primary production, might be expected to be amongst the more significant factors in seasonal breeding. Such effects tend to be accentuated in shallow inshore waters. Thus in temperate regions the majority of benthos and fishes inhabiting the shallower continental shelf waters will reproduce mainly in early spring and summer, so that the meroplanktonic larvae are usually concentrated over that period (cf. Thorson, 1950). A few species produce larvae over an extended period, in extreme cases stretching over almost the year (*vide infra*).

Many zooplankton surveys of inshore areas thus refer to meroplankton as being of significance over part of the main productive season. Bigelow (1926) found that in more neritic parts of the Gulf of Maine, cirripede nauplii became abundant during March or April, echinoderm and polychaete larvae and fish eggs somewhat later in spring, and medusae particularly about May. At Kiel, Lohmann (1908) found that rotifers and bivalve larvae made a contribution to the total population of zooplankton over summer, while the survey of Harvey *et al.* (1935) of the zooplankton of the western English Channel indicated only a small proportion as meroplankton, apart from cirripede larvae about March and occasional minor populations of rotifers and polychaete larvae. Wiborg's (1954) study of inshore Norwegian waters recorded small numbers of benthic invertebrate larvae for a short period of spring and summer. These results are generally confirmed by the investigations of Lie (1967) who agreed with previous workers on the outstanding importance of copepods in the overall seasonal cycle.

Eriksson (1973b) studying the zooplankton off Goteborg at two shallow stations (20 m and 40 m) over three years, commented on the great seasonal variation in abundance and in composition. While accepting the outstanding importance of the copepods which averaged 69 % of the population, with small contributions from cladocerans (6 %), chaetognaths (3 %) and appendicularians (2 %), the meroplanktonic larvae of invertebrates were important, averaging 19 %. Eriksson examined the meroplankton in some detail. Bivalve larvae made up the major meroplanktonic group (55 % of total meroplankton), with cirripede larvae next in importance (21 %). Hydromedusae, not listed with the meroplankton, were mainly represented by *Obelia* spp., *Rathkea octopunctata*, and *Aglantha digitalis*, but some hydromedusae were not adequately sampled. Rotifers and planulae, scyphomedusans and ctenophores were also not quantitatively captured. The importance of rotifers (mainly *Synchaeta* spp.) during summer, however, in the brackish waters of the Oresund dominated by copepods

(*Eurytemora affinis* and *Acartia bifilosa*) and cladocerans (*Bosmina*, *Podon polyphemoides*, *Evadne nordmanni*), was noted by Eriksson (1973c).

Of the benthic meroplankton off Goteborg, bivalve larvae occurred through the whole year, and were particularly abundant from June to December. Of the species identified, *Mytilus edulis* predominated, with *Mya*, *Cardium crassum*, *C. edule*, *Macoma* and *Anomia* (*Heteranomia*) *squamula* also present. Gastropod larvae occurred over much the same period but were far less abundant. Cirripede nauplii, also found through the year, were particularly plentiful in early spring, probably mainly *Balanus balanoides*, and from May to October probably *B. improvisus* (cf. Table 4.1). Polychaete larvae, especially spionids, occurred through the year in rather small numbers, and larval echinoderms were present in rather low densities, asteroid larvae mainly in summer, plutei from March to December. Decapod zoeas, though found in higher numbers in July, were never abundant. The occasional occurrence of a few other meroplanktonic larvae is shown in Table 4.1.

For very brief periods in temperate waters, a great outburst of meroplanktonic larvae, especially those arising from spawnings of intertidal or very shallow-water-abundant benthic species showing spawning periodicity, may overwhelm the normally abundant copepods. Rarely also in some shallow estuarine areas meroplankton may be significant over longer periods of the year. In Southampton Water, Raymont and Carrie (1964) demonstrated the dominance of cirripede nauplii (*Balanus balanoides*, *Balanus crenatus* and especially *Elminius modestus*) for a substantial part of the annual cycle. Copepods, on average, ranked second in importance numerically.

The timing of the appearance of different meroplanktonic species is a useful tool in investigating breeding patterns.

Orton (1920) stressed the importance of temperature in controlling the spawning periods of marine animals, the time necessary for the development of the gonad being the other important factor. Often the attainment of a critical temperature appears to be essential to trigger spawning. Many authors (e.g., Thorson), therefore, list temperature and the animal's life cycle as the significant factors. If the onset of spawning coincides with a recognizable seasonal change in water temperature (for example, breeding beginning when the water warms to a certain temperature), a fairly widely distributed species will presumably spawn in different geographical areas at different seasons (e.g., winter in warmer parts of the range; late spring or summer in colder areas). Early quoted examples illustrating this kind of spawning behaviour included the oyster, *Ostrea edulis*, with a wide distribution in European waters, *Crassostrea virginica*, with a large geographical range in North America and the drill, *Urosalpinx cinerea* (cf. Loosanoff, 1958). However, the capacity for adaptation in organisms suggests that marine species might adapt their spawning times to local temperature conditions, either as genetic races or phenotypic populations. Mileykovskiy (1968), for instance, demonstrated that *Asterias rubens*, ranging widely over Arctic and boreal regions of the Atlantic, including the Barents and White Sea, appears to be represented by three physiological races, spawning commencing at 3.5°–4.5°C off North and South Norway, in the Baltic Sound, Barents Sea, North Cape and Murman coast; at a slightly higher temperature (6.5–7°C) in the White Sea; at 8°–10°C in Copenhagen Harbour, Limfjord, the English Channel and Irish Sea; and at 13°–15°C in the low salinity Baltic waters off Kiel.

In Britain, which tends to be in a geographical position bordering between northern

Table 4.1. Periods of occurrence of certain taxa off Goteborg during 1968–1970 (Eriksson, 1973)

Month	J	F	M	A	M	J	J	A	S	O	N	D
Holoplankton												
Sagitta setosa	+	+	+	+	+	+	+	++	++	++	++	+
S. elegans	+	+	+	+	+	+		+	+	+	+	
Oikopleura dioica	+		+	+	+	++	++	++	++	++	++	+
Fritillaria borealis			+	++	++	+	+					
Hyperia sp.			+	+	+							+
Parathemisto sp.	+			+			+		+	+		+
Metopa sp.		+	+	+	+	+	+					
Euphausiid nauplii			+	+	+	+	+	+	+			
Euphausiid eggs					+							
Ostracods	+		+	+					+			
Spiratella retroversa	+						+	+				+
Clione limacina								+				
Tomopteris helgolandica				+						+	+	+
Benthos larvae												
Barnacle cyprids	+	+	+	++	+	+	++	++	+	+	+	+
Anomuran zoeae	+			+		+	+	+				
Brachyuran zoeae				+	+	+	+	+	+	+		+
Megalopa larvae									+			
Spionid larvae	+	+	+	++	+	+	++	+	++	++	+	+
Pectinaria larvae					+		+	+	+			
Mitraria larvae			+	+	++	+		+				
Magelona larvae					+		+	+				
Bipinnaria larvae				+	+	+	+					
Brachiolaria larvae					+	+	+	+				
Pluteus larvae			+	+	+	+		+	+	+	+	+
Large spatangoid plutei							+	+				
Ophiouroids (metamorph.)						+		+		+		
Pilidium larvae					+					+		+
Actinotrocha larvae				+			+	+		+	+	+
Ascidian larvae				+	+		+			+		
Cnidaria												
Sarsia tubulosa					+							
Ectopleura dumontieri				+								
Steenstrupia nutans							+					
Euphysa aurata						+		+				
Cladonema radiatum								+				
Podocoryne carnea							+			+		
P. borealis				+								
Rathkea octopunctata				+	+		+		+			+
Leuckartiara octona								+	+			+
Obelia spp.			+	+	+	+	+	+	+	+	+	+
Eirene viridula				+								
Eutonina indicans					+	+						
Tima bairdi		+	+		+		+					
Aglantha digitalis		+		+	+		+	+		+	+	+
Arachnactis larvae				+	+			+				

+ = Present; + + = numerous.

(boreal) and southern (warmer) forms, there is often a grading of the time of breeding of benthic species and fishes which seems to be largely temperature dependent (cf. Chapter 2). A similar pattern has been demonstrated for the spawning times of a number of marine animals, especially echinoderms, in Norway.

Farther south, although temperatures are considerably higher through the year along the northern Mediterranean coast, investigations at Villefranche by Fenaux (1969) have demonstrated a clear seasonal pattern in the occurrence of echinoid larvae. The plankton hauls were made close to the coast where echinoids were very common. It is remarkable that >60 % of the larval forms taken over a year were echinoplutei, the majority being plutei of regular echinoids but with a fair proportion of spatangoids. Some echinoplutei may be taken in any month. However, Fenaux states that two maxima occur in the plankton, the lesser in spring and the greater from the end of summer through the autumn. Some species (*Paracentrotus lividus*, *Sphaerechinus granularis*, *Echinocardium cordatum* and *Echinocardium mediterraneum*) are represented in almost every month, though showing greater abundance at certain periods. Other species may occur in winter/spring (*Psammechinus microtuberculatus*) or early summer to autumn (*Echinocyamus pusillus*). Fenaux found that the autumn period of abundance was characterized by a succession of species (Table 4.2). Seasonal breeding is frequently observed therefore in somewhat lower latitudes, though there often appears to be an increasing tendency in warmer waters for breeding to be more extended in time and certainly not confined to one season.

Although in temperate latitudes, some species are claimed to have greatly extended spawning periods, careful re-examination sometimes indicates more restricted times. Earlier reports of neritic plankton collections frequently classified meroplanktonic larvae in terms of comparatively large taxa (e.g., gastropod, lamellibranch, cirripede larvae), since specific identification was very difficult. Precise identification has sometimes indicated narrower spawning times for separate species. A combination of examination of ripeness of gonad and of larval abundance can be helpful.

An examination by Millar (1952, 1954) of the adults of certain ascidians for eggs and larvae demonstrated that, although several species such as *Dendrodoa* and *Ciona* have been described as breeding through most of the year in Britain, the period may be less extended. *Dendrodoa grossularia*, studied in two localities, Millport and Essex, was found to have a prolonged breeding period. Two maxima in breeding could be seen in Essex waters, the first about March/April, the second in late summer/autumn, with a short break. There was no breeding during a brief winter period. At Millport breeding began somewhat later but continued as late as November. Similarly, Millar found that *Ciona intestinalis* breeds from about May to August/September rather than continuously in Scottish (Millport) waters. On the other hand, at Plymouth the season is a little longer, from April till about November, but breeding ceases in the winter months. Millar quotes Runnstrom that *Ciona* apparently has three physiological races with somewhat different breeding times in west Norway according to the temperature range. Breeding occurs from May to August. Gulliksen (1972) more recently found that the spawning of *Ciona* in Norwegian waters could occur from May to October. The most marked settlement occurred in June and since the larvae are very limited in their few days of planktonic life spawning is presumably more intensive in that month. On the other hand, in the much warmer waters off Naples breeding is believed to be virtually continuous.

Table 4.2. Occurrence of echinoid larvae during the year at some Mediterranean stations (Fenaux, 1969)

Month	J	F	M	A	M	J	J	A	S	O	N	D
Species												
A. lixula												
Villefranche	..	..	..	..	..	+	++	++	++	+	+	+
Messine (Mortensen, 1898)	..	..	..	..	..	..	..	..	..	+	+	+
P. lividus												
Villefranche	+	+	+	+	+	++	+	+	++	++	++	+
E. acutus												
Villefranche	..	+	+	..	..	..	..	..	..	..	..	..
P. microtuberculatus												
Villefranche	+	+	+	++	+	+	..	..	+	+	..	..
Trieste (Mortensen, 1898)	..	..	..	+	+	+	..	..	..	..	..	..
S. granularis												
Villefranche	++	+	+	+	..	+	+	+	+	++	++	++
Messine (Mortensen, 1898)	..	..	..	..	..	..	..	..	..	+	+	+
Alger (Rose, 1926)	..	..	..	+	..	..	..	..	..	..	..	..
C. cidaris												
Villefranche	..	..	..	..	..	..	..	..	+	+	+	+
Alger (Rose, 1926)	..	..	+	..	..	..	..	..	..	..	..	..
G. maculata												
Villefranche	+	+	+	+	..	+	+	+	+	+	++	++
E. pusillus												
Villefranche	..	..	..	..	..	+	+	+	+	+	+	..
E. cordatum and												
E. mediterraneum												
Villefranche	+	++	++	++	+	+	+	+	+	+	+	..
E. flavescens												
Villefranche	+	++	++	+	..	..	..	..	..	..	..	..
S. purpureus												
Villefranche	..	+	+	+	..	..	..	..	..	++	+	..
Messine (Mortensen, 1898)	..	..	..	+	..	..	..	..	..	..	..	..
B. lyrifera												
Villefranche	++	++	+	+	..	..	..	..	+	++	++	++
Echinopluteus solidus (Mortensen)												
Villefranche	..	..	..	..	..	+	+	+	+	..	..	..

Millar finds that three other ascidians (*Ascidiella aspersa*, *Diplosoma listerianum*, *Botryllus schlosseri*), breed for a few months in the summer in British waters. The immigrant species *Styela clava* was also found by Holmes (1968) to breed only during June, July and August in Southampton Water. On the other hand, Millar (1954) found that the essentially Arctic circumpolar species, *Pelonaia corrugata*, which occurs only to a limited extent in Scotland, breeds in the very cold time of the year, about January–February. The development is direct.

Factors other than temperature may of course modify spawning times. Once the gonads are ripe, lunar periodicity appears to be significant for some species, and so perhaps is water quality, including possibly the composition of material released by phytoplankton. Barnes (1957) showed that the synchronization of spawning of the barnacle *Balanus balanoides* was particularly tuned to the main phytoplankton outburst and only indirectly to temperature. Even though liberation of larvae may be delayed by factors such as turbidity of the water, the main spring flowering of diatoms appears to be the major controlling factor (cf. Moyse and Knight-Jones, 1967).

Spawning periods of other taxa in western European (British) shallow waters have been considered in relation to geographical distribution and temperature.

Of bivalves, *Mytilus edulis* has a very long spawning period in British waters, approximately from about March/April to as late as September. Local variations in time appear around the coasts of north-west Europe. Savage (1956) states that spawning time in the Bay of Aiguillon, France, appears to be as early as February, but for British waters the earliest month is March. Most frequently in north-west European waters spawning is in May/June but in many localities larvae are present through the summer at least till August. At Millport, Scotland, the main settlement period is September and October and settlement as late as October also holds for Denmark and Wales. Savage reports Cole's findings that mussel settlement could occur in almost any month, though chiefly in March, April, July and August. While in Britain spring is the main time for spawning, it may occur anytime from March to September. Holmes (1970) states that coastal power stations round Britain recommend low level chlorination against mussel settlement from April to November.

At Woods Hole on the American coast *Mytilus* can also spawn from April to September. *Cardium* (*Cerastoderma*) *edule* is regarded as a spring–summer spawner at Plymouth, roughly from about April/May to the end of the summer. At Millport, July–August is quoted as the main spawning time, presumably reflecting the slightly later time of attaining the necessary temperature. *Macoma baltica*, a typical boreal species, is said to breed mostly about April or May, when temperatures are beginning to rise after the winter minimum. In contrast with these lamellibranchs, two species appear to be more limited in their spawning time to near the summer maximum. *Ostrea edulis*, the European oyster, will not spawn below 16°–17°C, approaching the yearly maximum for British waters. It is essentially a high summer spawner (cf. Rodhouse, 1977). A recent immigrant or, more likely, transplant to British (Southampton) waters is the American hard clam, *Mercenaria mercenaria*. This is quoted as requiring temperatures above 20°C for spawning in American waters where it typically occurs from Cape Cod to Florida. In Britain it appears able to spawn at a slightly lower temperature, but even so is restricted to peak summer months. The exact time appears to be largely determined by local and temporary conditions, such as a water temperature rising above 18°C (Mitchell 1974).

An example of a taxon where identification of spawning periods of species has modified early views is that of cirripedes. Barnacle larvae are said to be present in temperate waters such as those off Britain mainly in early spring. Pyefinch (1948) examined the times of release of nauplii in British waters and agreed that the spawning of *Balanus balanoides* occurred at that time. Pyefinch found that *Balanus balanoides* shows a very sharp burst of larvae from about the last week of February through the first two or three weeks of March. The cyprids appear at the beginning of April. Thereafter no *Balanus balanoides* normally appear in the plankton. *Balanus crenatus* shows a similar main outburst in March and early April, but this species also continues to liberate larvae at intervals throughout the summer months. *Verruca stroemia* shows a main outburst of larvae from mid-February, through March to early April, but very much smaller numbers of larvae may be liberated almost throughout the whole year.

Barnes (1975), however, discussing the variations in spawning times with latitude, especially for barnacles, suggests that synchronization of spawning with the main phytoplankton outburst is significant so that the timing may be only indirectly related to temperature. It is striking that at Woods Hole, where the phytoplankton outburst is as early as December, the spawning of *Balanus balanoides* is some two to three months earlier than in Britain (Barnes and Barnes, 1958). Barnes and Stone (1973), Barnes (1975) and Barnes and Barnes (1975) discuss also the factors concerned with the breeding of the barnacle *Verruca stroemia* in western Scottish waters. The major brood is in early spring and the production of nauplii is synchronous. Barnes believes that the synchrony is due to a gradual building of the gonads during the comparatively poorer nutritional conditions of late autumn and winter, so that by February/March the ovaries of all the population are ripe. In the absence of factors strongly promoting synchrony in population breeding, Barnes claims that reproductive patterns will tend to get out of phase, factors in the whole environment being subject to considerable variation. Thus, with *Verruca*, a second spawning takes place later in the season, involving about 70% of the population, but breeding is asynchronous and no reproduction peaks are discernible (Fig. 4.1). Breeding ends in late autumn, probably a reflection of the poor nutritional conditions: temperature is certainly far above that for the first spring breeding. Slow development of the gonad continues from autumn to achieve ripeness by February/March. In a species such as *Balanus balanoides*, the gonad is gradually built from food stores, but, following development of the embryos, release synchronous with the spring diatom increase is advantageous. As an Arctic–boreal species, there is an upper temperature limit to breeding in *Balanus balanoides* but release of nauplii below the limit is not related to temperature. Spring growth of diatoms is a result mainly of seasonal light changes, not of temperature variation.

Soares (1957) confirmed the sharp timing of the liberation of larvae of *Balanus balanoides* in Southampton Water. *Balanus crenatus* was restricted mainly to the cold months of the year, but some larvae occurred later in the summer. The immigrant Australian barnacle *Elminius modestus* also occurs in the area. Larvae appear in great numbers about May and often dominate the plankton populations throughout the summer until September. Small numbers of this barnacle can appear in any month, however, throughout the year.

Crisp and Davies (1955) studied the time of release of nauplii in *Elminius*. They found that the time for development of embryos depended on temperature only, but speed of regeneration of the ovary was a reflection of food supply. Although a warm

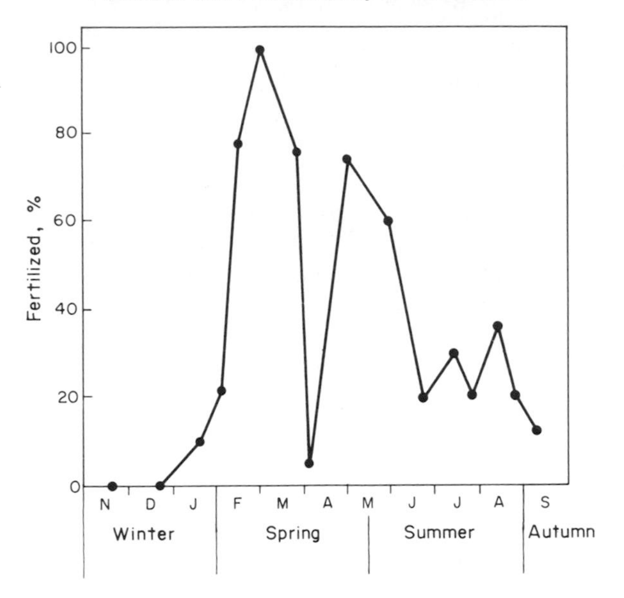

Fig. 4.1. Brood production in *Verruca stroemia*: the first brood, because of nutrient conditions, is formed almost in synchrony throughout the whole population; the second brood is only in partial synchrony due to the desynchronization by random variation; the summer broods are not synchronous (Barnes, 1975).

water species, *Elminius* was found to be remarkably eurythermal. Breeding is possible at temperatures exceeding *ca.* 8°C. A series of fairly rapid broods could occur over the summer, but although breeding could continue over a substantial part of the year, especially in south-west Britain, it is greatly diminished in colder, less productive periods of the year. Raymont and Carrie (1964) found several late spring/summer broods in Southampton Water with very small numbers even in late autumn. The contrast between the breeding of *Elminius* and that for the single brood of *Balanus balanoides* is very obvious. With other barnacles (*B. crenatus*, *B. improvisus*), despite more continuous breeding, the regeneration of the ovary is slow and seasonal.

Patel and Crisp (1960) comment on the breeding of three warmer-water barnacles. *Chthamalus stellatus* breeds at about 15°–16°C, from May to September in south-west Britain. *Balanus perforatus* has a similar temperature limit but breeds about July, possibly as a deeper-living species. *Balanus amphitrite* var. *denticulata* is an immigrant barnacle found in Europe only south of the Charente area of the Bay of Biscay. It has colonized certain specialized habitats in Britain where artificial warming of waters has occurred as in Queen's Dock, Swansea, Shoreham Harbour Canal and a few other areas (cf. Naylor, 1965). Patel and Crisp (1960) found the lower temperature limit was around 18°C. Successful breeding occurred in the Shoreham Harbour Canal, where temperatures frequently reached about 25°C from May to August. With comparatively high temperatures in Queen's Dock, Swansea, breeding was found to take place in both spring and autumn, but with subsequent engineering changes leading to a fall in environmental temperature, the *B. amphitrite* population declined and the colder species *B. crenatus* appeared. Patel and Crisp found that for all three barnacles, breeding increased with temperature to about 22°–25°C. The upper limit (28°–32°C) depended on species.

Another example of different spawning periods in one invertebrate taxon is provided by the crustacean decapods. The common shrimp, *Crangon vulgaris*, appears to have a very long breeding period. Allen (1966) suggests approximately two, or possibly three, broods a year, with eggs carried by the females as early as November/December to early autumn. On the whole, the two months October and November would appear as the most likely brief period without eggs. Therefore larvae can appear in the plankton all the year round except for two or three months of the late autumn or winter. Forster (1951) found that the common prawn *Leander serratus* has a very long breeding season with up to three broods during a year. The main breeding period appears to be January to June and since larvae can have a planktonic life of several weeks, they will occur at almost any time from February to August/September.

Though some other decapod species have a prolonged breeding period, only one brood is produced per year. According to Allen (1966) *Pontophilus spinosus* and *Spirontocaris* spp. are also annual but have a more limited breeding period during winter. Egg laying occurs about November/December and hatching follows in April. Mistakidis (1957) found, for *Pandalus montagui*, that eggs were carried by the female mainly from November to January or early February in the Thames Estuary. Hatching began in March and the larvae were found from April until July. Various records from other regions, indicated that larvae might be found until late summer, and Mistakidis suggests that this might be attributed to somewhat lower temperatures in other areas causing a slightly later season. Allen's (1966) records for the Northumberland coast, however, suggest that the species shows a fairly short period for egglaying (November/December) and that the larvae are chiefly present in April and May. The same breeding period appears to be true for the closely related *Pandalus borealis*, a species near the southern limit of its distribution, with larvae hatching in March/April. Both species of *Pandalus* are protandrous hermaphrodites.

Hippolyte is a spring/summer breeder off Northumberland, mainly from April to October, but has two broods annually. Forster (1951) found that breeding of *Palaemon serratus* off Plymouth began about December and continued for several months. Since the eggs are carried for about four months, the larvae appear in the plankton from about April/May to August/September. Off North Wales breeding appears to be a little later, hatching of larvae not beginning until June and the majority of larvae appearing in July. For the edible crab, *Cancer*, off the north-east coast of Britain, hatching is mainly over the period July to October, but off Plymouth breeding is somewhat earlier since zoeas and megalopas appear mostly from April to August. For *Carcinus maenas* the main period of abundance of larvae is quoted as July to October and for *Portunus* from September to November.

These observations indicate that the main breeding period for most decapods is limited to a period extending from spring to autumn, though different species show some separation of periods of abundance. Some variation in the breeding time of a species with geographical area is linked to local conditions, probably particularly to temperature. Allen (1966) points out that very few eggs are produced by any decapod in British waters after September and that the first batches of winter eggs hatch about the time of the spring phytoplankton increase. The last larvae for the year are present about the period of the autumn rise in phytoplankton. These results might be compared with those of Bodo *et al.* (1965) who give a fairly complete analysis of the occurrence of decapod larvae off Roscoff. Many species were recorded over almost every month (e.g., *Hippolyte* sp., *Crangon crangon*, *Pontophilus trispinosus*, *Galathea squamifera*, *Upogebia* sp., *Anapagurus hyndmani*, *Ebalia* sp., *Macropipus* sp., *Inachus* sp., *Macropodia* sp.). Others were more restricted (e.g., *Porcellana platycheles* May–October, *Pagurus bernhardus* January–June, *Pinnotheres pisum* July–October, *Pisa* July–October, *Corystes cassivelaunus* January–June), though marked abundance was often for one or two months (Table 4.3).

Table 4.3. Occurrence of decapod larvae off Roscoff (Bodo, Razouls and Thiriot, 1965)

Month	J	F	M	A	M	J	J	A	S	O	N	D	Total
DECAPOD LARVAE													
Natantia Caridea													
Pandalina brevirostris (Rathke)		+	++	+	+	+	+	+	+	+	+		200
Caridion stevenu Lebour			+	+		+				+			6
Hippolyte sp.	+	+	+	+	+	+	+	+	++	+	+	+	2500
Alphaeus macrocheles (Hailstone)	+	+	+		+	+	++	+	+	+	+	+	300
Athanas nitescens (Montagu)						+	+	++	+	+	+		400
Processa edulis (Risso)					+	+	+	+	++	+			1200
Processa canaliculata Leach	+	/	+		+	+				+		+	40
Palaemon sp.	+		+	+	+	++	+	+	+	+			150
Crangon crangon Fabricius	+	+	+	+	++	+	+	+	+	+			600
○ *Crangon allmani* Kinahan			/			+	+						4
Pontophilus sculptus (Bell)					+	+	+	+	+	+	+		50
Pontophilus trispinosus (Hailstone)	+		+	+	+	+	+	+	+	++	+		450
Pontophilus fasciatus (Risso)					+	+	++	+	+	++			500
Pontophilus bispinosus (Hail. West)				+		+	+	+	++		+		50
Pontophilus echinulatus (M. Sars)			/										1
Reptantia Macrura													
Palinurus elephas (Fabricius)					+	/							2
Scyllarus arctus (Linné)							+	+	+				10
Nephrops norvegicus Linné				/									3
Anomura													
Munida rugosa Fabricius			+										2
Galathea strigosa (Linné)	++	/	+		+	+	+	+	+	+	+	+	200
Galathea squamifera Leach	+	+	++	+	+	+	+	+	+	+	+	+	4500
Galathea sp.	+	/	+	+	++	+	+	+	+	+			500
Pisidia longicornis (Linné)	+	+	+	+	+	+	+	+	++	+	+	+	14000
Porcellana platycheles (Pennant)					+	++	+	+	+	+			900
Axius stirhynchus Leach						+	+	++	+				30
Callianassa tyrrhena (Petagna)								+	+	+			10
Upogebia sp.	/	+	+	/	+	+	++	+	+	+	+	+	11000
Clibanarius erythropus (Latreille)								/	+				2
Catapaguroides timidus (Roux)						+	+	+	+	+	+		18
Diogenes pugilator (Roux)						+	+	+	++	+			300

Table 4.3 (Contd.)

Month	J	F	M	A	M	J	J	A	S	O	N	D	Total
Pagurus bernhardus (Linné)	+	+	++	+	+	+							2300
Pagurus cuanensis (Bell)			+		+	+	++	+	+	+	+		300
Pagurus prideauxi (Leach)	+	+	++		+	+	+	+	+		+		100
Anapagurus laevis (Thompson)	+		+			+		/		+			7
Anapagurus hyndmani (Thompson)	+	+	++	+	+	+	+	+	+	+	+		3100
Dromia vulgaris M. Edw.								+					2
Brachyura													
Ebalia sp.	+	++	+	+	+	+	+	+	+	+	+	+	600
Corystes cassivelaunus (Pennant)	+	+	++	+	+	+							400
Pirimela denticulata (Montagu)	+	+	+	+	+	++	+	+	+	+			2000
Thia polita Leach								+	+				10
Cancer pagurus + *Atelecyclus* sp.	+	+	++	+	+	+	+	+	+	+			2900
Macropipus sp.	+	+	++	+	+	+	+	+	+	+	+	+	13300
Portumnus latipes (Pennant)						/	+	+	+				20
Carcinus maenas (Linné)	+	++	+	+	+	+	+	+	+	+		+	11400
Xantho sp.						++	+	+	+				1600
Pilumnus hirtellus (Linné)						+	+	+	+	++	+		3900
Pinnotheres pinnotheres (Linne)						/	+	+	++	+	+		300
Pinnotheres pisum (Pennant)							+	++	+	+			60
Brachynotus sexdentatus (Risso)							+		+				200
Maia squinado (Herbst)							+	++	+	+	+		3
Eurynome sp.	+	+	+	+	++	+	+	+	+	+	+		1000
Pisa sp.						+	+	++	+	+			200
Hyas coarctatus Leach		+	++	+	+								1700
Inachus sp.	+	+	+	+	+	++	+	+	+	+	+	+	400
Macropodia sp.	+	+	+	+	+	+	++	+	+	+	+	+	200
Achaeus sp.				+	+	+	++	+	+	+	+	+	80

O, species not recorded again at Roscoff.
/, 1963 occurrence only.

Some examples may be quoted of breeding times for other taxa and in other temperate regions. Kandler (1950) gives the time of maximal abundance for certain medusae in the waters of the Baltic as follows: *Sarsia tubulosa*—mainly April/June, *Rathkea octopunctata*—April/May, *Hybocodon prolifer*—April, *Phialidium hemisphaericum* and *Euphysia aurata*—August to November, ephyrae of *Aurelia* and *Cyanea* —February to April. Russell's (1953) summary of the seasonal distribution of medusae in the western English Channel (Table 4.4) shows that while a few species are present in all months, they are more plentiful over a limited period, and that many medusae are restricted to particular seasons. Very few species are present during December, January and February. Allewein (1968) also studied the times of occurrence of hydromedusae in the Oresund waters off Denmark. *Obelia* medusae were present almost all the year round and *Rathkea* for about 10 months. These two species were the most numerous. A peak in total numbers and in species of medusae occurred in May. Possibly a later summer (August) secondary peak may occur but thereafter numbers

Table 4.4. Seasonal distribution of medusae off Plymouth (Russell, 1953)

Month	J	F	M	A	M	J	J	A	S	O	N	D
Anthomedusae												
Rathkea octopunctata	√	√	√	×	×	.	.	.	.	.	√	√
Hybocodon prolifer	.	√	√	√	×	√	.	.	.	.	.	.
Euphysa aurata	.	.	√	√	√	√	.	.	√	.	.	.
Steenstrupia nutans	.	.	.	√	×	×	√	√	√	.	.	.
Podocoryne borealis	√	.	.	.	.	√	.	√	√	.	.	.
Podocoryne carnea	.	√	√	.	.	.	√	√	.	.	.	.
Bougainvillia principis	.	.	.	√	√	.	.	.	.	.	.	.
Sarsia tubulosa	.	.	√	√	√	√	√	.	√	√	√	.
Sarsia eximia	.	.	.	√	√	√	√	.	√	.	.	.
Bougainvillia britannica	.	.	.	√	×	×	√	.	.	.	.	.
Ectopleura dumortieri	.	.	.	√	√	.	.	.	√	√	√	.
Lizzia blondina	.	.	√	×	×	×	×	√	×	.	√	.
Leuckartiara octona	.	√	√	√	×	×	×	×	×	√	√	√
Bougainvillia ramosa	.	.	.	√	√	.	.	√	√	√	√	.
Sarsia gemmifera	.	.	.	.	√	×	×	.	√	.	√	.
Amphinema rugosum	.	.	.	√	×	×	×	√	√	√	.	.
Zanclea costata	.	.	.	.	√	.	√	√	√	.	.	.
Sarsia prolifera[a]	.	.	.	√	√	√	×	√	√	√	×	.
Dipurena halterata	.	.	.	√	√	√	√	√	√	.	.	.
Dipurena ophiogaster	.	.	.	.	.	√	√	.	.	.	.	.
Amphinema dinema	√	.	.	.	.	.	√	.	×	×	√	.
Turritopsis nutricula	√	.	.	.	.	.	.	√	√	×	√	√
Eucodonium brownei	.	.	.	.	.	.	.	.	√	.	.	.
Pochella polynema	.	.	.	.	√	√	.	.	.	.	.	.
Leptomedusae												
Tiaropsis multicirrata	.	.	√	√	.	.	.	.	.	.	.	.
Phialidium hemisphericum	√	√	×	×	×	×	×	×	×	×	√	√
Obelia spp.	√	√	√	×	×	×	×	×	×	×	×	√
Phialella quadrata	.	.	√	√	×	√	√	.	√	√	.	.
Cosmetira pilosella	.	.	√	×	×	×	×	×	√	√	.	.
Mitrocomella brownei	.	.	.	√	√	.	.	√	√	.	.	.
Eutima gracilis	√	√	√	√	×	×	×	√	√	√	√	√
Aequorea forskalea	.	.	√	√	√	√	√	√	√	√	√	√
Aequorea vitrina	.	.	.	.	.	√	.	.	.	.	.	.
Laodicea undulata	.	√[b]	√	√	√	√	√	×	×	√	√	.
Lovenella clausa	√	.	.	√	.	√	√	.	√	×	√	√
Helgicirrha schulzei	.	.	.	.	.	√	√	√	√	√	.	√
Eirene viridula	.	.	.	.	.	.	√	√	×	×	√	√
Octorchis gegenbauri	√	.	.	.	.	.	√	×	√	√	.	.
Agastra mira	.	.	.	.	.	.	√	√	.	.	√	.
Limnomedusae												
Proboscidactyla stellata	√	√[c]	√[d]	√	√	√	×	√	×	√	√	√
Gossea corynetes	.	.	.	.	.	.	√	√	√	√	√	√

Table 4.4 (Contd.)

Month	J	F	M	A	M	J	J	A	S	O	N	D
Trachymedusae												
Aglantha digitale var. *rosea*	√	√	√	√	√	×	×	×	×	√	√	√
Liriope exigua	√	√	.	.	√	√	√	×	×	×	×	×
Narcomedusa												
Solmaris corona	.	.	.	.	.	.	.	.	√	.	.	.

√recorded as present; × months of greatest abundance.
[a] Plymouth Marine Fauna gives Jan–Oct.; this is possibly a misprint for June–Oct.
[b] 21 February 1938 one specimen.
[c] Hydroid with medusa ready for liberation.
[d] Liberated from hydroid, 30 March 1936.

rapidly decline. *Obelia* hydroids can apparently release medusae in any month. *Rathkea octopunctata*, though described by Kramp as a spring species in British waters and the North Sea, and a summer species in colder waters, was not so restricted seasonally in the Oresund. Allewein suggests that its period of occurrence may be extended by gemmation.

Tima bairdi and *Phialidium hemisphaericum* medusae also appear to occur in outside Danish waters in most months. The occurrence of the latter species, however, appears to be due to its relative longevity; though mainly released in spring, the medusae may survive for about a year so that older animals may occur in autumn and winter. Of other medusae, *Euphysa* is described as a summer form and *Sarsia tubulosa* and *Hybocodon prolifer* as spring medusae (mainly April/May) (cf. Kandler, 1950). Allewein believes that the occurrence of various species in different months partly reflects their life cycle, but is strongly influenced by temperature.

Schram (1968) studying the meroplankton of the inner Oslo Fjord, found that although polychaete and lamellibranch larvae of a number of species were present, together with a few gastropod and echinoderm larvae, the meroplankton was dominated by larval polychaetes (90 % by number) almost through the year. One species, *Polydora ciliata*, was especially abundant (*ca.* 78 % of total) and was found throughout the year, though numbers were low in December, rising to a peak in May and with further high densities in late summer. *Polydora antennata* contributed on average 7 %, and showed little change in density over the year, but for all polychaete larvae combined the peak population was in spring. Apart from polychaetes, *Mytilus edulis* made up 8 % of the total meroplankton. The larvae occurred regularly from April to November, chiefly from June to October, often with maxima in June and August. In contrast, two species of bivalve larvae were restricted in time. *Mya arenaria* is a summer spawner and the related *Mya truncata* a winter spawner, possibly a reflection of their geographical distribution. Thorson (1950) lists *Mya truncata* as one of the very few species producing planktonic larvae in Arctic waters.

Larvae of nudibranchs and of *Limapontia* are recorded by Schram as being present, though in small numbers, all through the year. Thorson (1950) found larvae of *Limapontia capitata* and of *Zirphaea crispata* in all months in waters of the Sound

Table 4.5. Seasonal occurrence of *Polydora* larvae in Maine Waters, 1966–1968 (Blake, 1969)

Month	Jan	Feb	March	April	May	June	July	Aug	Sept	Oct	Nov	Dec
P. ligni												
P. websteri												
P. aggregata												
P. commensalis												
P. concharum												
P. socialis												
P. quadrilobata II												
P. caulleryi												

Rare – – –
Abundant ———

(Denmark) and larvae of *Akera* and *Philine* for 8–9 months. Echinoderm larvae, dominated by *Psammechinus*, were found only from March to September by Schram. It is of interest that Blake (1969) published an account of the occurrence of polydorid larvae in the Gulf of Maine just after Schram's study. Though the study did not include a true quantitative survey, relative abundance could be elucidated. No species was found throughout the year. Most species were seasonal and Blake related their occurrence mainly to water temperature. Most appeared to be abundant mainly between May and August.

Table 4.5 gives an indication of the different times of abundance of *Polydora* spp. and again emphasizes the importance of specific identification. An excellent example of this specific study concerns polychaete larvae belonging to another family, the Phyllodocidae. Cazaux (1975) found the reproductive period of *Phyllodoce laminosa* at Arcachon extended from July to February with a peak in August/September. He was able to differentiate the trochophores of this species from other meroplankton by the marked pink–orange pigment, and showed that the pelagic larval life has a duration of approximately 9 months.

Bodo, Razouls and Thiriot (1965) also found polychaete larvae, including *Pygospio elegans* and *Scolecolepis fuliginosa*, abundant in coastal waters off Roscoff early in the year, with other species important from April. *Spio martiniensis* was especially abundant from August to November (Fig. 4.2). Other meroplankton found were

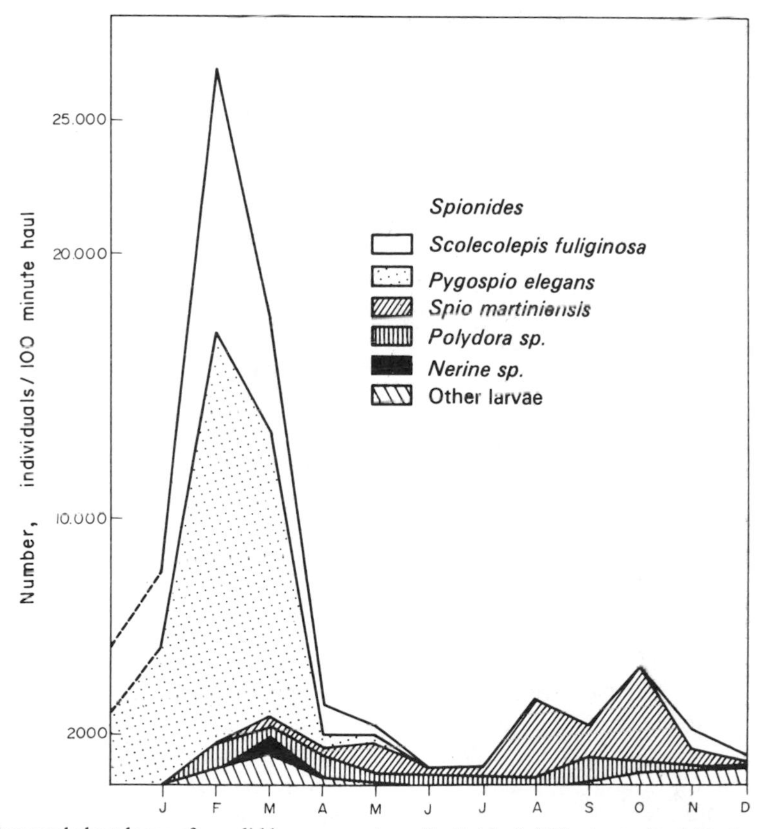

Fig. 4.2. Seasonal abundance of annelid larvae; number of individuals/100 minutes haul (Bodo *et al.*, 1965).

lamellibranch and gastropod larvae, especially in May. The greatest number of species of hydromedusae was recorded in June, but maximum abundance was in July, due particularly to *Obelia* which was common also in March and August. Other significant species were *Sarsia eximia* (June and October), *Phialidium* (October) and *Hybocodon* (May) (Fig. 4.3). Crustacean larvae included cirripede nauplii from January to March, dominating the meroplankton in February/March with a few occurring through the year. Decapod larvae were important early in the year and a second maximum occurred about June/July. Other species were important in the early autumn. Details of the occurrence of species have already been given (page 459). The meroplankton as a whole dominated the plankton early in the year, with a maximum in March but a second period of abundance occurred between June and October. Bodo *et al.* described the breeding cycles for many of the meroplanktonic species as similar to that found off Plymouth. Comparisons with species at Marseilles are less precise but many of them appear to have a more extended breeding period in the plankton in the Mediterranean waters.

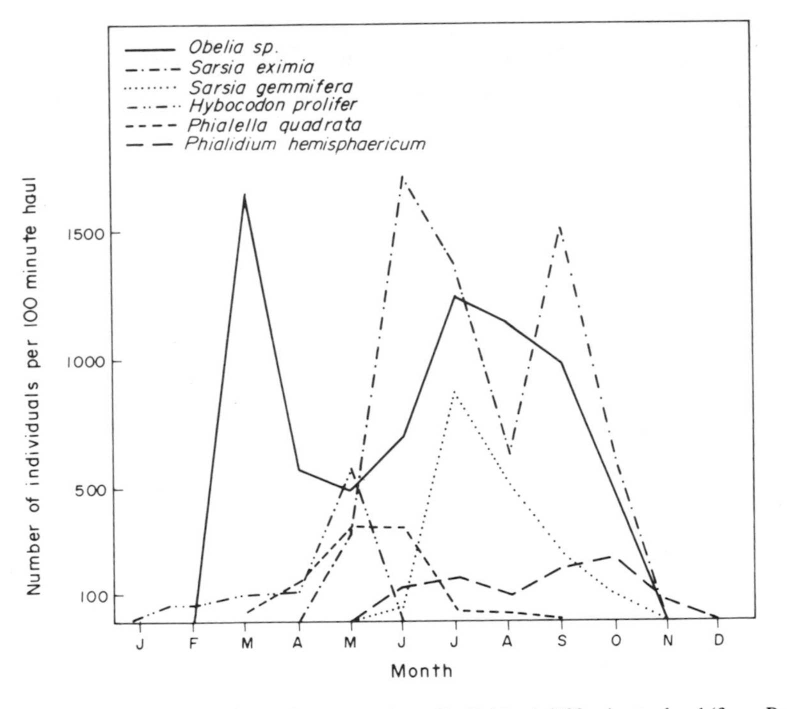

Fig. 4.3. Seasonal abundance of hydromedusae; number of individuals/100 minutes haul (from Bodo *et al.*, 1965).

Daro and Polk (1973) found larvae of *Polydora ciliata* occurring all the year at Ostend, Belgium. They appear in the plankton at the 3-setigerous segment stage and remain till the 16-setigerous segment stage—about a 3-week period at Whitstable, but just under 2 weeks at Ostend. Early larvae eat nanoplankton and detritus but older larvae take mainly small zooplankton (copepod and cirripede nauplii etc.) as well as detritus and phytoplankton. Development can be very rapid. About three weeks after settlement eggs can be laid, giving a total life cycle at the shortest of about six weeks.

Animals surviving the winter produce larvae in April; the second generation occurs about June–July but suffers heavy mortality and a few larvae about September represent a third generation. In at least one year larvae occurred as early as January/February so that four generations were found in that year.

Dorsett (1961) found major spawning and an abundance of larvae in plankton in April at Whitstable, Kent and again at the end of May, the spawnings arising from one homogenous population. After metamorphosis the worms grew up till late summer and gave rise to the over wintering stock. Some of the earlier spawned animals matured as early as January and produced larvae early. Others produced the main burst in April. The worms are thus annual.

A last example of variations in breeding times of meroplankton in temperate waters relates to Russell's (1935c) work on the spawning of fishes in British waters. Details are included in Chapter 2.

High Latitudes

So far we have been considering the contributions of meroplankton to the total neritic plankton in temperate regions. At very high latitudes there is a very marked tendency to omit larval stages and for development to be direct. In Arctic waters, Thorson (1950) calculated that only about 5% of benthic invertebrates had planktonic larvae, though the few with pelagic stages include some of the commonest species such as *Saxicava arctica*, *Mya truncata*, *Ophiocten sericeum* and *Strongylocentrotus droebachiensis*. The larval stages are also concentrated over the very brief summer months and are in the surface layer so that over a very short period the neritic plankton may show some contribution from meroplankton.

An example comes from a study by Mileikovskiy (1970a) on planktonic larvae of benthic invertebrates in an area of the western White Sea. The high latitude area experienced a very marked change in the water temperature in the inshore waters over summer. Mileikovskiy found very marked seasonal changes in larval abundance. Spawning times of the various species from several zoological groups were largely dependent on the zoogeographical origin of the species. Arctic species produced larvae in spring, Arctic–boreal forms from about June to August, and boreal species from summer to autumn (June to September). Some more cosmopolitan species including the pteropod holoplanktonic species (e.g. *Clione limacina*) had larvae throughout the whole sampling period and probably spawned throughout the year, though with greater intensity over spring/summer (cf. Table 4.6). The whole period of larval abundance was, however, shortened as compared with more temperate–boreal areas which may be correlated with the very shortened period of higher surface temperature and high phytoplankton production in the White Sea.

Lunar periodicity of spawning in some littoral and shallow-water invertebrates (several polychaetes, lamellibranchs, prosobranch gastropods, nudibranchs, 2 barnacles, bryozoans and 2 ophiuroids) was reflected in marked rhythm of abundance of larvae related to lunar tidal cycle. Daily changes in populations of meroplanktonic larvae also reflect daily spawning rhythms of a number of local invertebrates.

Very few investigations appear to have been carried out in estuarine waters at high latitudes but Chislenko's (1975) study of the Enisei River, flowing into the Arctic Ocean, emphasizes the profound effects of the tremendous run-off and of freezing and

Table 4.6. Reproduction periods of certain species of Loricata, Gastropoda, Bryozoa, Lamellibranchiata and Echinodermata with pelagic development in the Velikaya Salma Sound during 1957 (Mileikovskiy, 1970)

Species	Dates of occurrence of larvae in the plankton	Dates of main spawning period	Water temperature (°C) at beginning of main spawning period	Zoogeographical affinity
Tonicella marmorea	Jun–Aug	Jul	8–10	AB
Velutina velutina	15. May–20. Jun	15. May–5. Jun	0–1.5	PA
Lacuna divaricata	1. Jun–15. Sept	10. Jun–10. Jul	2.5–4.5	AB
Acmaea testudinalis	20. Jun–31. Jul	20. Jun–10. Jul	6–7	AB
Littorina littorea	15. Jun–15. Sept	25. Jun–10. Aug	7.5–8.5	B
Onchidoris muricatus *Acanthodoris pilosa* *Ancula cristata* *Palio dubia* *Coryphella rufibranchialis* *Aeolidia papillosa* *Eubranchus exiguus* *Cuthona* sp. *Cratena viridis*	Jun–Sept	25. Jul–15. Aug	6.5–7.5	AB and B
Diaphana minuta	1. Jul–15. Aug	1. Jul–1. Aug	8–9	AB
Philine aperta	15. Jul–20. Aug	15. Jul–20. Aug	9–10	B
Limaponlia capitata	May–Sept	Jan–Dec (?)	>0°	AB
Clione limacina	May–Sept	Jan–Dec (?)	>0°	C
Limacina helicina	May–Sept	Jan–Dec (?)	>0°	C
Mytilus edulis	1. Jun–15. Sept	25. Jun–15. Aug	7–8	AB
Macoma baltica *Mya arenaria*	1. Jun–15. Sept	25. Jun–15. Aug	7–8	B
Bryozoa	Jun–Sept	10. Jul–20. Aug	10–11	B
Ophipholis aculeata *Ophiura robusta*	15. Jun–31. Jul	25. Jun.–15. Jul	5.5–6	AB
Asterias rubens	1. Jul.–25. Aug	7. Jul.–25. Jul	8–9	B

AB—arctic–boreal.
PA—predominantly arctic.
B—boreal.
C—cosmopolitan.

thawing on the plankton. Meroplankton appears to have made a significant contribution only during the months of May and June of the brief productive season. In Antarctic waters it appears that the tendency for the benthos to omit planktonic stages is even more pronounced. However, Pearse (1969) found that one widespread Antarctic starfish *Odontaster validus* undergoes indirect development but the larvae are essentially demersal and the development is very slow. Of the few other Antarctic benthic species, including some polychaetes and crustaceans which do not have direct development, Pearse believes that a number of them may have feeding larvae which are mainly demersal, so that few truly planktonic larvae are captured. The same may be true of some deep-sea forms. Studying the drift of pelagic larvae in the Oyashio and Norwegian Current areas, Mileikovskiy (1968c) found a dramatic change in quantity of larvae with depth as they are carried away from the rich spawning areas on the continental shelf. He concludes that abyssal benthic populations are constantly replenished by such larvae settling out of the surface current.

Low Latitudes

Although, as Barnes (1975) has emphasized, in the tropics factors such as temperature, light and food approach more constant conditions, seasonal breeding cycles are often imposed on a more or less low level of reproduction. Examples will be quoted later, but Barnes instances an Indian limpet *Cellana radiata* which exhibits some breeding through the year with peaks in April and November. According to Barnes, even the deep sea, with its remarkable constancy, includes some invertebrates (isopods, echinoderms) which display greater breeding activity during part of the year. Moving away from the tropics, seasonal variations in such factors as temperature, food and light begin to become more obvious and regular. Breeding cycles tend towards a pattern such that larvae are more abundant when nutritional conditions are favourable. Typically, in the warm waters of low latitudes a maximum proportion of the benthos has meroplanktonic stages. Thorson (1950) suggests that 80% or more of the benthic invertebrate fauna produce planktotrophic larvae. Although the proposition has been often advanced that marine organisms breed more or less continuously in warm seas, there are indications that there are fairly specific breeding periods for tropical benthic animals though reproductive periods may be extended. The main characteristic of warm waters is that the different benthic species breed at different times, the period ranging over most months of the year, so that at any time a considerable abundance of meroplanktonic larvae may be found in the neritic waters—a contrast to the observations for temperate areas.

Thorson (1950) reviewed earlier literature, including Mortensen's observations that in tropical waters different echinoderms had distinct breeding periods. The results from the Great Barrier Reef Expedition for certain echinoderms indicated more active breeding in the warmer months. On the other hand, Thorson points out that off the Madras coast, polychaetes bred more actively in the winter months. It seems likely that monsoonal influences in such a region might affect breeding times. However, at least as regards brackish water polychaete species off Madras, some apparently bred all through the year, some at two or more restricted seasons, some more or less continuously but with one more intense period, and certain species were restricted to a limited period only.

The results of the investigations of Stephenson (1934) on breeding times of a number of invertebrates on the Great Barrier Reef were similar. She found that while some species bred throughout the year with more intense periods (e.g. *Thalamita*), others had a single period (e.g. *Hippopus*, *Ophiothrix*), while others again had two breeding times with an apparent quiescent period between (e.g. *Tripneustes*). *Acanthozostera* showed a burst of breeding every four weeks over about eight months (cf. Fig. 4.4). Stephenson points out that further studies would probably modify precise periods. For example, with *Cypraea annulus*, which breeds from October to March, there was the suggestion of a less active May/June stage and possibly no complete resting period occurred throughout the year. The important conclusion reached by Stephenson is that breeding was not confined to one part of the year. Some species were spawning in every month, though a majority appeared to breed mainly in summer.

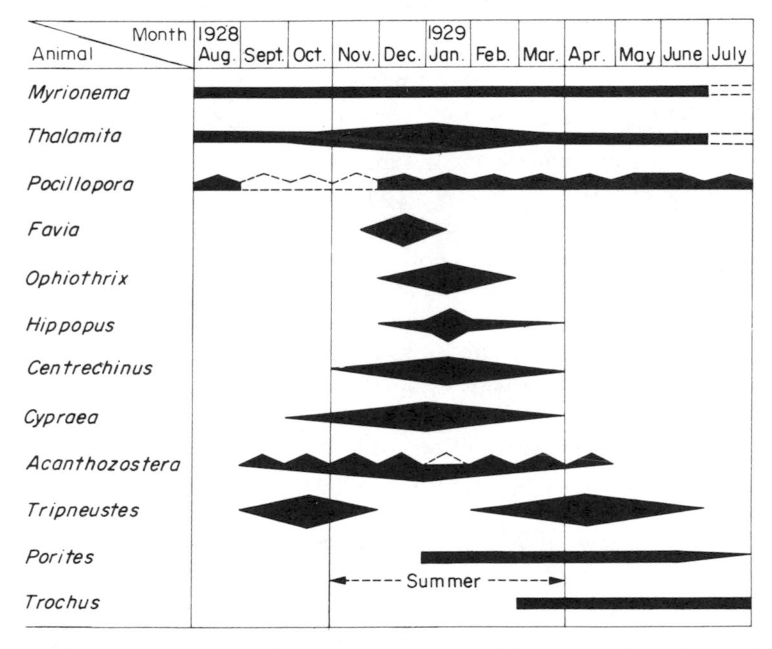

Fig. 4.4. Probable spawning periods of certain species at Low Isles. Broken lines show probabilities, in absence of data. Periodicity is indicated by a succession of peaks. Widening of figure indicates increased activity (*Thalamita* and *Hippopus* only) but diamond-shaped figures indicate duration of breeding only (not intensification at widest point) (Stephenson, 1934).

Apart from the extension of breeding periods in warm waters, where more than one cycle occurs annually, it is believed that for many species, especially those with a considerable geographical range, the length of the breeding cycle is shortened at higher environmental temperatures. Yamaguchi's (1970) work is an example. He studied the reproductive cycles of two ascidians off the coast of Japan and showed that apart from factors affecting the periodicity the duration of the cycle varied with water temperatures ranging from 12°–28°C. Thus the cycle varied from 1–2 months for *Ciona intestinalis* and from 2–5 months for *Styela plicata*. With the more continuous production of phytoplankton in inshore tropical waters, though at a somewhat lower level than in a typical temperate spring burst, some meroplankton may survive and feed successfully in almost any month in warm seas, so that competition between species can be thereby reduced.

In considering data on the breeding cycles of the benthos in warm waters, Pearse (1970, 1978) accepts that reproductive periodicity is less obvious, in part due to the more minor temperature fluctuations as compared with temperate regions. However, he suggests that a number of widespread tropical invertebrates inhabiting areas on the limits of or even outside the tropics, may show more restricted breeding periods, while some may exhibit more continuous breeding near the equator. He describes two widespread intertidal tropical molluscs, *Acanthopleura haddoni* and *Haliotis pustulata*, as spawning only in the warmest time of the year in the summer in the Gulf of Suez (which has considerable seasonal temperature variation), whereas in the Red Sea with its less seasonal temperature change, *Haliotis* spawns in winter and *Acanthopleura* probably has several summer spawnings. *Ornithochiton lyelli* has no strict periodicity in the Gulf of Suez but spawns more in the summer.

Pearse quotes results of Glynn for two tropical chitons from Puerto Rico. Both spawned in late summer, but one species in the more tropical east Panama region showed continuous reproduction. Pearse also found that tropical echinoids varied in their spawning times. In the Gulf of Suez, *Lovenia elongata* showed no clear periodicity, while *Diadema setosum* was restricted to summer. Kobayashi (1967) and Kobayashi and Nakamura (1967) found the latter species spawning from June to September in Japanese waters. The widespread *Echinometra mathaei* was restricted in the Gulf of Suez and in Japanese waters to summer spawning, but in the Red Sea and most other areas of its more tropical range, it reproduces throughout the year. Pearse suggests a critical temperature of 18°–20°C as necessary for the spawning of the species. Bauer (1976) observed that the major spawning of *Diadema antillarum* in regions such as Indian Key, Key West and Bermuda coincided with the time of decreased temperature (October/November to December/January). Lunar periodicity influences the release of gametes (cf. Kobayashi and Nakamura, 1967). Pearse (1972) found that there was periodicity in the reproduction of *Centrostephanus coronatus* and *Diadema setosum* which took place between March and September at Catalina, but that the periodicity was attuned to tidal rhythms.

Observations in the warm waters of South Florida also indicate variable spawning times for different benthic invertebrates. Moore and Lopez (1970) recorded two spawning periods, in January and October, for the bivalve *Dosinia elegans* whereas *Tellina alternata* spawns in spring and summer. Borkowski (1971) described the spawning times for seven warm water species of littorinids. Though four species showed some specimens with ripe eggs and sperm in every month, the great majority of individuals and all specimens of the remaining three species possessed spent gonads over the winter months between December and April. The gonads ripened in the spring and egg capsules were formed in the warmer months. Plankton tows showed the egg capsules mainly from June to October.

Despite the less obvious restriction of breeding to a limited time of the year in warm waters, neritic–tropical and sub-tropical zooplankton shows considerable changes in abundance during the year which are partly a reflection of meroplankton density. Herman and Beers (1969) examined changes in inshore plankton at a station in Bermuda having only very restricted exchange with outside waters. Another station was in open communication with the oceanic waters (cf. Chapter 3). While the mean zooplankton density was similar at the two stations ($2300/m^3$ and $1900/m^3$) and copepods were the dominant group at both sites, in the more restricted inner station at certain times during spring and summer, meroplanktonic molluscan larvae were predominant. Considerable fluctuations occurred at both stations. An autumn rise in phytoplankton at the inner station appeared to be correlated with a rise in zooplankton density in mid-winter or early spring, but summer populations built up to high levels in 1965. At the inshore station, Cladocera, especially *Evadne spinifera* with some *Penilia* and *Podon polyphemoides*, were important during the summer. Among the wide variety of larval forms, bivalve and gastropod larvae were the most important meroplankton and though numbers fluctuated, they seemed to occur almost throughout the year. Smaller numbers of polychaete and dacapod larvae were usually present and fish eggs and larvae again were few.

Reeve (1964) studying the meroplankton in Biscayne Bay, reported that medusae and decapod larvae occurred over most of the year. Porcellanids were frequently represented among the decapods. Cirripede nauplii showed larger changes in density. They were found mostly between September and April and formed 23% of the plankton in January. Reeve suggests that earlier studies by Moore and Frue (1959) in the same area, showing major settlement by barnacles in March and October, accord well with the main periods of abundance of cirripede nauplii. Polychaete larvae also varied greatly in density, the main appearance being from March to November with a peak exceeding 31% of the total zooplankton in September (cf. Fig. 4.5). Few gastropod veligers were encountered, though high densities were recorded during an earlier study. Fish eggs were abundant in July.

In a later study in the same area Reeve (1970b) found that meroplankton, chiefly decapod, mollusc and polychaete larvae, formed an important constituent of the zooplankton. The mean numbers of decapod larvae, presumably drawn mainly from the 50 species identified as adults in the Bay, amounted to $44/m^3$ (maximum density $294/m^3$) pagurid and brachyuran larvae, and $53/m^3$ (maximum $252/m^3$) caridean larvae. Rather lower densities farther out in the Bay might reflect the very moderate increase in water depth. Summer was a minimal time for decapod larvae. Crab larvae exhibited spring and autumn maxima, while shrimp larvae had a brief period of abundance in autumn and a more extended period over winter and spring. Of mollusc larvae (138 species of adult benthic molluscs have been identified from the Bay), bivalve veligers amounted to a mean density of $2600/m^3$ with one huge peak of $26{,}000/m^3$ in July; gastropod veligers had an average density of $4250/m^3$ with a maximum of $11{,}900/m^3$. There was no obvious seasonal pattern. The third main contribution to the meroplankton, polychaete larvae, averaged $600/m^3$ (maximum $3470/m^3$), with low numbers in early spring and over summer, but a peak density in April and rising numbers from September to December. There were few fish eggs and occasionally other meroplanktonic larvae were taken but some of the more delicate forms may have been lost in the preservation of samples. In general, the numbers of meroplankton were considerably higher than those recorded for earlier surveys in the area and this was attributable in part to the use of a very fine ($35\mu m$) net in the later work.

Woodmansee (1958) found gastropod veligers as a very important part of the zooplankton over most of the year in Biscayne Bay. While copepods were the major

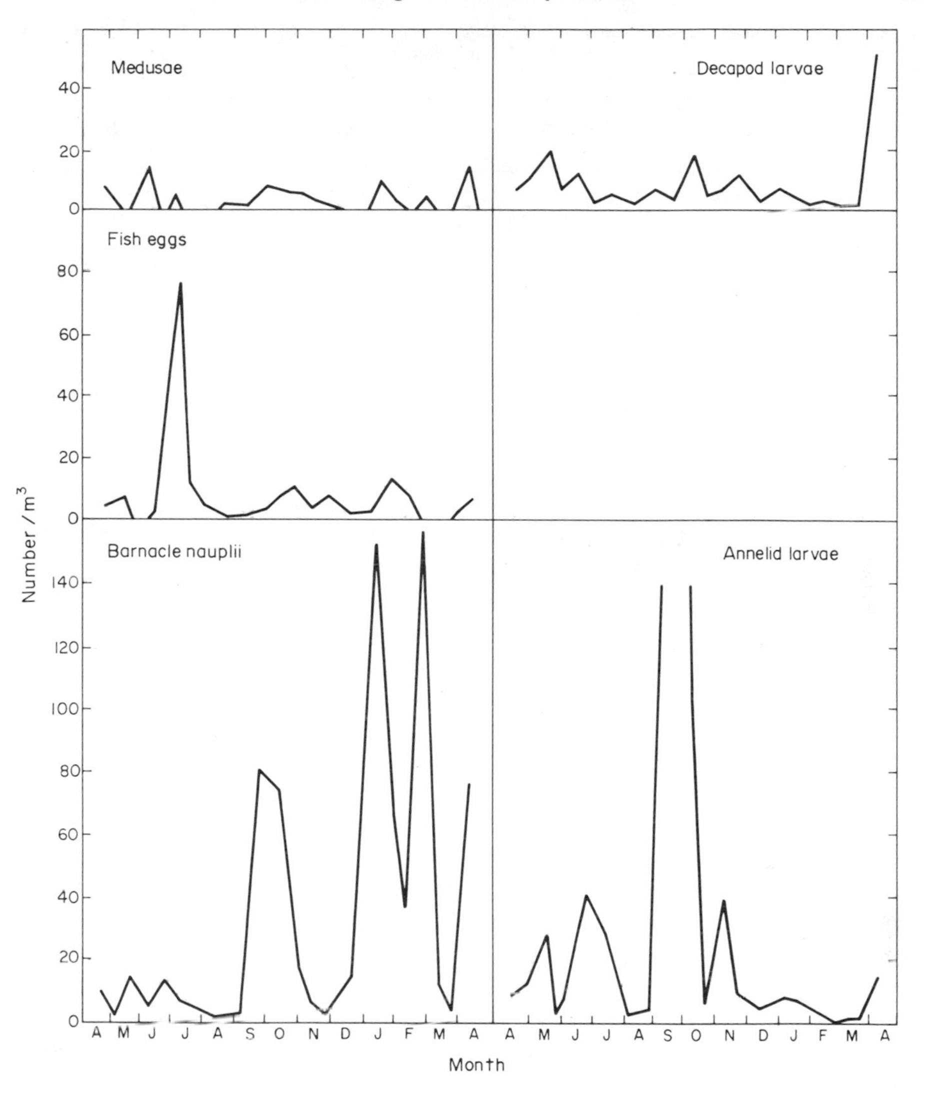

Fig. 4.5. Seasonal variation in the numbers of some meroplanktonic species (Reeve, 1964).

constituents, the veligers were often numerous, with peak percentage contributions in November/December and in June. Decapod zoeas were occasionally plentiful, chiefly during early spring and in the autumn, but numbers were far below those of gastropod larvae. Few other meroplankton were present. Hopkins (1977) sampled the zooplankton in the shallow, relatively enclosed waters of Tampa Bay, Florida, on a quarterly basis. Meroplankton was not identified beyond major groups, but formed a significant part of the population (19 % of total number; 8 % of biomass). The major taxa were lamellibranch, cirripede, polychaete and gastropod larvae, the bivalves on average being more plentiful (7973/m^3) than the other three groups together. Echinoderm, bryozoan and decapod larvae were generally far less abundant (cf. Table

Table. 4.7. Seasonal changes and yearly average of the main groups of meroplankton in Tampa Bay (modified from Hopkins, 1977)

Meroplankton groups	Autumn	Winter	Spring	Summer	Yearly average
Lamellibranch	4,200	1,200	15,500	5,300	7,973
Gastropod	360	40	2,000	3,300	1,585
Cirripede	1,500	990	4,600	1,600	2,175
Polychaete	1,000	370	670	3,800	1,616
Decapod	30	0	190	90	106
Echinoderm	60	0	220	1,000	448
Bryozoan	340	160	110	880	337
Total Meroplankton	7,800	3,300	25,200	17,200	15,358

All values are numbers/m^3.

4.7). Fish eggs were found mainly in spring and autumn but exceeded 500/m^3 only on two sampling dates. All the main groups of meroplankton were least plentiful in winter, except bryozoans which were less plentiful in spring (Table 4.7). Minor taxa also occurred more over the warmer seasons. Despite the lower density of meroplankton in winter, the presence of considerable numbers all through the year of a variety of taxa is evidence for the breeding of some species occurring in almost any month as in other warmer inshore environments.

Wickstead (1963) comments briefly on the greater abundance of larval meroplankton, including some forms apparently found only at night, in inshore zooplankton hauls in the Zanzibar Channel as compared with hauls only some 20 miles offshore in oceanic waters of around 750 m. In the Channel itself meroplankton was more important in the shallow (12 m) waters than somewhat farther offshore (40 m). Gapishko (1971a, b) found some periodic changes in certain meroplankton in the Gulf of Aden. The general zooplankton biomass is high during August and September, apparently due to upwelling in August. Echinoid larvae had been abundant (794/m^3 were recorded in March) and the numbers were maintained, although much of the zooplankton decreased from March onwards. However, during August, despite a marked increase in some holoplanktonic species (*vide infra*), echinoid larvae decreased. A decline also occurred over the same period in mollusc larvae.

With regard to the occurrence of meroplankton in other warm neritic waters, Lebour (1959) reported on considerable numbers of decapod larvae from tropical West Africa. Although many were larvae of holoplanktonic decapods, meroplanktonic larvae of benthic species were also present, many being Palaemonidae, though only *Palaemon elegans* was identifiable. Others included thalassinids (*Upogebia* and *Callianassa*) and alpheids. Anomuran larvae were represented by porcellanids and pagurids and there was an assortment of brachyuran zoeas and megalopas. Phyllosomas of *Palinurus* and *Scyllarus* occurred both near the coast and in the vicinity of the Cape Verde Islands. Since precise identification of many larvae was not attempted, it is not possible to discuss specific breeding cycles. Earlier, Lebour (1954) described decapod larvae, including several benthic species, from the Benguela Current. Gurney (1938) contributed descriptions of palaemonid and alpheid larvae from the Great Barrier Reef, and Lebour (1949) described decapod larvae from

Bermuda. However, there appears to be little or no information on breeding cycles.

Bainbridge (1972) observed less marked changes in zooplankton density during the period of upwelling, close to the coast than offshore in Nigerian waters and related differences largely to the greater numbers of meroplanktonic invertebrate larvae which partly masked the change in holoplankton.

Furnestin (1957) listed a number of meroplanktonic larvae for waters off Morocco. Identified larval decapod species included *Penaeus*, *Aristaeus*, *Palaemon*, *Lysamata*, and many pandalid and crangonid larvae. *Homarus* larvae and phyllosomas of *Palinurus vulgaris*, *Scyllarus arctus* and *Scyllarus latus* were also taken. Anomuran larvae were represented by *Porcellana*, *Pagurus* and galatheid zoeas. Brachyuran zoeas and megalopas included *Maia*, *Portunus*, *Inachus*, *Xantho* and *Carcinus*. Other meroplankton were represented by cyphonautes, echinoplutei (especially of *Paracentrotus lividus*), ophioplutei (chiefly *Ophiothrix fragilis*) and fish eggs and larvae. The identification of many of the larvae could be of great assistance in any future examination of breeding cycles.

Identification of breeding periods in a number of decapods in warm waters has been made by Seridji (1968), who examined the populations of decapod larvae in the Mediterranean (Bay of Algiers) close to the coast in depths from 40 to 100 m. There was a wide variety of species with some larvae occurring through the year, maximal numbers being found in September. Certain species (e.g. *Goneplax angulata*, *Ebalia* spp., *Processa* spp., *Philocheras* spp., *Anapagurus chiroacanthus* and *Alpheus glaber*) occur in all months but some of these show greater abundance at a particular season (e.g. *Anapagurus* and *Alpheus* in October/November; *Philocheras bispinosus* in spring). Other decapods have a limited breeding period, though the larvae have a fairly long planktonic life of more than three months (e.g. *Upogebia deltaura*—July–October; *Diogenes pugilator*—March–November). Certain species have larvae which remain in the zooplankton for periods of 1–3 months only (e.g. *Corystes cassivelaunus*, September/October; *Athanas nitescens*, May–July; *Pandalina brevirostris*, February/March). The number of species increases in spring to a maximum in summer. In general the decapod fauna resembles that off Marseilles and in the Adriatic, but there are differences in the breeding times of some of the species between Algiers and Marseilles (e.g. *Corystes* breeds in January–March off Marseilles).

Gosselck and Kuehner (1973) reported major seasonal changes in the holoplankton and an abundance of the larvae of the cephalochordate *Branchiostoma senegalense* from June to September off Cape Blanc off the Mauritanian coast, an area of pronounced seasonal upwelling off West Africa. Table 4.8 indicates that other meroplankton (lamellibranch and decapod larvae) could be plentiful in the region, though details of species are not available. The effect of upwelling on the incidence of meroplankton was confirmed by LeBorgne and Binet (1974) who examined vertical hauls over nine days during upwelling. They noticed a rapid increase in dry weight of zooplankton with synchronization of maxima. Thiriot (1977) states that off Cape Blanc larvae of Phoronidae and Cerianthidae and especially decapod larvae could be present in considerable densities during March and April, rivalling some of the holoplanktonic taxa. Different populations of meroplankton were reported as occurring in zones approximately parallel to the coast and varying in depth. Thiriot also quotes data from Santamaria showing that lamellibranch larvae could be equally important in the region, reaching maximum densities during March exceeding 400/m^3

Table 4.8. Qualitative and quantitative composition of plankton samples from which 1,200 Branchiostoma senegalense larvae were taken for investigation of gut contents (Gosselck and Kuehner, 1973)

Taxa	Number/12 m^3
Calanoida	
Below 1 mm	23400
1–2 mm	3900
Above 2 mm	1900
Cyclopoida	11500
Harpacticoida	present
Ostracoda	present
Mysidacea	100
Brachyura larvae	present
Malacostraca larvae	1500
Isopoda	present
Amphipoda	100
Nauplia	300
Foraminifera	100
Radiolaria	100
Noctiluca	present
Coelenterata	present
Chaetognatha	1400
Polychaeta	1400
Pteropoda	100
Branchiostoma senegalense larvae	7600
Bivalvia larvae	1200

over the continental slope. Gosselck and Kuehner's results indicate, as might be expected in a tropical environment, that abundance of meroplanktonic forms was not confined to that one season.

Neto and Paiva (1966) studying for one year an area close inshore (*ca.* 30 m) off Angola, found that although copepods were the most important animals they formed an annual average of only 44 % of the total zooplankton and were not dominant in every sample. Appendicularians (average 17 %) occurred in all samples and were dominant on six occasions. The meroplankton, however, made significant contributions. Cirripede and lamellibranch larvae were taken on every sampling date, the former were particularly plentiful in September and November and lamellibranch larvae were abundant in October and actually dominant in March. Cirripedes formed 9 % and lamellibranch 7 % of the total. The only other significant groups were pteropods present in all samples (5 %) and doliolids occurring through most of the year (2 %).

Though less important, other meroplankton, especially a variety of polychaete larvae and decapod larvae, some echinoderms, enteropneusts and medusae were represented. The importance of the meroplankton and the lower proportion of copepods recorded may be associated with the marked neritic character of the sampling area. This is borne out by the dominance of *Oithona nana* throughout the year as the chief copepod. The changes in salinity and temperature in such an inshore situation were associated with seasonal fluctuations in some other zooplankton species. In neritic areas of Zaire, Binet (1970) draws attention to the contributions of polychaete, lamellibranch and cirripede larvae, mostly during the major cooler upwelling season (May–September) and of decapod larvae chiefly in the warmer period of the year. Hydromedusae could also be important, though the periodicity was not obvious.

Seguin (1966) found a range of meroplanktonic larvae in shallow waters off Dakar. Bivalve veligers were common all through the year and some polychaete, particularly spionid, larvae were taken. Of echinoderm larvae, ophioplutei were most numerous and were found all through the year. Many larvae of benthic crustaceans were identified (Alpheidae, Palaemonidae, Processidae, and Hippolytidae). *Upogebia* and *Callianassa* larvae were among those identified, and amongst pagurids, *Pagurus*, *Diogenes* and *Albunea*. Zoeas of *Galathea* and Porcellanidae were also collected. Zoeas of Brachyura, which appeared to be represented in most months, include Xanthidae, Atelecylidae and Pinnotheridae. Cirripede nauplii were also numerous and more or less present throughout the year. In a study of the annual cycle of zooplankton off Abidjan, Seguin (1970) found that neritic components were low as expected off the edge of the continental shelf. Brachyuran zoeas were moderately abundant, however, particularly in November and May. Other major constituents of the meroplankton were ophioplutei, with some bivalve larvae and a few veligers of gastropods. Thiriot (1977) has emphasized that along the coast of West Africa meroplankton can be important on occasions, but particularly in periods of the year outside the time of maximal upwelling.

In other warm seas in the Atlantic off Venezuela Legare (1961) studying the zooplankton of the Gulf of Cariaco, found that bivalve larvae could occur in considerable abundance and in high frequency in the samples. Cirripedes were moderately frequently present and occurred in fair numbers. Echinoderm larvae, decapod larvae and larval polychaetes were also well represented, with varying frequencies and abundance (cf. Fig. 4.6). Decapod larvae included numerous brachyuran zoeas. Alpheid and thalassinid larvae were next in abundance and there were smaller numbers of a range of other decapod families. Meroplankton was much more abundant inside the Gulf of Cariaco than over the deep Trench region. A variety of meroplanktonic larvae extending over the whole period investigated appears to be typical of tropical waters. Investigations on the fluctuations in meroplankton in the Maracaibo estuary (Venezuela) by Rodriguez (1969) showed that such species as cirripede nauplii and *Uca* zoeas were present through the year but, as in other warm seas, they exhibited marked changes in density. Cirripedes reached a maximum ($>5000/m^3$) in December–February, while *Uca* zoeas achieved their maximum ($>3000/m^3$) in December. These crab larvae were present in significant numbers only in one area characterized by dense mangrove vegetation.

In other mangrove areas in South America, near Santos (Brazil), Teixeira, Tundisi and Kutner (1965) found that certain meroplankton made significant contributions to the total plankton population. Very close inshore, cirripede nauplii could be important, reaching maximal densities in July. Mollusc larvae were in greatest abundance in July and October. In very slightly deeper water, polychaete larvae were significant and showed seasonal variation, being at minimum density in April and at maximum density in October.

In the Caribbean area, Millar (1974) examined the breeding of some ascidians. Whereas in temperate waters ascidians tend to breed mainly in the warm season, in tropical regions some species have a long but fluctuating breeding period and others

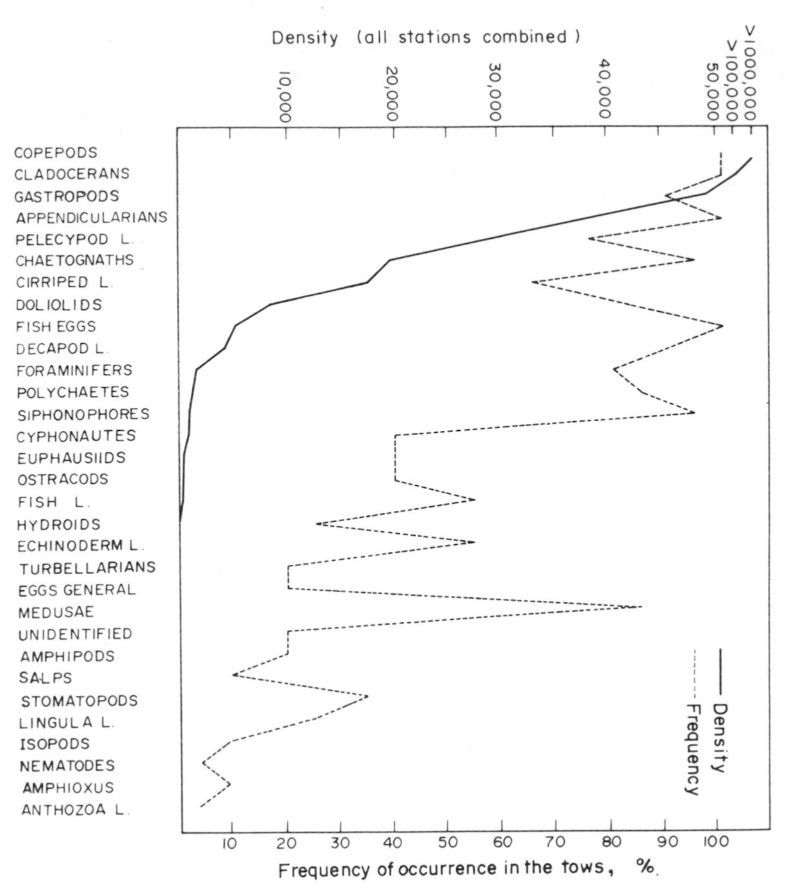

Fig. 4.6. Density (number of animals) and percentage frequency of zooplankton groups found in May 1960 in the Cariaco region (Legare and Zoppi, 1961).

have an interrupted breeding cycle. Millar quotes earlier data from Goodbody concerning three Jamaican species of ascidians, that settlement continued all through the year, though not at a uniform rate. Millar's investigations at Panama, using as a criterion of breeding the presence of embryos or larvae in the colony, demonstrated that *Eudistoma* had a well defined breeding period from June–November, *Tridemnum solidum* a period from June–December, and *Didemnum* bred almost continuously, though with more intense periods (cf. Fig. 4.7). Millar points out that it is doubtful whether temperature change could trigger breeding even in the first two species since temperature variation is so small in the region.

Studies in Indian coastal waters have also illustrated the role of the meroplankton. Goswami and Selvakumar (1977) found that certain meroplankton (decapod and polychaete larvae, veligers, fish eggs and larvae) made a fairly significant contribution to the estuarine zooplankton populations in the Mandovi–Zuari estuarine system at Goa, India. The zooplankton total biomass was considerably influenced by the monsoons, though not at exactly the same period in the two estuaries. In the Mandovi, polychaete larvae, mostly nereid, eunicid and terebellid larvae, were important. Fish

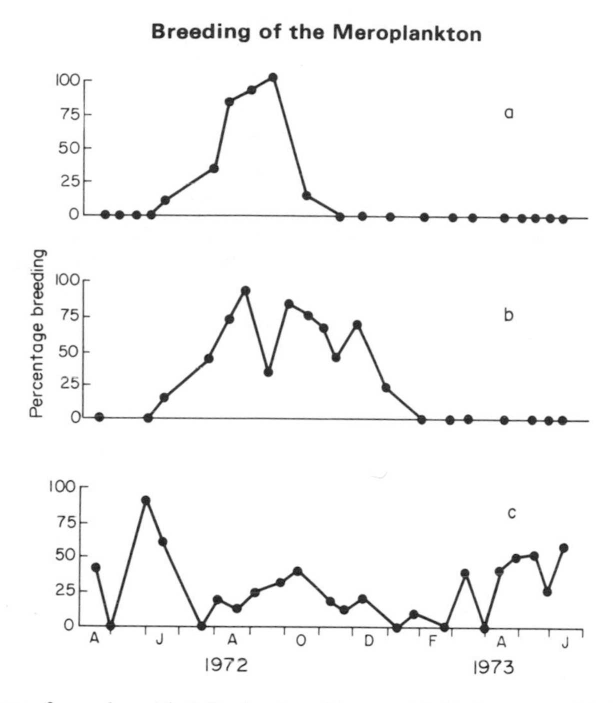

Fig. 4.7. Percentage of samples with fully developed larvae. (a) *Eudistoma* sp; (b) *Tridemnum solidum*; (c) *Didemnum* sp. (Millar, 1974).

eggs and the larvae of Clupeidae were especially plentiful in July. Brachyuran zoeas and the larval stages of *Metapenaeus dobsoni*, *M. monoceros* and *Penaeus merguiensis* were common at both stations. Veligers were present practically throughout the whole year and were dominant when salinity was low. Larvae of *Lucifer hanseni* were especially common during the month of July in the Zuari estuary.

A number of important contributions to the life histories of decapod crustaceans and other invertebrates involving meroplanktonic larvae were also reported during the Warm Water Symposium on Zooplankton (1977), but it is difficult to relate life histories to breeding cycles in warm seas. To some extent, however, examination of gonad condition might assist in elucidating breeding cycles. For example, Pillay and Nair (1971) used this approach to study the breeding cycle of three decapods near Cochin. Female *Uca annulipes* were found to have a high gonad index from July–March, *Portunus pelagicus* from August–March, and *Metapenaeus affinis* from August–April. The breeding season for all three species, while not continuous, extended over several months with peaks of more intense activity, particularly about December and January. No breeding occurred during the low salinity period of the monsoon. During the pre- and post-monsoon periods, the abundance of plankton food, as well as suitable environmental salinity, may favour breeding. Earlier reports for some other decapods suggest that the very low salinity on the west coast of India during the south-west monsoon may interefere with breeding cycles. On the east coast, though the salinity is not so sharply reduced, breeding could be triggered but not stopped by salinity changes. *Portunus pelagicus*, for instance, is said to breed continuously on the east coast, with several periods of more intense reproductive activity.

A further example of the effect of lowered salinity on breeding cycles can be drawn from the work of Vannucci, Santhakumari and Dos Santos (1970). Hydromedusae were sampled over two years near Cochin, India. Two stations were visited: one at Aroor in the Backwaters, where the south-west monsoon could lower salinity from about 32‰ to complete freshening of the water; the second at Fairway Buoy, a few miles outside the harbour, where during the south-west monsoon the surface salinity could fall to about 12‰ though at depth the water was of relatively high salinity. From about November to January some upwelling could bring high salinity water to the surface. Since the attainment of a suitable salinity at Aroor, following the fresh water flushing, did not occur until November/December, few medusae appeared before January. They were present from then until about May/June, with numbers depending on species. The reproductive cycles were obviously adapted to the seasonal change in salinity. With *Obelia* species, though they were found at both stations, the vast majority occurred at Aroor, where they were generally present until the extreme freshening due to the monsoon. At Fairway Buoy they were chiefly present from January to March, but only in smaller numbers.

Phialidium brunescens was found at Aroor in February/March and May. It appeared to be a normal inhabitant of the Backwaters, but occurred also at Fairway Buoy. *Blackfordia virginica* was abundant in January, February, May and June at Aroor, but was not found at Fairway Buoy. The species is known widely in tropical swamps and temperate estuaries. At Aroor there were at least two generations, one in January and the other in May/June, before the coming of the south-west monsoon. *Eirene ceylonensis*, found only at Aroor, was very common from March to June (Fig. 4.8a). *Eutima commensalis*, though found at both stations, was far more plentiful at Aroor,

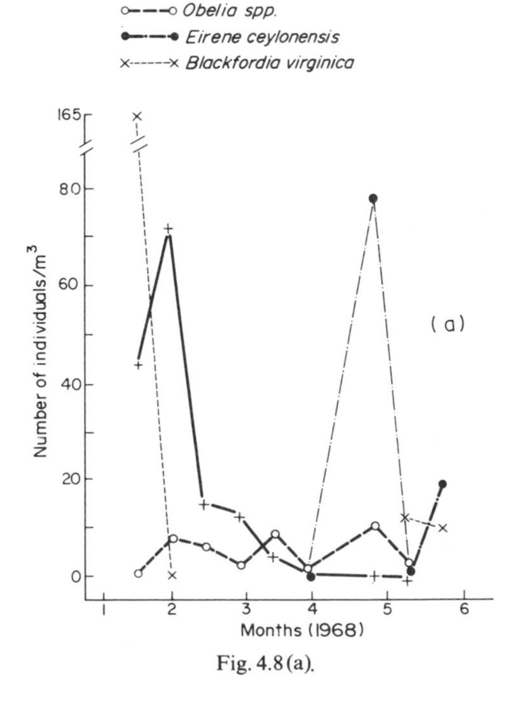

Fig. 4.8 (a).

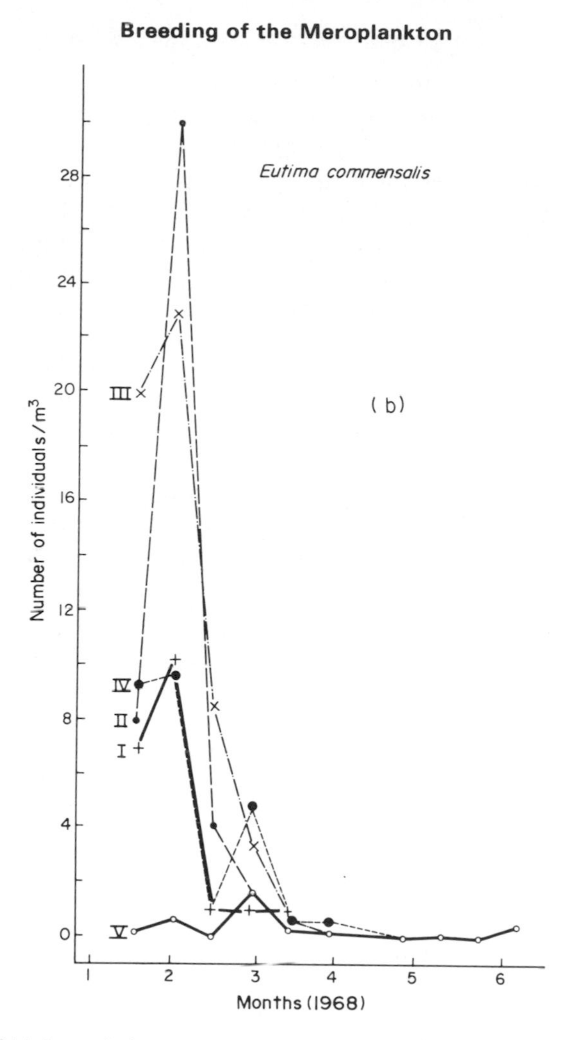

Fig. 4.8. Dynamics of (a) 4 populations at Aroor; (b) *Eutima commensalis* of stages I–V (Vanucci *et al.*, 1970).

and occurred from January to June with a peak in February (cf. Fig. 4.8b). Certain medusae (e.g. *Cytaeis*, *Pandea*) appeared to be carried into the Backwaters as dead specimens, presumably by the upwelling water off the mouth of the estuary. Certain medusae were found only at Fairway Buoy, one example being the cosmopolitan warm-water *Liriope tetraphylla*, from August to October.

Although no medusae occurred during the south-west monsoon inside the Backwaters, recolonization took place from resting stages such as the fixed forms of hydromedusae. Some form of resting stage or of encapsulation is essential to keep the species alive. With the restoration of suitable salinity, active medusae can be budded off during the inter-monsoon period. Trachymedusae and Narcomedusae, lacking a fixed stage, cannot be found inside the Backwater system, except as occasional visitors transported in the deeper waters. Although few species of Hydromedusae occur at Aroor, there can be extremely high populations with very rapid multiplication. Eight species accounted for nearly 10,000 specimens. At Fairway Buoy there appear to be two groups of medusae: from January–June, low salinity species brought out from the coast and the Backwaters, and from May–October, in the deeper higher salinity water, more open ocean visitors from the Arabian Sea. Santhakumari and Vannucci (1977) suggest, from sampling at Aroor at very short (48hr) intervals, that during the pre-monsoon period of rapid growth the attainment of sexual maturity and the turnover rate of the medusae may be exceedingly rapid. The whole cycle could perhaps take as little as three–four days. Salinity is the over-riding factor, limiting growth and reproduction.

Nair and Tranter (1971) also examined the zooplankton of the Backwaters in relation to the south-west monsoon. They showed that whilst during the pre-monsoon period (April) there was a high abundance and fair diversity of zooplankton, the summer freshening led to a reduction of the marine plankton. During the post-monsoon period (November) the biomass began to rise again. Apart from copepods, which were easily dominant at both pre- and post-monsoon seasons, cladocerans occurred, mostly towards the mouth of the Backwaters. Meroplankton included cirripede larvae, brachyuran zoeas and gastropod larvae during both pre- and post-monsoon periods, but there was some periodicity with other larvae. For example, bivalve larvae were common in April but were relatively sparse in November, polychaete larvae were present in both seasons but were more abundant in April, and penaeid larvae occurred mostly in November.

In sharp contrast to brackish estuaries and lagoons, another tropical marine environment where meroplankton may be significant is that of coral reefs and atolls. Emery (1968) described the different kind of zooplankton inhabiting a reef and showed that the population varied in different parts of the reef (e.g. reef top, caverns, etc.). While some widely distributed members of the general plankton (e.g. species of *Acartia*, *Oithona* and a few siphonophores) were part of the population, many epibenthic and tychopelagic species, especially harpacticoids, mysids and cumaceans were abundant, together with some meroplankton such as crab, shrimp and polychaete larvae.

Tranter and George (1972) studied the zooplankton of two atolls in the Laccadive Islands. Table 4.9 illustrates the marked differences between the plankton of the open

Table 4.9. Taxonomic composition at an ocean station compared with that of a lagoon station (Tranter and George, 1972)

Taxon	Density (number/m^3)		Percentage	
	Ocean	Lagoon	Ocean	Lagoon
Calanoid + Cyclopoid copepods	917	2	75	0.4
Pelagic Tunicates	48	0	4	0
Chaetognaths	28	0	2	0
Pelagic Polychaetes	7	0	0.6	0
Siphonophores	4	0	0.3	0
Harpacticoid copepods	2	366	0.1	65
Decapod larvae	3	53	0.3	9
Mysids	0	11	0	2
Gammarids	0	2	0	0.4
Ostracods	149	123	12	22
Miscellaneous	69	8	5.6	1.4
Total	1227	565	99.9	100.2

Indian Ocean, with calanoids, chaetognaths and salps as conspicuous components, and that of the coral lagoon, with harpacticoids, mysids and other epibenthic animals dominating. The contribution of meroplankton (decapod larvae) was notable. The biomass of zooplankton inside the atoll lagoons was higher than that of the open eastern Indian Ocean (Table 4.10), but the high biomass, particularly in the Kavaratti Lagoon, was considerably increased by the very dense swarms, especially at night, of the ostracod, *Cypridina*. However the plankton just outside the atoll was much richer than in the open ocean or even over the continental shelf (Table 4.10). This finding supports the conclusions of Russell (1934) that the zooplankton of coral reefs may resemble in richness that of temperate neritic areas.

Table 4.10. Nocturnal zooplankton biomass (surface) observed in the Laccadives, compared with similar observations from the N.S.W. continental shelf (Australia) and the eastern Indian Ocean (Tranter and George, 1972)

Location	Zooplankton Mean biomass (mg/m^3)	No. of observations
Ocean around Kavaratti	336	14
Ocean around Kalpeni	132	22
Eastern Indian Ocean	74	87
Lagoon, Kavaratti	189	6
Lagoon, Kalpeni	99	8
N.S.W. Continental shelf	136	12

Glynn's (1973) investigation of the plankton of a coral reef in south-western Puerto Rico indicated that the zooplankton was considerably depleted in streaming across the reef. Some 60 % of the zooplankton could be removed. The zooplankton was much richer at night, and during autumn the density varied with the phases of the moon. Though holoplankton organisms were most numerous, with copepod nauplii, especially *Acartia* and *Oithona*, forming the major part of the whole plankton, with some harpacticoids and appendicularians (Fig. 4.9a), a considerable fraction of the plankton was tychopelagic. Meroplanktonic animals were also present on occasions and sampling indicated that the breeding times of some groups of reef invertebrates were more intensive over certain periods. For example, bivalve larvae appeared all the year, but mainly in September/October, while gastropod veligers were more plentiful in July/August (cf. Fig. 4.9b). Zoeas occurred in early summer and autumn and echinoplutei in July to September. Generally the total biomass showed a seasonal fluctuation, with spring and especially autumn/winter as the main periods of abundance.

Sale, McWilliam and Anderson (1976) investigated the zooplankton in very close proximity to coral reefs on a part of the Great Barrier Reef, using a light trap for collecting the zooplankton. They compared the plankton of the shallow lagoon with that over the coral reef and also with that of the more open water in a channel between reefs. Fig. 4.10 illustrates differences in ten major groups of zooplankton between the three sites, and also compares samples taken close to the surface and close to the substratum on the reef and in the lagoon. There is considerable spatial and temporal patchiness in the faunal distribution. The fauna appears to be rich and abundant, though samples indicated fewer organisms for the open water. Open water samples

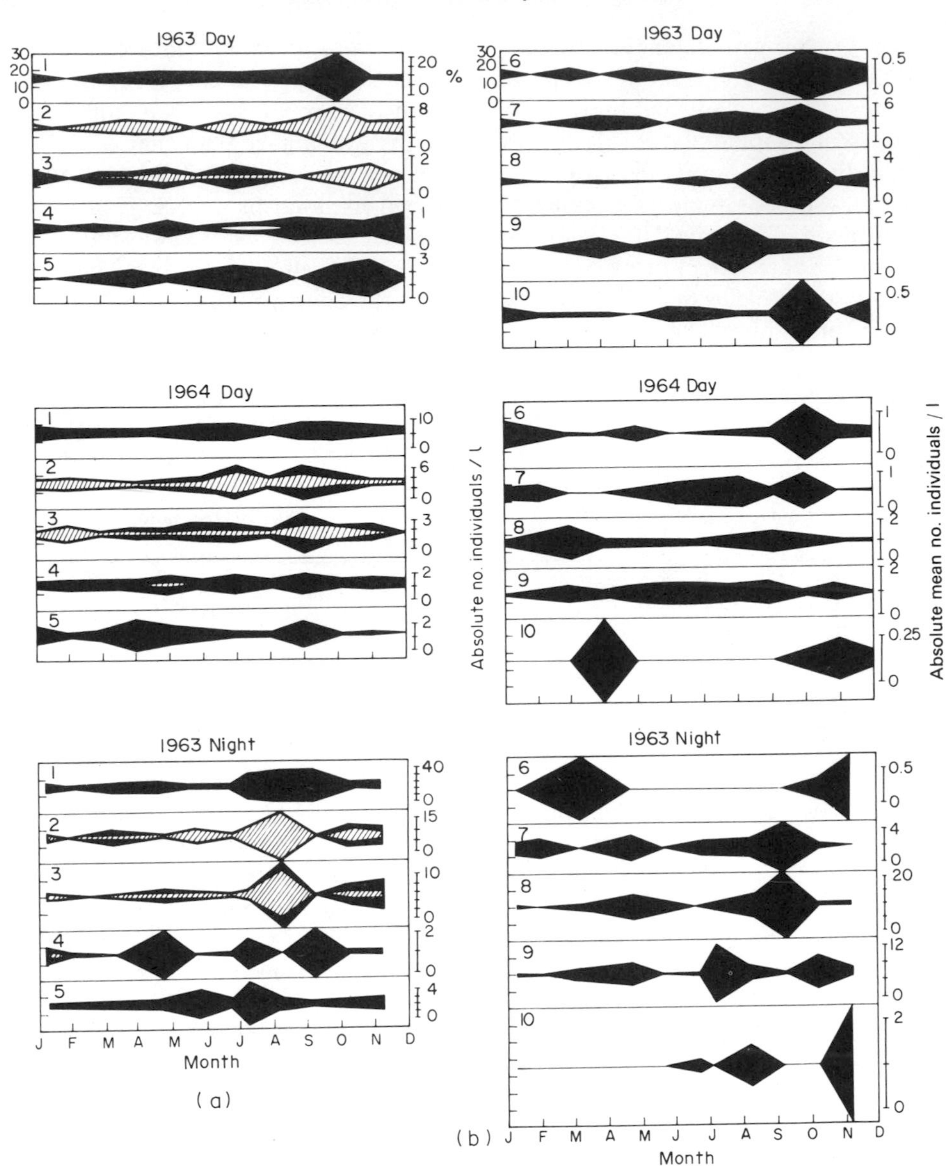

Fig. 4.9. Seasonal distribution of the dominant groups of (a) holoplankton and (b) meroplankton. Each value represents mean percentage abundance calculated from total number of individuals observed in all monthly collections at 3 stations. The night plot shows mean abundance at 3 stations for single monthly collections. Diagonal shading: relative abundance of young copepodid stages (1) copepod nauplii; (2) *Acartia* spp.; (3) *Oithona* spp.; (4) harpacticoids; (5) appendicularians; (6) cirripede nauplii and cyprii; (7) unidentified eggs and larvae; (8) bivalve larvae; (9) veliger larvae; (10) polychaete larvae (Glynn, 1973).

were also qualitatively different. They were dominated by calanoid copepods (36 %) and appendicularians (33 %) and had a lower species diversity. The near-reef samples had a smaller but considerable population of calanoids (24 %), but appendicularians were very few and other groups (cumaceans, amphipods, monstrilloid copepods and

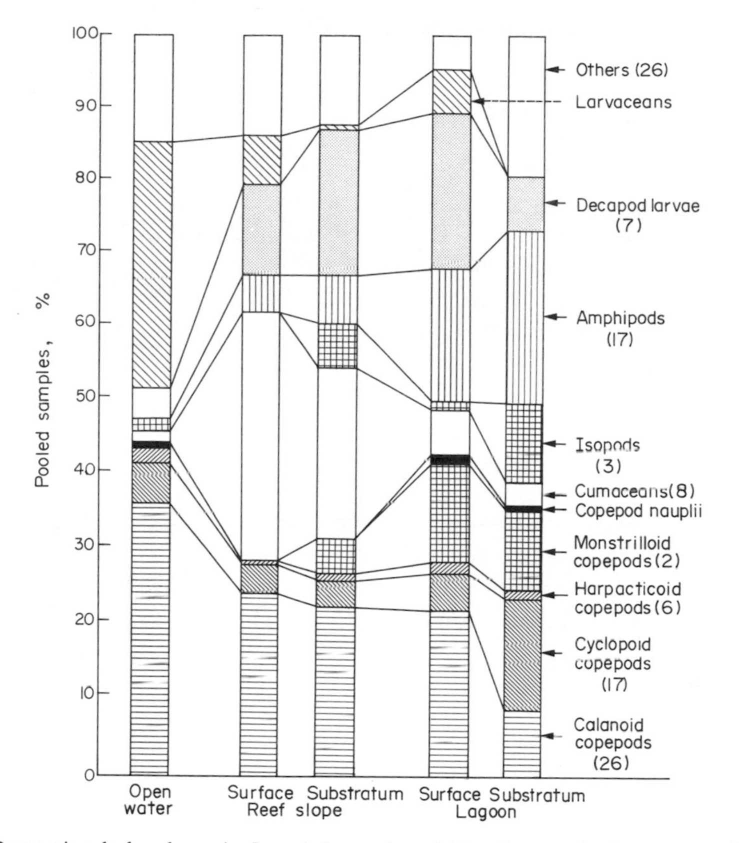

Fig. 4.10. Proportional abundance in 5 pooled samples of 10 major zooplankton groups. Numbers in parentheses under group names are numbers of taxa provisionally recognized in that group (larvaceans and copepod nauplii were each considered one taxon). Heavy lines across figure emphasize the changing importance of calanoid copepods, but also the relatively constant importance of total copepods and the changing importance of total arthropods as one proceeds from open water to the lagoon (Sale, McWilliam and Anderson, 1976).

Table 4.11. List of commonest copepod taxa, and percentage of all copepods belonging to each, for all samples from each station at Heron Reef (Sale *et al.*, 1976)

Station	Order	Species	Percentage of copepods
Open water	Calanoida	*Eucalanus crassus*	30.8
	Calanoida	*Undinula vulgaris*	12.8
	Calanoida	*Acartia* sp. "G"	9.4
			53.0
Reef slope	Calanoida	*Calanopia elliptica*	35.1
	Monstrilloida	*Thaumaleus* sp.	11.4
	Calanoida	*Acartia* sp. "G"	8.8
			55.3
Lagoon	Monstrilloida	*Thaumaleus* sp.	35.8
	Cyclopoida	*Oncea* sp. "F"	9.8
	Calanoida	*Calanopia elliptica*	8.9
			54.5

decapod larvae) formed a significant part of the fauna. Isopods, mysids and polychaetes could also be important (Fig. 4.10).

Although copepods were the dominant group at all three stations, the species differed. Table 4.11 shows that three species formed more than 50% of the total copepod population at each station, but there were marked differences between the near-reef and open water species. Many of the near-reef taxa, which are rare or absent in the open channel samples, may probably be regarded as truly resident animals. A number of them, as other investigators have suggested, are epibenthic rather than truly planktonic and are more plentiful in samples taken close to the bottom. Not only hydrological patterns but also the behaviour of much of this resident fauna tends to maintain it close to the reef itself.

Oceanic Meroplankton

So far this account of meroplanktonic larvae has been almost entirely confined to neritic waters with brief mention of atolls. Reference has been made to the unusually long larval life of *Emerita* which tends to be rather widely distributed off California. The phyllosoma larvae of *Palinurus interruptus* are also long-lived. Some details have already been given of their distribution off the coast of California. Comparatively long-lived larvae frequently exhibit wide dispersal over open oceans, but Johnson's (1960) investigation suggests that *Palinurus interruptus* remains a fairly near coastal form. The larvae hatched off California almost entirely between July and October, mainly during August and September. Only one brood was spawned per year, but the development included a fairly long larval period (nearly eight months) with eleven phyllosoma stages. Johnson (1960) quotes observations by Sheard that larval stages of the related species *Palinurus longipes* are planktonic for about five months. A number of species of phyllosomas are known to be long-lived and may therefore appear in open ocean plankton.

Robertson (1971) suggests from rearing of the phyllosomas and from field observations of *Scyllarus depressus* a minimum period for development of two and a half months at a temperature of about 25°C. A marked extension of the period can occur, however, especially with lower winter and spring conditions in the North Atlantic. Phyllosomas were observed in the Florida Straits from January–May. Late phyllosomas or post-larval stages recovered off the New England coast in early September might be derived from hatching in June off North Carolina, with transport by the Gulf Stream. Other meroplanktonic larvae that might be expected in ocean zooplankton are the leptocephali of eels, spawned over oceanic depths. Nauplii of cirripedes of the *Lepas* type, including *L. anatifera* and *L. fascicularis* larvae, characterized by a fairly prolonged existence, also occur in open sea plankton, beyond the edge of the continental shelf. Since the adult barnacles are typically surface (neuston) drifters, the larvae should not perhaps be classed as meroplankton, but it is convenient to mention their distribution here. For *L. anatifera*, the larval life has been confirmed as much longer than for shore cirripede species. The development appears to be limited to temperatures exceeding 15°C and normally the nauplii are distributed mostly south of latitude 49°N in the North Atlantic (cf. Bainbridge and Roskell, 1966). Moyse (1963) and Moyse and Knight-Jones (1967) showed that the larvae of Lepadidae (*L. anatifera* and *L. pectinata*), although small when first liberated, have a long larval life (57 days in culture for *L. anatifera*) and have large growth increments, resulting in relatively enormous late nauplii, aside from the great length of the spines. In the warmer North Atlantic *L. anatifera* nauplii occur from about July to October, probably being liberated early in July. They may have a life in nature approaching four months.

This brief examination of oceanic meroplankton would tend to show the very few taxa from which examples can be drawn and would suggest that such larvae are comparatively rare. While, in general, oceanic meroplankton are not plentiful, it has been known for some time that some phyllosomas appear to be able to delay metamorphosis when occurring over great depths. Such an ability would increase the chance occurrence of larvae. It is now generally recognized that this ability to delay metamorphosis is true of some species spread over a wide range of invertebrate taxa (cf. Thorson, 1950; Scheltema and Hall, 1975). Lack of suitable food may be a major factor in delaying metamorphosis but the earlier experiments of workers such as Wilson (1937, 1948) on the survival, particularly of polychaete larvae, suggested that minute differences in water quality and in characteristics of the bottom were important factors. Other investigators such as Crisp, Knight-Jones, Jagersten, Meadows, Gray and others have carried these studies further.

Physical, chemical and biological characteristics can all interact in a complex manner to accelerate or delay metamorphosis. An excellent recent review is that of Scheltema (1974). If, therefore, some meroplanktonic larvae can survive, even while growing only very slowly, they may be carried considerable distances, and contribute to the oceanic population. While it is not possible at this stage to give times and other details of breeding, some account of such larvae is required. Ekman (1953) believed that trans-oceanic transport of meroplankton larvae was unlikely in view of the short larval life. This view was supported by Thorson (1961), but his conclusions were mainly based on the average length of larval life for a large number (195) of species of mainly cold/temperate or boreal species. For the usual benthic coastal species dispersal across oceans would be almost an impossibility (cf. Mileikovskiy, 1968c). However, Thorson refers to certain groups (polychaetes, echinoids, asteroids, and particularly ophiuroids) where the length of drifting life may be extended by metamorphosis occurring away from the bottom and by possible larval reproduction. He instances also certain gastropods (*Cypraea*, *Tonna*, *Cymatium*, *Bursa*) where the larvae grow into relatively large veligers, with greatly hypertrophied lobes and which appear to be truly adapted for trans-ocean passage. Scheltema and Hall (1975) have also pointed out, as indeed Thorson (1950) accepted, that many tropical meroplanktonic species have much longer planktonic lives than colder-loving species. With the possible extension of the period before metamorphosis, Scheltema believes that occasional transport of a very few larvae can occur across oceans. Some tropical larvae, he finds, have an estimated life of four to eight months, some have been maintained longer in the laboratory, and he claims there is, therefore, sufficient time for transport.

Although most meroplankton will remain near the coast, of the "stragglers" swept offshore a very small percentage will survive. In the oceanic plankton some of the prosobranch larvae listed, and a sprinkling of larvae from various taxa adapted to prolong larval life, will occur together with the long-lived leptocephali and phyllosomas. Thus, especially, in warmer open sea plankton, there will be some meroplankton, though the proportion will inevitably be small. Some of the taxa having long-lived larvae include the sipunculids and hemichordates. Owre (personal communication) has collected many such larvae from open warm North Atlantic waters. Some arachnactis larvae may also be found very far offshore. Off the Californian coast, apart from the larvae of *Emerita* (*vide supra*), Longhurst (1968) has described larvae of *Pleuroncodes* far from shore. The larvae again appear to be long-lived. *Pleuroncodes planipes* is somewhat exceptional in that it is one of the very few galatheids which can have a pelagic existence for part of the adult life, though there are also substantial benthic populations. Some breeding, which occurs mostly from February to March, arises from pelagic adults, but larger larval concentrations over the continental shelf may probably be attributed to benthic crabs. A considerable proportion of late zoeas and megalopas are carried offshore and may metamorphose only after an extended time in more or less oceanic regions. There is no clear evidence whether megalopas are returned before metamorphosis, or whether some, still drifting over the narrow continental shelf, can metamorphose quickly there.

Some other examples of extended duration of meroplanktonic life may now be listed. Scheltema (1966) has claimed that a comparatively rapid development of some gastropods to the "creeping stage" can occur, which may be followed by a slower change with little growth, so conserving energy and prolonging the period before metamorphosis. For larvae of *Cymatium*, which, he believes, may even occasionally cross the Atlantic, 80% from tows in the North Atlantic were found in areas where surface temperature was 13°–21/23°C. Although there is a chance of survival of only one in two million due to mortality and predation, with the numbers of eggs per female there might be a one in forty chance of one female producing one transatlantic larva which survives. The larval life appeared to vary from less than one month to more than three.

Scheltema (1971) gives another example from shipworms (Teredinidae), some species of which are widely dispersed as planktonic, fairly long-lived, phytoplankton-feeding larvae. They occur in the temperate and tropical North Atlantic and adjacent seas, in surface waters. The most common is probably *Teredora malleolus* but other species occur widely. Teredinid veligers were found in 19% of the samples from the temperate and tropical North Atlantic. They are carried apparently for long distances along continental shores and probably across oceans, with numbers diminishing as the distance from the coast increases. It is well known experimentally that the rate of growth of many bivalve larvae is dependent largely on food supply. In the absence of rich food conditions, as in the open sea, Scheltema believes that some larvae may survive over long periods and show little growth. For shipworm larvae, the duration of the straight-hinged stage is highly dependent on food supply and may be considerably extended when food is scarce. The larvae of one shipworm species were kept alive for more than five weeks without the onset of metamorphosis.

A study by Allen and Scheltema (1972) concerning *Planktomya henseni* is also relevant to oceanic meroplankton. This is a pelagic bivalve, known in warm-temperate parts of the North and South Atlantic Ocean and now known also from the tropical Pacific. It probably has a roughly circumtropical distribution. The organism has generally been regarded as a neotenous species, i.e. as a holoplanktonic bivalve which could be broadly analogous to holoplanktonic pteropods or heteropods. Allen and Scheltema point out, however, that no specimen has ever been found with maturing reproductive organs and they believe that these animals are bivalve larvae, though long-lived ones. Indeed, they may be the larvae of a shallow water species of

Leptonacea, of which younger and smaller stages are found closer to land, with the numbers decreasing with distance from the coast. The concentration of larvae decreased 500-fold between the coast of Senegal and mid-Atlantic. This appears to be a particularly striking example of prolonged larval existence on the high seas.

A further example of oceanic larvae is *Planctosphaera*, a rarely found organism which has been taken in a few plankton hauls at very long distances from the coast in warm-temperate areas. Such captures suggest a very long pelagic development. Scheltema (1970) found only two *Planctosphaera* in about 800 plankton hauls in the tropical and warm-temperate Atlantic. The larvae are of the tornaria type and may be the larvae of a sub-tropical or tropical enteropneust. Bhaud (1972) described many larvae and juveniles of polychaetes belonging to the family Amphinomidae over deeper water along the continental slope off Nosy Bé (Madagascar). Though these young forms are derived from the neritic zone, their special buoyancy mechanisms permit an extended planktonic existence. If carried offshore, descent to the bottom is delayed and the larvae can apparently survive in open waters, though in reduced numbers. Bhaud (1975) also notes very large larvae (the largest *ca.* 2 mm) of Sabellariidae common in tows off Nosy Bé.

Mileikovskiy (1966), while emphasizing that the vast majority of meroplanktonic larvae keep fairly close to the coast, points out that a few species apparently travel considerable distances. For instance, the littoral gastropods *Littorina* and *Patella* apparently cross the North Sea from Scottish to Norwegian waters. The gastropod *Velutina plicatilis* is transported in the Irminger Current from Iceland to Greenland. Mileikovskiy's own observations suggest that certain larvae, mainly polychaetes, cyphonautes and particularly opisthobranch and lamellibranch larvae, may be carried very considerable distances, the latter >700 miles. With the longer larval life in tropical seas, he admits that even longer transport is possible. He quotes the prosobranch *Philippia krebsii* as probably drifting not infrequently right across the Atlantic. He also mentions larvae of *Ptychodera* (an enteropneust) as having long oceanic drift and some phyllosomas such as those of *Thenus orientalis* in warm oceans, though transport may not be at all regular.

While the density of meroplankton in oceanic samplings is generally acknowledged to be low, some recent data from Scheltema (1974), Scheltema and Hall (1975) and Hall and Scheltema (1975) are relevant. Scheltema (1974) found that certain genera of the polychaete family Chaetopteridae had long-lived planktonic larvae widely dispersed by the warm currents of the North and tropical Atlantic. In experiments, larvae of *Spiochaetopterus* were kept in some instances more than a year without metamorphosing. Some other polychaetes (Spionidae, Amphionidae, Sabellariidae, Oweniidae) may also have widely dispersed larvae. Although most of the larvae keep close to shore and only a fraction of a percentage may be dispersed across the ocean, Scheltema believes that the exchange of populations is possible. From very wide and numerous plankton hauls over the warm-temperate and tropical North Atlantic the collection of moderate numbers of pelagosphaera-type larvae have been recorded. (Hall and Scheltema, 1975). These are long-lived larvae of certain sipunculids (e.g. *Phascolosoma agassizii*, *Sipunculus nudus*), the larva being preceded by a trochophore stage. Other sipunculids may develop directly, include only a trochophore stage, or add a short-lived pelagosphaera larva after the trochophore (cf. Rice, 1967). The authors suggest that a number of tropical and sub-tropical forms exist with a widespread distribution (cf. many specimens obtained by Damas from the Dana collections). Some specimens brought back by Hall and Scheltema were maintained alive for over a year! Successful metamorphosis and settlement occurred with some larvae maintained in the laboratory.

Scheltema (1975) discusses the genetic continuity between sipunculid populations, separated by the Atlantic, through the dispersal of such long-lived larvae. For one larval type, one estimate suggests that 3.8×10^{12} larvae are dispersed across the Atlantic. From estimates of larval densities from plankton samples and from estimates of water transport, the chance of one larva surviving may approach 10^{-9}. Scheltema and Hall (1975) point out that the occurrence of sipinculid long-lived larvae is quite common. Small-sized, as well as the more conspicuous forms are known. Half of the hauls (692) from the area had such larvae, and they occurred in every major east–west and west–east current.

Chapter 5
Vertical Migration of Zooplankton

The migration of zooplankton from at least 200 m depth towards the surface at night was known from the results of the Challenger Expedition, and since that time the widespread nature of this diurnal vertical migration has been recognized as one of the most striking and characteristic aspects of the behaviour of marine zooplankton. Even the roughest comparisons between surface tow-nettings taken by day and at night show that far greater numbers of animals occur in the uppermost layers during the dark hours. Hardly a single major group of animals represented in the zooplankton does not show at least some species displaying diel movement. Earlier workers (Sewell, 1948; Farran, 1949; Zenkevitch and Birstein, 1956), in summarizing results of vertical distribution studies, emphasized that the maximum zooplankton biomass occurs above 1000 m. Thereafter numbers of species and of individuals drop sharply in most areas. The introduction of modern sampling gear has enabled workers to give a more accurate assessment of the various depths occupied by different species. Reference has already been made to vertical distribution of zooplankton biomass in Chapter 1, while seasonal and depth distribution of the various taxa is included in Chapters 2 and 3. The following pages therefore will deal mainly with ontogenetic and diurnal migration.

Ontogenetic Migration

It has been recognized for a long time that the behaviour of an animal can vary considerably both seasonally and in relation to generation. Moreover, almost every worker on vertical distribution of plankton has observed that frequently the various developmental stages of a species show different distribution patterns. In general, the youngest stages tend to be nearer the surface and appear to be living at a higher light optimum (Russell, 1935). These differences may apply also to meroplanktonic forms; Russell found that the younger larval stages of pagurid and galatheid larvae inhabited a shallower layer than the later post-larvae. Similarly, the distributions of the naupliar and cyprid stages of the same species of *Balanus* are stated by Pyefinch (1948) to occur at different levels.

Many of the deeper-living members of zooplankton groups, especially the Crustacea, undergo changes in depth distribution with age. Sekiguchi (1975a, b) reports that the females of *Calanus cristatus* lay eggs in deep water (> 500–1000 m) during winter in the North Pacific. The young Copepodites II and III ascend to surface layers in the spring, where feeding and development take place. Copepodite V appears about June, after which the descent to deeper water commences, and by November sexually mature adults predominate below 500 m. By contrast C V and adults of *Eucalanus bungii bungii* migrate to the near surface waters in spring, where spawning takes place, while in summer they inhabit the 400 m layer and descend to 600 m in

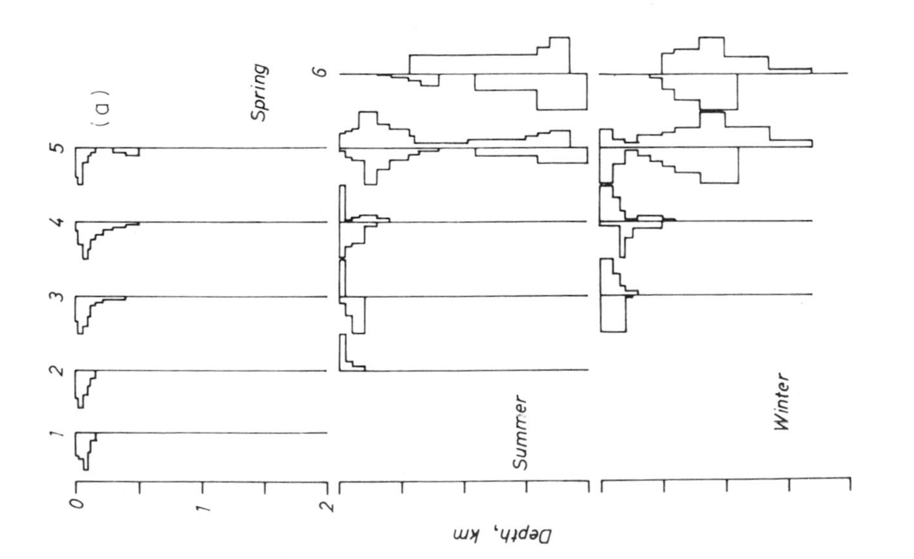

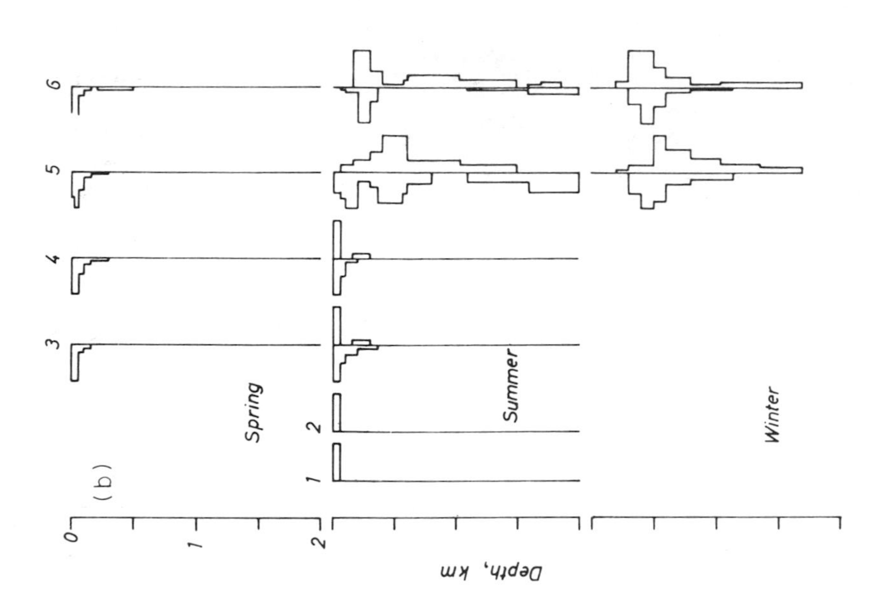

Fig. 5.1. Vertical distribution of (a) *Calanus cristatus* and (b) *Eucalanus bungii bungii*; daytime and night populations are shown to the left and right of the vertical line respectively. Numbers indicate the copepodite stages (Sekiguchi, 1975).

winter (Fig. 5.1a, b). Sekiguchi believes these migrations are associated with the feeding behaviour of the naupliar stages in contrast to species of *Pareuchaeta*, in which the early nauplii lack feeding appendages. Longhurst and Williams (1979) found at one station in the North Atlantic that adults of *Pleuromamma robusta* dominated a layer between 400–500 m, while C II, III and IV showed peak populations between 50–250 m.

For the Clyde Sea area, Nicholls (1933b), in his studies of *Calanus finmarchicus*, reported that the eggs and nauplii were nearly always in the upper 30 m and that there was little change in the distribution of the nauplii between day and night. Copepodite Stages I, II and III were abundant down to 45 m or so, and they showed a slight tendency to rise towards the uppermost layers during the hours of darkness. He suggests that it was C III which showed the main movement. The next stage exhibited a very definite diurnal vertical migration, most of the individuals being very deep (*ca.* 100 m) during the day, and moving right to the surface by nightfall (Fig. 5.2). On the other hand, Stage V copepodites seemed very unresponsive as regards vertical

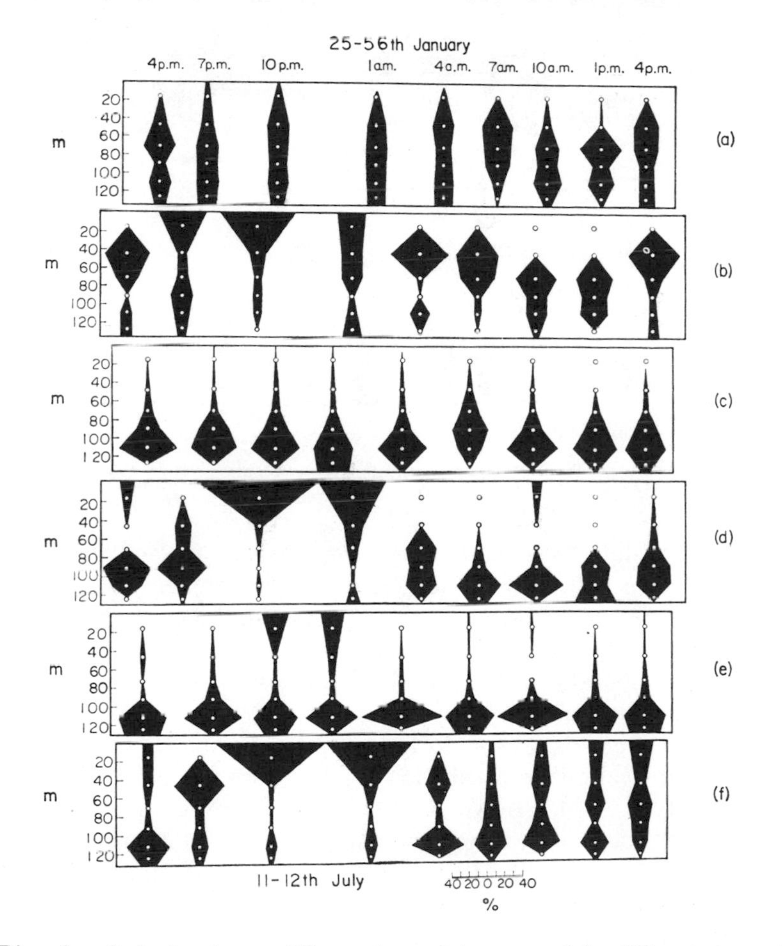

Fig. 5.2. Diurnal vertical migration at different times of the year and for different stages of *Calanus finmarchicus* in the Clyde Area; (a) males, January; (b) females, January; (c) Stage V, January; (d) females, July; (e) Stage V, July; (f) Stage IV, July (Marshall and Orr, 1955).

migration. Copepodite V was a deep-living stage occurring mostly from 80–100 m depth and there was little change between day and night. Nicholls found that male *Calanus*, which were mostly deep-living, were also largely unresponsive as regards vertical migration and contrasted sharply in this respect with females (Fig. 5.2). Farran (1947) also observed marked differences in behaviour of the various copepodite stages of *Calanus*. Females showed the clearest diurnal migration, with C V and IV being rather unresponsive (Fig. 5.3). Young copepodites were mainly in the upper layers by day. On the other hand, for *Metridia lucens* C I–IV were mainly below 50 m and females, and C V, though deep-living by day, moved into intermediate layers at night.

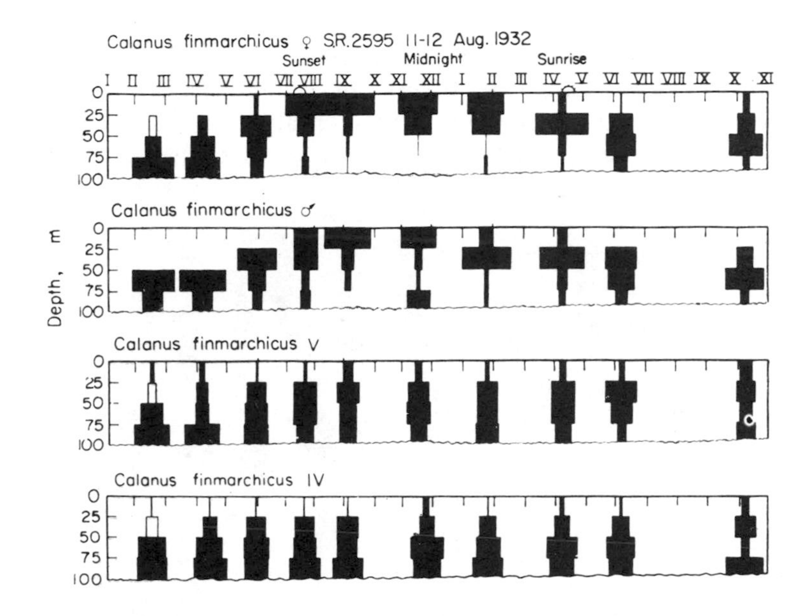

Fig. 5.3. Vertical distribution of different age groups of *Calanus finmarchicus* at a station south of Ireland (51°N; 8°W) (Farran, 1947).

Superimposed on differences in behaviour of the various stages there would also appear to be marked differences between the vertical distribution of generations. Many workers have noted daytime surface swarming of copepods, particularly *Calanus*, including adults and Stage V copepodites. This massing close to the surface often seems to be of a local and rather temporary nature. Russell has suggested that this behaviour, which is so different from the usual day depth distribution, may be due to physiological differences between generations. Off Plymouth, he showed that not only was there surface swarming, but that the distribution of adults followed definite changes with the season (Russell 1928, 1934c). In April, for example, the maximum for *Calanus* was at about 10 m depth, but the maximum sank gradually to about 20 m by June. This might be explained as an increase in depth with the gradually increasing penetration of light from spring to mid-summer. But about July, August and September there was a sharp rise towards the surface which clearly cannot be correlated with a light change. The individuals swarming in the later summer were exceptionally small and Russell believes that this generation produced late in the year had different physiological requirements, including a much higher light optimum. Stage V copepodites also vary considerably in their behaviour according to the particular brood. Thus Stage V of the overwintering generation may be very unresponsive to light changes and show little or no vertical migration. On the other hand, Stage V of the spring and early summer broods can perform more distinct vertical migrations. The unresponsiveness of the overwintering stage may be associated with the conservation of energy when food is relatively scarce.

Amongst others, Grainger (1959) has studied the life history of *Calanus hyperboreus* in the Arctic Ocean; he found that C II and III overwinter under ice at about 300–500 m depth and rise up to 150 m in summer only when food is more plentiful; adult males were always found slightly deeper than females (cf. Chapter 3). Miller (1974) reports that *Calanus cristatus*, *C. plumchrus* and *C. finmarchicus* show seasonal vertical migration and have an cxtended range in the Atlantic and Pacific (Chapter 3 page 425), while young *C. hyperboreus* migrate upwards 1–2 months after being spawned (Conover, 1962). Kosobokava (1978) found that the adult females of *C. glacialis* do not perform vertical migrations during winter, but remain in the less cold deeper layers.

Vinogradov (1970) refers to *Calanus propinquus* and *Calanoides acutus*, two Antarctic species in which spawning takes place near the surface where the young remain feeding actively; later stages descend to depths to overwinter. Andrews (1966) found the bulk of the population of the latter species below 500 m during July and August; from October a gradual rise takes place with animals mostly in the upper 250 m until March, when they descend to deep water (Fig. 5.4). Off tropical west Africa, *Calanus carinatus* produces eggs in the upper 10 m, where phytoplankton food is abundant for the young stages following upwelling (Binet and Suisse de Sainte Claire, 1975, cf. Chapter 3).

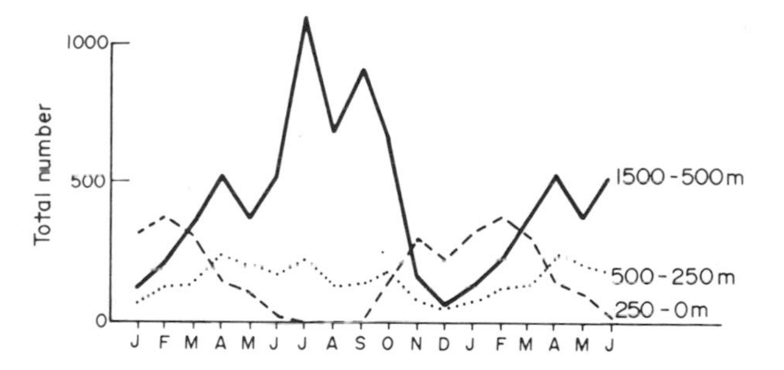

Fig. 5.4. Monthly seasonal variation in the average number of all stages of *Calanoides acutus* found at 3 depths (Andrews, 1966).

Amongst ostracods, Moguilevsky (1976) showed that ripe females and eggs of *Gigantocypris mulleri* were common in the 1200 m layer, while juveniles lived more shallowly at 800–1000 m. Angel (1979) records *Conchoecia daphnoides* adults at daytime depths of at least 600 m, with females showing active night migration up to 400 m and even above. The Stage IV juveniles of this large species remain at 50–200 m, whereas older Stages V and VI behave as adults, males being less active migrators.

The hyperiid amphipods *Primno macropa*, *Scina crassicornis* and *Vibilia armata*, exhibit the tendency for adults of both sexes to inhabit deeper water than juveniles

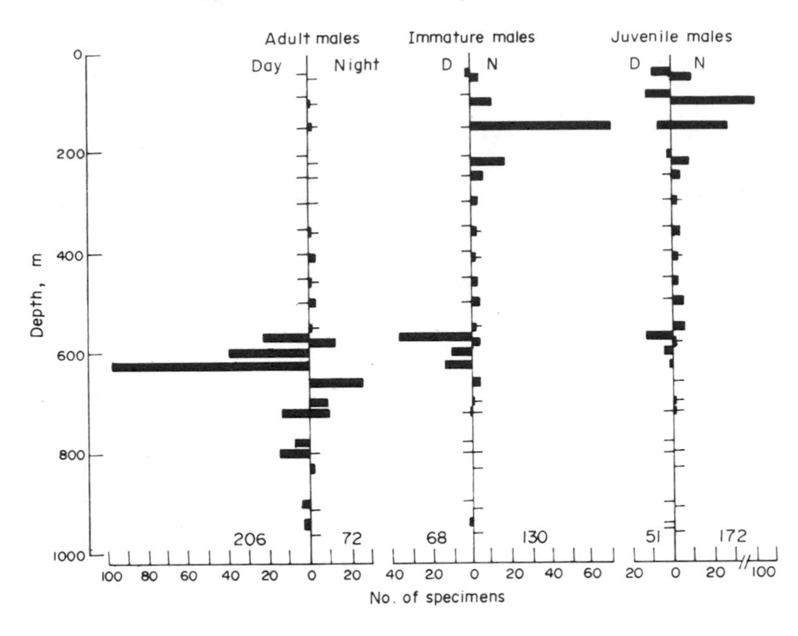

Fig. 5.5. Total catches of different age groups of male *Primno macropa* (Thurston, 1976).

(Thurston, 1976) (Fig. 5.5). This also applies to *Parathemisto japonica*, whose young of more than 5 mm in length remain in the top 100 m (Semenova, 1974).

Because of their importance as a zooplankton group, the euphausiids have been studied by many early workers (see Chapters 2 and 3) and also by Brinton, Mackintosh, Mauchline, Nemoto and Roger, amongst others. Komaki (1967) describes the pre-spawning swarming of euphausiids, which may coincide with the presence of cold water cores. The well-documented life history of *Euphausia superba* and of the northern krill *Meganyctiphanes norvegica* appears to show that eggs laid in the upper 150 m will sink in oceanic areas to a depth of 1500 m or near to the bottom in shelf areas before the nauplius hatches. Thereafter the development ascent commences. Mackintosh (1972) found that for *Euphausia superba* ascent occurred most commonly at the first calyptopis stage. There is clearly a large degree of variation between different species and localities. In the eastern tropical Pacific, the calyptopis stages of *E. diomedeae* occurred only in the upper 100 m by day and night; the large furcilias showed some degree of migration but did not reach the daytime 400 m levels of adults and large juveniles (Brinton, 1979). Vinogradov (1970) gives examples of the bathypelagic species *Bentheuphausia amblyops* and *Thysanopoda cornuta*, whose larvae have been caught more shallowly than 100 m, whereas the adults rarely penetrate up to the 700 m level and usually live more deeply (>2000 m–see Chapter 2).

In contrast, Omori (1974) quotes Heegaard (1969), who recorded that the large yolky eggs of the caridian decapod *Amphionides reynaudii* rise from the 2000-m spawning areas to the epiplanktonic layers, where larval development continues. The juveniles and adolescents descend as growth continues to the deep zone of the adults. The younger stages of *Hymenodora frontalis* and *H. glacialis* show the same pattern of descent with increase in body length (Vinogradov, 1970, Fig. 5.6). The ovigerous

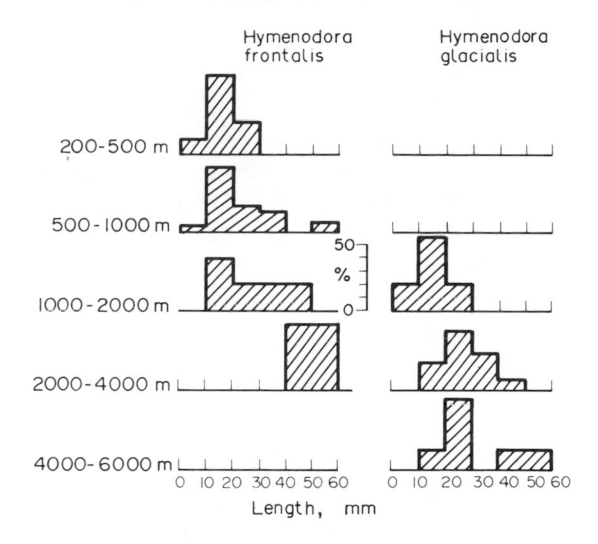

Fig. 5.6. Changes in the length of two species of *Hymenodora* with increasing depth (Vinogradov, 1970).

females of *Sergia lucens* ascend to spawn near the surface, where the larvae remain feeding on phytoplankton, and the return to deeper water commences at the zoeal stage (Omori, 1974, Fig. 5.7). As with many copepods and other zooplankton, the dietary regime appears to influence these ontogenetic migrations.

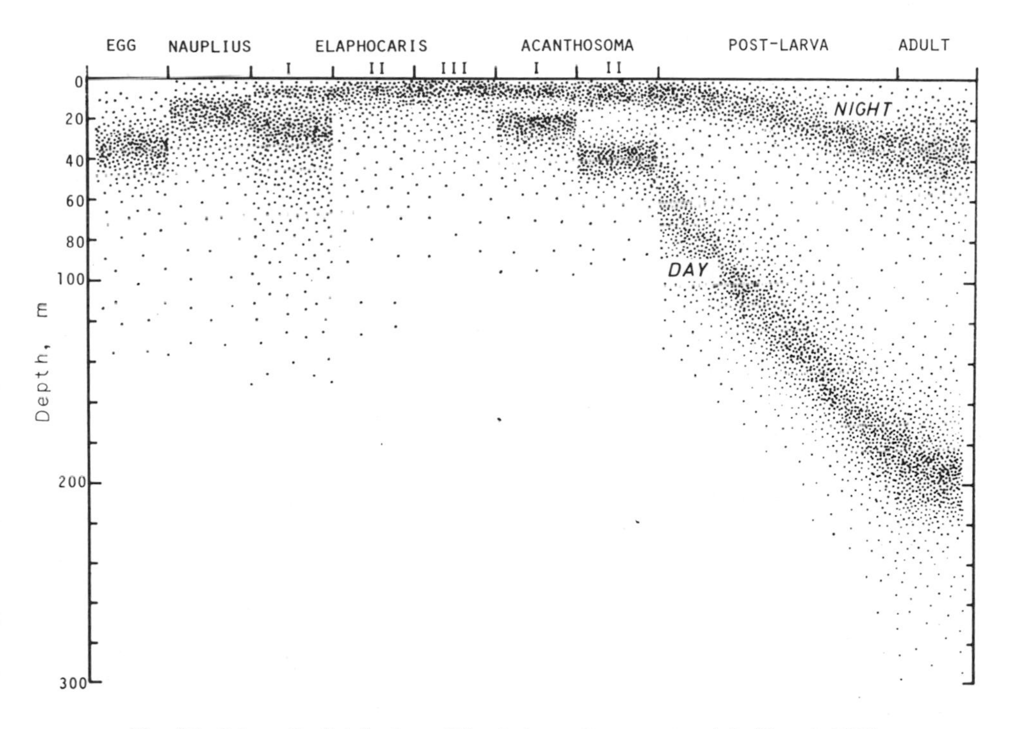

Fig. 5.7. Schematic distribution of *Sergia lucens* from egg to adult (Omori, 1974).

Several earlier workers have noted that for chaetognaths the younger stages live nearer the surface. For *Sagitta elegans*, Farran (1947) observed that the largest individuals were near the bottom by day and moved into the upper 30 m at night, with a tendency to spread down during darkness. Younger *Sagitta* were distributed nearer the surface by day and appeared to move right to the surface at nightfall. The classic work of Russell (1927, 1931, 1935) on the distribution of chaetognaths has already been quoted. In a shallow area of the Pacific at Suruga Bay, Japan, Nagasawa and Marumo (1975) investigated the populations of the neritic *Sagitta nagae* and found Stage I tended to frequent the surface by day, while Stages III and IV showed a higher percentage of individuals in the top 20 m at night. Vinogradov (1970) gives examples of *Eukrohnia hamata* and *Sagitta gazellae* spawning in deep waters and migrating upwards during the early developmental stages, while certain bathypelagic myctophid fish spawn at the uppermost limits of their migratory range. The behaviour of some young cephalopods has already been described in Chapter 3; Nesis (1972) recorded that the larvae of the mesopelagic *Abraliopsis affinis* occur in epiplanktonic layers and the larvae of the bathypelagic *Japetalla diaphana* are present throughout the water column.

Diurnal Vertical Migration

As early as 1912 Esterly showed that, off the Californian coast, *Calanus* would migrate from as great a depth as 400 m at its daytime level of concentration to near the surface at night. *Metridia lucens* migrated similarly from a daytime depth of about 300 m, approaching the surface at night. Other less abundant species of copepod also exhibited migration to the surface with the onset of darkness. Some six species, however, including *Euchirella galeata* and *Gaetanus unicornis*, were deep-living during the day, mostly below 400 m, and though there was clear evidence of a rise during the night to depths of less than 200 m, these species were never taken at the surface. The question of how far truly bathypelagic plankton was involved in diurnal migration was a difficult one, though the work of Murray and Hjort (1912) demonstrated that plankton down to 800 m would migrate towards the surface. Hardy and Gunther (1935) also showed that certain fairly deep-living copepods such as *Metridia* spp., *Scolecithricella minor* and especially *Pleuromamma robusta* in Antarctic waters, might migrate a distance of 400–500 m every day, moving towards the surface during the hours of fading daylight. Later work by Waterman, Nunnemacher, Chace and Clarke (1939) demonstrated a diurnal vertical migration by various decapods, mysids, euphausiids and amphipods living at depths down to at least 1000 m; it appeared that the range of movement for the different species exceeded 200 m and might amount to as much as 600 m.

More accurate sampling techniques using closing nets have enabled workers to assess the extent of migration in various groups, especially in oceanic areas. Rudyakov and Tseytlin (1976) estimate that about 13% of zooplankton biomass off the Canary Islands migrates more than 100 m, and cite copepods, ostracods, euphausiids with gonostomid and myctophid fish as the principal groups. Even amongst the microplankton of the neustonic layers, infusorians move down 10–20 m from the surface between 1400 and 1600 hours; tintinnids may also show slight changes in depth (Zaika and Ostrovskaya, 1972). It is, however, the migratory habits of copepods which have been most studied and documented by planktonologists.

Marshall and Orr (1955a), in their summary of the vertical migration and vertical distribution, have emphasized the complexity of the behaviour of *Calanus*. For example, in the Clyde Sea Area female *Calanus* are deep-living by day during late winter and early spring, though marked migrations occur at night. About April/May, however, there is a sudden rise and surface swarming occurs, after which the later brood tends to descend towards deeper levels. Thus the times of the year at which the generations approach the surface vary even between the Clyde and Plymouth (cf. page 492). In other areas the changes in depth pattern may be different from those observed in British waters. In polar waters Bogorov (1946) has observed *Calanus* swarming at the surface during the summer months of continuous light. Not only *Calanus* but all the dominant zooplankton species appeared to maintain themselves in the uppermost layers and to show no daily vertical movement to deeper water. In autumn, however, with alternating light and darkness all the major copepods, including *Calanus*, *Pseudocalanus*, *Microcalanus* and *Metridia*, performed pronounced diurnal vertical migrations, even from as deep as 100–200 m to near the surface. Only *Oithona* appeared unaffected. During the continuous darkness of winter the copepods maintained their level constant in deep water.

These results furnish yet another example of different behaviour patterns even in polar waters, for whereas Bogorov claims that *Metridia* spp. and *Microcalanus* were at the surface in summer, Ussing (1938), working off East Greenland, found that *Metridia longa* (and probably *Microcalanus*) occurred in the upper layers in winter but in deep water during continuous summer. He agrees that *Metridia* make extensive vertical migrations during later summer/autumn. Ussing's results indeed imply that *Calanus* on the one hand and *Metridia* (and *Microcalanus*) on the other represent two distinct types of behaviour. Digby (1954), however, suggests that a limited amount of vertical migration occurs with *Calanus* even during winter; both *Metridia* and *Calanus* exhibit migration, but the apparent difference in behaviour pattern arises from the fact that *Metridia* lives at a deeper level over the summer.

One of the most remarkable observations on differences in the vertical movements of a single species was that of Clarke (1934) in the Gulf of Maine. Clarke observed that *Calanus* and *Metridia* migrated from depths of over 120 m towards the surface at night, the migration for *Calanus* including adults and Stage IV and V copepodites. This pronounced movement occurred at a comparatively deep station in the Gulf where the depth of the bottom was of the order of 200 m, whereas at the same time over the shallow Georges Bank, Clarke found that *Calanus* was confined to approximately the upper 30 m and showed very little movement. *Metridia*, however, continued to perform well-marked vertical migrations. He attributed the difference in behaviour of *Calanus* to different generations occurring over the Georges Bank area and in the deeper part of the Gulf. With the remarkable circulation of the calanoid population in the Gulf of Maine which Redfield (1941) has investigated, it is quite possible that different broods could occur in the two regions. The variable nature of diurnal vertical migration is also borne out by the observations made by many workers (cf. Farran, Nicholls) that even within the same population of zooplankton, individuals belonging to the same species, sex and stage may exhibit different vertical migration patterns. It is likely that differences in physiological state due to such factors as ripening gonads and food are partly responsible for such variations in behaviour. Marshall and Orr (1960) have demonstrated the variability of diurnal vertical migration from year to year. The ripeness of the gonad is certainly a factor in migration, ripe females migrating more actively.

Marshall (1949), investigating the depth distribution of *Microcalanus* in Loch Striven (Clyde Area), found, as Ussing had, that all stages of the calanoid were absent from the surface during spring and summer, though it possibly approached the upper layers during winter. The other small copepods in Loch Striven all showed a similar depth distribution—from April until June they were in the upper layers, but they descended to a deeper level during July and August. *Pseudocalanus* and *Acartia* exhibited this change in level most clearly; in *Centropages* and *Temora* the descent was slightly less marked, and in *Oithona* less still. As with *Calanus*, the younger stages of all species tended to be nearer the surface.

In reports from the Discovery Expedition, Hardy and Gunther (1935) stated that some Antarctic zooplankton species, notably *Calanus acutus*, *Rhincalanus gigas* and *Eukrohnia* apparently do not perform diurnal vertical migrations, but show a clear seasonal change in depth distribution. Other Antarctic species take part in pronounced diurnal migrations; these include *Pleuromamma robusta*, *Metridia*, *Pareuchaeta antarctica*, *Scolecithricella minor*, *Euphausia triacantha* and *E. frigida*. In contrast, *Salpa fusiformis* and the amphipod *Vibilia antarctica* show some evidence of migration, though the vertical movement is slower and occurs later in the day. Again, *Drepanopus pectinatus* migrates very rapidly to the surface at night from a depth of at least 200 m and another copepod, *Calanus simillimus*, though not showing such extensive migrations, regularly and rapidly moves from a daytime depth of some 100–150 m right to the surface at night. The amphipod *Parathemisto* also appears to migrate from about 80–100 m to the surface, though the migration is somewhat slower. There appear to be all grades of migration in very high latitudes, and behavioural patterns are seasonally modified during the long periods of winter darkness. In the central Arctic Ocean *Calanus hyperboreus* and *C. glacialis* show little vertical change, though adult females of *C. glacialis* perform slight migrations in summer and autumn (Kosobokova, 1978). Similarly, Andrews (1966) found that diurnal movement of *Calanoides acutus* occurs only during the summer months in the Antarctic; the copepods remain below 500 m in winter (cf. Chapter 3, Fig. 5.8). Such seasonal variations in activity of copepods were noted by Moraitou-Apostolopoulou (1971). *Calanus tenuicornis*, a common species in the Saronica Gulf during winter, performed daily migrations from 100 m to the surface in the cold period, while animals remained at depths of 100–200 m in the summer and did not migrate. *Mecynocera clausi*, *Lucicutia flavicornis* and *Oncaea mediterranea* showed greatest activity in the autumn and remained below 100 m in the warmer months.

In deep water areas near the Sea of Japan *Metridia pacifica* shows a high biomass at about 2000 m with a long migratory range at night (Vinogradov and Sazhin, 1978). In a study of bathypelagic copepods in the tropical Pacific, Gueredrat and Friess (1971) list *Gaussia princeps* as a strong migrator over a 460-m range, with *Megacalanus princeps* and *Pareuchaeta hanseni* covering shorter distances (340 and 130 m respectively). In reviewing distributions of cyclopods in the upper 300 m of the west equatorial Pacific, Tsalkina (1972) found only two species of *Oncaea* (*O. mediterranea* and *O. conifera*) which showed pronounced vertical movement (*ca.* 125 m). In the north-east Indian Ocean, however, she records that the total population nucleus of Oncaeidae shifts upwards at night. Maximum migration occurs in the halistic zone of the central waters.

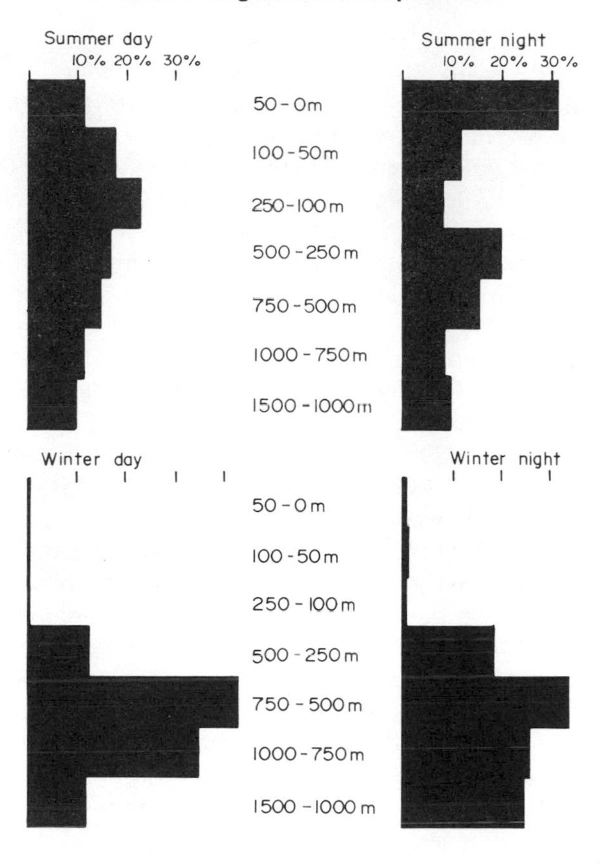

Fig. 5.8. The percentage distribution of *Calanoides acutus* in summer (December–March) and winter (June–September) months. The seasonal migration shows in the comparison of the winter and summer histograms; the diurnal migration is evident only in the summer histograms (Andrews, 1966).

In his studies of several groups off the Canaries, Rudyakov (1979) reported a variation in amplitude of migration for copepods: *Euchirella* (500 m), *Pleuromamma xiphias* (324 m) and *P. robustus* (190 m) were the main species noted.

It is interesting to find that change in depth distribution somewhat similar to that for copepods is seen for other zooplankton. Russell (1931), for example, investigated the distribution of *Sagitta*. He suggests that for both *Sagitta elegans* and *S. setosa* there is a tendency for the animals in daytime to move somewhat deeper from about April to June from a depth of some 15–20 m to well below 25 m. This may again be explained as a reaction to the increasing light intensities from spring to summer. In July/September, however, the sagittae rose considerably, entering zones of light intensity which they previously avoided, and it may be that this is another reflection of a difference in physiological state (*vide infra*). Bigelow and Sears (1939) and Pearre (1973) also noted that *S. elegans* frequented the surface at night and sank by day—a pattern which varied according to the time of year and age of animals and is repeated by *S. nagae* in shallow Japanese neritic waters (Nagasawa and Marumo, 1975). In the tropical Pacific, *S. enflata*, *S. bipunctata*, *S. regularis* and *S. serratodentata pacifica* all perform short migrations which Kolosova (1972) believes are influenced by light, though the first two species commence the upward movement as early as 1400 hours. Russell (1928) had earlier recorded the migration of *S. bipunctata.* Roe (1974) observed that off the Canary Islands *S. enflata* did not migrate actively, but *S. serratodentata* moved from the 250 m layer at night while juvenile *Eukrohnia* spp. came up to that depth.

Some of the records of the less frequently observed reverse migration concern neritic cladocerans. In shallow areas of the Japan Inland Sea dense surface populations of *Podon polyphemoides* and *Evadne tergestina* migrate at night to the bottom or just above it (Onbe 1977); this pattern also occurred for the former species in Chesapeake Bay (Bosch and Taylor, 1973b).

In common with copepods, euphausiids include some of the strongest migrators with great diversity in habit. Mauchline and Fisher (1969), Roger (1974b) and Brinton (1979), amongst many others, give excellent reviews and only a few examples will be quoted here. Some of the early records may now be of doubtful value because of the lack of closing nets. Because of swarming, Marr (1962) found it was not possible to define precisely the vertical movements of *Euphausia superba* but suggested that there was no obvious rhythm, though there was an accumulation of swarms within 5 m of the surface between 2200 and 0400 hours. By day, few catches were obtained below 160 m. Mauchline (1960) and Wiborg (1971) give details of the vertical movements of *Meganyctiphanes norvegica* and correlate these with size classes. From a large number of samples taken in three zones of the south tropical Pacific Ocean, Roger (1974b) was able to group some of the many species captured into three categories according to the distances migrated (Table 5.1) and emphasized that the range of migration was not influenced by the presence of a thermocline (*vide infra*).

Thurston (1976a, b) gives details of amphipod migrations in the Canary Islands area. The range was usually less than 100 m, but was greater amongst some of the deeper-living species; the gammarids *Cyphocaris anonyx* and *C. challengeri* were found about 250 m above their daytime depths of 500 and 800 m respectively. Of the hyperiids, *Scina crassicornis* and *Vibilia armata* might travel upwards for more than 400 m, although the average change was about 250–300 m. The dominant species for that area, *Primno macropa*, showed a range varying from 100–400 m, the deeper-living animals, especially males, moving farther from the daytime level of *ca.* 600 m up to 40 m (Fig. 5.5; cf. also Rudyakov, 1979).

The decapods have a wide vertical distribution from the surface to about 6000 m (Vinogradov, 1970; Omori, 1974), and include some extensive migrators. Amongst these are *Sergia robustus*, *Gennadas valens* and *Funchalia villosa*, with a change of depth from 450 m to 70 m at night (Foxton, 1970b) (Table 5.2; Fig. 5.9). Donaldson (1975) emphasizes the variation in migratory amplitude of sergestids from *S. splendens* (825 m) to *S. japonicus* (<100 m) and suggests that this is linked to a competition for food.

Amongst the thecosomes it is interesting to note that in a wide-ranging species such as *Limacina inflata*, large scale migrations of some 200–300 m across the pycnocline have been recorded by Haagensen (1976) for the Caribbean Sea. The temperature difference may be as much as 20°C, with about a 4‰ salinity variation. The juveniles, however, usually remained in the warm surface water. This was also noted off Madagascar by Frontier (1976). Adult *Limacina bulimoides* showed a similar pattern, with 97 % of the population rising from a daytime level of 225 m to about 50 m at night. Other migrators

Table 5.1. Mean estimated levels of maximum concentrations by day and night in three zones of the South Pacific. Euphausiids are classed into groups according to their depth distribution (Roger, 1974)

Long migrators (Amplitude of migration ⩾ 200 m)			Short migrators (Amplitude of migration 50–150 m)			Non-migrators (or amplitude of migration ⩽ 50 m)	
	Day	Night		Day	Night		
Thysanopoda tricuspidata	450	100	Mesopelagic			Epipelagic	
Thysanopoda monacantha	600	250					
Thysanopoda pectinata	600	300	*Stylocheiron abbreviatum*	300	200	*Styocheiron suhmii*	150
Thysanopoda obtusifrons	600	250				*Stylocheiron microphthalma*	150
Thysanopoda aequalis	550	100	Deep			*Stylocheiron carinatum*	150
Euphausia fallax	350	150	*Nematoscelis tenella*	500	350	*Stylocheiron affine*	200
Euphausia diomedae			*Nematobrachion flexipes*	500	400		
Euphausia brevis	450	50	*Nematobrachion sexspinosus*	500	400		
Euphausia mutica			*Nematobrachion boopis*	600	500	Mesopelagic	
Euphausia gibba	600	150	*Thysanopoda orientalis*	600	500	*Stylocheiron longicorne*	300
Euphausia tenera	450	100	*Thysanopoda cristata*	600?	400?	*Stylocheiron elongatum*	350
Nematoscelis microps	500	250	*Bentheuphausia amblyops*	900?	700?		
Nematoscelis gracilis	550	350					
Nematoscelis atlantica	500	300					

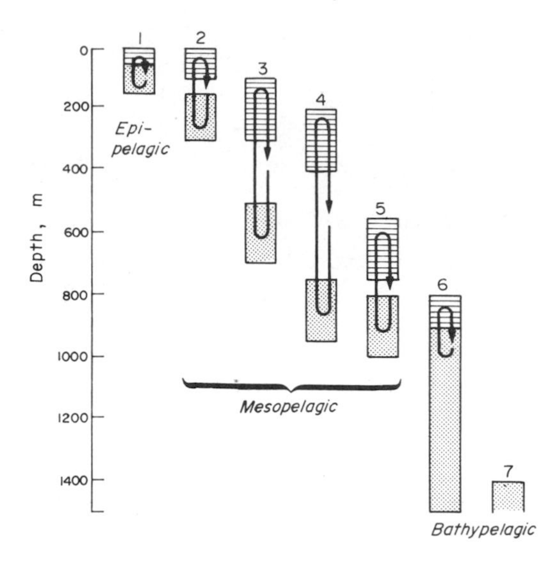

Fig. 5.9. Schematic illustration of the diurnal migration of pelagic shrimps. Dotted and hatched areas indicate the depths of the main day and night concentrations (Omori, 1974).

Table 5.2. Groups of decapods (adults and adolescents) in relation to their vertical migration (Omori, 1974)

(1) Main part of population in upper 150 m layer both night and day.

Neritic and epipelagic species including all members of the genera *Acetes*, *Peisos* and *Lucifer*.

(2) Surface–100 m (night) to 150–300 m (day).

Upper mesopelagic species: *Sergestes arcticus*, *S. similis*, *Sergia lucens*, *Pasiphaea sivado* (Risso).

(3) 100–300 m (night) to 500–700 m (day).

Lower mesopelagic species: *Gennadas incertus*, *G. propinquus* Rathbun, *Funchalia villosa*, *Sergestes armatus* Krøyer, *S. atlanticus* H. Milne Edwards, *S. corniculum* Krøyer, *S. erectus* Burkenroad, *S. orientalis* Hansen, *S. pectinatus* Sund, *S. sargassi* Ortmann, *S. vigilax* Stimpson, *Sergia prehensilis*, *S. scintillans* (Burkenroad), *Acanthephyra quadrispinosa*, *A. sexspinosa* Kemp, *Systellaspis debilis*, *Parapandalus richardi*, *P. zurstrasseni* Balss.

(4) 200–400 m (night) to 750–950 m (day).

Lower mesopelagic species (strong migrants): *Gennadas bouvieri* Kemp, *G. parvus*, *G. valens*, *Sergia inequalis* (Burkenroad), *S. gardineri* (Kemp), *S. splendens* (Sund), *Acanthephyra purpurea*.

(5) 550–750 m (night) to 800–1000 m (day).

Lower mesopelagic species (short-distance migrants): *Gennadas capensis* Calman, *G. elegans*, *Sergia robustus*, *Acanthephyra acanthitelsonis* Bate, *A. pelagica*.

(6) 800–1400 m (night) to 900–1500 m (day).

Bentheogennema borealis, *B. intermedius* (Bate), *Sergia japonicus*, *Petalidium foliaceum*, *P. obesum* (Krøyer), *Parapasiphae sulcatifrons* (Smith), *Acanthephyra prionota* Foxton, *Hymenodora frontalis*.

(7) Main part of population living lower than 1400 m both night and day.

Acanthephyra stylorostratis (Bate), *Hymenodora glacialis*, *H. gracilis*, *Physetocaris microphthalma* Chace, *Amphionides reynaudii*.

include adult *Limacina lesseurii*, *Cuvierina columnella*, *Diacria trispinosa*, *Cavolinia inflexa* and *Styliola subula*. *Creseis acicula* showed a slight upward movement from 65 m towards the surface at night, although some adults remained there over the 24 hours at two stations marked by upwelling near the surface. In contrast, both forms of *Creseis virgula* performed reverse migrations; for *C. virgula conica*, 55 % of adults inhabited the surface during the day, but this was reduced to 2.5 % at night.

Nesis (1972) gives a summary of the vertical movements of some cephalopods. The epiplanktonic squids *Symplectoteuthis oualaniensis* and *Onychoteuthis banksi banksi* migrate a short distance from 100–200 m to the surface at night, while their daytime level is occupied by mesopelagic species such as *Abrialopsis affinis*, *Leachia pacifica* and *Pterygoteuthis giardi*, rising from 600–700 m. A similar pattern, though from shallower maximum levels, was recorded for the last species and *Pyroteuthis margaritifera* in the Atlantic at Ocean Acre by Gibbs and Roper (1970). These data refer to adults; behaviour of larvae is described in Chapter 3. The mesopelagic squid *Histioteuthis heteropsis* migrates from 700 m to above 400 m at night off California (Belman, 1978).

Although some siphonophores have a fairly extensive vertical distribution, there is evidence of vertical migration. Barham (1963, 1966) suggests that the deep scattering layer is associated in part with physonectid siphonophores, essentially *Nanomia bijuga*. This siphonophore occurred at moderate depths of the order of 260–440 m off California and the concentrations, which were at times quite considerable, showed definite vertical migration patterns which appeared to be associated with the deep scattering layer. Barham comments on the strong active swimming of these *Nanomia*; with gas regulation for ascent and descent, their capabilities for vertical migration are marked. Two other Physonectids (one probably *Halistemma rubrum*) occurring at shallower depths also showed diurnal vertical migrations. In a study of zooplankton in the top 300 m off Bermuda and in the Florida Current, Moore (1949, 1953) demonstrated a correlation between the daytime depths of certain siphonophores and light intensity or temperature. Pugh (1975) comments on the long (*ca.* 200 m) migratory range of two species, *Amphicaryon acauli* and *Hippopodus hippopus*, having weak swimming ability, which may represent a continuous cycle associated with temperature because the distances recorded by Moore in the Caribbean are shorter than off the Canaries, where Pugh's observations were made.

The pattern of slow migration is shown in Fig. 5.10 for three species. *Chelophyes appendiculata* probably has a maximum daytime depth about 250 m, while *Hippopodus hippopus* may exist more deeply and reaches the top 50–100 m at night. *Vogtia glabra* and *Halistemma rubrum* move from a deeper area upwards through the 250 m zone, in which greatest numbers were found in hauls at sunrise and sunset. The eudoxid stage of *Ceratocymba sagittata* is a rapid and long distance migrant moving from 300–400 m depth up to the near surface (10–25 m) at night. Casanova (1977) found that *Abylopsis eschscholtzii* and *Lensia fowleri*, while occurring at the surface at all times, dispersed through the entire water column down to 700 m by day. He does not confirm Pugh's observations for the other species and attributes this possibly to the different types of net used in capture.

The movements of massive groups of colonial gelatinous organisms such as salps may contribute markedly to the food requirements of bathypelagic and benthic animals; Wiebe, Madin, Haury, Harbison and Philbin (1979) found large numbers of

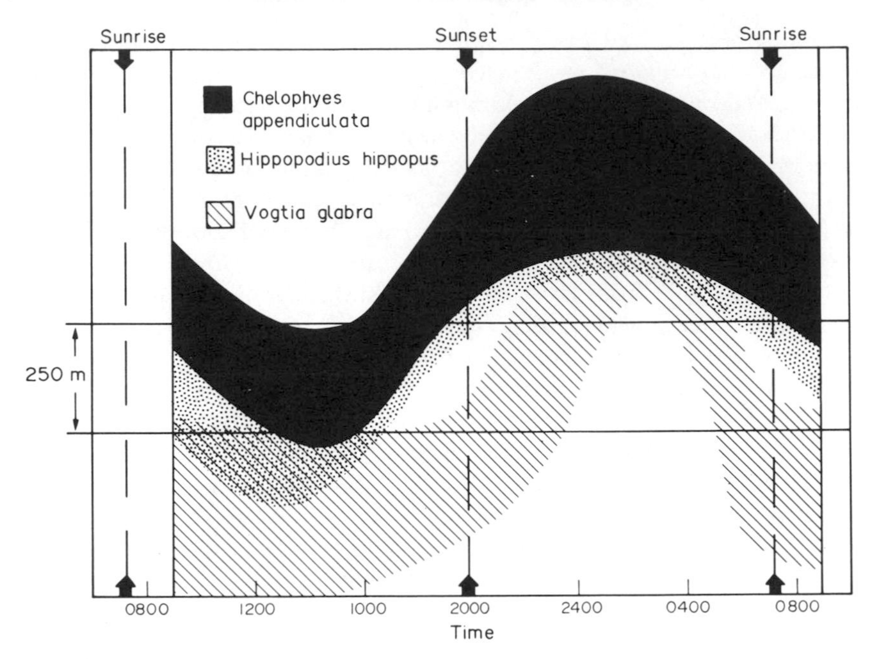

Fig. 5.10. Possible pattern of diel vertical migration in 3 species of siphonophore (Pugh, 1977).

Salpa aspersa travelling about 800 m between day and night locations and claim that this is one of the longest shifts recorded; further details are given in Chapters 1 and 6.

Kawaguchi (1969) describes how, amongst micronektonic fish, strong swimmers of genera such as *Tarletonbeania*, *Protomyctophum*, *Hygophum*, *Centrobrachus*, *Symbolophorus* and *Myctophum* come up to the surface layers at night, while hunting in the 200–400-m zone by day. Genera such as *Diaphus*, *Diogenichthys*, *Benthosema*, *Lampanyctus* and *Gonostoma* are examples of mesopelagic fish which rarely rise above 20–100 m at night.

It is necessary now to consider more precisely the mechanisms which may be responsible for diurnal vertical migration. Although it is generally conceded that light is probably the most important factor, the precise way in which the light exerts its effect and how far other factors may be concerned, has not yet been determined. One of the earlier suggestions was that of Loeb, who postulated that for nauplii of *Balanus* there was a movement towards light of low intensities but a reversal of phototactic sign with increasing light intensity occurred, so that the nauplii moved away from intense light. He thus explained vertical migration as a phototaxis with a reversal of phototactic sign. Later Loeb considered that geotaxis might be important, the animals being negatively geotactic in weak light or in darkness so that they would tend to move surfacewards at dusk and at night. Parker (1902), working on *Labidocera*, also considered that geotactic reactions were important.

In watching plankton swimming vertically in a column of water it is relatively easy to see how the animals sink as soon as the swimming movements cease and it must be remembered that constant upward swimming is necessary for the animals to maintain themselves at their selected depth. As a development of this observation, Ewald (1912), working on *Balanus* larvae, suggested that the vertical migration could be explained on the grounds that in weak diffuse light the larvae swam steadily upwards, but that in intense light swimming movements were inhibited and the animals sank to a greater depth. Thus, by constantly swimming up and then passively sinking, the animals were maintained roughly at a constant depth during the day, but rose towards the surface in the evening. Ewald believed that such vertical movements might apply fairly widely to crustacean plankton. This type of behaviour is a particular case of the well known kineses, behavioural patterns in which variations in the intensity of a non-directional stimulus cause variations in speed of movement.

Ewald observed that small increases in light intensity temporarily increased swimming speed. Many workers, therefore, have attempted to explain the movements of plankton on the grounds that the movements are rapid at low light intensities, so that the animals will tend to swim upwards. Ussing (1938), for example, suggests that copepods such as *Calanus* need the constant photokinetic stimulus of low light intensities. Cushing (1951) quotes several cases of a positive kinetic effect of light on zooplankton, though high intensities may cause inhibition, as may *rapid* increases in intensity. He also suggests from a later study (Cushing, 1955) on the behaviour of *Temora* and *Pseudocalanus* allowed to aggregate in tubes, that the results are more consistent with photokinesis being an important factor in migration. At the "optimum zone" the copepods apparently move most slowly; above and below this light intensity, movement increases. Although kinetic effects may play a part in vertical migration, as Cushing admits, other reactions are also involved. Even though some of the directionally orientated movements observed in laboratory experiments on zooplankton may prove to be artefacts, the movements including downward swimming would appear to be natural directed migrations.

Boden and Kampa (1967), studying the location of sound scattering, concluded that except in areas where a strong thermocline or oxygen deficiency layer occurred, light was the factor which influenced vertical movements. Russell (1931, 1933), Clarke (1933, 1934), Moore (1949, 1953), Vinogradov (1970) and Longhurst (1976) are some of the many planktonologists who have published data on migration in response to light changes, both under natural conditions and in the laboratory. According to Bogorov (1948), *Sagitta elegans* occupies a fairly definite depth stratum in the Barents Sea related to the optimum light intensity, but there is no vertical migration with the constant light in summer. It would be interesting to see if this species shows any migration during the brief spring and autumn seasons when there is a time of light day and dark night periods. He showed that at this time of year some species which tended to remain at one level during either the long summer or winter did migrate vertically. On the other hand, he shows that in the White Sea there is a migration of *Sagitta elegans* which he attributes to tidal flow rather than to light. From laboratory and field observations, Foxon (1940) interpreted that *Hemimysis lamornae* exhibits a negative geotaxis which is reversed in the presence of light; the mysid shows a stronger tendency to swim during darkness and remains on the bottom in the light.

Personal observations of other mysid species would appear to indicate that these animals continue swimming or foraging along the bottom under any type of illumination. Mauchline (1965, 1969) records the marked and rapid reaction shown by some larval stages of *Meganyctiphanes norvegica* and *Thysannoessa raschii* to changes in light intensity caused by bright sun followed by cloud cover, and quotes Lewis (1954), who believes that light intensity controls vertical migrations. The zooplankton of the deep scattering layer are maintained in regions of rich food supply in response to light; Isaacs, Tont and Wick (1974) found a correlation between displacement of organisms and transparency—the animals spend less time in higher regions and remain longer in regions of low transparency, which could be associated with high phytoplankton biomass.

The precise way in which the light stimulus acts on the animals is, however, very difficult to ascertain. It has already been noted that photokinetic effects may play some part, but orientated phototactic movements appear to be true of many zooplankton species.

There seems little doubt that true phototactic movements occur in some zooplankton. Rose (1925) considered that most zooplankton were negatively phototactic to high daylight intensities which they would encounter near the surface, but they became positively phototactic to lowered intensities deeper down. Thus one might think of a plankton animal as being more or less indifferent to light at its optimum intensity. If such an animal wandered upwards, encountering increasing intensities, it would become negatively phototactic, whereas if it descended into the deeper layers, it would become photopositive and would move under the influence of phototaxis towards its normal depth layer.

One of the simplest explanations of animals aggregating in an optimum zone is the reversal of phototactic sign with change of light intensity. Spooner (1933) observed that various species of the zooplankton moved in the direction of the light rays either positively or negatively, but this was true irrespective of any accompanying changes in intensity. Clarke (1933, 1934) considered that it was the rate of change of light intensity rather than the absolute magnitude of illumination which was the factor in inducing migration, a view also held by Johnson (1938) and Bary (1967). Any reduction in intensity tended to render the animals photopositive. In this connection, Johnson (1938) found that *Acartia clausi* was normally negatively phototactic and that under constant light intensity or during increasing light intensity the negativity was maintained; with a reduction in light intensity the animal became temporarily photopositive. This, however, contrasted with later results obtained by Johnson and Raymont (1939) on *Centropages*, in which no migration away from a light source could be produced either with increasing the light or with constant but relatively high light intensities. Many workers (e.g. Esterly, Clarke) have suggested that the real response is geotactic; that the changing light intensities effect geotactic responses of the animals rather than causing phototactic reactions. Negatively geotactic movements are induced by fading light intensities, whereas strong illumination or increasing light causes a positively geotactic movement, so inducing downward migration.

A negative geotaxis in darkness would, of course, be a useful mechanism for retaining zooplankton in the uppermost layers during the night. Many investigators have obtained results on diurnal vertical migrations which apparently show a departure of the zooplankton from the surface during darkness. This "midnight sinking", as it has been termed, has been explained by Russell as due to the random wanderings of the animals in the absence of any light stimulus. Cushing also believes that midnight sinking is common; on a photokinesis theory, movement is negligible and the animals may sink. Where sinking occurs and the plankton appears to become more or less evenly distributed during the night, a rise towards the surface may follow at, or just before, dawn. This "dawn rise" has been regarded as a light reaction by the animals to the very low dawn light intensity—"the plankton takes up its optimum". Many examples from plankton investigations seem to show midnight sinking and dawn rise (cf. Kikuchi, 1930). To some extent a fall in level at night may be due to an active downward migration. If animals migrate to the surface from great depths where the light intensity is extraordinarily low, the intensity at the surface in bright moonlight may be too high and they may go deeper during the time of most intense moonlight. At the same time, some species appear to remain truly aggregated at the surface during the dark, and there is no tendency to spread downwards. It is difficult to envisage any stimulus, other than a negative geotaxis, which is responsible for such a sharp distribution. With the coming of light, of course, the geotaxis is reversed or over-ridden by a negative reaction to light.

Numerous workers have attempted to elucidate the behaviour of zooplankton by means of laboratory experiment. One of the problems is that the results of such experiments are always complicated (sometimes almost vitiated) by the difficulty of maintaining zooplankton animals in good physiological condition in the laboratory. As well as the intrinsic factors of age and brood, which appear to cause marked variations in behaviour, animals brought into the laboratory probably do not show their normal reactions. A zooplankton species may show its main centre of vertical distribution at a certain depth but with a number of individuals scattered at other levels; thus the depth at which the animals are captured for experiment may modify the results.

Another complicating factor is that zooplankton species probably exhibit adaptation to light intensity. Field experiments by Russell indicated that older specimens of *Sagitta* were more sensitive to light in the early morning after a prolonged period of darkness; during the day they became adapted to higher light intensities. *Calanus* also appeared to show increasing adaptation throughout the hours of daylight. In laboratory experiments, therefore, the state of adaptation must be considered (cf. Clarke, 1930; Johnson, 1938). Often too, the light intensities employed in the laboratory are excessively high and outside the animals' normal range; plankton animals are frequently harmed by crowding in experimental vessels. Thus the results of laboratory investigations must be accepted with caution. The ingenious experiments of Hardy and Paton (1947), in which copepods were allowed to swim in closing chambers operated in the sea, and the further experiments by Hardy and Bainbridge (1954) with the continuous wheel, point the way to new types of experiments which may be more properly related to natural conditions.

As an example of the differences seen in behaviour in the laboratory and in the field, Schallek (1942, 1943) stated that *Acartia tonsa* in diffuse light exhibited movement towards the surface at sunset or in darkness, probably as the results of a geotactic response; with diffuse illumination of higher intensity they ceased swimming and tended to descend. However, in the laboratory, with highly directional light, *Acartia* was very strongly positively phototactic and swam constantly towards a light source. One of the many problems involved in experiments is to maintain the zooplankton feeding under more or less natural conditions. It seems very likely that the reactions of zooplankton to environmental factors such as light will be strongly conditioned by the amount and perhaps even by the type of food which they have taken.

Photosensitivity has been recorded for some Cnidaria; for example, *Nanomia*,

Halistemma rubrum and *Forskalia edwardsii* all react to strong illumination (Mackie and Boag, 1963), and although the calcyophoran *Hippopus* shows no co-ordinated response, aggregation of animals in light may indicate orthokinesis. In a study of the phototropic reaction of *Anamalocera patersoni*, Champalbert and Gaudy (1971) found that this was more pronounced in the mornings than evenings and that the sexes did not respond to the same extent. Variations in light intensity did not greatly affect orientation in males, but females showed a stronger positive reaction in reduced light.

Most of the earlier work on diurnal vertical migration concerned the more shallow-living plankton, but evidence has accumulated for such behaviour among deeper-living animals. Work on movements of the deep scattering layers of the oceans has shown that at least some of these layers are made up largely of planktonic animals. In discussing the effect of light on migration, it is necessary to consider how far light intensities can change in deep water. In Volume 1 it was noted that even in the clearest tropical waters, light is ineffective for photosynthetic activity below a maximum depth of about 150 m (Jerlov, 1968, 1976). However, light can penetrate much farther into the sea and may therefore be of great importance in relation to vision and to the photic reactions of animals generally, quite apart from photosynthesis (Fig. 5.11).

Some earlier observations using photographic plates suggested that an appreciable quantity of light was present at depths exceeding 500 m. For instance, Murray and Hjort showed that photographic plates would show some degree of blackening even at depths of 1000 m in oceanic waters, though they failed to detect any effect on plates exposed at depths of 1700 m. Observations from diving bells also showed that very small amounts of light could be detected at several hundred metres depth. Beebe (1935) found that complete darkness to the human eye was reached in Bermuda waters at a depth between 500 and 600 m. How far light produced by luminescence from marine animals could affect the general distribution and amount of light in the depths of the ocean was largely unknown.

Studies by Clarke and Wertheim (1956) and Clarke and Backus (1956) gave information on the penetration of light into deeper waters and on the role of luminescent light. Observations made by an ingenious bathyphotometer which combines a depth meter with a photomultiplier capable of detecting illumination at an intensity of about 10^{-12} that of full sunlight, suggest that the deeper ocean water between depths of 100 and 600 m has a relatively uniform and fairly high transparency. Some idea of the relatively great attenuation of light in the depths may be gained from Fig. 5.11.

Clarke and his co-workers found that below a depth of about 200–300 m the amount of light registered by the bathyphotometer was subject to relatively sharp fluctuations. These are believed to have been due to flashes of luminescent light emitted by marine animals. At greater depths the amount of bioluminescent light appears to have been greater than that due to the penetration of daylight. Further observations showed that this background of light in deep water could be resolved to some extent into individual flashes of relatively high intensity. These flashes may be obtained in relatively shallow water, but also at greater depths, down to at least 400 m. The record obtained during daylight hours shows a maximum at about 70 m depth in the euphotic zone and the intensity of the flashes at shallow depths is actually much higher than at deeper levels owing to the fact that these light intensities must be reckoned against the ordinary illumination due to daylight. There is a clear smaller maximum at a depth approaching

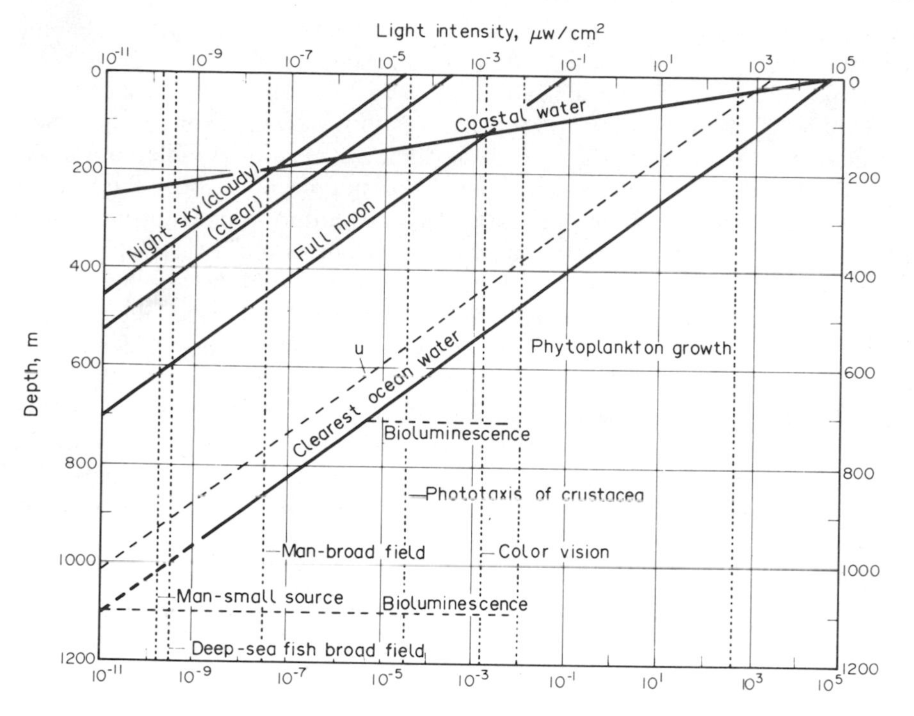

Fig. 5.11. Schematic diagram to show the penetration of sunlight into the clearest ocean water ($K = 0.033$) and into clear coastal water ($K = 0.15$) in relation to minimum intensity values for the vision of man and of certain deep sea fishes. U indicates approximate value of upward scattered sunlight (Clarke, 1970).

300 m which presumably is due to organisms of the deep scattering layer which occur at about this depth during the daytime.

Earlier observations from diving work suggested that bioluminescence could occur at all depths, but the work of Clarke and his collaborators indicates that variations in intensity of bioluminescence may occur with depth and with different types of water. Cnidarians, euphausiids (*E. pacifica*), mesopelagic decapods (*Oplophorus*, *Systellaspis*, *Sergestes*), salps (*Pyrosoma*) and myctophid fish may be some of the more important groups associated with bioluminescence. Later work by Clarke and Hubbard (1959) has revealed that apart from a relatively shallow zone exhibiting maximal bioluminescent flashing, another maximum zone occurred at a depth of about 900 m. This is of especial interest as this depth approximates to a zone of maximum zooplankton abundance found by Leavitt in the same oceanic area. Clarke and Hubbard found that flashing tended to diminish in greater depths, but they were able to obtain records of bioluminescent flashes even down to depths of 3750 m. Working in shallower waters off San Diego, Kampa and Boden (1956) were able to demonstrate that the frequency of bioluminescent flashing apparently varied with the diurnal vertical migration of the deep scattering layer.

Experiments quoted by Herring (1977) have shown that certain fish, squids, and crustaceans possess a photophore filter pigment which modifies the emission spectrum so that it approximates to that of ambient daylight. This would enable the animal to alter the intensity of its bioluminescence and to move up or down in the water column independently of ambient light changes. However, as has already been noted, many organisms follow a pattern close to that of a particular isolume.

The migrations of deeper-living plankton in relation to light intensity were examined by Clarke and Backus (1956). Observations were made on the movement of the isolumes with the changing incident daylight over a considerable portion of the day, and at the same time the pattern of the vertical migration of a deep scattering layer was also determined. The animals forming the deep scattering layer appeared to move with the changing light intensities during the day but resolved themselves into two different populations. During the afternoon these populations appeared to move upwards faster than the changing light intensities and in fact reached illuminations in one case more than a hundred times brighter than those they had experienced previously at midday.

These observations would suggest that while changing light intensities are of significance in migration, an optimum zone of light intensity is not the only factor determining the vertical movement of deep water animals of the scattering layer. Either the animals must adapt during the day to considerably different light intensities or the rate of change of light intensity may trigger off other behavioural patterns which cause the animals to move into higher light intensities during the latter part of the day. Bary (1967) believes that organisms move as a result of rate of change from ambient intensity and not optimal intensity. He found a linear relation between day depth scattering and speed of migration of the organisms. This proved faster in clear ocean water than in near-shore turbid areas.

Among earlier workers, Esterly (1917, 1919) suggested that internal rhythms of behaviour might control vertical migration. These could continue to some extent even under constant conditions, such as darkness. He obtained evidence for this pattern in *Acartia tonsa* and *A. clausi* and in some *Calanus*. More recently, Harris (1963) suggested the possibility of endogenous timing influencing the vertical movements of *Calanus finmarchicus*. Though experimental data did not give significant proof for the copepod, the results of experiments with the freshwater cladoceran *Daphnia magna* appeared more convincing. Enright and Hamner (1967) reviewed the behaviour of mixed zooplankton including copepods, amphipods, mysids, decapods, medusae, chaetognaths and mollusc larvae in a tank under a light/dark cycle regime of three days duration, followed by two days under constant dim light conditions. Positive migration was observed amongst the Crustacea only, and showed varying degrees of response to subsequent illumination. The authors have attempted to classify the full range of behaviour exhibited.

The amphipod *Nototropis* showed a well marked circadian rhythm conditioned by an imposed lighting cycle, while the response of some copepods such as *Acartia* spp. showed a slight trend towards rhythmic behaviour on cessation of the light cycle. In laboratory experiments with three species of pontellid copepods, Champalbert (1978) demonstrated a clear-cut circadian rhythm controlled by light and emphasized by pressure changes which conditioned responses causing an increase or decrease in swimming activity. Hydrostatic pressure was regarded as a factor in synchronizing the circadian rhythm. Two peaks of feeding activity coinciding with tidal phases and not connected with vertical migration were observed in Stage IV *Calanus glacialis* by Peruyeva (1977).

As regards temperature, there is evidence that high temperatures tend to induce or increase negative phototaxis; low temperatures, on the other hand, cause the animals to become positively phototactic. Rose (1925) considered that temperatures exceeding 20°C were of great importance in causing negative phototaxis. Despite the fact that there are clear cases of vertical migration right through a thermocline (cf. Esterly, 1912), there are examples such as Nikitin described where a temperature barrier may apparently affect the vertical distribution of plankton. Farran (1947) also suggested that the upward migration of *Metridia* in the sea off the south of Ireland was limited in one year by the higher temperatures prevailing in the upper layers. Hansen (1951), working in the Oslo Fjord during the summer, noted a marked discontinuity layer at about 10 m depth, with a fall in temperature approaching 10°C and a considerable rise in salinity. While some copepods (e.g. *Calanus* females, *Centropages hamatus*) were mostly between the discontinuity layer and the surface by day and rose towards the surface at night, others (*Calanus* males, *Temora longicornis*) appeared to remain in or near the discontinuity layer throughout the 24 hours. Yet other species (*Pseudocalanus* females, *Sagitta*, *Meganyctiphanes*) were mainly below the discontinuity layer during daytime, and although they moved upwards to the layer at night, they did not pass beyond it; presumably their upward movement was limited by the relatively high surface temperature. On the other hand, Banse (1957) observed only very weak vertical movements for the copepods *Temora longicornis* and *Centropages hamatus* in the stratified shallow waters near Kiel. For these and for other copepods (*Pseudocalanus*, *Oithona*) there was an obvious massing of the populations into narrow depth zones but there was no clear migration in relation to the movement of isolumes; day and night distributions were similar. Banse believes that movements of the water layers were mainly responsible for the changes in distribution of the zooplankton.

The laboratory experiments of Harder (1952, 1957), however, suggest that some zooplankton animals such as cladocerans and copepods actively seek discontinuity layers, both thermal and haline, when illuminated from above or when left in darkness. These animals are probably reacting to the density of the medium. Although polychaete larvae, mysids and chaetognaths were found to distribute themselves without reference to discontinuity layers and forms such as fish eggs and medusae might be massed at a boundary layer entirely passively on account of their body density, Harder claims that several copepod species (*Acartia*, *Pseudocalanus*, *Centropages*, *Temora*) migrate to a discontinuity layer. They may be sufficiently sensitive to react to layers differing by as little as 0.2‰ in salinity.

In more offshore waters such salinity gradients are unlikely to occur, but temperature changes may be sufficiently marked to modify diurnal vertical migration. Bogorov (1958), for example, suggests from results of Vinogradov that in Far Eastern waters the presence of a cold intermediate layer affected diurnal migration. Most of the zooplankton moved vertically only in the superficial layers between depths of about 25–50 m. *Metridia pacifica*, however, was not limited by the cold sub-surface stratum and migrated through the layer to reach a day depth of some 500 m. It may be as Moore (1950, 1952) and Moore and Corwin (1956) have suggested, that while many zooplankton species appear to be little affected by temperature, for some forms, both temperature and light may be important. This is perhaps borne out by the earlier results of Clarke (1934) on the migrations of two copepods in the Gulf of Maine. *Calanus* migrated from deep water right to the surface and in doing so crossed a marked thermocline, whereas *Metridia*, though showing a pronounced vertical migration, did not seem to rise higher than the discontinuity layer. Angel (1968, 1979) tentatively suggested that current shears, often found to occur at discontinuities, may be avoided by many species.

Mauchline and Fisher (1969) quote various authors as suggesting that a thermocline may act as a barrier to certain migrants or cause variation in speeds of ascent and descent. However, Mauchline did not find that *Meganyctiphanes norvegica* or *Thysanoessa raschii* were inhibited by the thermocline in the Clyde area and quotes Lacroix (1961), who recorded that a sharp thermocline caused *T. raschii* and *T. inermis* to ascend more slowly and descend more rapidly through the thermocline. Rudyakov (1972) postulates that swimming activity is at a minimum in the zone of optimum temperature and increases on either side so that a sharp drop in temperature at the thermocline causes animals to swim more rapidly until they become acclimatized, and the rate is then reduced. Roger (1974b), in his study of euphausiid migrations in the south tropical Pacific, maintains that even a very strong thermocline has little affect on ranges. In shallower waters of the Aegean Sea, Moraitou-Apostolopoulou (1971) suggests that psychrophilic copepods become much more active in their vertical movements from autumn onwards. *Haloptilus longicornis* and *Pleuromamma gracilis* are two species which frequent the 200 m level by day in summer and may reach 50 m at night; these, with *Calanus minor* and *C. tenuicornis* (50–100 m daytime depths), rise to the surface during cooler nights. Particularly in tropical areas, light may be the major influence in regulating diurnal migration in upper layers, while temperature may be important at greater depths (Moore and Foyo, 1963; Zalkina, 1970).

Amongst fish larvae, temperature appears to have more influence than pressure. Petersen, Miller and Hutchinson (1979) found a greater abundance of adults of *Merluccius productus*, *Leuroglossus* sp. and *Bathylagus* sp. within and below the thermocline, occurring at lower temperatures than the larvae in the uppermost layers.

Much of the information on the effects of oxygen concentration and of hydrostatic pressure has been gathered from laboratory experiment, but in a survey of the euphausiids from the equatorial tropical Pacific, Brinton (1979) has clearly demonstrated that oxygen concentration limits the location and vertical movement of certain species. *Euphausia distinguenda*, *E. diomedeae* and *Nematoscelis gracilis* are extensive migrators whose daytime depths is in the oxygen-deficient zone. Three mesopelagic species of Thysanopoda, *T. orientalis*, *T. pectinata* and *T. monocantha* extended to water of high oxygen content (2 ml O_2/l), but could not tolerate areas where concentrations fell below 0.1 ml O_2/l. Longhurst (1967) noted that the principal migrators inhabiting the oxygen-deficient layers below 150 m in the Californian Current area, e.g. *Euphausia eximia*, *Pleuromamma abdominalis* and certain myctophids, could adjust to a wide range of oxygen concentrations (from <0.5 to 5.0 ml O_2/l). He points out that several species of copepods tolerate low oxygen tensions either as resting stock for long periods or as vertical migrators for short periods. Examples are *Calanus helgolandicus* (California Current), *C. finmarchicus* (Black Sea) and *Pleuromamma indica* (Arabian Sea). It is interesting that two epiplanktonic species of *Pleuromamma* are sensitive to low oxygen tension.

Certain inshore surf plankton species are sensitive to pressure changes and wave actions which later impose a diurnal rhythm, causing them to swim upwards from the sand on a rising tide (Naylor and Atkinson, 1972). MacDonald (1972) was able to demonstrate varying responses to pressure and thinks that in mesopelagic animals this may stimulate vertical migration. The deep-living *Gigantocypris mulleri* showed a gradual recovery of swimming activity at pressures greater than 200 atm, although *Conchoecia* sp. failed to recover after similar treatment. The oxygen consumption of the decapod *Systellaspis debilis* increased with pressure, and *Acanthephyra pelagica* showed greater pleopod activity when subjected to a pressure increase of 50 atm per five minutes up to a total maximum of 200 atm. Champalbert (1978) also noted the increase in swimming activity with increased pressure and a corresponding decrease following decompression (or pressure reduction). Teal and Carey (1967) showed that pressure increases at temperatures normally encountered did not increase respiratory rates and therefore would not create a barrier to migrating euphausiids (cf. Chapter 6). Similarly, Belman (1978) found a lack of effect on metabolic rate at pressures up to 136 atm, although the squids (*Histioteuthis heteropsis*) reduced oxygen consumption in response to lower available oxygen. Other workers also report that pressure increase failed to stimulate metabolic response in Crustacea (Pearcy and Small, 1968; Quetin and Childress, 1976) and in pteropods (Smith and Teal, 1973). However, the mesopelagic migrant *Gaussia princeps* showed marked sensitivity even at comparatively low pressures, although this might be modified at ambient daytime depth temperatures (Childress, 1977). Details of pressure and temperature in relation to metabolic activities are given in Chapter 6.

It is convenient now to consider the rates at which migrants move during both ascent and descent. Many workers, in considering the sinking rates of planktonic animals, have suggested that these may play a part in the migratory movements. Ostwald and also Eyden (1923) maintained that, following temperature changes between day and night, a change in the viscosity of the water might take place which could be responsible for the animals sinking farther in the day. It is almost certain from our knowledge of the very slight depth of diurnal temperature change in the sea that viscosity changes would be too slight. Eyden attempted to explain vertical movements in the freshwater cladoceran *Daphnia pulex* partly on the assumption that the animals fed on the phytoplankton in the upper layers during the night and that the excess weight due to feeding caused their density to increase, so that they dropped into deeper levels by day.

Buoyancy mechanisms, some of which have been evolved in response to the varying atmospheric pressures encountered by meso- and bathypelagic species, will of course modify sinking rates. Gilpin Brown (1972) states that in certain cephalopods neutral buoyancy is achieved by the presence of liquid-filled chambers in the shell, while Biggs (1976) suggested that physonect siphonophores might control their buoyancy by varying the amount of gas in the pneumatophores. An earlier report by Pickwell (1967) substantiates this. Denton and Shaw (1962) found that many gelatinous plankters are neutrally buoyant. Later observations by Biggs (1976) point to ionic or biochemical regulation of the body fluids in the calycophoran *Rosacea cymbiformis* which might assist this slow-swimming organism in vertical migrations. Hamner's experiments, quoted by Pugh (1977), showed that some medusae could regulate their specific gravity when placed in varying densities of seawater, and thus retain neutral buoyancy. The ostracod *Gigantocypris mulleri* maintains its buoyancy by ionic regulation (cf. Denton, 1963) and some organisms may be supported by a mucus secretion.

Rudyakov (1970) suggests that in *Cypridina sinuosa* changes in density caused by physiological state will accelerate or retard sinking; recently-moulted individuals sink at less than half the rate of the more dense gravid females. The experiments of Gooday and Moguilevsky (1975) showed that both anaesthetized and normal *Conchoecia* sank at similar rates. Again, a density difference was apparent: animals with a full gut or a large egg mass descended more rapidly. Most of the untreated species were able to control their sinking rate by orientation of the body and the authors suggest a link between ability to control sinking velocity (*C. imbricata*) or rarely observed orientation (*C. haddoni*), with the range of vertical migration—in the former species it is about 500 m, while the latter is a non-migrant (cf. Rudyakov,1979). Further evidence that vertical movement is due to changes in locomotory rate balanced against sinking rate due to specific gravity is given by Rudyakov (1972b) and Rudyakov and Voronina (1973) for some copepods, and by Omori (1974) for the decapod *Sergia lucens*. Rudyakov (1973) gives calculated estimates of sinking velocity for *Cypridina sinuosa*, pointing out that downward movement is mitigated by bouts of locomotor activity. Angel (1970) argues that limbs must be used to accelerate motion of *Conchoecia spinirostris*. In general, ostracods have a shorter migration and higher sinking rate than copepods. Table 5.3 gives details of velocity of descent coupled with speed of active movements and their duration, taken by filming the animals. During the short phases of activity this approaches 10 cm/sec—indeed, faster than any predaceous fish, although this is due to the "leap and stop" type of movement described by Vlymen (1970), amongst others. He gives details of swimming speeds for *Labidocera trispinosa* (*ca.* 2–9 cm/sec) which are maintained for approximately quarter of a second.

In considering extensive vertical migrations such as are performed by euphausiids, speed of swimming is an important problem. Hardy and Gunther (1935) considered that in the migration of Antarctic species such as *Euphausia triacantha* and *Euphausia frigida*, which appeared to climb some 200 m in approximately two hours, some change in density possibly assisted the movement towards the surface. The difficulties involved in determining the speed of movements of zooplankton animals under natural conditions are considerable. Experiments such as watching their swimming movements in restricted tubes of seawater are often vitiated by the fact that the animals behave unnaturally or move against the side. The work of Hardy and Bainbridge (1954) is perhaps the most illuminating on this subject (cf. page 507). They employed an ingenious "plankton wheel" such that the zooplankton swam continuously as it were in an endless tube. The speed of movement of some zooplankton animals suggested by Hardy and Bainbridge's results is indeed remarkable: *Calanus* can move upwards at a rate of some 15 m/hr; *Centropages* at nearly double this speed; the small copepod *Acartia* at 9 m/hr; and the euphausiid *Meganyctiphanes norvegica* can swim upwards at over 90 m/hr. These speeds can be maintained for one hour. They have also shown that for short bursts the speed, both upwards and downwards, can be far higher. Thus over a period of two minutes *Calanus* swam downwards at over 100 m/hr, *Euchaeta* at 135 m/hr and *Meganyctiphanes* and the pelagic polychaete *Tomopteris* achieved speeds of more than 200 m/hr.

In most of the plankton animals tested, the movement under a strong light stimulus was a steady swimming, although species like *Acartia clausi* performed rapid zigzag movements. It appears, therefore, that previously we have underestimated the speed of swimming of plankton animals when they are stimulated to migrate. Hardy and Bainbridge believe that the speed achieved by zooplankton and the duration of swimming are such that the extensive vertical migrations observed are perfectly possible. Enright (1977a) showed in field studies off La Jolla (samples were taken in later

Table 5.3. Results of measuring the rate of active movement and the sinking rate of planktonic crustaceans, mm/sec (Zaikin and Rudyakov, 1976)

Indices	Species						
	Euchirella rostrata	*Pleuromamma xiphias*	*Candacia pachydactyla + C. acthiopica*	*Euchaeta marina*	*Conchoecia atlantica*	*Conchoecia concentrica*	*Cypridina sinuosa + C. acuminata*
Length of shell or cephahothorax, mm	2.6	3.0	2.0	2.2	3.6	1.8	1.7
Sinking rate							
Number of measurements	10	40	30	10	15	8	20
Average	10.4	6.0	5.3	5.4	34.6	18.1	17.7
rate of movement							
Number of tracks	63	994	118	378	26	368	1093
Duration of active phase	0.23	0.18	0.35	0.29	0.09	0.19	0.28
Maximum rate	174.0	155.2	94.0	22.2	128.8	120.0	82.0

afternoon to early evening at half-hourly intervals) that *Metridia pacifica* can maintain upward swimming speeds of 30–90 m/hr (equal to 3.5–10 body lengths/sec) while migrating from below 150 m to the surface. Such speeds could continue for at least an hour; later migrants appeared to accelerate from 150 m to the surface. Pearre (1973) described rapid and sporadic movements of *Sagitta elegans* towards food concentrations during winter, while Nagasawa and Marumo (1975) reported similar behaviour in *S. nagae*. Enright (1977b) suggested that the copepods he studied could maximize their energy gain by reaching the phytoplankton during peaking of stored photosynthetic products. He calculates data from Vinogradov (1970) for *Metridia pacifica* (19 m/hr) and from Clarke for *M. lucens* (50–60 m/hr) which agree with the estimations of Hardy and Bainbridge (1954) whose figures may be rather lower, possibly due to lack of stimuli provided for the animals. Reference has already been made to the amplitude of diel vertical change shown by some amphipods. Semenova (1974) gives details of these for *Parathemisto japonica*; for size groups of 10–14 mm females, upward speeds were approximately 67 m/hr and for males 54 m/hr over a range of approximately 350 m. Smaller individuals (above 5 mm in length) rose rapidly at the start, but then slowed down; descent was achieved at an average rate of 43 m/hr. Specimens less than 5 mm remained in the top 60 m and showed little vertical change. Zhuravlev and Neyman (1976) observed this species in a tank and found descent was hastened by withdrawal of appendages to give a streamlined body contour.

In field studies, whether by the use of vertical hauls at regular time intervals or measurements from sonic scattering layers, data obtained must relate to the average distance travelled by a population and its mean sinking rate. However, some workers have found a fair agreement between these and results from experiments. Thus Klyashtorin and Kuzmicheva (1975) report calculated and observed rates for *Neomysis mirabilis* to be consistent, and Omori's (1974) findings for downward migration of *Sergia lucens* (6.2 m/hr for zoeas, 108 m/hr for adults) correspond to the average velocity for freshly killed shrimps, although slightly lower (cf. also Rudyakov, 1972b; Gooday and Moguilevsky, 1975). While this may make a valuable contribution to the knowledge of energetics of migrators, experimental data should always be viewed with caution. In small containers, animals will not experience the change of temperature and pressure, and therefore viscosity of the medium, nor stimuli from the presence of any food concentration or population aggregations. Rudyakov and Voronina (1973) admit that observed behaviour of *Metridia gerlachei* in the Scotia Sea was not simulated in their laboratory experiments, where the maximum rate of descent of the copepod (36 m/hr) exceeded the passive sinking rate; they suggest that this may be due to lack of any marked temperature and pressure difference. Furthermore, it is not known to what extent anaesthesia may paralyse the receptor mechanisms of the animals, thus depriving them of, for example, visual and chemical stimuli.

Photographic studies such as those of Zaikin and Rudyakov (1976) could be most useful in recording the speeds of movement. They filmed copepods and then measured the sinking rates of anaesthetized animals; they state that migration proceeds at an average rate and that fast speeds apply only to escape reactions. They also stress the danger of relying on rates of individuals, as activity may increase with concentration of animals. Haury and Weihs (1976) describe the energy-efficient swimming behaviour of some negatively buoyant zooplankton and show by a series of equations that the energy thus saved is a function of the drag ratio and speeds of swimming and sinking.

An ideal method, now made possible by Scuba diving techniques, whereby *in situ* observations may be later compared with laboratory measurements has been employed. Madin (1974a,b) found that in the field *Salpa maxima* and *S. cylindrica* swam at around 9 m/min and for *S. fusiformis* a velocity of 1.2–3 m/min was recorded in the laboratory. He suggests that with their energy-efficient method of jet propulsion, salps may move continuously: water entering at the anterior end passes through a filter net before being ejected, thus enabling salps to combine the means of propulsion with that of feeding (cf. Chapter 6). The great amplitude of migration (up to 800 m) and vast numbers of salps sometimes occurring in swarms have an important effect on the ecosystem, where up to 90% of the total zooplankton biomass shifts from the upper 500 m by day, while faecal pellet production is estimated to contribute as much as 12 mg C/m²/day to deeper layers (Wiebe *et al.*, 1979). Another example of energy saving is given by Barham (1970), who suggests that myctophids possibly use respiratory water currents expelled from the gill cavities by rhythmic opercular pulses to move in a vertical plane. The rate of ascent may vary greatly even amongst members of the same genera; in the Atlantic, Roe (1974), analysing a 24-hour series of hauls at 250 m, found that *Nematoscelis microps* arrived later and left earlier than *N. tenella*, even though migrating a shorter distance.

There are many conflicting theories regarding assessment of energy required during vertical shift. Obviously the advantage gained, whether by rising to a region of optimal food concentration or sinking to allow a reduced metabolism at lower temperatures, must result in a net energy gain for the animal. Early calculations were misleadingly high and probably based on the continuously fast swimming rate which is reached only in the "leap" movements or acceleration at the commencement or cessation of the upward migratory phase. Thus, Ostrovskaya, using figures reported by Petipa (1967), calculated an energy requirement of 10–100 times the standard metabolic rate to maintain a speed of 15 cm/sec for crustaceans. Klyashtorin (1978) has drawn attention to these errors and suggests that two copepods studied (*Calanus finmarchicus* and *Metridia longa*) do not exceed 1–3 cm/sec during upward migration, while the rate for euphausiids is faster, at about 5–6 cm/sec (180–220 m/hr). He maintains that the extra energy required is relatively small and would not exceed 20–40% of the standard metabolic rate. This is confirmed by an earlier report of Klyashtorin and Yarzhombek (1973), who calculated energy requirements using results of respiratory rate measurements. Klyashtorin and Kuzmicheva (1975) observed *Neomysis mirabilis* to move at speeds of 8–10 cm/sec, which for a 10-mg animal would involve an increase of one and a half times that of basic metabolic rate, a figure which was consistent with calculated results. They believe that the relative amount of energy expended decreases with diminishing size; thus fish would need greater effort to move corresponding distances. An energy balance equation associating morpho-physiological data with ecological and migratory conditions was derived by Tseytlin (1977), who states that for movement at optimum speeds, 2–3 times the basic metabolic rate would be needed—that is, about 35% of the energy spent per day.

It is obvious that the need for food plays an important role in the responses of any animal, and many planktonologists believe that this will trigger the commencement of migration. An interesting description of the feeding habit related to rising and sinking of *Cypridina sinuosa* is given by Arashkevich (1977). At the onset of upward migration at about 1500–1700 hours, the ostracods defaecate and the gut becomes empty. They feed as they rise and remain at maximum height until 0100–0300 hours when, with the gut approximately 50–60% full, descent commences while feeding continues; the gut becomes full and grossly distended, with consequent increase in density. Digestion proceeds during the day when the middle layer of parenchyma cells becomes filled with fat, thus reducing body weight and creating greater buoyancy. *Parathemisto japonica* exhibits two types of behaviour in different regions of the Okhotsk Sea; little migration takes place in shallow (75 m) areas, but over the deeper zone (800 m) Zhuravlev and Neyman (1976) report maximum catches of these amphipods, most with full stomachs, at 2100 hours and a smaller peak at 0300 hours at 50 m below the surface. Fewer animals with near empty stomachs were taken at other times during night hauls. It is interesting that the same pattern was exhibited by their salmonid predators.

Although some micronektonic fish do not migrate, Kawaguchi (1969) found that *Cyclothone* and *Stenobrachius* occasionally occurred in the upper layers of the Pacific Kuroshio area at night, presumably in pursuit of their migrant prey. Nagasawa and Marumo (1975) found that *Sagitta nagae* closely followed the distribution pattern of its main prey, *Calanus pacificus*. Diel variation in feeding activity appears to interact with vertical movement. Mackas and Bohrer (1976) examined the feeding rate of planktonic herbivores *in situ* by fluorescent analysis. They suggest that full *Metridia lucens* descend rapidly, and the gut contents of these copepods shows a steady increase to a pre-dawn peak. McLaren (1969) and Sameoto (1976) are among those who have recorded a synchronization of feeding with high phytoplankton and chlorophyll levels; this would coincide with the theory expressed by Enright (1977b).

A relation exists between biochemical composition and vertical distribution of zooplankton. Although this subject is fully discussed in Chapter 7, it may be noted here that temporary changes in the levels of organic constituents, especially fats, are often associated with migratory movements. In his study of sergestid distribution, Donaldson (1975, 1976) suggests that those species with an extended amplitude of migration have a higher protein and lower lipid content, while the converse is true of the shorter-ranging species. This would appear to be consistent with greater muscle development and consequently improved swimming ability. In decapods also the level of wax esters in some species (e.g. *Acanthephyra purpurea*) is correlated with the rate at which the animal biosynthesizes these lipids and may be influenced by physiological requirements (Sargent, Morris and McIntosh, 1978). Thus a rapid interconversion of fatty acids to wax esters may be achieved and assist as a buoyancy device (Ackman, 1978). Omori (1974) thinks that such a facility may be an evolutionary adaptation to a migratory habit. Many workers accept the idea that lipid may displace neutrally buoyant body fluids (Childress and Nygard, 1974). The liver of the deep sea squid *Bathyteuthis* consisted of two oil-filled chambers which may function as buoyancy tanks; though this is not a migratory species, it is possible that a similar device exists amongst other cephalopods (Roper, 1970). Lee (1974) recorded that in *Calanus hyperboreus* the ratio of fatty acid to wax ester content depended on the depth or shallowness of habitat. A seasonal variation in lipid fractions was noted by Gatten and Sargent (1973), who showed that during early summer *Calanus finmarchicus* feeding actively in the surface layers at night have a faster rate of wax ester biosynthesis than those deeper-living calanoids having higher lipid levels and less active wax ester biosynthesis.

With so many species involved in different types of migrations, it is not surprising that there have been numerous theories postulated by planktonologists as to the reason for this complex but common type of behaviour. Clearly, there must be some overall advantage to the species and it would seem unlikely that it is merely a reflection of an orientating depth mechanism in response to the stimuli of light and gravity (Harris, 1953), though such factors will modify behaviour. Zooplankton animals are known to be adversely affected by high light intensities and presumably they are normally unable to remain near the surface during daylight hours. However, a movement towards the surface to the zone most thickly populated by the phytoplankton would be of survival value in that the animals could feed extensively in the uppermost layers during the night. They would have to move away during the day owing to the inimical influence of light. Whether factors other than light in the uppermost layers could play a part is uncertain.

Hardy suggested from work in the Antarctic and in the North Sea that very dense patches of phytoplankton are unfavourable to zooplankton, so that the animals cannot stay in the vicinity of such very rich phytoplankton for long. Perhaps active photosynthesis causes the production of harmful metabolites, though Bainbridge's results, as far as they go, hardly support this. Hardy (1956) has advanced a more general theory for the value of the diurnal vertical migration. He points out that in almost all seas the uppermost layers are moving at higher speeds and often even in different directions from the lower layers. The regular movement of plankton between the layers therefore allows the animals to be carried over greater distances at night, when moving in the uppermost layers, than during the day. The plankton population can be distributed over a much greater area of the ocean than if it continually moved with one body of water. By fortuitously drifting with water masses moving at different speeds the zooplankton is able, as it were, to sample a much wider area of the ocean.

Evans (1978) described a mathematical model of vertical current shear interacting with a nutrient–phytoplankton–herbivore ecosystem, and the possibility of patch-forming effects by the vertical movements of the plankton. Since horizontal currents in the sea may differ at various depths, nutrients and phytoplankton may be displaced by water movements, but zooplankton can move vertically and so remain in an area of optimal conditions. Figure 5.12 illustrates three layers and a reservoir which is a nutrient source and where the animals will be situated during daytime. On rising, migrants will enter an appropriate layer which provides the required food concentrations, then, in response to certain stimuli, for example light increase or full gut, they return to the reservoir. Where animals have become adapted to diel migration a stock may be retained in a favourable locale by using a current system. A response to light stimulus and interaction with currents can maintain organisms of the deep scattering layer in regions of high phytoplankton (Isaacs, Tont and Wick, 1974).

There are many instances whereby diel migration enables the zooplankton stock to remain resident in a certain locale. Figure 5.13 shows how some euphausiids maintain east–west equilibrium in the Eastern Equatorial Pacific (Brinton, 1979). In Chesapeake Bay populations of *Podon polyphemoides* exhibit a pattern of reverse migration which Bosch and Taylor (1973b) suggest is light-mediated; the daytime depth of these cladocerans is about 4 m where seaward currents exist. They descend before sunset, where a deeper landward flow maintains large stocks of this species. Off the Oregon coast, where dense phytoplankton patches occur some 20 km offshore, vertical migration by the younger copepodites of *Calanus marshallae* allows these stages to be returned by

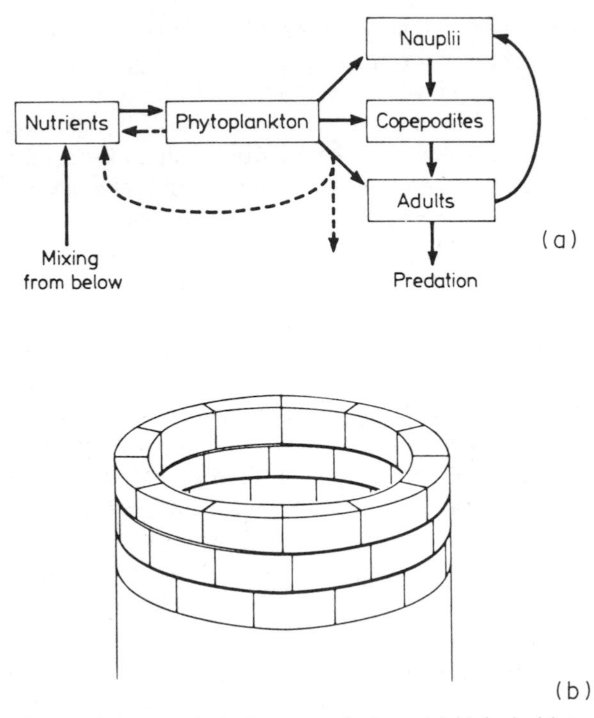

Fig. 5.12. A model of vertical shear and plankton populations: (a) biological interactions; (b) physical framework. The soluble and insoluble losses during grazing, shown as dashed lines for adults, occur also for the smaller classes (Evans, 1978).

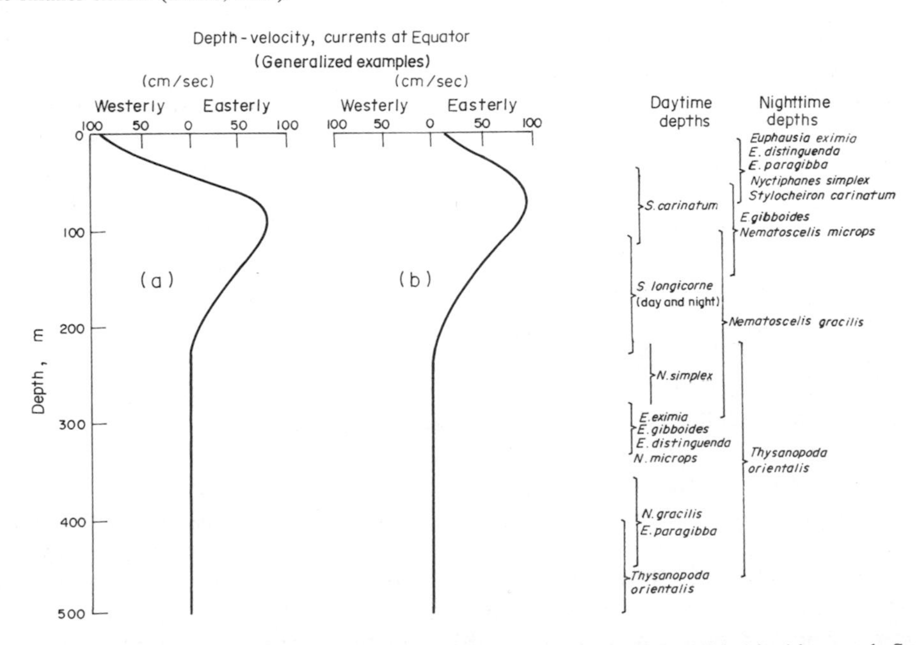

Fig. 5.13. Examples of current at the Equator as transport vectors for euphausiids (a) with westerly South Equatorial Current above easterly Undercurrent; (b) with easterly Undercurrent surfacing, shown in relation to day and night depths of occurrence of adults of the species occupying this regime (Brinton, 1979).

currents from the nearshore spawning ground to the upwelling zone (Peterson, *et al.* 1979, cf. Chapter 8). Similarly, in the neritic waters of the Gulf of Mexico, the decapod *Lucifer faxoni*, in the course of diel movement, achieves landward transport by tidal currents (Woodmansee, 1966), and Omori (1974) describes how *Sergia lucens* migrates obliquely between deeper offshore and shallower inshore layers in Suruga Bay (Fig. 5.14). *Sebastes* larvae, amongst other young fish, frequent the surface during four to five months of their pelagic life and may drift far, but older larvae descend to greater depth and thus avoid the surface currents (Bigelow, 1926; Kelley and Barber, 1961; cf. Chapter 2). This shift in horizontal position achieved by migrating organisms can thus confer a biological advantage to the species; David (1961) postulated that an interchange of populations could promote a gene flow through a species, although supporting evidence is slight.

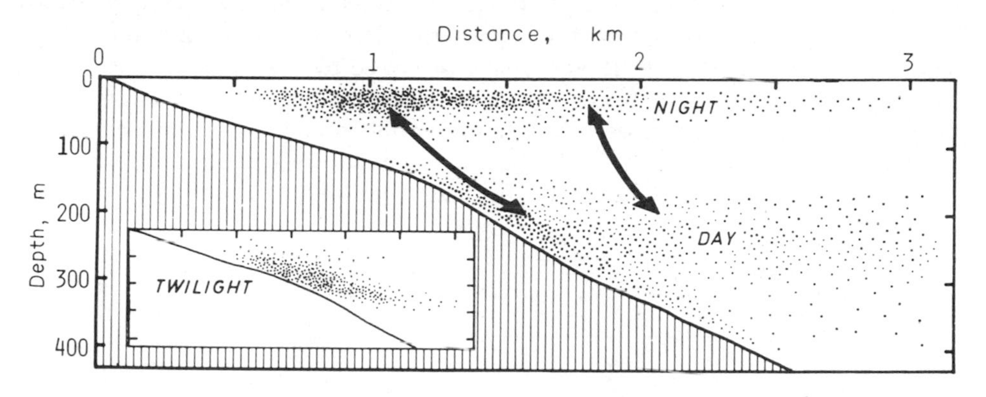

Fig. 5.14. Schematic diagram showing diurnal migration of *Sergia lucens* in Suruga Bay (Omori, 1974).

Another theory that prey species might avoid predators by remaining in deeper zones by day is supported by several authors. Vinogradov (1970), pointing out that by feeding actively on the rich phytoplankton of the epiplanktonic layer during darkness, herbivores may avoid their visual predators, noted that *Boreomysis* spp. descend with full stomachs. This view is supported by Zaret and Suffern (1976), who observed that in Gatun Lake, Panama, the copepod *Diaptomus gatunensis* appeared near the surface at night and thus lessened the likelihood of attack by its predator, a planktivorous fish; the nauplii of the species were not readily taken by this fish and were non-migratory. A similar observation by Robertson and Howard (1978), studying the eelgrass community off Victoria, would confirm that calanoid copepods and some decapod larvae achieve predator avoidance in this way: these migrants appeared to be smaller and almost transparent in contrast to the larger, opaque non-migratory plankters. Hairston (1976) believes that the degree of pigmentation of another *Diaptomus* sp. from Washington influences the vertical migration and behaviour pattern in varying light intensities. Experiments showed that marked pigmentation reduced photodamage to the copepod but made it easily visible to carnivores; to survive, the animals must remain at a depth where the light is too low for them to be seen by day, while those which are only lightly pigmented, though less vulnerable as prey, lack photoprotection and must also descend during the day. Omori (1974) found a correlation between pigmentation and day depth of shrimps. Angel (1979) discusses size and shape in relation to migratory behaviour and predator avoidance in halocyprid ostracods. Vinogradov and Sahzin (1975) recorded long migrations by *Metridia pacifica* down to 2000 m in the Sea of Japan, where there is an absence of deep water carnivores. In surveys of the tropical Pacific euphausiid populations, Roger (1972, 1973) comments that animals are eaten by micronektonic fish which in turn provide prey for larger fish; the euphausiids were taken at all stages of migration.

Enright (1977b) and Enright and Honeggar (1977) attempted to construct a metabolic model for obtaining maximum energy; the behaviour shift to upward swimming may be hunger-mediated, while a downward trend caused by satiation could mean that the herbivores spend a shorter time at the surface, thereby lessening the risk of predation while gaining a greater reduction in metabolic rate on reaching cooler deeper layers. The authors were not able to substantiate their theory experimentally; for a review of critical correspondence reference should be made to letters by Pearre (1979b) and others. Koslow (1979) disagrees that late afternoon migration avoids predators and points out that the light flashes produced by copepods such as *Gaussia* and *Metridia* may be an escape reaction to divert pursuing predators, while phytoplankton bioluminescence might influence the timing of migrations. Pearre (1979a) has reviewed some of the problems and puts forward a model of vertical migration with light and food as controlling factors. A diagram (Fig. 5.15) of

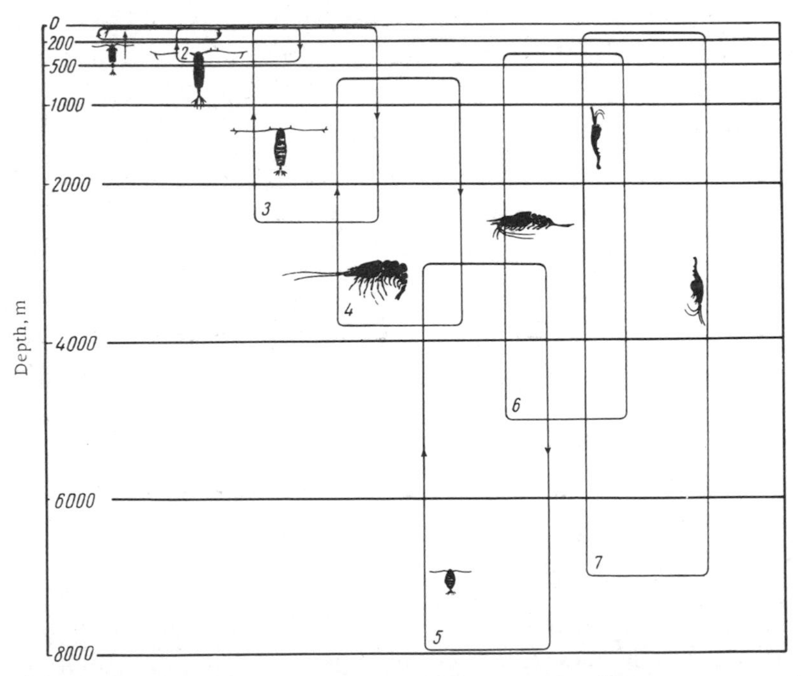

Fig. 5.15. Scheme of vertical migrations of oceanic plankton providing for active transport of organic matter from surface to deep-sea layers, with the example of the northwestern part of the Pacific Ocean: 1 – diurnal migrations encompassing surface zone; 2 – diurnal and seasonal migrations encompassing the surface zone and the intermediate layer; 3 – seasonal and ontogenetic migrations encompassing the surface zone and the bathypelagic; 4 – ontogenetic migrations encompassing the bathypelagic and abyssopelagic; 5 – migrations within the limits of the abyssopelagic; 6 – ontogenetic migrations encompassing the intermediate layer and a greater part of the deep-sea zone; 7 – short-term, irregular (?) food migrations covering almost the entire water column (Vinogradov, 1970).

distribution in different depth layers is given by Vinogradov (1970), who shows the strata at which the various stages and trophic groups are mainly found. Longhurst (1976) also gives a comprehensive review. McLaren (1963) claims that the advantage conferred by migration in thermally stratified waters is one of energy gain. When a zooplankter feeds in warmer surface layers and descends to cold depths where a lower metabolic rate reduces demand for food, this energy bonus would provide for increased food storage. If, by seasonal migration, animals were to remain in deep cold water, an additional selective advantage of retarded maturation with consequent increase in size and fecundity would accrue. Teal and Carey (1967) argue that for savings to be significant the difference between day and night temperature must be considerable in order to reduce metabolism sufficiently to offset the energy expended in migration. Some animals can tolerate near anoxic conditions for part of the day and then rise to more shallow layers where aerobic respiration enables them to burn off an oxygen debt (Childress, 1977; Belman, 1978). From a comparison of body size, clutch size and depth of occurrence in halocyprid ostracods, Angel (1979) finds no clear evidence of a demographic advantage arising from a diel migration.

Although the theory of Vinogradov (1962) that bathypelagic forms relied for nutrition on a rain of particles from the euphotic zone has now been modified, there is little doubt that a ladder-like series of migrating forms play an important role in the food chain (Raymont, 1970). Deeper-living planktivores will feed on herbivores from higher strata and in addition detritus and faecal pellets will add to the food available to omnivores. Thus Omori (1974) lists a group of mesopelagic decapods which migrate to the surface to feed, while the deeper-living species may move up at night to occupy the zones vacated by the surface migrants (Table 5.2; Fig. 5.9). The migrating trophic chain will also include the nekton where squid and planktivorous fish will act as secondary carnivores. Myctophids must adopt a migratory pattern and follow their prey food in order to obtain sufficient nutrition. Roger describes pelagic food chains with particular

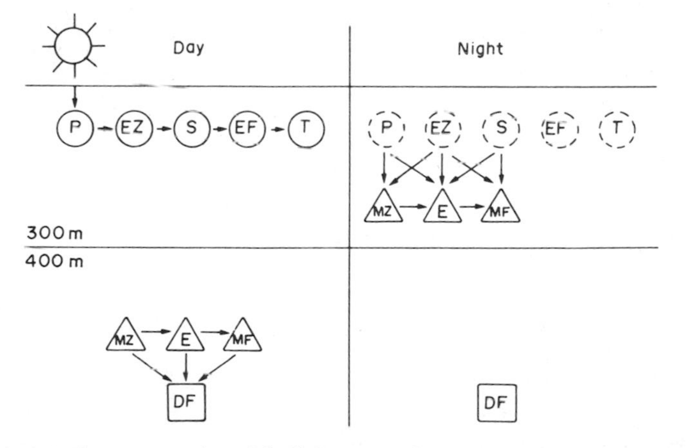

Fig. 5.16. Diagrammatic representation of the links among the pelagic communities of the tropical Pacific. Solid lines represent feeding activity. Dotted lines—non-feeding activity; circles—epipelagic species; triangles—vertically-migrating species; squares—deep sea species. Arrows indicate main paths of energy transfers through predator-prey relationship. P = phytoplankton; EZ = epipelagic zooplankton; S = *Stylocheiron* euphausiids; EF = epipelagic fish (e.g. Bramidae and Gempylidae); T = tuna; MZ = vertically-migrating zooplankton; E = *Euphausia* euphausiids; MF = vertically migrating fish (e.g. Myctophidae); DF = deep sea fish (Roger, 1977)

reference to euphausiids and stresses their importance in the transfer of energy to deeper layers (Fig. 5.16).

In a 24-hour study of hauls from 250 m in the Atlantic, Roe (1974) found that a continuous up and down movement of plankton took place, with fish, decapods and copepods especially dominating the night catches. He suggests that this sequentially ordered arrangement of migration could segregate species groups for part of the time and thus reduce competition. Amongst 12 common species of sergestids occurring off Bermuda at Ocean Acre, Donaldson (1975) found that a difference in feeding habits and in vertical distribution were important in this respect. Species which co-existed for part of the day achieved separation by migration at night; *Sergestes splendens* showed the greatest amplitude of movement (825 m), while *S. japonicus* travelled less than 100 m. At this stage no conclusive answer can be given to the question as to the value of diel vertical migration. In their excellent summary, Mauchline and Fisher (1969) point out that the advantages and causal factors are varied. Although some non-migrating species in a group appear to survive with as much success as the migrants, dispersion will reduce competition and allow each group to occupy an optimal ecological niche, thus exploiting to the full the resources of the environment. The tremendously wide occurrence of this phenomenon in the seas is one of the most challenging aspects of marine plankton study.

Chapter 6
The Food and Feeding and Respiration of Zooplankton

Food and Feeding

In the upper layers of the open ocean and in shallower inshore areas away from the sea floor the transfer of primary to secondary production is through the zooplankton. This might imply that the whole zooplankton population is herbivorous, consuming the phytoplankton algae. This is far from the truth; many groups of carnivorous zooplankton (medusae, siphonophores, chaetognaths, gymnosomes, amphipods and the great majority of fish larvae and planktonic decapods) have long been recognized. Planktonic groups that are normally regarded as mainly herbivorous may include carnivorous species. There are also numerous examples of zooplankton which are mixed feeders. Referring to copepods, Marshall (1973)* wrote that it is generally accepted that most species are neither purely herbivorous nor purely carnivorous. Essentially herbivorous zooplankton such as *Calanus hyperboreus* will take animal food (e.g. *Artemia* nauplii) offered experimentally (Mullin, 1963), and field observations indicate that such species will take some animal material. Species with sufficiently fine filtration mechanisms may also include detrital particles and bacterial aggregates in their diet. Within limits, many zooplankton species take what food is available.

The diet patterns of the carnivorous zooplankton often appear to be even more complex. Many of the carnivores, numerous species of chaetognaths or medusae for example, may feed directly on herbivorous copepods and are classified as primary carnivores. Similarly, gymnosomes can consume thecosomes, which are algal feeders and are also, therefore, primary carnivores. However, some zooplankton species feed on plankton animals which are themselves carnivorous. For example, the ctenophore *Beroe* feeds on other lobate or tentaculate carnivorous ctenophores; *Beroe* is, therefore, classified as a secondary carnivore. There may be a further link in the food chain; the planktonic gastropods *Glaucus* and *Phyllirhoe* feed on siphonophores, including *Porpita*, but this member of the pleuston feeds to a considerable extent on pontellid copepods which are themselves carnivorous. *Glaucus* and *Phyllirhoe* are, therefore, third stage (tertiary) carnivores.

Petipa, Pavlova and Mironov (1970) describe these different trophic levels with examples from the plankton fauna of the Black Sea. Animals from higher trophic levels, however, are rarely exclusively limited to one dietary path. Figure 6.1 illustrates the food web in the epiplanktonic community. Different size groups of herbivores are shown and are represented by *Oikopleura*, various copepodites of *Paracalanus*, larval

* Author's note: Some authors e.g. Marshall (1973) have continued to use the name *Calanus pacificus* for *C. helgolandicus*. The name specified by the particular author has been retained.

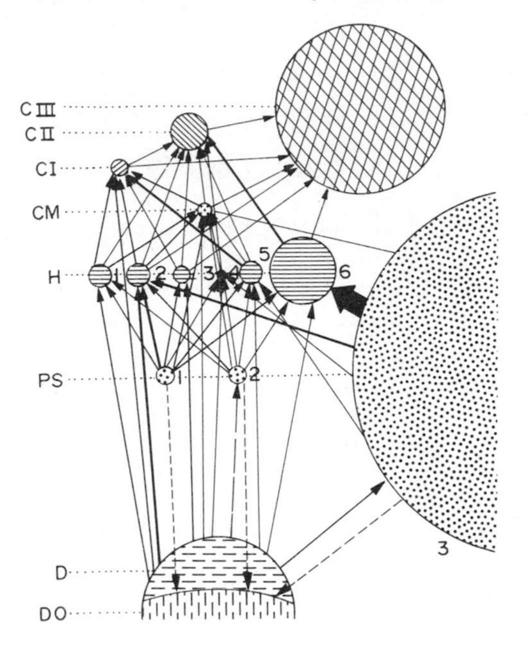

Fig. 6.1. Food webs in the epiplankton community. C III—tertiary carnivores; C II—secondary carnivores; C I—primary carnivores; CM—mixed-food (plant and animal) consumers; H—herbivorous organisms; PS—producers and saprophagous organisms (phytoplankton); D—detritus; DO—dissolved organic matter; 1, 2, 3 etc., written near the circles, are the ecological groups of the trophic level. The level of producers: 1—small-sized forms; 2—medium-sized forms; 3—large-sized forms. The level of herbivores: 1—*Oikopleura*; 2—II–III Copepodites; 3—IV–VI copepodites of *Paracalanus*; 4—larvae of molluscs and polychaetes; 5—nauplii; 6—IV–VI copepodites of *Pseudocalanus* and *Calanus*. Areas of circles are proportional to the average daily standing stock of the ecological groups. Thickness of solid pointers is proportional to daily specific rate of this or that food consuming. Dotted pointers mean the processes of dissolved organic matter excretion and the processes of detritus consumption, where no quantitative data were obtained (Petipa, Pavlova and Mironov, 1970).

molluscs and polychaetes, and different stages of *Pseudocalanus* and *Calanus*. The primary carnivores consisted of predatory Copepoda (*Oithona* and *pontellids*) which consumed small copepods and cladocerans. Mixed feeders were represented by some copepods, chiefly *Acartia clausi*. *Sagitta setosa* is typical of the epiplanktonic secondary carnivore category, its chief prey being herbivorous copepods, but *Oikopleura* and small sagittae are also taken. The bathypelagic group is represented mainly by large sagittae, which also consume copepods (Fig. 6.1). *Pleurobrachia* is regarded as almost the sole tertiary carnivore, since it was held to consume sagittae, but it also preyed directly on both herbivorous copepods and mixed feeders. The trophic relationships illustrated by Thiriot (1977) (Fig. 6.2) for equatorial African waters demonstrate different levels for several carnivorous zooplankton species from different taxa and also possible alternative pathways for some species (e.g. *Euphausia lucens*, *Nyctiphanes capensis*, *Parathemisto*, *Sagitta* spp.).

These alternative trophic patterns, typical of most species, lead to remarkable complexities. Many secondary or tertiary carnivores not only prey directly on herbivorous plankton, but a primary carnivore may consume a carnivore at the same or even at a higher trophic level. Thus a heteropod *Cardiopoda placenta* feeding on

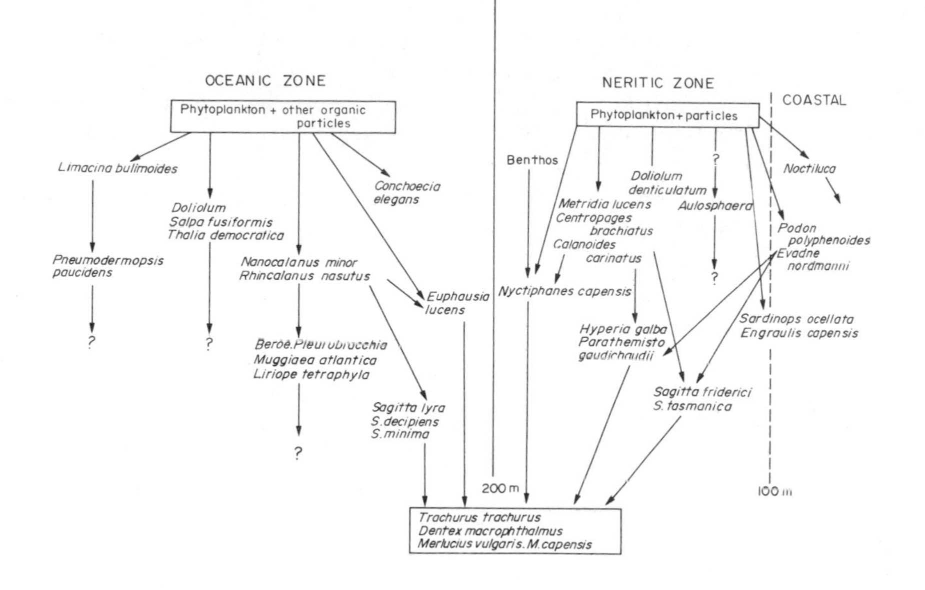

Fig. 6.2. Trophic relationship of different species in West African equatorial waters (Thiriot, 1977).

herbivorous salps will eat another heteropod (*Firoloidea*) or fish larvae or *Phyllirhoe* (Hamner, Madin, Alldredge, Gilmer and Hamner, 1975). Sagittae are generally regarded as primary or secondary carnivores since they can feed directly on herbivorous copepods and other planktonic species, as well as on carnivorous forms including fish larvae and, according to Bernard (1967), even medusae. Some post-larval fish, however, including herring, may take sagittae (cf. Hardy 1924) and even some planktonic forms such as medusae and copepods (cf. Wickstead (1959) for *Candacia*) can feed on sagittae.

The more correct term "food web" rather than food chain is, therefore, usually applied to the intricate trophic relationships of the planktonic ecosystem. The complexity tends to be increased by the capability of many species to supplement a predominantly carnivorous diet with the consumption of small particles, often of detrital material. In particular, in the deeper levels of the ocean, where food tends to be very scarce, whilst the carnivorous habit is widespread, the capacity to utilize a variety of dead material such as chitinous remains, decaying wood particles, dead or dying protozoans such as deep sea Phaeodaria (cf. Vinogradov), faecal pellets, sinking macroalgal particles (Rowe and Wolf, 1962) as well as the very occasional larger particles of disintegrating larger animals, may be of great survival value.

Paffenhofer and Knowles (1979) found for *Temora turbinata* and *Eucalanus pileatus* that the rate of faecal pellet production is a function of food concentration and size of copepod. They examined how far the considerable amount of organic material remaining in pellets may be available to other zooplankton. *Calanus helgolandicus* will remove faeces when phytoplankton is depleted, and although the value of the material has been a matter of debate, the authors find that the size and age (with reference to sinking rate) of faecal pellets is significant. Females of *Eucalanus* and *Temora* would graze on faecal pellets produced by nauplii of *Eucalanus* at about the same rate as a culture of

Rhizosolenia. Smaller, rounded pellets seemed to be of greater value; they may have relatively larger populations of micro-organisms. The organic content is also clearly related to the age of pellet. Vinogradov (1970, 1972) and other workers frequently employ the term euryphagous to describe this ability to use varied foods, and he points to the considerable proportion of euryphagous fauna at many deeper levels in the open ocean.

Food intake may also be substantially enhanced, qualitatively if not quantitatively, by the presence of bacterial colonies and even small saprozoic protozoans living on the surfaces of these particles. How far the relatively large amounts of dissolved organic matter might be used by such a bacterial microflora and thus contribute to the cycling of organic matter is still a subject of dispute. The role of the microzooplankton, especially small protozoans such as ciliates in providing food for particulate feeding metazoan plankton not only in the deeper oceans but in near surface layers undoubtedly needs emphasis. Heinbokel (1978a) has suggested that tintinnids and ciliates could be an important influence on energy flow. Many zooplankton can use alternative trophic pathways, using microzooplankton when larger food is not available (Parsons and LeBrasseur, 1970; Parsons, 1975) (cf. Volume 1). The part played by organic aggregates in providing particulate food from dissolved organic matter is still uncertain. "Marine snow" has been observed sometimes in considerable quantities in the marine environment. Hamner *et al.* (1975) observed amorphous aggregates even of sizeable dimensions in open waters and they draw attention to the abundant mucus produced by many invertebrates and which can be of considerable nutritional value to plankton. Mucus food webs used by many fine filter feeders and then discarded, as well as the empty "houses" of appendicularians, may be used as food (Alldredge, 1972, 1976b). The material serves also as a base for bacterial aggregates and small protozoans.

With the number of stages in the trophic cycle in the pelagic ecosystem and the complexity of the food web, it is not surprising that a significant loss of energy occurs in the many transferences in secondary production. Petipa *et al.* (1970) stated that some of the primary and secondary carnivores in the Black Sea community required up to 100 % of their body weight as food daily and *Pleurobrachia* was said to need more than 300 %. Estimates of overall fish production, even of pelagic fish which might be regarded as one typical trophic end product, are therefore not easy to relate to levels of primary production. While recent estimates of primary production may be reasonably accurate, the pathways from primary production to fish are complex and this may be responsible for considerable errors in attempts at quantification.

Cushing (1975) suggests that an ecological efficiency often advanced (*ca*. 10 %) may be too low for the link from primary to secondary production. He also refers to higher transfer efficiencies at the lower levels of food chains and refers to Steele's analysis for the North Sea, suggesting that ecological efficiency may be higher between primary and secondary production. However, an increased number of stages in a food web, especially in some demersal fishes, must lower overall efficiency. An ecosystem involving an approximately direct pathway: phytoplankton→zooplankton→pelagic fish, should give a higher transfer efficiency, but this could be lowered if excessive primary production and a time lag in the utilization occurred. In deep oceans also, a complication in trophic relationship may be introduced by the vertical migrations of zooplankton pertaining to different trophic habits (Vinogradov and Menshutkin,

1977). Such considerations and their relevance to the modelling of marine ecosystems underline the significance of accurate assessments of plankton food webs both quantitatively and qualitatively.

Methodology

As with other animals, an examination of gut contents is a well-tried method for attempting to determine the diet of zooplankton, though great care must be exercised in the interpretation of results, since digestion is often extremely rapid in zooplankton. Frequently, only a small percentage of the animals captured have recognizable food in the gut. Observations suggest that some herbivorous zooplankton may produce faecal pellets every 15–20 minutes; any food remaining may not give a proper picture of the diet. Examination of freshly-caught zooplankton on shipboard is preferable and almost essential in prolonged oceanographic cruises, especially since animals which are killed and preserved in formalin for later examination frequently void a considerable portion of the food from the gut due to the muscular contractions in dying.

Some spurious interpretations may be placed on the feeding of some of the carnivorous plankton caught with ordinary net methods, since while the animals are crowded in the cod end of the plankton net after capture, they often attack and feed on each other almost indiscriminately. Not only the contents of the gut but the faecal pellets produced by zooplankton may be used in some cases to identify the diet organisms. Many zooplankton produce discrete and characteristic faecal pellets (cf. Raymont and Gross, 1942; Martens, 1978) and as Marshall (1973) points out, it is possible to identify and even to count food organisms if these have indigestible skeletons remaining largely unbroken in the faeces. For example, diatom valves, thecae of dinoflagellates or acantharian skeletons may be recognizable. Marshall counted *Prorocentrum* skeletons in *Calanus* faeces. An obvious limitation not only in examining faecal pellets but also with gut contents is that any soft-bodied organisms in the food consumed (naked protozoans, ciliates, flagellates and the like) will be unrecognized.

Experiments to test the dietary preferences of zooplankton became feasible only after suitable techniques had been developed to grow plankton cultures in the laboratory. Even then, experiments were entirely confined to offering a few herbivorous zooplankton species (usually copepods) a diet of cultured unicellular algae. For many years very few algal species were in culture, so that the experiments were necessarily limited; many experiments were carried out offering the widely cultured diatom *Nitzschia* (*Phaeodactylum*) *closterium*. The development of algal culturing techniques has greatly extended the range of available phytoplankton species (diatoms and dinoflagellates) to include many nanoplankton organisms, so that a greater variety of food cultures, usually not bacteria-free, may now be offered. The great majority of species in regular culture, however, are coastal algae, some even estuarine or intertidal, so that results of testing oceanic zooplankton with such potential foods must be open to question. A much more serious limitation, however, is that all these experiments refer to herbivorous zooplankton. Almost the only animal food of suitable size that could be offered to potential carnivorous zooplankton was *Artemia* nauplii, which are easily grown in the laboratory. More recently, a few other cultured animals such as ciliates and rotifers have been offered as food. Such experiments merely answer the question whether animal food is consumed. Other tests of carnivorous feeding in the past have relied on capturing zooplankton in the field and offering this either unsorted or roughly sorted to the predator.

Over the last few years the experimental analysis of feeding carnivores, particularly chaetognaths, ctenophores and fish larvae by workers such as Reeve, Greve, Hirota and Howell, using graded sizes of rotifers, cirripede larvae, copepod nauplii and similar foods has greatly expanded our knowledge of food selection and feeding intensity, but there is still an urgent need to have a variety of zooplankton species in continuous laboratory culture to serve as standard food sources.

With algal cultures the criterion for assessing feeding is usually a cell count of the food culture. With animal food such as *Artemia* nauplii, counting is also usual. Direct observation of feeding on algae is obviously difficult, though Conover (1966) observed *Calanus hyperboreus* manipulating a large *Coscinodiscus* cell, mainly with the second maxillae and maxillipeds. After orientating the diatom, it was pushed against the mandibles and ingested. Apparently, the eliciting of the feeding response occurs when contact is made by a diatom cell with only a limited area of the mouth parts. When faecal pellets touched the mouth parts, they were rejected. Cushing (1968) also describes experiments by Petipa in which *Calanus* was observed to "locate" and capture a large diatom.

The use of algal cultures for feeding experiments was especially developed by Marshall and Orr (1955a, b) who demonstrated the considerable variety of phytoplankton ingested by *Calanus* (*vide infra*). Each experiment can demonstrate only whether a zooplankton species will accept a particular food item under the experimental conditions, when no other food is available. By carrying out a number of experiments—each using phytoplankton species from various taxa and differing, for example, in size, shape and even chemical composition—and then summarizing results of the series of experiments, the "preference" of the zooplankton animal for the algal food species could be determined. A problem is the wide variety of algal food which is accepted experimentally, but an analysis of the intensity of grazing will also help in assessing how far such foods are likely to be used in nature. Careful standardization of the conditions including food cultures is essential.

One improvement in the experimental analysis of food selection is to offer mixtures of two or more species of cultured algae and to deduce how far a zooplankton grazer will select one species (cf. Harvey, 1937; Marshall, 1973; Gaudy, 1974). Certain precautions are necessary. For example, if the algal species offered are of different mean size, the total volume of food represented by the different species should be approximately equal. In experiments of this type, Harvey (1937) found that with a mixture of *Ditylum* and *Lauderia* the former was strongly selected by *Calanus*. Similarly when mixtures of *Lauderia* and *Nitzschia* were offered, *Lauderia* was selected. There was a clear indication that larger cells are selected from a mixture, and the volume of water filtered was greater than when *Calanus* were fed on a culture of small cells only. Moreover, when a mixture of *Lauderia* and *Chaetoceros* was offered the former was markedly selected, suggesting that cell size may not be the sole factor in selection; shape and spikiness may be significant. Most experiments using mixtures of algal food species have indicated this tendency to select the larger cells.

Observations by Curl and McLeod (1961) also showed that with a natural tow netting rich in six species of diatoms, copepods (mostly *Pseudocalanus* and *Acartia tonsa*) fed actively on *Skeletonema* so that its density was markedly reduced, while the larger *Rhizosolenia* increased in numbers. Petipa (1965) claims that in nature copepods (*Calanus helgolandicus*) tend to feed on the species that are most plentiful rather than selecting larger cells, though this may be observed in laboratory experiments. For example, *Melosira*, *Cerataulina* and *Coscinodiscus* are selected in preference to *Nitzschia seriata* and *Prorocentrum*; *Chaetoceros curvisetus* is neutral. Yet *Prorocentrum* is consumed and is a satisfactory food in nature.

Mullin (1963) tested the food selection of several species of copepods regarded as being herbivorous, omnivorous or carnivorous taken from the Indian Ocean, using food mixtures consisting of *Artemia* nauplii and two or three species of phytoplankton (*Coscinodiscus perferatus*, *Thalassiosira fluviatilis* and *Cyclotella nana*). The difference in cell volume between the largest and smallest alga was nearly four orders of magnitude. Apart from finding fairly clear distinctions between predatory, omnivorous and particle feeders, Mullin observed that the Eucalanidae and Calanidae, which belong to the last category, also fed actively on *Artemia* nauplii. When offered the diatom food mixture, *Neocalanus gracilis* consumed all three species but selected *Coscinodiscus*; *Eucalanus attenuatus* appeared to feed only on the largest diatom and *Nannocalanus minor* showed no clear distinction between *Coscinidiscus* and *Thalassiosira*, but did not take *Cyclotella*. When offered only *Thalassiosira*, the Calanidae and Eucalanidae including *Rhincalanus* showed good production of faecal pellets. Most of the other copepods gave very poor production, indicating little particle grazing ability, but *Pleuromamma xiphias*, *P. abdominalis* and *Euchirella bella* exhibited a fair production of faeces. In all experiments with *Pleuromamma* spp., however, and in all with *Euchirella* (except one with C. III, *E. bella*) *Artemia* was selected completely in food mixtures including diatoms. Somewhat unexpectedly, *Chirundina* and *Lophothrix*, which are more generally regarded as being largely carnivorous, consumed some diatoms as well as *Artemia* from food mixtures. Some of the copepods certainly appear to be capable of varying their diets.

A modification of the method using one or more algal cultures as food was to label the algal cultures with an isotope, usually ^{32}P or ^{14}C. With single species the extent of grazing was determined by radioactivity counts of the cell suspension remaining after feeding, together with estimates of the activity of the bodies of grazers, faecal pellets or any other products (filtrate for excretory products, eggs) (*vide infra*). With food mixtures of two species Marshall (1973) also describes how two parallel experiments can be run with one of the algae labelled in one experiment and the alternative cells in the other. Assessment of the level of radioactivity in the two sets of grazing zooplankton would indicate whether either alga is selected.

Evaluation of the diet of zooplankton by exposing animals to a near natural population of food organisms has been achieved using Coulter Counter techniques, especially by Parsons and colleagues (e.g. Parsons, LeBrasseur and Fulton, 1967; Parsons and LeBrasseur, 1970). Samples of plankton, essentially phytoplankton, are analysed by the Coulter Counter so that a spectrum of the population is obtained, the density of various categories of phytoplankton being expressed in terms of size normalized to diameter (cf. Volume 1 for further details). A rapid microscopic examination of a plankton sample can also identify the major categories.

Relatively large samples of natural phytoplankton populations are obtained by means of containers such as large Van Dohrn or Niskin samplers. Zooplankton is collected independently by net hauls and carefully sorted into the main species. Known densities of a selected species are introduced to the natural samples of phytoplankton collected. Grazing is determined in terms of the quantities of each size category of phytoplankton consumed according to the change in the spectrum of the phytoplankton sample determined before and after the experiment. Poulet (1974), for example, studied food selection by the copepod *Pseudocalanus minutus* using essentially this technique (cf. also Boyd, 1976; Nival and Nival, 1976) (for results *vide infra*). Owing partly to the limitations of the Coulter Counter technique, it is not yet possible to use effectively a similar method to examine food selection and intensity of feeding for carnivorous zooplankton.

A slight variation of the method used by Parsons has been developed by Barlow and Monteiro (1979) and Greenwood (1971). Samples of plankton are obtained as before by the large bottle type of sampler, but part of the sample is carefully fractionated by filtering through plankton netting of selected mesh size. The "filtrate" contains the vast majority of the phytoplankton with bacteria and some small zooplankton, mainly protozoans and larvae. Of the "residue", essentially the zooplankton, a small sample is counted to ascertain the main species and their densities. Two similar cultures of the filtrate regarded mainly as phytoplankton, are then set up. To one a known volume of the zooplankton "residue" is added, thus introducing a grazing population of known density and composition; the other culture serves as a "phytoplankton" population control. After a suitable time interval the phytoplankton size spectrum is then determined for each culture using the Coulter Counter technique, the difference between the experimental and control culture indicating the grazing activity both qualitatively and quantitatively for the particular zooplankton population introduced.

Petipa, Pavlova and Sorokin (quoted by Vinogradov, 1973) employed a modification of the isotopically labelled food method to investigate the feeding in shipboard experiments of planktonic animals from the tropical Pacific. A variety of cultures of foods were grown and tagged with ^{14}C. These included bacteria (a known strain of *Pseudomonas* and natural planktonic bacteria) and several phytoplankton cultures, both dinoflagellates and diatoms. Labelled animal food was prepared by picking out various species of copepod from a natural plankton sample and feeding these with ^{14}C-tagged algae so that part of the radioactive carbon was assimilated. The species included *Acartia*, *Paracalanus*, *Scolecithrix* and *Temora* as well as the somewhat larger calanoids, *Eucalanus*, *Rhincalanus* and *Undinula*. The labelled bacteria, phytoplankton and copepods were subsequently offered as mixtures to various captured plankton animals to determine food selection. The relative quantities of the various foods consumed must be assessed, since many zooplankton, especially warm water species, are mixed feeders.

Detailed morphological examination of the feeding appendages and mechanisms of planktonic animals may be a useful guide to the diet. The investigations cannot determine the precise choice of diet but may suggest broad categories such as carnivorous or herbivorous feeding. Herbivorous or particle grazers tend to have mouth parts, especially the second maxillae, plentifully supplied with very fine setae, and mandibles with many blunt grinding cusps. Carnivorous feeders usually have sharp coarse-spined and grasping mouth parts and mandibles with large sharp, bladed edges, but this is not an absolute criterion (*vide infra*).

Few methods employed alone can yield accurate information on the diet of zooplankton. A combination of methods, including examination of mouthparts and of gut contents, with laboratory experiments using a variety of foods or natural phytoplankton may yield valuable results. The experiments should include some assessment of the rate of feeding and quantities of the different foods consumed. With particle feeding also, some further morphological determinations, for example, estimates of the inter-setular distance ("mesh size") on the main filter such as the second maxillae in calanoids may give an indication of the lower limit of size of food which can be retained effectively. Variations in the inter-setular distance and estimates of the main filtering areas can assist in providing a probability function for size of particle ingested (cf. Nival and Nival, 1976). Boyd (1976) has pointed out the need to allow for "leakage" of the filter in calculations. Although a number of factors may modify the filter feeding process, Nival and Nival obtained a good fit between the spectrum for particle retention and the mean inter-setular distance. Capture of large particles by the edges of the mouth parts and the seizing of larger items by raptorial feeding may occur in addition to filter feeding.

Whilst recognizing the paucity of information relating to the food of many zooplankton, especially of many open ocean inhabitants, and the difficulties in interpreting the results of investigations in diet, it is possible to distinguish broadly herbivorous, omnivorous and carnivorous types. There is general acceptance of the view that many species may also vary their diets as opportunity demands.

Zooplankton diets

The Protozoa is one of the least known taxa with regard to diet. Some indications of the food of particular groups of protozoans have already been indicated in Chapter 2. Many, especially deeper-living species, feed on bacteria and detrital material, but vast numbers of protozoans including foraminiferans, near-surface radiolarians and tintinnids are mainly herbivorous, feeding on phytoplankton. Heinbokel and Beers (1979) found that tintinnids could consume more than 20% of the total primary production, while earlier Beers and Stewart (1971) suggested that microzooplankton could account for 70% of the primary production. Thus with a shorter turnover time, the microzooplankton could prove a serious competitor to small crustaceans. It has been shown by Anderson, Spindler, Bé and Hemleben (1979) that the spined foraminiferan *Hastigerina pelagica* is carnivorous, but most non-spined foraminifera are omnivorous. Many radiolarians include symbiotic algae in their cell protoplasm. Most of the smaller naked ciliates are probably mainly detrital and bacterial feeders, but some of the larger ciliates are clearly carnivorous, and there is considerable evidence, albeit most of it circumstantial, that many Protozoa, especially small flagellates, live saprozoically, absorbing dissolved food materials from the water. How far they do this effectively at the normal concentration of organic matter in the sea is unknown; it may be that a limited number of Protozoa merely supplement their diets in this way.

Carnivorous groups

Among metazoan zooplankton, information concerning the diet of carnivorous forms is more scattered and limited than for herbivores, though groups such as medusae, siphonophores, chaetognaths and gymnosomes are generally acknowledged to be exclusively carnivorous. Part of the complexity in determining food items is undoubtedly related to the great variety of food which most predatory zooplankton will eat. Many species are opportunistic in their feeding habits. Copepods are, for example, regarded as a major item in the diet of species of chaetognaths. Indeed, in some areas such as Georges Bank and the North Sea, attempts have been made to model predator production on the consumption of *Calanus* by *Sagitta* (cf. also Petipa *et al.*, 1970, for the Black Sea). There is no doubt that sagittae prey very largely on copepods, but this is a reflection of the usual predominance of copepods in the marine plankton.

Chaetognaths will take a large range of other zooplankton. Reeve (1966) found that off Miami copepods, largely *Acartia* and *Paracalanus*, comprised 65–85% of the zooplankton and were the main food of *Sagitta hispida*, but this was supplemented by cirripede nauplii, decapod larvae and other crustacean and non-crustacean food. David (1955) reported that the Antarctic *Sagitta gazellae* consumed largely copepods including *Calanus propinquus*, *Rhincalanus*, *Metridia* and *Pleuromamma*, but other crustaceans (e.g. euphausiids and larvae, amphipods) and other chaetognaths were also eaten. Alvarino (1965) summarizes the food of chaetognaths as consisting of a range of crustaceans and fish larvae, other chaetognaths and medusae. She cites the results of Furnestin, which indicate that *Sagitta enflata* and *S. hispida* preyed on copepods, but also on tunicates, fish larvae, siphonophores, *Lucifer* and sagittae; *S. friderici* also consumed hyperiids and *S. hexaptera* was observed to eat *Candacia*. Since Wickstead (1959) found *Candacia* feeding on sagittae, this last observation is an example of the complex reversal of feeding relationships which can occur at higher trophic levels. Sagittae prey on fish larvae, but many fish larvae consume chaetognaths, giving another example of the complexities of the food web (cf. Lebour, 1922, 1923).

Reeve (1966) demonstrated that moving prey was selected by *Sagitta hispida* and that some size selection of prey was evident. *Sagitta* fed equally well in darkness and the light, and experiments by Figenbaum and Reeve (Chapter 2) suggested that artificial stimulation of non-motile cilia-like structures on the body of sagittae by vibration of the water, simulating movements of prey, elicited attack of the vibrating probe. In the natural environment copepods exceeding 1 mm in length were most often consumed. When offered *Artemia* nauplii of two sizes, *Sagitta* showed some preference for the larger prey and there was some evidence that the size of prey selected increased with size of the predator. Later work by Reeve and Walter (1972) confirmed this predator–prey size relationship. Whereas the youngest larvae chose prey of mean size (width) of about 88 μm, the largest older sagittae selected equally organisms of 250 and 400 μm size. They may also be cannibalistic to some extent.

Medusae, ctenophores and siphonophores are among the most characteristically carnivorous zooplankton. They take copepods and other common herbivores extensively, but this is a reflection of the abundance of copepods. Many are general zooplankton feeders, and since they may take both herbivores and carnivorous plankton, they are partly secondary carnivores. A more precise examination reveals, however, a restricted diet in some species. For instance, the ctenophore *Beroe cucumis* feeds on *Pleurobrachia*, and in Russian waters according to Kamshilov (1955) it feeds exclusively on other ctenophores, *Pleurobrachia*, *Mnemiopsis* and *Bolinopsis*. Swanberg (1974) has described the voracious attacks of *Beroe* on other ctenophores, the macrocilia of the lips cutting through the tissues. By preying on lobate ctenophores, it can reduce predation on general crustacean zooplankton and since some fish (cod, mackerel) eat *Beroe*, there is a complex "feed-back system". Swanberg found that *Beroe* is stimulated to swim by the presence of *Bolinopsis* (a chemokinesis), but it apparently can also locate its prey chemotactically. Dense patches of *Beroe* can decimate *Bolinopsis* populations. Although *Beroe* feeds almost exclusively on lobate ctenophores, Swanberg has seen it attack *Cestrum veneris*.

In contrast, as the investigations of Fraser (1970) have demonstrated, *Pleurobrachia* is a non-selective carnivore feeding mainly on copepods which together with some other crustaceans make up more than 80% of the diet in Scottish waters. Although *Pleurobrachia* eats fish eggs and larvae, these do not appear as a major fraction of the diet. Other accounts of the feeding of *Beroe* on ctenophores by Greve (1970, 1975), Zelickman (1972) and Burrell and Van Engel (1976) are discussed in Chapter 2. Though there is some argument as to how far species of *Beroe* are specific in their diet, there is general agreement, despite one observation of salp as prey, that they feed only on other ctenophores.

While the tentaculate and lobate ctenophores frequently seem to feed indiscriminately, Reeve and Walter (1978) report some conflicting observations from other workers. *Bolinopsis* and *Pleurobrachia* were held by Bishop to select small *Pseudocalanus*, although both Kamshilov and Lebour regarded the comparatively large *Calanus* as its major food. Kremer is also quoted as finding a reduced feeding rate with *Mnemiopsis* when offered zooplankton rich in cyclopoids and veligers, compared with abundant copepod and cladoceran prey. Reeve and Walter believe that cirripede nauplii are not readily captured by tentaculate ctenophores, though they are not discriminated against by *Mnemiopsis*. The relative activity of some potential zooplankton prey may also be significant in the amounts captured. Hirota (1972, 1974) found that a series of live copepod larvae, carefully graded in relation to the size of the ctenophore *Pleurobrachia bachei*, was required for its successful culture. Prey may be graded according to their active movement, which may break the fishing tentacles. Tentaculate ctenophores may show periods when feeding is discontinued and the tentacles even withdrawn, but lobate forms such as *Mnemiopsis* seem to feed almost continuously (Reeve and Walter, 1978), so that with rich zooplankton the gut can become gorged. Reeve and Walter (1977) also noted that plankton carnivores such as *Sagitta hispida* and *Mnemiopsis mccradyi* have the ability to ingest in a few minutes sufficient food to cover the daily requirement.

The patchiness of plankton is widely recognized and the successful nutrition of zooplankton is believed to be linked to encountering rich patches. The feeding behaviour of these carnivores appears to be especially well adapted to a patchy distribution of plankton (cf. *Acartia tonsa*, *vide infra*). The investigations of Harbison, Madin and Swanberg (1978) demonstrate that warmer water ctenophores at least show different feeding mechanisms, apart from the well known difference between lobate and tentaculate forms and that the prey may differ (cf. Chapter 2). The difference between *Lampea*, feeding only on salps, *Leucothea*, feeding on copepods and similar prey, and *Ocyropsis*, feeding on larger moving animals (euphausiids, small fish, hyperiids and even *Beroe*) is striking. The early observations of Lebour (1922) indicated that most hydromedusae fed on crustacean plankton, fish larvae and on other medusae. Zelickman *et al.* (1969) showed that *Tiaropsis* fed on *Rathkea*, though both medusae appeared to be rather unselective feeders. A number of hydromedusans and scyphomedusans (e.g. *Cosmetira*, *Aequarea*, *Cyanea* and *Chrysaora*) will eat *Pleurobrachia* and other ctenophores. Russell (1970) indicates some of the differences in the diet of three common Scyphozoa. *Aurelia* will eat almost any zooplankton organisms, copepods, amphipods and other crustaceans, eggs and a variety of meroplanktonic larvae, pteropods, medusae such as *Obelia* and *Phialidium* and a variety of young fish. Small ctenophores may also be eaten. *Chrysaora* is less omnivorous, not consuming copepods or other crustaceans, but eating hydromedusae, siphonophores and ctenophores. *Sagitta* and *Tomopteris* are eaten, but neither *Leuckartiara* nor *Beroe* are accepted. Miller (1974) described *Chrysaora* as feeding voraciously on *Mnemiopsis*.

The complexities of the planktonic food web are again obvious. *Cyanea lamarckii* appears to be even more restricted in its diet. Though the ephyrae will feed on copepods and eggs and medusae, the adults take only medusae (especially *Phialidium*, *Cosmetira*, *Laodicea*, *Obelia* and *Steenstrupia*) and some small *Beroe* and *Pleurobrachia*. Some medusae, such as *Neoturris pileata*, are not accepted, but according to some authorities ctenophores are very regularly consumed. The food of many medusae, especially those from more open waters, seems little known. Hamner *et al.* (1975) have called attention, however, to the great depredations made by medusae observed by Scuba diving methods. In more open sea conditions the trailing extended tentacles can sweep very large volumes of water. They report a large species of *Aequorea*, for example, with tentacles up to 2 m in length, ingesting large quantities of a swarm of salps. The food of the very large deep-sea medusans *Stygiomedusa* (Russell, 1960) and *Deepstaria* (Russell, 1967) is unknown. As carnivorous animals, they presumably feed on the very sparse zooplankton in the deep ocean. Possibly their growth rate and, even more significantly, their metabolic rate, may be very low.

Apart from these exceptionally large medusae, it would be interesting to know something of the food requirements of the deep sea *Trachylina* and of *Periphylla* and *Atolla* and of those siphonophores which inhabit ocean depths. Pugh (1977), reporting observations by Biggs, stated that siphonophores generally feed on a wide range of zooplankton from copepod nauplii to decapod crustaceans and small fish. Bieri (1966, 1970), however, demonstrated a remarkable separation of feeding niches with three siphonophores (*Velella*, *Porpita* and *Physalia*) occurring in the neuston. *Velella* fed mostly on fish eggs and on small larval crustaceans, and only some 10 % of the food consisted of copepods. The diet of *Porpita* differed sharply and was composed chiefly of carnivorous calanoids, including *Euchaeta marina* and large active pontellids such as *Labidocera acuta*. These predaceous copepods made up about 90 % of the diet, the remainder consisting of crab megalopas and fish. Separation of the food niches also applied to *Physalia*; Bieri concluded that *Physalia* fed essentially on fish. As we have earlier noted, certain pelagic gastropods feed on these siphonophores. Bieri found *Ianthina*, *Glaucus* and *Fiona* in the neuston feeding on the three species. They would thus rank at least as tertiary carnivores if feeding on *Porpita* or *Physalia*. Bayer (1963) confirmed that the three siphonophores were eaten by *Ianthina*; he also stated that *Glaucus* and *Fiona* preyed on *Velella*.

The economic importance of fish larvae is probably responsible for some considerable knowledge of the diet of these meroplanktonic animals. Though a little phytoplankton may be eaten in the very early stages of development, in the great majority of fish larvae the diet is carnivorous following absorption of the yolk sac (Lebour, 1921). Many larval species also appear to show little selection in the prey. Thus many fish larvae feed mainly on copepods, reflecting the general abundance of these crustaceans. But size of prey is often significant. For example, Wiborg has shown that cod larvae of different ages select different species and stages of copepods and that the breadth of mouth and gullet of the fish larva is an important factor in this selection. Cod larvae were also observed by Wiborg (1948a,b, 1949) to take euphausiids, decapod larvae and other crustaceans in addition to copepods. Occasionally the larvae of other fish were consumed. Any selection was apparently based mainly on size.

Lebour (1918, 1919), studying the food of several species of larval teleosts, noted that following the brief period during which a little phytoplankton was eaten, the main food consisted of the common copepods, chiefly *Pseudocalanus*, *Temora*, *Acartia*, *Calanus* and *Oithona*. Cladocera were readily eaten when available, and also balanid nauplii and larval molluscs. The diet was influenced largely by the abundance of the food species, but there was evidence of some selection, mainly with regard to size in relation to the size of gape and of the larva. There was some evidence of further selection. Few decapod larvae were eaten by young fish and the amphipod *Hyperia* was not taken, even though readily available. Marak (1960) concluded that there was very little selection apart from size in the feeding of larvae of the gadoids, cod, haddock and coalfish. Small larvae subsisted mainly on larval copepods; larger post-larvae on adult copepods, while the largest post-larvae could feed to some extent on young euphausiids and amphipods as well as on the copepods. *Centropages*, *Pseudocalanus*, *Paracalanus*, *Calanus*, *Temora* and *Tortanus* were all used as food provided they were suitable in size and the larval fish fed on the most abundant species. Russell's (1976) recent publication on British larval fishes indicates the food of a large number of species. Copepods usually rank as the major item, but other crustaceans as well as mollusc larvae, sagittae and other zooplankton are consumed, following the preliminary feeding on algae. Medusae are sometimes eaten. Wiborg (1960) reports that haddock try feeding on them.

That dietary requirements change with age in fish larvae is illustrated by the investigations of Hardy (1924) on the food of the herring. At a very early stage following the limited consumption of phytoplankton food, herring larve eat mostly larval molluscs and young copepods, especially nauplii. As larval molluscs become less important with growth of the herring larva, *Pseudocalanus* replace them as the chief food. Older post-larvae will take these and slightly larger copepods (e.g. *Temora*), but there is a considerable range of diet including sagittae, mysids, decapod larvae and other species (cf. Fig. 6.3).

Factors other than size of prey may be significant in food selection. The vulnerability of zooplankton may be important, especially in relation to developing predatory ability of the carnivore. Rosenthal and Hempel (1970) described the increasing abilities of herring larvae with age to locate suitable prey, essentially by sight, and the development of a rapid swimming capability. Greater manoeuvrability and co-ordination enabled larvae to capture sighted prey more effectively. There is distinction between the presence of an apparently sufficient plankton biomass and the ability of a fish larva to ingest enough at a given food density. Investigations by LeBrasseur (1969) and Parsons and LeBrasseur (1970) in which juvenile salmon were allowed to feed on zooplankton suggested that they tended to select ("elect") the larger copepod *Calanus plumchrus* over *Pseudocalanus*. Feeding was tested on crops of the two copepods and of a euphausiid. At a prey density equivalent to 20 g/m^3 for the three foods, juvenile pink salmon could obtain about 30 mg (wet wt) of food per hour from the *Calanus* crop, less than 20 mg/hr from the euphausiid and only around 0.3 mg/hr from *Pseudocalanus*. At a proper prey density the young fish could thus obtain a sufficiency of food for maintenance purposes from *Calanus*, whereas this was not attained with

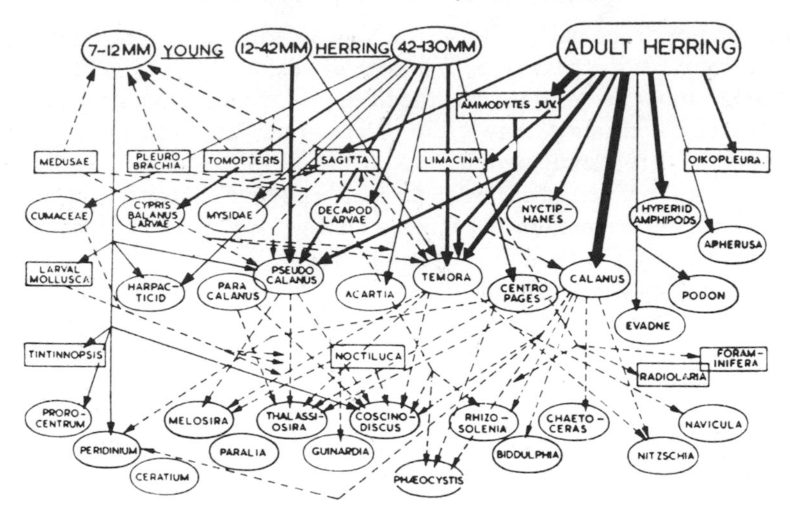

Fig. 6.3. The food of the herring (Hardy, 1924).

Pseudocalanus alone even when the prey density was increased to 90 g/m^3.

With the euphausiid *Euphausia pacifica*, it was just possible for sufficient food to be captured, though *Calanus* provided a more adequate diet (cf. Fig. 6.4). With *Microcalanus*, the fish also obtained the required food less rapidly at relatively high prey densities than when *Calanus* was offered, though a maximum ration could be obtained from *Microcalanus* at a reasonably low biomass. Parsons and LeBrasseur found that when offered a variety of zooplankton, juvenile salmon elected copepods of 2–4 mm length, were less selective to smaller copepods and were negatively selective to

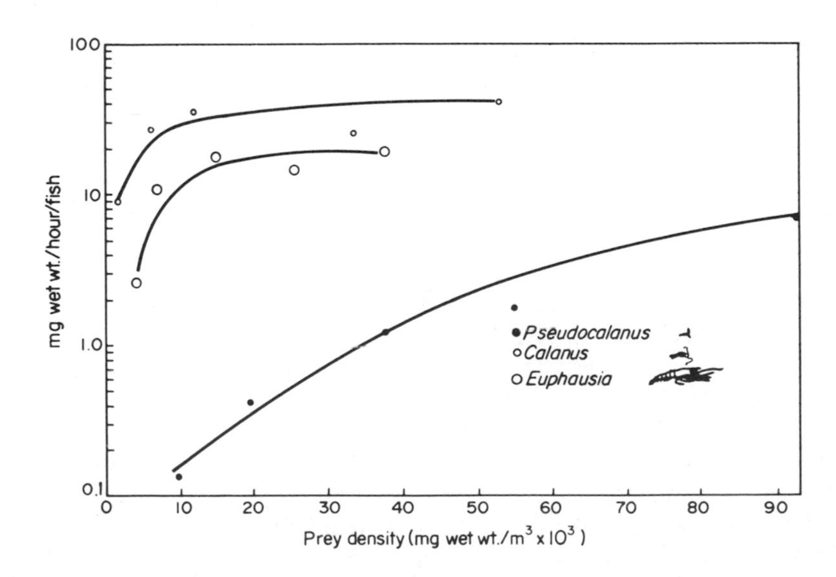

Fig. 6.4. Juvenile pink salmon rations at different prey concentrations (●) *Pseudocalanus*; (○) *Euphausia pacifica*; (○) *Calanus plumchrus* (Parsons and LeBrasseur, 1970).

other non-copepod prey. Although only preliminary tests were carried out with larval fish (*Ammodytes* and *Hexagrammus*), a size range between 0.5 and 1.5 mm of prey was selected, represented by the copepods *Microcalanus*, *Oithona* and *Pseudocalanus* and nauplii of larger species.

Some examples of greater selectivity among fish larvae are known, however. Plaice (*Pleuronectes platassa*) larvae take phytoplankton at an early stage and later become carnivorous and are believed to feed largely on copepod nauplii and mollusc larvae. Shelbourne (1953, 1957, 1962) demonstrated that although copepod nauplii were taken, *Oikopleura* was also selected. Investigations suggested that North Sea plaice larvae fed exclusively on *Oikopleura*. While smaller larvae were restricted to smaller prey (as shown by the size of *Oikopleura* in faecal pellets), older larvae took a wider size range and appeared to show some discrimination for larger prey. Ryland (1964) confirmed the great importance of *Oikopleura* in the diet of plaice larvae, but *Fritillaria* was also eaten. Even copepod nauplii were not significant as food. *Ammodytes* larvae, especially older stages, also take appendicularians, though they show a slightly wider range of diet. Despite the restricted diet of North Sea larval plaice, they can be reared on other animal foods. A mixture of barnacle larvae and *Artemia* nauplii has been successfully used for the younger stages, and larger *Artemia* for metamorphosing plaice (Shelbourne, Riley and Thacker 1963). Howell (1973) has fed plaice larva on the rotifer *Brachionus*.

Although a large number of fish species feed for a brief period on planktonic algae and then become carnivorous, a few pelagic fishes such as the Peruvian and Argentinian anchovies apparently feed on zooplankton in the youngest stages when the gill rakers are relatively coarse and change to a largely herbivorous diet of phytoplankton as adults. The adult gill raker is a finer filter (cf. Ciechomski, 1967; Rojas de Mendiola, 1971). Kuttalingam (1959) found the same change in diet with certain Indian fishes which are plankton feeders as adults.

There is little precise information on many of the generally carnivorous zooplankton taxa beyond the statement that the animals prey widely, probably feeding to a large extent opportunistically (e.g. pelagic polychaetes and nemertines). The widespread plankton polychaete *Tomopteris* may be ranked as a secondary carnivore; Suszczewski (1968) has observed its feeding, which includes a kind of sucking predation on *Sagitta*. Decapods are believed to be mainly carnivorous, though early stages are often herbivorous and later larvae progress to an omnivorous diet. Omori (1969, 1974) found that *Sergestes lucens*, whilst occasionally having detritus and diatoms in the gut, was essentially a carnivore. It fed mainly on copepods, including such carnivorous species as *Euchaeta marina*, and on euphausiids and their larvae, other decapods and some sagittae. *Sergestes similis* was observed to consume chiefly *Calanus pacificus* and *Metridia pacifica*, both dominant copepods in upper mesopelagic strata. Decomposing material was also taken.

Omori describes some differences in diet with depth distribution. For example, *Lucifer* ate some algae (*Trichodesmium*) in shallow coastal waters; *Sergestes* spp. and *Pasiphaea* at approximately 100 m fed chiefly on herbivorous copepods; *Acanthephyra* spp. and *Systellaspis debilis* living below 150 m consumed ostracods, decapods and *Cyclothone*. In addition, they take particulate material such as faecal pellets from upper levels. Bathypelagic species (e.g. *Bentheogennema borealis*) consume carnivorous copepods, chaetognaths, ostracods and medusae and detrital particles. The change in feeding types recalls the observations of Vinogradov (1973) for the north-western Pacific referring more generally to zooplankton. The uppermost layers are dominated by filterfeeders; a mixture of trophic types occurs to about 500 m; from 1500 to 3000 m there is relative abundance of carnivores; from 3000 to 4000 m many euryphagous animals occur with carnivores; and euryphages are dominant near the bottom. Vinogradov and Parin (1973) suggest that an alternation of layers of carnivorous groups occurs in the tropical Pacific.

Amphipods are also regarded as usually carnivorous. Kane (1963) suggested that *Parathemisto* might take dead as well as living tissue and Conover (1960) fed *Hyperia* and *Parathemisto* on living and dead plankton. Della Croce (1973) observed *Anchylomera* feeding on *Penilia*. Little is known of the food of the deep-water Lanceolidae, though Bowman and Gruner (1973) believe that they feed on the tissues of deep-water siphonophores and medusae. Although the precise relationship in the associations between near- and surface-living species of hyperiid and various "gelatinous" type zooplankton (medusae, ctenophores, siphonophores, salps) is uncertain, undoubtedly many of the hyperiids feed on the tissues of the animals with which they are associated. The degree of dependence varies from species to species and also with stage of development and other factors. Thus Bowman and Gruner (1973) quote observations from Laval that *Lestrigonus schizogeneios* and *Bougisia ornata*, both associated with *Phialidium*, shared food with the medusa when feeding was adequate, but fed on the medusa tissues when food was scarcer.

A *Vibilia* female transfers each young stage from the marsupium to a salp. The larva enters and feeds on the tissues until it can become planktonic. The young of *Phronima sedentaria* also feed to some extent on the tissues of the salp which also form the protective "barrels". According to Shih (1969), little is known about the diet of adult Phronimidae. Though phytoplankton is sometimes found in the gut, it may not have been taken directly and the structure of the limbs would not appear to support filter feeding. From an examination of the mouth parts and the general habit of Phronimidae, Shih believes that these hyperiids are probably omnivorous. Madin and Harbison (1977) found that *Vibilia* feeds on the food material collected by the salp with which it associates, but most hyperiids consume some of the host tissue (cf. Chapter 2). There is no doubt, however, that some hyperiids largely use medusae, ctenophores and other plankton as surfaces from which they can be carried in the plankton, occasionally moving free in the water. Forms such as *Parathemisto* and many oxycephalids are much more active swimmers, and many of the latter feed actively on ctenophores (Harbison *et al.*, 1978). Repelin (1970) also found siphonophores and medusae as a major part of the diet of amphipods (cf. Chapter 2).

Sheader and Evans (1975) confirmed the carnivorous feeding of *Parathemisto gaudichaudi*. A little phytoplankton occasionally found in the gut is believed to be derived from the contents of the alimentary canal of herbivorous prey. Unlike many hyperiids, *Parathemisto* is a very general carnivore. Larger copepods are especially acceptable, but other zooplankton, including smaller copepods, sagittae, decapod larvae, euphausiids and fish larvae are all readily taken. At times of the year predation on fish larvae can be considerable. A few species, e.g. *Clione* and *Tomopteris* were rejected. Hydromedusae of several species were also acceptable and from field observations at least one medusa, *Aequorea*, was fairly commonly eaten over certain seasons. *Parathemisto* often remained on the surface of a medusa after eating some of the tissues, a condition reminiscent of other hyperiid associations. Scyphozoa were not consumed. The type of prey was related to the body size of the hyperiid, but there was a seasonal difference in the feeding intensity. The usually smaller animals fed at about fifteen times the rate of the winter specimens, but only about one half of the summer population were actively feeding as against almost the whole of the winter population. Sheader and Evans suggest that the gut in *Parathemisto* is distinctly simplified compared with as other amphipods. The foregut especially lacks the cardiac masticatory chamber with chitinous ridges and spines and the pyloric filtering chamber. Instead there is an extensive digestive chamber with only one pair of digestive caeca. Apparently such a simple gut is typical of other hyperiids and may be related to their general diet of gelatinous zooplankton.

Heteropods are exclusive carnivores. Some species eat sagittae as well as copepods and other zooplankton, including fish larvae, ctenophores, siphonophores and other pelagic gastropods, but a favourite food appears to be salps (cf. Chapter 2). According to Hamner *et al.* (1975), the heteropod hunts its prey visually, apparently lying on its dorsal side and thus obtaining a silhouette of the prey.

Although most carnivorous zooplankton have a wide dietary range, a few examples have been quoted of species exhibiting considerable selection (e.g. *Beroe*, *Pleuronectes* larvae). An unusually clear example of selection appears from an examination of the feeding of gymnosomes. *Clione limacina* is a well known gymnosome, occurring widely in very cold polar seas and in boreal temperate seas. Ussing (1938) stated that in polar waters the prey of *Clione* consisted largely of shelled pteropods. According to Mileikovskiy (1970b), *Clione* feeds exclusively on *Limacina* (*Spiratella*), in polar waters on the Arctic species *L. helicina* and in boreal waters on the temperate species *L. retroversa*. The young veligers of *Clione* are herbivorous, feeding on phytoplankton. A marked specificity of thecosomes (shelled pteropods) as prey for particular species of gymnosomes (naked pteropods) appears to apply to other gymnosomes. Lalli (1970) suggests that gymnosomes are highly selective carnivores preying on thecosomes and she instances *Pneumoderma* consuming *Cavolinia* and quotes reports by Sentz-Braconnot that *Pneumodermopsis paucidens* is a predator on only two species of *Creseis*; even the thecosome *Cavolinia* was rejected. A further example is that *Spongiobranchaea australis* feeds solely on *Clio* (Conover and Lalli, 1972).

Examination of the behaviour and mechanisms of feeding by *Clione* indicate that an extreme selective feeding pattern has evolved. The whole feeding apparatus of the gymnosome is perfectly adapted to seize the prey and extract all the organic matter, leaving the shell. A precise method of orientating the prey when captured is included in the feeding process. Digestion and assimilation is remarkably efficient and *Clione* can withstand fairly long periods without food, but obtains maximum benefit from the abundant food when *Limacina* swarms. There is also selection for size of prey. The obvious disadvantage of the extreme prey specialization, that *Clione* (and other gymnosomes) cannot make use of the general zooplankton as food, thus appears to be outweighed by the remarkable efficiency of the nutritional pattern (*vide infra*). *Paedoclione*, also investigated by Lalli (1970) and Lalli and Conover (1973), appears to feed on the same food as *Clione*. The trochophore is herbivorous, but develops into a highly selective carnivorous adult. The adult is neotenous, however, and in size and time of abundance appears to avoid direct competition with *Clione* for the available smaller *Limacina*.

There is no very obvious distinction between the trophic relationships of the larger zooplankton (macroplankton) and the smaller micronektonic fishes and cephalopods. The majority of the micronektonic fishes live by day at 600 m or 800 m; they include small gonostomatids, hatchet fishes and above all myctophids. These fishes are plankton feeders, preying particularly on copepods and on euphausiids, amphipods and chaetognaths as well as on small squid and pteropods—also possibly on siphonophores and jellyfish. A substantial diurnal vertical migration occurs in many of these oceanic fishes, though the hatchet fishes move relatively little. Marshall (1954, 1960) claims that this pattern is essential for myctophids if they are to obtain the necessary food supply in the surface layers.

Amongst other nekton, some cephalopods appear to use migration and the "nutrition ladder" (cf. Chapter 5) to feed near the surface. Some small species such as *Cranchia* live in the upper zones together with the young of some deeper species, but others migrate from intermediate depths (cf. also migrations of *Pyroteuthis margaritifera* and *Pterygioteuthis giardi*; Gibbs and Roper, 1970). Many cephalopods consume zooplankton, including larger copepods, decapods and other crustaceans and pteropods. Nesis (1965) found that young squid of *Gonatus fabricii* fed on plankton (copepods, euphausiids, amphipods and sagittae) and that there was a clearly defined grading of size of prey with increased size of the squid predator. Milliman and Manheim (1968) refer to concentrations which were probably sergestid shrimps, squid and myctophids all showing vertical movements. The migrations may have a trophic relationship (cf. Chapter 5). Sergestids may act as intermediate carnivores in the food web. Omori (1969) described pronounced vertical migrations by *Sergestes lucens* from >200 m by day to near the surface at night. They fed on zooplankton (*vide supra*) and at least two fishes, *Diaphus caeruleus* and *Gephyroberyx japonicus*, accompanied the migration and fed very actively on them. The vertical migrations of herbivorous and carnivorous zooplankton and of mixed feeders may assist markedly in an efficient utilization of the basic primary production in all oceans (cf. Vinogradov, 1962; Foxton, 1964; Raymont, 1970; Vinogradov and Menshutkin, 1977).

In contrast to the outstanding carnivorous taxa of zooplankton, the predominantly herbivorous groups include the copepods, cladocerans, euphausiids, probably the ostracods and some other crustaceans together with the appendicularians, salps and the shelled pteropods. The larvae of most of the bottom invertebrates, such as veligers, trochophores, nauplii and the plutei and other larvae of echinoderms, feed on phytoplankton, though most decapod larvae are at least partly carnivorous at later stages. The younger stages of many holoplanktonic animals are also algal feeders. It is possibly better to describe this array of zooplankton as "particle feeders" rather than herbivores, since a large proportion of species, possibly almost all, may take non-living matter and small animals such as protozoans to a limited degree.

Predominantly herbivorous groups

Though the major taxa, particularly the copepods and euphausiids, are described as predominantly particle feeders, they include many species which are clearly omnivorous and numerous carnivorous forms. For convenience, these are described in the present section, especially since some species can vary their dietary habits.

Copepoda

The copepods are undoubtedly the most generally studied taxon of the herbivorous groups. In very early investigations the dependence of copepods on the blooms of diatoms, particularly in sub-polar regions, was widely recognized. Dakin, Esterley and Marshall were among the earlier investigators who showed that diatoms were the chief food of many copepods, but that a wide variety of phytoplankton might be consumed. Dakin (1908) recorded phytoplankton, chiefly diatoms, but also peridinians and unidentified microalgae, from the guts of *Calanus finmarchicus*. Esterly (1916), however, from an examination of *Gaidius* and *Undeuchaeta*, found that copepod remains as well as phytoplankton occurred in the gut contents, indicating a mixed diet. Beklemishev (1954) investigated the gut contents of several calanoids, *Calanus* spp. and *Eucalanus bungii*, in North Pacific waters. Diatoms predominated in the diet in more northern waters, many of the larger diatom species being of considerable importance. The larger cells are partly broken before being swallowed, but smaller diatoms are eaten whole. There was evidence that the comparatively deep-living copepods *Gaidius* sp. and *Gaetanus minor* also ate large diatoms such as *Coscinodiscus oculusiridis*. In warmer areas Beklemishev found that the copepods consumed a greater proportion of *Exuviaella*, coccolithophores and peridinians. Dinoflagellates and other flagellates are also of importance in providing nutriment over the summer in temperate waters, when diatoms are less abundant.

Competition for the phytoplankton available, especially in the more oligotrophic areas of the ocean such as the sub-tropics, may be partly offset by particle feeders differing in their selection for particle size. There may be some vertical separation of grazing faunas with similar demands on the phytoplankton. In the North Pacific, where comparatively rich crops of phytoplankton occur, Vinogradov (1973) suggests that the dominant copepods, *Calanus cristatus*, *C. plumchrus* and *Eucalanus bungii*, feed on virtually the same food. For example, of 31 diatom species acting as major food items, 15 were found in all three species and 10 diatoms in two copepod species. But direct competition was avoided to some degree by differences in the vertical distribution of the three copepods, marked in summer and especially in autumn when food was limited, but little observed during spring when abundant food was available. *Pseudocalanus*, according to Vinogradov, avoided competition to some extent since it tended to consume generally smaller diatom species.

There seems little doubt that *Pseudocalanus minutus* is almost entirely herbivorous, even though crustacean remains have very occasionally been found in the gut. Earlier studies listed not only diatoms (*Coscinodiscus*, *Thalassiosira*, *Chaetoceros*), but also flagellates, including coccolithophorids, as food. Corkett and McLaren's (1978) review indicates that long-spined *Ceratium* and spiny chain-forming diatoms such as *Chaetoceros* may be less readily eaten, though chains of *Chaetoceros* were consumed, according to Poulet's (1974) observations. Small algae such as *Exuviaella* are often an important part of the diet. Corkett and McLaren analysed a study by Zagorodnyaya on *Pseudocalanus* from the Black Sea which demonstrates that of 50 species of phytoplankton consumed, mostly diatoms and dinoflagellates, *Exuviaella*, *Coccolithus huxleyi* and the diatom *Cyclotella caspia* were the major diet items, but that diatoms tended to be more important than dinoflagellates in relation to their abundance in the water.

The main selection appeared to be based on particle size, however, and the general conclusion is that there is a selection of smaller algae, mostly < 10 μm, by copepodites and adult females. The extensive studies of Poulet (e.g. 1973, 1974, 1976) dealt with selection of particle size by *Pseudocalanus* and demonstrated that although a wide range of sizes can be filtered from < 4 μm to >100 μm diameter, smaller particles (<25 μm) were grazed when these amounted to about half of the total and larger particles when these were equally or more concentrated. Within limits, *Pseudocalanus* appeared to graze on those particles in greatest concentration. However, there are indications that the size range below 10 μm is less effectively grazed, though Poulet (1977) found that younger copepodites consumed particles < 10 μm more effectively than adults.

Beklemishev's (1954) observation that almost all the obvious herbivores in the Pacific could show some animal remains in the gut, especially in winter, agrees with older observations from the Atlantic that *Calanus finmarchicus*, though outstandingly herbivorous, may take some animal food in times of algal food scarcity. The same is probably true of *Pseudocalanus*, *Paracalanus*, *Calocalanus* and many others. Mullin (1966) describes Calanidae and Eucalanidae (*Eucalanus* and *Rhincalanus*) as herbivorous copepods which will take some animal food. Details of the diet of a particle feeding copepod are best known for *Calanus finmarchicus* and have been excellently summarized in the work of Marshall and Orr (1955a). *Calanus finmarchicus* (*sensu lato*) feeds on a wide variety of diatoms, including some of the largest forms like *Coscinodiscus* and *Ditylum brightwelli*. It will also readily graze on small species of *Chaetoceros* and especially on *Skeletonema costatum* which, in boreal waters such as the Clyde area, is one of the main spring diatoms. *Calanus* will take even the very small *Nitzschia closterium* var. *minutissima* (= *Phaeodactylum*). Harvey (1937b) suggested, however, that there is some selectivity, larger species being selected and a greater volume being swept clear. *Calanus* may also feed on dinoflagellates. Field examinations and laboratory experiments demonstrated that such species as *Prorocentrum*, *Peridinium*, *Gymnodinium* and *Syracosphaera* are readily utilized by *Calanus*, but some powers of selection may again be apparent.

Marshall found that silicoflagellates and coccolithophores did not form an important part of the diet, and *Ceratium*, though often very abundant in the seas, was rarely eaten by *Calanus*. Small flagellates such as *Dicrateria*, *Pyramimonas* and *Chlamydomonas* which were included in the nanoplankton were shown by Raymont and Gross (1942) to be grazed, although the quantity of food obtained from nanoplankton may be small. From an examination of the distance between the setules of the feeding appendages, Ussing (1938) claimed that organisms below a diameter of 4–5 μm might pass through the filtering appendages; Marshall and Orr's (1956) measurements gave the smallest distance as 2–3 μm. Marshall (1973) suggests that at the comparatively high food densities used in feeding experiments the setules may be partly clogged. Extremely small algae (e.g. *Nannochloris oculata*) may not be retained (Fig. 6.5) by *Calanus*, and probably even larger nanoplankton may not contribute very significantly to the food supply (cf. Berner, 1962) (Table 6.1).

Mullin (1963) tested four species of *Calanus*, including *C. finmarchicus* and *C. helgolandicus*, for food selectivity, using eight species of phytoplankton. Larger-celled forms gave higher grazing rates when small- and large-celled species were fed in a mixture, i.e. larger species were selected. From a mixture of five species, nanoplankton contributed < 6% to the diet.

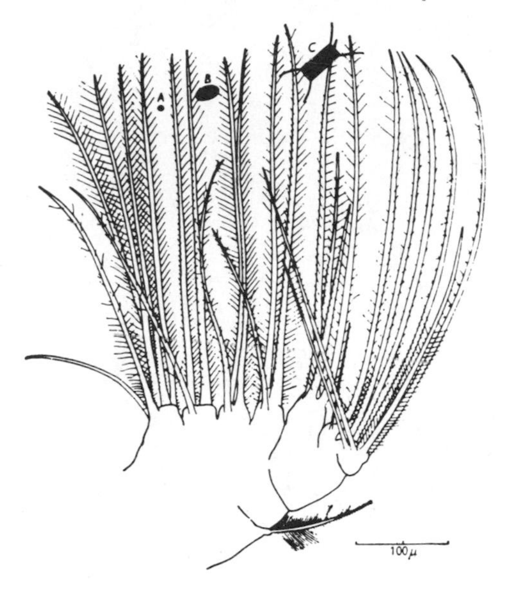

Fig. 6.5. The left maxilla of *Calanus helgolandicus* (A) *Nannochloris oculata*; (B) *Syracosphera elongata*; (C) *Chaetoceros decipiens* (Marshall and Orr, 1956).

Even in an important herbivorous taxon such as the copepods, of which many species are outstandingly dominant grazers over vast oceans, carnivorous species occur. Esterly (1916) described abundant copepod remains from the gut of the copepod *Scolecithrix* (*Scottocalanus*) *persecans*. Deep-living copepod families such as the Aeteidae, Arietellidae, Augaptilidae and Euchaetidae, on the evidence of mouth part structure, are believed to have a large number of omnivorous or carnivorous species (cf. Chapter 2). The shearing mandibles, coarse setae and piercing and holding spines in maxillae and prehensile maxillipeds in such deeper-living genera as *Pareuchaeta* are associated with a predatory habit (Bogorov, 1958c). Ussing (1938) also points out that species such as the deep-sea copepods *Bathycalanus*, *Megacalanus* and *Valdiviella* are almost certainly animal feeders. They live on other copepods, particularly those that tend to drift down from the upper layers. Harding (1974) refers to work by Russian investigators indicating that 15 out of 19 copepod species from the deep sea were carnivorous. There is little doubt, however, that such deep-sea forms will supplement their diet with carcasses and faecal material. The relatively coarse texture of the setae in such copepods implies also that they are unlikely to be able to retain much phytoplankton or other fine particle food (cf. Conover, 1960). Laboratory experiments confirm that *Euchaeta* will not feed on phytoplankton. Many other deeper-living and fairly widespread genera (e.g. *Metridia*, *Pleuromamma*, *Euchirella*) are also thought to be partly carnivorous. Haq (1967) fed *Metridia* with small and large diatom and flagellate food mixtures to which *Artemia* nauplii were added. Apart from some selection of the larger cells, *Metridia*, especially *M. lucens*, showed a preference for animal food (cf. also Mullin, 1966). A mixed diet appeared to satisfy nutritional requirements. Beklemishev also observed mixed food in gut contents of *Metridia pacifica* and *M. ochotensis*.

Table 6.1. Contribution of algae to food supply and clearance rates (Jorgensen, 1966a)

Species	Algae in suspension	Clearance, ml/24 hr/animal	Reference
Marine species			
Calanus finmarchicus, V + VI	*Ditylium*, 50 × 100 μm	170–240	Harvey, 1937
	Lauderia, 30 μm	50–100	
	Chaetoceros	0	
Calanus finmarchicus, V	*Chlamydomonas*, 7 × 12 μm	100	Gauld, 1951
Calanus finmarchicus, IV		50	
Calanus finmarchicus, III		29	
Calanus finmarchicus and *C. helgolandicus*, ♀	>10 μm flagellates	Max. 84	Marshall and Orr, 1961a,b
	>10 μm diatoms	Max. 43	
C. finmarchicus, Nauplius III	>10 μ algae	Max. 1.0	Marshall and Orr, 1956
C. finmarchicus, Nauplius IV		Max. 1.2	
C. finmarchicus, Nauplius V		Max. 1.7	
C. finmarchicus, Nauplius VI		Max. 2.0	
C. finmarchicus, Copepodite I		Max. 2.8	
C. finmarchicus, Copepodite II		Max. 6.4	
C. finmarchicus, Copepodite III		Max. 9.2	
Calanus finmarchicus, V		128–7,018, av. 2,850[a]	Cushing and Vucetic, 1963
Calanus finmarchicus, V	Natural phytoplankton	60	Cushing, 1958a
C. helgolandicus		16–50, av. 30; winter[b]	Cowey and Corner, 1963
		10–75, av. 50; summer	
Pseudocalanus minutus, ♀	>10 μm flagellates	9	Marshall and Orr, 1962
Centropages hamatus, VI	*Chlamydomonas*, 7 × 12 μm	13	Gauld, 1951
Centropages hamatus, ♀	>10 μ flagellates	Max. 15	Marshall and Orr, 1962
	>10 μ diatoms	Max. 2.7	
Temora longicornis, VI	*Chlamydomonas*, 7 × 12 μm	8	Gauld, 1951

Temora longicornis, ♀	>10 μm flagellates	Max. 16		Marshall and Orr, 1962
	>10 μm diatoms	Max. 18		
Temora longicornis, ♀	*Skeletonema costatum* chains	Max. 27, av. 6.5		Berner, 1962
Temora longicornis, ♀	*Skeletonema costatum* chains		150	Cushing, 1958a
Acartia clausi ♀	>10 μm flagellates	Max. 9		Marshall and Orr, 1962
	>10 μm diatoms	Max. 2.7		
Metridia lucens, ♀	>10 μm flagellates	Max. 1.3		
	>10 μm diatoms	Max. 0.4		
Oithona similis, ♀	>10 μm flagellates	Max. 0.02		
	>10 μm diatoms	Max. 0		
Freshwater species				
Diaptomus gracilis and *D. graciloides*	*Chlorococcus*, *ca.* 6 μm	4.1 (370 ml/mg dry wt.)		Malovitskaya and Sorokin, 1961
Diaptomus graciloides	Natural phytoplankton	1–3 (200 ml/mg dry wt.)		Nauwerck, 1959

[(a)] Estimates from field observations.
[(b)] Calculated as volumes of water that should be cleared of various essential amino acids in particulate form in order to cover requirements of starving copepods.

The experiments of Mullin (1966), carried out on shipboard in the Indian Ocean, have already been mentioned with regard to food selection. The investigation was supplemented by an examination of gut contents and mouthpart structure. A somewhat fuller account of the differences in the structure of mouthparts of various copepods feeding on plant and animal foods is given on pages 549–551. Most Calanidae and Eucalanidae had mouthparts adapted to herbivorous feeding; the gut contents consisted of diatom and radiolarian fragments, though some crustacean remains were found. In *Neocalanus* and *Rhincalanus*, however, the mandibular teeth are somewhat sharper than in other genera, the gut contents and the marked preference in these two copepods for *Artemia*, when offered, also suggested stronger carnivorous tendencies.

Pleuromamma spp. had mixed gut contents and mouthparts which appeared capable of omnivorous feeding. *Chirundina* mouthparts were of the carnivorous type and the gut contained abundant zooplankton fragments. *Euchirella* appeared to be more capable of feeding at least partly on phytoplankton; the mandibular teeth were blunter and the gut contained dinoflagellates. The two species of Scolecithricidae had mouthparts which might be interpreted as being adapted also to omnivorous feeding and though *Scolecithrix* did not take diatoms offered and the gut contents had crustacean remains, *Lophothrix* grazed on algae. Not only the absence of phytoplankton feeding but also the structure of the mouthparts indicated that species studied belonging to the families Euchaetidae, Candaciidae, Augaptilidae and Pontellidae were probably exclusively carnivorous. Arashkevich and Timonin (1970), from a combination of ^{14}C-labelled algal food experiments, and examination of gut contents and of mouthpart anatomy, also concluded that Calanidae, Eucalanidae, Pseudocalanidae and Temoridae are mainly filterers feeding on phytoplankton and some radiolarians. *Euchirella bella*, *Pleuromamma xiphias*, *Centropages elongatus* and *Scolecithrix danae* took a mixed diet. The assimilation index calculated for algae was generally lower. The mouthparts (maxilla 2) suggested some filtering ability, but predatory claw-like setae were also present.

Euchaeta, *Haloptilus* and *Candacia* appeared to be solely carnivorous (cf. also Petipa, Pavlova and Sorokin, 1973). As a result of these and other experiments and with data from other authorities, Timonin (1971) listed a series of copepods (*Neocalanus*, *Eucalanus*, *Paracalanus*, *Acrocalanus*, *Calocalanus*, *Clausocalanus* and *Undinula*), as major epiplanktonic forms found commonly in warmer seas, as being herbivores, though the first two genera are coarse filterers. Genera such as *Gaetanus*, *Euchirella*, *Undeuchaeta*, *Centropages*, *Pleuromamma* and *Chirundina*, many of which tend to live rather more deeply (cf. Chapter 2), are regarded as omnivorous, while *Euchaeta*, *Augaptilus*, *Haloptilus* and *Candacia*, as well as the cyclopoid genera *Oncaea* and *Oithona*, are carnivorous. Timonin finds that some carnivorous species have the mouthparts adapted for seizing and biting; others, especially cyclopoids and *Haloptilus* and *Candacia*, are of the piercing and sucking type.

Many of the Candaciidae and almost all pontellids are epiplanktonic. Their carnivorous habits demonstrate that although predation is very common among deep-sea copepods, the predatory habit is by no means confined to deeper-living forms. Wickstead (1959) reported *Candacia bradyi* feeding on chaetognaths. Many cyclopoids (*Oithona*, *Oncaea*, *Corycaeus*) live near the surface (Chapter 2) and appear to be largely predatory. Zalkina (1970) suggests that in warm Pacific waters there is a step-like distribution of the cyclopoids modified by the varying extent of diurnal vertical migration in the different species, and that this may serve to obtain maximal use of the food which is common to many cyclopoids. *Tortanus discaudatus*, a surface-living copepod off North America, is believed to be exclusively carnivorous (cf. Raymont, 1959; Landry, 1978). Lebour (1922) had earlier considered the pontellids *Anomalocera* and *Labidocera* as mainly predatory. A number of other shallow-living, frequently neritic copepods such as *Temora longicornis*, *Centropages* spp. and *Acartia* spp., though they fed actively in the laboratory on phytoplankton, appeared from the work of Lebour and others to live frequently on a mixed diet of phytoplankton and animal food under natural conditions. In plankton blooms they may graze almost entirely on algae.

Lance (1960) and others have shown that the copepods *Acartia* and *Centropages* may sometimes consume younger stages of other copepods, and some copepods will take protozoans, other microzooplankton and detritus. For many species, therefore, there is a marked tendency towards an omnivorous diet. Investigations due to Anraku and Omori (1963) have done much to elucidate the diets of some of these suspected omnivorous species. Six species of copepods were studied, each being offered either diatoms only, a mixture of diatoms and *Artemia* nauplii or as a third alternative, *Artemia* only. Anraku and Omori also examined the structure of the mouthparts in each of the calanoids, in relation to the food. (cf pages 544, 550). *Calanus finmarchicus* was confirmed as being essentially herbivorous, though it would take animal food, while *Tortanus discaudatus* was clearly exclusively carnivorous. The mouthparts of *Calanus finmarchicus*, a typical herbivore, and *Tortanus discaudatus*, a carnivorous copepod, are illustrated in Fig. 6.6. The most conspicuous differences are the numerous comparatively blunt teeth on the grinding molar-like mandible of *Calanus* and the sharp cutting large incisor-like teeth on the cutting edges in *Tortanus*. Maxilla 1 and especially maxilla 2 in *Calanus* have numerous long fine setae with many setules. *Tortanus* has the corresponding mouthparts with shorter stout, often hooked setae, few in number, forming prehensile appendages, and the maxilliped is somewhat similar.

The mouthparts in *Labidocera* are also of the carnivorous pattern (cf. Mullin, 1966). Anraku and Omori showed that the mouthparts of *Acartia* and *Centropages* spp. seemed to be intermediate in pattern. For example, maxilla 2 appeared to be partly adapted to filtering and partly prehensile. The edge of the mandible also suggested both grinding and biting tendencies, indicating omnivorous feeding. The results of the experiments on diet confirmed that *Acartia tonsa* was omnivorous. *Centropages hamatus* and *C. typicus* were also omnivorous, but there was some preference for animal food, especially by *C. typicus*. The pontellid, *Labidocera aestiva*, while predatory, exhibited very slight filtering ability, possibly due to a few finer setae on maxilla 2.

Anomalocera also is mainly carnivorous (cf. Gauld, 1964), but the structure of the second maxilla indicates that it is capable of some filter feeding. Landry (1978), observing the feeding behaviour of the copepod *Labidocera trispinosa*, found that the animal uses the second antenna and the maxillipeds to create a current funnelling prey towards the head of the predator. Prey is then captured by the second maxilla. The predator must sense the food, but only prey from in front of the head is sensed and subsequently actively captured. There is no threshold feeding behaviour at low food

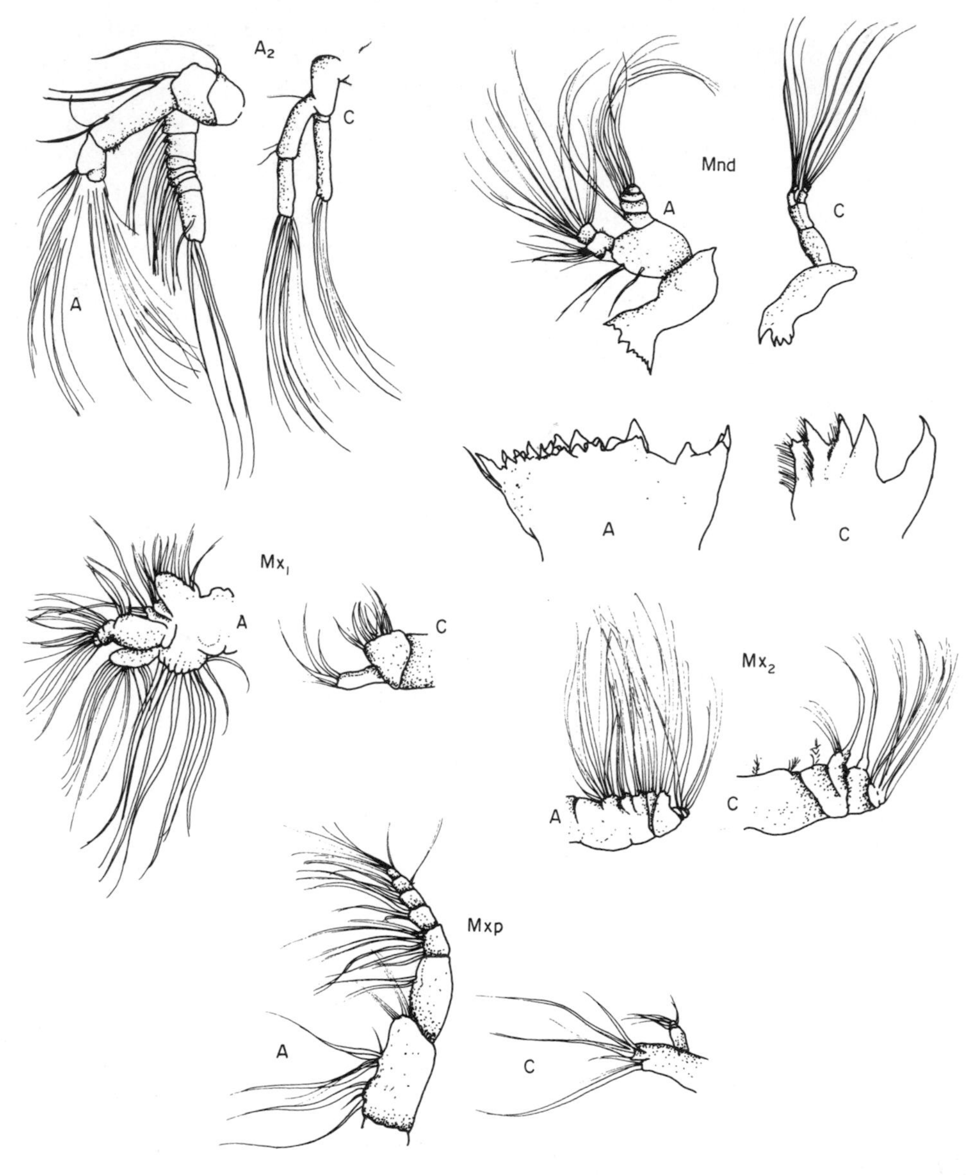

Fig. 6.6. The mouthparts of copepods, a typical herbivore (A) *Calanus finmarchicus*; and a typical carnivore (C) *Tortanus discaudatus*.

concentrations, but *Labidocera* continues to search and does not effect any saving of energy in food seeking. This imposes a very low tolerance to starvation in view of the constant high activity and lack of lipid reserves.

These differences in structure refer to calanoids belonging to several families. Vyshkvartseva (1975) studied the differentiation of mouthparts within a single genus, *Calanus*, examining the precise tooth structure on the masticatory edge of the mandible. Of the outer (ventral) group of teeth, a tooth named V2 is well developed in species of *Calanus* from high latitudes both north and south (e.g. *C. hyperboreus*, *C. glacialis*, *C. acutus*, *C. propinquus*), whereas species from more temperate latitudes, such as *C. plumchrus* and *C. helgolandicus* from the Black Sea, show V2 as weak or absent but with a diastema carrying a wide tubercle. In tropical and sub-tropical species (*C. robustior*, *C. gracilis*, *C. minor*, *C. patagoniensis*, *C. tonsus*) both V2 and the tubercle are lacking (Fig. 6.7). Vyshkvartseva suggests that the tooth structure is correlated with differences in the diet rather than with temperature. With diatoms predominating at high latitudes and dinoflagellates and coccolithophorids in warmer seas, she believes that a well developed V2 tooth may assist in holding larger diatoms such as *Coscinodiscus*, plentiful at high latitudes, before they are crushed. The majority of the larger cells are crushed and not swallowed whole.

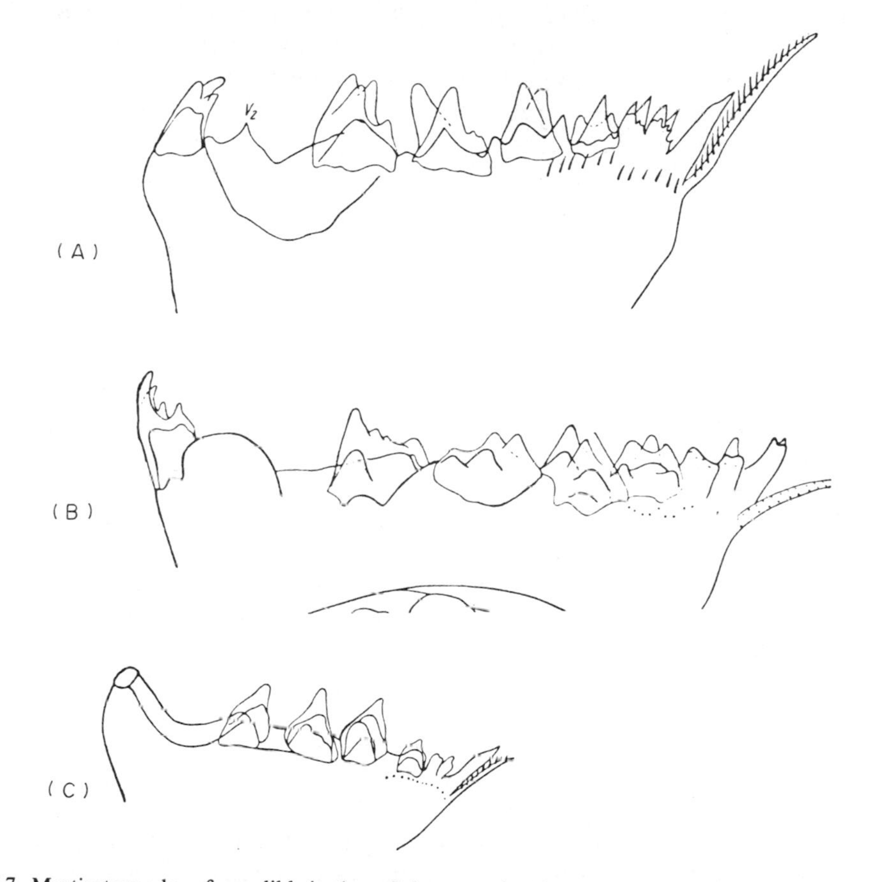

Fig. 6.7. Masticatory edge of mandible in three *Calanus* species (A) *C. glacialis*; (B) *C. Plumchrus*, Stage V; (C) *C. patagoniensis* (Vyshkvartseva, 1975).

Marshall and Orr (1966) have also used the distance between the setules as a guide to the size of food retained. The distance varies between setae and along the length of a seta from a minimum of *ca.* 2 μm to 7–20 μm or more in *Calanus* (Marshall, 1973). While *Pseudocalanus* and *Temora* have generally similar comparatively fine setules, those of *Centropages* and possibly *Acartia* are somewhat coarser. Conover found seasonal changes in the setule distances with specimens in the U.S.A. The setules of *Oithona* are very few and stout and the setae are longer. Marshall and Orr suggest that the observations agree with the results of feeding experiments using labelled foods. All the species except *Oithona* are capable of filter feeding but do take some animal food, especially *Centropages*. *Oithona* is mainly carnivorous, but *O. nana* supplements its diet when necessary with particulate matter, though this is seized and not filtered (Lampitt, 1978). According to Anraku and Omori (1963), *Acartia* appears to feed very actively for its size, and Conover (1956) suggested that its feeding setae are rather coarse, so that while taking a wide variety of food it tends to be a rather "wasteful" feeder. Heinrich (1963) also found variation in the inter-setular distance on maxilla 2 not only between species, but between setae of the appendage and in position on the seta. Age and sex could also influence the distance, which usually varied between 1.5 and 10 μm.

Tropical zooplankton included more of the very fine filterers with inter-setular distances of less than 4 μm; boreal waters had many particle feeders of medium coarseness, though younger copepodites of *Calanus plumchrus*, *C. cristatus* and *Pseudocalanus* were capable of fine filtration. While improved exploitation of the smaller algae is possible with the fine filtering types, Heinrich's data suggest that none of the copepods could filter bacteria successfully. Petipa, Pavlova and Sorokin (in Vinogradov, 1973), however, suggest that some of the "small copepods" and fine filterers in tropical waters may take bacteria as well as small algae to a very limited extent.

There have been repeated references to the importance of size of particle to many copepod grazers, though in the field, as Poulet's investigations especially have stressed, there is a marked tendency to feed (within limits) on the largest particles in greatest abundance. Mullin (1963) found not only that *Calanus* spp. selected larger particles, but he took auxospores from cultures of *Rhizosolenia* and then obtained comparatively large-celled cultures from the auxospores. When *Calanus* was fed with the larger- and smaller-celled *Rhizosolenia* the former were selected. Richman and Rogers (1969) cultured *Ditylum* and by adjusting the photoperiod and light intensity, they could influence that period when the percentage of cells in division ("pairs"), and thus of substantially greater volume, was high. *Calanus* showed higher filtration rates when there was a greater proportion of "pairs." Thus when cell division was mainly in darkness the filtering rate for "pairs" was markedly increased but was unchanged for single cells. There was also a more or less constant feeding rate when "pairs" were less than 20 %, an exponential rise in rate to a constant value with "pairs" increasing to 40 %. Richman and Rogers suggest that this implies that large "pairs" are actively hunted and seized; small cells are filtered.

Many authors believe that particle feeders may use both filtration and the seizing of particles, though Conover (1968) believes they are mutually exclusive in *Calanus hyperboreus*. Conover (1966) gives details showing how five strong setae of the second maxillae and other setae of the maxillipeds combine with the "normal" filtering setae of second maxillae to "handle" a large particle before this is pressed by the first maxillae against the mandible and crushed. Jorgensen (1966) and Gauld (1964) suggest that a mean size of about 40 μm may be an upper limit for filtration. Above this size, particles are individually seized. Jorgensen (1966) includes a useful analysis of the sizes of phytoplankton and the distance between setules for several species of marine copepod (cf. Table 6.2). Even such a herbivorous feeder as *Calanus finmarchicus* shows

Table 6.2. Distances (μm) between setules for two species of (a) *Calanus* and (b) sizes (μm^3) of phytoplankton (Jorgensen, 1966a)

(a)

			Coxa	Basipod			Endopod			
Calanus Species	Stage	External seta	Endite 1	Endite 2	Endite 3	Endite 4	Segment 1 (Endite 5)	Segments 2–4	Segment 5 "Feather"	Segment 5 Others
C. finmarchicus	♂	—	5–12	4–8	6–9	5–11	4–16	5–21	4–5	6–17
C. helgolandicus	♂	2	5–9	8–9	8–10	8–11	12–17	11–19	7–9	13–22
C. finmarchicus	♀	2	2–6	3–6	3–7	3–8	4–7	5–14	3–5	5–14
C. helgolandicus	♀	2	5–9	6–9	4–9	6–12	7–13	11–19	3–9	12–20
C. finmarchicus	V	1–3	2–4	3–4	2–4	2–4	3–6	4–8	2–5	4–10
C. helgolandicus	V	1–3	3–9	4–7	3–8	4–10	3–15	5–17	3	7–20
	IV[a]	2	3	3–4	3–6	3–5	3–5	4–7	3–4	5–9
	III[a]	2	2–4	3–4	3–4	2–5	4–7	4–8	3–4	4–11
	II[a]	—	2–3	3–4	2–4	3–4	3–5	3–7	3–4	3–7
	I[a]	—	2–4	2–4	2–5	2–5	3–6	4–7	2–4	4–8

[a] *C. finmarchicus* and *C. helgolandicus* cannot be distinguished until Copepodite V.

(b)

Phytoplankton	Cell volumes in μm^3
Diatoms	20–20,000,000
Dinoflagellates	500–100,000
Flagellates	14–500
Coccolithophores	14–500

different types of setae on maxilla 2, some of which are believed to assist the catching of larger particles before passing them to the mandibles. Copepods such as *Acartia* use the maxilla as a scoop for catching algae, and many other copepods apart from obvious carnivores make grasping movements with the maxillae (cf. Marshall, 1973 for a full discussion).

There is little information on feeding in relation to particle size for other omnivores. Robertson and Frost (1977) examined the feeding of *Aetideus divergens*, usually regarded as an omnivorous calanoid. They found that the copepod captured prey, represented by *Artemia* nauplii, and large algal cells. While large cells were grazed very effectively, with smaller particles, though the ration taken rose with increased cell concentration, the maximum ration was not attained. Feeding was especially inefficient on the smallest algae. The method of swimming in *Aetideus* and the comparatively large size of the maxillipeds suggested that the copepod probably searched for and seized particles. Other features of the structure of the mouthparts such as the varied patterns of the setae on the second maxillae and the comparative coarseness of the setules also indicated mixed feeding for the copepod. Robertson and Frost believed that feeding in the natural environment was opportunistic and that the copepod could not obtain sufficient food by filtering smaller particles, but that it relied on a mixed diet of larger items, including faecal pellets.

Mullin and Brooks (1967) examined the feeding of younger stages of *Rhincalanus* and showed that Nauplius II and III ate *Ditylum* from four species (*Coscinodiscus*, *Ditylum*, *Thalassiosira* and *Cyclotella*) in order of descending size, while older nauplii and copepodites selected the largest cells. Richman, Heinle and Huff (1977) also found that *Eurytemora*, *Acartia tonsa* and *A. clausi*, when feeding on natural food, select to some extent for size, but always on a size group of major abundance (cf. also Boyd, 1975; Nival and Nival, 1976; *vide supra*). Wilson (1973) tested the feeding of *Acartia tonsa* on particle size using plastic beads of 7–70 μm diameter. Copepodites I–VI showed a rise in the maximum size of bead ingested with increasing body size up to beads of about 59 μm diameter, but the minimum remained constant at 14 μm. Despite the comparatively few beads ingested, the ratio beads ingested/beads present for each size category permits the observation that although advanced copepodites (C V and VI) can still feed over the entire size range (14–59 μm), given narrow size, they select bands of increasing size to a maximum of around 55 μm diameter. Younger copepodites also select to the largest size band which is within their range of ingestion. Other scattered observations suggest that many copepods tend to take larger cells or longer chains of diatoms if these are available (e.g. Mullin, 1963; Martin, 1970). There are a few observations, such as those of Petipa, that little selection for size occurs when food is abundant. For *Euterpina*, Nassogne (1970) also found that algal food of intermediate size ranges (about 7–16 μm) gives successful culture from egg to adult; algae of smaller or larger size are not suitable or cannot be filtered for some stages.

Size is not the only factor in successful nutrition. The age of a culture may be important. Conover observed that grazing by *Acartia* was thus affected (cf. also Mullin, 1963). Paffenhofer (1970) found that *Calanus* nauplii were even killed if cultures of *Lauderia* (which are normally eaten) were of 12–21 days old. Results also often indicate that feeding on one species tends to favour continuation of grazing on the same alga. Size of food particle is to some extent related to body size and stage of the copepod. Reference has already been made to Mullin and Brook's experiment on

Rhincalanus, including successful rearing on *Ditylum*. *Calanus pacificus* could not be reared on *Ditylum* cultures; the nauplii are apparently unable to cope with the comparatively large cells (Mullin and Brooks, 1970). In *Calanus finmarchicus*, Marshall and Orr (1956) suggested that while N I and II did not feed, N III could eat phytoplankton up to approximately 20 μm diameter, and that copepodites fed on a wide variety of algae, with the filtering ability showing little change with size. Heinrich (1963), however, claimed considerable variation within a species between the copepodite stages in size filtration ability.

Many carnivorous copepods lay comparatively yolky eggs and the naupliar development may depend on the yolk as the sole food (e.g. *Chiridius armatus*, *Aetideus armatus*, probably *Xanthocalanus fallax*—Matthews, 1964). Bernard (1964) also described the six nauplius stages in *Euchaeta marina* and *Candacia armata* as being passed through rapidly (5–6 days) and without feeding, the yolk being entirely utilized for the development. *Euchaeta japonica* feeds on plant food at N III stage, however. The carnivorous copepod *Pontella meadi* also feeds on dinoflagellates in the nauplius stage (Gibson and Grice, 1977); other carnivorous forms may feed on animal food in the young stages. The stage at which feeding commences varies in both herbivorous and predatory copepods.

Though herbivorous copepods frequently use a wide variety of phytoplankton, not only size but the type of alga may be significant. Presumably the dietary requirements, particularly the necessary array and relative quantities of specific amino acids and possibly some lipid constituents, are not fully met by certain algae. Marshall and Orr (1952, 1955) found that egg-laying by ripe female *Calanus* was dependent on food supply and they therefore used egg production as a criterion of the suitability of specific algae as food. All diatoms tested (*Lauderia*, *Chaetoceros*, *Coscinodiscus*, *Skeletonema*, *Rhizosolenia*, *Ditylum*) were shown to be suitable food material causing oviposition, as were some dinoflagellates (*Peridinium* and *Gymnodinium*). Certain flagellates such as *Chlamydomonas* (*Dunaliella*), *Syracosphaera*, and an unidentified chrysomonad were satisfactory foods, but others (*Dicrateria*, *Hemiselmis* and *Chlorella*), though ingested, did not increase egg production over starved copepods. Indeed, *Chlorella* appeared to pass through the gut almost unchanged.

A few phytoplankton species are known to be toxic and are possibly avoided by herbivores. *Calanus* will filter the toxic *Gymnodinium veneficum* and *Prymnesium parvum* in culture, though Marshall and Orr (1955a) observed some mortality with *Gymnodinium*. Marshall (1973) also reports that *Dicrateria*, *Chlorella* and *Amphidinium* have a deleterious effect on *Pseudocalanus*. Even a cell-free filtrate of *Chlorella* culture appeared to reduce the life span. Some phytoplankton species in very high concentrations are avoided by zooplankton: Bainbridge (1953) observed this behaviour with very high concentrations of two species of flagellate, but the concentrations would hardly ever be encountered in nature. Paffenhofer (1970) tested the suitability of three algae, *Skeletonema*, *Lauderia* and *Gymnodinium*, as food for *Calanus pacificus*. He had previously been able to grow viable adults on each of the separate algal diets and also demonstrated that the food concentration affected the duration of life from egg to adult and the size of resulting females. Using a standard quantity of food (100 μg C/l), he showed that *Gymnodinium* gave the most rapid growth (18 days), *Lauderia* was intermediate (24 days) and *Skeletonema* gave the slowest growth to adult (36 days). Mortality was also greatest with *Skeletonema*. The amount of food ingested is highest in *Gymnodinium*, however, so that quantity as well as quality may be a factor.

Nassogne (1970) successfully cultured *Euterpina acutifrons* on a considerable number of unialgal cultures and on mixed species. For five algae, all of suitable size and in excess food concentration, he demonstrated differences in the value of the various diets using egg production and adult life span as criteria (cf. Table 6.3).

Table 6.3. Egg production and adult life span of *Euterpina acutifrons* fed on different algal species under conditions of excess food (Nassogne, 1970)

Algae	Culture (cells/ml)	(μg fresh weight/ml)	Total eggs per adult	Days of adult life
Prorocentrum micans	2.10^3	23.5	15 ± 11	14 ± 3
Platymonas suecica	1.10^5	26	84 ± 7	27 ± 2
Gymnodinium sp.	1.10^5	25	135 ± 26	37 ± 3
Phaeodactylum tricornutum	4.10^5	24	151 ± 42	32.6 ± 7.5
Chaetoceros danicus	8.10^4	24	151 ± 42	28 ± 8
Mixture	3.10^4	24	281 ± 17	38 ± 1

The investigations of Provasoli and his colleagues (e.g. Shiraishi and Provasoli, 1959) on the harpacticoid copepod *Tigriopus japonicus* might be of relevance to these discussions on the nutritional needs of marine calanoids. *Tigriopus* may be grown on many different bacteria-free algal cultures, but growth and the reproductive capacity of the copepod varies greatly. Some algal foods do not promote egg production; with some cultures, algae growth slows so that copulation and egg production is delayed. Other algal food permits a limited number of generations, whereas some food species permit apparently continuous reproduction (cf. Table 6.4). The presence of bacteria may modify the effect of an inadequate algal food, suggesting that vitamins or other growth factors are made available by the bacteria (Provasoli, 1966). Table 6.4 shows the lack of successful culture with *Dunaliella* spp., *Brachiomonas* etc., and the very few generations with *Syracosphaera* and *Hemiselmis*. *Monochrysis* appears to be a complete food, allowing continuous reproduction. A *Rhodomonas* and *Isochrysis* mixture is also a complete diet.

Gilat's (1967) less detailed results for *Tigriopus brevicornis* are somewhat similar in that although *Gyrosigma* and *Tetraselmis* can act as a food alone, better results were achieved on a mixed diet. Bacteria and detritus were also useful and Gilat holds that such material adds substantially to the nutrition of herbivorous plankton in oligotrophic areas. Heinle, Harris, Ustach and Flemer (1977) find that egg production in species such as *Eurytemora affinis* is affected by both quality and quantity of algal food available; it appeared that detrital food alone was not qualitatively adequate. They suggest that ciliated protozoans appear to be important in the transfer of energy from detritus to copepods (cf. also Chapter 7). The inadequacy of certain unialgal diets for the complete length of adult life of *Pseudocalanus* is also noted by Corkett and McLaren (1978). With regard to the marine zooplankton, the limited evidence suggests that a range of algae exists in the oceans, from harmful and neutral to beneficial species from a dietary standpoint. The latter are in the great majority but

Table 6.4. Utilization of food organisms by *Tigriopus japonicus* (Edmondson, 1966)

Food Organisms	Stage reached by first generation[a]	No. generations obtained
Chlorophyceae		
Dunaliella parva (brine)	L	
Dunaliella tertiolecta (marine)	L	
Tetraselmis tetrathele (La Jolla)	A	3
Tetraselmis maculata (No. 5)	A	8 → (b)
Nannochloris oculata	A	1
Stephanoptera sp.	L	
Stichococcus fragilis	L	
Brachiomonas pulsifera	L	
Pyramimonas inconstans	L	
Chrysophyceae		
Isochrysis galbana	A	9
Monochrysis lutherii	A	8 → (b)
Stichochrysis immobilis	A	2
Syracosphaera elongata	A	1
Cryptophyceae		
Chroomonas sp.	A	5
Hemiselmis virescens	A	1
Rhodomonas lens	A	6
Eugleninae		
Eutreptia sp.	A	2
Dinophyceae		
Gyrodinium cohnii	L	
Peridinium sp.	L	

[a] L, larval stages; A, adult.
[b] continuing culture.

differ in their precise nutritional value. The zooplankton normally satisfy their needs by feeding on several species.

Though detailed examinations of egg production or life span as a test of diet have not been carried out on many copepods, even some of the calanoids which are omnivorous show good survival on certain phytoplankton. Thus Gauld (1952) was able to keep several small copepods (*Temora*, *Pseudocalanus* and *Centropages*) on cultures of *Chlamydomonas* and *Skeletonema*. Ikeda (1977b) found that *Skeletonema costatum* did not maintain female *Paracalanus aculeatus* for more than six days. *Dunaliella salina*, however, promoted healthy cultures of *Temora turbinata*.

Raymont (1959) observed that flagellates, mostly *Dunaliella* and *Dicrateria*, were filtered extensively by several copepods (*Centropages hamatus*, *Eurytemora herdmani*, *Pseudocalanus minutus*, *Temora longicornis* and *Metridia lucens*). The unpublished results of investigations by Gross and Clarke on growth of plankton in enclosed tanks at WHOI indicated that smaller copepods flourished on a wide variety of diatom and other phytoplankton. The subsequent tank experiments of Raymont and Miller (1962) at Woods Hole demonstrated that following the growth of a variety of diatoms and μ-flagellates there was a very marked tendency for only one or two species of phytoplankton (*Exuviaella marina*, *Nannochloris* sp. and *Nitzschia closterium*) to dominate the population. Several small copepods, *Paracalanus crassirostris*, *Eurytemora hirundoides*, *Oithona brevicornis* and especially *Acartia tonsa*, flourished and passed through several reproductive cycles in the tanks. They must therefore have fed almost exclusively on these few species of phytoplankton and utilized them for egg production and naupliar development. The real suitability of these algae as food is apparent from the very high copepod densities achieved in these experiments. *Centropages*, in contrast, appeared not to breed successfully.

Parsons, LeBrasseur and Fulton (1967) found that *Calanus pacificus* had a wide range in its ability to filter plankton algae, including nanoplankton < 7–8 μm in size. On the other hand, only a relatively small amount of food was obtained (2–4 % body wt) from such nanoplankton.

Euphausiacea

Although differences in the diet of species of copepods have been most intensively studied, some information is available on the food requirements of other groups of zooplankton.

A few euphausiids are undoubtedly herbivorous; the Antarctic *Euphausia superba* appears to feed almost exclusively on the rich blooms of diatoms. Barkley (1940) concludes that there is some selection in the species of diatom used as food; *Fragilariopsis antarctica* is most generally consumed and other small- and smooth-celled species (minimum size *ca.* 7 μm) are also widely eaten. According to Barkley, no organisms apart from diatoms appear to be of any nutritional significance to *E. superba*, but Nemoto (1968a) suggested a rather wider variety (cf. Table 6.5). The persistence of the diatom valves may slightly exaggerate their importance. Nemoto describes the fineness of the filter formed by the thoracic legs of *E. superba* and the herbivorous pattern of the mandible and maxillae. A cluster of spines on one of the

Table 6.5. Main food of antarctic krill *Euphausia superba* (Nemoto, 1968)

Chrysophycea	*Fragilariopsis antarctica*
Phaeocystis sp.	*F. cylindricus*
Coccospheres	*F. rhombica*
Not dominant	*F. ritscheri*
Silicoflagellates	*F. curta*
Distephanus regularis	*Micropodiscus oliveranus*
Dinoflagellata	*Navicula criophila*
Not dominant	*Nitzschia barkleyi*
Diatoms	*Thalassiosira gracilis*
Asteromphalus heptactis	*T. antarctica*
A. parvulus	*Thalassiothrix antarctica*
A. regularis	*Tropidoneis belgicae*
A. hookeri	*T. glacialis*
A. hyalinus	Foraminifera
Biddulphia polymorpha	Not dominant
Charcotia actinochilus	Radiolaria
Coscinodiscus lentiginosus	Not dominant
C. minimus	Tintinnids
Cocconeis gaussi	Not dominant
	Crustacea
	Not dominant

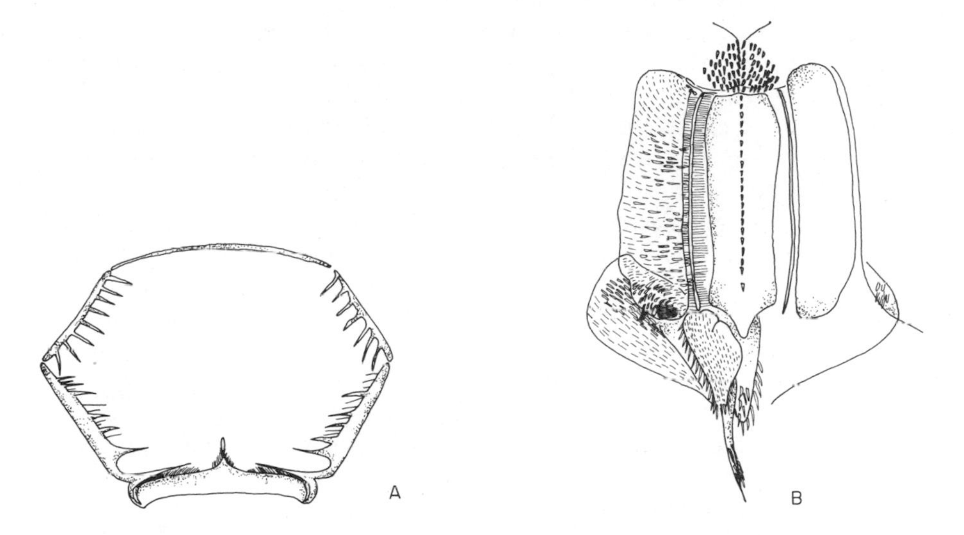

Fig. 6.8. Stomach of *Euphausia superba*, (A) Cross-section of stomach at middle; (B) base part of stomach (Nemoto, 1968).

stomach plates in the euphausiid is also regarded as assisting in crushing diatoms (Fig. 6.8). More details of the adaptations to herbivorous feeding are given later.

According to Lebour (1924), *Nyctiphanes couchi* fed on phytoplankton, mostly diatoms, but there was some unidentifiable debris which might have indicated detrital feeding; animal food was occasionally taken. Le Roux (1973) found that larval stages of *Nyctiphanes couchi* consumed both algae and animal food (*Artemia* nauplii). Reduced food supply increased the duration of the intermoult periods during development. This was also true when the larvae were fed on algal foods instead of *Artemia* nauplii, and it appeared that some algae were not completely digested. Hickling (1925), however, claimed that euphausiids (*Nyctiphanes*, *Meganyctiphanes*, *Thysanoessa*) were feeding very largely on detritus, though some investigators regarded them as being chiefly carnivorous.

Mauchline and Fisher (1969) have also emphasized the importance of detritus, especially for *Meganyctiphanes* and *Thysanoessa raschii*, and point out that some euphausiids living in sufficiently shallow water actively stir up the bottom deposit and feed on it. They also quote several other species (*Thysanoessa inermis*, *T. longipes*, *Thysanopoda tricuspidata*, *Stylocheiron* spp., *Euphausia pacifica*) in which, according to Ponomareva, detritus feeding is important. As Nemoto (1971/72) points out, the importance of detritus feeding in sub-surface levels of open oceans has probably been underestimated. Ussing (1938) described euphausiids occurring off east Greenland as carnivorous and Einarsson (1945) considered that for *Thysanoessa* spp., *Meganyctiphanes norvegica* and *Thysanopoda acutifrons*, phytoplankton food was unimportant, detritus and some zooplankton being eaten.

An excellent history of research on the food and feeding of euphausiids is given by Nemoto (1971/72). He earlier (1967) divided euphausiids into filter feeding types in highly productive areas (*E. superba*, *E. vallentini*, *E. mucronata*), mixed feeders (some *Thysanoessa* spp.), surface feeders in tropical waters—usually partly carnivorous, consuming copepods (e.g. *E. diomedeae*), and bathypelagic forms, such as the large *Thysanopoda*—usually eating zooplankton (copepods, sagittae and even micronekton). *Bentheuphausia*, though bathypelagic, was formerly believed to take phytoplankton and possibly microzooplankton, since greenish material was usually present in the gut. Nemoto (1968b) and Nemoto and Saijo (1968), however, demonstrated that *Bentheuphausia* contains mostly degraded chlorophyll in the gut and they believe that the euphausiid filters material with decomposing plant pigment, possibly faecal pellets from more surface-feeding zooplankton. Mesopelagic euphausiids may be largely carnivorous (e.g. *Nematobrachion*) but some species of *Nematoscelis* may contain detrital material which shows some degraded chlorophyll. Some species, including *E. pacifica*, *Thysanoessa longipes*, *Tessarabrachion oculatus* and *Thysanopoda tricuspidata*, feed mainly in the surface layers at night and show a fair proportion of chlorophyll in the gut. *E. similis* and *E. nana*, dominant species in certain Japanese waters, show even more obviously active feeding on phytoplankton, mainly at night, so that relatively large amounts of chlorophyll appear in the gut. Very great quantities of faecal pellets are produced (Nemoto, 1968b).

Macdonald (1927) investigated the feeding of *Meganyctiphanes norvegica* in the Clyde area. He observed that apart from the importance of detritus, diatoms and microplankton formed part of the food, particularly of the younger stages, but that the largest *Meganyctiphanes* were carnivorous to a considerable extent, feeding on copepods such as *Calanus* and *Euchaeta*. Fisher and Goldie (1959) confirmed that detritus was usually more important than phytoplankton in Clyde animals, though dinoflagellates and some diatoms are eaten; crustaceans, including copepods and Eucarida, were taken in significant amounts. Mauchline (1960) considered that the smallest *Meganyctiphanes* were mainly filter feeders, while the largest were chiefly carnivorous. Crustacean food was largely *Calanus*, *Euchaeta* and Eucarida, but *Sagitta* was also taken. The amount of animal food increased with the size of *Meganyctiphanes*.

Ponomareva (1955) investigated the food of *Thysanoessa inermis*, *T. raschii*, *T. longipes* and *Euphausia pacifica* in the Sea of Japan. All the species except *T. longipes* will feed on phytoplankton, usually filtering the most abundant species in spring. Feeding was more active at night and nearer the surface. The filtering ability of *Euphausia pacifica*, feeding on *natural* phytoplankton blooms was investigated by Parsons, LeBrasseur and Fulton (1967). *E. pacifica* filters phytoplankton very effectively, especially *Chaetoceros*, with a minimum size limit of about 12 μm and an upper limit of 60–80 μm. Although nanoplankton (*ca.* 8 μm) was taken, it was much less effectively filtered. A marked selection for size appears to be true of several euphausiid filter feeders. Thus, while *E. pacifica* feeds on chains of *Chaetoceros* in coastal Pacific waters off western Canada, it takes smaller forms offshore. The euphausiid also has the ability to switch to carnivorous feeding when sufficient zooplankton prey is available; Lasker's (1966) suggestion is that it obtains its required ration more readily on animal diet, if possible (*vide infra*). This may be an example of a change from filtering to a raptorial type of feeding. Jorgensen's (1966) division into the two types on the basis of particle size may hold good, as suggested for copepods.

The elucidation of different trophic types amongst the euphausiids is in part based on examination of the feeding appendages. Nemoto (1967) considers that the maxillae, which have comparatively well-developed setae in herbivorous types (e.g. *E. vallentini*,

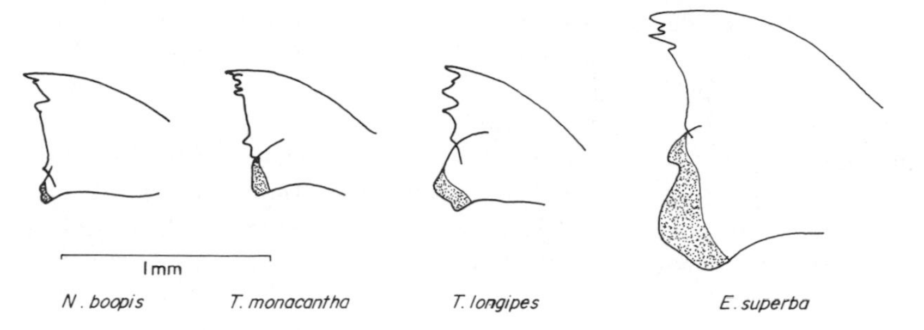

Fig. 6.9. Pars inisiva and pars molaris (shaded) in the mandibles of euphausiids (Nemoto, 1967).

E. superba) and which help to crush diatoms, have degenerated into simple, more grasping structures in carnivores (e.g. *Nematobrachion*). The mandible exhibits a much stronger and heavier pars molaris associated with crushing diatom food in herbivorous types such as *E. superba*; a reduction in the pars molaris is very evident in carnivorous species, for example, *Nematobrachion boopis* (Fig. 6.9). The thoracic legs, which serve as the main filtering mechanism, exhibit marked differences in relation to the predominant type of food. *E. superba*, *Thysanoessa raschii* and other herbivores have well developed setae with finely distributed barbs on the legs; *Thysanopoda* may lack filtering setae, or a limited number may be present on part of some thoracic appendages, but the barbs tend to be much coarser. Grasping spines are developed on certain joints. The carnivorous *Nematobrachion* and *Stylocheiron* have thoracic appendages of a similar type. Some species have very elongated second or third legs which may also be associated with the predatory habit (cf. Fig. 6.10). The spacing between the fine barbs of the setae varies from 5–7 μm in *E. superba* to 25–50 μm in *Thysanopoda monacantha*. Fairly shallow tropical species appear to be partly carnivorous, though some coarse filtering may occur. Roger (1975) suggests from

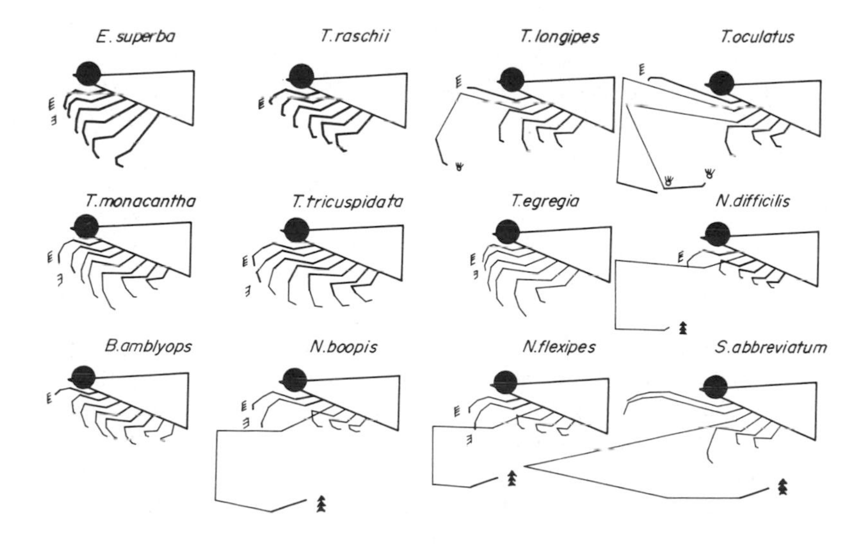

Fig. 6.10. Patterns of leg arrangements in euphausiids (Nemoto, 1967).

analysis of stomach contents of tropical euphausiids that the detailed structure of the thoracic legs may not always be a clear indication of the diet (*vide infra*). Modifications of the stomach with spines, ridges and armoured areas and of the size of the chamber are considered to be associated with differences in diet.

Ponomareva's observations that *Thysanoessa longipes* did not filter phytoplankton may be associated with the somewhat deeper habitat of the species. *T. raschii* was believed to take only some eggs and nauplii of copepods, and detritus, in addition to phytoplankton, but Mauchline (1966) found abundant crustacean fragments. The other three species are more obviously carnivorous on the zooplankton, in addition to their filter feeding. Molluscs, amphipods, *Oikopleura* and, at times of scarcity of larger zooplankton, tintinnids and *Globigerina*, were all consumed, but copepods formed the chief animal food. Euphausiids also appear to feed on each other to some degree. There was some selection with regard to the copepod diet: *Metridia pacifica* and *Calanus plumchrus* were most commonly eaten, though they are not always the most abundant in the plankton. While copepods such as *Eucalanus bungii* and *Pareuchaeta japonica* may be too large, it is surprising that the common small copepod *Oithona similis* was rarely taken. The mesopelagic and especially the bathypelagic euphausiids such as *Nematobrachion* and *Thysanopoda* are certainly mainly carnivorous.

So far detailed discussion of food and feeding in euphausiids has largely been concerned with temperate or high latitude species. In the warm Pacific, Roger (1975, 1973a, 1977) found that of 28 species of euphausiid, a great majority were largely carnivorous. Twelve species, including *Stylocheiron suhmii*, *S. affine*, *S. abbreviatum*, *Thysanopoda aequalis* and *Nematobrachion boopis*, were strictly carnivorous, feeding on copepods and microzooplankton such as tintinnids, radiolarians and foraminiferans. Twelve species were more omnivorous (e.g. *Nematoscelis gracilis*, *Thysanopoda monacantha*), with some (e.g. *Nematoscelis tenella* and *Thysanopoda cristata*) apparently able to change from carnivorous to herbivorous feeding according to the availability of food. Only four species were strictly herbivorous (e.g. *Euphausia paragibba* and *E. gibboides*), but a euphausiid such as *Thysanopoda tricuspidata* fed mainly on plant material but also took some animal food. This latter species possessed some filtering setae on the thoracic limbs and was thus somewhat unusual among species of the genus. Roger points out that while a number of surface-living euphausiids are carnivorous, some living at subsurface levels migrate to the surface to feed at night (e.g. *Euphausia diomedeae*, *E. tenera*). *Stylocheiron* spp. are non-migratory and feed only during daylight hours. *Thysanopoda cristata* is mesopelagic and does not migrate, but tends to feed only by day, since its potential prey migrate surfacewards in darkness.

Other zooplankton

A few general observations are included on food of certain other groups of zooplankton, many of which are herbivorous or mixed feeders.

Some species of mysids undoubtedly filter phytoplankton effectively (cf. Lucas, 1936; Ferguson, 1973), but Tattersall and Tattersall (1951) suggest that many may supplement their food with animal tissue, while others are clearly omnivorous or carnivorous. Zenkevitch and Birstein (1956) state that the deep-living mysid *Gnathophausia gigas*, while mainly carnivorous, may feed by filtering particles and can also utilize the faecal pellets of other plankton. *Eucopia* spp., which are an important part of the zooplankton at depths of *ca.* 1500–2500 m in the north-western Pacific, also appear to be mainly carnivorous. Vinogradov (1970, 1972) points out that at deeper levels in the north-western Pacific several species of *Boreomysis* perform relatively enormous vertical migrations to feed on phytoplankton in subsurface layers. *Boreomysis incisa* in particular contributes substantially to the zooplankton biomass in the Kurile–Kamchatka Trench at depths exceeding 4000 m. The stomachs of these animals are filled with greenish matter, abundant in large diatom fragments and cholorophyll, and with tintinnids. There is no evidence that the diatom remains had formed part of faecal pellets which had been consumed by the mysids and Vinogradov believes that the extensive vertical migration permits *B. incisa* to feed at subsurface levels. In the earlier summary of Vinogradov (1962) *B. spinifera* is listed among the few mysids feeding on detritus. In contrast to *Boreomysis*, many of the deeper-living mysids are carnivores. Vinogradov includes *Petalophthalmus armiger* and *Longithorax* which appear to suck out the contents of their prey.

The Ostracoda includes animals with varied diets. Vinogradov (1962) classes many of the somewhat deeper-living species as detritus feeders, but the gut often also contains remnants of radiolarians, foraminiferans, crustaceans and chaetognaths. As in many other zooplankton, the diet appears to combine carnivorous and particle feeding. In some shallower-living ostracods (e.g. *Conchoecia alata*) phytoplankton is consumed. Iles (1961) reviewed the very limited information on the food of *Conchoecia* spp. and indicated that ostracods were generally predatory, with copepod remains particularly predominant in the stomachs. However, fine matter (detritus and diatom material) was also frequently present and may be trapped by the sticky secretion of the marginal glands of the carapace or by the mandibular palps. Arashkevich (1977), studying the feeding habits of *Cypridina sinuosa*, also found that particulate food such as diatoms and flagellates can be taken with the assistance of secretions of the oral glands, but that the main diet is of copepods and that the ostracod is able to feed on prey of size equal to its own length.

Angel (1968) commented that stomach contents and faecal pellets of *C. spinirostris* and *C. curta* contained abundant diatoms and silicoflagellates although he had observed both ostracods attacking chaetognaths. Timonin (1971) classified *Conchoecia* spp. as omnivores. Apart from the Conchoeciidae, Cannon's (1931) investigations on *Cypridina levis* indicated that the animal fed by filtering particles, but his study (Cannon, 1940) of *Gigantocypris mulleri* showed that the ostracod was predaceous on sagittae, copepods and even fish larvae, despite its comparatively poor swimming powers. An example of an extremely specialized diet in a deep sea crustacean is drawn from the Leptostraca: *Nebaliopsis* appears to prey only on squid and fish eggs.

It is difficult to define the nutritional needs of decapod larvae, though considerable experience has now been gained on the rearing of commercially important species. The pioneer work of Lebour (1928) showed that various species of crab larvae would feed on *Nitzschia* and other algae, but needed some animal food to complete the necessary series of moults. Costlow and Bookhout (1959) used *Arbacia* eggs and *Artemia* nauplii supplemented by beef liver to rear *Callinectes sapidus* in the laboratory. Broad (1957) investigated the rearing of two species of *Palaemonetes* using algal cultures (*Nitzschia*, *Chlamydomonas*, *Thorocomonas*, *Nannochloris*, *Porphyridium* and *Pyramimonas*) and animal food including *Artemia* nauplii, chaetognaths and other zooplankton.

Differences in survival, frequency of moulting and rate of development were associated with the amount of food available, but as Lebour had found, some animal food was necessary. Experiments in Costlow's laboratory suggest that such animal food tends to have an all-or-nothing effect, i.e., if suitable, development will continue through all moult stages; if unsuitable, the larvae do not develop at all. A possible variation, however, is that sometimes a larval stage is "skipped". With algal food or mixed algal and animal food there may be a lengthening of development rather than the all-or-nothing result. Apparently, algal food alone does not permit development to extend to two months in some crab and some shrimp species (Edmondson, 1966). Provasoli (1966) reported a time difference in requirements in the semi-commercial cultivation of the penaeid shrimp, *Panaeus japonicus*. The nauplii would moult without feeding, zoeas could be fed with *Skeletonema*, the mysid stage also required zooplankton (given as *Artemia* nauplii and bivalve larvae) and later other animal tissue was required (cf. Edmondson, 1966).

Cladocerans are believed to be entirely herbivorous plankton. They often appear in dense numbers when phytoplankton becomes abundant. According to Sorokin (1971), *Penilia* and the appendicularian *Oikopleura* are among the very few zooplankton which are capable of very fine filtration, so that even bacteria may be eaten. Some copepods and even euphausiids are capable of feeding on bacteria aggregates, but the minimum particle diameter appears to be approximately 4 μm. According to Bainbridge (1958), however, the cladoceran *Evadne nordmanni* feeds on large algae and exhibits a considerable degree of selection. The crustacean appears to become abundant in the Clyde area mainly when certain types of phytoplankton, peridinians and tintinnids are plentiful. Examination of gut contents and of the food held by feeding appendages confirmed that *Evadne* was a very selective feeder. *Peridinium* spp., *Ceratium furca* and probably other peridinians were eaten, though even other species of *Ceratium* were apparently neglected (cf. Table 6.6). Tintinnids are also eaten, but diatoms tend to be unimportant. It seems likely that *Evadne* actively seizes particles rather than filtering the food. In contrast, the freshwater *Daphnia pulex* is reported by Horton, Rowan, Webster, and Peters (1979) to browse on the bottom detritus of its culture vessel when algal food is scarce; such switching may account for the sudden population outbursts observed amongst cladocerans.

The larvae of the cirripedes *Balanus balanoides*, *Verruca stroemia* and *Chthamalus stellatus* have been reared through a series of moults by Bassindale (1936) on a culture of *Nitzschia closterium*. Certain other algae appeared in some of the cultures, however, and growth was not followed by settlement. Lockhead (1936) claimed that cells exceeding 6 μm in diameter could not be swallowed by nauplii of *Balanus perforatus*, though the length of cells could be considerably greater. He included Protozoa, small diatoms, filamentous algae and especially small flagellates in the diet. Lockhead held that there was no selection of food apart from size, but that *B. perforatus* nauplii could not masticate larger diatoms. There seems little doubt, however, that a wide variety of food organisms can be utilized by cirripede larvae.

Soares (1957) showed that diatoms as large as *Coscinodiscus* can be broken and eaten by the nauplii of *Elminius modestus*, yet both *E. modestus* and *Balanus crenatus* larvae filtered flagellates of the size of *Dunaliella* and the small diatom *Nitzschia closterium*. Growth of cirripede larvae in the Clyde area is promoted by the spring phytoplankton increase, which is dominated by diatoms, especially *Skeletonema costatum* (Barnes,

Table 6.6. The percentage composition of organisms held by preserved *Evadne* (The percentages for 17 and 18 October 1951 are compared with the percentage composition of the microplankton at 1 m) (Bainbridge, 1958)

	Organisms held by *Evadne*	Microplankton
Phytoplankton		
Diatoms	—	0.03
Peridinium spp.	20.9	3.0
Peridinium spp. fragments	1.6	—
Ceratium furca	56.4	34.6
C. furca fragments	15.0	—
Ceratium spp. less *C. furca*	—	55.1
Other dinoflagellates	1.6	6.6
Zooplankton		
Tintinnopsis spp.	1.6	0.06
Tintinnopsis spp. fragments	—	—
Small copepod eggs	3.2	0.5
Copepod egg membranes	—	—
Copepod nauplii	—	—
Larval lamellibranchs	—	—
Unidentified debris	—	—
Number of microplankton organisms examined	62	—
Mean number of organisms per litre	—	12,760

1957). Hudinaga and Kasahara (1941) were also able to rear *B. amphitrite* var. *hawaiiensis* on *Skeletonema*. Costlow and Bookhout (1957, 1958) successfully reared nauplii of *Balanus eburneus* and of *B. amphitrite* var. *denticulata* to the cyprid stage, and through metamorphosis to the settled adult on *Chlamydomonas* sp. culture to which some developing *Arbacia* eggs were added. There was a suggestion that development was not successful on *Chlamydomonas* diet alone.

Moyse (1963) showed that various cirripede larvae need different foods. *Balanus balanoides* was successfully reared on diatoms (*Skeletonema*, *Ditylum*, *Asterionella*) but not on flagellates; even the widely-used flagellate *Dunaliella* was useless. Toxic dinoflagellates, *Amphidinium* and *Prymnesium* and to some extent *Chlorella* were not suitable for any cirripede tested. *Chthamalus* was reared successfully on several flagellates and on the dinoflagellate *Prorocentrum*, but diatoms were not adequate as a diet. *Elminius*, however, was successfully reared on diatoms and on some flagellates. Moyse and Knight-Jones (1967) found that *Lepas*, which has a relatively long larval life, successfully completed growth and metamorphosis to the cyprid stage on flagellate diets. Moyse suggests that the species of cirripede show dietary requirements associated to some extent with the predominant type of phytoplankton at their normal latitudes. There may be specific needs also.

Few studies have been made on special food requirements of zooplankton other than Crustacea. The larvae of several species of bivalves have been fed on algal cultures by Loosanoff and his associates. The larvae of *Venus mercenaria*, for example, have been grown on *Chlorella* and sulphur bacteria cultures as well as on detrital suspensions (cf. Loosanoff, Miller and Smith, 1951). The economic importance of the oyster focused attention on the dietary requirements of the larvae. Many workers (Cole, Korringa, Loosanoff, Nelson, Bruce, Knight and Parke) stressed the necessity of providing food of suitable size for the larvae. Algal cells exceeding 10 μm in diameter cannot be ingested. Bruce, Knight and Parke (1940), however, demonstrated that not all algae of suitable size were equally effective as food. Subsequent work by Davis (1953) on the food of the larvae of *Crassostrea virginica* confirmed that whilst size of food particle was of great importance, it was not the only significant factor in adequate diet. Investigations demonstrated that four species of bacteria and three flagellate species would not promote good growth and that detritus was also ineffective as a food source. Later, Davis showed that nine different species of bacteria and one species of chrysomonad failed to meet the dietary demands of oyster larvae, but five species of nanoplankton (*Dicrateria inornata*, *Hemiselmis rufescens*, *Isochrysis galbana*, *Chromulina pleiades*, *Pyramimonas grossii*) all gave very good growth, as did a "mixed *Chlorella*" culture.

There was some indication that the suitability of the various food organisms changed with growth of the larvae. The youngest oyster larvae could not utilize pure *Chlorella* cultures, whereas older larvae were capable of growth on *Chlorella* (cf. also Walne, 1963). Davis did not obtain strong evidence for interaction between different algal food species, but Bruce, Knight and Parke (1940) suggested that with *Ostrea edulis* larvae, *Pyramimonas* and *Chromulina* fed together were more effective as a food supply. Differences between the value of algae of suitable size as food suggest variation in the biochemical composition of the diet, but the few analyses carried out so far do not give ready support to clear differences in the composition of the diet (cf. Corner and Cowey, 1968). Larvae of *Venus* (*Mercenaria*) appear to be rather less selective than those of *Crassostrea*, being able to grow on *Chlorella* alone and on some bacteria which were unsuitable for oyster larvae.

Walne (1963) investigated the growth of oyster larvae using *Isochrysis* and bacteria as a standard. *Dunaliella* and *Phaeodactylum* were positive; bacteria alone and *Monochrysis* alone showed only limited, if any, positive effect. *Chlorella* gave a definitely negative effect and *Hemiselmis* was also slightly negative. Walne's (1974) later experiments suggest that *Chaetoceros calcitrans*, *Monochrysis* and *Skeletonema* may, however, be superior to *Isochrysis*. Loosanoff's extensive work indicated that a mixture of *Monochrysis* and *Isochrysis* was the most suitable food for many bivalve larvae.

Loosanoff and Davis (1963) suggest that apart from toxic forms (e.g. *Prymnesium*), even the poorest of "naked" flagellates are superior as food for oyster larvae to species with a cell wall. Pilkington and Fretter (1970) found some differences in suitable algal foods for the growth of veligers of the prosobranchs *Nassarius* and *Crepidula*. For example, *Cricosphaera carterae* was an excellent food; *Monochrysis* and *Pyramimonas* were reasonably satisfactory at fairly high densities, but the larvae did not metamorphose. *Exuviaella baltica* was as satisfactory as *Cricosphaera* for *Crepidula* but was not suitable for *Nassarius*, and another species of *Exuviaella* (*E. pusilla*) was not a satisfactory food for either veliger. Apart from variations in digestibility of some of the food algae, the difference between the two *Exuviaella* species is remarkable.

Appendicularians and thaliaceans rank among some of the most important essentially herbivorous grazers in the marine plankton. They may compete with other grazing zooplankton (cf. Chapter 2). The method of feeding in Appendicularia has already been described. Jorgensen (1966b) has emphasized the extreme fineness of the filtering apparatus, which may retain particles exceeding 0.1 μm. Paffenhofer (1973) demonstrated the culture of the appendicularian *Oikopleura dioica* through 19 generations without loss of fertility when fed on different unialgal foods.

Madin (1974) examined the feeding of six species of salp, including the widespread swarming form *Thalia democratica* (cf. Chapter 2), observing the animals in the open sea by Scuba diving. The circular body muscles of the salp effect a pumping of water through pharynx and atrium, the flow being filtered through a conical mucus net. The net is continuously secreted by the peripharyngeal bands, passed backwards and, when rolled into a cord, is passed to the oesophagus. The net fills the whole pharynx, which is bounded by peripharyngeal bands, branchial bar and oesophagus. This type of feeding had been previously reported, but another method was described where particles were trapped on isolated mucus strands. Madin believes that this second slower method is not practised under natural conditions. More than 90% of the salps observed were feeding, so that the process may be practically continuous and the renewal of the mucus net may be exceedingly rapid—from 10–30 seconds to a few minutes—depending on species and conditions.

There appears to be no selection for particle size and the fineness of the filtration mechanism is remarkable. Particles of around 1 μm can be retained; the upper limit appears to be as large as 1 mm. Though the majority of particles are of phytoplankton and nanoplankton, foraminiferans, radiolarians and tintinnids are also taken. The enormous grazing effect of such continuous feeding, especially with the high reproductive rate and dense swarming of salps, must be of great ecological significance. Some data on feeding rates are discussed in subsequent pages. Harbison and Gilmer (1976) confirmed that the mucus net of *Pegea confederata* could retain particles of 0.7 μm. Salps may thus be very important links in a food web from very small nanoplankton. Faecal pellets and organic debris such as dead tests from large numbers of migrating *Salpa aspersa* could contribute markedly to the food requirements of bathypelagic and benthic organisms (Wiebe, Madin, Haury, Harbison and Philbin 1979; cf. Chapter 5).

The quantity of food

Apart from specific food requirements, feeding of the zooplankton must involve the amount of food necessary for maintenance, growth and reproduction. Many particle feeders employ filtration as one of the major feeding mechanisms and many experiments on the amount of food required by zooplankton have thus been largely concerned with filtration rates, especially in copepods. Results were usually expressed as volume of water filtered ("swept clean") per unit of time, but in view of more recent knowledge of variation in the type of particle feeding, it is now more usual to quote data as the weight of food (or carbon or nitrogen equivalent) consumed per unit animal or per unit body weight. Feeding is often given as ingestion rate, which must be distinguished from filtering rate.

Filtration in calanoids, though it may appear largely automatic, may cease voluntarily. Its rate is dependent on physical and chemical factors; for example, it is generally agreed that the rate increases with temperature (cf. Fuller, 1937). Age of culture, which may reflect dissolved organic and inorganic substances in the medium as well as cell composition, may affect the rate, and the volume of water per animal (size of container) may be a significant factor (*vide infra*). Food concentration, according to earlier investigators, seemed to have little effect on filtering rate, though this view may now need modification; very high concentrations of food might depress filtration.

Fuller (1937) obtained some slight evidence of a diurnal rhythm in feeding and considered that light might lower the feeding intensity. Wimpenny (1938) also believed, as a result of field observations, that *Calanus* and some smaller calanoids showed some diurnal feeding rhythm. Gauld (1953), however, from observations of *Calanus* in the field and in the laboratory, claimed that copepods could have full guts at all hours of the day and could feed continuously over long periods. Differences in feeding between copepods living more in the surface layers at night and by day at greater depths might be attributed to the relative abundance of phytoplankton in the different strata.

Meaningful laboratory experiments on the feeding rates of copepods are not easy to achieve in view of such difficulties as the collection of animals of the same species and stage in healthy condition, and the provision of a proper volume and concentration of food culture in standard condition and growth phase. There are often serious difficulties in reconciling experimental data with field observations. Nevertheless experiments are essential.

Some earlier work, which relied on algal cell counts to estimate rates, suggested fairly low values. Fuller and Clarke (1936) obtained a rate for *Calanus finmarchicus* Stage V copepodites, grazing on *Nitzschia*, of 5–6 ml filtered/animal/day. Estimates suggested that for the average concentration of algae in waters off Cape Cod, at least 70 ml/day should be filtered for maintenance alone. Harvey (1937b), however, found much higher clearance rates with *Calanus finmarchicus*, though his experiments covered comparatively short periods. For *Lauderia* an average rate of 3 ml/hr/copepod, and for *Ditylum* rates of 7 and even 10 ml/hr were obtained. These values would correspond to approximately 70 ml filtered per day for *Lauderia* and a much larger volume for *Ditylum* if these rates continued for 24 hours. More probably, rates would decline with time. Filtration was probably depressed in the experiments of Fuller and Clarke by the small volumes of water employed, and *Nitzschia* may be less actively filtered than *Ditylum* in view of the large size of the latter. Gauld (1951), working with *Calanus* Stage V feeding on cultures of the flagellate *Chlamydomonas* (slightly larger than *Nitzschia*), obtained filtration rates varying from 40 to 100 ml/day, with a mean of around 80 ml/day. For experiments lasting 18–24 hr, rates were calculated according to the relationship

$$F = \frac{(\log_{10} C_0 - \log_{10} C_t)\, V}{t \log_{10} e},$$

where F = volume filtered
C_0 and C_t = initial and final cell concentrations
V = volume of water per animal
t = time

Gauld believed that the rate of filtration was independent of the food concentration, although the copepods possibly fed only intermittently. Comparison of filtering rates using *Chlamydomonas* and a diatom of approximately equivalent size gave much lower rates for the diatom, indicating that although selection for size may be important in filter feeding, other factors in the quality of the food may modify feeding rates.

Marshall and Orr (1955b) fed *Calanus* on a variety of unialgal cultures of diatoms, dinoflagellates and flagellates (Chlorophyceae, Chrysophyceae and Cryptophyceae) previously labelled with radioactive phosphorus (^{32}P). They estimated the volume of culture cleared from the radioactivity of the culture and that of the bodies of the copepods, due consideration being paid to the tracer in faeces and in any eggs produced. Although almost all cultures were freely filtered, there was a considerable range in feeding rates even when the same food organism was used. Rates were somewhat lower with the highest food concentrations, but there appeared to be little relation between food concentration and rate of filtration. Individual copepods were remarkably variable, however, in their feeding. In the same experiment some would feed very actively and others would cease feeding altogether, though all the animals appeared healthy. The state of maturity and other factors affecting the physiology of the animal probably influence the filtration rate. The time of the year as well as the age of food culture and of the grazers may also be significant.

Marshall and Orr (1955b) believe that *Calanus* is not an automatic feeder; the copepods may cease filtering or appear to reduce filtration at very high food concentrations. Marshall and Orr suggest that *Calanus* is then unable to ingest all the food offered, rather than experiencing any inhibitory effect from the algae.

The question of near automatic feeding was investigated by Reeve (1963), who worked with *Artemia* cultures in order to obtain more standardized filtering animals with respect to age and size. The results, though referring to an animal that is not planktonic, may give useful clues in considering zooplankton filtering. Reeve found that using Gauld's formula for calculating filtration, while the filtration rate (F.R.) was practically constant at comparatively low cell densities, the rate declined with further rise in cell concentration. The ingestion rate (I.R.), computed according to the equation

$$\text{I.R.} = \frac{V(C_0 - C_t)}{t},$$

shows an increase with cell density to a maximum, after which no further increase occurs (Fig. 6.11). *Artemia* disperses accumulated food which is filtered at concentrations above the maximum, indicating an approach to automatic filtration. Reeve also demonstrated that the maximum increased with body size in *Artemia*, and by using two unialgal cultures of different cell size (*Chlorella* and *Dunaliella*), showed that the maximum varied inversely with cell volume.

The investigations of Mullin (1963) on *Calanus* indicated that filtering rates were dependent to some degree on the duration of the experiment. With several algal cultures (*Asterionella*, *Ditylum*) rates were higher initially and subsequently declined. Over short term experiments (4–8 hr) the rates approached 150–350 ml/copepod/day for the two diatoms. Over a full 24-hour experiment *Calanus helgolandicus* gave much lower rates of *ca.* 20–55 ml/day. Mullin's experiments with *Calanus hyperboreus*, however, were particularly relevant to Reeve's results for *Artemia* in that there was a clear effect of food concentration upon filtering rate. Mullin also calculated ingestion rates. There was the expected rise in ingestion rate with increasing cell density at lower concentrations, but at higher cell densities an optimum rate was reached, followed by a decline at still higher cell densities. The optimum concentration varied with the size of diatom used (*Ditylum*, *Thalassiosira*), indicating that the maximum represents a volume of food as maximum ration. However, unlike Reeve's results, there appears to be no sharp change to a plateau at the maximum (cf. Fig. 6.12a,b). Though the filtration rate data

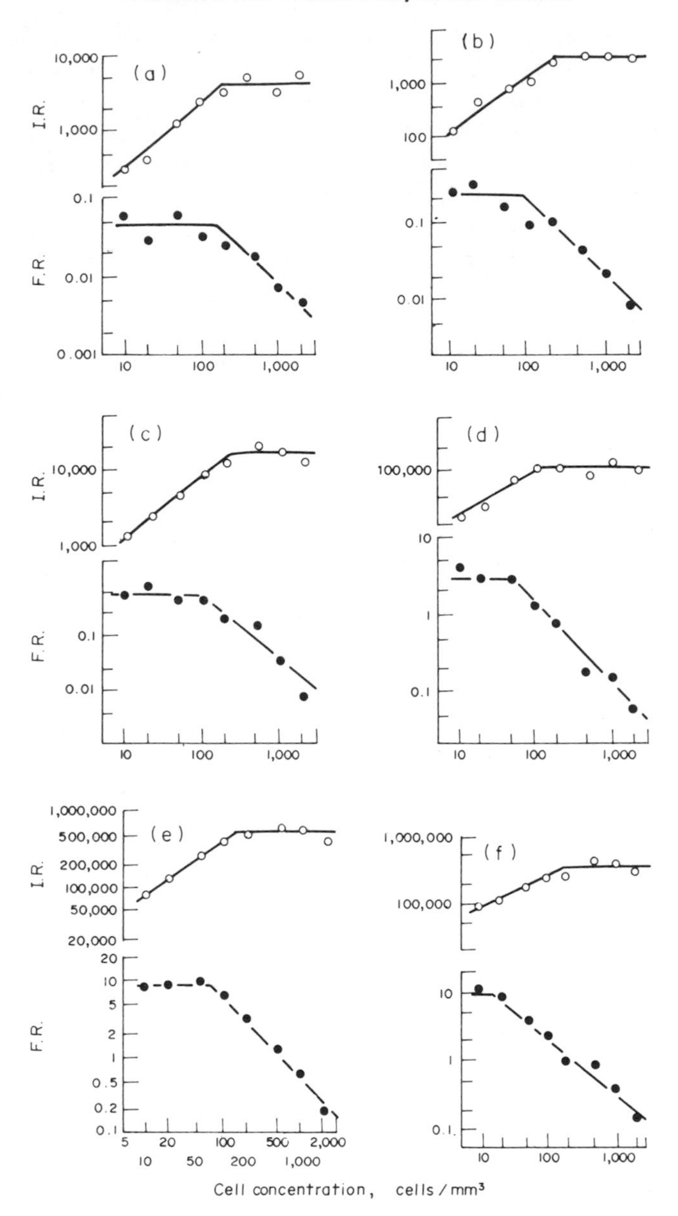

Fig. 6.11. Filtration (F.R.) and ingestion (I.R.) rates with increasing cell concentration for *Artemia* at six different ages (increasing from a–f) feeding on *Phaeodactylum*. Both axes logarithmic (Reeve, 1963).

show considerable scatter, there is generally a decline as particle concentration increases. It has been suggested that zooplankton species may show two types of relationship between filtration and food concentration. In many, filtration rate is unchanged at low cell densities but declines as the maximum ration is achieved (A); in others, filtration may be reduced at very low particle concentrations, but it rises as cell

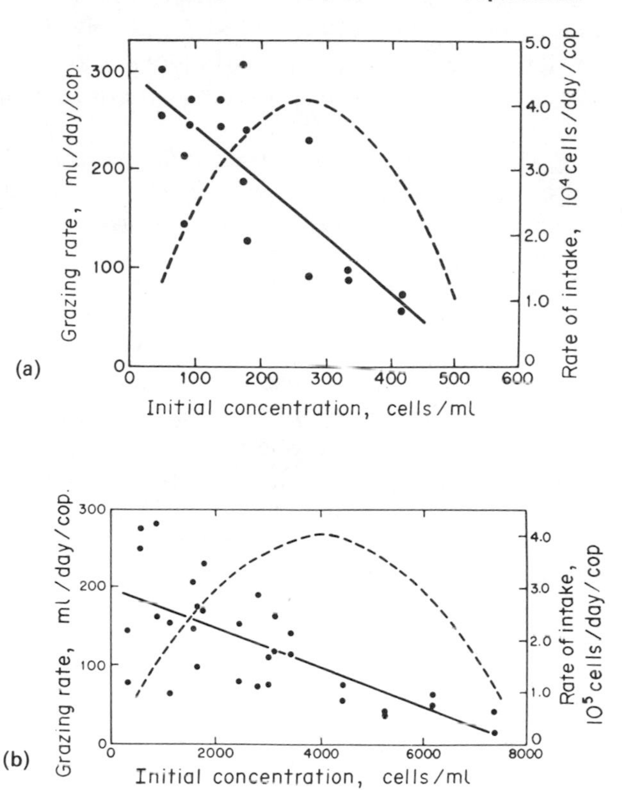

Fig. 6.12. Grazing by female *Calanus hyperboreus* on various food concentrations of (a) *Ditylum brightwellii* and (b) *Thalassiosira fluviatilis*. Solid line—grazing rate; broken line—intake rate (Mullin, 1963).

density increases to reach a maximum, after which further increase in particle concentration causes a decline in filtering rate (B) (Fig. 6.13).

Observations by Anraku (1964) on certain species of copepods (*Calanus*, *Pseudocalanus* and *Acartia*) grazing on algae, and on some carnivorous species (*Centropages*, *Tortanus*, *Labidocera*) fed on *Artemia*, gave generally similar results showing an optimum ingestion rate with increasing food concentration. That there is some relationship between the amount of food ingested and the size of copepod was demonstrated by Mullin (1963), who restricted comparison of feeding rates to species of

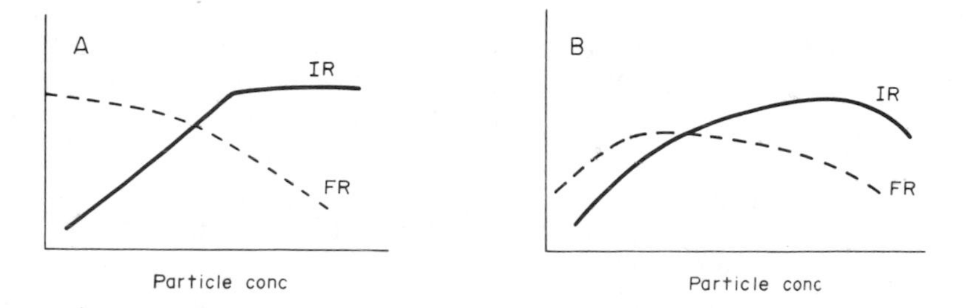

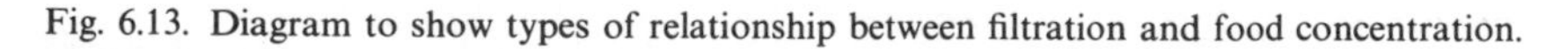

Fig. 6.13. Diagram to show types of relationship between filtration and food concentration.

the same genus, *Calanus*. Females of any one species were found to feed more than the Copepodite V stage and the amount of food declined in the order *C. hyperboreus*, *C. glacialis*, *C. finmarchicus* (*sensu stricto*), *C. helgolandicus*—the order of decreasing body size.

The studies of Haq (1967) on the feeding of *Metridia lucens* and *Metridia longa* filtering *Dunaliella* also showed an inverse relationship between filtering rate and food concentration. Ingestion increased to a maximum of 3 μg organic matter/copepod/day when the food concentration amounted to 1.2 μg/ml ($\equiv 32 \times 10^3$cells/ml) but gradually declined thereafter in *M. lucens*. In *M. longa* the ingestion rate was slightly lower but continued to increase to a maximum of 4.1 μg organic matter/copepod/day at a food concentration of 2.8 μg/ml ($\equiv 75 \times 10^3$cell/ml) (Fig. 6.14). With animal food (*Artemia* nauplii), which was preferred, ingestion also rose to a maximum and then declined with increasing food concentration. The maximum was 54 μg organic matter/copepod/day for *M. lucens* and 16 μg organic matter/copepod/day for *M. longa*, both substantially higher than with phytoplankton food. Apart from selection of animal food, especially by *M. lucens*, both species showed different maximum filtration rates and maximal quantities of ingested food when offered various mixtures of algal species (normally two species), with a tendency for larger cells to be more effectively filtered. Feeding rates also tended to be increased when feeding on mixed algae.

Gaudy (1974) reports similar findings from a study of grazing rates of four copepods. *Calanus helgolandicus* gave a wide range of maximal grazing rates (*ca.* 10–100 ml/copepod/day) when fed on a single species of alga (*Phaeodactylum*, *Skeletonema*, *Ditylum*, *Lauderia*, *Asterionella*), but these were considerably above those of the omnivorous species *Centropages typicus* (mostly <ml/day), *Temora stylifera* (5–30 ml/day) and *Acartia clausi* (<3 ml/day). Food intake above a minimum level rises with food concentration to a maximum and then is relatively constant. Filtration rate declined with high cell densities. With pluri-algal diets offered to *Centropages* and *Temora*, feeding rates could be higher, but there was considerable selection; *Lauderia* was preferentially consumed. Cell size appears to be one significant factor. Not only cell concentration but also cell shape and size may affect filtration rates (cf. Cushing, 1968).

As Conover (1964) has pointed out, there is now general acceptance that calanoids are not automatic filterers, and the control of a species such as *Calanus* over its grazing appears to be somewhat greater than for a semi-automatic filterer such as *Artemia*. There is evidence that the frequently observed decline in filtering rate with age of culture may not necessarily be attributable to any deleterious algal metabolite, but may result from a different organic content (calorific value) of the cells. Frost's (1972) experiments, using different sized diatoms of similar (pill box) shape and using rather longer periods, suggest that ingestion rate rises with cell concentration to a maximum, and he claims that the constant level is then maintained. With previously starved copepods, the plateau for maximal consumption is higher. The optimum obtained by Anraku, Mullin and others might then represent a type of overshoot with animals not sufficiently conditioned. Frost's data for filtration rates suggest a constant level at lower cell densities, falling at higher concentrations (cf. Fig. 6.15a,b).

It is important to note the marked effect of cell size. Filtration rates began to decline earlier with increasing cell densities for the largest diatom (a centric species) and least rapidly with the smallest, *Thalassiosira fluviatilis*. It is hardly surprising that the data for quantity of water filtered ("volume swept clear") and the amount of food ingested by a

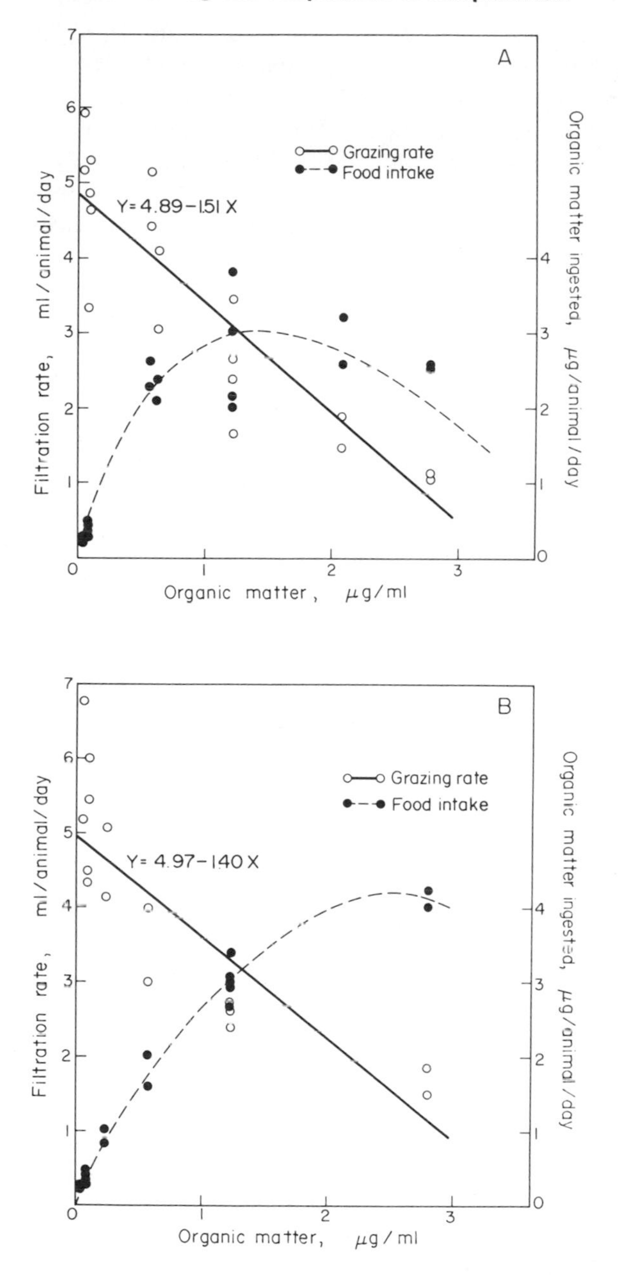

Fig. 6.14. Filtering rate and rate of ingestion of *Dunaliella* by (A) *Metridia lucens* and (B) *M. longa* at various concentrations (Haq, 1967).

species of copepod may show considerable variation, bearing in mind all the differences in experimental technique. Table 6.7 gives some of the information for marine species summarized by Marshall (1973), omitting some of the more carnivorous copepods. There has been considerable argument over the effect of the volume of water (container size) used in feeding experiments (cf. Marshall, 1973). Cushing (1959) believed that very large volumes were essential if the data were to be applicable to field conditions. Table 6.8 indicates the marked increase in volumes filtered by *Temora*, with the volume of water per copepod varying from 5 ml to 500 ml. One *Calanus* or *Anomalocera* was held to require a volume of several litres. Although many workers have not accepted the necessity of these very large quantities, too small a volume certainly limits feeding. Geen and Hargrave (1966), for example, showed that mixed populations of three calanoids, mostly *Acartia tonsa*, filtered 5–6 ml/day in laboratory containers, but increased their

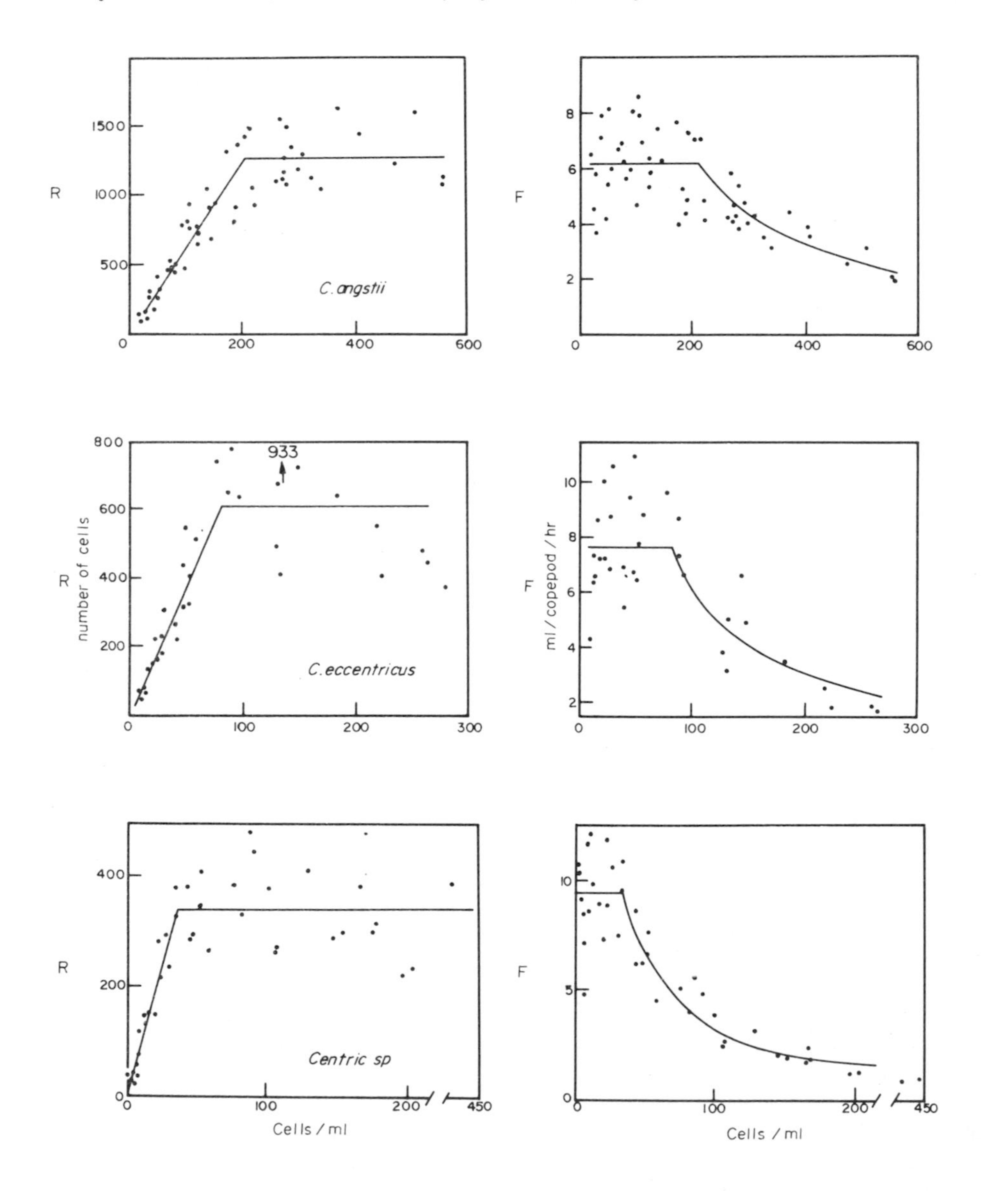

Fig. 6.15a. Effect of cell concentration on ingestion rate (R) (cell numbers) and volume swept clear (F) (ml/cop./hr) of adult females of *Calanus* feeding on *Coscinodiscus angstii*, *C. eccentricus* and centric species (Frost, 1972).

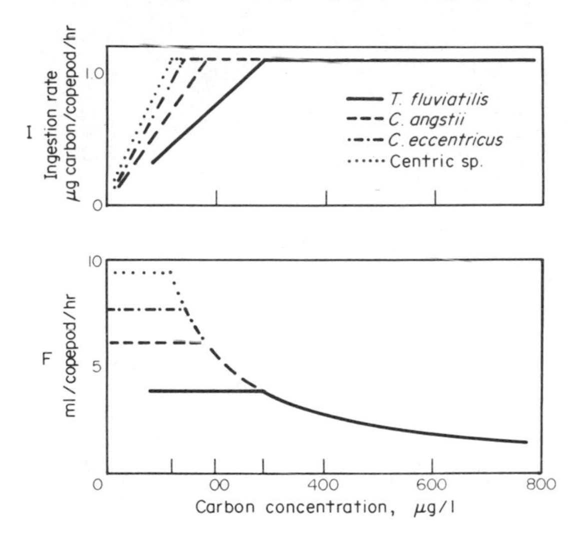

Fig. 6.15b. Effect of size, (species) and concentration (as carbon) of food particles on ingestion rates I and volume swept clear (F), of adult female *Calanus* (Frost, 1972).

filtration to 30 ml/day in large cylinders with nylon covers suspended in a lake. Corner (1961) obtained a filtration rate of 10–36 ml/day for *Calanus helgolandicus* using a constant flow method.

Cushing (1963) attempted to reconcile the sharp reduction of natural algal concentrations due to grazing by *Calanus*, as seen in an area of the North Sea, with the generally low feeding rates observed in laboratory experiments. Copepods such as *Calanus* and *Anomalocera* might clear more than 1000 ml/day, though not by the usual type of filtration feeding. He advanced an encounter theory of feeding to account for these much increased volumes (Cushing, 1968). Copepods are believed to search for and locate food particles and then seize them. Although the hairs on the antennules were regarded as important in locating algae, it was shown that capture was little impaired by removal of the appendages. Other sensory structures on the appendages may, however, be able to locate particles. The "perceptive range" has been suggested as 0.1–0.6 mm for *Calanus*, though this may be an underestimate (cf. Petipa, 1965). Cushing calculates the volumes of water filtered from the speed of swimming of the copepod in relation to its length and the time necessary to search out, capture and eat a cell in relation to the cell concentration. The volumes calculated, apart from the relation to perceptive range, depend on cell size and density. At high cell concentrations, algal size is the major factor.

The observation that copepods usually seem to increase their food intake with a rise in cell density led Parsons and his colleagues (e.g. Parsons *et al.*, 1967, 1969; Parsons and LeBrasseur, 1970) to employ a concept first used by Ivlev for fish feeding, relating dependence of a daily ration on food concentration. The relationship proposed by Ivlev is

$$r = R\,(1 - e^{-kp}),$$

where r = the amount of food ingested at a prey density p
R = is the maximum ration taken.

Table 6.7. Ingestion of food by a variety of copepods (Marshall, 1973)

Species	Total length (mm)	Food	Filtering rate		Method	Source
			(ml/copepod/day)	(ml/mg dry wt/day)		
Calanus finmarchicus	♀ (3–4)	Cells <10 μ diam.	0–4	—	^{32}P	Marshall and Orr, 1955a
C. finmarchicus	♀	Cells >10 μ diam.	0–84	—	^{32}P	Marshall and Orr, 1955a
C. finmarchicus	late stage	Nat. sea.	1–52	—	Chlor. a, at sea	Adams and Steele, 1966
C. finmarchicus	♀ 2.8–4.3	Thal.	35–190	300–2150	Cell count	Anraku, 1964a
C. finmarchicus	♀ 2.7–4	Var. phyto.	5–164	—	Cell count	Mullin, 1963
C. helgolandicus	♀ (*c.* 3)	Nat. sea.	10–36	—	Cascade expt.	Corner, 1961
C. pacificus	♀ —	Dit., Go.	68–123	—	Cell count	Mullin, 1963
C. pacificus	♀ —	L.	201–835	—	Cell count	Paffenhöfer, 1971
C. pacificus	♀ —	Gym.	316–1428	—	Cell count	Paffenhöfer, 1971
C. hyperboreus	♀ 6.3–8.5	Var. phyto.	0–329	—	Cell count	Mullin, 1963
C. glacialis	♀ 4.4–5.2	Var. phyto.	9–448	—	Cell count	Mullin, 1963
Neocalanus gracilis	♀ 3.0–3.7	Cos., Thal., Art.	20–222	—	Cell count	Mullin, 1966
Nannocalanus minor	♀ 1.8–1.9	Cos., Thal., Art.	18–22	—	Cell count	Mullin, 1966
Rhincalanus nasutus	♀ 3.8–4.3	Art.	17	—	Count	Mullin, 1966
R. nasutus	♀ —	Thal., Art.	98–669	—	Count	Mullin and Brooks, 1967
R. cornutus	♀ 3.1–3.4	Thal., Art.	16–120	—	Count	Mullin, 1966
Eucalanus attenuatus	♀ 3.8–5.8	Thal., Art.	15–101	—	Count	Mullin, 1966
Pseudocalanus minutus	♀ 1.2–1.6	Thal.	6–40	750–3500	Cell count	Anraku, 1964a
P. minutus	♀ —	Flag.	5–6	—	Cell count	Geen and Hargrave, 1966
P. elongatus	♀ (1.4)	Var. phyto.	0–12	—	^{32}P	Marshall and Orr, 1966
Chirundina indica	♀ 3.3–3.9	Thal., Cy., Art.	49–197	—	Count	Mullin, 1966
Euchirella curticauda	♀ 3.6–3.7	Art.	264	—	Count	Mullin, 1966
E. bella	♀ 3.1–3.9	Art.	232	—	Count	Mullin, 1966
Euchaeta acuta	♀ 3.2–3.9	Art.	85	—	Count	Mullin, 1966
Scolecithrix danae	♀ 1.7–2.0	Art.	118	—	Count	Mullin, 1966
Lophothrix latipes	♀ 2.9–3.1	Cos., Art.	31–58	—	Count	Mullin, 1966
Centropages typicus	♀ (1.3–1.8)	Thal.	5–50	30–540	Cell count	Anraku and Omori, 1963
C. typicus	—	Diatoms	0–11	—	Cell count	Gaudy, 1968
C. hamatus	♀ (1.4)	Thal.	1–15	65–1200	Cell count	Anraku and Omori, 1963
C. hamatus	—	Var. phyto.	0–6	—	^{32}P	Marshall and Orr, 1966

Diaptomus oregonensis		(1.3–1.5)	Nano.	<1–2.5	—	^{14}C	Richman, 1964
D. oregonensis		—	Flag. nat. phytopl.	0–13	—	Cell count	McQueen, 1970
Temora longicornis	♀	(1–1.5)	Skel.	1–27	—	^{32}P	Berner, 1962
T. longicornis		—	Var. phyto.	0–21	—	^{32}P	Marshall and Orr, 1966
T. longicornis	♀	—	Flag.	5–6	—	Cell count	Geen and Hargrave, 1966
Metridia lucens	♀	(2.5–3)	Ch. Cri.	0.4–1.3	—	^{32}P	Marshall and Orr, 1966
M. lucens		—	Mix. phyto.	1–15	—	Cell count	Haq, 1967
M. lucens		—	Mix. phyto., Art.	18–31	—	Count	Haq, 1967
M. longa		(4–4.5)	Mix. phyto.	1–5	—	Count	Haq, 1967
M. longa		—	Mix. phyto., Art.	3–7	—	Count	Haq, 1967
Pleuromamma xiphias	♀	4.2–5	Art.	520	—	Count	Mullin, 1966
P. abdominalis	♀	3–3.4	Art.	210	—	Count	Mullin, 1966
P. gracilis	♀	1.7–1.8	Art.	73	—	Count	Mullin, 1966
P. piseki	♀	1.7–1.8	Art.	110	—	Count	Mullin, 1966
Labidocera acutifrons	♀	3.3–3.9	Art.	80	—	Count	Mullin, 1966
L. aestiva	♀	1.8–2.0	Art.	—	20–1400	Count	Anraku, 1964a
Acartia tonsa	♀	(1–1.2)	Skel.	8–25	—	Cell count	Conover, 1956
A. tonsa	♀	—	Thal.	1–105	200–8500	Cell count	Anraku, 1964a
A. clausi	♀	(1.2–1.3)	Thal.	3–20	400–5000	Cell count	Anraku, 1964a
A. clausi	♀	—	Var. phyto.	0–11	—	^{32}P	Marshall and Orr, 1966
A. clausi		—	nat. phyto.	5–6	—	Cell count	Geen and Hargrave, 1966
Tortanus discaudatus		2–2.3	Art.	—	220–1700	Count	Anraku and Omori, 1963

Where total length is not given by the authors, it is taken from Sars (1903) or Wilson (1932) and enclosed in brackets.

Abbreviations for foods:

Nat. sea.	—natural seawater	L.	—*Lauderia borealis*
Var. phyto.	—various phytoplankton cultures	Skel.	—*Skeletonema costatum*
Mix. phyto.	—mixtures of phytoplankton cultures	Thal.	—*Thalassiosira fluviatilis*
Nano.	—nanoplankton	Ch.	—*Chaetoceros* sp.
Flag.	—flagellates	*Dit.*	—*Ditylum brightwellii*
Art.	—*Artemia* nauplii	Gym.	—*Gymnodinium splendens*
Cos.	—*Coscinodiscus* sp.	Go.	—*Gonyaulax* sp.
Cy.	—*Cyclotella* sp.	Cri.	—*Cricosphaera elongata*

Table 6.8. Increase in volume filtered by *Temora* in relation to container size (Cushing, 1959)

Volume of vessel (ml)	5	25	100	500
Volume (ml) swept clear by *Temora longicornis* in one day	6	13	26	150

Thus, the ration increases with food concentration to a maximum level. The experiments of Parsons *et al.* (1967) on grazing indicated that zooplankton would not feed below a minimum level of prey (p_0). Thus a modified equation was proposed, and, employing new symbols, is that used by McAllister (1970)

$$I = I_m \, (1 - e^{-\delta(P - P')})$$

where I = rate of ingestion of phytoplankton/unit grazer;
I_m = maximum rate of ingestion;
P = mean phytoplankton concentration during grazing period;
P' = minimum phytoplankton concentration required to induce filtering;
δ = a constant defining rate of change of ingestion with food concentration.

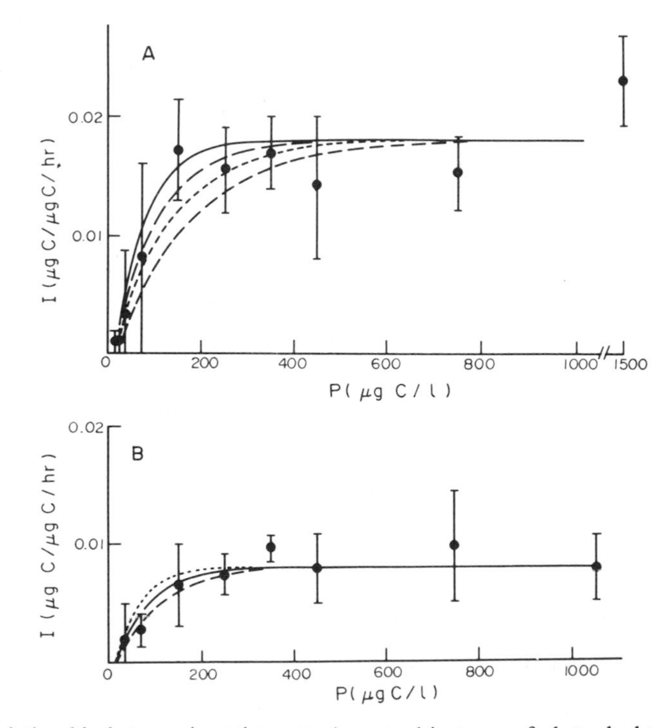

Fig. 6.16. (a) Relationship between ingestion rates (expressed in terms of phytoplankton carbon/unit or zooplankton carbon/hour) (A) by nocturnally feeding copepods and phytoplankton concentration. Circles with bars are means and standard deviations. Lines were fitted assuming $I_M = 0.0180$, $P' = 15$. Solid line $\delta = 0.0153$; long dashes $\delta = 0.0109$; dots $\delta = 0.0081$; short dashes $\delta = 0.0063$.
(B) by continuously feeding copepods. Circles with bars are means and standard deviations. Lines were fitted assuming $I_M = 0.0080$, $P' = 15$. Solid line $\delta = 0.0140$; dotted line $\delta = 0.0184$; dashed line $\delta = 0.0098$ (McAllister, 1970).

In Fig. 6.16 the results of grazing experiments with *Calanus pacificus* are illustrated for nocturnal and continuous (24-hour) grazing. McAllister obtained evidence that feeding did not commence below a minimum food concentration (15 μg C/l), but that ingestion increased with rising food concentrations to a maximum. Several values for the rate of change of ingestion (δ) are given (Fig. 6.16). Parsons *et al.* (1969) allowed *Calanus plumchrus* to graze on natural phytoplankton populations in the Strait of Georgia.

Figure 6.17 illustrates the different concentrations required for threshold feeding and to attain maximum ration. In some experiments the maximum ration was shown to be greater than the highest concentration of phytoplankton naturally available. The type of phytoplankton is obviously of significance in addition to concentration. Thresholds were computed to vary from 50–190 μg C/l with different phytoplankton populations; maximum ration was *ca.* 400 μg C/l. Table 6.9 gives an evaluation of the ration to body weight of the copepod. Also included are data on grazing using *Pseudocalanus*. Corkett and McLaren (1978), reviewing feeding in this calanoid and using various conversion factors to attempt to compare data on food concentration by different authors, comment on the very great variation in phytoplankton concentration necessary for maximum ingestion rate (Fig. 6.18). Settlement of cells in culture or inadequate filtration may be partly responsible (Table 6.10). Paffenhofer and Harris' (1976) technique of using large rotating vessels may have assisted in yielding low saturation levels. The daily ration was computed (cf. Table 6.11) and also shows large differences. *Chaetoceros* is almost certainly not eaten to any great extent. The differences in the significance of those phytoplankton species which are easily consumed is, however, substantial. Corkett and McLaren (1978) also compare the daily ration of *Pseudocalanus* in nature. Zagorodnyaya's data for an index of gut fullness, on the rate of passage of food through the gut (*ca.* 30 min) permitted the ration as a percentage of wet body weight to be calculated. For Copepodite V and adult females values were 5.7–12.1 %, compared with an average of 11 % obtained by Poulet. A wider

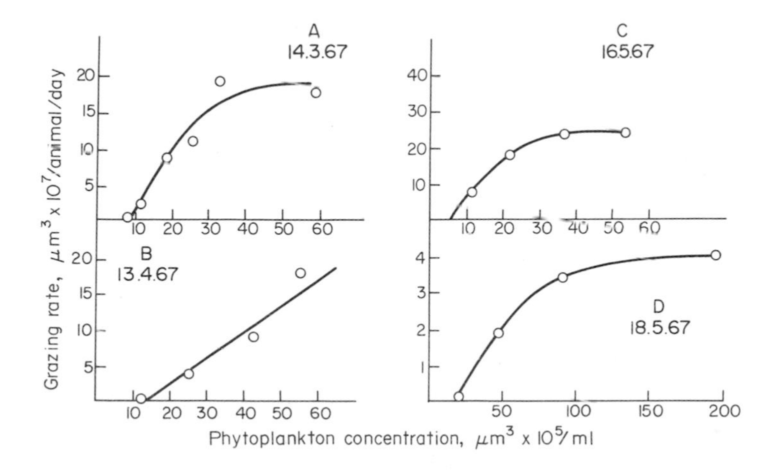

Fig. 6.17. Zooplankton grazing at different concentrations of phytoplankton. A. *Calanus pacificus* grazing on *Thalassiosira* spp.; B. *Calanus plumchrus* III and IV grazing on *Skeletonema costatum* and μ flagellates; C. *Calanus plumchrus* V grazing on *Skeletonema costatum* and μ flagellates; D. *Pseudocalanus minutus* and *Oithona* grazing on *Skeletonema costatum* and μ flagellates (Parsons, LeBrasseur, Fulton and Kennedy, 1969).

Table 6.9. Summary of grazing experiments indicating different phytoplankton rations required to obtain maximum rations (Parsons *et al.*, 1969)

Predominant zooplankton	Modal food size diam. (μ)	Predominant phytoplankter	Food density (μg C/l) min.	Food density (μg C/l) max.	Max. ration obtained (mg C/animal/day)	Mean zooplankter body weight (mg C)	Ration (% body wt/day)
Pseudocalanus minutus Krøyer	8 and 14	*Skeletonema costatum* and μ-flagellates	81	>163	0.0004	0.010	4.0
Calanus pacificus Brodsky	90	*Thalassiosira* spp.	142	>305	0.0168	0.100	16.8
Calanus pacificus	57	*Thalassiosira* spp.	85	380	0.0184	0.100	18.4
Calanus pacificus and *Calanus plumchrus* IV	45	*Thalassiosira* spp.	48	>205	0.0202	0.100	20.2
Calanus plumchrus III and IV	5 and 9	μ-flagellates	39	>108	0.0017	0.030	5.7
Calanus plumchrus III and IV	16	*Skeletonema costatum* and μ-flagellates	119	>520	0.0181	0.030	60
Calanus plumchrus V	14	*Skeletonema costatum* and μ-flagellates	62	285	0.0260	0.175	14.8
Pseudocalanus minutus and *Oithona* sp.	14	*Skeletonema costatum* and μ-flagellates	190	760	0.0045	0.010	45

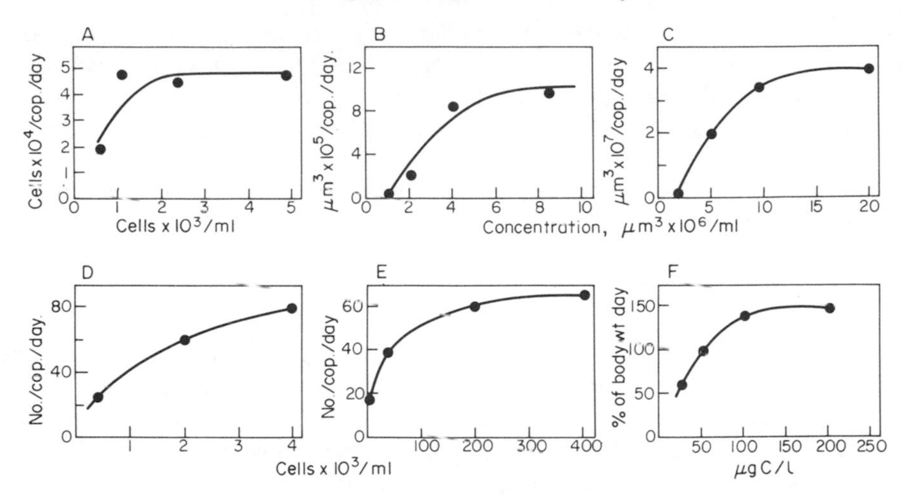

Fig. 6.18. Evidence of saturation of feeding rates of *Pseudocalanus* from various authors. (A) ingestion rate of the diatom *Thalassiosira fluviatalis*; (B) ingestion rate of mixed natural phytoplankton; (C) ingestion rate of a mixture of the diatom *Skeletonema* sp. and μ flagellates; (D) number of faecal pellets produced with the diatom *Lauderia borealis* as food; (E) number of faecal pellets with the chrysomonad—*Isochrisis galbana*; (F) ingestion rate (mean of all stages) of *Pseudocalanus* feeding on *Thalassiosira rotula* as a function of *nominal* food densities (Corkett and McLaren, 1978).

Table 6.10. Laboratory estimates of near-saturation food concentrations for *Pseudocalanus* (Corkett and McLaren, 1978)

Food species	Cell volume (μm^3)	Phytoplankton concentration at 90% of maximum ingestion rate		
		(cells/ml)	(μm^3/ml)	(μg C/ml)
Thalassiosira fluviatilis	1370	1700	3.9×10^6	0.32
Isochrysis galbana	100	160,000	16.0×10^6	2.72
Lauderia borealis	27,000	4300	116.0×10^6	5.14
Chaetoceros sp.	—	—	5.2×10^6	0.27
Skeletonema sp. and "μ-flagellates"	—	—	11.5×10^6	1.09
Thalassiosira rotula	—	—	1.1×10^6	0.09

Table 6.11. Estimates of near-saturation ingestion rates by *Pseudocalanus* (Corkett and McLaren, 1978)

Food species	Daily ration	
	(μm^3)	(μg C)
Thalassiosira fluviatilis	110×10^6	8.74
Chaetoceros sp.	0.8×10^6	0.04
Skeletonema sp. and "μ-flagellates"	40×10^6	3.8
Thalassiosira rotula	195×10^6	16.1

range of 2.2–55.5% obtained by Poulet allows for seasonal variations and younger copepodite grazing, but Corkett and McLaren point to the usually much greater saturation food levels normally obtained from laboratory experiments. Gaudy (1974), for example, gives maximal rates for four calanoids ranging from three to ten times their body weight.

In the open ocean, zooplankton must feed at considerably lower concentrations than the threshold values suggested for the Strait of Georgia. Reeve and Walter (1977) and Dagg (1977), amongst others, have recently emphasized the importance of patchiness. Dagg considered both spatial and temporal patchiness and examined the effects on four species of copepod, using as criteria survival under starvation, and egg production with discontinuous feeding. *Acartia tonsa* and *Centropages typicus* were believed to be more sensitive to small scale patchiness than *Pseudocalanus minutus* or *Calanus finmarchicus*. Reeve and Walter found that the grazing rate of *Acartia tonsa* increased with food concentration to a maximum at *ca.* 350 μg C/l, with a threshold at 7–18 μg C/l, but that ingestion rate could increase at greater food concentrations. The feeding behaviour is adjusted to high and low intensities rather than achieving a steady level, in conformity with the copepod encountering rich phytoplankton patches. They compare the type of feeding in some respects, therefore, to that of some carnivores such as *Sagitta hispida* and *Mnemiopsis mccradyi*, which probably meet rich prey densities only rarely, but can then feed at such a high rate as to fill the gut in a brief period and satisfy up to one day's food requirement.

Patchiness of plankton may be of the greatest significance in allowing zooplankton to graze intensively and discontinuously. There is some doubt, however, as to whether a threshold feeding concentration is real in zooplankton. It is also not clear whether the usual curve relating increasing rate of consumption with increasing food concentration is curvilinear or rectilinear (cf. Mullin, Stewart and Fuglister, 1975). The latter may describe the data best for well-fed copepods, but at low food concentration interpretation is difficult. Frost (1975) described experiments which indicated that clearance rates were constant above a certain limiting cell concentration, but that below this low density filtration was "impaired", though it did not necessarily cease.

Lam and Frost (1976) later suggested that a filtering copepod exhibits feeding behaviour which might tend to maximize its net rate of energy gain. From a considerable body of data on feeding in *Calanus pacificus*, covering a variety of size of food particle and range in concentration, they described three phases in filtering behaviour. At low food concentrations filtration is at a low rate; this rises with increasing particle concentration to a maximum which is maintained until at some higher critical food concentration filtering rate declines. The point at which reduced filtering rate is succeeded by the maximal rate is dependent on the size of food particle. Both the maximum rate and the concentration of food necessary to achieve maximum filtering depend on the body size of the copepod.

An equation was developed by Lam and Frost for energy gain to the copepod in terms of ingestion rate, deduced from food concentration, speed of water through the filter and the effective filtering area, assuming assimilation efficiency is constant. Loss of energy is due partly to drag forces in the filtering mechanism, a function of filtering speed, and other general losses due to oxidative metabolism, which might be regarded as proportional to length^2 of the animal. Assumptions made included the existence of a maximum speed for filtering and of a maximum rate of energy gain which cannot be exceeded. A net energy loss will be expected at low and very high filtering speeds. The model predicted the relationships between food concentration and filtering rate. At very low food concentrations whether filtering continues or ceases there is a net loss of energy. With rising food concentration the speed of filtering must be adjusted to maximize net gain. Assuming a limit to the rate at which energy can be delivered for filtering, the copepod will gain from food increase at this rate until a limit to ingestion is reached, for instance, related to the assimilative capability of the animal. The model also demonstrated that filtering rates increase with body size of the copepod at any given food concentration, and that below the maximum filtering rate, the rate rises with particle size. With a copepod offered a mixture of particle sizes, Lehmann (1976) demonstrated by a model which also simulated the feeding behaviour of a copepod attempting to maximize its net rate of energy gain, that the animal will harvest optimally if the rates of ingestion are maximal for particles slightly larger than those most numerous in the particle mixture. Regarding the microzooplankton, the Mediterranean tintinnid *Stenosemella ventricosa* feeds on nanoplankton particles. Although smaller particles (3–12 μm) are preferentially taken, the size choice is a function of particle concentration; the tintinnid can switch size categories and may consume food up to 27 μm (Rassoulzadegan and Etienne, 1981).

A minimum cell density below which filtration ceases has been suggested as being of importance in the field, since the low phytoplankton concentration is permitted to increase without depredation. This concept of a "refuge" for phytoplankton may, however, be met by a time variation in feeding. McAllister (1970) observed the growth of two parallel series of diatom cultures which under the conditions of the experiment were growing at an approximate doubling time of 36 hours. One culture was continuously grazed, the other subject to only nocturnal (12-hour) grazing by an equal population of copepods. The culture under nocturnal grazing attained much higher population levels; the algae could grow unhindered during the day, whereas with constant grazing the increase was poor. McAllister subsequently modelled changes in phytoplankton/zooplankton stocks using a computer programme which considered such factors as the initial stock of algae, their growth rate and respiration, the initial stock of zooplankton, their recruitment rate, assimilation efficiency, respiration, etc., and especially examining the differential effects of continuous and nocturnal grazing, the latter for varying time periods.

Since many investigations have suggested that feeding after a period without food intake is very intensive and declines with time (e.g. McAllister, 1970; Ikeda, 1971, 1977a), though some studies (e.g. Paffenhofer, 1971, 1976b) have not confirmed this, McAllister also examined the effect on phytoplankton crop of declining feeding intensities. While nocturnal grazing certainly does not always result in high algal stocks, the duration and changing intensity is obviously a most important factor in considering the relationship between food ration and phytoplankton concentration. A clear example of the difference between continuous and nocturnal grazing comes from an estimation of total monthly secondary production at Weather Station P in the north-eastern Pacific Ocean (Fig. 6.19), assuming different grazing schemes and different levels of zooplankton respiration. The algal growth rate and monthly changes in stock were corrected for loss due to plant respiration and the food ration thus available was corrected for assimilation efficiency and respiration of the zooplankton. The amount was then partitioned assuming continuous grazing, nocturnal grazing only, and

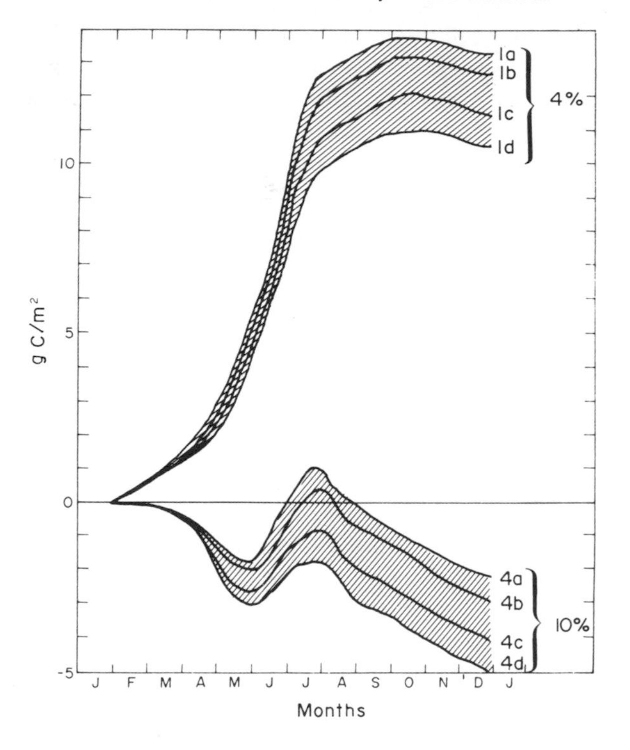

Fig. 6.19. Cumulative totals of monthly secondary production estimated from field data at Ocean Station "P" assuming different grazing schemes and zooplankton respiratory rates. Curves 1 and 4 show the production estimated on the assumption of zooplankton respiratory rates of 4% and 10% of the body weight/day respectively. Curves a and b were estimated assuming nocturnal grazing, each with a different rate of decline of the feeding rate. Curve c assumes nocturnal grazing at a constant rate equal to the mean of the variable rate. Curve d assumes continuous grazing (McAllister, 1970).

nocturnal but declining grazing rates. Some examples of the considerable variations in standing stock are shown in Fig. 6.19.

The quantity of food ingested varies amongst other factors with body size for a given species and with temperature. Mullin and Brooks (1970) observed with *Rhincalanus* and *Calanus helgolandicus* that feeding rate increased with body size (age). Ingestion rate also varied directly with temperature; both species ate and grew more slowly, for example, at 10°C than at 15°C, though the copepods did not necessarily achieve a larger size. Mullin and Brooks also found that the rate of ingestion per unit body weight decreased with body size (age), lending support to the well known concept that within limits the smaller the body size the higher the metabolic rate. Investigations by Paffenhofer (1971) carried out on *Calanus helgolandicus* reared in the laboratory gave further evidence for this view. The copepods were fed on phytoplankton cultures but at concentrations varying around those off La Jolla (*ca.* 50 μg C/l). Grazing and ingestion rates increased approximately linearly with body weight of the copepods, but there was a decline per unit body weight. Thus, while for Nauplius IV some 0.2–0.8 μg C/l was taken, corresponding to 4–21 ml/day filtered, 18–69 μg C/l was taken by adult female *Calanus* with a filtration rate of 286–773 ml/day. The daily ingestion rate was equivalent to

290–480 % of the body weight for Nauplius V and only 28–85 % of the body weight for adult females.

The fall in grazing rate per unit weight is clear from Fig. 6.20, which also indicates that Nauplius V had a higher unit grazing rate than all other stages. Ingestion rates were highest for *Calanus* grazing on the large-celled *Lauderia* (36 μm). Figure 6.21 demonstrates the relatively high grazing rates obtained by Paffenhofer, especially with the large-celled diatoms. Rates for *Thalassiosira* (12–17 μm) and *Chlamydomonas* (10–14 μm) are appreciably lower. As one example, for Copepodite III Marshall and Orr's data suggest a filtering rate of approximately 9 ml/day. Mullin and Brooks found 18 ml/day and Paffenhofer's data are 135–150 ml/day. Although volume of water (container size) appeared to be important for reared copepods, there was not a clearly demonstrable relationship for wild stocks. Paffenhofer emphasizes that older copepodites cannot graze so effectively on small phytoplankton and that growth rate decreased with older copepodites fed only on smaller diatoms (cf. also Mullin and Brooks).

Adams and Steele (1966) examined *Calanus* filtering rates on shipboard using natural densities of phytoplankton. While results showed considerable variation, for adult females clearance rates were usually 15–40 ml/day, values somewhat lower than

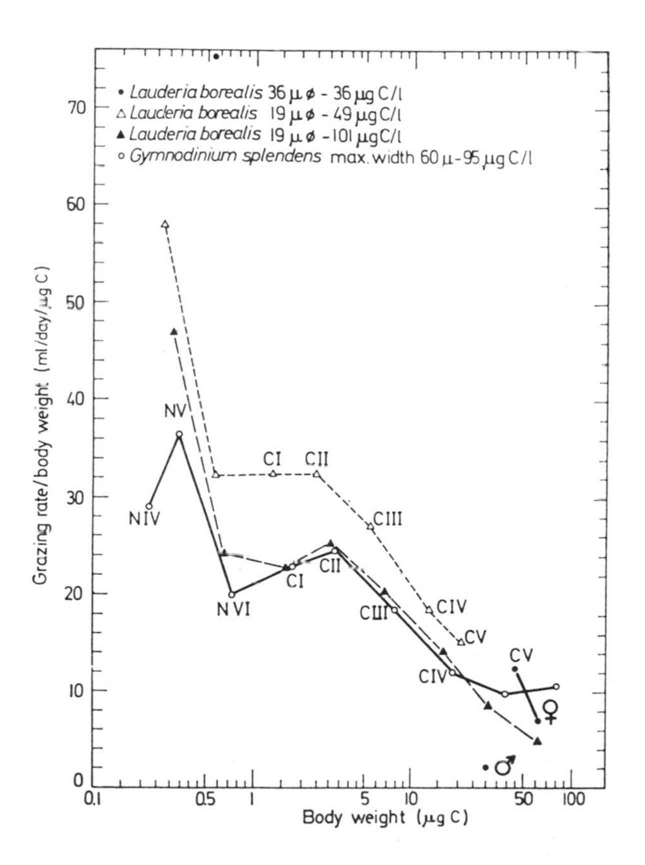

Fig. 6.20. Grazing rates (at 15°C) per body weight of copepod (*Calanus helgolandicus*). Data for N IV, V and VI feeding at 36 μg C/1 *Lauderia borealis*, 36 μm diameter (ϕ), are 91.5, 145.80, 75.7 ml/day/μg C of copepod, respectively (Paffenhofer, 1971).

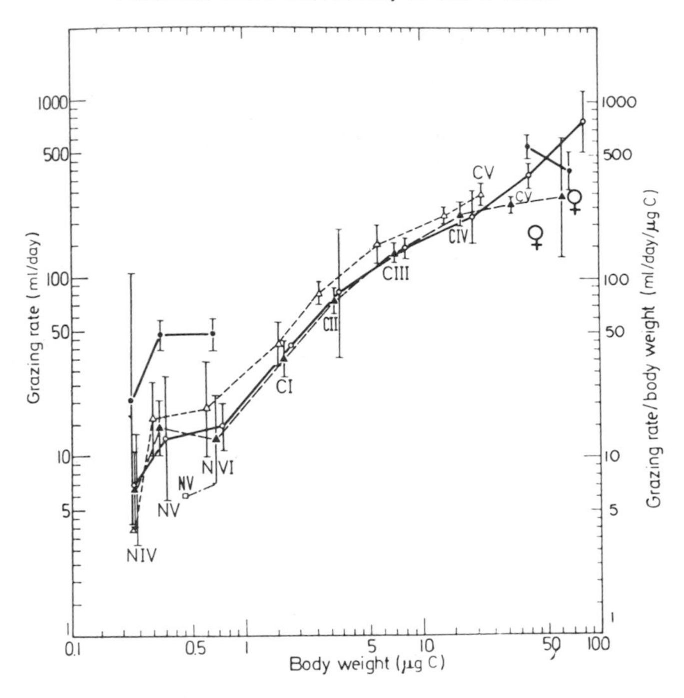

Fig. 6.21. Mean grazing rate per copepod for various stages and mean weights at 15°C as influenced by species, size and concentration of food organism presented. Vertical bars: 95% confidence limits around each mean. Symbols as for Fig. 6.20 (Paffenhofer, 1971).

Mullin's and very far below those of Paffenhofer. From the quantity of particulate matter in the North Sea, Adams and Steele estimated that *Calanus* would not acquire sufficient food to satisfy their maintenance needs when seasonal levels of particulate matter are low. Thus during winter *Calanus* might be forced to rely on microzooplankton, thus adopting a carnivorous diet. During periods of phytoplankton blooms, however, more food than is sufficient for maintenance will be ingested; the surplus would then be able to contribute to egg production.

There are comparatively few data for feeding rates for copepods other than *Calanus* spp. Gauld's (1951) early data indicate the generally lower values for smaller copepods and for earlier copepodites of *Calanus* (cf. Table 6.12) (cf. also Gaudy, 1968). Conover (1956) found clearance rates of 6–12 ml/day for *Acartia tonsa* and *A. clausi* at *ca.* 10°C. He believed that apart from the expected relationship between filtering rate and temperature, some seasonal acclimation might occur. Some scattered data do exist for zooplankton other than copepods. For example, Lasker and his colleagues found fairly high filtering rates for *Euphausia pacifica* feeding on flagellates (15–25 ml/hr), but low values for diatoms (cf. Raymont and Conover, 1961). Jorgensen (1962a, 1966a) quotes some very high filtering rates for *Penilia* (26.5 ml/day/mg dry wt).

A very useful contribution is an analysis of feeding rates on salps recently carried out by Harbison and Gilmer (1976). They examined feeding in three salps, chiefly *Pegea confederata.* Rates increased logarithmically with body length (and body carbon) and even rates of over 100 ml/mm were recorded, though aggregates and solitaries of *Pegea* showed different slopes for body length. For body carbon, however, the slopes were the

Table 6.12. Grazing rates of different species and stages of copepods (Gauld, 1951)

Species	No. of copepods per vessel	Volume of vessel (ml)	Duration (hr)	Temperature (°C)	Mean volume swept clear per copepod (ml)
Pseudocalanus minutus	1	10	24	10	4.28
Temora longicornis	10	150	24	10	8.38
Centropages hamatus	10	150	24	10	12.99
Calanus finmarchicus Stage V[a]	1	100	24	12.5	64.36
C. finmarchicus Stage V[a]	1	100	18	17	71.03
C. finmarchicus Stage IV	2	100	18	17	36.65
C. finmarchicus Stage III	3	100	18	17	22.24

[a] Two series of experiments.

same. The rate for *Pegea* (30–39mm size) was of the order of 48 l/hr/g (dry wt) or 804 l/hr/g C. There appears to be no change in filtration rate with particle concentration. The experiments mainly involved clearance rates for cultures of *Thalassiosira pseudonana*, but a few experiments used other algae. The slow swimming of *Pegea* led the authors to believe that these aquarium-obtained results may approximate to those under natural conditions. For faster swimming species the values are probably underestimates. Harbison and Gilmer point out that the high grazing rates of salps is especially significant in view of these animals including, in the wide range of particles filtered, small nanoplankton that are not normally taken by copepods. Griffiths and Caperon (1979) describe a method for determination of grazing rates of mixed zooplankton, using ^{14}C-labelled phytoplankton with high specific activity. The experiments, lasting a maximum of two hours, showed that grazing increased linearly with increased phytoplankton concentration up to four times the natural ambient levels. Ressoulzadegan and Etienne (1981) showed that *Stenosemella ventricosa* had a high feeding rate of 4×10^5 μm^3 wet volume food/tintinnid/day after 24 hours starvation, decreasing to 4.6×10^4 μm^3 after 20 hours.

More detailed investigations, however, have been conducted on two copepods, *Pseudocalanus* and *Temora*, cultured in the laboratory and the results compared with further work on *Calanus helgolandicus*, also cultured by Paffenhofer and Harris (1976) and Harris and Paffenhofer (1976a,b).

Paffenhofer (1976) examined the feeding of *Calanus helgolandicus* in continuous culture fed on *Lauderia*, *Gymnodinium*, *Gonyaulax* and *Prorocentrum* in concentrations from about 40–100 μg carbon per litre, approximating to concentrations experienced naturally in the Californian area. The grazing rates on the four foods were generally similar, but owing to the different carbon/volume ratios in the various algal species the amounts of carbon ingested by the copepods were very different. For example, the amount of *Prorocentrum* ingested was much higher than for *Lauderia*. All the food species tested were readily ingested and the quantity ingested varied directly with cell concentration, confirming earlier investigations. Most of the mortality occurred from hatching to the first copepodite stage. With *Lauderia*, markedly lower food concentration was shown to increase mortality. Both food concentration and the type of food affected the size (weight) of the resulting adult copepods. The amount of food and its specific type also influenced mortality, but mainly in the nauplius stages.

In the continuous culture of the inshore calanoid copepods *Pseudocalanus elongatus* and *Temora longicornis*, the diatom *Thalassiosira rotula* was used as food. (Paffenhofer and Harris, 1976; Harris and Paffenhofer, 1976a,b). The range of food concentration selected approximated to that experienced in inshore conditions. For both species mortality was mainly in the nauplius stages, especially in the last nauplius in *Temora*. Mortality in the copepodite stages was comparatively low; for the whole period of development mortality averaged 30 % for *Pseudocalanus* and 37 % for *Temora*. Food concentration did not bear any clear relationship to mortality.

In both copepods ingestion of food increased with food concentration. With *Pseudocalanus*, for example, the average daily ration was 63 % at a food concentration of 25 μg C/l, rising to a daily ration of 148 % at a food concentration of 200 μg C/l; for *Temora*, the comparable figures were: daily rations of 80 % and 146 % at food concentrations of 25 and 200 μg C/l, respectively. As expected, both grazing and ingestion rates increased linearly with increasing body size, estimated as body weight, from C I to C VI (cf. Figs 6.22, 6.23). For *Temora*, adults reared at the lowest food concentration (25 μg C/litre) were smaller than those at higher food concentrations, and grazing rates tended to be higher at lower food concentrations. The rate of feeding decreased as Nauplius VI *Temora* approached the moult to the copepodite stage. Harris and Paffenhofer cite a similar observation by Petipa of a decline in feeding rates by N VI when approaching the moult for the two calanoids *Acartia clausi* and *Calanus helgolandicus*.

For *Pseudocalanus* no clear correlation appeared between food concentration and the size (weight) of the adult female. On the other hand, *Pseudocalanus* males were smaller and lacked oil reserves when reared at the lowest food concentration. In neither species of calanoid were there statistically significant differences between the amounts of food ingested per unit body weight by the earlier nauplius and copepodite stages and by adults, although there was apparently some decline in ingestion per unit body weight at later copepodite stages. This contrasts with the earlier findings of Paffenhofer for *Calanus* and of Mullin and Brooks (1970) for the same species.

The volumes swept clear during grazing quoted by Paffenhofer and Harris are much higher than those of earlier workers. For example, with *Pseudocalanus* various investigations suggested clearance rates of *ca.* 10–40 ml/adult female copepod/day; Paffenhofer and Harris found rates of 50–80 ml/copepod/day for Copepodites IV and V and 80–150 ml/day for adult females. Average clearance rates for female *Temora* were 150–200 ml/copepod/day. These high values may be partly related to the uncrowded condition of the experimental animals and the comparatively low food concentrations; Paffenhofer also believes the size of container may be significant. Possibly these high grazing rates may be associated with the copepods having been cultured in the laboratory. For the three calanoid copepods *Calanus helgolandicus*, *Pseudocalanus elongatus* and *Temora longicornis* grazing rates increase linearly with body weight on a log scale. Grazing rates for the three species of the same body weight are also in the same range, for example, at 2 μg C body size—35–45 ml; at 4 μg C body size—60–90 ml swept clear/copepod/day. A general relationship of this type might be developed for temperate herbivorous calanoids.

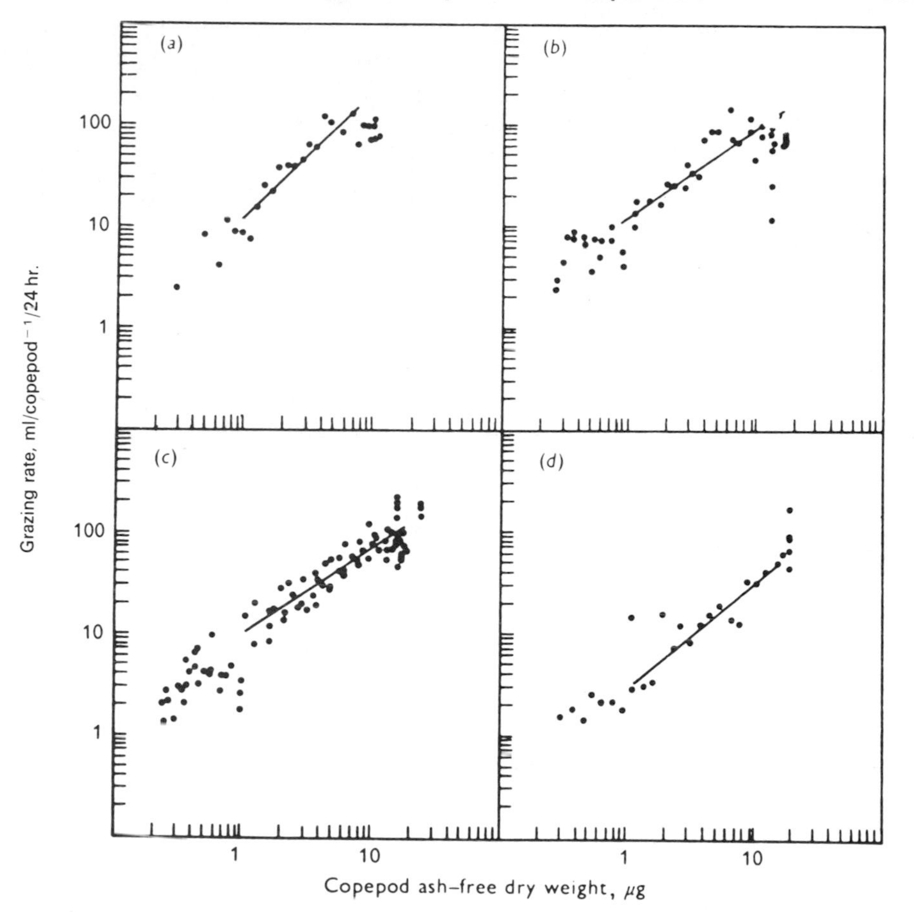

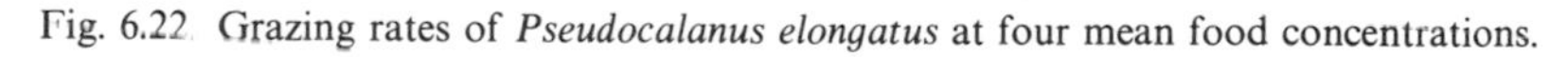

Fig. 6.22 Grazing rates of *Pseudocalanus elongatus* at four mean food concentrations.

(*a*) $25\,\mu g\,C\,l^{-1}$ $\log y = 1.2295 \log x + 1.0878$, $r = 0.9117$
(*b*) $50\,\mu g\,C\,l^{-1}$ $\log y = 0.8827 \log x + 1.0846$, $r = 0.8765$
(*c*) $100\,\mu g\,C\,l^{-1}$ $\log y = 0.8490 \log x + 0.9910$, $r = 0.9109$
(*d*) $200\,\mu g\,C\,l^{-1}$ $\log y = 0.9831 \log x + 0.5212$, $r = 0.9090$

Regression equations are calculated for the period of growth from copepodite I to copepodite V. (Paffenhofer and Harris, 1976)

Harris and Paffenhofer also quote previously determined ingestion rates for *Temora stylifera* for comparison with their results. A female ingested 50–75% of its body weight daily at a food concentration of about 500 μg C/litre (Gaudy, 1974). The same species ingested 19.4% of its body weight daily when feeding at a concentration of about 30 μg C/litre (cf. Petipa, Pavlova and Sorokin, 1971). Both values are considerably below rates for *Temora longicornis*, but there could be marked differences in experimental conditions.

Although most studies on feeding rates of zooplankton have been concerned with herbivorous zooplankton, some investigations have been conducted with carnivorous species. While it is possible to estimate feeding rates in terms of volume swept clear, the amount of food ingested, especially the daily ration, is more meaningful. The Ivlev

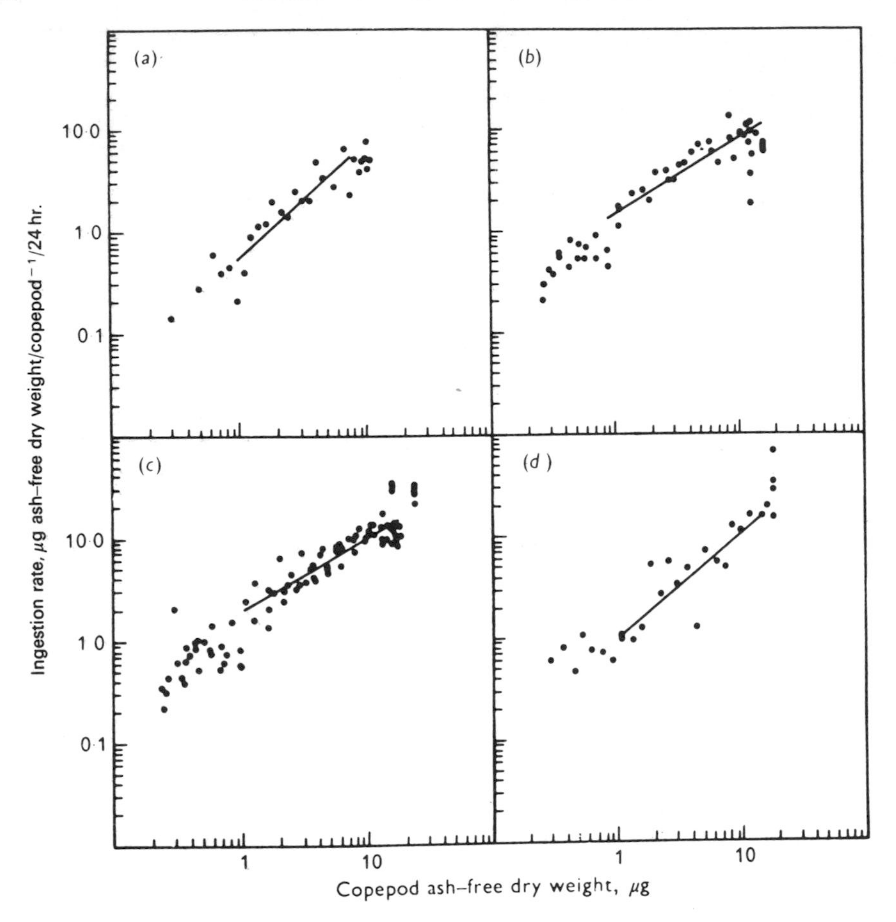

Fig. 6.23. Ingestion rates of *Pseudocalanus elongatus* at four mean food concentrations.

(*a*) 25 μg C l^{-1} $\log y = 1.1319 \log x - 0.2611$, $r = 0.8509$
(*b*) 50 μg C l^{-1} $\log y = 0.7494 \log x + 0.1514$, $r = 0.8818$
(*c*) 100 μg C l^{-1} $\log y = 0.7038 \log x + 0.3079$, $r = 0.8851$
(*d*) 200 μg C l^{-1} $\log y = 0.9799 \log x + 0.0488$, $r = 0.8439$

Regression equations are calculated for the period of growth from copepodite I to copepodite V. (Paffenhofer and Harris, 1976)

equation, as modified by Parsons and his colleagues (*vide supra*) to examine the daily food ration has been used by Parsons and LeBrasseur (1970) to investigate the feeding of juvenile salmon on zooplankton. They showed increased ingestion rates with increasing prey concentration up to a maximum ration, using several species of copepods (*Calanus plumchrus*, *Microcalanus*, *Pseudocalanus*) and the euphausiid *Euphausia pacifica*. At a given prey density (20 g/m^3), juvenile pink salmon (90 mm length) obtained very different quantities of food from various prey species viz. *ca.* 30 mg wet wt/hr from *Calanus plumchrus*, <20 mg/hr from *Euphausia pacifica* and 0.3 mg/hr from *Pseudocalanus*. From the daily requirements of the juvenile salmon (683 mg/day), it would appear that the predator could readily obtain the necessary daily ration feeding on *Calanus* (assuming continuous feeding); with *Euphausia* the requirement was barely possible; with *Pseudocalanus* there was very clear insufficiency. With the latter prey species, even prey densities exceeding 90 g/m³ (equivalent to > 670 individuals/litre) were insufficient.

Another comparison using smaller juvenile pink salmon (34 mm) demonstrated that while *Calanus* would fulfil the daily requirements at a prey density of 10 g/m³ (= 4 *Calanus*/l), and the smaller *Microcalanus* at a prey density of only 1.4 g/m³ (= 23 *Microcalanus*/l), the amount of food captured per hour was nearly three times as high when feeding on *Calanus*. Presumably a major factor would be the relative time required and the energy expended in the efficient capture of the two sizes of prey. The specific effect of different foods on growth rates of predator is another important consideration. A real problem in reviewing the results of all such experiments, however, is that the prey densities are very high in relation to the natural environment; predators may have to concentrate their feeding on aggregations of zooplankton in the sea.

Lasker (1966) examined the feeding rate of *Euphausia pacifica* using *Artemia* nauplii. The euphausiid could capture about 50 nauplii per day. While *E. pacifica* could "filter" this amount of prey successfully at concentrations of 50 nauplii/100 ml (i.e. 1 *Artemia*/2 ml) and there was little decline in capture rate at lower concentrations down to 1 *Artemia*/20 ml, further decreases in prey density lowered the capture rate. The euphausiid was probably sensing and then seizing prey rather than filtering the food. Lasker calculated that for total requirements (growth, respiration, etc.) a young euphausiid weighing some 0.24 mg would need to consume 12 *Artemia* per day and a 9-mg euphausiid about 340 *Artemia* nauplii/day. *Euphausia pacifica*, according to Lasker, is chiefly herbivorous on rich phytoplankton blooms inshore, though mainly carnivorous in open waters. The volume which would need to be swept clear when feeding herbivorously, assuming that all the particulate carbon in Pacific waters was utilized (equivalent to 0.16 mg C/l), to satisfy its needs, would amount to 43 ml/day for an animal of 0.24 mg, and 1170 ml/day for an animal of 9 mg.

Anraku and Omori (1963) fed several calanoids with *Artemia* nauplii. With the carnivorous copepod *Tortanus discaudatus* there was some indication of an increase in the amount of food ingested with the density of prey. *Tortanus* could eat >30 nauplii/day. Ikeda (1977b) estimated that *T. gracilis* would consume 20–30 *Artemia* nauplii per day, but at high food concentrations overkill took place. *Eucalanus subcrassus* would eat > 140 nauplii daily. Female *Tortanus discaudata* preferentially ate all stages of *Acartia*. Mullin (1979) found that when offered *Acartia* adults, copepodites and nauplii in addition to *Tortanus* nauplii, the larger sizes were selected. He points out that *Tortanus* is a marked predator, killing more than 30 % of its body weight in 5 hours, and emphasizes the importance of size as well as biomass of potential prey, and its ability to escape from active predators. Similarly Landry (1978) studied the feeding of *Labidocera trispinosa* on the nauplii of five species of calanoids (*Acartia clausi*, *Acartia tonsa*, *Paracalanus parvus*, *Calanus pacificus* and *Labidocera trispinosa* itself). Clearance rates were of the order of 0.5–1.0 l/copepod/day. For all prey species except *Labidocera* the clearance rates increased with size of nauplii to the oldest (largest) stage (cf. Fig. 6.24), but fell when copepodites were offered. With *Labidocera* the feeding rate increased to the copepodite (C I) stage. Clearance rates were greater for *Calanus* nauplii than for *Acartia* or *Paracalanus*, suggesting that selection for size is significant, but since the rates fell with copepodite prey it is presumed that copepodites are

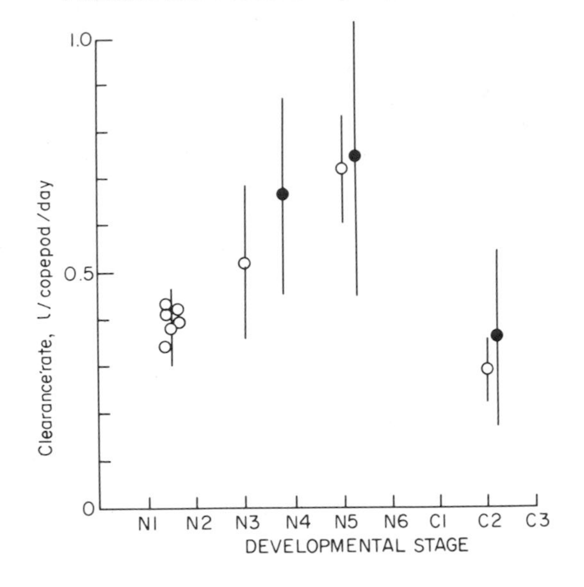

Fig. 6.24. Clearance rates of *Labidocera* on various developmental stages of *Acartia tonsa* (○) and *Acartia clausi* (●). Vertical lines represent 95 % confidence limits for mean of eight replicates. Points without confidence limits represent mean (8 replicates) clearance rates on *A. tonsa* N I in standard experiments with various mixtures of alternate prey (Landry, 1978).

generally sufficiently active to resist capture, though readily sensed. The larger sizes of nauplii may be more easily sensed than small prey. The feeding behaviour appears to be related only to detection and capture ability; there is no adaptation to prey density or any prey preference. *Labidocera* must therefore exert greater predatory pressure on the larger species of copepod nauplii (e.g. *Calanus*) than on the small forms (e.g. *Acartia*).

Some details have already been given of the amount of food consumed by the omnivorous calanoid *Aetideus*. Alvarez and Matthews (1975) observed the feeding of *Chiridius armatus* on *Artemia*. Ingestion rate increased with the prey concentration, though the rate fell at high prey levels. The maximum intake was 200 *Artemia*/mg *Chiridius*/day. Alvarez and Matthews point to a fairly comparable rate for *Lucifer chacei* in the laboratory, but state that the rate is some four times that found by Lasker for growth and maintenance in *Euphausia pacifica*.

Reeve (1966) demonstrated, by feeding *Sagitta hispida* with *Artemia* nauplii, that temperature markedly affected feeding rates. A maximum rate was observed at 25°C and feeding ceased at a lower limit of 11°C and a higher limit exceeding 30°C. The maximum number of *Artemia* nauplii eaten was about 70 per day. At maximum intake the food corresponded to about 70 % body wt/day; for maintenance requirements measurements of oxygen uptake suggested that *Sagitta* required some 24–59 % of body wt/day, depending on whether lipid or carbohydrate was oxidized. Reeve (1970) also studied the growth of *Sagitta hispida* at different food levels. When *Artemia* nauplii were offered at different food densities varying from 0 to 100 nauplii per day, growth was observed to be almost linearly related to ration up to *ca.* 20 nauplii/day, but there was little effect on growth of higher rations. However, when egg production was taken into account an increase in the percentage of *Sagitta* with mature eggs was observed with the amount of food up to 100 nauplii/day.

Reeve and Walter (1978), reviewing feeding rates in ctenophores, comment on the very wide range of prey concentration over which ingestion rate is proportional to food density. They list such ranges as 105–3600 copepods/l found by Bishop, 4–44/l by Miller and 1–100 food organisms/l by Kremer, all for the ctenophore *Mnemiopsis leidyi*. For *Mnemiopsis mccradyi*, Reeve and his colleagues found the ingestion rate to be proportional to the concentration over the entire very large range of 1–3000 copepods/l. This result, however, applied to two sizes of adult ctenophore. With the larval tentaculate stages and with the tentaculate ctenophore *Pleurobrachia bachei* the feeding rate became approximately constant and independent of prey concentration at densities exceeding 200 copepods/l or, according to experiments by Rowe, at densities greater than 400 nauplii/l. Adult lobate ctenophores do not appear to have a critical food concentration above which ingestion rate is constant. Reeve describes the formation of partly digested masses of prey surrounded by mucus when *Mnemiopsis* is allowed to feed at very high prey densities. These masses are later rejected. However, he points to experiments on starved *Mnemiopsis* showing that initial feeding rates are higher than in the fed ctenophores, indicating that feeding is not entirely automatic. Feeding activity can be expressed as volumes swept clear, though the quantities are variable and can be very large, e.g. 1–74 l/animal/day for *Mnemiopsis mccradyi* of size 5–70 mm length, at 26°C and approximately half these volumes for *M. leidyi* at 20–25°C. For both ctenophores the weight-specific clearance rates declined with increasing body size, as has been found with some other zooplankton species.

The observations of Corner, Head, Kilvington and Pennycuick (1976), when *Calanus helgolandicus* fed carnivorously on *Elminius* nauplii, clearly demonstrated a rapid rise in daily ration with increase in prey density to a maximum ration of *ca.* 80 nauplii/day at a food concentration of around 225 nauplii/l, equivalent approximately to the maximum for *total* microplankton found off Plymouth. Further increase in prey concentration even to 1300 nauplii/l had no obvious effect on ration.

Fish larvae have been used for studying carnivorous feeding rates. Stepien (1976) reared larvae of the sea bream, *Archosargus rhomboidalis*, in the laboratory. The fish larvae were fed on natural zooplankton. There was evidence of size selection with the growth of the fish larvae, and as expected there was an increase in feeding rate with age, the increase being exponential. Between temperatures of 23 and 29°C there was a small positive relationship between temperature and feeding rate. The daily ration, calculated as a percentage of body weight, declined with age over a 10-day period of larval development, but the average daily ration over the 10-day period was also markedly affected by temperature, rising from about 66% of the body weight at the lowest temperature to 151% at 29°C.

Stepien draws attention to similar studies by Rosenthal and Hempel (1970) and by Howell (1973) on the feeding of herring and plaice larvae, indicating that for all three species intake of food increased with age. According to Howell's investigation, the increase in feeding rate was exponential, as with the sea bream. *Archosargus* larvae appeared to be feeding much faster than either larval herring or plaice, but the data cannot be accurately compared since the food organisms employed were different and the size of fish larvae was not estimated precisely.

Houde (1978) fed three species of fish larvae found off Miami (*Anchoa mitchilli, Achirus lineatus* and *Archosargus rhomboidalis*) with natural zooplankton, mainly copepods, at varying concentrations. Some larvae of all three species survived at

Table 6.13. Reported concentrations of some potential larval fish food organisms from coastal and estuarine areas (Houde, 1978)

Reference	Place	Organisms	Concentration (no/l)
Burdick (1969, cited in May, 1974)	Kaneohe Bay, Hawaii	Copepod nauplii	50–100 common 200 sometimes present
Duka (1969)	Sea of Azov	*Acartia clausi* nauplii	62–65
		Other copepod nauplii and copepodids	>30
		Total	>90
Mikhman (1969)	Gulf of Taganrog, Sea of Azov	Early stages of copepoda	39–546
Hargrave and Geen (1970)	two eastern Canada estuaries	Copepod nauplii and copepodids	> 60
Reeve and Cosper (1973)	Card Sound, South Florida	Copepod stages 20–200 μm in breadth	range 23–209 mean for 28 collections 72
		Tintinnids	range 40–369
Heinle and Flemer (1975)	Patuxent River estuary	*Eurytemora affinis* nauplii and copepodids	>100 frequently >2000 occasionally
Houde (unpublished data)	Biscayne Bay, South Florida	Copepod nauplii and copepodids <100 μm in breadth	usually 50–100
		Tintinnids	frequently >100

minimum food levels (10–50 prey/l). Food levels were adjusted several times daily to maintain average concentrations. Approximately 10% of the larvae survived to metamorphosis at the following food concentrations: *Anchoa*, 107/l; *Achirus*, 130/l; *Archosargus* 34/l. However, increased food concentrations markedly influenced survival and growth. For example, 50% survival of *Anchoa* was achieved at about 1850 prey/l and for *Archosargus* at 216/l. Houde believes that some laboratory experiments have tended to exaggerate the food requirements of fish larvae. He quotes data indicating that the richness of microzooplankton, especially in many inshore areas, appears to be sufficient for the successful growth of many fish larval species (Table 6.13). The effects of predation may have been underestimated.

Respiration

The transfer of primary to secondary production within the plankton ecosystem involves not only the ingestion of suitable food material in adequate quantity but also its digestion and assimilation, aspects which will be discussed later. In considering the amount of food ingested and assimilated, reference has been already made to daily requirements or daily ration. The amounts of matter ingested by a zooplanktonic animal and the quantities utilized for different purposes may be expressed by the familiar equation which appears in a number of forms:

$$I = F + G + E(+R) + M$$

where I = matter ingested;
F = matter lost as faeces;
G = quantity used for growth;
E = amount excreted;
R = matter used for egg production in mature animals;
M = amount lost in oxidative metabolism.

Growth is discussed later, but clearly one of the factors influencing the amount of food ingested (the daily ration) is the amount used in oxidative metabolism, the mechanism for providing energy for the animal. It is also true that the respiratory (metabolic) demands on the food requirement in most zooplankton tend to form a large proportion of the total. Respiration of zooplankton has, therefore, received considerable attention, though until recently, investigations had concentrated on a very few species of mainly copepods. One of the earliest studies undertaken was by Putter and was mostly carried out on mixed small copepods. Though the values suggested were almost certainly far too great (cf. Krogh, 1931), they drew attention to the importance of respiratory demands and during the 1930s several significant investigations on respiration followed, mostly on *Calanus finmarchicus* (*sensu lato*) (Marshall, Nicholls and Orr, 1935; Clarke and Bonnet, 1939).

Methods and applicability

A serious difficulty in any experimental work on marine zooplankton is the capture and sorting of animals in healthy condition. A recent important advance has been the provision of animals from the culture of a few species (copepods, sagittae, appendicularians, ctenophores), but it is essential to compare the behaviour of field and cultured animals. Even with cultured stocks, the handling and transfer of animals to experimental vessels for studies on respiration can lead to abnormal responses. Moreover, most zooplankton are of small size and some degree of crowding is almost essential to obtain measurable results. Antibiotics have been employed, especially in longer term experiments to reduce errors due to the growth of bacteria.

With the relatively high concentrations of bicarbonate ion in seawater and the marked buffering action, comparatively few investigations have been concerned with monitoring carbon dioxide concentration as an estimate of respiration (but cf. Teal and Kanwisher, 1966; Raymont and Krishnaswamy, 1968). Changes in the concentration of dissolved oxygen are usually followed, indicating oxygen consumption and a respiratory quotient (R.Q.) assumed, or such a relationship as the oxygen/nitrogen ratio estimated to gain knowledge of respiratory substrate. The Winkler method is regularly employed to measure oxygen concentrations. Bryan, Riley and Williams (1976) have developed a very sensitive and accurate micro-Winkler modification, so that single zooplankton animals (e.g. small copepods) can be studied. Manometric methods have also been used (e.g. the Dixon–Haldane constant pressure manometer and the Warburg method). Zeuthen (1947) used the Cartesian diver technique so that single animals may be studied, but the volume of water in which the animal respires is very small. The oxygen electrode has been used with some success. Ikeda (1979) measured, by this method, respiration rates of tropical copepod larvae and of a ciliate. These varied from 9–2.8 μl O_2/mg wet body wt/hr for copepods and from 4.6–2.0 μl O_2/mg wet body wt/hr for the ciliates. He suggests that the relative respiration rate for microzooplankton is comparable to that of the net zooplankton, and should be considered as a factor in studying the energy flow through tropical seas. A different approach is to estimate electron transport system (ETS) activity, usually involving a colour intensity measurement of a tetrazolium salt product (cf. Krishnaswamy, Raymont and Tundisi, 1967; Raymont, Krishnaswamy and Tundisi, 1967; Mayzaud, 1978). Corkett and McLaren (1978) also cite the interesting approach by Muhammad who found a strong correlation between oxygen consumption and succinic dehydrogenase activity in mixed zooplankton, including the copepod *Pseudocalanus*.

As an instance of the problem of the applicability of experimental laboratory data on respiratory rates to natural populations, many workers (e.g. Marshall *et al.*, 1935; Raymont, 1959; Berner, 1962; Ikeda, 1970) have observed a raised respiratory rate at the beginning of an experiment, the rate falling off with time to an approximately steady lower level. Ikeda (1970, 1974, 1977a) observed some fluctuations even after several days although a lapse of time between capture and experiment had little effect on tropical microzooplankton (Ikeda, 1979). The high initial rate is usually believed to be artificial and due to a type of "hyper-excitement" after handling. Many workers therefore allow 24 hours before making estimates, though it is possible that the usual laboratory conditions depress respiration. It is uncertain in any event, what the "steady" level, usually accepted as the respiratory rate, really represents. It may approach a basal metabolic rate but is obviously not strictly comparable, since the animals are moving around in the experimental vessels or at least some appendages (e.g. pleopods, gills etc.) are active. A few trials of respiratory rates have been carried out with anaesthetized animals (e.g. *Metridia*).

Hyper-excitement is not always observed but some animals seem to take several days before "normal" activity and respiration is established (e.g. Conover, 1962, with

Calanus hyperboreus and Marshall's, 1973, report on male *Eurytemora affinis*). Another difficulty in assessing respiratory data relates to the increase in respiration accompanying locomotory activity. This has been applied particularly to vertical migration by zooplankton. Petipa (1965), for example, claimed an enormous increase in respiratory demand with vertical migration and further related this (Petipa, 1970) to very large demands as rations to cover migration (cf. also Mullin, 1969). Vlymen (1970) strongly argued against the estimates of excessive energy demands in vertical migration; he suggested that there might be a doubling of energy output. More recently, the energy demand in locomotion has been investigated (cf. Chapter 5). Where several animals are used in an experiment, their natural excitation leading to movement may be an important factor in the level of respiration. While excessive crowding can be obviously deleterious and, over a period of time, release of metabolites might affect zooplankton even in less crowded conditions, the evidence is somewhat conflicting, as Marshall (1973) points out. It may be, as Razouls (1972) has reported, that a certain fairly low density of animals gives the maximum respiration rate, since above this density, crowding is progressively harmful. Below this density, the few animals fail to stimulate one another. The degree of crowding and the duration of experiment are obviously linked (Corkett and McLaren, 1978).

Factors influencing respiratory rate

Although no strict separation of experiments designed to observe the effects of particular factors on respiration will be made it is convenient to single out certain parameters.

Temperature

Temperature has been widely investigated. The early experiments by Marshall *et al.* (1935) indicated that Copepodite V *Calanus finmarchicus* utilized 0.4–0.5 μl O_2/copepod/hr at 15°C. From about 0°–10°C the respiratory rate was approximately doubled, recalling the Q_{10} relationship that has been applied to several physiological processes, but the rise in oxygen uptake was somewhat steeper between 10° and 20°C (Fig. 6.25). Experiments with male and female *Calanus* not only showed a somewhat greater oxygen uptake than Copepodite V at any one temperature, an effect which may be attributed to body size, but demonstrated that Q_{10} varied somewhat for different temperature ranges and for the different stages. Clarke and Bonnet (1939) used mainly a manometric method but obtained a similar rise in respiratory rate of Copepodite V *Calanus* with temperature, though their rates are somewhat greater than those found by Marshall *et al.* (e.g. at 15–17°C, 0.5–1.0 μl O_2/copepod/hr). The difference in method is unlikely to have affected the result, since Raymont and Gauld (1951) obtained a very similar rate to that of Marshall *et al.* using the Dixon–Haldane manometric method. Possibly, differences in the stocks of *Calanus* are significant (*vide infra*).

Similar observations on the effect of temperature on respiration have been made on several other species of copepod. Gauld and Raymont (1953) demonstrated the expected relationship between respiration and temperature for several small copepods (*Temora longicornis*, *Acartia clausi*, *Centropages hamatus*), and Conover (1956) did likewise for *Acartia tonsa* and *Acartia clausi*. In the comparatively deep water inhabited by *Calanus hyperboreus*, where temperature varies little, Conover (1962),

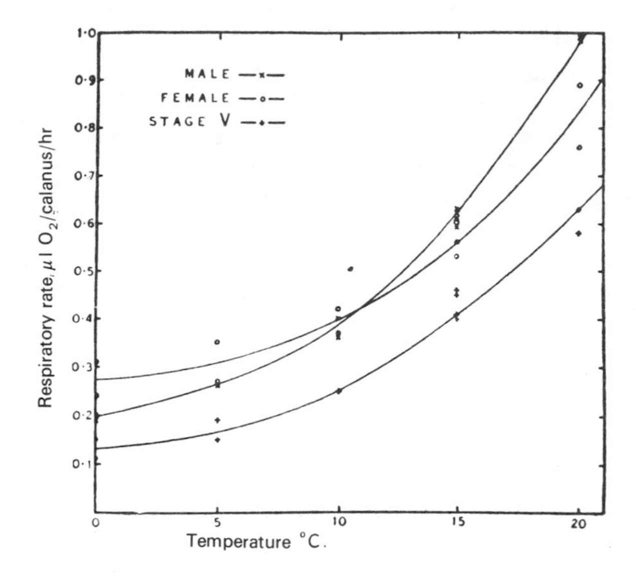

Fig. 6.25. Respiration of *Calanus* (Marshall and Orr, 1955).

examining the effect of temperature on oxygen consumption, confined experimental temperatures to narrow limits: 2°, 5° and 8°C. Stage V copepodites, kept at 3°–4°C for three weeks prior to experimentation, showed an increase in respiration with temperature, the Q_{10} being 1.76. When the animals were maintained for two weeks at the experimental temperature (2°, 5° or 8°C) before testing at that one temperature, however, these "conditioned" copepods regulated their respiratory rate, so that the Q_{10} was only 1.14 (Fig. 6.26).

Mullin and Brooks (1970) found that the Q_{10} differed substantially between the two copepods *Rhincalanus nasutus* and *Calanus pacificus* which were reared in the laboratory at two temperatures, 10° and 15°C. The copepods were tested at the same temperature at which they were reared and three copepodite stages (C IV, C V and C VI ♀) were studied. For *Calanus* the mean Q_{10} was 3.1 and for *Rhincalanus* 8.3, indicating that *Rhincalanus* is a poor regulator of respiratory rate with temperature, as compared with some temperate zooplankton species. Wild *Calanus pacificus*, after being retained in the laboratory overnight, were also tested. No significant difference was found in their respiratory rate at either temperature as compared with laboratory reared copepods. The expected increase in oxygen consumption with temperature to an optimum, above which there was a sharp decline, was observed by Anraku (1964b) for the copepod, *Pseudocalanus minutus*. Temperatures exceeding 20°C appeared to be harmful. Considerable differences were apparent in the relationship between respiration and temperature of this copepod at different seasons, with comparatively high oxygen uptake at low temperatures for animals taken in the colder months and a low respiratory rate at the same low experimental temperature for copepods caught in summer (Fig. 6.27). Anraku suggests a change in the Q_{10} value from 1.3 in February to 3.7 in August.

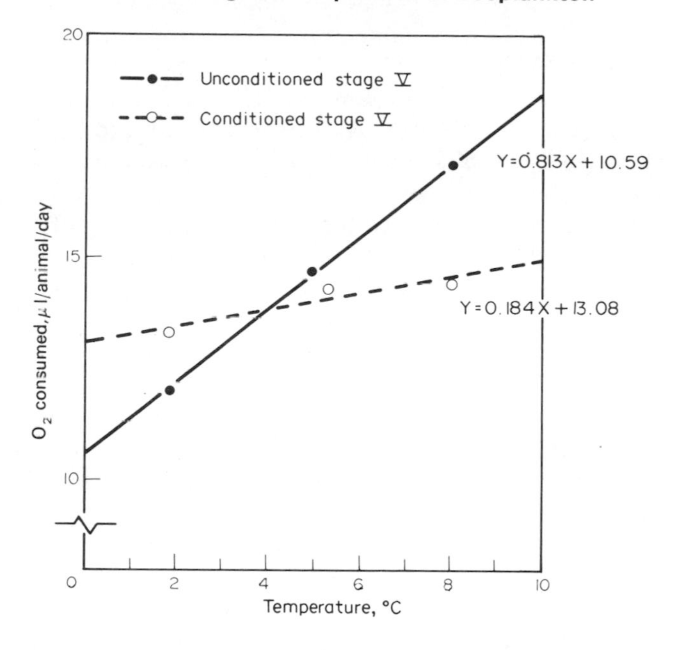

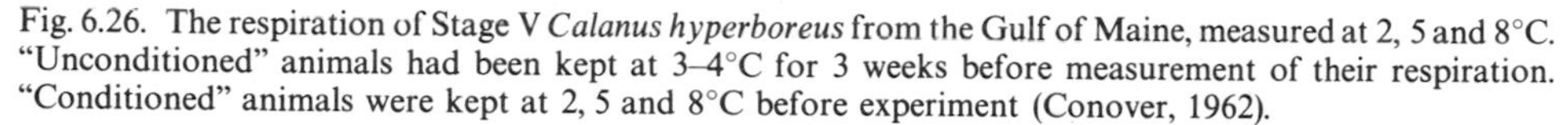
Fig. 6.26. The respiration of Stage V *Calanus hyperboreus* from the Gulf of Maine, measured at 2, 5 and 8°C. "Unconditioned" animals had been kept at 3–4°C for 3 weeks before measurement of their respiration. "Conditioned" animals were kept at 2, 5 and 8°C before experiment (Conover, 1962).

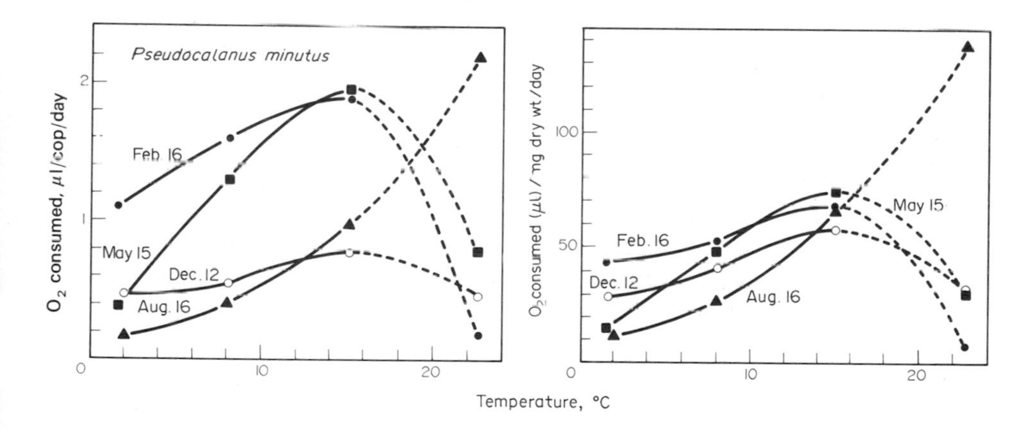

Fig. 6.27. Respiration rates of *Pseudocalanus minutus* plotted against temperature (Anraku, 1964).

There was a clear indication of different effects on the respiratory rate over various temperature ranges for colder and warmer water species of copepod. Thus *Labidocera aestiva*, a warmer-water form, exhibited the greatest rate of oxygen consumption at the highest temperature employed (22.5°C), while respiration was negligible at 2°C and many of the copepods did not survive at that temperature. *Acartia tonsa* also showed low oxygen uptake at 2° and 8°C, but marked increases in rate at 15° and 22.5°C. Winter stocks showed greater respiration rates at lower temperatures than summer stocks, and a less obvious rise with temperature. Respiratory rates for *Acartia clausi*, in contrast, fell between 15° and 22.5°C.

Champalbert and Gaudy (1972) examined the respiration of several species of copepods in the waters off north-west Africa (South Morocco and the Canaries) at temperatures between 10° and 24°C. *Anomalocera patersoni* and *Pontellopsis villosa* adults showed an increase in oxygen consumption up to 18°C. A decline occurred at higher temperatures, especially marked for *Anomalocera*, though for younger stages (Copepodite IV and V) the decline was not so obvious. For *Anomalocera*, the Q_{10} was 2.07 between 10° and 14.5°C, and 5.61 between 14.5° and 18°C. The species is limited to colder months of the year in the area. The Q_{10} was much smaller for *Pontellopsis villosa*, a more eurythermal species in the region. A few other hyponeustonic species also showed fairly high temperature optima for respiration.

Of epiplanktonic species, *Centropages typicus* and *Temora stylifera* appeared to have a maximal rate between 21° and 24°C, though *Centropages* showed little rise in oxygen consumption between 10° and 18°C. *Centropages typicus* is near its latitudinal limit off South Morocco. The sharp rise in metabolism at the higher temperature may reflect this geographical distribution. *Calanus helgolandicus*, while showing little change in oxygen consumption between 10° and 14°C, displayed a sharp rise to 18°C and at temperatures above 20°C. There was considerable scatter in values at the higher temperatures. The difference between this more southerly species and the relationship of respiration to temperature for *Calanus finmarchicus* is clear. A few experiments were carried out on mesopelagic and bathypelagic species. *Pleuromamma xiphias*, for example, showed a fairly steady increase in respiration with a rise in temperature from 10° to 24°C. *Undeuchaeta plumosa* also displayed a regular rise in metabolism with temperature. There was little evidence of adaptation by such deeper-living copepods to the temperature conditions of deeper strata. Possibly this is associated with the marked diurnal vertical migrations performed by species like *Pleuromamma xiphias*.

Champalbert and Gaudy (1972) summarize their findings for temperature/respiration relationships for copepods from a warmer region by emphasizing the variety of patterns of adaptation to temperature conditions shown by the different species. For species with some latitudinal spread in distribution they believe that some degree of adaptation is often shown, but that generally respiratory rates for copepods occurring off South Morocco and the Canaries are somewhat lower than for the same species in more boreal temperate areas. Champalbert (1973) compared the respiratory rates of *Anomalocera patersoni* sampled from a temperate and a sub-tropical area. Optimum temperatures were about 18°–19°C for the warm water area, while those from the colder region were about 14°–15°C. Champalbert believes that metabolic rate showed some adaptation in accordance with environmental temperatures.

Nival, Malara, Charra, Palazzoli and Nival (1974) carried out further experiments on four species of calanoids in the upwelling zone off Morocco. In the majority of experiments temperatures up to 20°–23°C caused a rise in respiratory rate for the species *Acartia clausi* and *Centropages typicus*, regarded as warm water copepods in the area of capture. *Temora stylifera* and *Calanus helgolandicus* however, showed that following an increase in oxygen uptake within the temperature range 13°–20°C, respiration rates decreased above 20°C.

The effect of temperature increase on oxygen consumption has been demonstrated for some zooplankton other than copepods. Jawed (1973) showed the rise in oxygen uptake for two species of mysid, *Archaeomysis grebnitzkii* and *Neomysis awatschensis*, between temperatures of 5° and 15°C (Table 6.14). For *Neomysis* it would appear that the optimum has been exceeded at 15°C. Quetin and Childress (1976) found for the pelagic crab, *Pleuroncodes planipes*, that oxygen consumption was a function of temperature. The Q_{10} values ranged for adjacent temperatures between 10° and 25°C from 1.9–2.5, tending to be greater at the lower temperature ranges. Rates compared well with other epiplanktonic crustaceans.

Table 6.14. Oxygen consumption in relation to temperature (Jawed, 1973)

Temperature (°C)	Oxygen consumption ($\mu l\, O_2$/mg/day)	
	Archaeomysis	*Neomysis*
5	21.25	27.43
10	32.71	45.37
15	59.34	38.70

Experiments have successfully been carried out on oxygen consumption of a few species of euphausiids in relation to temperature. Lasker (1966) found for *Euphausia pacifica* that the Q_{10} was approximately 2 between 5° and 12°C. Migration into warmer upper layers could increase the metabolism. Teal and Carey (1967b) found that the Q_{10} of several species of euphausiids (Table 6.15) varied from 2 to 3.5. Paranjape (1967) found that *E. pacifica had a* Q_{10} of 2.21 between 5° and 10°C, but that between 10° and 15°C this had risen to 2.55. As with other zooplankton, the Q_{10} would appear to vary with species and with temperature range. However, Teal and Carey found that the relationship between respiratory rate and temperature for a series of euphausiids (*Thysanopoda monacantha*, *T. tricuspidata*, *T. obtusifrons*, *Meganyctiphanes norvegica*, *Euphausia americana*, *E. recurva* and *E. hemigibba*) was represented by a straight line on a semi-log plot. Paranjape (1967) found respiratory rates of *E. pacifica*, *Thysanoessa raschii*, *T. spinifera*, *T. longipes* and *Tessarabrachion oculatum* averaged 0.53 $\mu l\, O_2$/mg dry wt/hr at 5°C, 0.79 at 10°C, and 1.26 at 15°C. Respiratory rates of *Euphausia pacifica* from off Oregon are adapted to a seasonal range in environmental temperatures between 5° and 10°C. Small and Hebard (1967) demonstrated that both acclimatized and non-acclimatized animals were adapted to the same range and that whether adapted to 5° or 10°C, the Q_{10} values were very similar (2.13 and 2.23).

The euphausiids which migrate deeply to 600 m, reaching a temperature of approximately 5°C, show slightly lower respiratory rates at all temperatures than those migrating to only *ca.* 250 m where the environmental temperature is 10°C. Mean rates are *ca.* 0.7 $\mu l\, O_2$/mg dry wt/hr at 5°C, and *ca.* 1.0 $\mu l\, O_2$/mg dry wt/hr at 10°C. Occasionally surface temperatures of 15°C may be experienced but there is some respiratory compensation so that oxygen uptake is not much increased. Gilfillan (1976) also noted that *E. pacifica* can withstand temperature changes up to 15°C. He found that the coastal animals which usually have a higher respiratory rate, appear better able to adapt to changes in environmental conditions.

Table 6.15. Oxygen consumption of euphausiids (Mauchline and Fisher, 1969)

Species	Temperature range (°C)	Oxygen consumption (μl O_2/mg wet weight/hr)	Authority
Thysanopoda monacantha	10	0.18	Teal and Carey (1967b)
	20	0.40	Teal and Carey (1967b)
	25	0.70	Teal and Carey (1967b)
T. tricuspidata	10	0.15	Teal and Carey (1967b)
	15	0.29	Teal and Carey (1967b)
	20	0.52	Teal and Carey (1967b)
T. obtusifrons	10	0.42	Teal and Carey (1967b)
	15	0.65	Teal and Carey (1967b)
	20	0.97	Teal and Carey (1967b)
Meganyctiphanes norvegica	8–10	0.08–0.31	Mauchline (unpublished)
	5	0.07	Teal and Carey (1967b)
	10	0.19	Teal and Carey (1967b)
	15	0.43	Teal and Carey (1967b)
Euphausia americana	25	0.90	Teal and Carey (1967b)
E. recurva	15	0.40	Teal and Carey (1967b)
E. superba	0–15	0.04–0.24	McWhinnie and Marciniak (1964)
E. pacifica	10	0.17–0.30	Lasker (1960
	10	0.20–0.30	Small and Hebard (1967)
E. hemigibba	10	0.21	Teal and Carey (1967b)
	20	0.55	Teal and Carey (1967b)
	25	1.00	Teal and Carey (1967b)
Thysanoëssa raschii	8–10	0.12–0.47	Mauchline (unpublished)
Nematoscelis megalops	4–10	0.11–0.14	Raymont and Conover (1961)
Unidentified euphausiids	5	0.13–0.26	Conover (1960)

An extensive series of experiments relating oxygen uptake and temperature was carried out by Ikeda (1970, 1974), who examined zooplankton from a wide variety of taxa and from boreal, temperate, sub-tropical and tropical environments. Altogether, some 77 species in the first investigation and 112 species in the second series were studied. Respiratory rate was determined at the environmental temperature for all zooplankton investigated, so for most forms there was a very limited temperature range for any one species studied. Nevertheless, Ikeda was able to demonstrate a general rise in respiratory rate with environmental temperature, so that, for example, oxygen consumption was lower for boreal than for tropical plankton (cf. Fig. 6.28). Moreover, the Q_{10} value increased from cold-water to warmer-water species.

Several references have already been made, for example in the studies of Anraku, Champalbert and Gaudy, and Small and Hebard, to adaptation of respiratory rate to temperature, especially with fairly marked seasonal environmental temperature changes, but also in relation to variations in temperature with geographical location and depth. Acclimatization of the rates of physiological processes in many animals to changed environmental conditions is well recognized. After observing metabolic rates from 13 taxa at different latitudes, temperatures and conditions, Musayeva and Shuskina (1978) believe that compensatory mechanisms are responsible for maintaining the optimum metabolic level for a particular species. With zooplankton,

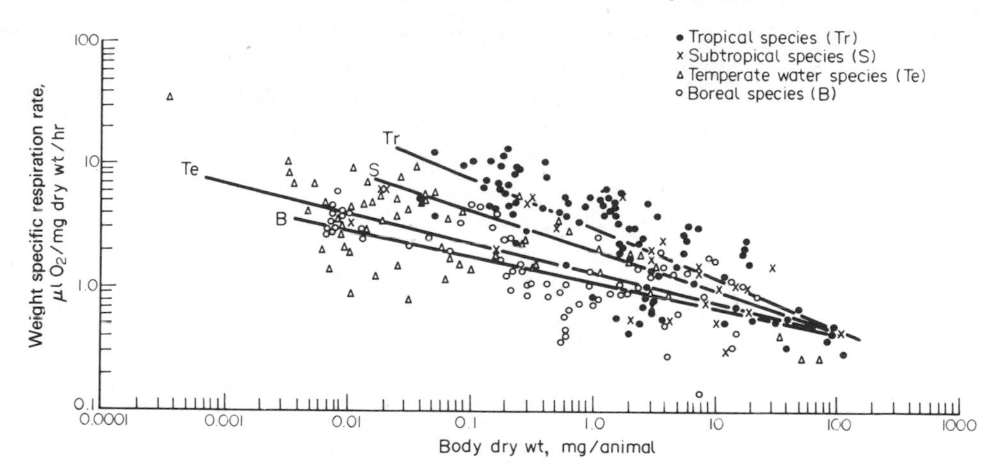

Fig. 6.28. Relationship between weight specific respiration rate and body dry weight for tropical, subtropical, temperate and boreal zooplankton (Ikeda, 1974).

acclimatization of metabolic rate to temperature is frequently demonstrable and this may lead to some difficulty in the critical interpretation of data. Thus, when zooplankton is captured in the field and transported to the laboratory, even on board ship, it inevitably experiences some changes in temperature and other environmental characteristics. Most investigators maintain the animals under standard conditions for a period, often 24 hours, before experimentation. They may then transfer the animals directly to the experimental temperature. Others "condition" the zooplankton for some hours to the experimental temperature before commencing an experimental run. While acclimatization ("acclimation") is usually looked upon as occupying a rather longer period, the varying times used for conditioning obviously may approach acclimatization. Indeed, experimental animals are sometimes described as "partly acclimatized" or "fully acclimatized".

Some experiments have been specifically performed to investigate acclimatization to temperature in zooplankton. Halcrow (1963), using the O_2-electrode technique, found for animals from the "spring" brood of *Calanus finmarchicus* (i.e. growing at *low* temperature) and maintained at 4°C, that they showed the usual O_2/temperature relationship in experiments between 4° and 10°C (perhaps also <4°C). But for experiments at temperatures exceeding 10°C there was a decline in respiratory rate, suggesting that such a temperature was beyond the physiological limit. However, if animals were maintained under good conditions in the laboratory at about 10°C and thus acclimatized to a higher temperature, experiments at temperatures even exceeding 10°C showed some increase in respiratory rate. Experiments which were similarly conducted, but with a "summer" brood of *Calanus*, developing at temperatures of around 15°C, showed only a small increase in respiratory rate at lower temperatures between about 4° and 10°C, but with experiments at 15°C, there was a very sharp rise in respiration. Little increase occurred with experiments at 20°C. There appears to have been a shift of the O_2/temperature curve between the spring and summer brood ("translocation"). Each curve generally demonstrates the well-known relationship of a rise in respiration with temperature to a maximum value (optimum), beyond which no further increase in rate occurs and at slightly higher temperatures there is usually a very sharp decline in respiration with a rapid approach to lethal conditions.

Other examples of acclimatization include observations by Moreira and Vernberg (1968) on males of the copepod, *Euterpina acutifrons*. The animals, which occur in comparatively warm waters, were acclimatized to two temperatures (15° and 25°C). When the respiratory rate was tested over this range, there was a clear difference in the response of the two sets of animals. Of considerable interest also was the occurrence of two different size groups of male *Euterpina*. These, when acclimatized to the two temperatures, showed different patterns of response. Seasonal modification of the respiration/temperature relationship was demonstrated by Anraku (1964b) for several calanoids, *Calanus finmarchicus*, *Pseudocalanus*, *Acartia tonsa* and *Acartia clausi*. Conover (1956) had earlier described some changes in the effect of temperature rise on respiration seen in summer and winter stocks of both species of *Acartia* (Fig. 6.29). A genuine adaptation of the metabolic rate with season also appears from the study of *Centropages typicus* by Gaudy (1968). Over most of the year oxygen uptake varied directly with temperature up to *ca.* 17°C, but the effect of further temperature rise varied with season. For *Centropages* obtained in March, for example, respiration fell at a temperature of 20°C and declined almost to zero at 25°C. On the other hand, for May and August copepods, the respiratory rate rose regularly to 20°C. At the end of the year, a temperature even of 17°C was excessive; maximum oxygen uptake occurred at *ca.* 10°C. Several studies indicate that especially where seasonal temperature changes are substantial, upper temperatures which were lethal to zooplankton over winter are satisfactory during summer.

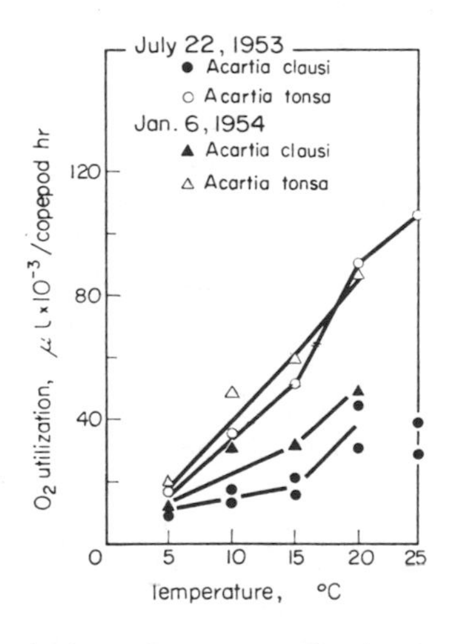

Fig. 6.29. Relationship between respiration and temperature for *Acartia* spp. on two dates during summer and winter (Conover, 1956).

For the chaetognath *Sagitta elegans* Sameoto (1972) demonstrated a rise in oxygen consumption with temperature and there was evidence of some seasonal acclimatization to temperature. Figure 6.30 illustrates the results of experiments on sagittae of different body size (expressed as body wt), all performed at an experimental temperature of 4°C. A generally lower respiratory rate was found in November and February as

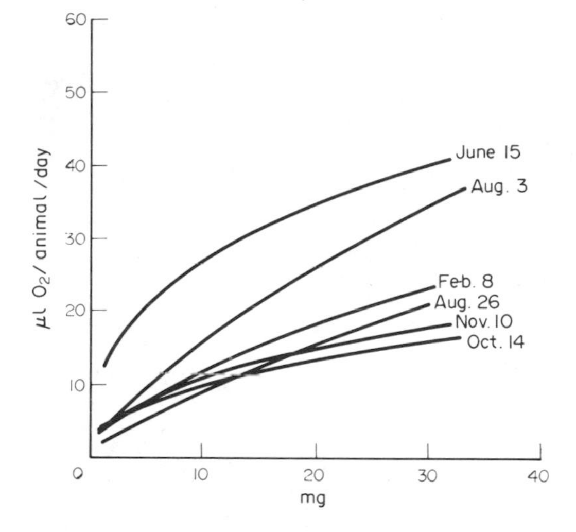

Fig. 6.30. Comparison of regression lines for Sagitta respiration experiments conducted at 4°C throughout the year (Sameoto, 1972).

compared with June. A complication in Sameoto's study is that *Sagitta* performs a marked diurnal vertical migration over much of the year, when the environmental temperature could range from *ca.* 4°C in deeper layers to as much as *ca.* 14°C near the surface in late summer.

A particularly striking example of adaptation of metabolic rate to low temperatures was found by McWhinnie and Marciniak (1964) in a study of the respiration of *Euphausia superba*. They measured oxygen uptake at temperatures from 0°–15°C within 30 min of capturing the euphausiids. Consumption of oxygen rose slightly for animals kept at 0° and 5°C but no further increase in oxygen consumption occurred between 10° and 15°C (Fig. 6.31a). The Q_{10} value for 0°–5°C was 1.1–1.3. They collected animals from the sea at a temperature ranging from 0.1° to 0.28°C, divided the catch and maintained one group at 0°C and the other group at 1°C for 36–40 hr before estimating their respiratory response through a temperature range (Fig. 6.31b). The individuals kept at the lower temperature consumed more oxygen than those maintained at the higher temperature, thus indicating metabolic adaptation to low temperature. Further evidence of this was obtained by maintaining two groups of individuals, caught in the sea at temperatures ranging from −0.05° to 0.9°C, in an experimental temperature of 0°C. The temperature response of one group was tested after 16–36 hr and the other after 4–5 days (Fig. 6.31c). The animals kept longer at 0°C had a higher oxygen consumption and a lower Q_{10} over a low temperature range than the animals kept at 0°C for the shorter period. Conversely, at a high temperature range, the animals kept for longer at 0°C had a lower oxygen consumption than the others.

This adaptation to low environmental temperature is seen on a more restricted scale in Conover's experiments on *Calanus hyperboreus*. Furthermore, at the very low temperatures of Antarctic waters, Rakusa-Suszczewski, McWhinnie and Cahoon (1976) measured the respiratory rate of female *Rhincalanus gigas* as 0.47 μl O_2/mg dry wt/hr at −1.8°C. The smaller *R. nasutus* from warm-temperate waters would consume

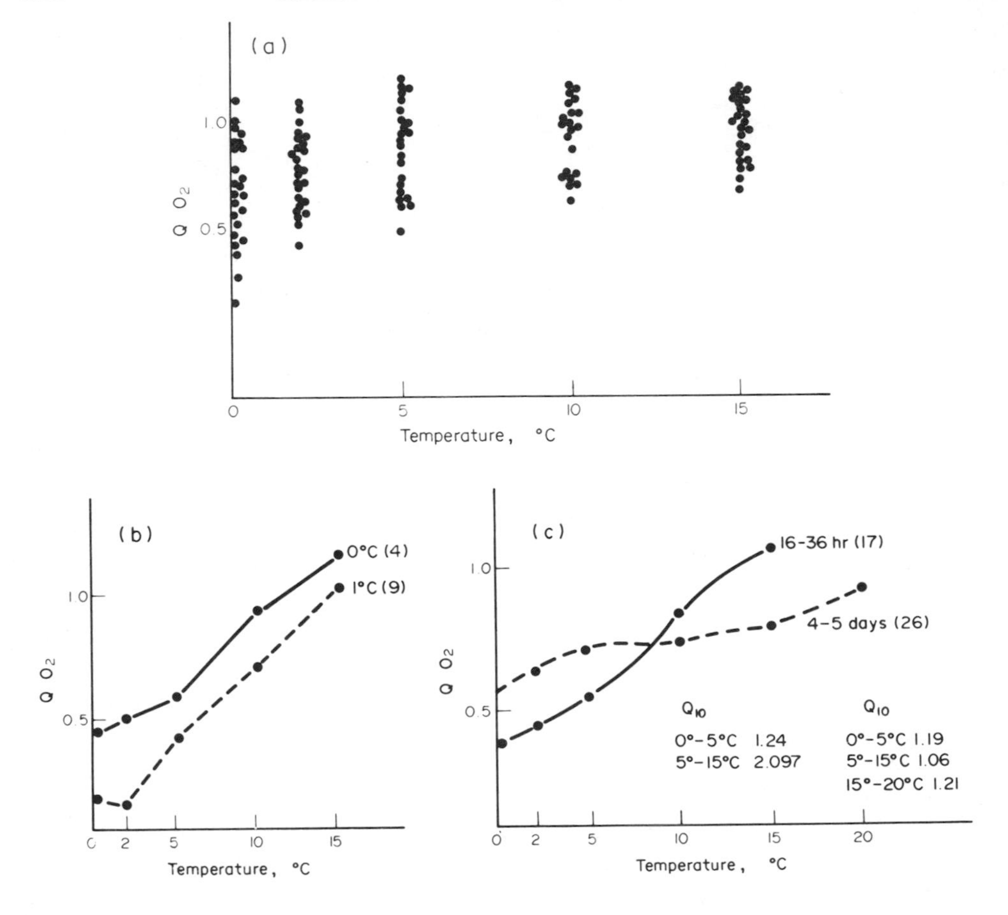

Fig. 6.31. (a) Distribution of responses of oxygen consumption of *Euphausia superba* to various temperatures immediately after collection from water of $-0.9°$ and $-1.27°C$; (b) Oxygen consumption 36–40 hr after introduction to 0°C and +1°C, number of animals in parentheses; (c) Influence of time upon the oxygen consumption-temperature responses when maintained at 0°C, number of animals in parentheses; values of Q_{10} on left are for animals kept 16–36 hr, those on right for animals kept 4–5 days (McWhinnie and Marciniak, 1964).

0.61 $\mu l\ O_2$/mg dry wt/hr at 6.5°C. For *R. gigas*, CO_2 production indicated R.Q. values of 0.44, well below the calculated R.Q. for pure lipid. This in addition to the high lipid body content (60.5% dry wt), suggests extensive oxidation of lipid.

Childress (1971a) has produced strong evidence for a very clear reduction in metabolic rate for deep sea zooplankton. It would be premature to attribute this effect to any one factor such as the low environmental temperature of the deep sea. Indeed, Childress suggests that the weak masculature and relatively large amounts of lipid tissue in bathypelagic animals are more likely to be related to the low metabolism. However, incomplete metabolic adaptation to low temperature might also be a factor (cf. Meek and Childress, 1973). Childress compared a variety of zooplankton (euphausiids, mysids, decapods, ostracods, etc.) from different depths and all experiments were carried out at 5.5°C, characteristic of a depth of *ca.* 600 m. Table 6.16 indicates the marked decline of metabolic rate with depth. Further experiments on the

Table 6.16. Relationship between depth of occurrence and respiratory rate in some pelagic fishes and crustaceans (Childress, 1971)

Depth (m)	Respiratory rate ($mg\ O_2$/kg dry wt/min) Range	Mean	SD	n	Dry wt range (g)
0–400	6.2–15.3	12.6	±3.7	19	0.0041–0.104
400–900	1.46–10	4.5	±2.5	53	0.0049–2.22
900–1300	0.65–1.63	1.2	±0.3	8	0.132–2.02

deep-living copepod *Gaussia* and the mysid *Gnathophausia* by Childress (1971b, 1975) also indicate that the respiratory rate is greatly reduced in these bathypelagic forms. Childress (1977) investigated the dependence of oxygen consumption on temperature in *Gaussia princeps*, a comparatively deep-living copepod, which performs daily vertical migrations off California from below 400 m, in the oxygen minimum layer, to depths of 200–300 m at night. Measurements of oxygen uptake showed a rise with temperature from 3.5°–10°C, but the effect was proportionally greater in the lower range; the Q_{10} between 3.5° and 7°C was 2.60 as against only 1.31 between 7° and 10°C. The effect of temperature on metabolic rate was also linked with hydrostatic pressure. Thus, whereas the Q_{10} for respiration for the range 3.5°–10°C was 1.46 at 1 atm pressure, this rose with pressure increase to reach 2.42 at 181 atm. Both factors will tend to reduce oxygen consumption at the lower temperatures and higher hydrostatic pressures typical of greater depths (cf. Fig. 6.32).

A possible industrial example of acclimatization in marine copepods is reported by Gaudy (1977) who obtained evidence of some modification of the raised respiratory rate at higher temperature in *Acartia clausi* which had experienced passage through the cooling system of electricity power station installations and from those copepods obtained from nearby areas which received heated effluent. A similar pattern of acclimatization was demonstrated for the respiration of the species previously subjected to temperature acclimatization in laboratory experiments between temperatures of 10° and 28°C. An example of temperature acclimatization in warm-water zooplankton comes from a study by Biggs (1977), who found that the siphonophore *Forskalia* could adjust its metabolic rate to a temperature range of at least 9°C. Short term exposures to temperatures only 5°C lower than the environmental temperature, however, were found to result in a marked lowering of oxygen uptake of between two-fold and five-fold. Biggs questions whether siphonophores and similar plankton might conserve metabolic demands by migrating to deeper levels after feeding at night, provided pressure changes had little influence on respiratory rates.

One or two species of zooplankton have been examined for the effect of temperature on respiration at different parts of the species' geographical range (cf. Gaudy for *Centropages* and *Calanus helgolandicus*). Results indicate that some degree of latitudinal acclimatization may occur. Conover (1956, 1959) found a significant difference in respiratory rate of *Acartia clausi* between Long Island Sound, U.S.A. and Southampton, England, that might reflect temperature acclimatization. With very widely distributed zooplankton, the range of environmental temperature thus being considerable, it would be of interest to examine the respiration/temperature relationship of a species from several regions. With such cosmopolitan species as *Oithona*

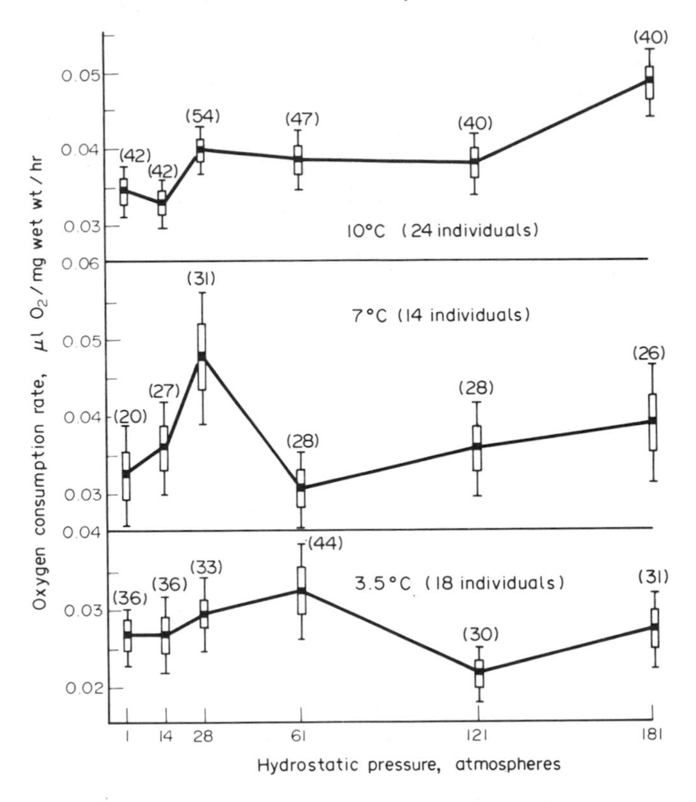

Fig. 6.32. Oxygen consumption rates of *Gaussia princeps* at different temperatures and hydrostatic pressures. Numbers in parentheses: numbers of observations; blocks: standard error of mean at each point; vertical lines: 95% confidence limits (*t* test) (Childress, 1977).

similis, *Beroe cucumis* and *Pleurobrachia pileus* or very widely distributed zooplankton (*Dimophyes arctica*, *Paracalanus parvus*) it is likely that different races exist which may be genetically distinct, but phenotypic acclimatization to different temperature regimes, with its reflection on oxygen uptake, might also be revealed.

Apart from acclimatization due to genetic variation, other types of acclimatization have been included in the examples already quoted. In some experimental techniques, individual groups of experimental animals have been conditioned by maintaining them at different temperatures before exposing them to the test. In contrast to these acclimatized groups are the populations of animals of a species which, by living in different areas, have been acclimatized to different temperature regimes. Many of the examples of acclimatization have, however, been drawn from seasonal differences in the metabolic response of species to temperature. In most cases this in turn relates to different reactions of generations of a species through the year. The problem of metabolic variation in different broods (generations) of a species is complex. Not only may generations of zooplankton develop under different temperature regimes, the total environment, including quantity and quality of food and organic and inorganic trace substances, may vary seasonally at least in more inshore habitats and in temperate and higher latitudes. Many species display size differences between broods, but there may be subtler differences such as those in biochemical composition. Some of the relationships between generation and respiratory rate will, therefore, be reviewed separately. In the discussion of variation in oxygen consumption with season and brood, the variable body size of a species must be considered.

The significance of relating body size to oxygen consumption is by no means to be confined, however, to considerations of brood and season. Intra-specific differences in size will apply to the developmental stages and may also be characteristic of individuals of species with a wide distribution drawn from different areas. Inter-specific differences in size of zooplankton within a taxon or even between taxa may also be considered in relating body size and oxygen consumption.

Body size

A considerable amount of data relates oxygen consumption to body size, usually dry weight, since the size of a single zooplanktonic animal may vary with season, lipid content, water content and other factors. In the study quoted by Sameoto (1972), the respiratory rate was shown to increase with body weight in *Sagitta elegans* at each experimental temperature from 1.5° to 15°C (Fig. 6.33a,b). Champalbert and Gaudy's (1972) data better relate to weight and similarly those of Anraku (1964b) are plotted as respiratory rate per copepod and per unit dry body weight. As Corkett and McLaren (1978) point out, the relationship is much clearer on a weight basis and it is now common practice to relate oxygen uptake to weight. Methods for dry weight determination need to be standardized, however, and the literature has many examples of discrepancies in data on dry weight.

For some gelatinous zooplankton dry weights can be misleading in view of the very high water and ash contents (cf. Chapter 1). Many workers, therefore, have attempted to find a relationship between body size and respiration. Raymont and Gauld (1951) and Gauld and Raymont (1953), from the results of oxygen uptake determinations on several small copepods and on *Calanus* and *Euchaeta*, suggested that respiratory rate was related to surface area. Since there is much variation in the proportion of urosome to cephalothorax in calanoids, surface area was calculated in relation to cephalothorax length, and for several species and developmental stages of calanoids respiration appeared to be proportional to (cephalothorax length)$^{2.2}$. There was good agreement using the equation:

$$\log R = 2.19 \log L - 0.928$$

where R = respiratory rate at 17°C;
L = cephalothorax length.

However, variations in the shape of calanoid species, especially departures from the approximately cylindrical shape of the cephalothorax (e.g. *Temora longicornis*), demand a correction to the length before using the equation.

Later investigations (Raymont, 1959) with the same species of copepods off the north-east coast of U.S.A. and with some other calanoids also gave good agreement using the relationship between respiration and surface area, allowing for differences in cephalothorax shape as, for example, in *Eurytemora herdmani*. One exception was the copepod, *Tortanus discaudatus*, in which respiratory rate appeared to be high (*vide infra*). Many workers including Conover (1956, 1959, 1960) have considered respir-

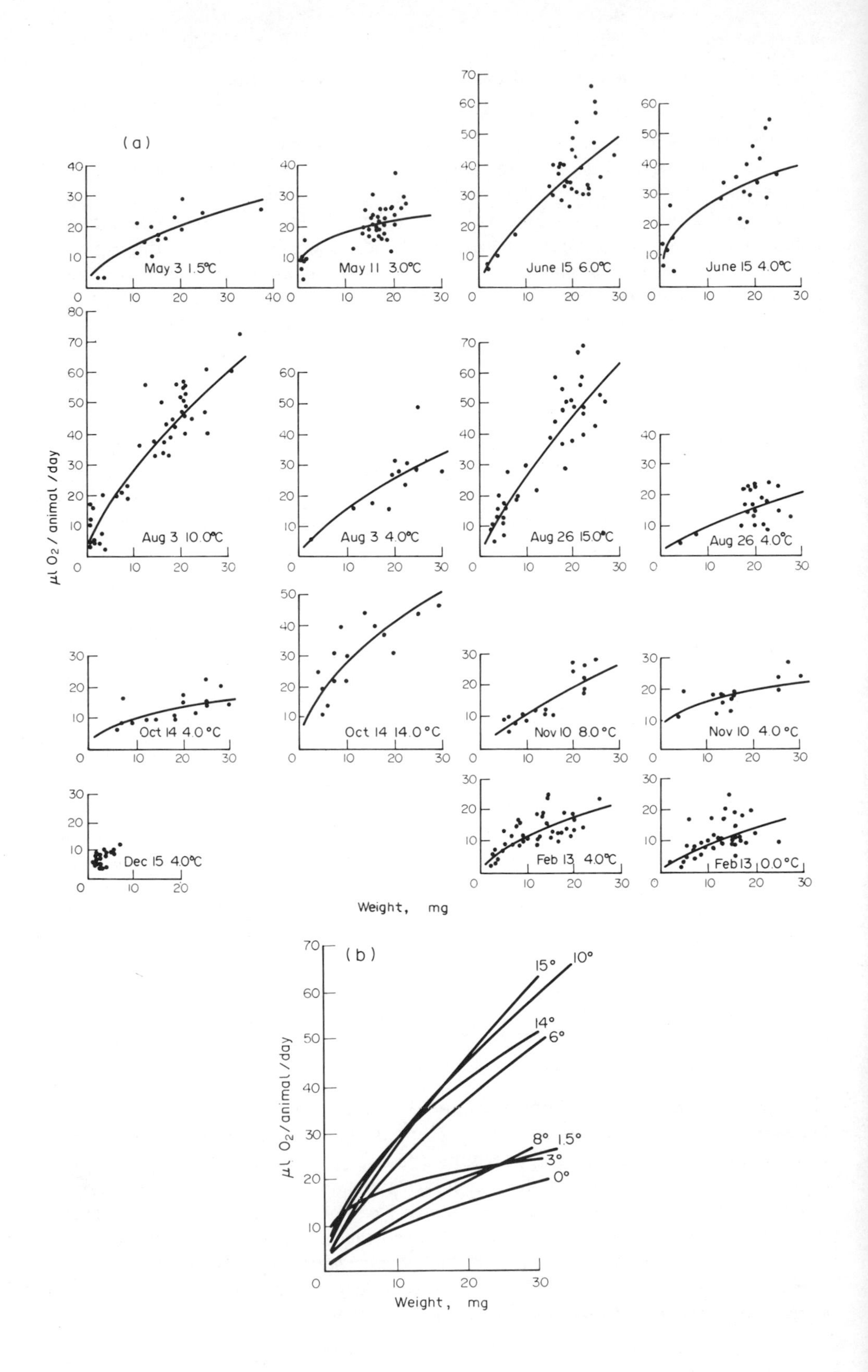
(a)
May 3 1.5°C
May 11 3.0°C
June 15 6.0°C
June 15 4.0°C
Aug 3 10.0°C
Aug 3 4.0°C
Aug 26 15.0°C
Aug 26 4.0°C
Oct 14 4.0 °C
Oct 14 14.0 °C
Nov 10 8.0 °C
Nov 10 4.0 °C
Dec 15 4.0°C
Feb 13 4.0°C
Feb 13 0.0°C
μl O2 / animal /day
Weight, mg
(b)
15° 10°
14° 6°
8° 1.5°
3°
0°
μl O2/animal /day
Weight, mg

ation as more closely related to body weight than to surface. A relationship is suggested of the form:

$$R = a\,W^b$$

or

$$\log R = b \log W + \log a$$

where R = respiratory rate, which can be expressed as $\mu l\ O_2$/animal/hr or as $\mu l\ O_2$/mg body wt/hr;

W = weight; a and b are constants.

The constant a reflects the metabolic level and apart from being affected by temperature, may be modified by other factors such as regional and seasonal differences, generation and food levels (Conover, 1968). The constant b has been found to vary in different investigations from 0.622 to 1.06. If surface area was directly proportional to respiratory rate the theoretical value of b would be 0.67; if it was proportional to weight, its value would be 1.0. As Marshall (1973) points out, most experimental results tend to give values between these two limits (e.g. Conover, 0.86—mean for seven neritic copepods; Raymont and Gauld (1951) 0.73 for four marine copepods). Champalbert and Gaudy (1972) also cite a series of equations for the various ecological groupings of copepods from off north-west Africa. Occasionally values for b are found well below 0.67. Small and Hebard's (1967) data and Paranjape's (1967) results for *Euphausia pacifica*, however, are very close to 1.0.

Perhaps the most extensive series of determinations covering a wide range of zooplankton is that due to Ikeda (1970, 1974). The relationship between respiration and body weight has been calculated for each latitudinal grouping, boreal, temperate and tropical species, and body weight has been considered as wet, dry and ash-free dry weight. The results indicate that metabolic rate is more closely related to total organic weight and the correlation is especially highly significant for boreal and temperate species (Fig. 6.34), with values for b of 0.865 and 0.679, respectively (Ikeda, 1970). The extensive data obtained subsequently by Ikeda (1974) gave regression equations with correlation coefficients ranging from 0.959 for boreal to 0.886 for tropical species. Correlations were highly significant for all geographical groups.

The increase in respiratory rate with habitat temperature, previously noted, is clearly demonstrated by plotting respiration against body weight. That respiration is more closely linked to organic weight is hardly surprising, but the variable amounts of some biochemical constituents, especially lipid may introduce complexities. Moreover, respiratory exchange, whether via special gills or through the general body, is essentially surface mediated and surface area is presumably therefore significant. Conover's (1960) examination of the respiration of a variety of zooplankton (amphipods, euphausiids, copepods) also demonstrated the effect of body size. In these and other investigations it was also possible to demonstrate that the finding, well authenticated for many animals, that respiration (metabolic) rate per unit body weight

Fig. 6.33. (a) Data points for oxygen consumption versus live body weight for respiration rate experiments on *Sagitta elegans* at different temperatures and dates. Data are fitted to power curves ($Y = aX^b$). (b) Comparison of relationship between respiration rate and live body weight at temperatures from 0°–15°C (excluding 4°C) (Sameoto, 1972).

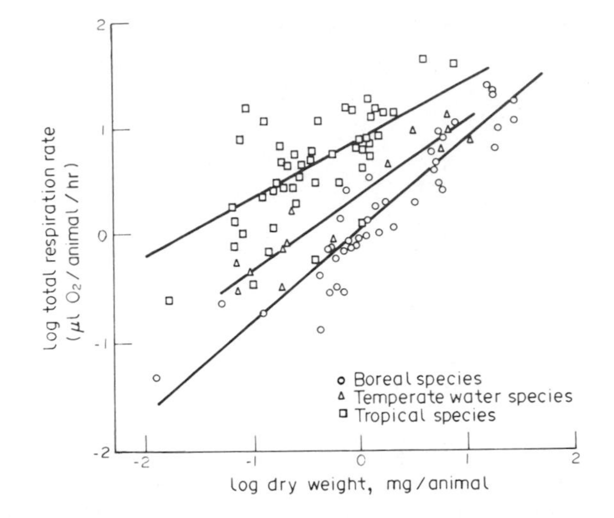

Fig. 6.34. Regression lines for the plankton animals from boreal, temperate and tropical waters, drawn for log total respiration rates (μl O_2/animal/hr) plotted against log dry weight (mg/animal) (Ikeda, 1970).

decreases with increase in body size, applies to zooplankton. Ikeda (1970) demonstrated the finding separately for plankton animals from boreal, temperate and tropical environments (Fig. 6.34), and he further concluded that the Q_{10} for respiration decreased with increasing body weight for the three groups (Table 6.17). Conover's (1960) data not only demonstrated the decrease in weight-specific respiration rate with increasing body size, but suggested that carnivorous zooplankton of a particular size group had higher respiratory rates than herbivores of the same group. He attributed this to a greater expenditure of energy in counteracting negative buoyancy, and active swimming in search of prey. The earlier indication of a much higher rate in the carnivorous *Tortanus* as compared with more herbivorous copepods (Raymont, 1959) would support this conclusion (cf. also Conover and Corner, 1968).

Ikeda (1970) found little difference between the two groups but agreed that more active zooplankton such as euphausiids and amphipods appeared to respire at a higher rate than more sluggish species. In his later investigation, Ikeda (1974) confirmed the decline in weight-specific respiration rate with increasing body weight for each of the four geographical groupings, and the general rise in metabolic rate of zooplankton

Table 6.17. The Q_{10} for respiration rate in plankton animals. The calculation is based on the experiments at different latitudes (Ikeda, 1970)

	Body dry weight (mg)		
	0.1	1.0	10
Boreal species (8°C)[a]			
Temperate water species	3.3	2.3	1.6
(17°C)	3.6	2.7	2.0
Tropical species (29°C)			

[a] The temperature in parentheses indicates mean experimental temperature.

with habitat temperature. He suggested that respiration rate (R: $\mu l\ O_2$/animal/hr) could be predicted if the habitat temperature (T: °C) and body weight of an animal (W: dry wt/mg) were known, from the equation:

$$\log R = (-0.01089T + 0.8918) \log W + (0.02438T - 0.1838)$$

He also found that active zooplankton groups, euphausiids, amphipods, and fish larvae, respired at a slightly higher rate, and gelatinous zooplankton (cnidarians, ctenophores and tunicates) at a slightly lower rate than other zooplankton of the same size. Thus, even such highly carnivorous taxa as cnidarians and ctenophores had a respiratory rate below that of some more obvious herbivores. Conover (1960) was mainly comparing species from the same taxon. A comparison of respiration rates for different species of some siphonophores and ctenophores might be relevant in view of recent findings (cf. Mackie, 1973; Harbison, Madin and Swanberg, 1978—Chapter 2) that some are much more active in the pursuit and capture of their prey. In this connection, the investigations of Biggs (1977) are of especial interest. He measured the respiration of many tropical and sub-tropical gelatinous zooplankton animals (siphonophores, medusae, ctenophores, heteropods, salps, etc), collected by Scuba divers and then incubated them on board ship in a flowing seawater bath (average temperature *ca.* 26°C). In view of the low ratio of carbon content to dry weight in most gelatinous zooplankton, measurements of respiration (and excretion) were standardized to protein content. There was no obvious short term increase in metabolism after capture although, as Biggs points out, the activity of some of the species is likely to have been inhibited in the experimental vessels. Respiration ranged from 2 to 192 $\mu l\ O_2$/mg protein/hr.

Apart from variation due to size, with smaller animals exhibiting higher respiratory rates, there were some taxonomic differences. For example, physonect siphonophores had higher respiratory rates than cystonect species, a difference which may be associated with the lack of swimming bells in cystonects and their consequent lesser activity. With calcycophore siphonophores, there was a wide range of metabolic rates which probably reflects the variable form, from slow swimming species such as *Hippopodius*, *Abyla* to the more active *Stephanophyes*. The herbivorous thaliaceans tended to have higher metabolic rates especially the small doliolids, presumably reflecting their active muscular pumping. The rates for aggregate salps were lower than for solitary forms of similar dimensions. Cydippid ctenophores and many medusae were characterized by low metabolic rates, but some lobate ctenophores, especially the active *Ocyropsis*, had higher respiratory rates. Ikeda's suggestion of the significance of the level of activity of plankton as an influence upon respiratory rate is not therefore contradicted, but Biggs clearly demonstrates the broad range of activity within the so-called gelatinous zooplankton.

The dependence of metabolic rate on body size is clearly relevant to variation in oxygen consumption with generation and with area within a zooplankton species (*vide supra*), since members of a species often show considerable differences in body dimensions.

Seasonal and geographical variations

In examples already quoted of seasonal and regional differences in oxygen uptake within a species, rates have usually been calculated in terms of body weight (e.g.

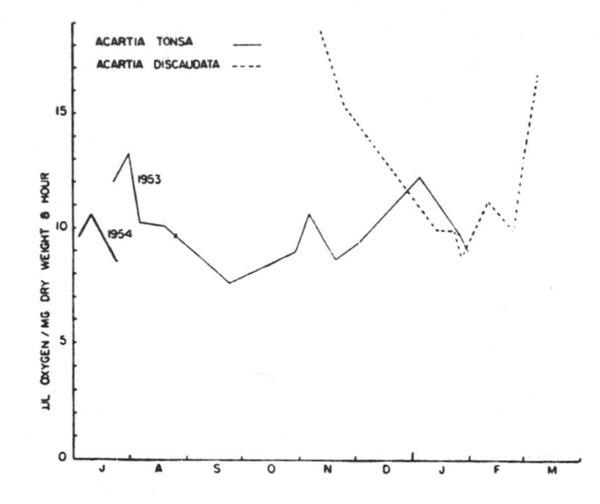

Fig. 6.35. Seasonal variation in respiratory rate for *Acartia tonsa* from Long Island Sound, U.S.A. and for *A. discaudata* from Southampton, U.K. (Conover, 1959).

Anraku, 1964b; Gaudy, 1968; Corkett and McLaren, 1978). Further illustrations include studies by Conover (1959) who obtained strong evidence for a seasonal decline in respiration calculated on a weight basis with several calanoids, especially *Acartia discaudata*, from the Southampton area (cf. Fig. 6.35). Raymont (unpublished) had also found a marked reduction over winter in the respiration of small calanoids from the same area, when due allowance was made for size differences of generations. Marshall and Orr (1958, 1966) reported seasonal changes in the oxygen consumption of several calanoids, *Calanus*, *Centropages hamatus*, *Acartia clausi*, *Pseudocalanus* and *Temora*, the main difference being a rise in spring and a decline in early summer. Though only the period March to July was studied, a peak about April/May was apparent (cf. Fig. 6.36).

Marshall (1973) points out that several factors could contribute to a spring rise in respiration: body size of the spring generation of copepods is usually maximal, food is abundant (*vide infra*) and ripe females are actively reproducing (though respiration even of ripe copepods declines later). The rising spring temperature might also trigger a response. Further studies, however, indicate that some physiological change with generation is in part responsible. Berner (1962), after correcting for size variation in *Temora*, found a spring peak of respiration. Gaudy's (1968) studies on *Centropages*, while allowing for size differences of generations, indicated a clear change in respiration with generation. In a further study in the Mediterranean, Gaudy (1973) found variations with season in the levels of respiration, calculated on a unit weight basis for four calanoids, *Centropages typicus*, *Temora stylifera*, *Acartia clausi*, and *Calanus helgolandicus*. The maximum rate was usually in spring, approximating to the time of the phytoplankton increase, but the main change was due to temperature acclimatization. Changes in the Q_{10} values, however, appeared to be related to the occurrence of different generations of the copepods.

Cowey and Corner (1963b) and Conover and Corner (1968) demonstrated that female *Calanus* showed a seasonal adaptation in respiratory rate, despite differences in size and in the degree of sexual maturity. The latter had been shown earlier by Marshall

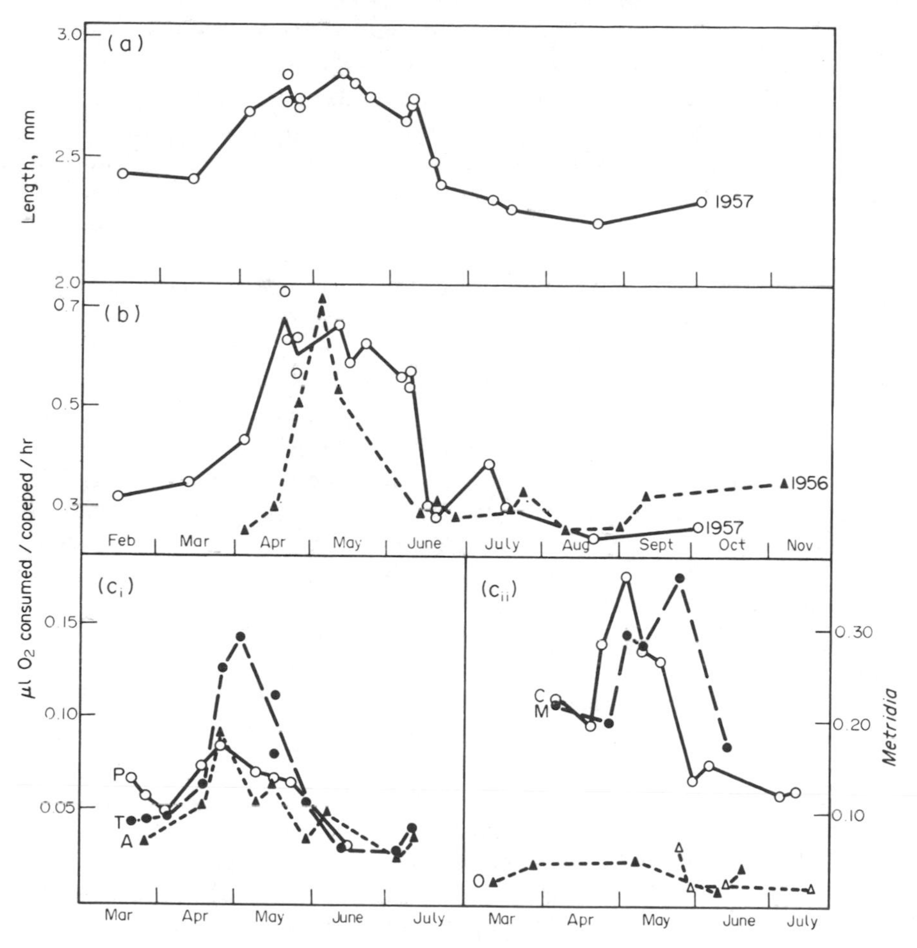

Fig. 6.36. Seasonal changes in oxygen consumption in various copepods: (a) Length of metasome in mm of ripe female *Calanus finmarchicus* in 1957; (b) Oxygen consumption of ripe female *C. finmarchicus* in 1956 and 1957; (c_i) oxygen consumption of *Pseudocalanus elongatus* (open circles), *Temora longicornis* (closed circles) and *Acartia clausi* (triangles) in 1956; (c_{ii}) *Centropages hamatus* (open circles) *Metridia lucens* (closed circles) and *Oithona similis* (closed triangles) all in 1956; *O. similis* (open triangles) in 1965 (Marshall, 1973).

and Orr (1956) to influence respiration. Ripe females had a distinctly higher oxygen uptake than unripe specimens of the same size. However, the finding that Copepodite V *Calanus* showed an obvious seasonal adaptation in metabolic rate was especially significant. Stage V *Calanus* normally overwinters in deeper water and is comparatively inactive, not undertaking diurnal vertical migrations over winter. It also appears to live largely on food (lipid) reserves (*vide infra*). Any reduction of metabolic rate during the overwintering period would be advantageous. Conover and Corner (1968) showed that the oxygen uptake of Stage V in winter was very markedly reduced from 79–31 μl O_2/mg/day. Similar reductions have been repeatedly confirmed in other studies. The reduction of respiration in other copepods agrees with these results and complements Marshall and Orr's finding of a peak in respiration in spring. An exception to this seasonal rise in respiratory rate in the spring observed by Marshall and Orr (1966) related to the cyclopoid, *Oithona*. Species of this genus are known to be markedly carnivorous and the absence of a spring rise in respiration may be associated with its carnivorous habit. According to Marshall and Orr its respiratory rate was high in relation to its body size as compared with other copepods. Whether it reduces its metabolic level over winter is not known. Lampitt (1978), however, reports a low metabolic rate for *O. nana* and states that despite the reduced daily requirements, it must rely on omnivorous feeding since predation on nauplii could supply only 25 % of daily needs.

With *Calanus hyperboreus*, Conover (1962) found evidence of an even more marked seasonal variation in metabolic rate, with a maximum in spring, coinciding with a period of activity and pronounced feeding. The peak occurred about April/May and after a comparatively short time, metabolic rate declined through the summer and autumn to low winter values. The seasonal changes applied to Copepodite V and adult females. Starved animals generally had a lower metabolic rate than fed copepods, though on occasions feeding Copepodite V copepods seemed to affect respiration very little. Ripe female *Calanus hyperboreus* had a higher oxygen uptake than unripe females and even by November/December, long before any increase in phytoplankton food was evident, adults with evidence of gonad maturation showed an average oxygen consumption of 20 $\mu l\ O_2$/copepod/day as compared with unripe females with an average consumption of *ca.* 15 $\mu l\ O_2$/copepod/day. When respiration levels were adjusted for weight differences, ripe females respired at nearly twice the rate of immature Copepodite VI. The period of active feeding and raised metabolism appears to occupy as little as 10–20 % of the year and over much of the twelve months *Calanus hyperboreus* reduces the metabolic level and lives comparatively economically on food reserves (*vide infra*). In the laboratory, animals offered food following some weeks of adjustment will become more active and show an increase in metabolic rate.

A number of examples of differences in level of respiration appear in species examined in different areas of their range. While some of the differences observed for *Calanus finmarchicus* may be attributed partly to laboratory technique and to partial acclimatization, different stocks appear to show variations not attributable merely to differences in size. Thus, Conover (1960) reported that *Calanus* from Georges Bank had a much higher oxygen uptake rate than the same species of copepod taken from the Gulf of Maine, though the sampling points were less than 100 miles apart. Redfield (1941) showed the presence of different broods in the two areas. Omori (1970) demonstrated differences in the oxygen consumption of *Calanus cristatus* (C V) in northern and southern parts of its range, after due allowance had been made for size difference. Ikeda's (1970) values for the same copepod are again considerably higher than either of Omori's. In later experiments Ikeda (1971) obtained an oxygen uptake approximately twice that of his earlier figure. Although the latter experiments were carried out with an open flow system on shipboard, so that the copepods were stimulated to swim and feed actively, there were wide differences in the respiratory rates, suggesting that amongst other factors, the experimental copepods were sampled from different stocks.

Not all zooplankton show these geographical variations. Conover and Lalli (1972, 1974) found that in the gymnosome *Clione limacina* though respiratory rate followed the normal relation to body size, there was no discernible relationship to region or to season.

Feeding and starvation

Although some experiments have demonstrated little effect of food intake on respiration, zooplankton which are starved generally exhibit a depressed metabolic rate. In general zooplankton naturally inhabiting an environment where the food supply appears to be inadequate also tend to show a lowered respiration as compared with the same population inhabiting an area when food is plentiful. The experiments of Conover with *Calanus hyperboreus* have already been noted. Other long term experiments (Conover, 1964) showed a definite reduction in oxygen uptake with starvation. However, the effects of feeding are not always as obvious with this copepod as with other species (cf. Conover and Corner, 1968). Ikeda (1971) claims that the doubling of oxygen consumption in *Calanus cristatus* maintained in an open flow system as compared with copepods kept in closed chambers without food is mainly due to the active feeding which he monitored during the experiments with the open flow system. Taguchi and Ishii (1972) also reported higher respiratory rates for *Calanus cristatus* and for *Calanus plumchrus* in experiments in which the copepods were placed in natural seawater on board ship immediately after capture. They attribute the increased metabolism to the presence of abundant food.

Corner, Cowey and Marshall (1965) measured oxygen consumption in feeding and starving female *Calanus*, making allowance for the respiration of the algae used as food. Respiration increased with feeding. For example, in one experiment as the food supply increased from zero to 78 μg N/l, oxygen uptake increased from 38 to 110 μl O_2/mg dry wt/day. In very preliminary experiments Raymont (1959) found no obvious effect on respiration of feeding and starving *Pseudocalanus*. With *Tortanus discaudatus*, however, a very sharp reduction in oxygen consumption followed starvation and with *Centropages hamatus*, which is at least partly carnivorous, starvation also appeared to reduce oxygen consumption. *Acartia* has been observed to increase its oxygen uptake with feeding (Conover, 1956). The effect obviously is partly dependent on species. Perhaps carnivorous species are more affected by starvation, and in particular, herbivorous species with large lipid stores may be less dependent.

Ikeda (1974) describes the differential effect of starvation on survival and respiration of different groups and body sizes of zooplankton. A discussion of his data is left until other aspects of metabolism are reviewed, but it is worth noting that a later study by Ikeda (1977a) on several herbivorous species (*Calanus plumchrus*, *Paracalanus parvus*, *Euphausia pacifica*) and on the carnivorous zooplankton, *Pleurobrachia* and *Parathemisto*, collected off British Columbia, showed that fed animals maintained or even raised their respiratory levels, whereas respiratory rate decreased progressively with starvation. Rates varied e.g. *Calanus* and especially *Euphausia* showed only a fairly slow drop in respiratory rate with starvation, *Paracalanus* and *Pleurobrachia* exhibited rapid falls. Another example from warmer waters is taken from an experiment by Nival, Malara, Charra, Palazzoli and Nival (1974) with *Temora stylifera* taken off Morocco. They found that starvation resulted in a rapid decrease of respiratory rate, together with a regular decline in excretion rate over seven days of experiment. Dry body weight showed a decrease after four days starvation, and the O/N ratio declined from approximately 17 to 7, suggesting the use of protein for survival after only about four days fasting. Ikeda also discusses changes in excretory rates and in body weight (*vide infra*). It is perhaps significant that Conover and Lalli (1972,

1974) found a very clear increase in respiration in *Clione limacina* with feeding; a relationship was established between metabolic rate and food ration.

Salinity

Over a vast extent of most oceans, the variation in salinity is almost certainly too slight to have any significant effect on respiration. For inshore areas and especially estuaries where the salinity changes can be substantial, very few data are available for the effect of these changes on the respiratory rate of zooplankton.

Marshall and Orr (1955) and Marshall (1973) reported a decline in oxygen uptake with reduced salinity with gradual transfer of *Calanus finmarchicus* to salinity as low as 17‰. Ankru (1946b) using dilutions of seawater to 21.5‰, confirmed the fall in respiratory rate with lowered salinity. He obtained the same result with *Centropages hamatus*, reducing salinity to as little as 13.5‰, when respiration had fallen to about a third of its original level. With the euphausiid, *Euphausia pacifica*, Gilfillan (1972) also found that reduction in salinity tended to cause a lowering of oxygen consumption. The effect is, however, temperature dependent. At a temperature of 5°C, there appears to be little effect on respiration for salinities down to 24‰. At higher temperatures the effect on oxygen uptake is variable, but at 15°C, for example, even a reduction to 27‰ lowers respiratory activity. Jawed (1973) also demonstrated some reduction in respiration for the two mysids, *Archaeomysis grebnitzkii* and *Neomysis awakschensis*, when salinity was reduced from 30‰ to 22.5‰, but while oxygen uptake decreased significantly at 10°C; no effect was observed for either species at 5°C.

Gilfillan regards the change with lowered salinity as a stress effect and this is probably true for other open sea zooplankton such as *Calanus*. With *Acartia tonsa*, however, Lance (1965) observed a rise in respiratory rate with salinity reduction. There was very little change in oxygen consumption between 36‰ and 33‰ salinity, but with a more substantial decline in salinity to 11‰, respiration was approximately double that at 36‰. *Acartia tonsa* is well known as a markedly euryhaline estuarine copepod and it would appear that the marked contrast in the effects of salinity on respiration exhibited by this copepod and typically open sea zooplankton may be partly associated with the osmotic regulation which *Acartia* performs. With open sea species, any lowering of salinity probably merely depresses all physiological functions.

Light

As Marshall (1973) states, sunlight is lethal to *Calanus* and to many other zooplankton and most experiments on oxygen uptake are carried out in darkness or in subdued light. There is very little information, therefore, on the effects of light intensity on respiration. Bright light has been observed to raise the oxygen uptake in *Calanus* but the exposure tends to damage the animals. Conover (1956) found very little effect of light on the respiration of *Acartia tonsa*. There is some indication of a greater oxygen uptake by male and female *Pseudocalanus* in diffuse light than in darkness; the effect may reflect greater locomotory activity (cf. Corkett and McLaren, 1978). *Calanus* has been observed to swarm near the surface but the effect of light on the respiration is unknown. The swarming might also be related to the stage of sexual maturity which itself affects respiratory rate. The response of neustonic zooplankton which are exposed to very high light intensities in the tropics, would be of interest.

Oxygen content

Though little information exists on the effect of lowered oxygen concentration on respiratory rate, there are substantial variations in oxygen content of seawater and, in particular, oxygen minimum zones exist in the oceans where oxygen concentrations can fall well below 1 ml O_2/l. Longhurst (1967) described dense populations of zooplankton west of Baja California, a region where a pronounced oxygen minimum zone is well known. A rich assemblage of epiplanktonic forms was present in the upper well oxygenated strata. Below 100–200 m a sharp decline in oxygen was observed, but although the variety of zooplankton was less, dense zooplankton populations were present in the oxygen-deficient layers. Oxygen content fell to less than 0.5 ml O_2/l, in some instances < 0.2 ml O_2/l, but there was no evidence that a decrease in oxygen was associated with a regular decline in zooplankton. Often the zooplankton was concentrated in comparatively thin strata. The plankton consisted partly of species (e.g. *Euphausia eximia*, *Pleuromamma abdominalis* and myctophid fishes) which migrated surfacewards at night, reaching a layer where the oxygen content was about 5 ml O_2/l, i.e. a daily range of ten-fold oxygen concentration was tolerated (cf. Brinton, 1979).

In the oxygen-deficient zone, other species of zooplankton which exhibited no diurnal vertical migration were also present. These were dominated by *Calanus pacificus* with much smaller numbers of *Rhincalanus nasutus* and *Eucalanus bungii californicus*. The *Calanus* stock consisted almost entirely of Copepodite V stages and Longhurst compares them with the deep-living overwintering stock of *Calanus finmarchicus* V which occur in the Atlantic. They may persist in the deep layers for about six months, but can rapidly take advantage of the rich algal blooms which occur with coastal upwelling. The presumably lowered metabolic rate of the copepods at reduced temperature and possibly low oxygen content would favour an overwintering stock.

Longhurst quotes observations from Vinogradov and Voronina for oxygen-deficient sub-surface waters of the northern Arabian Sea that no resting populations of copepods occurred in those areas nor were there species migrating daily from the oxygen-poor layers surfacewards. A marked reduction in numbers of zooplankton and of biomass was found in the oxygen-poor strata. *Rhincalanus nasutus* and *Pleuromamma indica* were amongst the few species tolerating the very low oxygen content which could be as little as < 0.15 ml O_2/l. Haq, Khan and Chugtai (1973) have confirmed the very great importance of *Pleuromamma indica* in the oxygen-deficient waters off Pakistan. Vannucci and Navas (1973) also point to the many species of hydromedusae in the Indian Ocean which can tolerate extremely low oxygen concentrations down to 0.2 ml O_2/l. In the Black Sea, Longhurst quotes experiments by Nitikin and Mahn demonstrating that *Calanus helgolandicus*, not previously adapted to low oxygen, could tolerate concentrations between 0.5 and 0.2 ml O_2/l. They live in the Black Sea close to the 0.2 ml O_2/l level, the same concentration as off California.

Although low oxygen concentrations may be tolerated by *Calanus*, Marshall and Orr (1955) found that below a critical concentration of around 3 ml O_2/l (at 15°C), respiratory rate was depressed. The effect was less obvious at lower temperatures and was less marked for Copepodite V copepods, underlining Longhurst's finding that the stock off California were in the C V stage.

A few observations by Zeuthen (1947), mostly on meroplanktonic larvae, suggested that depression of respiratory rate occurred only at considerably reduced oxygen tensions, but Morrison (1971), who studied meroplanktonic larvae of *Mercenaria mercenaria*, found a clear effect of oxygen content on growth rate. Between oxygen concentrations of 4.2 ml O_2/l and 2.4 ml O_2/l growth was markedly slowed, but the larvae would survive even at tensions as low as 1 ml O_2/l, and would continue normal growth when transferred to fully oxygenated seawater. An example of more marked tolerance to low oxygen was reported by Teal and Carey (1967a), who described large zooplankton populations both around and above the pronounced oxygen minimum zone off Chile.

Euphausia mucronata was especially abundant below the minimum layer and would migrate vertically through the layer, encountering almost zero oxygen tensions. Experiments showed that respiration was almost unchanged when the euphausiid was maintained in water of 50% oxygen saturation (at 20°C) and that the respiratory rate was little affected until the oxygen content reached almost zero value. This ability contrasted with that of the euphausiid, *Thysanopoda monacantha*, and the decapods *Sergestes*, *Hippolyte* and *Systellaspis* in the same region which showed a sudden fall in respiratory rate when the oxygen tension declined to a critical level. The decapods, however, showed a considerable degree of regulation of oxygen uptake, since they were able to maintain their respiratory rates to oxygen tensions of *ca.* 2–6% atm (≡ 0.4–1.2 ml O_2/l). The lower limit for the northern Pacific euphausiid, *Euphausia pacifica*, of *ca.* 0.7 ml O_2/l quoted by Lasker (1966) would be comparable (cf. Childress, 1975).

Childress (1971) in his studies of the deeper-living zooplankton which inhabit waters of extremely low oxygen content off California, has pointed to the generally very low metabolic rate of the animals. The pelagic crab *Pleuroncodes planipes* may, however, live at depths from the surface to some 300 m, so that the oxygen content of the water may range from saturation to approximately 0.1 ml O_2/l. The rate of oxygen consumption in *Pleuroncodes* in saturated seawater was similar to that of other epipelagic crustaceans and the metabolic rate is not lowered as in the deep-living animals. Quetin and Childress (1976) found that at all temperatures *Pleuroncodes* regulated its respiratory rate to the partial pressure of oxygen, so that a constant rate of consumption was maintained to a very low oxygen level (*ca.* 1.9–4.9 mm Hg O_2, depending on temperature). Therefore it can live aerobically even in very low oxygen environments. The oxygen uptake appears to be independent of hydrostatic pressure within the range (up to *ca.* 30 atm) normally encountered (*vide infra*).

One of the species frequently found in the oxygen minimum zone is the copepod, *Gaussia princeps*. The relationship between the respiratory rate of this copepod and temperature has already been described. Childress (1977) found that *Gaussia* was apparently unable to satisfy its metabolic requirements at the very low oxygen concentrations (*ca.* 0.2 ml O_2/l) in the deep environment. Experiments carried out at different temperatures from 5.5° to 10°C showed that there was little or no change in oxygen consumption with very large reductions in the oxygen concentration (quoted as partial pressures). But a critical level is reached at about 10 mm Hg O_2 partial pressure, below which *Gaussia* shows a clear decline in oxygen uptake. Childress claims that most other species which inhabit the oxygen minimum zone are able to regulate their oxygen consumption down to the lowest oxygen level at which they are found (*ca.* 6 mm Hg O_2 partial pressure). These species cannot live anaerobically; in the total absence of oxygen they survive for less than 30 min. In contrast, *Gaussia* can survive without oxygen for about 12 hours. In the natural environment it compensates at the deepest levels for its inability to regulate its aerobic respiration by living partly anaerobically, but it also migrates at night to depths of 200–400 m. At these levels, where the oxygen concentration is much higher, though there is little temperature difference, the comparatively low hydrostatic pressure has a marked effect on oxygen uptake. The copepod respires aerobically at a much higher rate (though this is still low compared with epiplanktonic species), is comparatively active and feeds and "burns off' the oxygen debt accumulated at depth. Because of the greatly lowered metabolism at depth, with the Q_{10} being more affected at lower temperature, anaerobic respiration has to account for only about one-third of the total respiration. The copepod, therefore, is not forced to seek the oxygenated shallower layers regularly each night.

Childress (1971b) found that the deep-living mysid *Gnathophausia ingens* has a remarkable ability to regulate its oxygen consumption at exceedingly low oxygen tensions. As with other deep fauna in the oxygen-depleted strata the metabolic rate is very low. The respiratory rate calculated per unit weight falls as the oxygen tension decreases until no oxygen can be detected. A precise critical level cannot be stated, since respiratory uptake is variable, above all with the animal's activity. Swimming ceases within 30 minutes after oxygen is exhausted, though the animals will revive in oxygenated water after exposure to anaerobic conditions for up to 6 hours. Its anaerobic capability is, however, not large and it lives successfully in the oxygen minimum zone (*ca.* 0.2 ml O_2/l) respiring aerobically but at a clearly much reduced metabolic rate, at which level it is capable of slow swimming. Childress suggests that anaerobic respiration occurs over brief periods during occasional bursts of activity. The gill area and the capacity to increase the ventilation volume in *Gnathophausia* are considerable and are significant in its ability to respire at very low oxygen levels.

A study of a large number of mid-water crustaceans obtained off California by Childress (1975) indicated that species regulated their oxygen consumption so that a comparatively constant level was maintained over a range of oxygen concentrations. The critical partial pressure of oxygen, below which oxygen consumption declined, was remarkably low for the deep-living species in the oxygen depleted layers. Childress points out that the critical level is close to that of the immediate environment indicating that aerobic respiration is predominant and most species have very limited anaerobic capacity. Exceptions are *Gaussia* and *Pleuroncodes* (cf. Table 6.18). The amphipod *Hyperia galba* resembles *Gaussia* in not fully regulating its oxygen consumption to the lowest partial pressure of oxygen where the species exists and like *Gaussia* also shows considerable capacity for anaerobic respiration. Belman (1978) also lists the squid *Histioteuthis heteropsis* occurring in the oxygen depleted layers as a good metabolic regulator. It can live for short periods under anoxic conditions and apparently respires anaerobically but must migrate surfacewards later to burn off the oxygen debt acquired. Childress suggests that while *Gaussia* differs from the great majority of the deep-living animals in showing pronounced vertical migrations, *Hyperia* has comparatively abundant food in its deep environment since it lives on medusae. The low food supply may preclude anaerobic respiration in other animals living in the oxygen-poor layers and apart from their very low metabolic rate, such animals must be either vertical migrators or be in a resting stage such as *Calanus* Copepodite V. A low protein content of the species typically inhabiting the oxygen minimum layers, in addition to the low temperature, will contribute to their low metabolic rate, but other unknown factors

Table 6.18. Vertical distribution of zooplankton respiration in the north-west Atlantic Ocean (Childress, 1975)

Depth (m)	Biomass (mg wet wt./m^3)	Zooplankton respiration rate[a] (ml O_2/mg/yr)	Zooplankton respiration (this study) calculated (ml O_2/l/yr)	Total respiration (Riley, 1951) calculated (ml O_2/l/yr)
200	11	1.40	0.015	0.21
400	5.3	0.46	0.0024	0.050
600	7.5	0.24	0.0018	0.033
800	10.4	0.18	0.0019	0.028
1000	3.9	0.15	0.0006	0.013
1200	2.1	0.14	0.0003	0.005
1400	2.0	0.12	0.0002	0.002

[a] Values from this study (respiration = $5.77X^{-0.564}$, X = minimum depth of occurrence in meters) using 20 m minimum depth (MDO) for 200 m depth, 100 m MDO for 400 m, 300 m MDO for 600 m, 500 m MDO for 800 m, 700 m MDO for 1000 m, 800 m MDO for 1200 m, 1000 m MDO for 1400 m.

must play a part. The effect of increased hydrostatic pressure on deep living fauna generally appeared to be rather slight.

Hydrostatic pressure

Although rather few experiments have been carried out on the effect of hydrostatic pressure on the respiration of zooplankton, for epiplanktonic species there appears to be little effect (cf. Quetin and Childress, 1976). Teal and Carey (1967b) investigated the effect of pressure changes on eight species of euphausiids taken from various areas. All except *Thysanopoda monacantha* were epiplanktonic. Four species showed some increase in respiration with increased pressure but at temperatures which could be experienced only at the surface where normal atmospheric pressure holds. Only *Meganyctiphanes* showed a fall in respiration with pressure rise. Generally the effects of pressure changes on the epiplanktonic euphausiids were slight. Temperature would appear to be the significant factor in vertical migration. The one mesoplanktonic species studied, *Thysanopoda monacantha*, showed a greater increase in respiration with pressure at lower than at higher temperatures. Mauchline and Fisher (1969) also quote the results of Pearcy and Small (1968) who examined the oxygen uptake of the euphausiids *Thysanoessa spinifera* and *Euphausia pacifica* under changing pressures. There was no significant difference in oxygen consumption between 1 and 50 atm. Meek and Childress (1973) showed respiration of a non-migratory mesopelagic fish living in the oxygen minimum zone was also unaffected by pressure changes and they believe this holds also for deep-sea forms and vertically migrating invertebrates as well as epipelagic species.

Investigations by Macdonald, Gilchrist and Teal (1972) were chiefly concerned with the tolerance of zooplankton to pressure changes. Activity tended to be reduced at high pressure (500 atm) and this might affect oxygen consumption, but deep sea species were generally less sensitive. A few experiments were carried out on the deep-sea ostracod, *Gigantocypris mulleri*, which indicated a reduction in the respiration of small specimens with increased pressure, but only a slight decline in oxygen uptake in larger animals.

The effect as the investigators point out, could have been a reflection of changes in locomotory activity. It is frequently held that deep-sea animals are little affected by pressure increases. The mesoplanktonic decapod, *Systellaspis debilis*, showed a more marked reaction. Pressures were increased by 150 atm and subsequently reduced by 50 atm, before another similar cycle of increased and decreased pressure. Respiration increased with each rise of hydrostatic pressure and declined when pressure fell. Meek and Childress quote the results of Napora on the same species, which demonstrated an increase in respiratory rate with increased pressure. This rise in metabolic rate is counteracted, however, by the decrease in respiration with the lowered temperature of greater depths, so that *Systellaspis* maintains a relatively constant respiratory rate throughout its range of vertical migration. Teal (1971) found that five species of North Atlantic decapods maintained a relatively constant metabolic rate when subjected to different pressure and temperature changes: the decrease in respiration rate due to temperatures is offset by the increase caused by rising pressure. He suggests that in contrast to the euphausiids studied earlier (cf. page 620), these predatory decapods maintain a higher metabolic rate which enables them to hunt prey by day or night.

Smith and Teal (1973) examined the respiration of four thecosomes. For *Diacria trispinosa*, *Cuvierina columnella* and *Clio pyramidata* there was a similar rise in respiratory rate with temperature, the Q_{10} ranging from 1.7 to 2.2 at the lower temperatures, and being slightly less at higher ranges. At each temperature, increased hydrostatic pressure caused a rise in oxygen uptake for pressures exceeding *ca.* 20 atm (75 atm for *Clio*). In relation to their normal depth of occurrence the respiration of *Diacria* and *Cuvierina* is thus influenced by temperature only: *Clio* has a somewhat deeper range but, in the natural environment, its respiration will be affected by temperature but not by pressure. This is similar to earlier findings for epiplanktonic euphausiids. The thecosome *Limacina helicoides*, differs in that the effect of temperature on respiration is slight at lower temperatures ($Q_{10}=1.2$ at 10°C) and hydrostatic pressure has very little effect on respiration. A small increase was seen only at pressures exceeding 150 atm and at a temperature of 10°C. The latter species is mesopelagic or bathypelagic and its metabolic rate appears to be very constant over its large depth range, varying slightly with the environmental temperature.

Developmental stages

The respiration of the developmental stages of zooplankton has been studied for few species. While it is possible that factors such as changing concentration of respiratory enzymes with age and the effect of dietary changes which take place with some animals, might influence respiration, probably little can be said beyond the obvious decline in weight specific respiration rate with age, which is a reflection of the increase in body size. There are a very few indications, however, of relatively greater changes in metabolic rate at certain growth stages. Marshall and Orr (1958), for example, found that, compared with a mean respiratory rate of *ca.* 7.6 μl O_2/copepod/day for adult females, the rate fell to 3.1 μl O_2/copepod/day for C IV, and to values as small as 0.08 and 0.05 μl O_2/copepod/day for the youngest nauplii. They suggested, however, a particularly sharp decline in respiration per animal at the C III stage (cf. Fig. 6.37).

Mullin and Brooks (1970) compared the respiration of *Rhincalanus nasutus* and *Calanus pacificus* and showed the expected increase in oxygen uptake with body size (age), which may be compared with the increase in food intake. For three develop-

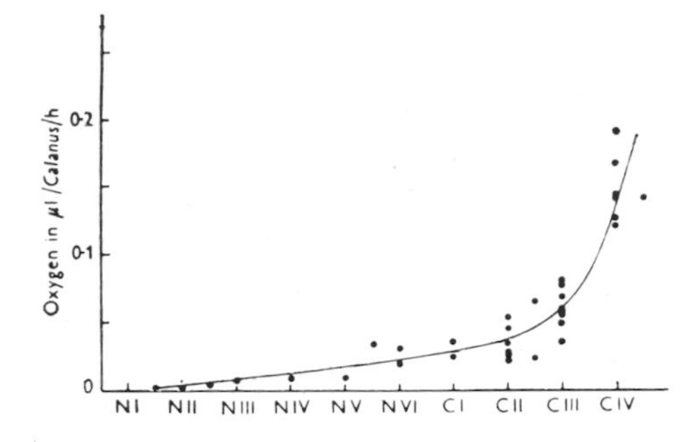

Fig. 6.37. Oxygen consumption of early stages of *Calanus* (Marshall and Orr, 1958).

mental stages, they calculated the respiratory rate on a unit weight basis and showed a decline with increasing body size (Table 6.19). It is interesting to note that for each stage, the respiration per unit weight is greater for *Calanus*, the smaller species. *Acartia clausi* is one of the very few other species for which data exist on the oxygen consumption of developmental stages. Table 6.20 (Marshall, 1973) gives the data from Petipa, and illustrates the marked increase in oxygen uptake on a unit weight basis with the decreasing body size of the younger stages.

A major object in determinations of respiratory rates in zooplankton was the calculation of maintenance requirements. While the early work of Putter suggested large amounts, mainly for small mixed populations of copepods, much of the later work on respiration indicated considerably lower requirements. Thus Marshall and Orr's (1958) experiments with *Calanus* indicated that 3.9% (as lipid) or 7.2% (as protein or carbohydrate) of the body weight would be required daily by females over summer at 10°C, and 2.8–6.7% body weight as the daily requirement in winter at 10°C. Ikeda (1971) calculated that *Calanus cristatus* (C V) required 7% of the body weight daily. The amount of oxygen consumed is used to estimate requirements:

$$\text{mg } O_2/\text{unit time} \times \frac{12}{32} \times \text{R.Q.} = \text{mg C/unit time required}$$

or

$$\text{ml } O_2(\text{N.T.P})/\text{unit time} \times \frac{12}{22.4} \times \text{R.Q.} = \text{mg C/unit time required}$$

Table 6.19. Respiratory rate of two copepol species calculated on a unit weight basis (Mullin and Brooks, 1970)

	Respiratory rate (µg C respired/µg body wt/day)	
	Rhincalanus	*Calanus*
Stage IV	0.026	0.095
Stage V	0.025	0.060
Stage VI ♀	0.017	0.037

Cultures grown on Thalassiosira.

Table 6.20. Oxygen consumption in young stages of *Acartia clausi* and *Calanus finmarchicus* (Marshall, 1973)

Species	Stage	Number used	Dry wt (μg) Range	Dry wt (μg) Mean	Temp. (°C)	Duration of expt (h)	Oxygen consumption (μl/copepod/day) Range	Oxygen consumption (μl/copepod/day) Mean	Oxygen consumption (μl/mg dry wt/day) Range	Oxygen consumption (μl/mg dry wt/day) Mean	Location	Source
Acartia clausi, large	♀+♂	212	4.3–10.2	8.8	15–26	Most 24	0.9–3.4	1.61	126–525	214		
	♀	45	4.3–5.9	5.0	23–25	7–24	0.4–2.7	1.28	89–595	267		
	♂	28	2.6–5.4	4.6	7–24	8–24	1.2–4.6	2.87	211–1295	753		
A. clausi, small	♀	95	1.4–1.9	1.7	20–26	16–24	0.3–2.0	0.91	153–1770	632		
	♂	63	1.4–1.6	1.5	24–26	24	0.7–1.4	0.98	483–950	676		
A. clausi, young	C V	13		1.8	24–26	24		2.02		1050	Black Sea	Petipa (1966)
	C IV(+III and V)	16		1.0	25	15		0.34		336		
	C IV and V	16		1.5	24–26	24		1.51		1008		
	C III	44		0.6	25–26	15–24	0.3–1.0	0.67		1108		
	C II	24		0.3	24	24		0.24		800		
	C I and II	25		0.6	25	17		0.27		446		
	N V and VI	24		0.1	24	24		0.10		1056		
Calanus finmarchicus	♀	386 (June–Mar.)		(176)	10	48	6.0–13.9	7.6		(43)		
	♀	191 (Apr.–May)		(242)	10	48	10.3–17.8	15.1		(62)		
	♂	213		(203)	10	48	6.8–12.4	10.4		(27)		
	C V	519		(240)	10	48	2.4–9.3	5.5		(23)		
	C IV	73			10	19–48	1.4–4.0	3.1				
	C III	70			10	19–48	0.9–1.9	1.4			Firth of Clyde	Marshall and Orr (1958)
	C II	36			10	19–48	0.5–1.3	0.9				
	C I	4			10	19–48	0.3–0.9	0.6				
	N VI	4			10	19–48	0.7–0.8					
	N V	2			10	19–48	0.2–1.9					
	N IV	4			10	19–48	0.2–0.8					
	N III	28			10	19–48	0.19	0.19				
	N II and II and III	874			10	19–48	0.07–0.09	0.08				
	N I–II	720			10	19–48	0.05	0.05				

Dry weights of *Acartia* calculated, according to Petipa, as 16% wet weight.
Calanus dry weights in brackets averaged from 112 samples taken throughout the years 1933 and 1961–64. The averages of the two sets of samples agreed well.

Table 6.21. Respiration and food requirements of female copepods (Marshall and Orr, 1962)

Species	Size (mm)	No. used	Month	Temp. (°C)	Respiratory rate (μl O_2/cop./day)	Respiratory rate (μl O_2/mg dry wt./day)	food needed/day (mg) carbo-hydrate	food needed/day (mg) lipid	Reference
Bathycalanus sp.	13	1	VIII	5.1	163.4	9.1	217.3	80.5	4
Calanus hyperboreus	(5.5–6)	6	VIII	<10	25.0	5.8	33.3	12.3	4
Pareuchaeta	5.1	11	VIII	<10	49.5	12.8	65.9	24.4	4
P. norvegica		10	IX	17	114.8		152.7	56.6	8
Rhincalanus nasutus	4.2–4.7	6	IV	6.5	14.4	14.9	19.2	7.1	4
Euchirella rostrata	2.9–3.1	9	IV	6.5	25.5	29.0	33.9	12.6	4
Pleuromamma robusta	(2.7)	6	VIII	8	17.3	60.4	23.0	8.5	4
Calanus		19	VIII	7.5	9.1	42.4	12.1	4.5	4
C. finmarchicus		75	VIII–IX	10	12	—	16.0	5.9	2
	2.3–2.8	783	I–XI	10	2.7–17.7	—	3.6–23.6	1.3–8.7	5
C. helgolandicus	2.6	192	X–IV	10	4.6–11.5	—	6.1–15.3	2.3–5.7	5
Calanus? f or *h*	2.5	44	VIII–IX	17	20.6	—	27.4	10.2	8
Metridia lucens	1.8	76	IV–VI	10	4.7–7.1	—	6.3–9.5	2.3–3.5	6
Centropages	0.7–1.5	269	IV–VII	10	1.7–3.4	—	2.3–4.5	0.8–1.7	6
C. hamatus	1.1	17	IX	17	3.0	—	4.0	1.5	8
	1.0	Many	V–VI	20	2.2–3.0	142–155	2.9–4.0	1.1–1.5	3
C. typicus	1.3	209	IX	17	4.3–6.9	—	5.7–9.2	2.1–3.4	8
Temora longicornis	0.5–1.3	468	III–VII	10	1.0–2.5	—	1.3–3.3	0.5–1.2	6
	0.7	137	V–IX	10	0.2–1.2	—	0.3–1.6	0.1–0.6	1
	0.8–1.1	Many	II–VI	20	2.4–4.7	190–328	3.2–6.3	1.2–2.3	3
Pseudocalanus	0.8–1.4	387	III–VI	10	0.7–2.2		1.0–2.9	0.4–1.1	6
P. minutus	0.8–0.9	Many	I–III	20	1.3–1.9	196–257	1.7–2.5	0.6–0.9	3
Acartia clausi	0.7–0.8	523	I–VII	5.6	0.2–0.7	43–96	0.3–0.9	0.1–0.3	3
	0.8–1.3	326	III–V	10	0.8–1.4		1.1–1.9	0.4–0.7	6
	0.8–1.0	Many	I–XII	20	0.7–2.2	138–304	0.9–2.9	0.3–1.1	3
A. tonsa	0.9–1.1	Many	I–XII	20	1.1–2.6	181–318	1.5–3.5	0.5–1.3	3
A. bifilosa	0.8–0.9	Many	II–V	20	1.5–2.8	319–394	2.0–3.7	0.7–1.4	3
A. discaudata	0.7–0.8	Many	XI–III	20	0.8–1.6	207–436	1.1–2.1	0.4–0.8	3
Eurytemora herdmani	0.9	248	IV–V	15	2.1	—	2.8	1.0	7
Oithona similis	0.5–0.7	157	V–VII	10	0.2–0.8	—	0.3–1.1	0.1–0.4	6

In the column for size, figures in brackets denote sizes from Sars, 1903.

1. Berner, unpublished.
2. Clarke and Bonnet, 1939.
3. Conover, 1959.
4. Conover, 1960.
5. Marshall and Orr, 1958.
6. Marshall and Orr, unpublished.
7. Raymont, 1959.
8. Raymont and Gauld, 1951.

Table 6.22. Respiratory rates and theoretical amounts of lipid metabolised by five copepod species and one euphausiid (Corner and Cowey, 1968)

Species	Dry body wt mg	Oxygen uptake (μl O_2/mg dry wt/day)	Lipid used daily (%)
Acartia clausi	0.0045	69	57.5
Metridia longa	0.25	35	10.5
Calanus finmarchicus	0.40	25	4.9
C. hyperboreus	2.30	12.5	1.7
Euchaeta norvegica	3.0	17.5	2.8
Euphausia pacifica	5.4	35	17.4

Usually the R.Q. is unknown and is assumed as lying between 0.7 (lipid) and 1.0 (carbohydrate). There is an extensive literature on calculated maintenance requirements of zooplankton. Table 6.21 gives an indication from some earlier data on copepod respiration of maintenance requirements, showing the fall with decreasing body size of copepod. If, however, the amount is related to the body weight of copepod and calculated on a percentage basis, the high demands for small species becomes clear. Table 6.22 illustrates this principle for five copepod species. Similarly younger developmental stages will tend to have greater daily demands on a unit weight basis. Corkett and McLaren calculate 7% and 12% for *Pseudocalanus* females and nauplii, averaging some ten-fold difference in body weight. Some of the differences in developmental stages are, however, much more striking (*vide infra*). Some values given by Petipa (1966) for the daily demands of *Acartia* may seem exceptionally large, but are partly due to the comparatively high environmental temperature in which the copepods live.

Chapter 7
Metabolism and Biochemical Composition

The Metabolism and Excretion of Zooplankton

Information on diet and the rates of feeding of zooplankton together with data on oxygen uptake have permitted some calculation of the food required by a species to fulfil its metabolic needs. However, such calculations must be related to the type of food utilized. Clearly there is the need to study metabolic pathways in zooplankton and to determine variations in the substrates utilized.

Much of the earlier work on metabolic studies in zooplankton has been concerned with excretion of phosphorus and nitrogenous products, though the interpretations of these experiments were necessarily limited due to inadequate knowledge of the relative nutritional status (detailed biochemical composition) of the animals used. These earlier studies indicated that there was a fairly rapid turnover of phosphorus; thus Gardiner (1937) showed that a planktonic animal could excrete significant quantities of phosphate in a matter of a few hours. Later investigations, including those of Conover, Marshall and Orr (1959) and Cushing suggested that there was a considerable turnover of soluble phosphorus when zooplankton grazed actively. Harris (1959) drew attention to the great turnover of phophorus by copepods such as *Acartia*. Approximately 12 μg of phosphorus were excreted daily per mg dry wt by this copepod, and as Corner and Cowey (1968) have pointed out, this approaches 150 % of the body phosphorus per day. Other zooplankton may exhibit relatively high rates of phosphorus excretion. For example, the chaetognath, *Sagitta hispida*, excretes nearly 40 % of its total body phosphorus per day. Pomeroy, Matthews and Min (1963) suggest that approximately 100 % of total body phosphorus is released per day by warm water zooplankton. The amount of phosphorus excreted by planktonic animals may be in part related to body size. Thus Corner and Cowey (1968) quote the small ciliate, *Euplotes*, as excreting the equivalent of its body phosphorus in 30 minutes, as compared with a larger animal such as *Calanus finmarchicus* which excretes its body phosphorus in approximately 20 hours.

It is generally now recognised, however, that not all the phosphorus released by a planktonic animal is truly excreted because not all this phosphorus has been incorporated into the tissues. Phosphorus released either as phosphate, or as organic material easily hydrolysed by phosphatases, or indeed as more stable organic phosphorus compounds, may have been released during the passage of food through the alimentary canal and thus have arisen from non-assimilated food; true excretion concerns metabolic waste alone. A substantial part of the phosphorus released by zooplankton can arise from non-metabolic sources. Thus Johannes (1964), distinguishing between truly excreted and merely liberated phosphorus, emphasizes that the amount of time necessary for a planktonic animal to excrete an amount of phosphorus equivalent to its whole body content is far less on an ecological basis—i.e. relating to total phosphorus liberated irrespective of source, than on a physiological basis—i.e. relating only to excretion of metabolized phosphorus.

There is no doubt, however, that phosphorus is exceptionally labile in many planktonic organisms. Earlier studies by Marshall and Orr (1955) using ^{32}P, showed that a considerable amount of ingested phosphorus was distributed relatively rapidly to body tissues. In females, for example, a high proportion passed to the ovary or was liberated in the eggs of actively reproducing females. In Stage V copepodites a large proportion passed to the fat. In contrast, males had a surprisingly small amount in the reproductive system but a fair quantity in the fat and slightly more in the musculature. Later work by Marshall and Orr (1961), in conjunction with Conover, attempted to investigate the stability of phosphorus in *Calanus* using ^{32}P fed over various time periods. A period of feeding of at least a week with labelled phosphorus (half-life 14 days) is essential to establish equilibrium in the body, but following this period the total body phosphorus would be turned over in some 20 days. This presumably represents the rate of true metabolic excretion. Excretion of phosphorus was more rapid in fed than in starved copepods, but a remarkably high amount (up to 55 % or more) of total body phosphorus may be lost in starving *Calanus*, apparently without injurious effects. Female *Calanus* may transfer a large proportion (up to 70 %) of the phosphorus assimilated as food to the eggs within a week.

Conover suggested that there might be two pools of phosphorus-containing substances in the copepod body, a labile pool with a half-life of less than half a day, and a stable pool with a half-life of about 13–14 days, as found by Marshall and Orr. The stable pool would amount to at least 94 % of the total phosphorus. It is impossible at this stage to identify the labile pool since phosphorus is present in many endogenous compounds in the body such as phospholipids, phosphoproteins, sugar-phosphates and nucleotides.

Although the high rates of phosphorus excretion originally described were artificially exaggerated by the release of non-metabolized phosphorus, zooplankton undoubtedly has a relatively high rate of metabolic activity. Hargrave and Geen (1968) investigated phosphorus excretion by zooplankton, mostly *Temora* and *Pseudocalanus*, from an isolated lake in Nova Scotia. They found that about two-thirds of the dissolved phosphorus excreted by adult and Stage V copepodite *Pseudocalanus* was in organic form. Mixtures of copepodites of *Temora* and *Pseudocalanus* tested in filtered and natural seawater showed substantially larger amounts of dissolved inorganic phosphorus (presumably phosphate) excreted when the copepods were living in unfiltered water. The amount of dissolved organic and inorganic phosphorus was also reduced when excretion by *Pseudocalanus* was tested in water passed through filters of 3–5 μm porosity. Presumably the lack of feeding in filtered water reduced excretion, though complications are introduced by the effects of faecal pellet production during experiments. However, Hargrave's earlier experiments with non-feeding copepods suggested that 7.7 % of the total body phosphorus might be excreted daily.

Butler, Corner and Marshall (1969) estimated that small zooplankton (mostly *Pseudocalanus*) in filtered seawater excreted 0.77 μg P/mg dry wt/day, equivalent to approximately 11 % of the body phosphorus. Corkett and McLaren (1978) indicate that apart from experimental artefacts which may be attributed to crowding and to length of experiment, feeding, temperature and body size would appear to be the most significant factors affecting phosphorus excretion in copepods. Butler *et al.* (1969) studied P and N excretion in *Calanus* removed from the sea, and kept in Millipore-filtered seawater so that leaching of nutrients from faecal pellets and possible uptake of nutrients due to the presence of algae was avoided. Animals were sorted rapidly and the duration of experiments limited to four hours so that, in the opinion of the investigators, there was no significant starvation factor. The copepods were, therefore, regarded as excreting at essentially the same rate as when feeding naturally in the sea. Since there was no defaecation during the experiment, any phosphate excreted would be truly metabolized phosphorus.

During April 1968, when phytoplankton food was abundant, female *Calanus* and Stage V copepodites lost approximately 20% and 27%, respectively, of body phosphorus daily; males lost only 15% per day. The excretion rate dropped after spring, and in late October when little phytoplankton was present the daily loss of phosphorus amounted to 10% of body P for females and 7% for Copepodite V.

Later work over a more extended period (Butler *et al.*, 1970) demonstrated that excretion of phosphorus, whether expressed as μg/mg dry wt/day or as percentage body loss, showed a seasonal pattern following the changing levels of plant food expressed as chlorophyll *a* (chlor. *a*). The highest excretion point was in spring (April); excretion was lower in summer, but another high level occurred in August/September, with excretion declining to its lowest level in January/February. Seasonally high food levels appear to induce high excretory rates in *Calanus*. In winter, when little natural food is present, freshly caught or starving animals showed a low rate of excretion. In spring, freshly caught *Calanus* excreted phosphate at a much higher rate than starved specimens. Although the temperatures of experiments varied between 6° and 18°C according to the natural environmental change, the Q_{10} value for excretion over that range was almost exactly 2. In summer, most (*ca.* 80%) of the excreted phosphorus appeared as phosphate, both for starved and fed animals, but during the spring diatom bloom a very high proportion (>70%) of phosphorus excreted was in organic form.

The high metabolic level of zooplankton, as earlier suggested by a study of phosphorus excretion in, mainly *Calanus*, is generally confirmed by examining nitrogen excretion. One of the earliest investigations, by Harris (1959), dealt with nitrogenous excretion by the zooplankton (mainly *Acartia*) of Long Island Sound. Excretion, essentially of ammonia, averaged 36.4 μg N/mg dry wt/day. If this is added to the nitrogen demands for growth, the total requirement for nitrogen apparently exceeded that available from phytoplankton. The daily excretion of nitrogen was equivalent to about 45% of the body protein—*Acartia* possessing 89 μg N/mg dry wt, of which some 90% may be assumed to be protein.

Webb and Johannes (1967) also investigated nitrogen excretion in a variety of zooplankton groups including microzooplankton. Though the values obtained, ranging from 2.4 to 30 μg N/mg dry wt/day, were lower they were still relatively large. The observations of Beers (1964) for *Sagitta Hispida* suggested that about 12.6 μg/mg/day were excreted, corresponding to about 14.5% of the total body nitrogen/day; Reeve, Raymont and Raymont (1970) obtained a similar figure (16 μg N/mg/day) for the same species. The more detailed studies by Corner, Cowey and Marshall (1965, 1967) and Corner and Cowey (1968) on nitrogen excretion of *Calanus* indicated that female *Calanus*, under the same temperature and food conditions used by Harris for *Acartia*, excreted some 10–16 μg N/mg dry wt/day, a value considerably lower than that for Acartia. Earlier observations had suggested that excretion was temperature dependent and probably also was affected by feeding. The experiments of Corner *et al.* confirmed that nitrogen excretion varied with a number of factors such as food—probably both quality and quantity—temperature and body size. A rise in temperature from 5° to 15°C almost doubled nitrogenous excretion in *Calanus*, though only a small increase in excretion occurred with a further rise of 10°C (i.e. to 25°C). Butler *et al.* (1970) tested the effect of temperature on nitrogen excretion in *Calanus* over the range 6°–19°C; the Q_{10} was close to two.

Body size is a major factor in nitrogenous excretion. Under constant temperature Corner *et al.* found that adult female *Calanus* excreted approximately 10 μg N/mg dry wt/day; the slightly smaller Stage V lost relatively less (7 μg/mg/day) which may reflect a somewhat lower metabolic rate; but a mixture of younger copepodites (Stages II–IV) excreted 21 μg N/mg/day, and younger copepodites and nauplii lost relatively even more—39 μg N/mg/day. This value may be compared with the excretion rate for the small copepod, *Acartia*. Corner and Davies (1971) calculated, from Harris's data, a rate equivalent to 43 μg N excreted/day (equivalent to 48% of the body nitrogen) for *Acartia*, a value close to that for younger stages of *Calanus*. Corner suggested that nitrogen excretion per unit weight is roughly halved for a four- or five-fold increase on body size (weight).

Corkett and McLaren (1978) reported on experiments by Christiansen carried out on nitrogenous excretion of *Pseudocalanus* in the same lake in Nova Scotia where Hargrave and Geen studied phosphorus excretion. They also stressed the complications introduced in experiments where faecal pellets are released by feeding of the copepods, the problems of crowding and effects of long term experiments. For example, in one experiment with mixed copepodites in unfiltered water, excretory rates were five-fold higher at 2 hr from the start of the experiment than they were 12 hr later. They attribute much of the decline in rate in longer term experiments to food depletion (but cf. Ikeda, 1977, *vide infra*). Certainly Christiansen demonstrated, with Copepodite V *Pseudocalanus*, that excretion rose from 0.046 to 0.068 μg NH_3-N/copepodite/hr for food levels increasing from 39 to 93 μg N/l, respectively.

Body size was also confirmed as an important factor by Christiansen. For several species of copepod and for individual developmental stages of *Pseudocalanus* the weight specific excretory rate fell with increasing body size, a similar pattern to respiratory rate (cf. Fig. 7.1). Although Corkett and McLaren suggest that the excretory rates of the nauplius stages may have been exaggerated and point to a lower rate calculated from an equation relating body size and excretion by Ikeda, the effect on excretion of size—both inter- and intra-specifically—is clear.

Christiansen suggests a considerably less marked effect of temperature rise on excretion with mixed populations of *Pseudocalanus* and *Temora* ($Q_{10} = 1.37$) than Butler *et al.* obtained for *Calanus*. Both low oxygen content and reduced salinity caused a clear rise in excretory rate. *Acartia tonsa*, in contrast, showed little effect of reduced salinity on excretion, but *Pseudocalanus* may be able to acclimatize to environments of lowered salinity and oxygen content. Corkett and McLaren calculate that an adult *Pseudocalanus*, non-feeding, might excrete around 5% of its body nitrogen daily. Taguchi and Ishii (1972) observed somewhat lower excretory rates for

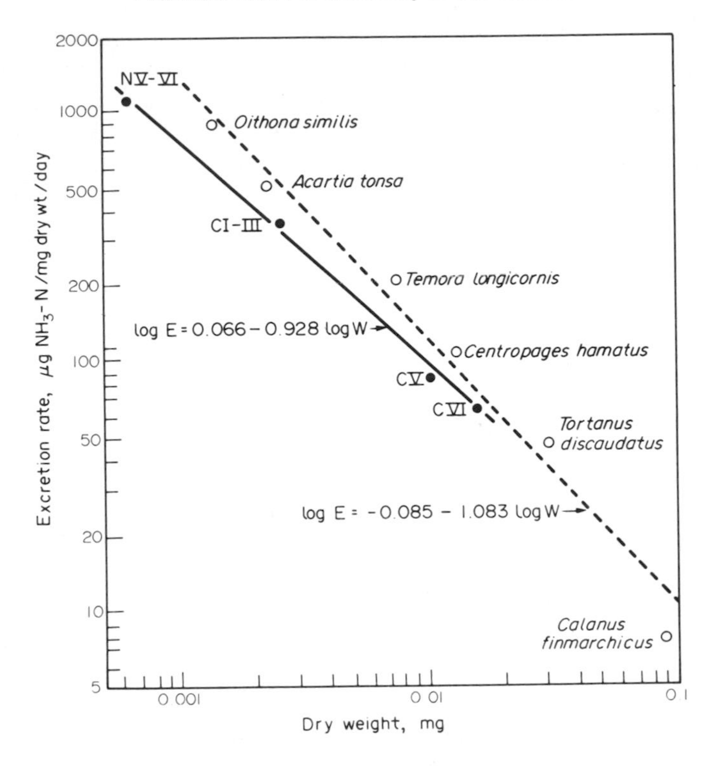

Fig. 7.1. Weight specific excretion rate for adults of different species of copepods (open circles, broken line) and for different stages of *Pseudocalanus* (closed circles, unbroken lines) (Corkett and McLaren, 1978).

nitrogen and phosphorus for *Calanus cristatus* and *Calanus plumchrus* as compared with more neritic zooplankton, when rates were calculated on a dry weight basis. The rate for *C. plumchrus* was greater than that for *C. cristatus*, presumably a reflection, in part, of body size.

Although excretion in copepods has been the most studied, Jawed (1969) investigated excretion in a mysid, *Neomysis rayii*, and the euphausiid *Euphausia pacifica*. He obtained rates of approximately 3.0 and 2.5 μg N/mg/ day respectively, for the two species (at 10°C). Rough calculation suggests that these species lose approximately 2.5 and 2.1 % of the body nitrogen per day. Ikeda (1974) gives a rate of *ca.* 5.3 μg/mg/day. Ferguson (1972) studied nitrogen excretion in *Neomysis integer* at 15°C and in 25 % seawater (i.e. conditions approximating to their environmental conditions near Southampton) and found that excretion rates varied seasonally. Thus in February/March the rate was approximately 11 μg N/mg dry wt/day; this was reduced by slightly more than 50 % in September/October. The excretion rate for February/March however, amounts to some 10 % of the body nitrogen per day, a value considerably higher than for *Neomysis rayii* though lower than for some small copepods. Table 7.1 gives an approximate relationship between nitrogen loss due to excretion and body size, and shows the reduction in loss with increasing body size even with crustaceans of different taxonomic groups.

Mayzaud's data (1973) confirm the relationship which has been advanced for

Table 7.1. Relationship between nitrogenous excretion and body size

Animal	Rate of excretion (% body N/day)
Acartia	40
Calanus	
(nauplii & young copepodites)	38
,, C II–IV	20
,, C VI ♀	8
Euphausia pacifica	2
Neomysis rayii	2.5
Neomysis integer	9

nitrogenous excretion rate and size. For the two larger zooplankton crustaceans *Meganyctiphanes* and *Phronima* this investigator obtained excretion rates of 8.0 and 5.5 μg N/mg dry wt/day, as compared with approximately 52 and 51 μg N/mg dry wt/day for the two smaller zooplankton species *Acartia* and *Sagitta setosa*. In other experiments on nitrogenous excretion, Mayzaud and Dallot (1973) examined the effect of temperature on *Phronima*. A substantial proportion of the nitrogen excreted was in organic form, but the expected rise in excretory rate was seen with temperature increasing from 5° to 19.5°C. Total nitrogen excreted rose from 5.1 to 12.5 μg N/mg dry wt/day; ammonia excreted rose from 2.4 to 9.6 μg N/mg dry wt/day. The Q_{10} values, both for respiratory rate and excretion, were greater in the lower temperature range (5°–10.5°C) than between 10.5° and 19.5°C. For *Sagitta setosa*, where also a considerable amount of organic nitrogen was lost, total nitrogen excreted increased with temperature (from 15° to 20°C) from 5.6 to 8.4 μg N/animal/day, but in terms of dry weight of animal this was four- or five-fold greater than rates determined by Beers (1964) and Reeve *et al.* (1970) for the neritic species, *Sagitta hispida*. Whether the greater rate and amount of organic nitrogen lost is associated with the animals having been starved previously is uncertain. The level of oxygen uptake appeared to be very low.

Of interest is the relationship between body size and loss of nitrogen during starvation in zooplankton. Corner and his colleagues found that loss of nitrogen due to starvation in *Calanus* could account for 2–4 % of the body nitrogen per day (cf. Corner and Davies, 1971). Mayzaud's experiments suggested that whereas for the two larger zooplankton forms, *Meganyctiphanes* and *Phronima*, the loss by starvation was not very different (8 % body protein/day), for the small species, *Acartia*, 68 % of the body protein was lost every day during starvation (Table 7.2). Ikeda (1977) has, however, referred to some complicating factors which may affect experiments on excretory rates. In his studies on the effects of feeding and starving on the respiration of several species of zooplankton collected in waters off British Columbia, excretory rates were also measured. Though with herbivorous species there was a decline in both nitrogenous and phosphorus excretion for the first few days of the experiment, fed animals had a higher excretory rate than starved specimens. With fed *Calanus plumchrus* a remarkable recovery of nitrogenous excretion was observed after about two weeks. The level of phosphorus excretion with feeding was relatively constant after the initial period, but in starved *C. plumchrus* phosphorus excretion was undetectable after

Table 7.2. Relationship between body size and protein loss during starvation (Mayzaud, 1973)

Animal	Rate of loss (% body protein/day)
Meganyctiphanes	8
Phronima	8
Acartia	68

twelve days. The decline in rate of both types of excretion was more rapid for starved specimens of *Paracalanus* than for *Calanus*. Fed *Paracalanus* had much higher rates than starved animals.

In *Euphausia*, the period of survival was again extended, though not as much as for *Calanus*. Phosphorus excretion was lower for starved animals, but there was very considerable scatter in the data for nitrogen excretion, and excretion in starved animals was not obviously reduced. With the carnivorous form, *Pleurobrachia*, there was a clear pattern of increased nitrogenous and phosphorus excretion with feeding, and a marked decline with starvation, without the complication of initial fluctuations. Ikeda believes that the variations in the excretory rates in the other animals cannot be attributed solely to the effect of feeding and starvation, though this is a very significant factor. The capturing and handling of animals and especially the repeated testing practised by Ikeda, with its physiological stress, appears to have more complex effects on excretion than on respiration. Another factor is the switching from one substrate to another when animals are transferred from their natural environment and fed in the laboratory.

Data for nitrogenous excretion in zooplankton are still relatively limited. Probably the most widespread investigations of ammonia excretion are the earlier studies of Ikeda (1974), involving 81 species including seven taxa of non-crustacean and seven of crustacean plankton, drawn from boreal, temperate, sub-tropical and tropical environments. As with oxygen uptake measurements, ammonia excretion showed considerable variation, and there was generally a decline in the rates with continued maintenance in the laboratory (cf. Fig. 7.2). A measure of standardization was achieved by maintaining animals after capture for one day at approximately environmental temperature before experimentation, but Ikeda's (1977) critical comments on the particular problems of laboratory experiments on excretion, discussed earlier, must not be forgotten.

Body size (weight) was strongly correlated with excretion rate, as suggested by earlier investigators, though Ikeda's studies were based on a much greater variety of species. A relationship of the same form as for O_2-uptake ($\log E = b \log W + \log a$) was calculated for each environmental (habitat) temperature (Table 7.3). The effect of habitat temperature on the relationship between the rate of ammonia–nitrogen excretion and body weight was similar to that for respiration. Weight specific excretion rate of ammonia–nitrogen increased with decrease in body weight and increase of habitat temperature (cf. Table 7.3 and Fig. 7.3). In dealing with respiration Ikeda noted that euphausiids, Amphipoda and fish larvae belonged to a higher rate group, and Cnidaria, Ctenophora, Tunicata and some benthos larvae formed a group with lower than average rates. This same tendency appears for ammonia excretion.

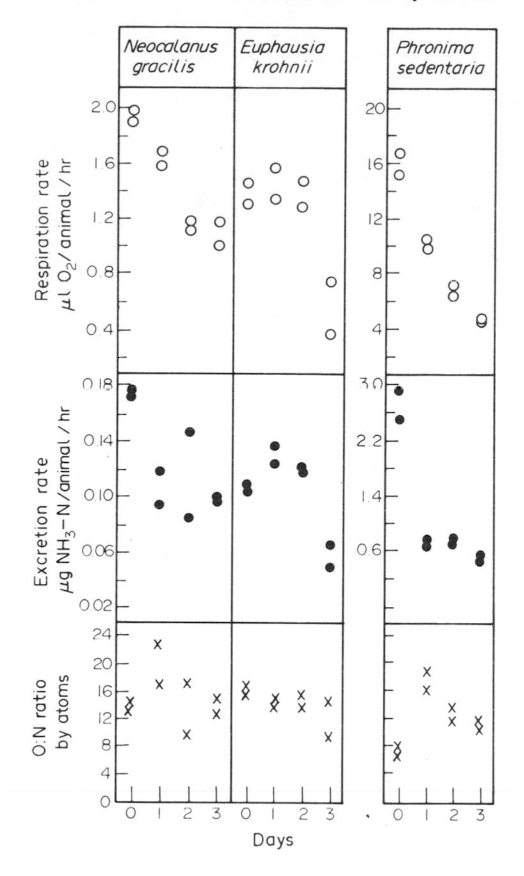

Fig. 7.2. Changes in rates of respiration and ammonia excretion and the O : N ratio of zooplankton during maintenance in the laboratory. Experiments conducted at 19.7°C (Ikeda, 1974).

Table 7.3. Relationship between total ammonia–nitrogen excretion rate (E: μg NH_3 N/animal/hr) and body dry weight (W: mg/animal) in zooplankton (Ikeda, 1974)

Zooplankton	Mean temperature and range (°C)	Equation of regression line
Boreal species	8.2 (4.5–14.3)	$\log E = 0.790 \log W - 1.198$
Temperate species	15.0 (13.8–16.0)	$\log E = 0.635 \log W - 0.857$
Subtropical species	20.2 (17.3–22.5)	$\log E = 0.654 \log W - 0.767$
Tropical species	27.3 (25.7–28.5)	$\log E = 0.591 \log W - 0.639$

These findings, however, should be compared with those of Biggs (1977), who determined ammonia excretion rates on gelatinous zooplankton which were collected by Scuba diving. Rates of nitrogenous (ammonium) excretion range from 0.1 to 3.3 μg NH_4^+/mg protein/hr, with the smaller animals showing higher rates per unit protein than larger forms. Differences in the metabolic activity of the various taxa of gelatinous zooplankton, however, also affected excretory rates in a comparable

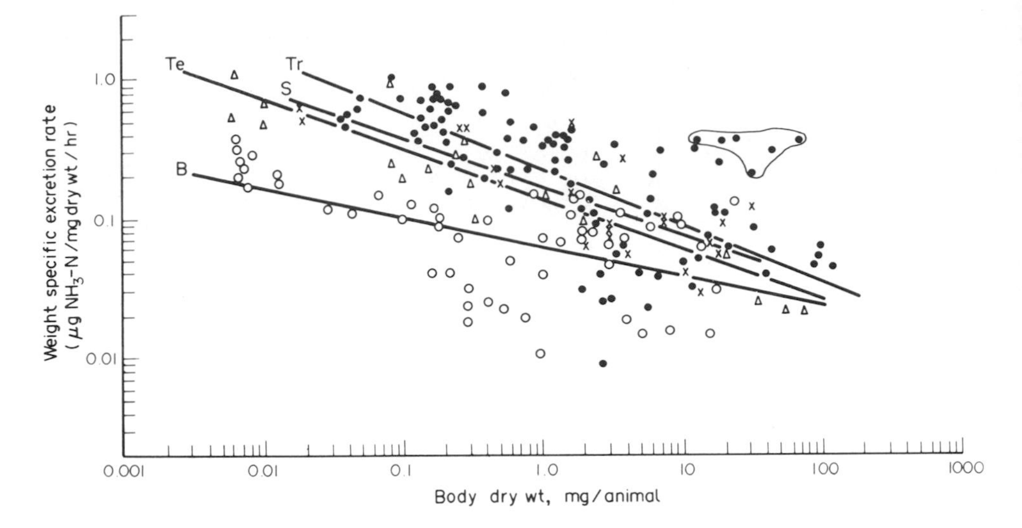

Fig. 7.3. Relationship between weight specific excretion rate of ammonia-nitrogen and body dry weight for tropical, sub-tropical temperate and boreal zooplankton. Points encircled denote fish larvae. ● = Tropical sp. (Tr); x = Sub-tropical sp. (S); Δ = Temperate water sp. (Te); ○ = boreal sp. (B) (Ikeda, 1974).

manner to the influence on the level of oxygen consumption previously noted (*vide supra*). For example, cystonect siphonophores showed lower rates of ammonium excretion than physonect siphonophores, and within the Physonectae rates were somewhat lower in species such as *Agalma rosacea* and *A. okeni* which are rather inactive and slow swimming. Thaliaceans, especially the smaller forms, had higher than average excretory rates and the less active cydippid ctenophores showed lower levels of nitrogenous excretion than actively swimming lobate forms such as *Ocyropsis maculata*. The high excretory rates in *Ocyropsis* and in the heteropod *Pterotrachea hippocampus* (2.9 and 3.0 $\mu g\ NH_4^+$/mg protein/h, respectively) caused the oxygen/nitrogen ratios to be low, whereas the very low rates of ammonium excretion in the pseudothecosomes *Corolla spectabilis* and *Gleba cordata* were reflected in very high ratios (89 and 54, respectively).

Although it is difficult to select examples from the great volume of Ikeda's results, especially since body size and temperature—both factors affecting metabolism—are variable, there are notably low values for ammonia excretion for *Beroe* (0.023 μg N/mg/hr), *Salpa* (0.10 μg N/mg/hr) and *Liriope* (0.031 μg N/mg/hr), especially since the last two were measured at relatively high temperatures (*ca.* 27°C) (Table 7.4). However, the rates found by Biggs (1977) for *Salpa maxima* and *S. cylindrica* at approximately the same temperature, are very much higher.

Ikeda's data indicate that moderately large boreal copepods could show relatively small ammonia excretion rates at their "normal" environmental temperatures if these are low. Thus *Calanus plumchrus* IV and V had rates of 0.028 and 0.031 μg N/mg/hr respectively at temperatures of 4.5°–7.0°C, while *Eucalanus bungii* ♀ excreted 0.057 μg N/mg/hr at approximately the same environmental temperatures. But a summer population of *Calanus plumchrus* V, of almost half the body size and maintained at a higher temperature (14°–16°C), gave a much enhanced excretion rate—0.23 μg N/mg/hr.

Table 7.4. Excretion rates of smaller-sized copepods (Ikeda, 1974)

Animal	Experimental temperature (°C)	Excretion (μg NH_3-N/mg dry wt/hr)
Acartia clausi	13.8–14.5	0.82
	15.0–15.7	0.58
Calanus minor V♀	26.6–27.1	0.53
		0.56
		0.47
Undinula vulgaris ♀	26.4	0.50
	27.8	0.69
	28.5	0.76
Labidocera acuta	28.5	0.65
Eucalanus attenuatus	27.3–27.4	0.43
	27.8	0.39

Smaller copepods (*Pseudocalanus*, *Acartia* spp.) show much higher weight specific excretion rates, especially with increased temperature (e.g. *Acartia clausi*, *ca.* 0.70 μg N/mg/hr at 14°–16°C). With further rise in environmental temperature typical herbivorous or carnivorous tropical copepods, mostly of medium size (*Labidocera acuta*, *Calanus minor*, *Undinula vulgaris*, *Eucalanus attenuatus*), show weight specific excretion rates of the same high order (0.39–0.76 μg N/mg/hr) (Table 7.4).

Apart from copepods *Phronima* of relatively large size, though tested at *ca.* 20°C, had only a relatively small ammonia output of < 0.1 μg N/mg/hr—a value distinctly lower than Mayzaud's findings (*ca.* 0.23 μg N/mg/hr). At 27°C, however excretion was trebled according to Ikeda. With relatively cold water species of smaller size (*Parathemisto japonica*) the excretion rate was 0.066 μg N/mg/hr at *ca.* 7°C.

The excretion rate recorded by Jawed (1969) for *Euphausia pacifica* (*ca.* 0.10 μg N/mg/hr) is not far different from Ikeda's (1974) value of 0.067 μg N/mg/hr at 10°C. But two tropical euphausiids show the expected increase in excretion rate with temperature—*Euphausia diomedeae*: 0.28 μg N/mg/hr and *E. tenera*: 0.42 μg N/mg/hr, at *ca.* 27°C. Two chaetognaths both show relatively high rates but whereas *Sagitta elegans* (a boreal species—5.5°–7.0°C) gives a rate of 0.12 μg N/mg/hr, *Sagitta serratodentata*, a warm water form, has the raised rate of 0.65 μg N/mg/hr at 27°C. Reeve *et al.* (1970) quote an almost identical rate for the warm water species *S. hispida*—0.67 μg N/mg/hr.

Despite experimental difficulties with the wide range of animals studied by Ikeda (1974) and the problem of taxonomic differences, certain generalizations are possible. While the direct effect of environmental temperature on excretion is clear, the Q_{10} tends to be higher for boreal than for warm water species. A decrease of the Q_{10} follows increase in body size. For animals of about 0.1 mg dry wt, irrespective of habitat, the mean Q_{10} is 2.4; this falls to 1.6 for animals of about 10 mg dry wt.

As Corner and Newell first showed, nitrogenous excretion in *Calanus* is predominantly in the form of ammonia, but whether excretion in most zooplankton is ammonotelic has been a matter of debate. For *Sagitta hispida*, Beers found approximately 70% of nitrogen excreted was ammonia, and Jawed showed an even higher percentage of ammonia for *Neomysis rayii* and *Euphausia pacifica*. Some 70–90% of the total nitrogen was excreted as ammonia, with traces of urea, by *E. pacifica* and small amounts of other forms of nitrogen were excreted by both species. Webb and Johannes (1967), however, claimed that a relatively large amount of nitrogen was excreted as amino acids. Mayzaud and Dallot (1973) reported that, in addition to ammonia, substantial quantities of organic nitrogen were excreted by two salps (*Thalia democratica*, *Salpa fusiformis*), and by *Phronima sedentaria* and *Sagitta setosa*; in the latter two species organic nitrogen amounted to nearly 50% and 30%, respectively. The condition of the plankton and occurrence of bacteria have been cited as factors affecting the proportion of organic nitrogen and ammonia. It seems likely that for most zooplankton excretion is predominantly ammonotelic, but that small amounts of amino acids and urea can be liberated by some species. From an examination of excretion in four zooplankton species, *Meganyctiphanes*, *Phronima*, *Acartia* and *Sagitta setosa*, Mayzaud (1973) found that the nitrogen excreted was ammonia but that some organic nitrogen was released which could be degraded to ammonia by bacteria.

While very few data are available on excretion in ctenophores, Reeve and Walter (1978) quote results obtained by Kremer on *Mnemiopsis* which are unusual for zooplankton in that a large fraction (46%) of the nitrogen was excreted in dissolved organic form. The remainder was made up of ammonia; urea was insignificant. Reeve also reports O/N ratios for both *Mnemiopsis* and *Pleurobrachia* of approximately 15. A lower value might have been expected for such active carnivores (*vide infra*). Eppley, Renger, Venrick and Mullin (1973) found that urea accounted for up to 50% of total nitrogen excreted by mixed zooplankton from the North Pacific. Corner, Head, Kilvington and Pennycuick (1976) examined nitrogen excretion in *Calanus* feeding carnivorously. Although the total excretion of nitrogen rose with increase in food ration (from 7.7 to 11.5% body N/day for a seven-fold increase in ration), the proportion of urea did not change significantly—the percentage amounting to 9 or 10% of total N. Possibly some zooplankton species are more predominantly urea-excreting, or the conditions of Eppley's experiment (temperature 20°C) may have favoured urea excretion.

The investigations of Butler *et al.* (1969, 1970), and Corner *et al.* (1974) referred to previously have demonstrated that excretion of nitrogen, as of phosphorus, by *Calanus finmarchicus* and *C. helgolandicus* is considerably influenced by season. Mayzaud (1973) also drew attention to a marked change in nitrogen excretion with season in *Meganyctiphanes*. He associated a decline in spring with use of lipid as substrate. The studies on *Calanus* suggested that nitrogen excretion was rapid in spring when phytoplankton food was plentiful. In April, 1968, for example, female *Calanus* excreted approximately 13% of body nitrogen daily, while the value for Stage V and males was 15% and 11%, respectively. Excretion fell during the summer and by late October, when little phytoplankton was available, had dropped to a daily loss of body nitrogen of 3.5% for females and nearly 4% for Stage V. The proportion of nitrogen lost, expressed as percentage of body nitrogen, was always less than the proportion of phosphorus excreted but the total *amount* of nitrogen was considerably greater. The N/P ratio, which was generally comparatively constant through the year, approximated to 5. A fuller investigation in the subsequent year confirmed the strong seasonal influence on nitrogen excretion. As with phosphorus, whether loss of nitrogen was expressed as percentage of body nitrogen or as weight lost per day, the seasonal pattern followed the variations in plant food available. A maximum excretory rate was found in spring and somewhat lower levels in summer, but with high levels coinciding with peaks of phytoplankton, as during August/September. A decline to a minimum occurred over mid-winter (January/February).

The change in the amount of nitrogen excreted with season is considerable; maximal values, in April, approached 12–13 μg N/mg dry wt/day; minimal winter values were of the order of 2 μg N/mg dry wt/day for female *Calanus*. Butler *et al.* confirmed that females had a higher metabolic rate than males and excreted more nitrogen. There was also a small but significantly higher rate for females than for Stage V. The positive effect of abundant food in the natural environment on excretory rate is evident from experiments on starvation. In winter both freshly caught and experimentally starved *Calanus* showed a low excretory rate, whereas in spring freshly caught animals had a much enhanced rate over starved specimens. The Q_{10} for excretion between 6° and 18°C was approximately 2, although some difference appeared between the sexes. Values found were as follows: females, 2.08; males, 2.12; Stage V, 2.21 (cf. Corner *et al.*, 1965).

In contrast to phosphorus excretion, in which there was a considerable proportion in organic form at certain periods (especially during the spring bloom), the vast proportion of nitrogen excreted (*ca.* 90 %) throughout the year was ammonia. Some earlier laboratory results suggested that the daily amount of nitrogen excreted varied when *Calanus* were fed on three different algae (*Brachiomonas*, *Cricosphaera*, *Skeletonema*), all at concentrations corresponding to the mean annual level of particulate nitrogen off Plymouth. Whether the digestibility of the different chemical compositions of the algal species is responsible was not clearly established. With each of the algae, however, experimental feeding clearly demonstrated the marked rise in excretion with increased food supply, and the reduction in rate with starvation. As Corner *et al.* point out, there must be an upper limit to the excretion rate dependent on biochemical pathways in the copepod and on enzyme concentrations, but feeding under natural conditions during the spring diatom bloom is reflected in a high excretion rate. The domination of protein in the phytoplankton diet and the conversion of the carbohydrate "skeletons", after deamination, to lipid storage products, results in an extensive liberation of ammonia.

For phosphorus, little is known about the composition of substances liberated other than phosphate, except that some phosphorus is in organic form (cf. Corkett and McLaren, 1978). Again, the proportion of body loss is always much higher for phosphorus. For example, over three peak diatom periods off Plymouth, female *Calanus* lost *ca.* 20 % (April, 1968), 28 % (October, 1968) and 20 % (April, 1969) of the body phosphorus daily. Corresponding values for daily nitrogen loss were 11 %; 9 % (August/September) and 6–7 %.

The experiments of Butler *et al.* (1970) permit some discussion of utilization of food during the year, whether nitrogen or phosphorus be considered. All three stages of *Calanus* (V, VI♂, VI♀) laid down extra nitrogen and phosphorus during the spring diatom bloom, and calculations show that approximately 77 % P and 62 % N of the food captured by the animals in spring was assimilated (i.e. digested and absorbed) (cf. Table 7.5). The daily requirements in terms of nitrogen and phosphorus for spring and winter seasons may also be estimated (Table 7.5). But assimilation efficiencies are only 62.4 % for N and 77 % for P (cf. pages 658 *et seq*). Thus, to sustain growth and metabolism in spring *Calanus* would require to capture 13.4 % of body N and 17.6 % body P per day. Butler *et al.* point out that such values are close to the earlier estimates by

Table 7.5. **Daily requirement (A) and utilization of food (B) (nitrogen and phosphorus) by *Calanus* (Butler *et al.*, 1970)**

A

Season	Daily requirement to sustain growth and metabolism Body N (%)	Body P (%)
Spring	8.4	13.6
Winter	1.9–2.8	6.2–7.2

B

Fate of total N or P taken each day in late April Total nitrogen	Total phosphorus
36 % excreted	60 % excreted
37 % lost as faeces	23 % lost as faeces
27 % for growth	17 % for growth

Harvey that *Calanus* required some 11–14 % of body wt daily; Cushing's estimate of approximately 50 % daily is much higher. With the average level of particulate matter in the Clyde Sea area, these calculations suggest that in spring, 29 ml/day would need to be swept clear by *Calanus*. With the lesser amount of particulate matter—but including detritus—in the sea during winter, about 37–59 ml must be swept clear per day. While the lower value is probably within the filtering abilities of *Calanus*, it is unlikely that all the organic particulate matter present during winter, when little living phytoplankton is available, is digestible and of real food value. The nutrition of overwintering copepods is still debatable.

Corner *et al.* (1965) include data showing the excretion per unit weight for various developmental stages of *Calanus* and for the amounts of nitrogen in the tissues of the stages. These permit a crude estimate of gross growth efficiency (*vide infra*), assuming an assimilation efficiency of around 60 %. The young copepodites (C II–IV) laid down as tissue approximately 0.9 μg N/day. Average nitrogen excreted for the same copepodites was about 0.6 μg N/day, so that a total of 1.5 μg N/day must be assimilated. But if percentage of food assimilated is around 60 %, 2.5 μg N food is required/day.

Then, **gross growth efficiency** $\left(\dfrac{\text{growth (weight)}}{\text{food (weight) ingested}}\right)$ is

$$\frac{0.9}{2.5} \times 100 = 36\,\%$$

and **nett growth efficiency** $\left(\dfrac{\text{growth (weight)}}{\text{assimilated food (weight)}}\right)$ is

$$\frac{0.9}{1.5} \times 100 = 60\,\%$$

These values may be compared with those of Conover for *C. hyperboreus* of 30% gross growth efficiency and 52% nett growth efficiency.

In a later study, Corner, Cowey and Marshall (1967) analysed all stages of *Calanus* and determined assimilation efficiencies by two methods: (i) N determinations (micro-Kjeldahl) on animals, faeces and uneaten cells separated by successive filtration, to give totals of food assimilated/food ingested; (ii) Conover's ratio method (cf. pages 662 *et seq*). Results suggested that the assimilation efficiency averaged 61.7% with *Skeletonema*, and as a first approximation this value was used for the complete development. Determination of the body nitrogen and excretion rate of the various naupliar and copepodite stages indicated that nitrogen content rose from 0.04 μg/animal for N I, to 0.37 μg/animal for C I and to 17.8 μg/ animal for C VI♀. Nitrogen excretion increased from *ca.* 0.01 μg N/animal/day for N I/II, to 0.4 μg N/animal/day for C III/IV and to 0.84 μg N/animal/day for C VI♀ (Fig. 7.4). These data permit calculation of the growth efficiency of *Calanus* for the pre-adult phase and of efficiency of egg production for adult females:

(a) *For Pre-Adult Phase* (35 days)
13.9 μg N/animal was used for metabolism (area under Fig. 7.4b). Assuming that moulting accounts for about 2%/day/body N, this is equivalent to *ca.* 0.64 μg for 11 moults.

Increase in body N = 17.8 − 0.04 = 17.76 μg.
Food captured (as N)

$$= (17.76 + 13.9 + 0.64) \times \frac{100}{61.7} \quad \text{(assimilation efficiency)}$$

$$= 52.3\ \mu g\ N$$

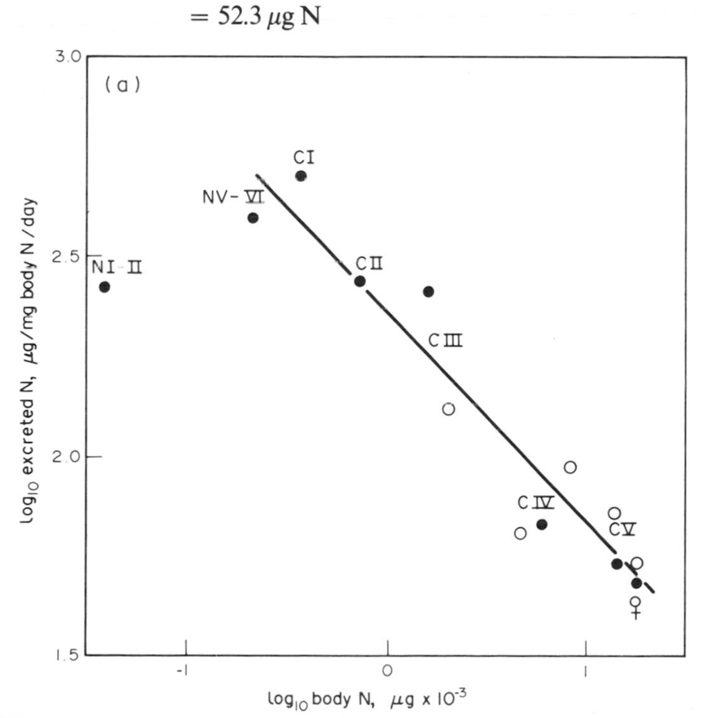

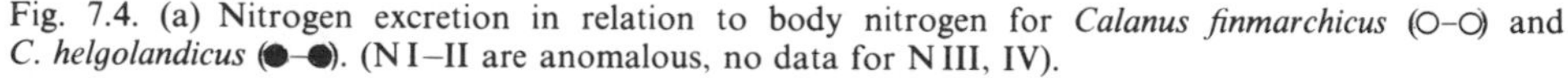
Fig. 7.4. (a) Nitrogen excretion in relation to body nitrogen for *Calanus finmarchicus* (O–O) and *C. helgolandicus* (●–●). (N I–II are anomalous, no data for N III, IV).

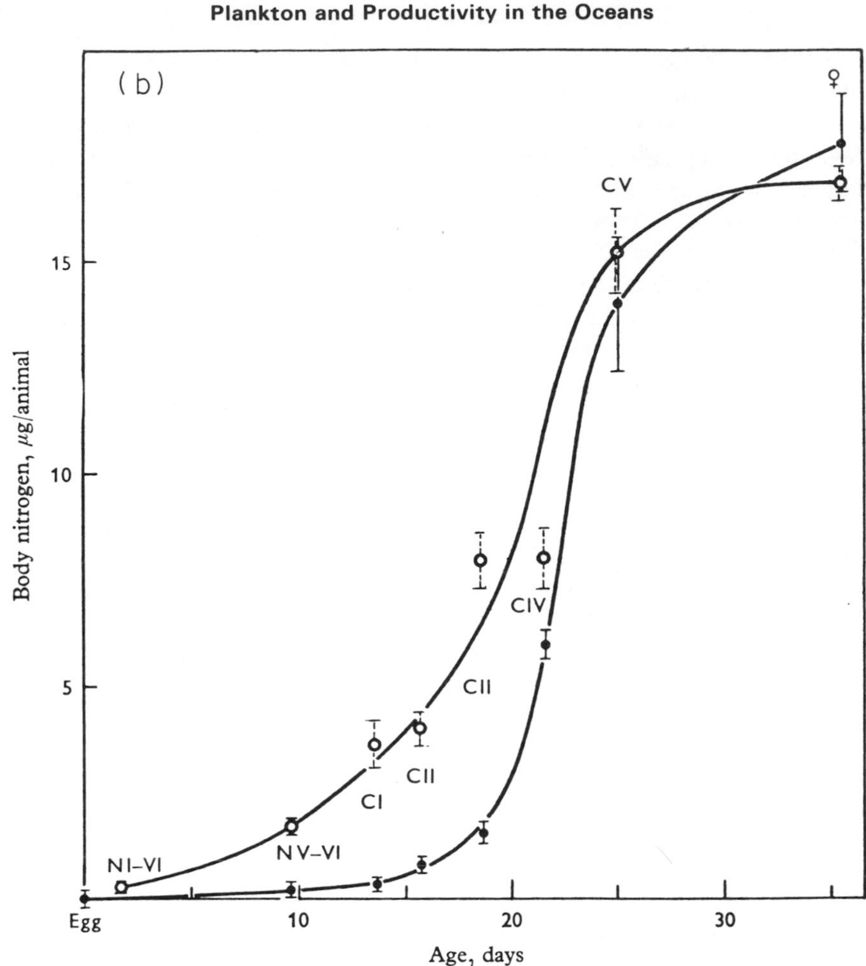

Fig. 7.4. (b) Nitrogen retained (●–●) and excreted (○–○) by different developmental stages of *C. finmarchicus* N I–VI, nauplius stages; C I–V, copepodite stages (Corner, Cowey and Marshall, 1967).

Therefore gross growth efficiency $\frac{\text{(growth)}}{\text{(food ingested)}} \times 100 = 34\%$

(b) *For Adult Phase* (*ca.* 35 days)

Only metabolism and egg production (mean 250 eggs/♀) must be considered.

N value of ingested food = Metabolism + eggs

$$= [(0.84 \times 35) + (0.035 \times 250)] \times \frac{100}{61.7} \quad \text{(assimilation efficiency)}$$

$$= 29.4 + 8.75 \times \frac{100}{61.7}$$

$$= 61.8\,\mu\text{g N}$$

Gross efficiency of egg production $= \frac{8.75}{61.8} \times 100 = 14\%$

For the whole life, a rough average efficiency is 24%, i.e. about a quarter of the food is used for growth and eggs and three-quarters for faeces, moulting and especially metabolic maintenance.

Corkett and McLaren (1978) include a calculation for *Pseudocalanus minutus* from data on nitrogen content and excretion rate to suggest that for the interval Nauplius V to adult, the gross growth efficiency was only 7.9%. From a mean value of 10% of the body weight as daily loss due to respiration, some 10% for excretion, an assimilation efficiency of 65% and a growth rate of 10 to 17% of body weight per day (from Paffenhofer and Harris, 1976), Corkett and McLaren suggest that the ingestion rate to balance the equation for intake of food must be of the order of 60% of body weight daily.

Although excretion falls in winter for *Calanus*, the body loss is still approximately 2% body N/day and 7% body P/day (cf. Table 7.6) and this loss must be compensated. Food sources other than phytoplankton should be considered. It might be relevant that Martin (1968) found autumn excretion levels much higher than spring (approximately three-fold for phosphorus and ten-fold for nitrogen) in Narragansett Bay zooplankton. The population was possibly then feeding extensively on detritus and microzooplankton, especially as the zooplankton was dominated by *Acartia*, recognized as omnivorous.

To examine possible food sources, Corner *et al.* (1974) fed *Calanus* on four diets: (1) suspended matter from Clyde water in winter; (2) particulate matter formed by bubbling seawater enriched with soluble extracts of plant cells; (3) living *Elminius* nauplii; (4) dead *Elminius* nauplii. Only diets (3) and (4) maintained both body nitrogen and phosphorus in *Calanus*. Dead nauplii gave 34% body N/day and 44.5% body P/day; live nauplii 25% body N/day and 47% body P/day. These rations are much higher than those deduced earlier from the spring diatom increase in the Clyde (about 13% body N/day and 17.5% body P/day), and while some of the increase may be attributed to the capture of "large" particles (i.e. *Elminius*) rather than feeding on "small" particles (diatoms), the better contribution is in part a reflection of the carnivorous diet. Carnivorous feeding might contribute considerably to the nutrition of overwintering *Calanus*. Particulate matter might help indirectly in maintaining the microzooplankton, and these in turn could contribute to the *Calanus* diet. Barnacle (*Elminius*) larvae are not likely themselves to be a source of food, since they are not

Table 7.6. Mean values for nitrogen and phosphorus excretion by Calanus in winter and spring (Butler, Corner and Marshall, 1970)

		Winter	Spring
N excretion	(μg/animal/day)	0.47	1.16
	(μg/mg dry wt/day)	2.65	5.43
P excretion	(μg/animal/day)	0.091	0.250
	(μg/mg dry wt/day)	0.483	1.176
% body N excreted daily		2.58	4.62
% body P excreted daily		6.45	10.30

Triplicate determination on females, males and C V's in winter and spring. Body N = 11.8% dry wt in winter and 12.6% in spring; corresponding values for body P, 0.80% and 0.97% respectively.

normally abundant over several winter months, but the microzooplankton available might well include protozoans. Whatever the food source, the winter demand as a daily ration would appear to be very high whether expressed as nitrogen or phosphorus, but this may have been increased by the experimental feeding, especially by the pure animal diet.

The relationship between reduction in metabolic rate, with starvation (usually demonstrated by reduced oxygen uptake) and period of survival, has been considered by Ikeda (1974) (cf. Fig. 7.5). Ikeda reports *Calanus plumchrus* as losing 1 % body weight (cf. *Calanus finmarchicus, vide supra*—2 %), *Acartia clausi* as losing 6 % and *Acartia longiremis* as losing 16 % over short starvation periods. Such losses will depend in part on body size and habitat temperature, both factors affecting metabolic rate. Generally, for tropical species, reduction of metabolic rate under poor nutritional conditions is not of high survival value, but for many boreal species, even a small decrease in metabolism may be highly effective. Many instances have been quoted of a

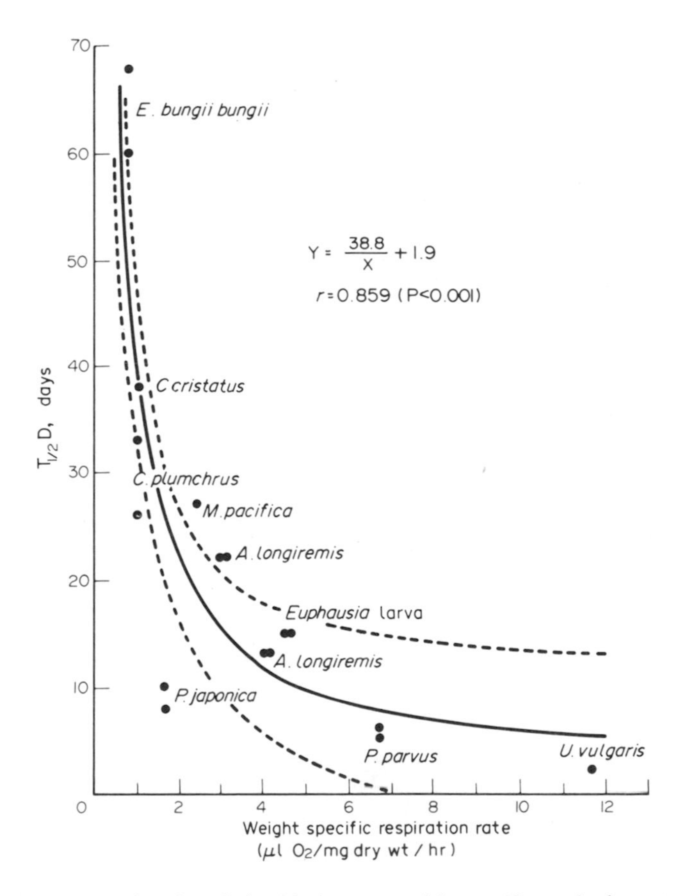

Fig. 7.5. Scatter diagram showing the relationship between weight specific respiration rate and $T_{\frac{1}{2}}D$ for tropical (*Undinula vulgaris*), temperate (*Paracalanus parvus*) and boreal (*Calanus cristatus* V, *C. plumchrus* V, *Eucalanus bungii bungii*, *Metridia pacifica*, *Acartia longiremis*, *Parathemisto japonica* and euphausiid larva) zooplankton species.
Dotted line represents 95 % confidence limits of the regression curve. $T_{\frac{1}{2}}D$ denotes time when half of initial (day 0) number of individuals died (Ikeda, 1974).

reduction in metabolism being usually reflected in a lessened liberation of excretory products. Thus, *Calanus finmarchicus* loses around 4% daily of body protein during starvation, whereas feeding *Calanus* lose at least 15% daily of body protein.

The relation between oxygen uptake and nitrogen excretion is a useful tool in investigating metabolic level. This relation, between the quantity of oxygen consumed and the amount of nitrogen excreted (the O/N ratio), is widely used as an indication of metabolic substrate. Relationships of this kind need not be restricted to O/N ratios; oxygen/phosphorus (O/P), carbon/nitrogen (C/N), or nitrogen/phosphorus (N/P) ratios might equally well be used. Butler *et al.* (1969) examined differences in the N/P ratio of phytoplankton and zooplankton. Earlier workers had obtained a value (in terms of weight) of *ca.* 7 for phytoplankton and 11 for zooplankton. The N/P ratios for several species of phytoplankton are generally lower than that found for the faecal pellets of grazing zooplankton forms. The excretory products of animals feeding on phytoplankton and apparently retaining excess nitrogen, must however show a N/P ratio distinctly less than that of their tissues. Harris (1959) obtained an N/P ratio of 4.4 for excretory products of mixed zooplankton in Long Island Sound. The values of Butler *et al.* are somewhat higher (mean 5.4), but are far below that for zooplankton tissue. The N/P ratio of the excretory products showed little variation over a substantial period of time, even from year to year, and there appeared to be no real difference between male, female and Stage V *Calanus*, although the actual levels of nitrogen and phosphorus excretion varied considerably through the year. This comparative constancy of the N/P ratio is useful, for if total excreted phosphorus is determined (a comparatively simple matter) a rough estimate of the total excretion of nitrogen may be computed.

The O/N ratio can be of especial value in indicating metabolic substrates. "Average" protein is held to contain about 45% carbon and 16% nitrogen; if pure protein is being metabolised, the O/N ratio should approximate to 8. The composition of the "average" living particulate organic matter in seawater is usually reckoned as 106 atoms C/16 atoms N/1 atom P (cf. Volume 1). Theoretically, for the complete oxidation of this particulate matter, roughly equivalent to a mixed diet, the O/N ratio should be approximately 17. If pure carbohydrate or pure lipid (except for the small amount of nitrogen present if lipid includes phospholipid) is metabolized, theoretically the O/N should be infinity. In practice, high values exceeding 50, even approaching 100, may be obtained when rich carbohydrate or, more usually, rich lipid diets are being utilized.

Ikeda (1977) gives values for mean O/N, N/P and O/P ratios (based on earlier data) for the average composition of organic matter, which are slightly different from those quoted. It is important also to recognize the variations which may occur in the chemical composition of protein and lipid, and that the ratios, while valuable, can give only limited indications of metabolic substrates. Generally, apart from the high O/N and O/P ratios characteristic of carbohydrate oxidation, low N/P and O/N ratios are typical of lipid- and protein-orientated metabolism, respectively. Table 7.7 gives an indication of the wide range of ratios found for species of zooplankton. Anomalous ratios should be critically examined. Damaged or moribund zooplankton included in experiments and the use of unfiltered seawater may greatly affect rates of excretion of nitrogen and phosphorus.

Harris (1959) was one of the first to examine O/N ratios in mixed zooplankton

Table 7.7. Oxygen : nitrogen (O : N), nitrogen : phosphorus (N : P), and oxygen : phosphorus (O : P) ratios from measurements of respiration, ammonia-nitrogen excretion and inorganic phosphate-phosphorus excretion rates reported by former workers (Ikeda, 1977)

Animals	O:N	N:P	O:P	Source
Mixed zooplankton (UFW)	7.7	7.0	54	Harris (1959)
Mixed zooplankton (UFW)	41	9.98	222	Martin (1968)
Calanus cristatus (UFW)	5.7	19	110	Taguchi and Ishii (1972)
Calanus plumchrus (UFW)	6.8	13	89	Taguchi and Ishii (1972)
Mixed zooplankton (UFW)	13.48	10.33	142.4	Le Borgne (1973)
Sagitta hispida (FW)	—	11.3	—	Beers (1964)
Mixed zooplankton (UFW)	—	—	72	Satomi and Pomeroy (1965)
Calanus helgolandicus (FW)	9.8–15.6[a]	—	—	Corner *et al.* (1965)
Boreal zooplankton (10 species) (UFW)	6–200[a]	—	—	Conover and Corner (1968)
Calanus finmarchicus (FW)	—	10.8	—	Butler *et al.* (1969)
Calanus finmarchicus (FW)	—	11.0[b]	—	Butler *et al.* (1970)
Sagitta hispida (FW)	6.8	—	—	Reeve *et al.* (1970)
Calanus helgolandicus (FW)	—	16.5[b]	—	Corner *et al.* (1972)
Temora stylifera (FW)	7–15[c]	—	—	Nival *et al.* (1974)
Phronima sedentaria	4.27	—	—	Mayzaud (1973)
Meganyctiphanes norvegica (FW)	4.77–12.13	—	—	Mayzaud (1973)
Acartia clausi (FW)	1.61	—	—	Mayzaud (1973)
Sagitta setosa (FW)	1.75	—	—	Mayzaud (1973)
Boreal, temperate, sub-tropical and tropical zooplankton (81 species) (FW)	4–115	—	—	Ikeda (1974)
Mixed zooplankton (UFW)	—	6.8	—	Mullin *et al.* (1975)

[a] Ninhydrin N.
[b] Total N : total P.
[c] Read from their figure 10.
The use of unfiltered (UFW) and filtered (FW) sea water for measurement is noted.
Ratios are by atoms.

populations (mainly *Acartia*). Though the ratio showed considerable variation from approximately 4–16, the mean (*ca.* 8) suggested that protein was extensively used by these copepods. Conover and Corner (1968) examined the O/N relationship in four species of calanoid from higher latitudes, *Calanus hyperboreus*, *Calanus finmarchicus*, *Metridia longa* and *Euchaeta norvegica*, with particular emphasis on considering changes in the ratio with species and with season. For *Calanus hyperboreus*, an Arctic copepod, the O/N ratio was relatively high, usually in excess of 30; on some occasions values exceeding 50 were obtained. The O/N maximum occurred in spring, about May, after the copepods had commenced feeding actively (*vide supra*) (cf. Fig. 7.6). The respiration level, subject to seasonal variation, showed a rise to a maximum about May. The rate of nitrogenous excretion, also seasonally variable, exhibited a more sudden increase from a low winter level to reach a maximum somewhat earlier, about April. With a marked fall in nitrogen excretion through late spring and summer, and a rather slower decline in oxygen uptake, the O/N ratios generally decreased during summer and autumn, fluctuating around 30 throughout autumn and early winter. The O/N ratio was still relatively high in autumn, presumably because animals were mainly using lipid as an energy source. The minimum O/N ratio occurred shortly before the phytoplankton bloom, although excretory rates reached low values in winter (cf. Fig.

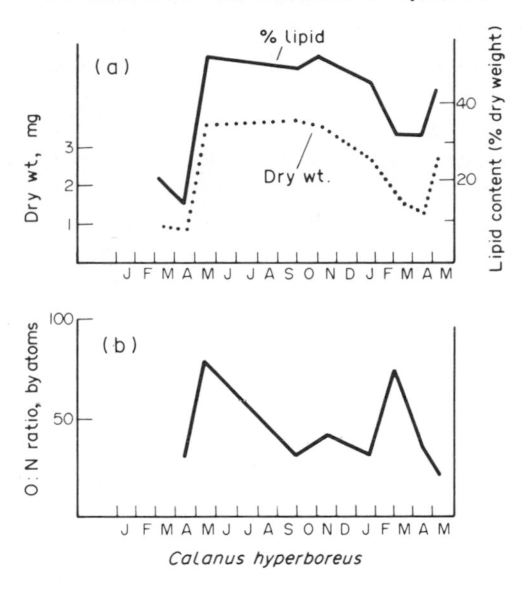

Fig. 7.6. Seasonal changes in (a) dry weight and fat content and (b) O : N ratio. Curves smoothed (Modified from Conover and Corner, 1968).

7.6b). Conover and Corner interpret the low O/N ratio as demonstrating that the animal had then so far depleted its lipid reserves that it was utilizing protein to some extent; O/N values were less than 20.

Even during the spring bloom, since respiration and excretion were both increasing, the O/N ratio was not very high. Rapid feeding at this season resulted in reserves, essentially lipid, being laid down. Much of the reserve would be derived from dietary protein which would need to be deaminated with loss of nitrogen. It is only just *after* the bloom, when the animals were feeding little but were using almost pure lipid for respiratory substrate, that the O/N ratio became maximal. These observations on *Calanus hyperboreus* were confirmed by the variations in dry weight and the proportion of lipid. Figure 7.6a suggests that both dry weight and percentage lipid reached low values immediately before the spring bloom, but a rapid rise in body weight with deposition of lipid followed with only a gradual reduction during summer and early autumn as the copepod used its reserves sparingly with the lowered metabolic rate.

Calanus finmarchicus also showed a pattern of seasonal change in respiration, excretion and O/N ratio which was fundamentally similar. A low O/N ratio was found in late winter and there was a rapid rise with values of *ca.* 50 in May. However, in the late summer and autumn the O/N ratio fell much more sharply, characteristic of a species which produces several generations and relies more on the food immediately available. The O/N ratio was frequently less than 20 over this period, indicating utilization of a mixed phytoplankton diet. Some rise in O/N ratio occurred with storage of lipid for the overwintering stage in autumn and relatively high levels of lipid (exceeding 40% of the dry body wt) were recorded at this time. In late winter (February/March), however, a sharp reduction in lipid and the relatively low O/N ratio (little more than 10) suggests that lipid reserves are then mainly exhausted and that the copepod must, to a considerable extent, be metabolizing protein. *Calanus finmarchicus* appears to be much more dependent than *C. hyperboreus* upon available food supplies.

Metridia longa tends to have somewhat lower O/N ratios than *Calanus finmarchicus*, possibly reflecting the partly carnivorous diet (cf. Haq, 1967). Although respiration and excretion rates showed an increase in spring and a decline in autumn, the O/N ratio showed relatively little change (<20 throughout) except just after the spring bloom. There was little change in dry body weight or lipid content, which was normally less than 20 % of the body weight. *Metridia* makes use of body protein if starved; normally it probably feeds omnivorously (cf. Chapter 6).

Limited data for the carnivorous copepod *Pareuchaeta norvegica* suggest that the respiration rate is relatively high but shows rather few fluctuations apart from a small increase in spring. Nitrogen excretion changes were also rather irregular and the O/N ratio did not show much variation, but was generally lower than for *Calanus*. The copepod presumably relies little on stored food material, but heavily on immediate food supply with its substantial amount of protein. Conover and Corner advance the view that the four copepods show different degrees of dependence on their environment, with the carnivorous *Pareuchaeta* and the omnivorous *Metridia* showing highest dependence and *Calanus hyperboreus*, with its marked ability to store lipid, achieving some degree of temporary independence.

Although Martin (1968) did not use O/N ratios to deduce metabolic substrates, his analysis of seasonal variations in excretion in zooplankton indicated that when phytoplankton food was abundant, nitrogen and phosphorus excretion were low and lipid reserves were laid down, a low level of phosphorus excretion usually being indicative of lipid deposition. Later in the season when phytoplankton was relatively scarce, nitrogen excretion rose, suggesting that dietary protein was used; the protein may come from a change to carnivorous diet. A rise in phosphorus excretion also indicated that lipid was not being deposited.

Jawed (1969) found for the mysid *Neomysis rayii* an O/N ratio of 8.4, and for the euphausiid *Euphausia pacifica* a rather similar value of 9.4. Both these animals were kept without food during the experiment; under these semi-starvation conditions there is a strong suggestion of protein metabolism. Assuming that protein is metabolized, at 10°C about 14 μg protein for *Neomysis* and 13 μg for *Euphausia* were used per mg dry weight per day, equivalent to about 2 % of the body protein per day. A study of the mysid *Neomysis integer* by Ferguson (1973) demonstrated that over late winter/spring (February and March) the O/N ratio was minimal, averaging 4 and 4.5, but that during September and October the values were 8.8 and 8.7, indicative of pure protein metabolism. The figure of around 4 is abnormally low and below that calculated for protein (cf. some of Harris's earlier data). Mayzaud also found low values, frequently <4, for O/N ratios in the zooplankton (*Meganyctiphanes*, *Phronima*, *Acartia*). He attributed these excessively low values to species with little or no reserve but with a high turnover, degrading body protein faster than the residues can be oxidized. In some experiments with the same species he found O/N ratios of 12, indicating fairly substantial protein metabolism. Taguchi and Ishii (1972) also determined ratios from their data on excretion and respiration. Whilst O/N and O/P ratios were variable, the values were low (means: 5.7 and 6.8 respectively for *C. cristatus* and *C. plumchrus* for O/N; 110 and 89 respectively for *C. cristatus* and *C. plumchrus* for O/P), suggesting that excretory losses might have exceeded respiratory loss somewhat.

An extensive series of determinations of O/N ratios for zooplankton is given by Ikeda (1974). A considerable range is sometimes observed for a single species and though this could be attributed in part to experimental error, since O_2 and NH_3 changes were on occasions very small, some of the variation may genuinely reflect change in metabolic substrate. For example, a marked range is observed for *Metridia pacifica* (16–93); since Haq found for *M. longa* and *M. lucens* a strong tendency to omnivorous or carnivorous feeding (a view supported by Conover and Corner), the range for *M. pacifica* may indicate a change from largely carnivorous to herbivorous diet. The data (cf. Table 7.8) also illustrates different O/N ratios for various zooplankton typical of herbivorous or carnivorous species. For example, *Calanus plumchrus* in boreal waters is quoted as showing ratios of 40 and 51, characteristic of herbivores, whereas *Beroe* (17), *Sagitta elegans* (10) and *S. serratodentata* (11) all appear to be essentially carnivorous. The low O/N ratios for *Acartia* (*A. longiremis*, 12; *A. clausi*, 6) are noteworthy since these copepods have frequently been described as omnivores. Likewise, though the O/N ratios are subject to considerable variation, *Euphausia pacifica* has a fairly high O/N ratio of 27, indicative of mainly herbivorous feeding, as against 15.5 and 13 for *E. diomedeae* and *E. tenera* which are presumably omnivorous tropical species (cf. Roger, 1973a) (Table 7.8).

Ikeda has plotted the relative percentage frequency of the O/N ratio for various environmental groups. He suggests that values of less than 24 denote mainly protein metabolism; values of more than 24 indicate a greater degree of lipid metabolism. Species from temperate and warm waters mostly show low O/N values, reflecting mainly protein metabolism. Boreal zooplankton exhibit a much greater range of ratios (7–93) but more than half have very high O/N ratios; this is held to denote widespread lipid metabolism since so little carbohydrate reserve is generally observed, (cf. page 696). Some O/N ratios exceeded 90. The high O/N ratios in boreal species tended to be characteristic of herbivores, but even single species showed a wide range, possibly

Table 7.8. Ratio of O : N for various zooplankton species (Ikeda, 1974)

Animal (copepods)	O:N ratio	Range in Experimental Temperature (°C)
Metridia pacifica V	28 (4)	4.5–9.7
M. pacifica ♀	93 (1)	8.2–10.5
Calanus plumchrus IV	51 (3)	4.5–7.0
C. plumchrus V	40 (3)	4.5–7.0
Acartia longiremis	12 (3)	8.9–11.0
A. longiremis	16 (3)	8.4–14.3
A. clausi	6	
Beroe cucumis	17 (3)	15.0–15.3
Sagitta elegans	10 (3)	5.5–7.0
S. serratodentata	11 (2)	27.4
Euphausia pacifica	37 (4)	7.0–10.7
E. tenera	13 (9)	26.6–27.5
E. diomedeae	15.5 (2)	27.2

Mean values given, number of determinations in parenthesis.

indicating variable nutrition. There appeared to be generally lower ratios for carnivorous than for herbivorous species from boreal regions, as Conover and others had suggested. On the other hand, temperate and warm water species did not show such considerable differences in O/N ratios; perhaps in the absence of massive phytoplankton outbursts a more omnivorous diet is common, and there is less reliance on lipid storage.

Biggs (1977) found that O/N ratios for the carnivorous gelatinous zooplankton studied were mostly in agreement with the range of 8–24 suggested by other workers, notably Ikeda. The metabolism of pure protein would give the lower values; oxidation of equal amounts of protein and lipid would yield the higher ratios. A particularly interesting determination on eudoxids of *Diphyes* and *Ceratocymba* gave very low ammonium excretion rates and remarkably high O/N ratios of 121 and 88 respectively. Biggs mentions that some eudoxids do not feed. The high ratios might be explained if the eudoxids relied on reserves of carbohydrate or (more likely) lipid during their brief existence. Herbivorous gelatinous zooplankton generally showed higher O/N ratios (18 to 89) than those for carnivores. Mayzaud and Dallot (1973) had reported very low ratios for *Thalia democratica* (0.6) and *Salpa fusiformis* (3.8) and for the amphipod *Phronima sedentaria* (2.0). The low values may have been partly associated with animals having been starved for 12 hours before the experiment, but the effect seems exceptionally marked in comparison with other studies. The oxygen uptake was very low. Since the animals were collected by nets, some damage may have been experienced apart from starvation effects.

Nival, Malara, Charra, Palazzoli and Nival (1974) determined rates of nitrogenous excretion, in addition to their studies on respiratory rate, for several calanoids from upwelling waters off the coast of Morocco. Rates, determined usually at 13° or 15°C, are quoted per unit dry weight and per unit of body protein. For three species, *Calanus helgolandicus* (and *finmarchicus*), *Acartia clausi* and *Centropages typicus*, the rates are of the same order as those found by other workers, provided the experimental temperature is taken into account (Table 7.9). The rate for *Acartia clausi*, however, is clearly lower off Morocco than off Villefranche. Temperature and excretion of ammonium appear to be directly related, with a Q_{10} value of 1.3–1.8 for temperatures between 13° and 23°C except for *Acartia*, which showed no increase in excretory rate with temperature rise. Nival *et al.* also found that excretion was related to body size, but especially to the amount of body protein, so that the copepod population off Morocco excreted some 14% of the body protein daily, more or less independent of temperature.

The effect of starvation was tested for the copepod *Temora stylifera* over a 7-day period. Respiration and excretion showed a clear and regular decline. Body weight remained approximately constant for the first 4 days, but then declined. Protein fell over the first 3 or 4 days and though Fig. 7.7 suggests levels were then maintained, other experiments indicated a continued decline. The O/N ratio at the beginning of the experimental was 15, corresponding to carnivorous or possibly partly omnivorous feeding. Though this ratio held for the first few days of starvation, the value then declined to about 7 (cf. Fig. 7.7). Nival *et al.* believe that over some 3 or 4 days of starvation *Temora* is able to maintain an approximately normal metabolism, but following that period body protein is utilized exclusively.

Effects of periods of experimental feeding and starvation on zooplankton may

Table 7.9. Nitrogen excretion per unit dry weight for adults of three species of copepods (Nival *et al.*, 1974)

Species	Location	Excretion (μg N/mg dry wt/day)	Experimental temperature (°C)	Reference
Calanus helgolandicus &	Atlantic			
C. finmarchicus	Clyde	11.0–12.1	15	Corner, Cowey and Marshall, 1965
	spring	14.94 (1.63)[a]	10–14	Butler, Corner and Marshall, 1969
	autumn	3.21 (1.23)		
	winter	2.65 (0.90)	7	Butler *et al.*, 1970
	spring	5.43 (2.58)	9	
	North American coast	0.5–3	4–6	Conover & Corner, 1968
	Moroccan coast	12.1–15.9	15	Nival *et al.*, 1974
Acartia clausi	Mediterranean			
	Villefranche	52.04 (4.25)	15	Mayzaud, 1973a
	Atlantic			
	North American coast[b]	30.0–81.0	16–18	Harris, 1959
	Moroccan coast	15.5	15	Nival *et al.*, 1974
Centropages typicus	Atlantic			
	North American coast	11.9	3–5	Conover and Corner, 1968
	Moroccan coast	18.02	15	Nival *et al.*, 1974

[a] Standard deviation of the mean is given in parenthesis.
[b] Plankton of more than 70% *Acartia clausi*.

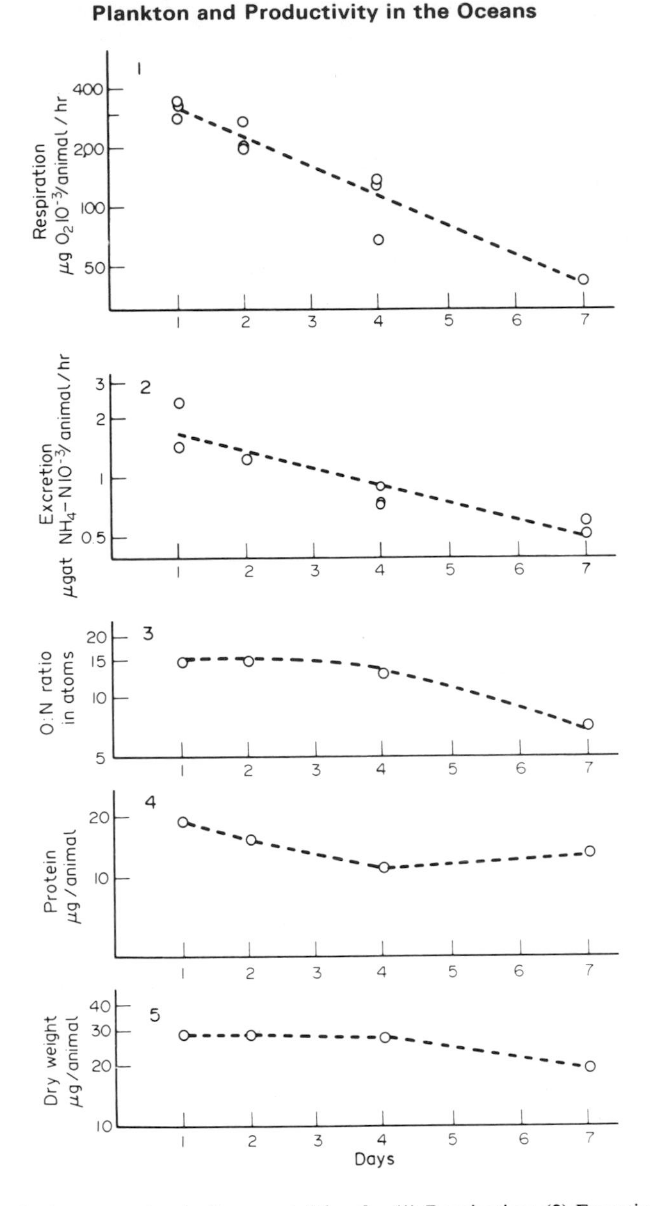

Fig. 7.7. Changes during starvation in *Temora stylifera* for (1) Respiration; (2) Excretion; (3) O : N ratio; (4) Protein; (5) Dry weight (Nival *et al.*, 1974).

therefore be of value in the consideration of metabolic substrate. Ikeda's (1977) data on respiration and excretion have been used to examine changes in O/N, O/P and N/P ratios. Changes in fed and starved animals are illustrated in Figs 7.8, 7.9, 7.10. Consideration of the data for *Calanus plumchrus* are difficult to interpret. In *Euphausia pacifica* there is a clear decrease in the O/N ratio, with starvation indicating an increasing use of body protein. In *Paracalanus* the low N/P and O/P ratios are interpreted as indicating the importance of lipid in metabolism. High O/N ratios would support this conclusion. The O/N ratios for the carnivorous *Pleurobrachia* and

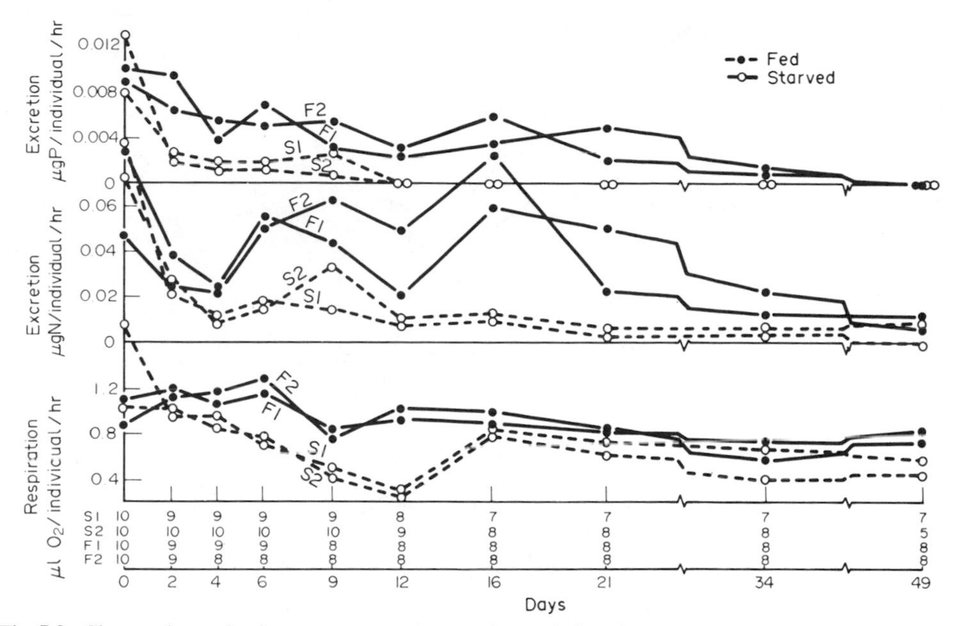

Fig. 7.8. Changes in respiration rates, ammonia excretion and phosphate excretion for *Calanus plumchrus*. Successive change in survival number of copepods in each fed (F1, F2) and starved (S1, S2) series is shown in lower part of figure (Ikeda, 1977).

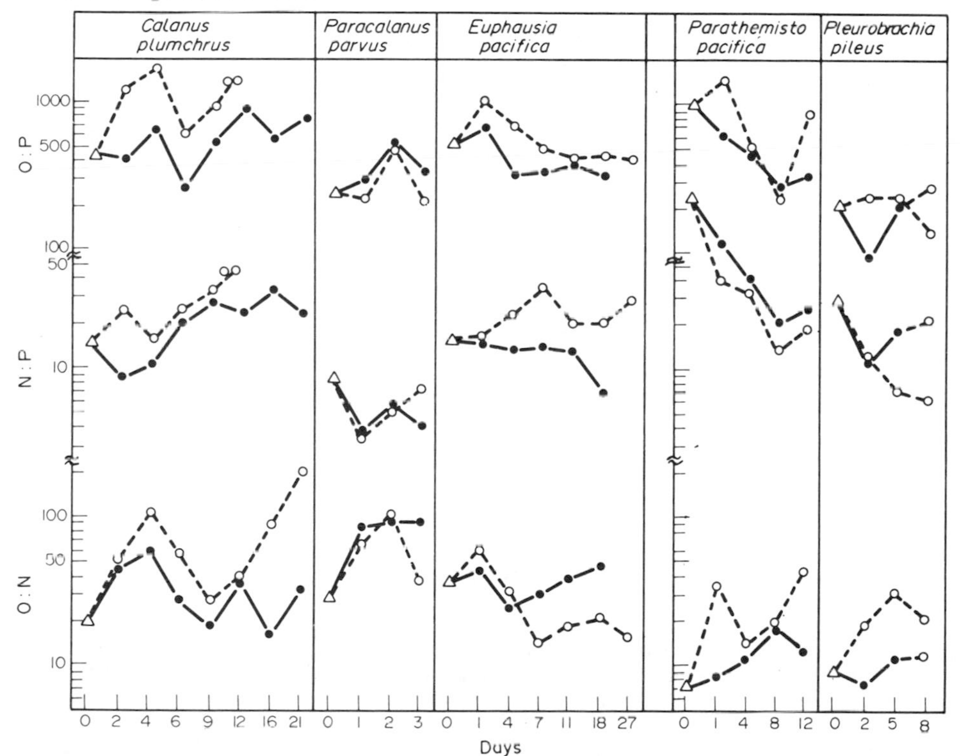

Fig. 7.9. Changes in averaged O : N, N : P and O : P ratios calculated from rates of respiration, ammonia excretion and phosphate excretion in zooplankton. Ratios of starved *P. pacifica* are from the only series (S2) in which no cannibalism occurred. Closed and open circles denote values on fed and starved individuals respectively. At start of experiments (Day O), pooled data on fed and starved specimens are indicated by open triangles (Ikeda, 1977).

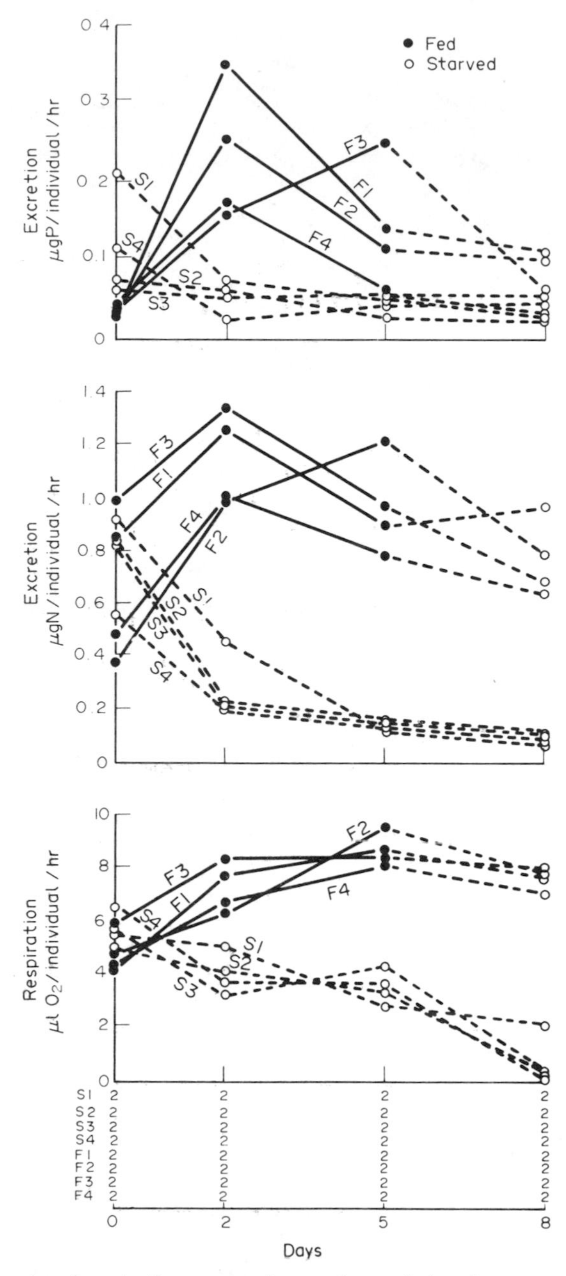

Fig. 7.10. Changes in rates of respiration, ammonia excretion and phosphate excretion for *Pleurobrachia pileus*. Successive change in survival number of ctenophores in each fed (F1, F2, F3, F4) and starved (S1, S2, S3, S4) series is shown in lower part of figure. (Ikeda, 1977).

Parathemisto were always low, especially in fed animals. Protein appears to be an important regular substrate for oxidation. During starvation, changes in the N/P ratio might indicate some use of lipid.

To a limited degree, differences between zooplankton species, including those from various latitudes, in the major biochemical constituents of the body (protein and lipid) may also give some indication of likely food reserves and substrates. Ikeda found a great range in nitrogen content among the zooplankton species investigated, with a maximum (*ca.* 13 %) in a number of taxa. Several major crustacean groups as well as chaetognaths and fish larvae all tended to have high nitrogen content. The minimum (as low as 0.4 %) occurred in a salp. There was no obvious variation in body nitrogen with latitude. On the other hand, as regards body carbon content, boreal species tended to be higher than tropical species with a maximum even greater than 65 % for *Pseudocalanus*. The C/N ratio in body composition may also assist in indicating major biochemical differences. Thus, high C/N ratios denote high lipid percentage and low C/N ratios (*ca.* 3) denote high protein percentage. Other authors have variously suggested a constant ratio (e.g. 6) for zooplankton or that the ratio varies with species from 4 to 11.

Ikeda observed ratios for many species of the order of 3 or 5, but there were fairly obvious latitudinal differences, with lower values indicating protein as the major body constituent, in warmer water and temperate species. While about 50 % of boreal species and those from deep water followed this pattern, in others ratios were higher (10 or more) presumably indicating a greater proportion of lipid, since very little carbohydrate appears to exist in zooplankton (Raymont *et al.* 1960, 1970, 1971). Ikeda quotes some other characteristics (e.g. the size of oil sacs in copepods) in support of the general view that although body composition may be subject to considerable change at high latitudes because of the highly variable food supply, there is a marked tendency for boreal and many deep-sea species to deposit considerable amounts of lipid. Rakusa-Suszczewski, McWhinnie and Cahoon (1976) comment on the high lipid content of zooplankton at high latitudes which may be used in metabolism and for reproduction. Deposition of lipid appears to be generally negatively correlated with environmental temperature (cf. chemical composition, *vide infra*). Low latitude zooplankton require to feed frequently and have little tolerance of starvation.

One example of the changes in various biochemical fractions with starvation for a boreal copepod, *Calanus cristatus*, is given in Table 7.10. Although protein is utilized extensively, the percentage loss of lipid is greater. Very little carbohydrate is present and there is no obvious change in carbohydrate. Starvation experiments for a number of species showed, apart from a general decrease in body weight, a loss of either lipid or protein partly dependent on whether the species was "fatty", and partly dependent on season in some species. In general, lipid is first extensively used and after prolonged starvation protein is lost (cf. Conover and Corner). In animals with little lipid reserve

Table 7.10. Changes in dry weight, total lipid, total protein, total chitin and total carbohydrate in *Calanus cristatus* (Stage V) from the Bering Sea during starvation (Ikeda, 1974)

Days of starvation	Dry weight	Ash weight	Lipid	Protein	Chitin	Carbohydrate
0	2203	308	474	1311 ± 143	94 ± 9	14 ± 1
3	1599	355	218	931 ± 125	82 ± 12	12 ± 3
7	1209	348	35	715 ± 103	99 ± 10	12 ± 1
13	1218	308	108	696 ± 129	91 ± 11	14 ± 2

numbers are μg/animal: mean of eight determinations ± standard deviation.

(e.g. *Acartia*) protein is metabolized from the beginning of starvation (cf. Table 7.10). Although a number of boreal species do not store any appreciable amount of lipid, they are mostly carnivorous (e.g. chaetognaths), relying on a more steady level of food at high latitudes. For some bathypelagic species lipid storage may be used as an energy reserve, especially for the younger stages before they have sufficient locomotor activity to catch their own prey. The production of relatively few eggs, well supplied with lipid-rich yolk, may tide the young over a brief juvenile period when vulnerability to starvation might be very real.

Relatively large zooplankton capable of storing considerable lipid reserves are particularly well adapted to high latitudes. The "high turnover" zooplankton species, more usually regarded as typical of lower latitudes, tend to show rapid growth and achieve only relatively small size at maturity; the body lays down comparatively little food reserve and the life cycles of the species tend to be brief (cf. Roger, 1974a). Animals with a "low turnover", more usually found in high latitudes, are frequently characterized by slower growth rate but greater body size at maturity, with more abundant metabolic reserves; the life cycles are lengthened.

Russian workers in particular have used a relationship between rate of production (P) and the biomass of organisms at one trophic level (B), namely the rate of production per unit biomass (P/B), to compare trophic levels. In warm waters material is generally rapidly cycled. Any food available is rapidly consumed. The zooplankton reproduces rapidly and life cycles are short; there is little food storage and the biomass tends to be low. P/B values thus tend to be high. In contrast, at high latitudes phytoplankton production is often excessive during spring/summer; the zooplankton feed and grow well and store substantial food reserves over the rich food period, but although the crop may thus increase, reproduction is typically rather low. P/B values thus may be low.

An assessment of the daily rate of production of matter (P) at one trophic level can be determined from the number of organisms eaten, mortality, the number of organisms transferring to the next trophic level per day and the difference in standing stock at the end of the day. Petipa, Pavlova and Mironov (1970) analysed the transfer of energy through the epiplankton community of the Black Sea. The reproductive rate of algae and respiration of zooplankton (copepods) were determined experimentally and environmental data (light, temperature, nutrients, etc.) were used to determine effects on rates. Total weight increase of zooplankton was assessed from difference in weight at the end and beginning of a stage (e.g. Stage IV to Stage III) divided by the average duration of the stage, taking temperature effects into account. Food composition and rations for zooplankton were determined from gut contents and rate of digestion. Both P and B tend to decrease at higher trophic levels and the P/B ratio generally declincs. Animals are generally slower-growing and spend an increasing amount of energy in capturing their prey. A decrease was observed in ecological efficiency (efficiency of transfer of energy from one trophic level to another at the expense of food) from *ca.* 70% (primary herbivores) to *ca.* 7% (tertiary carnivores).

Pasternak (1978), investigating growth and metabolism of the mysid *Neomysis mirabilis* in the laboratory and in nursery ponds, found that nett growth efficiency declined with age. The specific growth increment also showed a fairly steady fall with age, apart from the first few days of observation, when the young mysids were adapting to environmental conditions in the aquarium or pond and mortality tended to be high.

Shuskina, Kislyakov and Pasternak (1974) had earlier described a study of the rate of production of plankton in an approximately natural community of zooplankton isolated in 140-litre containers to which ^{14}C-labelled phytoplankton was added. After exposure overnight, the radioactivity of the phytoplankton and of each group of zooplankton (copepods, sagittae and an amphipod) was determined, together with their biomass. Assuming a percentage assimilation for each type of feeding, the productivity per unit of biomass (P/B) for the various groups was estimated.

Ikeda and Motoda (1978) comment on a high annual P/B ratio (0.035) on the continental shelf of the eastern Bering Sea where small zooplankton predominate, in contrast to the relatively low ratio in the central oceanic area and south of the Aleutian Chain (0.024 and 0.022 respectively). Although estimates for the annual zooplankton production in the Bering Sea range from 6.0–18.2 gC/m^2/yr, that of the Kuroshio Current region (11.3–20.8 gC/m^2/yr) appears similar; the high summer production in the Bering Sea (40–121/mg C/m^2/day) accounts for this.

Conover and Lalli (1974) cite the interesting example of the gymnosome *Clione limacina*, a species with a large geographical range which appears to be able to adapt its metabolism to some degree to low or high turnover in relation to the population's distribution in the southern or northern part of its range. They also suggest that this highly specialist carnivore (cf. page 687) may benefit from its ability to utilize almost the whole of its prey and to survive for several weeks without food, though these two characteristics are not limited to specialist carnivores. Mayzaud (1976), from a study of changes in O_2 uptake and in nitrogen (NH_3) excretion in four zooplankton species, has suggested that body size and amount of lipid reserve not only influence to a marked extent the period of survival under starvation, but also are related to flexibility in the animal's metabolism (cf. Ikeda, 1977). With two species (*Acartia* and *Sagitta*) under starvation, an initial fall in O_2 uptake and NH_3 excretion was followed by a more-or-less steady state. The C/N ratio was generally maintained by *Sagitta* during the starvation period; protein was the major component lost, but lipid was reduced at a fairly constant rate so that the protein/lipid ratio remained steady. Unusually, a major portion of the lost lipid appeared to be phospholipid. (The copepod *Eurytemora*, which was also subjected to a few experiments, apparently lost triglyceride rather than phospholipid). The pattern of change was probably similar in *Acartia*, although the picture was not as clear, as different developmental stages were used. Protein, which amounted to 60–70% dry body weight, suggested the greatest loss, though a steady reduction of lipid (*ca.* 25 µg/day) occurred over the 6-day starvation period.

In contrast to *Acartia* and *Sagitta*, *Calanus*, which survived starvation over a much longer period, showed an initial fall in O_2 uptake, followed by a rise to an intermediate level and in some experiments, a further rise later. Ammonia excretion fell initially, but then oscillated. Mayzaud believes that *Calanus*, which showed losses of both lipid and protein during starvation, practises an alternation of lipid and protein metabolism. Carbon was lost faster than nitrogen and the R.Q. was somewhat lower than for *Acartia* and *Sagitta*, suggesting that lipid was used in metabolism, but this could change to periods of mainly protein utilization. The copepod appeared to have some degree of flexibility to switch the type of metabolism, thus conserving body nitrogen as far as possible. Such an alternation might be relevant to the interpretation of the results of Ikeda (1977) with *Calanus plumchrus*, in which nitrogenous excretion seemed to fluctuate over extended periods of starvation. In Mayzaud's experiments, when lipid was used triglyceride and wax ester tended to be utilized rather than phospholipid.

Much of the data given so far on O/N ratios relate either to animals tested soon after capture in the field or to those subjected to starvation. A very few species have been maintained in the laboratory so that they can be tested experimentally. The results of Reeve, Raymont and Raymont (1970) for *Sagitta hispida* fed on more or less natural zooplankton confirmed an O/N ratio of about 7, typical of an animal feeding carnivorously. Ferguson (1973) fed *Neomysis integer*, a neritic mysid, on different foods (cf. page 648). Control animals collected direct from the field gave an O/N ratio of 10. With natural detritus, which included some live algae, O/N ratios were rather lower (10.5) than with crab hepatopancreas as food (16), but both results indicated that protein was being partly metabolized. There was evidence that with natural detritus live cells were selected; much of the carbohydrate remaining from the dead material was probably relatively resistant "fibre". Hepatopancreas would presumably include a proportion of carbohydrate reserve and some lipid in addition to protein. Ferguson used kaolin impregnated with three different food materials: glucose and starch (pure carbohydrate), oleic acid (pure lipid) and casein (pure protein) in preliminary experiments with artificial foods. The O/N ratios obtained for *Neomysis* fed on these diets were 91, 26.5 and 9.5 respectively. The very high O/N ratio with glucose-starch confirmed that the mysids were metabolizing carbohydrate. On a pure lipid diet the O/N ratio was lower than expected; possibly the only lipid material available, oleic acid, was not readily assimilated and the animals may therefore also have utilized some body protein. With the casein diet the very low O/N ratio confirmed the expectation of extensive use of protein.

Assimilation

Knowledge of the proportion of food ingested by zooplankton which is assimilated is important not only with reference to the physiology of particular species but also in relation to the transfer of food material through the marine ecosystem.
Assimilation efficiency is usually defined as

$$\frac{\text{Food assimilated (weight or calories)}}{\text{Food ingested (weight or calories)}} \times 100.$$

Earlier field studies such as those of Harvey, Cooper, Lebour and Russell (1935) and Riley (1947) suggested that the zooplankton eats perhaps one-half to one-third of its weight of food per day, but the much smaller amount required for maintenance and sustaining growth (10–15%) would indicate that assimilation efficiencies were relatively low (25–33%). It was widely believed that zooplankton herbivores frequently showed very incomplete digestion of the algal food. The rapid passage of food through the gut of such active grazers as copepods and the voiding of green faecal pellets lent support to these views of incomplete digestion. Marshall and Orr (1955) suggested that with rapidly feeding *Calanus* the time for passage of the food was as little as 20 minutes when algal cells were present at high densities. Marshall (1973) emphasizes that this is approximately the maximum speed, with some 12 faecal pellets produced per hour and Sekiguchi (1975) has suggested 30 minutes as the time for total passage of food for herbivorous copepods.

Although rate of production of faecal pellets is not a precise method for estimating feeding rates, the data of Urry (1965) and of Corkett and McLaren (1978) indicated some regulation of pellet production at very high food concentrations, arguing somewhat against excessive feeding, a view supported by Arashkevich and Tseytlin (1978). The intensity of grazing of planktonic herbivores which can take place, however, led Beklemishev (1957, 1962) to propound the theory of superfluous feeding. In effect, he proposed that reduced assimilation efficiencies followed peak densities of phytoplankton crop; with such levels of plant production the phytoplankton available greatly exceeded the requirements of the grazers for growth and maintenance. An approximate level of phytoplankton was proposed beyond which superfluous feeding occurs. Beklemishev suggested a plankton biomass of about 3 g/m^3, equivalent to about 390 μg C/l. While the variability of cell biomass and cell carbon with species was acknowledged, another estimate indicated a level for superfluous feeding of *ca.* 10^4–10^6 cells/l. Beklemishev's view rests on the supposition that filtration rate in zooplankton herbivores such as copepods is not affected by cell concentration, so that filtration continues unabated whatever the algal density; feeding is largely automatic. This view should be contrasted with more recent theories on the effects of cell concentration on filtration rate and knowledge of different methods of feeding (cf. Chapter 6).

Sekiguchi (1975a) commenting on the comparative abundance of phytoplankton, copepods and faecal pellets in waters of Ise Bay and the Inland Sea of Japan, suggests that the concentration of faeces increased with greater density of phytoplankton to about 5 μg/l (as chlorophyll) but then declined, suggesting that the ingestion by the zooplankton reached a maximum intake and that superfluous feeding did not occur (cf. Arashkevich and Tseytlin, 1978). The waters in the area are extremely eutrophic and very high phytoplankton concentrations occur, so that the maximal rations of small copepods (*Acartia*, *Oithona* and *Microsetella*) are easily satisfied. Thus the concentration of faeces in the water and the chlorophyll (phytoplankton) density were not directly correlated. Petipa (quoted in Marshall, 1973) also believed that superfluous feeding was an important factor in the marine ecosystem, since copepods were estimated from the rate of disappearance of the phytoplankton population to take 50–400 % of their body weight daily. Estimates of food intake were also made from the weight of food in the gut and rate of passage through the alimentary canal and, for migrating copepods, from the changes in size of the fat storage sac in the body, assuming that all the lipid comes from the food and was deposited in the sac.

Arashkevich (1977) also studied the feeding rhythms of the ostracod *Cypridina sinuosa* and found that a single ration, which will completely fill and distend the gut, amounts to 25–30 % of the body weight (including the shell). Lipid stores accumulate in the tissues following digestion of the ration during the day; they act as a buoyancy mechanism and may presumably be used partly to supply energy demands including the next upward migration (cf. Chapter 5). Although Petipa's estimates for the daily requirements of *Calanus* are extremely high, she believes that superfluous feeding may occur, particularly when *Calanus* begins to lay eggs. Food concentrations indicated as critical levels for superfluous feeding were 3 mg/l for *Prorocentrum* (wet weight) or 5–12 mg/l for *Nitzschia* or *Coscinodiscus*. However, Arashkevich and Tseytlin (1978) maintain that when food reserves are high, active feeding ceases in phytophagous copepods and superfluous feeding is unlikely to take place.

The importance of intensity of grazing was emphasized by Cushing (1959) (cf. also Cushing and Vucetic, 1963). A copepod such as *Calanus* was believed to have very high food requirements of the order of 50 % of the body weight daily to sustain growth and maintenance. The grazing capabilities and feeding intensities of planktonic herbivores were believed to be very high, but according to Cushing during periods of diatom abundance an animal such as *Calanus* might consume three times its body weight of food per day.

If superfluous feeding occurs and assimilation efficiency is low, this would imply that digestion in herbivorous plankton is incomplete. Is the range of digestive enzymes in herbivorous copepods likely to ensure the proper digestion of food? Very few investigations have been carried out, but Conover (1968), summarizing earlier work, points out that the enzyme complement is capable of digesting all three classes of food. Several esterases, a peptidase and an amylase are listed for copepods in later work. The results of Vaccaro suggest also that bacteria existing in the gut of herbivores are capable of digesting chitin and alginic acid; cellulose digestion is probably similarly effected. A study by Molloy (1958) on *Neomysis integer*, an omnivorous semi-planktonic mysid, might also give some indication of the range of digestive enzymes in some zooplankton. Molloy demonstrated the presence of proteases, lipases, a chitinase, cellulase, esterase and amylase in gut extracts. Boucher and Samain (1974) have reported amylase activity in the digestive enzymes of a variety of zooplankton, including a number of copepods. They describe a semi-automatic method for analysing amylase activity and protein content of zooplankton and correlate enzyme activity with the nitrogen fraction of the diet (Samain and Boucher, 1974). Samain, Daniel and Coz (1977) also demonstrated protease, (trypsin) and amylase activity in zooplankton and used the relative intensities of the two types of digestive enzymes as indicators of carnivorous and herbivorous feeding. Mayzaud and Poulet (1978) have identified several proteases and carbohydrases from copepod populations (*Pseudocalanus minutus*, *Acartia clausi*, *Temora longicornis*, *Oithona similis* and *Eurytemora herdmani*) from eastern Canadian waters.

There are strong indications, therefore, that digestion by zooplankton ought to be reasonably complete. Perhaps, however, the food moves through the gut during very heavy grazing too rapidly for adequate digestion, especially where the relatively large ash content of diatom food, in particular might tend to dilute or reduce the efficiency of the enzymes. Aside from the level of daily demand, Cushing suggests that the loss of algal tissue to the herbivorous zooplankton population may be due less to incomplete digestion than to spoilage of cells during the grazing process.

The extent of spoilage during grazing is still uncertain. Marshall and Orr (1955) refer to the possibility. Petipa (quoted in Marshall, 1973) observed that when large cells were broken by *Calanus* some of the contents were lost. Conover (1966) also described a loss of 15% of the particulate carbon in *Calanus hyperboreus* feeding on large cells. Vyshkvartseva (1975) confirms the observations of many earlier workers that species of *Calanus* eat smaller cells whole, whereas large algae, such as some species of *Coscinodiscus*, are broken up by the mandibles before the majority of the contents are swallowed. She illustrates cells of *Coscinodiscus* which had been damaged by the mandibles during the feeding of *Calanus glacialis*. Vyshkvartseva believes that *Calanus* species from high latitudes have a greater ability to break large cells, whereas those from lower latitudes tend to swallow most of the food whole. This might explain some differences that have been found in testing for spoilage of food. Conover held that some degree of breaking of large diatoms may occur for the Arctic/boreal species, *Calanus hyperboreus*; on the other hand, Corner, Head and Kilvington (1972), using

British species of *Calanus* and testing their feeding on large cells of *Biddulphia*, found very little evidence of spoilage, although assimilation was relatively poor.

Though some field observations have tended to support the theory of superfluous feeding, it has been difficult to match such observations with controlled laboratory experiments. Daily food requirements not only vary with species but are also related to the stage of the grazer. Paffenhofer (1971) observed in rearing *Calanus helgolandicus* in the laboratory that whereas adult *Calanus* had daily food requirements which were not very different from estimates suggested by earlier workers, young stages had much greater demands in terms of percentage body weight. For example, Nauplius V had daily requirements equivalent to at least 300% of the body weight (*vide supra*). Paffenhofer thus argues that calculations of food requirements for a grazing population must take account of the great variation in requirements of different developmental stages. The additional data now available on the daily food requirements of developmental stages of other species (*Temora*, *Acartia*, *Pseudocalanus*) may also assist in calculating demands of mixed populations. Tomasini and Mazza (1978) found for *Centropages typicus* and *Acartia clausi* that although ingestion rate increased with cell density to an optimal level and consumption increased at levels far in excess of that encountered in the sea, filtering rate often fell and was not automatic. They suggest that the copepod's feeding rate stabilizes in relation to a regulation of the animal's metabolic needs. Superfluous feeding would thus be unlikely.

Testing the theory of superfluous feeding in the laboratory is difficult, since experiments often tend to give a wide range of results. Variations in season and ambient temperature influence rate of digestive ability and consequently ingestion rate (Mayzaud and Poulet, 1978). The method employed in most early experiments consisted of placing a number of animals of a herbivorous species (e.g. a calanoid) in a concentration of a unialgal cell culture and subsequently carrying out an assessment of the cells consumed and the faeces produced. Such experiments are subject to complications such as the effects of crowding of animals and confinement in containers and the accuracy of cell counts, particularly at low concentration, as already discussed for grazing experiments. But an additional major difficulty is the total recovery of the faeces produced, an essential element in the experiment.

The use of a radioactive tracer, usually ^{32}P, by Marshall and Orr in grazing experiments provided another technique for estimating assimilation efficiency. Their results were of great interest since, though a considerable range occurred in the experimental data, the assimilation efficiencies recorded were generally high. This was true more or less irrespective of the feeding intensity, which though variable, could at times be very high. Efficiencies varied with the type of food. For diatoms it was nearly always high; four species, for example, gave efficiencies always greater than 50% and usually greater than 80% (Marshall and Orr, 1955b). Dinoflagellates showed similar high efficiencies—normally greater than 60%, usually above 80%, sometimes exceeding 90%. Some of the flagellates tested gave very high efficiencies, even exceeding 90%, but again a considerable range occurred.

A few species of flagellates (*Chromulina*, *Dicrateria* and *Chlorella*) gave very low assimilation efficiencies. Reference was made earlier to the observation that these species were shown to be ineffective as food for *Calanus*; indeed, *Chlorella* was largely undigested. With the diatom *Skeletonema*, which is so widely grazed during the spring increase, despite a wide range in efficiencies, generally very high values were found (62–87%). It is also of interest that with the diatom *Chaetoceros*, even when grazing was extremely intense and faecal pellets were being produced as frequently as every 5–6 minutes, a very high efficiency of assimilation (>80%) was still preserved. Despite doubts that some of the experiments using ^{32}P might have certain inherent errors, other experiments using labelled ^{14}C as tracer confirmed the generally high assimilation efficiencies. Presumably, however, any technique using biologically active tracer might be open to the criticism that a proportion of the isotope assimilated might be lost through respiration and excretion during the course of the experiment. Other complexities, particularly involving cycling and metabolic pools, are discussed by Conover and Francis (1973).

Despite these problems, several investigators have successfully used isotope methods for estimating assimilation efficiency. Thus Berner (1962) confirmed high efficiencies using the ^{32}P method with the copepod *Temora* (50–98%); ^{14}C is even more widely used. Some limited data are included by Corkett and McLaren (1978) which indicate that *Pseudocalanus* did not digest the dinoflagellates *Gymnodinium kowaleskii* and *Peridinium trochoideum* very efficiently; assimilation efficiency was only 38% and 24% respectively. Other foods, including detritus from *Platymonas*, showed much higher assimilation efficiencies (71%) and the range for most foods offered was 53–82%. However, in view of the generally low and variable food intake and the high proportion of radioactive carbon estimated as respired compared with the quantity retained in the tissues, the differences in the assimilation efficiency between various foods may not be significant.

The experiments of Corner (1961) had the advantage of using a constant flow method to estimate assimilation efficiencies as the ratio of total carbon removed to total carbon as faeces. In experiments on *Calanus helgolandicus*, Corner confirmed high efficiency levels (74–91%). Other experiments include those of Mullin (1963), who used essentially the total recovery method, but made estimates as carbon. He found a very wide range in efficiencies with some low values. Taguchi and Ishii (1972), using dry weight and carbon measurements, reported assimilation efficiencies of 75% and 85% for *Calanus cristatus* and *Calanus plumchrus* respectively.

An important alternative technique is that suggested by Conover (1962, 1964, 1966 a,b)—the so-called "ratio" method. This is a convenient way of measuring assimilation efficiencies, since only samples of food and of faeces need be recovered for subsequent determination of the proportion of organic matter and ash. Knowing these values, the assimilation efficiency (U) may be calculated according to the following equation:

$$U = \frac{F' - E'}{F'(1 - E')}$$

where F' = ratio of organic matter in food;
E' = ratio of organic matter in faeces.

i.e.

$$F' = \frac{(F - A_f)}{F} \quad \text{and} \quad E' = \frac{(E - A_e)}{E}$$

where F and A_f are the dry weights of food and of ash in the food, respectively; E and A_e are the dry weights of faeces and of ash in the faeces, respectively.

Certain assumptions must be made in using this ratio method: the organic matter and ash must be ingested in the same proportions as the food itself; there must be no loss of food by regurgitation or other occurrence except faecal production; the ash must be unaffected by digestive processes. The use of an inert material in the food (e.g. chromic oxide) has been suggested as a check on accuracy. Conover has proved that the ratio method for herbivorous zooplankton species gives a fairly accurate measurement; his (1966a) comparison of the direct method and the ratio method was satisfactory. Although Conover's earlier data for *Calanus hyperboreus* feeding on diatoms (cf. Conover, 1964) included low efficiencies, they were hardly ever less than 40 % and although the later data (Conover, 1966a,b) showed a considerable range with occasional low values, in general relatively high efficiencies were recorded, e.g. 60 %–65 % for three diatoms. With some foods (*Thalassiosira fluviatilis*, *Exuviaella*, *Dunaliella*) maximal efficiencies exceeded 80 %.

Different species of algal food gave different efficiencies, though Marshall and Orr (1955b) had earlier shown that many diatoms and dinoflagellates gave similar percentage assimilations. Conover suggests that digestibility is unlikely to be very variable unless particularly resistant materials occur in the food. In any case differences in exact assimilation efficiencies between foods were not entirely consistent. But, there was a clear relationship for percentage assimilation and ash content, even for the same food species; percentage ash was negatively correlated with assimilation efficiency. Conover suggests that the thicker cell wall may protect the food cells from the digestive enzymes or that the inert ash competes in some manner (perhaps by adsorption) for enzymes. Conover's results suggest that neither temperature, age of culture nor cell concentration of algae affected the assimilation efficiency in *Calanus hyperboreus*. The lack of any clear relationship between food concentration and percentage efficiency and the high level of assimilation efficiency recorded for many foods argues against the hypothesis of superfluous feeding.

Reference has already been made to investigations, in which assimilation calculated on a basis of nitrogen analysis and by the ratio method was shown to be relatively high (54–68 %; mean 61.7 %) for *Skeletonema* fed to *Calanus* (cf. Corner *et al.*, 1967). Calculations of assimilation efficiency by Butler *et al.* (1970) for *Calanus* captured in the field, based on the amounts of nitrogen and phosphorus laid down as tissue and the quantities excreted and lost with faeces, indicated a high assimilation efficiency (77 % as phosphorus; 62 % as nitrogen) during spring time. Although Corner *et al.* (1972) found a much poorer assimilation for *Calanus* fed on *Biddulphia* (40 % as phosphorus; 34 % as nitrogen) than when feeding on natural phytoplankton, there was no evidence of wasteful feeding on *Biddulphia*. Why digestion and assimilation should be so much poorer for *Biddulphia* is not clear: it appears not to be due to ash content. A possible explanation appears from a later study by Sargent, Gatten, Corner and Kilvington (1977), who examined the assimilation of lipid from *Biddulphia sinensis* by *Calanus*. Uptake appears to be substantial and rapid for Stage V and female copepods, with Stage V depositing most lipid as wax ester and females laying down triglyceride and wax ester equally.

Although assimilation efficiency in terms of lipid is not directly calculable since the amount of lipid lost in faeces was not measured, an estimate from the amount of lipid deposited suggests a minimum value for assimilation efficiency of 60 % for Stage V and 64 % for females, a value similar to that also quoted for a hydrocarbon (60 %). This value considerably exceeds that for nitrogen (34 %), but as Sargent *et al.* point out, this does not necessarily indicate that lipid is more efficiently absorbed from the gut than protein. Amino acids from dietary protein may be deaminated and the carbon skeleton subsequently used for the synthesis of lipid. Also lipids are probably formed from other substances (e.g. carbohydrates).

In summary, in spite of some conflicting evidence, there is fairly strong support for moderately high assimilation efficiencies for grazing calanoids. However, green faecal pellets are frequently recovered in experiments, and there are even reports of algal cultures successfully started from viable cells remaining in faecal pellets. The number of such viable cells is probably extremely small and they do not form a significant proportion of the algal food densities. The green faeces, Conover suggests, may also indicate partly digested chlorophyll (phaeophytins) rather than genuinely undigested cells.

Estimates of assimilation efficiency in zooplankton apart from copepods include those of Lasker (1966), who found in short-term experiments, feeding *Euphausia pacifica* with unicellular algae labelled with ^{14}C, very high assimilation efficiencies, usually exceeding 90 %. When the euphausiid was fed on ^{14}C-labelled *Artemia* nauplii, assimilation was also high, varying from 66 %–95 % (mean 84 %), though there was a suggestion that the available food was not sufficient in all experiments to supply all the metabolic needs. In food selection experiments *Euphausia* clearly favoured animal food, and there was evidence that flagellates and dinoflagellates were filtered more readily than some diatoms. Moulting depressed feeding.

Ferguson (1973) investigated assimilation efficiencies in the semi-planktonic mysid *Neomysis integer*. This animal will feed in the laboratory on a variety of foods including phytoplankton, animal food and detritus. Ferguson found that efficiencies were generally high, even exceeding 80 % with some diatoms, but that there were specific differences; for example, *Skeletonema* gave a lower efficiency than *Coscinodiscus*. Very high efficiencies were obtained on animal diets; with dead mysids assimilation efficiency was especially high—of the order of 90 %. It is perhaps significant that the animals are naturally cannibalistic and also scavengers. Low efficiencies were usually obtained with detritus; this included "artificial" detritus formed by growing unialgal cultures in the laboratory and then allowing these to degenerate in darkness, "natural" detritus obtained from the local environment and a mixed algal mat collected from aquaria which included dead detrital organic matter as well as living cells (cf. Figs 7.11, 7.12).

Assimilation efficiency was also studied in relation to the different major food components. Ferguson fed *Neomysis* on a selection of animal and plant food which had been analysed to determine the basic biochemical composition. Assimilation of the carbohydrate component in the diet was high (75–80 %) on both animal and plant foods, although dead *Neomysis* food, which had the lowest carbohydrate content, gave the highest assimilation efficiency. Protein assimilation was also high (75–90 %), with dead *Neomysis* food yielding the maximum efficiency. Assimilation of the lipid component was generally high for various foods (usually exceeding 80 %). The assimilation efficiency with the diatom *Skeletonema* was lower for all three components (*ca.* 65 %).

Salinity had little or no effect on assimilation efficiency in *Neomysis*. Variations in temperature were not followed by marked changes; the main effect of temperature was

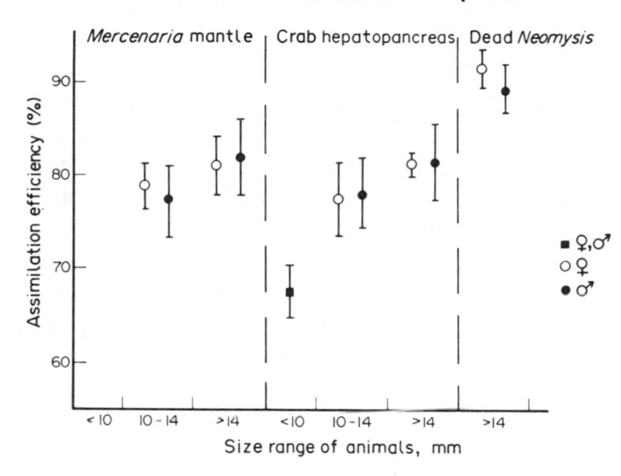

Fig.7.11. Assimilation efficiencies in *Neomysis integer* feeding on animal food (Ferguson, 1973).

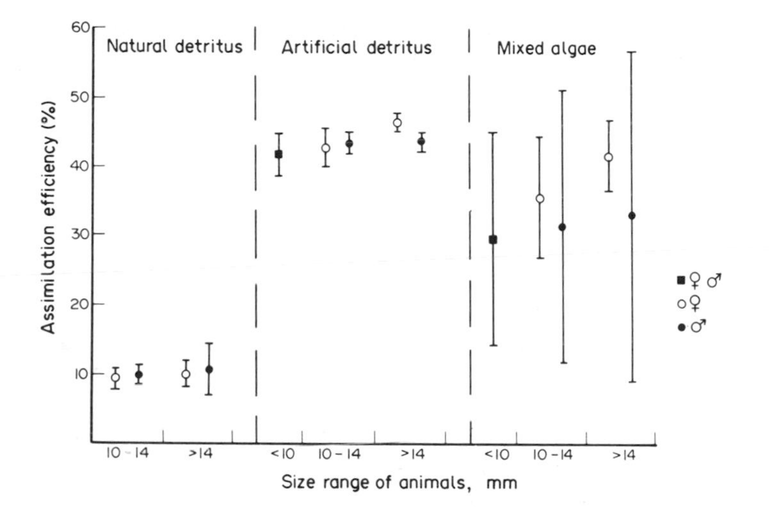

Fig. 7.12. Assimilation efficiencies of *Neomysis integer* feeding on detritus and mixed algae. Vertical lines represent ± 2 k standard error of the mean (Ferguson, 1973).

on ingestion rate. The release of young by mature female mysids, however, caused a temporary lowering of assimilation efficiency; moulting caused a similar effect (cf. Lasker's results with euphausiids).

The assimilation efficiency of the mainly carnivorous copepod *Chiridius armatus* was found to be high by Alvarez and Matthews (1975). The mean value was 90% for females fed on *Artemia* nauplii and 95% when fed on mixed zooplankton. The authors point out that such short-term experiments frequently do not take growth requirements nor uptake of inorganic matter into account. Thus in some of the experiments carried out by Conover on *Calanus hyperboreus* growth was considerable, although assimilation efficiency was rather low. In crustaceans which do not moult as adults, there will be a difference in considerations of assimilation efficiency between pre-adult and adult stages. With pre-adults, inorganic material may well be incorporated into the body from the food, especially in relation to the formation of exuviae. In the adult the main transfer of assimilated food will be organic matter only, chiefly to the ovary. Alvarez and Matthews believe that this may in part account for their assimilation efficiencies being generally higher than those quoted by earlier workers (*ca.* 70% for *Exuviaella* and 85% for *Skeletonema*).

Whether assimilation efficiencies tend to be greater with animal diet should also be considered, especially in view of the results of Ferguson and of Conover and Lalli (1974). The proportion of ash may be a significant factor (*vide infra*). Reeve and Walter (1978) found that digestive efficiency of the ctenophore *Mnemiopsis* feeding on *Acartia* was comparatively high (*ca.* 74%) and of the same order as for chaetognaths and herbivorous zooplankton. The experimental food concentrations must be restricted, however, to food of up to 100 *Acartia*/l. At higher densities (up to a concentration of 1000 *Acartia*/l) the ctenophores reject much of the prey, which is then only partly digested and wrapped in excess mucus. The efficiency was very variable but averaged as little as 20%.

Ferguson's experiments on *Neomysis* gave clear indications that ash content of the food is inversely related to assimilation efficiency, supporting Conover's (1966) suggestion for *Calanus*. The high efficiencies obtained with animal food, especially mysid flesh, suggest the "suitability" of the biochemical composition of an animal diet, rather than merely relative ease of digestion (cf. the hypothesis of "specific dynamic effect of food"). Ferguson's results on assimilation efficiency in relation to biochemical components support the suggestion.

How far external factors such as temperature, salinity, food concentration and quality of food, apart from "internal" factors, including stage of development, body size, moulting, sex and degree of maturity, affect assimilation efficiency in zooplankton generally is not altogether clear from the relatively few studies so far carried out. Most of the present evidence indicates that concentration of food, as Conover suggested earlier for *Calanus hyperboreus*, is not a major factor. Thus Sushchenya (1970) points out that while assimilation of food varies directly with food concentration until the maximal ration is attained, assimilation efficiency does not appear to be much affected by food concentration until the maximum ration is approached. Sushchenya found the relationship $A = 0.66R$ to hold for *Artemia*, where A = assimilation efficiency and R = ration (i.e. assimilation efficiency or percentage assimilation is practically constant at 66% until maximum assimilation is reached). In some zooplankton, lower percentage assimilation may appear at very high food concentrations.

Cosper and Reeve (1975) point out that estimates of gross growth efficiency (cf. page 668 *et seq.*) in zooplankton are often high and that a large proportion of assimilated food is used in metabolism and thus assimilation efficiency must necessarily be high. Values found generally for particle-feeding zooplankton confirm this. Data for assimilation efficiencies for carnivorous feeding of zooplankton are few. Apart from the high values just quoted from Ferguson and from Alvarez and Matthews, and Lasker's value of 84% for *Euphausia pacifica*, Cosper and Reeve (1975) found a value of around 80% for the chaetognath *Sagitta hispida* feeding on copepods. *Sagitta* apparently assimilates variable quantities of ash from its food so that the Conover ratio method is not reliable in studying this species and generally gave lower values. *Sagitta* feeds on fairly large food particles compared with a herbivore diet, but the assimilation efficiency is not affected by food size.

Conover and Lalli (1972, 1974) examined the feeding of the gymnosome *Clione limacina* on the thecosome *Limacina* (*Spiratella*) *retroversa*. They first established a relationship for the size (shell diameter) of *Limacina* and its carbon and nitrogen content, so that the number of *Limacina* consumed could be expressed as carbon or nitrogen. Very little faecal material was produced when *Clione* fed on *Limacina*, suggesting substantial utilization of the food. Carbon and nitrogen analyses of the small amount of faecal material produced enabled an assimilation efficiency to be calculated. This was surprisingly high—at least 95 % for carbon and approaching 100 % for nitrogen.

Corner *et al.* (1976) also studied assimilation efficiency in carnivorous feeding. They describe a method of determining efficiency in terms of nitrogen without collecting and analysing faecal pellets. The following equations would apply:

$$I = G + T + E$$

and

$$AI = G + T$$

or

$$A = \frac{G}{I} + \frac{T}{I}$$

where I = ingested daily ration as N
G = growth in body N
T = N excreted
E = N as egested faeces
A = assimilation efficiency.

If the gross growth efficiency (K_1) (*vide infra*), which is equal to $\left(\frac{G}{I}\right)$, is determined in terms of nitrogen and the excretion of nitrogen in relation to nitrogen intake $\left(\frac{T}{I}\right)$, the assimilation efficiency (A) may be calculated.

Using *Calanus helgolandicus* feeding carnivorously on *Elminius* nauplii, Corner *et al.* demonstrated that assimilation efficiency was very high, as with some other carnivorous zooplankton. When feeding maximally, efficiency was approximately 90 %. There was a just discernible effect of size of ration; assimilation varied overall between 80 % and almost 100 %. The much lower values previously quoted for assimilation efficiency in *Calanus* when grazing herbivorously (*ca.* 60 % or less) underline the "suitability" of animal food. The basic biochemical composition of animal flesh may perhaps lead to a lower energy requirement for digestion and absorption. The greater proportion of ash in phytoplankton may also cause digestion to be less efficient.

The possibility that certain zooplankton might vary their assimilative abilities is relevant when considering a change of diet from herbivorous to carnivorous feeding and variations in food concentration. The well recognized patchiness of plankton implies that plankton species could gain appreciably from an ability not only to vary their feeding rate to reap maximum advantage from a rich food patch but also from a capacity for raising or lowering their assimilative abilities to match the variable seasonal standing crops, particularly in neritic areas. Mayzaud and Poulet (1978) found a linear relationship between feeding rate and food supply for five species of neritic copepod common in the Bedford Basin (*Pseudocalanus minutus*, *Oithona similis*, *Acartia clausi*, *Temora longicornis* and *Eurytemora herdmani*). The animals adjusted their ingestion rate over comparatively long time periods (weeks or months), although short term experiments indicated the usual saturation of feeding with maximum ration. Several digestive enzymes (protease and carbohydrases) obtained from the copepod population showed seasonal adaptation to the type of food available. The levels of digestive enzymes were found to vary linearly on a seasonal basis. There was also adaptation to the biochemical nature of the food. In June, when protein content of the food was relatively high, the high affinity of the acidic protease enzyme (low half-saturation constant) indicated an adaptation by the copepods to the richer protein diet. Conversely, in late July the high half-saturation constant for the protease reflected a diet richer in carbohydrate. The variations in the initial slope of the saturation curve with time also suggest a strong reactivity for all the four enzymes studied.

Growth Rate and Growth Efficiency

If the total food ingested by an animal is designated as C, then:

$$C = G + M + R + E + F,$$

where G = growth;
M = respiratory metabolism;
R = reproduction;
E = soluble excretion;
F = faeces (growth in crustaceans must include moults).

In the relatively few studies of marine zooplankton energy budgets; E is sometimes not separated from F. Then:

$$C - F = G + M$$

for animals not reproductively active.
This emphasizes that, reproduction apart, the excess intake of food over metabolic requirement is available for growth. These energy budgets may be conveniently considered in terms of calories, though growth is more usually measured as change in weight (ΔW), frequently estimated from body size (length) measurements. Thus:

$$\frac{\Delta W}{t} = C - M$$

where t = time period.
A useful measure of the efficiency of an animal in converting food intake to growth is therefore:

$$\frac{\Delta W}{tC}.$$

This is usually termed the gross growth efficiency, frequently designated as K_1.

But if A = food assimilated, sometimes described as the "physiologically useful ration", then:

$$A = C - F$$

and

$$\frac{\Delta W}{tA}$$

is a measure of growth rate in relation to the amount of food assimilated, termed nett growth efficiency and often designated as K_2.

The nett growth efficiency must be higher than the gross efficiency for a given species, but growth efficiencies generally in zooplankton could be fairly low, since it has already been pointed out from work on *Calanus* that a large percentage of the food intake is utilized in oxidative metabolism. Nevertheless, growth rates in some zooplankton may be high given optimal environmental conditions, particularly a rich food supply.

Problems of growth rates and growth efficiencies in herbivorous zooplankton have been examined by Mullin and Brooks (1967, 1970a,b). They reared the copepods *Calanus pacificus* and *Rhincalanus nasutus* in the laboratory and made measurements of the food consumed and the body weight of developmental stages of the copepods in terms of carbon. Estimates were also made of respiratory rates of the two calanoids.

Growth was approximately described by the expression:

$$W_t = W_0 e^{kt},$$

where W_0 and W_t are initial and final body weights as carbon;
t is time in days.

Despite growth being discontinuous in copepods in so far as a discrete series of moults occurs, there is presumably increase in body weight between moults.

Earlier experiments (1967) on *Rhincalanus* indicated that the amount of carbon per unit dry weight increased with body size, probably reflecting storage of lipid as the copepod approached maturity. The rate of growth appeared to vary with temperature and food concentration (presumably to maximal food ration). Mullin and Brooks (1967) observed that growth rate as expressed by (k) declined with the period in the life history. For *Rhincalanus*, for three major periods of development, results were:

$$k = \begin{array}{l} \text{N I–C II} \simeq 0.24/\text{day} \\ \text{C II–C IV} \simeq 0.17/\text{day} \\ \text{C IV–C VI}♀ \simeq 0.12/\text{day}. \end{array}$$

These k values did not take account of the amounts of carbon lost in exuvia during moults, which were equivalent to about 10% carbon per moult. The results were not unexpected; it is characteristic of many animals that growth declines with age. Of the few zooplankton species studied, some appear to reach a clear maximum size; other species, e.g. *Sagitta hispida*, continue to exhibit some increase in body size even as relatively old animals (cf. Fig. 7.13). The relationships may be somewhat complicated, however, as some crustaceans are characterized by having a specific number of instars in development and not moulting as adults (e.g. copepods), whereas others moult throughout life (e.g. euphausiids). Even if body length does not alter in adult life, ripening of the gonads may involve body weight increase.

Mullin and Brooks (1970b) demonstrated in experimental culture of *Calanus*

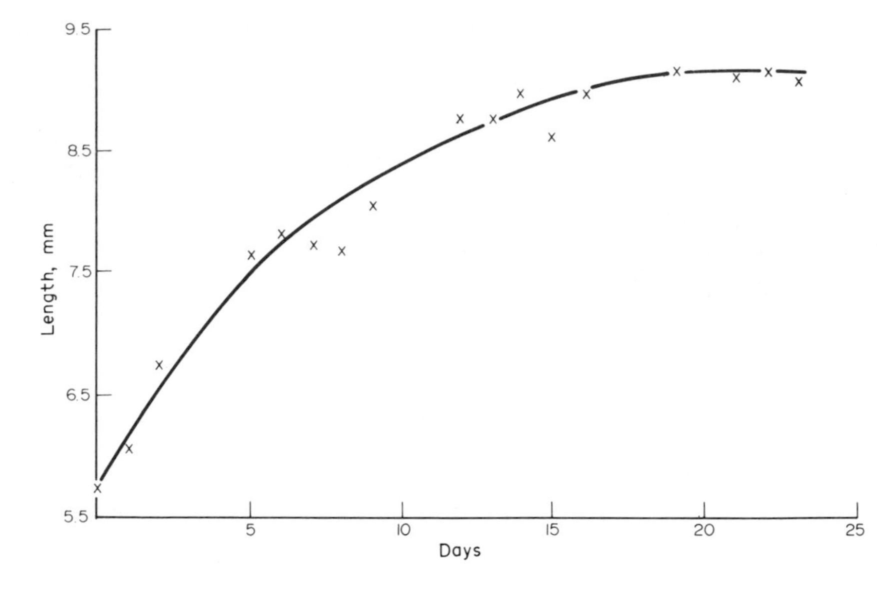

Fig. 7.13. Growth rate of *Sagitta hispida* in terms of length increase with time at 21°C (Reeve, 1970a).

pacificus on *Thalassiosira* and on *Gymnodinium*, from N I to C IV, an increase, as expected, in the duration of development with reduction in the food concentration especially below 200 μg C/l. However, the total amount of food ingested over the period declined with lower food densities below 300 μg C/l; the mean weight of the C IV also declined. There was no clear evidence for any decrease in gross growth efficiency at lower food concentrations, but there was an indication of an inverse relationship between total ingestion and gross growth efficiency (Fig. 7.14). The animals maintain a high efficiency at lower concentrations and achieve a given stage of development at a smaller size.

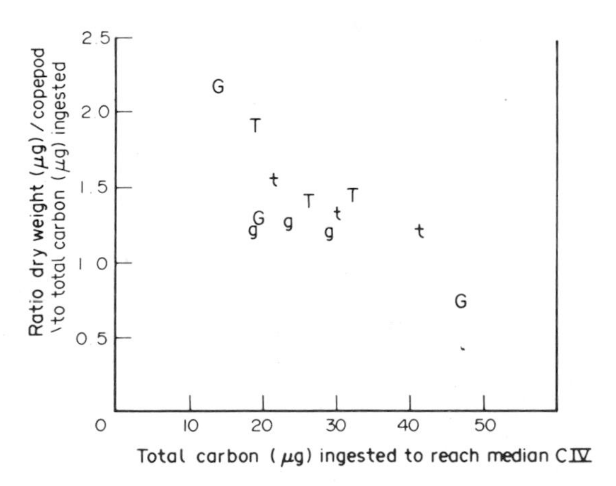

Fig. 7.14. Relationship between the cumulative amount of food ingested and ratio of the dry weight of a median C IV *Calanus*. G = Gymnodinium, 12°C; g = Gymnodinium, 17°C; T = Thalassiosira, 12°C; t = Thalassiosira, 17°C (Mullin and Brooks, 1970b).

Within limits the copepods appear to make metabolic compensations to achieve high nett and gross efficiencies at lower food densities. If assimilation efficiency is independent of food concentration, as indicated earlier, it is not a question of more efficient assimilation of food at lower concentrations. The main effects of lesser food availability appear to be slower development, smaller size and the likely effect of the latter on fecundity.

Using both *Rhincalanus nasutus* and *Calanus pacificus*, Mullin and Brooks (1970a) examined the effect of two foods, *Thalassiosira* and *Ditylum*, but with *Calanus* the data refer only to *Thalassiosira* since the nauplii will not eat *Ditylum*. Both copepods, however, showed a preference for larger food particles. Rate of growth decreased with age in both copepods (cf. Fig. 7.15) except for experiments in which *Rhincalanus* was fed *Thalassiosira* at 15°C (Table 7.11). *Thalassiosira* generally appeared to be less satisfactory as a food for *Rhincalanus*, the nauplii growing less rapidly and ingesting the food at lower rates. But neither temperature nor type of food has any clear effect on body size (expressed as weight of carbon) nor on the C/N ratio of the body. A comparison of the coefficient of exponential growth (k) in the equation:

$$W_t = W_0 e^{kt} \qquad (\textit{vide supra})$$

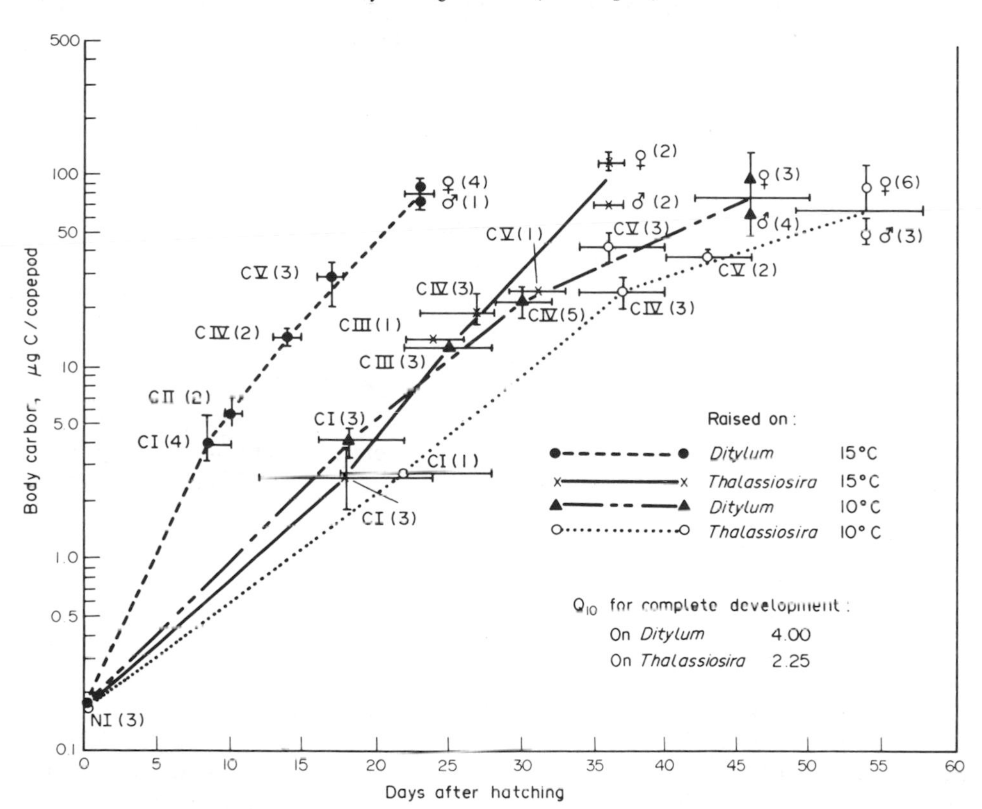

Fig. 7.15. Growth of *Rhincalanus* under different experimental conditions. Vertical and horizontal lines indicate the ranges around each point and the numbers in parentheses indicate the number of body carbon measurements. Lines have been drawn from NI–CI, CI–CIV and CIV to adult representing three developmental stages (Mullin and Brooks, 1970a).

Table 7.11. Comparison of coefficient of exponential growth (*k*) for three different diets and growth periods of calanoids (Mullin and Brooks, 1970a)

Diet	Growth coefficient (*k*/day)			Temperature (°C)
	NI–CI	CI–CIV	CIV–CVI	
Rhincalanus on	0.64	0.22	0.17	15
Ditylum	0.18	0.14	0.08	10
Rhincalanus on	0.15	0.22	0.18	15
Thalassiosira	0.13	0.15	0.06	10
Calanus on	0.30	0.27	0.22	15
Thalassiosira	0.18	0.17	0.08	10

may be made for the three major growth periods in development N I to C I; C I to C IV; C IV to C VI (see Table 7.11). The results indicate that *Calanus* apparently grew better than *Rhincalanus* on *Thalassiosira* diet. In both calanoids the rate of growth generally fell with age, confirming earlier results. The lower *k* values obtained for *Rhincalanus* on a *Thalassiosira* diet underline the classification of the latter as a less satisfactory food, though this did not affect final body size. Both copepods showed slower growth at 10°C as compared with 15°C.

Paffenhofer and Harris (1976) and Harris and Paffenhofer (1976a) studied the growth of two species of calanoid, *Pseudocalanus elongatus* (herbivorous) and *Temora longicornis* (omnivorous), maintained in continuous culture and feeding on *Thalassiosira rotula*. Detailed study of growth was limited mainly to the copepodite stages. From C I to C V, growth was exponential at all food concentrations. Data on newly hatched nauplii (N I) as well as on the copepodite states and on adults (C VI) permit the calculation of growth rates as *k* values for the range of development periods.

In *Pseudocalanus* the fastest growth was achieved from C I to C III/IV, falling slightly to C V/VI; after the moult to the adult, growth was almost zero. Lower food concentrations affected growth rates, particularly in the C III/IV to C V/VI phase. The highest *k* value was 0.38. Corkett and McLaren (1978), whilst pointing to laboratory evidence in addition to extensive field data for an effect of temperature on body size in *Pseudocalanus*, suggest that there appears to be no obvious effect of food concentration on body size, apart from decreased food supply causing a retardation of development (i.e. a slower growth rate—cf. Paffenhofer and Harris). Higher temperature is related to decrease in body size, but while this relationship may be seen in most of the development, the effect, from Thompson's results, seems to vary with developmental stage. Thus, from N V to C I there is some indication of a positive effect of temperature on size; the negative relationship is especially clear in later copepodite stages. Paffenhofer and Harris (1976b) found that both grazing and ingestion rates in *Pseudocalanus* increased linearly with body size (weight) from C I to C VI. Ingestion also increased with food concentration; at a concentration of 24 μg C/l the daily ration was equal to 63% of body weight; at 200 μg C/l the daily ration was 148%.

Temora nauplii had low growth rates and as for *Pseudocalanus* growth was fastest from C I to C III/IV, decreasing during the late copepodite (C V) stage. After the moult to the adult, however, a considerable increase in weight took place, especially in females, presumably owing to egg development. Growth rate was reduced at the lowest food concentration; the resulting adult copepods were also smaller, a result in contrast to that for *Pseudocalanus*. The k value did not exceed 0.3, except for one experiment (*ca.* 0.5). Grazing and ingestion rates increased linearly with body weight from C I to C VI; grazing also increased with size in the naupliar stages, but there appeared to be a depression of feeding as N VI approached the moult to C I. At a food concentration of 25 μg C/l the mean daily ration was 80 % of body weight; at 200 μg C/l the daily ration rose to 145 %.

As far as the data indicated for both calanoids, ingestion per unit body weight was more or less constant throughout the life history, in contrast to Paffenhofer's (1971) earlier findings for *Calanus helgolandicus*.

A later analysis by Paffenhofer (1976a) for the calanoid *Calanus helgolandicus*, comparing growth rate and efficiency on diatom (*Lauderia*) and dinoflagellate (*Gymnodinium, Gonyaulax, Prorocentrum*) diets, indicated that growth coefficients were relatively high and approximately similar for stages N I to C III; there was a decrease for later copepodites, with the lowest value typical of young adults (C VI). During the faster-growing younger stages, growth coefficients were generally higher on dinoflagellates than on *Lauderia*. On the diatom diet greater body weights were obtained for all developmental stages with increased food concentrations, a result different from that of Mullin and Brooks. However, this relationship was not always true for *Calanus* reared on a dinoflagellate (*Prorocentrum*) diet. Some effect of specific diet on adult size was apparent; adults fed on *Prorocentrum* were heavier than those reared on *Lauderia*. The general decline in growth rate with age, confirmed in these experiments, agrees, as Paffenhofer points out, with results obtained by Petipa, who found a fall in daily (calorific) growth after the early copepodite stages in *Acartia clausi*. Petipa's results with *Calanus*, however, were somewhat different; growth rates were generally lower and nauplii showed a relatively slow growth (cf. Table 7.12).

Table 7.12. Mean daily increment of one copepod expressed as percent of the caloric content of one copepod (Paffenhofer, 1976)

Copepod stage	*C. helgolandicus*	*A. clausi*
Nauplii (N III to N VI)	6.3	20.3
C I	21.0	21.3
C II	22.5	18.2
C III	21.7	24.3
C IV	21.5	13.7
C V	5.9	8.3
F	3.2	1.2 (eggs)
M	—	—

Although the earlier experiments of Conover (1964) with *Calanus hyperboreus* were confined to late copepodite stages, so that discussion of growth rates is hardly justified, the data permit calculation of gross and nett growth efficiencies in terms of weight and calorific value for copepods fed on diatoms in the laboratory. Gross growth efficiency (K_1) ranged from *ca.* 13—36 %, mean 21 % (determined on a dry weight basis); on a calorific basis the values were higher (18—50 %; mean 30 %). Nett growth efficiency

(K_2) ranged from 25—64 % (dry wt); from 34—89 % (on calorific values). The higher values were influenced by *C. hyperboreus* laying down fat. Conover (1964, 1968), whilst pointing out that few attempts had been made to determine growth efficiencies in zooplankton, gave some data for freshwater zooplankton and other animals (Table 7.13). Among the data was Reeve's report of very high gross growth efficiencies (more than 60 %) for *Artemia*. Factors such as temperature, developmental stage and food concentration might affect growth efficiency; *Artemia* showed decreased efficiency with increase in ration and a higher growth efficiency for animals between 15 and 30 days of age.

The investigations of Mullin and Brooks (1970a) permit the calculation of growth efficiencies for *Calanus* and *Rhincalanus*. The gross growth efficiencies (K_1) varied considerably over the different growth periods (Table 7.14). For the whole development, K_1 ranged from 30–45 %, which is of the same order as earlier data on *Calanus*. Mullin and Brooks, however, did not find the clear decline in gross growth efficiency with age typical of vertebrates. For *Rhincalanus* there was no significant difference in K_1 between copepods reared on the two diets, nor any obvious difference in K_1 between *Calanus* and *Rhincalanus* reared on the same diet. Temperature also had no effect on gross growth efficiency for either copepod. They suggest that temperature probably acted approximately equally on ingestion and respiration so that the same fraction of the food consumed would be available for growth.

Temperature appears to have little effect on body size and total food ingested and mainly affects rates of ingestion, respiration and growth. Since assimilation efficiency apparently does not vary with food concentration (cf. page 666), even if *Calanus* and *Rhincalanus* ingest the same amounts of C/day in the field as laboratory reared animals, their gross growth efficiency (K_1) will probably be lower since they must spend more energy searching for food and in diurnal vertical migration. Marshall (1973) has summarized some of the growth efficiency data for *Calanus* spp. and *Rhincalanus*, which have already been described, and has included data from Petipa for *Acartia* for comparison (Table 7.14). As she points out, the data for *Calanus* and *Rhincalanus* obtained by different investigators are reasonably comparable, especially when several developmental stages are grouped together. But, Paffenhofer's data show the much lowered efficiency of N IV and N V *Calanus*.

Petipa's results demonstrate the generally lower growth efficiency of *Acartia* compared with *Calanus*. In Petipa's data, K_1 for *Calanus* N V shows a remarkable fall which she suggests is due to the incidence of vertical migration, actually in C IV, as Marshall points out. The very low values quoted for adult females are puzzling, except that growth should be considered only as regards ovary development and egg deposition, which may be greatly variable with time.

Harris and Paffenhofer (1976b) used their data on growth of the laboratory-reared calanoids *Pseudocalanus* and *Temora*, to study growth efficiency (K_1) in terms of ash-free dry weight. Except during the naupliar stages, *Temora longicornis* gave better weight increments than *Pseudocalanus elongatus*. For the earlier stages to about C III, the maximum appeared to be at food concentrations of about 100 μg C; for late copepodites the maximum increased somewhat with advancing age.

The gross growth efficiency (K_1) was higher for *Temora* throughout (except at the lowest food concentration for the naupliar stages). Values of K_1 showed no relationship to food concentration for *Pseudocalanus*. A negative correlation was

Table 7.13. Efficiency of growth in different planktonic crustaceans (Conover, 1968)

Species	Age or stage	Basis for calculation	K_1	K_2	Source
Daphnia pulex	5 days	calories	—	42.3	Ivlev (1938); cited by Pavlova (1964)
	15 days	calories	—	30.3	Ivlev (1938); cited by Pavlova (1964)
Penilia avirostris	juveniles	calories	—	53.4	Pavlova (1964)
	females	calories	—	21.1	Pavlova (1964)
	production of young	calories	—	52.9	Pavlova (1964)
Daphnia pulex	pre-adult	calories	3.87–13.22	55.36–58.64	Richman (1958)
	adult	calories	0.43–1.25	2.99–3.95	Richman (1958)
	production of young	calories	10.02–16.52	52.08–70.46	Richman (1958)
	40 days' growth	calories	10.01–17.43	55.99–72.98	Richman (1958)
Daphnia pulex	young stages	calories	>30–>60		Pechen and Kuznetsova (1966)
	adults		4.5–29.0		
Artemia salina	young stages	dry weight	<20–>60		Reeve (1963*b*)
	reproduction	dry weight	15–25		Reeve (1963*b*)
Artemia salina	1–9 days		2.8–5.6		Mason (1963)
	4–8 days	carbon[14]	5.4–8.3		Mason (1963)
	first 17 days		4.3–5.2		Mason (1963)
	5 mm length		5.8–7.3		Mason (1963)
Artemia salina	1–9 days	dry weight	11–53		Gibor (1957)
Artemia salina	12 days' growth	calories	9.02–18.52	23.65–27.44	Sushchenya (1964)
Calanus hyperboreus	IV	dry weight	13.0	24.7	Conover (1964)
	V	dry weight	14.6–36.4	28.6–64.3	Conover (1964)
	IV	calories	18	34	Conover (1964)
	V	calories	20–50	40–89	Conover (1964)
Calanus finmarchicus	sub-adult	nitrogen	34		Corner, Cowey and Marshall (1967)
	adult (egg production)	nitrogen	14		Corner, Cowey and Marshall (1967)
Diaptomus siciloides	females (egg production)	calories	7	4.18–19.8	Comita (1964)
Euphausia pacifica	adult	carbon[14]	32.4		Lasker (1960)
	adult	carbon[14]		17.1	Lasker (1966)
	including molts	carbon[14]		25.2	Lasker (1966)
	including molts & eggs (field populations)	carbon[14]		33.6	Lasker (1966)

$K_1 = \Delta W/R/\Delta t$ or the fraction of ingested food or energy used in growth;
$K_2 = \Delta W/A' R \Delta t$ or the fraction of assimilated food or energy used in growth.

Table 7.14. Growth efficiencies of *Calanus*, *Rhincalanus* and *Acartia* on different foods (Marshall, 1973)

Species	*C. pacificus*[a]		*Rhincalanus nasutus*[a]				*Calanus pacificus*[b]		*Calanus pacificus*[b]		*C. helgolandicus*[c]	*Acartia clausi*[c]	
Food	*Thalassiosira*		*Ditylum*		*Thalassiosira*		*Lauderia*	*Gymnodinium*	*Lauderia*	*Gymnodinium*	Natural seawater		
μg C/litre	226	177	200	148	352	196	101	95	101	95			
Temperature (°C)	10	15	10	15	10	15	15	15	15	15			
Stage							Newly moulted body wts		Medium body wts				Stage
N I													N I
N II													N II
N III													N III
N IV	21	18	39	39	22	21			7.6	9.8			N IV
N V							17.3	20.1	14.7	14.1	34	14	N V
N VI									29.8	36.7			N VI
C I							20.9	29.6	22.0	22.0	50	17	C I
C II	72	26	31	32	55	32			17.6	21.2	39	16	C II
C III							18.6	34.7	22.4	27.2	28	23	C III
C IV									15.7	25.3	21	16	C IV
C V	30	37	34	48	25	40		27.6	19.6	22.2	5	11	C V
♀													♀
Total N I–C VI	35	34	34	45	30	37							

Petipa also gives figures of 2% for efficiencies of adult females of *Calanus* and *Acartia*.
[a] from Mullin and Brooks, 1970.
[b] from Paffenhöfer, 1971.
[c] from Petipa, 1967.
Marshall uses the alternative specific name *pacificus*, although the authors quoted refer to *helgolandicus*.

demonstrated between K_1 and food concentration for *Temora* during the period C III to C VI. This could also be seen over the total N I to C VI period. Similarly, a significant correlation was found between log K_1 and ration for *Temora*. With further data an inverse relationship between gross growth efficiency and ration might be demonstrable also for *Pseudocalanus* (Table 7.15a,b). Corkett and McLaren (1978) point out that the daily ingestion rates in Paffenhofer and Harris's data which continued to rise beyond those food concentrations at which growth rates were maximal, should be reflected in lower growth efficiency, probably due to reduced assimilation efficiency. Further, the regressions of growth efficiency on food concentration for all stages were all negative, though none were significant. Obviously, such a relationship cannot apply for either species below a limiting food concentration. As the quantity of food ingested approaches that required for metabolism alone, growth efficiency must decline rapidly. Harris and Paffenhofer believe that growth efficiency is affected not only by food concentration and developmental stage but probably also by other factors (e.g. temperature and salinity).

Values for K_1 from hatching to adult were *ca.* 17–27 % for *Temora* and 13–18 % for *Pseudocalanus*, emphasizing again the lower value for *Pseudocalanus* under the conditions of the experiment. The value of 25 % for *Pseudocalanus* cited by Corkett and McLaren (1978) is quoted erroneously for *Temora*. If calorific values were used as a basis for comparing K_1, *Pseudocalanus* might show higher values since fat droplets found in the copepod were not taken into account in the dry weight analysis.

These values tend to be somewhat lower than those found by previous workers for *Calanus* spp. Efficiencies measured by Paffenhofer (1976a) in studying *Calanus helgolandicus* are also slightly lower. Gross growth efficiencies varied with stage of development; for different diets, including various dinoflagellates and the diatom *Lauderia*, K_1 rose to a maximum at about the C I to C III stage, falling again to the adult (C VI). For the first few days of adult life, efficiencies varied from *ca.* 4 %–10 % (see Fig. 7.16). The highest values for any stage were around 33 %. Over the whole development from nauplius to adult K_1 varied from 18.5 % to 30 % on various diets. On *Lauderia*, there was evidence for higher efficiencies at lower food concentrations (Fig. 7.17). At food concentrations approximately equivalent to those occurring off La Jolla, California, K_1 varied from 20–24 %. Only small differences appeared whether efficiencies were measured as dry organic weight, carbon or calories.

Paffenhofer quotes results of Petipa for comparison; her data suggest that *Calanus helgolandicus* reached a peak of growth efficiency at the C I stage (nearly 50 %); very low values (5 % and 2 %) were found for C V and C VI. For *Acartia clausi*, C III had the highest K_1 value, exceeding 20 %, C VI having very low values. Different species of grazing calanoids appear to have specific developmental patterns for growth (Table 7.16 gives summary). Heinle, Harris, Ustach and Flemer (1977) found a reduction in numbers of eggs per brood and in the number of broods with lowered food (algal culture) levels for *Eurytemora affinis* and a decline in average body weight of females produced. The assimilation efficiency, however, was unchanged with varied food concentration. Food chain efficiency was much higher for lowest food concentration; gross growth efficiency (K_1) was also substantially greater at the lowest food concentration (Table 7.17).

Data on the growth of herbivorous copepods at approximately environmental temperatures and food concentrations were obtained by Taguchi and Ishii (1972) in the

Table 7.15a. Data for cumulative ingestion, weight increase and gross growth efficiency for *T. longicornis* at 12.5°C (Harris and Paffenhöfer, 1976b)

	Experimental food concentrations (μg C/l)	N I to C I			C I to C III			C III to 50 % adult			Hatching to 50 % adult		
		$\bar{x}$	s	n	$\bar{x}$	s	n	$\bar{x}$	s	n	$\bar{x}$	s	n
Food concentration (μg ash-free dry wt/l)	25	47.0	—	1	41.6	—	1	47.6	—	1	45.1	—	1
	50	101.4	14.6	2	96.6	43.1	2	88.7	15.5	3	104.0	22.9	3
	100	228.8	18.5	5	177.6	18.4	5	158.6	33.6	5	194.9	16.7	5
	200	358.0	50.9	2	376.6	63.6	2	345.6	24.2	2	354.8	5.8	2
Cumulative ingestion (μg ash-free dry wt)	25	3.0	—	1	8.0	—	1	31.6	—	1	42.6	—	1
	50	3.9	1.9	2	23.6	7.0	2	56.4	13.2	3	83.1	14.5	3
	100	4.0	1.9	5	15.0	7.3	5	74.6	10.8	5	93.6	10.9	5
	200	5.9	2.8	2	27.5	4.6	2	88.8	40.8	2	121.6	42.6	2
Weight increment (μg ash-free dry wt)	25	0.8	—	1	3.2	—	1	6.8	—	1	10.7	—	1
	50	0.9	0.1	2	6.0	3.5	2	14.8	0.4	3	22.0	1.9	3
	100	0.8	0.1	5	3.3	0.8	5	15.1	3.0	5	19.3	3.0	5
	200	1.2	0.2	2	7.0	2.2	2	11.6	4.8	2	19.7	6.7	2
Gross growth efficiency (%)	25	24.7	—	1	39.7	—	1	21.5	—	1	25.1	—	1
	50	25.8	20.2	2	24.7	7.2	2	27.1	6.6	3	26.8	3.2	3
	100	24.4	7.4	5	24.8	7.0	5	20.1	2.0	5	20.6	2.3	5
	200	25.6	18.9	2	25.8	12.3	2	14.5	12.0	2	17.3	11.5	2

$\bar{x}$ = mean, s = 95 % confidence limits, n = number of determinations.

Table 7.15b. Data for cumulative ingestion, weight increase and gross growth efficiency for *P. Elongatus* at 12.5°C (Harris and Paffenhöfer, 1976b)

	Experimental food concentrations (μg C/l)	N I to C I			C I to C III			C III to 50 % adult			Hatching to 50 % adult		
		$\bar{x}$	s	n	$\bar{x}$	s	n	$\bar{x}$	s	n	$\bar{x}$	s	n
Food concentration (μg ash-free dry wt/l)	25	55.8	—	1	48.4	—	1	37.2	—	1	47.0	—	1
	50	96.3	8.5	2	96.0	22.2	2	98.8	4.0	2	96.8	11.6	2
	100	181.4	21.9	4	179.1	38.7	4	164.2	21.8	4	176.0	26.0	4
	200	357.4	—	1	365.1	—	1	364.0	—	1	362.1	—	1
Cumulative ingestion (μg ash-free dry wt)	25	2.5	—	1	10.9	—	1	39.9	—	1	53.3	—	1
	50	6.3	2.4	2	33.7	0.9	2	53.1	4.8	2	93.1	3.3	2
	100	9.6	5.1	4	27.5	8.2	4	63.3	20.3	4	100.4	8.2	4
	200	6.7	—	1	28.4	—	1	92.6	—	1	127.7	—	1
Weight increment (μg ash-free dry wt)	25	1.1	—	1	1.9	—	1	6.4	—	1	9.4	—	1
	50	1.1	0.0	2	7.7	1.1	2	4.4	2.8	2	13.1	5.5	2
	100	1.8	1.5	4	4.2	1.0	4	10.4	1.1	4	16.4	3.4	4
	200	1.2	—	1	3.5	—	1	12.6	—	1	17.3	—	1
Gross growth efficiency (%)	25	45.5	—	1	17.0	—	1	16.0	—	1	17.6	—	1
	50	17.8	6.7	2	22.7	3.6	2	8.0	6.7	2	14.0	5.3	2
	100	17.1	5.1	4	15.9	4.3	4	17.8	6.3	4	16.0	3.7	4
	200	17.4	—	1	12.4	—	1	13.6	—	1	13.5	—	1

$\bar{x}$ = mean, s = 95 % confidence limits, n = number of determinations.

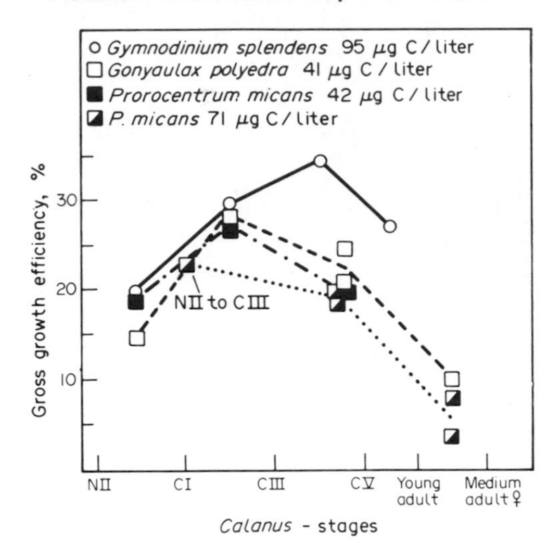

Fig. 7.16. Gross growth efficiency of *Calanus helgolandicus* feeding for various periods of lifetime on different dinoflagellate diets (Paffenhofer, 1976a).

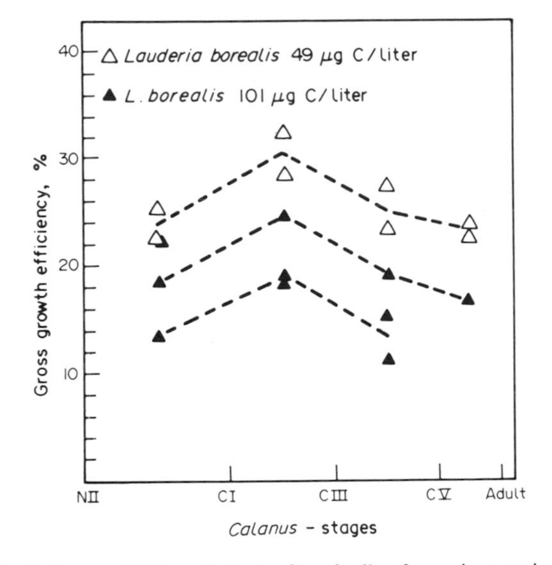

Fig. 7.17. Gross growth efficiency of *Calanus helgolandicus* feeding for various periods of its juvenile life on *Lauderia borealis* at different food concentrations. The data are based on organic carbon (μg C/litre) (Paffenhofer, 1976a).

northern North Pacific using shipboard experiments. Unfortunately their data for growth, expressed as micrograms of carbon assimilated per animal per day, cannot be so readily compared with those of other workers and are confined to late copepodites. *Calanus plumchrus* was found to ingest only about 20% of the quantity of food taken by *Calanus cristatus*, but showed a higher respiratory metabolism in relation to assimilation. This was reflected in a lower growth rate, *viz.* 1.57 μg C/animal/day for

Table 7.16. Gross growth efficiencies of *Calanus finmarchicus*, *Calanus helgolandicus*, *Calanus hyperboreus*, and *Rhincalanus nasutus* (Paffenhöfer, 1976)

Species and stages	Temp (°C)	Food concentration (μg C/litre)	Food sp.	Gross growth efficiency (%)	Basis	Reference
C. hyperboreus mostly C V	2–5	6.4–6.7[a]	*T. fluviatilis* (laboratory)	13.0–17.3	ash-free dry wt	Conover 1964
C. hyperboreus C V	2–5	1.7–1.8[a]	*T. fluviatilis* (laboratory)	18.6–36.4	ash-free dry wt	Conover 1964
C. helgolandicus *C. finmarchicus* C II to C V	10	Approx 210	*Skeletonema costatum*	35.7	Nitrogen	Corner et al. 1965
C. finmarchicus hatching to adulthood	10	Not given	Not given (sea)	34	Nitrogen	Corner et al. 1967
C. finmarchicus *C. helgolandicus* C V and adults	12	Not given	Not given (sea)	18.9–34.6 21.4–37.7	Phosphorus Nitrogen	Butler et al. 1969
C. finmarchicus *C. helgolandicus* C V and adults	6–18	Approx 490 (sea)	*Thalassiosira* sp. *S. costatum*	17.2 26.8	Phosphorus Nitrogen	Butler et al. 1970
Rhincalanus nasutus hatching to adulthood	10 15 10 15	352 196 200 148	*T. fluviatilis* *T. fluviatilis* *Ditylum brightwellii* *D. brightwellii*	30 37 34 45	Carbon Carbon Carbon Carbon	Mullin and Brooks 1970a

[a] Values expressed as mg ash-free dry wt per litre.

Table 7.17. ***Eurytemora affinis*. Energetic statistics from experiment with three levels of algal food (Heinle *et al.*, 1977)**

Statistic	Ration		
	High	Medium	Low
Average female weight (μg)	8.35	4.90	5.23
Food provided (μg/copepod/day)			
Mean	19.872	9.936	5.244
Range	6.998–30.840	3.542–18.836	0.346–13.824
% of body weight/day	238	203	100
Percent of food actually ingested	76	78	81
Egg production (μg/female/day)	1.955	0.911	0.978
Mean number of eggs/brood	22.37	15.80	11.27
Mean number of broods produced	5.6	4.7	3.5
Productivity (per day)			
Mean	0.231	0.213	0.183
Range	0.073–0.614	0.085–0.362	0.028–0.362
Food-chain efficiency (%)	7.34	6.84	13.92
Gross growth efficiency, K_1 (%)	9.66	8.77	17.18

Assumption was made that all dry weights = 2 (carbon content). Food provided was calculated from daily cell counts and an established regression between cell counts and content of particulate carbon. Statistics were calculated only for time that females were actively producing eggs. Food-chain efficiency = egg biomass produced/biomass of food provided, gross-growth efficiency = egg biomass produced/biomass of food ingested.

C. plumchrus, compared with 15.5 μg C/animal/day for *C. cristatus*, equivalent to *ca.* 25% of the assimilated food for *C. plumchrus* and approximately double that for *C. cristatus*. The authors compare their growth data with values of 6.3 μg C assimilated in growth per day for the euphausiid *Euphausia pacifica*, as found by Lasker, and an estimate of *ca.* 3 μg C per day for *Calanus finmarchicus* and *C. helgolandicus*, from Corner's experiments. Gross growth efficiencies were 39% for *C. cristatus* and 22% for *C. plumchrus*; possibly these values should be slightly reduced to allow for excretion losses or moults (cf. Corner's value of *ca.* 35% and Paffenhofer's recent estimates for other *Calanus* spp.).

Studies on the growth of zooplankton other than copepods include that of Lasker (1966) for the euphausiid *Euphausia pacifica*. Euphausiids moult as adults and moulting may be frequent. Lasker found for *E. pacifica* that the inter-moult period varied between three and eight days (mean five days), apparently depending on temperature rather than food supply. Each exuvium corresponded to approximately 10% dry weight, containing about 17% organic carbon. In estimating growth, measurements of uropods of exuviae were used, since uropod size was regularly related to total length and weight. In the laboratory, juvenile euphausiids achieved a maximum growth rate of 0.048 mm/day, more than twice that recorded by Ponomareva (1966) from field data; the smaller stages showed faster rates. Lasker attributed the acceleration mainly to the availability of an abundant food supply in the form of *Artemia* nauplii. *E. pacifica* could incorporate as much as 30% of the assimilated carbon as growth, though in nature Lasker believes that a more realistic estimate would be about 10%. Respiratory metabolism accounted for the largest fraction of the carbon assimilated. Lasker calculated that of the total carbon available to *E. pacifica*, the animal utilized the following proportions:

6–30 % as growth, 6–11 % as moults and 62–87 % as respiration.

For field populations, he believed that a likely energy budget would be 9 % as growth, 15 % as moults, 67 % in respiration and 9 % in reproduction.

With short term experiments using ^{14}C-labelled algae, Lasker estimated growth efficiency in *Euphausia*: this varied from 11–74 % (mean 32 %). In another series of experiments where ^{14}C-labelled *Artemia* were offered, efficiency was lower (mean 25.6 %), but Lasker comments that the prey were probably too large to be captured in sufficient quantity for maximal growth. These values, expressed as percentages of ingested carbon, would correspond to gross growth efficiencies. A comparison of the utilization of food by *Euphausia* and *Calanus* is:

	% utilization of assimilated food	
	Calanus (as N) for 10 weeks	*Euphausia* (as C) for 20 months
Growth	25	10
Metabolism	61	72
Moults	1	17
Egg Production	12	1[a]

The values are slightly altered from those given in the original paper
[a] This value is a revised figure according to Corner *et al.* (1967).

Outstanding is the high proportion of food used for metabolism in both species.

An example of remarkably rapid growth amongst marine zooplankton is that of the widely distributed salp *Thalia democratica*. Heron (1972 a,b) found that under optimal conditions, growth in the laboratory, at least for up to about 5 hr, approached that in a sea area experiencing bloom conditions of phytoplankton. After that time growth rate in the laboratory declined, probably associated with a reduction in the density of food organisms which leads to a much reduced rate of feeding. Field and laboratory data indicate that over much of the life cycle of the salp the growth rate at best approximated to 10 % of the length per hour. The life history of the salp includes a regular alternation of generations (cf. Chapter 3). If zero age is taken as the commencement of the sexual stage in the life cycle, the first birth of the next generation was about 42 hours later, under optimal conditions, but Heron suggests that since birth and fertilization are both to some extent timed to a diurnal cycle, the generation time at best is probably of the order of two days. Under somewhat less favourable conditions the generation times for the completion of asexual and sexual generations increases to about 4–14 days. Temperature, apart from food supply, is among the factors affecting growth rates.

Assuming a two-day generation time, and from calculated age specific mortality, it appears that the population could be increased by between 1.6 and 2.5 times per day during the initial stage of colonization of a phytoplankton bloom. Such extraordinary outbursts of population appear to be consistent with the sudden dense swarming of *Thalia*, especially in tropical and sub-tropical waters. Heron states that the rate of increase appears to be higher than that of any multicellular organism yet measured. The very rapid growth, he believes, may be associated with the relatively simple type of construction of the salp body, the barrel-like form and the pumping action permitting ease of feeding, locomotion and respiration. Growth of the embryo stage is also presumably accelerated by the direct maternal nutrition *via* the placenta. It is of interest that the copepod *Sapphirina angusta*, living in and on the salp *Thalia democratica*, can consume the salp tissues at a remarkable rate and grow rapidly. Heron (1973) found that a female copepod feeding on *Thalia* in the laboratory could feed, produce and shed eggs, which hatched to the Nauplius I stage, all within two days.

Oikopleura provides another example of very rapid growth among zooplankton. Fenaux (1977) demonstrated that *Oikopleura dioica* reared at a temperature of 22°C hatches in 2–3 hr from fertilization of the egg. After only 8 hr the tail has shifted to its proper position and by this time cell division is practically completed. Apart from maturation of the gametes, further increase occurs largely in the volume of the somatic cells. The secretion of the first "house" soon follows, a mucus covering which swells and begins to separate from the trunk. At 14°C the times are extended to 6 and 18 hr respectively. The generation time at 22°C, as indicated by the first egg release, may be as little as three days (Fig. 7.18). Paffenhofer (1973, 1976c) found similar generation times for *O. dioica* fed on algal food concentrations of 25–80 μg C/l (9.5 days at 13°C). With animals fed on natural detritus, the times were 5.5 days at 18°C and 24 days at 7°C. He also believes that generation time is temperature-dependent.

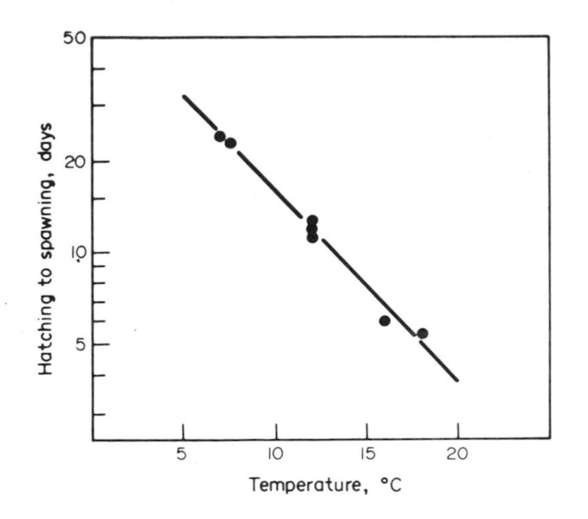

Fig. 7.18. Generation time of *Oikopleura dioica* reared in natural seawater at different temperatures (Paffenhofer and Harris, 1979).

The growth of the tail of the oikopleuran is particularly remarkable, apparently mainly due to increase in cell size. Fenaux found that at 22°C the length of tail increased from 195 μm at 1 hr after hatching, to 620 μm 24 hr after hatching. Over the first day the increase was 590 %. Although the rate of growth of the tail declines subsequently, it is still impressive, the length achieving 2300 μm after 4 days.

Growth studies on carnivorous zooplankton are few. Reeve (1970) examined the growth of the chaetognath *Sagitta hispida* fed on *Artemia* nauplii. Growth rate was directly related to temperature over the normal range for the species; the rate was also dependent on food concentration. Reeve and Cosper (1975) quote earlier work confirming that higher temperature causes accelerated maturity, although the mature animals would be smaller in size and dry weight (263 μg at 7°C, 119 μg at 31°C). Gross growth efficiency, estimated as dry weight and as total nitrogen, ranged for immature *Sagitta* from 22 % to 50 % on a dry weight basis; from 19 % to 48 % as nitrogen; the overall mean value was 34.5 %. With mature *Sagitta* (including egg production in the estimates) efficiency ranged from 19 % to 54 % as total nitrogen. Efficiency appeared to be inversely related to temperature. Over the entire size range studied, efficiency was 36 % in terms of nitrogen. Efficiency was therefore of the same order as that found by Lasker for *Euphausia*, by Corner for *Calanus* and by Mullin and Brooks for juvenile marine copepods.

Reeve calculates from oxygen uptake values that a minimum of 24–59 % body weight/day is required as food; this may be compared with a *maximum* uptake when feeding on *Artemia* corresponding to approximately 70 % body wt/day. Part of the ration was, therefore, available for growth and reproduction. Reeve and Walter (1972) point out, however, that daily ration as a percentage of body weight declines with age, as in many other zooplankton species. Whereas young *Sagitta hispida* daily consumed their own weight of food, the ration declined progressively to about 10 % daily for the adult. Heinbokel (1978a) found that the gross growth efficiencies for two tintinnid species fed on algal cultures (25–470 μg C/l) could exceed 50 %.

Among other studies on growth of zooplankton, Bamstedt (1976) observed a seasonal variation for the euphausiid *Meganyctiphanes norvegica*. Growth (increase in weight) occurred mostly in summer/autumn and ceased in mid-winter. Estimates showed a fairly high value in August (1.2 % per day), falling to 0.3 % per day in November and negative values in December/January. By March/April the average growth was around 0.5 % per day, reaching 1.1 % daily by June (Fig. 7.19). There was some evidence, when two generations of the euphausiid were present, that the younger

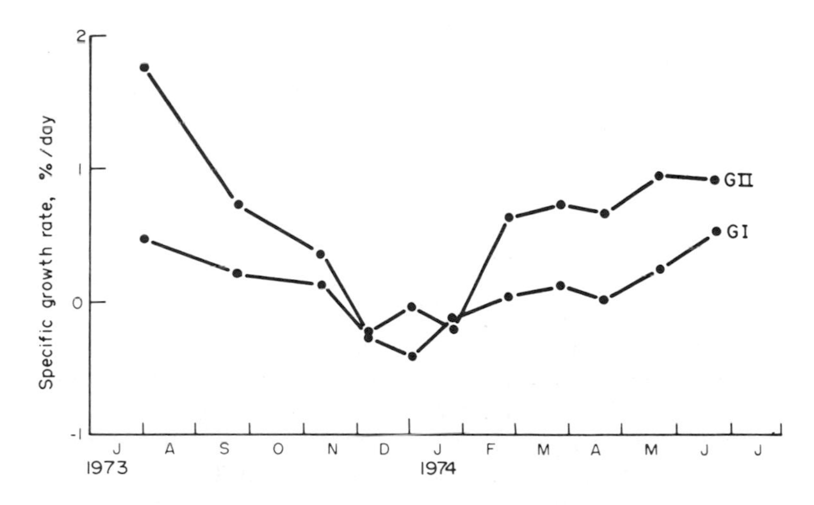

Fig. 7.19. Seasonal changes in specific growth rate of an average individual of *Meganyctiphanes norvegica*, from an old (G I) and a young (G II) generation. Values based on dry weight changes as shown in Fig. 7.20 (Bamstedt, 1976).

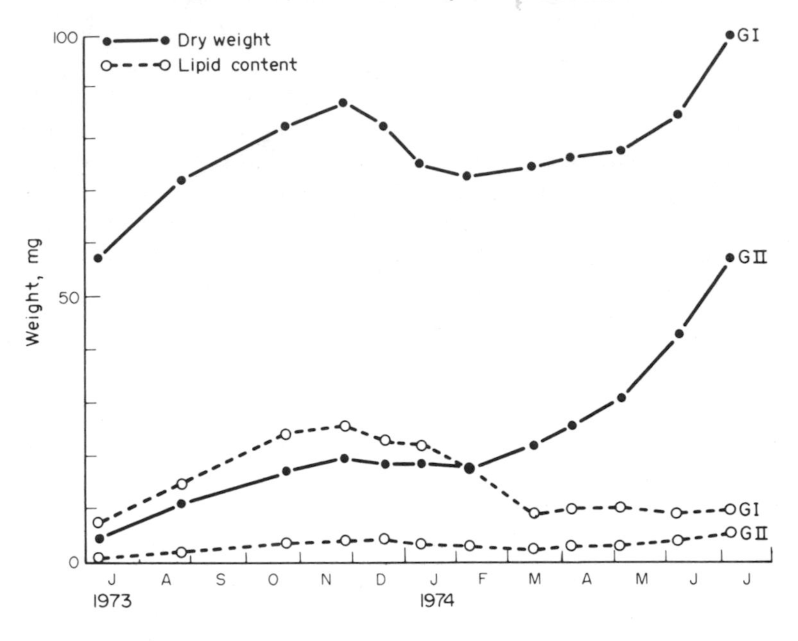

Fig. 7.20. Seasonal changes in dry weight and lipid content of an average individual *Meganyctiphanes norvegica* from an old (G I) and a young (G II) generation, based on the curve for growth in carapace length of Jorgensen and Matthews (1975) and other data (Bamstedt, 1976).

generation showed a higher growth rate, indicating some decline of growth with age (cf. Fig. 7.20). Raymont, Srinivasagam and Raymont (1971) found a higher proportion of protein in younger *Meganyctiphanes*, indicating more active growth.

In studies on the carnivorous copepod *Euchaeta norvegica* Bamstedt and Matthews (1975) and Bamstedt (1975) also deduced mean specific growth rates as percentages of the dry body weight for the later copepodite stages. For growth from C IV to C V the daily rate was calculated as 3.3 %, for C V to C VI♂ as 1.2 % and for C V to C VI♀ as 2.2 %. However, adult females increased in weight at a rate of only 0.5 %/day, i.e. C IV grew seven-fold faster than females. This decline in growth rate with age is confirmed also in studies by Dagg and Littlepage (1972) on *Euchaeta elongata*; growth measured as increase in dry body weight was 15 % daily for C III and 12.8 % for C IV, but fell to only 1.9 % per day for C V. They draw attention to the similarity of the results obtained by Petipa for *Calanus*, showing the marked slowing of growth in C V and VI stages, even though their growth rates are generally somewhat higher than Petipa's. Bamstedt (1975) also compares the growth rates of the chaetognath *Sagitta hispida*, which he estimates in terms of dry body weight increase, as 5–10 % per day, the rate being proportional to temperature between 17° and 31°C. He suggests that the growth rates estimated for *Euchaeta* are not unreasonable in the light of the much lower environmental temperature. A study of growth rate in another outstandingly carnivorous phylum, the Ctenophora, would be of interest, but Reeve and Walter (1978) allude to some of the difficulties involved in assessing growth in ctenophores, particularly measurements of lobate forms.

Few estimates have been made in the laboratory and there are often great variations, though some very high growth rates are reported. Using the coefficient of exponential growth (k) to compare results for different species, Reeve and Walter point to Hirota's data for *Pleurobrachia*, which suggest a period of slow growth (up to 2 mm diameter), then an accelerated phase (6.5 mm), followed by another slower period. Values of k varied between 0.12 to 0.17, 0.21 to 0.47 and 0.04 to 0.17 for the three periods respectively. They also report data from Greve for *Bolinopsis* and *Pleurobrachia* with maximum growth of 10–40 mm in 20 days ($k = 0.2$), and a faster rate of 5–15 mm in 8 days for *Beroe* ($k = 0.4$). Reeve and Walter found higher growth rates for *Pleurobrachia*, but the same three phases of growth were evident. In the middle period k was approximately 0.47 (i.e. animals were increasing their biomass by nearly 50 % per day) and an even higher coefficient (0.76) was achieved at increased temperature (20°C). In *Mnemiopsis mccradyi* the fastest growth was obtained at a temperature of 26°C. The highest growth rate over 10 days gave a growth coefficient of 0.78; this fell to 0.23 over the following 10 days and to 0.07 over the next 20 days. At the highest growth rate for young animals, Reeve and Walter found a daily doubling of biomass, a rate of growth which they point out might approach that recorded for *Thalia democratica*.

Another study of growth in carnivorous zooplankton is that of Conover and Lalli (1972, 1974) for the gymnosome *Clione limacina*. By establishing relationships between body dimensions in *Clione* and dry weight and calorific value and computing similar values for the prey (*Limacina*), Conover and Lalli were able to examine growth rates and growth efficiencies. Rate of growth and ultimate size of *Clione* seemed to be regulated more by the size of prey than by its abundance and larger prey was clearly selected, but feeding rate, as influenced by size of prey, markedly affected growth.

Growth was exponential and could be described by the same equation as for herbivorous species, but determination of growth rate was complicated by the very variable behaviour of individual *Clione* in the laboratory; some would grow rapidly from the start of an experiment, but would then slow down, whereas others might not grow initially but thereafter would show low growth rates; some individuals lost weight. Ikeda (1974) also describes decrease in body size (length) with shortage of food for a carnivorous species, the chaetognath *Sagitta elegans*, and cites similar reports for *Sagitta hispida* (Reeve) and *Clione* (Lalli). These effects are not apparent for animals with a firm exoskeleton; such animals experience the depletion of reserves with food shortage.

Growth in *Clione*, was calculated as instantaneous rates:

$$\left(\frac{1}{t_i+1^{-t_i}}\ln.\frac{W_{i+1}}{W_i}\right),$$

where W_i = weight at time t_i,
W_{i+1} = weight at a later time t_{i+1}.

Despite great variation, values could be very high under near-optimal temperature, reaching 0.4 on a dry weight basis. Gross growth efficiency (K_1) could also be very high—0.7 on a dry weight basis and a maximum of 0.8 on calorific values. Since, however, nett efficiencies (K_2) were *not* unusually high, the large values for K_1 are probably related to the extraordinarily efficient food assimilation—very little faecal material is egested by *Clione* when feeding (cf. Chapter 6).

No significant relationship was found which suggested a decreased growth efficiency with body size or with increased ration for *Clione*, according to Conover and Lalli. Provided *Clione* receives enough food of the right size, growth efficiency does not decline as the gymnosome grows larger and consumes more. Undoubtedly the metabolic rate of the animal is dependent on ration; tests showed a clear increase in respiration with feeding. Some variation in respiratory rate with season and area was also probably mainly related to abundance of food. Over the animal's normal very wide geographical range, temperature exerts relatively little effect on the metabolic rate. *Clione* appears to adjust its "standard" rate to the habitat temperature.

The study also illustrates a useful method of relating growth rate to metabolism and food consumption (cf. Conover, 1968; Parsons and Takahashi, 1973; Conover and Lalli, 1974). If R = ration; p = food assimilated/ingested; T = metabolism; G = growth; W = weight, then:

$$G = \frac{\Delta W}{\Delta t} \quad T = \alpha W^{t}$$ (the normal relationship for metabolism and weight)

But

$$pR = T + \frac{\Delta W}{\Delta t}$$

$$K_1 = \frac{\Delta W}{\Delta t R}$$

Thus

$$\frac{\Delta W}{\Delta t\, R} = e^{-a-bR}$$ (assuming ration is described by Ivlev equation),

or

$$\frac{\Delta W}{\Delta t} = R\,e^{-a-bR}$$

Therefore

$$T = R(p - e^{-a-bR}).$$

Thus if ration, assimilation efficiency and growth are known, metabolism can be calculated and tested by experiment.

Another energy budget may be estimated from investigations of Corner, Head, Kilvington and Pennycuik (1976), who examined aspects of the growth and metabolism of *Calanus helgolandicus* feeding carnivorously on nauplii of *Elminius*. The daily ration increased with the concentration of food according to the Ivlev equation and there was no evidence of a critical ("threshold") food concentration. The maximum ration in terms of nitrogen was equivalent to 22.4 % body N (cf. Fig. 7.21). Intake of food was balanced by egestion of faeces and excretion when *Calanus* ingested the equivalent of 7.8 % of body N. The number of *Elminius* nauplii consumed over a 24-hour experiment was measured, and the gross growth efficiency (K_1) estimated from the intake of nitrogen in terms of *Elminius* (their mean nitrogen content having been previously estimated) and the increase in body nitrogen of *Calanus* over the same period. With maximal ration, K_1 was 0.49 (i.e. 49 %). At a daily ration equivalent to 7.8 % body N, K_1 is zero, since *Calanus* is then in balance (the total food assimilated being used in oxidative metabolism). As the ration rises, K_1 increases to reach approximately 0.5 at maximum ration (cf. Fig. 7.22).

Corner *et al.* point out that this gross growth efficiency is considerably higher than those found by Butler, Corner and Marshall (1970) for *Calanus* feeding during the

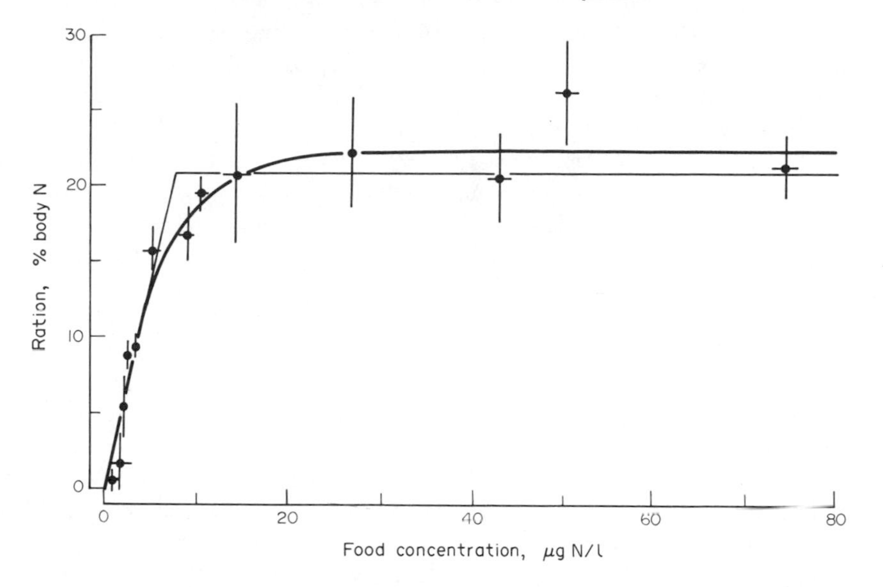

Fig. 7.21. Rations as percentage body nitrogen ingested by adult female *Calanus* feeding on different concentrations of *Elminius* nauplii. Best fit to the data (thicker line) is given by the Ivlev (1955) type of equation. Horizontal and vertical bars through the points show 95 % confidence limits (Corner, Head, Kilvington and Pennycuik, 1976).

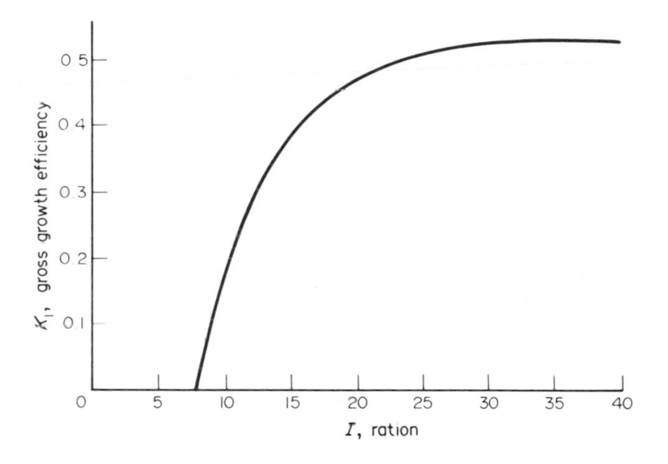

Fig. 7.22. Gross growth efficiency K_1, calculated as the G/I ratio for different I values. (I = daily ration) (Corner, Head, Kilvington and Pennycuik, 1976).

natural spring diatom increase. Reference has already been made (page 643) to the balance sheet drawn up in terms of nitrogen and phosphorus for *Calanus* in that earlier investigation. Over two years (1969, 1970) gross growth efficiencies were 33 % and 27 % for nitrogen; 28 % and 17 % for phosphorus. The lower gross growth efficiencies in the second year might be related to growth being measured then only for stages C V and C VI, whereas in 1969 estimates were made over the whole life span. It is generally accepted that K_1 values are lower for later copepodites.

Corner *et al.* (1976) also point out with regard to the generally lower growth efficiencies found by Butler *et al.* that when the copepods were feeding in the natural environment they would be receiving only some 13 % body nitrogen as daily ration. At the same food intake, the more recent investigations of Corner *et al.* suggest a similar K_1 value of 0.33. As with other studies of energy budgets, the proportion of ingested food used in metabolism appears to be high. Nitrogen excretion rises with food intake and at maximal ration (equivalent to 22.5 % body nitrogen), excretion amounts to around 11.5 % of body nitrogen, i.e. metabolism must be equivalent to approximately 50 % of total nitrogen intake.

Detritus has been suggested as a supplementary food, for example, during winter scarcity, by a number of investigators. Indications are that assimilation efficiency is lower for detritus than for living food material. Ferguson (1973) has also given some preliminary results for growth efficiency using detritus. For the mysid *Neomysis integer* she has compared growth efficiencies when *Neomysis* (8–10 mm length) was fed on an animal diet and on "artificial" detritus (dead algal culture—cf. page 664). With artificial detritus larger quantities of egesta were found in relation to the amount of food ingested and the gross growth efficiency was very low (mean 7.5 %) compared with 27.5 % (mean) for animal diet. The growth efficiency on animal diet might reasonably be compared with the findings of Clutter and Theilacker (1971), who reported for the mysid *Metamysidopsis elongata* K_1 varying from 19 % to 29 % when fed *Artemia*. Growth was calculated from uropod measurements, relationships having previously been established between uropod size and length of body and between length/weight of body. Respiratory metabolism was again a major part of the energy budget. Clutter and Theilacker suggested an energy budget (cf. Table 7.18). Ferguson attempted an energy budget for the mysid fed on a mixed diet; the results can be regarded as preliminary only and excretory losses were not directly measured. The relatively high demands for metabolism are, however, in agreement with those for other zooplankton (cf. Table 7.18).

Growth efficiency studies in other carnivorous zooplankton include Stepien's (1976) investigation of larvae of the sea bream, *Archosargus rhomboidalis*. The amount of food consumed was estimated from the number of prey taken and the mean dry weight of food organisms in the gut; daily growth increments of the fish larvae were calculated from measured samples of larvae taken from the cultures and established length/weight relationships. Over a 10-day period the mean values for K_1 at three experimental

Table 7.18. Energy budget for two mysid species (Clutter and Theilacker, 1971)

Function	Percent of energy used	
	Metamysidopsis elongata	*Neomysis integer* (12–14 mm length)
Growth	19	13
Metabolism	55	52
Moults	7	
Egg production	19	
Lost as faeces		21
Presumed lost in excretory activity		14

temperatures (23–29°C) were 23.9 %–30.6 %. Though K_1 appeared to decrease slightly with an increase in temperature, the difference was not significant.

Reeve and Walter (1978) have pointed out that the high water and ash content of ctenophores may introduce serious errors in estimating growth efficiencies. The amount of organic carbon may be very low—only 2 % of the dry weight, compared with >40 % for the food (copepods). On a carbon basis, Reeve and his colleagues estimated growth efficiency in *Mnemiopsis* to range from 7 % at lowest food densities (3 copepods/l) to 2 % at the highest food concentrations (300 copepods/l). With *Pleurobrachia*, efficiencies ranged from 11 % to 3 % with increasing food concentration.

Even at food concentrations where digestive efficiency is known to be high, growth efficiency is low compared with other zooplankton. However, ctenophores grow as fast as zooplankton which have high growth efficiencies. *Pleurobrachia* appears to have high energy requirements, however. According to Reeve and Walter, they require about 63 % of their food intake to satisfy their metabolic activities, even where digestive efficiency is high. The amount of living material providing the energy required for movement of the ctenophore is relatively very small and thus a very large proportion of the ingested food must be directed to energy demands.

Biochemical Composition and Metabolic Pathways

General considerations

In investigating the significance of O/N ratios in zooplankton, Conover and Corner (1968) analysed changes in lipid content as an indication of its use as metabolic substrate (*vide supra*). The study illustrates the importance of accurate knowledge of the basic biochemical composition of zooplankton species in contributing to investigations on secondary production. More recent investigations of Ikeda (1974), Mayzaud (1976), Bamstedt (1975, 1978) and Raymont and colleagues also focus attention on the value of precise data on biochemical composition.

Early studies on basic composition tended to concentrate on lipid because of its likely role as a food reserve; by contrast, direct analysis of carbohydrate was rare. Thus the earlier data on the basic biochemical composition of zooplankton are relatively scanty and incomplete. However, a review of the older literature shows that the assumption was usually made that either carbohydrate or lipid or a mixture of these was used as respiratory substrate. Unfortunately, much of the earlier work on basic biochemical composition involved analysing mixed zooplankton hauls or mixtures of species from one taxon (e.g. copepods). In the few studies where selected species were chosen, frequently only protein or lipid was analysed, often not on the same haul of zooplankton. Moreover, precise values for protein may be open to question, since in most of the earlier work nitrogen was analysed by Kjeldahl determinations, the protein content being subsequently calculated on the assumption that protein contained 16 % nitrogen. Protein may, however, have a nitrogen content varying from 12 % to 20 %. Though some criticisms of direct estimates of protein by both biuret and Folin methods may be justified, such direct estimates are more accurate than data based on nitrogen analyses. A summary of some of the earlier data on biochemical composition of zooplankton is included in Table 7.19.

Although it is desirable that biochemical analysis is carried out on fresh zooplankton,

Table 7.19. Major biochemical constituents of marine zooplankton (Raymont, 1972)

Species	Carbohydrate[a]	Lipid[a]	Protein[a]	Chitin[a]	Ash[a]	Author
Copepods	20	7	59	4.7	9.3	Brandt, 1898
Copepods	—	—	—	4.0	14.8	Brandt and Raben, 1919–1922
Euchaeta norvegica	—	18–36	31–44	3.1–5.0	3.6–4.4	Orr, 1934a
Calanus finmarchicus	—	10.5–47	30–77	3.0	4.0	Orr, 1934b
Copepods	0–4.4	4.6–19.2	70.9–77	—	4.2–6.4	Krey, 1950
Sagittae	13.9	1.9	69.6	—	16.3	
Copepods (mainly *Acartia* sp.)	—	9.5	—	12.6	—	Lafon *et al.*, 1955
Calanus helgolandicus	—	11.0	75.2	—	2.8	Nakai, 1955
Calanus plumchrus	—	53.3	34.8	—	2.4	
Paracalanus parvus	—	19.1	70.1	—	2.9	
Pseudocalanus elongatus	—	17.3	71.5	—	2.3	
Euchaeta japonica	—	33.7	51.8	—	2.1	
Acartia clausi	—	5.8	82.6	—	3.3	
Euphausia pacifica	—	2.7	79.3	—	7.1	
Euphausia superba	6–28	11–26	55–61	—	8.0	Vinogradova, 1960
Mysis microphthalma[b]	13	32	52	—	3.0	Vinogradova, 1964
Paramysis loxolepis[b]	13	27	52	—	8.0	
Calanus helgolandicus	9	46	43	—	2.0	

Note: Apart from the analyses of Krey, the few data for carbohydrate appear to have been obtained by difference rather than by direct analysis.
[a] Percentage of dry weight.
[b] Caspian sea.

this is not always possible or convenient. Tests carried out in our laboratory clearly demonstrated that analyses of material preserved in such standard fixatives as formalin and alcohol are entirely unreliable (cf. also Fudge, 1968). If it is not possible to use fresh plankton, analysis should be confined to freeze-dried material or to zooplankton which has been deep frozen (−20°C). Bamstedt (1974) describes certain modifications in the analytical techniques for use with dried zooplankton material.

In any examination of major biochemical constituents there is a need to study seasonal variations through the year. Very few studies of seasonal changes in all the biochemical fractions have been carried out on any marine crustaceans, even benthic species. An exception is the work of Barnes, Barnes and Finlayson (1963). Scarcely any investigations were carried out on zooplankton until comparatively recently, but the work of Orr (1934a,b) is outstanding in that changes in protein and lipid for two species of copepods were followed over a considerable period. Orr's data suggested that protein in *Calanus* ranged from about 30–78 % dry weight; lipid content could change four-fold (cf. Table 7.19). *Euchaeta* also showed large changes, though the ranges were not quite as great. The studies of Nakai (1955) (see Table 7.19) demonstrated that lipid and protein could vary considerably with the species of zooplankton. Nakai was one of the first investigators to suggest that the amounts of lipid and protein might be complementary.

Unfortunately, neither Orr nor Nakai included direct estimates of carbohydrate. Krey (1950), one of the earliest investigators to determine carbohydrate, obtained very low values for copepods though somewhat higher values were obtained for *Sagitta*. Vinogradova gives some data for carbohydrate (Table 7.19), but the values were apparently obtained by difference as opposed to direct analysis. Preliminary studies by Raymont and Krishnaswamy (1960) and Raymont and Conover (1961) indicated that total carbohydrate in certain zooplankton species (*Calanus* spp., *Neomysis*, *Meganyctiphanes*, *Thysanoessa*, *Euchaeta*) was very low. It was essential to obtain information on seasonal variations in each of the biochemical components in one species. An investigation of the brackish water semi-planktonic mysid *Neomysis integer* (Raymont, Austin and Linford, 1966) confirmed that throughout the year carbohydrate was remarkably low, averaging only 2 % of dry body weight. Though variations occurred, the amount of carbohydrate was always small, the maximum not exceeding 3 % dry weight. Ash and chitin both averaged *ca.* 7–8 % of the body weight throughout the year; lipid was variable ranging from below 7 % to a maximum of 14 % in different months (mean value 13 % dry body weight). The major fraction was protein averaging 71 %; the monthly means varied from *ca.* 60 % to 73 % of dry body weight.

Analyses carried out on other neritic mysids yielded very similar data. *Leptomysis lingvura*, an inshore Mediterranean species studied over a few weeks only, gave average values of protein 70 %, lipid 11 % and carbohydrate 2 % (Raymont and Linford, 1966). Seguin (1968) found an almost precisely similar basic composition for the British inshore mysid *Praunus flexuosus*. Lee and Chin (1971), studying *Neomysis awatchensis* from South Korea, quote results which are remarkably similar. Preliminary results (Raymont—unpublished) for the mysid *Acanthomysis sculpta* from the Californian coast suggest a generally similar composition.

While this basic biochemical composition may apply to a number of inshore mysids, that of open ocean zooplankton from different taxa may be very different. Biochemical analyses of zooplankton obtained during oceanographic cruises, stored deep frozen and freeze dried and later studied have yielded information on open ocean plankton. For a number of planktonic decapods (e.g. *Acanthephyra*, *Gennadas*, *Sergestes*, *Systellaspis*, *Funchalia*, *Oplophorus*) protein is the major fraction, normally exceeding 50 % or even 60 % of dry body weight. Lipid is variable (10–>20 %) but carbohydrate is again very low, usually around 2 % of dry body weight. Some very preliminary analyses for the well known bathypelagic mysids *Eucopia* and *Gnathophausia* suggested that lipid might be distinctly higher, particularly in *Eucopia*, but protein was still the major fraction and carbohydrate was as low as in other zooplankton (cf. Childress and Nygaard, 1974).

Figure 7.23 indicates the relative proportions of carbohydrate, lipid and protein as percentage of total organic matter for several different species. Values for certain decapods have been generally confirmed by Donaldson (1976), analysing sergestids taken off Bermuda. Protein varied from 65.7 % dry weight for *Sergestes splendens* to 58.6 % for *S. grandis*; *S. japonicus* was much lower (*ca.* 47 %), but many of the

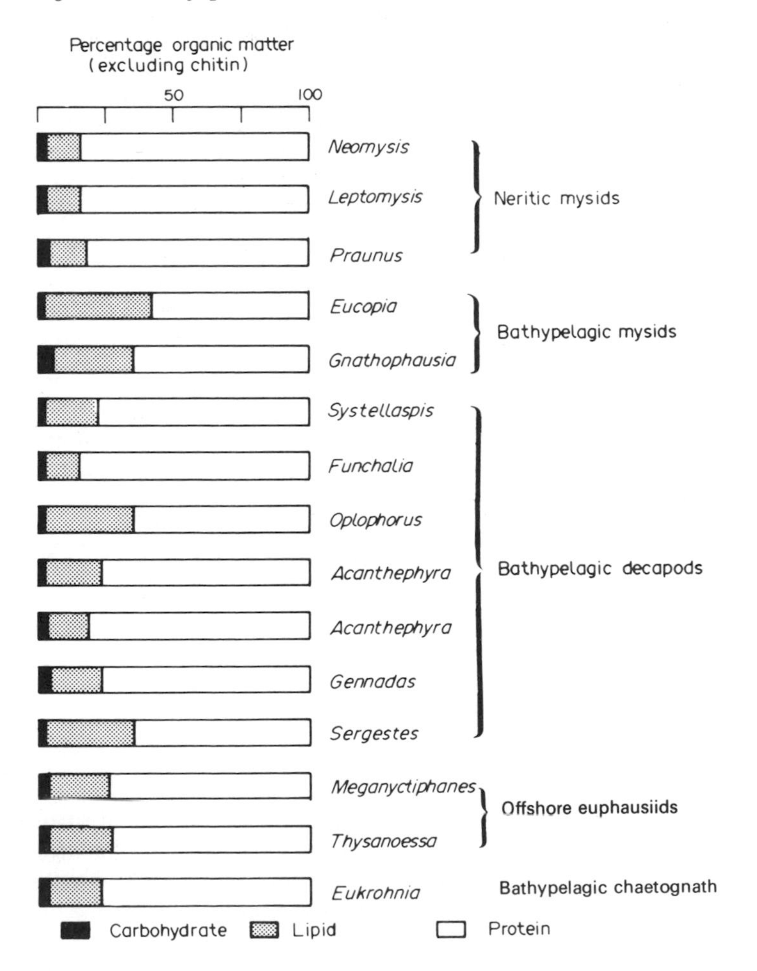

Fig. 7.23. A comparison of the carbohydrate, lipid and protein content in different zooplankton species, expressed as percentage of total organic matter (excluding chitin). Data from different sources (Raymont, Srinivasagam and Raymont, 1969b).

individuals were in relatively poor condition. Some seasonal change (71.5–62.7 %) was observable in the protein content of *S. splendens*. Mean lipid content was moderate (the maximum for the five species was 12.7 % dry weight in *S. robustus*) and seasonal fluctuations were comparatively small. Carbohydrate was low in all sergestids, not exceeding 2.5 % dry weight. Donaldson suggests that those species exhibiting the larger ranges of diurnal vertical migration tend to show higher protein content, perhaps indicating greater opportunities for increased feeding in richer upper water layers, and greater muscular development.

It is frequently suggested that zooplankton from high latitudes may be particularly high in lipid; the data given by Conover and his colleagues, Ikeda, Rakusa-Suszczewski *et al.* and by others confirm this suggestion. An investigation of the seasonal changes in biochemical composition of the boreal euphausiid *Meganyctiphanes norvegica*, taken over six months from Norwegian waters and from Scottish waters over 11 months, showed similar patterns of change over the two areas (Raymont, Srinivasagam and Raymont, 1969a, 1971). Protein was the major fraction, amounting to about 57 % dry body weight, lipid was variable (10–29 %), and carbohydrate very low (2 %). The variation in lipid suggested that it might be laid down when food was plentiful and utilized partly as a metabolic substrate when food was scarce and in part for gonad formation.

Roger's (1978) analysis of *Meganyctiphanes* from the Mediterranean gives a value for total nitrogen, which on the assumption that it all represents protein, would give a content ranging from 51 % to 69 % dry weight. No data were given for lipid. Bamstedt (1976) has more recently analysed *Meganyctiphanes* from Korsfjorden, Norway and found a very similar average biochemical composition, although ash content was much less variable over the year. Chitin and the very small amount of carbohydrate were also fairly constant. Protein was by far the major component but lipid was most variable. Both Raymont *et al.* and Bamstedt find a build-up in lipid to a maximum of 19–25 % dry body weight by October and a subsequent decline to a minimum in March (*ca.* 10 %). Bamstedt finds lipid levels remain fairly low till July and a steady increase then follows. In the material analysed by Raymont *et al.* some minor fluctuations in lipid were observed before the rise in late summer/autumn. Bamstedt found a strong correlation between lipid concentration and dry body weight from July to December, suggesting a close relationship between storage and utilization of lipid and body weight. Although a build-up of lipid when food is comparatively abundant may be associated with gonad development and considerable loss of lipid will occur at spawning, both Raymont *et al.* and Bamstedt attribute some of the marked reduction in lipid in winter/early spring to metabolic demands of the animal when food supply is very limited.

A somewhat comparable pattern appears to hold for changes in the biochemical components in the Antarctic krill, *Euphausia superba*, though sampling is far less complete. Raymont, Srinivasagam and Raymont (1971b) found that for krill captured during the Antarctic summer, protein was the major biochemical fraction, averaging 50 % dry body weight, and carbohydrate was very low (<5 %). Ash comprised about 17 % and chitin about 4 % dry weight in December.

Lipid was the most variable fraction; individuals showing as great a range as 6–33 % dry weight. There also appeared to be a seasonal change. Early in the season (December) the mean value was relatively low (13 %), but as the season advanced

(January) the amount of lipid almost doubled. Later samples showed even higher amounts, with some animals exceeding 40 % of dry body weight by the end of February (Ferguson and Raymont, 1974). Fisher (1962) found lower values for *E. superba*, ranging from 3 % to 21 % dry body weight for animals caught over 2 months. Vinogradova (1960) obtained a maximum lipid content of 26 % for *Euphausia superba*, with juvenile specimens having much lower lipid—2.5 % of dry body weight.

The marked increase in lipid found by Ferguson and Raymont during the Antarctic spring/summer might suggest that with active feeding on the rich phytoplankton the krill lay down lipid as food reserve, but use it during the period of scarcity. Support for this interpretation comes from earlier observations by Littlepage (1964) on the high Antarctic species *Euphausia crystallorophias*. He found 9–35.5 % of the dry body weight as lipid, a range comparable to that found for *E. superba* and also for *Meganyctiphanes*. Littlepage's study suggests that after the lipid peak in late summer/autumn, amounting to some 35 % of dry body weight, a more or less steady decline followed during the winter to reach the lowest lipid content in early spring, just before the start of the phytoplankton increase. The same author, investigating lipid changes in the Antarctic copepod *Euchaeta antarctica*, observed large fluctuations but of a different pattern. The body lipid, mainly in a dorsal oil sac, increased from mid-June (18 % body weight) to an August peak of 46 %, but this peak was associated with egg production. Extremely yolky eggs are produced and the body lipid fell sharply as the eggs were released. Apart from this change associated with egg production, lipid varied relatively little during the rest of the year (34–39 %).

Bamstedt and Matthews (1975) and Bamstedt (1975) examined seasonal changes in body weight and biochemical composition in the related boreal species *Euchaeta norvegica*, taken from Korsfjorden, Norway. The major components for the later copepodites studied were protein and lipid at all seasons. Carbohydrate rarely exceeded 2 % dry body weight; ash amounted to 6–7 % (rarely >10 %) for adult females, males and Copepodite V and chitin was only about 5 %. Ash and chitin were not specifically determined for Copepodite IV stages, but by difference analysis it appeared that the amounts were somewhat higher. Protein varied between 43 and 53 % and lipid averaged about 21 % of the body weight for C IV, so that together these two components made up about two-thirds of the weight, whereas in the older stages protein and lipid could amount to as much as 80 %. Protein varied from 36–47 % for females; 41–55 % for males and 39–50 % for C V. Lipid ranged from 26–47 % for females, 29–43 % for males and 24–43 % for C V (cf. Fig. 7.24).

Although protein fluctuated to some extent with body weight, lipid, which was more variable, influenced body weight to a greater degree. Body weight rose from a low level for the later copepodites in spring (about April) to a maximum about July and to a second maximum in November. This was particularly marked for females. The build-up, particularly of lipid, may be associated with maturation of the gonads. Two spawnings occur in the fjord, the first beginning in June/July the second and greater one in November/December. Bamstedt and Matthews also calculated the ratio protein/lipid and observed that seasonal changes in the ratio were generally similar for males, females and C V, reflecting a common environmental effect. The changes were most obvious, however, for females, probably indicating the effect, especially on lipid reserves, of the maturation of gonads and subsequent spawning (cf. Littlepage for *Euchaeta antarctica*). The eggs of *Euchaeta norvegica* are higher in lipid than the female

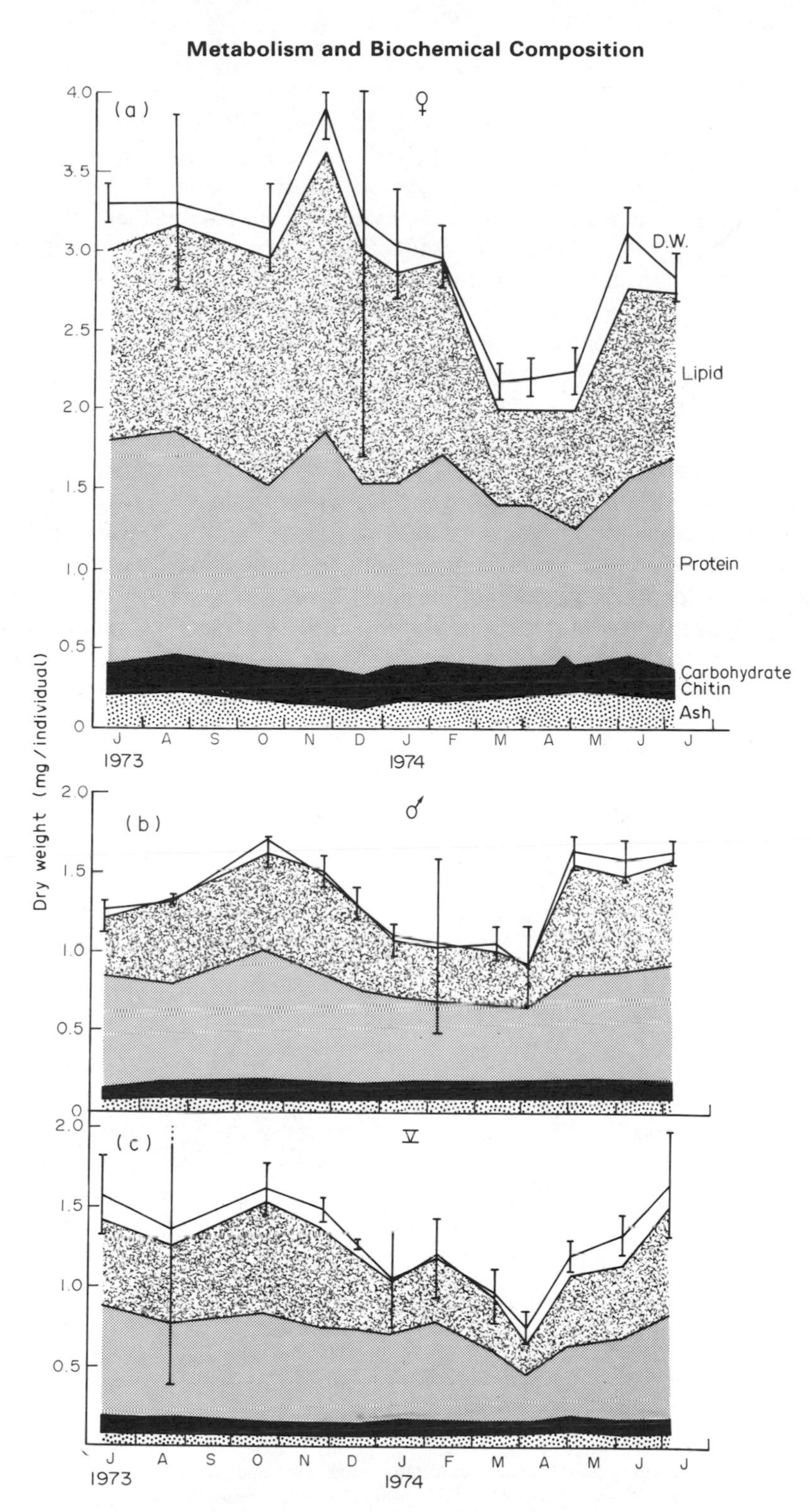

Fig. 7.24. Mean values of dry weight and biochemical composition of *Euchaeta norvegica* over 13 months (a)—female; (b)—male; (c)—Stage V (Bamstedt and Matthews, 1975).

tissues (38–49 % protein; 50–58 % lipid; *ca.* 2 % carbohydrate).

Bamstedt pointed out that since male *Euchaeta* cannot feed, the animal must rely on reserves for its oxidative metabolism. The chief reserve is lipid, mainly present as wax ester (*vide infra*). The lowest amount corresponds to about 21 % body weight, presumably the minimum for survival. The maximum may be of the order of 43 %. Assuming a respiration rate similar to adult females, Bamstedt calculates that male *Euchaeta* use approximately 6 μg lipid/mg dry weight daily for their metabolic needs and that some 50 % of the male stock might then survive (assuming no predation) for some 29 days; none could survive for a period exceeding 41 days on his calculations.

Variations in the biochemical composition with season in some other zooplankton inhabiting the Korsfjorden have also been studied. Bamstedt (1978) attempted to relate the changes to two main factors, food supply and the reproductive cycle. For females of the carnivorous deep-living copepod *Chiridius armatus* protein averaged 46 % and lipid 24 % of the dry weight over the year. Carbohydrate was low (2 %), as apparently in all zooplankton. The seasonal variations (Fig. 7.25) showed lipid to be the most variable. Dry weight was maximal in summer and minimal in winter, with lipid showing a similar periodicity, but with the changes occurring slightly later. Breeding begins in September and continues until March, but it is not clear whether the changes in lipid are linked to the reproductive cycle. Bamstedt observed, however, that several zooplankton species from Korsfjorden showed very low lipid content about late winter/spring (*vide infra*), and that this might be associated with the continued metabolic demands when food is very scarce.

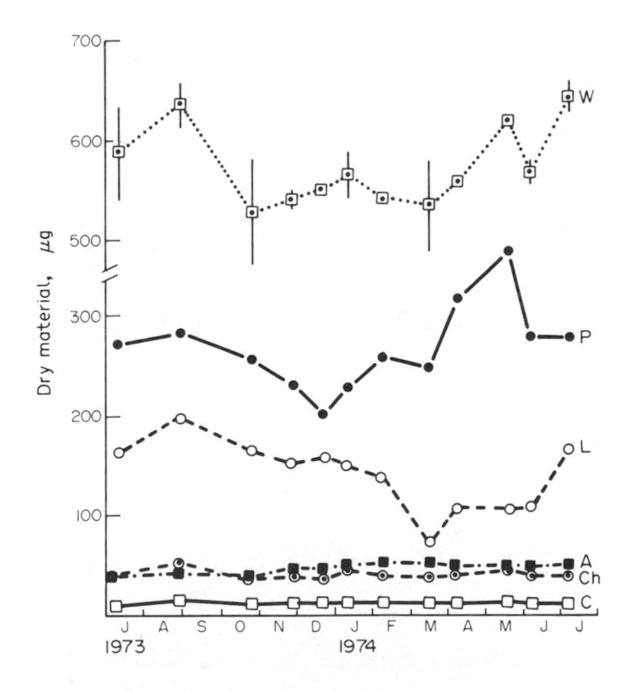

Fig. 7.25. Seasonal variation in the average individual dry weight (W), and the average individual content of protein (P), lipid (L), ash (A), chitin (Ch), and carbohydrate (C) in adult females of *Chiridius armatus*. Mean values for Ch and C proportions estimated where some determinations were lacking. Vertical bars denote ± one standard deviation (Bamstedt, 1978).

For *Boreomysis arctica* from Korsfjorden protein averaged about 38 % and lipid a little less (34 %); carbohydrate was again very low (2 %). During autumn and winter lipid became the major component (Fig. 7.26) but decreased from mid-winter to summer. Dry body weight was highest in autumn, when lipid is high. Low values in early spring may be associated with poor nutritional conditions. For females with and without marsupia, differences in biochemical composition were slight, suggesting that maturation of gonad and egg production was not the most important factor affecting the composition.

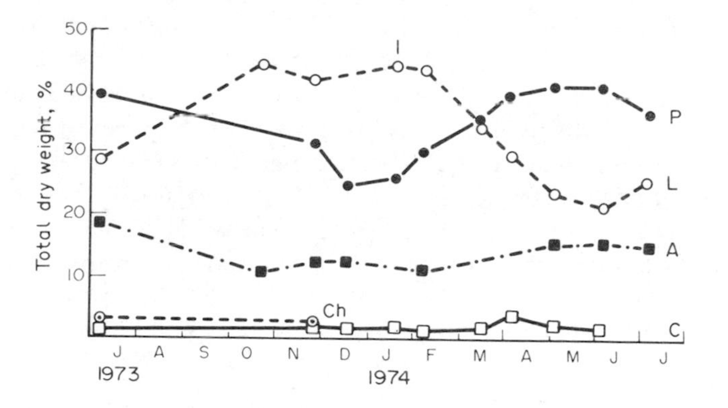

Fig. 7.26. Seasonal variation in the average proportion of protein, lipid, ash, chitin and carbohydrate in juvenile *Boreamysis arctica*. Symbols as in Fig. 7.25 (Bamstedt, 1978).

Eukrohnia hamata showed the main component as protein (39 %) with substantial lipid (32 %), very little carbohydrate (1.5 %), but considerable ash (*ca.* 18 % dry wt), probably reflecting the high water content (Fig. 7.27). Lipid was again the most variable, with maximal proportions in autumn/winter and minimum in winter/spring. The trend is similar in immature and mature individuals and Bamstedt believes that this reflects a deposition of lipid reserve during summer and early winter when food is abundant and a reduction of lipid when food is scarce in winter/spring. Some lipid is drawn upon for reproduction.

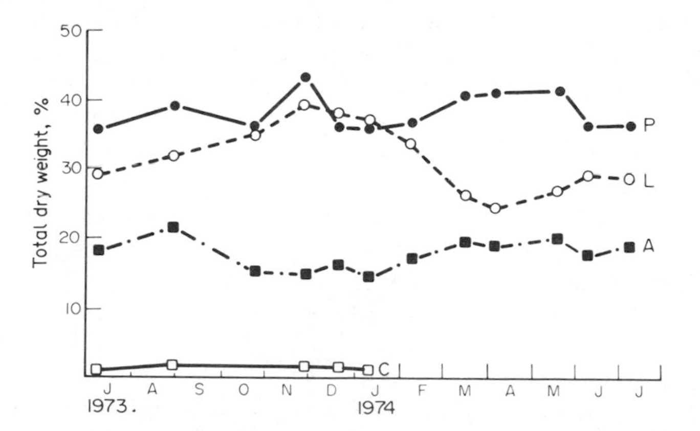

Fig. 7.27. Average proportion of protein, lipid, ash and carbohydrate of *Eukrohnia hamata* (6 mg dry weight) Symbols as in Fig. 7.25 (Bamstedt, 1978).

Childress and Nygaard (1974) examined the biochemical composition of several mid-water zooplankton species off southern California. Whilst confirming the very low carbohydrate content and the tendency for deeper-living plankton to have a greater lipid content than shallow species, their lipid fractions were more substantial (6–20% of *wet* wt for species between 400 and 800 m). In terms of ash-free dry weight, Childress and Nygaard quote values such as 40% for *Gnathophausia ingens* and *G. gracilis*, 56% for *Notostomus*, 49% for *Hymenodora* and 35% for *Bathycalanus princeps*, though several species show smaller contents (e.g. *Sergestes* spp. 16–24%, *Gaussia* 22%). The lipid and other components need to be related to wet weight, however, since water content and total organic matter show great variation, e.g. *Notostomus* reduces skeletal and organic matter and is actually negatively buoyant. *Gigantocypris agassizii* has an extremely high water content. Calorific content increased at intermediate depths, but both lipid and calorific content apparently declined at the greatest depths. Protein showed a marked trend to decrease with increasing depth of occurrence. *Notostomus* and *Hymenodora*, for example, had only 24% and 29% (ash-free dry wt), *Gnathophausia* spp. contained 36%, but *Gaussia* reached 51%. They suggest that the lower protein levels which they record may be associated with the greater depth of habitat of the plankton.

Childress and Nygaard relate the protein content to respiratory rate and suggest that a substantial part of the decrease in respiration is associated with less protein, reduction of actively metabolizing tissue and lower density. Relative buoyancy decreased with depth (i.e. tended to become very small or even negative). Lipid may displace neutrally buoyant body fluids. Lower relative buoyancy may conserve energy by allowing quiescent periods, although quite high respiratory activity may occur. With deep-living species, however, in which the lipid and calorific content is lower than in mid-water forms, Childress and Nygaard believe relatively large size is of greater significance than neutral buoyancy. The species are weak in musculature, and low in calorific content but are able to grow to comparatively large size in the deep sea habitat with very low food supply. The amount of energy stored is high relative to the low respiratory demand and the apparent lack of marked vertical migration.

Mayzaud and Martin (1975) point to the lack of information on the mineral composition and its significance in zooplankton, apart from studies specifically concerned with radioactive elements and despite the recent upsurge of interest in organic components. They demonstrate the wide variability of inorganic constituents amongst different species, although the causes are not obvious. Though zooplankton does not show such very marked capacities for concentrating elements as phytoplankton, there is still considerable ability to do so. Of the four abundant metals in seawater, potassium is concentrated and sodium, calcium and magnesium discriminated against by both *Calanus* and *Sagitta*. For the four trace metals iron, manganese, copper and zinc, *Sagitta* shows almost equal concentration factors (exceeding 70) for the first three, but zinc is concentrated more than a thousand-fold. *Calanus* shows a concentration factor of 630 for iron, *ca.* 1500 for copper and a remarkable value (*ca.* 2500) for zinc. Though certain of these metals are part of enzyme systems, such concentrations cannot be explained on these grounds. Some of the metals may be adsorbed on body surfaces rather than incorporated in tissues. This factor is considered by Martin (1970). His analysis of zooplankton collections demonstrated the great variation between the amounts of different elements; lead, iron, cadmium and zinc, for example, were among the most concentrated metals, though such data do not take account of the large interspecific differences in the animals' ability to concentrate particular elements. Martin emphasizes the increased amounts of some metals which are adsorbed on cast exuviae, especially of such abundant zooplankton as copepods.

Small and Fowler (1973) reported average quantities of zinc in the euphausiid *Meganyctiphanes norvegica* from Mediterranean waters, amounting to 73 μg Zn/g dry body wt. An approximate calculation, based on the range of concentration of zinc in seawater (0.01–1.0 μg/l—Volume 1, Chapter 2) suggests a concentration factor for *Meganyctiphanes* of 1.5×10^4–1.5×10^6 dependent on the amount of zinc in Mediterranean waters. The factor is much greater than that of Martin for zooplankton in general. Small and Fowler discuss the flux of zinc and its route to deeper waters. They believe that faecal pellets produced by the euphausiid are responsible for the largest proportion of the zinc flux (36–98 % of total body zinc, dependent on food availability). Excretion of zinc and loss through moults make only a comparatively minor contribution.

Work on the turnover of heavy metals by zooplankton and the toxic effects of these elements has been reviewed by Davies (1978). While it is not within the scope of this text to give a detailed account of pollution, some heavy metals have been shown to affect growth and development of certain zooplankton. For instance, after initial exposure to 5 μg Cu/l for four days, *Acartia tonsa* released more eggs than did controls but higher concentrations depressed egg production, which ceased at 50 μg Cu/l, while exposure for more than 10 days lowered fecundity (Reeve, Walter, Darcy and Ikeda, 1977). Two other copepods, *Pseudocalanus* and *Paracalanus*, also showed signs of stress on exposure to mercury and copper.

The molecular forms and biochemical associations of trace metals in zooplankton have received little investigation. Johnson and Braman (1975) showed that very small fractions of the total arsenic in *Lepas* and in an unspecified shrimp from the pelagic *Sargassum* community were present in methylated forms, particularly as dimethylarsinic acid. There is more evidence concerning the occurrence of methylated arsenic compounds in organisms of higher trophic levels and in macroalgae and phytoplankton; for example, arsenobetaine has been isolated from *Palinurus longipes cygnus* Edmonds, Francesconi, Cannon, Raston, Skelton and White, 1977). Leatherhead, Burton, Culkin, McCartney and Morris (1973) report that more than 90 % of the mercury in the decapod crustacean *Oplophorus* was present in the methylmercuric form.

An understanding of the metabolic pathways in zooplankton demands some knowledge of the detailed chemical composition of the main organic components. Little information is available on constituent carbohydrates, in part no doubt owing to the generally extremely low total carbohydrate in zooplankton. Preliminary analysis by Raymont, Austin and Linford (1968) on *Neomysis integer* suggested that the majority (nearly 50 %) of the total carbohydrates was probably ribose from nucleic acids. Glycogen amounted to about 25 % and a very small quantity (5 %) of free pentoses was present. Glucose and galactose were not detected in the uncharacterized remainder. Mayzaud and Martin (1975) agree that most of the small amount of carbohydrate which they identified in *Calanus* and *Sagitta* was probably ribose derived from nucleic acid. They identified traces of glucose.

Lipid

Substantial information is available on the lipid composition of zooplankton. This increased knowledge is largely due to advances in thin layer and gas liquid chromatography. Lipid is important in plankton both as a reserve and as structural material. The major lipid components recognized in zooplankton are triglycerides, regarded chiefly as reserve storage material, phospholipids, mainly structural components of cells, sterols and to a very much smaller extent, hydrocarbons. Traces of free fatty acids may frequently be identified and very small amounts of diglycerides and monoglycerides. Wax esters may be a major constituent of the total lipid, but are by no means universal in zooplankton. In *Neomysis integer* the main constituents are triglycerides (34 %), phospholipids (50 %) and sterols (13 %) (cf. Raymont *et al.*, 1968). Traces of free fatty acids, mono- and diglycerides were identified. Morris (1973) also reported traces of these lipid materials and hydrocarbons in *Neomysis*; no wax ester was detected in either investigation.

Lee, Hirota and Barnett (1971) found that of the copepod population in an area of the sub-tropical Pacific, the species to about 250 metres depth generally had low total lipid (mean 18 % dry weight). The lipid was made up of relatively large amounts of triglyceride (*ca.* 14 % total lipid), but very little wax ester (*ca.* 2.5 % total lipid). A second group of lower mesopelagic or bathypelagic copepod species, usually below 600 m, had a much higher lipid content (mean 40 % dry wt, with some species exceeding 50 %). Comparatively little triglyceride was present in these deeper-living species (mean 4 % total lipid); the major part of the lipid was made up of wax ester (mean >60 %). Some copepods inhabiting a zone approximately 300–600 m depth included a number of species which migrated towards the surface at night. In this apparently transitional zone the total lipid content of the copepods was approximately the same as for the surface-living group; the proportion of wax ester was highly variable and triglyceride formed a distinctly larger proportion of the total than in surface forms.

Calanoid copepods collected from temperate regions (the upper 200 m) and from polar regions (upper 500 m) also had high total lipid (25– >70 % dry wt), with wax ester a very important component and triglyceride of relatively little significance. This was especially true of the polar copepods; wax ester exceeded 60 % and triglyceride was usually less than 10 %. Wax ester appears, therefore, to be the major lipid reserve in copepods living in deeper and colder waters. To some degree this distinction between proportions of wax ester and triglyceride may be seen in some families of calanoids. All genera examined of the Euchaetidae, Lucicutiidae, Heterorhabdidae and Augaptilidae, generally typical of the colder, deeper water, had more than 20 % lipid as wax ester; in contrast, genera from the largely surface and warmer water families, the Pontellidae and Candaciidae, had less than 10 % wax ester.

Triglyceride and wax ester together, however, make up approximately 70 % of the total lipid for copepods drawn from various families from different areas. These two components appear to constitute the chief reserve lipid, which may show some loss on starvation. In contrast, phospholipid and sterols, which make up about 25 % of the total, are believed to be essentially structural components. Mayzaud and Martin (1975) believe that there is considerable evidence for some variation in the chemical composition of a zooplankton species and quote as factors influencing composition the stage of development, the season and geographical location. In their analysis of *Calanus* taken in November from the Nova Scotia shelf, total lipid is somewhat higher (44 % dry wt) than found in most investigations and equals the protein content. The phospholipid component of the total lipid is, however, remarkably low (*ca.* 4 %). Lee Nevenzel and Paffenhofer (1971) found, for example, 44 % of total lipid in *Calanus helgolandicus*. In contrast, Mayzaud and Martin, while finding only moderate levels (9 %) of total lipid in *Sagitta elegans*, agreeing with Reeve, Raymont and Raymont's (1970) observations on *Sagitta hispida*, obtained very high phospholipid (88 % of lipid). The much higher amounts of lipid in the deeper-living *Eukrohnia* recorded by Fisher (1962), by Raymont *et al.* (1969b) and espccially by Lee, Hirota and Barnett (1971), however, suggest that a genuine specific difference may be true of various chaetognaths. *Eukrohnia* may make use of lipid in metabolism to a greater extent than at least some species of *Sagitta* which predominantly metabolize protein (cf. Reeve and colleagues). Bamstedt's (1978) observations on *Eukrohnia hamata* also indicate substantial lipid which he believes is used in metabolism and reproduction.

The lipids of marine zooplankton include a great range of fatty acids—saturated, mono-unsaturated and polyunsaturated acids. Certain fatty acids are relatively common in most species, for example, the saturated acid $C_{16:0}$, the mono-unsaturated acids $C_{16:1}$ and $C_{18:1}$ and the polyunsaturated acids $C_{20:5}$ and $C_{22:6}$*. Frequently the saturated acids $C_{14:0}$ and $C_{18:0}$ and the mono-unsaturated acids $C_{20:1}$ and $C_{22:1}$ are also present in reasonable quantities. Lee, Hirota and Barnett's investigations confirmed that the same five fatty acids were the major ones in the copepods analysed. But, whilst $C_{16:0}$ was important in triglyceride and phospholipid fractions, the two mono-unsaturated fatty acids were of chief importance in triglyceride and wax ester. The polyunsaturated acids, especially $C_{22:6}$, were of greater significance in phospholipids.

Morris (1972, 1973) extended the observations of Lee *et al.* (1971) on the amount of lipid and proportion of wax ester in relation to depth more widely to planktonic crustaceans—copepods, euphausiids, mysids and decapods. In the warm-temperate North Atlantic, Morris found that upper water crustaceans, extending to about 600 m, were generally low in lipid (1–2 % wet wt), the lipid occurring mainly as triglyceride and phospholipid; wax ester made up less than 10 % of the total. Mid-water species, mainly exceeding 500 m, showed much higher total lipid (3–12 % wet wt). A large proportion of the lipid (30–80 %) was wax ester, the remainder being chiefly triglyceride. Deep-water near-bottom crustaceans generally showed lower lipid levels (1–3 % wet wt), with wax ester being markedly reduced. Three species of *Acanthephyra* exemplify this pattern: *A. purpurea*, an upper water species, yielded 5–10 % wax ester; *A. pelagica*, from mid-water, contained 44–65 %; *A. eximia*, the deepest-living, had 5–10 % wax ester. The vertical ranges of the first two species, however, overlap to some extent.

Morris also found relatively high amounts of wax ester in calanoid copepods from near surface waters of higher latitudes; calanoids appear, therefore, to be somewhat exceptional among the epipelagic zooplankton. Confirmation of the conclusions on the distribution of lipids with depth and latitude also comes from investigations by Lee and Hirota (1973). In a tropical area of the South Pacific, 32 species of copepod

* In this nomenclature the number before the colon denotes the number of carbon atoms, the number after the colon, the number of double bonds.

showed a great range in total lipid (<5 %–60 % dry wt) and in the proportion of wax ester (trace to 70 %). Generally, surface forms showed low lipid with triglyceride as the major neutral lipid; deep-living species had high lipid and wax ester. Comparison of tropical, sub-tropical and higher latitudes showed that high lipid content and substantial amounts of wax ester (>20 %) were typical of copepods living deeper than 500 m at any latitude and of those in the upper 500-m layers in latitudes exceeding 50° (both north and south). For copepods living shallower than 500 m, temperate and polar species had higher lipid and wax ester contents but less triglyceride than warm water forms. For species living deeper than 500 m, sub-tropical species have more lipid than tropical species but the proportions of wax ester to triglyceride are similar. The two genera *Calanus* and *Eucalanus* are exceptional in that shallow-living species even in the tropics have some wax ester, but the various species of these genera show a progressive rise in wax ester from tropical, through sub-tropical and temperate to polar latitudes.

A number of planktonic decapods and euphausiids collected in sub-tropical waters showed triglyceride as the main storage lipid; wax ester occurred only in trace quantities, but the species analysed were mostly taken above 1000 m depth. Only *Gnathophausia*, a cranchiid squid and a chaetognath had substantial amounts of wax ester. Herring (1973) found that species of pelagic Caridea may have very high lipid levels. While most species ranged from 1–5 % of *wet* weight, some were much higher (e.g. 18.4 % in *Systellaspis cristata*, 18.5 % in *Hymenodora gracilis* and a maximum of 29.6 % in *Meningodora miccylus*). There was some indication of greater lipid content with increase in depth. Penaeidea did not show the high lipid contents; the range was 1.5–4.5 % for most species. Herring confirmed that wax ester was low or absent in shallow pelagic decapods, in which triglyceride was the important fraction, but wax ester was a major constituent of the lipid of deeper species, though there were a few exceptions (e.g. *Gennadas valens* had little).

The hepatopancreas contained a very substantial proportion of the body lipid in decapods, but the components differed. Thus *Acanthephyra sexspinosa* hepatopancreas had wax ester but triglyceride was present in greater quantity. In *Systellaspis cristata* wax ester was by far the major component. The eggs of all pelagic Caridea analysed had large quantities of lipid (6–39 % wet wt). Some other data are given by Herring (1974). Eggs of Penaeidea were not investigated. Triglyceride was always the dominant lipid component even in those species which as adults had high wax ester.

Only four species, including *Systellaspis* spp. and *Notostomus auriculatus*, had significant quantities of wax ester in the egg. Phospholipid was comparatively important in the eggs, especially those of small size. In a study of the eggs of *Acanthephyra*, Herring and Morris (1975) described the deep red colour of the eggs as being duc to a carotenoid associated with a lipoprotein in the yolk. The lipoprotein contained phospholipid and triglyceride in relatively large proportions. Triglyceride was the major fraction in the remaining lipid. Total lipid was remarkably reduced during the development from 20 % to 5 % (*wet* wt) (Fig. 7.28), but the relative composition of the lipid showed little change. Herring and Morris found very little wax ester in the embryo, although adult *Acanthephyra pelagica* have large quantities.

Although some five fatty acids are common in most zooplankton, there appear to be taxonomic differences. In *Neomysis integer* about a third of the total fatty acids occur as saturated fatty acids, chiefly $C_{16:0}$; about 40 % is represented by polyunsaturated

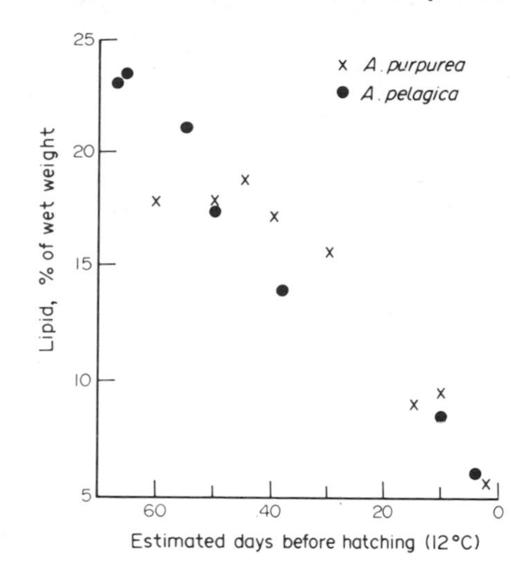

Fig. 7.28. Lipid content of the eggs of *Acanthephyra purpurea* and *A. pelagica* (Herring and Morris, 1975).

acids, chiefly $C_{20:5}$; mono-unsaturated acids, mainly $C_{18:1}$, are relatively low. Morris (1973) found that upper-water oceanic copepods had almost equal quantities of saturated and polyunsaturated acids but that mono-unsaturated fatty acids were very low. He described a somewhat similar pattern for oceanic euphausiids. On the other hand, mono-unsaturated acids were relatively much more abundant in pelagic decapods. Depth range, however, may also affect the pattern; deeper water species of copepods, euphausiids and especially mysids (*Gnathophausia* and *Eucopia*) have much greater quantities of mono-unsaturated acids than their upper water relatives (cf. Fig. 7.29a–f).

Preliminary analyses of a few plankton species, including euphausiids, decapods and a copepod, taken at different stations over a fairly wide geographical area of the north-east Atlantic suggested that there was little variation in fatty acid pattern over different parts of the area (Morris, 1973). The pattern of fatty acids may, however, be modified by certain environmental factors, including temperature. Not only, it is suggested, is total lipid much increased in zooplankton from waters of lower temperature (cf. Sheard, 1953), but plankton from colder waters also contains a higher proportion of the long chain polyunsaturated fatty acids. Belloni, Cattaneo and Pessani (1976) found that the lipid content of *Meganyctiphanes norvegica* taken from the Ligurian Sea was *ca.* 7% dry weight. Important fatty acids were $C_{16:0}$, $C_{18:1}$, $C_{20:5}$, $C_{22:6}$; but $C_{16:0}$ and $C_{18:1}$ were higher than in the samples from the north Atlantic (Morris, 1971a, 1972). They suggest that the warmer ambient temperature and the geographical location at the southern limit of distribution influenced not only lipid content but also specific composition. Farkas and Herodek (1964) demonstrated an increase in the quantity of C_{20} and C_{22} polyunsaturated acids in certain copepods at lower temperatures. Morris (1971b) also claimed some increase in polyunsaturates, especially $C_{22:6}$, when *Neomysis* was subjected to lower temperatures, though he found no effect of salinity on fatty acid composition.

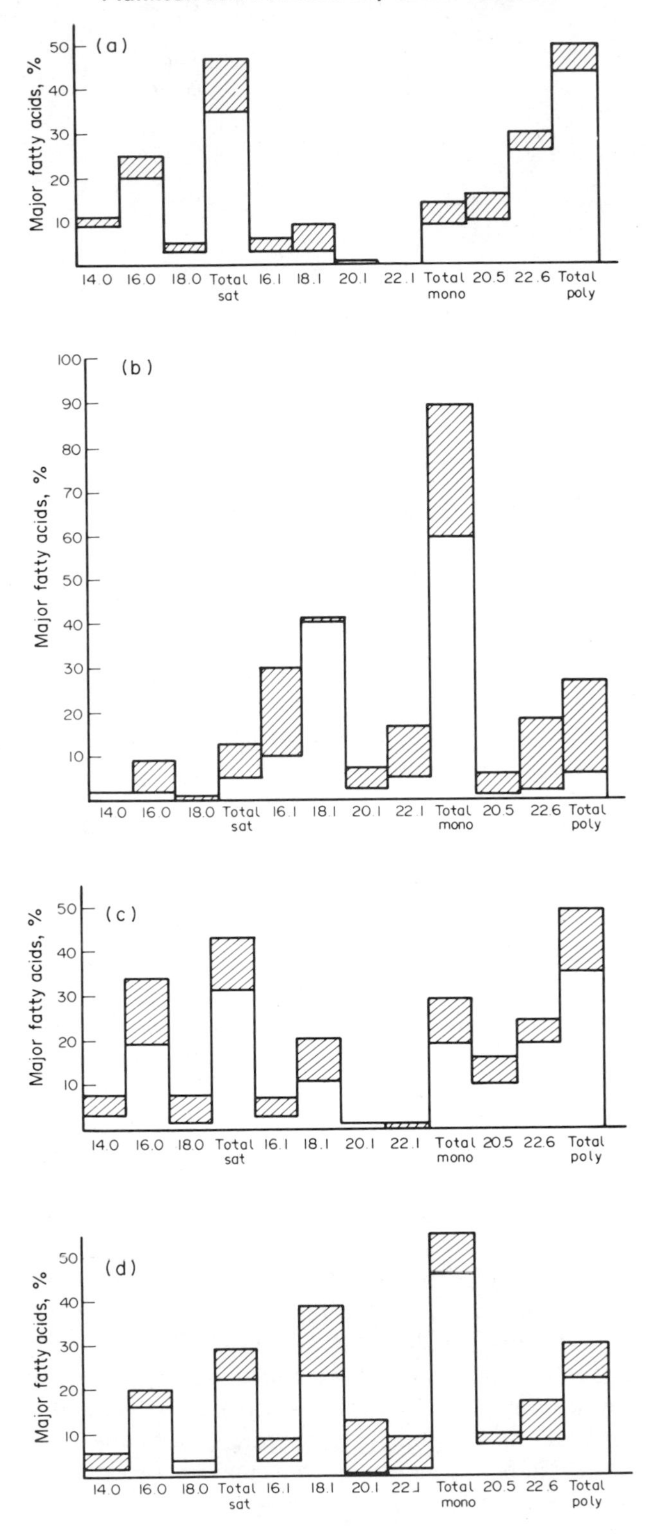

(a)
(b)
(c)
(d)
Major fatty acids, %
14.0
16.0
18.0
Total sat
16.1
18.1
20.1
22.1
Total mono
20.5
22.6
Total poly

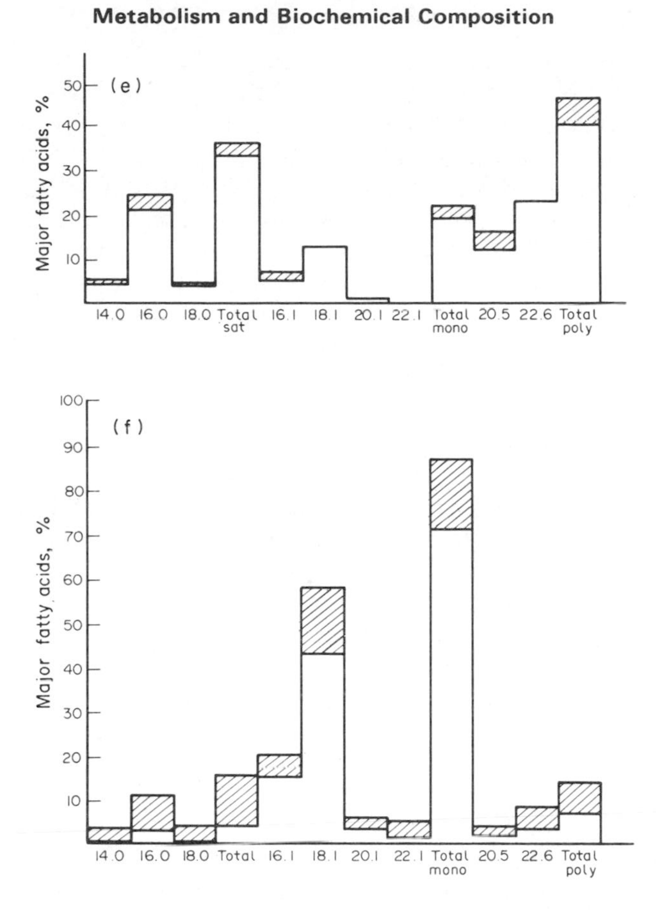

Fig. 7.29. Variations in the major fatty acids found in oceanic zooplankton. Hatched lines indicate variation in individual fatty acid as percentage of total. (a) upper water copepods; (b) deep water copepods; (c) eight species of upper water euphausiids; (d) Six species of upper water oceanic decapods; (e) two species of upper water oceanic mysids; (f) two species of deep water oceanic mysids (Morris, 1973).

The fatty acids of the two Antarctic euphausiids *Euphausia superba* and *E. crystallorophias* examined by Bottino (1974) showed the same five acids as major constituents as in other zooplankton, but there were some well marked differences between the two species. Of the saturated acids, *E. superba* had rather more $C_{14:0}$ than *E. crystallorophias*, $C_{16:0}$ being relatively more abundant in the latter, but the chief distinction was in the amount of mono-unsaturated acids. The chief fatty acid in *E. crystallorophias* was $C_{18:1}$, forming an average 44 % of the total, approximately twice that for *E. superba* (21 %). The difference was mainly attributable to the importance of wax fatty acids in *E. crystallorophias*, oleic acid ($C_{18:1}$) being overwhelmingly abundant among the wax fatty acids (83 %). There was a general similarity in the polyunsaturated acids of the two species, $C_{20:5}$ being more abundant (*ca.* 15–18 %) than $C_{22:6}$ (5–10 %). Bottino suggests that the greater degree of unsaturation in the total fatty acid spectrum of *E. crystallorophias* might be associated with its very high latitude neritic habitat; *E. superba* is less frequently covered by ice (cf. Chapter 2). He is considering the suggestions already referred to, that low temperature is associated with a greater degree of unsaturation. The fatty acid pattern of the euphausiids is fairly comparable to that of Antarctic phytoplankton, but the importance of waxes in *E. crystallorophias* is distinctive. Although Bottino indicates that the small quantities of polyunsaturates and of wax esters in the phytoplankton might be selectively retained by the grazing euphausiids, synthesis is also probably active (*vide infra*).

The proportions of the lipid components may change during the development of a species. Lee, Nevenzel and Lewis (1974) investigated variations in the wax ester contents of different stages of the copepod *Euchaeta japonica* using both wild and laboratory reared animals. While wax ester was present in all stages of development, triglyceride was identified only in the eggs, nauplius II and adults. Total lipid decreased as a percentage of dry body weight from egg to copepodite II, though there was some increase in absolute quantity of total lipid from N V. Development from the C II stage, however, included a marked increase in the proportion of lipid, with levels reaching 30 % dry weight in C IV and 50 % dry wt in C VI (cf. Bamstedt and Matthews, 1975). In adult female *Euchaeta* there was a reasonable quantity of triglyceride (18 % of total lipid) compared with 60 % as wax ester. In starving females lipid tended to be used conservatively; triglyceride was used first, but when nearly exhausted wax ester was called upon and was utilized slowly. Mature females transfer substantial amounts of lipid to the eggs, including considerable quantities of triglyceride. This appears to be utilized preferentially in development, so that little remains by the N III stage and all triglyceride is used up by N VI. By this time wax ester is also being used heavily. A large lipid store may be of significance in relation to the ability of *Euchaeta* to reproduce throughout the year, though reproduction tends to be maximal in the spring. It is significant that the young stages apparently feed little.

For all stages of development about 20 % of total lipid was present as sterol and phospholipid, 80 % as wax ester and triglyceride. However, differences were apparent in the relative importance of certain fatty acids in the composition of wax ester and triglyceride. For example, $C_{16:0}$ and $C_{18:1}$ were the major fatty acids of triglyceride, with $C_{16:1}$ and the polyunsaturate $C_{20:5}$ also present in reasonable amount. $C_{16:0}$ was of no importance in wax ester; $C_{18:1}$ and $C_{16:1}$ with the polyunsaturates $C_{20:5}$ and $C_{22:6}$ were the major fatty acids. The dominant alcohol of the wax esters was $C_{16:0}$, with considerable amounts, especially in C V, of $C_{14:0}$. Phospholipids resembled triglycerides to some extent in that both $C_{16:0}$ and $C_{18:1}$ were important, but there were fairly substantial contributions of the polyunsaturates $C_{20:5}$ and $C_{22:6}$ especially in the eggs and the C VI stage.

Another example of differences in lipid fractions in different developmental stages is the observation by Sargent and Lee (1975) that whereas the amphipod *Cyphocaris challengeri* had the major portion (*ca.* 75 %) of its lipid as wax ester in the adult female, the eggs contained only triglyceride.

Among factors which might be expected to modify lipid components, including fatty acid patterns, are feeding and starvation. In preliminary experiments Lee, Hirota and Barnett (1971) found that the copepod *Gaussia*, when subjected to long periods of starvation, used triglyceride first but subsequently used wax ester quite extensively. Morris (1972, 1973) observed that the proportions of total lipid and wax ester varied directly in the oceanic mysid *Eucopia*, a finding which he suggested indicated that wax ester was used as a food reserve. Gatten and Sargent (1973) described wax ester as the major lipid, amounting to more than 80 % of the total in *Calanus finmarchicus*, whereas triglyceride was present only in very small amounts. The small calanoid *Acartia* also had substantial amounts of wax ester, though in smaller proportion than in *Calanus*. Gatten and Sargent observed some indications of increased biosynthesis of wax ester at night by these near-surface calanoids. They believe that this reflects a storage of lipid with feeding (cf. Arashkevitch, 1977). In experiments on the copepod *Euchaeta*, using cell-free enzyme preparations, they found that wax ester could be formed *de novo* from all three major biochemical constituents—protein, lipid and carbohydrate. Lee *et al.* had also shown that *Calanus* could form wax ester from a diet lacking this material. There is evidence, therefore, for wax ester being used as an energy food store. Gatten and Sargent suggest that it may be used especially by zooplankton for laying down lipid quickly during periods of intensive feeding on a comparatively rich food supply, such periods being followed by times of relative food shortage. Zooplankton migrating to the surface at night to feed on the more dense phytoplankton may be particularly adapted to this form of lipid storage.

The calanoid *Calanus hyperboreus* is a well known Atlantic Arctic species (cf. Chapter 2). Studies by Lee (1974) off an ice island in the Arctic Ocean (83°–88°N) demonstrated a build-up of lipid, chiefly as wax ester, during the brief summer Arctic phytoplankton bloom, with lipid reaching >70% of the dry weight in August/September, declining to about 66% by October, and to 51% by December (mean weight of lipid, mg/copepod: August 1.7; September 2.1; December 0.9). The decline in lipid to December was paralleled by the fall consequent on starvation of *Calanus* maintained at 3°C in the laboratory. Thereafter, however, the decline in lipid content was much more severe in the laboratory animals, so that only 0.3 mg/copepod (22% dry wt) remained after 90 days. In the field, *Calanus hyperboreus* showed only a very slow fall from December, so that by June 0.4 mg lipid/copepod (equivalent to 29% dry wt) remained, after which the amount began to increase again, presumably with feeding.

Lee found that wax ester exceeded 80% of the total lipid in *Calanus*, while triglyceride varied between negligible amounts and 7–8%. The laboratory-starved animals suffered a fall in wax ester after 90 days to 32% total lipid. The specific fatty acids which were important in the phytoplankton appeared to be incorporated into the wax esters only during summer and only trace amounts were identified in the more structural phospholipids of *Calanus*. Although wax ester is obviously utilized when food is not available and the typical phytoplankton fatty acids disappear, the polyunsaturated fatty acids in *Calanus* are apparently conserved (*vide infra*). The marked use of lipid confirms the earlier observations of Conover and Corner on *Calanus hyperboreus*. In mid-water species wax ester may function partly as an energy store, especially for animals which live in zones of comparative food paucity, but wax ester may also function in transferring lipid to eggs and may have a buoyancy function (cf. Lee *et al.*, 1971; Morris, 1973).

Other observations on variations in lipid patterns with feeding include those of Morris (1971b), who followed changes with season in *Neomysis integer*. He found total lipid to be relatively high in spring, presumably with feeding on phytoplankton and at this time $C_{16:1}$ was particularly abundant. Throughout the late summer and autumn total lipid tended to fall; $C_{16:1}$ fell markedly, $C_{14:0}$ to a lesser extent. Over the same period the polyunsaturated acids, particularly $C_{22:6}$, increased, indicating that polyunsaturated acids could possibly be built up from $C_{16:1}$ and similar dietary acids by a process of chain elongation. Lee, Nevenzel and Paffenhofer (1971) studied the effects of diet on zooplankton lipids. They analysed wild stocks of *Calanus helgolandicus* and laboratory-reared stocks fed on precise diets, namely three species of diatoms (*Skeletonema*, *Lauderia* and *Chaetoceros*) and the dinoflagellate *Gymnodinium*. Total lipid in *Calanus* ranged from 12% to 24% dry body weight. Of this a substantial portion was wax ester (25–41%) and phospholipid (28–59%); smaller quantities of triglyceride and sterols were present. None of the three diatoms used as food had a detectable quantity of wax ester. They possessed moderate amounts of triglyceride and sterols, but the majority of lipid (about 50%) was phospholipid; appreciable amounts of free fatty acids and hydrocarbons were present. The amount of food consumed showed a direct relationship with the total lipid in the copepods; the composition of the wax ester laid down was also considerably affected by the amount, though not the species of diatom available as food.

With starvation or a very much reduced diet, there was a considerable increase in the proportion of short-chain wax esters, especially those with total chain length of C_{30}. These were probably esters of saturated and mono-unsaturated acids and alcohols of C_{14} and C_{16} chain length. On the other hand, with richer feeding there was a greater proportion of longer chain wax esters. Considerable similarity was also evident in the range and proportion of fatty acids in the phytoplankton food and in the fatty acids of both wax esters and triglycerides. Lee, Nevenzel and Paffenhofer (1971) therefore suggest that with more concentrated feeding the food fatty acids are to a large extent incorporated into the fatty acids of the reserve lipids, both wax ester and triglyceride recalling observations on *Calanus hyperboreus* in the Arctic summer. On the other hand, when food is very limited, the copepods degrade the food fatty acids to small molecules, even to acetate. The specific fatty acids of the copepod are subsequently built up *de novo*, but only relatively short chain fatty acids are synthesized. This gives rise to the C_{14} and C_{16} type.

As regards the long chain alcohols that make up the wax esters, these are believed to be fairly specific. They consist chiefly of $C_{11:0}$, $C_{16:1}$, $C_{20:1}$, $C_{20:4}$ and $C_{22:6}$ and are believed to be either synthesized *de novo* or are modified from dietary material. Chain elongation and increase in unsaturation, especially in the formation of the polyunsaturated alcohols, $C_{20:4}$ and $C_{22:6}$, appear to be particularly active transformations. The phospholipid fatty acids seem to be of an essentially conservative pattern and are not affected to any marked extent by changes in the amount or composition of the diet. This emphasizes the structural stability of phospholipids. A relatively large amount of the fatty acids in phospholipids is polyunsaturates, particularly $C_{22:6}$. This fatty acid was either not detected or was found only in trace amounts in the three diatoms used as food, though some other polyunsaturated fatty acids were identified. Presumably, therefore, chain elongation must have occurred, particularly in the production of $C_{22:6}$, probably derived from the fatty acid $C_{18:3}$. The long chain alcohols of wax esters included polyunsaturates; earlier results had suggested that wax ester alcohols were mostly saturated or mono-unsaturated and of C_{16} or C_{18} chain length. The fatty acids of wax esters were mostly $C_{20:5}$, $C_{20:4}$, $C_{16:3}$, $C_{16:1}$ and $C_{16:0}$ when animals were fed on a rich diet. The dietary $C_{14:0}$ was not used; it was either oxidized or used to form triglyceride.

Among the sterols found in *Calanus* was cholesterol, also earlier identified in *Euphausia* and in some neritic plankton species. Cholesterol, however, was not identified from any of the food algae, so that it must be synthesized by *Calanus* from the assimilated lipids. Some of the sterols in phytoplankton are mentioned in Volume 1, Chapter 4. Morris (1972, 1973) has indicated some of the possible pathways in lipid metabolism in planktonic crustaceans (cf. Fig. 7.30).

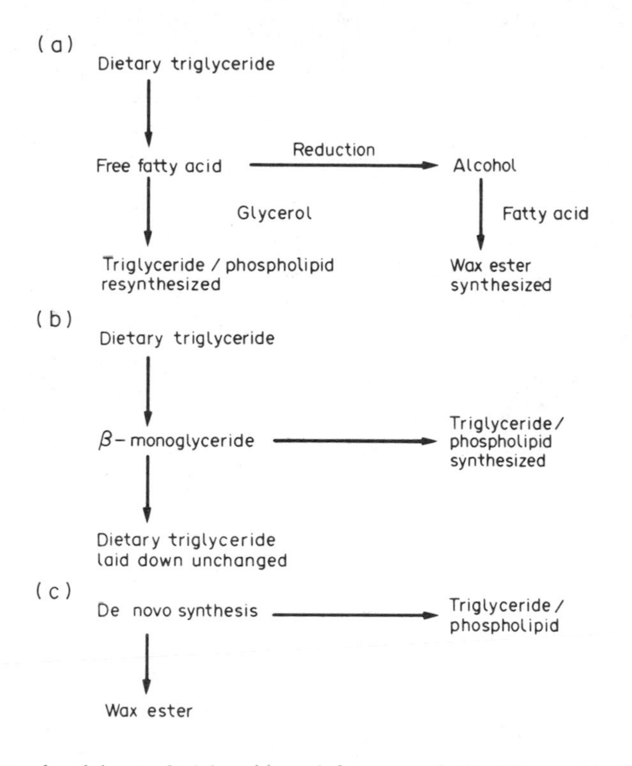

Fig. 7.30. Dietary breakdown of triglyceride and *de novo* synthesis of fatty acids (Morris, 1973).

Comparatively small quantities of hydrocarbons occur in marine zooplankton. The investigations of Blumer and his colleagues (e.g. Blumer, Mullin and Thomas, 1964; Blumer, Mullin and Guillard, 1970) have made outstanding contributions to our knowledge of the naturally occurring hydrocarbons. Corner (1978), reviewing their occurrence, confirms the importance of pristane (cf. Volume 1, Chapter 4), though only calanoid copepods appear to have comparatively substantial quantities (0.86–2.9% of total lipids for Gulf of Maine calanoids). The relatively high quantity in *Calanus hyperboreus* is maintained, the level even showing a slight increase, with prolonged starvation and considerable loss of total lipid. Pristane and the other hydrocarbons present in much smaller quantities in zooplankton are probably all derived from the phytol part of chlorophyll (Fig. 7.31) (cf Volume 1, Chapter 4). Corner indicates some metabolic pathways. There is also a suggestion that some hydrocarbons, such as HEH (a $C_{21} H_{32}$ olefinic hydrocarbon) may be more easily accumulated by some zooplankton (e.g. *Rhincalanus nasutus*) from certain algae. Possible biological functions of naturally-occurring hydrocarbons include buoyancy and an influence on sex ratio in calanoids.

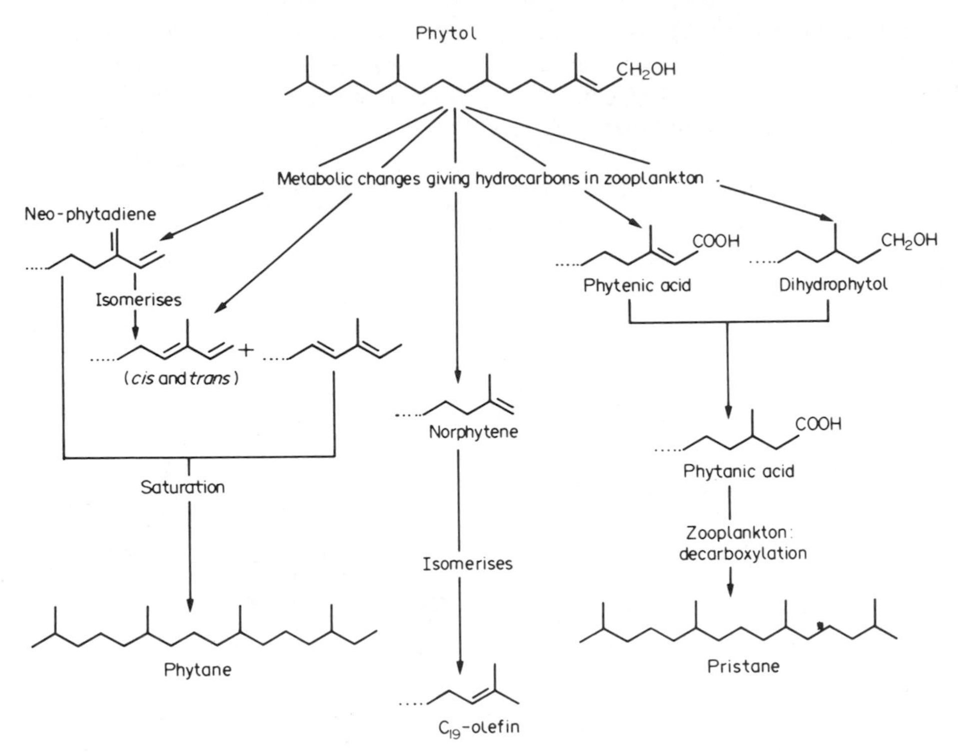

Fig. 7.31. Inter-relationships between phytol-derived hydrocarbons (Corner, 1978).

An important feature of the hydrocarbons, as first stressed by Blumer, is their comparative stability in the marine food web. Corner (1978) has also discussed the occurrence and effects of exogenous hydrocarbons in zooplankton. No attempt will be made here to review this subject, but certain aspects are of great interest from the point of view of the metabolism of zooplankton. Some of the studies by Corner himself with colleagues have emphasized the very slow release of small residues of hydrocarbons, possibly again reflecting their stability. Corner also reports a comparison of the rate of release of a ^{14}C-labelled naphthalene by *Calanus helgolandicus* when the hydrocarbon is taken up from solution alone and when feeding on cells of *Biddulphia* which had incorporated the hydrocarbon. Loss was much slower when the hydrocarbon was taken up with food (cf. Fig. 7.32).

Laboratory experiments including feeding plankton species with labelled fatty acids and other lipid materials might assist in the study of metabolic pathways. In experiments with the mysid *Neomysis integer*, ^{14}C-labelled fatty acids were fed using pure starch as a carrier and the animal subsequently sacrificed and analysed for fatty acid composition. *Neomysis* fed labelled $C_{16:0}$ and $C_{18:1}$ fatty acids converted them into a series of fatty acids including polyunsaturates of the C_{20} and C_{22} series. Apparently only relatively small amounts were degraded. Conversions included chain shortening, increase in unsaturation and chain elongation (Morris, Ferguson and Raymont, 1973). Acids of the C_{20} and C_{22} series can therefore be produced by these crustaceans without specific dietary demand for acids such as linoleic and linolenic

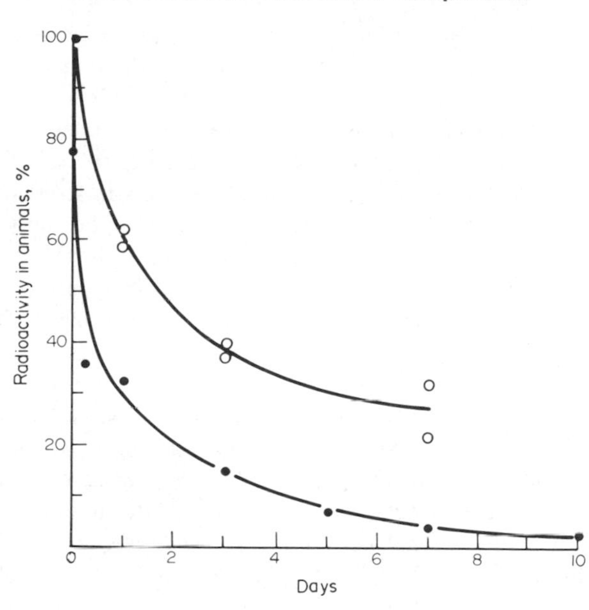

Fig. 7.32. Release of radioactivity by adult female *Calanus helgolandicus* that had accumulated ^{14}C-1-naphthalene from a sea water solution (solid circles) or from a diet of *Biddulphia* cells (open circles). Levels expressed as percentage of the radioactivity originally present in the animals (Corner, 1978).

($C_{18:3}$) which are apparently required by mammals. Linoleic acid, $C_{18:2}$, for example, is essential to mammals for the production of arachidonic acid, $C_{20:4}$.

In *Neomysis* the labelled fatty acids synthesized from the dietary acids appeared in both the triglyceride and phospholipid fractions—less in the phospholipid, though the proportion increased with duration of feeding. Phospholipids appear to be the more conservative component, confirming earlier indications. Experiments by Morris and Sargent (1973) have also demonstrated that the oceanic mysid *Gnathophausia*, the euphausiid *Nematobrachion* and the decapod *Acanthephyra*, when injected or fed with labelled $C_{16:0}$ fatty acid, showed chain elongation and extensive transformations, ^{14}C-labelled fatty acids $C_{16:1}$, $C_{18:0}$, $C_{18:1}$, $C_{20:5}$ and $C_{22:6}$ appearing (as with *Neomysis*) in triglyceride and phospholipid fractions. But wax esters are also relatively plentiful in *Gnathophausia*. The wax ester incorporated the synthesized labelled fatty acids, both as the C_{16} alcohol moiety and as the $C_{18:1}$ fatty acid. Other investigations with marine copepods have also demonstrated that labelled fatty acid (e.g. ^{14}C-palmitic acid, $C_{16:0}$) can be incorporated into triglyceride and wax ester fractions, production being stimulated by the addition of ATP and coenzyme A (cf. Gatten and Sargent, 1973). In further experiments Sargent, Gatten and McIntosh (1974), using *Euchaeta* extract, demonstrated that addition of the alcohol $C_{18:1}$ stimulated the formation of wax ester; of the lipid synthesized, generally more ^{14}C was incorporated preferentially into wax ester than into triglyceride.

It has been generally assumed that lipid may be synthesized from other dietary components. Gatten and Sargent (1973) demonstrated this synthesis in cell-free extracts of *Euchaeta* (*vide supra*).

Morris, Ferguson and Raymont (1973) demonstrated that *Neomysis* fed pure starch labelled with ^{14}C synthesized carbon-labelled lipid materials. A series of labelled fatty acids were identified and though triglyceride was the usual product, other lipid fractions were produced, including phospholipid. Labelled protein was also produced to a very limited extent, presumably by transamination processes. Further experiments (Morris, Armitage, Raymont, Ferguson, and Raymont, 1977) feeding pure labelled starch to *Neomysis* demonstrated that though the carbohydrate was converted to lipid it was unable to maintain the body lipid at normal levels. The triglyceride fraction of the total lipid decreased much more obviously than phospholipid or sterol. A substantially greater proportion of the labelled starch was converted into saturated and mono-unsaturated acids of the triglyceride fraction, but when ^{14}C-labelled starch was fed over a longer period, the proportion of radioactive polyunsaturated fatty acids in the phospholipid fraction increased. With an inadequate diet, therefore, phospholipid and the polyunsaturated fatty acids appear to be conserved as far as possible.

Morris *et al.* (1977) also compared laboratory feeding of *Neomysis* on pure non-radioactive starch with feeding on pure kaolin. *Neomysis* will filter kaolin fairly effectively and produce faecal pellets but, since the material is inert, though feeding activity is maintained the animal is in effect starved. *Neomysis* fed on pure starch exhibited a decline in total lipid; protein also declined very sharply. With such an inadequate starch diet the phospholipid fraction was to a considerable extent retained, whilst triglyceride was very largely metabolized. These effects were more pronounced in the starved (i.e. kaolin-fed) animals. In general, the quality of the fatty acid composition was fairly well maintained on the starch diet. Although, therefore, pure starch is an inadequate diet for *Neomysis*, feeding with the pure carbohydrate results in a considerable saving of the body lipid.

These experiments do not assist in elucidating wax ester synthesis, since *Neomysis* possesses none. The investigations of Sargent and Lee (1975), specifically aimed at studying wax ester synthesis, suggested that a wide variety of zooplankton can synthesize wax ester *de novo* and that these materials are transmitted only to a limited extent through the food chain. Sargent and Lee found very different proportions of wax ester in four species of zooplankton. The amphipod *Cyphocaris challengeri* was richest in total lipid (nearly 50 % dry wt) and in wax ester (75 % of total); *Calanus pacificus*, with 26 % total lipid, had more triglyceride (55 %) than wax ester (20 %); *Thysanoessa raschii* and *Sagitta elegans* had 22–23 % total lipid, the great majority (*ca.* 70 %) as phospholipid (cf. Mayzaud and Martin, 1975), but of the neutral lipid present *Sagitta* possessed considerably more wax ester than *Thysanoessa.*

Although fatty acids $C_{14:0}$, $C_{16:0}$, $C_{18:1}$ and the polyunsaturated acids $C_{20:5}$ and $C_{22:6}$ were important in all four species in triglyceride and phospholipid fractions, the wax ester fatty acids showed species differences. *Cyphocaris* had mainly $C_{16:0}$ and $C_{18:1}$, whereas *Calanus* had chiefly $C_{20:1}$ and $C_{22:1}$ (cf. also Bottino's observation on *Euphausia crystallorophias*—*vide supra*). The major alcohols in both species were $C_{20:1}$ and $C_{22:1}$.

Sargent and Lee observed that all the zooplankton species incorporated labelled glucose and labelled palmitic acid ($C_{16:0}$) present in filtered seawater into their lipids. Phospholipid was the major fraction synthesized from glucose, but neutral lipid, both triglyceride and wax ester, was produced from palmitic acid. Although phospholipid is believed, therefore, to be turned over more rapidly than neutral lipid, its level is maintained by synthesis, whereas neutral lipid is a more static, essentially storage constituent. *Calanus* and *Sagitta* showed both fatty acids and fatty alcohols carrying the ^{14}C label from $^{14}C_{16:0}$ and since palmitic acid can be synthesized *de novo*, these two plankton species must be able to synthesize wax ester *de novo*. In other species, however, the fatty alcohol was not readily synthesized. Only species normally rich in wax ester appear to effect the transformation to fatty alcohol, although all species examined will readily incorporate ^{14}C-labelled fatty alcohol into a wax ester. Wax ester, when formed in substantial quantities in the marine food chain, is probably produced *de novo* mainly by certain zooplankton species by the production of the necessary fatty alcohols from dietary fatty acids, protein or carbohydrates.

Proteins and amino acids

Zooplankton animals have a relatively high proportion of protein and usually some non-protein nitrogen mostly present as free amino acids in the body (cf. Cowey and Corner, 1963; Srinivasagam, Raymont and Raymont, 1969; Childress and Nygaard, 1974; Mayzaud and Martin, 1975). Although many species have relatively large quantities of lipids which may be used as metabolic substrate, protein may be used at times even by these "low turnover" plankton species (cf. Ikeda, 1974) and "high turnover" species appear to use protein extensively. The lower protein content of deep-sea plankton has already been noted (cf. also Childress and Nygaard, 1974).

Amongst the few analyses carried out on the proteins of zooplankton animals, Manwell, Baker, Ashton and Corner (1967) demonstrated that starch gel electrophoresis could be used for differentiating between species even of such small size as zooplankton copepods. Their studies dealt with three species of *Calanus* and were mainly concerned with esterases and dehydrogenases. They showed that *Calanus finmarchicus* and *Calanus helgolandicus* had only fifteen enzyme zones in common and differed by 24–28 zones, suggesting very strongly that the two forms were separate species. Table 7.20 shows a comparison of the electrophoretic mobilities of the two species. Even though the precise identification of protein in zooplankton is difficult, this method of examining closely related zooplankton species appears to be of considerable value. Another preliminary study by Blakey-Marshall and Wade (unpublished data) used electrophoresis to examine possible differences between swarms of *Euphausia superba* captured in the Antarctic. Though only a few experiments could be carried out on board ship, zymograms for general protein indicated considerable differences in the patterns indicative of genetic variations among the different stocks. However, as some *Euphausia* were maintained unfed on board before analysis, some variation might be attributable to partial starvation. Useful patterns were obtained for esterases and for lactic dehydrogenase in *Euphausia*. A difference in pattern, especially of esterases, between young and older *Euphausia superba* might have been associated with age group distribution or might have indicated genetic variation in krill patches. Lin and Zubkoff (1973) have demonstrated by electrophoretic techniques that the malate dehydrogenase isozyme patterns of the three scyphozoans *Aurelia aurita*, *Chrysaora quinquecirrha* and *Cyanea capillata* are distinct. Tetrazoliumoxidase bands can also be regarded as being distinct in number or mobility. Although Scyphistomas were used in these determinations, similar identification of some enzymes might be profitably employed in separating races of planktonic medusae and other taxa.

Table 7.20. Comparison of the electrophoretic mobilities of some enzymes in *C. helgolandicus* and *C. finmarchicus* (Manwell *et al.*, 1967)

Enzyme	Zones in common	Zones distinct
Esterases		
Region I (includes eserine-inhibited zone)	1	6
Region II (includes 2 eserine-inhibited zones)	1	2
Slower moving esterase	2	0
EDTA-inhibited	0	1
PMSF-inhibited	0	1
E-600-insensitive	1	0
β-Galactosidase	1	0
Acid phosphatases	0	3
"Peptidase"	2	3–4
Malate dehydrogenases		
NAD-dependent	2	1
NADP-dependent	0	3–4
Glucose-6-phosphate DH	1	0
6-Phosphogluconate DH	1	0
Triose-phosphate DH	3	2–3
Aldolase	0	1–2
Amylase	0	1
Total	15	24–28

Not all isozymic differences are tabulated because in such cases the actual number of distinct polypeptide chain types is usually less than the number of zones.

In the studies by Manwell *et al.* (1967) an amylase was identified. Using biochemical techniques, Boucher and Samain (1973, 1975) and Samain and Boucher (1974) have studied amylase and protease in species of marine zooplankton and have also estimated the protein content (*vide supra*). The more specific identification of enzymes in zooplankton may follow.

The study by Barnes and Blackstock (1975a) of seasonal changes in acid and alkaline phosphatases, though carried out on benthic animals (adult cirripedes), demonstrated a maximum in alkaline phosphatase related to the spring phytoplankton bloom and changes in acid phosphatases, possibly associated with metabolic changes in the moulting cycle. This may indicate a fruitful field of investigation on zooplankton. Moreover, as Barnes and Blackstock (1975b) emphasize, there is a great need to study seasonal changes in the enzymes of marine crustaceans since the carbohydrate content of crustaceans is accepted now as generally being low and the emphasis is on lipid and protein metabolism.

In consideration of the protein composition and possible metabolic patterns in marine zooplankton, another type of analysis has involved an examination of the spectrum of amino acids in total protein hydrolysates of plankton species. Cowey and Corner (1963a,b) investigated the protein amino acid composition and the free amino acids in *Calanus finmarchicus*. In general the amino acids from protein hydrolysates resembled the amino acid composition of the phytoplankton and other suspended material in seawater, but there were some differences, for instance, in the proportion of tyrosine. They further compared (1966) the amino acid composition of *Calanus* and its food with the faeces produced during feeding, but found little evidence for the selective absorption of specific amino acids. Of the non-protein nitrogen in *Calanus* (some 10% of total nitrogen), a very large proportion (*ca.* 90%) was made up of free amino acids, with small amounts of betain, trimethylaminoxide and other materials. Amino acids made up a high proportion of the total organic nitrogen in zooplankton (cf. Corner and Cowey, 1968). Mixed zooplankton, mainly small copepod species, have 70–80% as amino acids; *Calanus finmarchicus* has 90% (76% as protein; 14% as free amino acids) and *Calanus helgolandicus* has 80%. Jeffries (1969) also examined the free amino acids in zooplankton (mostly species of *Acartia*) but the samples were not sorted species.

Several authors (Suyama *et al.*, 1965; Burkholder *et al.*, 1967; Moiseev, 1970; Sidhu *et al.*, 1970) Srinivasagam, Raymont, Moodie and Raymont, (1971) have examined the amino acid composition of body protein in *Euphausia superba*. The last study compared the composition of *E. superba* with that of the euphausiid *Meganyctiphanes norvegica* and the mysid *Neomysis integer*. Five amino acids were dominant in the protein hydrolysates of the three crustaceans: glutamic acid was present in greatest concentration, followed in descending order of importance by aspartic acid, lysine, leucine and arginine, the order being identical in the three species (Table 7.21). The total percentage contribution of the five major amino acids ranged in the three species from 51% to 54%. Even the amino acids in lesser concentration were present in generally similar relative proportions. Thus alanine, valine, isoleucine and phenylalanine formed a secondary group next in order of importance in the three species. Methionine and cystine were always present in smaller quantities, but the analytical method employed (acid hydrolysis followed by ion-exchange chromatography) does not estimate these

Table 7.21. The amino acid composition of *Meganyctiphanes norvegica*, *Neomysis integer* and *Euphausia superba* (Srinivasagam *et al.*, 1971)

Amino acids	Amino acids in protein hydrolysate[a]			Free amino acids[a]		
	Meganyctiphanes norvegica	*Neomysis integer*	*Euphausia superba*	*Meganyctiphanes norvegica*	*Neomysis integer*	*Euphausia superba*
Cysteic acid	—	—	—	3.7	0.8	1.8
Taurine	—	—	—	10.6	23.1	15.9
Aspartic acid	11.5	10.9	10.7	7.7	1.9	0.9
Threonine	4.5	4.3	4.2	5.7	1.3	3.8
Serine	4.1	3.9	3.8	4.5	1.4	2.1
Glutamic acid	18.0	17.9	15.2	0.8	5.5	1.0
Proline	3.1	2.8	3.3	4.5	4.2	8.3
Glycine	4.1	3.9	4.6	8.9	30.9	14.0
Alanine	5.5	5.3	5.6	3.7	3.8	10.2
Valine	5.2	5.4	5.2	1.6	1.0	4.1
Cystine	1.1	0.7	1.2	0.4	0.6	—
Methionine	2.4	1.6	2.8	0.8	0.2	0.5
Isoleucine	4.8	5.3	5.0	1.2	0.6	3.1
Leucine	7.5	7.7	7.7	1.6	0.7	2.6
Tyrosine	4.3	3.9	4.3	3.3	0.6	3.1
Phenylalanine	4.7	5.0	5.0	1.2	0.4	0.9
Ornithine	0.1	0.3	0.4	7.3	0.1	7.5
Lysine	9.6	10.2	10.0	11.0	1.5	8.4
Histidine	3.0	3.3	3.5	3.3	1.5	1.9
Tryptophan	—	—	—	—	0.7	2.0
Arginine	6.8	7.5	7.5	18.3	19.1	8.1

[a] Values as percentages of total amino acids in protein hydrolysates and of total free amino acids.

quantitatively. Tryptophan is destroyed by the method. The results of the investigation by Srinivasagam *et al.* (1971) indicated, however, however, that the *total* body protein composition was similar in the three species of zooplankton despite the marked difference in habitat. A few analyses of *Neomysis* by Armitage (1979) using a slightly different preliminary extraction, gave generally comparable results, though glycine occurred in greater and arginine in lesser concentrations.

In contrast, the free amino acids, as extracted by hot distilled water from lipid-free tissue, showed considerable differences between the three zooplankton species. Free amino acids formed about 20 % of total amino acids in *Euphausia superba*; for *Calanus*, Cowey and Corner (1963a) found the free fraction to be about 16 %. In both *Neomysis* and *Meganyctiphanes* the free amino acids formed about 8 % and 15 % respectively, but Kjeldahl estimations for *Neomysis* suggested that the value should be approximately doubled. Extraction techniques may be partly responsible for such differences.

In *Neomysis* three free amino acids, glycine, taurine and arginine, were dominant, making up more than 70 % of the free pool. Alanine, glutamic acid, proline and aspartic acid are of secondary importance. A much more detailed study by Armitage (1979) has confirmed the importance of the three amino acids in *Neomysis*. Proline was present in somewhat larger amounts in the second group. Extraction with ethanol did not result in any substantial changes. In *Euphausia superba* taurine and glycine were fairly important, together forming 30 % of the total (cf. Table 7.21). In *Meganyctiphanes norvegica* the free amino acid present in greatest concentration was arginine, followed by lysine, taurine and glycine (Table 7.21).

These species differences in free amino acids may to some degree be associated with the pronounced differences in habitat. *Neomysis*, as an estuarine animal, shows considerable osmoregulatory ability (cf. Ralph, 1965) and glycine and taurine especially appear to play an important role in osmoregulatory functions (but cf. Armitage, 1979). *Euphausia superba* and *Meganyctiphanes norvegica* are oceanic species, living at high southern and northern latitudes respectively.

Amino acids of the protein hydrolysate of the estuarine mysid, *Mesopodopsis slabberi*, closely resembled those of *Neomysis* (Raymont, Ferguson and Raymont, 1973). The neritic chaetognath *Sagitta setosa* also showed overall similarity in the protein amino acids, but of the major amino acids, alanine replaced leucine in order of importance. Of the less abundant amino acids, glycine and histidine were present in somewhat larger proportions than in the mysids (Table 7.22).

Although the ctenophore *Pleurobrachia pileus* has a relatively very low protein content, the protein hydrolysate amino acid composition shows a considerable resemblance to that of the other two neritic zooplankton species. For example, glutamate is most plentiful, aspartate only slightly less abundant and leucine, lysine and arginine are present in significant quantities (cf. Table 7.22). In both mysids, glycine, taurine and arginine are the dominant free amino acids. *Sagitta setosa* shows a considerably different pattern. Glutamate is most abundant and though glycine is fairly rich, other acids are equally important; taurine was not identified (cf. Table 7.23). Corkett and McLaren (1978) note that the free amino acids of *Pseudocalanus* included the normal variety, but that glycine amounted to about one-third, and proline one-fifth of the total. Taurine, alanine and arginine were well represented.

The comparatively few data available on the amino acid composition of oceanic zooplankton prompted other analyses of acid hydrolysates of lipid-free residues,

Table 7.22. The amino acid composition of protein hydrolysates in four inshore species expressed as percentages (Raymont *et al.*, 1973)

Amino acids	*Neomysis integer* Ovigerous females	*Neomysis integer* Less mature	*Mesopodopsis slabberi*	*Sagitta setosa*	*Pleurobrachia pileus*
Aspartic acid	11.2	10.6	11.0	8.8	11.7
Threonine	4.1	3.9	4.4	3.5	5.3
Serine	3.8	3.5	4.6	4.5	5.6
Glutamic acid	17.2	19.1	14.8	12.0	12.2
Proline	3.0	2.4	3.2	3.6	5.0
Glycine	3.6	4.1	4.0	6.8	7.6
Alanine	5.2	5.4	5.8	8.4	5.8
Valine	5.2	5.4	5.6	5.9	5.6
Cystine	1.1	0.7	1.4	1.8	0.5
Methionine	2.3	1.7	2.4	2.7	0.5
Isoleucine	5.3	5.3	5.3	5.4	4.8
Leucine	7.7	8.4	7.8	6.0	7.4
Tyrosine	4.0	3.9	4.3	3.2	2.8
Phenylalanine	5.1	5.0	4.6	4.7	4.3
Lysine	10.2	10.2	10.4	9.4	8.6
Histidine	3.5	2.4	3.6	5.0	3.3
Arginine	7.3	7.8	6.7	8.4	7.1
Taurine	—	—	—	—	—
Ornithine	0.2	0.2	0.2	0.3	—

Table 7.23. A comparison of the major free amino acids in three planktonic species Each concentration is expressed as a percentage of the total (Raymont *et al.*, 1973)

Amino acids	*Neomysis integer* Ovigerous females	*Neomysis integer* Less mature	*Mesopodopsis slabberi*	*Sagitta setosa*
Aspartic	3.1	2.9	4.7	12.0
Glutamic	6.0	5.6	7.9	17.8
Proline	5.9	2.4	—	8.9
Glycine	27.4	29.6	22.5	10.0
Alanine	4.8	5.2	2.9	6.7
Lysine	1.3	2.2	9.0	11.0
Arginine	17.6	12.0	25.5	10.0
Taurine	26.4	26.6	14.7	—

mostly protein, of several oceanic species—the copepod *Labidocera acutifrons*, the euphausiid *Nematobrachion sexspinosum*, the decapods *Acanthephyra purpurea* and *A. pelagica*, the mysid *Gnathophausia* sp. and the tunicate *Pyrosoma* sp. (Raymont, Morris, Ferguson and Raymont, 1975). Results confirmed an earlier suggestion that the amino acid composition of the total body protein for a variety of neritic and oceanic zooplankton is closely similar. Thus, except for *Labidocera*, the same five amino acids first identified for *Neomysis*, *Meganyctyphanes* and *Euphausia* were again dominant, with glutamate always in greatest concentration in all species. Changes in the order of the succeeding amino acids were comparatively insignificant. In

Labidocera total protein, leucine was less plentiful than glycine and alanine. The compositions of the two species of *Acanthephyra* which inhabit different depths in the ocean were especially similar. The proportions of the amino acids in the body protein of many marine zooplankton species appear, therefore, to be little affected by difference in degree of maturity, sex, environmental depth and geographical area. On the other hand, the much more limited data indicate that the free amino acids may exhibit quite wide specific variation, especially in offshore zooplankton.

Following the report by Jeffries (1969) of some seasonal change in the free amino acids of field populations of mixed species of *Acartia*, Armitage (1979) examined gravid females, adult and juvenile *Neomysis integer* and found some differences in the free amino pool, especially in gravid females. A 12-month study of variations in the free amino acids was, therefore, confined to mixed (non-gravid) adults and showed some seasonal differences, for example, a greater total concentration in February than in August. Certain individual acids (e.g. glycine and lysine) were also higher in late winter/early spring than in late spring/summer. Apart from salinity changes, seasonal variation in environmental temperature and diet may affect the free amino acid pattern. Cowey and Corner (1963b), however found that the total amino acid composition of *Calanus helgolandicus* was remarkably constant over a year.

Armitage (1979) studied the effect of changes in salinity on the amino acid patterns of adult *Neomysis integer*. With salinity experimentally varying continuously from *ca.* 27 to *ca.* 2‰ in about 3 hours, simulating part of the natural tidal cycle in Southampton Water, no significant change in total free amino acids or in any of the amino acids was detected during the simulated ebb and flood phases of the tidal cycle. This was true for those free amino acids usually regarded as important in osmoregulation. The result is in contrast to earlier studies by Ralph (1965) and Austin (1970) but their work concerned steady state conditions after sudden transfer to a new substantially different salinity. Armitage found, however, that during the periods of comparatively constant salinity at high and low tide, especially marked with the "double stand" in Southampton Water, changes occurred in the level of free amino acids in *Neomysis*. For example, with mysids retained for increasing periods at approximately 27‰ salinity ("high tide") after the simulated flood phase, the concentration of glycine, alanine, proline, glutamate and valine increased and the level of total free amino acids rose. *Neomysis* however, subjected to the simulated ebb phase and maintained for increasing time at *ca.* 2‰, showed a fall in the same amino acids to regain approximately their original levels (Fig. 7.33). Farmer and Reeve (1978) found that the free amino acid pool in *Acartia tonsa* was reduced in proportion to an experimental salinity decrease; ammonia excretion and oxygen consumption also rose. A corresponding response did not occur with rise in salinity, but again, these changes were rapid, unlike those in Armitage's experiments. Taurinc, although in relatively high concentration and thought to have osmoregulatory functions in the benthic mollusc *Mya* (Garrett and Allen, 1972), did not show any change in concentration.

Although the protein amino acid composition of marine plankton appears to be comparatively constant, the continual breakdown and synthesis of protein as a dynamic component, with the selective absorption, assimilation, interconversion and degradation of particular amino acids might lead to changes in their relative proportions. Further, since some zooplankton species such as *Neomysis* have a very high proportion of protein and it appears to be a significant substrate for oxidative

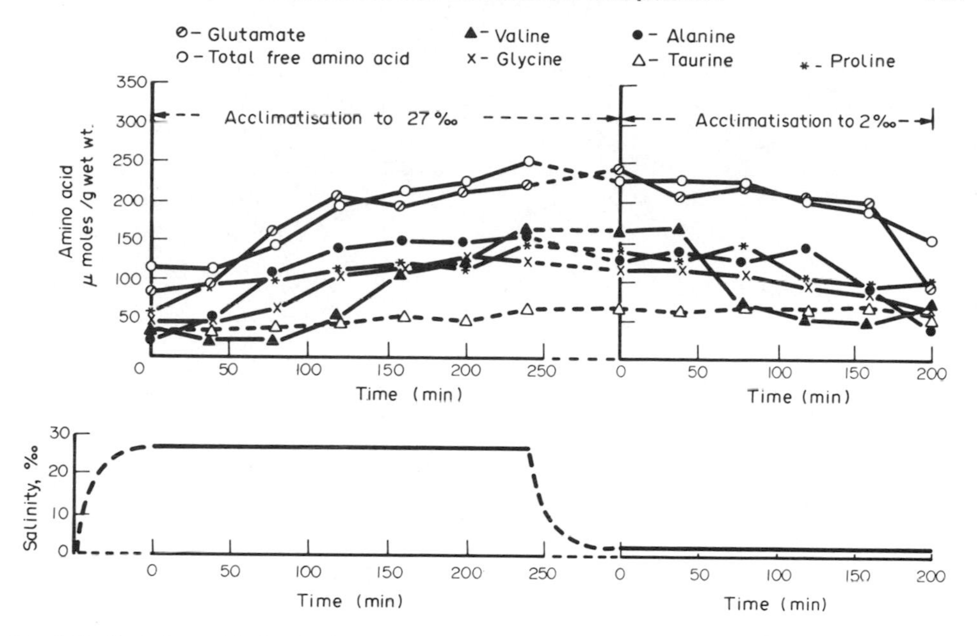

Fig. 7.33. Variations with time in the concentration of total free amino acid *Neomysis integer* exposed to salinities of 27 ‰ and 2‰ (○), glycine (x), taurine (△), glutamate (ϕ), valine (▲), alanine (●), and proline (*) in whole body extracts of (Armitage, 1979).

metabolism (Raymont, Austin and Linford, 1968), studies on feeding and starvation might indicate some aspects of its metabolism.

Feeding on pure carbohydrate (starch) diet and, in some experiments, with kaolin (equivalent to starvation) (*vide supra*) for varying periods up to 12 days showed the expected fall in total protein compared with control animals. The loss of protein varied with duration of the experiment and with different stocks of mysid but could even exceed 60 % of total protein. The reduction was more severe in starved animals (cf. Armitage, Raymont and Morris, 1977). The results suggest that feeding on pure carbohydrate allows *Neomysis* to meet some of its energy requirements, but that some degradation of protein occurs to meet essential demands for amino acids and other specific metabolic processes. Differences in the amino acid composition of the protein hydrolysates of starch-fed and control animals were small. This is not surprising since the comparison is between total amino acids comprising the considerable variety derived from the hydrolysis of the whole body proteins. Nevertheless, statistical analysis demonstrated that there were differences. Aspartic acid, serine and histidine suffered the largest reductions in the starch-fed mysids; the smallest changes occurred in alanine and valine. Figure 7.34 illustrates these differences as determined by Friedman Rank Sum analysis.

In contrast to the marked reduction in total protein in carbohydrate-fed mysids, the total concentration of free amino acids showed only an insignificant decrease. The major free amino acids (taurine, glycine, aspartate and arginine) showed no significant reduction. In contrast, others, especially tyrosine, but also leucine, isoleucine, valine and lysine were considerably reduced (cf. Figs 7.34, 7.35). Of the four amino acids

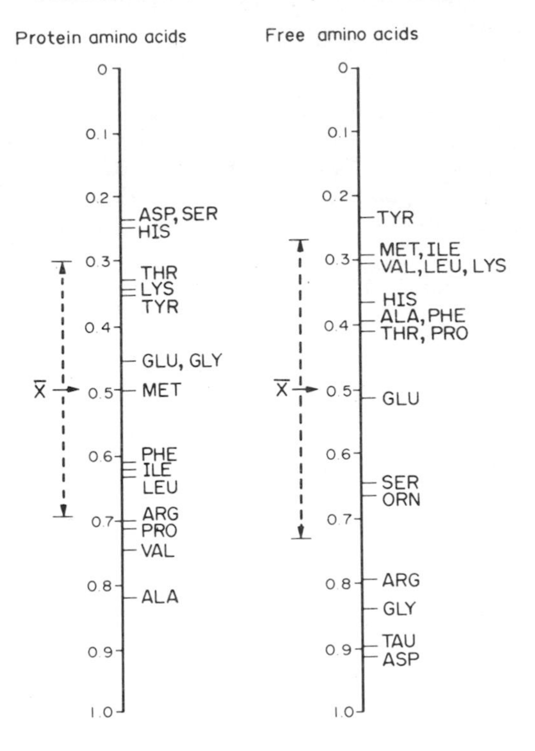

Fig. 7.34. Mean rank values of the change in protein and free amino acid composition of *Neomysis integer* fed pure carbohydrate (Armitage, 1979).

showing least change, glycine is believed to be concerned in osmoregulation in *Neomysis*; a high level of taurine may also be essential. Aspartate is important because of its central position in the metabolic paths of several keto and amino acids. Arginine, as arginine phosphate, is important in energy transfer. The results might imply that these four free amino acids are conserved, perhaps at the expense of body protein.

Experiments were performed (Armitage *et al.*, 1978) on *Neomysis*, in which animals were fed on protein in the form of lipid-free egg albumen over periods of about one week. After sacrifice, their amino acid composition was compared with control animals and with those fed on kaolin (starved). The starved animals, as expected, lost about 40 % of body protein but, somewhat surprisingly, the protein-fed mysids lost an almost equal amount, though their survival was better. Despite the high nitrogen input, the demand for energy caused the rate of degradation to exceed the rate of protein synthesis. There was also possibly a demand for some amino acid in the free pool present only in limited amount in the diet, or the lack of some cofactor in the purified diet may have reduced the rate of protein building. Statistical analysis showed that significant changes occurred in the protein hydrolysate amino acids from protein-fed and starved animals compared with controls (Fig. 7.36). For instance, glutamate and lysine were markedly reduced in both protein-fed and starved animals. However, there were also differences between the protein-fed and starved mysids. For example, histidine was markedly reduced in starved animals but not in albumen-fed mysids; proline was reduced in protein-fed but not in starved animals.

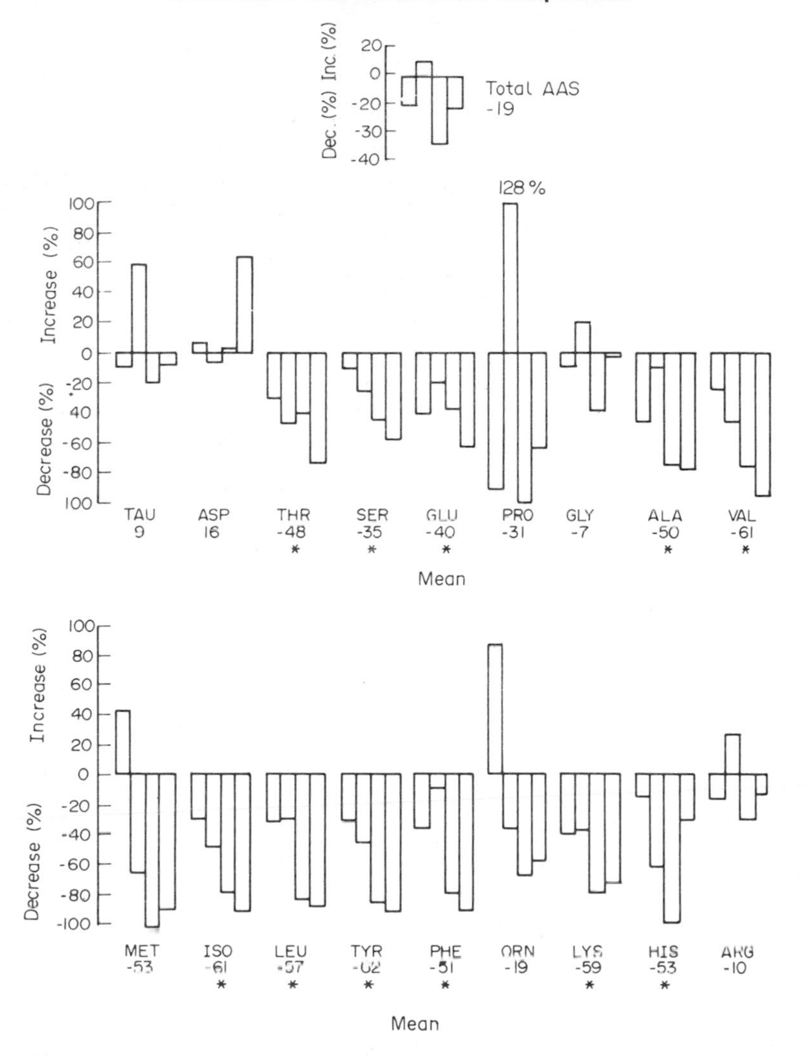

Fig. 7.35. Percentage change in free amino acids of *Neomysis integer* fed pure carbohydrate. The mean value is given below each histogram and an asterisk denotes a significant change (Armitage, 1979).

With regard to the free amino acids, starvation caused a reduction of total free acids of 28 % and although most were reduced, taurine, aspartate and arginine were conserved and glycine only very slightly reduced (cf. results on carbohydrate feeding). The albumen-fed animals showed total free amino acid contents which were virtually unchanged, but there was considerable variability in the changes in individual acids between different experiments, presumably reflecting the simultaneous transfer of some amino acids from degraded body tissues and some taken from the digested albumen. Arginine and aspartate, however, were markedly conserved; glycine and taurine were almost unchanged. Proline and alanine were most reduced in both fed and starved animals. The results of both starch and albumen feeding experiments suggest that the body protein in *Neomysis* may very broadly be divided into that which can be

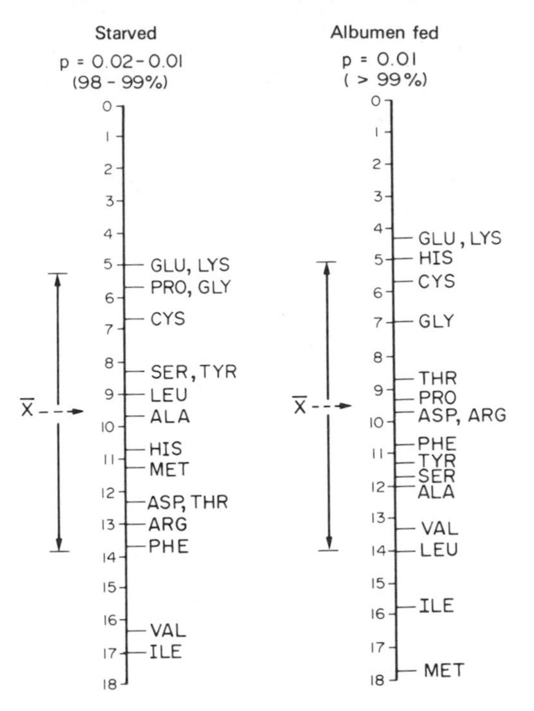

Fig. 7.36. Mean rank values of the change in protein amino acids of *Neomysis integer*, either starved or fed pure albumen (Armitage, 1979).

readily utilized for metabolic requirements in times of nutritional deficiency and that which is essentially structural protein (presumably enzyme systems, membranes, etc.) which must be conserved at all costs. Protein may also be degraded when the diet is insufficient to supply the essential pool of free amino acids. Cowey and Sargent (1972) have suggested that in fish the particular amino acid composition of the dietary protein largely influences whether it is used for oxidative metabolism or for synthetic purposes. In zooplankton, with marked protein turnover, the same might apply to body as well as to dietary protein.

Although the protein hydrolysates of starch-fed and albumen-fed *Neomysis* show marked differences due to the input of nitrogen in protein-fed animals, the levels of individual amino acids in the diet appear to have no direct influence on the composition of the body protein which is synthesized (cf. Cowey and Corner (1963b) for *Calanus*). Similarly, the composition of the free amino acid pool does not appear to reflect dietary composition. Protein from the body (and if obtainable from the diet) is mobilized, but levels of particular acids, especially aspartate, glycine, taurine and arginine, are maintained, presumably because of their special significance.

These investigations indicate that there are "essential" and "non-essential" amino acids in marine zooplankton, important not only to an appreciation of metabolic mechanisms and pathways *per se*, but also of significance in understanding ecological food relationships in the sea. The lack of a technique for feeding zooplankton using a formulated diet of known amino acid composition followed by the successive removal of individual acids limits the study. Feeding with carbohydrate and protein diets can give only indications of the importance of particular amino acids. A useful complementary approach is to feed or inject animals with radioactively labelled materials. Apart from possible shock effects and overloading by injection, this latter technique is almost impossible to use with zooplankton, owing to their small size. Only a very few relatively large species with comparatively firm exoskeletons can be successfully injected.

Feeding with labelled compounds, ^{14}C-glucose, ^{14}C-aspartate and ^{14}C-glutamate, has, however, been employed with *Neomysis* (Armitage, 1979). In each case, the isotope was incorporated into two kinds of diet base, pure starch and a mixture of egg–albumen–lipid–starch, which might prove to be a more adequate diet. Experiments lasted for a maximum of 13 days and after sacrificing of the animals radioassay of the free and protein amino acids was carried out. Despite difficulties of interpretation of data, for example, due to the generally low level of ^{14}C incorporation into the amino acids, it was possible to identify those which had been significantly labelled. Table 7.24 summarizes the data for free and protein amino acids from experiments with the three labelled substrates. Armitage suggests from the frequency of significant labelling that taurine, glycine, aspartate, threonine, glutamate, proline, arginine, tyrosine and lysine are probably non-essential. The labelling of others was less extensive but some synthesis appears to have occurred. At least two acids, histidine and methionine (possibly cystine) appear to be unlabelled and presumably are essential. A marked difference between the free and the protein amino acids is that even when aspartate or glutamate was used as the labelled substrate, these amino acids were never significantly labelled in the free pool but always in the protein.

A very few similar experiments have been carried out using the same labelled substrates fed to the oceanic deep-living mysid *Gnathophausia*. The radioactivity of the chitin fraction after hydrolysis and of the lipid were determined in addition to the free and protein amino acids. Table 7.25 indicates that although a range of amino acids was significantly and frequently labelled, glycine, arginine and tyrosine, in addition to methionine and histidine, were not significantly labelled either in the free pool or protein fraction. Although they can occur in fair quantities in the free amino acids they were apparently not synthesized from the substrates offered. The difference between the estuarine *Neomysis*, in which glycine production appears to be very prominent, and the oceanic *Gnathophausia*, which did not synthesize glycine from the three substrates, is striking. It is curious that tyrosine was not labelled in *Gnathophausia* in view of the ready incorporation of ^{14}C into phenylalanine. There was substantial transfer of ^{14}C from labelled glucose to chitin in *Gnathophausia*, very little transfer from glutamate and none from labelled aspartate. For chitin synthesis in *Gnathophausia* it is suggested that glucose rather than amino acids is a significant source. Some ^{14}C from labelled glucose and, in one experiment from labelled glutamate, was transferred to lipid. In *Neomysis* there appeared to be a more substantial synthesis of ^{14}C-labelled lipid from all the substrates, especially from glucose.

The limited information obtained on the essential amino acids of crustaceans applies mainly to larger benthic species. Of the ten acids regarded as essential by Hartenstein (1970), six appeared to be labelled in the experiments with *Neomysis*. Only histidine and methionine of those listed by Hartenstein were apparently not synthesized. It would be premature to attempt to define "essential" and "non-essential" amino acids from these experiments on *Neomysis* and *Gnathophausia*. The generally low levels of radioactivity in the amino acids with feeding as opposed to injection and other

Table 7.24. Summary of extent to which amino acids were labelled following feeding of *Neomysis* with ^{14}C-glucose, ^{14}C-aspartate and ^{14}C-glutamate (Armitage, 1979)

Free amino acids Mean concentration (μmoles/g wet wt)		Instances of significant labelling (a)		Protein amino acids Mean concentration (μmoles/g wet wt)		Instances of significant labelling		Total number of instances of significant labelling out of maximum of 12	
1. GLY	54.22	TAU	+ + + + + −	1. LYS	162.99	GLU	+ + + + + +	GLY	7
2. TAU	35.19	GLY	+ + + ± ± ±	2. GLU	160.15	ASP	+ + + + + +	ASP	6
3. ARG	3.14	PRO	+ + ± ± − −	3. ASP	128.79	THR	+ + + + + +	THR	6
4. ALA	2.91	ARG	+ + ± ± − −	4. ALA	105.43	ARG	+ + + + ± ±	GLU	6
5. PRO	2.83	ORN	+ + ± − − −	5. GLY	93.83	TYR	+ + + + ± ±	PRO	6
6. ORN	1.54	ALA	+ ± ± ± − −	6. PRO	75.77	GLY	+ + + + ± −	ARG	6
7. SER	1.01	LYS	+ ± ± ± − −	7. LEU	73.63	PRO	+ + + + ± −	TYR	5
8. GLU	0.49	SER	+ ± ± − − −	8. ARG	54.65	LYS	+ + + + − −	LYS	5
9. LEU	0.40	PHE	+ − − − − −	9. VAL	53.28	ILE	+ + + ± ± −	PHE	4
10. LYS	0.39	TYR	+ − − − − −	10. SER	51.82	VAL	+ + + ± − −	ILE	3
11. THR	0.33	(b)CYS^c	+ − − − − −	11. THR	47.25	PHE	+ + + ± − −	LEU	3
12. ASP	0.28	LEU	± ± − − − −	12. ILE	46.64	LEU	+ + + − − −	VAL	3
13. PHE	0.27	THR	± ± − − − −	13. PHE	41.25	SER	+ ± ± ± ± −	ALA	2
14. CYS^c	0.26	GLU	± − − − − −	14. TYR	36.06	ALA	+ ± ± ± − −	SER	2
15. ILE	0.25	ASP	± − − − − −	15. HIS	26.06	CYS^c	± ± ± − − −	CYS^c	1
16. MET	0.25	HIS	− − − − − −	16. MET	13.85	MET	± − − − − −	MET	0
17. VAL	0.25	ILE	− − − − − −	17. (c)CYS^N	3.61	HIS	± − − − − −	HIS	0
18. TYR	0.19	MET	− − − − − −	18. CYS^C	2.73	CYS^N	− − − − − −		
19. HIS	0.16	VAL	− − − − − −						

(a) + = >95%; ± = 68–95%; − = <68% confidence.
(b) CYS^C refers to cysteic acid.
(c) CYS^N refers to cystine.

Table 7.25. Summary of extent to which amino acids were labelled following feeding of *Gnathophausia* sp. with ^{14}C-glucose, ^{14}C-aspartate and ^{14}C-glutamate (Armitage, 1979)

Free amino acids Mean concentration (μmoles/g wet wt)		Instances of significant labelling (a)		Protein amino acids Mean concentration (μmoles/g wet wt)		Instances of significant labelling		Total number of instances of significant labelling out of maximum of 12	
1. PRO	26.08	TAU	+ + + + + –	1. GLY	61.95	GLU	+ + + + ± –	GLU	6
2. GLY	8.78	PHE	+ + + – – –	2. GLU	60.60	VAL	+ + + ± ± –	ALA	5
3. TAU	8.10	PRO	+ + ± ± ± –	3. LYS	54.75	ASP	+ + + ± – –	VAL	4
4. ALA	6.67	ALA	+ + ± – – –	4. ASP	52.92	ALA	+ + + ± – –	PRO	4
5. LEU	6.05	GLU	+ + ± – – –	5. ALA	49.45	PRO	+ + – – – –	ASP	3
6. VAL	3.83	(b)CYS^C	+ + ± – – –	6. PRO	48.12	SER	+ ± ± – – –	PHE	3
7. ILE	3.57	VAL	– ± ± – – –	7. SER	31.98	THR	+ ± ± – – –	SER	2
8. LYS	2.92	ILE	+ ± – – – –	8. VAL	30.42	GLY	± ± ± – – –	CYS^C	2
9. ARG	2.17	LYS	+ ± – – – –	9. LEU	30.35	ARG	± ± ± – – –	THR	1
10. ORN	2.05	LEU	+ – – – – –	10. TYR	29.67	(c)CYS^N	± ± ± – – –	ILE	1
11. MET	1.68	SER	+ – – – – –	11. ILE	23.05	HIS	± ± – – – –	LEU	1
12. PHE	1.65	THR	± ± ± ± – –	12. ARG	18.90	CYS^C	± ± – – – –	LYS	1
13. THR	1.27	GLY	± ± – – – –	13. PHE	18.17	MET	± – – – – –	GLY	0
14. SER	1.18	ARG	± ± – – – –	14. THR	16.38	LYS	– – – – – –	ARG	0
15. TYR	1.15	MET	± ± – – – –	15. HIS	15.22	LEU	– – – – – –	TYR	0
16. GLU	0.87	TYR	± ± – – – –	16. MET	4.95	TYR	– – – – – –	HIS	0
17. HIS	0.73	ASP	± ± – – – –	17. CYS^N	1.36	ILE	– – – – – –	MET	0
18. ASP	0.38	ORN	± – – – – –	18. CYS^C	1.02	PHE	– – – – – –		
19. CYS^C	0.09	HIS	± – – – – –						

(a) + = >95%; ± = 68–95%; – = <68% confidence.
(b) CYS^C refers to cysteic acid.
(c) CYS^N refers to cystine.

complexities in the experiments pose problems in interpretation. It must also be appreciated that micro-organisms in the gut of the mysids may be responsible for some of the synthesis, though this does not greatly alter relevance to considerations of the ecological food chains in the sea.

Nevertheless, investigations of this type may eventually lead to a better appreciation of nitrogen metabolism in zooplankton. With the pronounced protein metabolism true of many species, interconversions would be expected as part of protein synthesis and breakdown. Raymont, Austin and Linford (1968) reported the preliminary identification of transaminase activity in *Neomysis*. Activity of α-ketoglutarate transaminase activity was strong, especially with aspartate, leucine and alanine. Glyocylate transaminase activity appeared to be very weak. Armitage (1979) has considered possible pathways for the biosynthesis of amino acids by *Neomysis*. She also obtained evidence for a glutamate–oxaloacetate transaminase more active than the glutamate–pyruvate system. In *Gnathophausia* the frequent labelling of alanine and glutamate indicates that the glutamate–pyruvate reaction may be more active.

The successful culture of several species of zooplankton in the laboratory is a major contribution to further investigation on the physiology and biochemistry of zooplankton. Very few analyses on zooplankton species reared on different diets have so far been attempted. Corner and Cowey (1968) refer to preliminary experiments indicating that juvenile clams (*Mercenaria mercenaria*) fed on "poor" (*Chlamydomonas coccoides*) and "good" (*Tetraselmis suecica*) phytoplankton showed little difference in body protein composition, though the amino acid composition of the two foods was generally rather similar. The rearing of other zooplankton on "poor" and "good" foods has unfortunately not been followed by amino acid analysis of the zooplankton. In any event, the variety of lipid material and range of amino acids in feeding zooplankton on unialgal food cultures or on specific animal diets is so great that it is usually impossible to elucidate the effects of particular lipid or amino acid components. There is need now for the laboratory culture of zooplankton on defined diets. Guerin and Gaudy (1977) and Gaudy and Guerin (1977) have indicated such an approach to culturing using defined artificial diets, some used in commercial fish culture, others adapted from medical formulae, for growing the harpacticoid *Tisbe holothuriae*. One diet among several listed, selected as a high nitrogen feed, includes casein, cod liver oil and yeast.

Preliminary studies on *Neomysis integer* (Raymont and Raymont, unpublished) maintained in the laboratory at 15°C and fed an egg albumen–glucose powder, desiccated ox liver powder and other artificial foods showed that juvenile mysids feed actively and exhibit excellent growth, especially on a high nitrogen diet, e.g. achieving growth of 4–5 mm in 10 days.

Physiological and biochemical investigations carried out on such laboratory-reared "standard" animals would not suffer from the problems attending the use of different experimental stocks, although due consideration must be paid to differences already suggested between "wild" and "laboratory" zooplankton. In particular, if zooplankton species could be reared on a defined diet of known amino acid composition, investigations involving the successive removal of amino acids as practised in work on mammals would go far in contributing to our understanding of protein metabolism in zooplankton, supplementing the comparatively little information now available from the use of other techniques.

Chapter 8
Water Masses and Zooplankton Population—Indicator Species

The qualitative and quantitative characteristics of zooplankton populations are greatly influenced by the breeding patterns of holoplankton and in some areas by meroplankton breeding cycles. While seasonal vertical migration may markedly affect zooplankton populations, other factors such as upwelling and various hydrological changes, with the effects of mixing and stabilisation, may produce pronounced changes in the phytoplankton (cf. Volume 1) and in turn on the characteristics of the zooplankton populations. Since the deep significance of vertical water movements on plankton production is well recognised, the horizontal transfer and exchange of superficial water layers and the movements of water masses may profoundly modify both the abundance and the quality of the zooplankton. Such drifting of water masses parallel to, towards or away from a coast, especially where there are conspicuous seasonal varying strengths of warmer and colder currents with their endemic plankton populations, are particularly obvious. They can be responsible for fluctuations in the density of the zooplankton and for marked changes of fauna, with the appearance of assemblages of species characteristic of a body of water foreign to the area. Frequently, especially near coasts, such horizontal flows accompany changes in vertical circulation patterns. The drifting of superficial water masses with their endemic zooplankton is as typical for open oceans as for coastal areas, though it is often easier to study near shore.

Knowledge of the faunal composition of water masses and of change in the plankton due to drifting of water layers is thus of the utmost significance and often cannot be separated from changes in populations with breeding. Pavshtiks (1975) emphasized that seasonal changes in the boundaries of species cannot be understood without accurate knowledge of the water masses and currents. He defines a water mass in relation to this statement as a comparatively large volume of water where the distribution of physical, chemical and biological characteristics remain nearly constant for a considerable period. As a comparatively simple example of changing boundaries of zooplankton species, where a pronounced difference in climate exists between winter and summer, the varying seasonal intensity of the warmer and colder currents not only results in varying abundance of the zooplankton, but may lead to the advance and retreat of the distributional limits of species. It has become increasingly clear that many plankton animals have their main centres of distribution within a specific water mass. A species inhabiting a water mass will tend to have an overall distribution beyond the area where it reproduces and maintains itself successfully and it will be carried to a greater or lesser degree outside its own area by drift. Closely similar species inhabiting water masses which adjoin and are not too dissimilar will therefore compete and a seasonal succession is likely, but may be associated either with slow changes in the characteristics of the same water mass owing to climatic change, or with seasonal variations in current patterns or frequently with a combination of the two.

The horizontal (geographical) distribution of many species of zooplankton belonging to different taxa has been described in Chapter 2. Since many species are centred in a particular water mass, an assemblage of species often appears to be typical of a particular water mass and is then often termed a plankton community. It is not necessarily the presence of certain plankton species, but more frequently the proportion of a number of species, which characterizes a specific water mass or current. It would appear that the physical and chemical features of the environment, such as temperature and salinity are important controlling factors in a community, and examination of T–S–P diagrams (temperature–salinity–plankton) can be very useful in correlating plankton populations with water masses. This approach was used particularly by Bary (e.g. Bary, 1964), who found that a proper relationship could be established between plankton species and surface water masses, as delimited by T–S diagrams, in spite of seasonal temperature fluctuations.

Although temperature and salinity are useful factors in defining a water mass, more subtle influences such as inorganic micronutrients, organic growth factors and inter specific growth-promoting and growth-inhibiting substances which are typical of a water mass may be the really significant factors affecting the proportion and distribution of the assemblage of species. Such characteristics, sometimes referred to as the "quality" of the water mass, cannot be defined. As Bary (1963b, 1964) suggests, the T–S diagram is rightly used to describe the water mass, but temperature (and salinity) does not of itself have a clear effect on the distribution of a species. When mixing of water masses takes place, the populations of species must adjust to the unspecified properties of the mixture of water; the degree of success in active reproduction under such conditions determines whether the population flourishes or declines. Bary (1963b) divided some 17 species of zooplankton from the north-east Atlantic according to similarities of distribution. The six groupings could be related to near-surface water masses. For instance, warm (southern), transitional and colder (northern) oceanic groups could be distinguished, certain species showing some ability to spread across boundaries. All three oceanic groups tended to keep outside the continental shelf, i.e. there was little capacity for extending into mixed waters. Other species were more specifically coastal.

Since species remain mainly within their particular water mass despite seasonal changes in temperature and salinity, Bary proposes that some unique property is the critical factor for the species reaction (tolerance). Tolerance will become critical at some point along a concentration gradient of the factor when two water masses mix. For many species it will be the actual boundary of the two water bodies. Some species (e.g. *Metridia lucens*, *Temora longicornis*) appear to have considerable capacity to cross boundaries, i.e. they have a fairly high tolerance. In further discussion, Bary (1964) divided the zooplankton of the north-east Atlantic primarily into southern, northern and coastal waters and noted the comparative constancy of the assemblages of species despite seasonal temperature fluctuations. While at the border of two water masses a species experiencing a "foreign" water quality may be intolerant of such a characteristic, the mixing of two water masses may affect the species differently than either factor when present alone. The effect on a zooplankton population must be within the existing T–S limits; temperature and salinity may indeed modify the reaction of a population.

In the development of a zooplankton population within a water mass, we know very little of the biological inter-relationships between phytoplankton and zooplankton and between species of zooplankton. These, however, may be all-important, especially in combination with more obvious factors such as temperature, in the degree of success of the reproduction of one species in competition with another. Unfortunately, we have little information apart from some data on feeding relationships, and very sparse knowledge of the effects of some toxic or inhibitory species in limiting distribution. A few examples of the occurrence of certain species of phytoplankton which appear to promote the presence of others were discussed in Volume 1. In speaking of communities in the marine zooplankton, the term cannot normally indicate the degree of biological inter-relationships which have been elucidated for some land communities. In a less strict sense, however, the term "zooplankton community" will be employed.

It is inappropriate to attempt to describe the considerable variety of zooplankton communities ranging over cold high latitudes, temperate, sub-tropical and tropical waters and neritic and oceanic areas. In any event, considerable detail has already been given for some communities in describing the distribution of various taxa (Chapter 2) and in accounts of the breeding patterns of plankton (Chapter 3). However, it is relevant to give a few examples of communities in order to illustrate the effect of changes in the boundaries of species in relation to hydrological, including seasonal, changes, to examine such changes on species abundance and to describe some examples of plankton indicators.

Before considering communities it is necessary to discuss the existence of cosmopolitan species. A few species of zooplankton appear to have an enormously wide distribution, sometimes stretching almost throughout the world oceans (cf. Chapter 1), some even apparently capable of flourishing in neritic and oceanic waters and in the coldest and warmest waters. Some forms, which Ekman (1953) called "cold water cosmopolites", usually occur in high densities in cold superficial waters at higher latitudes both in the northern and southern hemisphere, but are found in smaller numbers in sub-tropical and tropical areas, usually in deeper layers, so that they avoid warmer water. Cosmopolitan species and cold water cosmopolites appear to be the least useful in attempting to delimit communities. However, populations of such very widespread species frequently show slight variations in body characteristics in different parts of their range and are often regarded as separate races.

Brodsky (1965) emphasized that precise taxonomic studies can also reveal the existence of closely allied species (polytypic species), each having its own separate distribution. The division by Brodsky (1975) of the species *Calanus finmarchicus* (*sensu lato*) is an example (cf. Chapter 2). *Pseudocalanus minutus* may be divisible into three closely allied species. With several zooplankton species, though different races or sub-species are apparent, it is not clear whether these are phenotypic or genotypic. Examples are *Calanus plumchrus* (Kos, 1975), *Oithona similis* and other *Oithona* species (Shuvalov, 1975), *Liriope tetraphylla*, *Eukrohnia hamata*, *Sagitta planctonis* and *Beroe cucumis*. Other widespread species exhibiting racial differences are included in Chapter 2.

McGowan (1963) discusses possible mechanisms for the evolution of genetically distinct populations of a zooplankton species, where isolation of separate populations appears to be unlikely or incomplete. He examines two types of the pteropod *Limacina helicina*—the variant having a shell with a much flattened spine and lacking the usual vertical striations found in North Pacific sub-Arctic waters, including the Alaskan gyre and the Californian Current. The two populations, termed "A" and "B", show intergrades where the two overlap but the distribution patterns differ for the two types, with "A" characteristic of *pure* sub-Arctic waters and "B" typical of *mixed* sub-Arctic water. The distributions, however, agree with the circulation pattern, provided it is assumed that variant "B" is derived from "A". McGowan believes that the types are genetically rather than phenotypically distinct, though proof is still awaited. The morphological differences, however, cannot be readily correlated with temperature variations or other environmental factors. Variant "B" may be a special genotype of "A" regularly carried into the different (mixed water) environment to which it is better adapted.

There is every likelihood that inter-specific variation occurs in oceanic zooplankton from other taxa, despite the exchanges of water in the open ocean. Closely allied species or sub-species may, therefore, then be restricted to their particular area by adaptation to the precise ecological conditions obtaining in a specific water mass. Currents may be significant in maintaining a population in the particular area, vertical migrations of the animals in combination with currents at deeper levels being part of the mechanism. The biogeographical boundaries of closely allied species and even of sub-species of so-called cosmopolitan forms may thus agree with that of a particular water mass, although in attempting to follow changes in the boundaries of species it is more difficult to elucidate the distribution of closely allied forms. Such investigations usually require statistical analysis of morphometric characters.

Arctic Zooplankton

Certain species of zooplankton appear to be typical of Arctic waters, especially in the Atlantic sector. While the precise limits of some species may still be a matter for debate, the following appear to be fairly typical: *Calanus glacialis*, *C. hyperboreus*, *Pseudocalanus minutus*, *Microcalanus pygmaeus*, *Metridia longa*, *Pareuchaeta glacilis*, *Oithona similis*, *Oikopleura vanhoeffeni*, *Themisto libellula*, *Themisto abyssorum*, *Pseudalibrotus* spp., *Ptychogastria polaris*, *Mertensia ovum*, *Crossota norvegica* and *Limacina helicina*.

Grice (1962) mentions other moderately common copepods in Arctic waters (*Scaphocalanus magnus*, *Chiridius obtusifrons*, *Ectinosoma finmarchicum*, *Spinocalanus magna* and *Oncaea borealis*). Many of these are common to the Pacific sector of the Arctic, but *Calanus finmarchicus*, *C. hyperboreus*, *Chiridius obtusifrons*, *Heterorhabdus norvegicus* and *Pareuchaeta glacialis* do not occur in the Pacific. Johnson (1963) adds to these species of copepods *Gaidius tenuispinus* and *G. brevispinus*, *Microcalanus pygmaeus* and other microcalanoids as juveniles, *Scaphocalanus brevicornis*, *Spinocalanus abyssalis* and *Temorites brevis* and some members of the ostracod genus *Conchoecia* as other components of the Arctic zooplankton. A few other species may avoid the highest latitudes (e.g. *Thysanoessa inermis*, *T. raschii*), but occur in Arctic and sub-Arctic areas with a number of the Arctic copepods such as *Calanus hyperboreus* and *Metridia longa*. *Aglantha digitale*, *Meganyctiphanes norvegica* and the chaetognaths *Eukrohnia hamata*, *Sagitta arctica* and *Sagitta maxima* are also found in Arctic/sub-Arctic waters.

The range of Arctic and sub-Arctic species is also considerably modified in relation to the vertical distribution of the plankton. The very cold low density superficial layer in the Arctic Ocean has the greatest density of zooplankton, but species diversity is greater in the intermediate warmer Atlantic water layer. Cold, deep Arctic water comprises a third stratum (cf. Grainger, 1965). Although Dunbar and Harding (1968) accept the existence of the cold, superficial Arctic water layer with its limited fauna, they doubt the identity of the intermediate Atlantic water as a distinct water mass with a specific fauna. Grainger (1965) describes eight typically Arctic copepods as forming >99 % of the zooplankton in the upper 50 m in the Arctic Ocean and they appear to be able to breed in the surface waters. While the same species occur in the deeper warmer layer, they are present in smaller numbers but are accompanied by other less typical Arctic species (e.g. *Spinocalanus magnus*, *Gaidius tenuispinus*, *Gaidius brevispinus*, *Heterorhabdus*, *Aetidiopsis rostrata*).

Grainger also found that coastal waters in the Canadian Arctic had the same species of surface-living Arctic copepods, together with species such as *Aglantha digitale*, *Aeginopsis laurenti*, *Oikopleura vanhoeffeni*, *Fritillaria borealis* and *Sagitta elegans*. In some areas meroplanktonic medusae and species such as *Eurytemora herdmani*, *Acartia longiremis*, *Limnocalanus grimaldii* and *Hyperoche* were also found. In western parts of the region the sub-Arctic copepods *Eucalanus bungii bungii* and *Calanus cristatus* can spread from the Pacific Ocean, but they apparently do not breed at such very high latitudes, an observation confirmed by Dunbar and Harding (1968).

Calanus finmarchicus (*sensu stricto*) is the most typical representative of sub-Arctic (boreal) waters, together with *Euchaeta norvegica* and many species, especially *Pseudocalanus* and *Metridia*, already listed for high latitudes. *Calanus finmarchicus* does not appear to extend to the very highest latitudes, at least in the superficial waters, where it is replaced by *Calanus glacialis* and *Calanus hyperboreus*. Grainger (1963) distinguishes between areas such as the Labrador Sea, Davis Strait and Hudson Strait which have boreal Atlantic waters dominated by *Calanus finmarchicus* and areas with Arctic water (Arctic Ocean, most of Hudson Bay, etc.), which are dominated by *C. glacialis*. Where a mixture of water occurs both species are present. *C. glacialis* is predominant over much of Baffin Bay west of Davis Strait and part of Hudson Strait, i.e. areas with a greater proportion of Arctic Water. *C. hyperboreus* follows *C. glacialis* over most of its distribution.

A distinction must be made between the normal range of the species in which it is actively reproducing and the extreme extension of a few specimens, especially where a branch current may lead to inflow of expatriate species. Tidmarsh (1973), studying an area between north-west Greenland and Ellesmere Island (79°N), found that *C. finmarchicus* occurred regularly in the epiplankton, although this was dominated by Arctic species (*Metridia longa*, *C. glacialis*, *C. hyperboreus*). *Euchaeta norvegica* and *Scolecithricella ovata* were also present in the intermediate Atlantic water, apparently because of the penetration of some mixed Atlantic water as part of the West Greenland current. However, little reproduction appeared to occur in these species, so that they may properly be regarded as immigrant forms.

Dunbar (1942) describes a somewhat similar pattern in the Canadian Arctic. He points to several gammarids as being typically Arctic, of which a few (e.g.

Pseudalibrotus glacialis, *P. nanseni*, *Apherusa glacialis*) are said to be truly pelagic and can serve as good Arctic plankton indicators. The hyperiid *Themisto libellula* occurs in greatest abundance in high Arctic areas. In Disko Bay there is some upwelling of slightly warmer, more saline Atlantic water and in this region *Themisto libellula* is comparatively scarce. It is replaced by *Thysanoessa* spp. and some *Meganyctiphanes*, both normally regarded as sub-Arctic species (Chapter 2). In deeper hauls *Boreomysis nobilis*, a true Arctic species, was found. Thus some mixture of Arctic water was presumably present. Dunbar points to the difference between the plankton of Disko Bay with its preponderance of euphausiids and the high Arctic plankton typical of waters off Baffin Island.

Boreal Zooplankton

The domination of *Calanus finmarchicus* in sub-Arctic waters led Bigelow (1926) to describe the widespread plankton community typical of the boreal parts of the North Atlantic as the *Calanus* community. As so often found at higher latitudes, the community has comparatively few dominant species. Apart from the overwhelming importance of *Calanus finmarchicus*, the most typical species are *Metridia lucens*, *Euchaeta norvegica*, *Pseudocalanus*, *Microcalanus*, *Themisto compressa*, *Thysanoessa* spp., *Clione limacina*, *Limacina retroversa*, *Eukrohnia hamata*, *Oithona similis* and under slightly less oceanic conditions, *Meganyctiphanes* and *Sagitta elegans*.

A small proportion of Arctic species particularly the copepods, *Calanus hyperboreus* and *Metridia longa*, may be included in boreal waters, depending on latitude and on current patterns. Bigelow (1926) found the community typical of more open waters of the Gulf of Maine and of the waters of the North Atlantic, but as more coastal areas of the Gulf are approached the numbers of neritic species increase. This community is limited on the American side of the Atlantic to the south by the warmer, higher salinity water of the Gulf Stream (cf. Bigelow and Sears, 1939; Clarke, 1940), but to the north and east it spreads across the Atlantic over oceanic regions round Iceland and the south of Greenland to areas off the British Isles and the Norwegian Sea.

Jespersen (1940, 1944) demonstrated the presence of the *Calanus* community off the Faroes and off Iceland, and Wiborg (1954, 1955) found it in Norwegian waters. Rae and Fraser (1941) and Rae and Rees (1947) have shown a similar zooplankton to exist in the North Sea, though as the coasts and banks are approached, just as in American waters the number of neritic species becomes greater. This type of zooplankton community is also recorded widely over the Barents and Kara Seas (cf. Zenkevitch, 1963). *Calanus finmarchicus* was usually dominant in the biomass of plankton of the Kara Sea, though *Pseudocalanus* was more abundant in some areas. Medusae and other neritic zooplankton species became more plentiful in coastal areas. Grainger (1959), working in a coastal section of the Canadian Arctic, also found medusae, ctenophore fragments and sagittae abundantly represented in the shallow-water plankton, as well as the more oceanic species such as *Calanus finmarchicus*, *C. hyperboreus*, *Pseudocalanus*, *Oithona* and *Aglantha*. In general, over the North Atlantic, while the density of *Calanus finmarchicus* is very large at higher latitudes the proportion of the extreme cold-loving species such as *Calanus hyperboreus* and *Metridia longa* and *Aglantha* tends to increase (cf. Jespersen, 1939, 1940).

Seasonal breeding

The examination of Arctic and sub-Arctic (boreal) communities in the North Atlantic has given some evidence of seasonal change in the distribution of species in relation to water masses. The advance and retreat of warmer and colder species with season in the Barents Sea was described by Zenkevitch (1963). He also instances the Kara Sea as an area where biological indicators can be employed particularly usefully. Water masses that contribute to the Kara Sea include those from the Barents Sea in the west, from the central Arctic basin, which may contribute cold deep Arctic water or warmer intermediate water, and cold brackish inflow from the land mass. Similar changes in the very cold boreal waters off the Norwegian coast and in the Norwegian Sea have also been described by Wiborg, Hansen and Ostvedt. Much seasonal change is due to the reproduction of endemic species and to vertical migration, but Ostvedt (1955), for example, mentions certain changes such as the appearance of slightly less cold species (*Metridia lucens*, *Limacina retroversa*) during late summer, which may be associated with incursions of water of Atlantic origin.

The analysis by Pavshtiks (1975) of seasonal changes in the zooplankton of Davis Strait illustrates the marked effects of current patterns and water masses in a fairly typical *Calanus* community. For a large part of the year Davis Strait is icebound; spring for plankton development follows the ice melt. Contributions to the waters of Davis Strait include a portion of the Irminger Current with its Atlantic water mixing with the cold East Greenland current, the very cold Canada and Labrador Currents flowing southwards along Baffin Land and run-off from the land masses. A very complex mixing of waters thus occurs, especially in the southern part of the Strait. *Calanus finmarchicus* and *Oithona similis* are usually the dominant zooplankton species, but whereas *Calanus* breeds mainly from May/June and is dominant until July, *Oithona* becomes more important from July onwards. By September *Oithona* is the more important copepod. Both species live throughout the winter in the area, but in low density.

The very great changes in biomass (approaching 100-fold) and even greater amplitude of seasonal change in numerical density of the zooplankton are profoundly affected by the current patterns and ice melt. In the southern less cold part of the Strait the biological winter lasts from October until about March/April. *Calanus*, partly brought in by the Irminger Current and partly overwintering in the area, can start to spawn about May, and only slightly later (June/July) in deeper parts of the Labrador Sea. Farther north, however, the ice delays development so that the biological winter lasts until about July; numbers of *Calanus*, though the dominant species, are low in the northern area and in the region of the very cold Labrador Current at this time. The so-called biological spring (i.e., the period over which species are actively reproducing) in the northern area and the cold Canada Current may, therefore, be as late as August or September. *Calanus glacialis*, a truly Arctic species, also occurs in this water mass and commences spawning. Grainger (1963) believes that a possible second late summer breeding may occur in the Labrador Sea but that breeding of *C. finmarchicus* does not normally occur north of Davis Strait. *C. glacialis* breeds once throughout its Arctic range: breeding may be in late summer in some areas. In these very cold areas Pavshtiks found that *Metridia longa*, *Calanus hyperboreus*, *Limacina helicina* and other Arctic forms contribute substantially to the biomass of the zooplankton, even though *Calanus finmarchicus* and *Oithona* are still numerically important. The appearance of these Arctic species with the ctenophore *Mertensia* reflects the boundary of the Canada Current against the warmer West Greenland Current where the typical *Calanus finmarchicus* community is dominant.

Other species (e.g. *Aglantha*) may appear in the more southern part of the Davis Strait, especially later in the season, about August or September. Where the Atlantic component is more marked, *Limacina retroversa*, *Tomopteris* and *Physophora* can be found and in the most southern regions of Davis Strait there is an abundance of *Salpa fusiformis*. In the extreme southern region the equivalent of a biological summer occurs, though it is very late (about August/September). It is characterized not only by the importance of *Oithona* but by the relative abundance of more boreal species already listed, by many medusae contributed by more coastal water, and by a greater diversity of the fauna. A biological summer never occurs in the cold northern part of the Strait. Changes in abundance of zooplankton and the extension of species in Davis Strait thus strongly reflect hydrological seasonal changes.

The composition of the plankton and the changes in Davis Strait generally resemble those of the northern Norwegian Sea. Davis Strait is less rich and less diverse in fauna, however—possibly a reflection of the longer and more severe ice cover (cf. Pavshtiks, 1975). Hansen (1960) found that the biomass of zooplankton in the upper 50 m in the open waters of the southern Norwegian Sea varied approximately thirty-fold between June and November/December, when it was at its minimum. While much of the change was due to vertical migration and to massive reproduction during the spring/summer, there were differences associated with various water masses. For example, a lower biomass was typical of the central part of the Arctic/sub-Arctic waters of the East Iceland Current and also of fully coastal waters, whereas mixed Atlantic/sub-Arctic waters and inflowing Atlantic water had a richer biomass.

Hansen distinguishes between the Atlantic water and mixed waters, which are dominated by *Calanus finmarchicus* on the one hand and a different population in the East Iceland Current water on the other, with larger sized *Calanus* having slower development which probably included some *Calanus glacialis*. The Atlantic water had the richer variety of species, including many copepods such as *Eucalanus*, *Euchirella*, *Pleuromamma*, *Candacia*, *Rhincalanus* and *Euchaeta norvegica*, the chaetognaths *Sagitta elegans* and *S. serratodentata* and other forms—*Aglantha*, *Themisto gaudichaudi* and *Limacina retroversa*. A seasonal effect on the composition of the population in Atlantic and mixed waters was obvious. Whereas in June *Calanus* formed the vast percentage of the population, in November/December *Calanus* was reduced in numbers and the Atlantic element of the population became more important.

In more Arctic waters, although the three common copepods *Calanus finmarchicus*, *Pseudocalanus minutus* and *Oithona similis* were still present, the stronger the Arctic component the smaller the combined percentage of these three forms. Other typically Arctic species, *Calanus hyperboreus*, *Metridia longa*, *Themisto libellula*, *Themisto abyssorum* and *Eukrohnia hamata* were present, but as the proportion of *Calanus finmarchicus* decreased, *Pseudocalanus minutus*, a species widely distributed in Arctic waters, became increasingly important. In the coastal neritic water apart from some prominence of meroplankton, *Calanus* was important over summer, though other species such as *Temora longicornis* and cladocerans played a significant part. During the winter *Calanus* was of decreased importance and the neritic species *Acartia* and

Pseudocalanus elongatus became dominant. Some extension of *Evadne* into more open Norwegian Sea waters occurred over summer; cladocerans were absent during winter. The changing water masses and currents in the Norwegian Sea thus again play a significant part in faunistic seasonal changes and abundance of species.

Plankton Indicators

Several examples have already been quoted of the occurrence of certain species of zooplankton acting as indicators of the presence of specific water masses and of current movements. The development of the study of plankton indicators was greatly stimulated by the work of Russell (1934, 1935b, 1936, 1939) on the zooplankton populations appearing off Plymouth. To an observer at the entrance to the English Channel it would appear likely that water entering from the Atlantic Ocean should have a great mixture of oceanic planktonic species, some of which (such as *Thysanoessa*, *Aglantha*, *Meganyctiphanes* and *Clione limacina*) would be of boreal origin, whereas others (such as *Agalma elegans* and *Sagitta serratodentata*) would be typical of warmer Gulf Stream water. The occurrence of both groups of species would indicate that water from the Atlantic reaching the English Channel was of mixed origin. Russell's work on the zooplankton appearing off Plymouth has shown that such a mixture may indeed occur. However, the meroplanktonic medusa *Cosmetira pilosella* (Fig. 8.1) may also appear in the Channel. This medusa breeds off south-west Ireland and not in the English Channel itself, but can enter the Channel with Atlantic water which has already circulated off the Irish coast. *Sagitta elegans* appears in great numbers in this incoming Atlantic water, which is mixed with some water of more coastal origin.

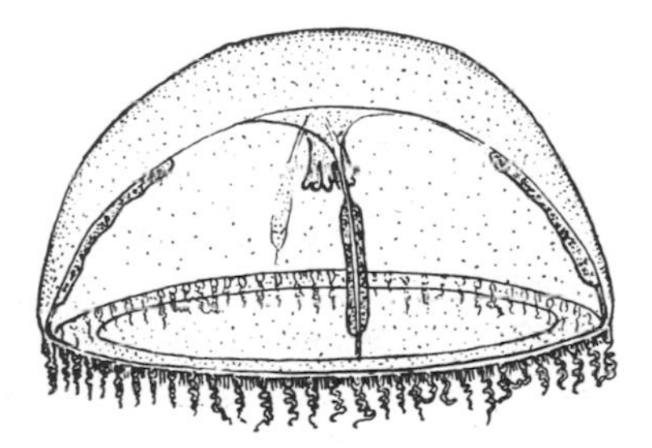

Fig. 8.1. *Cosmetira pilosella* (Russell, 1963).

At times, salps and *Doliolum* may appear at the mouth of the English Channel and since no species of salp or of *Doliolum* is boreal or Arctic in distribution, but all North Atlantic forms are sub-tropical or tropical, their presence indicates a flow of warmer Atlantic water. Russell called this water of warm origin bringing salps and *Doliolum* off Plymouth "south-western water". Accompanying these forms are *Muggiaea* spp. (cf. Fig. 2.16), *Physophora borealis*, *Arachnactis* larvae and *Lepas fasicularis*, sometimes with such warmer-water copepods as *Euchaeta hebes* and *Rhincalanus*.

Some species of plankton are typically brackish forms (e.g. *Eurytemora affinis*) while others, though not quite so brackish in distribution, such as *Acartia bifilosa* and

Oithona nana, are normally confined to rather low salinity water; *Centropages hamatus* is a typically neritic species. Although the majority of meroplanktonic forms tend to be most plentiful under coastal conditions, there are some holoplanktonic species apart from the copepods listed (e.g. *Sagitta setosa*) which appear to be confined to certain neritic waters. The appearance of some zooplankton species normally found offshore in the neritic English Channel may therefore be indicative of waters of more distant origin flowing into the Channel. These species may be regarded as plankton indicators, that is, plankton organisms which may be of value in indicating the movements of particular masses of water. Many species are too widespread to be useful as indicators; cosmopolites, such as *Pleurobrachia* and *Calanus finmarchicus*, are obvious examples. Russell (1935b) has pointed out that plankton indicators may be selected from any zoological group, but often it will be the association of several species from different groups which will indicate the flow of a particular mass of seawater. He has emphasized that an indicator must be practical, that is, it must be of reasonable size and fairly easily recognized, if it is going to be of value.

Sometimes individuals of species outside their normal area of distribution occur as secondary constituents of the plankton together with indigenous species which are much more numerous and the water may then be regarded as of mixed origin, as in the North Sea. Local conditions, however, need full study before any indicator can be accepted. In other words, a species which is a true indicator of water movements in one region may not be effective in another sea area. Few indicators may be regarded as universal, though forms such as *Cyanea* and *Aurelia* are typical of coastal conditions. Some other indicators are of very wide use; for example, the Arctic species *Limacina helicina* (cf. Fig. 2.46), if it appears off the New England Coast, off the south of Ireland or off the coasts of north-west Europe, may be regarded as indicating a flow of Arctic water. *Calanus hyperboreus* also appears to be a fairly good indicator of Arctic water to the north of the British Isles.

On the other hand, Bigelow (1926) demonstrated that off the north-east coast of America, although most *Calanus hyperboreus* are brought in by an Arctic current flowing south in the spring, the copepod apparently has a small centre of local reproduction in the waters north of Cape Cod. Similarly, in some of the Norwegian fjords *Calanus hyperboreus* occurs in the deeper layers. This copepod is therefore useless as an Arctic indicator off Norway and off the north-east American coasts. The distribution of *Metridia longa*, another Arctic form, in the Gulf of Maine appears to be similar. *Clione limacina* is normally thought of as an Arctic form, but it appears to be able to live fairly successfully, perhaps as a slightly different variety, in the more temperate parts of the North Atlantic. Thus when it appears in the English Channel it can be of use as an indicator of Atlantic water, but cannot demonstrate a flow of Arctic water.

The use of some indicators in determining the movements of particular water masses may now be followed in more detail. In the western English Channel area off Plymouth, Russell showed that there are three types of water: (1) the warm south-west Channel or Biscay water, in which the characteristic indicators are salps, *Doliolum*, *Muggiaea* spp. and the medusa *Liriope exigua*, probably included in this water as it sweeps round the French coast. Other warm water species such as *Rhincalanus* may appear with this flow of more southern water. (2) Western water which occurs flowing past the south of Ireland. (3) True Channel water. The elucidation of the last two

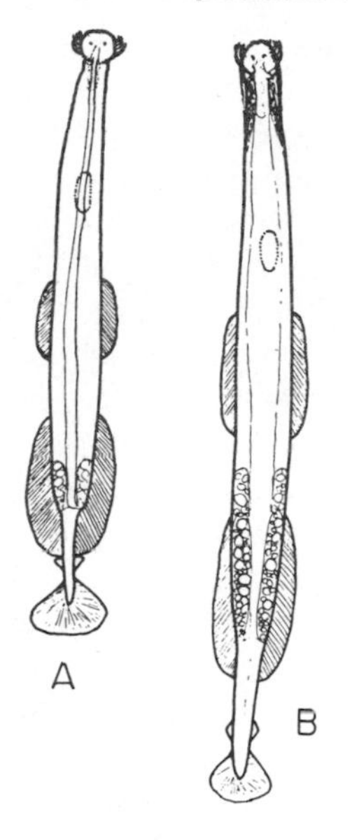

Fig. 8.2. (A) *Sagitta setosa*; (B) *Sagitta elegans* (Fraser, 1957).

bodies of water was based mainly on an analysis of *Sagitta* populations off the Plymouth coast (Fig. 8.2).

Russell showed that *Sagitta elegans* dominated the catches off Plymouth to the exclusion of *Sagitta setosa* up to the autumn of 1930, but that thereafter *Sagitta setosa* occurred and very few *S. elegans* were taken. It is possible that both species lived in competition off Plymouth, but Russell observed that the medusa *Cosmetira pilosella*, a species found in deeper waters off the south-west coast of Ireland, was always associated with large populations of *Sagitta elegans*, whereas it did not occur with *S. setosa*. The waters to the south of Ireland are also rich in *S. elegans*, but *S. setosa* does not occur there. It would appear, therefore, that when western or Atlantic water, containing such boreal species as *Aglantha*, *Clione limacina*, *Meganyctiphanes* and *Thysanoessa* as well as the medusa *Cosmetira* flowed past Plymouth, there was an incursion of *Sagitta elegans*. On the other hand, when only small populations of *S. elegans* were found off Plymouth, the boreal Atlantic species occurred only in small numbers. The boreal species together indicate a strong flow of mixed Atlantic water, of which *S. elegans* is so characteristic that Russell has applied the term "elegans water" to such a water mass. Occasionally the oceanic and somewhat warmer water species *S. serratodentata* occurs off Plymouth at the same time as *S. elegans*, probably indicating a flow of water from rather farther out in the Atlantic Ocean.

Russell linked these changes with the cyclonic circulation in the Celtic Sea and suggested that the water increases in nutrient content with upwelling off the Atlantic slope. In this mixing of oceanic Atlantic with more coastal seawater *Sagitta elegans* flourishes. The production of zooplankton is very high in this mixed ("elegans") water and includes high densities of such typical widespread copepods as *Calanus*, *Centropages typicus*, *Paracalanus* and *Pseudocalanus* when the flow of Atlantic western water into the Channel is strong. These four calanoids are normally present in the

Channel and therefore cannot be used as indicators. Nevertheless, the abundance of these species is obvious. Apart from the high nutrient content of the inflowing mixed "elegans" water, mixing of water masses frequently appears to have a beneficial effect upon production. It is likely that the supply of other trace growth-promoting substances is increased.

In contrast, when *Sagitta setosa* is dominant off Plymouth the water is generally low in phosphate as well as being of somewhat lower salinity, typical of Channel water. "Channel" or "setosa" water in general shows a marked poverty of zooplankton. *S. setosa* occurs widely in inshore waters (cf. Chapter 2). When "setosa" water occurs off Plymouth, though such typically widely distributed copepods as *Calanus*, *Pseudocalanus*, *Oithona*, *Centropages*, *Paracalanus*, *Temora* and *Acartia* (as well as the cladocerans *Podon* and *Evadne*) occur and meroplanktonic forms may also be observed sometimes in fair numbers, the density of the whole zooplankton tends to be low. Russell (1939) later suggested the medusa *Turritopsis* (Fig. 8.3) as another indicator of Channel water.

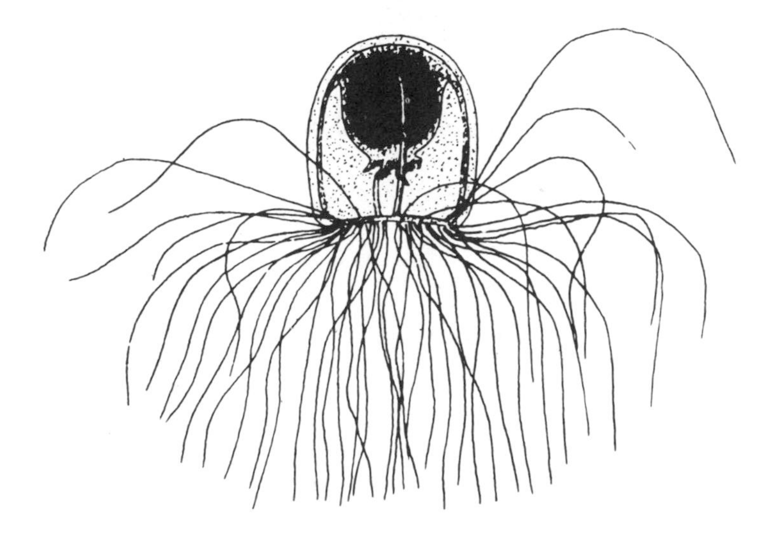

Fig. 8.3. *Turritopsis nutricula* (Russell, 1939).

Biscay water is also characterized by having very little zooplankton, but at the time of Russell's investigations there was considerable green flocculent matter present, so that it was possible to distinguish the three types of water even without detailed microscopic examination. At times the south-western water from Biscay, whilst mainly oceanic, can vary in its flow and be replaced by rather more coastal water of lower salinity entering round Ushant.

Russell has correlated the occurence of "elegans" or "setosa" water off Plymouth with the strength of flow of water from the English Channel into the North Sea. When the circulation off south-west Ireland results in a fairly strong inflow of mixed Atlantic "elegans" water, an increased easterly flow of water occurs through the Dover Straits. To some extent a strong flow of Atlantic water round the north of Scotland may be accompanied by a lessened current through Dover Straits, causing Channel water

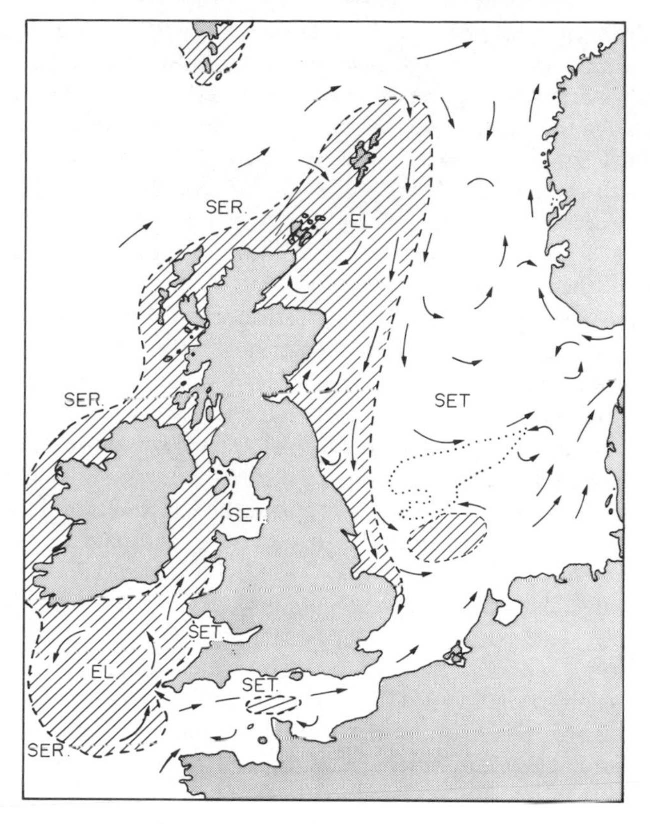

Fig. 8.4. Generalized picture of the distribution of plankton around the British Isles in an autumn when the influx of Atlantic water into the North Sea from the north was strong. SET = *Sagitta setosa*; EL = *S. elegans* SER = *S. serratodentata* (adapted from Russell, 1939).

dominated by *Sagitta setosa* to occur off Plymouth. At times, also, when the flow through the Straits is weak, south-west Biscay water may creep past Plymouth (cf. Fig. 8.4).

Whilst the study of these water masses off Plymouth is in itself of great fundamental interest, the results are not without practical significance. Russell (1930–1947) has shown that on the whole the nutrient-rich water typical of *S. elegans* off Plymouth was accompanied by relatively large numbers of young fish, excluding clupeids, in the area. There was also a good winter herring fishery over this period. Relatively high phosphate of the order of 0.67 μg PO_4-P/l (winter maximum) were demonstrated up to about 1930, but following that year the winter maximum of phosphate fell to about 0.48 μg PO_4-P/l, the numbers of young fish fell very considerably and the herring fishery collapsed (cf. Cooper, 1938).

The distribution of indicator organisms in the western Channel area has also been studied by Southward (1961, 1962). Dealing with the macroplankton, Southward suggests that there are north-westerly species (e.g. *Aglantha*, *Sagitta elegans*) which are approximately equivalent to the forms previously named as western species. Many of these are northerly species verging on their southern limits of distribution. They tend, however, to extend their distribution across the entrance to the English Channel from May to July. In contrast, what may be described as south-westerly forms (e.g. *Euchaeta hebes*) are mostly warm water species near their northerly limits of distribution. Whilst they are restricted in their area of abundance to the south-west of Ushant, they tend to extend across the Channel entrance from July to January. These movements appear to be correlated with the anti-clockwise swirl off the entrance to the Channel, south-westerly forms becoming dominant as water moves from the south-west as the swirl contracts and north-westerly species becoming more abundant as lower salinity water moves south with the expansion of the swirl.

The appearance of particular indicator species off Plymouth will be further complicated by the changing pattern of the surface currents in the western Channel between summer and winter. Southward's analysis suggests that the changes in the Channel approaches observed in earlier years may be partly a result of generally rising surface temperature in the area and thus a shift in the boundaries of distribution of some indicator species. Comparable analyses of plankton and nutrients off Plymouth since the war have demonstrated that from about the late 1960s an increase in winter maximum phosphate concentration has occurred, accompanied by richer spring phytoplankton and more abundant zooplankton, with a reasonable density of species which had virtually disappeared in the catches during the 1930s. Young teleosts increased over the same period and there are signs of some return of herring (cf. Russell, Southward Boalch and Butler, 1971). There are indications also of a reduction over the same period in the intensity of spawning of the pilchard, a more southerly distributed fish. These fluctuations may be associated with hydrological changes involving a retreat of the more boreal plankton northwards during times of climatic warming, and a return southwards over a period of worsening climate. The cycling of nutrients may be affected by these fluctuations in the ecosystem, as outlined in Volume 1.

A study of plankton indicators rather similar to that of Russell was made by Fraser (1939) off the Scottish coast. He demonstrated that *Sagitta elegans* was absent from truly oceanic water entering the Scottish area. The species was not found in the Faroe Channel, but was more common in the sea lochs of the west coast and in the channel between the islands and the mainland. It was very abundant, however, in the northern North Sea off the east coast of Scotland. The distribution suggests that *Sagitta elegans* is a species which flourishes in a mixture of oceanic and coastal water, as Russell showed off Plymouth. Fraser finds that this mixed water includes species such as *Thysanoessa*, *Themisto* and *Aglantha*, which are Atlantic forms, as well as *Calanus finmarchicus* and *Limacina retroversa*. There are also fair numbers of medusae like *Cosmetira* and *Laodicea* from deeper offshore areas, as well as other species such as *Neoturris* and *Bougainvillia* from more coastal regions of north Scotland. This mixed water seems to favour the breeding of Crustacea, so important as fish food, and indeed the whole zooplankton tends to be rich. Fraser believes that the mixing of water on the edge of the continental shelf causes a considerable enrichment of nutrients, promoting phytoplankton growth and in turn promoting zooplankton production.

Fraser found that there were various types of water flowing round the north of Scotland. The oceanic water coming in from the Atlantic contained a number of species, some of which could be recognized as typically warm oceanic species. *Sagitta arctica*, *Calanus hyperboreus* and *Metridia longa* may well be regarded as very cold water forms indicating sub-Arctic inflow, and north of the Shetlands there is a greater proportion of these cold water forms. *Pleuromamma robusta* is an Atlantic form and *Sagitta maxima* is another cold Atlantic but deeper-living species. On the other hand, warm water species such as *Sagitta hexaptera* and the more warm-temperate *Sagitta serratodentata* can occur, together with such typical warm water indicators as salps, doliolids, *Rhincalanus*, *Euchirella* and *Physophora*. The incoming Atlantic oceanic water thus contains a mixture of cold and warmer-water zooplankton.

Further studies by Fraser (1952, 1955) have demonstrated that the Atlantic zooplankton, especially some of the warmer-water forms, include species derived from the comparatively deep warm outflow from the Mediterranean Sea. Part of this outflow, after leaving the Mediterranean at a depth of about 1250 m, proceeds northwards as a current at around 950 m depth through the Bay of Biscay. Thence it appears to pass even farther north at somewhat shallower depths to the west of Ireland, though the extent of the flow probably varies from year to year. The fauna, modified by the addition of plankton from the deeper waters between the Azores and the Bay of Biscay, has been designated by Fraser as the "Lusitanian" fauna. It includes *Pelagia noctiluca*, *Sagitta lyra*, doliolids, *Vibilia* spp., *Lensia* spp., *Chuniphyes multidentata*, *Vogtia* spp., *Sapphirina* spp., *Praya cymbiformis* and *Rosacea plicata*.

West of Scotland this deep warm flow may emerge on the surface to some extent by upwelling against the continental slope. "Lusitanian" species may, therefore, appear in the incoming Atlantic water round the north of Scotland. The Lusitanian flow tends to keep to the more southern parts of the Faroe Channel. The contribution of Mediterranean species and of warm Atlantic zooplankton to the total Lusitanian fauna is discussed in Chapter 2. The strength of flow of the Lusitanian water varies considerably from year to year; in some years it may reach little farther than the west of Ireland; in years of strong flow it may stream between the Shetland and Orkney Islands and so reach the North Sea (Fig. 8.5). Even in years of strong flow Lusitanian water does not appear to pass north of the Shetlands. By contrast, western North Atlantic

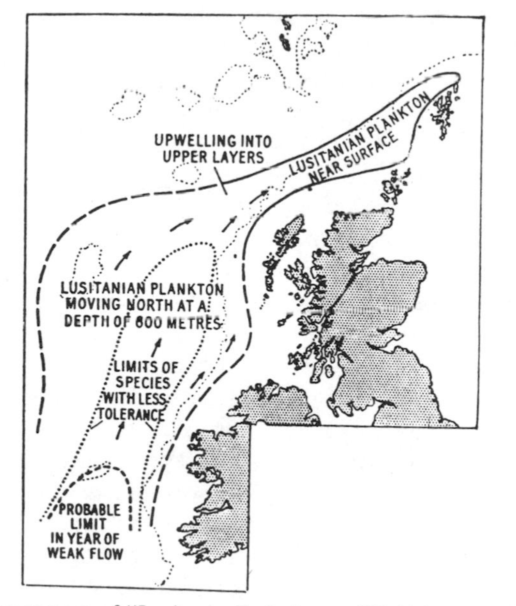

Fig. 8.5. The movement of "Lusitanian" plankton off British coasts (Hardy, 1956).

water enters the North Sea to an appreciable extent by this route. This water is much richer in *Calanus* and other crustaceans and is marked by the presence of other species such as *Clione limacina*, *Dimophyes arctica*, *Agalma elegans*, *Laodicea* and *Eukrohnia*.

There are several zooplankton species, however, which are common to both the Lusitanian and Atlantic water—*Aglantha*, *Clio*, *Rhincalanus*, *Eucalanus*, *Thysanoessa longicaudata*, *Sagitta planctonis* and to some extent *S. lyra* and *S. maxima*. As the islands and mainland of the Scottish coast are reached the Atlantic water, including any Lusitanian flow, grades into the mixed "elegans" water. Another body of water which may be regarded as of North Sea origin is characterized by *Sagitta setosa*. As with the "setosa" water of the English Channel, the zooplankton population is relatively poor and there is a reasonably clear demarcation between the rich "elegans" and the poor "setosa" water (Fig. 8.4). The farther the Atlantic water flows round the north of Scotland into the North Sea, the more the exotic species typical of the warm Atlantic and Lusitanian waters tend to die out. More cold water inflows containing zooplankton may be added between the Faroes and the Shetlands, but as the water moves farther southwards into the North Sea proper these cold water forms are also killed. The fauna will therefore change relatively rapidly with entry into the North Sea, though under exceptional conditions, with a greater degree of inflow, a few of the original Atlantic species may survive and may even breed for a short time before being killed off (e.g. asexual breeding of salps) (cf. Chapter 2).

Fraser (1965) has described eight areas in the North Sea, the boundaries of which are very distinct and vary with season, but which have different plankton communities. The differences depend, amongst other factors, on the extent of the North Sea inflow. Rakusa-Suszczewski (1967) defines three different water masses in the Fladen Ground area using population age structures of *Calanus finmarchicus* and *Sagitta elegans* as indicators of current patterns in the North Sea. Fraser (1969), however, points to the significance of minute differences in the characteristics of water masses in determining the success of a species. Although a species may be tolerant of the environmental conditions, a limiting factor could be its food organisms, which may not be so tolerant. The total organic content, which includes the biological history, of a water mass may be far more important than other physical and chemical characteristics.

Fraser illustrates the complexity of the problem of survival of species outside their usual area with reference to the inflow of water into the North Sea in different years. In 1947 the maximum extension of *Dolioletta gegenbauri* occurred, whereas 1955 was the year for greatest abundance and penetration of *Salpa fusiformis*. Both species are common and widespread in the north-east Atlantic and appear to be fairly tolerant species, though both die out over winter in the North Sea. Yet the water quality was presumably sufficiently different in the two years to favour one of the two immigrants. Even more striking, however, is Fraser's observation that a number of exotic siphonophores normally distributed more to the south as Lusitanian species did not penetrate at all during the marked 1947 inflow into the North Sea, but showed a big advance during 1953 and 1954, continuing in most years until 1964 or 1965, when there were indications of a retreat. Their success was presumably therefore not solely dependent on the volume of inflow, especially since the volume of Lusitanian water is small in comparison with the amount from the western North Atlantic. Fraser (1965) lists some of the species characteristic of each area of the North Sea and also the oceanic species entering from the north of Scotland.

Similarly, zooplankton found in the western Baltic and carried in with the North Sea inflow and species found in the northern part of the North Sea but which demonstrate a flow from southern or central areas of the North Sea are identified. Many of the species in the different communities may at times, however, be found in any of the areas and indicators may be used only where adjacent communities are very different. Fraser suggests that indicator species could be of three kinds:

(1) Organisms transported by currents; no recruitment occurs.
(2) Organisms thriving only in mixed waters, thus indicating an inflow of water of mixed origin or an inflow leading to a mixture of water. These species may live in reduced numbers when conditions are not ideal, so that recruitment is not entirely dependent on outside sources.
(3) Planktonic larvae of sessile animals transported to areas where the adults are not found.

The numbers of an indicator depend on several factors including the volume of inflow, distance of transport and the elapsed time during which a change of environmental characteristics could occur. Clearly, if there is a considerable volume and little environmental change, species may continue in active reproduction even after long transport. The number of exotic species must also be influenced by the initial density of the zooplankton and whether any reproduction has occurred *en route*. Fraser points out, however, that as a very broad rule numbers will decline progressively with distance, so that numerical pattern of an indicator species may sometimes be used as a guide to the pattern of inflow, e.g. the extent of inflow from the north into the North Sea.

Briggs (1974), questioning the use of plankton indicators for water masses, considers that the arguments for believing that each water body is a distinct environment with its own characteristic community of organisms are rather weak. Few species even from taxa recognized as useful appear to qualify as indicators. Most investigators, however, have recognized the necessity of a very critical appraisal of individual indicator species. Fraser (1965) states that individual indicators may prove to be unreliable and that a larger number, and above all, a community of species, is a much more reliable index of a water mass. The value of employing numerical balances of many species is also discussed later.

The danger of using uncommon species which have become adapted to the local conditions as indicators is also instanced by Fraser. Examples are *Apolemia* and *Pneumodermopsis* in the North Sea. Cooper and Forsyth (1963) found *Pneumodermopsis paucidens* moderately abundant over comparatively shallow waters of the European and American continental shelves. From 1956 to 1958, though fairly common off the Atlantic coast of Ireland, the species was never found east of a line from Iceland to Scotland. During 1960 and 1961 there was a marked change, with the species becoming more widespread; low numbers were found off Ireland, but it was comparatively abundant over the northern and central North Sea where it had not been taken before. Moreover, it had a rather longer season of abundance in those years. The gymnosome *Clione limacina*, found from 1948 to 1952 over deep Atlantic waters and over the continental shelf and in the northern North Sea, also subsequently declined in abundance over the deep ocean and from 1956 to 1959 it appeared to be confined to the continental shelf regions and northern North Sea. This change, while somewhat similar to that described for *Pneumodermopsis paucidens*, did not occur over the same period. While the fluctuations in the distribution of *Pneumodermopsis* might in part be due to hydrological changes, the species may have become adapted to slight variations in the environment. Clearly, it cannot be used as an indicator in Atlantic water flowing into the North Sea.

Fraser (1965) has emphasized the value of chaetognaths as plankton indicators (cf. also Alvarino, 1965). Some taxa have been little used, possibly due to the occurrence of races and varieties of species which are not easily identified. Comparatively rapid adaptability to environmental conditions is obviously disadvantageous to the employment of indicators. Although a few siphonophores are restricted to particular water masses (e.g. *Muggiaea atlantica* and *M. kochi* in more neritic waters), Alvarino (1967) suggests that most warm-water siphonophores have a very broad distribution. Margulis (1972), while accepting a wide range for most species, believes that their distribution may be associated with the characteristics of water masses, especially in the Atlantic. Dealing with the tropical and sub-tropical Atlantic forms, Margulis suggests that they may be divided into three groups according to the extent of spread into cooler, higher latitudes. Details of the distribution are given in Chapter 2. Although the detailed physico-chemical characteristics of the water masses were not identified, Margulis believes that the productivity of the water mass in association with the hydrological conditions may be relevant to the distribution of the species.

Pugh (1975, 1977) in considering siphonophore distributions, draws attention to the significance of species assemblages and the proportions of the various forms rather than to the presence of indicator species. While the depth of certain isotherms may be significant in siphonophore distribution, as suggested by Alvarino (1967), more complex inter-relationships may exist between the plankton fauna and the physico-chemical characteristics of the water mass. From oblique plankton samplings across the North Atlantic near latitude 32°N, Pugh (1975) studied 19 species of siphonophore. When subjected to factorial analysis, 13 species showed a clear east–west trend which might be associated with the different water masses of the Sargasso Sea and of the eastern North Atlantic. The commonest species, *Eudoxoides mitra*, while being found in all samplings, had its maximum distribution in the west, and the second most abundant siphonophore, *Chelophyes appendiculata*, its maximum distribution in the east. Thus though many species of siphonophores are cosmopolitan or at least widely distributed, trends in distribution and association of groups of species may still be discerned. "Western" species, according to Pugh's analysis, included *Lensia conoidea*, *Eudoxoides spiralis*, *Diphyes bojani* and *Eudoxoides mitra*; "eastern" species were *Lensia multicristata*, *Vogita spinosa* and *Clausophyes ovata*, the last living somewhat deeper (*ca.* 800 m).

In a further study, Pugh (1977) examined the vertical distribution of species in zones and water masses. He concludes that apart from the two major siphonophore assemblages, one in the Sargasso Sea and one in the eastern North Atlantic, the application of multivariable statistical analysis to the data demonstrated the existence of faunal zones. The fauna is made up of certain species with characteristic relative numerical abundance rather than individual faunal assemblages. Often a species occurs in more than one zone but is then associated with other species and in different numerical proportions. The faunal zones can often be identified with water masses, but the depth distribution of water masses under a water column is a very significant factor.

A single station may include several water masses at different depths and thus different faunal zones.

The results are in many respects comparable to those of Angel and Fasham (1975) and Fasham and Angel (1975), which deal with the ostracod fauna of the North Atlantic. They examined the vertical distribution of the water masses and faunal groups in addition to the horizontal distribution. The faunal zones separated on the basis of species abundance were examined in relation to water masses. The zones were also related to T–S diagrams in order to attempt to identify the component water masses (e.g. North Atlantic Central Water, South Atlantic Central Water, Antarctic Intermediate Water, North Atlantic Deep Water, Mediterranean Water). The authors also emphasize the importance of recognizing that different faunal boundaries may be present at different depths at any one station. Although some water masses have an associated faunal assemblage, this is not true for every water mass and some may be subdivided faunistically, e.g. the North Atlantic Central Water divides into northern and southern zones. Very few of the ostracods are restricted to a single zone. As regards the relevance of these findings to the identification of indicator species, only one species, *Conchoecia macromma*, appears to be a true indicator in that it was restricted to one zone or water mass, but *Conchoecia oblonga* approaches indicator status in that its main abundance was restricted to one zone. Fasham and Angel (1975) suggest that indicator species, at least for ostracods, are rare (cf. Briggs, 1974). Haagenson (1976) pointed out that in assigning species to a specific water mass, note should be taken of their ability to perform diurnal vertical migrations. Many of the Caribbean pteropod species moved between a daytime habitat in the sub-tropical underwater to tropical surface water at night (cf. Chapter 5).

Some indication has been given of the use of certain hydromedusans in characterizing water masses especially from the investigations of Kramp and of Russell. More recently, the study of Vannucci and Navas (1973b) of Hydromedusae from the Indian Ocean collections suggests how these animals can be employed in discussions of distribution. Widespread epiplanktonic warm water oceanic Trachymedusae, especially *Liriope tetraphylla* and *Aglaura hemistoma*, therefore made up nearly half the species collected. However, there was an obvious difference between the Bay of Bengal, with its water of lowered salinity due to the run-off from the major river systems and its generally greater proportion of neritic components, and the fauna of the Arabian Sea. The Bay of Bengal had a considerable proportion of Leptoline medusae, all believed to have a hydroid benthic stage, but the life span of the medusae was relatively long, permitting their drift offshore. Apart from the abundant *Liriope*, *Aglaura hemistoma* and *Rhopalonema velatum* were very plentiful trachymedusans, while the narcomedusan *Cunina tenella*, a euryhaline eurythermal species tolerant of low oxygen content, was also characteristic of these waters. In the Arabian Sea *Aglaura hemistoma* and the narcomedusan *Solmundella bitentaculata* were characteristic.

In the Indian Ocean Central Water, with slightly lower temperature and salinity than the Arabian Sea surface waters, there were many species common to sub-surface waters in the Arabian Sea and the generally widely distributed cooler-water species were prevalent. In areas of upwelling water *Ectopleura sacculifera* appeared to be typical. A number of species inhabited the deeper layers and although several showed a considerable depth range through deep and intermediate layers and were living essentially in cold waters, they were reasonably eurythermal. Other species were more or less confined to low temperature waters. A number of the medusae could also withstand a remarkable lowering of oxygen tension. Thus *Aegina citrea* and *Aeginura grimaldii* are stenohaline species but fairly eurythermal, *A. citrea* occurring mainly in colder waters and *A. grimaldii* in deep cold waters. Both species, especially *A. grimaldii*, are markedly tolerant of lowered oxygen. While species such as *Amphogona apicata* are not Antarctic in origin and are commonly found at the boundary layers of Antarctic Intermediate and Indian Ocean Central Water, many of the deep-living medusae are Antarctic forms and appear to be transported into the Indian Ocean, particularly with the Antarctic Intermediate Water. These include *Botrynema brucei*, *Halicreas minimum*, *Colobonema sericeum*, *Crossota alba*, *C. brunnea* and *Euphysora furcata*. Vannucci and Navas (1973a,b) point to the latter as being a good indicator of Antarctic Intermediate or Antarctic Deep Water in the Indian Ocean; the number of medusae increases near the Antarctic Convergence at depths of 500–1000 m. By contrast, *Euphysora annulata* is limited to warm surface layers with a narrow range of temperature and salinity and *Euphysora bigelowi* is regarded as a more widely distributed warm-temperate species.

Species which are believed to be immigrants from the Mediterranean into the Arabian Sea include the anthomedusans *Koellikerina fasciculata* and possibly *Bougainvillia maniculata*. *Aequoria* and several species of *Koellikerina* (*K. constricta*, *K. elegans*, *K. multicirrata*, *K. octonemalis*) are species tolerant of low salinity and the latter also of low oxygen.

In considering the plankton fauna of water masses, however, it is necessary to return to consider communities and not only representatives of a single taxon. The investigations of the plankton communities of the English Channel and North Sea were amongst the earliest studies of this kind.

Zooplankton of the North Pacific

Another region where the water masses and circulation are well known and where the distribution of several taxa has been carefully described is the North Pacific Ocean. The overall distribution of the major water masses (cf. Volume 1) indicates a broad zone of sub-Arctic waters marked by a cyclonic gyre and separated from the warmer central Pacific waters by a transitional zone approximating to 45°–40°W, with its extension as the Californian Current along the west coast of North America. The anticyclonic gyre marks off the central mainly sub-tropical waters, which tend to divide into an eastern and larger western central water mass. Tropical waters lie north and south of the equator as the equatorial water mass. There is a general east–west flow in this region. This distribution refers essentially to surface or near-surface layers, to a depth of 200 m or more.

Apart from the obvious differences in volume of zooplankton, with very large biomass but few species in the sub-Arctic waters, very large numbers of species and very low biomass in the central (sub-tropical) gyres and somewhat higher biomass and numerous species in the equatorial zone, certain species can be listed as being characteristic of different water masses (cf. Johnson and Brinton, 1963; Reid, Brinton, Fleminger, Venrick and McGowan, 1978). Some have a wider distribution, however; for example, a number of "warm" species inhabit both central and equatorial waters. In delimiting a species it is essential to establish that it forms an active reproducing population. The younger stages of planktonic open ocean species are frequently less tolerant of changed environmental conditions in a different water mass. Neritic forms also, though often regarded as somewhat more tolerant than oceanic species, may not form a breeding population.

Of the major water masses, the whole sub-Arctic zone has a fairly characteristic fauna which to some degree extends into the Californian Current (cf. Chapter 2). The euphausiid *Euphausia pacifica* is characteristic and *Sagitta elegans* has a generally similar distribution (cf. North Atlantic). The widespread copepods *Calanus cristatus*, *Calanus plumchrus* and *Eucalanus bungii bungii* are also typical of the sub-Arctic Pacific, though *C. plumchrus* extends at least to 35°N in the western Pacific at somewhat deeper levels under the influence of the cold Oyashio Current. Kos (1975) also describes separate populations of this polytypic species, which is typical of different water masses (cf. Chapter 2). The southward extension of *Calanus plumchrus*, *Sagitta elegans* and some other sub-Arctic species (e.g. *Calanus cristatus*) at deeper levels is well recognized (cf. Omori, 1965, 1967, 1969; Marumo, 1966).

The meeting of the Kuroshio and Oyashio Currents off Japan leads to a complex distribution of species. For example, Motoda and Marumo (1963), while describing the abundance of the boreal copepods *Calanus pacificus* and *Calanus plumchrus* and the chaetognath *Sagitta elegans* in the cold waters north of the main Kuroshio off Japan and also the presence of warm water copepods (e.g. *Calanus minor*, *Undinula darwini*) and chaetognaths (*Sagitta hexaptera*, *S. enflata*) to the south of the Current, point out that these warm species can be found in some areas north of the Current. Isolated masses of Kuroshio water may apparently be present surrounded by cold Oyashio waters (cf. pockets of Gulf Stream water and their fauna in the North Atlantic—Volume 1). The distribution area of the warm species also extends north over summer with a shift in the path of the Kuroshio Current and retreats south in winter.

Neritic plankton including meroplankton extends from the Japanese coast eastwards to a considerable distance with the flow of the Kuroshio Current, especially along its northern border. Motoda and Marumo list several plankton species from different taxa as useful indicators of Kuroshio water. Furuhashi (1966) lists copepods at various depths in the Oyashio and Kuroshio regions; he claims that the presence of cold water copepods in the Kuroshio deep water must indicate their recent arrival from Oyashio areas and states that they are unlikely to survive beyond normal life span. Hida (1957) has also drawn attention to the comparatively abrupt change from sub-Arctic to Transition Zone in mid-western Pacific compared with the more extended change in the more western Pacific. He refers to *Limacina helicina*, *Sagitta elegans* and *Eukrohnia hamata* as species abundant in the sub-Arctic zone, in addition to the few species of copepods and euphausiids already noted as being typical in high density in that water mass. He emphasizes the high biomass of the zooplankton of the zone and the low species diversity.

In the more extreme northern Arctic/sub-Arctic waters *Calanus glacialis* occurs, as in the coldest areas of the North Atlantic and *Calanus marshallae* (Frost, 1974) is found in the eastern northern North Pacific (cf. Chapter 2). *Limacina helicina* is another sub-Arctic representative, as are two euphausiids with a more northerly distribution than the widespread *Euphausia pacifica*—*Thysanoessa longipes* (both "spined" and "unspined" forms) and *Thysanoessa inermis*, which is even more limited to a northerly distribution (cf. Chapter 2). Bowman (1960) regards *Parathemisto pacifica* as an indicator of the cold low salinity Sub-Arctic water with a distribution pattern somewhat similar to *Euphausia pacifica* (cf. Chapter 2).

The gyres in the eastern and western North Pacific separate water masses which may be correlated with the partial separation of zooplankton populations between east and west central Pacific waters ("sub-tropical waters"), from the equatorial Pacific water mass. Comparable water masses are identifiable in the South Pacific Ocean. Brinton (1962) and others have studied the distribution of euphausiids in relation to these water masses (Table 8.1). To a lesser extent, populations in the North Atlantic may similarly be related to gyres. Haury, McGowan and Wiebe (1978) noted that the presence of the cool transition water species *Nematoscelis megalops* in isolated patches in the Sargasso Sea was related to the occurrence of Gulf Stream cold core rings.

In the Transition zone there is considerable movement of water in an easterly direction; the generally southward flow of the Californian Current also tends to transport epiplanktonic species. It is not entirely clear, therefore, how stocks of zooplankton are maintained. Some species live in sub-surface layers and though they may migrate vertically they rarely reach the surface, thus avoiding major drift. Other forms appear to retain at least a portion of the stocks in deeper water during the season of less active reproduction. There appear to be species which are truly endemic to this type of water as well as others which occur in higher densities here than in adjoining water masses.

Thus the Californian Current might be expected to have sub-Arctic species, forms from the sub-tropical gyre central water mass and a few forms even from lower latitudes transported by the counter-current and some endemic coastal species. Some maintain themselves as an actively breeding population; others will be represented only in so far as they are brought in by currents and may rely solely on immigration; some may reproduce, but only to a limited extent. The fauna includes *Sagitta scrippsae*, which enters the Alaskan gyre and the Californian Current, where it decreases in density and also lives more deeply as the waters are transported southwards. Alvarino (1965) regards it as a particularly good indicator of the Current; fluctuations in population along the Californian coast reflect the variations in the strength of current with season and from year to year.

Nematoscelis difficilis is also typical. According to Gopalokrishnan (1973), its limits in the North Pacific are about 35°–45°N and it spreads southwards in the Californian Current even to about 20°N. *Thysanoessa gregaria* has a similar distribution and is also found in transitional waters of the sub-Antarctic. *Limacina inflata* and *Calanus pacificus* are other characteristic species. Hida (1957) regards *Limacina inflata* as being of increased abundance in transition waters: *Euclio balantium* is also important and *Peraclis apicifulva* is said to be confined to the Transition Zone. Brodsky (1965) proposes that three subspecies of *Calanus pacificus* may be distinguished, with *C. pacificus pacificus* being confined mainly to the coast of Oregon and California. Since this is an actively reproducing population, there are presumably certain mechanisms (e.g. seasonal depth migrations) which maintain the population near the coast (cf. Chapter 3).

Peterson, Miller and Hutchinson (1979) found an abundance of zooplankton in upwelling areas off Oregon up to 20 km from shore between 0–20 m depth. *Acartia clausi* and *Centropages abdominalis* are confined to the inshore 5 km and *Pseudocalanus* sp. lives up to 15 km offshore, but returns to the 5-km zone to breed.

Table 8.1. Distribution of some typical euphausiid species in the Pacific Ocean (Brinton, 1962)

Species	Depth (m)
Sub-arctic epipelagic:	0–280
Thysanoessa raschii	
T. inermis	
T. spinifera	
T. longipes	
Euphausia pacifica	
Transition-zone epipelagic:	0–280
Nematoscelis difficilis N.	
N. megalops S.	
Euphausia pacifica	
Thysanoessa gregaria N,S.	0–700
Thysanopoda acutifrons	
Central epipelagic:	0–700
Thysanopoda obtusifrons	
T. aequalis, T. subaequalis	
Euphausia brevis	
E. mutica	
E. recurva 20°–42°N,S	
E. hemigibba (North Pacific)	
E. gibba (South Pacific)	
Nematoscelis atlantica	
N. microps	
Stylocheiron carinatum	
S. abbreviatum	
S. suhmii	
S. affine "Central Form"	
Nematobrachion flexipes	
Equatorial epipelagic:	0–280
Euphausia tenera	
E. distinguenda	
Stylocheiron microphthalma	
Thysanopoda tricuspidata	0–700
Euphausia diomediae	
E. eximia	
E. lamelligera	
E. fallax	
Nematoscelis gracilis	
Stylocheiron affine	
"West Equatorial Form"	
"East Equatorial Form"	
"Indo-Australian Form"	
Central-equatorial mesopalagic:	140–1000
S. elongatum	
Thysanopoda pectinata	
T. orientalis	
T. monacantha	
Nematoscelis tenella	
Nematobrachion boopis	
Central mesopelagic:	280–1000
Nematobrachion sexspinosus	
Stylocheiron robustum	
Thysanopoda cristata	

Acartia longiremis and older copepodites of *Calanus marshallae* frequent the offshore 15–20 m layer with dense phytoplankton blooms, but the females of the latter migrate to shallower inshore water to spawn, with nauplii and young copepodite stages appearing within the 5-km area. It is likely that ontogenetic vertical migrations allow the animals to be returned by currents to the upwelling zone. In contrast, Reid, Brinton, Fleminger, Venrick, and McGowan (1978) suggest that *Eucalanus californicus*, a transition species with the same distribution as *Thysanoessa gregaria*, has a population in the Sea of Japan and could be transported to the Transition Zone from that source.

Patterns of breeding and changes in abundance may thus be intimately connected with distribution. Alvarino (1971) describes the siphonophore *Muggiaea atlantica* as extending along the Transition Zone and being present all the year in the Californian Current. It is claimed, however, that its centre of reproduction might be far distant in the west Pacific and that it might be transported by the easterly flow. The problem of a return of a proportion of the population of such species by westward flowing currents, perhaps at greater depths farther south, is still unresolved. Alvarino, however, reports on another siphonophore, the temperate species *Chelophyes appendiculata*, which she says occurs all the year and also breeds more or less continuously in the region of the Californian Current off La Jolla. The warm-water species *Chelophyes contorta*, however, displaces *C. appendiculata* during the season of inflow of the warm water Davidson Current; the species could, therefore, be regarded as an indicator.

In the sub-tropical or central waters, *Euphausia brevis* and *Nematoscelis atlantica*, both also found in the South Pacific sub-tropical waters, and *Euphausia hemigibba*, found in the north-east, are typical. Reid *et al.* (1978) refer also to the copepod *Clausocalanus lividus*, which has a sub-tropical distribution north and south of the equator. Sherman (1963) suggested that some pontellids might be characteristic of water masses in the North Pacific. *Labidocera acutifrons* and *L. detruncata* were associated with North Pacific Central Water and could be useful indicators over the summer in regions such as Hawaii. *Pontella tenuiremis* and *Pontellina plumata* also occurred mainly in the same water mass, but small numbers extended to mixed equatorial/central waters. Over winter, with the narrowing of the area between central Pacific and equatorial Pacific waters, the distribution of *L. acutifrons* and *L. detruncata* was more extensive. Hida (1957) comments on the high species diversity and low biomass of the sub-tropical zone. The northern limit with Transition waters, however, shifts seasonally and the change varies with longitude. He regards *Pterosagitta draco* as a particularly good indicator of sub-tropical waters, this chaetognath being accompanied especially by *Sagitta serratodentata* and *S. hexaptera*, which reach only to a limited extent into the Transition Zone. On the other hand, *Sagitta minima* is most characteristic of the mixed waters of the Transition Zone and *S. lyra* is also found there. In the sub-tropical zone, Hida also found *Creseis acicula* to be comparatively abundant, though it is markedly reduced towards the Transition Zone. *Cavolinia inflexa* apparently occurs more or less equally in both regions.

According to Bieri (1959) a chaetognath typical of the north central Pacific water is *Sagitta pseudoserratodentata*, but it appears to be excluded from Equatorial Water. *Sagitta bipunctata*, whilst occurring north and south of the equator, is also very rare in the equatorial zone. Bieri points to several chaetognaths, however, which are found in equatorial and central Pacific water masses (e.g. *Sagitta enflata*, *S. hexaptera*, *S. pacifica* and *Pterosagitta draco*). *S. robusta*, *S. regularis* and *S. ferox* also inhabit equatorial and west central waters, but tend to avoid eastern Pacific central waters.

Three chaetognaths, *S. bedoti*, *S. neglecta* and *S. pulchra*, are absent from the wide central area of the equatorial Pacific but occur in the western Pacific (south-east Asia, East Indies, Australia) and off the coasts of central America. Although there is some tendency for these populations to occur nearer land masses, Bieri points out that any such "coastal" effect is experienced even for hundreds of miles offshore. The exclusion of the chaetognath *Sagitta pseudoserratodentata* from the equatorial zone apparently does not hold for the extreme western Pacific. The ecological conditions in the western sector permit a breeding population in the equatorial region. A similar pattern appears to be true of the euphausiid *Stylocheiron abbreviatum*. *Thysanopoda aequalis* occurs in the slightly more fertile eastern central waters, but less abundantly in the central Pacific. The western Pacific waters are occupied by *Thysanopoda subaequalis*. Other zooplankton limited to the sub-tropical gyres of the Pacific include the pteropods *Limacina leseurii*, *Styliola subula* and *Cavolinia inflexa* (but cf. Haagensen, 1976).

Prominent among the copepods in the waters of the Pacific equatorial zone are species of the genus *Pontellina* which are typically surface-living. Fleminger and Hulsemann (1974) point out that *P. sobrina*, which is strictly confined to the east Pacific, and *P. morii*, which is found widely in the tropical Indian and Pacific Oceans, are both characteristic of the more eutrophic equatorial upwelling waters and that seasonal changes in the winds to some extent maintain the circulation of the section of equatorial waters inhabited by the copepod, so that breeding stocks are retained in a suitable area. *P. plumata*, on the other hand, has a very wide range in warmer seas; in the Pacific its approximate limits are from 40°N to 40°S, i.e. in the equatorial and central Pacific waters. According to Fleminger and Hulsemann, it appears to be less plentiful in the tropical waters where the other species of *Pontellina* are common and is more typical of the oligotrophic warm seas.

Fleminger and Hulsemann (1973) also report on differing distributions of the warm-water species group *Clausocalanus arcuicornis*. Whereas *C. arcuicornis* itself occurs in central Pacific waters and is absent from true equatorial regions, *C. paululus* spreads across the whole sub-tropical and tropical region. *C. minor* and *C. farrani* are confined to lower latitude areas, *C. minor* particularly being found in equatorial waters only. In the species group *Eucalanus elongatus*, *E. inermis* is a tropical species, again apparently confined to the eastern Pacific. Similar distributions which can be identified for some other zooplankton taxa emphasize the remarkably fine adaptation of a species to the exact environmental conditions of a water mass, the need for mechanisms to isolate closely allied species genetically and the need for current patterns changing with season or depth to retain a proportion of breeding stock or for more permanent currents to supply immigrants from an outside centre of reproduction.

Of the euphausiids, *Euphausia diomedeae* and *Nematoscelis gracilis* are fairly restricted to the equatorial Pacific, whilst *Euphausia distinguenda* appears to be limited to the eastern Equatorial Zone and according to Reid *et al.* (1978) migrates daily in and out of the low oxygen zone, which is well known in that region. *E. paragibba* is another equatorial species, but is absent from part of the eastern equatorial North Pacific. *E. eximia* and *E. lamelligera* also occur in the east Pacific, especially nearer coasts (cf. Table 8.1).

The pteropod *Limacina trochiformis*, although present in very low density in other parts of the tropical Pacific, occurs in comparatively high numbers in the eastern tropical region. As previously noted for copepods (e.g. *Pontellina plumata*, *Clausocalanus paululus*), sagittae, euphausiids (e.g., *Euphausia tenera*) and for some other zooplankton taxa, a very wide distribution is typical through the tropical and sub-tropical belts. With the large numbers of purely sub-tropical species, this leads to a very large species diversity in central Pacific waters. In the equatorial zone the number of species is somewhat lower. An example from the euphausiids of separation of truly tropical and sub-tropical forms is that of *Stylocheiron*: *S. suhmii* occurs in central waters both in the North and South Pacific, but the two populations are separated by the equatorial waters to which *S. microphthalma* is confined.

Increase in precise knowledge of the distribution of species, including the elucidation of spatial separation of sub-species and of distribution patterns of polytypic species, may materially assist in characterizing water masses. Sufficiently distinctive and large zooplankton species may also be available to rank as indicators. Several of the euphausiids, chaetognaths and larger copepods are examples. Thus *Sagitta scrippsae*, *Nematoscelis difficilis* and *Thysanoessa gregaria*, being typical of Transitional waters, could well indicate the strength of flow of the Californian Current off the American coast, though numerical abundance and the presence of several such species together is important in relating the occurrence of indicator species to strength of flow of a water mass. The occurrence of some of the warm-water species off California may equally relate to the strength of flow of the Davidson Current.

An example related to the Transitional Zone of the value of precise analysis of species groups in relation to water masses comes from Fleminger's (1973) study of the copepod genus *Eucalanus*. In the *E. attenuatus* group he has described a large and distinctive form separated as a new species, *Eucalanus parki*. All the records are from the North Pacific and occur in the Transition Zone. A few records, however, extend south along longitude 155°W to about 32°N latitude, a finding which supports the restriction of this species to a particular water mass, since the extension marks the separation between the eastern and western water masses of the central Pacific. In the South Pacific, Geynrikh (1973) records two cold-temperate calanoids, *Metridia lucens* and *Calanus tonsus*, penetrating as far as 23°S off the coast of Chile in the Peru Current.

Aside from the description of indicator species, it is important to recognize that temporary changes in major current flows, induced to some extent by seasonal or even year-to-year climatic variations, influence the disposition of water masses. The effects on water masses of the fluctuating meanderings of the Gulf Stream in the Atlantic and of the Kuroshio in the Pacific are obvious examples. From a biological point of view, such variations must have pronounced effects on the spread and successful breeding of the indigenous zooplankton populations, especially bearing in mind that the species (particularly closely related species) will be in strong competition. The relationships between the hydrological conditions and the spread and reproduction of the species are again, therefore, of paramount significance. An investigation by Gueredrat (1971) in the region already under review (the tropical Pacific) may illustrate such considerations.

Earlier studies indicated a decrease in abundance of zooplankton from east to west in the tropical Pacific. Thus Mais and Jow (1960) describe, from oblique plankton tows through the upper 300 m in the eastern tropical Pacific, higher zooplankton volumes (59–622 ml/1000 m^3) in certain areas (off central America and between Ecuador and the Galapagos) than generally occur in the region. They cite data from earlier cruises indicating that zooplankton is generally more abundant in the eastern tropical Pacific (mean values 51–165 ml/1000 m^3) than in the central Pacific (27 ml/1000 m^3). Numerically copepods were most abundant in their collections, with chaetognaths, tunicates, euphausiids, siphonophores, ostracods, amphipods and decapods in descending order of importance. Copepods were also dominant in volume. Gueredrat (1971), accepting this decrease from east to west, suggested that certain comparatively large copepods showed distinct meridional distributions along the equatorial Pacific. *Eucalanus subtenuis*, which is confined to the eastern tropical Pacific (*vide supra*), where it may dominate the copepod population, forming up to 45 % of the total, is associated with that region primarily in relation to the more abundant phytoplankton consequent on the stronger upwelling. *Euchaeta marina* is also comparatively plentiful in the eastern region, but reaches its maximum somewhat farther westwards than *Eucalanus subtenuis* and then declines in density towards the western Pacific. Its abundance, it is suggested, is associated with that of the herbivorous copepods which form its prey.

Another group of copepods (*Rhincalanus cornutus*, *Eucalanus attenuatus* and *Eucalanus subcrassus*) is more abundant in the western Pacific, possibly in relation to the higher temperature of the superficial layers and the deeper thermocline. Fleminger's (1973) suggested division of the *E. attenuatus* group also describes *E. attenuatus* itself as an Indo-Pacific tropical species most abundant downstream of eutrophic regions along the equator. Gueredrat groups these and other copepods in east–west distribution patterns, largely in relation to upwelling and surface currents, surface temperature and thermocline depth, phytoplankton crop and other factors. Presumably the successful spread and maintenance of stocks and breeding of the various species is largely controlled by the hydrological instability with upwelling and the intensity of flow of the cold Peru Current in the eastern Pacific and by some climatic variation, including wind strength, over the whole region.

Zooplankton of the Morocco Coast, South-West Africa and Mediterranean

The effects of changing water masses on the occurrence and abundance of zooplankton species is well exemplified by Furnestin's (1957, 1964) investigations off the coast of Morocco. Reference has already been made (Chapter 3) to the breeding of some species, particularly in relation to upwelling. Although the coast is generally open, a band of neritic water of moderate salinity exists as a superficial layer overlying the narrow continental shelf.

A group of zooplankton species is fairly typical of this neritic water. Certain hydromedusae, especially *Obelia*, with two Limnomedusae and in the south, *Phialidium hemisphaericum*, are found, with a few essentially littoral mysid species (except *Siriella thompsoni*, which occurs in offshore waters) which emphasize the neritic character of the plankton. A characteristic species is the chaetognath *Sagitta friderici*, which forms 90 % of the total chaetognath population (Fig. 8.6). Not only is the species relatively abundant, but the population is fairly stable, reproducing more or less throughout the year and the chaetognath appears to be an excellent indicator of coastal water. Siphonophores generally avoid inshore waters, but *Muggiaea atlantica* appears to be truly neritic, maintaining its position fairly close to the coast and only

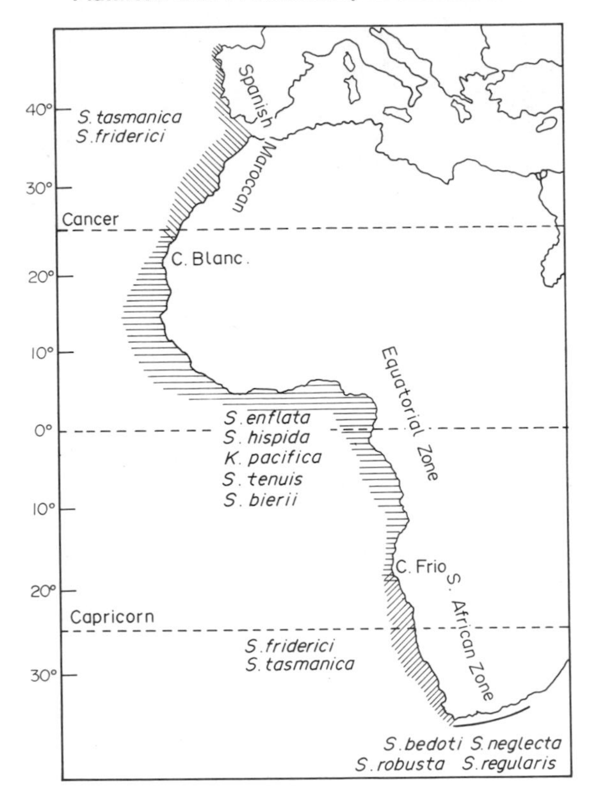

Fig. 8.6. Diagram of distribution of surface and sub-surface chaetognaths along the West coast of Africa (Furnestin, 1970).

the strength of upwelling. This species has already been noted as a coastal indicator off California (cf. page 752). Cladocerans (*Podon polyphemoides*, *Evadne nordmanni* and *Evadne spinifera*, with a few *Podon intermedius*) are also typical of the inshore waters, where they may occur at certain seasons in very large numbers. Only one species of appendicularian, *Oikopleura dioica*, occurs frequently in nearshore samples: it can become very abundant.

Although a species such as *Sagitta friderici* is itself a good indicator of coastal water off Morocco, it is the assemblage of species which characterizes the water mass. The species form an actively reproducing population, even though their density may change markedly with season.

It is characteristic of the area that deeper water, particularly from the continental slope, rises towards the surface. This upwelling is particularly strong at certain places along the coast (cf. Fig. 8.7) and is markedly seasonal, beginning in spring, reaching maximum intensity in summer and declining in autumn. Upwelling is, however, complicated by a movement of warm high salinity offshore surface water during summer which invades the coastal zone, particularly from the southwest (Furnestin, 1964). While the movements may be partly associated with upwelling, they are in part independent and of considerable amplitude, moving more superficial oceanic water to and from the coast. Thus, greater numbers of offshore oceanic species characteristi-

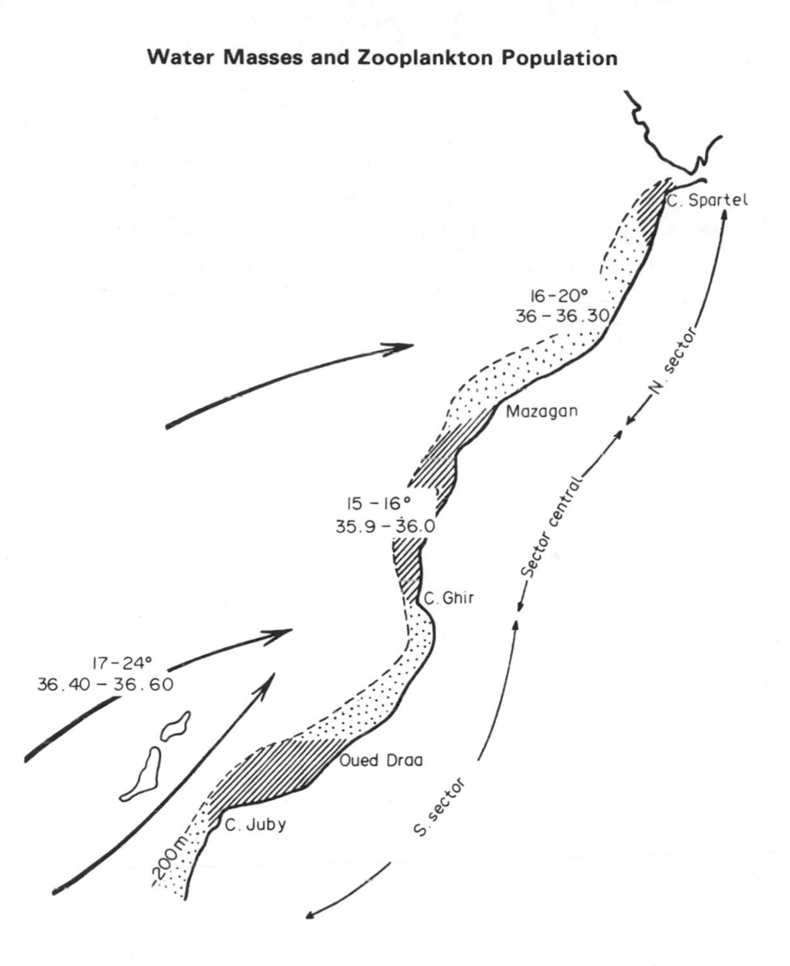

Fig. 8.7. Diagram of principal hydrological formations in the Spanish–Moroccan Bay with mean values for temperature and salinity. Dotted area = coastal water; hatched area = slope water; arrows indicate oceanic water (Furnestin, 1964).

cally appear in coastal areas mainly during summer when the offshore waters advance from the south-west. They include offshore warm-water hydromedusans such as *Rhopalonema velatum*, *Solmaris corona*, *Solmundella bitentaculata* and various siphonophores, a group (apart from *Muggiaea atlantica*) which is typical of the open high seas, including *Lensia conoidea*, *Chelophyes appendiculata*, *Bassia bassensis*, *Diphyes dispar*, *Abyla trigona*, *Eudoxoides spiralis*, *Abylopsis tetragona* and many others. The occurrence of several of these species, although not in high density, is a good indication of inflow of more ocean water.

Not all of the siphonophores are equally useful as indicators. According to Furnestin the best is *Chelophyes appendiculata*, which reflects very precisely the seasonal changes in the movements of the more oceanic offshore waters. Thus in spring *Chelophyes* appears in small numbers inside the 200-m line; by summer, they occur in considerable numbers well inshore, particularly to the south; during autumn they retire towards the edge of the continental shelf and are over deeper water for the winter. Warm-water thecosomatous pteropods, including *Limacina inflata*, *Creseis virgula*, *Styliola subula* and *Cavolinia*, also enter coastal areas during the inflow of the warmer offshore water. However, their movement is mainly towards the south of Morocco, where the influence of the inflow is stronger. According to Furnestin *Cavolinia inflexa* is an especially good indicator because it occurs in fairly high densities.

A few heteropods (*Atlanta* spp., *Carinaria lamarcki*, *Pterotrachea*) which are open ocean warm-water species can appear in very small numbers in southern areas of the Moroccan coast. In contrast to the domination of the appendicularians by *Oikopleura dioica* in the neritic waters, several other appendicularians, particularly *Oikopleura longicauda*, are typical of the inflowing offshore waters. Certain chaetognaths, *viz. Sagitta bipunctata* and *Sagitta serratodentata* (which are found in more open oceans) and *Sagitta tasmanica*, which is perhaps less oceanic in distribution, are also characteristic. Furnestin (1957) points out that although some juvenile *Sagitta bipunctata* are found off Morocco, the species subsequently seems to disappear, and no adults are found in summer. Presumably their virtual disappearance is due to the influx of colder upwelling water. The warm-water species *Sagitta enflata* is encountered in small numbers and although common in mixed warm waters, the individuals may represent part of a tropical or sub-tropical population.

Two Trachymedusae, *Liriope tetraphylla* and *Aglaura hemistoma*, which are widely distributed in warmer oceans, are claimed by Furnestin to be typical of the upwelling waters, presumably owing to the mixing of ocean water at the edge of the continental slope. Certain chaetognaths are also characteristic. Though *Sagitta hexaptera* and *S. lyra* are rather deeper-living, juveniles may spend some time in the more superficial waters. These two species appear in waters upwelling on to the continental shelf off Morocco and are accompanied to some extent by *S. minima* and *Pterosagitta draco*. Although not as deep-living as *S. lyra*, these species appear to be sub-surface and *S. minima* occurs especially in mixed waters.

Most characteristic of upwelled waters along the Morocco coast is the salp *Thalia democratica*, which may be present over summer in enormous numbers. Although inhabiting more superficial layers offshore, it appears to be carried down in the slope waters and onto the shelf with the upwelling. Furnestin emphasizes that although some species, particularly the chaetognaths, are useful as plankton indicators, it is frequently an assemblage of species in reasonable density which indicates a flow of water into the area.

Aside from these examples of specific investigations of water masses and indicators, observations during other types of study on zooplankton have frequently shown the effects of exchange of bodies of water on the zooplankton. Several references were made in Chapter 3 to approximately horizontal movements of water masses which led to marked quantitative and qualitative changes in the plankton population. Some of the observations dealt with plankton fairly near the coast where, in some instances, the horizontal exchange of water masses was associated with upwelling. Such hydrological movements are more or less regular, occurring at particular seasons, but others can be more long-term and irregular. The extension southwards of warm tropical waters along the coast of Peru, the well-known phenomenon known as El Niño, which brings a different plankton population along the coast, is seen only in certain years. Irregular hydrological movements, which may be short-term and are more particularly observable near coasts, may sometimes be correlated with climatic changes such as fluctuations in wind direction and strength.

An example of a change which could not readily be associated with climatic variation was observed by Stander and de Decker (1969) off south-west Africa (cf. Chapter 3). During one year (1963) an invasion of comparatively warm saline water from the north-west covered much of the northern and coastal regions of Walvis Bay. A comparison of the environmental conditions averaged over the previous ten years with those during 1963 showed that surface water temperature, instead of having a small seasonal decline from about May, rose to a comparatively high value which was maintained for one or two months. Near normal levels, as indicated by the results for the previous 10 years, were reached only by about October. Surface salinity was also raised from April to September. There were clear indications of the advance of surface warm saline water from the north and west beginning about May, with a subsequent retreat commencing in July. The advance of the atypical water mass was accompanied by a lowering of oxygen values.

With the unusual water flow there were marked changes in the zooplankton. Swarms of copepods are normally found in the inner part of Walvis Bay. These were drastically reduced, the populations of *Paracalanus parvus* being especially badly affected. A coastal population of neritic copepods (*Paracalanus crassirostris*, *Paracartia africana*) and cladocerans (*Podon polyphemoides*, *Evadne nordmanni*) also underwent drastic reduction in density as the warm inflow pushed along the coast. The zooplankton fauna more typical of the deeper waters in the region was scarce and less diversified during the incursion of warm water. Apart from these zooplankton forms three copepods (*Calanoides carinatus*, *Metridia lucens* and *Centropages brachiatus*) which are dominant in the Bay and especially common during upwelling periods were very greatly reduced. By contrast, a copepod which greatly increased in density with the extension of the warm saline waters into Walvis Bay was *Nannocalanus minor*: it was accompanied by *Centropages chierchiae*, *Centropages bradyi*, *Calanus tenuicornis* and *Acartia danae*. (Some further details of changes in zooplankton are included in Chapter 3.)

Casanova (1977), analysing the zooplankton characteristic of different areas of the Atlanto-Mediterranean, also points to the significance of the inflow of superficial Atlantic waters into the Mediterranean and the outflow of deeper waters through the Straits of Gibraltar, though there is restriction due to the limited depth at Gibraltar. Although a regular seasonal change in inflow is not apparent, variation in the easterly passage of superficial water will strongly affect the zooplankton transported. Many species, including some deeper-living medusae, siphonophores, euphausiids, decapods and copepods present in nearby Atlantic waters do not appear in the Mediterranean. While this may be in part due to the comparatively shallow depth of the Straits, Casanova believes that the relatively high salinity and temperature typical of Mediterranean waters precludes the establishment of flourishing populations of many would-be immigrants. For example, *Periphylla periphylla* is the only species among four Atlantic bathypelagic medusae to penetrate the Mediterranean and although *Colobonema sericeum* can cross the sill, it fails to establish itself.

Of siphonophores, the common *Chuniphyes multidentata* occasionally penetrates eastwards, but apparently dies out rapidly; *Dinophyes arctica*, despite its remarkable eurythermal characteristics, is excluded (cf. Bigelow and Sears, 1937); *Nectopyramis* spp. are rare (Fig. 8.8). Some of the more superficial siphonophores, however, flourish in Mediterranean waters (e.g. *Diphyes dispar* appears to follow the influence of the Atlantic current) (Fig. 8.9). Of the species of *Gennadas* in the Atlantic, only *G. elegans*

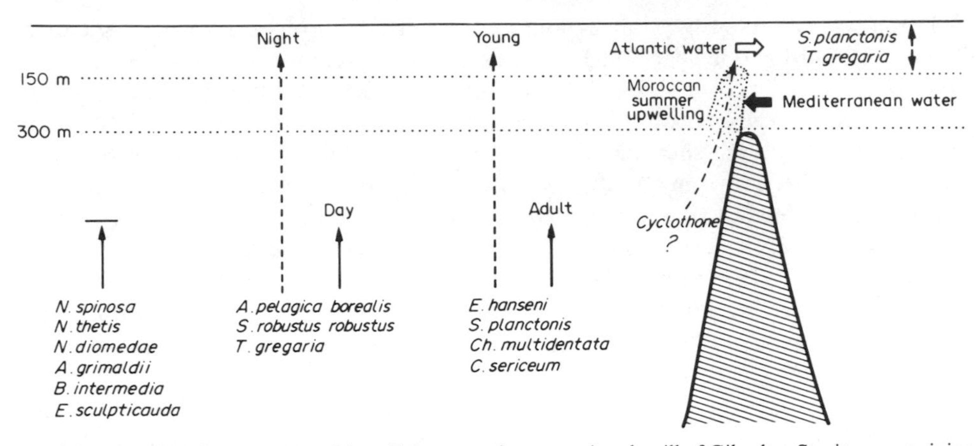

Fig. 8.8. Diagram of movements of deep-living organisms crossing the sill of Gibraltar Strait, or remaining in the Atlantic (Casanova, 1977).

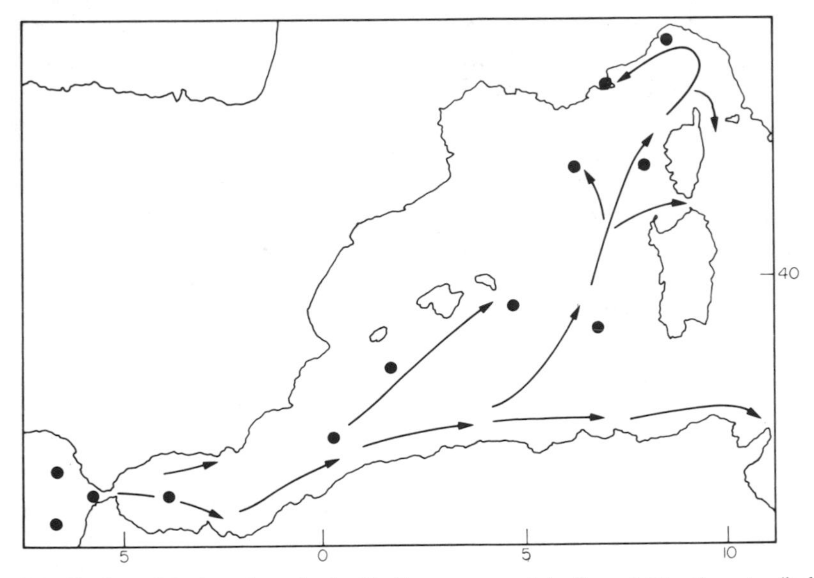

Fig. 8.9. Distribution of *Diphyes dispar* in the Mediterranean and the flow of Atlantic water (indicated by arrows). Various authors (Casanova, 1966).

occurs in the Mediterranean. *Acanthephyra pelagica* occurs but not *A. purpurea*, though Casanova suggests that *A. pelagica* is represented by a sub-species and that its presence is not due to the Atlantic inflow. A number of mainly shallow-living species are, of course, transported, and while some are only temporary immigrants not competing successfully with the local fauna, others flourish and indeed tend to contribute to substantially more dense populations with the mixture of waters.

The outflow of the deeper Mediterranean water is marked by two species, *Cymbulia peroni* and the Mediterranean (small) variety of *Periphylla periphylla*. Other species from the Mediterranean, such as endemic decapods, presumably do not find conditions favourable on being transported to the Atlantic. However, a great diversity of plankton find the particular conditions of the Lusitanian outflow (i.e.

Mediterranean water flowing at depth in the Atlantic) especially favourable. This applies more particularly to some tropical/sub-tropical species, which are probably adapted to high salinity. This rich fauna, therefore, can appear even as far north as off Scottish coasts, as described by Fraser (cf. page 743).

Much of the discussion of water masses and indicator species in the Mediterranean, however, has been concentrated on the spread of the superficial Atlantic water. Gaudy (1962), describing zooplankton off Marseilles, lists no fewer than eleven species typical of different areas of the Atlantic (tropical/sub-tropical, temperate, neritic) which could serve as indicators of Atlantic inflows at different seasons. The strength of flow of superficial water is all-important. Bernard (1967) states that the inflow into the Alboran Sea not only brings an abundance of Atlantic species but also a comparatively rich zooplankton. *Paracalanus parvus*, the most abundant species, is superseded by *Clausocalanus arcuicornis* over winter. In the Gulf of Lyons the shallow and lower salinity waters permit the flourishing of neritic species, e.g. *Sagitta setosa* and *Penilia*, but the more saline offshore zone is characterized by *Sagitta bipunctata*, *Sagitta serratodentata* and *Aglaura hemistoma* (i.e. warm open water species).

The fauna of the Catalonian Sea is mainly of Atlantic origin in the eastern part of the area, but is poor in species elsewhere, with *Centropages typicus* dominant. Off Corsica in Calvi Bay *Clausocalanus arcuicornis*, normally regarded as an oceanic species, was the most important copepod, with peak populations in March, April and May, while the more neritic *Temora stylifera* appeared for only a short period in autumn. Dauby (1980) states that productivity in general is low and more typical of oceanic areas, despite the neritic situation; he considers this is due largely to lack of land run-off. In the Northern Tyrrhenian Sea *Clausocalanus arcuicornis*, *Temora stylifera*, *Centropages typicus* and *Acartia clausi* are the dominant copepods, but different communities can be identified in respect of the copepod fauna. The inflow of Atlantic water off Algiers, which may extend as far as Tunisia, leads to fairly regular seasonal changes in the zooplankton population. Aside from regular fluctuations, Bernard (1970) observed an unusual flow of colder surface waters in the Bay of Algiers during October/November 1956. The inflow permitted a marked production of the copepods *Acartia* and *Oithona* in place of the usual zooplankton fauna, which normally at that season is characterized by an abundance of *Temora stylifera*.

Zooplankton of Other Areas

More irregular changes can be significant. Legare and Zoppi's (1961) observations of changes in the zooplankton in the region of the Cariaco Gulf were shown to be largely dependent on the direction and strength of local winds causing an inflow or outflow of surface waters to and from the Gulf (cf. Chapter 3). The local water exchanges not only influenced the density of the zooplankton but also its qualitative composition, as Rao and Urosa (1974) demonstrated.

On the eastern coast of Venezuela the strength of the Amazon outflow, which is dependent on seasonal climatic conditions, affects the composition and richness of the zooplankton along the coast and even to some distance in the open ocean. Many other instances are given of hydrological variations causing changes in the plankton, even though the factors inducing the water movements are not known. Tundisi (1970) refers to variations from year to year in the zooplankton off Santos. On the Great Barrier

Reef occasional incursions of ocean waters lead to differences in the zooplankton, for instance, the introduction of oceanic copepods such as *Acartia negligens* and *Acartia australis*. Dessier and Laurec (1978) analysed the seasonal changes in various zooplankton taxa in the Pointe-Noire (west African coast) area and confirm that hydrological changes are significant in influencing the presence or absence of species. Examples are *Paracalanus aculeatus*, *Lucifer* sp. and *Centropages furcatus* which have peak abundance in the warm season (surface temperature 25°–28°C), *Paracalanus scotti* and *Oithona nana*, having high densities during the main cold season when the lowest surface temperature is 20°C. *Temora turbinata* is common around January—the transition period between the minor cold season and the main hot season.

The alternation of wet and dry seasons, particularly in certain tropical regions of the world may lead to marked hydrological changes close to shore, resulting in enrichment of surface waters without obvious upwelling. Such enrichment not only causes changes in the density of the zooplankton but leads to considerable change in the faunistic composition. At Madagascar, Bour and Frontier (1974) studied seasonal changes in the zooplankton close inshore at Nosy-Bé and towards the edge of the continental shelf. They associated the changes particularly with the alternation of the wet and dry seasons (cf. also Frontier 1973, 1974). Inshore, from January to April, they found the waters markedly stratified, with the surface layer sharply reduced in salinity by the heavy rainfall. The waters became homogeneous during the succeeding months (the dry season), but a sudden run-off in December, consequent on the heavy rains, caused a very sudden blooming of phytoplankton and after a very short delay, a burst of herbivorous zooplankton followed.

As shown in Fig. 8.10 there was a sharp fall in total zooplankton and in copepods with the beginning of the rains until mid-December and a subsequent rapid rise in

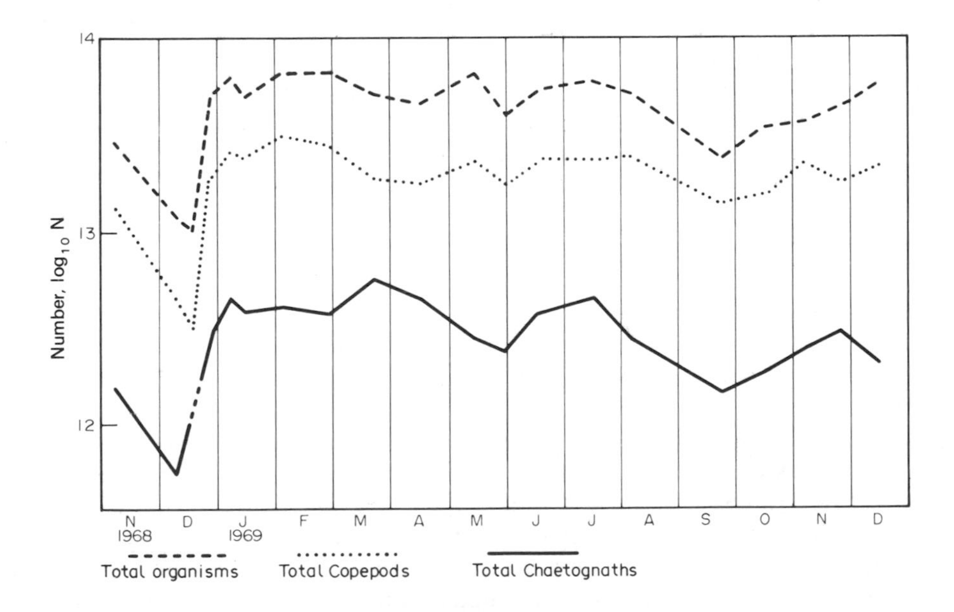

Fig. 8.10. Annual cycle in the Bay of Ambaro: variations in total number of zooplankton (hatched line); copepods (dotted line); chaetognaths (unbroken line) (Bour and Frontier, 1974).

population. Of the herbivorous plankton which was closely linked in time of abundance to the phytoplankton increase, cladocerans (chiefly *Penilia* and *Evadne tergestina*) were most important and towards the outer continental shelf region the euphausiid *Pseudeuphausia latifrons* followed the same cycle. Thecosomatous pteropods (*Creseis acicula*, *Creseis chierchiae*, *Atlanta gaudichaudi*) appeared in considerable densities near the coast at the end of the dry season and *Atlanta gaudichaudi* juveniles reached a maximum in the wet season in the more inshore areas with the abundance of other herbivores.

A detailed study of the chaetognaths showed that the total chaetognath population followed the overall changes, with a fall in density in December and a later increase. The chief species, *Sagitta enflata*, was distributed both inshore and across the shelf, though populations were richer inshore. Indeed, the inshore zone was generally rich in carnivorous zooplankton, which from about May tended to spread across the whole shelf. The most typical neritic chaetognath, *Sagitta neglecta*, though not so abundant as *S. enflata*, followed the same seasonal pattern but was mostly coastal and reached its maximum density close inshore in August. More typically, oceanic chaetognaths could increase in abundance over the shelf during the dry season, but disappeared with the rains (e.g. *Sagitta bedfordi*).

Local changes in climate in other tropical areas can greatly influence the zooplankton and obscure more regular patterns by causing local flows of surface waters. Panikkar and Rao (1973) point out that over much of the coast of India, although seasonal effects on the zooplankton due to the monsoon climate can be clearly recognized, marked changes in density and composition are often associated with brief local climatic changes and exchanges of local water masses.

Jillett's (1971) observations on the changes in the sub-tropical zooplankton in Hauraki Gulf, New Zealand, show the effects of inflows of more oceanic waters, leading to variations in the zooplankton of more inshore waters which may overshadow the more regular seasonal pattern.

At higher latitudes irregular and short-term variations in the movements of water masses with their effects on the zooplankton are not easily recognized. The generally strong seasonal effect at higher latitudes tends to dominate the picture. In some very shallow inshore waters, however, more local climatic effects, particularly in estuarine flows, may sometimes be observed and in certain seas variations in the exchange of water masses, sometimes on a very long-term basis, can be recognized. Eriksson (1973a) refers to hydrological changes in the shallow waters off the west coast of Sweden. The considerable variation in the zooplankton apart from the major seasonal change is probably partly a reflection of water exchanges. Variations on a very long-term basis in the inflow of Atlantic and mixed waters into the English Channel and North Sea are also in part associated with plankton changes as described in earlier work on plankton indicators.

The inflow of deeper saline water into the Baltic is another example of a moderately long-term though irregular event strongly affected by the superficial run-off from the Baltic, which is in turn largely influenced by climatic variations. Radziejewska, Chojnacki and Maslowski (1973) observed during 1972 a number of zooplankton species in the southern Baltic not previously recorded from the region. This incursion of unusual species was accompanied by an influx of higher salinity water. Thirteen species from the North Sea were identified in the inflow, eight being first records. These included *Tomopteris kefersteini* and *Aetideus armatus*, normally oceanic residents in the North Atlantic, *Rosacea plicata* and *Eucalanus elongatus*, probably of "Lusitanian" origin, and *Centropages typicus* and *Metridia lucens* of mixed oceanic–neritic character. The species could be regarded as satisfactory indicators of North Sea water. The study illustrates the well-known principle that while a species occurring widely in a region cannot be used as an indicator (e.g. *Metridia lucens* or *Centropages typicus* over much of the North Atlantic), it may be acceptable as such in an adjacent area.

Chaetognaths have been particularly useful in characterizing water masses in several parts of the world. Apart from the earlier discussion of the classic work around British coasts, in the North Pacific and off Morocco (*vide infra*), several other studies on plankton indicators have centred on chaetognaths. Murakami (1959) studied the chaetognaths of the Seto Sea, Japan, where fifteen species representing almost all the fauna in the local seas were identified. *Sagitta crassa*, *S. enflata*, *S. bedoti* and *S. minima* were the dominant species. Three regions were distinguished:

(1) An outer region with the fauna richest in abundance and variety but with oceanic chaetognath species only.
(2) An inner region, poorest in variety and quantity, with *S. crassa* and *S. enflata* only.
(3) An intermediate region, with *S. minima* and *S. bedoti*, described as "sub-oceanic" presumably mixed slope species, in addition to *S. crassa* and *S. enflata*.

About half of the oceanic species spawned in the Seto Sea. Murakami concluded that some of the chaetognath species could be successfully used as indicator species. In the Strait leading into the Sea *Sagitta bipunctata* as an oceanic form was typical of the highly saline ocean inflowing along the bottom; *S. minima* occurred in water of somewhat lower salinity including water of neritic origin overlying the bottom water; *S. crassa* was characteristic of superficial low salinity water diluted by river drainage from inner parts of the Seto Sea. As other workers have suggested, *S. enflata* is slightly neritic but certainly cannot flourish in dilute waters. The changing abundance of *S. enflata* in the Seto Sea was thus found to indicate the seasonal influence of the inflowing saline Kuroshio Current. Typical species of chaetognaths for the Kuroshio Current apart from *Sagitta enflata* include *S. hexaptera*, *S. serratodentata pacifica* and *Pterosagitta draco* according to Motoda and Marumo (1963). The two forms of *S. enflata* found in other areas were identified in the Seto Sea, the smaller race being typical of less oceanic water. In the brackish waters of the inlets where *S. crassa* flourished, Murakami suggests that since the species exists in several morphologically distinct forms, these may be used to distinguish different brackish water masses.

Marumo and Nagasawa (1973) also refer to the influence of the Kuroshio Current on chaetognath populations in Japanese waters of Sugami and Suruga Bays. During summer, with the influx of Kuroshio water, the chaetognath populations, dominated by *Sagitta nagae*, become denser but the inflow also transports the oceanic *Sagitta ferox* into the Bays.

Species of chaetognaths have also been employed to characterize water masses off the south-east Atlantic coast of America. Neritic species along part of the Florida coast include *Sagitta hispida* and *S. helenae*. Pierce (1953) found that *S. helenae* was practically restricted to the shelf waters off North Carolina and extended south along the Florida coast and northwards as far as Cape Hatteras (*ca.* 35°N). *S. tenuis*, another neritic species, was common inshore but was rarely found beyond the continental shelf. *S. enflata* was widely distributed over the shelf but was also abundant offshore. Along the edge of the shelf *S. minima* was plentiful, but was not found over the inner shelf region and it was uncommon beyond the continental shelf. It thus appears to be typical of slope waters, which might be regarded as largely "mixed" (oceanic and coastal) water. Offshore, in typical warm ocean waters of the Florida Current, Pierce found *S. serratodentata* very plentifully represented, though it was rarely taken over the shelf. *S. bipunctata* also was typically present in oceanic waters but was very scarce even in waters over the outer region of the shelf. Two other chaetognaths, *Pterosagitta draco* and *Krohnitta pacifica*, occurred most plentifully in Florida Current waters, though they were found more widely, extending into continental shelf waters, as was even more characteristic of *Sagitta enflata*. Pierce suggests that the distribution of the chaetognaths in the area would appear to be typical of three zones: inshore shelf, slope and ocean, though these do not have precise limits, and their extent varies from time to time with variations in the flow of the Florida Current, irregularities of flow and changing strength of the movement of water offshore.

Farther north, off the New England coast, a similar population of chaetognath species may be found offshore in the main Gulf Stream, but their numbers decrease rapidly in slope waters; over the edge of the continental shelf increasing populations of *Sagitta elegans* are encountered. Inside the shelf, especially in the Gulf of Maine, *S. elegans* is the characteristic and endemic species (cf. Bigelow, 1926). A portion of the population may be transported outside the Gulf, but the density declines sharply towards the edge of the continental shelf. The warm waters of the Gulf Stream are an impassable barrier. Bigelow (1926) contrasts the distribution of *Sagitta elegans* with that of *Sagitta serratodentata*, the species characteristic of the warm tropical Gulf Stream waters, which is carried into the Gulf of Maine during the summer but never forms a reproducing population.

Redfield and Beale (1940) also examined the distribution of *Sagitta elegans* in the Gulf of Maine. As well as reproducing inside the Gulf (cf. Chapter 3), where at times some of the population may be swept outside and lost, a separate population is maintained by a subsidiary anticyclonic eddy over Georges Bank and large populations can occur there over much of the year.

By contrast, *Sagitta serratodentata* is an immigrant species which flows in with the surface waters mainly via the Eastern Channel approach to the Gulf of Maine. As previously noted for other parts of the Atlantic, *Sagitta serratodentata*, though a warmer water chaetognath, can survive quite extensive cooling. It comes in as a surface immigrant in summer only, since water of a rather warmer character seems to enter at this time of the year. The appearances of *Sagitta serratodentata* are therefore highly seasonal and it does not reproduce at all in the colder conditions of the Gulf of Maine. It may be classed as a "terminal immigrant". This chaetognath occurs rather more commonly near the area of main inflow into the Gulf, though it does spread out to some extent more widely before dying off.

Three other chaetognaths, *Eukrohnia hamata*, *Sagitta maxima* and *Sagitta lyra*, are found in the Gulf waters but these all enter via the deeper inflow, whereas *Sagitta serratodentata* is brought in mainly in the superficial water layers. *Eukrohnia* is reasonably abundant but *Sagitta maxima* and *S. lyra* are both very scarce indeed. All three species appear to be terminal immigrants in that, although they come in with the inflowing water, there is no evidence of any breeding in the Gulf. They persist for a time, but in the case of *S. lyra* and *S. maxima* Redfield and Beale (1940) have shown fairly clearly that their numbers diminish rapidly the farther one goes from their main point of entry. On the other hand, *Eukrohnia hamata* appears in more or less the same density in the deeper layers all over the Gulf of Maine area. Far greater numbers of *Eukrohnia* are undoubtedly brought in from outside waters, but this chaetognath also appears to persist longer. It probably withstands the lowered temperatures of the Gulf waters better and the individuals have a longer life expectancy.

The loss of a portion of the *Sagitta elegans* population by drift southwards to the open ocean from the Gulf of Maine emphasizes the significance of some mechanism which either maintains a breeding population of zooplankton within the area to which it is specifically adapted or which returns a proportion of breeding stock to the area. There are numerous examples of the regular transport of a species in considerable numbers to a region where, with the slow change in environmental conditions, breeding is no longer successful, though a reasonably sized population is present.

Problems of transport, immigration, maintenance of stock and breeding are well exemplified by other studies in the Gulf of Maine area. Redfield (1939) demonstrated that the pteropod *Limacina retroversa* enters the Gulf of Maine, where it is widely transported by the prevailing (mainly anti-clockwise) circulation. Immigration appears to be mainly responsible for the *Limacina retroversa* populations in the Gulf.

Redfield (1941) has also shown that the general great anti-clockwise eddy which is typical of the water movements of the Gulf of Maine, with a time period of approximately three months, is largely responsible for the changes in distribution and abundance of the calanoid population of the area (cf. Fig. 8.11). He shows that, in the northern parts of the Gulf of Maine off the Mount Desert area the *Calanus* community of the upper water layers has its maximum abundance in the autumn period. But as the winter approaches there is a shift in maximal population south to the coast of Massachusetts, and then in the spring and early summer months, the maximum area of abundance of calanoids changes to just north of Georges Bank, east of Cape Cod. The late summer and early autumn populations reach their peak again in the more

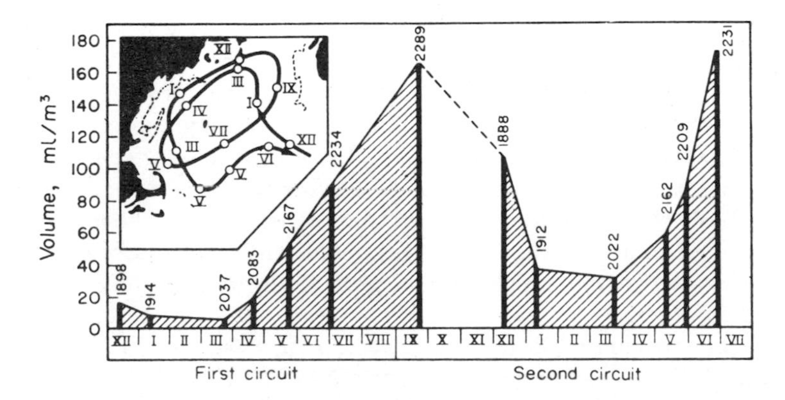

Fig. 8.11. The circulation of the calanoid community in the Gulf of Maine. The growth of the population in a mass of water assumed to move along a course indicated in inset. Zooplankton as volumes ml/m^3. Black bars indicate the volumes caught at selected stations (Redfield, 1941).

northern areas of the Gulf of Maine to give the distribution seen in the previous year. Redfield interprets these results on the grounds that in the late spring and summer the inflow through the eastern channel of waters off the Nova Scotia area into the Gulf of Maine is lessened and a fairly rich zooplankton is established in the northern areas of the Gulf.

Later in the year, with the general anti-clockwise circulation, this zooplankton tends to be carried south towards the Massachusetts coast and this movement is reinforced by new water flowing in round the coast of Nova Scotia. This main inflow is maximal over the winter period, but it tends to carry a poor zooplankton population, so that over the winter only low densities of zooplankton are present in the northern sectors of the Gulf. The general flow of the older water, which has already been for some time circulating in the Gulf of Maine and which has continued to support a fair zooplankton population, is towards the southern regions of the Gulf near the Georges Bank area. With the strong flow of new water into the Gulf during winter a considerable loss of plankton-rich water must occur and this is expelled mainly off Georges Bank and is passed to the waters of the Atlantic. In the late spring and summer, however, with the considerable lessening of inflow into the Gulf of Maine, a part of this water circulating as a great eddy can turn north again instead of being driven out of the Gulf. This water, rich in zooplankton, thus adds to the relatively dense populations of zooplankton which accumulate about September in the more northerly regions (cf. Fig. 8.11). To a small extent there may also be some reinforcement at this time by some zooplankton drifting from the Nova Scotia banks, but the general overall circulation is mainly responsible for maintaining a rich calanoid population in the Gulf, with the sharpest seasonal effects occurring in the more northern areas.

Antarctic Zooplankton

On a larger scale, the problem of the transport and possible return of part of a zooplankton stock in the North Pacific has already been discussed. A particularly clear and large scale transport of zooplankton has been described by Mackintosh (1937) for Antarctic waters. Changes in vertical distribution of certain common Antarctic species (*Calanoides acutus*, *Eukrohnia hamata* and *Rhincalanus gigas*) have been described with particular reference to breeding cycles (cf. Chapter 3).

During the Antarctic summer large populations are established in the upper layers (to *ca.* 200 m), and the superficial waters tend to drift northwards towards the Antarctic Convergence, where much of the stock might be carried farther north in the Antarctic Intermediate Water and would be lost. Mackintosh demonstrated that with the onset of winter the plankton migrated vertically into the deeper layers. For example, late in the Antarctic winter (September), although the total quantity of zooplankton was greatly reduced, it was also spread out in depth to a great extent, with the majority of the animals between 500 and 750 m and fewest in the surface layers (Fig. 8.12). Both *Rhincalanus* and *Eukrohnia* showed a maximum in winter at about 500 m, with a reasonable concentration below 1000 m, but *Calanus acutus* had an even more remarkable distribution, being completely absent from the surface during the winter period and being most abundant from 750 m to more than 1000 m. Below 200 m or so, all this plankton would be living in the rather warmer waters which are drifting southwards towards the Antarctic continent.

Mackintosh observed that in early spring (October) the plankton was generally

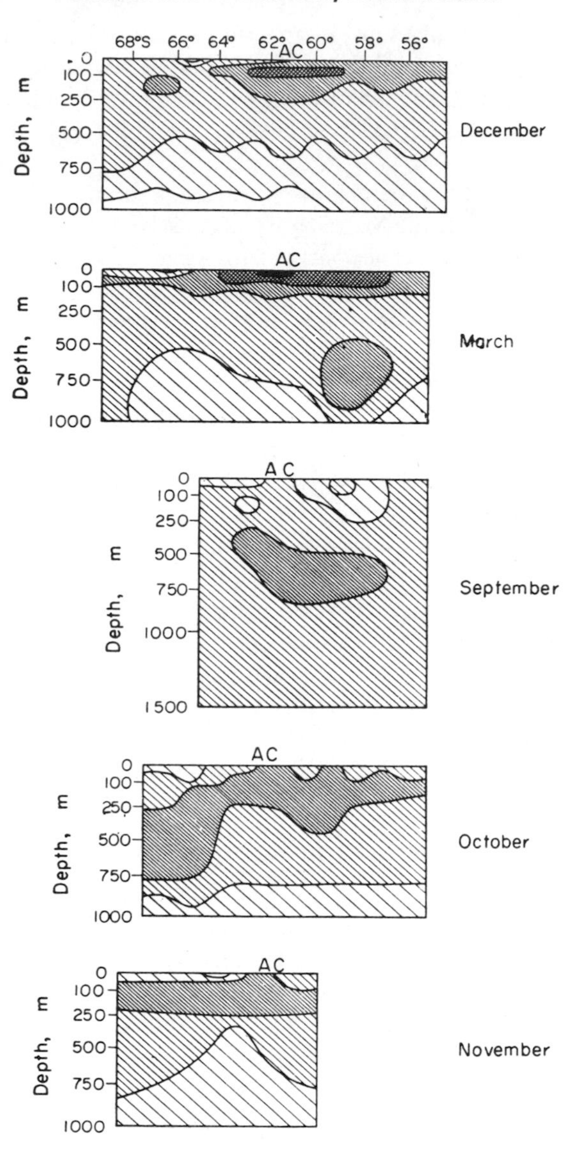

Fig. 8.12. Changes in the vertical distribution of the zooplankton throughout the year in the Antarctic. Heaviest shading denotes highest density (Mackintosh, 1937).

higher in the water, mostly lying at a depth between 100 and 300 m. With the advance of spring there was a progressive rise, though the surface was still poor in numbers (Fig. 8.12). By December most of the zooplankton was near the surface and during the following two or three months the great burst of reproduction of the zooplankton occurred in the surface waters (Fig. 8.12). Thus from about December to March the zooplankton was mainly in the surface waters drifting northwards, whereas over the winter the zooplankton though reduced in numbers, was chiefly in the deeper-water layers drifting south towards the Antarctic again. This is indeed a very large scale circulation, amounting in depth to at least 500 m and horizontally to some hundreds of miles.

The phenomenon of diurnal vertical migration is considered in some detail in Chapter 5, but it may be said here that the three species which dominated the zooplankton in Mackintosh's catches showed no obvious daily vertical migrations. On the other hand, certain zooplankton species which appeared in Mackintosh's collections in smaller numbers did not seem to show the marked changes in seasonal depth distribution which would have returned them to the Antarctic continent as we have postulated. Such species included the copepod *Pleuromamma robusta* and euphausiids such as *Euphausia triacantha* and *E. frigida*. However, other studies on these three zooplankton species have shown that they and many others display a pronounced diurnal vertical migration. *Pleuromamma robusta*, for example, lives in comparatively deep water at a depth of about 500–700 m by day, but it performs a diurnal vertical migration on an enormous scale, passing at night to the upper 50–100 m (cf. Hardy and Gunther, 1935). The euphausiids also migrate daily through at least 200 m. It appears, likely, therefore, that these species, instead of compensating for surface drift by a seasonal change in depth, are alternately drifting a little north in the surface layers during the night-time, and drifting southwards in the warmer deeper layers during their day-time sojourn.

Not only this circulation pattern but the general westerly drift of Antarctic waters under the influence of the east Antarctic Current (East Wind Drift) and the easterly transport due to the West Wind Drift on either side of the Antarctic Divergence at high latitudes (*ca.* 68°S) are undoubtedly partly responsible for the delimitation of the zooplankton species, most of which tend to be circumpolar in distribution (cf. Foxton, 1956). The complex circulation pattern in the Weddell Sea is also responsible for maintaining a zooplankton community mainly within that very large region. Zooplankton communities in the Antarctic have been described by Mackintosh (1934), and the limits of some species are indicated in Fig. 8.13. The different ranges of several southern species belonging to different taxa (euphausiids, copepods, chaetognaths) are described by Mackintosh (1964).

As Fig. 8.14 indicates, it is important to distinguish between the extreme latitudinal limit of a species and the range within which the species normally maintains an active breeding population. The fairly clear pattern of major current flows in the Southern Ocean delimits water masses within which a species is able to compete with related species as a reproducing stock. Examples of some of these distribution patterns include Brodsky's (1962) description of the well defined distribution of two circumpolar Antarctic copepods, *Calanus propinquus* and *Calanoides acutus*. For both, the limits are from the Antarctic Convergence to the continental coast. *Calanus simillimus* has a greater range; from the Antarctic Divergence to 45° or 50°S or even to the Sub-Tropical Convergence (*ca.* 40°S).

Calanus australis and *Calanus tonsus* occupy sub-Antarctic waters at approximately 50° or 55°S to 35° or 40°S latitude. Another example relates to the chaetognaths of the Southern Ocean. David (1959, 1962) points to the Sub-Tropical Convergence as separating the sub-tropical species from those of higher latitudes. He suggests that the Southern Ocean fauna has seven exotic species which penetrate the colder waters to some extent but which are normally found in sub-tropical waters. Of these, only two regularly appear in the sub-Antarctic waters (*Sagitta serratodentata* and *S. decipiens*) and they never penetrate the Antarctic. Four chaetognaths are shared with other

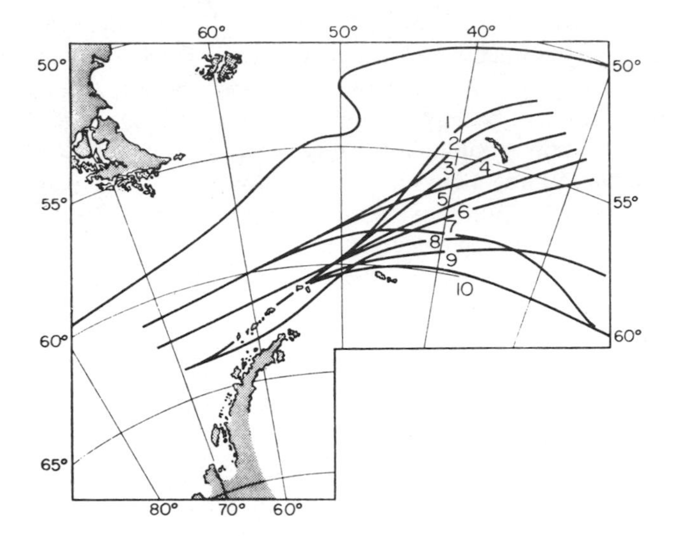

Fig. 8.13. Distributional limits for certain species of Antarctic macroplankton.

1. Northern limit of *Diphyes antarctica* and normal northern limit of *Euphausia superba*.
2. Northern limit of *Sibogita borchgrevinki*.
3. Northern limit of pelagic *Eusirus antarcticus*.
4. Southern limit of *Euphausia vallentini*.
5. Northern limit of *Auricularia antarctica*.
6. Southern limit of *Euphausia triacantha*.
7. Southern limit of *Candacia* sp.
8. Southern limit of *Limacina balea*.
9. Southern limit of *Calanus simillimus*.
10. Southern limit of *Pleuromamma robusta* and northern limit of *Haloptilus ocellatus*.

(Mackintosh, 1934).

regions: *S. macrocephala*, a deep-living form; *Heterokrohnia mirabilis*, also a deep-living but rare species; *S. maxima*, found at higher northern latitudes at intermediate depths; and *Eukrohnia hamata*, a mainly surface form at high latitudes and truly cosmopolitan. Only three chaetognaths are peculiar to the Southern Ocean. *S. gazellae*, a circumpolar form, is found near the surface, extending from the Sub-Tropical Convergence to the continental slope, but mainly in the sub-Antarctic. At intermediate depths *S. marri* is also circumpolar, with its maximum abundance in the Antarctic. *Eukrohnia bathyantarctica* is a deep-living species (cf. Fig. 8.14b).

The Southern Ocean zooplankton yields some examples of apparently widely distributed species which, however, show distinct geographical ranges associated with different races. David (1955) described two races of *Sagitta gazellae* north and south of the Antarctic Convergence. *Eukrohnia hamata* may also be polytypic (cf. Brodsky, 1965). Elsewhere in the oceans the existence of subspecies or races of a so-called cosmopolitan species has been noticed, especially from the investigations of Brodsky, Shavalov, Fleminger and others (cf. pages 731, 754). While it might be unwise to state that cosmopolitan species do not exist, they are few and further work may elucidate subspecific populations in many wide-ranging forms. Investigations such as those of Fleminger (1973) on *Eucalanus* illustrate how work may reveal previously unrecognized differences in populations which are genetically heterogeneous.

Species Groupings and Plankton Recorder Data

It is nevertheless true that widespread species may appear to separate into more or

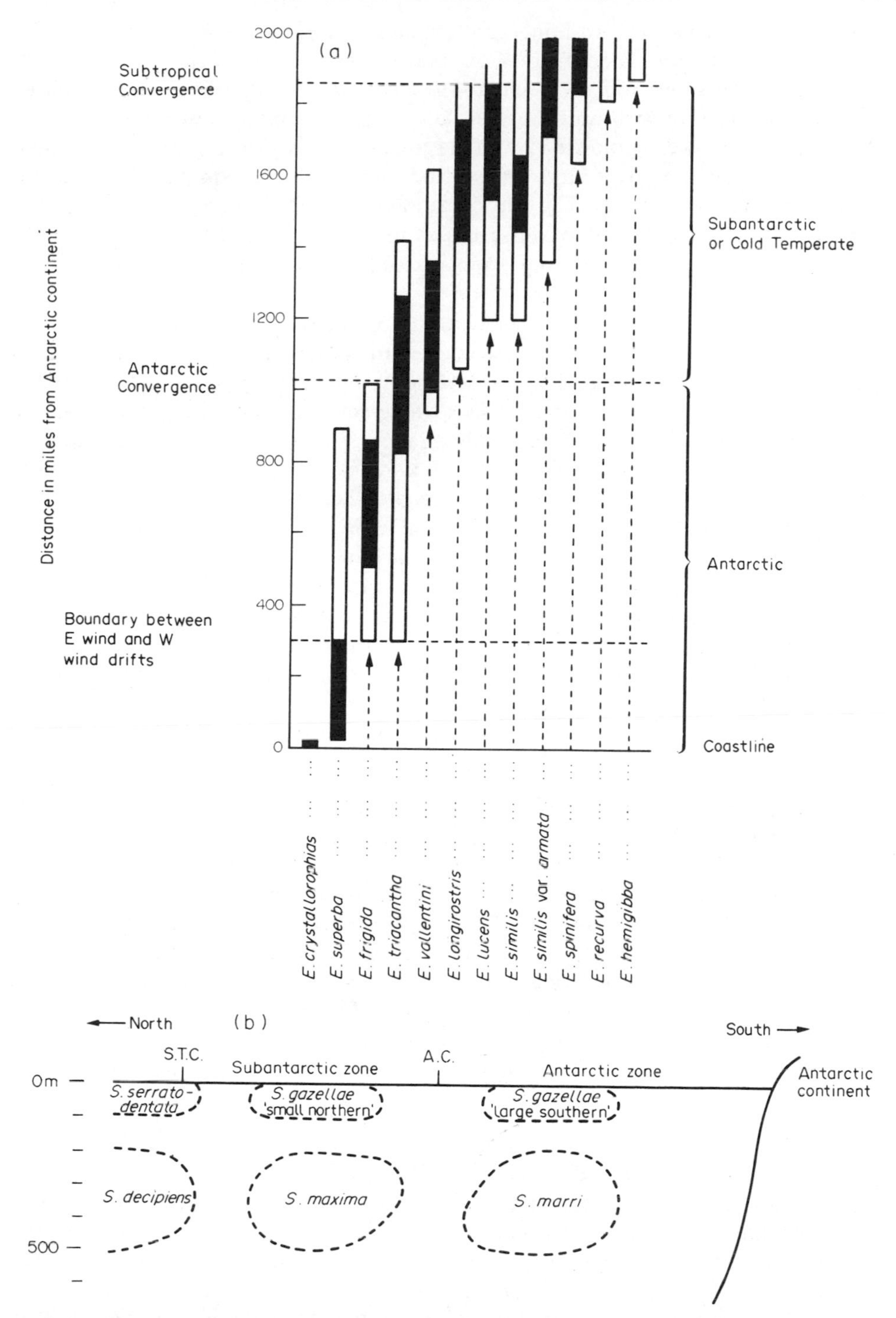

Fig. 8.14. (a) Succession of *Euphausia* spp. from the Antarctic Continent northwards; each species occupies a circumpolar zone, black parts indicate belt of concentration within outlined range. (b) Diagram to show relative positions of centres of concentration in summer for southern species of *Sagitta* (Mackintosh, 1964).

less distinct populations without any apparent morphological distinction. The extensive data accumulated from Continuous Plankton Recorder investigations furnish some examples and also relate to the complex problem of species boundaries and their association with water masses. Glover (1967) emphasizes the value of the Recorder type of investigation in view of the great practical difficulty of the continuous zooplankton sampling over large areas of ocean spread over long periods which is essential to this type of study. The time of sampling, and the large areas which must be sampled in relation to the drift of water masses and their populations, in addition to the problem of selection in capturing zooplankton, can introduce many artefacts in analysis.

The true variability of plankton populations in space and time is difficult to assess. Glover describes the methods employed in examining Recorder data accumulated over many years and across extensive ocean traverses. Using rectangles of 1° latitude and 2° longitude as a standard base for analysis, an assessment of plankton populations can be obtained giving information on the abundance of different species, data which are superior to presence/absence records. The data reflect species balance between major plankton constituents rather than emphasizing the occurrence of those species which may be regarded as "rarities". Despite seasonal variations in the abundance of species and fluctuations from year to year, examination of the distribution atlases for the plankton species suggests that the organisms can be grouped according to their similarity or dissimilarity of distribution (cf. Fig. 8.15). Major groupings indicate an oceanic and a neritic assemblage of species, with an intermediate group. The oceanic species may also be arranged according to their southern (warmer) or more northern (colder) distribution. Thus for the copepods, which comprise a large proportion of the Recorder samples, the following distributions may be suggested:

(1) *Oceanic group*, from south to north: species such as *Centropages bradyi*, *Sapphirina*, *Calanus gracilis*, *Euchaeta acuta*, *Calanus minor*, four species of *Pleuromamma*, *Rhincalanus*, *Pleuromamma robusta*, *Pareuchaeta norvegica*, *Calanus hyperboreus*, *Metridia longa*. The largest number of these species occurred outside the 200-m level, i.e. beyond the continental shelf.
(2) *Intermediate Group*: especially *Calanus finmarchicus* in the coastal Atlantic and the northern North Sea but passing into the Norwegian Sea. Several species (*Candacia*, *Metridia lucens* and *Centropages typicus*) were found farther offshore. *Paracalanus*, *Pseudocalanus* and *Temora* were typically farther inshore, grading into *Centropages hamatus*, which is more truly neritic.
(3) *Neritic Group*: *Labidocera* and *Isias*, grading into *Centropages hamatus* and *Temora*, already described.

Colebrook (1964) developed a method using a form of multi-variate analysis to examine the data more critically for degrees of similarity and species distribution. The method used principal component analysis and has been applied to 22 species of zooplankton from the eastern North Atlantic and North Sea. As Glover (1967) emphasizes, the analysis is based on samples from approximately 10 m depth and relates to yearly means; seasonal variations are not considered. Correlation coefficients between all pairs of species were calculated using standardized data and the data then arranged in a correlation matrix. In examining the data in relation to geographical distribution of possible groups, the copepod *Acartia clausi* showed low correlations

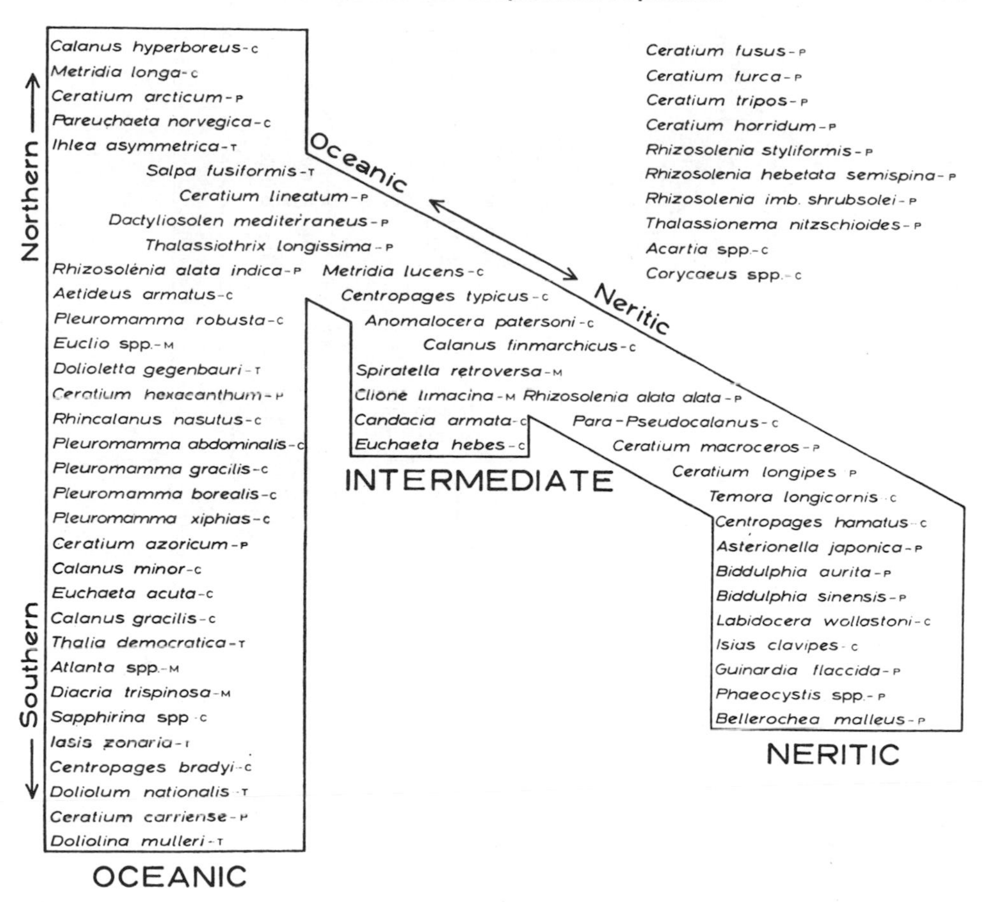

Fig. 8.15. A distribution series based on a visual examination of the charts given in the plankton atlas. The species are arranged in such a way that the distribution of each organism is most similar to those of neighbouring organisms in the list. P = phytoplankton; C = Copepoda; M = Mollusca; T = Tunicata (Colebrook, 1964).

with species in the matrix, but two populations were indicated, inhabiting the North Sea and Atlantic respectively. These have been separately analysed and included in the matrix (Fig. 8.16).

In a somewhat similar manner, *Calanus finmarchicus* (*sensu lato*) has been separated into *C. finmarchicus* (*sensu stricto*) and *C. helgolandicus* populations. Component analysis suggests that five groups of species exist with similar distributions within a group and with fairly obvious differences between groups. Figure 8.17 shows the geographical distribution of these groups and reference to Fig. 8.15 indicates how this geographical disposition might be related to the earlier distribution of species in the plankton atlas. The "intermediate" group has been divided latitudinally.

Colebrook's (1964) analysis showed that three components were responsible for 65% of the variance. He suggests that the components might respectively be identified with surface salinity, a function of temperature possibly involving mean temperature and range over the plankton season and its influence on vertical stability, and with the

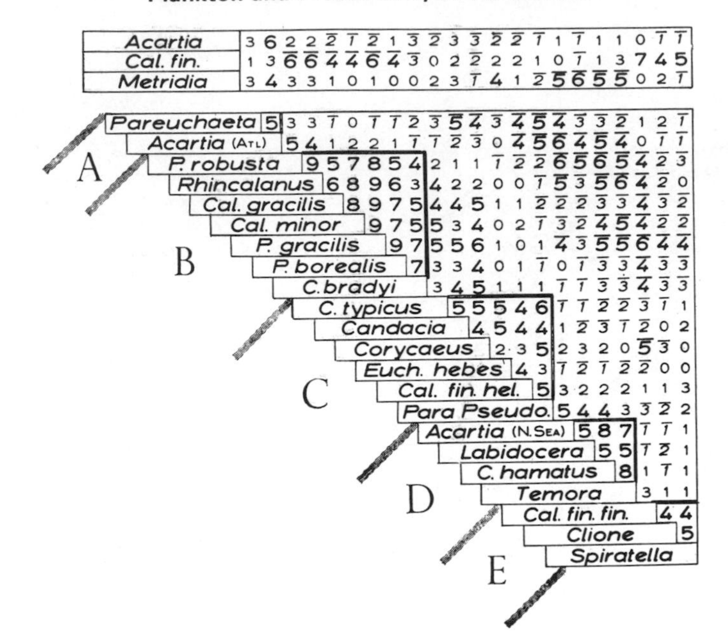

Fig. 8.16. The correlation matrix of the geographical distribution of the species listed along the principal diagonal of the matrix. The correlation coefficients are given to one place of decimals multiplied by ten. Two sizes of numbers have been used, the larger size for correlation of 0.4 and over, and the smaller size for the range +0.3 to −0.3. Negative correlations are shown by sloping numbers with a bar.

The groups of species indicated by vectors are identified by the letters A to E and the correlations within each group are marked off in the matrix by heavy lines.

The correlations of species not included in the component analysis are shown in a separate rectangular block above the triangular matrix (Colebrook, 1964).

mixing of oceanic and neritic waters. The last component could embrace a number of factors. Species of the northern oceanic group tend to have high numbers in cold waters with little temperature range; the southern oceanic group is abundant in waters of high salinity. Both intermediate groups occur in mixed waters, but show different relationships to temperature and salinity factors. The neritic group occurs abundantly in warm coastal waters of low salinity. Colebrook suggests that the second component was not given sufficient weighting in the earlier plankton analysis, nor in Russell's (1939) study of plankton distribution. The separation of geographically distinct but morphologically indistinguishable populations of *Acartia clausi* and possibly of other zooplankton species (*Corycaeus anglicus*, *Metridia lucens*) may be of interest in relation to earlier observations on separate races and varieties of zooplankton. The populations presumably represent more or less isolated groups specifically adapted to the somewhat different environmental conditions. A study of whether such populations are genetically distinct would be of obvious significance.

In a further analysis of Recorder data, Colebrook (1964) calculated the mean monthly variations in abundance of 18 zooplankton species for 1948–1962 for all areas, thus eliminating regional and annual fluctuations. After standardizing the seasonal fluctuations to eliminate the variability due to some species being always more abundant, correlation coefficients were calculated for all possible pairs of seasonal

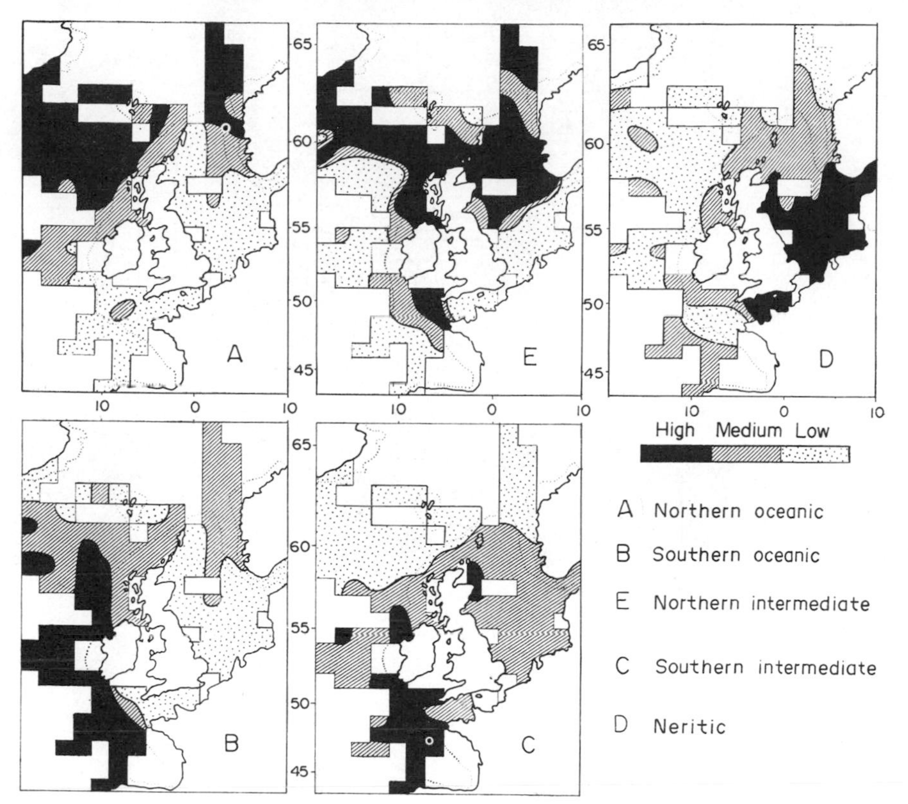

Fig. 8.17. Charts of the geographical distributions of the species groups. The distributions are shown by means of contours on an arbitrary scale indicating high, medium and low numbers (Colebrook, 1964).

cycles. Arrangement in a correlation matrix suggested that three overlapping groups were discernible, with high correlations within a group and low correlations between groups. The seasonal cycle for each group shows differences in timing (cf. Fig. 8.18): Group A (see diagram) peaks in about May or June; Group B somewhat later (about July/August) and Group C latest of all (about early autumn (September)). Colebrook and Robinson (1965) demonstrated similarity of the geographical patterns in the timing and duration of phytoplankton and copepod cycles, but the seasonal cycles of phytoplankton were much more variable. The abundance of copepods is probably largely determined by the timing of the spring phytoplankton increase.

Some zooplankton species show major differences in their seasonal cycles in different areas, and while this may be partly due to regional environmental variations, there would appear to be considerable support for the view that separate populations of the species are present (*vide supra*). It is desirable, however, to study fluctuations in a species both seasonally and from year to year over a considerable period and in the different areas. A study of the pteropod *Limacina* (*Spiratella*) *retroversa* suggested that separate populations existing in two areas, C2 (North Sea) and C5 (Atlantic), near the British Isles, were not very different in abundance from about 1948–1952, though their seasonal timing was not identical. From 1954 onwards, however, major distinctions

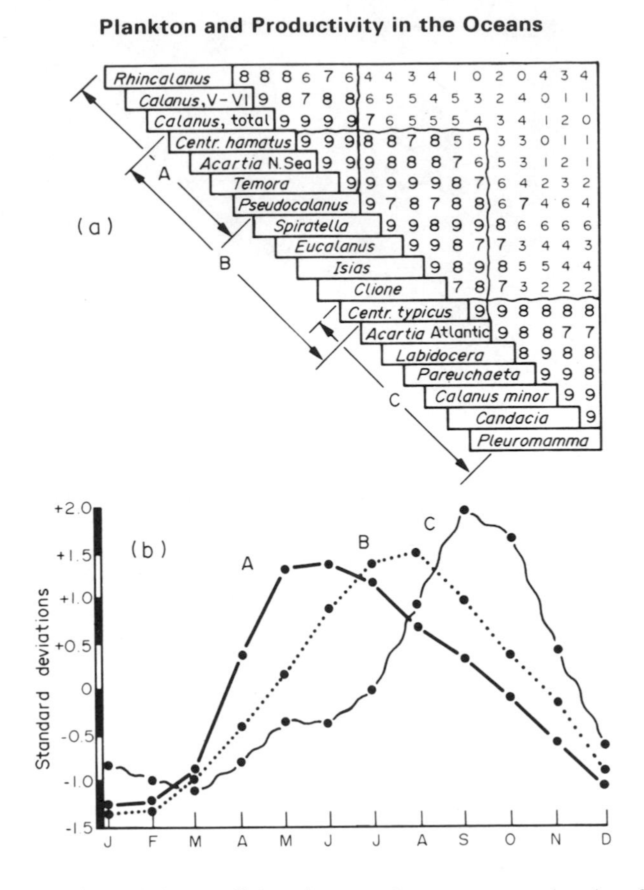

Fig. 8.18. (a) A matrix of correlation coefficients between the mean seasonal cycles of abundance of 18 zooplankton species. The numbers are correlation coefficients multiplied by ten. (b) Graphs of the standardized mean seasonal cycles of abundance of each of the three groups of species shown in the matrix A. (Glover, 1967, after Colebrook, 1965).

appeared between the two populations. The Atlantic Population became progressively smaller; in the North Sea *Limacina* maintained its time of main appearance and its abundance, though diminishing about 1962. Whilst they do not elucidate the reasons for the observed differences, such studies highlight the need for long-term study of fluctuations. Have the environmental conditions changed in the Atlantic and has the species become so precisely adapted to the conditions in Area C5 that it is no longer able to reproduce effectively?

Recorder samples may also be used for comparison with the results of investigations on water masses and indicators. Bainbridge (1963) summarized results on the chaetognaths in the north-east Atlantic. Figure 8.19 demonstrates that *Sagitta setosa* is an obviously neritic species especially common in the southern North Sea. *Sagitta elegans* might be classed as an intermediate type occurring over the shelf area and dividing into one group in the approaches to the English Channel and another in the northern North Sea, spreading southwards to some extent. Two species can be regarded as oceanic—*Sagitta serratodentata* occurs essentially outside the continental shelf, south-west and west of the British Isles and *Eukrohnia hamata* mainly in the Irminger Sea.

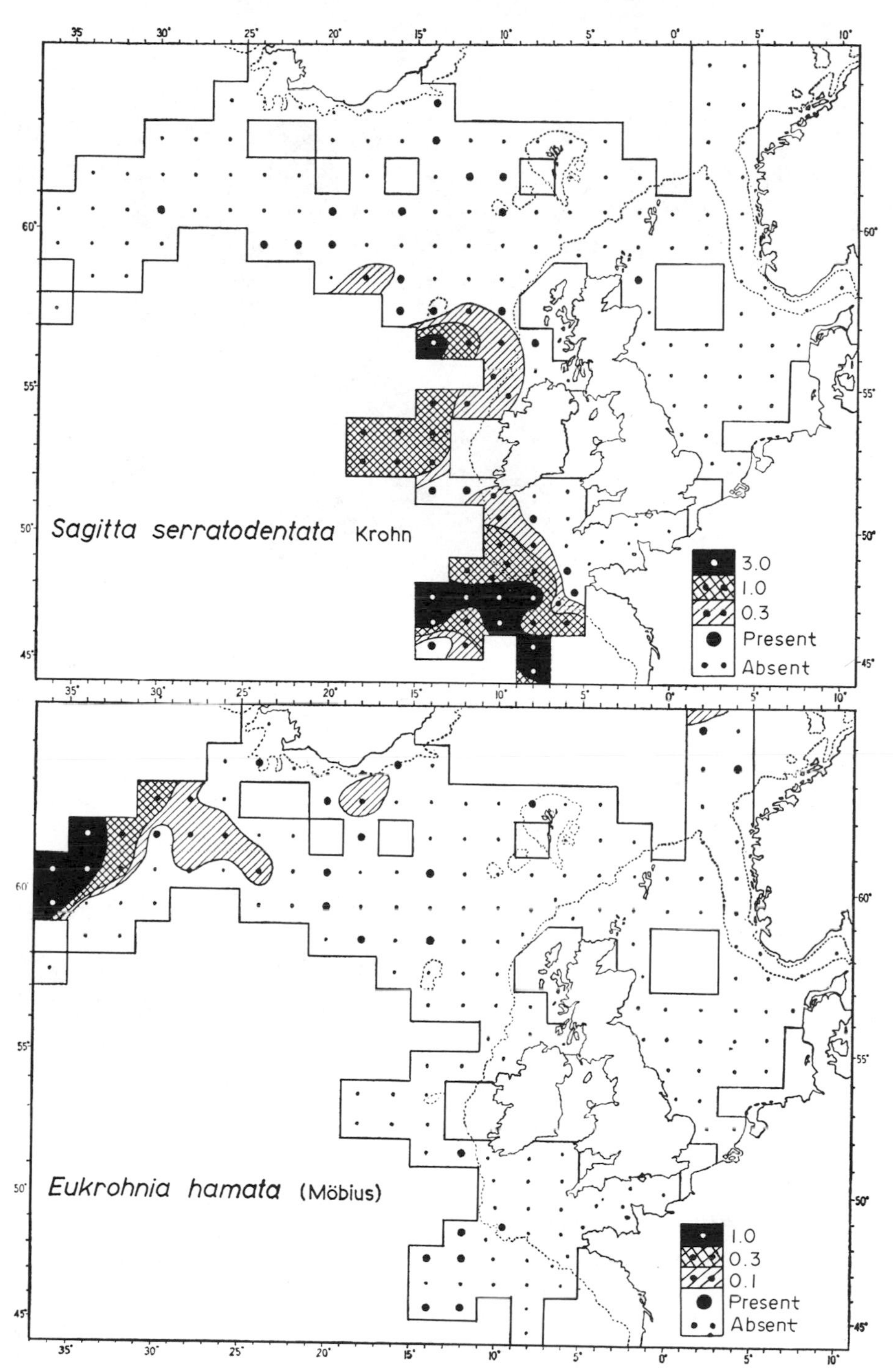

Fig. 8.19. Distributions, at a depth of 10 m of four species of chaetognaths in the main area sampled by the Continuous Plankton Recorder from 1958–60 (3 contour levels from the approximately logarithmic series are given, values indicate mean numbers per sample (*ca.* 3 m^3) (Bainbridge, 1963).

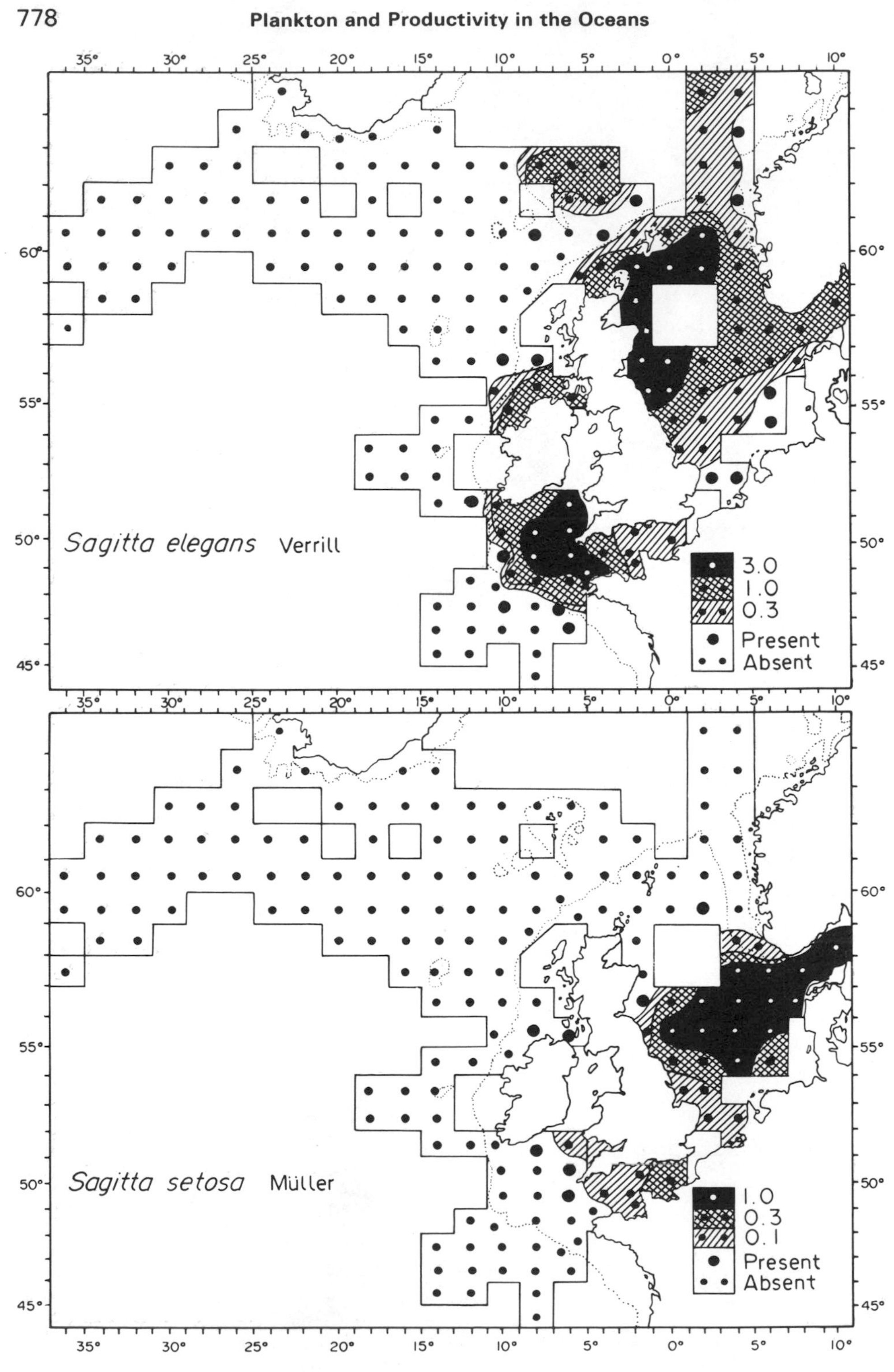

"Fig. 8.19 continued"

Comparatively rare species, all oceanic, include *Sagitta maxima* (south of Iceland) and *S. lyra* and *S. hexaptera*, south-west of the British Isles.

Matthews (1967, 1969) has used Recorder data to examine the problem of the spatial separation of species of *Calanus* in the North Atlantic. He suggests that biometric study shows that two species, *C. finmarchicus* (*sensu stricto*) and *C. helgolandicus* exist with morphological discontinuity despite some overlap in breeding distribution. The separation of *C. glacialis*, according to Matthews, is less obvious as a separate species, since some of the suggested morphological distinctions appear to be size-dependent and he instances rearing experiments using different temperatures to illustrate the size/character relationship. However, this in no way argues against the very successful use of *C. glacialis* as an indicator of Arctic water.

Matthews shows how an examination of Recorder data can be used to elucidate the geographical distribution of *C. finmarchicus*, *C. helgolandicus* and *C. glacialis*, even though only older copepodite stages (C V and VI) are used. From Fig. 8.20 it appears that although *C. finmarchicus* occurs throughout almost the whole survey area, its main centre of abundance is the area south of Greenland and east of Labrador and Newfoundland, with another centre in the Norwegian Sea and northern North Sea. It is more limited in the south-east Atlantic (off the Bay of Biscay). *C. glacialis*, in the western area, is found mainly off south-east Labrador and the northern slope of the Grand Banks, i.e. mainly under the influence of the inflowing Labrador Current and its eddies. This recalls its use as an Arctic indicator. *C. helgolandicus* occurs mainly east and south-east of the British Isles, with numbers decreasing west of Ireland and Scotland as *C. finmarchicus* increases. The species also appears in oceanic regions in the western Atlantic off the edge of the Gulf Stream.

Variations in the timing of the seasonal appearance and main period of increase of the three species are also related to geographical area. *C. helgolandicus* and *C. glacialis* populations would appear to come fairly close together only south-east of the Grand Banks, where waters of entirely different origin approach each other. *C. finmarchicus* and *C. helgolandicus*, however, obviously overlap where seasonal changes provide an environment first for one species and then for the other. Since any species may be regarded as having a total distribution greater than the area where it reproduces satisfactorily and maximally, in the North Sea and off the west coasts of the British Isles there would appear to be a pattern of varying success of reproduction of *C. finmarchicus* and *C. helgolandicus* with the slight change in environmental conditions with season, i.e. a true seasonal succession.

Figure 8.21 indicates the changing proportions of the two species with season and thus illustrates change in zooplankton population with local climatic (seasonal) change rather than with variations in water mass, except in so far as changes in water flow may be affected by season. Over the years, however, not only did the two species show differences in the timing of the spring increase but the duration of the maximum period of abundance varied. *C. helgolandicus* tended to have a shorter period and was also more variable.

Hopner Petersen and Curtis (1980) attempted a comparative estimation of biological production in sub-Arctic (west Greenland), temperate (North Sea) and tropical (Phangnga Bay, Thailand) areas (Fig. 8.22). They pointed out that the relative contribution of zooplankton and benthos to the ecosystem varied with latitude; zooplankton was the major food link in the tropics, while benthos became equally

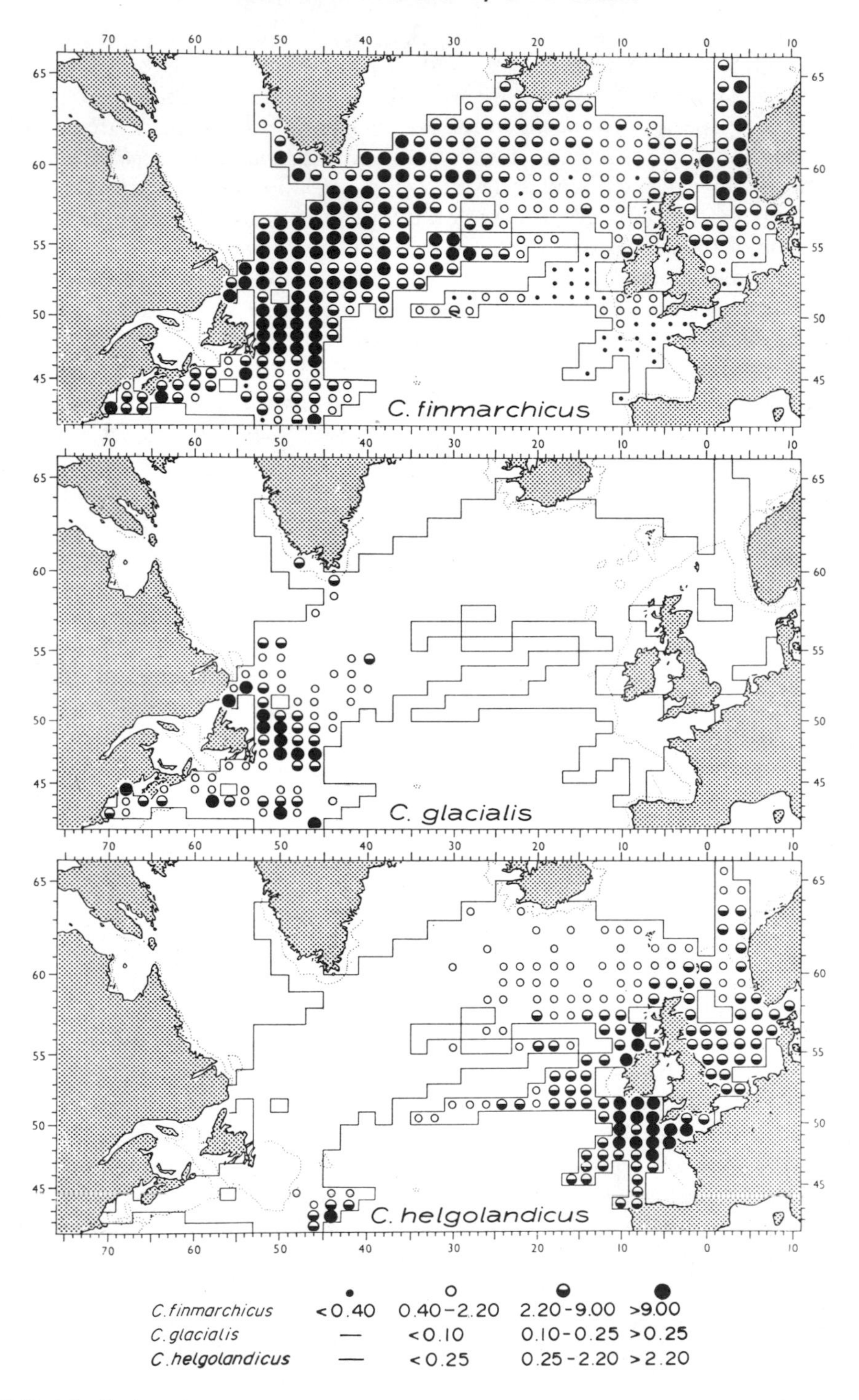

Fig. 8.20. Distribution of adults and Stage V *Calanus finmarchicus sensu stricto*, *C. helgolandicus* and *C. glacialis* at a depth of 10 m. Symbols represent mean numbers per sample (*ca.* 3 m^3) (Matthews, 1969).

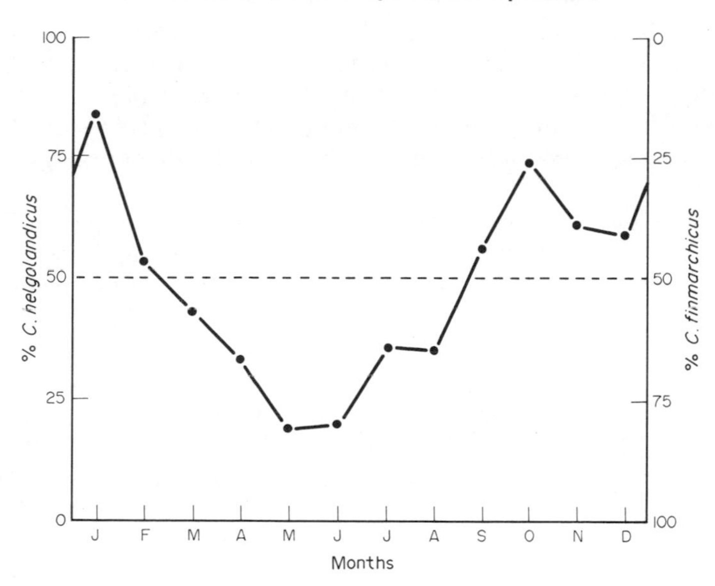

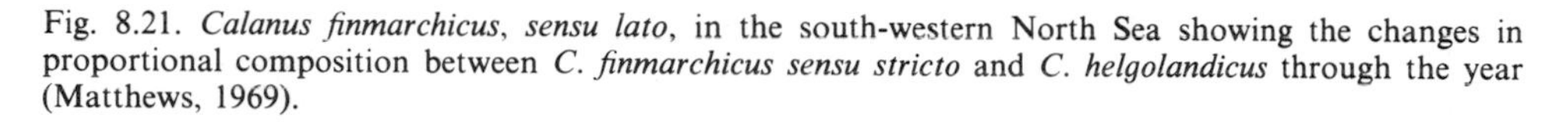

Fig. 8.21. *Calanus finmarchicus, sensu lato,* in the south-western North Sea showing the changes in proportional composition between *C. finmarchicus sensu stricto* and *C. helgolandicus* through the year (Matthews, 1969).

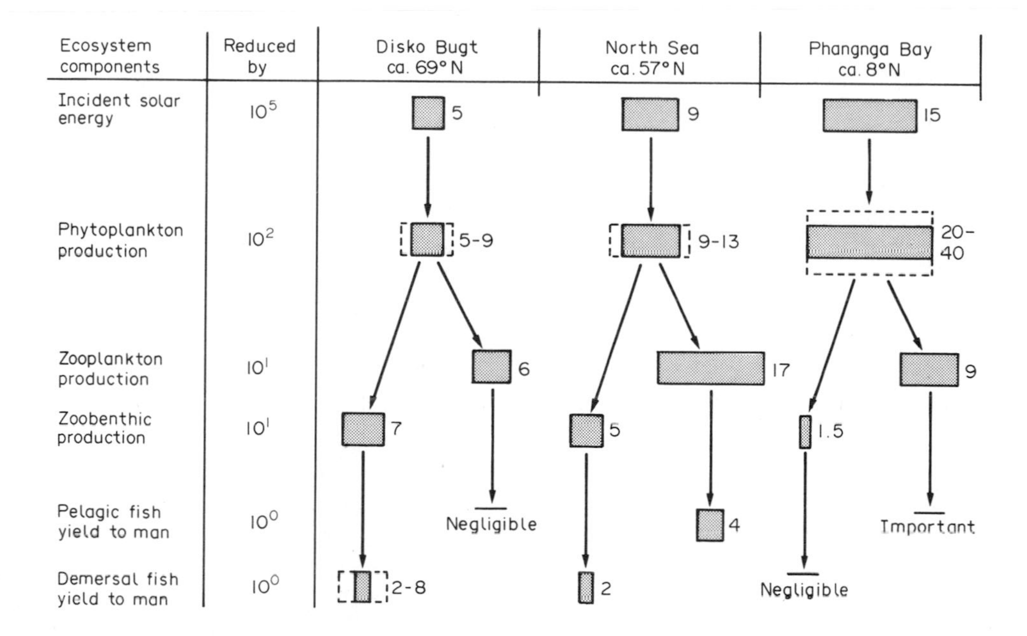

Fig. 8.22. Incident solar energy, phytoplankton production, secondary planktonic and benthic production, and fish yield to man for food webs from Disko Bugt, the North Sea and Phangnga Bay. All values expressed as kcal/m^2/yr. In fishery statistics "demersal" means "caught near the bottom"; in biology it means "feeding at or in the bottom" (Hopner Petersen and Curtis, 1980).

important at high latitudes. They believe that the high metabolic rates of tropical phytoplankton and of herbivorous zooplankton may dissipate the assimilated energy and thus lower the amounts available for transfer to other trophic groups. The slower turnover rate at higher latitudes provides a more stable food pool for benthic filter and detrital feeders, thus increasing their importance in the transfer of energy to the higher trophic levels.

The identification of factors responsible for long-term changes in the zooplankton fauna in any area are clearly difficult to determine. Aside from any long period climatic changes and their possible indirect effects on currents as well as direct influence on the annual population, other variations in water masses, genetic and phenotypic variations in the zooplankton itself and changes in the biological environment ("quality" of the water), including its biological history, may all be factors affecting the quantitative and qualitative characteristics of the plankton.

The complexity of physical and biological interactions and their influence on the pelagic ecosystem is obvious. Continued study with the accumulation of world-wide data allied to small scale experiments may help to reinforce the predictions of mathematical or computer model simulations. Steele (1978) suggests that records of plankton patchiness may provide some explanation of spatial and temporal variability in the occurrence of zooplankton communities.

The need for continued long-term investigations, using agreed standard sampling techniques wherever possible, is paramount.

References

ACKEFORS, H. (1965) *Ophelia* **2**, 269–280.
ACKEFORS, H. (1971) *J. Exp. Mar. Biol. Ecol.* **7**, 51–70.
ADAMS, J. A. and J. H. STEELE (1966) In H. Barnes (Ed.), *Some contemporary studies in Marine Science.* George Allen & Unwin; London, pp. 19–35.
ADLER, G. and P. JESPERSON (1920) *Meddr. Komm. Danm. Fisk -og Havunders, Serie Plankton* **2**, 1–45.
AHLSTROM, E. H. (1959) *Fishery Bulletin*, **60**.
AHLSTROM, E. H. (1961) *Rapp. P-v. Réun. Cons. Perm. Int. Explor. Mer* **150**, 169–176.
AHLSTROM, E. H. (1965) *CALCOFI Repts.* **10**, 31 pp.
AHLSTROM, E. H. (1966) *U.S. Fish Wildl. Serv. Spec. Sci. Rep. Fish*, **534**, 71 pp.
AHLSTROM, E. H. (1969) *CALCOFI Atlas*, **11**, pp. 1–187.
AHLSTROM, E. H. (1972) *CALCOFI Atlas*, **17**, pp. 1–396.
ALI KHAN, J. (1976) *Mar. Biol.* **37**, 305–324.
ALLDREDGE, A. L. (1972) *Science* **177**, 885–887.
ALLDREDGE, A. L. (1976a) *Mar. Biol.* **38**, 29–39.
ALLDREDGE, A. L. (1976b) *Limnol. Oceanogr.* **21**, 14–23.
ALLEN, J. A. (1966) *Oceanogr. Mar. Biol. Ann. Rev.* **4**, 247–265.
ALLEN, J. A. and M. R. GARRETT (1972) *Comp. Biochem. Physiol.* **41A**, 307–317.
ALLEN, J. A. and R. S. SCHELTEMA (1972) *J. Mar. Biol. Ass. U.K.* **52**, 19–31.
ALLEWEIN, J. (1968) *Ophelia* **5**, 207–214.
ALVAREZ, V. and J. B. L. MATTHEWS (1975) *Sarsia* **58**, 67–78.
ALVARINO, A. (1962) *Bull. Scripps Inst. Oceanogr.* **8**, 1–50.
ALVARINO, A. (1965) *Oceanogr. Mar. Biol. Ann. Rev.* **3**, 115–194.
ALVARINO, A. (1967) *Pacific Science* **21**, 474–485.
ALVARINO, A. (1968) *An. Inst. Biol. Univ. Nac. Autón. Mexico* **39**, 41–76.
ALVARINO, A. (1971) *Bull. Scripps Inst. Oceanogr.* **16**, 1–432.
ANDERSON, O., R. M. SPINDLER, A. W. H. BÉ and CH. HEMLEBEN (1979) *J. Mar. Biol. Ass. U.K.* **59**, 791–799.
ANDREWS, K. J. H. (1966) *Discovery Rep.* **34**, 117–162.
ANGEL, M. V. (1968) *Sarsia* **34**, 299–312.
ANGEL, M. V. (1969) *J. Mar. Biol. Ass. U.K.* **49**, 513–515.
ANGEL, M. V. (1970) *J. Mar. Biol. Ass. U.K.* **50**, 731–736.
ANGEL, M. V. (1979) *Progr. Oceanogr.* **8**, 3–124.
ANGEL, M. V. and M. J. R. FASHAM (1975) *J. Mar. Biol. Ass. U.K.* **55**, 709–737.
ANGELINO, M. I. and N. DELLA CROCE (1975) *Cah. Biol. Mar.* **16**, 551–558.
ANRAKU, M. (1964a) *Limnol. Oceanogr.* **9**, 46–60.
ANRAKU, M. (1964b) *Limnol. Oceanogr.* **9**, 195–206.
ANRAKU, M. and M. OMORI (1963) *Limnol. Oceanogr.* **8**, 116–126.
ARASHKEVICH, Y. G. (1977) *Oceanology* **17**, 466–469.
ARASHKEVICH, Y. G. and A. G. TIMONIN (1970) *Dokl. Akad. Nauk. S.S.S.R.* **191**, 935–938.
ARASHKEVICH, Y. G. and V. B. TSEYTLIN (1978) *Oceanology* **18**, 347–351.
ARAVINDAKSHAN, P. N. (1977) *Proc. Symp. Warm Water Zooplankton*, NIO – UNESCO, Goa, 137–145.
ARMITAGE, M. E. (1979) Ph.D. Thesis, University of Southampton.
ARMITAGE, M. E., J. E. G. RAYMONT and R. J. MORRIS (1977) In M. Angel (Ed.) *A Voyage of Discovery: George Deacon 70th Anniversary volume.* Pergamon Press; Oxford. pp. 471–481.
ARMITAGE, M. E., J. E. G. RAYMONT and R. J. MORRIS (1978) *J. Exp. Mar. Biol. Ecol.* **35**, 146–163.
ARON, W., E. H. AHLSTROM, B. MCK. BARY, A. W. H. BÉ and W. D. CLARKE (1965) *Limnol. Oceanogr.* **10**, 333–340.
AURICH, H. J. (1966) *Helgolander Wiss. Meeresunters* **13**, 246–274.
AUSTIN, T. S. and V. E. BROCK (1959) *Int. Ocean. Congress Preprints* A.A.A.S., Washington, 130–131.

BAAN, S. M. Van der and L. B. HOLTHUS (1969) *Nethl. J. Sea Res.* **4**, 354–363.
BAAN, S. M. Van der and L. B. HOLTHUS (1971) *Nethl. J. Sea Res.* **5**, 227–239.
BAILEY, R. S. (1974) *Mar. Res. Dept. Agric. Fish. Scotland* **1**, 1–29.
BAINBRIDGE, R. (1953) *J. Mar. Biol. Ass. U.K.* **32**, 385–445.
BAINBRIDGE, R. (1958) *J. Mar. Biol. Ass. U.K.* **37**, 349–370.
BAINBRIDGE, V. (1963) *Bull. Mar. Ecol.* **6**, 40–51.
BAINBRIDGE, V. (1972) *Bull. Mar. Ecol.* **8**, 61–97.
BAINBRIDGE, V. and J. ROSKELL (1966) in H. Barnes (Ed.) *Some Contemporary Studies in Marine Science*, Allen & Unwin, London, pp. 67–81.
BAKER, A. de C. (1959) *Discovery Rep.* **29**, 309–340.
BAKER, A. de C. (1970) *J. Mar. Biol. Ass. U.K.* **50**, 301–342.
BAKKE, J. L. W. (1977) *Sarsia* **63**, 49–55.
BAKKE, J. L. W. and V. A. VALDERHAUG (1978) *Sarsia* **63**, 247.
BAKKER, C. and N. de PAUW (1975) *Nethl. J. Sea Res.* **9**, 145–165.
BAMSTEDT, U. (1974) *Sarsia* **56**, 71–86.
BAMSTEDT, U. (1975) *Sarsia* **59**, 31–46.
BAMSTEDT, U. (1976) *Sarsia* **61**, 15–30.
BAMSTEDT, U. (1978) *Sarsia* **63**, 145–154.
BAMSTEDT, U. and J. B. L. MATTHEWS (1975) *Proc. 9th Europ. Mar. Biol. Symp.* H. Barnes (Ed.) Aberdeen University Press, pp. 311–327.
BANSE, K. (1964) in *Progress in Oceanography*, 2. Pergamon Press, Oxford. pp. 52–125.
BANSE, K. (1975) *Int. Revue Ges. Hydrobiol.* **60**, 439–447.
BARGMANN, H. E. (1945) *Discovery Rep.* **23**, 105–176.
BARHAM, E. G. (1957) *Hopkins Mar. Stat., Stanford Univ. Tech. Rep.* **1**, 1–182.
BARHAM, E. G. (1963a) *Science* **140**, 826–828.
BARHAM, E. G. (1963b) *Proc. XVI Int. Cong. Zool.* **4**, Washington, D.C. 296–300.
BARHAM, E. G. (1970) In R. Brooke-Farquhar (Ed.) *Proc. Int. Symp. Biological Sound Scattering in the Ocean.* Maury Centre Report 005, Washington, D.C. pp. 100–118.
BARKHATOV, V. A., A. F. VOLKOV, V. N. DOLZHENKOV and Y. P. KAREDIN (1973) *Oceanology* **13**, 559–563.
BARKLEY, E. (1940) *Dana Z. Fisch.* **1**, 65–156.
BARNARD, K. H. (1930) *Brit. Antarctic Terra Nova Exped. Nat. Hist. Rep. Zool.* **8**, 307–454.
BARNARD, K. H. (1932) *Discovery Rept.* **5**, 1–326.
BARNARD, K. H. (1937) *John Murray Expdn. Sci. Rept.* **4**, 130 pp.
BARNES, H. (1957) *Année Biologique* **33**, 67–85.
BARNES, H. (1975) *Publ. Staz. Zool. Napoli* **39**, Suppl. 8–25.
BARNES, H. and M. BARNES (1958) *Limnol. Oceanogr.* **3**, 29–32.
BARNES, H. and M. BARNES (1975) *J. Exp. Mar. Biol. Ecol.* **19**, 227–232.
BARNES, H. and J. BLACKSTOCK (1975a) In H. Barnes (Ed.) *Proc. 9th Europ. Mar. Biol. Symp.* Aberdeen University Press. pp. 299–310.
BARNES, H. and J. BLACKSTOCK (1975b) *J. Exp. Mar. Biol. Ecol.* **19**, 59–79.
BARNES, H., M. BARNES and D. M. FINLAYSON (1963) *J. Mar. Biol. Ass. U.K.* **43**, 185–211.
BARNES, H. and R. L. STONE (1973) *J. Exp. Mar. Biol. Ecol.* **12**, 279–297.
BARNES, J. H. (1966) In W. J. Rees (Ed.) *The Cnidaria and their Evolution. Symp. Zoo. Soc. London.* pp. 307–332.
BARNES, R. D. *Barnes Invertebrate Zoology*. 4th Edn W. B. SAUNDERS, London 1089 pp.
BARLOW, J. P., and J. D. C. MONTEIRO (1979) *Mar. Biol.* **53**, 335–344.
BARY, B. MCK. (1963a) In M. J. Dunbar (Ed.) *Marine Distributions. Roy. Soc. Canada Spec. Publ.* **5**, 51–67.
BARY, B. MCK. (1963b) *J. Fish. Res. Bd. Canada* **20**, 789–827.
BARY, B. MCK. (1964) *J. Fish. Res. Bd. Canada* **21**, 183–202.
BARY, B. MCK. (1967) *Deep. Sea Res.* **14**, 29–33.
BASSINDALE, R. (1936) *Proc. Zoo. Soc. London* **106**, 57–74.
BAUER, J. C. (1976) *Bull. Mar. Sci.* **26**, 273–277.
BAYER, F. M. (1963) *Bull. Mar. Sci. Gulf Caribb.* **13**, 454–466.
BAYER, F. M. and H. B. OWRE (1968) *The Free-living Lower Invertebrates* Macmillan & Co.: London and New York.
BÉ, A. W. H. (1962) *Deep Sea Res.* **9**, 144–151.
BÉ, A. W. H. (1966) *Abstracts of papers 2nd Int. Oceanogr. Congr.* Publ. Nauka, Moscow, p. 26.
BÉ, A. W. H. (1967) *Fich. Ident. Zooplancton* **108**.
BÉ, A. W. H. (1968) in *Zooplankton Sampling, Monographs on Oceanographic Methodology* **2**, UNESCO, Paris 173–174.
BÉ, A. W. H. (1969) *Antarctic Map Folio Series 11*. Amer. Geog. Soc. pp. 9–11.
BÉ, A. W. H., J. M. FORNS and D. A. ROELS (1971) In J. D. Costlow (Ed.) *Fertility of the Sea*, vol. 1, Gordon & Breach; New York. pp. 17–50.

Bé, A. W. H., S. M. Harrison and L. Lott (1973) *Micropaleontology* **10**, 150–192.
Bé, A. W. H. and D. S. Tolderlund (1971) In B. M. Funnel and W. R. Riedel (Ed.) *Micropalaeontology of Oceans*. Cambridge University Press; London. pp. 105–149.
Bé, A. W. H., G. Vilks and L. Lott (1971) *Micropaleontology* **17**, 31–42.
Beebe, W. (1935) *Half Mile Down*. John Lane, The Bodley Head; London. 344 pp.
Beers, J. R. (1964) *J. Cons. Perm. Int. Explor, Mer* **29**, 123–129.
Beers, J. R. and G. L. Stewart (1969) *Mar. Biol.* **4**, 182–189.
Beers, J. R. and G. L. Stewart (1970) *Bull. Scripps Inst. Oceanogr.* **17**, 67–87.
Beers, J. R. and G. L. Stewart (1971) *Deep-Sea Res.* **18**, 861–883.
Beklemishev, K. V. (1954) *Zool. J. Inst. Oceanol. Acad. Sci. U.S.S.R.* **33**, 1210–1229.
Beklemishev, K. V. (1957) *Trans. Inst. Oceanol. Acad. Sci. U.S.S.R.* **20**, 253–278.
Beklemishev, K. V. (1962) *Rapp. P-v. Réun. Comm Int. Explor. Scient. Mer Medit.* **16**, 165–176.
Belloni, S., R. Cattaneo and D. Pessani (1976) *Bull. Mus. Ist. Biol. Univ. Genova.* **44**, 103–112.
Belman, B. W. (1978) *Limnol. Oceanogr.* **23**, 735–739.
Belyayeva, N. V. (1969) *Oceanology* **9**, 854–860.
Berger, W. H. (1969) *Deep-Sea Res.* **16**, 1–24.
Bernard, M. (1962) *Fich. Ident. Zooplancton* **36**.
Bernard, M. (1963) *Pelagos* **1**, 35–48.
Bernard, M. (1964) *Pelagos* **2**, 51–71.
Bernard, M. (1967) *Oceanogr. Mar. Biol. Ann. Rev.* **5**, 231–255.
Bernard, M. (1970) *Pelagos* **11**, 196.
Berner, A. (1962) *J. Mar. Biol. Ass. U.K.* **42**, 625–640.
Berner, L. D. (1967) *CALCOFI Atlas* **8**, pp.
Berrill, N. J. (1950) *The Tunicata*. Ray Society; London, 354 pp.
Beyer, F. (1962) *Rapp. P-v. Réun. Cons. Perm. Int. Explor. Mer.* **153**, 79–85.
Bhaud, M. (1972) *Cah. ORSTOM Sér. Océanogr.* **10**, 203–216.
Bhaud, M. (1975) *Cah. ORSTOM Sér. Océanogr.* **13**, 68–77.
Bieri, R. (1959) *Limnol. Oceanogr.* **4**, 1–28.
Bieri, R. (1966a) *Publ. Seto Mar. Biol. Lab.* **14**, 161–170.
Bieri, R. (1966b) *Publ. Seto Mar. Biol. Lab.* **14**, 24–26.
Bieri, R. (1970) *Publ. Seto Mar. Biol. Lab.* **17**, 305–307.
Bigelow, H. B. (1926) *Bull. U.S. Bur. Fish.* **40**, II, 1–507.
Bigelow, H. B. and M. Sears (1937) *Rept. Danish Oceanogr. Exped. Mediterranean* **2**, (Biol.) 144.
Bigelow, H. B. and M. Sears (1939) *Mem. Mus. Comp. Zool. (Harvard)* **54**, 183–378.
Bigelow, H. B. and W. W. Welsh (1925) *Bull. U.S. Bur. Fish.* **40**(I), 1–567.
Biggs, D. C. (1977) *Limnol. Oceanogr.* **22**, 108–117.
Binet, D. (1970) *Doc. Scient. Pointe Noire, ORSTOM* **8**, 107 pp.
Binet, D. (1972) *Doc. Scient. Centre Rech. Oceanogr. Abidjan* **3**, 60–93.
Binet, D. (1973) *Doc. Scient. Centre Rech. Oceanogr. Abidjan* **4**, 77–90.
Binet, D. and E. Suisse de Sainte Claire (1975) *Cah. ORSTOM Sér. Océanogr.* **13**, 15–30.
Birstein, Y. A. and M. E. Vinogradov (1972) In B. G. Bogorov (Ed.) *Fauna of the Kurile-Kamchatka Trench and its Environment*. Israel Program for Scientific Translation, Jerusalem, pp. 419–438.
Blackbourn, D. J., F. J. R. Taylor and J. Blackbourn (1973) *J. Protozool.* **20**, 286–288.
Blackburn, M. (1968) *Fish. Bull. U.S. Fish and Wildlife Service* **67**, 71–115.
Blackburn, M. (1973) *Limnol. Oceanogr.* **18**, 552–563.
Blackburn, M. (1979) *Deep-Sea Res.* **26**, 41–56.
Blackburn, M., R. M. Laurs, R. W. Owen and B. Zeitzschel (1970) *Mar. Biol.* **7**, 14–31.
Blake, J. A. (1969) *Ophelia* **7**, 1–63.
Blumer, M., M. M. Mullin and R. R. L. Guillard (1970) *Mar. Biol.* **6**, 226–235.
Blumer, M., M. M. Mullin and D. W. Thomas (1964) *Helgolander Wiss. Meeresunters.* **10**, 187–201.
Bock, K. J. (1967) *Fich. Ident. Zooplancton* **110**.
Boden, B. P. (1955) *Discovery Rept.* **27**, 337–386.
Boden, B. P., M. W. Johnson and E. Brinton (1955) *Bull. Scripps Inst. Oceanogr.* **6**, 287–397.
Boden, B. P. and E. M. Kampa (1967) In N. B. Marshall (Ed.) *Aspects of Marine Zoology*. Proc. Symp. Zool. Soc. London. Academic Press. pp. 15–26.
Bodo, F., C. Razouls and A. Thiriot (1965) *Cah. Biol. Mar.* **6**, 219–254.
Boely, T. and C. Champagnat (1970) *Rapp. P-v. Réun. Cons. Perm. Int. Explor Mer* **159**, 176–181.
Bogorov, B. G. (1934) *J. Mar. Biol. Ass. U.K.* **19**, 585–612.
Bogorov, B. G. (1946) *J. Mar. Res.* **6**, 25–32.
Bogorov, B. G. (1958) In A. A. Buzzati-Traverso (Ed.) *Perspectives in Marine Biology*, Univ. California Press, Berkeley. pp. 145–158.
Bogorov, B. G. (Ed.) (1972) *Fauna of the Kurile-Kamchatka Trench and its Environment*. Israel Program for Scientific Translations, Jerusalem, 573 pp.

BONE, Q., and A. PULSFORD (1978) *J. Mar. Biol. Ass. U.K.* **58**, 565–570.
BONNETT, R., E. J. HEAD and P. J. HERRING (1979) *J. Mar. Biol. Ass. U.K.* **59**, 565–573.
BONNEVIE, K. (1913) *Rep. Sci. Res. 'Michael Sars' Expdn.* **3**, 69 pp.
BONNEVIE, K. (1931) *Rep. Sci. Res. 'Michael Sars' Expdn.* **5**, 1–9.
BORKOWSKI, T. V. (1971) *Bull. Mar. Sci.* **21**, 826–840.
BOSCH, H. F. and W. R. TAYLOR (1973a) *Mar. Biol.* **19**, 161–171.
BOSCH, H. F. and W. R. TAYLOR (1973b) *Mar. Biol.* **19**, 172–181.
BOSCH, H. F. and W. R. TAYLOR (1976) *Tech. Rep. Chesapeake Bay Inst.* **66**, 77.
BOTTAZZI E. MASSERA, K. VIJAYAKRISHNAN NAIR and M. C. BALANI (1967) *Arch. Oceanogr. Limnol.* **15**, 63–67.
BOTTAZZI E. MASSERA, and G. NENCINI (1969) *Fich. Ident. Zooplancton* **114**.
BOTTAZZI E. MASSERA, B. SCHREIBER and V. T. BOWEN (1971) *Limnol. Oceanogr.* **16**, 677–684.
BOTTAZZI E. MASSERA, and A. VANNUCCI (1964) *Arch. Oceanogr. Limnol.* **13**, 316–385.
BOTTAZZI E. MASSERA, and A. VANNUCCI (1965a) *Arch. Oceanogr. Limnol.* **14**, 1–68.
BOTTAZZI E. MASSERA, and A. VANNUCCI (1965b) *Arch. Oceanogr. Limnol.* **14**, 153–257.
BOTTINO, N. R. (1974) *Mar. Biol.* **27**, 197–204.
BOUCHER, J. and J. F. SAMAIN (1974) *Tethys* **6**, 179–188.
BOUCHER, J. and J. F. SAMAIN (1975) In H. Barnes (Ed.) *Proc. 9th Europ. Mar. Biol. Symp.* Aberdeen Univ. Press. pp. 329–341.
BOUCHER, J. and A. THIRIOT (1972) *Mar. Biol.* **15**, 47–56.
BOUR, W. and S. FRONTIER (1974) *Cah. ORSTOM Sér. Océanogr.* **12**, 207–219.
BOUSFIELD, E. L. (1954a) Ph.D. Thesis, Harvard Univ.
BOUSFIELD, E. L. (1954b) *Bull. Nat. Museum Canada* **132**, 112–154.
BOUSFIELD, E. L. (1955) *J. Fish. Res. Bd. Canada* **12**, 762–767.
BOWMAN, T. E. (1960) *Proc. U.S. National Mus. Smithsonian Inst.* **112**, 343–392.
BOWMAN, T. E. and H. E. GRUNER (1973) *Smithsonian Contrib. Zool.* **146**, 64 pp.
BOYD, C. M. (1976) *Limnol. Oceanogr.* **21**, 175–180.
BRACONNOT, J. C. (1974) *Rapp. P-v. Mer Medit.* **22**, 87–99.
BRIGGS, J. C. (1974) *Marine Zoogeography*, McGraw Hill; New York, 475 pp.
BRINKMANN, A. (1917) *Michael Sars Deep Sea Exped. 1910* **3**, 1–18.
BRINTON, E. (1962) *Bull. Scripps Inst. Oceanogr.* **8**, 51–269.
BRINTON, E. (1979) *Prog. Oceanogr.* **8**, 125–189.
BRINTON, E. and K. GOPALAKRISHNAN (1973) In B. Zeitzschel (Ed.) *Biology of the Indian Ocean.* Chapman & Hall, London; pp. 357–382.
BROAD, A. C. (1957) *Biol. Bull. Woods Hole* **112**, 162–170.
BRODSKY, K. A. (1961) *New Zealand Oceanographic Inst. Contribution* **95**, 21 pp.
BRODSKY, K. A. (1962) In R. Carrick, M. Holdgate and J. Prevost (Eds.) *Biologie Antarctique Proc. Symp. S.C.A.R.* Hermann; Paris. pp. 257–258.
BRODSKY, K. A. (1965) *Oceanology* **4**, 1–12.
BRODSKY, K. A. (1967) *Calanoida of the Far Eastern Seas and Polar Basin of the U.S.S.R.* Israel Program for Scientific Translations; Jerusalem. 440 pp.
BRODSKY, K. A. (1975) In Z. A. Zvereva (Ed.) *Geographical and Seasonal Variability of Marine Plankton.* Israel Program for Scientific Translations; Jerusalem. pp. 1–127.
BRUCE, J. R., M. KNIGHT and M. W. PARKE (1940) *J. Mar. Biol. Ass. U.K.* **24**, 337–374.
BRYAN, J. R., J. P. RILEY and P. J. LEB. WILLIAMS (1976) *J. Exp. Mar. Biol. Ecol.* **21**, 191–197.
BSHARAH, L. (1957) *Bull. Mar. Sci. Gulf Caribb.* **7**, 201–251.
BUCKMANN, A. (1970) *Mar. Biol.* **5**, 35–56.
BURKENROAD, M. D. (1936) *Bull. Bingham Oceanogr. Coll.* **5**(2) 1–151.
BURKILL, P. H. (1978) Ph.D. Thesis, University of Southampton.
BURKHOLDER, P. R., E. F. MANDELLI and P. CENTENO (1967) *J. Agr. Food Chem.* **15**, 718–720.
BURRELL, V. G. and W. A. VAN ENGEL (1976) *Estuar. Coast. Mar. Sci.* **4**, 235–242.
BURUKOVSKIY, R. N. (1967) *U.S. Dept. Commerce Service*, Joint Publications Research, Washington, D.C. pp. 37–54.
BUTLER, E. I., E. D. S. CORNER and S. M. MARSHALL (1969) *J. Mar. Biol. Ass. U.K.* **49**, 977–1001.
BUTLER, E. I., E. D. S. CORNER and S. M. MARSHALL (1970) *J. Mar. Biol. Ass. U.K.* **50**, 525–560.
CALEF, G. W. and G. D. GRICE (1967) *J. Mar. Res.* **25**, 84–94.
CAMPBELL, A. S. (1954) In R. C. Moore (Ed.) *Treatise on Invertebrate Palaeontology*, Part D, **3**, Geol. Soc. Am. and University Kansas Press.
CANNON, H. G. (1931) *Discovery Rep.* **2**, 435–482.
CANNON, H. G. (1940) *Discovery Rep.* **19**, 185–224.
CARRÉ, C. (1967) *Cah. Biol. Mar.* **8**, 185–193.
CARRÉ, D. (1967) *Cah. Biol. Mar.* **8**, 233–251.
CASANOVA, J.-P. (1968) *Rapp. P-v Comm. Int. Mer Medit.* **19**, 451–454.

CASANOVA, J.-P (1977) D.Sc. Thesis, Univ. of Provence (Aix-Marseille).
CASEY, R. E. (1971) In B. M. Funnell and W. R. Riedel (Eds.) *Micropalaeontology of Oceans*, Cambridge University Press, pp. 331–341.
CAZAUX, C. (1973) *Bull. Ecol.* **4**, 257–275.
CAZAUX, C. (1975) *Cah. Biol. Mar.* **16**, 541–549.
CHACE, F. A. (1940) *Zoologica* **25**, 117–209.
CHAMPALBERT, G. (1971) *J. Exp. Mar. Biol. Ecol.* **6**, 23–33.
CHAMPALBERT, G. (1973) *Mar. Biol.* **19**, 315–319.
CHAMPALBERT, G., and R. GAUDY (1972) *Mar. Biol.* **12**, 159–169.
CHAPMAN, D. M. (1966) In W. J. Rees (Ed.) *The Cnidaria and their Evolution.* Symp. Zoo. Soc. London, pp. 51–75.
CHEKUNOVA, V. L. and T. I. RYNKOVA (1974) *Oceanology* **14**, 434–440.
CHEN, C. and A. W. H. BÉ (1964) *Bull. Mar. Sci.* **14**, 185–220.
CHENG, C. (1947) *J. Mar. Biol. Ass. U.K.* **26**, 551–561.
CHENG, L. (1973) *Oceanogr. Mar. Biol. Ann. Rev.* **11**, 223–235.
CHENG, L. (1975) *Fich. Ident. Zooplancton* **147**, 4.
CHEVREUX, E. (1935) *Res. Camp. Sci. Prince Albert Monaco* **90**, 1–230.
CHEVREUX, E. and L. FAGE (1925) *Faune de France* **9**, 1–488.
CHILDRESS, J. J. (1971a) *Limnol. Oceanogr.* **16**, 104–106.
CHILDRESS, J. J. (1971b) *Biol. Bull.* **141**, 109–121.
CHILDRESS, J. J. (1975) *Comp. Biochem. Physiol.* **50**, 787–799.
CHILDRESS, J. J. (1977) *Mar. Biol.* **39**, 19–24.
CHILDRESS, J. J. and M. NYGAARD (1974) *Mar. Biol.* **27**, 225–238.
CHISLENKO, L. L. (1975) In Z. A. Zvereva (Ed.) *Geographical and Seasonal Variability of Marine Plankton.* Israel Program for Scientific Translations, pp. 263–274.
CIECHOMSKI, J. D. de (1967) *CALCOFI, Rep.* **11**, 72–81.
CIECHOMSKI, J. D. de (1971) In J. D. Costlow (Ed.) *Fertility of the Sea*, Gordon Breach; New York, pp. 89–98.
CIFELLI, R. and K. N. SACHS (1966) *Deep-Sea Res.* **13**, 751–753.
CLARKE, G. L. (1933) *Biol. Bull. Woods Hole* **65**, 402–436.
CLARKE, G. L. (1934) *Biol. Bull. Woods Hole* **67**, 432–448.
CLARKE, G. L. (1940) *Biol. Bull. Woods Hole* **78**, 226–255.
CLARKE, G. L. (1970) In G. Brooke Farquhar (Ed.) *Proc. Int. Symp. Biological Sound Scattering in the Ocean.* Maury Centre Report 005, Washington, D.C. pp. 41–50.
CLARKE, G. L. and R. H. BACKUS (1956) *Deep-Sea Res.* **4**, 1–14.
CLARKE, G. L. and D. D. BONNETT (1939) *Biol. Bull.* **76**(3) 371–383.
CLARKE, G. L. and C. J. HUBBARD (1959) *Discovery Rep.* **19**, 1–20.
CLARKE, G. L., E. L. PIERCE and D. F. BUMPUS (1943) *Bull. Mar. Biol. Lab. Woods Hole* **85**, 201–226.
CLARKE, G. L. and G. K. WERTHEIM (1956) *Deep-Sea Res.* **3**, 189–205.
CLARKE, M. R. (1966) *Adv. Mar. Biol.* **4**, 91–300.
CLUTTER, R. I. and M. ANRAKU (1968) *Zooplankton sampling. Monographs on Oceanographic Methodology 2*, UNESCO Paris, pp. 57–76.
CLUTTER, R. I. and G. M. THEILACKER (1971) *Fish. Bull. U.S.* **69**, 93–115.
COE, W. R. (1935) *Zool. Anz.* **111**, 315–317.
COE, W. R. (1945) *Zoologica, N.Y.* **30**, 145–168.
COE, W. R. (1954) *Bull. Scripps Inst. Oceanogr.* **6**, 225–286.
COE, W. R. (1956) *Fich. Ident. Zooplancton* **64**, 5 pp.
COLEBROOK, J. M. (1964) *Bull. Mar. Ecol.* **6**, 78–100.
COLEBROOK, J. M., R. S. GLOVER and G. A. ROBINSON (1961) *Bull. Mar. Ecol.* **5**, 67–80.
COLEBROOK, J. M., D. E. JOHN and W. W. BROWN (1961) *Bull. Mar. Ecol.* **5**, 90–97.
COLEBROOK, J. M. and G. A. ROBINSON (1965) *Bull. Mar. Ecol.* **6**, 123–139.
COLMAN, J. S. and F. SEGROVE (1955) *J. Anim. Ecol.* **24**, 445–462.
CONOVER, R. J. (1956) *Bull. Bingham Oceanogr. Coll.* **15**, 156–233.
CONOVER, R. J. (1959) *Limnol. Oceanogr.* **4**, 259–268.
CONOVER, R. J. (1960) *Biol. Bull. Woods Hole*, **119**, 399–415.
CONOVER, R. J. (1962) *Rapp. P-v Réun. Cons. Perm. Int. Explor. Mer* **153**, 190–196.
CONOVER, R. J. (1964) *Proc. Symp. Exp. Mar. Ecol.*, Univ. Rhode Island Occas. Publ. **2**, 81–91.
CONOVER, R. J. (1966a) In H. Barnes (Ed.) *Some Contemporary Studies in Marine Science.* George Allen & Unwin; London, pp. 187–194.
CONOVER, R. J. (1966b) *Limnol. Oceanogr.* **11**, 346–354.
CONOVER, R. J. (1968) *Am. Zoologist*, **8**, 107–118.
CONOVER, R. J. and E. D. S. CORNER (1968) *J. Mar. Biol. Ass. U.K.* **48**, 49–75.
CONOVER, R. J., S. M. MARSHALL and A. P. ORR (1959) W.H.O.I. Unpubl. MS. (Ref. No. 59–32) pp. 1–12.

CONOVER, R. J. and V. FRANCIS (1973) *Marine Biology*, **18**, 272–283.
CONOVER, R. J. and C. M. LALLI (1973) *J. Exp. Mar. Biol. Ecol.* **9**, 279–302.
CONOVER, R. J. and C. M. LALLI (1974) *J. Exp. Mar. Biol. Ecol.* **16**, 131–154.
COOPER, G. A. and D. C. T. FORSYTH (1963) *Bull. Mar. Ecol.* **6**, 31–38.
COOPER, L. H. N. (1938) *J. Mar. Biol. Ass. U.K.* **23**, 181–195.
CORKETT, C. J. and I. A. MCLAREN (1978) *Adv. Mar. Biol.* **15**, 1–231.
CORNER, E. D. S. (1961) *J. Mar. Biol. Ass. U.K.* **41**, 5–16.
CORNER, E. D. S. (1978) *Adv. Mar. Biol.* **15**, 289–380.
CORNER, E. D. S. and C. B. COWEY (1964) *Oceanogr. Mar. Biol. Ann. Rev.* **2**, 147–167.
CORNER, E. D. S. and C. B. COWEY (1968) *Biol. Rev.* **43**, 393–426.
CORNER, E. D. S. and A. G. DAVIES (1971) *Adv. Mar. Biol.* **9**, 102–204.
CORNER, E. D. S., C. B. COWEY and S. M. MARSHALL (1965) *J. Mar. Biol. Ass. U.K.* **45**, 429–442.
CORNER, E. D. S., C. B. COWEY and S. M. MARSHALL (1967) *J. Mar. Biol. Ass. U.K.* **47**, 259–270.
CORNER, E. D. S., R. N. HEAD and C. C. KILVINGTON (1972) *J. Mar. Biol. Ass. U.K.* **52**, 847–861.
CORNER, E. D. S., R. N. HEAD, C. C. KILVINGTON and S. M. MARSHALL (1974) *J. Mar. Biol. Ass. U.K.* **54**, 319–331.
CORNER, E. D. S., R. N. HEAD, C. C. KILVINGTON and L. PENNYCUICK (1976) *J. Mar. Biol. Ass. U.K.* **56**, 345–358.
CORNER, E. D. S., and B. S. NEWELL (1967) *J. Mar. Biol. Ass. U.K.* **47**, 113–120.
COSPER, T. C. and M. R. REEVE (1975) *J. Exp. Mar. Biol. Ecol.* **17**, 33–38.
COSTLOW, J. D. and C. G. BOOKHOUT (1957) *Biol. Bull. Woods Hole*, **112**, 313–324.
COSTLOW, J. D. and C. G. BOOKHOUT (1958) *Biol. Bull. Woods Hole* **114**, 284–295.
COSTLOW, J. D. and C. G. BOOKHOUT (1959) *Int. Ocean. Congress Preprints* A.A.A.S.; Washington. pp. 228–229.
COWEY, C. B. and E. D. S. CORNER (1963a) *J. Mar. Biol. Ass. U.K.* **43**, 485–493.
COWEY, C. B. and E. D. S. CORNER (1963b) *J. Mar. Biol. Ass. U.K.* **43**, 495–511.
COWEY, C. B. and E. D. S. CORNER (1966) In H. Barnes (Ed.) *Some Contemporary Studies in Marine Science*. George Allen & Unwin; London. pp. 225–231.
COWEY, C. B. and J. R. SARGENT (1972) *Adv. Mar. Biol.* **10**, 383–492.
CRISP, D. J. and P. A. DAVIES (1955) *J. Mar. Biol. Ass. U.K.* **34**, 357–380.
CRONIN, L. E., J. C. DAIBER and E. M. HULBERT (1962) *Chesapeake Science*, **3**, 63–93.
CURL, H. (1962) *Rapp. P-v Réun. Cons. Perm. Int. Explor. Mer* **153**, 183–189.
CURL, H. and G. C. MCLEOD (1961) *J. Mar. Res.* **19**, 70–88.
CUSHING, D. H. (1951) *Biol. Rev.* **26**, 158–192.
CUSHING, D. H. (1955) *J. Anim. Ecol.* **24**, 137–166.
CUSHING, D. H. (1959a) *Fish. Invest. Lond. Series II*, **22**, No. 6, 1–40.
CUSHING, D. H. (1959b) *J. Cons. Perm Int. Explor. Mer* **24**, 455–464.
CUSHING, D. H. (1963) *J. Mar. Biol. Ass. U.K.* **43**, 327–389.
CUSHING, D. H. (1968) *J. Cons. Perm. Int. Explor. Mer* **31**, 70–82.
CUSHING, D. H. (1975) *Marine Ecology & Fisheries*. Cambridge University Press; London. 278 pp.
CUSHING, D. H., G. F. HUMPHREY, K. BANSE and T. LAEVASTU (1958) *Rapp. P-v Réun. Cons. Perm. Int. Explor. Mer* **144**, 15–16.
CUSHING, D. H. and T. VUCETIC (1963) *J. Mar. Biol. Ass. U.K.* **43**, 349–371.
CUSHING, D. H. and J. J. Walsh (Eds.) (1976) *The Ecology of the Seas* Blackwell Scientific Publications; Oxford. 467 pp.
DAGG, M. (1977) *Limnol. Oceanogr.* **22**, 99–107.
DAGG, M. J. and J. L. LITTLEPAGE (1972) *Mar. Biol.* **17**, 162–170.
DAKIN, W. J. (1908) *Int. Rev. Ges. Hydrobiol. and Hydrogr.* **1**, 772–782.
DALES, R. P. (1953) *Proc. Zool. Soc. Lond.* **122**, 1007–1015.
DALES, R. P. (1957) *Bull. Scripps Inst. Oceanogr.* **7**, 95–167.
DALES, R. P. and G. PETER (1972) *J. Nat. Hist.* **6**, 55–92.
DALLOT, S. (1968) *Rapp. Commun. Int. Explor Mer Medit.* **19**, 521–523.
DARO, M. H. and P. POLK (1973) *Nethl. J. Mar. Res.* **6**, 130–140.
DAUBY, P. (1980) *Oceanol. Acta.* **3**, 403–407.
DAVID, P. M. (1955) *Discovery Rep.* **27**, 235–278.
DAVID, P. M. (1958) *Discovery Rep.* **29**, 199–228.
DAVID, P. M. (1959) *Rep. B.A. N.Z. Antarct. Exped. Ser. B* **8**, 73–79.
DAVID, P. M. (1962) In R. Carrick, M. Holdgate and J. Provost (Eds.) *Biologie Antarctique Proc. Symp. S.C.A.R.* Hermann; Paris. pp. 253–256.
DAVID, P. M. (1965) *Endeavor* **24**, 95–100.
DAVID, P. M. (1967) *Proc. Zool. Soc. Lond.* **19**, 211–213.
DAVIES, A. G. (1978) *Adv. Mar. Biol.* **15**, 381–508.

DAVIES, H. C. (1953) *Bull. Mar. Biol. Lab. Woods Hole* **104**, 334–350.
DAWSON, J. K. (1978) *Limnol. Oceanogr.* **23**, 950–957.
De DECKER, A. (1973) In B. Zeitzschel (Ed.) *The Biology of the Indian Ocean*. Chapman & Hall; London: Springer-Verlag; Berlin, 189–219.
DEEVEY, G. B. (1952) *Bull. Bingham Oceanogr. Coll.* **13**, 65–119.
DEEVEY, G. B. (1956) *Bull. Bingham Oceanogr. Coll.* **15**, 113–155.
DEEVEY, G. B. (1960) *Bull. Bingham Oceanogr. Coll.* **17**, 5–53.
DEEVEY, G. B. (1962) *AEC Rept. Contr. AT (30–1) 2646, Bermuda Biological Stn.*
DEEVEY, G. B. (1964) *J. Mar. Biol. Ass. U.K.* **44**, 589–600.
DEEVEY, G. B. (1968) *Peabody Mus. Nat. Hist. Yale Univ.* **26**, 1–125.
DEEVEY, G. B. (1971) *Limnol. Oceanogr.* **16**, 219–240.
DEEVEY, G. B. and A. L. BROOKS (1971) *Limnol. Oceanogr.* **16**, 927–943.
DEEVEY, G. B. and A. L. BROOKS (1977) *Bull. Mar. Sci.* **27**, 256–291.
DELLA CROCE, N. (1974) *Fich. Ident. Zooplancton* **143**, 4 pp.
DELLA CROCE, N. and P. VENOGOPAL (1972) *Mar. Biol.* **15**, 132–138.
DELLA CROCE, N. and P. VENOGOPAL (1973) *Int. Revue Ges. Hydrobiol.* **58**, 713–721.
DESSIER, A. and A. LAUREC (1978) *Oceanol. Acta.* **1**, 285–304.
DIGBY, P. S. B. (1950) *J. Mar. Biol. Ass. U.K.* **29**, 393–438.
DIGBY, P. S. B. (1953) *J. Anim. Ecol.* **22**, 289–322.
DIGBY, P. S. B. (1954) *J. Anim. Ecol.* **23**, 298–338.
DOMANEVSKY, L. (1970) *Rapp. P-v Réun. Cons. Perm. Int. Explor. Mer* **159**, 223–226.
DONALDSON, H. A. (1975) *Bermuda Mar. Biol.* **31**, 37–50.
DONALDSON, H. A. (1976) *Mar. Biol.* **38**, 51–58.
DORSETT, D. A. (1961) *J. Mar. Biol. Ass. U.K.* **41**, 383–396.
DUCRET, F. (1975) *Cah. Biol. Mar.* **16**, 287–300.
DUNBAR, M. J. (1940) *J. Anim. Ecol.* **9**, 215–226.
DUNBAR, M. J. (1942) *Canada J. Res. D* **20**, 33–46.
DUNBAR, M. J. (1946) *J. Fish. Res. Bd. Can.* **6**, 419–434.
DUNBAR, M. J. (1957) *Can. J. Zool.* **35**, 797–819.
DUNBAR, M. J. (1962) *J. Mar. Res.* **20**, 76–91.
DUNBAR, M. J. (1963) *Fich. Ident. Zooplancton* **103**, 4 pp.
DUNBAR, M. J. (1964) *Serial Atlas of Marine Environment Folio 6*, Am. Geograph. Soc.; New York.
DUNBAR, M. J. and G. C. HARDING (1968) In J. E. Sater (Ed.) *Arctic Drifting Stations*. Arctic Inst. N. America. pp. 315–326.
DUSSART, B. H. (1965) *Hydrobiol.* **26**, 72–74.
EDINBURGH PLANKTON RECORDS (1973) *Bull. Mar. Ecol.* **7**, 1–174.
EDMONDS, J. S., K. A. FRANCESCONI, J. R. CANNON, C. L. RASTON, B. W. SKELTON and A. H. WHITE (1977) *Tetrahedron Lett.* **18**, 1543–1546.
EDMONDSON, W. T. (Ed.) (1966) *Marine Biology III*. New York Acad. Sci., 313 pp.
EDWARDS, C. (1978) *J. Mar. Biol. Ass. U.K.* **58**, 291–311.
EFFORD, I. E. (1970) *Crustaceana* **18**, 293–308.
EINARSSON, H. (1945) *Dana Rept.* **27**, 1–185.
EKMAN, S. (1953) *Zoogeography of the Sea*. Sidgwick & Jackson Ltd. London. 417 pp.
EMERY, A. R. (1968) *Limnol. Oceanogr.* **13**, 292–303.
ENRIGHT, J. T. (1977a) *Limnol. Oceanogr.* **22**, 118–125.
ENRIGHT, J. T. (1977b) *Limnol. Oceanogr.* **22**, 856–872.
ENRIGHT, J. T. (1979) *Limnol. Oceanogr.* **24**, 788–791.
ENRIGHT, J. T. and W. M. HAMNER (1967) *Science* **157**, 937–941.
ENRIGHT J. T. and H. W. HONEGGER (1977) *Limnol. Oceanogr.* **22**, 873–886.
ERIKSSON, S. (1973a) *Zoon* **1**, 37–68.
ERIKSSON, S. (1973b) *Zoon* **1**, 113–123.
ERIKSSON, S. (1973c) *Zoon* **1**, 95–111.
EPPLEY, R. W., E. H. RENGER, E. H. VENRICK and M. M. MULLIN (1973) *Limnol. Oceanogr.* **18**, 534–551.
ESTERLY, C. O. (1912) *Univ. Calif. Public. Zool.* **9**, 253–340.
ESTERLY, C. O. (1916) *Univ. Calif. Public. Zool.* **16**, 171–184.
EVANS, G. T. (1978) In J. H. Steele (Ed.) *Spatial Pattern in Plankton Communities*. Plenum Press; London. pp. 157–179.
EWALD, W. F. (1912) *J. Exp. Zool.* **13**, 591–612.
EYDEN, D. (1923) *Proc. Camb. Phil. Soc. Biol. Sci.* **1**, 49–55.
FAGE, L. (1941) *Dana Rep.* **19**, 1–52.
FAGE, L. (1942) *Dana Rep.* **23**, 1–67.
FAGER, E. W. (1957) *Ecology* **38**, 586–595.

FAGER, E. W. and J. A. McGOWAN (1963) *Zooplankton Species Groups in the North Pacific. Science*, (N.Y.) **140**, 453–466.
FAGETTI, E. (1972) *Mar. Biol.* **17**, 7–29.
FARKAS, T. and S. HERODEK (1964) *J. Lipid Res.* **5**, 369–373.
FARMER, L. and M. R. REEVE (1978) *Mar. Biol.* **48**, 311–316.
FARRAN, G. P. (1929) *British Antarctic Terra Nova Exp. Zool.* **8**, No. 3.
FARRAN, G. P. (1936) *Gt. Barrier Reef Exped. Sci. Rept.* **5**, 73–142.
FARRAN, G. P. (1947) *Proc. Roy. Irish Acad.* **51 B**, 121–136.
FARRAN, G. P. (1948) *Fich. Ident. Zooplancton* **11–17**.
FARRAN, G. P. (1949) *Gt. Barrier Reef Exped. Sci. Rept.* **2**, 291–312.
FARRAN, G. P. (1951) *Fich. Ident. Zooplancton* (revised by W. Vervoort) **32**, **33**, **34**, **35**, **36**, **37**, **38**, **39**, **40**.
FASHAM, M. J. R. and M. V. ANGEL (1975) *J. Mar. Biol. Ass. U.K.* **55**, 739–757.
FAUVEL, P. (1916) *Res. Camp. Sci. Monaco* **48**, 1–152.
FEIGENBAUM, D. and M. R. REEVE (1977) *Limnol. Oceanogr.* **22**, 1052–1058.
FENAUX, L. (1969) *Bull. Inst. Oceanogr. Monaco* **69**, (No. 1394), 1–28.
FENAUX, R. (1967) *Faune de l'Europe et du Bassin Mediterranean.* Masson et Cie; Paris. 116 pp.
FENAUX, R. (1973) In B. Zeitzschel (Ed.) *The Biology of the Indian Ocean.* Chapman & Hall; London: Springer-Verlag; Berlin. pp. 409–414.
FENAUX, R. (1976a) *Ann. Inst. Oceanographique* **52**, 89–191.
FENAUX, R. (1976b) *Mar. Biol.* **34**, 229–238.
FENAUX, R. (1977) *Proc. Symp. Warm Water Zooplankton.* NIO – UNESCO, Goa, pp. 487–510.
FERGUSON, C. F. (1973) Ph.D. Thesis, University of Southampton.
FERGUSON, C. F. and J. K. B. RAYMONT (1974) *J. Mar. Biol. Ass. U.K.* **54**, 719–725.
FEVOLDEN, S. E. (1979) *Sarsia* **64**, 189–198.
FISH, A. G. (1962) *Bull. Mar. Sci. Gulf Carib.* **12**, 1–38.
FISH, C. J. (1936a) *Bull. Mar. Biol. Lab. Woods Hole* **70**, 118–142.
FISH, C. J. (1936b) *Bull. Mar. Biol. Lab. Woods Hole* **70**, 193–215.
FISH, C. J. (1954) *Symp. Marine and Freshwater Plankton in the Indo-Pacific.* Bangkok, pp. 3–9.
FISH, C. J. and M. W. JOHNSON (1937) *J. Biol. Bd. Canada* **3**, 189–321.
FISHER, L. R. (1962) *Rapp. P-v. Réun. Cons. Perm. Int. Explor. Mer* **153**, 129–135.
FISHER, L. R. and E. H. GOLDIE (1959) *J. Mar. Biol. Ass. U.K.* **38**, 291–310.
FIVES, J. M. and F. I. O'BRIEN (1976) *J. Mar. Biol. Ass. U.K.* **56**, 197–211.
FLEMINGER, A. (1972) *Trans. Amer. Microsp. Soc.* **91**, 86–87.
FLEMINGER, A. (1973) *Fishery Bulletin* **71**, 965–1010.
FLEMINGER, A. (1975) *Estuarine Research* **1**, 392–419.
FLEMINGER, A. and K. HULSEMANN (1973) In B. Zeitzschel (Ed.) *The Biology of the Indian Ocean.* Chapman & Hall; London: Springer-Verlag; Berlin. pp. 339–347.
FLEMINGER, A. and K. HULSEMANN (1974) *Fishery Bulletin* **72**, 63–120.
FLEMINGER, A. and K. HULSEMANN (1977) *Mar. Biol.* **40**, 233–248.
FLORES, M. and G. J. BRUSCA (1975) *Bull. S. Calif. Acad. Sci.* **74**, 10–15.
FORSTER, G. R. (1951) *J. Mar. Biol. Ass. U.K.* **30**, 333–367.
FORSYTH, D. C. T. and L. T. JONES (1966) *Nature, Lond.* **212**, 1467–1468.
FOXON, G. E. H. (1940) *J. Mar. Biol. Ass. U.K.* **24**, 89–97.
FOXTON, P. (1956) *Discovery Rep.* **28**, 193–235.
FOXTON, P. (1961) *Discovery Rep.* **32**, 1–32.
FOXTON, P. (1964) *Crustaceana* **6**, 235–237.
FOXTON, P. (1966) *Discovery Rep.* **34**, 1–115.
FOXTON, P. (1969) *J. Mar. Biol. Ass. U.K.* **49**, 603–620.
FOXTON, P. (1970a) *J. Mar. Biol. Ass. U.K.* **50**, 939–960.
FOXTON, P. (1970b) *J. Mar. Biol. Ass. U.K.* **50**, 961–1000.
FOXTON, P. (1971) *Discovery Rep.* **35**, 179–201.
FRADE, R. and E. POSTEL (1955) *Rapp. P-v Réun. Cons. Perm. Int. Explor. Mer.* **137**, 33–35.
FRASER, J. H. (1936) *Discovery Rep.* **14**, 3–192.
FRASER, J. H. (1937) *J. Cons. Perm. Int. Explor. Mer* **12**, 311–320.
FRASER, J. H. (1939) *J. Cons. Perm. Int. Explor. Mer* **14**, 25–34.
FRASER, J. H. (1947) *Fich. Ident. Zooplancton* **10**, 4 pp.
FRASER, J. H. (1952) *Mar. Res. Scot. 1952*, **2**, 1–52.
FRASER, J. H. (1955) *Mar. Res. Scot. 1955*, **1**, 1–12.
FRASER, J. H. (1961) *Mar. Res. Scot. 1961*, **4**, 1–47.
FRASER, J. H. (1962) *Nature Adrift.* Foulis; London. 178 pp.
FRASER, J. H. (1965) *Ser. Atlas Mar. Environment* Folio *8*. Am. Geograph. Soc. New York. 2 pp.
FRASER, J. H. (1969) *Progr. Oceanogr.* **5**, 149–159.

FRASER, J. H. (1970) *J. Cons. Perm Int. Explor. Mer* **33**, 149–168.
FRONTIER, S. (1973) *Cah. ORSTOM Ser. Océanogr.* **11**, 259–290.
FRONTIER, S. (1974) D.Sc. Thesis. Univ. Provence, (Aix-Marseille).
FROST, B. W. (1972) *Limnol. Oceanogr.* **17**, 805–815.
FROST, B. W. (1974) *Mar. Biol.* **26**, 77–99.
FROST, B. W. (1975) *Limnol. Oceanogr.* **20**, 263–266.
FUDGE, H. (1968) *Nature*, **219**, 380–381.
FULLER, J. L. (1937) *Bull. Mar. Biol. Lab. Woods Hole* **72**, 233–246.
FULLER, J. L. and CLARKE, G. L. (1936) *Bull. Mar. Biol. Lab. Woods Hole* **70**, 308–320.
FULTON, J. (1968) *Technical Report 55, Fish. Res. Bd. Canada*, 141 pp.
FURNESTIN, M. L. (1955) *Rapp. P-v Réun. Cons. Perm. Int. Explor. Mer* **137**, 26–28.
FURNESTIN, M. L. (1957) *Rev. Trav. Inst. Pêches Marit.* **21**, 1–356.
FURNESTIN, M. L. (1964) *Rev. Trav. Inst. Pêches Marit.* **28**, 257–264.
FURNESTIN, M. L. (1970a) *Rapp. P-v Réun. Cons. Perm. Int. Explor. Mer.* **159**, 90–115.
FURNESTIN, M. L. (1970b) *Dana Rept.* **79**, 1–51.
FURUHASHI, K. (1966) *Publ. Seto Mar. Biol. Lab.* **14**, 295–322.
GAARDER, K. R. (1946) *Rep. Scient. Results Michael Sars N. Atlantic Deep Sea Exped.* **2**, 1–37.
GAMO, S. (1967) *Publ. Seto Mar. Biol. Lab.* **15**, 133–163.
GANAPATI, P. N. and D. V. RAO (1958) *Proc. Indian Acad. Sci. B.* **48**, 189–209.
GAPISHKO, A. I. (1971a) In M. Uda (Ed.) *The Ocean World Proc. Joint Ocean Assembly Tokyo.* Japan Soc. Promotion Science. pp. 434–436.
GAPISHKO, A. T. (1971b) *Oceanology* **11**, 399–403.
GARDINER, A. C. (1937) *J. Cons. Perm. Int. Explor. Mer.* **12**, 144–146.
GATTEN, R. R. and J. R. SARGENT (1973) *Nethl. J. Sea Res.* **7**, 150–158.
GAUDY, R. (1962) *Rec. Trav. St. Mar. Endoume* **27**, 93–184.
GAUDY, R. (1968) *Rapp. Com. Int. Mer Medit.* **19**, 517–519.
GAUDY, R. (1973) *Netherl. J. Sea Res.* **7**, 267–279.
GAUDY, R. (1974) *Mar. Biol.* **25**, 125–141.
GAUDY, R. (1975) *Mar. Biol.* **29**, 109–118.
GAUDY, R. (1977) *Mar. Biol.* **39**, 179–190.
GAUDY, R. and J. P. GUÉRIN (1977) *Mar. Biol.* **39**, 137–145.
GAUDY, R. and G. SEGUIN (1964) *Rec. Trav. St. Mar. Endoume* **34**, 211–217.
GAULD, D. T. (1951) *J. Mar. Biol. Ass. U.K.* **29**, 695–706.
GAULD, D. T. (1953) *J. Mar. Biol. Ass. U.K.* **31**, 461–474.
GAULD, D. T. (1964) In D. J. Crisp (Ed.) *Grazing in Terrestrial and Marine Environments.* Blackwell Scientific Publications; Oxford. pp. 239–245.
GAULD, D. T. and J. E. G. RAYMONT (1953) *J. Mar. Biol. Ass. U.K.* **31**, 447–460.
GEEN, G. H. and B. T. HARGRAVE (1966) *Int. Ass. Theor. Appl. Limnol.* **16**, 333–340.
GEORGE, J. (1967) *Proc. Symp. Indian Ocean, Bull. N.I.S. India* **38**, 641–648.
GEYNRIKH, A. K. (1968) *Oceanology* **8**, 231–239.
GEYNRIKH, A. K. (1973) *Oceanology* **13**, 94–103.
GHIRARDELLI, E. (1968) *Adv. Mar. Biol.* **6**, pp. 271–375.
GIBBS, R. H. and C. F. E. ROPER (1970) In R. Brooke Farquhar (Ed.) *Int. Symp. Biological Sound Scattering in the Ocean.* Maury Centre. Rept. 005, Washington, D.C. pp. 119–133.
GIBSON, V. R. and G. D. GRICE (1977) *J. Fish. Res. Bd. Can.* **33**, 847–854.
GIESKES, W. W. C. (1971) *Nethl. J. Sea Res.* **5**, 342–381.
GILFILLAN, E. (1972) In A. Y. Takenouti (Ed.) *Biological Oceanography of the Northern North Pacific Ocean.* Idemitsu Shoten; Tokyo, pp. 443–463.
GILFILLAN, E. (1976) *Mar. Biol.* **38**, 305–313.
GILAT, E. (1967) *Sea Fish. Res. Sta. Haifa Bull.* **45**, 79–95.
GILMER, R. W. (1972) *Science* **176**, 1239–1240.
GILMER, R. W. (1974) *J. Exp. Mar. Biol. Ecol.* **15**, 127–144.
GLOVER, R. S. (1967) *Symp. Zool Soc. London* **19**, 189–210.
GLYNN, P. W. (1973) *Mar. Biol.* **22**, 1–21.
GODEAUX, J. E. A. (1962) *Result Scient. Exped. Oceanogr. Belge Eaux Cot. Afr.* **3**, 1–33.
GODEAUX, J. E. A. (1977) *Atlantide Report* **12**, 7–24.
GOLD, K. (1968) *J. Protozool.* **15**, 193–194.
GOLD, K. (1969) *J. Protozool.* **16**, 507–509.
GOLD, K. (1970) *Helgolander Wiss. Meeresunters* **20**, 264–271.
GOODAY, A. J. and A. MOGUILEVSKY (1975) *J. Exp. Mar. Biol. Ecol.* **19**, 105–116.
GOPALAKRISHNAN, K. (1973) *Bull. Scripps Inst. Oceanogr.* **20**, 87 pp.
GOSSELCK, F. and E. KUEHNER (1973) *Mar. Biol.* **22**, 67–73.

GOSWAMI, S. and R. A. SELVAKUMAR (1977) *Proc. Symp. Warm Water Zooplankton NIO – UNESCO*, Goa, pp. 226–241.
GRAINGER, E. H. (1959) *J. Fish. Res. Bd. Canada* **16**, 453–501.
GRAINGER, E. H. (1962) *J. Fish. Res. Bd. Canada* **19**, 377–400.
GRAINGER, E. H. (1963) In M. J. Dunbar (Ed.) *Marine Distributions*, Roy. Soc. Canada, Spec. Publ. 5.
GRAINGER, E. H. (1965) *J. Fish. Res. Bd. Canada* **22**, 543–564.
GREVE, W. (1970) *Helgolander Wiss. Meeresunters* **20**, 304–317.
GREVE, W. (1972) *Helgolander Wiss. Meeresunters* **23**, 141–164.
GREVE, W. (1975a) *Fich. Ident. Zooplancton* **146**, 6 pp.
GREVE, W. (1975b) *Aquaculture* **6**, 77–82.
GREVE, W. (1977) *Helgolander Wiss. Meeresunters* **30**, 83–91.
GRICE, G. D. (1961) *Bull. Mar. Sci. Gulf Carib.* **10**, 217–226.
GRICE, G. D. (1962) *J. Mar. Res.* **20**, 97–109.
GRICE, G. D. (1963) *Bull. Mar. Sci. Gulf Carib.* **13**, 493–501.
GRICE, G. D. and V. R. GIBSON (1975) *Mar. Biol.* **31**, 335–337.
GRICE, G. D. and A. D. HART (1962) *Ecol. Monogr.* **32**, 287–309.
GRICE, G. D. and K. HULSEMANN (1965) *J. Zool.* **146**, 213–262.
GRIFFITHS, F. B. and J. CAPERON (1979) *Mar. Biol.* **54**, 301–309.
GUEREDRAT, J. A. (1971) *Mar. Biol.* **9**, 300–314.
GUEREDRAT, J. A. (1974) D. Sc. Thesis, ORSTOM Univ. Paris.
GUEREDRAT, J. A. and R. FRIESS (1971) *Cah. ORSTOM* **9**, 187–196.
GUÉRIN, J.-P. and R. GAUDY (1977) *Mar. Biol.* **44**, 65–70.
GULLIKSEN, B. (1972) *Sarsia* **51**, 83–96.
GUNTER, G. (1967) In G. H. Lauff (Ed.) *Estuaries*. A.A.A.S.; Washington, D.C. pp. 621–638.
GURNEY, R. (1938) *Gt. Barrier Reef Exped. Sci. Rep.* **6**(1), 1–60.
GURNEY, R. (1942) *Larvae of Decapod Crustacea*. Ray Society, London. 306 pp.
HAAGENSEN, D. A. (1976) *Caribbean Zooplankton*, Part II. O.N.R. Washington, D.C. pp. 551–712.
HADA, Y. (1972) In A. Y. Takenounti (Ed.) *Biological Oceanography of the Northern North Pacific Ocean*. Idemitsu Shoten, Tokyo. pp. 173–188.
HAECKEL, E. (1879) *Sytem der Medusen*. Gustav Fischer; Jena. 205 pp.
HAECKEL, E. (1881) *Challenger Rep. Zool.* **4**.
HAECKEL, E. (1887a) *Challenger Rep. Zool.* **18**.
HAECKEL, E. (1887b) *Die Radiolaria Eine Monographie* II.
HAECKER, V. (1907) *Archiv. fur Protistenkunde* **10**, 114–126.
HAIRSTON, N. G. (1976) *Proc. Nat. Acad. Sci. USA* **73**, 971–974.
HALCROW, K. (1963) *Limnol. Oceanogr.* **8**, 1–8.
HALL, J. R. and R. S. SCHELTEMA (1975) *Proc. Int. Symp. Biol. Sipuncula and Echiura I.* Kotor, pp. 183–197.
HAMBURGER and V. BUDDENBROCK (1911) *Nordisches Plankton Zoologie* **7**, 1–152.
HAMNER, W. M. (1977) *Proc. Symp. Warm Water Zooplankton*, NIO – UNESCO Goa, pp. 284–296.
HAMNER, W. M., L. P. MADIN, A. L. ALLDREDGE, R. W. GILMER and P. P. HAMNER (1975) *Limnol. Oceanogr.* **20**, 907–917.
HANSEN, J. H. (1922) *Res. Camp. Sci. Albert Ier Monaco*, **64**, 1–232.
HANSEN, V. K. (1960) *Meddr. Komm. Danm. Fisk. -og Havunders NS 2*, **23**, 1–53.
HAQ, S. M. (1967) *Limnol. Oceanogr.* **12**, 40–51.
HAQ, S. M., J. ALI KHAN and S. CHUGTAI (1973) In B. Zeitzschel (Ed.) *The Biology of the Indian Ocean*. Chapman & Hall; London: Springer-Verlag; Berlin. pp. 257–272.
HARBISON, G. R., D. C. BIGGS and L. P. MADIN (1977) *Deep-Sea Res.* **24**, 465–488.
HARBISON, G. R. and R. W. GILMER (1976) *Limnol. Oceanogr.* **21**, 517–528.
HARBISON, G. R., L. P. MADIN and N. R. SWANBERG (1978) *Deep-Sea Res.* **25**, 233–256.
HARDER, W. (1952) *Kurze Mitt. Fischereibiol. Abt. Max-Plank Inst. Meeresbiol. Wilhelmshaven* **1**, 28–34.
HARDER, W. (1957) *Ann. Biol.* **33**, 227–232.
HARDING, G. C. H. (1974) *J. Mar. Biol. Ass. U.K.* **54**, 141–155.
HARDY, A. C. (1924) *Fish. Invest. London Ser. II*, **7**, 1–53.
HARDY, A. C. (1956) *The Open Sea, Its Natural History: The World of Plankton*. Collins; London. 355 pp.
HARDY, A. C. and R. BAINBRIDGE (1954) *J. Mar. Biol. Ass. U.K.* **33**, 409–448.
HARDY, A. C. and E. R. GUNTHER (1935) *Discovery Rep.* **11**, 1–456.
HARDY, A. C. and W. N. PATON (1947) *J. Mar. Biol. Ass. U.K.* **26**, 467–526.
HARGRAVE, B. T. and G. H. GEEN (1968) *Limnol. Oceanogr.* **13**, 332–342.
HARRIS, E. (1959) *Bull. Bingham Oceanogr. Coll.* **17**, 31–65.
HARRIS, J. E. (1953) *Quart. J. Micr. Sci.* **94**, 537–550.
HARRIS, R. P. and G. A. PAFFENHOFER (1976a) *J. Mar. Biol. Ass. U.K.* **56**, 675–690.
HARRIS, R. P. and G. A. PAFFENHOFER (1976b) *J. Mar. Biol. Ass. U.K.* **56**, 875–888.

HARTENSTEIN, R. (1970) In J. W. Campbell (Ed.) *Comparative Biochemistry of Nitrogen Metabolism*, Vol. *1* Academic Press; London. pp. 299–385.
HARTMANN, J. (1970) *J. Cons. Perm. Int. Explor. Mer* **33**, 245–255.
HARTMANN, J. (1972) *Int. Revue Ges. Hydrobiol.* **57**, 559–571.
HARVEY, H. W. (1937) *J. Mar. Biol. Ass. U.K.* **22**, 97–100.
HARVEY, H. W. (1950) *J. Mar. Biol. Ass. U.K.* **29**, 97–137.
HARVEY, H. W., L. H. N. COOPER, M. V. LEBOUR and F. S. RUSSELL (1935) *J. Mar. Biol. Ass. U.K.* **20**, 407–441.
HARVEY, G. W. (1971) *Abstract Proc. Joint Oceanogr. Assembly* (Tokyo) S7–6, p. 275.
HAURY, L. R. (1973) *Limnol. Oceanogr.* **18**, 500–506.
HAURY, L. R., J. A. McGOWAN and P. H. WIEBE (1978) In J. H. Steele (Ed.) *Spatial Pattern in Plankton Communities.* Plenum Press; London. pp. 277–327.
HAURY, L. R. and D. WEIHS (1976) *Limnol. Oceanogr.* **21**, 797–803.
HAYS, J. D. (1965) *Biology of the Antarctic Seas II, Antarctic Research Series* **5**, 125–184.
HEATH, H. (1930) *J. Morph. Physiol.* **49**, 223–249.
HECHT, A. D., A. W. H. BÉ and L. LOTT (1976) *Science* **194**, 422–424.
HEDIN, H. (1974) *Zoon* **2**, 123–133.
HEDIN, H. (1975) *Zoon* **3**, 125–140.
HEEGARD, P. (1969) *Dana* **77**, 1–82.
HEINBOKEL, J. F. (1978a) *Mar. Biol.* **47**, 177–189.
HEINBOKEL, J. F. (1978b) *Mar. Biol.* **47**, 191–197.
HEINBOKEL, J. F. and J. R. BEERS (1979) *Mar. Biol.* **52**, 23–32.
HEINLE, D. R. (1966) *Chesapeake Sci.* **7**, 59–74.
HEINLE, D. R., R. P. HARRIS, J. F. USTACH and D. A. FLEMER (1977) *Mar. Biol.* **40**, 341–353.
HEINRICH, A. K. (1962) *J. Cons. Perm. Int. Explor. Mer* **27**, 15–24.
HEINRICH, A. K. (1963) *Translation RTS 7347 Trudy Inst. Oceanol.* **71**, 61–71.
HEINRICH, A. K. (1971a) *Abstract, Proc. Joint Oceanogr. Congress Tokyo.* pp. 432–433.
HEINRICH, A. K. (1971b) *Mar. Biol.* **10**, 290–294.
HEMPEL, G. and H. WEIKERT (1972) *Mar. Biol.* **13**, 70–88.
HENDERSON, G. T. D. (1964a) *Annls. Biol. Copenh.* **19**, 57–60.
HENDERSON, G. T. D. (1964b) *Annls. Biol. Copenh.* **20**, 80–87.
HENDERSON, G. T. D. (1964c) *Annls. Biol. Copenh.* **21**.
HENDERSON, G. T. D. (1966) *Annls. Biol. Copenh.* **23**, 21–63.
HERMAN, S. S. and J. R. BEERS (1969) *Bull. Mar. Sci.* **19**, 483–503.
HERON, A. C. (1968) in *Zooplankton Sampling. Monographs on Oceanographic Methodology 2.* UNESCO, Paris. pp. 19–25.
HERON, A. C. (1972a) *Oceanologia* **10**, 269–293.
HERON, A. C. (1972b) *Oceanologia* **10**, 294–312.
HERON, A. C. (1973) *J. Mar. Biol. Ass. U.K.* **53**, 429–435.
HERRING, P. J. (1967) *Symp. Zool. Soc. London* **19**, 215–235.
HERRING, P. J. (1971) *Comp. Biochem. Physiol.* **39B**, 739–746.
HERRING, P. J. (1972) *Nature* **238**, 276–277.
HERRING, P. J. (1973) *J. Mar. Biol. Ass. U.K.* **53**, 539–562.
HERRING, P. J. (1974) *Deep-Sea Res.* **21**, 91–94.
HERRING, P. J. (1977) *Nature* **267**, 788–793.
HERRING, P. J. (1978) *Comp. Biochem. Physiol.* **61B**, 391–393.
HERRING, P. J. and R. J. MORRIS (1975) In H. Barnes (Ed.) *Proc. 9th Europ. Mar. Biol. Symp.* Aberdeen Univ. Press. pp. 299–310.
HICKLING, C. F. (1925) *J. Mar. Biol. Ass. U.K.* **13**, 735–745.
HIDA, T. S. (1957) *Spec. Sci. Rep. U.S. Fish Wild Serv. Ser. Fish* **215**, 13 pp.
HIROTA, J. (1972) In A. Y. Takenouti (Ed.) *Biological Oceanography of the Northern North Pacific Ocean.* Idemitsu Shoten; Tokyo pp. 465–484.
HIROTA, J. (1974) *Fishery Bull.* (*U.S. Dept. Commerce*) **72**, 295.
HIROTA, R. (1961) *J. Sci. Hiroshima Univ.* **B 20**, 83–145.
HIMMELMAN, J. H. (1975) *J. Exp. Mar. Biol. Ecol.* **20**, 199–214.
HOBSON, E. S. and J. R. CHESS (1976) *Fishery Bull.* **74**, 567–598.
HOENIGMAN, J. (1963) *Rapp. P-v. Comm. Int. Mer. Medit.* **17**, 603–616.
HOLM-HANSEN, O., F. J. R. TAYLOR, and R. J. BARSDATE (1970) *Mar. Biol.* **7**, 37–46.
HOLMES, N. J. (1968) Ph.D. Thesis, Univ. of Southampton.
HOLMES, N. J. (1970) *C.E.R.L. Lab. Report RD/L/R 1672*, 1–20.
HOLT, E. W. L. and W. M. TATTERSALL (1905) *Ann. Rep. Fish. Ireland*, 99–152.
HOLTHUIS, L. B. (1951) *Atlantide Rep.* **2**, 7–187.

HOPKINS, T. L. (1969) *Limnol. Oceanogr.* **14**, 80–85.
HOPKINS, T. L. (1977) *Bull. Mar. Sci.* **27**, 467–478.
HOPNER PETERSEN, G., and M. A. CURTIS, (1980) *Dana* **1**, 53–64.
HORTON, P. A., M. ROWAN, K. E. WEBSTER and R. H. PETERS (1979) *Can. J. Zool.* **57**, 206–212.
HOTTA, H. and J. NAKASHIMA (1971) *Bull. Seikai Reg. Fish. Res. Lab.* **39**, 33–50.
HOUDE, E. D. (1978) *Bull. Mar. Sci.* **28**, 395–411.
HOWELL, B. R. (1973) *J. Cons. Perm. Int. Explor. Mer* **35**, 1–6.
HUDINAGA, M. and H. KASAHARA (1941) *Zool. Map. (Tokyo)* **54**, 108–118.
HUNT, H. G. (1968) *Bull. Mar. Ecol.* **6**, 225–249.
HYMAN, L. H. (1940) *The Invertebrates: Protozoa through Ctenophora* McGraw-Hill; London. 726 pp.
HYMAN, L. H. (1959) *The Invertebrates: Phylum Chaetognatha* McGraw-Hill; London. 783 pp.
ICANBERRY, J. W. and R. W. RICHARDSON (1973) *Limnol. Oceanogr.* **18**, 333–335.
IKEDA, T. (1970) *Bull. Fac. Fish. Hokkaido Univ.* **21**, 91–112.
IKEDA, T. (1971) *Bull. Fac. Fish. Hokkaido Univ.* **21**, 280–298.
IKEDA, T. (1974) *Mem. Fac. Fish. Hokkaido Univ.* **22**, 1–97.
IKEDA, T. (1977a) *Mar. Biol.* **41**, 241–252.
IKEDA, T. (1977b) *J. Exp. Mar. Biol. Ecol.* **29**, 263–277.
IKEDA, T. (1979) *J. Oceanogr. Soc. Japan* **35**, 1–8.
IKEDA, T. and S. MOTODA, (1975) *Pac. Sci. Ass. Spec. Symp. Mar. Sci. Dec. Hongkong Sess.* **2**, 24–28.
IKEDA, T. and S. MOTODA (1978) *Marine Science Communications* **4**, 329–346.
ILES, E. J. (1961) *Discovery Rep.* **31**, 299–326.
ISAACS, J. D., S. A. TONT and G. L. WICK (1974) *Deep-Sea Res.* **21**, 651–656.
IVANOV, B. G. (1970) *Mar. Biol.* **7**, 340–351.
JASCHNOV, W. A. (1961) *Translation U.S.S.R. Acad. Sci. Zool. Journ.* **40**, 1314–1334.
JASCHNOV, W. A. (1970) *Int. Revue Ges. Hydrobiol.* **55**, 197–212.
JAWED, M. (1969) *Limnol. Oceanogr.* **14**, 748–754.
JAWED, M. (1973) *Mar. Biol.* **21**, 173–179.
JEFFRIES, H. P. (1964) *Limnol. Oceanogr.* **9**, 348–358.
JEFFRIES, H. P. (1967) In G. H. Lauff (Ed.) *Estuaries* A.A.A.S.; Washington, D.C. pp. 500–503.
JEFFRIES, H. P. (1969) *Limnol. Oceanogr.* **14**, 41–52.
JERLOV, N. G. (1968) *Optical Oceanography*, (1st Edn) Elsevier; Amsterdam. 194 pp.
JERLOV, N. G. (1976) *Marine Optics*, (2nd Edn) Elsevier; Amsterdam. 231 pp.
JESPERSON, P. (1924) *Int. Rev. Ges. Hydrobiol. Hydrogr.* **12**, 102–115.
JESPERSON, P. (1935) *Dana Rept.* **7**, 1–44.
JESPERSON, P. (1939) *Meddr. Gronland.* **121**(3), 1–66.
JESPERSON, P. (1940) *Meddr. Komm. Danm. Fisk.-og Havunders. Serie Plankton* **3**(5), 1–77.
JESPERSON, P. (1944) *Meddr. Komm. Danm. Fisk.-og Havunders. Serie Plankton* **3**(7), 1–44.
JILLETT, J. B. (1968) *Aust. J. Mar. Freshwat. Res.* **19**, 19–30.
JILLETT, J. B. (1971) *New Zealand Oceanogr. Inst. Mem.* **53**, *New Zealand D.S.I.R. Bull.* **204**, 103 pp.
JITTS, H. R. (1969) *Aust. J. Mar. Freshwat. Res.* **20**, 65–75.
JOHANNES, R. E. (1964) *Limnol. Oceanogr.* **9**, 224–234.
JOHNSON, D. L. and R. S. BRAMAN (1975) *Deep-Sea Res.* **22**, 503–507.
JOHNSON, M. W. (1940) *J. Mar. Res.* **2**, 236–245.
JOHNSON, M. W. (1958) *J. Mar. Res.* **17**, 272–281.
JOHNSON, M. W. (1960) *Bull. Scripps Inst. Oceanogr.* **7**, 413–461.
JOHNSON, M. W. (1963) *Limnol. Oceanogr.* **8**, 89–102.
JOHNSON, M. W. and E. BRINTON (1963) In M. N. Hill (Ed.) *The Sea* Vol. 2 Wiley Interscience Publishers; London and New York pp. 381–414.
JOHNSON, W. H. (1938) *Biol. Bull. Woods Hole* **75**, 106–118.
JOHNSON, W. H. and J. E. G. RAYMONT (1939) *Biol. Bull. Woods Hole* **77**, 200–215.
JONES, N. S. (1950) *Biol. Rev.* **25**, 283–313.
JONES, N. S. (1955) *Discovery Rep.* **27**, 279–291.
JONES, N. S. (1956) *J. Anim. Ecol.* **25**, 217–252.
JONES, N. S. (1957) *Fich. Ident. Zooplankton* **73–76**.
JORGENSEN, C. B. (1962) *Rapp. P-v Réun. Cons. Perm. Int. Explor. Mer* **153**, 99–107.
JORGENSEN, C. B. (1966a) *Biology of Suspension Feeding*. Pergamon Press; Oxford. 357 pp.
JORGENSEN, C. B. (1966b) In W. T. Edmondson (Ed.) *Marine Biology III*. New York Acad. Sci., New York. pp. 69–133.
JORGENSEN, E. (1924) *Dana Rept. Exped. Mediter.* **2**, 101 pp.
KAMPA, E. M. and B. P. BODEN (1956) *Deep-Sea Res.* **4**, 73–92.
KAMSHILOV, M. M. (1955a) *Doklady Akad Nauk S.S.S.R.* **102**, 399–405.
KAMSHILOV, M. M. (1955b) *Tr.* Murmanskoy Biol. St. **2**, seen in abstract only.

KANDLER, R. I. (1950) *Ber. Dtsch. Komm. Meeresforsch* **12**, 47–85.
KANE, J. E. (1963) *Trans. Roy. Soc. N.Z.* **3**(5), 35–45.
KANE, J. E. (1966) *Discovery Rep.* **34**, 163–198.
KANWISHER, J. W. (1959) *Limnol. Oceanogr.* **4**, 210–217.
KASAHARA, S. and S. UYE (1979) *Mar. Biol.* **55**, 63–68.
KASAHARA, S., S. UYE and T. ONBE (1974) *Mar. Biol.* **26**, 167–171.
KASTURIRANGAN, L. R., M. SARASWATHY and T. C. GOPALAKRISHNAN (1973) In B. Zeitzschel (Ed.) *The Biology of the Indian Ocean.* Chapman & Hall; London: Springer-Verlag; Berlin. pp. 331–333.
KAWAGUCHI, K. (1969) *Bull. Plankton Soc. Japan* **16**, 63–66.
KELLY, G. F. and A. M. BARKER (1961) *Rapp. P-v. Réun. Cons. Perm. Int. Explor. Mer* **150**, 220–233.
KIELHORN, W. V. (1952) *J. Fish. Res. Bd. Can.* **9**, 223–264.
KILIACHENKOVA, V. A. (1970) *Rapp. P-v Réun. Cons. Perm. Int. Explor. Mer* **159**, 194–198.
KINZER, J. (1966a) *Deep-Sea Res.* **13**, 473–474.
KINZER, J. (1966b) *Proc. Symp. Oceanogr. Fish. Res. Trop. Atlantic* p. 45.
KING, J. E. and J. DEMOND (1953) *U.S. Fish and Wildlife Service Fish Bull.* **54**, 111–144.
KING, J. E. and T. S. HIDA (1957) *Pt. II U.S. Fish and Wildlife Service, Fish Bull.* **57**, 365–395.
KING, J. E. and R. T. B. IVERSON (1962) *Fish Bull. U.S. Wildlife Ser.* **210**, 271–321.
KLIE, W. (1943) *Fich. Ident. Zooplancton* **4**, 4 pp.
KLIE, W. (1944a) *Fich. Ident. Zooplancton* **5**, 4 pp.
KLIE, W. (1944b) *Fich. Ident. Zooplancton* **6**, 4 pp.
KLYASHTORIN, L. B. (1978) *Oceanology* **18**, 91–94.
KLYASHTORIN, L. B. and A. A. YARZHOMBEK (1973) *Oceanology* **13**, 575–580.
KOBAYASHI, N. (1967) *Publ. Seto Mar. Biol. Lab.* **15**, 403–414.
KOBAYASHI, N. and K. NAKAMURA (1967) *Publ. Seto Mar. Biol. Lab.* **15**, 173–184.
KOFOID, C. A. and A. S. CAMPBELL (1929) *Univ. Calif. Publs. Zool.* **34**, 1–403.
KOFOID, C. A. and A. S. CAMPBELL (1939) *Bull. Mus. Comp. Zool. Harvard* **84**, 473.
KOLLMER, W. E. (1963) *Invest. Rep. Mar. Res. Lab. S.W. Afr.* **8**, 1–78.
KOLOSOVA, Y. G. (1972) *Oceanology* **12**, 105–113.
KOLOSOVA, Y. G., K. A. KOKIN and A. N. VISHENSKIY (1977) *Oceanology* **17**, 705–709.
KOMAKI, Y. (1967) *Pacific Science* **21**, 433–448.
KONOVALOVA, G. V. and L. A. ROGACHENKO (1974) *Oceanology* **14**, 561–566.
KOS, M. S. (1975) In Z. A. Zvereva (Ed.) *Geographical and Seasonal Variability of Marine Plankton.* Israel Program for Scientific Translations, Jerusalem. pp. 128–168.
KOSLOW, J. A. (1979) *Limnol. Oceanogr.* **24**, 783–784.
KOSOBOKOVA, K. N. (1978) *Oceanology* **18**, 476–480.
KOTORI, M. (1972) In A. Y. Takenouti (Ed.) *Biological Oceanography of the Northern North Pacific Ocean.* Idemitsu Shoten, Tokyo. pp. 291–308.
KOZASA, E. (1974) *Bull. Seikai Reg. Fish. Res. Lab.* **45**, 15–21.
KRAEUTER, J. N. and E. M. SETZLER (1975) *Bull. Mar. Sci.* **25**, 66–74.
KRAMP, P. L. (1957) *Discovery Rep.* **29**, 1–128.
KRAMP, P. L. (1959a) *Mem. Inst. Sci. Nat. Belg. Galathea Rept.* **3**, 1–33.
KRAMP, P. L. (1959b) *Dana Rept.* **8** (No. 46) 1–283.
KRAMP, P. L. (1961) *J. Mar. Biol. Ass. U.K.* **40**, 1–469.
KRAMP, P. L. (1965) *Dana Rept.* **11** (No. 63).
KRAMP, P. L. (1968a) *Videnk Medd. Dansk. Natur. og Faun.* **131**, 67–98.
KRAMP, P. L. (1968b) *Videnk Medd. Dansk. Natur. og Faun.* **131**, 199–208.
KREMER, P. and S. NIXON (1976) *Estuar. Coastal Mar. Sci.* **4**, 627–639.
KREY, J. (1950) *Kieler Meeresforsch.* **7**, 58–75.
KRISHNASWAMY, S., J. E. G. RAYMONT and J. TUNDISI (1967) *Int: Revue Ges. Hydrobiol.* **52**, 447–451.
KROGH, A. (1931) *Biol. Rev.* **6**, 412–442.
KUKENTHAL, W. and T. KRUMBACH (1933) *Handbuch der Zoologie* **5**(2) de Gruyter, Paris.
KUTHALINGAM, M. D. K. (1959) *Current Science* **28**, 75–76.
LAGARDERE, J. P. (1978) *Fich. Ident. Zooplancton* **155**, **156**, **157**. 15 pp.
LALLI, C. M. (1970) *J. Exp. Mar. Biol. Ecol.* **4**, 101–118.
LALLI, C. M. (1972) *Biol. Bull. Mar. Biol. Lab. Woods Hole* **143**, 392–402.
LALLI, C. M. and R. J. CONOVER (1973) *Mar. Biol.* **19**, 13–22.
LAM, R. K. and B. W. FROST (1976) *Limnol. Oceanogr.* **21**, 490–500.
LAMPITT, R. S. (1978) *Limnol. Oceanogr.* **23**, 1228–1231.
LANCE, J. (1960) Ph.D. Thesis, University of Southampton.
LANCE, J. (1965) *Comp. Biochem. Physiol.* **14**, 155–165.
LANDRY, M. R. (1978) *Limnol. Oceanogr.* **23**, 1103–1113.
LASKER, R. (1966) *J. Fish. Res. Bd. Can.* **23**, 1291–1317.

LAURSEN, D. (1953) *Dana Rept.* **38**, 40 pp.
LAVAL, P. (1965) *Compte Rendu Séances Acad. Sci.* **260**, 6195–6198.
LEATHERLAND, T. M., J. D. BURTON, F. CULKIN, M. J. MCCARTNEY and R. J. MORRIS (1973) *Deep-Sea Res.* **20**, 679–685.
LEBORGNE, L. P. (1973) *Mar. Biol.* **19**, 249–257.
LEBORGNE, R. and D. BINET (1974) *Tethys* **6**, 321–322.
LEBOUR, M. V. (1918a) *J. Mar. Biol. Ass. U.K.* **11**, 433–469.
LEBOUR, M. V. (1918b) *J. Mar. Biol. Ass. U.K.* **12**, 22–47.
LEBOUR, M. V. (1919) *J. Mar. Biol. Ass. U.K.* **12**, 261–324.
LEBOUR, M. V. (1921) *J. Mar. Biol. Ass. U.K.* **12**, 458–467.
LEBOUR, M. V. (1922) *J. Mar. Biol. Ass. U.K.* **12**, 644–677.
LEBOUR, M. V. (1923) *J. Mar. Biol. Ass. U.K.* **13**, 70–92.
LEBOUR, M. V. (1924) *J. Mar. Biol. Ass. U.K.* **13**, 402–420.
LEBOUR, M. V. (1928) *Proc. Zool. Soc. London* **98**, 473–556.
LEBOUR, M. V. (1949) *Proc. Zool. Soc. London* **20**(2), pp.
LEBOUR, M. V. (1954) *Discovery Rep.* **27**, 219–234.
LEBOUR, M. V. (1959) *Atlantide Rept.* **5**, 119–143.
LEBRASSEUR, R. J. (1965) *Fish. Res. Bd. Can. Manuscript Rep. Ser.* **202**, 153 pp.
LEBRASSEUR, R. J. (1969) *J. Fish. Res. Bd. Can.* **26**, 1631–1645.
LEBRASSEUR, R. J. and O. D. KENNEDY (1972) *Fishery Bull.* **70**, 25–36.
LEE, R. F. (1974) *Mar. Biol.* **26**, 313–318.
LEE, B. D. and P. CHIN (1971) *Publ. Mar. Lab. Pusan Fish. Coll.* **4**, 1–8 (In Korean; English Abstract).
LEE, R. F. and J. HIROTA (1973) *Limnol. Oceanogr.* **18**, 227–239.
LEE, R. F., J. HIROTA and A. M. BARNETT (1971) *Deep-Sea Res.* **18**, 1147–1165.
LEE, R. F., J. C. NEVENZEL and A. G. LEWIS (1974) *Lipids* **9**(11), 891–898.
LEE, R. F., J. C. NEVENZEL and G. A. PAFFENHOFER (1971) *Mar. Biol.* **9**, 99–108.
LEGALL, J.-Y and M. L'HERROUX (1971) *Rapp. Sci. Tech. Centre Nat. Expl. Océans* **1**, 1–31.
LEGAND, M. (1969) *Aust. J. Mar. Freshwat. Res.* **20**, 85–103.
LEGAND, M., P. BOURRETT, P. FOURMANOIR, R. GRANDPERRIN, J. A. GUEREDRAT, A. MICHEL, P. RANCURIEL, R. REPELIN and C. ROGER (1972) *Cah. ORSTOM Sér. Océanogr* **10**, 303–393.
LEGARE, J. E. H. (1961) *Bol. Inst. Oceanogr. Univ. Oriente* **1**, 191–218.
LEGARE, J. E. H. and E. ZOPPI (1961) *Bol. Inst. Oceanogr. Univ. Oriente* **1**, 149–171.
LEHMAN, J. T. (1976) *Limnol. Oceanogr.* **21**, 501–516.
LELOUP, E. (1955) *Rep. Sci. Res. 'Michael Sars' Expedition* **5**, 1–24.
LENZ, J. (1973a) *J. Cons. Perm. Int. Explor. Mer* **35**, 32–35.
LENZ, J. (1973b) In B. Zeitzschel (Ed.) *The Biology of the Indian Ocean.* Chapman & Hall; London: Springer-Verlag; Berlin. pp. 239–241.
LE ROUX, A. (1973) *Mar. Biol.* **22**, 159–166.
LEWIS, J. B. (1954) *Bull. Mar. Sci. Gulf Carib.* **5**, 161–189.
LEWIS, J. B., J. K. BRUNDRITT and A. G. FISH (1962) *Bull. Mar. Sci. Gulf Carib.* **12**, 73–94.
LIE, U. (1967) *Sarsia* **30**, 49–74.
LIGHTHART, B. (1969) *J. Fish. Res. Bd. Can.* **26**, 299–304.
LIN, A. L. and P. L. ZUBKOFF (1973) *Helgolander Wiss, Meeresunters.* **25**, 206–213.
LITTLEPAGE, J. L. (1964) *Actual. Scient. Ind.* **1312**, 463–470.
LOCHHEAD J. H. (1936) *J. Linn. Soc. (Zool.)* **39**, 429–441.
LOHMANN, H. (1908) *Wiss. Meeresuntersuch. Abt. Kiel N.F.* **10**, 129–370.
LOHMANN, H. (1933–34) In W. Kukenthal and T. Krumbach (Eds.) Handbuch der Zoologie **5** (2) de Gruyter, Paris. pp. 1–202.
LONGHURST, A. R. (1966) *Deep-Sea Res.* **13**, 213.
LONGHURST, A. R. (1967a) *Deep-Sea Res.* **14**, 51–63.
LONGHURST, A. R. (1967b) *Deep-Sea Res.* **14**, 393–408.
LONGHURST, A. R. (1968) *Limnol. Oceanogr.* **13**, 143–155.
LONGHURST, A. R. (1976) In D. H. Cushing and J. J. Walsh (Eds.) *The Ecology of the Seas.* Blackwell; Oxford. pp. 116–137.
LONGHURST, A. R. and V. BAINBRIDGE (1964) *Bull. Inst. Fr. Afr. Noire* **26**, Ser A(2) 337–402.
LONGHURST, A. R., A. D. REITH, R. E. BOWER and D. L. R. SEIBERT (1966) *Deep-Sea Res.* **13**, 213–222.
LONGHURST, A. and R. WILLIAMS (1979) *J. Plankton Res.* **1**, 1–28.
LOOSANOFF, V. L. (1958) In A. A. Buzzati-Traverso (Ed.) *Perspectives in Marine Biology*, Univ. California Press, pp. 483–495.
LOOSANOFF, V. L. and H. C. DAVIS (1963) *Adv. Mar. Biol.* **1**, 1–136.
LOOSANOFF, V. L., W. S. MILLER and P. B. SMITH (1951) *J. Mar. Res.* **10**, 59–81.
LOPEZ, M. T. (1966) *Bolm. Inst. Oceanogr. Sao Paulo* **16**, 47–54.

LOVEGROVE, T. (1956) *Fich. Ident. Zooplancton* **63**, pp.
LUCAS, C. E. (1936) *J. Cons. Perm. Int. Explor. Mer* **11**, 343–361.
LYSHOLM, B. O. and K. F. NORDGAARD (1945) *Rep. Sci. Res. Michael Sars Deep-Sea Expn.* **5**, 1–60.
MCALLISTER, C. D. (1970) In J. H. Steele (Ed.) *Marine Food Chains* Oliver & Boyd, Edinburgh. pp. 419–457.
MACDONALD, R. (1927) *J. Mar. Biol. Ass. U.K.* **14**, 753–784.
MACDONALD, A. G. and I. GILCHRIST (1978) *Mar. Biol.* **45**, 9–21.
MACDONALD, A. G., I. GILCHRIST and J. M. TEAL (1972) *J. Mar. Biol. Ass. U.K.* **52**, 213–223.
MCGOWAN, J. A. (1960) *Deep-Sea Res.* **6**, 125–139.
MCGOWAN, J. A. (1963) *Systematics Assoc. Pub.* **5**. Speciation in the Sea. 109–128.
MCGOWAN, J. A. (1967) *CALCOFI Atlas* **6**, 1–218.
MCGOWAN, J. A. (1968) *Veliger* **3**, Supplement, 103–130.
MCHUGH, J. L. (1967) In G. H. Lauff (Ed.) *Estuaries*. A.A.A.S.; Washington, D.C. pp. 581–620.
MACKIE, G. O. (1964) *Pro. Roy. Soc. B.* **159**, 266–291.
MACKIE, G. O. (1973) *Publ. Seto. Mar. Biol. Lab.* **20**, 745–756.
MACKIE, G. O. and D. A. BOAG (1963) *Publ. Staz. Zool. Napoli* **33**, 178–196.
MACKINNON, D. L. and R. S. J. HAWES (1961) *An Introduction to the Study of Protozoa*. Clarendon Press, Oxford. 506 pp.
MACKINTOSH, N. A. (1934) *Discovery Rep.* **9**, 67–158.
MACKINTOSH, N. A. (1937) *Discovery Rep.* **16**, 367–412.
MACKINTOSH, N. A. (1962) *Biologie Antarctique Proc. SCAR Symp.* Hermann; Paris. pp. 29–38.
MACKINTOSH, N. A. (1964) *Proc. Roy. Soc. A.* **281**, 21–38.
MACKINTOSH, N. A. (1972) *Discovery Rep.* **36**, 1–94.
MCLAREN, I. A. (1963) *J. Fish Res. Bd. Can.* **20**, 685–727.
MCLAREN, I. A. (1966) *Ecology* **47**, 852–855.
MCWHINNIE, M. A. and P. MARCINIAK (1964) *Antarctic Res. Series 1 Biology of the Antarctic Seas*. Am. Geophysical Union, pp. 63–72.
MACKAS, D. and R. BOHRER (1976) *J. Exp. Biol. Ecol.* **25**, 77–85.
MADHUPRATAP, M. and P. HARIDAS (1975) *Indian J. Mar. Sci.* **4**, 77–85.
MADIN, L. P. (1974) *Mar. Biol.* **25**, 143–147.
MADIN, L. P. and G. R. HARBISON (1977) *Deep-Sea Res.* **24**, 449–463.
MADIN, L. P. and G. R. HARBISON (1978a) *Bull. Mar. Sci.* **28**, 335–344.
MADIN, L. P. and G. R. HARBISON (1978b) *Bull. Mar. Sci.* **28**, 680–687.
MAHNKEN, C. V. W. (1969) *Bull. Mar. Sci.* **19**, 550–567.
MAHNKEN, C. V. W. and J. W. JOSSI (1967) *J. Cons. Perm. Int. Explor. Mer* **31**, 38–45.
MAIS, K. F. and T. JOW (1960) *Calif. Dept. Fish Game Fish. Bull.* **46**, 117–150.
MAKAROV, R. R. (1969) *Oceanology* **9**, 251–261.
MAKAROV, R. R. (1977) *Oceanology* **17**, 208–213.
MAKAROV, R. R. (1979) *Mar. Biol.* **52**, 377–386.
MANRIQUE, F. A. (1977) *Proc. Symp. Warm Water Zooplankton* NIO – UNESCO, Goa pp. 242–249.
MANWELL, C., C. M. A. BAKER, P. A. ASHTON and E. D. S. CORNER (1967) *J. Mar. Biol. Ass. U.K.* **47**, 145–169.
MARAK, R. R. (1960) *J. Cons. Perm. Int. Explor. Mer* **25**, 147–157.
MARE, M. F. (1940) *J. Mar. Biol. Ass. U.K.* **24**, 461–482.
MARGALEF, R. (1963) *Rapp. P-v Comm. Int. Mer Medit.* **17**, 511–512.
MARGALEF, R. (1968) *Rapp. P-v Comm. Int. Mer Medit.* **19**, 565–566.
MARGALEF, R. (1969) *Publ. Stat. Zool. Napoli* **37**, Suppl. 40–61.
MARGINEANU, C. (1963) *Rapp. P-v Comm. Int. Mer Medit.* **17**, 523–530.
MARGULIS, R. Y. (1972) *Oceanology* **12**, 420–425.
MARINARO, J. Y. (1968) *Pelagos* **10**, 133–137.
MARR, J. (1962) *Discovery Rep.* **32**, 33–464.
MARSHALL, N. B. (1954) *Aspects of Deep Sea Biology*. Hutchinson; London. 380 pp.
MARSHALL, N. B. (1960) *Discovery Rep.* **31**, 1–122.
MARSHALL, S. M. (1934) *Gt. Barrier Reef Exped. Sci Rept.* **4**, 623–664.
MARSHALL, S. M. (1949) *J. Mar. Biol. Ass. U.K.* **28**, 45–122.
MARSHALL, S. M. (1969) *Fich. Ident. Zooplancton* **117–127**.
MARSHALL, S. M. (1973) *Adv. Mar. Biol.* **11**, 57–120.
MARSHALL, S. M., A. G. NICHOLLS and A. P. ORR (1933–34) *J. Mar. Biol. Ass. U.K.* **19**, 823.
MARSHALL, S. M., A. G. NICHOLLS and A. P. ORR (1935) *J. Mar. Biol. Ass. U.K.* **20**, 1–28.
MARSHALL, S. M. and A. P. ORR (1952) *J. Mar. Biol. Ass. U.K.* **30**, 527–547.
MARSHALL, S. M. and A. P. ORR (1955a) *The Biology of a Marine Copepod Calanus finmarchicus (Gunnerus)* Oliver & Boyd, Edinburgh. 188 pp.

MARSHALL, S. M. and A. P. ORR (1955b) *J. Mar. Biol. Ass. U.K.* **34**, 495–529.
MARSHALL, S. M. and A. P. ORR (1956) *J. Mar. Biol. Ass. U.K.* **35**, 587–603.
MARSHALL, S. M. and A. P. ORR (1958) *J. Mar. Biol. Ass. U.K.* **37**, 459–472.
MARSHALL, S. M. and A. P. ORR (1960) *J. Mar. Biol. Ass. U.K.* **39**, 135–147.
MARSHALL, S. M. and A. P. ORR (1961) *J. Mar. Biol. Ass. U.K.* **41**, 463–488.
MARSHALL, S. M. and A. P. ORR (1962) *Rapp. P-v Réun. Cons. Perm. Int. Explor. Mer* **153**, 92–98.
MARSHALL, S. M. and A. P. ORR (1966) *J. Mar. Biol. Ass. U.K.* **46**, 513–530.
MARTENS, P. (1978) *Fich. Ident. Zooplancton* **163**, 4 pp.
MARTIN, J. H. (1968) *Limnol. Oceanogr.* **13**, 63–71.
MARTIN, J. H. (1970) *Limnol. Oceanogr.* **15**, 756–761.
MARUMO, R. (1966) *J. Oceanogr. Soc. Japan* **122**, 7–15.
MARUMO, R. and S. NAGASAWA (1973) *J. Oceanogr. Soc. Japan* **129**, 267–275.
MATSUZAKI, M. (1975) *Oceanogr. Mag.* **26**, 57–62, 63–72.
MATTHEWS, J. B. L. (1964) *Sarsia* **16**, 1–46.
MATTHEWS, J. B. L. (1966) In H. Barnes (Ed.) *Some Contemporary Studies in Marine Science*. Allen & Unwin; London. pp. 479–492.
MATTHEWS, J. B. L. (1967) *Bull. Mar. Ecol.* **6**, 159–179.
MATTHEWS, J. B. L. (1969) *Bull. Mar. Ecol.* **6**, 251–273.
MATTHEWS, J. B. L. (1973) *Sarsia* **54**, 75–90.
MATTHEWS, J. B. L. and J. L. W. BAKKE (1977) *Helgolander wiss. Meeresunters* **30**, 47–61.
MATTHEWS, J. B. L. and L. HESTAD (1977) *Sarsia* **63**, 57–63.
MATTHEWS, J. B. L., L. HESTAD and J. L. W. BAKKE (1978) *Oceanologica Acta* **10**, 277–284.
MATTHEWS, J. B. L. and S. PINNOI (1973) *Sarsia* **52**, 123–144.
MAUCHLINE, J. (1959) *Proc. Zool. Soc. Lond.* **132**, 627–639.
MAUCHLINE, J. (1960) *Proc. Roy. Soc. Edinb. B.* **67**, 141–179.
MAUCHLINE, J. (1966) In H. Barnes (Ed.) *Some Contemporary Studies in Marine Science*, Allen & Unwin; London. pp. 493–510.
MAUCHLINE, J. (1972) *Deep-Sea Res.* **19**, 753–780.
MAUCHLINE, J. (1973) *J. Mar. Biol. Ass. U.K.* **53**, 801–817.
MAUCHLINE, J. and L. R. FISHER (1967) *Ser. Atlas Mar. Environ.* **13**, 2 pp.
MAUCHLINE, J. and L. R. FISHER (1969) *Adv. Mar. Biol.* **7**, 1–454.
MAYER, A. G. (1910) *Medusae of the World I, II and III* Carnegie Inst; Washington. pp. 499–735.
MAYER, A. G. (1912) *Ctenophores of the Atlantic Coast of North America*. Carnegie Inst; Washington Publ. **162**, 58 pp.
MAYZAUD, P. (1973) *Mar. Biol.* **21**, 19–28.
MAYZAUD, P. (1976) *Mar. Biol.* **37**, 47–58.
MAYZAUD, P. and S. DALLOT (1973) *Mar. Biol.* **19**, 307–314.
MAYZAUD, P. and J. L. M. MARTIN (1975) *J. Exp. Mar. Biol. Ecol.* **17**, 297–310.
MAYZAUD, P. and S. A. POULET (1978) *Limnol. Oceanogr.* **23**, 1144–1154.
MEEK, R. P. and J. J. CHILDRESS (1973) *Deep-Sea Res.* **20**, 1111–1118.
MENON, M. D. and K. C. GEORGE (1977) In *Proc. Symp. Warm Water Zooplankton*. NIO – UNESCO, Goa, pp. 205–213.
MENON, M. K., P. G. MENON and V. T. PAULINOSE (1967) *Proc. Symp. Indian Ocean Bull. NISI.* **38**, pp. 753–757.
MENON, P. G. and V. T. PAULINOSE (1973) *IOBC Handbook* **5**, 163–171.
MENSAH, M. A. (1966) *Proc. Symp. Oceanogr. Fish. Res. Tropical Atlantic*. pp. 50–51.
MENZEL, D. W. and J. H. RYTHER (1961) *J. Cons. Perm. Int. Explor. Mer* **26**, 250–258.
MICHAEL, E. L. (1911) *Univ. Calif. Publ. Zool.* **8**, 21–186.
MICHEL, H. B. and M. FOYO (1976) *Caribbean Zooplankton Part. 1* ONR., Washington, D.C. 549 pp.
MILEIKOVSKIY, S. A. (1966) *Oceanology* **6**, 396–404.
MILEIKOVSKIY, S. A. (1968a) *Oceanology* **8**, 553–562.
MILEIKOVSKIY, S. A. (1968b) *Mar. Biol.* **1**, 161–167.
MILEIKOVSKIY, S. A. (1968c) *Sarsia* **34**, 209–216.
MILEIKOVSKIY, S. A. (1969) *Oceanology* **9**, 549–557.
MILEIKOVSKIY, S. A. (1970a) *Mar. Biol.* **5**, 180–194.
MILEIKOVSKIY, S. A. (1970b) *Mar. Biol.* **6**, 317–334.
MILLAR, R. H. (1952) *J. Mar. Biol. Ass. U.K.* **31**, 41–61.
MILLAR, R. H. (1954) *J. Mar. Biol. Ass. U.K.* **33**, 33–48.
MILLAR, R. H. (1974) *Mar. Biol.* **28**, 127–129.
MILLAR, R. H. (1974) *Chesapeake Sci.* **15**, 1–8.
MILLIMAN, J. D. and T. F. MANHEIM (1968) *Deep-Sea Res.* **15**, 505–507.
MINODA, T. (1967) *Recent Oceanog. Works Japan* **9**, 161–168.

MINODA, T. (1972) In A. Y. Takenout (Ed.) *Biological Oceanography of the Northern North Pacific Ocean. Idemitsu Shoten*; Tokyo. pp. 323–333.
MISTAKIDIS, M. N. (1957) *Fish. Invest. Ser. II*, **21**, 1–52.
MITCHELL, R. (1974) Ph.D. Thesis. Univ. of Southampton.
MOGUILEVSKY, A. and M. V. ANGEL (1975) *Mar. Biol.* **32**, 295–302.
MOGUILEVSKY, A. and A. J. GOODAY (1971) In H. Löffler and D. Danielopol (Eds.) *Proc. 6th Int. Symp. on Ostracods.* W, Junk; The Hague. pp. 263–270.
MOISEEV, P. A. (1970) In M. W. Holdgate (Ed.) *Antarctic Ecology 1.* Academic Press; London. pp. 213–216.
MOLLOY, F. M. E. (1958) Ph.D. Thesis, Univ. of London.
MONROE, C. C. A. (1930) *Discovery Rep.* **2**, 1–222.
MONTEIRO, J. D. C. (1970) M.Sc. Thesis. Univ. of Southampton.
MOORE, E. and F. SANDER (1976) *Estuarine and Coastal Mar. Sci.* **4**, 589–607.
MOORE, E. and F. SANDER (1977) *Ophelia* **16**, 77–96.
MOORE, H. B. (1949) *Bull. Bingham Oceanogr. Coll.* **12**, 1–97.
MOORE, H. B. (1950) *Biol. Bull. Woods Hole* **99**, 181–212.
MOORE, H. B. (1952) *Bull. Mar. Sci. Gulf Carib.* **1**, 278–305.
MOORE, H. B. and E. G. CORWIN (1956) *Bull. Mar. Sci. Gulf Carib.* **6**, 273–287.
MOORE, H. B. and A. C. FRUE (1959) *Bull. Mar. Sci.* **9**, 421–440.
MOORE, H. B. and N. N. LOPEZ (1970) *Bull. Mar. Sci.* **20**, 980–986.
MORAITOU-APOSTOLOPOULOU, M. (1969) *Mar. Biol.* **3**, 1–3.
MORAITOU-APOSTOLOPOULOU, M. (1971) *Mar. Biol.* **9**, 92–97.
MORAITOU-APOSTOLOPOULOU, M. (1978) *Thalassographica* **2**, 155–172.
MORAITOU-APOSTOLOPOULOU, M. and V. KIORTSIS (1973) *Mar. Biol.* **20**, 137–143.
MORAITOU-APOSTOLOPOULOU, M. and V. KIORTSIS (1977) *Thalassographica* **1**, 205–213.
MOREIRA, G. S. and W. B. VERNBERG (1968) *Mar. Biol.* **1**, 282–284.
MORIOKA, Y. (1972) In A. Y. Takenouti (Ed.) *Biological Oceanography of the Northern North Pacific Ocean.* Idemitsu Shoten; Tokyo. pp. 309–321.
MORRIS, R. J. (1971a) *Deep-Sea Res.* **18**, 525–529.
MORRIS, R. J. (1971b) *J. Mar. Biol. Ass. U.K.* **51**, 21–31.
MORRIS, R. J. (1972) *Mar. Biol.* **16**, 102–107.
MORRIS, R. J. (1973) Ph.D. Thesis. Univ. of Southampton.
MORRIS, R. J., M. E. ARMITAGE, J. E. G. RAYMONT, C. F. FERGUSON and J. K. B. RAYMONT (1977) *J. Mar. Biol. Ass. U.K.* **57**, 181–189.
MORRIS, R. J., C. F. FERGUSON and J. E. G. RAYMONT (1973) *J. Mar. Biol. Ass. U.K.* **53**, 657–664.
MORRIS, R. J. and J. R. SARGENT (1973) *Mar. Biol.* **22**, 77–83.
MORRISON, G. (1971) *J. Fish. Res. Bd. Can.* **28**, 379–381.
MORTON, J. E. (1954) *J. Mar. Biol. Ass. U.K.* **33**, 297–312.
MORTON, J. E. (1957) *Fich. Ident. Zooplancton* **79**, 4 pp. and **80**, 4 pp.
MORTON, J. E. (1958) *J. Mar. Biol. Ass. U.K.* **37**, 287–297.
MOTODA, S. and R. MARUMO (1963) *Proc. Symp. Kuroshio* pp. 40–61.
MOYSE, J. (1963) *J. Cons. Perm. Int. Explor. Mer* **28**, 175–187.
MOYSE, J. and E. W. KNIGHT-JONES (1967) *Proc. Symp. on Crustacea MBA India* **2**, 595–611.
MULLIN, M. M. (1963) *Limnol. Oceanogr.* **8**, 239–240.
MULLIN, M. M. (1966) In H. Barnes (Ed.) *Some Contemporary Studies in Marine Science.* Allen & Unwin; London. pp. 545–554.
MULLIN, M. M. (1969) *Oceanogr. Mar. Biol. Ann. Rev.* **7**, 293–314.
MULLIN, M. M. (1979) *Limnol. Oceanogr.* **24**, 774–777.
MULLIN, M. M. and E. R. BROOKS (1967) *Limnol. Oceanogr.* **12**, 657–666.
MULLIN, M. M. and E. R. BROOKS (1970a) In J. H. Steele, (Ed.) *Marine Food Chains.* Oliver & Boyd; Edinburgh. pp. 74–95.
MULLIN, M. M. and E. R. BROOKS (1970b) *Limnol. Oceanogr.* **15**, 748–755.
MULLIN, M. M. and E. R. BROOKS (1970c) *Bull. Scripps Inst. Oceanogr.* **17**, 89–103.
MULLIN, M. M., E. F. STEWART and F. J. FUGLISTER (1975) *Limnol. Oceanogr.* **20**, 259–263.
MURAKAMI, A. (1959) *Nakai Reg. Fish Res. Lab.* **12**, 1–5.
MURRAY, J. and J. HJORT (1912) *Depths of the Ocean.* Macmillan & Co.; London. 821 pp.
MUSAYEVA, E. I. and E. A. SHUSHKINA (1978) *Oceanology* **18**, 343–346.
MUUS, B. J. (1953) *Fich Ident. Zooplancton* **52**, 6 pp. and **53**, 5 pp.
MUUS, B. J. (1963) *Fich Ident. Zooplancton* **94**. 5 pp.; **95**, 3 pp.; **96**, 6 pp.; **97**, 5 pp.; **98**, 4 pp.
NAGASAWA, S. and R. MARUMO (1975) *Bull. Plankton Soc. Japan* **21**, 23–40.
NAIR, K. K. C. (1972) *Current Science* **41**, 185–186.
NAIR, K. K. C. (1977) *Proc. Symp. Warm Water Zooplankton.* NIO – UNESCO, Goa, pp. 155–167.
NAIR, K. K. C. and D. J. TRANTER (1971) *J. Mar. Biol. Ass. India* **13**, 203–210.

NAIR, K. K. C., P. G. JACOB and S. KUMARAN (1973) In B. Zeitzschel (Ed.) *The Biology of the Indian Ocean.* Chapman & Hall; London; Springer-Verlag; Berlin. pp. 349–356.
NAIR, V. R. (1973) *IOBC Handbook* **5**, 87–96.
NAIR, V. R. (1977) *Proc. Symp. Warm Water Zooplankton*, NIO – UNESCO, Goa, pp. 168–195.
NAIR, V. R. and T. S. S. RAO (1973) In B. Zeitzschel (Ed.) *The Biology of the Indian Ocean.* Chapman & Hall, London; Springer-Verlag; Berlin. pp. 293–317.
NAKAI, Z. (1955) *Tokai Reg. Fish. Res. Lab. Spec. Publ.* **5**, 12–24.
NASSOGNE, A. (1970) *Helgolander wiss. Meeresunters* **20**, 333–345.
NAYLOR, E. (1955) *J. Anim. Ecol.* **24**, 255–269.
NAYLOR, E. (1957a) *J. Mar. Biol. Ass. U.K.* **36**, 599–602.
NAYLOR, E. (1957b) *Fich. Ident. Zooplancton* **77** and **78**.
NAYLOR, E. (1965) *Adv. Mar. Biol.* **3**, pp.
NELLEN, W. (1973) In B. Zeitzschel (Ed.) *The Biology of the Indian Ocean.* Chapman & Hall; London; Springer-Verlag; Berlin. pp. 415–430.
NELLEN, W. and G. HEMPEL (1970) *Wiss. Komm. Meeresforsch.* **21**, 311–348.
NEMOTO, T. (1957) *Sci. Rep. Whales Res. Inst., Tokyo* **12**, 33–89.
NEMOTO, T. (1967) *Inf Bull. Plankt. Japan. (Matsue Commem. Vol.)* 158–171.
NEMOTO, T. (1968a) *Symp. on Antarctic Oceanography.* pp. 240–253.
NEMOTO, T. (1968b) *J. Oceanogr. Soc. Japan* **24**, 253–260.
NEMOTO, T. (1970) In J. H. Steele (Ed.) *Marine Food Chains.* Oliver & Boyd, Edinburgh, pp. 241–252.
NEMOTO, T. (1971/72) *Proc. Roy. Soc. Edinb. B.* **73**, 259–265.
NEMOTO, T. and Y. SAIJO (1968) *J. Oceanogr. Soc. Japan* **24**, 46–48.
NEMOTO, T., K. KAMADA and K. HARA (1972) *Mar. Biol.* **14**, 41–47.
NESIS, K. N. (1965) *Oceanology* **5**, 102–108.
NESIS, K. N. (1972) *Oceanology* **12**, 416–437.
NETO, T. S. (1970) *Rapp. P-v Réun. Cons. Perm. Int. Explor. Mer.* **159**, 118.
NETO, T. S. and I. PAIVA (1966) *Notas Mimeogr. Centro. Biol. Aquat. Trop. (Lisbon)* **2**, 51 pp.
NICHOLLS, A. G. (1933a) *J. Mar. Biol. Ass. U.K.* **19**, 83–110.
NICHOLLS, A. G. (1933b) *J. Mar. Biol. Ass. U.K.* **19**, 139–164.
NIVAL, P. and S. NIVAL (1976) *Limnol. Oceanogr.* **21**, 24–38.
NIVAL, P., G. MALARA, R. CHARRA, I. PALAZZOLI and S. NIVAL (1974) *J. Exp. Mar. Biol. Ecol.* **15**, 231–260.
NOUVEL, H. (1943) *Res. Camp. Sci. Monaco* **105**, 1–128.
NOUVEL, H. (1950) *Fich. Ident. Zooplancton*, **18–27**.
OKUTANI, T. and J. A. MCGOWAN (1969) *Bull. Scripps Inst. Oceanogr.* **14**, 1–90.
OMMANNEY, F. D. (1936) *Discovery Rep.* **13**, 277–384.
OMORI, M. (1965) *J. Oceanogr. Soc. Japan* **21**, 212–220.
OMORI, M. (1967) *Deep-Sea Res.* **14**, 525–532.
OMORI, M. (1969) *Bull. Ocean Res. Inst. Univ. Tokyo* **4**, 1–83.
OMORI, M. (1970) In J. H. Steele (Ed.) *Marine Food Chains.* Oliver & Boyd, Edinburgh. pp. 113–126.
OMORI, M. (1974) *Adv. Mar. Biol.* **12**, 233–324.
OMORI, M. (1977) *Proc. Symp. Warm Water Zooplankton*, NIO – UNESCO, Goa, pp. 1–12.
ONBE, T. (1977) *Proc. Symp. Warm Water Zooplankton*, NIO – UNESCO, Goa, pp. 383–398.
ORR, A. P. (1934a) *Proc. Roy. Soc. Edinb. B.* **54**, 51–55.
ORR, A. P. (1934b) *J. Mar. Biol. Ass. U.K.* **19**, 613–632.
ORTON, J. H. (1920) *J. Mar. Biol. Ass. U.K.* **12**, 339–366.
OSTVEDT, O. J. (1955) *Hval. Skrifter* **40**, 1–93.
OWRE, H. B. (1960) *Bull. Mar. Sci. Gulf Carib.* **9**, 255–322.
OWRE, H. B. (1962) *Bull. Mar. Sci. Gulf Carib.* **12**, 489–495.
OWRE, H. B. (1972) *Bull. Mar. Sci.* **22**, 483–521.
OWRE, H. B. and M. FOYO (1964) *Bull. Mar. Sci. Gulf Carib.* **14**, 359–372.
PAFFENHOFER, G. A. (1970) *Helgolander Wiss. Meeresunters* **20**, 346–359.
PAFFENHOFER, G. A. (1971) *Mar. Biol.* **11**, 286–298.
PAFFENHOFER, G. A. (1973) *Mar. Biol.* **22**, 183–185.
PAFFENHOFER, G. A. (1976) *Limnol. Oceanogr.* **21**, 39–50.
PAFFENHOFER, G. A. and R. P. HARRIS (1976) *J. Mar. Biol. Ass. U.K.* **56**, 327–344.
PAFFENHOFER, G. A. and R. P. HARRIS (1979) *Adv. Mar. Biol.* **16**, 211–308.
PAFFENHOFER, G. A. and S. C. KNOWLES (1979) *J. Mar. Res.* **37**, 35–49.
PANIKKAR, N. K. and T. S. S. RAO (1973) *IOBC Handbook* **5**, pp. 111–162.
PARANJAPE, M. A. (1967) *J. Fish. Res. Bd. Can.* **24**, 1229–1240.
PARIN, N. V. (1967) *Oceanology* **7**, 115–121.
PARIN, N. V., N. N. GORBUNOVA and M. CHUVASOV (1972) *Oceanology* **12**, 894–899.
PARKER, G. H. (1902) *Bull. U.S. Fish. Comm.* **21**, 103–123.

PARSONS, T. R. (1976) In D. H. Cushing and J. J. Walsh (Eds.) *The Ecology of the Seas*. Blackwell, Oxford. pp. 81–97.
PARSONS, T. R. and D. J. BLACKBOURN (1968) *Nethl. J. Sea Res.* **4**, 27–31.
PARSONS, T. R. and R. J. LEBRASSEUR (1968) *Calif. Mar. Res. Comm. CALCOFI Rept.* **12**, 54–63.
PARSONS, T. R. and R. J. LEBRASSEUR (1970) in J. H. Steele (Ed.) *Marine Food Chains*. Oliver and Boyd, Edinburgh, pp. 325–343.
PARSONS, T. R. and M. TAKAHASHI (1973) *Biological Oceanographic Processes*. Pergamon Press, Oxford. 186 pp.
PARSONS, T. R., R. J. LEBRASSEUR and W. E. BARRACLOUGH (1970) *J. Fish. Res. Bd. Can.* **27**, 1251–1264.
PARSONS, T. R., R. J. LEBRASSEUR and J. D. FULTON (1967) *J. Oceanogr. Soc. Japan* **23**, 10–17.
PARSONS, T. R., R. J. LEBRASSEUR, J. D. FULTON and O. D. KENNEDY (1969) *J. Exp. Mar. Biol. Ecol.* **3**, 39–50.
PATEL, B. and D. J. CRISP (1960) *Physiol. Zool.* **33**, 104–119.
PAULINOSE, V. T. (1973) *IOBC Handbook* **5**, pp. 97–110.
PAULINOSE, V. T. and P. N. ARAVINDAKSHAN (1977) *Proc. Symp. Warm Water Zooplankton*. NIO – UNESCO. Goa, pp. 132–136.
PAULSEN, O. (1906) *Meddr. Komm. Danm. Fisk. -og Havunders. Serie Plankton* **1**, 1–21.
PAVSHTIKS, E. A. (1975) in Z. A. Zvereva (Ed.) *Geographical and Seasonal Variability of Marine Plankton*. Israel Program for Scientific Translations, Jerusalem. pp. 200–247.
PEARCY, W. G. and C. A. FORSS (1969) *Limnol. Oceanogr.* **14**, 755–765.
PEARCY, W. G. and L. F. SMALL (1968) *J. Fish. Res. Bd. Can.* **25**, 1311–1316.
PEARRE, S. (1973) *Ecology* **54**, 300–314.
PEARRE, S. (1979a) *J. Plankton Res.* **1**, 29–44.
PEARRE, S. (1979b) *Limnol. Oceanogr.* **24**, 781–782.
PEARSE, J. S. (1969) *Mar. Biol.* **3**, 110–116.
PEARSE, J. S. (1970) *Bull. Mar. Sci.* **20**, 697–720.
PEARSE, J. S. (1972) *J. Exp. Mar. Biol. Ecol.* **8**, 167–186.
PEARSE, J. S. (1978) *Bull. Mar. Sci.* **28**, 92–101.
PERMITIN, Y. E. (1970) In M. W. Holdgate (Ed.) *Antarctic Ecology 1*, SCAR, Academic Press; London. pp. 177–182.
PERUYEVA, Y. G. (1977) *Oceanology* **17**, 719–725.
PETER, G. (1972) *Rec. Zool. Surv. India*, **67**, 343–356.
PETER, G. (1973) *IOBC Handbook* **5**, pp. 76–86.
PETER, G. (1974) *Mahasagar. Bull. Natn. Inst. Oceanogr.* **7**, 95–86.
PETER, G. (1977) *Proc. Symp. Warm Water Zooplankton*. NIO – UNESCO. Goa, pp. 87–92.
PETERSON, W. T., C. B. MILLER and A. HUTCHINSON (1979) *Deep-Sea Res.* **26**, 467–494.
PETIPA, T. S. (1965) *Transl. N.S.* *72*, Acad. Sci. U.Kr.S.S.R., 102–110.
PETIPA, T. S. (1966) In *Physiology of Marine Animals*. Akad. Nauk. S.S.S.R. pp. 60–81.
PETIPA, T. S., E.V. PAVLOVA and G. N. MIRONOV (1970) In J. H. Steele (Ed.) *Marine Food Chains*. Oliver and Boyd, Edinburgh. pp. 142–167.
PETIPA, T. S., E. V. PAVLOVA and Y. SOROKIN (1973) In M. E. Vinogradov (Ed.) *Life Activity of Pelagic Communities in the Ocean Tropics*. Israel Program for Scientific Translations, Jerusalem. pp. 135–165.
PETRUSHEVSKAYA, M. G. (1971a) In B. N. Funnell and W. R. Riedel (Eds.) Micropalaeontology of Oceans. Cambridge Univ. Press. pp. 319–329.
PETRUSHEVSKAYA, M. G. (1972b) *Oceanology* **12**, 534–545.
PHILIPPON, M. R. (1972) Doctorate Thesis. Univ. de Provence 213 pp.
PIANKA, H. D. (1974) In A. V. Giese and J. S. Pearse (Eds.) *Reproduction of Marine Invertebrates* Vol. 1, Academic Press; London. pp. 201–265.
PIERCE, E. L. (1941) *J. Mar. Biol. Ass. U.K.* **25**, 113–124.
PIERCE, E. L. (1953) *J. Mar. Res.* **12**, 75–92.
PIKE, R. B. and D. I. WILLIAMSON (1958) *Fich. Ident. Zooplancton* **81**.
PILKINGTON, M. C. and V. FRETTER (1970) *Helgolander Wiss. Meeresunters* **20**, 576–593.
PILLAI, N. K. (1973) *IOBC Handbook* **4**, 1–125.
PILLAY, K. K. and N. B. NAIR (1971) *Mar. Biol.* **11**, 152–166.
PIRLOT, J. M. (1929) *Mem. Soc. Roy. Sci. Liege, Ser. 3* **15**, 1–196.
PIRLOT, J. M. (1939) *Res. Camp. Sci. Prince Albert, Monaco*, **102**, 1–64.
POMEROY, L. R., H. M. MATHEWS and H. S. MIN (1963) *Limnol. Oceanogr.* **8**, 50–55.
PONOMAREVA, L. A. (1955) *Zool. J.* **34**, 85–97.
PONOMAREVA, L. A. (1966) *Euphausiids of the North Pacific, their Distribution and Ecology*. Israel Program for Scientific Translations, Jerusalem. 154 pp.
POSTA, A. (1963) *Cah. Biol. Mar.* **4**, 201–210.
POSTEL, E. (1955) *Rapp. P-v Réun. Cons. Perm. Int. Explor. Mer* **137**, 14–16.

POULET, S. A. (1973) *Limnol. Oceanogr.* **18**, 564–573.
POULET, S. A. (1974) *Mar. Biol.* **25**, 109–123.
POULET, S. A. (1976) *Mar. Biol.* **34**, 117–125.
POULET, S. A. (1977) *J. Fish. Res. Bd. Can.* **34**, 2381–2387.
POULET, S. A. and MARSOT (1978) *Science* **200**, 1403–1405.
POULSEN, E. M. (1962) *Dana Rept.* **57**, 414 pp.
POULSEN, E. M. (1969a) *Dana Rept.* **75**, 1–100.
POULSEN, E. M. (1969b) *Fich. Ident. Zooplancton* **116**, 7 pp.
POULSEN, E. M. (1973) *Dana Rept.* **84**, 1–224.
PROKHOROV, V. S. (1968) *Rapp. P-v Réun. Cons. Perm. Int. Explor. Mer* **158**, 23–31.
PROVASOLI, L. (1966) In W. T. Edmondson (Ed.) *Marine Biology III* New York Acad. Sci.; New York. pp. 205–242.
PROVASOLI, L., K. SHIRAISHI and J. LANCE (1959) *Ann. N.Y. Acad. Sci.* **77**, 250–261.
PRYGUNKOVA, R. V. (1968) *Dokl. Akad. Nauk. SSSR* **182**(6) 1447–1450. (Translation).
PUGH, P. R. (1974) *J. Mar. Biol. Ass. U.K.* **54**, 25–90.
PUGH, P. R. (1975) *J. Exp. Mar. Biol. Ecol.* **20**, 77–97.
PUGH, P. R. (1977) *Proc. Symp. Warm Water Zooplankton*, NIO – UNESCO, Goa, pp. 362–378.
PYEFINCH, K. A. (1948) *J. Mar. Biol. Ass. U.K.* **27**, 464–503.
QASIM, S. Z. (1956) *J. Cons. Perm. Int. Explor. Mer* **21**, 144–155.
QUETIN, L. B. and J. J. CHILDRESS (1976) *Mar. Biol.* **38**, 327–334.
RADZIEJEWSKA, T., J. CHOJNACKI and J. MASLOWSKI (1973) *Mar. Biol.* **23**, 111–113.
RAE, K. M. and J. H. FRASER (1941) *Hull Bull. Mar. Ecol.* **1**(4) 171–238.
RAE, K. M. and C. B. REES (1947) *Hull Bull. Mar. Ecol.* **2**(11) 95–132.
RAKUSA-SUSZCZEWSKI, S. (1967) *J. Cons. Perm. Int. Explor. Mer.* **31**, 46–55.
RAKUSA-SUSZCZEWSKI, S. (1968) *J. Cons. Perm. Int. Explor. Mer.* **32**, 226–231.
RAKUSA-SUSZCZEWSKI, S., M. A. MCWHINNIE and M. O. CAHOON (1976) *Limnol. Oceanogr.* **21**, 763–765.
RALPH R. (1965) Ph.d. Thesis, Univ. of Southampton.
RAMIREZ, F. C. (1977) *Proc. Symp. Warm Water Zooplankton*, NIO – UNESCO, Goa, pp. 65–68.
RAO, T. S. S. (1973) In B. Zeitzschel (Ed.) *The Biology of the Indian Ocean.* Chapman & Hall; London: Springer-Verlag; Berlin. pp. 243–255.
RAO, T. S. S. and S. KELLY (1962) *J. Zool. Soc. India*, **14**, 219–225.
RAO, T. S. S., and L. J. UROSA (1974) *Bol. Inst. Oceanogr. Univ. Oriente* **13**, 67–78.
RASSOULZADEGAN, F. and M. ETIENNE (1981) *Limnol. Oceanogr.* **26**, 258–270.
RAYMONT, J. E. G. (1959) *Limnol. Oceanogr.* **4**, 479–491.
RAYMONT, J. E. G. (1966) In J. B. Cragg (Ed.) *Advances in Ecological Research.* Academic Press; London. pp. 117–205.
RAYMONT, J. E. G. (1970) In R. Brook Farquhar (Ed.) *Proc. Symp on biological Sound Scattering in the Ocean.* Maury Centre Rept. 005, Washington, D.C., pp. 134–146.
RAYMONT, J. E. G. (1972) In R. B. Clark and R. J. Wootton (Eds.) *Essays in Hydrobiology.* Univ. of Exeter. pp. 83–91.
RAYMONT, J. E. G. and B. G. A. CARRIE (1964) *Int. Revue. Ges. Hydrobiol.* **49**, 185–232.
RAYMONT, J. E. G. and R. J. CONOVER (1961) *Limnol. Oceanogr.* **6**, 154–164.
RAYMONT, J. E. G. and D. T. GAULD (1951) *J. Mar. Biol. Ass. U.K.* **29**, 681–693.
RAYMONT, J. E. G. and F. GROSS (1942) *Proc. Roy. Soc. Edinb. B*, **61**, 267–287.
RAYMONT, J. E. G. and S. KRISHNASWAMY (1960) *J. Mar. Biol. Ass. U.K.* **39**, 239–248.
RAYMONT, J. E. G. and S. KRISHNASWAMY (1968) *Int. Revue. Ges. Hydrobiol.* **53**, 563–572.
RAYMONT, J. E. G. and E. LINFORD (1966) *Int. Revue. Ges. Hydrobiol.* **51**, 485–488.
RAYMONT, J. E. G. and R. S. MILLER (1962) *Int. Revue. Ges. Hydrobiol.* **47**, 169–209.
RAYMONT, J. E. G., J. AUSTIN and E. LINFORD (1966) In H. Barnes (Ed.) *Some Contemporary Studies in Marine Science.* Allen & Unwin Ltd. London. pp. 597–605.
RAYMONT, J. E. G., J. AUSTIN and E. LINFORD (1968) *J. Mar. Biol. Ass. U.K.* **48**, 735–760.
RAYMONT, J. E. G., C. F. FERGUSON and J. K. B. RAYMONT (1973) *Spl. Publ. Mar. Biol. Ass. India*, 91–99.
RAYMONT, J. E. G., S. KRISHNASWAMY and J. TUNDISI (1967) *J. Cons. Perm. Int. Explor. Mer* **31**, 164–169.
RAYMONT, J. E. G., R. J. MORRIS, C. F. FERGUSON and J. K. B. RAYMONT (1975) *J. Exp. Mar. Biol. Ecol.* **17**, 261–267.
RAYMONT, J. E. G., R. T. SRINIVASAGAM and J. K. B. RAYMONT (1969a) *Deep-Sea Res.* **16**, 141–156.
RAYMONT, J. E. G., R. T. SRINIVASAGAM and J. K. B. RAYMONT (1969b) *Int. Revue. Ges. Hydrobiol.* **54**, 357–365.
RAYMONT, J. E. G., R. T. SRINIVASAGAM and J. K. B. RAYMONT (1971a) *Deep-Sea Res.* **18**, 1167–1178.
RAYMONT, J. E. G., R. T. SRINIVASAGAM and J. K. B. RAYMONT (1971b) *J. Mar. Biol. Ass. U.K.*, **51**, 581–588.
RAZNIEWSKI, J. (1970) *Rapp. P-v Réun. Cons. Perm. Int. Explor. Mer* **159**, 199–201.

RAZOULS, S. (1972) *J. Exp. Mar. Biol. Ecol.* **9**, 145–153.
REDFIELD, A. C. (1939) *Bull. Mar. Biol. Lab. Woods Hole* **76**, 26–47.
REDFIELD, A. C. (1941) *Bull. Mar. Biol. Lab. Woods Hole* **80**, 86–110.
REDFIELD, A. C. and A. BEALE (1940) *Bull. Mar. Biol. Lab. Woods Hole* **79**, 459–487.
REES, W. J. (1966) (Ed.) *The Cnidaria and their Evolution Symp. Zool. Soc. London*, pp. 199–222.
REEVE, M. R. (1963) *J. Exp. Biol.* **40**, 195–205.
REEVE, M. R. (1964) *Bull. Mar. Sci. Gulf Carib.* **14**, 103–122.
REEVE, M. R. (1966) In H. Barnes (Ed.) *Some Contemporary Studies in Marine Science*. Allen & Unwin Ltd.; London. pp. 613–630.
REEVE, M. R. (1970a) In J. H. Steele (Ed.) *Marine Food Chains* Oliver & Boyd; Edinburgh. pp. 168–189.
REEVE, M. R. (1970b) *Bull. Mar. Sci.* **20**, 894–921.
REEVE, M. R. (1977) *Proc. Symp. Warm Water Zooplankton* NIO – UNESCO, Goa, pp. 528–537.
REEVE, M. R. and T. C. COSPER (1975) In A. C. Giese and J. S. Pearse (Eds.) *Reproduction of Marine Invertebrates* Vol. 2. Academic Press; London. pp. 157–184.
REEVE, M. R. and M. A. WALTER (1972a) *J. Exp. Mar. Biol. Ecol.* **9**, 191–200.
REEVE, M. R. and M. A. WALTER (1972b) *Biol. Bull. Woods Hole* **143**, 207–214.
REEVE, M. R. and M. A. WALTER (1977) *J. Exp. Mar. Biol. Ecol.* **29**, 211–221.
REEVE, M. R. and M. A. WALTER (1978) *Adv. Mar. Biol.* **15**, 249–287.
REEVE, M. R., J. E. G. RAYMONT and J. K. B. RAYMONT (1970) *Mar. Biol.* **6**, 357–364.
REEVE, M. R., M. A. WALTER, K. DARCY and T. IKEDA (1977) *Bull. Mar. Sci.* **27**, 105–113.
REID, D. M. (1955) *Atlantide Rept.* **3**, 7–40.
REID, J. L., E. BRINTON, A. FLEMINGER, E. L. VENRICK and J. A. MCGOWAN (1978) In H. Charnock and G. Deacon (Eds.) *Proc. Gen. Symp. Joint Oceanogr. Ass., Edinburgh* Plenum Press; London. pp. 65–130.
REID, P. C. and A. W. G. JOHN (1978) *J. Mar. Biol. Ass. U.K.* **58**, 551–557.
RENZ, G. W. (1976) *Bull. Scripps Inst. Océanogr.* **22**, 1–267.
REPELIN, R. (1970) *Cah. ORSTOM. Sér. Océanogr.* **8**, 65–109.
RESHETNJAK, V. V. (1966) *Faunna SSSR N.S.* **94**, Nauka, Moscow.
RESHETNJAK, V. V. (1971) In B. M. Funnell and W. R. Riedel (Eds.) *Micropalaeontology of Oceans.* Cambridge University Press. pp. 343–349.
RICE, A. L. (1967) *Fich. Ident. Zooplancton* **112**.
RICE, M. E. (1967) *Ophelia* **4**, 143–171.
RICE, M. E. (1975) In A. C. Giese and J. S. Pearse (Eds.) *Reproduction of Marine Invertebrates 2.* Academic Press; London. pp. 67–127
RICHMAN, S. and J. M. ROGERS (1969) *Limnol. Oceanogr.* **14**, 701–709.
RICHMAN, S., D. R. HEINLE and R. HUFF (1977) *Mar. Biol.* **42**, 69–84.
RIEDEL, W. R. (1971) In B. M. Funnell and W. R. Riedel (Eds.) *Micropalaeontology of Oceans.* Cambridge Univ. Press. pp. 567–594.
RILEY, G. A. (1947) *J. Mar. Res.* **6**, 104–113.
RILEY, G. A., H. STOMMEL and D. F. BUMPUS (1949) *Bull. Bingham. Oceanogr. Coll.* **12**, 1–169.
ROBERTSON, A. I. and R. K. HOWARD (1978) *Mar. Biol.* **48**, 207–213.
ROBERTSON, P. B. (1971) *Bull. Mar. Sci.* **21**, 841–865.
ROBERTSON, S. B. and B. W. FROST (1977) *J. Exp. Mar. Biol. Ecol.* **29**, 231–244.
RODHOUSE, P. G. (1977) Ph.D. Thesis, Univ. of Southampton.
RODHOUSE, P. G. (1978) *J. Exp. Mar. Biol. Ecol.* **34**, 1–22.
RODRIGUEZ, G. (1969) *Mem. Simp. Intern. Lagunas Costeras.* UNAM–UNESCO, pp. 591–600.
ROE, H. S. J. (1972a) *J. Mar. Biol. Ass. U.K.* **52**, 315–343.
ROE, H. S. J. (1972b) *J. Mar. Biol. Ass. U.K.* **52**, 525–552.
ROE, H. S. J. (1972c) *J. Mar. Biol. Ass. U.K.* **52**, 1021–1044.
ROE, H. S. J. (1974) *Mar. Biol.* **28**, 99–113.
ROGER, C. (1973a) *Mar. Biol.* **18**, 312–316; 317–320; and 321–326.
ROGER, C. (1973b) *Mar. Biol.* **19**, 54–60; 61–65; 66–68.
ROGER, C. (1974a) *Mem. ORSTOM* **71**, 1–265.
ROGER, C. (1974b) *Cah. ORSTOM Sér. Océanogr.* **12**, 221–239.
ROGER, C. (1975) *Mar. Biol.* **32**, 365–378.
ROGER, C. (1977) *Proc. Symp. Warm Water Zooplankton.* NIO–UNESCO, Goa, pp. 309–318.
ROGER, C. (1978) *J. Exp. Mar. Biol. Ecol.* **33**, 57–83.
ROJAS de MENDIOLA, B. (1971) In J. D. Costlow (Ed.) *Fertility of the Sea 2.* Gordon & Breach; New York. pp. 417–440.
ROMAN, M. R. (1978) *Limnol. Oceanogr.* **23**, 1245–1248.
ROSE, M. (1929) *Res. Camp. Sci. Monaco*, **78**.
ROSE, M. (1933) *Faune de France 26. Copépodes Pélagiques* Lechavalier; Paris. 374 pp.

ROSENTHAL, H. and G. HEMPEL (1970) In J. H. Steele (Ed.) *Marine Food Chains*. Oliver and Boyd; Edinburgh. pp. 344–364.
ROSSIGNOL, M. (1955) *Rapp. P-v Réun. Cons. Perm. Int. Explor Mer* **137**, 17–20.
RUDYAKOV, Y. A. (1970) *Mar. Biol.* **6**, 98–105.
RUDYAKOV, Y. A. (1972a) *Oceanology* **12**, 886–890.
RUDYAKOV, Y. A. (1972b) *Oceanology* **12**, 773–775.
RUDYAKOV, Y. A. (1973) In M. E. Vinogradov (Ed.) *Life Activity of Pelagic Communities in the Ocean Tropics*. Israel Program for Scientific Translations, Jerusalem. pp. 240–255.
RUDYAKOV, Y. A. (1979) *Oceanology* **19**, 196–199.
RUDYAKOV, Y. A. and V. B. TSEYTLIN (1976) *Oceanology* **16**, 184–187.
RUDYAKOV, Y. A. and N. M. VORONINA (1973) *Oceanology* **13**, 423–426.
RUSSELL, F. S. (1927) *Biol. Rev.* **2**, 213–256.
RUSSELL, F. S. (1928) *J. Mar. Biol. Ass. U.K.* **15**, 429–454.
RUSSELL, F. S. (1931a) *J. Mar. Biol. Ass. U.K.* **17**, 391–414.
RUSSELL, F. S. (1931b) *J. Mar. Biol. Ass. U.K.* **17**, 767–775.
RUSSELL, F. S. (1932) *J. Mar. Biol. Ass. U.K.* **18**, 131–146 and 147–160.
RUSSELL, F. A. (1933) *J. Mar. Biol. Ass. U.K.* **18**, 555–558 and 559–574.
RUSSELL, F. S. (1934a) *J. Mar. Biol. Ass. U.K.* **19**, 555–558.
RUSSELL, F. S. (1934b) *Gt. Barrier Reef Exped. Sci. Rept.* **2**, 176–201.
RUSSELL, F. S. (1934c) *J. Mar. Biol. Ass. U.K.* **19**, 569–584.
RUSSELL, F. S. (1935a) *Rapp. P-v Réun. Cons. Perm. Int. Explor. Mer* **95**, 5–30.
RUSSELL, F. S. (1935b) *J. Mar. Biol. Ass. U.K.* **20**, 309–332.
RUSSELL, F. S. (1935c) *J. Mar. Biol. Ass. U.K.* **20**, 147–179.
RUSSELL, F. S. (1936) *J. Mar. Biol. Ass. U.K.* **20**, 507–522.
RUSSELL, F. S. (1939) *J. Cons. Perm. Int. Explor. Mer.* **14**, 171–192.
RUSSELL, F. S. (1953) *The Medusae of the British Isles*. Cambridge Univ. Press. 530 pp.
RUSSELL, F. S. (1967) *J. Mar. Biol. Ass. U.K.* **47**, 469–473.
RUSSELL, F. S. (1970) *The Medusae of the British Isles, II*, Cambridge Univ. Press, 284 pp.
RUSSELL, F. S. (1976) *The Eggs and Planktonic Stages of British Marine Fishes*. Academic Press; London. 584 pp.
RUSSELL, F. S. and J. S. COLMAN (1934) *Gt. Barrier Reef Exped. Sci. Rept.* **2**, 159–176.
RUSSELL, F. S. and J. S. COLMAN (1935) *Gt. Barrier Reef Exped. Sci. Rept.* **2**, 203–276.
RUSSELL, F. S. and W. J. REES (1960) *J. Mar. Biol. Ass. U.K.* **39**, 303–317.
RUSSELL, F. S., A. J. SOUTHWARD, G. T. BOALCH and E. I. BUTLER (1971) *Nature*, **234**, 468–470.
RUUD, J. T. (1929) *Rapp. P-v Réun. Cons. Perm. Int. Explor. Mer* **56**, 1–84.
RYLAND, J. S. (1964) *J. Mar. Biol. Ass. U.K.* **44**, 343–364.
SAIDOVA, K. M. (1972) In B. G. Bogorov (Ed.) *Fauna of the Kurile-Kamchatka Trench and its Environment*. Israel Program for Scientific Translations, Jerusalem. pp. 174–176.
SAKTHIVEL, M. (1973) In B. Zeitzschel (Ed.) *The Biology of the Indian Ocean* Chapman & Hall; London; Springer-Verlag; Berlin. pp. 383–397.
SALE, P. F., P. S. MCWILLIAM and D. T. ANDERSON (1976) *Mar. Biol.* **34**, 59–66.
SAMAIN, J.-F. and J. BOUCHER (1974) *Ann. Inst. Océanogr., Paris* **50**, 199–205.
SAMAIN, J.-F., J. Y. DANIEL and J. R. LECOZ (1977) *J. Exp. Mar. Biol. Ecol.* **29**, 279–289.
SAMEOTO, D. D. (1972) *J. Fish. Res. Bd. Can.* **29**, 987–996.
SAMEOTO, D. D. (1977) *Fish. and Marine Service Tech. Rept. 742*. Bedford Institute; Canada. 25 pp.
SANDERS, H. L. (1956) *Bull. Bingham Oceanogr. Coll.* **15**, 345–414.
SANTANDER, H. and O. S. de CASTILLO (1977a) *Bol. Inst. Mar. Peru* **3**, 73–94.
SANTANDER, H. and O. S. de CASTILLO (1977b) In *Proc. Symp. Warm Water Zooplankton*. NIO–UNESCO, Goa, pp. 105–123.
SANTHAKUMARI, V. and M. VANNUCCI (1977) In *Proc. Symp. Warm Water Zooplankton*. NIO–UNESCO, Goa, pp. 354–361.
SARASWATHY, W. (1973) *IOBC Handbook* **6**, 190–195.
SARGENT, J. R. and R. F. LEE (1975) *Mar. Biol.* **31**, 15–23.
SARGENT, J. R., R. R. GATTEN, E. D. S. CORNER and C. C. KILVINGTON (1977) *J. Mar. Biol. Ass. U.K.* **57**, 525–533.
SARGENT, J. R., R. R. GATTEN and R. MCINTOSH (1974) *Comp. Biochem. Physiol.* **47B** 217–227.
SARGENT, J. R., R. G. MORRIS and R. MCINTOSH (1978) *Mar. Biol.* **46**, 315–320.
SARS, G. O. (1902) *An Account of the Crustacea of Norway* **4**, Bergen, 171 pp.
SARS, G. O. (1918) *An Account of the Crustacea of Norway* **6**, Bergen, 225 pp.
SARS, G. O. (1921) *An Account of the Crustacea of Norway* **7**, Bergen, 32 pp.
SARS, G. O. (1925) *Res. Camp. Sci. Monaco*, **69**.
SAVAGE, R. E. (1956) *Fish. Invest. Ser. II* **20**, 1–22.

SCHALLEK, W. (1932) *Biol. Bull. Woods Hole* **82**, 112–126.
SCHALLEK, W. (1943) *Biol. Bull. Woods Hole* **84**, 98–105.
SCHELTEMA, R. S. (1966) *Deep-Sea Res.* **13**, 83–95.
SCHELTEMA, R. S. (1970) *Mar. Biol.* **7**, 47–48.
SCHELTEMA, R. S. (1971) *Mar. Biol.* **11**, 5–11.
SCHELTEMA, R. S. (1972) In B. Battaglia (Ed.) *5th European Mar. Biol. Symp.* Piccin Editore; Padua. pp. 101–114.
SCHELTEMA, R. S. (1974) *Thalassia Jugoslavica* **10**, 263–296.
SCHELTEMA, R. S. (1975) *Estuarine Res.* **1**, 372–391.
SCHELTEMA, R. S. and J. R. HALL (1975) *Proc. Int. Symp. Biol. Sipincula & Echiura I*, Kotor. pp. 103–115.
SCHRAM, T. A. (1968) *Ophelia* **5**, 221–243.
SEAPY, R. R. (1974) *Mar. Biol.* **24**, 243–250.
SEGUIN, G. (1966) *Bull. Inst. Franc. Afriq. Noire* **28A**, 1–90.
SEGUIN, G. (1968) *Deep-Sea Res.* **15**, 491–492.
SEGUIN, G. (1970) *Bull. Inst. Fond. Afriq. Noire*, **32A**, 607–663.
SEKI, H. (1973) *La mer, Bull. Soc. Franco-Jap d'Océanogr.* **2**, 153–158.
SEKIGUCHI, H. (1975a) *Bull. Fac. Fish. Mie Univ.* **2**, 1–10.
SEKIGUCHI, H. (1975b) *Bull. Fac. Fish. Mie Univ.* **2**, 29–38.
SEMENOVA, T. N. (1974) *Oceanology* **14**, 272–276.
SENTZ-BRACONNOT, E. (1965) *Cah. Biol. Mar.* **6**, 191–194.
SERIDJI, R. (1968) *Pelagos* **10**, 91–107.
SEWELL, R. B. S. (1947) *John Murray Expdn. Sci. Rept.* **8**, 1–303.
SEWELL, R. B. S. (1948) *John Murray Expdn. Sci. Rept.* **8**, 321–592.
SEWELL, R. B. S. (1953) *John Murray Expdn. Sci. Rept.* **10**, 1–90.
SHEADER, M. (1974) *Proc. Challenger Soc.* **4**, 247.
SHEADER, M. (1977) *J. Mar. Biol. Ass. U.K.* **57**, 943–954.
SHEADER, M. and F. EVANS (1975) *J. Mar. Biol. Ass. U.K.* **55**, 641–656.
SHEARD, K. (1953) *Rep. BANZ. Antarctic Res. Exped. Ser. B.* **8**, 1–72.
SHELBOURNE, J. E. (1953) *J. Mar. Biol. Ass. U.K.* **32**, 149–159.
SHELBOURNE, J. E. (1957) *J. Mar. Biol. Ass. U.K.* **36**, 539–552.
SHELBOURNE, J. E. (1962) *J. Mar. Biol. Ass. U.K.* **42**, 243–252.
SHELBOURNE, J. E., J. D. RILEY and G. T. THACKER (1963) *J. Cons. Perm. Int. Explor. Mer* **28**, 50–69.
SHERMAN, K. (1963) *Limnol. Oceanogr.* **8**, 214–227.
SHERMAN, K. and E. G. SCHANER (1968) *Limnol. Oceanogr.* **13**, 618–625.
SHIH, C. T. (1969) *Dana Rept.* **74**, 1–100.
SHIH, C. T. and M. J. DUNBAR (1963) *Fich. Ident. Zooplancton* **104**, 6 pp.
SHIRAISHI, K. and L. PROVASOLI (1959) *Int. Oceanogr. Congress* A.A.A.S; Washington. pp. 951–952.
SHOEMAKER, C. R. (1945) *Zoologica, New York*, **30**, 185–266.
SHUSHKINA, E. A., Y. Y. KISLYAKOV and A. F. PASTERNAK (1974) *Oceanology* **14**, 259–265.
SHUVALOV, V. S. (1965) *Oceanology* **5**, 116–123.
SHUVALOV, V. S. (1975) In Z. A. Zvereva (Ed.) *Geographical and Seasonal Variability of Marine Plankton.* Israel Program for Scientific Translations, Jerusalem. pp. 169–185.
SIDHU, G. S., W. A. MONTGOMERY, G. L. HOLLOWAY, A. R. JOHNSON and D. M. WALKER (1970) *J. Sci. Fd. Agric.* **21**, 293–296.
SIVERTSEN, E. and L. B. HOLTHUIS (1956) *Rep. Sci. Res. Michael Sars Expedn.* **5**(12) 1–54.
SKOGSBERG, T. (1920) *Zool. Bidr. Upps. Suppl.* **1**, 784 pp.
SMALL, L. F. and S. W. FOWLER (1973) *Mar. Biol.* **18**, 284–290.
SMALL, L. F. and J. F. HEBARD (1967) *Limnol. Oceanogr.* **12**, 272–280.
SMAYDA, T. J. (1966) *Int. American Tropical Tuna Comm. Bull.* **11**, 355–612.
SMIDT, E. L. B. (1951) *Meddr. Komm. Danm. Fisk, -og Havunders Serie Fiskeri* **11**(6), 1–151.
SMITH, K. L. and J. M. TEAL (1973) *Deep-Sea Res.* **20**, 853–858.
SOARES, M. (1957) M.Sc. Thesis. Univ. of Southampton.
SOMME, J. D. (1934) *Rep. Norway Fish and Mar. Invest.* **4**(9), 1–163.
VAN SOEST, R. W. M. (1973) *Bijdragen tot de Dierkunde* **43**, 119–125.
SOROKIN, J. I. (1971) *Int. Revue Ges. Hydrobiol.* **56**, 1–48.
SOROKIN, J. I. (1977) *Mar. Biol.* **41**, 107–117.
SOURNIA, A. (1969) *Mar. Biol.* **3**, 287–303.
SOUTHWARD, A. J. (1961) *J. Mar. Biol. Ass. U.K.* **41**, 17–35.
SOUTHWARD, A. J. (1962) *J. Mar. Biol. Ass. U.K.* **42**, 275–375.
SPOEL, S. van der (1967) *Noorduyn en Zoon*, Gorinchem, Netherlands. 375 pp.
SPOEL, S. van der (1972) *Fich. Ident. Zooplancton, 140–152.* 12 pp.
SPOONER, G. M. (1933) *J. Mar. Biol. Ass. U.K.* **19**, 385–438.

SRINIVASAGAM, R. T., J. E. G. RAYMONT, C. F. MOODIE and J. K. B. RAYMONT (1971) *J. Mar. Biol. Ass. U.K.* **51**, 917–925.
STANDER, G. H. and A. H. B. de DECKER (1969) *Investl. Rep. Div. Sea S. Africa* **81**, 1–46.
STEBBING, T. R. R. (1888) Challenger Rept. **29**, 1–1737.
STEELE, J. H. (1978) *Spatial Pattern in Plankton Communities*. Plenum Press; London. pp. 1–20.
STEEMANN NIELSEN, E. (1937) *J. Cons. Perm. Int. Explor. Mer* **12**, 147–153.
STEPHEN, A. C. (1931) *J. Mar. Biol. Ass. U.K.* **17**, 277–300.
STEPHEN, A. C. (1938) *J. Anim. Ecol.* **7**, 130–143.
STEPHENSEN, K. (1923) *Vidensk. Medd. Dansk Naturhistorisk Foreng Copenhagen* **76**, 5–20.
STEPHENSEN, K. (1924) *Rept. Danish Oceanogr. Exped. Mediterranean* **2**, D4 71–149.
STEPHENSEN, K. (1925) *Rept. Dan. Oceanogr. Exped. 1908–1910. 2 Biology*, 151–252.
STEPHENSEN, K. (1935) *Meddr. Grønland (Godthaab Expedn. 1928)* **80**, 1–94.
STEPHENSON, A. (1934) *Gt. Barrier Reef Exped. Sci. Rept.* **3**, 247–272.
STEPIEN, W. P. (1976) *Mar. Biol.* **38**, 1–16.
STIASNY, G. (1934) *Discovery Rep.* **8**, 329–396.
STIASNY, G. (1940) *Dana Rept.* **4**, No. 18.
STØP-BOWITZ, C. (1948) *Rep. Sci. Results 'Michael Sars' N. Atlantic Deep Sea Exped. 1910.* **5**(8), 1–91.
SUDA, A. (1973) In B. Zeitzschel (Ed.) *The Biology of the Indian Ocean.* Chapman & Hall; London: Springer-Verlag; Berlin. pp. 431–449.
SUND, O. (1932) *Sci. Rept. Michael Sars N. Atlantic Exped.* **3**, Part II.
SUSHCHENYA, L. M. (1970) In J. H. Steele (Ed.) *Marine Food Chains.* Oliver and Boyd, Edinburgh pp. 127–141.
SUYAMA, M., K. NAKAJIMA and J. NONAKA (1965) *Bull. Jap. Soc. Scient. Fish.* **31**, 302–306.
SWANBERG, W. (1974) *Mar. Biol.* **24**, 69–76.
TAGUCHI, S. and H. ISHII (1972) In A. Y. Takenouti (Ed.) *Biological Oceanography of the Northern North Pacific Ocean.* Idemitsu Shoten; Tokyo. pp. 419–431.
TATTERSALL, O. S. (1955) *Discovery Rep.* **28**, 1–190.
TATTERSALL, W. M. (1911) *Nordisches Plankton* **6**, 181–313.
TATTERSALL, W. M. (1936a) *Gt. Barrier Reef Expedn. Sci. Rept.* **2**, 277–290.
TATTERSALL, W. M. (1936b) *Zoologica* **21**, 95–96.
TATTERSALL, W. M. (1939) *John Murray Expedn. Sci. Rept.* **5**, 203–240.
TATTERSALL, W. M. (1951) *Bull. U.S. Nat. Mus.* **201**, 1–292.
TATTERSALL, W. M. and O. S. TATTERSALL (1951) *The British Mysidacea* Ray Soc. London. 460 pp.
TAVARES, D. Q. (1967) *Bolm. Inst. Oceanogr. Sao Paulo* **16**, 87–97.
TEAL, J. M. (1971) *Am. Zoologist.* **11**, 571–576.
TEAL, J. M. and F. G. CAREY (1967a) *Limnol. Oceanogr.* **12**, 548–550.
TEAL, J. M. and F. G. CAREY (1967b) *Deep-Sea Res.* **14**, 725–733.
TEAL, J. M. and J. KANWISHER (1966) *J. Mar. Res.* **24**, 4–14.
TEBBLE, N. (1960) *Discovery Rep.* **30**, 161–300.
TEBBLE, N. (1962) *Bull. Brit. Mus. Nat. Hist.* **7**(9), 373–492.
TEMPLEMAN, W. (1961) *Rapp. P-v Réun Cons. Perm. Int. Explor. Mer* **150**, 154–156.
TEMPLEMAN, W. (1968) *Rapp. P-v Réun Cons. Perm. Int. Explor. Mer* **158**, 41–53.
TESCH, J. J. (1946) *Dana Rept.* **5**, 82 pp.
TESCH, J. J. (1947) *Fich. Ident. Zooplancton* **8**, 6 pp.
TESCH, J. J. (1948) *Dana Rept.* **5**, 1–35.
TESCH, J. J. (1949) *Dana Rept.* **34**, 1–53.
TESCH, J. J. (1950) *Dana Rept.* **36**, 1–55.
TEIXEIRA, C., J. TUNDISI and M. B. KUTNER (1965) *Bolm. Inst. Oceanogr. Sao Paulo* **14**, 13–42.
THAMDRUP, H. M. (1935) *Meddr. Komm. Danm. Fisk -og Havunders, Serie Fiskeri* **10**(2), 1–125.
THIEL, H. (1966) In W. J. Rees (Ed.) *The Cnidaria and their Evolution* Symp. Zool. Soc. London. pp. 77–117.
THIRIOT, A. (1972) *Vie et Milieu* **25**, 243–295.
THIRIOT, A. (1977) *Doc. Scient. Centre Rech. Oceanogr. Abidjan* ORSTOM. **8**, 1–72.
THIRIOT-QUIEVREUX, C. (1973) *Publ. CNEXO, Ser. Res. Camp. Mer* **06**, 159–262.
THOMPSON, H. (1948) *Pelagic Tunicates of Australia.* CSIRO Australia. Melbourne. 196 pp.
THORSON, G. (1950) *Biol. Rev.* **25**, 1–45.
THORSON, G. (1957) In J. W. Hedgpeth (Ed.) *Treatise on Marine Ecology and Paleoecology 1. Ecology, Geo. Soc. Amer. Memoir* **67**, 461–534.
THORSON, G. (1961) In M. Sears (Ed.) *Oceanography A.A.A.S. Publ.* **67**, 455–474.
THORSON, G. (1964) *Ophelia* **1**, 167–208.
THURSTON, M. H. (1976a) *J. Mar. Biol. Ass. U.K.* **56**, 359–382.
THURSTON, M. H. (1976b) *J. Mar. Biol. Ass. U.K.* **56**, 383–470.
THURSTON, M. H. in M. Angel (Ed.) *A Voyage of Discovery.* Pergamon, Oxford, pp. 499–536.

TIDMARSH, W. G. (1973) *Arctic Inst. N. Am., Baffin Bay-North Water Project, Sci. Rept.* **3**, 181 pp.
TIMONIN, A. G. (1968) *Oceanology* **8**, 702–709.
TIMONIN, A. G. (1971) *Mar. Biol.* **9**, 281–289.
TJULEVA, L. S. (1971) In M. Uda (Ed.) *The Ocean World.* Proc. Joint Oceanogr. Assembly Tokyo. Japan Soc. Promotion Science. pp. 437–438.
TJULEVA, L. S. (1973) *Oceanology* **13**, 268–273.
TOKIOKA, T. (1960) *Publ. Seto Mar. Biol. Lab.* **8**, 352–443.
TOKIOKA, T. (1964) *Sci. Rep. Jap. Antarct. Res. Exped. E.* **21**, 16 pp.
TOKIOKA, T. (1965) *Publ. Seto Mar. Biol. Lab.* **13**, 231–242.
TOMASINI, J. A. and J. MAZZA (1978) *J. Cons. Perm. Int. Explor. Mer* **38**, 154–179.
TOTTON, A. K. (1954) *Discovery Rep.* **27**, 1–162.
TOTTON, A. K. and H. E. BARGMANN (1965) *A Synopsis of the Siphonophera.* Brit. Mus. Nat. Hist.; London. 230 pp.
TOTTON, A. K. and J. H. FRASER (1955) *Fich. Ident. Zooplancton* **55**, 4 pp., **61**, 4 pp. **62**, 4 pp.
TRANTER, D. J. (1962) *Aust. J. Mar. Freshw. Res.* **13**, 106–142.
TRANTER, D. J. (1963) *Nature, Lond.* **198**, 1179–1180.
TRANTER, D. J. (1973) In B. Zeitzschel (Ed.) *The Biology of the Indian Ocean.* Chapman & Hall; London; Springer-Verlag; Berlin. pp. 487–520.
TRANTER, D. J. and ABRAHAM, S. (1971) *Mar. Biol.* **11**, 222–241.
TRANTER, D. J. and J. GEORGE (1972) *Proc. Symp. Corals and Coral Reefs, 1964* Mar. Biol. Ass. India. pp. 239–256.
TRANTER, D. J. and J. D. KERR (1969) *Aust. J. Mar. Freshw. Res.* **20**, 77–84.
TRANTER, D. J. and P. E. SMITH (1968) In *Zooplankton Sampling.* Monographs on Oceanographic Methodology **2**, UNESCO, Paris. pp. 27–56.
TRAVERS, A. and M. TRAVERS (1970) *Tethys* **2**, 639–646.
TREGOUBOFF, G. (1963) *Rapp. P-v Comm. Int. Explor. Mer. Medit.* **17**, 531–538.
TREGOUBOFF, G. and M. ROSE (1957) *Manuel de Planctonologie Mediterranéenne* Centre National Rech. Sci. Paris 1, 587 pp.
TROOST, D. G. (1975) *Bull. Zool. Mus. Univ. Amsterdam* **4**, 201–211.
TROOST, D. G. A. B. SUTOMO and L. F. WENNO (1976) *Marine Research in Indonesia* **16**, 31–44.
TSALKINA, A. V. (1972) *Oceanology* **12**, 566–576.
TSEYTLIN, V. B. (1967) *Oceanology* **16**, 77–507.
TSEYTLIN, V. B. (1977) *Oceanology* **17**, 351–355.
TSYBAN, A. (1971) *Abstract, Proc. Joint Oceanogr. Assembly (Tokyo) 37–7*, pp. 276–277.
TUNDISI, T. (1970) *Bolm. Inst. Oceanogr. Sao Paulo* **19**, 131–144.
UNESCO (1968) *Zooplankton Sampling.* Monographs on Oceanographic Methodology **2**, UNESCO, Paris. 174 pp.
UNTERUBERBACHER (1964) *Invest. Rept. Div. Sea Fish* **11**, 1–42.
UROSA, L. J. and T. S. S. RAO (1974) *Bol. Inst. Oceanogr. Univ. Oriente* **13**, 53–66.
URRY, D. L. (1965) *J. Mar. Biol. Ass. U.K.* **45**, 49–58.
USSING, H. H. (1938) *Meddr. Grønland* **100**, 1–108.
VANE, F. R. and J. M. COLEBROOK (1962) *Bull. Mar. Ecol.* **5**, 247–253.
VANNUCCI, M. (1968) In *Zooplankton Sampling.* Monographs on Oceanographic Methodology **2**, UNESCO, Paris. pp. 77–86.
VANNUCCI, M. and D. NAVAS (1973a) In B. Zeitzschel (Ed.) *The Biology of the Indian Ocean.* Chapman & Hall; London: Springer-Verlag; Berlin. pp. 273–281.
VANNUCCI, M. and D. NAVAS (1973b) *IOBC Handbook* **5**, pp. 1–54.
VANNUCCI, M., V. SANTHAKUMARI and E. P. DOS SANTOS (1970) *Mar. Biol.* **7**, 49–58.
VENKATARAMANUJAM, K. and K. RAMAMOORTHI (1977) In *Proc. Symp. Warm Water Zooplankton*, NIO–UNESCO, Goa, pp. 474–485.
VENTER, G. E. (1969) *Invest. Rept. S.W. Afr. Mar. Res. Lab.* **16**, 73 pp.
VERVOORT, W. (1951a) *Verh. Acad. Wet. Amst. Afd. Nat.* **2**, 1–156.
VERVOORT, W. (1951b) *Fich. Ident. Zooplancton* (Farran revised) **32–39**.
VERVOORT, W. (1952) *Fich. Ident. Zooplancton*, **41–49**.
VERVOORT, W. (1963) *Atlantide Rep.* **7**, 77–194.
VERVOORT, W. (1965) *Atlantide Rep.* **8**, 9–216.
VINOGRADOV, M. E. (1962a) *Deep-Sea Res.* **8**, 251–258.
VINOGRADOV, M. E. (1962b) *Rapp. P-v Réun Cons. Perm. Int. Explor. Mer* **153**, 114–119.
VINOGRADOV, M. E. (1970) In *Vertical Distribution of the Oceanographic Zooplankton.* Israel Program for Scientific Translations. Jerusalem. 339 pp.
VINOGRADOV, M. E. (1972a) In A. Y. Takenouti (Ed.) *Biological Oceanography of the Northern North Pacific Ocean.* Idemitsu Shoten; Tokyo. pp. 333–340.

VINOGRADOV, M. E. (1972b) In B. G. Bogorov (Ed.) *Fauna of the Kurile-Kamchatka Trench and its Environment*. Israel Program for Scientific Translations, Jerusalem. pp. 104–123.
VINOGRADOV, M. E. (1973) *Life Activity of Pelagic Communities in the Ocean Tropics*. Israel Program for Scientific Translations, Jerusalem. 298 pp.
VINOGRADOV, M. E. and V. V. MENSHUTKIN (1977) In E. D. Goldberg, S. N. McCave, J. J. O'Brien and J. H. Steele (Eds.) *The Sea*. Wiley Interscience; New York. pp. 891–921.
VINOGRADOV, M. E. and N. V. PARIN (1973) *Oceanology* **13**, 104–113.
VINOGRADOV, M. E. and A. F. SAZHIN (1978) *Oceanology* **18**, 205–209.
VINOGRADOVA, Z. A. (1960) *Dokl. Acad. Nauk. SSSR.* **133**, 680–682.
VITIELLO, P. (1964) *Pelagos* **2**, 5–42.
VLADIMIRSKAYA, Y. V. (1976) *Oceanology* **15**, 359–362.
VLYMEN, W. J. (1970) *Limnol. Oceanogr.* **15**, 348–356.
VORONINA, N. M. (1970) *Antarctic Ecology* **1**, 160–172.
VOSS, G. L. (1971) *Bull. Mar. Sci.* **21**, 1–34.
VYSHKVARTSEVA, N. V. (1975) In Z. A. Zvereva (Ed.) *Geographical and Seasonal Variability of Marine Plankton*. Israel Program for Scientific Translations. Jerusalem. pp. 186–199.
WALNE, P. R. (1963) *J. Mar. Biol. Ass. U.K.* **42**, 767–784.
WATERMAN, T. H., R. F. NUNNEMACHER, F. A. CHACE and G. L. CLARKE (1939) *Biol. Bull. Woods Hole* **76**, 256–279.
WEBB, L. K. and R. E. JOHANNES (1967) *Limnol. Oceanogr.* **12**, 376–382.
WELLS, J. B. J. (1970) *Fich. Ident. Zooplancton* **133**, 4 pp.
WERNER, B. (1971) *Nature, Lond.* **232**, 582–583.
WESENBERG-LUND, E. (1939) *Rep. Danish Oceanogr. Exped. 1908–10 to the Mediterranean Adjacent Sea. Biology* **2**, 1–46.
WHEELER, A. (1969) *The Fishes of the British Isles and North-West Europe* Macmillan; London. 613 pp.
WHEELER, E. H. (1970) *Atlantic Deep-Sea Calanoid Copepoda. Smithsonian Contrib. Zool.* **55**, 1–31.
WHITELEY, G. C. (1948) *Ecol. Monogr.* **18**, 234–264.
WIBORG, K. F. (1948a) *Rep. Norweg, Fish and Mar. Invest.* **9**(3) 1–27.
WIBORG, K. F. (1948b) *Rep. Norweg. Fish and Mar. Invest.* **9**(4) 1–17.
WIBORG, K. F. (1949) *Rep. Norweg. Fish and Mar. Invest.* **9**(8) 1–27.
WIBORG, K. F. (1954) *Rep. Norweg. Fish and Mar. Invest.* **11**(1) 1–246.
WIBORG, K. F. (1955) *Rep. Norweg. Fish and Mar. Invest.* **11**(4) 1–66.
WIBORG, K. F. (1960) *Rep. Norweg. Fish and Mar. Invest.* **12**(8) 1–18.
WIBORG, K. F. (1971) *Fisk Dir. Skr. Ser. Havunders.* **16**, 10–35.
WICKSTEAD, J. H. (1958) *J. Cons. Perm. Int. Explor. Mer* **23**, 341–353.
WICKSTEAD, J. H. (1959) *J. Anim. Ecol.* **28**, 69–72.
WICKSTEAD, J. H. (1963) *Proc. Zool. Soc. London* **141**, 577–608.
WICKSTEAD, J. H. (1968) *J. Zool. London* **155**, 253–269.
WIEBE, P. H. (1971) *Limnol. Oceanogr.* **16**, 29–38.
WIEBE, P. H., S. BOYD and J. L. COX (1975) *Fish Bull. US.* **73**, 777–786.
WIEBE, P. H., L. P. MADIN, L. R. HAURY, G. R. HARBISON and L. M. PHILBIN (1979) *Mar. Biol.* **53**, 249–255.
WILLIAMSON, D. I. (1957) *Fich. Ident. Zooplancton* **67** and **68**.
WILLIAMSON, D. I. (1967) *Fich. Ident. Zooplancton* **109**.
WILLIAMSON, D. I. (1969) *Crustaceana,* **16**, 210–213.
WILSON, C. B. (1932) *Bull. 158*. U.S. National Museum (Smithsonian Inst.) Washington. 623 pp.
WILSON, C. B. (1942) *Carnegie Inst. Washington Publ.* **536**. 237 pp.
WILSON, C. B. (1950) *Bull. 100* U.S. National Museum (Smithsonian Inst.) Washington, pp. 141–441.
WILSON, D. P. (1937) *J. Mar. Biol. Ass. U.K.* **22**, 227–243.
WILSON, D. P. (1948) *J. Mar. Biol. Ass. U.K.* **27**, 723–760.
WILSON, D. S. (1973) *Ecology* **54**, 909–915.
WIMPENNY, R. S. (1937) *Fish. Invest. Lond. Ser. II* **15**(3) 1–53.
WIMPENNY, R. S. (1938) *J. Cons. Perm. Int. Explor. Mer* **13**, 323–326.
WOLF, T. (1962) *Galathea Report* **6**, 1–320.
WOODMANSEE, R. A. (1958) *Ecology* **39**, 247–262.
WOODMANSEE, R. A. (1966) *Ecology* **47**, 847–850.
WYATT, T. (1973) *Mar. Biol.* **22**, 137–158.
YAMAGUCHI, M. (1970) *Rec. Oceanogr. Works Japan* **10**, 147–155.
YONGE, C. M. (1926) *J. Linn. Soc.* **36**, 417–429.
YONGE, C. M. (1928) *Biol. Rev.* **3**, 21–76.
ZAIKA, V. Y. (1972) *Oceanology* **12**, 408–414.
ZAIKA, V. Y. and T. Y. AVERINA (1968) *Oceanology* **8**, 843–845.
ZAIKA, V. Y. and N. A. OSTROVSKAYA (1972) *Oceanology* **12**, 725–729.

Zaikin, A. N. and Y. A. Rudyakov (1976) *Oceanology* **16**, 517–519.
Zaikin, A. N. and Y. A. Rudyakov (1977) *Oceanology* **17**, 88–89.
Zaitsev, Y. P. (1971) *Marine Neustonology*. Israel Program for Scientific Translation, Jerusalem. 207 pp.
Zalkina, V. A. (1970) *Mar. Biol.* **5**, 275–282.
Zei, M. (1967) *Proc. Symp. Oceanogr. & Fisheries Resources of the Tropical Atlantic*. UNESCO, p. 69.
Zeitzschel, B. (1966) *Veroff. Inst. Meeresforch. Bremerh.* **2**, 293–300.
Zeitzschel, B. (1967) *Helgolander Wiss. Meeresunters.* **15**, 589–601.
Zelickman, E. A. (1969) *Oceanology* **9**, 558–564.
Zelickman, E. A. (1972) *Mar. Biol.* **18**, 256–264.
Zelickman, E. A., V. I. Gelfand, and M. A. Shifrin (1969) *Mar. Biol.* **4**, 167–173.
Zenkevitch, L. A. (1963) *Biology of the Seas of the U.S.S.R.* Allen & Unwin; London. 955 pp.
Zenkevitch, L. A. and J. A. Birstein (1956) *Deep-Sea Res.* **4**, 54–64.
Zeuthen, E. (1947) *C-r. Trav. Lab. Carlsberg, Ser. Chim.* **126**, 17–161.
Zhuravlev, V. M. (1977) *Oceanology* **17**, 81–83.
Zhuravlev, V. M. and M. Y. Neyman (1976) *Oceanology* **16**, 193–195.
Zilanov, V. K. (1968) *Rapp. P-v Réun Cons. Perm. Int. Explor. Mer* **158**, 116–125.
Zoppi, E. (1964) *Bol. Inst. Oceanogr. Univ. Oriente* **3**, 15–81.
Zo, Z. (1973) *Limnol. Oceanogr.* **18**, 750–756.

Index